The Big Book of Real Analysis

Syafiq Johar

The Big Book of Real Analysis

From Numbers to Measures

Springer

Syafiq Johar
Department of Mathematical Sciences
National University of Malaysia
Bangi, Malaysia

ISBN 978-3-031-30831-4 ISBN 978-3-031-30832-1 (eBook)
https://doi.org/10.1007/978-3-031-30832-1

Mathematics Subject Classification: 26-01, 28-01, 40-01

This Springer imprint is published by the registered company Springer Nature Switzerland AG
The registered company address is: Gewerbestrasse 11, 6330 Cham, Switzerland

Paper in this product is recyclable.

Preface

Understanding the methods of calculus is vital to the creative use of mathematics ... Without this mastery the average scientist or engineer, or any other user of mathematics, will be perpetually stunted in development, and will at best be able to follow only what the textbooks say; with mastery, new things can be done, even in old, well-established fields.

— Richard Hamming, mathematician

Upon my return to Malaysia after finishing my DPhil, apart from the usual teaching and administrative roles, I was asked for an opinion for a curriculum and course structure redesign for a mathematics degree program at the university. This is a very important job, at least in my eyes. What I noticed was, amongst other issues, in the previous years the real analysis course at this university was optional for mathematics majors and spans only one semester.

Since it was optional, a majority of the mathematics majors here choose to avoid taking this analysis course. On average, roughly only 10–20% of the cohort per year chooses to take the course because it is perceived to be one of the most difficult courses in undergraduate mathematics study. This is due to the rigour the course demands, the abstract content, and the stark difference to the mathematics that they have seen pre-university and other mathematics courses offered. Even so, due to the short amount of time allocated for it, this course only covers the topics of real sequences, real series, limits, and continuity of functions very hastily.

Since the mathematics program here is mainly based on computations and number-crunching, most of the students take up to three calculus courses, but were not properly exposed to the proofs, origins, and intuition of the concepts within. Understandably, the level of the calculus courses here are not as rigorous as a mathematician would require since they are also offered to non-mathematics majors (such as actuarial science and statistics majors).

For mathematics majors, these calculus courses are very important for the applications and as a precursor to applied mathematics courses. However, for them, these calculus courses should also be paired with a robust course on analysis to build their appreciation and understanding from the ground up. In fact, this is even welcome for non-mathematics majors as remarked in a quote by Richard Hamming at the beginning of the preface.

A not-so-surprising outcome to this mainly calculations-based system is that most of the mathematics students struggled with interpreting mathematical statements and the basics in proof-writing. They also were not exposed to in-depth discussions on mathematics and calculus such as: what are irrational numbers, what is infinity, what is a limit, how to give meaning to an infinite sum, when can we use L'Hôpital's rule, what is the difference between integration and antiderivatives, etc. This shaky mathematical foundation gave some of the students (and graduates) a rather warped idea of what mathematics is about.

Seeing these worrying issues, I proposed to expand the real analysis course to two semester-long compulsory courses since no mathematics major should graduate without learning how to prove mathematical statements, having a glimpse of what analysis is about, and understand the motivation for the concepts in calculus. Moreover, in my opinion, one semester is not enough to cover a myriad of topics in analysis with proper depth.

Overview

The contents of this book form a guide on these new courses that I designed. The text roughly follows the courses *Foundations of Analysis* and *Analysis 1–2* at Imperial College London, as well as *Introduction to University Mathematics*, *Analysis 1–3*, and *Integration* at the University of Oxford, where I was a tutor and stipendiary lecturer in 2015–2019. The books [46, 48, 63], and [74] as well as the lecture notes by Hilary Priestley and Zhongmin Qian of the University of Oxford also serve as important inspirations.

The contents of this book are presented in a mostly linear delivery since I believe it is the best way to provide motivation as well as justify some constructions or definitions. The discovery and history of mathematics have never been linear, but I leave that to the mathematical historians to discuss. However, minor historical anecdotes, people involved, and quotes are included in this book for a bit of a human interest and light humour. It is hoped that this could enhance intuition, appreciation, and enjoyment amongst the readers.

Despite the generally linear presentation, there are plenty of pauses within the text hinting at what the readers can expect in the later chapters and suggesting some relevant exercise questions. Some of the proofs for the results are left as exercises for the readers. At these points, it would be a good idea for the readers to take a break from reading, turn to the exercise page, and attempt the proofs themselves. This is a good opportunity for the readers to test their grasp on the material. As Pál Halmos (1916–2006) used to say: *"The only way to learn mathematics is to do mathematics."*

The text is also punctuated with many concrete examples, counterexamples, and important remarks. These are necessary to solidify, clarify, and motivate the abstract definitions and results. Some of the proofs and examples are written with a rather stream-of-consciousness detail to demonstrate the thought process and ideas behind them. This may cause some mild discomfort and tedium amongst seasoned mathematicians (as I have been told) since most experts prefer straight-to-the-point proofs. However, since this book is mainly aimed for beginner students in real analysis, the elaborations are intentionally kept to help the students develop their own mathematical intuition.

There are more than 600 exercise questions in total in this book and they are presented at the end of each relevant chapters. Hints for some of the exercises are provided at the end of the book to assist independent learners. The solutions to most of these problems are available for lecturers and instructors via the Springer Nature Extramaterial online platform. Some of the solutions can be rather sparse (for economical purposes) but they provide the general ideas on how to approach them. It is hoped that the lecturers, instructors, and readers can fill in the details and write them down rigorously in their own way.

The essential problems are labelled with $(*)$ and the interesting or difficult problems are labelled with $(\diamond)$. It is highly recommended for the readers to attempt all the $(*)$ problems because they are either the results which are stated (but not proven) in the text or the results that will be referred to again and used in the future chapters and exercises.

Some of the $(\diamond)$ problems can be rather long so they are suitable for homework questions or mini-projects. These questions may also offer a peek into applications, historic problems, or advanced topics. Readers who are keen for more exercise questions should consult these problem books: [2, 31, 38, 39, 40], and [60].

Course Structure Plan

As mentioned earlier, this book came about while I was designing the courses and curriculum structure at a local university. The design also takes into account of the other mathematical courses outside of basic real analysis to give a natural flow and synergetic feel to the overall learning experience.

The material of this book is enough to cover two to three semesters of real analysis courses ideally in the following timeline:

1. **Real Analysis 1: Numbers and Sequences.** In the first semester, the students could be taught with the material from Chaps. 1 to 6. This would provide a foundational knowledge on how to write proofs and the basics of naïve set theory. Chapter 5 is the most important part of this course as it lays down the concept of real sequences, which would appear everywhere in the later chapters. The course would then end with Chap. 6 which is an open-ended chapter showing where real sequences crop up in mathematics, providing an excellent discussion point or project topics for the class.

   Two good books that may be used as reference for this course are [28] and [43] in which the fundamentals of proofs, set theory, and the construction of the number systems are discussed in greater detail. It is also important for me to point out the books [34] and [50] which I myself have used as a first year undergraduate student.

It is recommended to pair this course with introductory courses on linear algebra, abstract algebra, discrete mathematics, combinatorics, and graph theory (similar to the content in [8,52,54,66], and [75]) since the basics of proof-writing are also required in these courses. Some algebraic structures such as rings, fields, and vector spaces are also mentioned in Chaps. 2–4, which tie in with the contents of generic introductory abstract and linear algebra courses. Having analysis and algebra courses side-by-side also allows the students to compare the structures and ideas in both fields.

From this course and the suggested pair courses, students can be led directly into an elementary number theory course since we have touched upon topics such as prime numbers, Bézout's identity, and modular arithmetic in Chap. 2. A reference for a suitable number theory course would be [37].

2. **Real Analysis 2: Series, Continuity, and Differentiability.** In the second semester, the course continues with Chaps. 7–14. The students will be reminded on what is a real sequence via the concept of real series in Chaps. 7 and 8. The students are then exposed to analysis of functions via limits and continuity. The topic of sequences continues with sequences and series of functions. A chapter is then devoted to the power series since there are many interesting results that can be obtained from them. Moreover, this topic will be important in the study of Taylor and Maclaurin series in Chap. 17 later. This course ends with the concept of differentiation and some of its applications.

   The two final chapters provide a good entry point into a course on ODEs, multivariable calculus, and complex analysis. In particular, the idea of differentiation in this course can then be expanded to other types of derivatives in these further courses such as partial derivatives, directional derivatives, full derivatives, and complex derivatives. Readers can refer to [3, 15, 18, 20], and [59] for references to these subjects.

   On the other hand, the bisection method, Newton-Raphson root-finding method, and Bernstein polynomials provide some introductory ideas to the study of numerical analysis. A good reference for this course would be [22].

   The topic on continuity also provides a first step into a course on metric spaces and topology, for which I recommend [17] and [73]. The latter, which is another book that I have used as an undergraduate, paints an important transition from the concept of metric spaces to more abstract topological spaces.

3. **Real Analysis 3: Integration.** Finally, in the third semester, the students are exposed to the concept of integration via Chaps. 15–20. This book presents four different types of integration (for other types of integration, see [61]).

   The first two concepts are the Riemann and Darboux integrals. In many literature, these are treated as the same construction, but here we strictly distinguish them by carrying out their distinct constructions and carefully comparing them. This section could be run in tandem with the topic of numerical quadratures in a course on numerical analysis (refer to [28] for material) to discuss other ways of approximating the area of a subgraph for a function. In the exercises, we present the construction of Riemann–Stieltjes integration as a generalisation to the Riemann integral.

In Chap. 16, we introduce and prove the fundamental theorem of calculus and its applications. This section pairs well with a course on multivariable calculus and complex analysis, in particular for the topic of path integrals.

Finally, the course ends with an introduction to measure theory, Lebesgue integration, and double integrals. There are many discussions on the motivation and application for these constructions within the text. For example, in the exercises, there are questions relating these topics to probability theory and functions spaces. Therefore, this third course on real analysis segues well into courses on functional analysis, functions spaces, distribution theory and Fourier analysis, PDEs, dynamical systems, and probability theory in the later years of studies. For more details, readers may refer to [4, 9, 12, 47, 56, 57, 64, 65], and [71].

The plan above can be summarised by the following table, where n is either $1, 2, 3, 4$.

Semester	Courses	Suggested paired and further courses
n	Real analysis 1 (Chaps. 1–6)	Abstract algebra, linear algebra, discrete mathematics
$n+1$	Real analysis 2 (Chaps. 7–14)	Abstract algebra 2, linear algebra 2, elementary number theory
$n+2$	Real analysis 3 (Chaps. 15–20)	Multivariable analysis and calculus, topology, ODEs, complex analysis, numerical analysis
$n+3$		PDEs, geometry of curves and surfaces, dynamical systems, measure and probability, functional analysis, functions spaces, distribution theory and Fourier analysis

In addition, but not less important, these courses should also be complemented with courses in applied mathematics (such as classical and quantum mechanics, fluids and waves, elementary physics, mathematical biology, financial mathematics, econometrics, modelling, and optimisation), statistics, data science, as well as history of mathematics to provide further applications, examples, and background.

On top of that, some exercises in this book encourage the readers to craft solutions using computer programs, so knowing some basic computer programming could possibly enhance your experience with this book.

There is a huge selection of literature for these complementary courses, but some excellent books and course material that I would like to highlight are [7, 13, 21, 26, 44], and [49]. A selection of references for history of mathematics, calculus, and analysis are [11, 19, 42, 68], and [69].

Alternative Course Structure Plans

An alternate plan is to cover Chap. 1 in a separate short course (similar to the arrangement in University of Oxford via a two-weeks course *Introduction to University Mathematics* taught at the very beginning of the undergraduate studies before branching off to all the other courses). The remaining chapters can then be covered in the following manner:

1. **Real Analysis 1: Numbers, Sequences, and Series.** Chapters 2–8.
2. **Real Analysis 2: Continuity and Differentiation.** Chapters 9–14.
3. **Real Analysis 3: Integration.** Chapters 15–20.

I really like this system since any other courses in the first row of the table of plans above also require the fundamental ideas in logic and proofs presented in Chap. 1. Therefore, having it taught in a separate foundational course is important to ensure uniformity in the baseline requirement of proofwriting.

Moreover, this arrangement has the advantage of naturally separating the topics on sequences and series (which are discrete in nature) from the topics on continuity and differentiation (which are continuous in nature). The set of real analysis problem books [38, 39], and [40] also splits these topics into the three groups accordingly.

For a shorter plan for these courses, many subtopics from this book can be omitted, such as the in-depth construction of real numbers via Dedekind cuts, additional topics and applications of sequences and series, the theory of measure, and Lebesgue integration. An outline for a speedy two-semester course on real analysis based on this book would be:

1. **Real Analysis 1: Numbers, Sequences, and Series.** Chapters 1–8 but omitting Sects. 3.7, 3.8, 4.2, 4.3, 4.6, 5.7, 6.3, 6.4, 8.1, and 8.2.
2. **Real Analysis 2: Continuity, Differentiation, and Integration.** Chapters 9–17 but omitting Sects. 9.6, 14.1, 14.4, and 16.2.

These omitted topics can then be assigned for outside reading or project topics.

Final Words

I am very thankful for finally finishing writing this book. I have done so many iterations of going through this book, each time fixing any errors and adding extra content which I think are relevant, so hopefully it converges to a fixed point of an ideal textbook. Alas, I am a human being and all this writing overwhelms me at times, so errors are inevitable. Therefore, if the readers have found any errors, mistakes, typos, or anything you are unhappy about in this book, please contact me at msyajoh@gmail.com.

Thank you to Austin Fuller, Alexis Hatto, Edward Hookway, and Richard Kruel for being constant companions during the writing of this book and for the numerous words of encouragement and suggestions; to my friends Kevin Pan, Claire Rebello, Sam Brzezicki, Ryan Roberts, Aman Pujara, Wael Al Jishi, Laura Abbott, Alex Holmes, Wei Sum Cheung, Cissy Chan, Jordan Noble, Ben Ashby, Ashara Peiris, and Ed Boyd for going through all this mathematics (and more) with me as young and carefree undergraduates; and to my DPhil friends Miles Caddick, Matthew Schrecker, Alexander Klimek, Kevin Schlegel, Luca Alasio, Francesco Della Porta, Bogdan Raita, Guangyu Xi, Seungchan Ko, Tabea Tscherpel, Ellya Kawecki, Sophia Koepke, Matt Rigby, Nikolaus Kolliopoulos, and Filip Zivanovic for the various important and unimportant mathematical discussions.

Special thanks are due to my mentors Darryl Holm for teaching me how to enjoy mathematics, David Gauld for teaching me how to communicate mathematics, and Andrew Dancer for teaching me how to do mathematics; to Luc Nguyen, Oliver Riordan, Chris Breward, and Kevin McGerty for the priceless lectureship opportunities in Oxford; to my students at Christ Church and St. Edmund Hall, Oxford (years 2015–2019) for teaching me the joys of teaching; and to my parents for not understanding why I would even write this really alien book yet understanding why it is important to me.

Finally, of course, this book is for you readers. I really hope you like it.

Bangi, Malaysia
May, 2023

Syafiq Johar

Contents

List of Figures

Logic and Sets 1

In high school, I wanted to study logic, which I thought would be useful in political debates or in legal battles against evil once I fulfilled my dream of becoming a solicitor. Unfortunately, I became neither a lawyer nor a politician, and I have since come to understand that logic is not a very useful tool in these areas in any case.

— Ariel Rubinstein, economist

Before we venture into the topic of analysis, we need to be sufficiently well-versed with the language of mathematical logic and proofs. The most common misconception among non-mathematicians is that mathematics is all about numbers and equations. Whilst they are important, they are not the main features of mathematics. Mathematicians view numbers as one of the languages or tools to communicate mathematics because they allow us to objectively quantify certain properties, structures, or act as convenient labels. Equations allow mathematician to compare or relate some concepts together.

In general, mathematics is so much more than numbers and equations. It is difficult to actually describe what mathematics is about but for me mathematics is a study of properties, structure, patterns, and relationships of abstract objects. From observed patterns within these abstract objects and structures, mathematicians come up with some general statements or claims. Some of these statements may be true and some could be false.

Therefore, mathematicians need to provide a rigorous proof to establish the truth of these statements, aided by numbers and equations as the language and logical arguments to bind them together neatly. The statements which have been proven to be true are then called propositions or theorems. Scattered throughout this book, we shall see some of these terminologies: definition, axiom, lemma, proposition, theorem, and corollary. The first two are very important when we are doing mathematics.

S. Johar, *The Big Book of Real Analysis*,
https://doi.org/10.1007/978-3-031-30832-1_1

Roughly speaking, definitions are declarations for the meaning of concepts and conventions. These declarations are important to ensure that everyone are well-informed and are on the same page with the assumptions and conventions used. Definitions are also used to state some conditions or concepts in a concise and condensed manner by giving it a name or a label.

Usually, what puts people off from mathematics is the amount of definitions, notations, and symbols involved in them. These can be very intimidating to some but they are actually your very dependable friends if you sit down and take some time to get to know them better! They are simply shorthand names or notations which can greatly condense the writing and communication of complicated ideas.

Example 1.0.1 As an example, consider the description of how one can find a solution to the quadratic equation $ax^2 + bx = c$ by Brahmagupta (c. 598–668) in the treatise *Brāhmasphuṭasiddhānta* (Correctly Established Doctrine of Brahma) [69]:

> To the absolute number multiplied by four times the [coefficient of the] square, add the square of the [coefficient of the] middle term; the square root of the same, less the [coefficient of the] middle term, being divided by twice the [coefficient of the] square is the value.

It is worth noting that the notations used for the equation $ax^2 + bx = c$ themselves were not familiar to Brahmagupta as mathematicians in older times used to describe their mathematical problems using words rather than symbols. Indeed, the addition and equality symbols $+$ and $=$ as well as the exponentiation notation were only introduced for use in mathematics in the 14th, 16th, and 17th century by Nicole Oresme (1323–1382), Robert Recorde (1510–1558), and René Descartes (1596–1650) respectively.

After many centuries of mathematical development in various different cultures, the idea by Brahmagupta can be rewritten in standard modern notation and symbols as:

A solution x for the equation $ax^2 + bx = c$ is $x = \frac{\sqrt{4ac+b^2}-b}{2a}$.

Even though it carries the exact same meaning as Brahmagupta's description, this is a much more concise sentence with fewer words and possibly fewer ambiguities. Of course, in order for this modern notation to make sense, we need to declare what a, b, c, and x are and define what the symbols $=, +, -$, and $\sqrt{}$ mean.

Without knowing what these symbols mean, it can be agreed that the sentence above can be very alien and intimidating, to the extreme point that we might be profiled as a terrorist if we were seen with such symbols. Indeed, in 2016, an American Airlines flight was delayed after a woman raised alarm after seeing her fellow passenger, economics professor Guido Menzio, writing some suspicious-looking cryptic codes on a notepad. Turns out it was a bunch of differential equations...

On the other hand, axioms (derived from the Greek word *axíoma* which means "that which is thought worthy or fit" or "that which commends itself as evident") are some notions which one assumes to be true or require to be true as a framework for some theory. As a result, these are very important in mathematics as they form the starting points or foundation upon which we establish our ideas and arguments.

As an example, let us refer to a very important set of axioms in mathematics, namely the Euclid's axioms in geometry. On top of some common notions, these axioms were systematically proposed by Euclid of Alexandria (c. 325B.C.-265B.C.) in a mathematical treatise known as Euclid's *Elements* [23] based on his observations of geometrical constructions on a plane. They are given as:

Definition 1.0.2 (Euclid's Axioms) The Euclid's axioms is a set of five postulates in planar geometry, which are:

1. It is possible to draw a straight line from any point to any other point.
2. It is possible to extend a line segment continuously in both directions.
3. It is possible to describe a circle with any centre and any radius.
4. It is true that all right angles are equal to one another.
5. Parallel postulate: It is true that, if a straight line falling on two straight lines make the interior angles on the same side less than two right angles, the two straight lines (produced indefinitely) intersect on that side on which the angles are less than two right angles (Fig. 1.1).

Remark 1.0.3 We make several remarks here.

1. Notice the usage of the term "Definition" at the beginning of the Euclid's axioms above. With the definition above, we have properly declared what we mean when we say "Euclid's axioms" and in the future we can simply quote "Euclid's axioms" without needing to repeat all of the points above.
2. But one has to be careful: different people or literature might use different definitions, so it is important to clarify or check what do they mean beforehand.
3. One funny example for this confusion is the term "distribution" which appears in the mathematical branches of probability, differential geometry, analysis, and number theory. In each of these mathematical branches, the term "distribution" have completely different meanings. So if you ask a probabilist or statistician, a

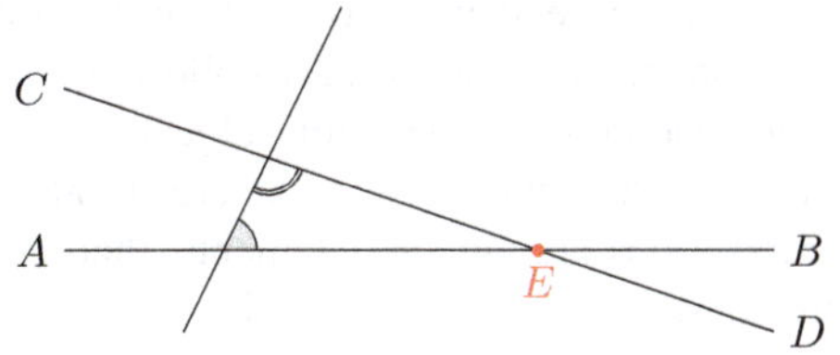

Fig. 1.1 Parallel postulate. The two marked angles add up to less than two right angles, so the lines AB and CD must meet somewhere (in this case at the point E)

geometer, an analyst, and a number theorist, you might get different meanings for this term!

4. A memorable passage from the book *Through the Looking-Glass, and What Alice Found There* by Lewis Carroll (1832–1898) regarding definitions and meaning is:

> "When I use a word," Humpty Dumpty said in rather a scornful tone, "it means just what I choose it to mean - neither more nor less."
> "The question is," said Alice, "whether you can make words mean so many different things."
> "The question is," said Humpty Dumpty, "which is to be master - that's all."

It is notable that, apart from being a celebrated author, Carroll is in actuality Charles Lutwidge Dodgson, a mathematician. This explains why his writings are often rife with logical and mathematical undertones.

The five Euclid's axioms or assumptions in Definition 1.0.2 form the basis for the study of planar geometry (also known as Euclidean geometry) from which we can build numerous results in geometry. However, if we instead propose a different postulate or system of axioms from Definition 1.0.2, we could get very different results.

For example, by removing or replacing the parallel postulate in Euclid's axioms with a different kind of postulate, we could get other geometrical framework such as the elliptic geometry or hyperbolic geometry (which are grouped under the umbrella term non-Euclidean geometry). Under these new axioms, geometry can behave in a very different way!

Some of the numerous major consequences of the Euclid's axioms are the similarity of triangles, Pythagorean theorem, trigonometric functions (sin, cos, tan,...), the Platonic solids, and even elementary number theory. Some of these were discussed in an elaborate manner in the 13 volumes of Euclid's *Elements*.

But how can we get all of these consequences from just the five points in the the Euclid's axioms? Obtaining these can be done by working from the axioms through a series of logical implications, deductions, and hard work. Quoting Thomas Huxley (1825–1895):

> The mathematician starts with a few [axioms], the proof of which is so obvious that they are called self-evident, and the rest of his work consists of subtle deductions from them. The teaching of languages, at any rate as ordinarily practised, is of the same general nature: authority and tradition furnish the data, and the mental operations are deductive.

These consequences come in the form of lemmas, propositions, theorems, and corollaries, which are usually collectively referred to as results. There is very little difference between these four objects and sometimes the distinction between them can be rather blurry. Lemmas are usually some preliminary statements or facts, propositions are bigger statements, theorems are the major results, and corollaries are consequences of these results.

These results are presented in the form of mathematical statements which have been rigorously proven to be true. They are used to explain mathematical phenomenon, as remarked by John B. Conway (1939-):

> To many, mathematics is a collection of theorems. For me, mathematics is a collection of examples; a theorem is a statement about a collection of examples and the purpose of proving theorems is to classify and explain the examples.

Remark 1.0.4 We end this section with some interesting historical remarks regarding Euclid's axioms:

1. The Euclid's axioms is one of the earliest examples of an axiomatic system in mathematics. However, due to its primitive nature, there are several gaps that were overlooked or assumed to be obvious in the proofs by Euclid from these axioms.
2. For example, the first ever proposition in the first volume of Euclid's *Elements* assumes that the two circles constructed in the proof intersect at a point. However, the existence of this intersection cannot be justified by any of the five axioms and common notions, thus creating a gap in the proof. This caused a great deal of criticism on the sufficiency of these axioms in developing the rest of *Elements*.
3. Euclid's axioms were fixed by David Hilbert (1862–1943) in his 1899 book *Grundlagen der Geometrie* (The Foundations of Geometry). In this book, Hilbert introduced an expanded system of 21 (and later reduced to 20 since one of them was proven to be redundant) axioms to include the additional assumptions used by Euclid and make the axiomatic system and *Elements* properly self-contained.

1.1 Introduction to Logic

In this book, we are mostly concerned with the usage of informal logic. To describe it broadly, informal logic is the study of correct reasoning via some set of rules or conventions using everyday language. The end goal is usually to deduce conclusions from some given information. These information are given in the form of mathematical statements.

What are mathematical statements? Roughly speaking, a mathematical statement (also known as proposition) is a sentence that is either true or false, but not both. Let us take a look at some examples and non-examples:

1. The sum of the angles in a triangle is equal to two right angles.
2. If it is raining, then Mr. X will carry an umbrella.
3. The letters A and B are both vowels.
4. Good morning, Mr. Magpie. How are we today?
5. Thank you for reading this book.
6. The weather is nice today.

Notice that the first three of the sentences above can be decidedly verified as true or false, but not both at the same time. On the other hand, sentences 4 and 5 are not mathematical statements since the former is a greeting/question and the latter is an imperative sentence.

The final sentence is interesting. It seems like a mathematical statement, but its truth is subject to some personal opinion of what a "nice weather" is. If you ask me, I like rainy and cloudy weather. But many people prefer the sunny weather. Therefore, unless we agree on a convention or define what "nice weather" objectively means, this is not a mathematical statement.

Remark 1.1.1 We make several remarks regarding truth and falsehood.

1. This dichotomy of being "true" or "false" is the main feature of a mathematical statement. This allows for no ambiguity in its meaning and interpretation.
2. Moreover, these truth or falsehood are "mathematical truths" as opposed to absolute truths. To elaborate, these truths depend on what we assume or agree with (in the form of axioms) at the beginning. This is apparent in the final statement above: the statement "The weather is nice today" can only be given a truth/false value once we declare what "nice" means.
3. Note that in Euclid's axioms in Definition 1.0.2, we can see clearly how these mathematical statements are presented. In particular, the fourth axiom says: "It is true that all right angles are equal to one another". The statement here is "All right angles are equal to each other" with the assumption that this statement is accepted or declared to be true.

The work of a mathematician is to verify whether a given mathematical statement is true or false, conditional to some definitions or axioms which are accepted to be true. The way to do this is via mathematical proofs, which we shall see throughout this book.

Next, we are going to present some basics in logic to help us understand how this can be done. Before we determine their truth or falsehood, we need to be able to read and interpret these statements.

And, Or, Not

Given some mathematical statements, we can combine them or manipulate them to create new statements. First, we can negate a statement by writing its opposite. Moreover, if we have two statements P and Q, we can combine them with logical connectives "and" or "or". The former connective is called a logical conjunction and the latter is called logical disjunction.

The resulting sentence obtained by combining multiple statements together is called a compound statement. Depending on the truth of each statements P and Q, we can also deduce the truth of the compound statement. It is easier to look at these via some examples.

Example 1.1.2 Consider two mathematical statements:

$$P : \text{A is a vowel,} \quad \text{and} \quad Q : \text{B is a vowel.}$$

We know for a fact that the statement P is true whereas the statement Q is false.

1. The negations of the statements P and Q, denoted as $\neg P$ and $\neg Q$ and read as "not P" and "not Q" respectively, are:

$$\neg P : \text{A is not a vowel,} \quad \text{and} \quad \neg Q : \text{B is not a vowel.}$$

   Now the statement $\neg P$ is false whereas the statement $\neg Q$ is true. Thus negation switches the truth of the statement, namely if we started with a true statement, its negation is false and vice versa.
2. Now let us look at the logical connectives "and" and "or":
   (a) The "and" connective is denoted by the symbol $\wedge$. The statement $P \wedge Q$ says "A is a vowel and B is a vowel". This is equivalent to "A and B are vowels". We know this is false since B is not a vowel. In short, the compound statement $R \wedge S$ can only be true when both the statements R and S are true.
   (b) The "or" connective is denoted by the symbol $\vee$. The statement $P \vee Q$ is "A is a vowel or B is a vowel" or equivalently "A or B is a vowel". This statement is true. In short, the compound statement $R \vee S$ can only be true when at least one of the statements R and S is true.
   We note that the "or" connective above is quite different to the conventional daily life language. In daily life, usually "or" is taken to mean that exactly one of the two options is true. In mathematical logic, this is instead called the "exclusive or" connective! We shall see this connective in Exercise 1.8.

Remark 1.1.3 Note that, even though we did not explicitly write it as a definition, in Example 1.1.2 we have defined what the three symbols $\neg$, $\wedge$, and $\vee$ mean. These definitions help us set the stage for future work by declaring what we mean when we use the symbols.

Example 1.1.4 Consider two mathematical statements:

$$P : \text{Lucy likes coffee,} \quad \text{and} \quad Q : \text{Lucy likes tea.}$$

If the compound statement $P \wedge Q$ is true, this means Lucy likes both coffee and tea. On the other hand, if the compound statement $P \vee Q$ is true, then at least one of the statements P or Q is true, namely Lucy likes coffee, tea, or both! Again, the latter is different to the usual daily life convention for the usage of the word "or" as remarked in Example 1.1.2(2)(b).

For the two connectives $\wedge$ and $\vee$ above, the order in which we apply the connectives do not matter. In other words, $P \wedge Q$ is the same statement as $Q \wedge P$ and $P \vee Q$ is the same statement as $Q \vee P$. We have a name for this situation.

Definition 1.1.5 (Logically Equivalent Statements) We say two statements P and Q are logically equivalent if their truth or falseness are the same. In other words, if either one is true, the other must be true as well. We write this as $P \equiv Q$.

Therefore, we write $P \vee Q \equiv Q \vee P$ and $P \wedge Q \equiv Q \wedge P$. These phenomenon are called the symmetry of the $\vee$ and $\wedge$ connectives.

Moreover, if we have three statements P, Q, and R, we can iteratively create the statements $(P \wedge Q) \wedge R$ and $P \wedge (Q \wedge R)$. In Exercise 1.9, the readers will show that both of these compound statements are true exactly when all three P, Q, and R are true. So we have $(P \wedge Q) \wedge R \equiv P \wedge (Q \wedge R)$. Likewise, $(P \vee Q) \vee R \equiv P \vee (Q \vee R)$. In both cases, we say $\wedge$ and $\vee$ are associative connectives, so we can unambiguously write them without the brackets as $P \wedge Q \wedge R$ and $P \vee Q \vee R$ respectively.

Example 1.1.6 From Example 1.1.4, consider three mathematical statements:

$$P : \text{Lucy likes coffee,} \quad \text{and} \quad Q : \text{Lucy likes tea,} \quad \text{and} \quad R : \text{Lucy likes juice.}$$

If the compound statement $P \wedge Q \wedge R$ is true, this means Lucy likes all of coffee, tea, and juice. On the other hand, if the compound statement $P \vee Q \vee R$ is true, then Lucy likes at least one of coffee, tea, or juice.

Let us look at another example of equivalent statements:

Example 1.1.7 Suppose that ξ is some unknown letter in the Latin alphabet. Consider two mathematical statements:

$$P : \text{The letter } \xi \text{ is a vowel,} \quad \text{and} \quad Q : \text{The letter } \xi \text{ is not a consonant.}$$

We claim that the statements P and Q are logically equivalent. We check:

1. Suppose that P is true, namely ξ is a vowel. Then ξ cannot be a consonant, and hence Q is true.
2. On the other hand, if P is false (namely, ξ is not a vowel), then ξ must be a consonant. This means Q is also false.

Thus the two statements are equivalent and we write $P \equiv Q$.

We can also combine the negation and the "and" and "or" connectives above. Usually, we study them by using truth tables, which is a gadget used to help logicians deduce the truth of some composite statements and also the equivalence of statements.

A truth table is a table which is filled with T and F where T denotes true and F denotes false (sometimes these are substituted with the symbols 1 and 0 respectively). The table has one column for each statements involved, where the rows denote all possible combinations of truth/false of the statements, and a final column to denote the desired compound statement and its truth. For example, given the statements P and Q, the truth table of the negation, "and", and "or" connectives are:

P	$\neg P$
T	F
F	T

P	Q	$P \wedge Q$
T	T	T
T	F	F
F	T	F
F	F	F

P	Q	$P \vee Q$
T	T	T
T	F	T
F	T	T
F	F	F

To read these tables, using the final table as an example, we say that the compound statement $P \vee Q$ can be true only when at least one of the statements P or Q is true. Truth tables also help us determine which statements are equivalent. Two mathematical statements are equivalent if they have identical columns/tables.

Example 1.1.8 Suppose that we have two statements P and Q. We want to find a logically equivalent statements to the compound statements $\neg(\neg P)$, $\neg(P \wedge Q)$ and $\neg(P \vee Q)$.

1. Suppose first that $\neg(\neg P)$ is true. This means $\neg P$ is false, which then means P is true. On the other hand, if $\neg(\neg P)$ is false, then $\neg P$ is true and therefore P is false. Hence we have the equivalence of statements $\neg(\neg P) \equiv P$. This equivalence is called double negatives.
2. (a) Suppose that $\neg(P \wedge Q)$ is true, which means $P \wedge Q$ is false. From Example 1.1.2(2), both of P and Q cannot be true at the same time (or otherwise $P \wedge Q$ is true). Thus at least one of the statements P or Q is false. In other words, at least one of the negation statements $\neg P$ or $\neg Q$ is true. Hence, the statement $(\neg P) \vee (\neg Q)$ is also true.
   (b) Now suppose that $\neg(P \wedge Q)$ is false. By an identical argument as in part (a), we can deduce that $(\neg P) \vee (\neg Q)$ is also false.

   Combining the observations above, we have the equivalence $\neg(P \wedge Q) \equiv (\neg P) \vee (\neg Q)$.

We can also show this equivalence of statements in a quick way by using a truth table. We construct, column by column, the following table:

P	Q	$P \wedge Q$	$\neg(P \wedge Q)$	$\neg P$	$\neg Q$	$(\neg P) \vee (\neg Q)$
T	T	T	F	F	F	F
T	F	F	T	F	T	T
F	T	F	T	T	F	T
F	F	F	T	T	T	T

We can see that the columns for $\neg(P \wedge Q)$ and $(\neg P) \vee (\neg Q)$ shaded above are identical depending on the truth values of P and Q. Hence, the two statements are equivalent.

3. Likewise, in Exercise 1.5, we can show that $\neg(P \vee Q) \equiv (\neg P) \wedge (\neg Q)$.

Conditional Statement

An important connective is the material implication or conditional statement. Given two statements P and Q, this compound statement is written symbolically as $P \Rightarrow Q$ (or, in many literature on logic, $P \rightarrow Q$). This statement can be read or written as:

1. P implies Q (short for: The truth of P implies the truth of Q).
2. P therefore/thus/so/hence Q (short for: P is true, therefore/thus/so/hence Q is true).
3. P is sufficient for Q (short for: The truth of P is sufficient for the truth of Q).
4. If P, then Q (short for: If P is true, then Q is true).
5. Q if P (short for: Q is true if P is true).
6. Q because P (short for: Q is true because P is true).

All of the above say that whenever P is true, the statement Q must also be true. Therefore, P cannot be true at the same time as Q is false. This means P can only be true when Q is true. As a result of this observation, additionally, the statement $P \Rightarrow Q$ can also be read as:

7. P only if Q (short for: P is true only if Q is true).
8. Q is necessary for P (short for: Q is necessarily true when P is true).

One needs to be familiar with the various ways we read the statement $P \Rightarrow Q$ above. Even though they are all written differently, they mean the exact same thing! This confused me so much when I first started doing mathematics since English is not my first language. But once we have done a lot of mathematics and practice these in our daily life thinking consistently, it simply becomes second nature to us.

Remark 1.1.9 We make two notational remarks from the observations above.

1. In the conditional statement $P \Rightarrow Q$, we sometimes call the statement P a sufficient condition for Q and the statement Q a necessary condition for P. The statement P is also called an assumption or an antecedent whereas the statement Q is called the consequence.
2. As the name suggest, in this compound statement, the truth of one of the statements (the consequence) is conditional on the truth of the other (the antecedent).
3. We also introduce two additional shorthand symbols which we shall see from time to time: $\because$ and $\therefore$ which stand for "because" and "therefore" respectively.

Remark 1.1.10 An important note is that the truth of the statement $P \Rightarrow Q$ does not tell us anything about the truth of Q when P is false. When P is false, the statement Q could either be true or false. This is a crucial point to take note of as it is an extremely common logical pitfall! The truth table for the material implication is given as:

P	Q	$P \Rightarrow Q$
T	T	T
T	F	F
F	T	T
F	F	T

Thus, we can say that the statement $P \Rightarrow Q$ is true only when both P and Q are true or when P is false (and in this case Q could be true or false). Now let us illustrate this connective with some examples.

Example 1.1.11 Note that the parallel postulate of the Euclid's axioms in Definition 1.0.2 utilises the feature of conditional statement. The parallel postulate asserts that:

> It is true that, if a straight line falling on two straight lines make the interior angles on the same side less than two right angles, the two straight lines (produced indefinitely) intersect on that side on which the angles are less than two right angles.

To see this clearly, let us break the parallel postulate into smaller pieces. We label the constituent statements as:

P: A straight line falling on two straight lines make the interior angles on the same side less than two right angles.

Q: The two straight lines (if produced indefinitely) intersect on that side on which the angles are less than two right angles.

Thus, in short, the parallel postulate simply says "It is true that $P \Rightarrow Q$" or equivalently "$P \Rightarrow Q$ is true".

Example 1.1.12 Apart from the mathematical example in Example 1.1.11, we use the material implication connective all the time in our daily life through decision-making. Suppose that we work as a concierge at an apartment complex where Mr. X, a man who leads a clockwork life, lives in. We have two statements:

$$P : \text{It is raining,} \quad \text{and} \quad Q : \text{Mr. X will carry an umbrella.}$$

From our observation as a dedicated concierge, we can see that when it is raining, Mr. X will always carry an umbrella with him as he leaves the building. In other words, "If P is true, then Q is true". So the statement $P \Rightarrow Q$ is observed to be true. Hence the truth table for this is:

P	Q	$P \Rightarrow Q$
T	T	T
T	F	F
F	T	T
F	F	T

The shaded cells are the statements which we know to be true.

1. On Monday morning, we look out the window and see that it is raining. So P is true. Since the statement $P \Rightarrow Q$ is true, we can guarantee that Q must be true as well, namely Mr. X will come downstairs with his umbrella. This is because there is only one row in the truth table above for which both P and $P \Rightarrow Q$ are true.
2. On Tuesday morning, we notice that it sunny so we know P is false. What can we say about Mr. X based on the statement $P \Rightarrow Q$ being true?
   Nothing! We only know that if it is raining, he will carry his umbrella. But we do not know anything about his umbrella-habit in other weather. So it could go either way: he might bring his umbrella down if he prefers the shade, or he could leave his umbrella at home if he wants some vitamin D from the sunlight.
   So when P is false, Q could either be true or false, which is what we noted in Remark 1.1.10. This can be seen in the truth table above: there are two rows for which P is false and $P \Rightarrow Q$ is true, so the truth of Q could go either way.
3. On Wednesday morning, we did not look out of the window to see the weather. We saw Mr. X coming down the stairs without his usual umbrella. This means Q is false. What can we deduce?
   The statement $P \Rightarrow Q$ is true says that having P is true and Q is false at the same time is impossible. We have also observed that Q is false. So P must be false as well. This can also be seen in the final row of the truth table above. Thus it must not be raining outside. If we are basing on intuition, this is also true: if it

is indeed raining outside, Mr. X would be leaving without his umbrella, which is not his usual habit!

4. On Thursday morning, we did not look out the window to see the weather. We saw Mr. X coming down the stairs with his trusty umbrella, namely Q is true. What can we deduce about the weather?
In this case, again, nothing! There are two different rows in the truth table for which the statements $P \Rightarrow Q$ and Q are true, so the statement P could have been true or false. His purpose of carrying the umbrella with him today could be because of rain, because of the sun, because of the snow, or other reasons. Therefore, knowing that Q is true does not tell us anything about the weather outside, so the truth of statement P remains unknown in this scenario.

From Example 1.1.12(4), we can see that assuming that the statement $P \Rightarrow Q$ is true, the truth of P guarantees the truth of Q, but not vice versa. Thus, unlike the connectives "and" or "or", the material implication connective $\Rightarrow$ is not symmetric, namely the two statements $P \Rightarrow Q$ and $Q \Rightarrow P$ are not equivalent. Therefore, we have a special name for this other statement: the statement $Q \Rightarrow P$ (also written as $P \Leftarrow Q$) is called the converse implication of $P \Rightarrow Q$.

Example 1.1.13 Suppose that Γ is a polygon. Consider the following statements:

$$P : \Gamma \text{ is a square,} \quad \text{and} \quad Q : \Gamma \text{ is a rectangle.}$$

From these statements, we can create two conditional statements, namely $P \Rightarrow Q$ and $Q \Rightarrow P$.

1. The former says "If Γ is a square, then Γ is a rectangle", which is true since Γ has four sides and each interior angle is a right angle, fulfilling what a rectangle is.
2. On the other hand, the converse $Q \Rightarrow P$ says "If Γ is a rectangle, then Γ is a square". This is false since the sides of Γ may not be of equal lengths for it to be a square.

This example illustrates that for any two conditional statements $P \Rightarrow Q$ and $Q \Rightarrow P$, it may be possible that one is true whilst the other is false.

We can combine the statements $P \Rightarrow Q$ and $Q \Rightarrow P$ using the connective "and" as $(P \Rightarrow Q) \land (Q \Rightarrow P)$, written succinctly as $P \Leftrightarrow Q$. This statement is called the biconditional statement and is read as "P if and only if Q" or "P iff Q" or "P is necessary and sufficient for Q". What is special about this statement is it is true only when both P and Q have the same truth or falsehood. Indeed, we can create a truth table for both of them as follows:

P	Q	$P \Rightarrow Q$	$Q \Rightarrow P$
T	T	T	T
T	F	F	T
F	T	T	F
F	F	T	T

From the table above, we have these cases:

1. If both P and Q are true, then $P \Rightarrow Q$ and $Q \Rightarrow P$ are both true, and so the statement $(P \Rightarrow Q) \wedge (Q \Rightarrow P)$ is true.
2. Similarly, if P and Q are both false then both $P \Rightarrow Q$ and $Q \Rightarrow P$ are true. This means $(P \Rightarrow Q) \wedge (Q \Rightarrow P)$ is true.
3. Now suppose that exactly one of P or Q is true. WLOG (without loss of generality), suppose that P is true and Q is false. Then, the statement $P \Rightarrow Q$ is false while the statement $Q \Rightarrow P$ is true. Thus, the "and" statement $(P \Rightarrow Q) \wedge (Q \Rightarrow P)$ is false since only one of the constituent statements is true.

Therefore, the statement $P \Leftrightarrow Q$ is true exactly when $P \equiv Q$. Hence, this is why sometimes the statement $P \Leftrightarrow Q$ is also read as "P is equivalent to Q".

Remark 1.1.14 Note the acronym WLOG above, which stands for "without loss of generality". This acronym is useful when we work on a problem with many cases, but the cases are symmetrical. In the example above, we assumed that exactly one of P or Q is true. So we have two possible cases, namely: P is true, Q is false and P is false, Q is true.

We could study these two cases separately but, by symmetry, the study for the two cases are similar and one can be obtained from the other by swapping the labels P and Q. Therefore, both of the cases can be studied by just looking at one of them since the other one can be done in an identical manner. So we are not losing any generality by focusing at only one of the cases in detail, hence the usage of WLOG.

Modus Ponens and Modus Tollens

The series of logical thinking process that we have demonstrated in Example 1.1.12 are called *modus ponens* (Latin for "method of affirming") and *modus tollens* (Latin for "method of denying"). This depends on one (or more) conditional statement that we assume or know to be true together with an additional observation. Indeed, the truth of the compound statement $P \Rightarrow Q$ alone does not tell us anything about the truth of any one of the statements P and Q.

As an example, consider the statements:

P : The Earth is flat.

Q : The globe is not an accurate representation of the Earth.

The conditional statement $P \Rightarrow Q$ simply says if the Earth is flat, then globe is not an accurate representation of the Earth. Even if we agree that the statement $P \Rightarrow Q$ is true, this does not assert that the Earth is flat! But from this conditional statement, we do know for sure that if the Earth is indeed flat, then all the globes must not be accurate. In short, a conditional statement $P \Rightarrow Q$ can be true even if P is false or absurd. This is a very common logical pitfall for many people, so please be wary of it!

Recall the table in Remark 1.1.10: knowing $P \Rightarrow Q$ is true does not pinpoint the exact truth/falsehood of the statements P and Q since there are three possible combinations of their truth that could lead to the statement $P \Rightarrow Q$ being true. Therefore, having an extra information could help us figure this out.

Definition 1.1.15 (*Modus Ponens*, *Modus Tollens*) Let P and Q be two mathematical statements and $P \Rightarrow Q$ is a conditional statement that is assumed or known to be true. Then:

1. *Modus ponens*: If P is true, then Q is true. In symbols $((P \Rightarrow Q) \wedge P) \Rightarrow Q$.
2. *Modus tollens*: If Q is false, then P is false. In symbols $((P \Rightarrow Q) \wedge \neg Q) \Rightarrow \neg P$.

A truth table may also be used to deduce *modus ponens* and *modus tollens* above. This was done extensively in Example 1.1.12. As we have mentioned before, we use this kind of logical reasoning all the time in our lives. Here is an example:

Example 1.1.16 Let P and Q be the statements:

P : Prof. Z is unwell today, and Q : Lecture by Prof. Z today is cancelled.

At the beginning of the term, Prof. Z announced that if he is feeling unwell on a particular day, his lecture on that day will be cancelled. So we know that the statement $P \Rightarrow Q$ is true axiomatically.

1. *Modus ponens* tells us that if P is true, namely Prof. Z is unwell today, then his lecture today is cancelled (Q is true).
2. *Modus tollens* tells us that if Q is false, namely the lecture by Prof. Z is running today, then Prof. Z is feeling well today (P is false).

Remark 1.1.17 On the other hand, if we know that the statement $P \Rightarrow Q$ is false, from its truth table, we can immediately infer that P is true and Q is false since there is only one row that has $P \Rightarrow Q$ as false.

Via *modus tollens* in Definition 1.1.15, if $P \Rightarrow Q$ is true and $\neg Q$ is true, then $\neg P$ is true. Note also that Q cannot be false when P is true, which means $\neg Q$ cannot be true at the same time as $\neg P$ is false. These two observations tell us that the conditional statement $\neg Q \Rightarrow \neg P$ is also true. This statement is called the contrapositive of $P \Rightarrow Q$ and they are in fact equivalent statements. Indeed, using the truth table, we can see that the shaded columns are identical:

P	Q	$P \Rightarrow Q$	$\neg Q$	$\neg P$	$\neg Q \Rightarrow \neg P$
T	T	T	F	F	T
T	F	F	T	F	F
F	T	T	F	T	T
F	F	T	T	T	T

Example 1.1.18 From Example 1.1.16, these following two material implications, which are contrapositives of each other, are the exact same statement:

Prof. Z is unwell today $\Rightarrow$ Lecture by Prof. Z today is cancelled.

Lecture by Prof. Z today is not cancelled $\Rightarrow$ Prof. Z is feeling well today.

Another useful observation is that the statement $P \Rightarrow Q$ is logically equivalent to the statement $\neg P \vee Q$. We shall prove this in Exercise 1.5 by using truth tables. In fact, with this identity, by using the fact that the connective "or" is symmetric and double negation, we have:

$$(P \Rightarrow Q) \equiv \neg P \vee Q \equiv Q \vee \neg P \equiv \neg(\neg Q) \vee \neg P \equiv (\neg Q \Rightarrow \neg P),$$

which is exactly what we claimed above regarding contrapositives being equivalent to each other.

Another equivalence of statements that we can obtain is the follows:

$$\neg(P \Rightarrow Q) \equiv \neg(\neg P \vee Q) \equiv P \wedge (\neg Q), \tag{1.1}$$

which is useful if we want to disprove a conditional statement.

To summarise this section:

Definition 1.1.19 Let P and Q be mathematical statements. We have:

1. Negation: $\neg P$.
2. And (conjunction): $P \wedge Q$.
3. Or (disjunction): $P \vee Q$.
4. Material implication/conditional statement: $P \Rightarrow Q$. This is equivalent to $\neg P \vee Q$.
5. Converse implication: the converse to $P \Rightarrow Q$ is $Q \Rightarrow P$.
6. Biconditional/equivalence: $P \Leftrightarrow Q$. This is equivalent to $(P \Rightarrow Q) \wedge (Q \Rightarrow P)$.
7. Contrapositive: the contrapositive to $P \Rightarrow Q$ is $\neg Q \Rightarrow \neg P$. This is equivalent to $P \Rightarrow Q$.

1.2 Proofs

As we have mentioned earlier, these mathematical statements are important in mathematics and we encounter them all the time. They could be some declarative sentences or assertions such as "Squares are polygons" or conditional statements such as "If Γ is a square, then Γ is a rectangle". Some of these statements are simply declared to be true as axioms: for example the five statements in Euclid's axioms.

But other times, mathematicians want to prove the truth or falsehood of some other more complicated statements. To start with, these statements are simply observations, guesses, or claims, which mathematicians call conjectures.

A conjecture could either be true or false. By a series of logical reasoning, only the truth or falsehood of the statement could be established and the conjecture becomes a mathematical result. Once this has been proven to be true (correctly and rigorously), its truth remains so for eternity. This process can be encapsulated in a quote attributed to Augustin-Louis Cauchy (1789–1857).

> First, it is necessary to study the facts, to multiply the number of observations, and then later to search for formulas that connect them so as thus to discern the particular laws governing a certain class of phenomena. In general, it is not until after these particular laws have been established that one can expect to discover and articulate the more general laws that complete theories by bringing a multitude of apparently very diverse phenomena together under a single governing principle.

Now let us demonstrate this process with a rather simple example.

The contrapositive to the parallel postulate in Euclid's axioms says that the sum of the interior angles to one side of a line transversal to a pair of parallel lines is exactly equal to two right angles. From Fig. 1.2, if CD is parallel to AB, this says $\angle BXF + \angle DYE$ is equal to two right angles or a straight line. So we expect that $\angle BXF = \angle CYE$.

We can put this guess as a conjecture:

Conjecture 1.2.1 Let AB, CD, and EF be three lines such that EF intersects the lines AB and CD at the points X and Y respectively. If AB is parallel to CD, then $\angle BXF = \angle CYE$.

The conjecture above is a conditional statement of the form $P \Rightarrow Q$ where:

$$P : AB \text{ is parallel to } CD, \quad \text{and} \quad Q : \angle BXF \text{ is equal to } \angle CYE.$$

Fig. 1.2 The configuration for Conjecture 1.2.1. The arrows denote that the lines CD and AB are parallel

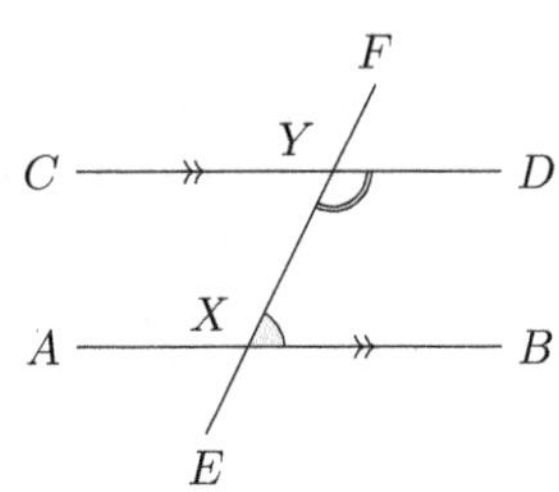

We want to decide whether the proposed statement $P \Rightarrow Q$ is true or false. In above, we have guessed that it is true, so let us aim to prove that it is indeed true. There are three ways we could do this:

1. We prove that when P is true, Q must be true as well directly.
   We start with assuming that P is true, namely AB is parallel to CD. Using the parallel postulate, we know that the angles $\angle DYE$ and $\angle BXF$ add up to two right angles. Moreover, the angles $\angle DYE$ and $\angle CYE$ also add up to two right angles (or a straight line).
   Since these two quantities are the same, they must be equal. Moreover, since they have a common angle $\angle DYE$, the remaining parts must be equal, namely $\angle BXF = \angle CYE$. This means Q is true. Therefore, we have proven that when P is true, Q must be true. So $P \Rightarrow Q$ is true.
2. Another way to do this is by using contrapositive. We note from Definition 1.1.19 that the statement $P \Rightarrow Q$ is equivalent to the contrapositive statement $\neg Q \Rightarrow \neg P$. Since they are equivalent, proving $P \Rightarrow Q$ is true is the same as proving $\neg Q \Rightarrow \neg P$ is true. So let us aim for this second goal instead.
   The negations of Q and P are:
   $$\neg Q : \angle BXF \text{ is not equal to } \angle CYE, \quad \text{and} \quad \neg P : AB \text{ is not parallel to } CD.$$
   Now we assume that $\neg Q$ is true, namely $\angle BXF$ is not equal to $\angle CYE$. We note that $\angle CYE$ and $\angle DYE$ add up to two right angles. Since $\angle CYE$ is not equal to $\angle BXF$, the angles $\angle BXF$ and $\angle DYE$ cannot add up to two right angles or otherwise $\angle BXF = \angle CYE$ would be true.
   Therefore the angles must add up to either an angle smaller or bigger than two right angles. Either way, by the parallel postulate, if we extend the lines AB and CD, they would intersect somewhere. Thus, AB and CD are not parallel to each other and hence $\neg P$ is true. This means $\neg Q \Rightarrow \neg P$ is true, which is equivalent to $P \Rightarrow Q$ being true.
3. Note also that from Definition 1.1.19 the statement $P \Rightarrow Q$ is equivalent to $\neg P \vee Q$. Thus, showing the statement $P \Rightarrow Q$ is true is equivalent to showing that $\neg P \vee Q$ is true. The latter is the same as showing its negation $\neg(\neg P \vee Q) \equiv P \wedge (\neg Q)$ is not true. To prove this, we assume on the contrary that $P \wedge (\neg Q)$ is true and get an absurd statement.
   Since $P \wedge (\neg Q)$ is assumed to be true, then both of the following statements are true:
   $$P : AB \text{ is parallel to } CD, \quad \text{and} \quad \neg Q : \angle BXF \text{ is not equal to } \angle CYE.$$
   Using the parallel postulate, since P is true, the angles $\angle DYE$ and $\angle BXF$ add up to two right angles. Moreover, the angles $\angle DYE$ and $\angle CYE$ also add up to two right angles. Hence, we have $\angle DYE + \angle BXF = \angle DYE + \angle CYE$. Since they have a common angle $\angle DYE$, the remaining parts must be equal, namely $\angle BXF = \angle CYE$.

On the other hand $\neg Q$ is also true, namely $\angle BXF \neq \angle CYE$. Putting the two equations together, we have $\angle BXF = \angle CYE \neq \angle BXF$, which is absurd! Thus $P \wedge (\neg Q)$ cannot be true or otherwise we end up with an impossible statement. As a result, we conclude that $P \Rightarrow Q$ must be true.

The conjecture has been proven to be true in three different ways above. Since its truth is now established, we can now upgrade Conjecture 1.2.1 by calling it a proposition or a theorem as such:

Proposition 1.2.2 *Let AB, CD, and EF be three lines such that EF intersects the lines AB and CD at the points X and Y respectively. If AB is parallel to CD, then $\angle BXF = \angle CYE$.*

Remark 1.2.3 We make some remarks here regarding the proposition and its proof.

1. Of course, we only need one valid proof to establish the truth of the proposition. So writing down all three, in practice, is not necessary. However, it is always good to have many different kinds of proofs as different proofs would give different insights to a mathematical phenomenon. In the words of Michael Atiyah (1929–2019):

   > I think it is said that Gauss had ten different proofs for the law of quadratic reciprocity. Any good theorem should have several proofs, the more the better. For two reasons: usually, different proofs have different strengths and weaknesses, and they generalise in different directions - they are not just repetitions of each other.

2. A lemma, proposition, or theorem is usually written in the form "P" or "$P \Rightarrow Q$" (or some other compound statement). The declaration that it is a lemma, proposition, or theorem asserts that it has been proven to be true, so they are to be read as "P is true" or "$P \Rightarrow Q$ is true" respectively.
3. However, mathematicians are usually very skeptical people: they always insist on seeing the proof of a statement for themselves to check its validity! This is usually a very good practice in mathematics and daily life.

Since Proposition 1.2.2 has been proven to be true, we can use them as we like in the future. This is how mathematics is built: from a set of basic axioms, we prove the truth of more statements using a chain of lemmas, propositions, theorems, and logical arguments. As long as the foundations of the framework (namely the axioms) are accepted to be true, the truth of propositions which are built on top of them remain true. As we have mentioned in Remark 1.1.1, their "mathematical truth" are based on the accepted axioms. In the words on Charles Steinmetz (1865–1923):

> Mathematics is the most exact science, and its conclusions are capable of absolute proof. But this is so only because mathematics does not attempt to draw absolute conclusions. All mathematical truths are relative, conditional.

The three strategies for proofs we have shown for Proposition 1.2.2 are some of the common strategies to prove a conditional statement of the form $P \Rightarrow Q$. For some proofs, one of the methods might be easier to follow than another. Therefore it is important to think of a strategy and approach to use when attempting to prove a result.

The three different approaches above come from the different but equivalent ways of writing the conditional statement, namely $P \Rightarrow Q$, $\neg Q \Rightarrow \neg P$, and $\neg P \vee Q$. They are called:

Definition 1.2.4 Suppose that P and Q are mathematical statements and we want to prove that the conditional statement $P \Rightarrow Q$ is true.

1. Direct proof: Assume P is true. Show Q must be true as well.
2. Contrapositive: Show instead $\neg Q \Rightarrow \neg P$ is true. To do this, assume that $\neg Q$ is true. Then show that $\neg P$ must be true as well.
3. Contradiction or *reductio ad absurdum*: Show instead $\neg P \vee Q$ is true. This is the same as showing that $\neg(\neg P \vee Q)$, or equivalently $P \wedge (\neg Q)$, is false. To show this, assume P and $\neg Q$ are both true. Then, deduce a false, contradictory, absurd, or impossible statement.

Remark 1.2.5 We make several remarks regarding the method of proofs above.

1. To prove a statement of the form "$P \Rightarrow Q$ is true", we do not have to worry about the case when P is false. This is because when P is false, Q is allowed to be either true or false for $P \Rightarrow Q$ to be true, as mentioned in Remark 1.1.10 and in the truth table for $P \Rightarrow Q$. Thus we only have to worry about the situation when P is true, for which we need to establish that Q must be true as well.
2. The phrase *reductio ad absurdum* is Latin for "reduce to absurdity". This is essentially the aim of proof by contradiction: we assume that the statement P is true whilst Q is false and proceed to reduce these two assumptions together to an absurd or impossible scenario. Hence, P being true and Q being false cannot occur simultaneously.
   We thus conclude that if P is true, necessarily Q is also true (and hence $P \Rightarrow Q$ is true). As the detective Sherlock Holmes said in the novel *The Sign of the Four* by Arthur Conan Doyle:

   > When you have eliminated the impossible, whatever remains, however improbable, must be the truth.

3. G.H. Hardy (1877–1947) also mentioned how powerful this method is in an excerpt from his book *A Mathematician's Apology*:

   > *Reductio ad absurdum* ... is one of a mathematician's finest weapons. It is a far finer gambit than any chess gambit: a chess player may offer the sacrifice of a pawn or even a piece, but a mathematician offers the game.

4. When proving a statement via contradiction, one may see the following notations used to signify that one has reached a contradiction: $\lightning$, $\divideontimes$, $\Rightarrow\Leftarrow$, or $\perp$.

Sometimes, we may have to prove that a biconditional statement $P \Leftrightarrow Q$ is true. To do this, we first note that, by definition, this is just the compound statement $(P \Rightarrow Q) \wedge (Q \Rightarrow P)$. So to prove its truth, we have to establish that both the statements $P \Rightarrow Q$ and $Q \Rightarrow P$ are true. We could prove them separately using the approaches in Definition 1.2.4 or, in some cases, simultaneously in one go. Let us show an example of this to prove the following proposition.

Proposition 1.2.6 *Let Γ be a polygon. Γ is a triangle if and only if the sum of the interior angles of Γ is equal to two right angles.*

The statements in this proposition are:

$$P : \Gamma \text{ is a triangle.}$$
$$Q : \text{Sum of the interior angles of } \Gamma \text{ is equal to two right angles.}$$

We want to prove that $P \Leftrightarrow Q$ is true. We shall prove the forward implication $P \Rightarrow Q$ using direct proof and the converse implication $P \Leftarrow Q$ using contradiction.

Proof We prove the implications separately.

($\Rightarrow$): Assume P is true, namely Γ is a triangle with vertices ABC. Draw a line DE parallel to BC at A so that D is on the same side of the triangle as B and E is on the same side of the triangle as C. Using Proposition 1.2.2, we have $\angle CBA = \angle BAD$ and $\angle BCA = \angle CAE$. Refer to Fig. 1.3 for this configuration.
Then, the sum of the angles in the triangle is $\angle BAC + \angle CBA + \angle BCA = \angle BAC + \angle BAD + \angle CAE$. The RHS (right-hand side) is equal to two right angles since the points D, A, and E lie on a straight line. Thus, Q is true.

($\Leftarrow$): We want to prove $Q \Rightarrow P$. Aiming for a proof by contradiction, we assume that Q and $\neg P$ are both true. Namely, assume that the sum of the interior angles of a polygon Γ is equal to two right angles and the polygon Γ is not a triangle.

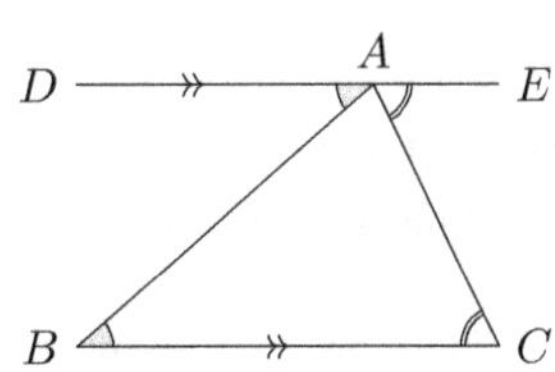

Fig. 1.3 Diagram for the proof of the forward implication in Proposition 1.2.6

Fig. 1.4 The polygon Γ is split into two smaller polygons Γ' and Γ''

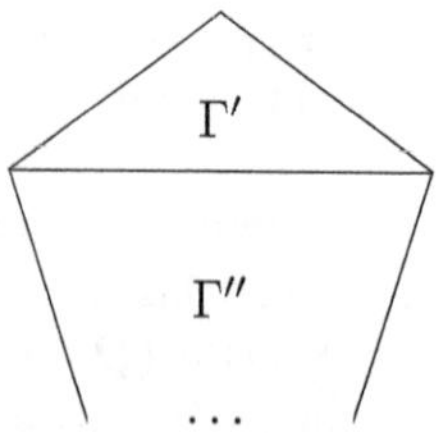

The latter means Γ has at least 4 vertices. Pick any three consecutive vertices of the polygon Γ and join them up to form a triangle Γ' and another polygon Γ'' with at least three sides. This can be seen in Fig. 1.4.
From the forward implication that we have proven, we know that the sum of the angles in the triangle Γ' is exactly equal to two right angles. This is already equal to the sum of all the angles in Γ by assumption. Therefore the sum of the angles in polygon Γ'' must be 0, an absurd statement. This provides us with the desired contradiction. □

Remark 1.2.7 Let us make some remarks from the proof above:

1. Note the acronym RHS in the proof. This is short for right-hand side. Likewise, we write LHS for left-hand side.
2. We usually draw □, called the Halmos square after Pál Halmos, to signify that a proof has ended. Sometimes the acronym QED (short for *quod erat demonstrandum*, which is "what was to be shown" in Latin) is also used to denote this.

With some basic ideas on logical statements and proofs, we now look at the most fundamental objects in mathematics, which are sets.

1.3 Sets

In this book, we are going to work with what modern mathematicians call naïve set theory as opposed to axiomatic set theory. In the naïve school of thought, a set is simply a collection of objects. Nothing more, nothing less. The members of a set are called elements or points of a set. We write this as:

Definition 1.3.1 (Set Membership) If x is an element of the set X, we write $x \in X$. Otherwise, if x is not an element of the set X, we write $x \notin X$.

There are various ways to describe sets. The easiest way to write down a set is by listing the elements one by one. This way of describing a set is called a roster notation. In this notation, the elements of a set are listed one by one with a comma separating them and enclosed in curly brackets.

Example 1.3.2 For the set A of integers between 1 and 3 inclusive, we write $A = \{1, 2, 3\}$. The number 1 is an element of A, so we write $1 \in A$ to denote this. The number 4 is not a member of the set A, so we write $4 \notin A$.

However, some sets have too many elements to be listed this way, so we use the set builder notation. As an example, the set B of natural numbers greater than or equal to 10 cannot be listed down in the roster notation since there are too many of them. In this case we may use the set builder notation $B = \{x \in \mathbb{N} : x \geq 10\}$ which is read as "the set of natural numbers x such that x is bigger than or equal to 10". The colon : symbol in the set builder notation is read as "such that". In other literature, a vertical bar | symbol may be used in place of the : symbol.

Example 1.3.3 Let us look at some examples on how to use the set builder notation.

1. The set A in Example 1.3.2 can also be written in set builder notation as $A = \{x \in \mathbb{N} : 1 \leq x \leq 3\}$. This is read as "The set of natural numbers x such that x is bigger than or equal to 1 and smaller than or equal to 3".
2. The set builder notation is also useful to specify the defining property of the elements of the set. For example, the set of even natural numbers C can be written as $C = \{x \in \mathbb{N} : x \text{ is even}\} = \{2x : x \in \mathbb{N}\}$.
3. Moreover, if the elements in a set is indexed by some indexing label, we can also write the indexing set as a subscript. For example, the set C above can be written as $C = \{x \in \mathbb{N} : x \text{ is even}\} = \{2x\}_{x \in \mathbb{N}}$.
4. More generally, given a statement $P(x)$ whose truth depends on $x \in X$, we can build sets of elements for which P is true or false, namely $\{x \in X : P(x) \text{ is true}\}$ or $\{x \in X : P(x) \text{ is false}\}$.
5. The study of naïve set theory can seem very elementary, but sometimes one might arrive at contradictory statements or impossible constructions. One of these contradictory statements is due to Bertrand Russell (1872–1970) via the construction of the following set: S is the set of all sets that do not contain themselves, namely $S = \{s : s \notin s\}$.
   Upon inspecting this seemingly innocent and naïvely constructed set, we can arrive at the conclusion $S \in S \Leftrightarrow S \notin S$, a paradox! A more concrete interpretation of this is given by the following question, which is called the barber paradox:

   > You can define the barber as 'one who shaves all those, and those only, who do not shave themselves'. The question is, does the barber shave himself?

   Due to these possible contradictions, philosophers and mathematicians spend a lot of time pondering on such paradoxes to fill in the gaps and formalise the study of set theory. This leads to the birth of axiomatic set theory in the late 19th century. Via these theories, mathematicians declare what set constructions are allowed axiomatically so that no paradoxes or contradictions shall arise.

Remark 1.3.4 A set can also be void of any elements. We call this set the empty set (or in some literature, null set) and denote it as $\{\}$ or $\varnothing$.

If we have two sets and one of the sets also has all the elements of the other set, we call the former set the subset and the latter set the superset.

Definition 1.3.5 (Subset, Superset) Let X and Y be sets such that for every $x \in X$, we have $x \in Y$. Then the set X is called a subset of Y and the set Y is called a superset of X. This is denoted as $X \subseteq Y$.

If $X \subseteq Y$, we also say that "The set X is contained in the set Y" or "The set Y contains the set X". Trivially, any set is a subset of itself by definition, so we also have $X \subseteq X$. Another important convention is that the empty set is always the subset of any set, namely $\varnothing \subseteq X$ for any set X. This is tautologically true because every element in $\varnothing$ (which is nothing) is also in the set X.

Example 1.3.6 The elements of the set $A = \{1, 2, 3\}$ are also contained in the set $D = \{x \in \mathbb{N} : x \geq 1\}$, so the set A is a subset of D and the set D is a superset of A. We also say that the set A is contained in the set D and the set D contains the set A. We write this symbolically as $A \subseteq D$.

Two sets are the same set or equal if every element of the first set is in the other set and vice versa. In axiomatic set theory, this is called the axiom of extensionality. We define:

Definition 1.3.7 (Equality of Sets) Two sets X and Y are called equal if $X \subseteq Y$ and $Y \subseteq X$. In other words, they have exactly the same elements. We denote this as $X = Y$.

Example 1.3.8 Consider the sets $A = \{1, 2, 3\}$ and $E = \{1, 1, 2, 3\}$. It seems like the set E has more elements than the set A. However, they are exactly the same set. Indeed:

1. For every $x \in A$ (so x could either be $1, 2$, or 3) we have $x \in E$. This implies $A \subseteq E$.
2. On the other hand, for every $y \in E$ (so y could either be $1, 2$, or 3) we have $y \in A$. Thus we have the inclusion $E \subseteq A$.

Hence by Definition 1.3.7, we have the equality $E = A$, namely $\{1, 1, 2, 3\} = \{1, 2, 3\}$. This means that duplicates of the same element in a set notation are counted only once.

Remark 1.3.9 Let us introduce some more notations.

1. Sometimes it may be useful to have the notation $X \subsetneq Y$ to mean that X is a subset of Y but X is not equal to Y. In other words, there is an element $y \in Y$ which is not in X. In this case, the set X is called a proper subset of Y. From Example 1.3.6, we have $A \subsetneq D$ since there are other elements of D which are not in A.
2. Another notation that we usually use is $X \nsubseteq Y$, which is short for X is not a subset and not equal to Y. This means that there are elements of X which are not in Y. In other words, $X \nsubseteq Y$ is the negation to the statement $X \subseteq Y$.

Set Algebra

The biggest set that contains all the possible elements that one is considering is called the universe, usually denoted as U. This is usually declared beforehand to avoid ambiguity.

Example 1.3.10 For example, if we declare our universe U to be the natural numbers, namely $U = \mathbb{N}$, then the set $D = \{\text{red, white, blue}\}$ is not a well-defined object since none of the elements in D is contained in our universe. However, the set $A = \{1, 2, 3\}$ is a well-defined object in our universe $U = \mathbb{N}$.

Remark 1.3.11 In fact, we have been using the concept of universe rather unconsciously before: recall the set builder notation that we used as a way to describe a set. Using this notation, in Example 1.3.6 we have constructed a set $D = \{x \in \mathbb{N} : x \geq 1\}$. The implicit universe in this construction is $\mathbb{N}$ and D is a subset of it.

In axiomatic set theory, one of the set axioms imply that the set builder notation can only be used for the construction of subsets of a larger set/universe. This is one of the most important axioms of set theory as it ensures that the construction of Russell's set in Example 1.3.3(5) is forbidden.

The universe is an important object required for the following definition:

Definition 1.3.12 (Complement) Let U be a universe and $X \subseteq U$. The complement of X in U, denoted as X^c, is the set of all elements in U that is not contained in X. Namely:

$$X^c = \{x \in U : x \notin X\}.$$

The complement of a set X in the universe U can be represented as the Venn diagram, named after John Venn (1834–1923), in Fig. 1.5. Essentially a Venn diagram shows the logical relationship between some collection of sets which can be useful as a visual aid.

Sometimes, if the universe is made known before, we simply call X^c the complement of X without referring to the universe. Clearly have $(X^c)^c = X$. We also have the following:

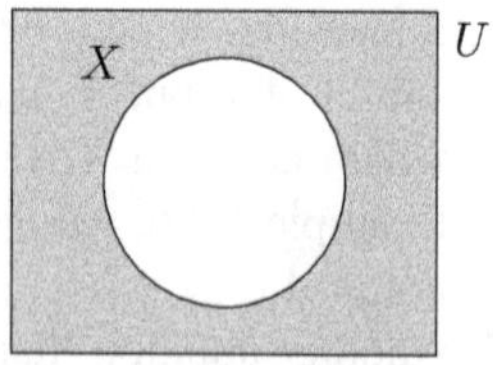

Fig. 1.5 If the disc is the set X, the shaded region is X^c

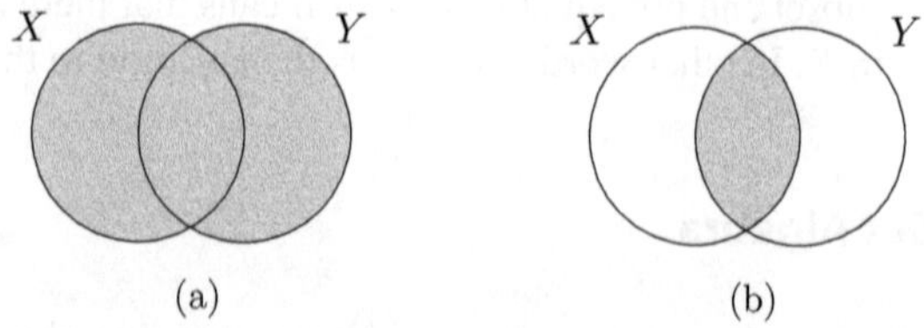

Fig. 1.6 The shaded region is $X \cup Y$ and $X \cap Y$ respectively. (**a**) $X \cup Y$. (**b**) $X \cap Y$

Proposition 1.3.13 *If X and Y are sets in a universe U with $X \subseteq Y$, then $Y^c \subseteq X^c$.*

Proof To show $Y^c \subseteq X^c$, pick any element $y \in Y^c$. By definition, $y \notin Y$. Since $X \subseteq Y$, we must also have $y \notin X$. This means $y \in X^c$. Since y is an arbitrarily chosen element in Y^c, all of the elements in Y^c must be in X^c and we conclude that $Y^c \subseteq X^c$. □

The next operations we are going to introduce are called union and intersection. As the names suggest, the union of two sets is the set obtained by combining the elements of the two sets together and the intersection is the collection of common elements from both sets.

Definition 1.3.14 (Union, Intersection) Let X and Y be sets.

1. The union or join of the sets X and Y, denoted $X \cup Y$, is the set of all elements that are members of either X or Y. Namely $X \cup Y = \{x : x \in X \text{ or } x \in Y\}$.
2. The intersection or meet of the sets X and Y, denoted $X \cap Y$, is the set of all elements that are members of both X and Y. Namely $X \cap Y = \{x : x \in X \text{ and } x \in Y\}$.

In other words, the union $X \cup Y$ is the smallest superset containing both X and Y and the intersection $X \cap Y$ is the largest subset contained in both X and Y. We can represent the sets $X \cup Y$ and $X \cap Y$ via the Venn diagrams in Fig. 1.6.

From the definition, the set operations $\cup$ and $\cap$ are symmetric, meaning that $X \cup Y = Y \cup X$ and $X \cap Y = Y \cap X$. Unions and intersections are also associative by unpacking their definitions. This means if we have three sets X, Y, and Z, then:

$$(X \cup Y) \cup Z = X \cup (Y \cup Z) \quad \text{and} \quad (X \cap Y) \cap Z = X \cap (Y \cap Z). \tag{1.2}$$

This can also be seen by drawing their respective Venn diagrams (however, one must be aware that "picture proofs" may not be regarded as a valid mathematical proof to some people!). Thus, by notational convention, we can remove the brackets in each of the set Eqs. (1.2) and denote them as $X \cup Y \cup Z$ and $X \cap Y \cap Z$ unambiguously.

Furthermore, we also have the following rules, which we leave as Exercise 1.19:

Proposition 1.3.15 *Let X and Y be sets in a universe U.*

1. $X \cap X^c = \varnothing$.
2. *Idempotent laws:* $X \cap X = X$ *and* $X \cup X = X$.
3. $X \cup U = U$ *and* $X \cap \varnothing = \varnothing$.
4. *Absorption laws:* $X \cup (X \cap Y) = X$ *and* $X \cap (X \cup Y) = X$.
5. $X \cup Y = U$ *and* $X \cap Y = \varnothing$ *if and only if* $X = Y^c$.

We can also extend the definition of unions and intersections to higher number of sets involved. We define:

Definition 1.3.16 (Indexed Union, Intersection) Let X_j be sets where j is contained in an indexing set J.

1. The union of the sets X_j, denoted $\bigcup_{j \in J} X_j$, is the set of all elements that are members of at least one of the X_j. Namely:

$$\bigcup_{j \in J} X_j = \{x : x \in X_j \text{ for at least one } j \in J\}.$$

2. The intersection of the sets X_j, denoted $\bigcap_{j \in J} X_j$, is the set of all elements that are members of all of the X_j. Namely:

$$\bigcap_{j \in J} X_j = \{x : x \in X_j \text{ for all } j \in J\}.$$

The indexing set J is also called the parameter set. We note that the indexing sets in the definitions above J could either be finite or infinite. If J is finite, say we are working with n sets, namely $X_1, X_2, \ldots, X_n$, we write their union and intersection respectively as:

$$\bigcup_{j=1}^{n} X_j = \{x : x \in X_j \text{ for at least one } j = 1, 2, \ldots, n\},$$

$$\bigcap_{j=1}^{n} X_j = \{x : x \in X_j \text{ for all } j = 1, 2, \ldots, n\},$$

where the former is the set of elements which are in at least one of the X_j's and the latter is the set of elements which are in all of the X_j's.

The union and intersection operations interact with each other via a distributive property, namely:

Proposition 1.3.17 *For any sets X, Y, and Z, we have:*

1. $X \cup (Y \cup Z) = (X \cup Y) \cup (X \cup Z)$.
2. $X \cap (Y \cap Z) = (X \cap Y) \cap (X \cap Z)$.
3. $X \cup (Y \cap Z) = (X \cup Y) \cap (X \cup Z)$.
4. $X \cap (Y \cup Z) = (X \cap Y) \cup (X \cap Z)$.

Proof We shall prove the third assertion only as the others can be done in a similar way.

3. ($\subseteq$): Pick any element $x \in X \cup (Y \cap Z)$. This means $x \in X$ or $x \in Y \cap Z$. Then, we have two possibilities:
   (a) If $x \in X$, then $x \in X \cup Y$ and $x \in X \cup Z$.
   (b) Otherwise, if $x \in Y \cap Z$, this means $x \in Y$ and $x \in Z$. Thus, $x \in Y \cup X$ and $x \in Z \cup X$.
   Either way, we have $x \in (X \cup Y) \cap (X \cup Z)$. Since x was arbitrarily chosen, this means $X \cup (Y \cap Z) \subseteq (X \cup Y) \cap (X \cup Z)$.

   ($\supseteq$): To show the reverse inclusion, we pick any element $x \in (X \cup Y) \cap (X \cup Z)$. Then, $x \in X \cup Y$ and $x \in X \cup Z$. From this, we have two cases:
   (a) $x \in X$.
   (b) Otherwise, if $x \notin X$, necessarily $x \in Y$ and $x \in Z$. Therefore, $x \in Y \cap Z$.
   Either way, we have $x \in X \cup (Y \cap Z)$. This shows the reverse inclusion.

Putting the two inclusions together, we obtain the equality of sets. □

Remark 1.3.18 To prove the equality of sets in Proposition 1.3.17 above, we used Definition 1.3.7 by showing the inclusions in both directions. This method is called the double inclusion method.

To relate the complements, unions, and intersection of sets, we have De Morgan's laws, which was named after Augustus De Morgan (1806–1871).

Theorem 1.3.19 (De Morgan's Laws) *Let X and Y be sets contained in the universe U. Then:*

1. $(X \cup Y)^c = X^c \cap Y^c$.
2. $(X \cap Y)^c = X^c \cup Y^c$.

Proof We prove the first assertion only as the second can be done in a similar way.

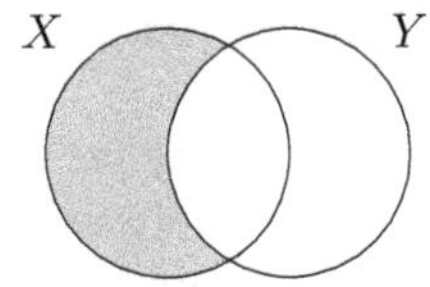

Fig. 1.7 The shaded region is $X \setminus Y$

1. Note that we have the following equivalences:

$$\begin{aligned} x \in X^c \cap Y^c \quad &\Leftrightarrow \quad x \in X^c \text{ and } x \in Y^c \quad \Leftrightarrow \quad x \notin X \text{ and } x \notin Y \\ &\Leftrightarrow \quad x \notin X \cup Y \\ &\Leftrightarrow \quad x \in (X \cup Y)^c. \end{aligned}$$

Going forward we have $X^c \cap Y^c \subseteq (X \cup Y)^c$ and going backwards we have $(X \cup Y)^c \subseteq X^c \cap Y^c$. Putting the two inclusions together, we get the equality $(X \cup Y)^c = X^c \cap Y^c$. □

The De Morgan's laws can be easily extended to an arbitrary number of sets using Definition 1.3.16. The proof is similar to the case of two sets above.

Another set operation that we can define is the set difference (Fig. 1.7):

Definition 1.3.20 (Difference) Let X and Y be sets. The difference of the sets X and Y, denoted $X \setminus Y$, is the set of all elements that are members of X but not Y. Namely:

$$X \setminus Y = \{x : x \in X \text{ and } x \notin Y\} = X \cap Y^c.$$

Thus the complement operation can also be written as a set difference. If U is the universe and X is contained in U, then we have the equality $X^c = U \cap X^c = U \setminus X$.

Furthermore, if U is the universe that contains the sets X and Y, by definition alone, the set difference can be written in terms of complement and intersection as $X \setminus Y = X \cap Y^c = X \cap (U \setminus Y)$. By using De Morgan's laws and distributivity of set operations, we can prove the following set identities:

Proposition 1.3.21 *Let U be a universe that contains the sets X, Y, and Z. Then:*

1. $X \cap Y = X \setminus (X \setminus Y)$.
2. $Z \setminus (X \cup Y) = (Z \setminus X) \cap (Z \setminus Y)$.
3. $Z \setminus (X \cap Y) = (Z \setminus X) \cup (Z \setminus Y)$.
4. $(X \setminus Y) \cap Z = (X \cap Z) \setminus Y = X \cap (Z \setminus Y) = (X \cap Z) \setminus (Y \cap Z)$.
5. $(X \setminus Y) \cup Z = (X \cup Z) \setminus (Y \setminus Z)$.
6. $Z \setminus (X \setminus Y) = (Y \cap Z) \cup (Z \setminus X)$.
7. $(Z \setminus X) \setminus Y = Z \setminus (X \cup Y)$.

Proof We prove assertions 1, 2, and 4 only. The others are left as Exercise 1.23.

1. We can use the method of double inclusion here. However, it is easier to show this equality using set algebra now that we have seen many useful results on these operations. By using definitions, De Morgan's laws, and distributivity of set operations, we have:

$$X \setminus (X \setminus Y) = X \setminus (X \cap Y^c) = X \cap (X \cap Y^c)^c = X \cap (X^c \cup Y)$$
$$= (X \cap X^c) \cap (X \cap Y) = X \cap Y.$$

2. By using definitions and De Morgan's laws, we have:

$$Z \setminus (X \cup Y) = Z \cap (X \cup Y)^c = Z \cap (X^c \cap Y^c) = (Z \cap X^c) \cap (Z \cap Y^c)$$
$$= (Z \setminus X) \cap (Z \setminus Y).$$

4. Using set algebra, by symmetry of intersections, we have:

$$(X \setminus Y) \cap Z = (X \cap Y^c) \cap Z = X \cap Z \cap Y^c, \tag{1.3}$$

which is both $(X \cap Z) \cap Y^c = (X \cap Z) \setminus Y$ and $X \cap (Z \cap Y^c) = X \cap (Z \setminus Y)$.
Finally, $(X \cap Z) \setminus (Y \cap Z) = (X \cap Z) \cap (Y \cap Z)^c = (X \cap Z) \cap (Y^c \cup Z^c) = (X \cap Z \cap Y^c) \cup (X \cap Z \cap Z^c) = X \cap Z \cap Y^c$, which is also (1.3). □

We now define symmetric difference operation of two sets as:

Definition 1.3.22 (Symmetric Difference) Let X and Y be sets. The symmetric difference of the sets X and Y, denoted $X \triangle Y$, is the set of all elements that are members of X or Y but not both. Namely:

$$X \triangle Y = (X \cup Y) \setminus (X \cap Y) = (X \setminus Y) \cup (Y \setminus X).$$

The Venn diagram for $X \triangle Y$ is given in Fig. 1.8 below.

Similar to unions and intersections, the symmetric difference is symmetric and associative, namely for sets X, Y, and Z, we have $X \triangle Y = Y \triangle X$ and $(X \triangle Y) \triangle Z = X \triangle (Y \triangle Z)$. Proving these are left as an exercise to the readers in Exercise 1.21.

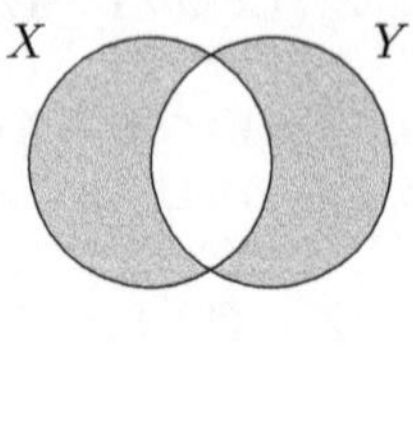

Fig. 1.8 The shaded region is $X \triangle Y$

Power Sets and Cartesian Product

Finally, from old sets, we can create new ones as well. From a given set, we can create a set of all its subsets. This is called a power set:

Definition 1.3.23 (Power Set) Let X be a set. The power set of X, denoted as 2^X or $\mathcal{P}(X)$, is the set of all subsets of X, including $\varnothing$ and X itself.

Example 1.3.24 Recall the set $A = \{1, 2, 3\}$. The power set of A is given by:

$$\mathcal{P}(A) = \{\varnothing, \{1\}, \{2\}, \{3\}, \{1, 2\}, \{1, 3\}, \{2, 3\}, \{1, 2, 3\}\}.$$

Here we note a distinction that $\{1\} \in \mathcal{P}(A)$ but $1 \notin \mathcal{P}(A)$. The former means the set containing 1 as its element and the latter is simply the element 1, so they are two completely different objects.

Next, we define the Cartesian product (named after René Descartes) of two sets as follows:

Definition 1.3.25 (Cartesian Product) Let X and Y be sets. The Cartesian product of the sets X and Y, denoted $X \times Y$, is a set defined as the set of ordered pairs of elements in X and Y. Namely $X \times Y = \{(x, y) : x \in X, y \in Y\}$.

We note that from this definition, it is clear that the Cartesian product is not commutative, namely $X \times Y \neq Y \times X$ in general. Moreover, if one of the sets X or Y is empty, the Cartesian product would also be empty.

Proposition 1.3.26 *Let X and Y be sets. Suppose that $A, B \subseteq X$ and $C, D \subseteq Y$. Then:*

1. $A \times (C \cap D) = (A \times C) \cap (A \times D)$.
2. $A \times (C \cup D) = (A \times C) \cup (A \times D)$.
3. $A \times (C \setminus D) = (A \times C) \setminus (A \times D)$.

We also have the symmetric identities:

4. $(A \cap B) \times C = (A \times C) \cap (B \times C)$.
5. $(A \cup B) \times C = (A \times C) \cup (B \times C)$.
6. $(A \setminus B) \times C = (A \times C) \setminus (B \times C)$.

Proof We prove the first assertion only as the others can be proven in a similar manner.

1. We prove the equality via the following sequence of equivalences which are obtained by the respective set operation definitions:

$$\begin{aligned}(x, y) \in A \times (C \cap D) &\Leftrightarrow x \in A \text{ and } y \in C \cap D \\ &\Leftrightarrow x \in A,\ y \in C, \text{ and } y \in D \\ &\Leftrightarrow (x, y) \in A \times C \text{ and } (x, y) \in A \times D \\ &\Leftrightarrow (x, y) \in (A \times C) \cap (A \times D).\end{aligned}$$

Going forward we have $A \times (C \cap D) \subseteq (A \times C) \cap (A \times D)$ and going backwards we have $(A \times C) \cap (A \times D) \subseteq A \times (C \cap D)$. Hence we have the required set equality. □

We also have the following results:

Proposition 1.3.27 *Let X and Y be sets. Suppose that $A, B \subseteq X$ and $C, D \subseteq Y$. Then:*

1. $(A \cup B) \times (C \cup D) = (A \times C) \cup (B \times C) \cup (A \times D) \cup (B \times D)$.
2. $(A \cap B) \times (C \cap D) = (A \times C) \cap (B \times D)$.

Proof We prove the first assertion only as the second is left as Exercise 1.24.

1. By using Proposition 1.3.26(2) and (5), we have:

$$\begin{aligned}(A \cup B) \times (C \cup D) &= ((A \cup B) \times C) \cup ((A \cup B) \times D) \\ &= (A \times C) \cup (B \times C) \cup (A \times D) \cup (B \times D),\end{aligned}$$

which is the desired equality. □.

We shall see more of set algebras on Cartesian products in Exercise 1.24.

More generally, the construction of Cartesian can be carried out to include finitely many sets. If the number of sets is n, this construction is called the n-fold Cartesian product.

Definition 1.3.28 (n-fold Cartesian Product) Let X_j for $j = 1, 2, \dots, n$ be sets. The n-fold Cartesian product of the sets X_j, denoted $X_1 \times X_2 \times \dots \times X_n$, is defined as the set of ordered pairs of elements $X_1 \times X_2 \times \dots \times X_n = \{(x_1, x_2, \dots, x_n) : x_j \in X_j \text{ for } j = 1, 2, \dots, n\}$.

1.4 Quantifiers

In the previous section, we have seen many statements that hold true for a set of elements $x \in X$. For example, in Definition 1.3.5, we say $X \subseteq Y$ if for all $x \in X$, we have $x \in Y$. For each $x \in X$, let us denote the mathematical statement $P(x)$ as "The element x is a member of Y". Using this, we have the following definitions:

1. $X \subseteq Y$ if and only if for all $x \in X$ the statement $P(x)$ is true.
2. $X \cap Y \neq \varnothing$ if and only if there exists an $x \in X$ such that $P(x)$ is true
3. $X \cap Y = \varnothing$ if and only if for every $x \in X$ the statement $P(x)$ is false.

In all of the examples above, we have used the phrases "for all", "there exists", and "for every". These are called quantifiers because they are used to express or quantify for how many of the elements $x \in X$ the statement $P(x)$ is true. The set X is called domain or universe of discourse and we say the statement $P(x)$ is parametrised by elements of X.

Remark 1.4.1 We make some pedantic remarks here:

1. Note that the object $P(x)$ by itself is not a mathematical statement if the variable x is left unspecified.
2. In fact, strictly speaking, in formal mathematical logic, the object P is called a predicate. The variable x is called a free variable, a placeholder, or unknown since it could be any of the elements in X. Once we have fixed or bound x to be one element in X, say $x = b$ for a fixed $b \in X$, then the resulting sentence $P(b)$ becomes a mathematical statement for which we can ascertain truth or falsehood.
3. Since we are not doing formal logic in this book, we may be sloppy with the notation and terminology. For readers who are keen to read more on this, an excellent reference for an in-depth introduction to formal logic is [33].

Example 1.4.2 Consider X to be the set of months in the Gregorian calendar, namely $X = \{\text{Jan}, \text{Feb}, \ldots, \text{Dec}\}$. We define $P(x)$ as the sentence "The month x has 30 days in it." At the moment, $P(x)$ by itself is not a mathematical statement since it cannot be definitively true or false: x is an unknown and could be any month!

However, we can turn it into a mathematical statement by fixing x, say $x = \text{Jan}$. Now $P(\text{Jan})$, which says "The month January has 30 days in it", is a mathematical statement since it is unambiguously true. In fact, January has exactly more days, but it definitely has 30 days in it.

Another way of getting rid of the free variable is via quantifiers. We define:

Definition 1.4.3 (Universal, Existential Quantifiers) Let X be a non-empty set and $\{P(x) : x \in X\}$ be a set of mathematical statements with domain X.

1. Universal quantifier: A universal quantifier is a symbol $(\forall x \in X)$, where the statement $(\forall x \in X), P(x)$ is true when $P(x)$ is true for every $x \in X$.
2. Existential quantifier: An existential quantifier is a symbol $(\exists x \in X)$: where the statement $(\exists x \in X) : P(x)$ is true when $P(x)$ is true for at least one $x \in X$.

Remark 1.4.4 We make several remarks here:

1. Universal quantifier:
   (a) The universal quantifier is reminiscent to the "and" connective where we require all the statements involved to be true for the compound statement becomes true.
   (b) The symbol $\forall$ is read as "for all" or "for every" or "for each" or "for any" or "for arbitrary". It is symbolically an upside-down letter A, which stands for "all".
   (c) The statement $(\forall x \in X), P(x)$ can be read or written in standard English as "For all $x \in X$, $P(x)$ is true" or "$P(x)$ is true for all $x \in X$".
2. Existential quantifier:
   (a) The existential quantifier is reminiscent to the "or" connective where we require at least one of the statements involved to be true for the compound statement to be true.
   (b) The symbol $\exists$ is read as "there exists" or "there are some" or "there is at least one" or "for some" or "for at least one". It is symbolically a reflected letter E, which stands for "exists".
   (c) The statement $(\exists x \in X) : P(x)$ can be read or written in standard English as "There exists an $x \in X$ such that $P(x)$ is true" or "$P(x)$ is true for some $x \in X$".
3. Similar to the colon : symbol in the set builder notation, the colon symbol : in the statement $(\exists x \in X) : P(x)$ is read as "such that". This colon symbol is not really required in the existential quantifier. Similarly, the comma symbol in the universal quantifier $(\forall x \in X)$, is not necessary. However, we include them with the quantifiers for more obvious readibility.
4. Most of the time, to declutter, we remove the brackets around the quantifier symbols when we are writing a statement, namely we write $\forall x \in X, P(x)$ and $\exists x \in X : P(x)$ instead. However, for this chapter, we keep the brackets for us to get used to them and to see some of their properties more clearly.

Example 1.4.5 Let us look at some examples of the usage of quantifiers.

1. Suppose that X and Y are sets. At the beginning of this section, we have defined for each $x \in X$ the mathematical statement $P(x)$ as "The element x is a member of Y". Using the quantifier symbols, we can write:
   (a) $X \subseteq Y \Leftrightarrow (\forall x \in X), P(x)$.
   (b) $X \cap Y \neq \varnothing \Leftrightarrow (\exists x \in X) : P(x)$.
   (c) $X \cap Y = \varnothing \Leftrightarrow (\forall x \in X), \neg P(x)$.

By using the quantifier symbols, even though they seem scary at first, wordy statements can be written more succinctly. Therefore, for many definitions in this book, apart from the usual English language text, we will also state them using the logical symbols.

2. Let X be the set of months in the Gregorian calendar, namely $X = \{\text{Jan}, \text{Feb}, \ldots, \text{Dec}\}$. We define the statements $P(x)$ as "The month x has 30 days in it."
   (a) Now we consider the statement $(\forall x \in X), P(x)$. This reads as "For every month x in X, the month x has 30 days". This statement is false because there is a month that has at most 29 days, which is February. Therefore the statement $(\forall x \in X), P(x)$ is false. Logically, we can see that this is false because we can find at least one $x \in X$ such that the statement $P(x)$ is not true.
   (b) On the other hand, consider the statement $(\exists x \in X) : P(x)$ which we read as: "There exists a month x in X such that the month x has 30 days in it". In other words, this statement claims that there is at least one month which has 30 days. This is true, since January has 30 days. Thus the statement $(\exists x \in X) : P(x)$ is true. The moral here is: to establish the truth of an existential statement, it is enough to show that there is one $x \in X$ such that $P(x)$ is true.
3. Let Γ be the set of all polygons. For each $\gamma \in \Gamma$, we define the following statements:

$$P(\gamma) : \gamma \text{ is a square,} \quad \text{and} \quad Q(\gamma) : \gamma \text{ is a rectangle.}$$

   (a) The statement $(\forall \gamma \in \Gamma), P(\gamma)$ is false. Indeed, this reads as "For any polygon $\gamma \in \Gamma$, γ is a square" which is clearly false since γ could be a triangle or a pentagon.
   (b) The statement $(\exists \gamma \in \Gamma) : P(\gamma)$ reads as "There exists a polygon $\gamma \in \Gamma$ such that γ is a square". This is true since we can find at least one polygon γ in Γ which is a square.
   (c) The statement $(\forall \gamma \in \Gamma), (P(\gamma) \Rightarrow Q(\gamma))$ is read as "For all polygons $\gamma \in \Gamma$, if γ is a square, then γ is a rectangle." This statement is true.
4. Consider the set of birds B and the family of statements $\{P(b) : b \in B\}$ where $P(b)$ is "The bird b can fly". Consider the statement $(\forall b \in B), P(b)$ which is "For each bird b, the bird b can fly". Clearly this statement is false since there are birds that cannot fly, for example penguins and kiwibirds who do not know how to operate an aircraft. Therefore the statement $(\forall b \in B), P(b)$ is false.
   This means the negation $\neg((\forall b \in B), P(b))$ must be true. But what is the negation of this statement? The negation should be "There is at least one bird b such that the bird b cannot fly". In symbols, this is $(\exists b \in B) : \neg P(b)$. So we have the equivalence:

$$\neg((\forall b \in B), P(b)) \equiv (\exists b \in B) : \neg P(b).$$

5. Let X and Y be sets. For each $x \in X$, define $P(x)$ to be the statement "$x \in Y$" and Q to be the statement "$X \cap Y \neq \varnothing$".
   (a) Q being true means that there is at least one element in X that is also in Y. We saw in the first example that $Q \equiv (\exists x \in X) : P(x)$.
   (b) The negation of Q, namely $\neg Q$, is then $X \cap Y = \varnothing$. We have also seen in the first example that this is equivalent to $\neg Q \equiv (\forall x \in X), \neg P(x)$.

   Thus, we have the equivalence:

$$(\forall x \in X), \neg P(x) \equiv \neg Q \equiv \neg((\exists x \in X) : P(x)).$$

In Examples 1.4.5(4) and (5), notice that when we apply a negation to a statement with quantifiers, we can commute the negation symbol with the quantifier by flipping the quantifier symbol from universal quantifier to existential quantifier or vice versa. Namely, if $P(x)$ is some statement parametrised by $x \in X$, we have the following rules:

1. $\neg((\exists x \in X) : P(x)) \equiv (\forall x \in X), \neg P(x)$.
2. $\neg((\forall x \in X), P(x)) \equiv (\exists x \in X) : \neg P(x)$.

The rules above are also called the De Morgan's laws in formal logic, which is similar in form to Theorem 1.3.19 if one thinks of the negation of statements as complement of sets.

The quantifiers can also be used to remove the variables in a predicate that depends on higher number of variables.

Example 1.4.6 Define two sets Γ and Δ where Γ is the set of letters in Latin alphabet and Δ is the set of all the words in *The Oxford English Dictionary*. Now we define a mathematical statement that depends on two variables, namely $(\gamma, \delta) \in \Gamma \times \Delta$. We denote the statement "The word δ begins with the letter γ" as $P(\gamma, \delta)$.

1. For any fixed $\gamma \in \Gamma$, we can create a mathematical statement $Q(\gamma) \equiv (\exists \delta \in \Delta) : P(\gamma, \delta)$. Varying γ, we get a family of mathematical statements $\{Q(\gamma) : \gamma \in \Gamma\}$ which is parametrised by $\gamma \in \Gamma$. Therefore, we can append this statement with a quantifier for γ to create a mathematical statement. We could have:
   (a) The statement:

$$(\forall \gamma \in \Gamma), Q(\gamma) \equiv (\forall \gamma \in \Gamma), (\exists \delta \in \Delta) : P(\gamma, \delta),$$

   which reads as "For each letter $\gamma \in \Gamma$, there exists a word $\delta \in \Delta$ such that the word δ begins with the letter γ". This statement is clearly true because for every letter from A to Z, we can always find a word that begins with that letter in *The Oxford English Dictionary*. This is a simple exercise which one can do if one has the dictionary at hand.

(b) Another statement that we can make is:

$$(\exists \gamma \in \Gamma) : Q(\gamma) \equiv (\exists \gamma \in \Gamma) : (\exists \delta \in \Delta) : P(\gamma, \delta),$$

which says "There exists a letter $\gamma \in \Gamma$ such that there exists a word $\delta \in \Delta$ such that the word δ begins with the letter γ". This statement is also true since for at least one letter γ, say $\gamma = \text{A}$, we can find at least one word that begins with γ. A word that would work is $\delta = \text{Analysis}$. One can check that this word is indeed in *The Oxford English Dictionary*.

2. Now let us look at some other combinations of quantifiers that we can construct for this statement.
   (a) $(\forall \delta \in \Delta), (\exists \gamma \in \Gamma) : P(\gamma, \delta)$ reads as "For every word $\delta \in \Delta$, there exists a letter $\gamma \in \Gamma$ such that the word δ begins with the letter γ". For each fixed word δ, it must start with one of the letters in Γ. So there is a $\gamma \in \Gamma$ for which the statement $P(\gamma, \delta)$ is true. Thus this statement is true.
   (b) $(\exists \delta \in \Delta) : (\forall \gamma \in \Gamma), P(\gamma, \delta)$ is read as "There exists a word δ such that for every letter $\gamma \in \Gamma$, the word δ begins with γ". Now this statement is false since there are no words δ that begins with every letter in the alphabet.
   (c) $(\exists \delta \in \Delta) : (\exists \gamma \in \Gamma) : P(\gamma, \delta)$ says "There exists a word $\delta \in \Delta$ such that there exists a letter $\gamma \in \Gamma$ such that the word δ begins with the letter γ". As an example, the word Analysis in the dictionary begins with the letter A in the alphabet. Thus this statement is true.

Remark 1.4.7 Let us make some remarks here regarding Example 1.4.6 above.

1. We have seen that if we have more than one quantifier in a statement, in general we could not move them around. We saw that the statements $(\forall \gamma \in \Gamma), (\exists \delta \in \Delta) : P(\gamma, \delta)$ and $(\exists \delta \in \Delta) : (\forall \gamma \in \Gamma), P(\gamma, \delta)$, where the order of the same quantifiers are swapped, are not equivalent since the former is true whereas the latter is false.
2. However, switching the order of two adjacent quantifier statements of the same type, namely either both are universal quantifier or both are existential quantifiers, is allowed. For example, we have seen that $(\exists \gamma \in \Gamma) : (\exists \delta \in \Delta) : P(\gamma, \delta)$ and $(\exists \delta \in \Delta) : (\exists \gamma \in \Gamma) : P(\gamma, \delta)$ are the same. This is because both of these statements can be read as $\exists(\gamma \in \Gamma) \wedge (\delta \in \Delta) : P(\gamma, \delta) \equiv (\exists(\gamma, \delta) \in \Gamma \times \Delta) : P(\gamma, \delta)$. So we have:

$$\begin{aligned}(\exists \gamma \in \Gamma) : (\exists \delta \in \Delta) : P(\gamma, \delta) &\equiv (\exists(\gamma, \delta) \in \Gamma \times \Delta) : P(\gamma, \delta)\\ &\equiv (\exists \delta \in \Delta) : (\exists \gamma \in \Gamma) : P(\gamma, \delta).\end{aligned}$$

   Similarly we have the equivalence of statement with nested universal quantifiers as such:

$$\begin{aligned}(\forall \gamma \in \Gamma), (\forall \delta \in \Delta), P(\gamma, \delta) &\equiv (\forall(\gamma, \delta) \in \Gamma \times \Delta), P(\gamma, \delta)\\ &\equiv (\forall \delta \in \Delta), (\forall \gamma \in \Gamma), P(\gamma, \delta).\end{aligned}$$

3. If we have more than one quantifier in a statement, we have seen that we can treat them as nested statements. This allows us to define negations on these statements more systematically. Recall that we can switch the order of negation and a quantifier by flipping the quantifier from universal to existential or vice versa. Using this rule, for example, we have the equivalence of statements:

$$\begin{aligned}\neg((\exists\gamma\in\Gamma):(\exists\delta\in\Delta):P(\gamma,\delta)) &\equiv (\forall\gamma\in\Gamma),\neg((\exists\delta\in\Delta):P(\gamma,\delta))\\ &\equiv (\forall\gamma\in\Gamma),(\forall\delta\in\Delta),\neg P(\gamma,\delta),\end{aligned}$$

where we moved the negation symbol inwards one quantifier at a time.

Example 1.4.8 Let X and Y be some sets and consider the two families of statements $\{P(x) : x \in X\}$ and $\{Q(y) : y \in Y\}$. Suppose that we have a statement $(\forall x \in X), (\exists y \in Y) : P(x) \Rightarrow Q(y)$ and we want to find its negation. Using Remark 1.4.7(3), by moving the negation inwards along one quantifier at a time, we have:

$$\neg((\forall x\in X),(\exists y\in Y):P(x)\Rightarrow Q(y))\equiv(\exists x\in X):(\forall y\in Y),\neg(P(x)\Rightarrow Q(y)).$$

Applying the equivalence in (1.1), we thus have:

$$\neg((\forall x\in X),(\exists y\in Y):P(x)\Rightarrow Q(y))\equiv(\exists x\in X):(\forall y\in Y),P(x)\wedge(\neg Q(y)).$$

Finally, we note that the existential quantifier simply denotes that there is at least one element $x \in X$ for which the statements $P(x)$ is true. The number of elements for which $P(x)$ is true is not specified otherwise. Sometimes, this quantity is crucial to a statement or definition, so we may also need the unique existential or non-existential quantifier.

Definition 1.4.9 (Unique, Non-existential Quantifier) Let X be a non-empty set and $\{P(x) : x \in X\}$ be a set of mathematical statements with domain X.

1. Unique existential quantifier: A unique existential quantifier is a symbol $(\exists! x \in X) :$ where the statement $(\exists! x \in X) : P(x)$ is true when $P(x)$ is true for exactly one $x \in X$.
2. Non-existential quantifier: A non-existential quantifier is a symbol $(\nexists x \in X) :$ where the statement $(\nexists x \in X) : P(x)$ is true when $P(x)$ is true for none of $x \in X$ (or, in other words, $P(x)$ is false for all $x \in X$).

Remark 1.4.10 We make a few remarks here:

1. The symbol $\exists!$ is read as "there exists a unique" or "there exists exactly one".
2. The symbol $\nexists$ is read as "there does not exist" or "there are no".

3. We note that the non-existential quantifier is just the negation of the existence quantifier. Since the non-existential quantifier requires all of the $P(x)$ to be false, therefore it requires all of the $\neg P(x)$ to be true. Hence we have the equivalence of the following three statements:

$$(\nexists x \in X) : P(x) \equiv (\forall x \in X), \neg P(x) \equiv \neg((\exists x \in X) : P(x)).$$

Example 1.4.11 Let X be the set of planets in the solar system and $P(x)$ be the sentence "Humans can live on planet x" for $x \in X$.

1. The statement $(\exists! x \in X) : P(x)$ is read as "There exists one and only one planet in the solar system that humans can live on". We do not have enough technology to determine that this statement is true since we have not tried living on a planet other than the Earth.
2. The statement $(\nexists x \in X) : P(x)$ says "There are no planets in the solar system that humans can live on". This statement is false since we can live on at least one planet of the the solar system, namely the Earth.

There are many other quantifiers in the mathematical language such as quantifier with cardinalities or quantifiers with measures (for the latter, see Definition 18.9.14). But for now, the quantifiers $\forall$, $\exists$, $\exists!$, and $\nexists$ are sufficient for us.

1.5 Functions

After sets, the next fundamental object in mathematics are functions. The concept of functions originates from the idea of how one quantity (dependent variable) changes when we change another quantity (independent variable). In other words, it is how one set transforms into another.

Intuitively, a function is a "machine" that produces an output when we feed a single input into it at a time. As a result, the concept of functions is used widely in natural sciences, computer science, technology, and also social sciences.

Early renditions of functions were given by Gottfried Wilhelm von Leibniz (1646–1716) and Johann Bernoulli (1667–1748) to describe a quantity related to points of a curve, such as a coordinate or a curve's slope, in their work on calculus. Leonhard Euler (1707–1783) gave a rough definition of what a function is in his text book *Institutiones calculi differentialis* (Foundations of Differential Calculus) in 1755:

> Those quantities that depend on others ... namely those that undergo a change when others change, are called functions of these quantities. This definition applies rather widely and includes all ways in which one quantity can be determined by others.

So all the early definitions of functions are quantitative, numerical, and algebraic in nature. As time progresses, Euler's definition is fine-tuned to remove ambiguities and contradictions that may arise. Johann Peter Gustav Lejeune Dirichlet (1805–1859) introduced an all-encompassing definition of functions between general sets. In modern terms, this is given by:

Definition 1.5.1 (Function) A function $f : X \to Y$ is a correspondence between two sets X and Y which assigns each element $x \in X$ a single element $f(x) \in Y$. In symbols, this is:

$$(\forall x \in X), (\exists! y \in Y) : f(x) = y.$$

Remark 1.5.2 We make a few important remarks here.

1. The sets X and Y are called the domain and codomain of the function respectively.
2. The correspondence f is usually called a map, mapping, assignment, or simply as function. Be warned that the latter is a widely accepted abuse of terminology since, technically, a function is the triple (X, Y, f) and each one of the components is important to specify beforehand. Although he was talking about something else at the time, the words of Georg Cantor (1845–1918) seems befitting here:

   > In order for there to be a variable quantity in some mathematical study, the domain of its variability must strictly speaking be known beforehand through a definition. However, this domain cannot itself be something variable, since otherwise each fixed support for the study would collapse.

3. The element $f(x) \in Y$ is called the image of the element x under f or after applying f. One needs to distinguish between the objects $f(x)$ and f, where the former is an element in Y and the latter is a mapping.
4. Hence, usually when we specify a function, we write it in full as "$f : X \to Y$ such that $x \mapsto f(x)$" where the latter is read as "the element x (in X) is mapped to the element $f(x)$ (in Y)".
5. The notation $f(x)$ for function acting on an element x in its domain was introduced by Euler in his book *Introductio in analysin infinitorum* (Introduction to the Analysis of the Infinite).

Example 1.5.3 Let us look at some examples of functions. Let $X = \{a, b, c\}$ and $Y = \{\clubsuit, \diamondsuit, \spadesuit, \heartsuit\}$ be two sets. Some mappings between them are given in Figs. 1.9 and 1.10.

In Fig. 1.9, we can see that both f and g are functions from the set X to Y as each element in the domain X is assigned to only one element in the codomain Y. Note that for the function f, the elements b and c in X are both assigned to the same element in the codomain. This is not a problem at all! The definition for functions does not specify that distinct elements in X need to be assigned to distinct images in Y.

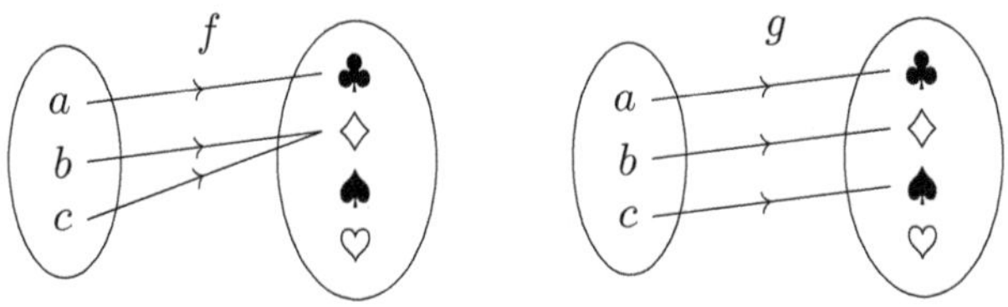

Fig. 1.9 f and g are functions between these sets

Fig. 1.10 f and g are not functions between these sets

On the other hand, in Fig. 1.10, both f and g are not functions from the set X to Y. The reasons are: for the map f, the element $c \in X$ is assigned to two distinct elements in the codomain and for the map g, the element $c \in X$ is not assigned to any element in the codomain. Therefore, both of these mappings violate the requirements needed in order to be called a function.

How do we represent functions? If the domain is finite and small, the most obvious way of doing it is by representing it pictorially as in Fig. 1.9.

We can also represent the function $f : X \to Y$ by listing down the pairs of elements and their images individually by the collection of pairs $\{(x, f(x)) : x \in X\}$ in the Cartesian product $X \times Y$. For example, in Fig. 1.9, the function $f : X \to Y$ can be represented by the set of pairs $\{(a, \clubsuit), (b, \diamondsuit), (c, \diamondsuit)\}$ whereas the function $g : X \to Y$ is represented by the pairs $\{(a, \clubsuit), (b, \diamondsuit), (c, \spadesuit)\}$. This representation is called a graph:

Definition 1.5.4 (Graph) Let $f : X \to Y$ be a function between two sets X and Y. The graph of the function f is given by the collection of pairs $G_f = \{(x, f(x)) : x \in X\} \subseteq X \times Y$.

Image and Preimage

For a function $f : X \to Y$, the sets X and Y are called the domain and codomain of the function f respectively. We may write the domain of the function as $X = \mathrm{Dom}(f)$. We also define the set $f(X) = \{f(x) : x \in X\} \subseteq Y$ as the image of the function f.

Remark 1.5.5 Sometimes the image of the function $f : X \to Y$ may be denoted as $\mathrm{Im}(f)$, but this notation may cause confusion when we deal with complex numbers (but it should not be confusing in context since f is a mapping, not a number).

If $Z \subseteq X$, then we can also define the set $f(Z)$ as the image of the subset $Z \subseteq X$ under the mapping f. Namely:

$$f(Z) = \{f(x) : x \in Z\} \subseteq f(X).$$

Clearly, if $W \subseteq Z \subseteq X$, then we must have $f(W) \subseteq f(Z)$. Moreover, the image of a function $f(X)$ may or may not be equal to the codomain Y. Note that in Fig. 1.9, both of the images $f(X)$ and $g(X)$ are proper subsets of the codomain. If the image coincides with the codomain, namely $f(X) = Y$, the function is called a surjective function or a surjection. We shall look at this kind of functions in more detail later.

For every element in the image of the function f, say $y \in f(X)$, it must be mapped from at least one element in the domain. We call the collection of such elements as the preimage of the element y, which we write as:

$$f^{-1}(\{y\}) = \{x \in X : f(x) = y\}.$$

If y is not in the image of the function (that is $y \notin f(X)$), we then write $f^{-1}(\{y\}) = \varnothing$ since there are no elements in X that are mapped to y by f. We can also define the preimage on subsets of the codomain Y. Namely, if $W \subseteq Y$, then we define the preimage of the subset W under the map f as all the elements in the domain X which are mapped to any element in W, namely:

$$f^{-1}(W) = \{x \in X : f(x) \in W\} \subseteq X.$$

Example 1.5.6 Recall the function $f : X \to Y$ defined in Fig. 1.9. We have:

1. $f^{-1}(\{\diamondsuit\}) = \{x \in X : f(x) = \diamondsuit\} = \{b, c\}$.
2. $f^{-1}(\{\spadesuit\}) = \{x \in X : f(x) = \spadesuit\} = \varnothing$.
3. $f^{-1}(\{\clubsuit, \diamondsuit\}) = \{x \in X : f(x) \in \{\clubsuit, \diamondsuit\}\} = \{a, b, c\} = X$.

The preimage sets satisfy the following results, which we leave for the readers to prove in Exercise 1.25.

Proposition 1.5.7 *Let $f : X \to Y$ be a function with $A \subseteq X$ and $B, C \subseteq Y$.*

1. *If $B \subseteq C$, then $f^{-1}(B) \subseteq f^{-1}(C)$.*
2. *$f^{-1}(f(X)) = X$.*
3. *$f(f^{-1}(Y)) = f(X) \subseteq Y$.*
4. *$f(X \setminus A) \supseteq f(X) \setminus f(A)$.*
5. *$f^{-1}(Y \setminus B) = X \setminus f^{-1}(B)$ or in other words $f^{-1}(B^c) = (f^{-1}(B))^c$.*

Proposition 1.5.7(5) says that the preimage operation preserves complements. This is not true for the image operations as strict inclusion may still occur for some functions and sets in Proposition 1.5.7(4). Readers are invited to come up with an example in Exercise 1.25.

Moreover, we have the following results on images and preimages of intersections and unions:

Proposition 1.5.8 *Let $f : X \to Y$ be a function, $V_i \subseteq X$ be a collection of subsets of X for each $i \in I$, and $W_j \subseteq Y$ be a collection of subsets of Y for each $j \in J$ where I and J are some indexing sets. Then:*

1. $f(\bigcap_{i\in I} V_i) \subseteq \bigcap_{i\in I} f(V_i)$.
2. $f(\bigcup_{i\in I} V_i) = \bigcup_{i\in I} f(V_i)$.
3. $f^{-1}(\bigcap_{j\in J} W_j) = \bigcap_{j\in J} f^{-1}(W_j)$.
4. $f^{-1}(\bigcup_{j\in J} W_j) = \bigcup_{j\in J} f^{-1}(W_j)$.

Proof We prove only the first two assertions. The others are left as Exercise 1.26.

1. Pick $y \in f(\bigcap_{i\in I} V_i)$. Then, by definition:

$$\begin{aligned}\exists x \in \bigcap_{i\in I} V_i \text{ such that } f(x) = y \quad &\Rightarrow \quad \exists x \in V_i \text{ for all } i \in I \text{ such that } f(x) = y\\ &\Rightarrow \quad y \in f(V_i) \text{ for all } i \in I\\ &\Rightarrow \quad y \in \bigcap_{i\in I} f(V_i),\end{aligned}$$

   and since y is arbitrary, we obtain the inclusion $f(\bigcap_{i\in I} V_i) \subseteq \bigcap_{i\in I} f(V_i)$.
2. We use double inclusion to prove this equality.

   ($\subseteq$): Pick an element $y \in f(\bigcup_{i\in I} V_i)$. Then, by definition:

$$\begin{aligned}\exists x \in \bigcup_{i\in I} V_i \text{ such that } f(x) = y &\Rightarrow \exists i \in I \text{ such that } x \in V_i \text{ with } f(x) = y\\ &\Rightarrow y \in f(V_i) \text{ for some } i \in I\\ &\Rightarrow y \in \bigcup_{i\in I} f(V_i),\end{aligned}$$

   which proves the first inclusion.

($\supseteq$): To show the reverse inclusion, pick an arbitrary $y \in \bigcup_{i\in I} f(V_i)$. Then, by definition:

$$\begin{aligned} \exists i \in I \text{ such that } y \in f(V_i) \quad &\Rightarrow \quad \exists i \in I \text{ such that } \exists x \in V_i \text{ with } f(x) = y \\ &\Rightarrow \quad \exists x \in \bigcup_{i\in I} V_i \text{ such that } f(x) = y \\ &\Rightarrow \quad y \in f\left(\bigcup_{i\in I} V_i\right), \end{aligned}$$

which shows the reverse inclusion.

By putting the two inclusions together, we obtain the desired equality of sets. □

Remark 1.5.9 We note that from Proposition 1.5.8, the preimage operation f^{-1} preserves union and intersection. However, the image operation f only preserves unions. Intersections may not be preserved under f. An example of this is the function $f : X \to Y$ in Fig. 1.9. If we set $U = \{b\}$ and $V = \{c\}$, we immediately get $U \cap V = \varnothing$ and thus $f(U \cap V) = f(\varnothing) = \varnothing$. However $f(U) = f(V) = \{\diamondsuit\}$ and so $f(U) \cap f(V) = \{\diamondsuit\} \neq \varnothing$. So this is an example for which $f(U \cap V) \subsetneq f(U) \cap f(V)$.

In fact, the preimage operations also satisfy the following:

Proposition 1.5.10 *Let $f : X \to Y$ be a function and $A, B \subseteq Y$. Then:*

1. $f^{-1}(B \setminus A) = f^{-1}(B) \setminus f^{-1}(A)$.
2. $f^{-1}(B \Delta A) = f^{-1}(B) \Delta f^{-1}(A)$.

We leave the proof of the above as Exercise 1.27.

So we conclude that the preimage operations preserves (finite and arbitrary) union, (finite and arbitrary) intersection, complement, set difference, and symmetric difference. On the other hand, the image operations do not necessarily satisfy this.

Injection, Surjection, Bijection

Now if each element in the image of a function has exactly one preimage, then we call the function an injective function or an injection. In other words, an injective function maps distinct elements in the domain to distinct elements in the codomain.

Example 1.5.11 In Fig. 1.9, the function $f : X \to Y$ is not injective because there are distinct elements in X which are mapped to the same element in Y, namely both b and c are mapped to the same element $\diamondsuit$. On the other hand, the function g is injective as each element in the image has exactly one preimage.

Together with surjection that we have mentioned earlier, we state the following definitions:

Definition 1.5.12 (Injection, Surjection, Bijection) Let $f : X \to Y$ be a function.

1. The function f is called an injective function or an injection if for each element $y \in f(X)$, there exists exactly one element $x \in X$ such that $f(x) = y$. In other words, whenever $f(x) = f(z)$, necessarily $x = z$. In symbols, this is:

$$(\forall y \in f(X)), (\exists! x \in X) : f(x) = y.$$

2. The function f is called a surjective function or a surjection if for every $y \in Y$, there exists an $x \in X$ such that $f(x) = y$. In other words, the image of the function coincides with the codomain, namely $f(X) = Y$. In symbols, this is:

$$(\forall y \in Y), (\exists x \in X) : f(x) = y.$$

3. The function f is called a bijective function or a bijection if it is both injective and surjective. In other words, every element in the codomain is mapped from exactly one element in the domain via f. Combining the above, in symbols this is:

$$(\forall y \in Y), (\exists! x \in X) : f(x) = y.$$

Remark 1.5.13 Let $f : X \to Y$ be a function. We make a couple of remarks here.

1. If f is an injection, we say that f injects into Y. Sometimes f is called a one-to-one function, but in this book, we avoid this terminology since it might cause some confusion with the third remark later.
2. If f is a surjection, we say f surjects onto Y. We also call f an onto function.
3. If f is a bijection, we call f a one-to-one correspondence between X and Y. This may be confusing with the first remark above, so we avoid using the one-to-one terminologies altogether.

Composite, Inverse, Restriction Functions

In Definition 1.5.1, we have seen functions as a general object. From Definition 1.5.12, injections, surjections, and bijections are special kinds of functions with additional constraints attached. Why are these kinds of functions special?

Bijective functions are nice because it gives a correspondence between the elements in the domain and elements in the codomain. This means that we can invert them. What does invert means?

Let $f : X \to Y$ be a bijective function. Since the function f is surjective, for each $y \in Y$, we can find at least one $x \in X$ such that $y = f(x)$. Furthermore, since the function f is injective, this x is unique. So we can create a new mapping from Y to X defined as:

$$\begin{aligned} g : Y &\to X \\ y &\mapsto x \quad \text{which satisfies } f(x) = y, \end{aligned} \tag{1.4}$$

which we can check to be a well-defined function: every element in the domain Y has exactly one image in X. This unique function g is called the inverse function of f.

It is important to note that this new function can only be defined uniquely if and only if the original function f is bijective. Indeed:

1. If the function f is not surjective, then there would be some $y \in Y$ which is not mapped from any $x \in X$. As a result, the function g in (1.4) cannot be defined for this y.
2. If f is not injective, then there exists an element $y \in Y$ that would be mapped to more than one element in X. Thus there are more than one possible choice for the image of y for the function (1.4).

So inverting a function roughly means switching the domain and codomain of a function whilst preserving the original correspondence between the elements in these sets. Before we give the proper definition of an inverse function, we would like to define the composition of functions.

Definition 1.5.14 (Composite Function) Suppose that we have the sets X, Y, W, and Z with functions $f : X \to Y$ and $g : W \to Z$. If $f(X) \subseteq \text{Dom}(g) = W$, we can define the composite function $g \circ f : X \to Z$ as $x \mapsto g(f(x))$.

The condition $f(X) \subseteq \text{Dom}(g) = W$ is necessary because we want to ensure that the image of every point $x \in X$ under f can be further acted upon by the function g.

Example 1.5.15 Consider the functions depicted in Fig. 1.11.

1. The image of the function f is the set $\text{Im}(f) = \{\clubsuit, \diamondsuit\}$ whereas the domain of the functions g and h are $\text{Dom}(g) = \{\diamondsuit, \heartsuit, \spadesuit\}$ and $\text{Dom}(h) = \{\diamondsuit, \heartsuit, \clubsuit\}$ respectively. Note that $\text{Im}(f)$ is not contained in $\text{Dom}(g)$. Therefore, the composition $g \circ f$ does not make sense. Indeed, we do not have a meaning for the object $(g \circ f)(a) = g(f(a)) = g(\clubsuit)$.

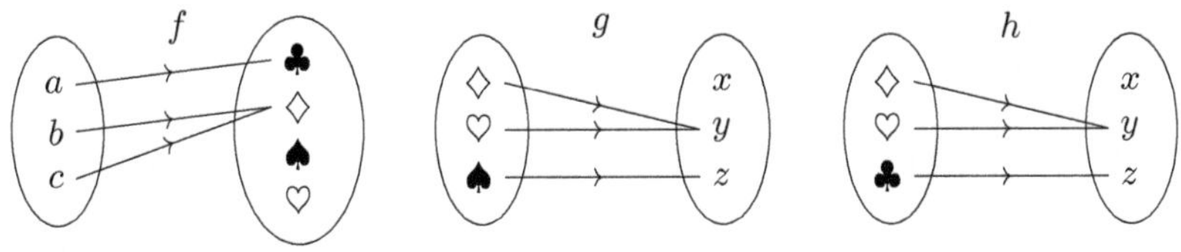

Fig. 1.11 f, g, and h are functions between some sets

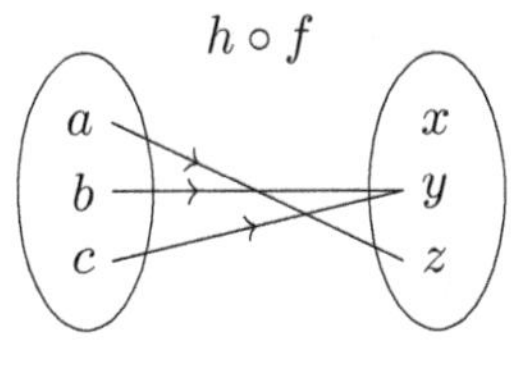

Fig. 1.12 The composite function $h \circ f$

2. On the other hand, we have the inclusion $\mathrm{Im}(f) \subseteq \mathrm{Dom}(h)$ and thus the composition $h \circ f$ can be defined. The image of a under this composition is $(h \circ f)(a) = h(f(a)) = h(\clubsuit) = z$. Likewise, $(h \circ f)(b) = y$ and $(h \circ f)(c) = y$. This describes the composite function $h \circ f$ fully, as shown in Fig. 1.12

Now suppose that $f : X \to Y$ is a bijective function with inverse $g : Y \to X$ as constructed in (1.4). Note that since $f(X) = Y = \mathrm{Dom}(g)$, we can compose these two functions. Fix an $x \in X$ and let us find its image under the composition $g \circ f$.

We have $(g \circ f)(x) = g(f(x))$ but recall that the image of a point $y \in Y$ under the inverse function g is the unique point $z \in X$ such that $y = f(z)$. In the case above, we want to find the image of the point $f(x)$ under g, which is the unique point $z \in X$ such that $f(z) = f(x)$. By injectivity of the function f, there is only one such point, namely $z = x$. Hence $g(f(x)) = x$. Thus, we conclude that for any $x \in X$, we have $(g \circ f)(x) = x$.

So the composition $g \circ f : X \to X$ maps every point to itself, which we call the identity map on X and denote as the function $\mathrm{id}_X : X \to X$ where $\mathrm{id}_X(x) = x$ for all $x \in X$. Using the same argument, the opposite composition, namely the composition $f \circ g : Y \to Y$, is also the identity map id_Y. Using this observation, we can now define:

Definition 1.5.16 (Inverse Function) Let $f : X \to Y$ be a bijective function. A function $g : Y \to X$ is the inverse function to f if $f(g(y)) = y$ for all $y \in Y$ and $g(f(x)) = x$ for all $x \in X$. In other words, we have $f \circ g = \mathrm{id}_Y$ and $g \circ f = \mathrm{id}_X$.

We call the function f an invertible function and g the inverse function of f which we usually denote as $g = f^{-1}$. We stress here that a function has an inverse if and only if it is bijective.

We also have a weaker versions of the above each of which satisfying only one of the two conditions, namely:

Definition 1.5.17 (Left-Inverse, Right-Inverse) Let $f : X \to Y$ be a function.

1. A function $g : Y \to X$ is said to be a right-inverse of the function f if $f(g(y)) = y$ for every $y \in Y$. In other words, $f \circ g = \text{id}_Y$.
2. A function $g : Y \to X$ is said to be a left-inverse of the function f if $g(f(x)) = x$ for every $x \in X$. In other words, $g \circ f = \text{id}_X$.

In each of the cases above, we say f is right-invertible or left-invertible respectively.

Necessarily, an invertible function is both left- and right-invertible. Moreover, we have the following characterisation:

Proposition 1.5.18 *Let $f : X \to Y$, where X is a non-empty set, be a function.*

1. *f is injective if and only if f is left-invertible.*
2. *f is surjective if and only if f is right-invertible.*

The readers are invited to prove this in Exercise 1.32.

Remark 1.5.19 We note that there is an ambiguity in the notation for f^{-1}. First, it was defined as the preimage $f^{-1}(\{y\})$ of a singleton set in Y, which can either be empty, unique, or multiple. Secondly it may be used to denote inverse function of f. Therefore, it is important to distinguish these two notations based on context.

Finally, we define the restriction of function to a subset of the domain:

Definition 1.5.20 (Restriction of Function) Let $f : X \to Y$ be a function and $Z \subseteq X$ be a subset of X. The restriction of the function f to the subset Z is the function $f|_Z : Z \to Y$ defined as $f|_Z(x) = f(x)$ for all $x \in Z$.

The restriction $f|_Z$ of a function is simply the function f with its domain restricted to the subset $Z \subseteq X$. This is useful when we need to do algebra on functions and to make sure that composition of functions or finding inverse functions work nicely.

Example 1.5.21 From Example 1.5.15, we have seen that the composite function $g \circ f$ cannot be defined since there are elements in $\text{Im}(f)$ which are not contained in $\text{Dom}(g)$, namely $\clubsuit \in \text{Im}(f)$ but $\clubsuit \notin \text{Dom}(g)$. If we insist that we want to compose these two functions, we need to restrict the function f to a smaller domain so that the offending element $\clubsuit$ is no longer contained in the image of the restricted function.

Let $X = \{b, c\} \subseteq \text{Dom}(f)$. The restriction $f|_X$ has image $\text{Im}(f|_X) = \{f(x) : x = b, c\} = \{\diamondsuit\} \subseteq \text{Dom}(g)$. Thus, the composition $g \circ f|_X$ exists and is given by the constant function $(g \circ f|_X)(x) = y$ for all $x \in X$.

Exercises

1.1 Write the following statements in symbols and write their negations.
  (a) The grapes are seedless and sweet.
  (b) The test is difficult but I got an A.
  (c) I got an A in the test because I worked hard.
  (d) If the train is full, then I will arrive late for the meeting.
  (e) You have to study analysis if you want to be a mathematician.
  (f) I will go to class only if I am feeling well.
  (g) I will be in my office or the cafeteria, but not both.

1.2 Interpret the following quotes (purportedly credited to the appropriate persons) using logical connectives and quantifiers. Or just enjoy them.
  (a) *"Life is either a daring adventure or nothing at all."* - Helen Keller (1880–1968), author and disability activist.
  (b) *"I've learned that people will forget what you said, people will forget what you did, but people will never forget how you made them feel."* - Maya Angelou (1951–2014), poet and civil rights activist.
  (c) *"If you set your goals ridiculously high and it's a failure, you will fail above everyone else's success."* - James Cameron (1954-), filmmaker.
  (d) *"I am not sick. I am broken. But I am happy to be alive as long as I can paint."* - Frida Kahlo (1907–1954), painter.
  (e) *"All our dreams can come true if we have the courage to pursue them"* - Walt Disney (1901–1966), film producer and entrepreneur.
  (f) *"A person who never made a mistake never tried anything new."* - Albert Einstein (1879–1955), theoretical physicist and Nobel laureate.
  (g) *"If your actions inspire others to dream more, learn more, do more, and become more, you are a leader."* - John Quincy Adams (1767–1848), president of USA.
  (h) *"If you judge people, you have no time to love them."* - Mother Teresa (1910–1997), Catholic nun and Nobel laureate.
  (i) *"There is no story that is not true."* - Chinua Achebe (1930–2013), novelist and poet.
  (j) *"I'm not offended by all the dumb blonde jokes because I know I'm not dumb... and I also know that I'm not blonde."* - Dolly Parton (1946-), singer-songwriter.
  (k) *"Everybody understood that if the proof is correct, then no other recognition is needed."* - Grigori Perelman (1966-), mathematician and Fields medalist.
  (l) *"I'm a feminist, because I see all women as smart, gifted, and tough."* - Zaha Hadid (1950–2016), architect and Pritzker laureate.

(m) *"If you know the enemy and know yourself, you need not fear the result of a hundred battles."* - Sun Tzu (544B.C.-496B.C.), military general.
(n) *"One cannot think well, love well, sleep well, if one has not dined well."* - Virginia Woolf (1882–1941), writer.
(o) *"You have enemies? Good. That means you've stood up for something, sometime in your life."* - Winston Churchill (1874–1965), prime minister of UK.
(p) *"Just 'cause you live in the ghetto doesn't mean you can't grow."* - Tupac Shakur (1971–1996), rapper.
(q) *"If there is no God, everything is permitted."* - Fyodor Dostoyevsky (1821–1881), novelist.
(r) *"They laugh at me because I'm different; I laugh at them because they're all the same."* - Kurt Cobain (1967–1994), musician.
(s) *"I raise up my voice – not so that I can shout, but so that those without a voice can be heard"* - Malala Yousafzai (1997-), education activist and Nobel laureate.
(t) *"If heaven had granted me five more years, I could have become a real painter."* - Katsushika Hokusai (1760–1849), painter.

The point of this exercise is to demonstrate that mathematical statements and logical connectives are used by everyone everyday. It is thus important for us to be able to understand, interpret, and appreciate them, not just in the mathematical setting! To quote Timothy Gowers (1963-):

> It is therefore good for the health of a country if its population has high standards of mathematical literacy: without it, people are swayed by incorrect arguments, make bad decisions and are happy to vote for politicians who make bad decisions on their behalf.

1.3 (∗) Let X be the set of cities in the world and Y be the set of countries in the world. For each $x \in X$ and $y \in Y$, define the statements:

$$P(x, y) : x \text{ is in } y, \quad \text{and} \quad Q(x, y) : x \text{ is the capital city of } y.$$

Determine the truth of the following compound statements:
(a) P(Kuala Lumpur, Malaysia).
(b) Q(Rio de Janeiro, Brazil).
(c) $\neg Q$(Nairobi, Kenya).
(d) P (London,United Kingdom) $\wedge$ P (Boston, United States of America).
(e) P(Seattle, United States of America) $\wedge$ P(Canberra, Azerbaijan).
(f) P(Paris, France) $\vee$ P(Beijing, Italy).
(g) P(Sapporo, Japan) $\Rightarrow$ Q(Sapporo, Japan).
(h) Q(Tokyo, Japan) $\Rightarrow$ P(Tokyo, Japan).
(i) Q(Riyadh, India) $\Rightarrow$ P(Addis Ababa, Ethiopia).
(j) Q(Buenos Aires, Egypt) $\Leftrightarrow$ Q(Stockholm, Sweden).

1.4 (∗) Write the following statements in symbols. Using *modus ponens* and *modus tollens*, answer the following questions.

(a) All whales are mammals. An animal named Lucy is not a mammal.
Is Lucy a whale?

(b) A fish is not a mammal. Lucy is not a mammal.
Is Lucy a fish?

(c) If you are a mathematician, you are clever. Lucy is clever.
Is Lucy a mathematician?

(d) A student is allowed to take Analysis 2 only if the student passed Analysis 1. I am taking Analysis 2.
Did I pass Analysis 1?

(e) If I do not do my assignments, I will get low marks in my course. If I get low marks in my course, I will not graduate. I did my assignments. Will I graduate?

(f) If I get an A in Analysis 1, I will go to town to celebrate. When I am celebrating in town, I always get a milkshake. I am not getting a milkshake. Did I get an A for Analysis 1?

1.5 (a) Complete the following truth tables:

P	Q	$\neg P$	$\neg Q$	$\neg P \wedge \neg Q$
T	T	F	F	F
T	F			
F	T			
F	F			

P	Q	$P \vee Q$	$\neg(P \vee Q)$
T	T	T	F
T	F		
F	T		
F	F		

Hence deduce that the statements $\neg P \wedge \neg Q$ and $\neg(P \vee Q)$ are equivalent.

(b) Using truth tables, show that the statements $\neg P \vee \neg Q$ and $\neg(P \wedge Q)$ are equivalent.

(c) Complete the following truth table:

P	Q	$\neg P$	$\neg P \vee Q$
T	T	F	T
T	F		
F	T		
F	F		

Hence deduce that all the statements $P \Rightarrow Q$, $\neg P \vee Q$, and $\neg Q \Rightarrow \neg P$ are equivalent.

(d) Show that the statements $P \Rightarrow Q$ and $Q \Rightarrow P$ are not equivalent.

(e) Using parts (a) and (c), show that the statements $\neg(P \Rightarrow Q)$ and $P \wedge (\neg Q)$ are equivalent.

1.6 (⋄) Suppose that P, Q, and R are mathematical statements. In the following, state whether Q is true, false, or cannot be determined.

(a) P is false and $P \vee \neg Q$ is true.
(b) $(P \wedge \neg Q) \Rightarrow R$ is false.
(c) $P \Rightarrow (R \wedge Q)$ is false and R is true.
(d) $(P \Rightarrow Q) \vee (P \Rightarrow R)$ is false.
(e) $(Q \Rightarrow P) \vee (P \Rightarrow R)$ is false.
(f) $(P \Leftrightarrow Q) \wedge (R \Rightarrow P)$ is true and R is true.

1.7 Suppose that P and Q are mathematical statements. Create a truth table for $P \Leftrightarrow Q$ with five columns P, Q, $P \Rightarrow Q$, $Q \Rightarrow P$, and $(P \Rightarrow Q) \wedge (Q \Rightarrow P)$.
Explain when can the statement $P \Leftrightarrow Q$ be true.

1.8 (∗) In this question, we define a connective "exclusive or" which we denote as $\veebar$. We have seen this in Exercise 1.1(g). For mathematical statements P and Q, the statement $P \veebar Q$ is defined as $(P \vee Q) \wedge \neg(P \wedge Q)$.

(a) Using truth tables, prove that $P \veebar Q \equiv \neg(P \Leftrightarrow Q)$.
(b) Deduce that $\veebar$ is symmetric.
(c) Using truth tables, show that $\veebar$ is associative, namely for statements P, Q, and R we have $(P \veebar Q) \veebar R \equiv P \veebar (Q \veebar R)$.

1.9 (∗) Let P, Q, and R be mathematical statements. Using truth tables, prove the following equivalences of statements. The truth table for the first question has been set up for you.

(a) $(P \wedge Q) \wedge R \equiv P \wedge (Q \wedge R)$.

P	Q	R	$P \wedge Q$	$Q \wedge R$	$(P \wedge Q) \wedge R$	$P \wedge (Q \wedge R)$
T	T	T	T	T	T	T
T	T	F				
T	F	T				
T	F	F				
F	T	T				
F	T	F				
F	F	T				
F	F	F				

(b) $(P \vee Q) \vee R \equiv P \vee (Q \vee R)$.
(c) $P \wedge (Q \vee R) \equiv (P \wedge Q) \vee (P \wedge R)$.
(d) $P \vee (Q \wedge R) \equiv (P \vee Q) \wedge (P \vee R)$.

1.10 (∗) Write the following statements in symbols and quantifiers and write their negations.

(a) All the students in the class got an A for the test.
(b) If it is 10pm, then every children in the village is asleep.
(c) I will go out only if I can find a friend who can watch after my cat.
(d) There exist polygons which are triangles.

(e) Every natural number greater than one can be divided by some prime number.
(f) Every even natural number is a sum of two prime numbers.
(g) There exists a solution to the equation $x^3 + y^3 = z^3$ in natural numbers.
(h) For each week in the year 2023, there exists a day in that week that is rainy.

1.11 (∗) Let X be the set of cities in the world and Y be the set of countries in the world. For each $x \in X$ and $y \in Y$, define the statements:

$$P(x, y) : x \text{ is in } y, \quad \text{and} \quad Q(x, y) : x \text{ is the capital city of } y.$$

Determine the mathematical truth of the following statements:
(a) $\forall y \in Y, \exists x \in X : Q(x, y)$.
(b) $\exists x \in X : \forall y \in Y, P(x, y)$.
(c) $\forall x \in X, \forall y \in Y, (Q(x, y) \Rightarrow P(x, y))$.
(d) $\forall x \in X, \forall y \in Y, (P(x, y) \Rightarrow Q(x, y))$.

1.12 (∗) Let X be a set and $\{P(x), Q(x) : x \in X\}$ be a set of mathematical statements parametrised by X. Let $X' = \{x \in X : P(x) \text{ is true}\} \subseteq X$. Explain why the following three statements are equivalent to each other.
1. $\forall x \in X : (P(x) \Rightarrow Q(x))$.
2. $\forall x \in X' : Q(x)$.
3. $x \in X' \Rightarrow Q(x)$.

1.13 (∗) Let X, Y, and Z be sets in a universe U. Draw the Venn diagrams of the following sets:
(a) $X \cap (Y \cup Z)$.
(b) $X \cup (Y \cap Z)$.
(c) $(X \cap Z) \cup (Y \setminus X)$.
(d) $(X \setminus Z)^c$

1.14 Let X, Y, and Z be sets.
(a) Draw the Venn diagram of the set $W = (X \setminus Y) \cup (Y \setminus Z) \cup (Z \setminus X)$.
(b) From the diagram in part (a), write down a conjecture for an expression of W with only one difference symbol $\setminus$ used.
(c) Prove your conjecture in part (b).

1.15 (◇) Let X, Y, and Z be subsets of the universe U. The set $X \cap Y \cap Z$ are all the points in the universe U that is contained in all three of X, Y, Z.
(a) Find an expression for the set of elements in U which is contained in exactly two of the sets X, Y, and Z.
(b) Find an expression for the set of elements in U which is contained in exactly one of the sets X, Y, and Z.

1.16 For sets X, Y, and Z, prove that if $X \subseteq Y$ and $Y \subseteq Z$, then $X \subseteq Z$.

1.17 (∗) Let X, Y, and Z be sets. Prove that:
(a) If $X \subseteq Y$ then $X \cup Z \subseteq Y \cup Z$.
(b) If $X \subseteq Y$ then $X \cap Z \subseteq Y \cap Z$.
(c) $X \subseteq Z$ and $Y \subseteq Z$ if and only if $X \cup Y \subseteq Z$.
(d) $Z \subseteq X$ and $Z \subseteq Y$ if and only if $Z \subseteq X \cap Y$.

1.18 Let X and Y be sets. Prove that:

(a) $X \cup Y = X$ if and only if $Y \subseteq X$.
(b) $X = X \cap Y$ if and only if $X \subseteq Y$.
(c) $X \cap Y = X$ if and only if $Y \cup X = Y$.
(d) $X \setminus Y = \varnothing$ if and only if $X \subseteq Y$.
(e) $X \setminus Y = X$ if and only if $Y \subseteq X^c$.
(f) $X \cap Y = \varnothing$ if and only if $X \subseteq Y^c$.

1.19 Prove Proposition 1.3.15, namely:
Let X and Y sets in a universe U. Prove:
(a) $X \cap X^c = \varnothing$.
(b) Idempotent laws: $X \cap X = X$ and $X \cup X = X$.
(c) $X \cup U = U$ and $X \cap \varnothing = \varnothing$.
(d) Absorption laws: $X \cup (X \cap Y) = X$ and $X \cap (X \cup Y) = X$.
(e) $X \cup Y = U$ and $X \cap Y = \varnothing$ if and only if $X = Y^c$.

1.20 ($*$) Prove the first two assertions in Proposition 1.3.17, namely:
For any sets X, Y, and Z, we have $X \cup (Y \cup Z) = (X \cup Y) \cup (X \cup Z)$ and $X \cap (Y \cap Z) = (X \cap Y) \cap (X \cap Z)$.

1.21 ($*$) Let X, Y, and Z be some sets.
(a) Show that the symmetric difference is symmetric and associative, namely $X \Delta Y = Y \Delta X$ and $(X \Delta Y) \Delta Z = X \Delta (Y \Delta Z)$.
(b) Show that $X \cap (Y \Delta Z) = (X \cap Y) \Delta (X \cap Z)$.
(c) Show that $X \Delta Y \subseteq (X \Delta Z) \cup (Z \Delta X)$.

1.22 Let X and Y be sets in a universe U.
(a) Prove that $(X \Delta Y)^c = (X^c \cup Y) \cap (X \cup Y^c)$.
(b) Show that $X \cup Y = (X \setminus Y) \cup (X \cap Y) \cup (Y \setminus X)$ where any two pairs of sets in the RHS are disjoint.
Hence, deduce that if $Y \subseteq X$, then $X = (X \cap Y) \cup (X \setminus Y)$ where the sets on the RHS are disjoint.

1.23 ($*$) Prove the remaining assertions in Proposition 1.3.21, namely:
For sets X, Y, and Z, we have:
(a) $Z \setminus (X \cap Y) = (Z \setminus X) \cup (Z \setminus Y)$.
(b) $(X \setminus Y) \cup Z = (X \cup Z) \setminus (Y \setminus Z)$.
(c) $Z \setminus (X \setminus Y) = (Y \cap Z) \cup (Z \setminus X)$.
(d) $(Z \setminus X) \setminus Y = Z \setminus (X \cup Y)$.

1.24 ($*$) Let X and Y be sets. Suppose that $A, B \subseteq X$ and $C, D \subseteq Y$. Prove that:
(a) $(A \cap B) \times (C \cap D) = (A \times C) \cap (B \times D)$.
(b) $(A \times C)^c = (A^c \times C) \cup (A \times C^c) \cup (A^c \times C^c)$ where the sets on the RHS are pairwise disjoint.
(c) $(A \times C) \cup (B \times D) = ((A \setminus B) \times C) \cup ((A \cap B) \times (C \cup D)) \cup ((B \setminus A) \times D)$ where the sets on the RHS are pairwise disjoint.
(d) $(A \times C) \setminus (B \times D) = (A \times (C \setminus D)) \cup ((A \setminus B) \times C)$.

1.25 ($*$) Prove Proposition 1.5.7, namely:
Let $f : X \to Y$ be a function with $A \subseteq X$ and $B, C \subseteq Y$. Prove the following statements:
(a) If $B \subseteq C$, then $f^{-1}(B) \subseteq f^{-1}(C)$.
(b) $f^{-1}(f(X)) = X$.

(c) $f(f^{-1}(Y)) = f(X) \subseteq Y$.
Give an example for which $f(f^{-1}(Y)) \subsetneq Y$.
(d) $f(X \setminus A) \supseteq f(X) \setminus f(A)$.
Give an example for which $f(X \setminus A) \supsetneq f(X) \setminus f(A)$.
(e) $f^{-1}(Y \setminus B) = X \setminus f^{-1}(B)$ or in other words $f^{-1}(B^c) = (f^{-1}(B))^c$.

1.26 $(*)$ Prove the remaining assertions in Proposition 1.5.8, namely:
Let $f : X \to Y$ be a function and $W_j \subseteq Y$ be a collection of subsets of Y for each $j \in J$ where J is an (finite or infinite) indexing set. Then:
(a) $f^{-1}(\bigcap_{j\in J} W_j) = \bigcap_{j\in J} f^{-1}(W_j)$.
(b) $f^{-1}(\bigcup_{j\in J} W_j) = \bigcup_{j\in J} f^{-1}(W_j)$.

1.27 $(*)$ Prove Proposition 1.5.10, namely:
Let $f : X \to Y$ be a function and $A, B \subseteq Y$. Then:
(a) $f^{-1}(B \setminus A) = f^{-1}(B) \setminus f^{-1}(A)$.
(b) $f^{-1}(B \Delta A) = f^{-1}(B) \Delta f^{-1}(A)$.

1.28 $(\diamond)$ Let $f : X \to Y$ be a function. Suppose that $A \subseteq X$ and $B \subseteq Y$. Prove that $A \cap f^{-1}(B) = \varnothing$ if and only if $f(A) \cap B = \varnothing$.

1.29 $(*)$ Let $f : X \to Y$ be a function. Prove that:
(a) If $A \subseteq X$, then $f^{-1}(f(A)) \supseteq A$.
In addition, show that if f is injective, then equality occurs.
(b) If $B \subseteq Y$, then $f(f^{-1}(B)) \subseteq B$.
In addition, show that if f is surjective, then equality occurs.

1.30 Let $f : X \to Y$ and $g : Y \to Z$ be functions and $A \subseteq Z$ be a subset. Prove that $(g \circ f)^{-1}(A) = f^{-1}(g^{-1}(A))$.

1.31 $(*)$ Let $f : X \to Y$ and $g : Y \to Z$ be functions. Prove that:
(a) If f and g are both injective, then $g \circ f$ is injective.
(b) If f and g are both surjective, then $g \circ f$ is surjective.
(c) If $g \circ f$ is injective, then f is injective.
Is it necessarily true that g is also injective? Give a proof or a counterexample.
(d) If $g \circ f$ is surjective, then g is surjective.
Is it necessarily true that f is also surjective? Give a proof or a counterexample.

1.32 Prove Proposition 1.5.18, namely:
Let $f : X \to Y$, where X is a non-empty set, be a function. Prove that:
(a) f is injective if and only if f is left-invertible.
(b) f is surjective if and only if f is right-invertible.

Integers

2

There was a single one, then there were ten. When ten made a hundred, and a hundred million.

— David Longstreth, musician

In Chap. 1, we have defined what sets are (naïvely, but still fits our purposes) and what we can do with them. However, sets are simply just a collection of objects. Nothing more, nothing less. The elements in a set do not interact with each other and the set is said to have no structure.

Now we would like to endow some sets with additional structures to make them more interesting to study. This is analogous to thinking the set as a pile of sand and we imbibe it with additional structures by adding clay or mud (to clump them together), cement and water (to give rigidity), heat (to melt them together and make glass), or scissors (if you listen to Echo and the Bunnymen) so the sand grains interact with each other in some way. By putting this additional structure, we have more interesting properties to manipulate and study.

Roughly speaking, a structure on a set is some additional features or constraints we declare on the set. These structures come in various types, for example: relations, order, algebraic, geometric, topological, differential, and measure. These structures are declared on a set via some rules or axioms that the elements in the set need to satisfy.

Sometimes, we may even place more than one structure on a set and study the interactions between them. If we have more than one structure on a set, we may require the structures to be compatible with each other in a certain way. An example of this is the compatibility of algebraic and order structure that we shall see in Definitions 2.6.1 and 3.3.1. Other times, one structure may induce or give rise to another structure on the set. Thus, having structures on a set can lead to a very intricate and rich mathematical study!

S. Johar, *The Big Book of Real Analysis*,
https://doi.org/10.1007/978-3-031-30832-1_2

2.1 Relations

In this section, we would like introduce a set structure called relation. As the name suggests, this structure is used to relate, compare, or connect two elements in one or more sets. A relation is defined by declaring whether the elements satisfy some specified condition or not. Pairs of elements that satisfy this condition are collected together and this collection is called a relation. We define:

Definition 2.1.1 (Relation) A relation $\mathcal{R}$ on the sets X and Y is a subset of the ordered pairs in a Cartesian product $X \times Y$, namely $\mathcal{R} \subseteq X \times Y$. We say an element $x \in X$ is related to an element $y \in Y$ iff $(x, y) \in \mathcal{R}$. We also denote $x \sim_{\mathcal{R}} y$ or simply $x \sim y$ iff x is related to y.

Example 2.1.2 Let us look at some examples of relations.

1. Let X be the set of countries in the world and Y be the set of capital cities in the world. We can define a relation $\sim_{\mathcal{R}}$ on the sets X and Y where $x \sim y$ iff y is the capital city of x. Therefore, we have Austria $\sim$ Vienna, Vietnam $\sim$ Hanoi, Nepal $\sim$ Kathmandu, and Bolivia $\sim$ La Paz. But Canada $\not\sim$ Dublin.
2. Another example of a relation that we have seen in Chap. 1 is a function. A function $f : X \to Y$ is a relation on the sets X and Y defined as $x \sim_{\mathcal{R}} y$ iff $f(x) = y$. The graph of this function is precisely the relation $\mathcal{R}$ since:

$$G_f = \{(x, f(x)) : x \in X\} = \{(x, y) : x \in X, y = f(x)\} = \mathcal{R}.$$

In this chapter, we are interested on how elements in a set relate to another within the same set.

Definition 2.1.3 (Binary Relation) A relation $\mathcal{R}$ on the set X with itself is called a binary or homogeneous relation. Namely, $\mathcal{R} \subseteq X \times X$.

For brevity, binary relations on $X \times X$ are simply referred to as relations on the set X.

Example 2.1.4 Suppose that $X = \{\text{father, mother, daughter, son}\}$ is a set of family members, which we abbreviate as $\{F, M, D, S\}$ in the obvious way. Let us define a relation $\mathcal{R}$ on this family for which $x \sim y$ means "x is a child of y".

Using traditional familial convention as an axiom, we can determine which of the family members are related via $\mathcal{R}$: each of the son and daughter are related to both the father and mother. Namely, we have $S \sim F$, $S \sim M$, $D \sim F$, and $D \sim M$. We do not have any other ordered pairs in this relation, so $\mathcal{R} = \{(S, F), (S, M), (D, F), (D, M)\} \subseteq X \times X$.

Note that in Example 2.1.4, $x \sim y$ does not necessarily mean $y \sim x$. Relations that satisfy this condition are called symmetric relations and it is one of many special types of relations. Here are some relevant special types of binary relations that we are going to look at.

Definition 2.1.5 (Some Types of Binary Relations) Let $\mathcal{R} \subseteq X \times X$ be a binary relation on the set X. We call the relation $\mathcal{R}$:

1. Reflexive: If $x \sim x$ for all $x \in X$.
2. Irreflexive: If $x \not\sim x$ for all $x \in X$.
3. Symmetric: If $x \sim y$, then $y \sim x$.
4. Antisymmetric: If $x \sim y$ and $y \sim x$, then $x = y$.
5. Transitive: If $x \sim y$ and $y \sim z$, then $x \sim z$.
6. Strongly connected: For any $x, y \in X$, either $x \sim y$ or $y \sim x$.
7. Trichotomous: For all $x, y \in X$, exactly one of $x = y$, $x \sim y$, or $y \sim x$ holds.

There are many other special types of relations such as asymmetric, dense, and connected, but we do not need to know them for this book.

Remark 2.1.6 When we plot a relation on X in a table or on a Cartesian diagram, some of these types of relations may be indicated by some visual features of the plot. For example, reflexivity means that the whole diagonal is plotted, irreflexivity means that the whole diagonal is not plotted, and symmetric relation means that the plot is symmetric about the diagonal.

We shall see some important relations on a set later on, namely strict and partial orders.

Equivalence Relation

A very important kind of relation that we can define now is:

Definition 2.1.7 (Equivalence Relation) Let $\mathcal{R} \subseteq X \times X$ be a relation on a set X. The relation $\mathcal{R}$ is called an equivalence relation on X if it is reflexive, symmetric, and transitive.

Example 2.1.8 Let us look at some examples:

1. Recall Example 2.1.4 earlier. The relation $\mathcal{R}$ on this family given by "is a child of" is not an equivalence relation. This is very easy to check because it is, for example, not a reflexive relation. This is clear because every family member is not a child of themselves and so this relation is not an equivalence relation.
2. If we define a new relation $\mathcal{S}$ on this family for which $x \sim y$ iff "x has the same gender as y", we can check that this is an equivalence relation. Indeed, we can list down the relation $\mathcal{S} = \{(S, F), (F, S), (S, S), (F, F), (D, M), (M, D), (D, D), (M, M)\}$ to check that it is reflexive, symmetric, and transitive.

A curious thing to note here is this relation splits the set of family members into two disjoint subsets in which the members are related only to each other, namely $\{F, S\}$ and $\{M, D\}$. We shall investigate this feature in more generality in Theorem 2.1.12.

If $\mathcal{R}$ is an equivalence relation on a set X, for any $x \in X$ we can group together elements which are related to the element x. This group is non-empty since every element must be related to itself. We call each of these subsets an equivalence class for the set element x:

Definition 2.1.9 (Equivalence Class) Let X be a non-empty set and $\sim$ is an equivalence relation on X. For an element $x \in X$, the equivalence class of x, denoted as $[x]$, is the subset:

$$[x] = \{y \in X : x \sim y\} \subseteq X.$$

In other words, the equivalence class $[x]$ is the subset of X consisting of all the elements in X which are related to x.

The set of all equivalence classes of X under the equivalence relation $\sim$ is called a quotient set and is denoted as:

$$X/\!\sim\, = \{[x] : x \in X\}.$$

Remark 2.1.10 An important distinction to take note here is that for any $x \in X$, the object $[x]$ is a subset in the set X, but is a point in the quotient set $X/\!\sim$. More succinctly, $[x] \subseteq X$ but $[x] \in X/\!\sim$. Therefore context is important!

Example 2.1.11 In Example 2.1.8(2) we have $X/\!\sim_{\mathcal{S}} \; = \; \{[F], [M], [D], [S]\}$. However we saw that there are only two equivalence classes for the relation $\mathcal{S}$, namely $\{F, S\}$ and $\{M, D\}$. Thus we have the equality of the equivalence classes $[F] = [S] = \{F, S\}$ and $[M] = [D] = \{M, D\}$. Because of the repetition, we can discard any repeated classes and the quotient set can be written as $X/\!\sim_{\mathcal{S}} = \{[F], [M]\}$.

Note that the choice for which repeated class that we want to discard is arbitrary. We call the choice of element $x \in X$ used to denote an equivalence classes as a representative of the class. In the above, the representatives chosen for each class are F and M respectively.

We could equivalently write the quotient set as $X/\!\sim_{\mathcal{S}} = \{[D], [S]\}$ by choosing different representatives from each class. The choice of representative for each class is not that important: they are used simply as a label.

We have noted in Example 2.1.8(2) that the equivalence relation $\mathcal{S}$ splits the set of family members $\{F, M, S, D\}$ into two disjoint subsets in which every element in every subset is only related to elements in the same subset as itself. This is true in general: for an equivalence relation $\sim$ on a set X, the equivalence classes of $\sim$ partitions the set X. A partition here means that the whole set X can be written as a union of non-intersecting subsets of X. In fact, the opposite is also true: any partition of the set X induces an equivalence relation on X. This is an important feature of equivalence relations.

Theorem 2.1.12 (Fundamental Theorem of Equivalence Relations) *Let X be a set. Then:*

1. *If $\mathcal{R} \subseteq X \times X$ is an equivalence relation on a set X, then the equivalence classes either coincide or are disjoint. In other words, the equivalence classes partition the set X, namely:*

$$X = \bigcup_{x \in X} [x],$$

   with either $[x] \cap [y] = \varnothing$ or $[x] = [y]$ for any $x, y \in X$.
2. *Any partition of the set X induces an equivalence relation on the set X.*

Proof We prove the assertions one by one:

1. First, by reflexivity of the relation, each element $x \in X$ is necessarily contained in an equivalence class $[x]$. Thus, by double inclusion, the union of all the equivalence classes $\bigcup_{x \in X} [x]$ is the whole of X.
   Next, we want to show that any two of these classes are disjoint or coincide. Let $[x]$ and $[y]$ be two equivalence classes in X. Suppose that they are not disjoint. We aim to show that $[x] = [y]$ via double inclusion. Since they are not disjoint, there exists an element $z \in [x] \cap [y]$. Pick any $w \in [x]$. Since $z \in [x]$, we must have $w \sim z$. Moreover, since $z \in [y]$, we must have $z \sim y$. By transitivity of the equivalence relation, we then have $w \sim y$. So $w \in [y]$ as well and thus $[x] \subseteq [y]$. By reversing the roles, we can show that $[y] \subseteq [x]$ and thus we conclude that $[x] = [y]$.
2. For a given partition of X, define a relation $\sim$ on X as $x \sim y$ iff x and y belong to the same partition subset. This is an equivalence relation since it is:
   (a) Reflexive: Any point $x \in X$ is related to itself naturally.
   (b) Symmetric: If $x \sim y$, then x and y are contained in the same subset of the partition and thus $y \sim x$.
   (c) Transitive: If $x \sim y$ and $y \sim z$, then x and y are in the same partition subset $A \subseteq X$ and y and z are in the same partition subset $B \subseteq X$. Since the subsets partition X and both contain a common element y, the subsets A and B must coincide. We conclude that $x \sim z$.

   Thus, this relation is an equivalence relation. □

An equivalence relation is used to treat different but related elements in a set X as the same element in the quotient set $X/\sim$ by collapsing the (possibly many) related elements in X into one single element in the quotient set $X/\sim$. As a result, it is a useful construction in many different areas of mathematics such as group theory, ring theory, number theory, and topology. We shall see in Sect. 2.4 how we can utilise this construction.

2.2 Natural Numbers $\mathbb{N}$

We have defined some basic mathematical concepts from set theory. Now we ask ourselves: what is analysis? It is a mathematical area that deals with limits, and infinity. This includes properties of real numbers, infinite sequences, infinite sums, continuous functions, and the fundamentals of calculus.

Students have seen these concepts in school and calculus classes as black boxes, but where do these concepts come from? These concepts are the product of hundreds (thousands, according to [68]) of years of mathematical ingenuity, study, and formalisations. But for practical reasons, these concepts are usually condensed into one or two years of classes in school. Therefore, naturally we expect that the finer details of the mathematical content are omitted, skipped over, or simplified.

Analysis is where we put these concepts under a magnifying glass and study them rigorously from first principles to understand where they come from. This would give us more insight on the techniques and more general problems that may arise from them. Real analysis is a subset of a wider subject of analysis in which we look at analysis related to real numbers.

Analysis on complex numbers can be extended easily from the analysis of real numbers, but is more rigid in nature. The rigid nature of complex analysis gives it a different flavour compared to real analysis, but let us hold that for a later exploration. We need to first understand real analysis before we move on to complex analysis!

First of all, what is a number? In order to understand its property, we need to first be able to define what the set of numbers is. This is a very difficult question; it is easy to think about numbers, but describing it or giving it a definition is a very difficult philosophical question! For a brief and enjoyable read on the nature and various interpretation of numbers, readers are directed to [27].

It is usually said that numbers are (hidden) in nature around us. It is hugely debated whether they are discovered or invented. John Fraleigh and Raymond Beauregard profoundly wrote in their book *Linear Algebra*:

> Numbers exist only in our minds. There is no physical entity that is number 1. If there were, 1 would be in a place of honor in some great museum of science, and past it would file a steady stream of mathematicians gazing at 1 in wonder and awe.

The concept of numbers were devised by our ancestors to make sense of the world as well as simplify their (and our) lives. Historically, numbers come about as tools to count, measure, and label. As time progresses, they are used in trade, commerce, record-keeping, taxation, travel and seafaring, time and calendars, astronomy, and science before they are studied in their own right.

As mentioned above, the first incarnation of the idea of numbers is probably via counting and labelling. We can count discrete objects using natural numbers, as we have done many times before in this book: for example when we enumerate pages, chapters, theorems, and label mathematical objects using these numbers as their indices. Using the Hindu-Arabic numeral notation, we denote this set of counting numbers as the set of symbols:

$$\mathbb{N} = \{1, 2, 3, 4, 5, 6, 7, 8, 9, 10, 11, 12, 13, 14, 15, \ldots\}.$$

Remark 2.2.1 We make several remarks here regarding the notation above.

1. It is important to note that natural numbers exist independently of notations. We can still count objects by pointing to each object one by one, counting down using closed fingers, using tallies such as bars and crosses, or separate the objects into piles without attaching a name to the counters.
2. Attaching a symbol, notation, or name to the counters allows us to communicate these numbers so that they are can be better understood and recorded. Ancient civilisations use various different notations to denote the natural or counting number. Some ancient civilisation, such as the Mayans and the Phoenicians, used tallies to denote the numbers. Others use more complicated symbolic notation, such as the Chinese or Roman numerals. A historical discussion of numerals in various civilisation can be found in [14].
3. Here we adopt the Hindu-Arabic positional numeral system, for which every natural number can be represented by a combination of ten distinct symbols $1, 2, 3, 4, 5, 6, 7, 8, 9$, and 0 arranged in a string. These are just a choice of symbols; we can use any different ten distinct symbols such as any of the ones in Fig. 2.1 if we like.
4. This notation is also called the base-10 system since we are using ten distinct symbols to represent them. This is one of the most favoured numeral system as it is widely used, more intuitive, and easily adapted for arithmetic. And most of all, it is because many of us were already taught (or indoctrinated) to use this system since pre-school that any attempt to move away from this system would be a chore.
5. In many literature, the number 0 is also regarded as a natural number. Even mentioning this usually sparks a heated debate, so we will just declare (for our purposes) that 0 is not a natural number here in this book. There, I said it!

Algebra of Natural Numbers

Natural numbers are also called whole numbers, because they represent whole discrete quantities. At the moment, the set $\mathbb{N}$ consisting of strings of the squiggly symbols in Fig. 2.1 is just a set. Nothing more, nothing less. Let us add an algebraic structure on this set by declaring how the elements interact with each other via algebraic operations.

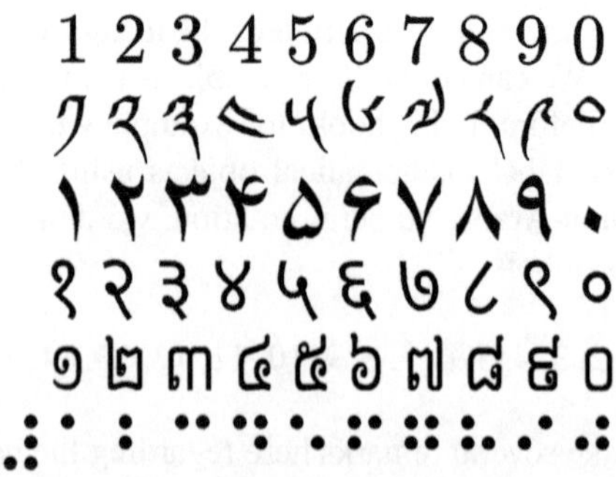

Fig. 2.1 Some examples of numeral symbols from various cultures which use the base-10 positional numeral system. From above: Hindu-Arabic, Tibetan, Persian, Devanagari, Khmer, and Braille numerals. Note that the first symbol in the Braille numerals list indicates that the symbols following it are treated as numerals rather than alphabets (these symbols are also used to denote the Latin letters A to J in Braille)

Remark 2.2.2 Here are some remarks regarding the term algebra:

1. The term algebra comes from the Arabic word *al-jabr* which means "reunion or rejoining of broken parts". This fits with our objective: to combine numbers together via some kind of interaction/operation between them. This term comes from the title of a book *al-Kitab al-Mukhtasar fi Hisab al-Jabr wal-Muqabalah* (The Compendious Book on Calculation by Completing and Balancing) by Muhammad ibn Musa al-Khwarizmi (c. 780–850).
2. The verbs completing and balancing in the title of the book by Khwarizmi refer to the methods of solving equations of real numbers in the form of $A = B$. We complete and balance the terms on each side by doing some sequence of operations on both sides of the equation. These operations are called algebraic operations.
3. In fact, algebra is a very wide terminology. It can also be used to describe ways to combine other abstract objects apart from numbers. Recall from Sect. 1.3 that we have the algebra of sets as ways to combine or interact sets. Later on we shall see other types of algebra such as algebra of limits and algebra of functions.

Here, we are going to define two basic algebraic operations that we can do on the set $\mathbb{N}$ for these completion and balancing processes. We can add and multiply natural numbers by thinking of them as counting numbers.

1. Addition is defined as follows: if we have m objects in one hand and n objects in the other, if we combine them together, we would have $m + n$ objects.
2. For multiplication: if we have m objects each in n bags, combining them all by adding the m objects n times, we can show that there are $\underbrace{m + m + \ldots + m}_{n \text{ times}} = m \times n$ objects in total. Alternatively, we can also view it as follows: if we have

n bags each containing m objects, we have a total of $\underbrace{n + n + \ldots + n}_{m \text{ times}} = n \times m$ objects.

This also means that multiplication is simply repeated additions.

The set $\mathbb{N}$ is called closed under addition and multiplication because whenever we combine two elements of $\mathbb{N}$ via addition or multiplication, the outcome is also an element of $\mathbb{N}$, namely $m + n, m \times n \in \mathbb{N}$ for all $m, n \in \mathbb{N}$. Moreover, from these intuitive definitions and by an easy check, both of these operations are:

1. Commutative: $m + n = n + m$ and $m \times n = n \times m$ for all $m, n \in \mathbb{N}$.
2. Associative: $(m + n) + p = m + (n + p)$ and $(m \times n) \times p = m \times (n \times p)$ for all $m, n, p \in \mathbb{N}$.
3. $\times$ is distributive over $+$: $p \times (m + n) = (p \times m) + (p \times n)$ for all $m, n, p \in \mathbb{N}$.

With all of these clarified, we can now define the natural numbers as an infinite set $\mathbb{N}$ endowed with the algebraic structures of addition and multiplication.

In modern day mathematics, the concept of natural numbers are axiomatised via Peano's axioms, after the Italian mathematician Giuseppe Peano (1858–1932). The Peano's axioms formalise the structure on the set $\mathbb{N}$ by the following axioms or rules:

Definition 2.2.3 (Peano's Axioms) The natural numbers $\mathbb{N}$ is a set such that:

1. $1 \in \mathbb{N}$.
2. If $n \in \mathbb{N}$, then $s(n) \in \mathbb{N}$. The operation $s : \mathbb{N} \to \mathbb{N}$ is called the successor operation and $s(n)$ is called the number succeeding n.
3. $1 \neq s(n)$ for any $n \in \mathbb{N}$.
4. For any $m, n \in \mathbb{N}$, $s(m) = s(n)$ if and only if $m = n$.
5. Axiom of induction: If $A \subseteq \mathbb{N}$ is such that $1 \in A$ and $n \in A \Rightarrow s(n) \in A$, then $A = \mathbb{N}$.

Principle of Mathematical Induction

An important note is that the final Peano's axiom formalises the principle of mathematical induction over the natural numbers that one would have seen before in school. Indeed, suppose that we have a series of statements $P(n)$ indexed by the natural numbers. If we set $A = \{n \in \mathbb{N} : \text{statement } P(n) \text{ is true}\}$ and we can show the two conditions:

1. Base case: $1 \in A$ (namely $P(1)$ is true), and
2. Inductive step: $k \in A \Rightarrow s(k) \in A$ (namely if $P(k)$ is true then $P(s(k))$ is also true),

are true, then the axiom of induction says that $A = \mathbb{N}$, namely $P(n)$ is true for all $n \in A = \mathbb{N}$. The assumption $P(k)$ is true in the inductive step is called the inductive hypothesis. In fact, we can also adjust this axiom to start from another base case but the inductive step stays the same.

Remark 2.2.4 In many introductory mathematical literature, the axiom of induction is stated with an analogy of dominoes falling on top of each other. Imagine we have an infinite number of dominoes arranged so that they stand on their smallest side, one after another in an infinite line. The n-th domino represents the statement $P(n)$ and toppling it over is analogous to proving that the statement $P(n)$ is true. The axiom of induction says:

1. (Base case) if we can topple the first domino, and
2. (Inductive step) toppling the k-th domino also topples the successor $s(k)$-th domino,

then all the dominoes in the line are toppled, namely all the statements $P(n)$ are true.

Indeed, by toppling the first domino and using the inductive step, we can topple the second domino. Since the second domino is toppled, by the inductive step, the third domino is toppled. This is repeated so that, eventually, any domino far along the sequence will be toppled (Fig. 2.2).

This axiom of induction is important because it allows us to define addition and multiplication from the rest of the Peano's axioms. Indeed, Peano defined addition and multiplication for all $a, b \in \mathbb{N}$ recursively as:

$$a + 1 = s(a) \quad \text{and} \quad a + s(b) = s(a + b), \tag{2.1}$$

$$a \times 1 = a \quad \text{and} \quad a \times s(b) = a + (a \times b). \tag{2.2}$$

Thus, the algebraic facts that we have laid out for natural numbers, namely the addition and multiplication prior to Definition 2.2.3, can be derived from the five axioms of Peano. Therefore, the natural numbers can be axiomatised formally.

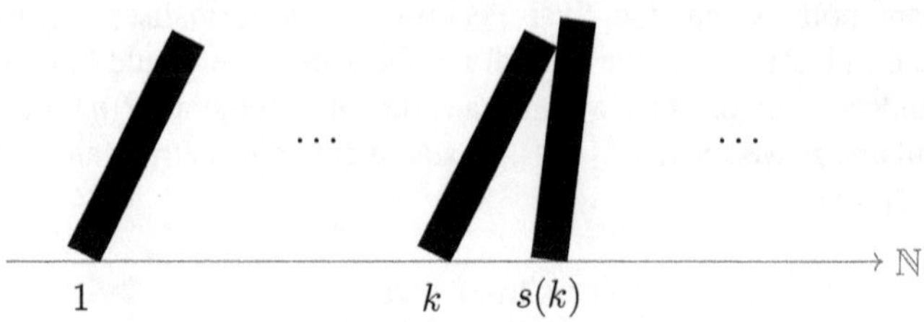

Fig. 2.2 Principle of mathematical induction using the dominoes analogy

However, we are not going to approach the constructive formalisation of natural numbers according to Peano and accept them, well, naturally. This agrees with the naturalism philosophy of Leopold Kronecker (1823–1891) [27] who stated that:

> God made the natural numbers; all else is the work of man.

Of course, this seems like a leap of faith to some people, so we invite the readers to build the natural numbers using Peano's axioms by checking that the addition and multiplication operations defined in (2.1) and (2.2) are well-defined and agree with our earlier definition. This checking is set for the readers as Exercise 2.9.

For simplicity, we now define some shorthand notations when writing down the operations. For $m, n \in \mathbb{N}$ we define the following simplifications:

1. $mn = m \times n$.
2. $m^n = \underbrace{m \times m \times \ldots \times m}_{n \text{ times}}$. This is called the exponentiation. In this notation, m is called the base and n is called the exponent.

Example 2.2.5 Now let us look at how the axiom of induction can be used to prove a property that we expect to hold for every natural number. This is especially useful when we want to prove a statement involving the quantifier $\forall n \in \mathbb{N}$.

We claim that the statement $\forall n \in \mathbb{N}, (1 + 3 + 5 + \ldots + (2n + 1) = (n + 1)^2)$ is true. To prove this via induction, we have to show two things: the base case and the inductive step. For any $n \in \mathbb{N}$, we denote the statement $1+3+5+\ldots+(2n+1) = (n + 1)^2$ as $P(n)$. We write LHS and RHS as the left-hand side and right-hand side of the equation respectively.

1. Clearly, $P(1)$ true since LHS $= 1 + 3 = 4$ and RHS $= (1 + 1)^2 = 2^2 = 4$.
2. Now assume that $P(k)$ is true, namely $1 + 3 + 5 + \ldots + (2k + 1) = (k + 1)^2$. We aim to show that this assumption implies that $P(k + 1)$ is also true, namely $1 + 3 + 5 + \ldots + (2k + 1) + (2k + 3) = (k + 2)^2 = k^2 + 4k + 4$ is also true. To achieve this, we use direct method, namely we start with the LHS and try to get to the RHS of the equation using the assumption that $P(k)$ is true. We start with:

$$\begin{aligned}\text{LHS} &= 1 + 3 + \ldots + (2k + 1) + (2k + 3)\\ &= (k + 1)^2 + (2k + 3) \quad (\because P(k) \text{ is true})\\ &= k^2 + 2k + 1 + 2k + 3\\ &= k^2 + 4k + 4 = \text{RHS},\end{aligned}$$

which is what we wanted to prove. Therefore the truth of $P(k)$ implies that $P(k+1)$ is true.

By the principle of mathematical induction, we can conclude that $P(n)$ is true for all $n \in \mathbb{N}$.

2.3 Ordering on $\mathbb{N}$

Apart from counting, measuring, and labelling, the natural numbers are also used to order a collection of objects. The ordering of elements of $\mathbb{N}$ occurs in a natural way: informally, the longer we need to successively count to a number, the greater it is.

Before we attempt to formalise this, let us first define what an ordering should mean. Generally, a strict total order structure on a set X is a binary relation that satisfies some special conditions:

Definition 2.3.1 (Strict Total Order) A strict total order $<$ on a set X is a binary relation on X that is:

1. Irreflexive: $a \not< a$ for all $a \in X$,
2. Transitive: If $a < b$ and $b < c$, then $a < c$, and
3. Trichotomous: For any $a, b \in X$ exactly one of the following holds: $a < b$, $a = b$, or $b < a$.

In Definition 2.3.1 we use the symbol $<$ instead of the usual notation $\sim$ used to denote the strict total order relation. This is to emphasise that this relation is a strict total order (and thus must fulfil the three conditions above) on the set X. Moreover, it is an instructive notation: the smaller end of $<$ points towards the "smaller" element in the order.

Remark 2.3.2 The term "total" (as opposed to "partial" order) in the name refers to the fact that any two arbitrarily picked distinct elements in X are related to each other in some way via the trichotomy condition. This means the strict total order $<$ allows us to totally compare any two distinct elements in the set X.

An example of a strict total order on a set is the following:

Example 2.3.3 Let $\mathcal{L}$ be the set of all the words in a simplistic fictional language with only two letters $\{x, y\}$. Every word in this language consists of either 1 or 2 letters, so there are only 6 different words in this language. Suppose that we want compile all of these words in a small dictionary. We can order the words in a dictionary according to a lexicographical ordering, defined as follows:

1. If $V, W \in \mathcal{L}$ are such that the first letter of the words V and W are x and y respectively, then $V < W$.
2. If $V, W \in \mathcal{L}$ share the first letter but V has fewer letters than W, then $V < W$.
3. Supposing that $V, W \in \mathcal{L}$ have two letters and share the first letter, if the second letter of V is x and the second letter of W is y, then $V < W$.

We can show that this defines a total order in the list of all words in this language by checking it fulfils Definition 2.3.1. Using the rules above, we can order all the words in the dictionary as follows:

$$x < xx < xy < y < yx < yy. \tag{2.3}$$

In fact, any finite set at all can be endowed with some kind of total order. One just has to list them down and declare axiomatically that the order is increasing as we go along the list, as in the list (2.3) above.

Back to natural numbers. Earlier in this section, we gave a very informal and intuitive way of defining an order on the natural numbers via successive counting. Let us provide a formal definition here:

Definition 2.3.4 (Natural Strict Total Order on $\mathbb{N}$) The natural strict total order on $\mathbb{N}$ is $m < n$ iff there exists an $x \in \mathbb{N}$ such that $m + x = n$.

This ordering satisfies all the strict total order axioms in Definition 2.3.1. We check:

1. Irreflexive: Clearly $n \not< n$ since $n + x \neq n$ for any $x \in \mathbb{N}$.
2. Transitive: If $m < n$ and $n < r$, then there are $x, y \in \mathbb{N}$ such that $m + x = n$ and $n + y = r$. Adding y to the first equation, we get $m + (x + y) = n + y = r$ which means $m < r$.
3. Trichotomous: If $m = n$, then clearly neither $n < m$ nor $n < m$ holds since there are no $x \in \mathbb{N}$ such that $m + x = n$ or $n + x = m$. Now suppose that $m \neq n$. We prove that both $m < n$ and $n < m$ cannot be true simultaneously. Assume for contradiction that they are simultaneously true. Then, there exist $x, y \in \mathbb{N}$ such that $m + x = n$ and $n + y = m$. Thus $m + (x + y) = n + y = m$ and so $m < m$, which contradicts irreflexivity. Hence $m < n$ and $n < m$ cannot both be true at the same time. This shows that for any $m, n \in \mathbb{N}$, at most one of $m = n$, $m < n$, or $n < m$ holds.
   Now we prove that at least one of $m = n$, $m < n$, or $n < m$ is true by induction on n with a fixed m. For the base case $n = 1$, if $m = 1$, then $m = n$. Otherwise if $m \neq 1 = n$, we must have $m = m \times 1 = \underbrace{1 + 1 + \ldots + 1}_{m \text{ times}}$ which means $m = 1 + (1 + 1 + \ldots + 1) = 1 + x = n + x$ for some $x \in \mathbb{N}$. This implies $m > n = 1$ is true. Thus, for the base case, either $m = n$ or $m > n$ is true.
   For the inductive step, assume that for $n = k$, at least one of $m = k, m < k$, or $k < m$ is true. We now show that at least one of $m = k + 1, m < k + 1$, or $k + 1 < m$ is true.
   (a) If $m = k$ is true, then $m + 1 = k + 1$, which implies $m < k + 1$ is true.
   (b) If $m < k$ is true, by definition of the order, there exists an $x \in \mathbb{N}$ such that $m + x = k$. This implies $m + x + 1 = k + 1$ and so $m < k + 1$ is true.

(c) If $k < m$ is true, then there exists an $x \in \mathbb{N}$ such that $k + x = m$. We have two cases for the value of x:
   i. If $x = 1$, then $m = k + 1$ is true.
   ii. Otherwise, $x \neq 1$ and, by the base case, there is a $y \in \mathbb{N}$ such that $x = y + 1$. Substituting this in the equation $k + x = m$, we have $k + y + 1 = m$ or $(k + 1) + y = m$. Therefore $k + 1 < m$ is true.

This completes the inductive step. Hence, at least one of $m = n$, $m < n$, or $n < m$ is true for any $m, n \in \mathbb{N}$.

Putting the two results together, we conclude that for any $m, n \in \mathbb{N}$, exactly one of $m = n$, $m < n$, or $n < m$ is true.

For simplicity, we denote $x \leq y$ if either $x = y$ or $x < y$. The symbol $\leq$ is called a weak inequality and the symbol $<$ is called a strict inequality. In contrast to the strict total order in Definition 2.3.1, the weak inequality is a relation called a total order which satisfies:

Definition 2.3.5 (Total Order) A total order $\leq$ on a set X is a binary relation on X that is:

1. Reflexive: $a \leq a$ for all $a \in X$,
2. Antisymmetric: If $a \leq b$ and $b \leq a$, then $a = b$,
3. Transitive: If $a \leq b$ and $b \leq c$, then $a \leq c$, and
4. Strongly connected: For any $a, b \in X$, either $a \leq b$ or $b \leq a$.

Based on this order, the set $\mathbb{N}$ does not have a largest element because we can theoretically just keep on counting forever by successively creating larger elements since $n < n + 1$ for all $n \in \mathbb{N}$. However, any non-empty subset of it must have a smallest element. This result is called the well-ordering principle:

Lemma 2.3.6 (Well-Ordering Principle) *Let $\mathbb{N}$ be the set of natural numbers equipped with the strict total order $<$ in Definition 2.3.4. Then, any non-empty subset of the natural numbers $\mathbb{N}$ must have a smallest element.*

Proof For $n \in \mathbb{N}$, let $P(n)$ be the statement "Any subset of $\mathbb{N}$ containing a natural number smaller than or equal to n has a least element". We prove that $P(n)$ is true for all $n \in \mathbb{N}$ via mathematical induction.

1. $P(1)$ is true since any subset of $\mathbb{N}$ which contains 1 has a least element, which is 1.
2. Now assume that $P(k)$ is true for some $k \in \mathbb{N}$, namely any subset of $\mathbb{N}$ that contains any natural number smaller than or equal to k contains a least element. We aim to show that $P(k+1)$ is also true. Pick any subset $S \subseteq \mathbb{N}$ which contains a natural number smaller than or equal to $k + 1$. We have two cases:
   (a) If S does not contain any natural number strictly smaller than $k + 1$, then S must contain $k + 1$ and this is the smallest element of S.

(b) Otherwise, if S contains a natural number strictly smaller than $k + 1$, then it contains a natural number smaller than or equal to k. By the inductive hypothesis, the set S contains a least element.

Thus, in both cases, $P(k + 1)$ is also true.

Hence, we conclude that the statement $P(n)$ is true for all $n \in \mathbb{N}$.

Now pick any non-empty subset $X \subseteq \mathbb{N}$. This set X must contain some $n \in \mathbb{N}$. Since $P(n)$ is true, the set X has a least element. □

For a non-empty subset $X \subseteq \mathbb{N}$, we call the smallest element of X the minimum of X, denoted as $\min(X)$. In other words, for all $x \in X$ we have $x \geq \min(X)$. Moreover, if X has a largest element, we call this element the maximum of X, denoted as $\max(X)$. In other words, for all $x \in X$ we have $x \leq \max(X)$.

However, the well-ordering principle does not guarantee that $\max(X)$ exists for any subset $X \subseteq \mathbb{N}$. Indeed, as we have mentioned above, the set $\mathbb{N}$ itself does not have the largest element. Therefore, it is convenient and useful to introduce a symbol to represent this boundless or endless concept if the maximum does not exist.

Definition 2.3.7 (Infinity) The infinite symbol ∞ is used to denote a boundless quantity.

Remark 2.3.8 Infinity is a scary concept that makes many people, including seasoned mathematicians, uneasy. Let us make some remarks regarding infinity:

1. Infinity does not represent a unique concept. Due to its ethereal and mysterious nature, this depends on various interpretation and context ranging from mathematical, philosophical, theological, and cultural. One thing for sure, as remarked by Steven Strogatz [72], it unnerved a lot of people:

   > Infinity lies at the heart of so many of our dreams and fears and unanswerable questions: How big is the universe? How long is forever? How powerful is God? In every branch of human thought, from religion and philosophy to science and mathematics, infinity has befuddled the world's finest minds for thousands of years. It has been banished, outlawed, and shunned. It's always been a dangerous idea.

2. The infinity symbol in Definition 2.3.7 is credited to John Wallis (1616–1703). Be aware that this symbol does not denote a natural number. As Karl Friedrich Gauss (1777–1855) vehemently disapproves:

   > I protest against the use of an infinite quantity as an actual entity; this is never allowed in mathematics. The infinite is only a manner of speaking, in which one properly speaks of limits to which certain ratios can come as near as desired, while others are permitted to increase without bound.

   In a sense, for Gauss ∞ is a "destination" rather than a "place".

3. In his works *Physics* and *Metaphysics*, Aristotle (c. 384B.C.-322B.C.) classified infinity into two different groups: potential infinity and actual infinity.
   (a) Any process that can be extended indefinitely (such as listing down the natural numbers) is called a "potential infinity". This process is never complete since we can never get to the end of it, similar to the interpretation of infinity by Gauss. Another example of this is the principle of mathematical induction. In the domino analogy, we topple the first domino, which topples the second domino, which topples the third domino, and so on and so forth, indefinitely going towards infinity.
   (b) In contrast, "actual infinity" is a definite infinity. This is a complete and definite infinite. It sounds like an oxymoron that even Aristotle rejected the possibility of an actual infinity. However, actual infinities do exist (in our minds, at least). An example of this is when we say: the set of natural numbers is infinite. This infinite is taken as a whole rather than an indefinite process.
4. To utilise the concept of actual infinity above, modern mathematicians and philosophers introduced number systems that treat infinity as actual numbers (opposing Gauss's protest) rather than a destination that one can never reach. In fact, these infinities may also satisfy some algebraic operations in the new number systems.
   These number systems are the cardinal and ordinal numbers and were introduced by Cantor. They are used to measure the size of an abstract set and order the elements in an abstract set respectively, thus they require Aristotle's concept of actual infinity. We shall encounter cardinal numbers in Definition 3.4.4 and observe that infinity also exists in many different variants.
5. Moreover, we shall also see the extended real number system (see Definition 18.1.1) in which ∞ is allowed to interact algebraically with the natural numbers (and most of the real numbers) meaningfully.

As we have noted, any subset of $\mathbb{N}$ has the smallest element but is not guaranteed to have the largest element. However, if a subset of $\mathbb{N}$ is finite (namely the set has n elements, where $n \in \mathbb{N}$), we can always find one:

Lemma 2.3.9 *Let $X \subseteq \mathbb{N}$ be a finite subset of $\mathbb{N}$. There exist elements $a, b \in X$ such that $\max(X) = a$ and $\min(X) = b$.*

Proof The minimum exists by well-ordering principle. For the maximum, this is proven via induction on the size of X. The base case for the induction is when the size of the set X is 1 so that $X = \{x_1\}$. Clearly $\max(X)$ exists and is equal to x_1.

Now assume that the maximum exists for the case of sets with size k. We want to prove that the maximum also exists for any set of size $k+1$. Suppose that $X = \{x_1, x_2, \dots, x_{k+1}\}$. We can write X as the union $X = \{x_1, x_2, \dots, x_k\} \cup \{x_{k+1}\}$ which is a union of a set of size k and a set of size 1. By inductive hypothesis, there exists a maximal element in the former set since it is of size k. Let x the maximal element of the set $\{x_1, x_2, \dots, x_k\}$. Then, there are two cases:

1. If $x_{k+1} \leq x$, then $x_j \leq x$ for all $j = 1, 2, \dots, k+1$ and so $\max(X) = x$.

2. If $x_{k+1} > x$, then $x_{k+1} > x \geq x_j$ for $j = 1, 2, 3, \ldots, k$. This also means $x_{k+1} \geq x_j$ for all $j = 1, 2, \ldots, k+1$ which then implies $\max(X) = x_{k+1}$.

Either way, the maximum of the set X is attained by some element in X. □

The readers shall also prove the converse of Lemma 2.3.9 in Exercise 2.14, namely any subset of $\mathbb{N}$ with a maximum must be a finite set.

Combining the algebraic operations $+$ and $\times$ with ordering in Definition 2.3.4, we obtain these important properties:

Proposition 2.3.10 *Suppose that $m, n, x \in \mathbb{N}$.*

1. *If $m < n$, then $m + x < n + x$.*
2. *If $m < n$, then $xm < xn$.*
3. *$m \leq mn$.*

Proof We prove the assertions one by one.

1. This can be proven by induction on x. Fix any $m, n \in \mathbb{N}$ with $m < n$. For the case $x = 1$, suppose for contradiction that $m + 1 \geq n + 1$. We would then have $m + 1 \geq n + 1 > n$ which then implies $m < n < n + 1 \leq m + 1$. But there are no natural numbers n strictly in between any consecutive natural numbers m and $m + 1$, which is a contradiction. Thus, we must have $m + 1 < n + 1$.
   Now assume that this is true for any $m, n \in \mathbb{N}$ and $x = k \in \mathbb{N}$. By the inductive case, we have $m + k < n + k$. Repeating the same argument for the base case, we can deduce $m + (k + 1) < n + (k + 1)$, which gives us the conclusion.
2. We prove this via induction on x. Clearly, the base case for $x = 1$ is true by assumption. Assume that the inequality is true for $x = k$, namely $km < kn$. To prove the case for $x = k + 1$, we apply the previous assertion to get $km + m < kn + m < kn + n$ which then implies $(k + 1)m < (k + 1)n$.
3. We note that $n \geq 1$. If $n = 1$, then $m = m \times n$. Otherwise, if $n > 1$, by applying the previous assertion, we have $m = m \times 1 < m \times n$. □

From Proposition 2.3.10, we also deduce the cancellation laws on the natural numbers:

Proposition 2.3.11 (Cancellation Laws on $\mathbb{N}$) *Suppose that $m, n, x \in \mathbb{N}$.*

1. *If $m + x = n + x$, then $m = n$.*
2. *If $m + x < n + x$, then $m < n$.*
3. *If $mx = nx$, then $m = n$.*
4. *If $mx < nx$, then $m < n$.*

Proof We shall prove the first two claims only. The other claims are left as Exercise 2.16.

1. By contrapositive, this is equivalent to proving $m \neq n$ implies $m + x \neq n + x$. Assume that $m \neq n$. WLOG, suppose that $m < n$. From Proposition 2.3.10, we have $m + x < n + x$ which means $m + x \neq n + x$.
2. Since $m+x < n+x$, there exists a $k \in \mathbb{N}$ such that $m+x+k = n+x$. Applying the first assertion, we have $m + k = n$ and hence $m < n$. □

Factors and Divisors

From the cancellation law on multiplication in Proposition 2.3.11(3), we can ask an important question: given two natural numbers $m, n \in \mathbb{N}$, does there exist another natural number x for which $m \times x = n$?

If there is such a number, we call m and x factors or divisors of n and we write $x = \frac{n}{m} \in \mathbb{N}$. Obviously, this question does not have an answer if $n < m$ since $mx \geq m > n$ for any $x \in \mathbb{N}$. Even if $n \geq m$, this question may not have an answer all the time.

Example 2.3.12 We can check for a small case where $m = 2$ and $n = 5$: we have $2 \times 1 = 2, 2 \times 2 = 4, 2 \times 3 = 6$, and $2 \times x > 6$ for any $x > 3$. Therefore there can never be an $x \in \mathbb{N}$ for which $2 \times x = 5$. In fact, for any $1 \leq m \leq 5$, the only two possible numbers m that allow us to do this are $m = 1, 5$.

On the other hand, if $n = 6$, we can write $1 \times 6 = 6, 2 \times 3 = 6, 3 \times 2 = 6$, and $6 \times 1 = 6$.

Let us define:

Definition 2.3.13 (Factors, Divisors) Let $n \in \mathbb{N}$. The number $m \in \mathbb{N}$ is a factor or divisor of n if there exists an $x \in \mathbb{N}$ such that $mx = n$. In this case, we write $x|n$ and $m|n$ which are read as "x divides n" and "m divides n" respectively.

We call the set of factors or divisors of n as $D(n) = \{m \in \mathbb{N} : \exists x \in \mathbb{N} \text{ such that } mx = n\}$. For any $n \in \mathbb{N}$, then $D(n)$ has at least one element, namely 1, since $1 \times n = n$. Moreover, by the same reasoning, if $n \neq 1$, $D(n)$ has at least two distinct elements, namely 1 and n. If $x \in D(n)$, we must have $1 \leq x \leq n$, so the set $D(n)$ is finite for any $n \in \mathbb{N}$. We define:

Definition 2.3.14 (Prime, Composite Numbers) Let $n \in \mathbb{N}$.

1. The number n is called a prime number if there are exactly two distinct elements in $D(n)$, namely 1 and n.
2. The number n is called a composite number if there are more than two distinct elements in $D(n)$.

Note that Definition 2.3.14 does not cover the number 1. The number 1 is neither prime nor composite. We call 1 a unit.

Example 2.3.15 From Example 2.3.12, we say that 5 is a prime number and 6 is a composite number since $D(5) = \{1, 5\}$ and $D(6) = \{1, 2, 3, 6\}$.

We also define:

Definition 2.3.16 (Coprime Numbers) Let $m, n \in \mathbb{N}$. We call m and n are coprime to each other if $D(m) \cap D(n) = \{1\}$. In words, the numbers m and n do not share any factors with each other apart from 1.

Example 2.3.17 Note that since $D(5) \cap D(6) = \{1, 5\} \cap \{1, 2, 3, 6\} = \{1\}$, the numbers 5 and 6 are coprime to each other. On the other hand, the numbers 9 and 6 are not coprime to each other since $D(9) = \{1, 3, 9\}$ and $D(6) \cap D(9) = \{1, 3\}$.

For any two numbers $m, n \in \mathbb{N}$, the set $D(m) \cap D(n)$ is necessarily a finite subset of $\mathbb{N}$ because each of them are finite sets. Hence, by Lemma 2.3.9, we can find the largest element in this set, which we call the greatest common divisor of m and n, denoted as $\gcd(m, n)$. This notation gives us another characterisation of coprime numbers, namely: the numbers m and n are coprime if and only if $\gcd(m, n) = 1$.

Prime numbers are very important in the study of mathematics as they are seen as the building blocks of the set $\mathbb{N}$. We shall see this important result, called the fundamental theorem of arithmetic, in Exercise 2.30.

Due to this, prime numbers are widely studied in algebra and number theory. In fact, prime numbers are the main defense in information technology in the form of public key security and cryptography. They are also studied on their own and are the main characters in some of the major unsolved mathematical problems such as the Goldbach conjecture, the twin prime conjecture, and the Riemann hypothesis.

2.4 Integers $\mathbb{Z}$

From the addition and multiplication operations that we have defined on the set of natural numbers $\mathbb{N}$, ideally we would like them to be reversible. This is reminiscent to the cancellation laws in Proposition 2.3.11 in which we can cancel the common component coming from addition or multiplication from either side of an equation or an inequality.

However, these cancellation laws in Proposition 2.3.11 stipulate very specific situations for the cancellations to be allowed. In this section, we want to investigate how we can reverse the addition process.

Reversing the addition operation is natural when we want to take away some quantities from another. For example, if $n > m$, from Definition 2.3.4, there is a number $x \in \mathbb{N}$ such that $x+m = n$. Moreover this number is unique. Indeed, if there are two distinct such number, say $x, y \in \mathbb{N}$ with $x < y$, then $n = x+m < y+m = n$

which is absurd! We can thus define x as the difference between n and m when $n > m$ and denote it as $x =$ "$n - m$". This operation is called subtraction or difference.

This concept of subtraction dates back to Diophantus of Alexandria (c. 200–284) and even way before that. But the case of Diophantus is notable because according to Florian Cajori (1859–1930) in his book *A History of Mathematics* [11]:

> Diophantus had no notion whatever of negative numbers standing by themselves. All he knew were differences, such as $2x - 10$, in which $2x$ could not be smaller than 10 without leading to an absurdity.

This is precisely the case that we have discussed above, namely we can subtract a natural number m from another natural number n provided that $n > m$.

In fact, Diophantus rejected the idea of negative numbers on the basis that they are absurd. He referred to the equation $4x+20 = 4$ in his series of books *Arithmetica* as absurd because it would lead to a meaningless answer. Of course it is meaningless if we require x to be in $\mathbb{N}$. This necessitates a new set of numbers which would then allow us to view this absurd equation in a meaningful way.

In order to solve the equation $4x + 20 = 4$ for x, we need to extend our current number system $\mathbb{N}$ into something that could accommodate this. We have seen above that subtraction of natural number m from n can be defined if $n > m$. Let us look at the other cases that Diophantus rejected.

1. What if $n = m$? Say $m = n = 1$. There are no natural number x for which $x + 1 = 1$ since $1 = x + 1 > 1$. Therefore, we need to extend our set of numbers which allows this. We want such a number x that satisfies $x + 1 = 1$, which we denote symbolically as $x =$ "$1 - 1$" or the ordered pair $(1, 1) \in \mathbb{N}^2$, whatever this is.
   However, this number, if it does exist, must also be the same as "$2 - 2$", "$3 - 3$", and "$n - n$" for any $n \in \mathbb{N}$. This is true since $x + 1 = 1$ is equivalent to $x + n = n$ by repeated additions/cancellations. So $x =$ "$n - n$" as well for all $n \in \mathbb{N}$. Therefore the number x can also be described by "$n - n$" for any $n \in \mathbb{N}$. We represent this new number x as all the pairs $(n, n) \in \mathbb{N}^2$.
2. Next, let us consider $n < m$. For concreteness, say $n = 1$ and $m = 2$. We want to define a new number y such that $y + 2 = 1$. As qualmed by Diophantus, y cannot be a natural number so it must be a new kind of number. Let us denote it as the symbol $y =$ "$1 - 2$". So y can be represented by the pair $(1, 2) \in \mathbb{N}^2$.
   Again, this representation is not unique. The equality $y + 2 = 1$ is equivalent to $y + (n + 1) = n$ for any $n \in \mathbb{N}$, so the number y can also be represented by the pairs $(n, n + 1) \in \mathbb{N}^2$ for any $n \in \mathbb{N}$.

Since we have many different pairs of elements in $\mathbb{N}^2$ that can be used to describe the same quantity, we would like to identify or group together these pairs together as one object. As we have discussed at the beginning of this chapter, this can be done by using some kind of equivalence relation and quotient set.

To this end, let us try and define a suitable equivalence relation on the Cartesian product of natural numbers $\mathbb{N}^2$. From the examples above, we want, amongst others:

1. the pairs $(1, 1)$ and any (n, n) where $n \in \mathbb{N}$ to be related, and
2. the pairs $(1, 2)$ and any $(n, n + 1)$ where $n \in \mathbb{N}$ to be related,

and in this new number system, we want the related elements to be treated as the same element.

Using the above observations, let us define a relation on $\mathbb{N}^2$ via $(a, b) \sim (c, d)$ iff $a + d = b + c$. This is a well-defined relation as we know perfectly well how to add and equate natural numbers. We now show that this is in fact an equivalence relation on the set $\mathbb{N}^2$:

1. Reflexive: $(a, b) \sim (a, b)$ since $a + b = b + a$.
2. Symmetric: If $(a, b) \sim (c, d)$, then $a + d = b + c$. By symmetry of the additions and equality, we have $c + b = d + a$ and so $(c, d) \sim (a, b)$.
3. Transitive: If $(a, b) \sim (c, d)$ and $(c, d) \sim (e, f)$, then $a + d = b + c$ and $c + f = d + e$. Adding f to the first equation and b to the second equation, we can equate them to get $a + d + f = c + b + f = d + e + b$. Using cancellation law for addition on $\mathbb{N}$, we get $a + f = b + e$. Thus $(a, b) \sim (e, f)$.

Therefore, $\sim$ is an equivalence relation on the set $\mathbb{N}^2$. Thus we can define the quotient set $\mathbb{N}^2/\sim = \{[(a, b)] : (a, b) \in \mathbb{N}^2\}$, where the related elements in $\mathbb{N}^2$ are now the same object. We call this the set of integers $\mathbb{Z}$. The points in $\mathbb{N}/\sim$ are depicted in Fig. 2.3 as the sets of lines.

Remark 2.4.1 At this point, it may seem a bit strange for the readers. How can numbers be the set of lines in Fig. 2.3? We just have to remember that the set of numbers is simply, well, a set. They are made up abstract intangible objects. Even the natural numbers $\mathbb{N}$ was defined via an abstract infinite set of strange symbols $\{1, 2, 3, \ldots\}$ first before we put algebraic and ordering structure on it.

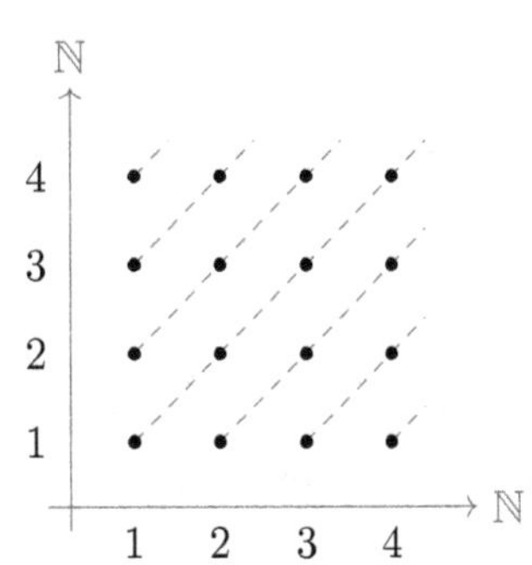

Fig. 2.3 The set of integers $\mathbb{Z}$ is given by the set of the equivalence classes of points in $\mathbb{N}^2$ which lie on the same dashed red line in the lattice above

One of my favourite quotes regarding this abstract nature of numbers is by Kate Owens [27] in which she joked:

> When you finish a PhD in mathematics, they take you to a special room and explain that i [the imaginary unit which will be defined in Exercise 3.24] isn't the only imaginary number - turns out that ALL the numbers are imaginary, even the ones that are real.

We can put more meaning to this abstract set $\mathbb{Z}$ by defining some algebraic operations and ordering on them, which we will do next.

2.5 Algebra on $\mathbb{Z}$

Now we would like to put an algebraic structure on this set $\mathbb{Z}$ by defining addition and multiplication operations on this set. We note that we have a copy of the set $\mathbb{N}$ within the set $\mathbb{Z}$ in the form of $z_n = [(n+1, 1)]$ for each $n \in \mathbb{N}$. Thus, we would like to define addition $\oplus$ and multiplication $\otimes$ operations on the set $\mathbb{Z}$ that coincides or compatible with the corresponding operations $+$ and $\times$ on the natural numbers $\mathbb{N}$.

To find a candidate for $\oplus$ that is compatible with the addition $+$ on $\mathbb{N}$, for any $m, n \in \mathbb{N}$ we require the copy of m and the copy of n in $\mathbb{Z}$ adds up under $\oplus$ to give the copy of $m + n$ in $\mathbb{Z}$. In other words, we require $z_m \oplus z_n = z_{m+n}$. This means:

$$[(m + 1, 1)] \oplus [(n + 1, 1)] = [(m + n + 1, 1)] = [(m + 1 + n + 1, 1 + 1)].$$

From this observation, we propose $\oplus$ on more general elements of $\mathbb{Z}$ as:

$$[(a, b)] \oplus [(c, d)] = [(a + c, b + d)]. \tag{2.4}$$

Likewise, for the operation $\otimes$ to be compatible with $\times$, we require $z_m \otimes z_n = z_{mn}$, namely:

$$\begin{aligned}[(m + 1, 1)] \otimes [(n + 1, 1)] &= [(mn + 1, 1)] \\ &= [(mn + m + n + 1, 1 + m + n)] \\ &= [((m + 1)(n + 1) + 1, (m + 1) + (n + 1))].\end{aligned}$$

Thus, we propose $\otimes$ on more general elements of $\mathbb{Z}$ as:

$$[(a, b)] \otimes [(c, d)] = [(ac + bd, bc + ad)]. \tag{2.5}$$

Clearly, these operations are symmetric just like $+$ and $\times$. However, these definitions hinge on the choice of representative for each class in order for us to write them down explicitly. As we have noted earlier, a class representative of any equivalence class is merely a label. Thus, we need to check that these operations $\oplus$ and $\otimes$ are well-defined, namely they are independent of the class representative choice.

By symmetry, it is sufficient to show that the operations in (2.4) and (2.5) are independent of the chosen equivalence class representative in the first argument. In other words, we need to check that if $(a,b) \sim (m,n)$, then we must also have:

$$[(a,b)]\oplus[(c,d)] = [(m,n)]\oplus[(c,d)] \quad \text{and} \quad [(a,b)]\otimes[(c,d)] = [(m,n)]\otimes[(c,d)].$$

1. To show the first one, note that $(a,b) \sim (m,n)$ means $a+n = b+m$. Adding $c+d$ on both sides and using the symmetry and associativity of $+$, we get $(a+c)+(n+d) = (b+d)+(m+c)$, and so $(a+c, b+d) \sim (n+d, m+c)$. Thus:
$$[(a,b)] \oplus [(c,d)] = [(a+c, b+d)] = [(m+c, n+d)] = [(m,n)] \oplus [(c,d)].$$
2. For multiplication, by equivalences, we have:
$$\begin{aligned} & [(a,b)] \otimes [(c,d)] = [(m,n)] \otimes [(c,d)] \\ \Leftrightarrow\quad & [(ac+bd, bc+ad)] = [(mc+nd, nc+md)] \\ \Leftrightarrow\quad & (ac+bd, bc+ad) \sim (mc+nd, nc+md) \\ \Leftrightarrow\quad & ac+bd+nc+md = bc+ad+mc+nd \\ \Leftrightarrow\quad & (a+n)c+(b+m)d = (b+m)c+(a+n)d. \end{aligned}$$
Thus, $[(a,b)] \otimes [(c,d)] = [(m,n)] \otimes [(c,d)]$ is true if and only if $(a+n)c+(b+m)d = (b+m)c+(a+n)d$ is true. We know that the latter is true by the equality $a+n = m+b$ from the assumption that $(a,b) \sim (m,n)$. Hence, $\otimes$ is also independent of choice of representative for the equivalence classes.

Therefore, $\oplus$ and $\otimes$ are well-defined operations on the set $\mathbb{Z}$. Now we note that there are special elements in $\mathbb{Z}$ which do not alter other elements under $\oplus$ and $\otimes$. These elements are called the additive and multiplicative identities and are given by $[(1,1)]$ and $[(2,1)]$ respectively. Indeed, for any $[(a,b)] \in \mathbb{Z}$, we compute:

$$\begin{aligned} [(1,1)] \oplus [(a,b)] &= [(a+1, b+1)] = [(a,b)], \\ [(2,1)] \otimes [(a,b)] &= [(2a+b, 2b+a)] = [(a,b)], \end{aligned}$$

and, by symmetry, the operations done in opposite order also yield the same result. Moreover, for every $[(a,b)] \in \mathbb{Z}$, we can find an element in $\mathbb{Z}$ such that its addition with $[(a,b)]$ is the additive identity. We claim that $[(b,a)]$ does the job. Indeed we have:

$$[(a,b)] \oplus [(b,a)] = [(a+b, a+b)] = [(1,1)].$$

We call the element $[(b,a)]$ the additive inverse to $[(a,b)]$. The readers are invited to show that the additive inverse to any element is unique in Exercise 2.22.

In fact, the set $\mathbb{Z}$ with these operations is called a commutative ring since it satisfies all of the axioms below, some of which we have proven above. In Exercise 2.21, one can check that our definition for the integers $(\mathbb{Z}, \oplus, \otimes)$ satisfy all of the following axioms.

Definition 2.5.1 (Commutative Ring) A commutative ring is a set X along with $+$ and $\times$ operations such that $a + b \in X$, $a \times b \in X$ for all $a, b \in X$ and satisfies the following axioms:

1. Associativity of $+$: $(a + b) + c = a + (b + c)$ for all $a, b, c \in X$.
2. Commutativity of $+$: $a + b = b + a$ for all $a, b \in X$.
3. Existence of identity for $+$: There exists a $0 \in X$ such that $0 + a = a + 0 = a$ for all $a \in X$.
4. Existence of inverses for $+$: For each $a \in X$, there exists a $-a \in X$ such that $a + (-a) = 0$.
5. Associativity of $\times$: $(a \times b) \times c = a \times (b \times c)$ for all $a, b, c \in X$.
6. Commutativity of $\times$: $a \times b = b \times a$ for all $a, b \in X$.
7. Existence of identity for $\times$: There exists a $1 \in X$ such that $1 \times a = a \times 1 = a$ for all $a \in X$.
8. Distributivity of $\times$ over $+$: $a \times (b + c) = (a \times b) + (a \times c)$ for all $a, b, c \in X$.

Other than $\mathbb{Z}$, there are many other examples of rings studied in mathematics. One example is the ring of integers modulo r, which will be looked at in Exercise 2.25. On the other hand, the set $(\mathbb{N}, +, \times)$, even though it is closed under addition and multiplication (namely any sum and product of two natural numbers is still a natural number), does not form a commutative ring because of many reasons. One of them is the lack of additive identity and additive inverse in the set $\mathbb{N}$.

Remark 2.5.2 The ring axioms in Definition 2.5.1 specified the properties of $+$ and $\times$ for 2 or 3 terms only. However, by induction, we can extend these properties to higher (but finitely many) number of terms. We set two notations for these:

1. Given finitely many elements $x_1, x_2, \dots, x_n \in X$ in a ring X, we define their sum and product using the following shorthand notation:
   (a) $\sum_{j=1}^{n} x_j = x_1 + x_2 + \ldots + x_n$, and
   (b) $\prod_{j=1}^{n} x_j = x_1 \times x_2 \times \ldots \times x_n$.
2. Due to the associativity and commutativity of $+$ and $\times$, we have omitted the brackets from each operations in the finite sum and product as the order on which we carry these operations out does not matter. As a result, each of the sum and the product above has a well-defined value in X independent of how we carry out the order of the $+$ and $\times$.
3. The symbols $\sum$ and $\prod$ in the above are derived from the Greek letters for S (sum) and P (product) respectively. Notice that in the notations, the index j is a dummy variable: they are simply used to parametrise the collection of summands or factors. As a result, the final value is independent of the index j. In fact, we

can also change this dummy variable to another one, as long as it does not cause confusion. For example:

$$\sum_{j=1}^{n} x_j = \sum_{k=1}^{n} x_k = \sum_{\spadesuit=1}^{n} x_{\spadesuit},$$

all mean $x_1 + x_2 + \ldots + x_n$.

4. If we have infinitely many terms in X, we may still write $\sum_{j=1}^{\infty} x_j$ or $\prod_{j=1}^{\infty} x_j$, but these are undefined for now. These infinite sum and product are simply symbolic with no meaning here, since we do not know how to add infinitely many terms at the moment. This is akin to gibberish words, such as "covfefe", which have no meaning despite us being able to write them down.
   The reason being is that the ring axioms allows us to add or multiply two or (by induction) finitely many number of terms of X and this sum or product is guaranteed to be in X. However, the axiom does not guarantee this if we carry out the summation or product indefinitely.
   Despite the lack of algebraic meaning, infinite sums are useful as a formal notation, as we shall see later in Sect. 4.4. Moreover, in Chap. 7 we shall discuss on how we can make sense of an infinite sum and give it a meaning beyond the ring axioms.
5. Roughly speaking, this is what sets the topics of algebra and analysis apart: algebra deals with finite structures and symmetries whereas analysis forays into the realm of infinity (whether infinitely large or infinitely small) and how to make sense of it. But do not quote me on this because the distinction between these two topics can be blurry sometimes! Moreover, these topics intermingle in many modern areas of mathematics such as functional analysis and Lie groups.

Due to the cumbersome notation for $\mathbb{Z} = \{[(a, b)] : (a, b) \in \mathbb{N}^2\}$ above, we want to simplify the notation. Before we do that, let us first observe the following lemma:

Lemma 2.5.3 *Let $[(a, b)] \in \mathbb{Z}$. There is a unique element (c, d) in the class $[(a, b)]$ such that $c = 1$ or $d = 1$.*

Proof If $a = b$, then $[(a, b)] = \{(n, n) : n \in \mathbb{N}\}$. There is only one element in this class with 1 in its coordinates, namely $(1, 1)$.

Now suppose that $a \neq b$. There are no elements of the form $(1, c)$ and $(d, 1)$ in the same class $[(a, b)]$ simultaneously. Indeed, if there is, then $(1, c) \sim (d, 1)$ implies $2 = d + c$. Since $c, d \in \mathbb{N}$, we must have $c = d = 1$. This means $(a, b) \sim (1, 1)$, namely $a + 1 = b + 1$ and thus contradicting the assumption that $a \neq b$. If $a > b$, then there exists a unique $x \in \mathbb{N}$ such that $a = b + x$. This means $a + 1 = b + x + 1$ and so $(a, b) \sim (x + 1, 1)$. Likewise, if $a < b$ then there exists a unique $x \in \mathbb{N}$ such that $b = a + x$ and so $(a, b) \sim (1, x + 1)$. □

Remark 2.5.4 Pictorially, the statement in Lemma 2.5.3 is clear from Fig. 2.3. Indeed, in any equivalence class, we can trace the line to find a unique point in the class such that one of its coordinate points has the value 1.

Lemma 2.5.3 allows us to introduce a simplified notation or symbols for elements in the set $\mathbb{Z}$. In fact, this is the notation that we all know and love from school. This notation arises by choosing the unique class representative for each class with a 1 in its coordinate. We write these numbers down using the following notations:

$$[(a,b)] = \begin{cases} n & \text{if } (a,b) \sim (n+1,1), \\ 0 & \text{if } (a,b) \sim (1,1), \\ -n & \text{if } (a,b) \sim (1,n+1), \end{cases}$$

where $n \in \mathbb{N}$. For each $n \in \mathbb{N}$, we read $-n$ as "negative n". The first kind of numbers are called positive integers, the middle is called zero, and third kind of numbers are called negative integers.

Note that the notation is independent of the choice of representative for the class $[(a, b)]$. This is because in each class $[(a, b)]$ there is exactly one element which has at least one 1 in its coordinate expression, as guaranteed by Lemma 2.5.3. In this notation, the additive and multiplicative identities $[(1, 1)]$ and $[(2, 1)]$ are denoted as 0 and 1 respectively.

Using this notation, we can see clearly that the integers $\mathbb{Z}$ can be divided into three classes: the positive integers, zero, and the negative integers. In Fig. 2.3, zero is represented by the diagonal line, the positive integers are represented by the lines below the diagonal, and the negative integers are represented by the lines above the diagonal. We denote these non-zero sets as $\mathbb{Z}_+$ and $\mathbb{Z}_-$ respectively. For brevity, we sometimes denote $\mathbb{Z}_{\geq 0} = \mathbb{N}_0 = \mathbb{Z}_+ \cup \{0\}$ as the non-negative integers and $\mathbb{Z}_{\leq 0} = \mathbb{Z}_- \cup \{0\}$ as the non-positive integers. Thus, we have the disjoint union $\mathbb{Z} = \mathbb{Z}_+ \cup \mathbb{Z}_- \cup \{0\}$.

From the representation above, we note that the positive integers coincide with the set of natural numbers and the negative integers are their mirror images; they are the same as the natural numbers, except for the negative symbol "$-$" at the front. These numbers annihilate each other in the sense that $n \oplus (-n) = 0$ for any $n \in \mathbb{N}$. Indeed, we have:

$$n \oplus (-n) = [(n+1,1)] \oplus [(1,n+1)] = [(n+2,n+2)] = [(1,1)] = 0,$$

or, in other words, $-n$ is the additive inverse of n.

Since the addition and multiplication operations $\oplus$ and $\otimes$ were defined by design to coincide with the addition and multiplication operations $+$ and $\times$ on $\mathbb{N} = \mathbb{Z}_+$, we may relabel the operations $\oplus$ and $\otimes$ as $+$ and $\times$ respectively to emphasise that they are merely the extensions of $+$ and $\times$ to a bigger set.

Back to the conundrum of Diophantus: the negative symbol allows us to extend the subtraction operation that we have defined partially on the set $\mathbb{N}$. Recall that we (and Diophantus) could only define the subtraction "$a - b$" as the natural number x such that $x + b = a$ if $a > b$ in $\mathbb{N}$. We note that in the extended number system $\mathbb{Z}$, the above is simply $x = a + (-b)$ because:

$$\begin{aligned} x = [(x+1, 1)] &= [(x+b+1, b+1)] \\ &= [(a+1, b+1)] \\ &= [(a+1, 1)] + [(1, b+1)] = a + (-b). \end{aligned}$$

Therefore we can extend this operation of subtraction to any pair of natural numbers in $\mathbb{N}$ and also any pairs of integers in $\mathbb{Z}$ via the shorthand notation $a - b = a + (-b)$ for any $a, b \in \mathbb{Z}$. Namely, to subtract b from a, we add $-b$ to a.

Another important lemma that we can prove is the following:

Lemma 2.5.5 *Let $a, b \in \mathbb{Z}$.*

1. *If either a or b is 0, then $ab = 0$.*
2. *If $a, b \in \mathbb{Z}_+$, then $ab \in \mathbb{Z}_+$.*
3. *If $a, b \in \mathbb{Z}_-$, then $ab \in \mathbb{Z}_+$.*
4. *If $a \in \mathbb{Z}_+$ and $b \in \mathbb{Z}_-$, then $ab \in \mathbb{Z}_-$.*
5. *If $ab = 0$, then $a = 0$ or $b = 0$.*
6. *For any $a \in \mathbb{Z}$, $a \times (-1) = (-1) \times a = -a$.*
7. *$-0 = 0$.*
8. *$1 \neq 0$.*

Proof We prove the first five assertions and the others are left as Exercise 2.23.

1. WLOG, assume that $b = 0$. Suppose that $ab = c$. Since $b = 0 = 0 + 0 = b + b$, we have $c = ab = a(b + b) = ab + ab = c + c$, which means $c = 0$.
2. Since $a, b \in \mathbb{Z}_+ = \mathbb{N}$, we have $ab \in \mathbb{N} = \mathbb{Z}_+$ by the property of multiplication in $\mathbb{N}$.
3. Let $a = -m$ and $b = -n$ for some $m, n \in \mathbb{N}$. By adding a 0 term, factorising, and using the first assertion, we have:

$$\begin{aligned} ab = (-m)(-n) &= (-m)(-n) + m(n + (-n)) \\ &= (-m)(-n) + mn + m(-n) \\ &= (-m + m)(-n) + mn \\ &= 0(-n) + mn = mn \in \mathbb{N} = \mathbb{Z}_+. \end{aligned}$$

4. Let $b = -n$ for some $n \in \mathbb{N}$. From the second assertion, $an \in \mathbb{Z}_+$. Consider the quantity $ab + an = a(-n) + an = a(-n + n) = 0$. Adding $-(an)$ on both sides, we get $ab = -(an) \in \mathbb{Z}_-$.
5. We prove this by contrapositive. Suppose that both a and b are non-zero. Then, by assertions 2, 3, and 4, we either have $ab \in \mathbb{Z}_+$ or $\mathbb{Z}_-$. So $ab \neq 0$. □

An important consequence of the above is that Lemma 2.5.5(5) gives us the cancellation law in $\mathbb{Z}$.

Corollary 2.5.6 (Cancellation Law on $\mathbb{Z}$) *If $a, b, c \in \mathbb{Z}$ with $a \neq 0$ and $ab = ac$, then $b = c$.*

Proof Note that $ab = ac$ is equivalent to $a(b - c) = 0$. By Lemma 2.5.5, we either have $a = 0$ or $b - c = 0$. Since $a \neq 0$, we must have $b - c = 0$ and hence the result. □

From the cancellation law of multiplication on $\mathbb{Z}$, we can also extend the idea of factors and divisors in Definition 2.3.13 from $\mathbb{N}$ to $\mathbb{Z}$ as follows:

Definition 2.5.7 (Factors, Divisors) Let $n \in \mathbb{Z}$. The number $m \in \mathbb{Z}$ is a factor or divisor of n if there exists an $x \in \mathbb{Z}$ such that $mx = n$. In this case, we write $x|n$ and $m|n$ which are read as "x divides n" and "m divides n" respectively.

2.6 Ordering on $\mathbb{Z}$

Let us now present a way to order elements in $\mathbb{Z}$. We define:

$$a \prec b \quad \text{iff} \quad b + (-a) \in \mathbb{Z}_+ = \mathbb{N}.$$

We first check that this is indeed a strict total order on $\mathbb{Z}$:

1. Irreflexive: $a \nprec a$ since $a + (-a) = 0 \notin \mathbb{N}$.
2. Transitive: If $a \prec b$ and $b \prec c$, then $a + (-b), b + (-c) \in \mathbb{N}$. Thus, their sum is also in $\mathbb{N}$, namely $a + (-b) + b + (-c) = a + (-c) \in \mathbb{N}$. This implies $a \prec c$.
3. Trichotomous: We check the following cases:
   (a) Suppose that $a = b$. Then, $a + (-b) = b + (-a) = 0 \notin \mathbb{N}$ and so $a \nprec b$ and $b \nprec a$.
   (b) Next, assume $a \neq b$. Then, $a + (-b) \neq 0$ and so $a + (-b)$ is in exactly one of $\mathbb{Z}_+$ or $\mathbb{Z}_-$. The former means $a \prec b$. The latter implies $-(a + (-b)) = b + (-a) \in \mathbb{Z}_+$ and so $b \prec a$. Thus, exactly one of $a \prec b$ or $b \prec a$ holds.

   This means exactly one of $a = b$, $a \prec b$, or $b \prec a$ holds.

Hence, $\prec$ is a strict total order. To elaborate this strict order, comparing the integers follow these hierarchy:

$$a \prec b \quad \text{iff either} \quad \begin{cases} a \in \mathbb{Z}_-, b \in \mathbb{Z}_+, \\ a = 0, b \in \mathbb{Z}_+, \\ a \in \mathbb{Z}_-, b = 0, \\ a, b \in \mathbb{Z}_+ \text{ and } a < b, \text{ or} \\ a, b \in \mathbb{Z}_- \text{ and } -a > -b. \end{cases}$$

Note that when $a, b \in \mathbb{N} = \mathbb{Z}_+$, the order $\prec$ that we have defined coincides with the strict total order $<$ on $\mathbb{N}$. Thus, we can view $\prec$ as an extension of the ordering $<$ on $\mathbb{N}$ to a larger set $\mathbb{Z}$. To simplify notations and emphasise this fact, we can relabel $\prec$ with $<$.

Moreover, this ordering on $\mathbb{Z}$ is compatible with the ring structure on $\mathbb{Z}$. What does "compatible" mean here? Compatibility of the two structures here means addition with any integer and multiplication with a positive integer do not alter the ordering. This compatibility condition is called the ordered ring axioms, given as follows.

Definition 2.6.1 (Ordered Ring Axioms) A ring $(X, +, \times)$ with a relation $<$ is an ordered ring if:

1. $<$ is a strict total order from Definition 2.3.1,
2. $<$ is compatible with $+$: If $a < b$ and $c \in X$, then $a + c < b + c$, and
3. $<$ is compatible with $\times$: If $0 < a$ and $0 < b$, then $0 < ab$.

We can check that the ring $\mathbb{Z}$ with the strict total order $<$ forms an ordered ring by showing it satisfies all the axioms in Definition 2.6.1:

1. The ordering $\prec$ was defined as a strict total order on $\mathbb{Z}$.
2. For any $m, n, x \in \mathbb{Z}$ we have:

$$\begin{aligned} m < n \quad &\Leftrightarrow \quad n + (-m) \in \mathbb{N} \quad \Leftrightarrow \quad n + (-m) + x + (-x) \in \mathbb{N} \\ &\Leftrightarrow \quad n + x + (-(m + x)) \in \mathbb{N} \\ &\Leftrightarrow \quad m + x < n + x. \end{aligned}$$

3. If $m, n \in \mathbb{Z}$ are such that $0 < m, n$, then $m, n \in \mathbb{N}$ and so $0 < mn$.

Similar to the natural numbers, we have the following cancellation laws:

Proposition 2.6.2 (Cancellation Laws on $\mathbb{Z}$) *Suppose that $m, n, x \in \mathbb{Z}$.*

1. *If $m + x < n + x$, then $m < n$.*
2. *If $mx < nx$ and $x > 0$, then $m < n$.*

Proof The first assertion is a direct consequence of the ordered ring axioms by adding $-x$ to both sides of the inequality. We thus prove the second assertion only.

2. Suppose for contradiction that $m \geq n$. Then, there are two cases:
   (a) If $m = n$, then $mx = nx$.
   (b) If $m > n$, since $m - n > 0$ and $x > 0$, by the ordered ring axiom, we have $0 < (m - n)x = mx - nx$ and so $mx > nx$.

   In either case, we get a contradiction. Hence $m < n$. □

However, unlike the natural numbers, we do not have the well-ordering principle for the set of integers $\mathbb{Z}$. This is because there are subsets of $\mathbb{Z}$ which does not have a minimal element. For example $\mathbb{Z}_- \subseteq \mathbb{Z}$ does not have the smallest element. Indeed, if it does, say $a \in \mathbb{Z}_-$ is the smallest element of $\mathbb{Z}_-$, we can always find an even smaller element in $\mathbb{Z}_-$ namely $a - 1 \in \mathbb{Z}_-$, thus contradicting our claim that a is the smallest element in $\mathbb{Z}_-$.

Remark 2.6.3 However, one can define a different strict total order on the set of integers that is well-ordered. For example, define a different strict total order $\prec$ on $\mathbb{Z}$ as:

$$m \prec n \quad \text{iff} \quad \begin{cases} |m| < |n|, \text{ or} \\ |m| = |n| \text{ and } m < 0 < n, \end{cases}$$

where $|n|$ is the absolute value or modulus of n, defined as:

$$|n| = \begin{cases} n & \text{if } n \geq 0, \\ -n & \text{if } n < 0. \end{cases}$$

One can check that this is a strict total order. In comparison to the order $<$ that we have defined earlier, the ordering $\prec$ tells us that if the absolute value of a number is smaller, they are regarded as smaller according to the order $\prec$ without taking into account of the signs of the two numbers. On the other hand, if the absolute value of two numbers are the same, the negative number is considered smaller. For example, using this ordering we have $-1 \prec 2$, $1 \prec -2$, and $-1 \prec 1$. So we can arrange the integers $\mathbb{Z}$ using this order as:

$$0 \prec -1 \prec 1 \prec -2 \prec 2 \prec -3 \prec 3 \prec \ldots$$

In contrast to the total order $<$ that we have defined earlier, this ordering is well-ordered, meaning that any non-empty subset of it has a smallest element. However, the drawback of this order is that it is not compatible with the ring structure that we have on $\mathbb{Z}$. As an example, in this ordering we have $-2 \prec 2$. Now notice that when we add -1 on both sides, we get $-2-1 = -3 \nprec 1 = 2-1$. Thus the order $\prec$ is not compatible with addition $+$ as in Definition 2.6.1. Hence $(\mathbb{Z}, \prec)$ is not an ordered ring.

Exercises

2.1 Write the conditions for all the types of binary relations in Definition 2.1.5, namely reflexive, irreflexive, symmetric, antisymmetric, transitive, strongly connected, and trichotomous, in logical symbols.

2.2 ($*$) Determine whether the following relations $\sim$ on $\mathbb{Z}$ are equivalence relations. If $\sim$ is an equivalence relation, describe the equivalence classes of $\sim$.

(a) $a \sim b$ iff $a < b$.
(b) $a \sim b$ iff $a \leq b$.
(c) $a \sim b$ iff $a + 1 = b + 1$.
(d) $a \sim b$ iff $a = -b$ or $a = b$.
(e) $a \sim b$ iff $\gcd(a, b) = 2$.
(f) $a \sim b$ iff $ab = 0$.
(g) $a \sim b$ iff $a + b = 2m$ for some $m \in \mathbb{Z}$.
(h) $a \sim b$ iff $a = mb$ for some $m \in \mathbb{Z}$.
(i) $a \sim b$ iff $a + b = m^2$ for some $m \in \mathbb{Z}$.
(j) $a \sim b$ iff $3|(2a + b)$.

2.3 (a) Let $X = \{1, 2, 3, 4\}$ and $\mathcal{R} = \{(1, 1), (1, 3), (2, 2), (2, 4), (3, 1), (3, 3), (4, 2), (4, 4)\} \subseteq X \times X$. Show that $\mathcal{R}$ is an equivalence relation.
(b) Let $\sim$ be a relation on the set of pairs $\mathbb{N} \times \mathbb{N}$ defined as $(a, b) \sim (c, d)$ iff $a = c$ or $b = d$. Determine whether $\sim$ is an equivalence relation.

2.4 Let $n \in \mathbb{N}$.
(a) How many different relations are there on the set $\mathcal{N} = \{1, 2, \ldots, n\}$?
(b) How many different relations are there on the set $\mathcal{N} = \{1, 2, \ldots, n\}$ which is reflexive?
(c) How many different relations are there on the set $\mathcal{N} = \{1, 2, \ldots, n\}$ which is symmetric?
(d) How many different relations are there on the set $\mathcal{N} = \{1, 2, \ldots, n\}$ which is reflexive and symmetric?
(e) How many different equivalence relations on the set $\{1, 2, 3\}$ are there?
(f) How many different equivalence relations on the set $\{1, 2, 3, 4\}$ are there?

2.5 Let X be a set and $\mathcal{R} \subseteq X \times X$ be a relation on X which is symmetric and transitive.
(a) Suppose that $(x, y) \in \mathcal{R}$ for some $y \in X$. Prove that $(x, x) \in \mathcal{R}$ as well.
(b) Explain and provide an example why being symmetric and transitive is not enough to guarantee that $\mathcal{R}$ is an equivalence relation.

2.6 Let $\mathcal{P} = \{P : P \text{ is a mathematical statement}\}$ be the set of all mathematical statements. Define a relation $\sim$ on the set $\mathcal{P}$ as $P \sim Q$ iff they are logically equivalent, namely $P \equiv Q$. Show that $\sim$ is an equivalence relation.

2.7 Let X be a set and $\mathcal{R}_1, \mathcal{R}_2 \subseteq X \times X$ be two equivalence relations on X.
(a) Prove that $\mathcal{R}_1 \cap \mathcal{R}_2$ is also an equivalence relation.
(b) Is $\mathcal{R}_1 \cup \mathcal{R}_2$ an equivalence relation? Provide a proof or a counterexample.

2.8 Let X be a set and $\mathcal{R} \subseteq X \times X$ be a relation on X. Define a relation $\mathcal{R}^{-1} = \{(b, a) : (a, b) \in \mathcal{R}\}$. Prove that $\mathcal{R}$ is an equivalence relation if and only if $\mathcal{R}^{-1}$ is an equivalence relation.

2.9 ($\diamond$) Recall Peano's axioms in Definition 2.2.3 which characterises the natural numbers. We can define the addition operation on $\mathbb{N}$ recursively via $a + 1 = s(a)$ and $a + s(b) = s(a + b)$ for all $a, b \in \mathbb{N}$ whereas multiplication is defined recursively using addition via $a \times 1 = a$ and $a \times s(b) = a + (a \times b)$ for all $a, b \in \mathbb{N}$. Using the Peano's axioms, show that for any $a, b, c \in \mathbb{N}$ we have:
(a) $a + b = b + a$.
(b) $a \times b = b \times a$.

(c) $a + (b + c) = (a + b) + c$.
(d) $a \times (b \times c) = (a \times b) \times c$.
(e) $a \times (b + c) = (a \times b) + (a \times c)$.
We have also defined ordering on $\mathbb{N}$ as $a < b$ if and only if there exists a $c \in \mathbb{N}$ such that $a + c = b$. Show that this ordering is compatible with the addition and multiplication above, namely if $a < b$ and $c \in \mathbb{N}$, then we have:
(f) $a + c < b + c$.
(g) $c \times a < c \times b$.

2.10 (∗) State, with justifications, whether the following statements are true or false.
(a) $\forall n \in \mathbb{N}, \exists m \in \mathbb{N} : m < n$.
(b) $\forall n \in \mathbb{N}, \exists m \in \mathbb{N} : m \leq n$.
(c) $\forall n \in \mathbb{N}, \exists m \in \mathbb{N} : n < m$.
(d) $\exists n \in \mathbb{N} : \forall m \in \mathbb{N}, m < n$.
(e) $\exists n \in \mathbb{N} : \exists m \in \mathbb{N} : n + m = 1$.
(f) $\forall n \in \mathbb{N}, \nexists m \in \mathbb{N} : n + m = 1$.
(g) $\forall m \in \mathbb{N}, \forall n \in \mathbb{N}, \exists p \in \mathbb{N} : m + p = n$.
(h) $\forall m \in \mathbb{N}, \forall n \in \mathbb{N}, \exists p \in \mathbb{N} : m + p = n$ or $n + p = m$.
(i) $\forall m \in \mathbb{N}, \exists n \in \mathbb{N}, \exists p \in \mathbb{N} : m + p = n$.
(j) $\nexists n \in \mathbb{N} : \forall m \in \mathbb{N}, nm = m$.
(k) $\forall m \in \mathbb{N}, \forall n \in \mathbb{N}, \exists! p \in \mathbb{N} : m = n + p$.
(l) $\forall m \in \mathbb{N}, \forall n \in \mathbb{N}, (n < m \Rightarrow (\exists! p \in \mathbb{N} : m = n + p))$.

2.11 Write down the principle of mathematical induction using quantifiers and connectives.

2.12 (∗) Using the principle of mathematical induction, prove the following statements:
(a) $\forall n \in \mathbb{N}, 5 | 6^n - 1$.
(b) $\forall n \in \mathbb{N}, 16 | (5^n - 4n - 1)$.

2.13 (∗) Using the principle of mathematical induction, prove the following inequalities:
(a) $\forall n \in \mathbb{N}, 2^{n+1} \geq 2^n + 2^{n-1} + 1$.
(a) $\forall n \in \mathbb{N}, (n \geq 2 \Rightarrow n^2 > n + 1)$.
(b) $\forall n \in \mathbb{N}, (n \geq 2 \Rightarrow n^3 > 2n + 1)$.
(c) $\forall n \in \mathbb{N}, (n \geq 4 \Rightarrow n! > n^2)$ where $n! = n \times (n-1) \times (n-2) \times \ldots \times 1$ is the factorial operation.
(d) $\forall n \in \mathbb{N}, (n \geq 7 \Rightarrow n! > 3^n)$.
(e) $\forall n \in \mathbb{N}, (n \geq 10 \Rightarrow 2^n > n^3)$.

2.14 Let $X \subseteq \mathbb{N}$ be a subset of the natural numbers. Prove that if the set X has a maximum, then it must be finite.

2.15 (◇) The Fibbonaci sequence (f_n) is a sequence or list of natural numbers that are defined recursively via $f_1 = f_2 = 1$ and $f_n = f_{n-1} + f_{n-2}$ for all $n \geq 3$. The first few numbers of this sequence are given by: $1, 1, 2, 3, 5, 8, 13, 21, 34, 55, 89, 144, \ldots$.
Leonardo Bonacci (1170–1240), also known as Fibonacci, discussed this sequence in the book *Liber abaci* (The Book of Calculations) which arises from a hypothetical question on the population of rabbits in a field. Today,

there are many applications to this sequence in mathematics and science in general, ranging from combinatorics to computer science and biology.
Due to its recursive definition, we can define many interesting properties for this sequence. Via induction, prove that for any $n \in \mathbb{N}$:
(a) f_n and f_{n+1} are coprime.
(b) $\sum_{j=1}^{n} f_j = f_{2n} - 1$.
(c) $\sum_{j=1}^{n}(-1)^{j+1} f_j = (-1)^{n+1} f_{n-1} + 1$.
(d) $\sum_{j=1}^{n} f_{2j-1} = f_{2n}$.
(e) $\sum_{j=1}^{n} f_{2j} = f_{2n+1} - 1$.
(f) $\sum_{j=1}^{n} f_j^2 = f_n f_{n+1}$.
(g) $\sum_{j=1}^{2n+1} f_j f_{j+1} = f_{2n}^2$.
(h) $f_{n+1}^2 - f_n f_{n+2} = (-1)^n$.
(i) For any fixed $m \in \mathbb{N}$ and any $n \geq 2$, $f_{m+n} = f_{n-1} f_m + f_n f_{m+1}$.
Using part (i), prove that:
(j) f_{2n} is divisible by f_n.
(k) More generally, for any $p \in \mathbb{N}$, f_{pn} is divisible by f_n.

2.16 (∗) Prove the remaining assertions in Proposition 2.3.11, namely:
For $m, n, x \in \mathbb{N}$, prove that:
(a) If $mx = nx$, then $m = n$.
(b) If $mx < nx$, then $m < n$.

2.17 Find the factors and the greatest common divisor of the following pairs of numbers. Determine which pair of numbers are coprime to each other.
(a) 38 and 18.
(b) 24 and 60.
(c) 15 and 64.
(d) 63 and 120.

2.18 (∗) Let $a, b, d, m, n \in \mathbb{N}$. Prove that if $d|a$ and $d|b$, then $d|(am + bn)$.

2.19 Let $a, b \in \mathbb{N}$. Let $\text{lcm}(a, b)$ be the lowest common multiple for a and b. In other words, $\text{lcm}(a, b)$ is the smallest natural number for which a and b both divide $\text{lcm}(a, b)$. For any pair $a, b \in \mathbb{N}$, explain why their lcm exists in $\mathbb{N}$. Show:
(a) $a|b \Leftrightarrow a = \gcd(a, b)$.
(b) $a|b \Leftrightarrow b = \text{lcm}(a, b)$.

2.20 (⋄) Let $a, b \in \mathbb{N}$ and $g = \gcd(a, b)$ so that $a = gp$ and $b = gq$ for some $p, q \in \mathbb{N}$.
(a) Show that $\gcd(p, q) = 1$ (in other words, p, q are coprime).
(b) Show that $g|\text{lcm}(a, b)$.
(c) Show that $\text{lcm}(p, q) = pq$.
(d) Prove that $\gcd(a, b) \times \text{lcm}(a, b) = ab$.

2.21 (∗) Prove that the set $\mathbb{Z} = \mathbb{N}^2/\sim = \{[(a, b)] : (a, b) \in \mathbb{N}^2\}$ with the addition and multiplication operations $\oplus$ and $\otimes$ defined in (2.4) and (2.5) satisfy the commutative ring axioms in Definition 2.5.1.

2.22 (∗) Let X be a non-trivial ring. Show that:
(a) The additive identity 0 in X is unique.

(b) The multiplicative identity element 1 in X is unique.
(c) For each element $a \in X$, its additive inverse $-a$ is unique.

2.23 ($*$) Prove the remaining assertions in Lemma 2.5.5, namely:
Let $a, b \in \mathbb{Z}$.
(a) For any $a \in \mathbb{Z}$, $a \times (-1) = (-1) \times a = -a$.
(b) $-0 = 0$.
(c) $1 \neq 0$.

2.24 We define odd and even integers as follows:

Definition 2.7.4 (Odd, Even Integer) Let $n \in \mathbb{Z}$. The number n is called:
1. even if $n = 2m$ for some $m \in \mathbb{Z}$, and
2. odd if $n = 2m + 1$ for some $m \in \mathbb{Z}$.

Prove carefully that:
(a) $\nexists n \in \mathbb{Z} : (n \text{ is even} \wedge n \text{ is odd})$.
(b) $\forall n \in \mathbb{Z} : (n \text{ is even} \veebar n \text{ is odd})$.
Now suppose that $a, b \in \mathbb{Z}$. Prove that:
(c) ab is odd $\Leftrightarrow$ a and b are odd.
(d) ab is even $\Leftrightarrow$ a or b is even.

2.25 ($\diamond$) Let $r \in \mathbb{N}$ be a natural number. Define a relation $\sim$ on the set of integers $\mathbb{Z}$ where $a \sim b$ iff $a - b$ is divisible by r.
(a) Show that $\sim$ is an equivalence relation. If $a \sim b$, we write this as $a \equiv b \pmod r$.
(b) Hence, show that there are r equivalence classes in $\mathbb{Z}/\sim$.
We denote this quotient set as $\mathbb{Z}/r\mathbb{Z} = \{[0], [1], [2], \ldots, [r-1]\}$ called the integer classes modulo r. We now define addition and multiplication in this set as:

$$[a] \oplus [b] = [a + b] \quad \text{and} \quad [a] \otimes [b] = [ab].$$

(c) Prove that these operations $\oplus$ and $\otimes$ are well-defined addition and multiplication operations. In other words, show that the definitions above are independent of the choice of class representatives for $[a]$ and $[b]$. What are the additive and multiplicative identity elements? Show that these identities are unique.
(d) Prove that the set $\mathbb{Z}/r\mathbb{Z}$ and the operations $\oplus$ and $\otimes$ satisfy the ring axioms.

The study of such classes are called modular arithmetic. It was developed by Gauss in his book *Disquisitiones arithmeticae* (Arithmetical Investigations). Today it is a very important object in mathematics, especially in the study of number theory, group theory, and cryptography.
Notably, modular arithmetic is the main concept used in the Diffie–Hellman key exchange by Whitfield Diffie (1944-) and Martin Hellman (1945-) as well as the RSA public-key cryptosystem developed by Ron Rivest (1947-), Adi Shamir (1952-), and Leonard Adleman (1945-), both of which are used in

modern-day secure data transmission. We use these everyday without even realising it!

2.26 (*) Let $a, b \in \mathbb{N}$ with $a \geq b$.

(a) Show that we can uniquely write $a = q_1 b + r_1$ where $q_1, r_1 \in \mathbb{N}$ and $0 \leq r_1 < b$.

The process above is called the long division process. The numbers q_1 and r_1 above are called the quotient and remainders respectively. Since $r_1 < b$, we can carry out the argument as above repeatedly to get a list:

$$\begin{aligned}
a &= q_1 b + r_1, \\
b &= q_2 r_1 + r_2, \\
r_1 &= q_3 r_2 + r_3, \\
r_2 &= q_4 r_3 + r_4, \\
&\vdots
\end{aligned}$$

(b) Prove that r_j is strictly decreasing and hence $r_n = 0$ eventually for some $n \in \mathbb{N}$.

This process is called the Euclidean algorithm.

(c) Show that r_{n-1} divides both a and b.

(d) Let d be any common divisor of a and b. Show that d divides r_{n-1}. What can we say about r_{n-1}?

(e) Hence, prove Bézout's identity, named after Étienne Bézout (1730–1783):

Theorem 2.7.5 (Bézout's Identity) *Let $a, b \in \mathbb{N}$ be two positive integers.*

1. If $am + bn = d$ for some $m, n, d \in \mathbb{N}$, then $gcd(a, b)|d$.

2. There exist some integers $x, y \in \mathbb{Z}$ such that $ax + by = gcd(a, b)$ where $gcd(a, b)$ is the greatest common divisor of the numbers a and b.

(f) Write a computer program that takes a two positive integers a and b as an input and expresses the $\gcd(a, b)$ as a combination of a and b.

2.27 (a) Let $a, b, c \in \mathbb{N}$ such that $\gcd(a, b) = \gcd(a, c) = 1$. Using Bézout's identity from Exercise 2.26(e), prove that $\gcd(a, bc) = 1$.

(b) Suppose that $a, b \in \mathbb{N}$ such that they are coprime to each other. Using part (a) and induction, show that for any $n \in \mathbb{N}$, a^n and b^n are also coprime to each other.

2.28 Let $a, b \in \mathbb{Z}$. Prove that the equivalence $ax \equiv b \pmod{r}$ has a solution $x \in \mathbb{Z}$ if and only if $\gcd(a, r)|b$.

2.29 Let $a, b, c \in \mathbb{N}$.

(a) Prove that if a and c are coprime to each other, then $c|ab$ implies $c|b$.

(b) Prove that if c is a prime number, then $c|ab$ implies $c|a$ or $c|b$.

2.30 (◇) We have seen the principle behind mathematical induction from the Peano's axioms. Here, we present a variant of it which is the strong mathematical induction. The only difference between the strong induction and regular induction is that the induction hypothesis used is stronger by assuming all the statements $P(n)$ for $n = 1, 2, \ldots, k$ are true in order to prove the statement $P(k+1)$. Despite this, the two induction principles are logically equivalent to one another.

Definition 2.7.6 (Strong Mathematical Induction) Suppose that we have a series of statements $P(n)$ indexed by $n \in \mathbb{N}$. If $A = \{n \in \mathbb{N} :$ statement $P(n)$ is true$\}$ and the following two conditions hold:
1. Base case: $1 \in A$ (namely $P(1)$ is true), and
2. Inductive step: $1, 2, 3 \ldots, k \in A \Rightarrow s(k) \in A$ (namely if $P(1), P(2), \ldots, P(k)$ are true, then $P(k+1)$ is also true),

then $A = \mathbb{N}$. In other words, $P(n)$ is true for all $n \in \mathbb{N}$.
(a) Using the strong mathematical induction, show that any natural number $n \geq 2$ can be written as a product of powers of primes. In other words, there exist prime numbers p_j and natural numbers a_j for $j = 1, 2, \ldots, m$ for some $m \in \mathbb{N}$ such that $n = p_1^{a_1} p_2^{a_2} \cdots p_m^{a_m}$.
(b) Show that the prime number decomposition of a natural number in part (a) is unique up to rearrangement of the factors.
The above proves the fundamental theorem of arithmetic:

Theorem 2.7.7 (Fundamental Theorem of Arithmetic) *Any natural number $n \geq 2$ can be written as a unique product of powers of primes. In other words, there exist prime numbers p_j and natural numbers a_j for $j = 1, 2, \ldots, m$ for some $m \in \mathbb{N}$ such that $n = p_1^{a_1} p_2^{a_2} \ldots p_m^{a_m}$ and this expression is unique up to rearrangement of the factors.*

(c) Prove that there are infinitely many prime numbers.

2.31 For any natural number $a \in \mathbb{N}$, we declare $a^0 = 1$. Suppose that $a, b \in \mathbb{N}$ with prime decomposition $a = p_1^{a_1} p_2^{a_2} \cdots p_m^{a_m}$ and $b = p_1^{b_1} p_2^{b_2} \cdots p_m^{b_m}$ where $a_j, b_j \in \mathbb{N}_0$ for $j = 1, 2, \ldots, m$. Prove that:
(a) $\gcd(a, b) = p_1^{c_1} p_2^{c_2} \ldots p_m^{c_m}$ where $c_j = \min\{a_j, b_j\}$ for all $j = 1, 2, \ldots, m$.
(b) $\operatorname{lcm}(a, b) = p_1^{d_1} p_2^{d_2} \ldots p_m^{d_m}$ where $d_j = \max\{a_j, b_j\}$ for all $j = 1, 2, \ldots, m$.

2.32 Prove Exercise 2.20(d) using Exercise 2.31.

2.33 (∗) Let $X \subseteq \mathbb{N}$ be an infinite subset of the natural numbers and $Y = \{x_1, \ldots, x_n\} \subseteq \mathbb{N}$ be a finite subset of $\mathbb{N}$. Show that the set $X \setminus Y$ is non-empty and there exists an $x \in X \setminus Y$ such that $x > x_j$ for all $j = 1, 2, \ldots, n$.

Construction of Real Numbers

3

The method of 'postulating' what we want has many advantages; they are the same as the advantages of theft over honest toil.

— Bertrand Russell, mathematician

In Chap. 2, we have seen how we can reverse the addition operation on $\mathbb{N}$ by extending the set $\mathbb{N}$ to include zero and the negative numbers. This set forms an ordered ring which we call $\mathbb{Z}$. In this chapter, we shall continue with figuring out how can we reverse the multiplication process on the number set $\mathbb{Z}$.

3.1 Rational Numbers $\mathbb{Q}$

From the integers, the next natural extension of the number system is by the definition of fractions or rational numbers as ratios of integers. Again, this is a natural concept in real life as they represent parts of a whole object or ratios of two quantities for comparison.

We write these quantities as pairs of two integers (p, q), read as "p out of q", where $q \neq 0$. As how it is taught in schools, this is also more commonly written as $p : q$. Our notation (p, q) simply emphasises that these elements belong in the Cartesian product $\mathbb{Z} \times (\mathbb{Z} \setminus \{0\})$.

However, there are some pairs of these integers which are equal to each other. As shown in Fig. 3.1, the pair $(1, 2)$ represents the same quantity as the pairs $(2, 4)$ and $(5, 10)$ because they all represent the quantity of "a half of the whole".

Therefore, we need to identify these three pairs of numbers in the set $\mathbb{Z} \times (\mathbb{Z} \setminus \{0\}) = \{(p, q) : p, q \in \mathbb{Z}, q \neq 0\}$ together into one equivalence class. More

S. Johar, *The Big Book of Real Analysis*,
https://doi.org/10.1007/978-3-031-30832-1_3

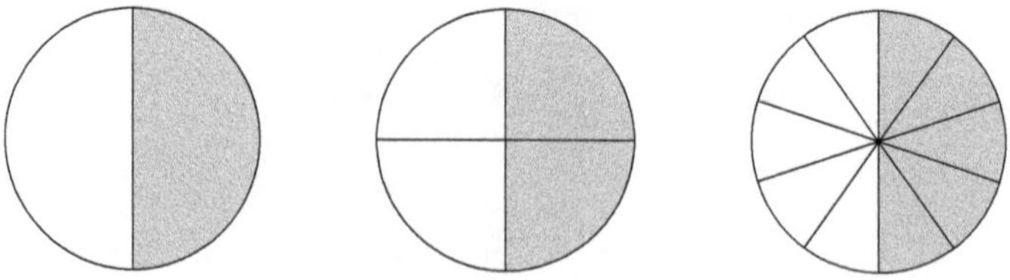

Fig. 3.1 Three different representations of "one half"

generally, we say that two pairs in this set are related, denoted as $(p, q) \sim (r, s)$, iff $ps = qr$ as integers. We can check that this is an equivalence relation:

1. Reflexive: $(p, q) \sim (p, q)$ since $pq = pq$.
2. Symmetric: If $(p, q) \sim (r, s)$, then $ps = rq$. By symmetry of the equality, this implies that $(r, s) \sim (p, q)$.
3. Transitive: If $(p, q) \sim (r, s)$ and $(r, s) \sim (m, n)$, then $ps = rq$ and $rn = ms$ as integers. Multiplying the first equation with n and the second equation with q, we have $psn = rqn = msq$. By cancellation law on $\mathbb{Z}$, since $s \neq 0$, we must have $pn = mq$. Thus, by definition, we have $(p, q) \sim (m, n)$.

Thus, we do not have to worry about two distinct pairs of integers representing the same quantity as these pairs are all identified together under this equivalence relation. Concretely, the rational numbers is the quotient set or the set of equivalence classes defined as:

$$\mathbb{Q} = (\mathbb{Z} \times (\mathbb{Z} \setminus \{0\}))/\sim = \{[(p, q)] : p, q \in \mathbb{Z}, q \neq 0\},$$

where $\sim$ is the equivalence relation defined above. These classes can be seen in Fig. 3.2. Usually, we denote the class $[(p, q)]$ as $\frac{p}{q}$ to simplify the notation. This number is called a fraction, a quotient, a ratio, or a rational number.

For a rational number $\frac{p}{q}$, the number p is called the numerator whereas the number q is called the denominator. As seen in Fig. 3.1, the denominator is thought of the number of divisions in a whole object and the number p is how many of such division represents the fraction. Note also that the integers $\mathbb{Z}$ are also contained in the set of rational numbers in the form of classes which can be represented by some $\frac{p}{q}$ with $q = 1$.

If we recall the discussion on equivalence classes in the previous chapter, there is not a unique representative for each class. However, we can choose the simplest representative where the numerator and the denominator are coprime to each other and the denominator is non-negative. Namely, we can find a representative for the class $\frac{p}{q}$ such that $\gcd(p, q) = 1$ and $q > 0$. This representative is called an irreducible representation or fraction in lowest terms.

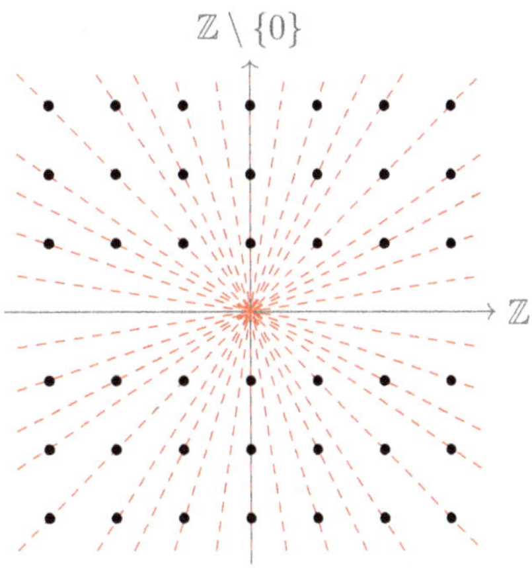

Fig. 3.2 Numbers in lines as numbers again! The set of rational numbers $\mathbb{Q}$ is given by the set of the equivalence classes of points in $\mathbb{Z}\times(\mathbb{Z}\setminus\{0\})$ which lie on the same dashed red line in the lattice above

So, for example, the numbers $\frac{1}{-2}, \frac{-2}{4}, \frac{5}{-10}, \ldots$ all could represent the same class, but the unique representative for this class that satisfy the requirements above would be $\frac{-1}{2}$. If we insist this rule of choosing the representative, then every rational number class can be represented uniquely by one single pair of integers.

The rational numbers form an object what modern mathematicians call a field. A field is a special commutative ring for which every non-zero element has a multiplicative inverse. More formally, a field is defined as follows:

Definition 3.1.1 (Field) A field is a set X along with $+$ and $\times$ operations such that $a+b\in X, a\times b\in X$ for all $a,b\in X$ and satisfies the following axioms:

1. $(X,+,\times)$ is a commutative ring as defined in Definition 2.5.1, and
2. Existence of inverse for $\times$: for each $a\in X\setminus\{0\}$, there exists a $\frac{1}{a}\in X$ such that $a\times\frac{1}{a}=1$.

Usually, we denote the set X with the operations $+$ and $\times$ as a whole by $\mathbb{F}$. The field axioms are what one takes for granted and uses in arithmetic and algebra at school level. For simplicity, for any $a,b\in\mathbb{F}$, let us introduce the following shorthand notations and conventions:

1. $ab=a\times b$.
2. $\frac{a}{b}=a\times\frac{1}{b}$.
3. $a-b=a+(-b)$.
4. If $n\in\mathbb{N}$, then $a^n=\underbrace{a\times a\times a\times\ldots\times a}_{n\text{ times}}$.
5. $a^0=1$ for any $a\neq 0$.

We shall justify the final convention in Sect. 2.4.

Example 3.1.2 Let us look an example and a non-example:

1. We have seen that the set $\mathbb{Z}$ with $+$ and $\times$ forms a commutative ring. However, not all non-zero elements in $\mathbb{Z}$ has a multiplicative inverse in $\mathbb{Z}$ and so $(\mathbb{Z}, +, \times)$ does not form a field. This is clearly true because, for example, the number 2 does not have a multiplicative inverse in $\mathbb{Z}$ since there are no integer $x \in \mathbb{Z}$ such that $2x = 1$. So the integers $\mathbb{Z}$ do not form a field.
2. As a simple example, the field made up of one element $\mathbb{F} = \{0\}$ is called a trivial field. This is a very boring field as it has only one element. We shall see more interesting examples of fields as we go along in this chapter.

From its definition, fields must satisfy the following lemma:

Lemma 3.1.3 *Let $\mathbb{F}$ be a number field.*

1. *For any $a \in \mathbb{F}$ we have $0 \times a = a \times 0 = 0$.*
2. $-0 = 0$.
3. *If $\mathbb{F}$ is a non-trivial field, then $1 \neq 0$.*

Since fields are also rings, the proof of Lemma 3.1.3 is similar to the proof of Lemma 2.5.5.

3.2 Algebra on $\mathbb{Q}$

In Example 3.1.2 we have seen that the set of integers $\mathbb{Z}$ is not a field. On the other hand, we claimed earlier that the set of rational numbers $\mathbb{Q}$ forms a field with a suitable addition and multiplication operations. We first need to define how we add and multiply rational numbers together.

First, we try to define addition. Similar to the natural numbers, we would like to add rational numbers by combining them together. For concrete cases, let us look at how we can add $\frac{2}{3}$ to $\frac{1}{4}$ before we declare a meaningful general addition operation. Representing these numbers as fractions of a disc, we can combine them with an addition operation which we denote by $\oplus$ as depicted in Fig. 3.3.

However, the addition above does not tell us what the sum is explicitly in terms of the constituent fractions $\frac{1}{4}$ and $\frac{2}{3}$. To get an explicit representation, we subdivide

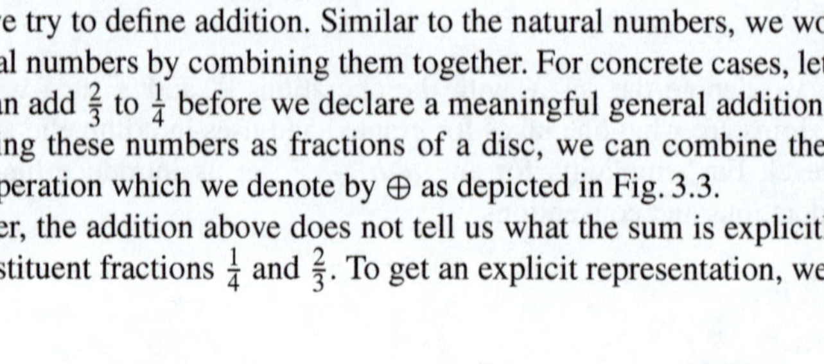

Fig. 3.3 Adding two fractions

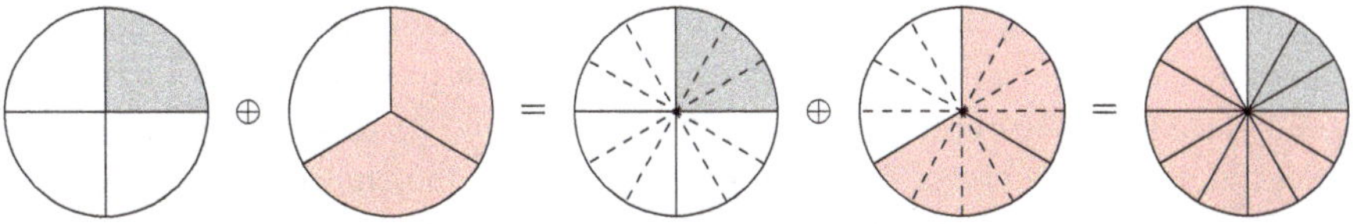

Fig. 3.4 Adding two fractions after subdividing them

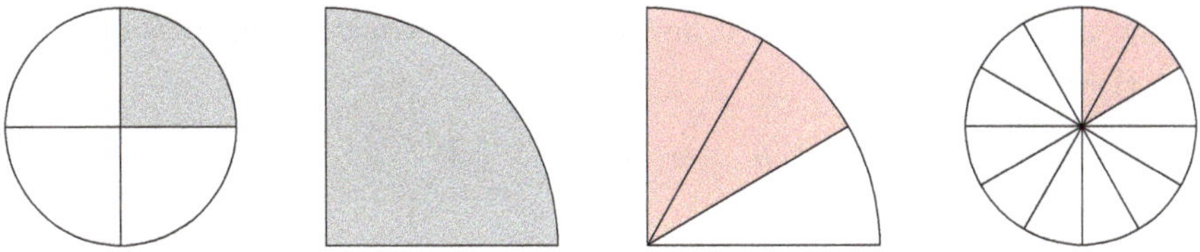

Fig. 3.5 Multiplying rational numbers. **(a)** $\frac{1}{4}$ of 1. **(b)** $\frac{1}{4}$. **(c)** $\frac{2}{3}$ of $\frac{1}{4}$. **(d)** $\frac{2}{3} \otimes \frac{1}{4}$ of 1

the fractions so that they have the same size divisions first as in Fig. 3.4 before combining them.

So, in our case of adding $\frac{1}{4}$ and $\frac{2}{3}$ together, we have $\frac{1}{4} = \frac{3}{12}$ and $\frac{2}{3} = \frac{8}{12}$. Therefore, their sum, as in Fig. 3.4, is $\frac{11}{12}$. In general, if $\frac{p}{q}$ and $\frac{r}{s}$ are two rational numbers, then we propose an addition $\oplus$ operation on $\mathbb{Q}$ by finding representatives of each rational number with the same denominator first and adding the numerators as follows:

$$\frac{p}{q} \oplus \frac{r}{s} = \frac{ps}{qs} \oplus \frac{rq}{qs} = \frac{ps + rq}{qs}.$$

Note that this operation can be done using our prior knowledge of the integers since we know how to add and multiply the integers which appear in the numerator and denominator.

For multiplication, we follow the same idea by representing the rational numbers pictorially and propose a meaningful multiplication operation. Using the numbers $\frac{1}{4}$ and $\frac{2}{3}$ again, we note that multiplication means we want $\frac{2}{3}$ (copies) of $\frac{1}{4}$. So if we represent a whole disc as 1 again, we can visualise multiplying these two rational numbers step by step as in Fig. 3.5. We denote this multiplication operation as $\otimes$.

Thus graphically, we can see that $\frac{2}{3} \otimes \frac{1}{4}$ is the same as the fraction $\frac{2}{12}$. We note that this is obtained by multiplying the numerators and the denominators of the two rational numbers respectively. In general, if $\frac{p}{q}$ and $\frac{r}{s}$ are two rational numbers, then we propose a multiplication operation $\otimes$ on $\mathbb{Q}$ as follows:

$$\frac{p}{q} \otimes \frac{r}{s} = \frac{pr}{qs}.$$

One can check that the operations $\oplus$ and $\otimes$ are well-defined regardless of which representative of the equivalence classes is chosen. For example, since $\frac{1}{2} = \frac{2}{4}$ and $\frac{5}{2} = \frac{15}{6}$, we naturally expect $\frac{1}{2} \oplus \frac{5}{2} = \frac{2}{4} \oplus \frac{15}{6}$ and $\frac{1}{2} \otimes \frac{5}{2} = \frac{2}{4} \otimes \frac{15}{6}$. The proof for the independence of the algebraic operations $\oplus$ and $\otimes$ on the choice of representative of rational numbers is left as an exercise for the readers in Exercise 3.1.

In fact, since the integers $\mathbb{Z}$ are contained in the rationals $\mathbb{Q}$, we can show that the restriction of the operations $\oplus$ and $\otimes$ on $\mathbb{Q}$ to the elements of $\mathbb{Z}$ agree with the $+$ and $\times$ operations on $\mathbb{Z}$. Indeed, any $m, n \in \mathbb{Z}$ can be represented by $\frac{m}{1}$ and $\frac{n}{1}$ in $\mathbb{Q}$ respectively. Thus:

$$\frac{m}{1} \oplus \frac{n}{1} = \frac{m(1) + n(1)}{1} = \frac{m+n}{1},$$
$$\frac{m}{1} \otimes \frac{n}{1} = \frac{mn}{1} = \frac{m \times n}{1}.$$

Thus, we can simply use the symbols $+$ and $\times$ in place of $\oplus$ and $\otimes$ since these are just extensions of the operations $+$ and $\times$ on $\mathbb{Z}$ to a larger set of numbers $\mathbb{Q}$.

With these operations defined, via a routine check, we can show that $\mathbb{Q}$ along with these operations satisfy the commutative ring axioms in Definition 2.5.1 with the additive and multiplicative identities $\frac{0}{1}$ and $\frac{1}{1}$ respectively.

With this, we can immediately see that an additive inverse for a rational number $\frac{p}{q}$ is $\frac{-p}{q}$ since $\frac{p}{q} + \frac{-p}{q} = \frac{p+(-p)}{q} = \frac{0}{p} = \frac{0}{1}$. We write this inverse as $-\frac{p}{q}$. Moreover, if $p \neq 0$, a multiplicative inverse of it is the number $\frac{q}{p}$ since $\frac{p}{q} \times \frac{q}{p} = \frac{pq}{pq} = \frac{1}{1}$. Hence $(\mathbb{Q}, +, \times)$ is a field.

The next question is: are there any more additive and multiplicative inverse of the number $\frac{p}{q}$? The answer is no, those are the only ones. In general, we have:

Lemma 3.2.1 *Let $\mathbb{F}$ be a non-trivial number field.*

1. *The additive identity* 0 *is unique.*
2. *The multiplicative identity element* 1 *is unique.*
3. *For each element $a \in \mathbb{F}$, its additive inverse $-a$ is unique.*
4. *For each element $a \in \mathbb{F} \setminus \{0\}$, its multiplicative inverse $\frac{1}{a}$ is unique.*

Proof The readers were asked to prove the first three assertions in Exercise 2.22. Here we shall prove the final assertion only.

4. Suppose that we have two multiplicative inverses for $a \in \mathbb{F}$, namely $\frac{1}{a^\circ}$ and $\frac{1}{a^*}$. Then, $a\frac{1}{a^\circ} = 1$ and $\frac{1}{a^*}a = 1$. Multiplying the former with $\frac{1}{a^*}$ and using the associativity of multiplication in a field, we have:

$$\frac{1}{a^*}\left(a\frac{1}{a^\circ}\right) = \frac{1}{a^*} \quad \Rightarrow \quad \left(\frac{1}{a^*}a\right)\frac{1}{a^\circ} = \frac{1}{a^*} \quad \Rightarrow \quad \frac{1}{a^\circ} = \frac{1}{a^*},$$

which means these two inverses are in fact identical. □

Using the results that we have so far, we can prove the following field properties, which are left as Exercise 3.2.

Lemma 3.2.2 *Let $\mathbb{F}$ be an ordered number field and $a, b, x, y \in \mathbb{F}$.*

1. *If $a + x = a$ for all $a \in \mathbb{F}$, then $x = 0$.*
2. *Cancellation property: If $a + x = a + y$, then $x = y$.*
3. $-(-a) = a$.
4. $-(a + b) = (-a) + (-b)$.
5. *If $ab = 0$, then $a = 0$ or $b = 0$ (in the language of abstract algebra, $\mathbb{F}$ is an integral domain).*
6. *If $ax = a$ for all $a \neq 0$, then $x = 1$.*
7. *Cancellation property: If $ax = ay$ for all $a \neq 0$, then $x = y$.*
8. *If $a \neq 0$, then $\frac{1}{\frac{1}{a}} = a$.*
9. $(-b)a = a(-b) = -(ab)$. *In particular,* $(-1)a = a(-1) = -a$.
10. $(-1)(-1) = 1$. *In particular,* $(-a)(-a) = a^2$.
11. *If $a, b \neq 0$, then $\frac{1}{a}\frac{1}{b} = \frac{1}{ab}$.*

3.3 Ordering on $\mathbb{Q}$

Now suppose that we have two rational numbers $\frac{m}{n}$ and $\frac{p}{q}$ where m, n, p, q are integers such that n and q are non-zero. How do we order these rational numbers? We know how to order integers so let us use the ordering $<$ on the integers $\mathbb{Z}$ as a reference point. We propose an ordering on $\mathbb{Q}$ as $\prec$ which is defined thus:

$$\frac{m}{n} \prec \frac{p}{q} \quad \text{iff either} \quad \begin{cases} nq > 0 \text{ and } mq < np, & \text{or} \\ nq < 0 \text{ and } mq > np. \end{cases}$$

We can check that this definition is independent of the choice for class representative in $\mathbb{Q}$. Suppose that $\frac{m}{n} = \frac{x}{y}$ and $\frac{p}{q} = \frac{r}{s}$ (which, by definition of the relation $\sim$, mean $my = nx$ and $ps = rq$), then we have several cases. Assume first that $nq > 0$, $ny > 0$, and $qs > 0$. This means all of n, y, q, s are of the same sign so that $ys > 0$. Then, we have:

$$\begin{aligned} \frac{m}{n} \prec \frac{p}{q} \quad \Leftrightarrow \quad mq < np \quad \Leftrightarrow \quad mqys < npys \quad &\Leftrightarrow \quad nxqs < nyrq \\ &\Leftrightarrow \quad 0 < nq(yr - xs), \end{aligned}$$

and since $nq > 0$, by the cancellation law in Proposition 2.6.2, this implies:

$$0 < yr - xs \quad \Leftrightarrow \quad xs < yr \quad \Leftrightarrow \quad \frac{x}{y} \prec \frac{r}{s}.$$

The other cases for other possible sign combinations of nq, ny, and qs can also be done using a similar argument. Therefore, regardless of the representative fraction used for the rational number, the ordering remains the same. Thus this relation is well-defined.

Next, we verify that $\prec$ is a strict total order on $\mathbb{Q}$. Since $\prec$ is independent of the choice of representative for elements in $\mathbb{Q}$, we can assume WLOG that the following rational numbers are chosen such that their representatives have positive denominators. We check:

1. Irreflexive: $\frac{a}{b} \nprec \frac{a}{b}$ since $ab = ba$.
2. Transitive: Suppose that $\frac{a}{b} \prec \frac{c}{d}$ and $\frac{c}{d} \prec \frac{e}{f}$. By definition of $\prec$ and the assumption that the denominators are positive, we have $ad < bc$ and $cf < de$. Since $\mathbb{Z}$ is an ordered ring and $b, f > 0$, we have $adf < bcf$ and $bcf < bde$. Thus, by transitivity of the ordering on $\mathbb{Z}$, we have $adf < bde$. Finally, since $d > 0$, by cancellation law on $\mathbb{Z}$ in Proposition 2.6.2, we get $af < be$ and conclude that $\frac{a}{b} \prec \frac{e}{f}$.
3. Trichotomous: Suppose that $\frac{a}{b}, \frac{c}{d} \in \mathbb{Q}$. By trichotomy of the ordering $<$ on $\mathbb{Z}$, exactly one of $ac = bd, ac < bd$, or $bd < ac$ holds. Thus, exactly one of $\frac{a}{b} = \frac{c}{d}, \frac{a}{b} \prec \frac{c}{d}$, or $\frac{c}{d} = \frac{a}{b}$ can be true.

Therefore, we conclude that $\prec$ is a strict total order on $\mathbb{Q}$.

Now that we have a well-defined strict ordering on the set on rational numbers, we want to combine the ordering structure and field structure together. The idea is similar to the ordered ring axioms in Definition 2.6.1. If the ordering structure and the field structure are compatible with each other in the following sense, we call the field an ordered field.

Definition 3.3.1 (Ordered Field Axioms) A field $(\mathbb{F}, +, \times)$ with a relation $<$ is an ordered field if:

1. $<$ is a strict total order from Definition 2.3.1,
2. $<$ is compatible with $+$: if $a < b$ and $c \in \mathbb{F}$, then $a + c < b + c$, and
3. $<$ is compatible with $\times$: if $0 < a$ and $0 < b$, then $0 < a \times b$.

Remark 3.3.2 An interesting note here is that, similar to rings, not every field can have an ordered field structure. For example, the complex numbers $\mathbb{C}$ or the modulo p integers $\mathbb{Z}/p\mathbb{Z}$ (as seen in Exercise 2.25) for some prime number p are both number fields. Moreover, we can define some strict total order on them. However any strict total order on the fields $\mathbb{C}$ and $\mathbb{Z}/p\mathbb{Z}$ could never be compatible with their field structures. We shall see this later in Exercises 3.29 and 3.31.

Back to the rational numbers. The strict order $\prec$ that we have defined on $\mathbb{Q}$ above can be shown to satisfy the ordered field axioms. We check this one by one.

1. $\prec$ is defined as a strict total order.
2. Pick $\frac{m}{n}, \frac{p}{q}, \frac{x}{y} \in \mathbb{Q}$ and, WLOG, suppose that $n, q, y > 0$. Suppose that $\frac{m}{n} \prec \frac{p}{q}$. Then:

$$
\begin{aligned}
\frac{m}{n} \prec \frac{p}{q} \quad \Leftrightarrow \quad mq < pn \quad &\Leftrightarrow \quad mqy^2 < pny^2 \\
&\Leftrightarrow \quad mqy^2 + xnqy < pny^2 + xnqy \\
&\Leftrightarrow \quad qy(my + xn) < ny(py + xq) \\
&\Leftrightarrow \quad \frac{my + xn}{ny} \prec \frac{py + xq}{qy} \\
&\Leftrightarrow \quad \frac{m}{n} + \frac{x}{y} \prec \frac{p}{q} + \frac{x}{y}.
\end{aligned}
$$

3. Let $\frac{m}{n}, \frac{p}{q} \in \mathbb{Q}$ and, WLOG, suppose that $n, q > 0$. Assume that $0 \prec \frac{m}{n}$ and $0 \prec \frac{p}{q}$ so that $m, p > 0$. Then, we have $pm > 0$ and $qn > 0$ which means $0 \prec \frac{mp}{nq} = \frac{m}{n}\frac{p}{q}$.

Again, the ordering $\prec$ on $\mathbb{Q}$ is exactly similar to the ordering $<$ upon restriction to $\mathbb{Z}$. Therefore, to emphasise this fact, we reuse the symbol $<$ in place of $\prec$ for the ordering on $\mathbb{Q}$. Inheriting the property from its subsets $\mathbb{N}$ and $\mathbb{Z}$, the field $\mathbb{Q}$ does not have a maximal element.

Moreover, the ordered field axioms allow us to deduce additional properties for the rational numbers. The following properties were taken for granted or as "obvious" in school arithmetic and algebra. However, they all can be deduced by carefully applying the field and ordering axioms. The readers will prove these properties in Exercise 3.2.

Lemma 3.3.3 *Let $\mathbb{F}$ be an ordered number field and $a, b, c \in \mathbb{F}$.*

1. $0 < 1$.
2. $a \leq b$ *if and only if* $-b \leq -a$.
3. *If* $a \leq b$ *and* $c > 0$, *then* $ac \leq bc$.
4. *If* $a \leq b$ *and* $c < 0$, *then* $bc \leq ac$.
5. $a^2 \geq 0$ *with equality if and only if* $a = 0$.
6. *If* $a > 0$, *then* $\frac{1}{a} > 0$.
7. *If* $a, b > 0$ *and* $a \leq b$, *then* $\frac{1}{b} \leq \frac{1}{a}$.
8. *If* $ab > 0$, *then either* $a > 0, b > 0$ *or* $a < 0, b < 0$.
9. *If* $ab < 0$, *then either* $a > 0, b < 0$ *or* $a < 0, b > 0$.

Example 3.3.4 In particular, Lemmas 3.2.2 and 3.3.3 hold for the ordered field $\mathbb{Q}$. These properties allow us to manipulate equalities and inequalities in $\mathbb{Q}$. Let us look at some examples on how to use them:

1. Suppose that we want to find all the pairs of rational numbers $x, y \in \mathbb{Q}$ that satisfy $x^2 + y^2 = 0$. We can add $-y^2$ on both sides of the equation to get $x^2 = -y^2$. But now, notice that from Lemma 3.3.3(5), we must have $x^2 \geq 0$ and $y^2 \geq 0$. The latter means $-y^2 \leq 0$. So the only possibility for $x^2 = -y^2$ is when $x^2 = y^2 = 0$. Again, by Lemma 3.3.3(5), this can only be achieved when $x = y = 0$. We can easily check that indeed the pair $x = y = 0$ satisfies the equation above. Hence the only pair of rational numbers (x, y) that satisfy the equation $x^2 + y^2 = 0$ is $(x, y) = (0, 0)$.
2. Suppose that we want to find all the rational numbers that satisfy the equation $2x + 3 = 0$. Using the assertions in Lemma 3.2.2, by applying the denoted operations on both sides of the equation, we have:
$$2x + 3 = 0 \quad \overset{-3}{\Longrightarrow} \quad 2x = -3 \quad \overset{\times\frac{1}{2}}{\Longrightarrow} \quad x = -\frac{3}{2}.$$
Furthermore, since all the implication arrows above are reversible, this is a solution to the equation and it is unique.
3. Now suppose that we want to find all rational numbers x that satisfy the inequality $x^2 + 3x > 4$. By the ordered field axioms, we can add -4 on both sides of the inequality to get $x^2 + 3x - 4 > 0$. Now we express the LHS as a product of two numbers, namely $(x - 1)(x + 4) > 0$. Applying Lemma 3.3.3(8), both factors are positive or both factors are negative. We investigate both cases:
   (a) If $x - 1 > 0$ and $x + 4 > 0$, then x must satisfy $x > 1$ and $x > -4$. So putting them together, we have $x > 1$.
   (b) If $x - 1 < 0$ and $x + 4 < 0$, then x must satisfy $x < 1$ and $x < -4$. Combining these two inequalities, we get $x < -4$.
   We now check that both of the cases above are valid. Indeed:
   (a) If $x > 1$, then $x^2 > 1$ and $3x > 3$. So $x^2 + 3x > 1 + 3 = 4$.
   (b) If $x < -4$, then $x + 3 < -1$. By applying Lemma 3.3.3(4) twice, we have $x^2 + 3x = x(x + 3) > (-4)(x + 3) > (-4)(-1) = 4$.
   Thus, all the $x \in \mathbb{Q}$ that satisfy the inequality $x^2 + 3x > 4$ are precisely $\{x \in \mathbb{Q} : x < -4\} \cup \{x \in \mathbb{Q} : x > 1\} = \{x \in \mathbb{Q} : x < -4 \text{ or } x > 1\}$.
4. Now suppose that we have a very simple equation $x - 3 = 2$. Clearly $x = 5$ is the only solution to this. But let us try a different route to get there. We carry out the following algebra by first squaring both sides of the equation as thus:
$$\begin{aligned} x - 3 = 2 \quad \Rightarrow \quad (x - 3)^2 = 2^2 \quad &\Rightarrow \quad x^2 - 6x + 9 = 4 \\ &\Rightarrow \quad x^2 - 6x + 5 = 0 \\ &\Rightarrow \quad (x - 1)(x - 5) = 0 \\ &\Rightarrow \quad (x - 1) = 0 \text{ or } (x - 5) = 0 \\ &\Rightarrow \quad x = 1 \text{ or } 5, \end{aligned}$$
so we get an extra solution of $x = 1$ using this method.

The above sequence of implications, in short, simply says $x-3=2 \Rightarrow x=1$ or 5, which is true since the x that satisfies the equation $x-3=2$ is either 1 or 5. Both of these are candidate solutions to the equation. However, $x=1$ does not satisfy the equation since $1-3=-2 \neq 2$ and so the converse implication $x=1$ or $5 \Rightarrow x-3=2$ is false.

If we were to carefully check the implication arrows above one by one, we can see that all the arrows are reversible, except for the very first one. The very first implication is not reversible since $(x-3)^2=2^2$ could either mean $x-3=2$ or $x-3=-2$.

The moral of the story is after carrying out a sequence of algebraic operations, it is important to check the solution candidates that we have obtained since we may pick up some sneaky false ones. These false solutions are called extraneous solutions.

Archimedean Property of $\mathbb{Q}$

An important property of the rational numbers is the Archimedean property, named after Archimedes of Syracuse (c. 287B.C.-212B.C.). This property says each positive rational number is strictly smaller than some natural number.

Proposition 3.3.5 (Archimedean Property of Rational Numbers) *For every positive rational number $r \in \mathbb{Q}_+$, there exists a natural number $n \in \mathbb{N}$ such that $n > r$.*

Proof Let $r=\frac{p}{q}$ where $p, q \in \mathbb{N}$. We claim that $n=2p \in \mathbb{N}$ is sufficient. Indeed, since $q \geq 1$ and hence $0 < \frac{1}{q} \leq 1$, by multiplying with $p > 0$, we have $\frac{p}{q} \leq p < 2p = n$. □

In analysis, we use the above property in many instances, so it is important to be familiar with it. Two consequences of the Archimedean property are the following:

Corollary 3.3.6 *For every positive rational number $r \in \mathbb{Q}_+$, there exists a natural number $n \in \mathbb{N}$ such that $0 < \frac{1}{n} < r$.*

Proof Since $r > 0$, Lemma 3.2.2 says $\frac{1}{r}$ is also a positive rational number. By the Archimedean property, we can find a natural number $n \in \mathbb{N}$ such that $\frac{1}{r} < n$. Finally, by algebra, this implies that $0 < \frac{1}{n} < r$. □

Corollary 3.3.7 *Let $a, b \in \mathbb{Q}$ be such that $a < b$. Then, there is an $r \in \mathbb{Q}$ such that $a < r < b$.*

Proof Since $b-a > 0$, Corollary 3.3.6 says there exists a natural number $n \in \mathbb{N}$ such that $0 < \frac{1}{n} < b-a$. Adding a on all sides of the inequalities, we obtain the desired result with $r = a + \frac{1}{n}$. □

In fact, we have a stronger result. We first state two lemmas, which the readers will prove in Exercises 3.14 (in more generality) and 3.16 respectively.

Lemma 3.3.8 (Binomial Expansion for $\mathbb{Q}$) *Let $j, n \in \mathbb{N}_0$ with $0 \leq j \leq n$. The quantity $\binom{n}{j} = \frac{n!}{(n-j)!j!}$ is called the binomial coefficient with $r! = r \cdot (r-1) \cdot (r-2) \cdot \ldots \cdot 1$ for any $r \in \mathbb{N}$ and $0! = 1$. For any $x, y \in \mathbb{Q}$ we have:*

$$(x+y)^n = \sum_{j=0}^{n} \binom{n}{j} x^j y^{n-j}.$$

Lemma 3.3.9 *Let $a, b \in \mathbb{Q}$ and $p \in \mathbb{N}$. For any $x \in \mathbb{Q}$, if $m \in \mathbb{N}$ is such that $m > \frac{2^p x^{p-1}}{b-a}$ and $m \geq \frac{2}{x}$, then $0 < (x + \frac{1}{m})^p - x^p < b - a$.*

Now we state and prove the following result which generalises Corollary 3.3.7:

Proposition 3.3.10 *Let $a, b \in \mathbb{Q}$ be such that $0 < a < b$ and $p \in \mathbb{N}$. Then, there is a positive rational number $r \in \mathbb{Q}_+$ such that $a < r^p < b$.*

Proof By Proposition 3.3.5 and Corollary 3.3.6, we can find some positive rational numbers $y, z \in \mathbb{Q}_+$ such that $0 < y^p < a < b < z^p$. For these y and z, using Proposition 3.3.5 again, we can find $m_1, m_2 \in \mathbb{N}$ such that $m_1 \geq \frac{2}{y}$ and $m_2 > \frac{2^p z^{p-1}}{b-a}$. Set $m = \max\{m_1, m_2\}$ so that for any $x \in \mathbb{Q}_+$ satisfying $y \leq x \leq z$, we have $\frac{2}{x} \leq \frac{2}{y} \leq m_1 \leq m$ and $\frac{2^p x^{p-1}}{b-a} \leq \frac{2^p z^{p-1}}{b-a} < m_2 \leq m$.

Hence, from Lemma 3.3.9 for all $x \in \mathbb{Q}_+$ with $y \leq x \leq z$, we have:

$$0 < \left(x + \frac{1}{m}\right)^p - x^p < b - a. \tag{3.1}$$

Increasing y by multiples of $\frac{1}{m}$, we define the set $S = \{k \in \mathbb{N} : (y + \frac{k}{m})^p > a\}$. This set is non-empty because for any $k \geq \frac{am}{py^{p-1}}$, we have $(y + \frac{k}{m})^p \geq (y + \frac{a}{py^{p-1}})^p > py^{p-1}\frac{a}{py^{p-1}} = a$ using the binomial expansion in Lemma 3.3.8. Therefore, due to well-ordering principle, S has a minimum element, say $k = t \in S$.

We claim $r = y + \frac{t}{m} \in \mathbb{Q}$ is the rational number that we are looking for. By definition of the set S, necessarily $r^p > a$. On the other hand, we have $r - \frac{1}{m} = y + \frac{t-1}{m} \geq y > 0$. Therefore, by minimality of r, we have $(r - \frac{1}{m})^p \leq a < z^p$ which then implies $r - \frac{1}{m} < z$. Since $y \leq r - \frac{1}{m} < z$, we can set $x = r - \frac{1}{m}$ in inequality (3.1) to get:

$$r^p - \left(r - \frac{1}{m}\right)^p < b - a \quad \Rightarrow \quad r^p < b - a + \left(r - \frac{1}{m}\right)^p \leq b - a + a = b.$$

Thus we can conclude that $a < r^p < b$. $\square$

Remark 3.3.11 Now that we have a total ordering on the rational numbers, let us make some important remarks on the ordering of rational numbers:

1. Does the set $\mathbb{Q}$ has a largest element? No, this set keeps on going forever, similar to the set $\mathbb{N}$. Indeed, supposing for a contradiction that there is a maximum element $r \in \mathbb{Q}$. Then, by Proposition 3.3.5, there is a natural number $n \in \mathbb{N} \subseteq \mathbb{Q}$ such that $r < n$, which contradicts the maximality of r. Likewise, using the same argument as the above, we can show that $\mathbb{Q}$ does not have a minimum element. Thus the integers $\mathbb{Z}$ and the rational number $\mathbb{Q}$ are unbounded above and below.
2. We have seen that we used the symbol ∞ in Definition 2.3.7 to denote the unbounded concept of $\mathbb{N}$. For $\mathbb{Q}$ and $\mathbb{Z}$, we have are two variants for the symbol of infinity, namely $+\infty$ (which we usually write as ∞ only) and $-\infty$. They denote the direction in which we follow: either positive or negative boundless quantity.
3. Another interesting property of the rational numbers is that there is no smallest positive rational number due to Corollary 3.3.6. More generally, given a rational number, there is no "next" rational number that comes after it. This behaviour is unlike the integers where for each integer n there is a succeeding integer $n + 1$ that comes after it via Peano's axioms.
   This can be seen via Corollary 3.3.7. Indeed, suppose that a is a rational number and we claim that $b > a$ is the succeeding rational number after a. Then we can always find another rational number r strictly in between them. But then this means the rational numbers a and b are not next to each other as there is at least one other rational number in between them.
   In fact, we can continue with this construction indefinitely. This property is called the density of rational numbers: for each pair of rational numbers, there are infinitely many rational numbers in between them.

Even though there are infinitely many rational numbers, the set of rational numbers are countable: we can count the rational numbers $\mathbb{Q}$ just as how we would count the natural numbers $\mathbb{N}$. This seems absurd at first: it looks like there are more rational numbers than the natural numbers and the integers since $\mathbb{N} \subsetneq \mathbb{Z} \subsetneq \mathbb{Q}$! In fact, we have noted in Remark 3.3.11(3) that in between 0 and 1 there are already infinitely many rational numbers.

However, the concept of infinity can be a strange and remarkable thing and we have to deal with infinite sets in a very delicate manner. As warned by Bernard Bolzano (1781–1848) in his posthumously-published book *Paradoxien des Unendlichen* (Paradoxes of the Infinite):

> Certainly most of the paradoxical statements encountered in the mathematical domain ... are propositions which either immediately contain the idea of infinite, or at least in some way or other depend on that idea for their attempted proof.

But we assure you here: the fact that the size of set of natural numbers and the set of rational numbers are the same is not a paradox! To demonstrate this, of course, we need to first clarify what do we mean by the "size of a set" via cardinality.

3.4 Cardinality

Roughly speaking, the cardinality or size of a set quantifies how many elements are there in the set. For a set X, we denote $|X|$ as its cardinality or size. Finite sets have well-defined sizes: we can just count the number of elements in the sets using $\mathbb{N}_0$. Moreover, by the ordering in $\mathbb{N}_0$, we can compare their sizes; if we have two finite sets X and Y with number of elements $m, n \in \mathbb{N}_0$ respectively, we say X has smaller cardinality compared to Y if $m < n$, greater cardinality compared to Y if $m > n$, or equal cardinality to Y if $m = n$. Easy.

For non-finite sets, we have to use a different approach since we cannot quantify their sizes using any element in $\mathbb{N}_0$. As mentioned in Remark 2.3.8, infinity has always been a divisive topic in mathematics. Thus, naturally, assigning a size to an infinite set has always been problematic and inconsistent.

As an example, the set of natural numbers $\mathbb{N}$ and the set of even natural numbers, denoted as $2\mathbb{N} = \{2n : n \in \mathbb{N}\}$, both have infinite "sizes". However, since $2\mathbb{N} \subsetneq \mathbb{N}$, instinctively we expect that the set $2\mathbb{N}$ is strictly smaller than $\mathbb{N}$ in terms of "size", even though they have the same infinite "size"! So is there a hierarchy within these infinite sizes? Some infinity is smaller than others? Instinct usually fails us when it involves infinity...

It was the genius of Georg Cantor who came up with how we can quantify the size of infinite sets consistently. For finite sets, our instinctive method is to count the number of elements in different sets using $\mathbb{N}_0$ which would then allow us to compare their sizes using the order on this set. For infinite sets, Cantor proposed to do it backwards: we compare the sizes of sets first and then assign their sizes based on this comparison. In fact, this system is also consistent with finite sets.

To demonstrate this, let us return to the finite sets X and Y above and determine how else can we compare their sizes without counting the number of elements beforehand.

Pick the first element $x_1 \in X$. After this selection, pick an element $y_1 \in Y$. Thus, we can pair up the element x_1 with y_1. Next, pick an element $x_2 \in X \setminus \{x_1\}$ and pair it with another element $y_2 \in Y \setminus \{y_1\}$. Refer to Fig. 3.6 for a demonstration.

We repeat this pairing construction until either the elements in X runs out, the elements in Y runs out, or elements in both sets run out at the same time. In each of the cases, we have:

Fig. 3.6 Pairing elements of the finite sets X and Y

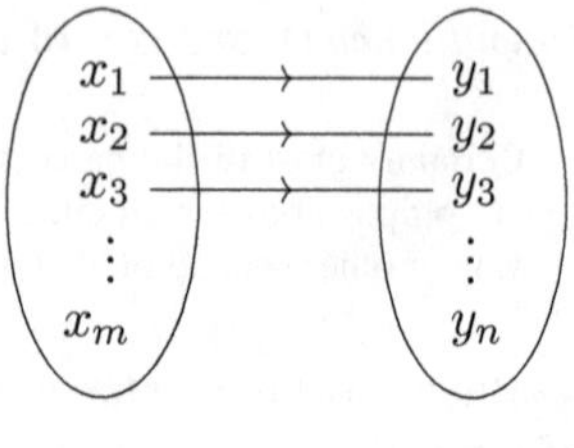

1. For the first case, the size of X is smaller since we have ran out of elements in X first. By the pairing that we have carried out, we have an injective function $f : X \to Y$ mapping distinct elements of X to distinct elements in Y.
2. For the second case, since there are still elements of X left over, we have to match the remaining elements in X with any of the elements in Y that has been mapped to. Thus, we have a surjective function $f : X \to Y$.
3. Finally, if both sets run out of elements at the same time, each $x \in X$ has a unique and distinct pair element $y \in Y$. This defines a bijection between the two sets.

So if the size of X is smaller than Y, we can find an injection $f : X \to Y$ and if the size of X is larger than Y, we can find a surjection $f : X \to Y$. Using this idea of comparing sizes of sets via functions, for more general sets, we can define:

Definition 3.4.1 (Cardinality Comparison) Let X and Y be two sets.

1. If there is an injection from X to Y, we say that the cardinality of X is smaller than or equal to Y and we write $|X| \leq |Y|$.
2. If there is a surjection from X to Y, we say that the cardinality of X is larger than or equal to Y and we write $|X| \geq |Y|$.
3. If there is a bijection between them, we say that the cardinality of X is equal to Y and we write $|X| = |Y|$.

Remark 3.4.2 We extend our notations above to strict comparison. We say $|X| < |Y|$ if there is an injection from X to Y but no surjection between the two. Likewise, we say $|X| > |Y|$ if there is a surjection from the set X to Y but no injection between the two.

We can relate the first two cases in Definition 3.4.1 to the third via the Cantor-Bernstein-Schröder theorem named after Cantor, Felix Bernstein (1878–1956), and Ernst Schröder (1841–1902). This theorem seems very obvious and was first stated by Cantor without proof.

Theorem 3.4.3 (Cantor-Bernstein-Schröder Theorem) *Let X and Y be sets such that there are functions $f : X \to Y$ and $g : Y \to X$ which are injective. Then, there must exist a function $h : X \to Y$ which is bijective.*
In terms of the language of cardinality in Definition 3.4.1, if $|X| \leq |Y|$ and $|Y| \leq |X|$, then $|X| = |Y|$.

Proof Suppose that there are injections $f : X \to Y$ and $g : Y \to X$. Since g is an injection, if we restrict the codomain of the function g to $\mathrm{Im}(g) = g(Y) \subseteq X$, this new function $\bar{g} : Y \to \mathrm{Im}(g) \subseteq X$ is a bijection and hence has an inverse $\bar{g}^{-1} : \mathrm{Im}(g) \to Y$.

Let $A, B \subseteq X$ be subsets of X defined as $A = \bigcup_{j=0}^{\infty}(g \circ f)^j(X \setminus \mathrm{Im}(g))$ and $B = X \setminus A$. By the latter, we have $X = A \cup B$ with $A \cap B = \varnothing$. Now we wish to construct a bijection from X to Y by using this disjoint decomposition of X into the sets A and B.

We claim that $B \subseteq \text{Im}(g)$. Indeed, if $x \in B$, then $x \notin A = \bigcup_{j=0}^{\infty}(g \circ f)^j(X \setminus \text{Im}(g))$. In particular, $x \notin (g \circ f)^0(X \setminus \text{Im}(g)) = X \setminus \text{Im}(g)$ which means $x \in \text{Im}(g)$. Thus we can construct a well-defined function $h : X \to Y$ described as:

$$h(x) = \begin{cases} f(x) & \text{if } x \in A, \\ \bar{g}^{-1}(x) & \text{if } x \in B. \end{cases}$$

By straightforward arguments, we can show that this function is injective and surjective which is left to the readers in Exercise 3.6. □

Theorem 3.4.3 is a very useful theorem to have when we are dealing with infinite sets. Given two infinite sets, it is usually tricky to construct an explicit bijection between them in order to conclude that their cardinalities are the same. However, thanks to the Cantor-Bernstein-Schröder theorem, instead of constructing a bijection between the two sets, we can just find an injection from one space to the other and vice versa to conclude that the spaces have the same cardinality.

Cardinality of a Set

Intuitively, any finite sizes can be classified according to the number of elements in them. Indeed, suppose that we have two sets X and Y such that the number of elements in them are $m, n \in \mathbb{N}_0$ respectively. By Definition 3.4.1, the two sets are said to have the same cardinality if there is a bijection between them. If $f : X \to Y$ is a bijection, then it must be surjective and injective. From Exercise 3.4, the first means $m \geq n$ and the latter means $m \leq n$. Thus $m = n$.

However, a set with an unbounded number of elements has a different cardinality to any of these finite sets because we cannot define a surjection from a finite set to this unbounded set. Thus, they must be strictly bigger than all finite sets and we call them infinite sets.

Moreover, even among infinite sets, there are infinite sets which is not bijective to some other infinite set. This means not all infinities are the same. So we extend the concept of natural numbers to cardinal numbers in order to be able to distinguish these infinities by comparing them using Definition 3.4.1. Using this idea, infinite sets can be classified into two smaller classes: countable and uncountable.

The intuition here is that we can count or label the elements of a finite set explicitly in finite time, no matter how big the size of the set is. It might take several lifetimes to do so, but it will end eventually. For example, the numbers 10^{100} and $10^{10^{100}}$, which are called the googol and googolplex respectively, are both finite even though they are very very huge. We can count them and eventually (theoretically) we will get to the end. This is not advisable though; even if a person can write two digits per second, writing a googolplex without rest would take about 1.51×10^{92} years!

On the other hand, if the set is infinite, we are not able to finish counting it and there are two distinct cases on how we cannot do this. In the first case where the set is countable, we can still theoretically write down all the elements of this set if we are given infinite time and energy to do so. The process will not end, but we can write the elements down one by one and eventually not miss any of them. In the second case where the set is uncountably infinite, even if we have an infinite amount of time and energy, we would not be able to list down all the elements of this set one by one. Therefore, we need to distinguish the two kinds of infinite sets. This can be encapsulated via the following definition:

Definition 3.4.4 (Cardinality of a Set) Let X be a set.

1. If $X = \varnothing$, then we set $|X| = 0$ and say it has size 0.
2. If there is a bijection between the set X and the set $\{1, 2, \ldots, n\} \subseteq \mathbb{N}$ for some $n \in \mathbb{N}$, we call X a finite set. The set X then has size n and is written as $|X| = n$.
3. Otherwise, the set X is called infinite if there exists no such $n \in \mathbb{N}$. Furthermore, we can classify them further as follows:
   (a) The set X is called countably infinite if it is an infinite set and there is an injection from X to the set of natural numbers $\mathbb{N}$. The set X is said to have size $\aleph_0$ (pronounced aleph-nought) and is written as $|X| = \aleph_0$.
   (b) Otherwise, the set X is called uncountably infinite if it is an infinite set and there are no injection from X to $\mathbb{N}$. This is written as $|X| = \aleph_1$ (pronounced aleph-one).

Remark 3.4.5 We make some remarks here:

1. We call the finite sets and countably infinite sets collectively as countable sets. Being countable (finite or infinite) means there is an injection from the set to some subset of $\mathbb{N}$.
2. For countably infinite set in Definition 3.4.4(3), the definition used in some literature is that there is a bijection from the set X to $\mathbb{N}$. However, by virtue of the Cantor-Bernstein-Schröder theorem, the existence of an injection from the infinite set X to $\mathbb{N}$ is an equivalent definition (an injection from $\mathbb{N}$ to X already exists since X is an infinite set).
3. The collection of uncountable sets can also be classified even further according to higher cardinality numbers. This is true because of Cantor's theorem, which we shall prove in Exercise 3.8. Cantor's theorem says for any set X, the set of all subsets of X or the power set given as $\mathcal{P}(X) = \{U : U \subseteq X\}$ satisfies the cardinality comparison $|X| < |\mathcal{P}(X)|$.
   So the power set of an uncountably infinite set X would have a strictly bigger size than X itself, which then classifies the uncountably infinite sets into smaller classes. We can repeat this again indefinitely to get even larger sets. These infinite cardinal numbers starting from $\aleph_0, \aleph_1, \aleph_2, \ldots$ are collectively known as transfinite numbers. However, the classification of these transfinite numbers according to Definition 3.4.4 is enough for the purposes of this book.
4. In the obvious way, the cardinality $\aleph_0$ is considered the smallest kind of infinity.

The beauty of the classification for infinite sets in Definition 3.4.4 is its consistency with Definition 3.4.1. For infinite sets, suppose that X and Y are both infinite sets which are bijective to each other, namely $|X| = |Y|$. If one is countably infinite, then the other must be countably infinite as well. Indeed, suppose that X is countably infinite. Then, there is an injection $f : X \to \mathbb{N}$. Furthermore, since $|X| = |Y|$, there is a bijection $g : Y \to X$. Thus, the composition $f \circ g : Y \to \mathbb{N}$ is an injection from Y to $\mathbb{N}$ by Exercise 1.31(a). This implies that Y is also countably infinite.

Example 3.4.6 Let us look at some examples to clarify the distinctions in Definition 3.4.4.

1. The set $X = \{\heartsuit, \diamondsuit, \clubsuit, \spadesuit\}$ is finite with size 4: you have just mentally counted the elements in the set "one... two... three... four..." in your head. This mental counting defines a function $f : X \to \{1, 2, 3, 4\}$ with $f(\heartsuit) = 1, f(\diamondsuit) = 2, f(\clubsuit) = 3$, and $f(\spadesuit) = 4$. Moreover, this function is a bijection and thus $|X| = 4$.
2. The set of even natural numbers smaller than or equal to 10 is also finite. This set is $X = \{2, 4, 6, 8, 10\}$ and it has a clear matching or labelling with the set $\{1, 2, 3, 4, 5\}$ via a bijection $f : X \to \{1, 2, 3, 4, 5\}$ defined as $f(x) = \frac{x}{2}$. Thus, this set has size 5.
3. Clearly, the set of natural numbers $\mathbb{N}$ is an infinite set since one can never find a bijection between $\mathbb{N}$ and the set $\{1, 2, \dots, n\}$ for any $n \in \mathbb{N}$. Furthermore, since we have an injective identity map from $\mathbb{N}$ to itself, the set $\mathbb{N}$ is countably infinite. Thus $|\mathbb{N}| = \aleph_0$
4. How about the set of even natural numbers $2\mathbb{N} = \{2n : n \in \mathbb{N}\}$? This set cannot be finite. If it is finite, then by Lemma 2.3.9, there is a biggest such even natural number, say m. However, $m + 2$ is also an even natural number but it is not in the list since the biggest element in the set is m, a contradiction. So the set of even natural numbers must be infinite.
   Now we ask ourselves: is it countable? Any even natural number must be of the form $2n$ for some $n \in \mathbb{N}$. Then, we can define a map from $2\mathbb{N}$ to the set of natural numbers $\mathbb{N}$ via the function $f : 2\mathbb{N} \to \mathbb{N}$ as $f(2n) = n$.
   This function is well-defined since every element in the domain gets mapped to one and only one element in the codomain. To show that it is injective, suppose that we have $f(x) = f(y)$ for two elements $x, y \in 2\mathbb{N}$. We can write $x = 2m$ and $y = 2n$ for some $m, n \in \mathbb{N}$. By definition of the function, we have $m = f(2m) = f(x) = f(y) = f(2n) = n$. So $m = n$ and therefore $x = 2m = 2n = y$. This means the function is injective and we conclude that the set of even natural numbers is also countably infinite.
   Here we can see our first example of two sets with one strictly contained in the other, but having the same size, namely $2\mathbb{N} \subsetneq \mathbb{N}$ but $|2\mathbb{N}| = |\mathbb{N}| = \aleph_0$.
5. The set of non-negative integers $\mathbb{N}_0$ is an infinite set because it contains $\mathbb{N}$, which is already an infinite set. But can we count it? Certainly. We simply have to find an injection from the set $\mathbb{N}_0$ to $\mathbb{N}$. At first, this may seem absurd since it looks

like there are more elements in the set $\mathbb{N}_0$ than there is in $\mathbb{N}$. But we can actually do this!

Define a function $f : \mathbb{N}_0 \to \mathbb{N}$ as $f(n) = n + 1$. This is a well-defined function and it is injective. Indeed, suppose that $f(m) = f(n)$ for some $m, n \in \mathbb{N}_0$. Then $m+1 = f(m) = f(n) = n+1$ which implies $m = n$, so the function is injective. Thus, the set of non-negative integers is countably infinite. Again, we encounter this funny behaviour, namely $\mathbb{N} \subsetneq \mathbb{N}_0$ but $|\mathbb{N}| = |\mathbb{N}_0| = \aleph_0$.

6. The set of integers $\mathbb{Z}$ is also countably infinite. Clearly it is an infinite set as it contains $\mathbb{N}$. But how can it be countable? We have to find an injective function from the integers to the natural numbers. Again, this seems impossible since the set $\mathbb{Z}$ seems so much bigger than the set $\mathbb{N}$. But this is actually possible!

   The idea behind this is to map the set of non-negative integers to the odd $\mathbb{N}$ and the negative integers to even $\mathbb{N}$. More concretely we define a function $f : \mathbb{Z} \to \mathbb{N}$ by:

$$f(n) = \begin{cases} 2n + 1 & \text{if } n \geq 0, \\ -2n & \text{if } n < 0. \end{cases}$$

   This is a well-defined function and is injective. Indeed, suppose that $f(m) = f(n)$ for some $m, n \in \mathbb{Z}$. Since they are equal, they must have the same parity: either both are odd or both are even. WLOG, let us assume that they are both odd. Then $f(m) = f(n)$ implies that $2m + 1 = 2n + 1$ and so $m = n$. The even case is also done similarly. Thus, the function is injective and $|\mathbb{Z}| = \aleph_0$ as well.

The constructions in Examples 3.4.6(5) and (6) stem from a thought experiment called the Grand Hilbert Hotel. David Hilbert proposed the following scenario: imagine that $X = \{x_j : j \in \mathbb{Z}\}$ are hotel guests and there are $\mathbb{N}$ hotel rooms in the Grand Hilbert Hotel. We want to check in the guests in the hotel so that each room has only one occupant at most.

1. The guests $\{x_j : j \in \mathbb{N}\} \subseteq X$ can be checked into the hotel immediately by assigning these guests to their matching room numbers. In other words, for each $j \in \mathbb{N}$, we assign the guest x_j to the room j as in Fig. 3.7. The hotel is now fully occupied.

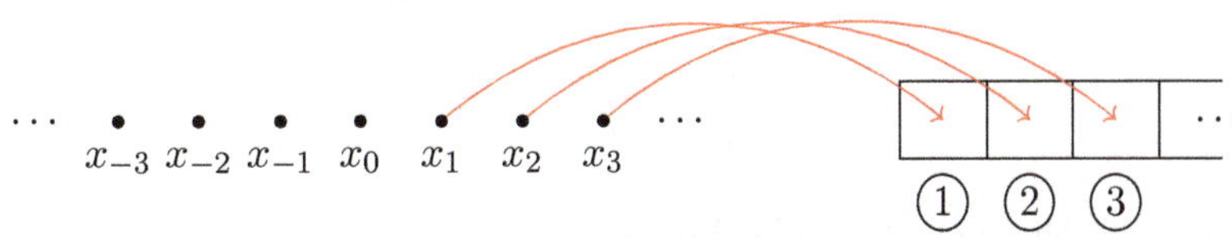

Fig. 3.7 Checking in the guests $\{x_j : j \in \mathbb{N}\}$

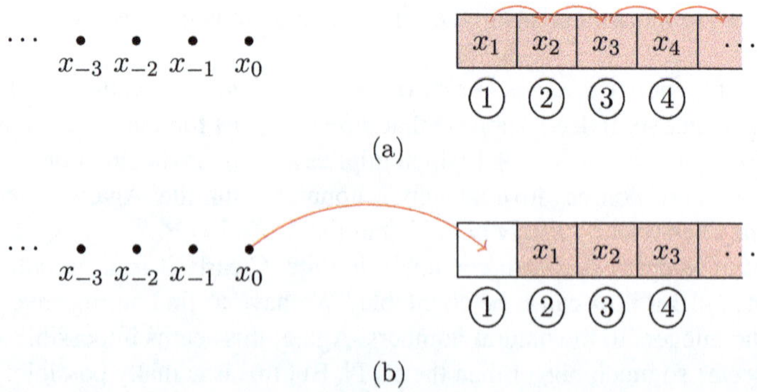

Fig. 3.8 Checking in the guest x_0. **(a)** Shift the occupant in room n to room $n+1$ first.... **(b)** ... then check in x_0 in the vacant first room

2. But what about the remaining guests $\{x_j : j \in \mathbb{Z}_{\leq 0}\}$ outside? The hotel is already fully occupied by all the guests $\{x_j : j \in \mathbb{N}\}$ so they cannot be checked in. Or can they?
   Actually, they can! Let us first check in guest x_0. The manager of the hotel can cleverly shift the checked-in guests along by one room, namely move the guest in room 1 to room 2, guest in room 2 to room 3, guest in room 3 to room 4, and so on, leaving the first room vacant as in Fig. 3.8. The hotel manager can then check in guest x_0 in the now vacant room 1. This is similar to the construction in Example 3.4.6(5) above.
3. So far we can accommodate one extra guest in the full hotel, but what about the other infinitely many guests $\{x_j : j \in \mathbb{Z}_-\}$ outside? The manager can move the guest in room n to the room $2n$ for every $n \in \mathbb{N}$, so that every other room is now vacant as demonstrated in Fig. 3.9. Now all the guests $\{x_j : j \in \mathbb{Z}_-\}$ can all be checked in comfortably into the odd-numbered rooms.

In the end, all the guests X can occupy each of the $\mathbb{N}$ hotel rooms.

We can even determine who is in which room for the hotel record-keeping purposes. The assignment of rooms above can be written explicitly as the injection $r : X \to \mathbb{N}$ given by:

$$r(x_j) = \begin{cases} -2j-1 & \text{if } j < 0, \\ 2 & \text{if } j = 0, \\ 2j+2 & \text{if } j > 0. \end{cases}$$

Now let us determine the cardinality of unions of sets.

Lemma 3.4.7 *The union of two disjoint countable (finite or infinite) sets is countable.*

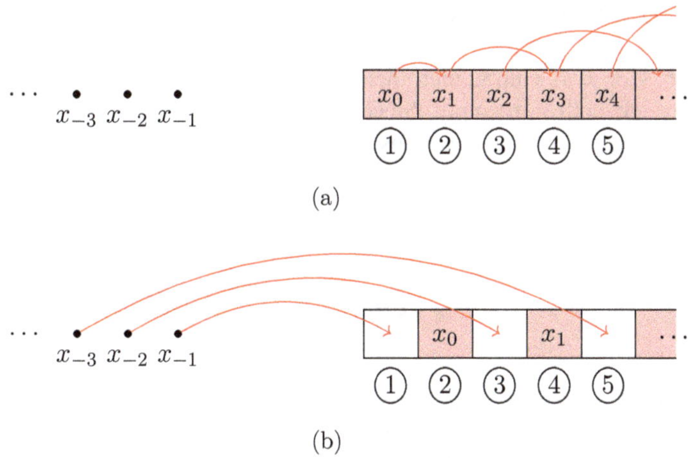

Fig. 3.9 Checking in everyone else. (**a**) Shift the occupant in room n to room $2n$ first.... (**b**) ... then check in everyone else in the vacant odd-numbered rooms. Enjoy your stay

Proof Let A and B be two disjoint countable sets. If either one is empty, then there is nothing to prove. Otherwise, suppose that they are both non-empty. Then, there are injective maps $f : A \to \mathbb{N}$ and $g : B \to \mathbb{N}$. Now we construct an injective map from the set $A \cup B$ to $\mathbb{N}$. To do this, we map the elements in the set A injectively to the set of odd natural numbers and the elements in the set B injectively to the set of even natural numbers. More precisely, we define a mapping from the set $A \cup B$ to $\mathbb{N}$ as:

$$h : A \cup B \to \mathbb{N},$$
$$x \mapsto \begin{cases} 2f(x) + 1 & \text{if } x \in A, \\ 2g(x) & \text{if } x \in B. \end{cases}$$

This is a well-defined function. Moreover, it is injective. Indeed, if $h(x) = h(y)$, then they have the same parity. If $h(x) = h(y)$ are even, then $2g(x) = h(x) = h(y) = 2g(y)$ and so $g(x) = g(y)$. Since g is injective, we must have $x = y$. Similar argument holds for the odd case. Thus, we conclude that $A \cup B$ is also countable. □

For finite sets, we know exactly the size of the union. Namely, we have the following lemma:

Lemma 3.4.8 *Let A and B be finite sets.*

1. *If $A \cap B = \varnothing$, then $|A \cup B| = |A| + |B|$.*
2. *In general, we have $|A \cup B| = |A| + |B| - |A \cap B|$.*

Proof We prove the assertions one by one.

1. If either A or B is empty, then there is nothing to prove. Otherwise, suppose that they are both non-empty. Since A and B are finite, say of sizes m and n respectively, there are bijections $f : A \to \{1, 2, \dots, m\}$ and $g : B \to \{1, 2, \dots, n\}$. Define a function:
$$h : A \cup B \to \{1, 2, \dots, m + n\},$$
$$x \mapsto \begin{cases} f(x) & \text{if } x \in A, \\ g(x) + m & \text{if } x \in B. \end{cases}$$
This is a well-defined function since every element in the domain $A \cup B$ is mapped to exactly one image in the codomain. Moreover, it is straightforward to prove that it is a bijection. Hence $|A \cup B| = m + n = |A| + |B|$.
2. By Exercise 1.22, we can write $A \cup B = (A \setminus B) \cup (B \setminus A) \cup (A \cap B)$ where the sets on the RHS are pairwise disjoint. By using the first assertion, we have $|A \cup B| = |A \setminus B| + |B \setminus A| + |A \cap B|$.
On the other hand, we can also express A and B are a union of disjoint sets, namely $A = (A \cap B) \cup (A \setminus B)$ and $B = (A \cap B) \cup (B \setminus A)$. Thus by using the first assertion again, we have $|A| = |A \cap B| + |A \setminus B|$ and $|B| = |A \cap B| + |B \setminus A|$. Combining the three equalities of sizes, we get the result. □

By induction, we can extend Lemmas 3.4.7 and 3.4.8 to finitely many collection of sets:

Lemma 3.4.9 *Let $\{A_j\}_{j=1}^n$ be a collection of n sets.*

1. *If A_j are all countable, then the union $\bigcup_{j=1}^n A_j$ is also countable.*
2. *If A_j are all finite, then the union $\bigcup_{j=1}^n A_j$ is also finite.*

Remark 3.4.10 The actual size of the union of two finite sizes has a closed form in Lemma 3.4.8. However, if we have more than two finite sets, the size of the union can be much more complicated to compute. The formula is given by the principle of inclusion-exclusion in Exercise 19.24.

Furthermore, we can also extend Lemma 3.4.9(1) to countably infinite union of countable sets. The readers will prove this result later in Exercise 4.24.

Lemma 3.4.11 *The countably infinite union of countable (finite or infinite) sets is countable.*

In the analogy of Hilbert Hotel, this happens when countably infinitely many buses each containing countably infinitely many guests arriving at the hotel simultaneously and need to be sorted out into their own rooms. Lemma 3.4.11 guarantees that they all can be checked in into their own rooms. Plenty of room at the Hotel Hilbert... any time of year, you can find it here.

Cardinality of $\mathbb{Q}$

Now we go back to the original question posed earlier before we defined cardinality, namely: how big is the set of rational numbers? We have seen in Example 3.4.6(6) that the sets $\mathbb{N}$ and $\mathbb{Z}$ are both countable. We can now show that the set of rational numbers is actually countably infinite as well.

Proposition 3.4.12 *The set of rational numbers $\mathbb{Q}$ is countably infinite.*

Proof We split the rational numbers into three disjoint subsets: $\mathbb{Q}_- \cup \{0\} \cup \mathbb{Q}_+$, namely the negative rational numbers, zero, and the positive rational numbers. We first show that the positive rational numbers is countably infinite.

Clearly the set of positive rational numbers contains the natural numbers, so it must be at least as big as the set of natural numbers. Hence the positive rational numbers is an infinite set. To show that it is countable, we find an injective map from $\mathbb{Q}_+$ to $\mathbb{N}$. For every $r \in \mathbb{Q}_+$, we may write $r = \frac{p}{q}$ uniquely in reduced form, namely $p, q \in \mathbb{N}$ are coprime to each other. To show this set is countable, we define the following map from $\mathbb{Q}_+$ to $\mathbb{N}$:

$$\begin{aligned} f : \mathbb{Q}_+ &\to \mathbb{N}, \\ r = \frac{p}{q} &\mapsto 2^p 3^q. \end{aligned}$$

This is a well-defined function because, by uniqueness of the reduced form of positive rational numbers, the positive rational number $r = \frac{p}{q}$ is mapped to one and only one element in the codomain $\mathbb{N}$. Now we check that this map is injective. Suppose that there are two positive rational numbers $r = \frac{p}{q}$ and $s = \frac{m}{n}$ that are mapped to the same element in $\mathbb{N}$, namely $2^p 3^q = f(r) = f(s) = 2^m 3^n$. We aim to show that $p = m$ and $q = n$.

Suppose for contradiction that $p \neq m$. WLOG, say $p > m$. By cancellation law, we have $2^{p-m} 3^q = 3^n$. Since 2^{p-m} is even, the LHS is even. But the RHS is odd for any $n \in \mathbb{N}$, which gives us the required contradiction. Thus, we must have $p = m$ and this leaves us with $3^q = 3^n$. Again, assume for contradiction that $q \neq n$. WLOG, say $q > n$. Thus $3^{q-n} = 1$. The LHS is strictly bigger than the RHS, again giving us another contradiction. This implies $q = n$. Since $p = m$ and $q = n$, we must have $r = s$. Thus, the map f is injective and hence the set of positive rational numbers is countably infinite.

The set of negative rational numbers is also countably infinite. Indeed, there is an obvious bijection $g : \mathbb{Q}_+ \to \mathbb{Q}_-$ given by $g(r) = -r$ which shows that $|\mathbb{Q}_+| = |\mathbb{Q}_-|$. Thus the set $\mathbb{Q}^* = \mathbb{Q}_- \cup \mathbb{Q}_+$ is countably infinite since it is a union of two countably infinite set as stated in Lemma 3.4.7. Taking the union of this resulting set $\mathbb{Q}^*$ with the finite set $\{0\}$ proves that the set of all rational numbers $\mathbb{Q} = \mathbb{Q}^* \cup \{0\}$ is also countably infinite. □

So if the set of rational numbers, the biggest set that we have seen so far, is still countable, what are examples of an uncountable set in Definition 3.4.4? Do they even exist? We digress to the discussion of irrational numbers for now before we provide an answer to that question in Chap. 4.

3.5 Irrational Numbers $\bar{\mathbb{Q}}$

So far, the natural numbers, the integers, and the rational numbers appear naturally in the real world. Thus, their concepts are well-understood and accepted readily. On the other hand, other types of numbers appear later in the scene. The concept of irrational numbers date back to the Greeks.

Pythagoras of Samos (c. 570B.C.–495B.C.) was a famous Greek mathematician and philosopher. One of his biggest contributions to mathematics is the Pythagorean theorem which relates the lengths of the sides of a right-angled triangle and is a staple in any modern high-school mathematics curriculum. The proof of this result hinges solely on the work of Euclid via similar triangles.

Pythagoras is also the founder of a devoted philosophical and intellectual school known as the Pythagoreans. The central tenet of this commune is "All is Numbers" signifying how sacred mathematics is as a unifying subject for science, law, music, philosophy, religion, and life in general. Another interesting doctrine held by them is that broad beans are forbidden in this commune due to Pythagoras's revulsion towards them.

From the works of Pythagoras, the Pythagoreans tried computing the length of the hypotenuse of a right-angled triangle with sides of length 1 unit as in Fig. 3.10. Using the Pythagorean theorem, they conclude that the length of the hypotenuse is a number x that satisfies the equation $x^2 = 1^2 + 1^2 = 2$. However, this is baffling: this number is not a rational number that they claim to encompass all the numbers in the world.

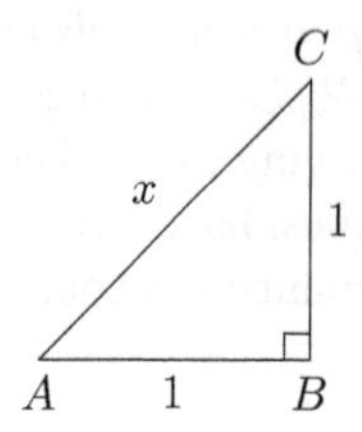

Fig. 3.10 Pythagoras theorem says $AB^2 + BC^2 = AC^2$

Indeed, if it is, we can write $x = \frac{m}{n}$ in the lowest terms where m and n are coprime non-negative integers with $n \neq 0$. Thus, we have $\frac{m^2}{n^2} = 2$ which implies $m^2 = 2n^2$. By Exercise 2.24(d), m is even so $m = 2p$ for some other natural number p. Putting this in the equation above, we get $2n^2 = m^2 = (2p)^2 = 4p^2$ which implies $n^2 = 2p^2$. However, this also implies that n is even by an exact similar argument. This gives us a contradiction as we have assumed that m and n are coprime natural numbers! So the assumption that the number x is rational must be false. This new kind of number is clearly bigger than 1 by the Pythagorean theorem and smaller than 2 by triangle inequality, so $1 < x < 2$.

This was a huge crisis for the Pythagoreans because it either meant that their geometrical theorems are wrong or their number system is wrong, both of which are equally devastating. Furthermore, since they considered mathematics as sacred, this discovery is nothing short of sacrilegious. Legend has it that this new number was discovered by Hippasus of Metapontum (c. 530B.C.-450B.C.) and as a punishment for this blasphemy, he was thrown overboard a ship and left to drown. To explain this anomaly, they then termed these quantities as incommensurable or inexpressible.

A way to explain this is that the length of the hypotenuse is a member of some new set of numbers. The existence of these non-rational numbers came about from the continuous nature of geometry rather than the discrete construction of integers and rational numbers.

In contrast to the rational numbers $\mathbb{Q}$ that can be expressed as a ratio of two integers, these numbers cannot be expressed as a ratio of two integers as we have seen above. Hence they are called the irrational numbers which is denoted as $\bar{\mathbb{Q}}$. By dichotomy, these two types of numbers are disjoint.

3.6 Bounds, Supremum, and Infimum

Before we address this issue of the existence of the new kind of numbers outside of $\mathbb{Q}$, let us define some basic terminologies. For a general ordered number field $\mathbb{F}$ we define:

Definition 3.6.1 (Maximum, Minimum) Let $X \subseteq \mathbb{F}$ be a non-empty subset of an ordered number field $\mathbb{F}$.

1. A number $a \in X$ is called a maximum of X if $x \leq a$ for all $x \in X$. We denote $a = \max(X)$.
2. A number $b \in X$ is called a minimum of X if $x \geq b$ for all $x \in X$. We denote $b = \min(X)$.

Essentially, the maximum of X is the largest element contained in X and the minimum of X is the smallest element contained in X. From the definition, a maximum and a minimum, if they exist, can be used to provide bounds for the subset in question.

Example 3.6.2 Let us look at some examples:

1. Consider the set $X = \{x \in \mathbb{Q} : x \in \mathbb{Z}, x \leq 10\} = \{\dots, 0, 1, 2, 3, 4, 5, 6, 7, 8, 9, 10\}$. This set has a maximum in $\mathbb{Q}$ which is 10. However, it does not have a minimum. Indeed, if it has a minimum, say $a \in X$, then a is an integer and is negative. However, the number $a - 1$ is also in X but is smaller than a, contradicting the assumption that a is smaller than any other element in X.
2. The set $X = \{x \in \mathbb{Q} : x < 10\}$ does not have a minimum by a similar argument as the above. Moreover, it does not have a maximum in $\mathbb{Q}$. Indeed, suppose for contradiction that $a \in \mathbb{Q}$ is a maximum of this set, namely $x \leq a$ for all $x \in X$. The rational number $\frac{a+10}{2}$ satisfies $a < \frac{a+10}{2} < 10$ and so it must be in X as well. However, it is greater than a and thus contradicts the assumption that $a = \max(X)$.
3. On the other hand, the set $X = \{x \in \mathbb{Q} : x \leq 10\}$ has a maximum of 10 since $10 \in X$ and $x \leq 10$ for any $x \in X$.

Bounds

We can see from Example 3.6.2 that not all subsets of $\mathbb{Q}$ have a maximum or a minimum. So we need to come up with a more robust quantification of bounds that would potentially work for any subset of $\mathbb{Q}$. For a set to have a maximum, it must necessarily be non-empty and is bounded from above. Likewise, for a set to have a minimum, the set must be non-empty and bounded from below. We first define:

Definition 3.6.3 (Upper, Lower Bound) Let $X \subseteq \mathbb{F}$ be a non-empty subset of an ordered number field $\mathbb{F}$.

1. A number $a \in \mathbb{F}$ is called an upper bound of the set X if $x \leq a$ for all $x \in X$. If such a number a exists, then the set X is called bounded from above in $\mathbb{F}$. In symbols:

$$X \text{ is bounded above} \quad \text{if} \quad \exists a \in \mathbb{F} : \forall x \in X, x \leq a.$$

2. A number $b \in \mathbb{F}$ is called a lower bound of the set X if $x \geq b$ for all $x \in X$. If such a number b exists, then the set X is called bounded from below in $\mathbb{F}$. In symbols:

$$X \text{ is bounded below} \quad \text{if} \quad \exists b \in \mathbb{F} : \forall x \in X, x \geq b.$$

3. If the set X is bounded from both above and below, then the set X is called bounded in $\mathbb{F}$. In other words, the set X is bounded if there exists an $M > 0$ such that $-M \leq x \leq M$ for all $x \in X$.

If no such upper or lower bounds exist, we call the sets unbounded sets:

Definition 3.6.4 (Unbounded Sets) Let $X \subseteq \mathbb{F}$ be a non-empty subset of an ordered number field $\mathbb{F}$.

1. The set X is called unbounded above if for any $a \in \mathbb{F}$, there exists an $x \in X$ such that $a < x$. In symbols:

$$X \text{ is unbounded above} \quad \text{if} \quad \forall a \in \mathbb{F} : \exists x \in X, x \geq a.$$

2. The set X is called unbounded below if for any $b \in \mathbb{F}$, there exists an $x \in X$ such that $b > x$. In symbols:

$$X \text{ is unbounded below} \quad \text{if} \quad \forall b \in \mathbb{F} : \exists x \in X, x \leq b.$$

3. If the set X is unbounded above or unbounded below, then the set X is called an unbounded set.

Example 3.6.5 Note that the upper and lower bounds of a set $X \subseteq \mathbb{F}$, if they exist, might not necessarily lie within the set X. Let us look at some examples. Suppose that $\mathbb{F} = \mathbb{Q}$.

1. From Example 3.6.2(2), the set $X = \{x \in \mathbb{Q} : x < 10\}$ does not have a maximum. However, it has an upper bound. For example, 20 is an upper bound for this set since it is bigger than every element in X. Indeed, if $x \in X$, then $x < 10 < 20$. This upper bound is not contained in X.
2. The set $Y = \{x \in \mathbb{Q} : x^2 \leq 2\}$ is bounded from above in $\mathbb{Q}$. The number $4 \in \mathbb{Q}$ is an upper bound for this set. Indeed, if it is not, by Definition 3.6.4, we can find some $x \in Y$ such that $x > 4$. Hence $x^2 > 16 > 2$ which means $x \notin Y$, giving us a contradiction. The upper bound 4 for the set Y is not contained in the set Y.
   Similarly, this set is bounded from below. A lower bound would be -4, but there are many other lower bounds, for example -3. Again, this is a lower bound because if it is not, then we can find an $x \in Y$ such that $x < -3$. By squaring both sides, this means $x^2 > 9 > 2$ so that $x \notin Y$, another contradiction.
3. The set $Z = \{x \in \mathbb{Q} : x \geq 0\}$ is not bounded above in $\mathbb{Q}$. Indeed, suppose for contradiction that $a \in \mathbb{Q}$ is an upper bound of this set, namely $x \leq a$ for all $x \in Z$. However, $a+1 \in Z$ as well but $a < a+1$. So there is at least one element in Z that is bigger than this upper bound, which is a contradiction. However, it is bounded from below. 0 is an obvious lower bound for the set Z since every $x \in Z$ satisfies $x \geq 0$ by definition. This lower bound is contained in the set Z.

Supremum and Infimum

An upper bound or a lower bound for a set, if they exist, may not be unique. Looking at Example 3.6.5(1), we can find infinitely many upper bounds for the set $X = \{x \in \mathbb{Q} : x < 10\}$. Therefore, we want to try and find the best upper bound of the set X which is its smallest possible upper bound. Likewise, for a set which is bounded from below, we want to find the best lower bound as the greatest possible lower bound for the set. We define these best bounds as:

Definition 3.6.6 (Supremum, Infimum) Let $X \subseteq \mathbb{F}$ be a non-empty subset of an ordered number field $\mathbb{F}$.

1. We call a number $a \in \mathbb{F}$ the supremum of the set X, denoted as $a = \sup(X)$, if a is an upper bound of the set X and it is the least upper bound of X. Namely, $x \leq a$ for all $x \in X$ and if a^* is another upper bound of X, then $a \leq a^*$.
2. We call a number $b \in \mathbb{F}$ the infimum of the set X, denoted as $b = \inf(X)$, if b is a lower bound of the set X and it is the greatest lower bound of X. Namely, $x \geq b$ for all $x \in X$ and if b^* is another lower bound of X, then $b \geq b^*$.

This would be a good candidate for the extension for the concept of maximum and minimum. Indeed, if a set has a maximum, then it has a supremum which coincides with the value of the maximum. Likewise holds true for minimum and infimum. These are given in the following proposition:

Proposition 3.6.7 *Let $X \subseteq \mathbb{F}$ be a non-empty subset of an ordered number field $\mathbb{F}$.*

1. *If $a = \max(X)$, then $\sup(X) = a$.*
2. *If $\sup(X) = a$ and $a \in X$, then $\max(X) = a$.*
3. *If $b = \min(X)$, then $\inf(X) = b$.*
4. *If $\inf(X) = b$ and $b \in X$, then $\min(X) = b$.*

Proof We shall prove the first two assertions only as the others can be done in a similar way.

1. Since $a = \max(X)$, by definition $x \leq a$, for all $x \in X$ and $a \in X$. The former means a is an upper bound for the set X. The latter implies that it is the smallest upper bound. Indeed, if there exists a smaller upper bound a^* for the set X, then $a^* < a \in X$, which means a^* cannot be an upper bound for the set X since it is strictly smaller than at least one element of the set X (namely a). Thus, a is the smallest upper bound for the set X, which implies $\sup(X) = a$.
2. Since $\sup(X) = a \in X$, we have $x \leq a$ for all $x \in X$. Moreover, since $a \in X$, by Definition 3.6.1, we have $a = \max(X)$. □

So any maximum is a supremum, but a supremum may not be a maximum, which we shall see in Example 3.6.9(1) later. From Definition 3.6.6, we have a useful duality of supremum and infimum:

Lemma 3.6.8 *Let* $X \subseteq \mathbb{F}$ *be a non-empty subset of an ordered number field such that* $\sup(X)$ *exists. Define a new set* $-X = \{-x : x \in X\}$. *Then,* $\inf(-X)$ *exists and* $\inf(-X) = -\sup(X)$.

Proof Denote $a = \sup(X)$. The set X is bounded from above and hence the set $-X$ would be bounded from below. Indeed, every element in $-X$ is of the form $-x$ for some $x \in X$. Since $a = \sup(X)$, we must have $x \leq a$ for all $x \in X$. Hence, $-x \geq -a$ for all $x \in X$. So every element in $-X$ is bigger than $-a$. Therefore, $-a$ is a lower bound for $-X$.

We now prove that $-a$ is the infimum of the set $-X$. To show that it is the greatest lower bound, we assume for contradiction that there exists another lower bound a^* for the set X which is greater than $-a$, namely $-a < a^*$. By definition of lower bound, we must have $-x \geq a^*$ for all $-x \in -X$. In other words, for all $x \in X$ we have $x \leq -a^* < a$. This means $-a^*$ is a smaller upper bound for the set X, but this is clearly false since a is, by assumption, the smallest upper bound for X. Thus, we have reached a contradiction. □

Example 3.6.9 Let us look at some examples:

1. We have seen in Example 3.6.2 that the set $X = \{x \in \mathbb{Q} : x < 10\}$ does not have a maximum in $\mathbb{Q}$. However, its supremum exists. Clearly this set is bounded from above, so its set of upper bounds is non-empty. Now we want to determine its smallest upper bound.
A good guess would be $x = 10$. This is an upper bound for the set X by definition of the set X. Furthermore, we can show that it is the smallest possible upper bound. For contradiction, suppose that there is a smaller upper bound for the set X and we call it $r < 10$. Since r must also be an upper bound of the set X, any element of X must be smaller than or equal to r. However, we note that $r < \frac{r+10}{2} < 10$. So $\frac{r+10}{2} \in X$ but it is strictly bigger than the upper bound r, which is a contradiction. Therefore there are no smaller upper bound for the set X in $\mathbb{Q}$ and we conclude that $\sup(X) = 10$.
2. Sometimes the supremum for a subset of $\mathbb{Q}$ may not exist even if the set is bounded from above. Thus, the set does not have the best upper bound in $\mathbb{Q}$.
Consider the set $Y = \{x \in \mathbb{Q} : x^2 \leq 2\}$ in $\mathbb{Q}$. This set is bounded from above as we have seen in Example 3.6.5(2). However, it does not have a supremum in $\mathbb{Q}$, which we are going to prove now.
Suppose for contradiction that there is a supremum for this set, which we call $r \in \mathbb{Q}$. Then there are three possibilities: either $r^2 = 2$, $r^2 < 2$, or $r^2 > 2$. Necessarily $r > 0$ since Y contains the number 1 and so $r \geq 1 > 0$. We note that the first case cannot happen since $r^2 = 2$ does not have a solution in $\mathbb{Q}$ as demonstrated by the Pythagoreans. We examine the other two cases:

(a) Suppose that $r^2 < 2$ and $r \in \mathbb{Q}$. Since $\frac{2-r^2}{2r+1} \in \mathbb{Q}_+$, by Corollary 3.3.6, we can find a natural number $n \in \mathbb{N}$ such that $0 < \frac{1}{n} < \frac{2-r^2}{2r+1}$. Set $\delta = \frac{1}{n}$. Then, $r + \delta \in \mathbb{Q}$ and:

$$\begin{aligned}(r+\delta)^2 = r^2 + 2r\delta + \delta^2 &\leq r^2 + 2r\delta + \delta \qquad (\because 0 < \delta \leq 1)\\ &= r^2 + \delta(2r+1)\\ &< r^2 + (2 - r^2) = 2 \qquad (\because \delta < \tfrac{2-r^2}{2r+1}),\end{aligned}$$

which means $r + \delta \in Y$ as well. However, $\sup(Y) = r < r + \delta \in Y$, which contradicts the fact that r is an upper bound for the set Y.

(b) If $r^2 > 2$ and $r \in \mathbb{Q}$, pick $\delta = \frac{r^2-2}{2r} > 0$. Note that $r - \delta = r - \frac{r^2-2}{2r} = \frac{r^2+2}{2r}$ and so $r - \delta \in \mathbb{Q}_+$. Also:

$$(r-\delta)^2 = r^2 - 2r\delta + \delta^2 > r^2 - 2r\delta = r^2 - (r^2 - 2) = 2.$$

This means $r - \delta$ is an upper bound for the set Y. Indeed, if there exists an $x \in Y$ such that $x > r - \delta$, this would then imply $x^2 > (r-\delta)^2 > 2$, so $x \notin Y$ which is a contradiction.
However, we have $r - \delta < r$ which means r is not the smallest possible upper bound for the set Y, namely $r \neq \sup(Y)$. This gives us yet another contradiction.

Since all of the cases lead to contradictions, we conclude that the set Y does not have a supremum in $\mathbb{Q}$.

Remark 3.6.10 The argument in Example 3.6.9(2) above seems very contrived. How were we supposed to know what such specific δ to pick so that we get the desired contradiction? This is actually a very common method in analysis. It seems like we have plucked the number δ out of thin air here, but in actuality, the process requires a lot of guesswork and strategy. The final presentation seems neat but this is just because we did not present all the rough work here! As in the quote by Richard Feynman (1918–1988):

> We have a habit ... to make the work as finished as possible, to cover all the tracks, to not worry about the blind alleys or to describe how you had the wrong idea first, and so on.

It seems like magic, but it is really a lot of hidden hard work! We shall explain more about this process and demonstrate it in more details in Chap. 5.

Completeness Axiom

Ideally, we would like for any subset of a number field which is bounded from above to have a least upper bound or supremum. However, the number field $\mathbb{Q}$ does not satisfy this requirement as evidenced by the subset Y in Example 3.6.9(2).

So in order to fulfill this criterion, we need to put an extra assumption on our number field. Of course, we would like the number set to be a field and also satisfies the ordering axioms. On top of those, we want to add an extra condition on our ordered number field, namely the completeness axiom.

Definition 3.6.11 (Completeness Axiom) Let $\mathbb{F}$ be an ordered number field. The number field $\mathbb{F}$ is called complete if for any non-empty subset $X \subseteq \mathbb{F}$ that is bounded from above, the quantity $\sup(X)$ exists in $\mathbb{F}$.

This axiom is also called the least upper bound property, the supremum property, or Dedekind completeness. The axiom can also be stated in terms of Cauchy sequences, but we will only state the definition of Cauchy sequences in Chap. 6 later. Therefore, it is a good idea to revisit this axiom and its consequences after we have defined Cauchy sequences later.

Remark 3.6.12 An equivalent statement of the completeness axiom is that every non-empty subset $X \subseteq \mathbb{F}$ which is bounded from below has an infimum in $\mathbb{F}$. This is due to Lemma 3.6.8.

Therefore, axiomatically, we want to have a special field of numbers which has a field-compatible order and satisfies completeness. We call such field, if it exists, the real numbers.

Definition 3.6.13 (Real Numbers) The set of real numbers $\mathbb{R}$ is a number set which satisfies:

1. the field axioms in Definition 3.1.1,
2. the ordered field axioms in Definition 3.3.1, and
3. the completeness axiom in Definition 3.6.11.

The set $\mathbb{Q}$ already satisfies the first two axioms in Definition 3.6.13, but not the third. Thus, the main distinction between the rationals $\mathbb{Q}$ and the real numbers $\mathbb{R}$ is the completeness axiom. What this axiom roughly mean is that there are no "gaps" in between the numbers in $\mathbb{R}$.

For example, we have seen earlier that in $\mathbb{Q}$ there is a gap somewhere in between 1 and 2, which can only be filled up by a number x which squares to 2, which is most definitely not a rational number as discovered by the Pythagoreans. Therefore, to complete the rational number field, we have to introduce these non-rational numbers to fill in or complete the gap. We call such numbers the irrational numbers, denoted as $\bar{\mathbb{Q}}$.

In fact, by dichotomy, the real numbers $\mathbb{R}$ is made up exactly from the rationals and the irrationals, namely $\mathbb{R} = \mathbb{Q} \cup \bar{\mathbb{Q}}$.

Now, we can just accept that there is a set of numbers that satisfy Definition 3.6.13 axiomatically and happily skip to Chap. 4. This set of real numbers consists of the rational numbers $\mathbb{Q}$ that we have constructed and the irrational num-

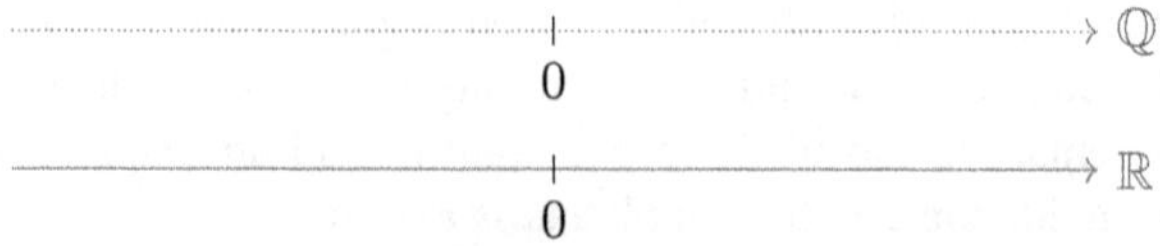

Fig. 3.11 The set of rational numbers $\mathbb{Q}$ and the number line $\mathbb{R}$. Lots of gaps in the set $\mathbb{Q}$ and no gaps in the set $\mathbb{R}$

bers $\bar{\mathbb{Q}}$ which "complete" them. This is what the general mathematical community up til the nineteenth century chose to do: the existence of these irrational numbers and the algebraic operations on them were simply accepted as a given and assumed to follow the operations that we already have on the rationals.

Moreover, via completeness, unlike the set $\mathbb{Q}$, there are no gaps in between the elements in $\mathbb{R}$. Thus the set of real numbers can be represented by an endless continuous straight line where each element of $\mathbb{R}$ is a point on the line. This line is called the real number line (Fig. 3.11).

Alternatively, we can take a constructive approach and show that such a set of numbers exist. This is more favoured by Russell, according to his quote at the beginning of this chapter. Explicit construction of the real numbers enables us to concretely see how the basic algebraic operations of addition and multiplication as well as ordering work on them. These would then allow us to define some new operations on the real numbers such as exponentiation and logarithm which we shall see in Chap. 4.

But to define the algebraic operations, we first need some concrete presentation of the real numbers. This is what we are going to do in the next section via a construction called the Dedekind cuts. In fact, there are many other equivalent construction of real numbers that one can carry out. One example is via Cauchy sequences, which the readers are invited to explore in Exercise 6.27 later.

3.7 Dedekind Cuts

We have stated the desirable things that we want in the set of real numbers in Definition 3.6.13. But how do we know such a field exists? If they do exist, how do they look like and how do we construct them?

We know how the rational numbers $\mathbb{Q}$ look like since they can be represented explicitly by pairs or ratios of integers. Let us now fill in these gaps between rational numbers with some new numbers using a construction called Dedekind cuts. This construction was devised by Richard Dedekind (1831–1916) as a way to rigorously define the real numbers. We define:

Definition 3.7.1 (Dedekind Cuts) Let $\mathbb{F}$ be an ordered number field. A Dedekind cut (L, U) on $\mathbb{F}$ is a partition of the set $\mathbb{F}$ into two disjoint sets $L, U \subseteq \mathbb{F}$ with $L \cup U = \mathbb{F}$ such that:

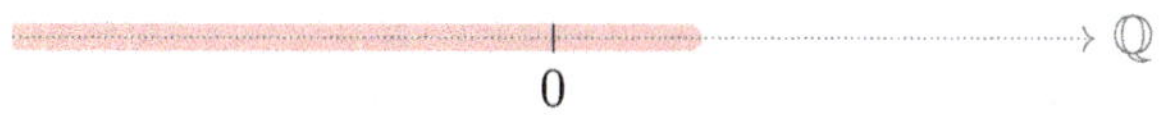

Fig. 3.12 Example of a Dedekind cut in $\mathbb{Q}$

1. $L \neq \varnothing, \mathbb{F}$,
2. L is closed downwards: for any $x, y \in \mathbb{F}$ with $x < y$, if $y \in L$, then $x \in L$ as well, and
3. L does not have a maximal element: for any $x \in L$ there exists a $y \in L$ such that $x < y$.

Remark 3.7.2 Let us make some remarks on Definition 3.7.1.

1. From definition, we can see that specifying L alone is enough to define the Dedekind cut pair (L, U). Therefore, for brevity, instead of the pair (L, U), we usually refer to a Dedekind cut as just L with the implicit knowledge that $U = \mathbb{F} \setminus L$.
2. Note that the second condition in Definition 3.7.1 is also equivalent to saying every element of U is greater than any element of L.

Now let us set $\mathbb{F} = \mathbb{Q}$ and call the set of all Dedekind cuts on $\mathbb{Q}$ as:

$$\mathcal{C} = \{L : (L, U) \text{ is a Dedekind cut of } \mathbb{Q}\}.$$

Example 3.7.3 Here are some examples of Dedekind cuts on $\mathbb{Q}$.

1. For any $q \in \mathbb{Q}$, define $L_q = \{x \in \mathbb{Q} : x < q\}$ and $U_q = \mathbb{Q} \setminus L_q = \{x \in \mathbb{Q} : x \geq q\}$. A graphical example of a Dedekind cut L_q is given in Fig. 3.12. We check:
   (a) Clearly L_q is neither $\varnothing$ nor $\mathbb{Q}$.
   (b) L_q is closed downwards since if $r < s$ and $s \in L_q$, then $r < s < q$ and hence $r \in L_q$ as well.
   (c) This set does not have a greatest element since for any $r \in L_q$, we can always find a greater element than it in L_q via the density of rationals. For example, we have $r < \frac{r+q}{2} < q$ with $\frac{r+q}{2} \in \mathbb{Q}$. Thus, L_q does not have a maximal element.

   We conclude that L_q is a Dedekind cut.
2. Recall that the set $Y = \{x \in \mathbb{Q} : x^2 \leq 2\}$ in Example 3.6.5(2) does not have a greatest element in $\mathbb{Q}$. This is not a Dedekind cut as it is not closed downwards since, for example, -10 is smaller than any element in Y but itself is not in Y. Therefore, we set $L = Y \cup \{x \in \mathbb{Q} : x < 0\} = \{x \in \mathbb{Q} : x^2 \leq 2 \text{ or } x < 0\}$ and $U = \mathbb{Q} \setminus L$. We can check that this resulting set is a Dedekind cut:
   (a) Clearly L is non-empty and is not the whole of $\mathbb{Q}$.

(b) Pick any $x \in L$ and $y < x$. If $y < 0$, then clearly $y \in \{x \in \mathbb{Q} : x < 0\} \subseteq L$. Otherwise, suppose that $0 \leq y < x$. We then have $0 \leq y^2 < x^2 < 2$ since $x \in L$. Therefore, $y \in Y \subseteq L$ as well. Thus, the set L is closed downwards.
(c) Suppose for contradiction that the set L has a maximal element. Necessarily, the maximum is positive and is thus contained in $Y \subseteq L$. By Proposition 3.6.7, since the set Y has a maximum, it must also have a supremum. However, we have seen in Example 3.6.9(2) that the supremum cannot exist in $\mathbb{Q}$, giving us a contradiction

So this L is another example of a Dedekind cut on $\mathbb{Q}$.

From Example 3.7.3(1), for each $q \in \mathbb{Q}$, we can see that the supremum of the cut $L_q = \{x \in \mathbb{Q} : x < q\}$ exists in $\mathbb{Q}$ and is exactly q. In fact, we have the following characterisation:

Lemma 3.7.4 *A Dedekind cut $L \in \mathcal{C}$ has a supremum in $\mathbb{Q}$ if and only if $L = L_q = \{x \in \mathbb{Q} : x < q\}$ for some $q \in \mathbb{Q}$.*

Proof We prove the implications separately.

($\Leftarrow$): If $L = L_q = \{x \in \mathbb{Q} : x < q\}$ for some $q \in \mathbb{Q}$, then clearly q is an upper bound for the set L. We claim that q is the smallest upper bound for L. Suppose for contradiction that there is a smaller upper bound $r < q$ of L. Then, we have $r < \frac{r+q}{2} < q$. But this is a contradiction since $\frac{r+q}{2} \in L$ but it is greater than the supposed upper bound r of L.

($\Rightarrow$): Suppose that $\sup(L) = q \in \mathbb{Q}$. We claim that $L = \{x \in \mathbb{Q} : x < q\}$. We prove this by using double inclusion:

($\subseteq$): Pick any $x \in L$. Then, $x \leq \sup(L) = q$. However, we must have $x \neq q$. If it does, Proposition 3.6.7 implies that $q = \max(L)$, contradicting the assumption that L is a Dedekind cut which should not have a maximal element. So $x < q$ and hence $L \subseteq \{x \in \mathbb{Q} : x < q\}$.

($\supseteq$): Pick any $x \in \{x \in \mathbb{Q} : x < q\}$. Then, $x < q = \sup(L)$ and hence $x \in L$. Indeed, suppose for contradiction that $x \notin L$. Since L is a Dedekind cut and is closed downwards, each $y \in \mathbb{Q}$ with $x < y < q$ must not be in L as well. This means x is a smaller upper bound for the set L, contradicting $\sup(L) = q$. Hence $\{x \in \mathbb{Q} : x < q\} \subseteq L$.

Thus, we conclude that $L = L_q$. □

A corollary of this, which the readers are asked to prove in Exercise 3.9, is:

Corollary 3.7.5 *Let (L, U) be a Dedekind cut on $\mathbb{Q}$. Then, $L = L_q$ if and only if U has a minimal element.*

Now that we have seen some examples of Dedekind cuts, how can we use them to extend our rational numbers to the real numbers? Each rational number $q \in \mathbb{Q}$ can be represented uniquely by a Dedekind cut L_q in $\mathcal{C}$. Conversely, using Lemma 3.7.4, we can use cuts L with a supremum in $\mathbb{Q}$ to represent each of the rational numbers $\mathbb{Q}$. Therefore, there is a correspondence between the set of rational numbers and the set of cuts L with a supremum.

However, in general, for a Dedekind cut (L, U), the set L may not have a supremum in $\mathbb{Q}$ and hence the set U may not have a minimum. For example, the supremum of the cut $L = \{x \in \mathbb{Q} : x^2 \leq 2 \text{ or } x < 0\}$ that we saw in Example 3.7.3(2) does not exist. In fact there are many more such cuts. So we can use these cuts to represent a new set of numbers which extends our rational numbers. Altogether, the set of cuts $\mathcal{C}$ could be our set of real numbers.

Remark 3.7.6 Again, this could be surprising to some of the readers at first because now we are treating a subset of the numbers $\mathbb{Q}$ as a number. But if we recall Remark 2.4.1, this is similar to our construction of the integers where the integers are the sets of points in $\mathbb{N}^2$ lying on a line in Fig. 2.3. We have also constructed the rational numbers $\mathbb{Q}$ using the subsets in $\mathbb{Z} \times (\mathbb{Z} \setminus \{0\})$ in Fig. 3.2. Nothing new here!

To give the set $\mathcal{C}$ the usual characteristics of a number set similar to $\mathbb{N}$, $\mathbb{Z}$, and $\mathbb{Q}$, we need to define algebraic operations and ordering on it. Moreover, these operations must also satisfy the field and ordering axioms in Definitions 3.1.1 and 3.3.1. So how do we define them?

3.8 Algebra and Ordering of Dedekind Cuts

Since the rational numbers are represented in the set of Dedekind cuts $\mathcal{C}$ by the cuts $\mathcal{C}_{\mathbb{Q}} = \{L_q : q \in \mathbb{Q}\} \subseteq \mathcal{C}$, let us first find a good candidate definitions for $\prec$, $\oplus$, and $\otimes$ on the set $\mathcal{C}_{\mathbb{Q}} \subseteq \mathcal{C}$ so that they agree with the operations $<$, $+$, and $\times$ that we already know on $\mathbb{Q}$. Before we do that, let us define some arithmetic operations on subsets of $\mathbb{Q}$.

Definition 3.8.1 Let $X, Y \subseteq \mathbb{Q}$ be non-empty sets of rational numbers and $\lambda \in \mathbb{Q}$. We define:

$$\lambda X = \{\lambda x : x \in X\},$$

$$X + Y = \{x + y : x \in X, y \in Y\},$$

$$XY = \{xy : x \in X, y \in Y\}.$$

Now let us try and define suitable ordering, addition, and multiplication on the set $\mathcal{C}_{\mathbb{Q}}$ first.

1. We define an ordering on the cuts $\mathcal{C}_{\mathbb{Q}}$ as $L_p \prec L_q$ if and only if $L_p \subsetneq L_q$ as subsets of $\mathbb{Q}$. This would then agree with the ordering on $\mathbb{Q}$ because:

$$p < q \quad \Leftrightarrow \quad L_p \subsetneq L_q \quad \Leftrightarrow \quad L_p \prec L_q.$$

Likewise, we say $L_p \preceq L_q$ if and only if $L_p \subseteq L_q$. We call a Dedekind cut $L_p \in \mathcal{C}_{\mathbb{Q}}$ positive if and only if $L_p \succ L_0$ (or equivalently $0 \in L_p$). This agrees with the convention on $\mathbb{Q}$ since:

$$p \text{ positive in } \mathbb{Q} \;\Leftrightarrow\; 0 < p \;\Leftrightarrow\; L_0 \subsetneq L_p \;\Leftrightarrow\; L_0 \prec L_p \;\Leftrightarrow\; 0 \in L_p.$$

Analogously, we call an element L_p of $\mathcal{C}_{\mathbb{Q}}$ negative if and only if $L_p \prec L_0$ (or equivalently $0 \notin L_p$). Similar ideas allow us to define non-negative and non-positive elements in $\mathcal{C}_{\mathbb{Q}}$. We call the cut L_0 the zero Dedekind cut.

2. For consistency with $+$ on $\mathbb{Q}$, we want to define addition $\oplus$ on the set $\mathcal{C}_{\mathbb{Q}}$ so that $L_p \oplus L_q = L_{p+q}$. One obvious way to define the addition operation is via the set addition as in Definition 3.8.1, namely $L_p \oplus L_q = \{x + y \in \mathbb{Q} : x < p, y < q\}$. We have to make sure that this resulting set is also a Dedekind cut by checking the conditions in Definition 3.7.1.
   (a) Clearly this set is non-empty. Moreover, this set is not the whole of $\mathbb{Q}$ as it is bounded from above by $p + q$.
   (b) This set is closed downwards: Fix any $a + b \in \{x + y \in \mathbb{Q} : x < p, y < q\}$ where $a < p$ and $b < q$. For any $z < a + b$, we have $z - a < b$. We can then write $z = x + y$ where $x = a < p$ and $y = z - a < b < q$ and thus $z \in \{x + y \in \mathbb{Q} : x < p, y < q\}$ as well.
   (c) This set does not have a maximal element: Fix any $a+b \in \{x+y \in \mathbb{Q} : x < p, y < q\}$ where $a < p$ and $b < q$. Since L_p and L_q do not have maximal elements either, we can find $c \in L_p$ and $d \in L_q$ such that $a < c < p$ and $b < d < q$. This implies $a + b < c + d \in \{x + y \in \mathbb{Q} : x < p, y < q\}$.

   Thus $\{x + y \in \mathbb{Q} : x < p, y < q\}$ is also a Dedekind cut and hence the proposed addition definition above is a perfectly well-defined binary operation on $\mathcal{C}_{\mathbb{Q}}$. To show that the resulting cut $L_p \oplus L_q = \{x + y \in \mathbb{Q} : x < p, y < q\}$ is exactly equal to $L_{p+q} = \{x \in \mathbb{Q} : x < p + q\}$, we use double inclusion.

   ($\subseteq$): Pick any $z \in \{x + y \in \mathbb{Q} : x < p, y < q\}$. Then $z = a + b$ is a rational number where $a < p$ and $b < q$. This implies $z < p+q$ and so $z \in L_{p+q}$.

   ($\supseteq$): Conversely, pick any $z \in L_{p+q}$. By definition, z is a rational number with $z < p+q$. Let $t = \frac{p+q-z}{2} > 0$. Then, we would have $z+t < p+q$ which implies $z + (t - p) < q$. Let $y = z + (t - p) < q$ so that $z = (p - t) + y$. Thus, we can write $z = x + y$ where $x = p - t < p$ and $y < q$. Hence, $z \in \{x + y \in \mathbb{Q} : x < p, y < q\}$.

Therefore, we have a satisfactory addition operation on $\mathcal{C}_{\mathbb{Q}}$ which satisfies the desired equality $L_p \oplus L_q = L_{p+q}$ for all $p, q \in \mathbb{Q}$. Clearly the additive identity here is $L_0 = \{x \in \mathbb{Q} : x < 0\}$.
The additive inverse for L_q, which we denote as $\ominus L_q$, can then be uniquely defined by using the fact that $(-q) + q = 0$. We then have $L_q \oplus L_{-q} = L_0$ which implies the additive inverse of L_q is:

$$\ominus L_q = L_{-q} = \{x : x < -q\} = \{-y : y > q\}.$$

The final term above looks almost like $-U_q = \{-y : y \geq q\}$ except with the point $-q = -\min(U_q)$ taken out. This is necessary because we want the resulting set to not have a maximal element and hence be a Dedekind cut. So $\ominus L_q = -U_q \setminus \{-q\}$. To recap, we have defined:

$$\begin{aligned} L_p \oplus L_q &= \{x + y : x \in L_p, y \in L_q\}, \\ \ominus L_q &= \{-x : x \in U_q, x \neq \min(U_q)\}. \end{aligned}$$

3. Supposing $p, q \geq 0$, for consistency with $\times$ in $\mathbb{Q}$, we require $L_p \otimes L_q = L_{pq} = \{z \in \mathbb{Q} : z < pq\}$. Using this idea, we want to define a suitable multiplication operation on $\mathcal{C}_{\mathbb{Q}}$ which satisfies this requirement. First, if $p, q > 0$, we define $L_p \otimes L_q$ as:

$$\begin{aligned} L_p \otimes L_q &= \{x \in L_p : 0 \leq x\}\{y \in L_q : 0 \leq y\} \cup \{x \in \mathbb{Q} : x < 0\} \\ &= \{xy \in \mathbb{Q} : 0 < x < p, 0 < y < q\} \cup \{x \in \mathbb{Q} : x \leq 0\} \end{aligned} \tag{3.2}$$

By introducing $z = xy$, we can write the above as:

$$\begin{aligned} L_p \otimes L_q &= \{xy \in \mathbb{Q} : 0 < x < p, 0 < y < q, z = xy\} \cup \{x \in \mathbb{Q} : x \leq 0\} \\ &= \left\{z \in \mathbb{Q} : 0 < x < p, 0 < \frac{z}{x} < q\right\} \cup \{x \in \mathbb{Q} : x \leq 0\} \\ &= \{z \in \mathbb{Q} : 0 < x < p, 0 < z < qx\} \cup \{x \in \mathbb{Q} : x \leq 0\} \\ &= \{z \in \mathbb{Q} : 0 < z < pq\} \cup \{x \in \mathbb{Q} : x \leq 0\} = L_{pq}. \end{aligned}$$

On the other hand, if at least one of p or q is 0, we define:

$$L_p \otimes L_0 = L_0 \otimes L_p = \{x \in \mathbb{Q} : x < 0\} = L_0.$$

Note that this is also equal to the definition for multiplication for positive p, q if we declare that the product of sets in Definition 3.8.1 as $XY = \varnothing$ if at least one of X or Y is $\varnothing$. To recap, we have defined the multiplication operation on non-negative cuts (namely $p, q \geq 0$) via:

$$\begin{aligned} L_p \otimes L_q &= \{x \in L_p : 0 \leq x\}\{y \in L_q : 0 \leq y\} \cup \{x \in \mathbb{Q} : x < 0\} \\ &= \{xy : x \in L_p, y \in L_q, x, y \geq 0\} \cup \{x \in \mathbb{Q} : x < 0\}, \end{aligned}$$

where the former set is empty if one p or q is 0.
This proposed definition is a bit strange and unnatural (unlike for $\oplus$) since we first truncated the sets L_p and L_q to the non-negative parts before multiplying them term-wise together. However, this is necessary here or otherwise, if we follow the construction for $\oplus$ by directly using Definition 3.8.1, the resulting set is not a Dedekind cut. The readers will show this and justify the definition (3.2) in Exercise 3.10.
We can now define multiplication for cuts corresponding to rational numbers of other signs: if $p < 0$ and $q \geq 0$, we define $L_p \otimes L_q = \ominus(L_{-p} \otimes L_q)$. Since $-p, q \geq 0$ we can then proceed with the multiplication operation for non-negative numbers above to show that it is still consistent with the multiplication of rational numbers. Indeed:

$$\begin{aligned} L_p \otimes L_q = \ominus(L_{-p} \otimes L_q) &= \ominus(\{z \in \mathbb{Q} : 0 \leq z < -pq\} \cup \{x \in \mathbb{Q} : x < 0\}) \\ &= \ominus(L_{-pq}) = L_{pq}. \end{aligned}$$

Likewise, multiplication for $p \geq 0$ and $q < 0$ is defined as $L_p \otimes L_q = \ominus(L_p \otimes L_{-q})$ and multiplication for $p, q \leq 0$ is defined as $L_p \otimes L_q = L_{-p} \otimes L_{-q}$. From these definitions, we can deduce that the multiplicative identity is $L_1 = \{x \in \mathbb{Q} : x < 1\}$. Notice that, by using the definition above, we have $L_q \otimes L_{\frac{1}{q}} = L_1$. Thus, we can define the multiplicative inverse of L_q for $q > 0$, denoted $\frac{1}{L_q}$, as $L_{\frac{1}{q}}$ which is given by:

$$\frac{1}{L_q} = L_{\frac{1}{q}} = \left\{x \in \mathbb{Q} : x < \frac{1}{q}\right\} = \left\{\frac{1}{y} \in \mathbb{Q} : y > q\right\} \cup \{x \in \mathbb{Q} : x \leq 0\}.$$

For $q < 0$, from definitions, we have $L_1 = L_q \otimes L_{\frac{1}{q}} = L_{-q} \otimes L_{-\frac{1}{q}} = L_{-q} \otimes (\ominus L_{\frac{1}{q}})$. This means L_{-q} is the multiplicative inverse of $\ominus L_{\frac{1}{q}}$ which then implies $\ominus L_{\frac{1}{q}} = \frac{1}{L_{-q}} = \frac{1}{\ominus L_q}$. Finally, since this means $\frac{1}{\ominus L_q}$ is the additive inverse of the Dedekind cut $L_{\frac{1}{q}}$, we have $L_{\frac{1}{q}} = \ominus\left(\frac{1}{\ominus L_q}\right)$.

Writing them all down together, we have defined the multiplication operation as:

$$L_p \otimes L_q$$
$$= \begin{cases} \{xy : x \in L_p, y \in L_q, x, y \geq 0\} \cup \{x \in \mathbb{Q} : x < 0\} & \text{if } L_p, L_q \succeq L_0, \\ \ominus(L_{-p} \otimes L_q) = \ominus((\ominus L_p) \otimes L_q) & \text{if } L_p \prec L_0, L_q \succeq L_0, \\ \ominus(L_p \otimes L_{-q}) = \ominus(L_p \otimes (\ominus L_q)) & \text{if } L_p \succeq L_0, L_q \prec L_0, \\ L_{-p} \otimes L_{-q} = (\ominus L_p) \otimes (\ominus L_q) & \text{if } L_p, L_q \prec L_0, \end{cases}$$

and the multiplicative inverse for non-zero Dedekind cut is:

$$\frac{1}{L_q} = \begin{cases} \left\{\frac{1}{y} : y \in U_q, y \neq \min(U_q)\right\} \cup \{x \in \mathbb{Q} : x \leq 0\} & \text{if } L_q \succ L_0, \\ \ominus\left(\frac{1}{L_{-q}}\right) = \ominus\left(\frac{1}{\ominus L_q}\right) & \text{if } L_q \prec L_0. \end{cases}$$

We have thus defined ordering and algebraic operations $\prec$, $\oplus$, and $\otimes$ on $\mathcal{C}_{\mathbb{Q}} \subseteq \mathcal{C}$ so that they are consistent to the ordering and algebraic operations on $\mathbb{Q}$. Now we simply extend these definitions to other cuts in $\mathcal{C}$ so that we have an ordered field structure on $\mathcal{C}$. Generalising from above, for any Dedekind cuts $L, M \in \mathcal{C}$ we define:

1. Order: $L \prec M$ if and only if $L \subsetneq M$ as shown in Fig. 3.13. Likewise, $L \preceq M$ if and only if $L \subseteq M$. We call an element L of $\mathcal{C}$ positive if and only if $L \succ L_0$ (or equivalently $0 \in L$). Likewise, we call an element L of $\mathcal{C}$ negative if and only if $L \prec L_0$ (or equivalently $0 \notin L$).
2. Addition:

$$L \oplus M = \{x + y \in \mathbb{Q} : x \in L, y \in M\}.$$

Some examples of additions of Dedekind cuts are shown in Fig. 3.14. The additive identity is denoted as $L_0 = \{x \in \mathbb{Q} : x < 0\}$ and the additive inverse for $L \in \mathcal{C}$ is:

$$\ominus L = \{-x : x \in L^c, x \neq \min(L^c)\},$$

where $\min(L^c)$ may or may not exist. Note that, by Corollary 3.7.5, this minimum exists if and only if $L \in \mathcal{C}_{\mathbb{Q}}$.

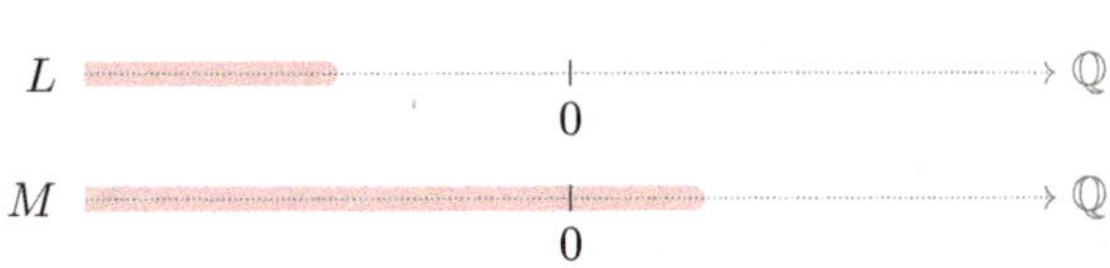

Fig. 3.13 Ordering of Dedekind cuts. Here we have $L \prec M$ since $L \subseteq M$

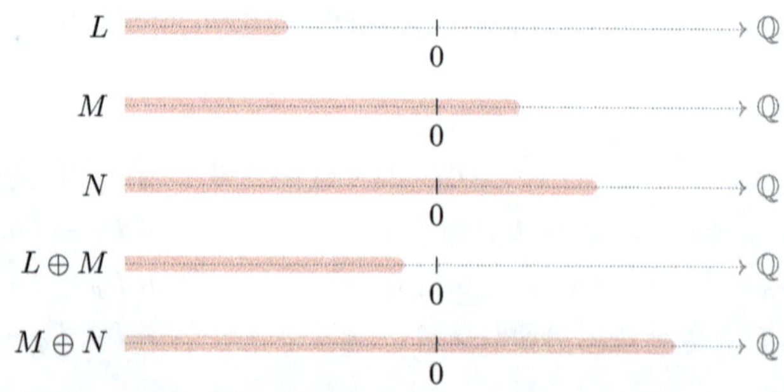

Fig. 3.14 Addition of some Dedekind cuts

3. Multiplication:

$$L \otimes M$$
$$= \begin{cases} \{xy : x \in L, y \in M, x, y \geq 0\} \cup \{x \in \mathbb{Q} : x < 0\} & \text{if } L, M \succeq L_0, \\ \ominus((\ominus L) \otimes M) & \text{if } L \prec L_0, M \succeq L_0, \\ \ominus(L \otimes (\ominus M)) & \text{if } L \succeq L_0, M \prec L_0, \\ (\ominus L) \otimes (\ominus M) & \text{if } L, M \prec L_0. \end{cases}$$

The multiplicative identity is denoted as $L_1 = \{x \in \mathbb{Q} : x < 1\}$ and the multiplicative inverse for non-zero Dedekind cuts $L \neq L_0$ is:

$$\frac{1}{L} = \begin{cases} \left\{\frac{1}{y} : y \in L^c, y \neq \min(L^c)\right\} \cup \{x \in \mathbb{Q} : x \leq 0\} & \text{if } L \succ L_0, \\ \ominus\left(\frac{1}{\ominus L}\right) & \text{if } L \prec L_0. \end{cases}$$

One can then check that these operations on $\mathcal{C}$ satisfy all the necessary field and ordered field axioms in Definitions 3.1.1 and 3.3.1. We have shown some of these (namely the existence of identities and inverses), but we leave the others for the readers to check in Exercise 3.11.

Therefore, the set of Dedekind cuts equipped with the ordering $\prec$ and algebraic operations $\oplus$ and $\otimes$ forms an ordered field. Furthermore, this field contains a copy of $\mathbb{Q}$ with consistent ordering and algebraic operations in the form of the cuts $\mathcal{C}_{\mathbb{Q}} \subseteq \mathcal{C}$.

Finally, we are going to prove that this ordered field also satisfies the completeness axiom in Definition 3.6.11. Recall that the field $\mathbb{Q}$ does not have this property, so this new extended field is an upgraded version of the rational numbers which satisfies the completeness axiom.

Proposition 3.8.2 *The ordered field of Dedekind cuts $\mathcal{C}$ with ordering $\prec$ and algebraic operations $\oplus$ and $\otimes$ satisfies the completeness axiom.*

Proof Let $\mathcal{D} \subseteq \mathcal{C}$ be a subset of the Dedekind cuts which is bounded from above. Define a subset $X \subseteq \mathbb{Q}$ as the union of all the elements in each of the cuts in $\mathcal{D}$, namely:

$$X = \{x : x \in D \text{ for some } D \in \mathcal{D}\} = \bigcup_{D \in \mathcal{D}} D.$$

It is easy to check that X is also a Dedekind cut by showing that it is neither $\varnothing$ nor $\mathbb{Q}$, is closed downwards, and does not have a greatest element. Furthermore, for every $D \in \mathcal{D}$, we have $D \subseteq X$ as sets in $\mathbb{Q}$ and so $D \preceq X$ as elements in $\mathcal{C}$. This means X is an upper bound for the set $\mathcal{D}$.

To show that X is the least upper bound of the set $\mathcal{D}$, pick any arbitrary upper bound Y of the set $\mathcal{D}$. We aim to prove that $X \preceq Y$ as Dedekind cuts or equivalently $X \subseteq Y$ as sets in $\mathbb{Q}$. For any $x \in X$, by definition, $x \in D$ for some $D \in \mathcal{D}$. Since Y is an upper bound of the set $\mathcal{D}$, we must have $D \preceq Y$. As sets in $\mathbb{Q}$, this means $D \subseteq Y$ and so $x \in D \subseteq Y$. Since $x \in X$ is arbitrary, we have the inclusion $X \subseteq Y$.

The cut X defined above is then called the supremum of $\mathcal{D}$, written as $\sup(\mathcal{D}) = X = \bigcup_{D \in \mathcal{D}} D$. □

Thus, since the Dedekind cuts is an ordered field that is complete, this would be our real numbers $\mathbb{R}$ that we have defined in Definition 3.6.13. Note that this field contains the rational cuts $\mathcal{C}_{\mathbb{Q}}$ which can be distinguished as the cuts which has a supremum in $\mathbb{Q}$ as per Lemma 3.7.4. The remaining cuts are called the irrational cuts, written as $\mathcal{C}_{\bar{\mathbb{Q}}}$. We now know that the irrational numbers exist and the addition and multiplication operations on them is just an extension of these operations from the rationals.

Example 3.8.3 Recall the Dedekind cut $L = \{x \in \mathbb{Q} : x^2 \leq 2 \text{ or } x < 0\}$ from Example 3.7.3(2). We have seen that this set does not have a supremum in $\mathbb{Q}$, so it cannot be in $\mathcal{C}_{\mathbb{Q}}$. Therefore, the real number corresponding to this cut is an irrational number.

1. This irrational number can also be seen as a supremum of some set of rational cuts in $\mathcal{C}$. We define a subset of rational cuts $\mathcal{D} = \{L_p : p \in \mathbb{Q}, p^2 \leq 2\} \subseteq \mathcal{C}_{\mathbb{Q}} \subseteq \mathcal{C}$.

This set is clearly bounded from above since, for example, $D \prec L_{10}$ for every $D \in \mathcal{D}$. From Proposition 3.8.2, the supremum of this set is:

$$\begin{aligned}
\sup(\mathcal{D}) &= \bigcup_{\substack{p^2 \le 2 \\ p \in \mathbb{Q}}} L_p = \bigcup_{\substack{p^2 \le 2 \\ p \in \mathbb{Q}}} \{x \in \mathbb{Q} : x < p\} \\
&= \bigcup_{\substack{p^2 \le 2 \\ p \in \mathbb{Q}}} \{x \in \mathbb{Q} : x^2 < p^2\} \cup \{x \in \mathbb{Q} : x < 0\} \\
&= \{x \in \mathbb{Q} : x^2 < 2\} \cup \{x \in \mathbb{Q} : x < 0\} \\
&= \{x \in \mathbb{Q} : x^2 \le 2\} \cup \{x \in \mathbb{Q} : x < 0\} = L,
\end{aligned}$$

where we added the empty set $\{z \in \mathbb{Q} : z^2 = 2\} = \varnothing$ in the final line.

2. Now let us compute $L \otimes L$. Clearly $L \succ L_0$ since $0 \in L$, so by using the appropriate definition of $\otimes$ here, we have:

$$\begin{aligned}
L \otimes L &= \{x \in \mathbb{Q} : x^2 \le 2 \text{ or } x < 0\} \otimes \{y \in \mathbb{Q} : y^2 \le 2 \text{ or } y < 0\} \\
&= \{xy : x, y \in L, x, y \ge 0\} \cup \{x \in \mathbb{Q} : x \le 0\} \\
&= \{xy : x, y \in \mathbb{Q}, 0 < x, 0 < y, x^2 \le 2, y^2 \le 2\} \cup \{x \in \mathbb{Q} : x \le 0\}.
\end{aligned} \tag{3.3}$$

Let us simplify the former set. By introducing $z = xy$, we have:

$$\begin{aligned}
&\{xy : x, y \in \mathbb{Q}, 0 < x, 0 < y, x^2 \le 2, y^2 \le 2\} \\
&= \{xy : x, y, z \in \mathbb{Q}, 0 < x, 0 < y, 0 < z, x^2 \le 2, y^2 \le 2, z^2 = x^2y^2 \le 4\} \\
&= \left\{z : x, z \in \mathbb{Q}, 0 < x, 0 < \frac{z}{x}, 0 < z, x^2 \le 2, \frac{z^2}{x^2} \le 2, z^2 \le 4\right\} \\
&= \{z : x, z \in \mathbb{Q}, 0 < x, 0 < z, x^2 \le 2, z^2 \le 2x^2, z^2 \le 4\} \\
&= \{z : z \in \mathbb{Q}, 0 < z, z^2 \le 4\},
\end{aligned}$$

and so, putting this in (3.3), we get:

$$\begin{aligned}
L \otimes L &= \{xy : x, y \in \mathbb{Q}, 0 < x, 0 < y, x^2 \le 2, y^2 \le 2\} \cup \{x \in \mathbb{Q} : x \le 0\} \\
&= \{z : z \in \mathbb{Q}, 0 < z, z^2 < 2^2\} \cup \{x \in \mathbb{Q} : x \le 0\} \\
&= \{z : z \in \mathbb{Q}, z < 2\} = L_2.
\end{aligned}$$

Therefore, unlike in $\mathbb{Q}$, there are Dedekind cuts in $\mathcal{C}$ that equates to L_2 when multiplied to itself. Now the Pythagoreans do not have anything to worry about!

From Example 3.8.3(1), we saw that the cut L can be represented by a supremum of some set of rational cuts. In general, any irrational cuts can be represented as such.

Proposition 3.8.4 *Let L be any irrational Dedekind cut. If we define the set of all rational cuts smaller than L as $\mathcal{D} = \{L_p : p \in L\}$, then:*

$$L = \sup(\mathcal{D}) = \bigcup_{L_p \in \mathcal{D}} L_p.$$

Proof Let us show that this is true by double inclusion:

($\subseteq$): Pick any $x \in L$. Since L is a Dedekind cut and does not have a maximum value, there is a $q \in L$ such that $x < q$. Then, $L_q \in \mathcal{D}$ and thus $x \in L_q \subseteq \bigcup_{L_p \in \mathcal{D}} L_p$. Since $x \in L$ is arbitrary, we then have the inclusion $L \subseteq \bigcup_{L_p \in \mathcal{D}} L_p$.

($\supseteq$): Pick any $x \in \bigcup_{L_p \in \mathcal{D}} L_p$. Then, $x \in L_p$ for some $p \in L$. This means $x < p$. Since L is a Dedekind cut which is closed downwards and $p \in L$, we must have $x \in L$ as well. Since $x \in \bigcup_{L_p \in \mathcal{D}} L_p$ is arbitrary, we conclude that $\bigcup_{L_p \in \mathcal{D}} L_p \subseteq L$. □

The Real Numbers

From now on, unless necessary, we refer to the the set of Dedekind cuts as the set of real numbers $\mathbb{R}$ and the Dedekind cuts as real numbers. To reconcile that the real numbers is simply an extension of the rational numbers (and mostly to declutter our writing), instead of using the notations $\prec$, $\oplus$, and $\otimes$, we revert to the original notations for ordering and algebraic operations that we used on $\mathbb{Q}$, namely $<, +$, and $\times$.

Moreover, for any $a \in \mathbb{R}$ and $n \in \mathbb{N}$, by induction, the n-fold multiplication $\underbrace{a \times a \times \ldots \times a}_{n \text{ times}}$ is well-defined. This is called the n-th power of a, denoted as a^n.

We also now write the numbers simply as symbols instead of cuts or subsets of $\mathbb{Q}$. However, unlike the integers or the rational numbers, these symbols may not be explicitly written with the digits in base-10 notation.

Some of the symbols used for these irrational numbers will be informative. For example the irrational number L in Example 3.8.3 is written as the symbol $\sqrt{2}$ or $2^{\frac{1}{2}}$, which is read as "square root of 2". With this notation, we have the implicit knowledge that this number is equal to 2 when multiplied with itself.

We shall see in Exercise 3.17, that for any $p \in \mathbb{N}$, the square root operation above can be extended to any p-th root uniquely on any non-negative real number $a \in \mathbb{R}_{\geq 0}$ as $\sqrt[p]{a}$ or $a^{\frac{1}{p}}$. We shall also show that at least countably infinitely many of such numbers are irrational. However, in spite of their irrationality of the number $a^{\frac{1}{p}}$ above, we know that their p-th power is the number a which allows us to roughly know their quantitative behaviour.

On the other hand, there are also symbols for irrational numbers which are not quite as instructive. For example the Euler-Napier constant e, the Archimedes constant π, and the golden ratio φ. All of these symbols do not tell us any quantitative (or even sign) information about these numbers, but they are understood as universal symbols of some important irrational numbers within the mathematical community. We shall see later why these irrational numbers are special, important, and deserve their own symbol.

To close this chapter, we state a proposition that was proven using Dedekind cuts in Proposition 3.8.4. This characterisation of real numbers allows us to describe irrational numbers using rational numbers. This is very important for algebraic purposes in the next chapter.

Proposition 3.8.5 *Let $x \in \mathbb{R}$ be any real number. This number can be expressed as the supremum of the set $x = \sup\{q \in \mathbb{Q} : q \leq x\}$.*

Exercises

3.1 (∗) Recall the definition of addition and multiplication operations on $\mathbb{Q}$:

$$\frac{p}{q} + \frac{r}{s} = \frac{ps + qr}{qs} \quad \text{and} \quad \frac{p}{q} \times \frac{r}{s} = \frac{pr}{qs}.$$

Show that these operations are well-defined for any representative of the rational numbers used. In other words, show that if $\frac{p}{q} = \frac{a}{b}$ and $\frac{r}{s} = \frac{c}{d}$ via the relation $\frac{p}{q} = \frac{a}{b}$ iff $pb = aq$, then $\frac{p}{q} + \frac{r}{s} = \frac{a}{b} + \frac{c}{d}$ and $\frac{p}{q} \times \frac{r}{s} = \frac{a}{b} \times \frac{c}{d}$.

3.2 Prove Lemmas 3.2.2 and 3.3.3 carefully by using the field and ordered field axioms.

3.3 Find the cardinalities of the following sets:

(a) $A = \{(2n, 3n) : n \in \mathbb{N}\}$.
(b) $A = \{(2n, 3n) : n \in \mathbb{N}, n^2 \leq 25\}$.
(c) $A = \{(2n, 3n) : n \in \mathbb{N}, n^2 \geq 25\}$.
(d) $A = \{(2n, 3n) : n \in \mathbb{Z}\}$.
(e) $A = \{(2m, 3n) : m, n \in \mathbb{N}\}$.
(f) $A = \{(2m, 3n) : m, n \in \mathbb{N}, n^2 \leq 25\}$.

3.4 (∗) Let A and B be finite sets and $f : A \to B$. Suppose that $|A| = m$ and $|B| = n$.

(a) Prove that if f is a surjection, then $m \geq n$.

(b) Prove that if f is an injection, then $m \leq n$.
Now suppose that $m = n$.
(c) Prove that if f is an injection, then it is a surjection.
(d) Prove that if f is a surjection, then it is an injection.
Give examples of infinite sets A and B for which both of the above statements are false.

3.5 (a) Prove that the Cartesian product $\mathbb{N} \times \mathbb{N}$ is countable.
(b) More generally, prove that for two countable sets A and B, the Cartesian product $A \times B$ is countable.

3.6 ($\diamond$) In Cantor-Bernstein-Schröder theorem in Theorem 3.4.3, X and Y are sets such that there are functions $f : X \to Y$ and $g : Y \to X$ which are injective. We have defined a function $h : X \to Y$ in the proof. Prove that this function is a bijection.

3.7 (a) Let A, B, and C be some sets such that $A \subseteq B \subseteq C$. If A and C are countably infinite, prove that B is also countably infinite.
(b) Let A and B be sets such that $A \subseteq B$. If A is uncountably infinite, then prove that B must also be uncountably infinite.

3.8 ($*$) In this question, we prove Cantor's theorem:

Theorem 3.9.6 (Cantor's Theorem) *Let A be a set and $\mathcal{P}(A)$ is the power set of A, which is the set of subsets of A. Then there are no surjective function $f : A \to \mathcal{P}(A)$.*

In the words of cardinality, we must have $|A| < |\mathcal{P}(A)|$.
(a) Suppose first that A is a finite set $n \in \mathbb{N}_0$. Prove that $|\mathcal{P}(A)| > n$.
(b) Now suppose that A is infinite. Suppose for contradiction that $|\mathcal{P}(A)| \leq |A|$ so that there exists a surjection $f : A \to \mathcal{P}(A)$. Thus, every $X \in \mathcal{P}(A)$, namely $X \subseteq A$, has a preimage under f. In other words, there exists an $x \in A$ such that $f(x) = X$. Define a subset $Y \subseteq A$ as:

$$Y = \{a \in A : a \notin f(a)\} \in \mathcal{P}(A).$$

Show that this leads to a contradiction.
Hence, for any set A at all we must have $|A| < |\mathcal{P}(A)|$. In particular, the power set of natural numbers $\mathcal{P}(\mathbb{N})$ is an uncountable set.

3.9 ($*$) Prove Corollary 3.7.5, namely:
Let (L, U) be a Dedekind cut on $\mathbb{Q}$. Show that $L \in \mathcal{C}_{\mathbb{Q}}$ if and only if U has a minimal element.

3.10 We have defined the $\otimes$ operation on the rational Dedekind cuts $\mathcal{C}_{\mathbb{Q}}$ for $p, q > 0$ as:

$$L_p \otimes L_q = \{x \in L_p : 0 \leq x\}\{y \in L_q : 0 \leq y\} \cup \{x \in \mathbb{Q} : x < 0\}.$$

Now we are going to justify this definition.

Suppose $p, q > 0$. Denote $P = \{x \in \mathbb{Q} : x < p\}$ and $Q = \{y \in \mathbb{Q} : y < q\}$.

(a) Show that the pairwise multiplication set $PQ = \{xy \in \mathbb{Q} : x, y \in \mathbb{Q}, x < p, y < q\}$ is unbounded above and below.

(b) Hence, show that $PQ = \mathbb{Q}$.
Explain how the presence of negative numbers in each of the sets P and Q caused this.

Let us remove the negative numbers from both of the sets P and Q. Define:

$$P' = P \cap \mathbb{Q}_{\geq 0} = \{x \in \mathbb{Q} : 0 \leq x < p\},$$
$$Q' = Q \cap \mathbb{Q}_{\geq 0} = \{x \in \mathbb{Q} : 0 \leq x < q\}.$$

(c) Show that the product of sets $P'Q' = \{x \in \mathbb{Q} : 0 \leq x < p\}\{y \in \mathbb{Q} : 0 \leq y < q\}$ is not closed downwards.

(d) Finally, to close the set $P'Q'$ downwards, we add in $\{x \in \mathbb{Q} : x < 0\}$, resulting in $P'Q' \cup \{x \in \mathbb{Q} : x < 0\}$. Show that this set does not have a maximal element.

Thus $P'Q' \cup \{x \in \mathbb{Q} : x < 0\} = \{x \in \mathbb{Q} : 0 \leq x < p\}\{y \in \mathbb{Q} : 0 \leq y < q\} \cup \{x \in \mathbb{Q} : x < 0\}$ is a Dedekind cut and this definition above gives us a well-defined multiplication operation on $\mathcal{C}_{\mathbb{Q}_+}$. In fact, we have proven that it coincides with L_{pq}, which is a bonus since this multiplication $\otimes$ agrees with the multiplication $\times$ on $\mathbb{Q}_+$ that we already know of.

3.11 Now we want to completely show that the set of Dedekind cuts satisfy the field and ordering axioms.

(b) Show that the $\oplus$ and $\otimes$ operations on $\mathcal{C}$ are commutative and associative.

(c) Show that $\otimes$ operation is distributive over $\oplus$ in $\mathcal{C}$.

(d) Show that the $\prec$ ordering on $\mathcal{C}$ satisfies the transitive and trichotomy laws.

(e) Show that the $\prec$ ordering on $\mathcal{C}$ is compatible with $\oplus$ and $\otimes$.

3.12 Using Dedekind cuts, show that if $L, M \in \mathcal{C}$ such that L is rational and M is irrational, then $\ominus M$ and $L \oplus M$ are both irrational.

3.13 (∗) Let $a, r, d \in \mathbb{R}$. By induction, prove for all $n \in \mathbb{N}$ that:

(a) $a + ar + ar^2 + ar^3 + \ldots + ar^n = \frac{a(1-r^{n+1})}{1-r}$ if $r \neq 1$.

(b) $a + (a + d) + (a + 2d) + \ldots + (a + nd) = \frac{n+1}{2}(2a + nd)$.

3.14 (∗) We are going to prove an extension of Lemma 3.3.8. Let $j, n \in \mathbb{N}$ with $0 \leq j \leq n$. The quantity $\binom{n}{j} = \frac{n!}{(n-j)!j!}$ is called the binomial coefficient with $r!$ (read as r factorial) defined as the product $r! = r \cdot (r-1) \cdot (r-2) \cdot \ldots \cdot 1$ for any $r \in \mathbb{N}$ and $0! = 1$.

(a) Show that for any $j, n \in \mathbb{N}$:

i. For $0 \leq j \leq n$ we have $\binom{n}{j} = \binom{n}{n-j}$.

ii. For $1 \leq j \leq n$ we have $\binom{n}{j-1} + \binom{n}{j} = \binom{n+1}{j}$.

Using the identities above, we can compile the binomial coefficients in a triangular array as follows: the j-th term in the n-th row is the number $\binom{n}{j}$ (we start with the 0-th row as a convention). The following are the first seven rows of the triangular array.

$n = 0$							1						
$n = 1$						1		1					
$n = 2$					1		2		1				
$n = 3$				1		3		3		1			
$n = 4$			1		4		6		4		1		
$n = 5$		1		5		10		10		5		1	
$n = 6$	1		6		15		20		15		6		1

The triangular array above is called Pascal's triangle after Blaise Pascal (1623–1662). However, this triangle has been studied long before Pascal's time by, amongst others, Omar Khayyam (1048–1131) and Yang Hui (1238–1298).

Notice that the property in part (a)(ii) implies that the sum of any two entries next to each other in the array equals the number below them. An example in the array above is highlighted as $6 + 4 = 10$. This property then allows us to build the triangle layer by layer easily.

One can deduce many other interesting results from this triangle. We are going to prove some of them here.

(b) Let $n \in \mathbb{N}$.
   i. If n is even, prove that for $0 \leq j < k \leq \frac{n}{2}$ we have $\binom{n}{j} < \binom{n}{k}$.
      Conclude that for any $\frac{n}{2} \leq k < j \leq n$ we have $\binom{n}{k} > \binom{n}{j}$.
   ii. If n is odd, prove that for $0 \leq j < k \leq \lfloor \frac{n}{2} \rfloor$ we have $\binom{n}{j} < \binom{n}{k}$.
      Moreover, prove that $\binom{n}{\lfloor \frac{n}{2} \rfloor} = \binom{n}{\lceil \frac{n}{2} \rceil}$.
      Conclude that $\binom{n}{k} > \binom{n}{j}$ for any $\lceil \frac{n}{2} \rceil \leq k < j \leq n$.

(c) For any $x, y \in \mathbb{R}$ and $n \in \mathbb{N}$, show by induction that:

$$(x + y)^n = \sum_{j=0}^{n} \binom{n}{j} x^j y^{n-j}.$$

   This is called the binomial expansion.

(d) Deduce that for $n \geq 1$, we have $\sum_{j=0}^{n} \binom{n}{j} = 2^n$ and $\sum_{j=0}^{n} (-1)^j \binom{n}{j} = 0$.

(e) Using part (d), show that for $n \geq 1$ we have:

$$\sum_{\substack{j=0 \\ j \text{ odd}}}^{n} \binom{n}{j} = \sum_{\substack{j=0 \\ j \text{ even}}}^{n} \binom{n}{j} = 2^{n-1}.$$

(f) Prove that $\sum_{j=0}^{n} \binom{n}{j}^2 = \binom{2n}{n}$.

(g) We want to show that for any fixed $n \in \mathbb{N}$, we have the following inequalities:

$$\frac{4^n}{\sqrt{4n}} \leq \binom{2n}{n} \leq \frac{3 \cdot 4^n}{4\sqrt{n+1}}. \tag{3.4}$$

The quantity $\binom{2n}{n}$ is called the central binomial coefficient because it is the middle term in the $2n$-th row of Pascal's triangle.

i. Show first that $\frac{1}{4^n}\binom{2n}{n} = \frac{1}{2n}\prod_{j=1}^{n-1}\left(1+\frac{1}{2j}\right) = \prod_{j=1}^{n}\left(1-\frac{1}{2j}\right)$.

ii. Show that $\prod_{j=1}^{n-1}\left(1+\frac{1}{2j}\right)^2 \geq n$.
Prove that $\left(\frac{1}{4^n}\binom{2n}{n}\right)^2 \geq \frac{1}{4n}$ and hence deduce the first inequality in (3.4).

iii. Prove that $\prod_{j=1}^{n}\left(1-\frac{1}{2j}\right)\left(1+\frac{1}{2j}\right) \leq \frac{3}{4}$.
Hence, deduce the second inequality in (3.4).

3.15 (*) In this question, we are going to prove the Bernoulli's inequality for natural number exponents. This inequality is named after Jacob Bernoulli (1655–1705), one of the mathematicians in the Bernoulli family.

(a) Using the binomial expansion in Exercise 3.14(c), prove that for any $x \geq 0$ and $n \in \mathbb{N}_0$ we must have $(1+x)^n \geq 1+nx$.

(b) Using mathematical induction, prove the Bernoulli's inequality which says for any $x \geq -1$ and $n \in \mathbb{N}_0$ we must have $(1+x)^n \geq 1+nx$. We shall extend this result to other exponents in Exercises 4.7 and 4.15.

3.16 (a) Let $x \in \mathbb{Q}_+$ and $m \in \mathbb{N}$. Prove by induction that if $xm \geq 2$, then for all $n \in \mathbb{N}$ we have $\left(1+\frac{1}{xm}\right)^n \leq 1+\frac{2^n}{xm}$.

(b) Using part (a), prove Lemma 3.3.9.

3.17 (∗) In this question, we are going to prove the existence of p-th roots of positive real numbers using Dedekind cuts.

(a) Let $M \succ L_0$ be a Dedekind cut. If $p \in \mathbb{N}$, show that:

$$L = \{x \in \mathbb{Q} : x \geq 0, x^p \in M\} \cup \{x \in \mathbb{Q} : x < 0\},$$

is also a Dedekind cut and $L \succ L_0$.

(b) By double inclusion, show that $\underbrace{L \otimes L \otimes L \otimes \ldots \otimes L}_{p \text{ times}} = M$.

(c) For any integer $p > 1$ and $x, y \in \mathbb{R}$, show that $x^p - y^p = (x-y)(x^{p-1} + x^{p-2}y + \ldots + xy^{p-2} + y^{p-1})$.

(d) Hence, conclude that if $a \geq 0$, then there exists a unique $b \geq 0$ such that $b^p = a$. We write this as $b = \sqrt[p]{a}$ or $b = a^{\frac{1}{p}}$ and call it the p-th root of a.

3.18 Using algebra, prove that the numbers $\sqrt{3}$, $\sqrt{3}+\sqrt{2}$, and $\sqrt{3}-\sqrt{2}$ are all irrational.

3.19 (a) Show that for any $n \in \mathbb{N}$, the number $\sqrt{n}$ is either an integer or an irrational number.

(b) More generally, show that for any $n, k \in \mathbb{N}$ with $k \geq 2$, the number $\sqrt[k]{n}$ is either an integer or an irrational number.

(c) Show that for any integer n greater than 1, the number $\sqrt[n]{n}$ is irrational.

(d) Hence, deduce that the set of irrational numbers $\bar{\mathbb{Q}}$ is at least countably infinite.

3.20 A polynomial of degree $n \in \mathbb{N}_0$ over $\mathbb{R}$ is a function $P : \mathbb{R} \to \mathbb{R}$ of the form $P(x) = \sum_{j=0}^{n} a_j x^j$ for $a_j \in \mathbb{R}$ for $j = 0, 1, \ldots, n$ and $a_n \neq 0$. The number a_n is called the leading coefficient of P and the terms $a_j x^j$ are called monomials.

By extension, a non-zero constant function may be considered a 0-th degree polynomial. This does not apply to the zero function for which we cannot meaningfully assign a degree to. So we leave the degree of a zero function as undefined.

We denote the degree of a (non-zero) polynomial P as $\deg(P)$. Suppose that P and Q are non-constant polynomials of degrees m and n respectively. Thus, we can write them as $P(x) = \sum_{j=0}^{m} a_j x^j$ and $Q(x) = \sum_{j=0}^{n} b_j x^j$ with $a_m, b_n \neq 0$. Prove that:

(a) $P + Q$, $P \times Q$, and $P \circ Q$ are all polynomials as well.

(b) If $P + Q$ is non-zero, then $\deg(P + Q) \leq \max\{m, n\}$.

(c) $\deg(P \times Q) = m + n$.

(d) If $P \circ Q$ is non-zero, $\deg(P \circ Q) = mn$.

(e) The non-zero conditions in parts (b) and (d) are necessary. Give examples of the polynomials P and Q for which the degree of either polynomial $P + Q$ or $P \circ Q$ is undefined.

(f) Show that if $P \times Q$ is zero identically, then either P or Q is zero identically.

3.21 $(*)$ A real polynomial P is called monic if its leading coefficient is 1.

(a) Show that for any monic real polynomial P of degree $n \geq 1$, if there is a number $c \in \mathbb{R}$ such that $P(c) = 0$, then we can write $P(x) = (x-c)Q(x)$ where Q is another polynomial of degree $n - 1$. We call the number c a root of P.

(b) Hence, deduce that we can write any monic real polynomial in the form $P(x) = (x - c_1)(x - c_2)\ldots(x - c_k)Q(x)$ where $0 \leq k \leq n$, $c_j \in \mathbb{R}$ for $j = 1, 2, \ldots, k$, and $Q(x)$ is a polynomial of degree $n - k$ such that $Q(x) = 0$ does not have a real solution.

Thus, after collecting similar linear factors $(x - c)$ together, we can always write a polynomial P as $P(x) = \prod_{j=1}^{k}(x - d_j)^{m_j} Q(x)$ where d_j are some real numbers, m_j are natural numbers such that $\sum_{j=1}^{k} m_j \leq n$, and Q is a polynomial with no real roots. The real numbers d_j are called roots of P and the exponent m_j is called the multiplicity of root d_j.

However, if we extend the number field that we are working with from $\mathbb{R}$ to $\mathbb{C}$ (which we shall explain in Exercise 3.24), we can always find complex roots of the polynomial Q. Thus, over $\mathbb{C}$, we can factorise the original polynomial P further into n linear factors. This result is called the fundamental theorem of algebra, first rigorously proven by Jean-Robert Argand (1768–1822).

3.22 (a) Prove that a polynomial P of degree n has at most n distinct real roots.
(b) Hence, deduce that for any constant $k \in \mathbb{R}$ the polynomial P of degree n attains the value k at most n times.

3.23 ($*$) Let $P(x) = ax^2 + bx + c$ be a quadratic (degree 2) real polynomial where $a, b, c \in \mathbb{R}$ and $a \neq 0$.
(a) Show that if $b^2 - 4ac < 0$, then there are no real solutions to $P(x) = 0$.
(b) Otherwise, show that the solutions to $P(x) = 0$ are $x = \frac{-b \pm \sqrt{b^2-4ac}}{2a}$.
(c) Deduce that P has two distinct roots if $b^2 - 4ac > 0$ and one root with multiplicity 2 if $b^2 - 4ac = 0$.
The quantity $b^2 - 4ac$ is called the quadratic discriminant for the polynomial P since it allows us to distinguish the cases for how many roots the polynomial P has.
(d) Suppose that the roots of P are r_1 and r_2 (not necessarily distinct). Show that $r_1 + r_2 = -\frac{b}{a}$ and $r_1 r_2 = \frac{c}{a}$.
These are called the Viéte's formulas which relates the coefficients of a polynomial with its roots. This family of formulas, which were discovered by François Viéte (1540–1603), can be extended to polynomials of higher degrees and also if the roots r_j of the polynomial are not real.

3.24 ($*$) The complex numbers is a set of numbers which is obtained by appending the real field with an imaginary unit i which satisfies $i^2 = -1$. This comes about along the similar line of thinking with the conundrum faced by Diophantus with the equation $4x + 20 = 4$ in $\mathbb{N}$ in Sect. 2.4. If we now focus on the equation $x^2 = -1$, this has no solution in $\mathbb{R}$ because we know that the square of any real number must be non-negative by Lemma 3.2.2, so this equation is absurd in $\mathbb{R}$!
However, if we really insist that we want a solution to this equation, we need to extend the real number system with some new kind of number, which enables us to square a number and get a negative result. We denote the solution to this equation as $x = \sqrt{-1}$, whatever this means.
This is exactly what Gerolamo Cardano (1501–1576) was led to do in his book *Ars Magna* (The Great Art). In this book, he studied the formula for finding the roots of cubic polynomials of the depressed form $x^3 + px + q = 0$ originally by Scipione del Ferro (1465–1526) and Niccolò Tartaglia (1500–1557). The formula by Cardano for a solution to the depressed cubic equation is given by:

$$x = \sqrt[3]{-\frac{q}{2} + \sqrt{\frac{q^2}{4} + \frac{p^3}{27}}} + \sqrt[3]{-\frac{q}{2} - \sqrt{\frac{q^2}{4} + \frac{p^3}{27}}},$$

which has a real solution only for $4p^3 + 27q^2 \geq 0$. This necessitates an explanation for the case of $4p^3 + 27q^2 < 0$ by allowing us to take the square root of a negative real number.

However, many mathematicians were dismissive of this new concept. Cardano himself described it as subtle and useless. Descartes derisively gave the number $\sqrt{-1}$ is current name: "imaginary number" or "imaginary unit". Few, such as Rafael Bombelli (1526–1572), gave it some attention. But largely this concept was ignored.

It was later on in the eighteenth century that the complex/imaginary numbers gained traction via proponents such as Argand, Cauchy, Abraham de Moivre (1667–1754), Euler, Gauss, and Caspar Wessel (1745–1818). Today, the complex numbers found extensive applications everywhere and is a staple concept in mathematics.

Due to the importance of the imaginary unit $\sqrt{-1}$ in algebra, Gauss proposed for it to be called the "lateral unit" instead. This name is chosen by Gauss so that we have positive unit $+1$, negative unit -1, and lateral unit $\sqrt{-1}$ where the latter is orthogonal/lateral to the other two units on the Argand diagram for the representation of complex numbers (see Fig. 4.6). However, the term imaginary number coined disparagingly by Descartes remains stuck.

The imaginary unit is denoted as the symbol $i = \sqrt{-1}$. The set of complex numbers can then be written as the combination:

$$\mathbb{C} = \{a + ib : a, b \in \mathbb{R}, i^2 = -1\}.$$

For any two elements $z_1 = a_1 + ib_1$ and $z_2 = a_2 + ib_2$ of $\mathbb{C}$, we define an addition $\oplus$ and multiplication $\otimes$ on $\mathbb{C}$ as:

$$z_1 \oplus z_2 = (a_1 + ib_1) \oplus (a_2 + ib_2) = (a_1 + a_2) + i(b_1 + b_2),$$

$$z_1 \otimes z_2 = (a_1 + ib_1) \otimes (a_2 + ib_2) = (a_1a_2 - b_1b_2) + i(a_1b_2 + a_2b_1).$$

Show that the set $\mathbb{C}$ with these operations forms a field.

Since the operations $\oplus$ and $\otimes$ are simply extensions of $+$ and $\times$ from $\mathbb{R}$ to $\mathbb{C}$, we denote them using $+$ and $\times$ as well.

3.25 (∗) Using Lemmas 3.2.2 and 3.3.3, find all the $x \in \mathbb{R}$ that satisfy each of the following inequalities:

(a) $2x + 5 > 0$.
(b) $x^2 - 6 < -x$.
(c) $2x^2 + 9x \geq 5$.
(d) $x^3 - x > 0$.
(e) $\frac{1}{x} < x^2$.
(f) $\sqrt{x}(x + 1) > 0$.

3.26 (∗) Let $a_1, a_2, \ldots, a_n, b_1, b_2, \ldots, b_n \in \mathbb{R}$. Prove the Cauchy-Schwarz inequality that says:

$$\left(\sum_{j=1}^{n} a_j b_j\right)^2 \leq \left(\sum_{j=1}^{n} a_j^2\right)\left(\sum_{j=1}^{n} b_j^2\right).$$

This inequality is named after Cauchy and Hermann Schwarz (1843–1921). Sometimes it is also called Cauchy–Bunyakovsky–Schwarz inequality to include Viktor Bunyakovsky (1802–1889) who proved the integral version of it (see Exercise 15.6).

3.27 (∗) In this question, we are going to prove the AM-GM inequality.

(a) Prove the arithmetic-geometric mean (AM-GM for short) inequality for two positive real numbers, namely for any real numbers $a, b \geq 0$, we have:

$$\frac{a+b}{2} \geq \sqrt{ab},$$

with equality if and only if $a = b$. The LHS is called the arithmetic mean and the RHS is called the geometric mean.

(b) The inequality above can be expanded to include n real numbers $a_1, a_2, \ldots, a_n \geq 0$ which produces:

$$\text{AM} = \frac{a_1 + a_2 + \ldots + a_n}{n} \geq \sqrt[n]{a_1 a_2 \ldots a_n} = \text{GM}.$$

Prove this statement by mathematical induction.

3.28 (◇) The AM-GM inequality can be extended to include two other means: the harmonic mean HM and the quadratic mean QM. For real numbers $a_1, a_2, \ldots, a_n > 0$, we define:

$$\text{HM} = \frac{n}{\sum_{j=1}^{n} \frac{1}{a_j}} \quad \text{and} \quad \text{QM} = \sqrt{\frac{\sum_{j=1}^{n} a_j^2}{n}}.$$

Show that for any collection of n positive real numbers, we have the ordering $0 < \text{HM} \leq \text{GM} \leq \text{AM} \leq \text{QM}$.

3.29 Consider the complex number field $\mathbb{C}$ from Exercise 3.24. For any two elements $z_1 = a_1 + ib_1$ and $z_2 = a_2 + ib_2$, define the lexicographic order $\prec$ on $\mathbb{C}$ as:

$$z_1 \prec z_2 \quad \text{iff either} \quad \begin{cases} a_1 < a_2, \text{ or} \\ a_1 = a_2 \text{ and } b_1 < b_2, \end{cases}$$

where $<$ is the usual strict total order on real numbers.

(a) Show that lexicographic ordering $\prec$ is a strict total order.

(b) Prove that the field $\mathbb{C}$ with the lexicographic order $\prec$ is not an ordered field.

(c) In general, prove via contradiction that any strict total order $\lhd$ at all on the field $\mathbb{C}$ cannot be compatible with the field structure.

3.30 ($\diamond$) Denote the set $\mathbb{Q}[\sqrt{2}] = \{a + b\sqrt{2} : a, b \in \mathbb{Q}\}$. Define addition and multiplication operations on this set as the restriction of the corresponding operations on $\mathbb{R}$.

(a) Show that for every $x, y \in \mathbb{Q}[\sqrt{2}]$, we have $x + y, x \times y \in \mathbb{Q}[\sqrt{2}]$.

(b) Show that for every $x \in \mathbb{Q}[\sqrt{2}]$ with $x \neq 0$, there exist $y, z \in \mathbb{Q}[\sqrt{2}]$ such that $xy = 1$ and $x + z = 0$.

(c) Conclude that $\mathbb{Q}[\sqrt{2}]$ with the algebraic operations $+$ and $\times$ is a field which is strictly contained in $\mathbb{R}$.

(d) Let $<$ be the usual strict total ordering on the field $\mathbb{Q}$. For $x, y \in \mathbb{Q}[\sqrt{2}]$ with $x = a_1 + b_1\sqrt{2}$ and $y = a_2 + b_2\sqrt{2}$, we define the lexicographic order $\prec$ on $\mathbb{Q}[\sqrt{2}]$ as follows:

$$x \prec y \quad \text{iff either} \quad \begin{cases} a_1 < a_2, \text{ or} \\ a_1 = a_2 \text{ and } b_1 < b_2. \end{cases}$$

Show that this is a strict total order on $\mathbb{Q}[\sqrt{2}]$.

(e) Prove that $\prec$ is not compatible with the field structure on $\mathbb{Q}[\sqrt{2}]$.

(f) What strict total order which is compatible with the given field structure can we define on $\mathbb{Q}[\sqrt{2}]$?

The set $\mathbb{Q}[\sqrt{2}]$ is an example of a field extension from the field $\mathbb{Q}$, which is studied in abstract algebra, particularly Galois theory [58] which is a field pioneered by Évariste Galois (1811–1832) before his untimely death in a duel.

3.31 ($\diamond$) Recall the integer classes modulo $r \in \mathbb{N}$ in Exercise 2.26 which we denoted as the set $\mathbb{Z}/r\mathbb{Z} = \{[0], [1], [2], \ldots, [r]\}$ with algebraic operations $\oplus$ and $\otimes$. We have shown that this is a ring. Suppose that r is a prime number.

(a) For each non-zero element $[a] \in \mathbb{Z}/r\mathbb{Z}$, show that there is an non-zero element $[b] \in \mathbb{Z}/r\mathbb{Z}$ such that $[a] \otimes [b] = [1]$.

(b) Hence, conclude that $\mathbb{Z}/r\mathbb{Z}$ is a field.

(c) Can we equip this field with a strict total order $\prec$ so that it forms an ordered field?

3.32 ($\diamond$) If $r \in \mathbb{N}$ is a composite number instead, would the integer classes modulo r, namely $\mathbb{Z}/r\mathbb{Z}$ be a field? Provide a proof or a counterexample.

Real Numbers 4

You know this boogie is for real.

— Jay Kay, musician

In Chap. 3, we have successfully defined and constructed the set of real numbers, which is a complete ordered number field, using the Dedekind cuts. So now we know such a number system exists. Figure 4.1 is a diagram which charts our quest on constructing the real numbers in the previous chapters.

Of course, we can extend the number system further according to what we want as long as we are careful and consistent with the construction. For example, in Exercises 3.21 and 3.24, we extended the real numbers to the complex numbers $\mathbb{C}$ if we wish that all real polynomials to have roots. Likewise, we can continue further with the construction of quaternios $\mathbb{H}$, octonions $\mathbb{O}$, and sedenions $\mathbb{S}$.

Also, there are other extensions of number systems such as the extended real numbers $\bar{\mathbb{R}}$ from $\mathbb{R}$ (see Definition 18.1.1), the p-adic numbers $\mathbb{Q}_p$ from $\mathbb{Q}$ (see Exercises 6.21, 6.22, and the end of Exercise 6.27), as well as the cardinal and ordinal numbers from $\mathbb{N}$ due to Cantor.

Examples of other exotic number system extensions that one can carry out include Clifford's dual numbers, tessarines, coquaternions, and biquaternions. The point that we would like to make here is number systems are human creations and products of imaginative minds. We can create them as we wish as long as they are well-defined and consistent.

Remark 4.0.1 A caveat of these extensions is that, even though we can get a system with some new desired properties, we may lose some other properties of the preceding number system. For example:

1. The extension $\mathbb{N}$ to $\mathbb{Z}$ gives us a ring structure, namely we introduce the inverse and identity of addition operation. But we lose the well-ordering principle.

S. Johar, *The Big Book of Real Analysis*,
https://doi.org/10.1007/978-3-031-30832-1_4

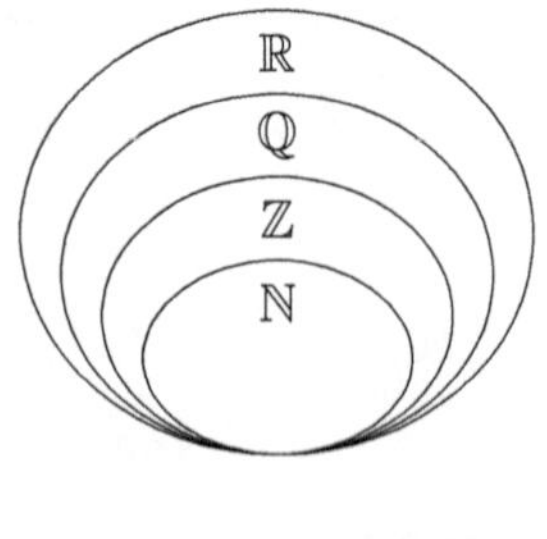

Fig. 4.1 Inclusion of the constructed number systems in Chaps. 2 and 3. Note that the algebraic operations $+$ and $\times$ and strict total order $<$ in $\mathbb{R}$ are consistent all the way down to the operations in $\mathbb{N}$

2. The extension $\mathbb{Q}$ to $\mathbb{R}$ gives us completeness. But we lose explicit representation for every element and countability which we shall see in Sect. 4.4.
3. The extension $\mathbb{R}$ to $\mathbb{C}$ allows us to fully factorise any real polynomials into linear factors (see Exercise 3.21). But we lose the ordered field structure (see Exercise 3.29).
4. $\mathbb{C}$ has a commutative multiplication operation, but $\mathbb{H}$ does not.
5. $\mathbb{H}$ has an associative multiplication operation, but $\mathbb{O}$ does not.

In this chapter, we shall focus on the set of real numbers $\mathbb{R}$ that we have constructed in Chap. 3 and study some of its properties, including how to represent its elements, and its cardinality. We shall also define new algebraic operations and special subsets of these numbers. Finally, we shall see some other extensions of these numbers, namely the complex numbers $\mathbb{C}$ and the real n-space $\mathbb{R}^n$.

Before we move on into the mathematics, we address a very common question that I get as a mathematician: Why are real numbers called real numbers? What makes them real? The answer is quite underwhelming actually. The term "real numbers" was coined by Descartes in the seventeenth century. In *La Géométrie* (The Geometry), he wrote:

> Moreover, the true roots as well as the false [roots] are not always real; but sometimes only imaginary [quantities]; that is to say, one can always imagine as many of them in each equation as I said; but there is sometimes no quantity that corresponds to what one imagines, just as although one can imagine three of them in this [equation], $x^3 - 6x^2 + 13x - 10 = 0$, only one of them however is real, which is 2, and regarding the other two, although one increase, or decrease, or multiply them ... one would not be able to make them other than imaginary [quantities].

Descartes came up with the term "real numbers" to distinguish these numbers from the "false" imaginary numbers that he dismissed as we have discussed in Exercise 3.24. Over time, the term real numbers stuck even though, as we have noted before, all number systems are as real (or as imaginary) as each other!

4.1 Properties of Real Numbers $\mathbb{R}$

Using Dedekind cuts, we know that the elements in the set $\mathbb{R}$ can be added and multiplied together and these operations coincide with what we know about the rational numbers. Furthermore, there is an ordering on the set $\mathbb{R}$, which again coincides with the ordering on $\mathbb{Q}$. Thus, we hope that we can extend the Archimedean property of rational numbers in Proposition 3.3.5 to the real numbers. Indeed we can!

Proposition 4.1.1 (Archimedean Property of Real Numbers) *For every positive real number $r \in \mathbb{R}_+$, there exists a natural number $\mathbb{N}$ such that $n > r$.*

Proof Suppose for contradiction that there exists a positive real number r such that no such $n \in \mathbb{N}$ exists. In other words, $r \geq n$ for every $n \in \mathbb{N}$. Since $\mathbb{N}$ is a non-empty subset of $\mathbb{R}$ and $r > n$ for every $n \in \mathbb{N}$, the set $\mathbb{N}$ is bounded from above in $\mathbb{R}$. By the completeness axiom of the real numbers, since the subset $\mathbb{N} \subseteq \mathbb{R}$ is bounded, it has a least upper bound $s = \sup(\mathbb{N}) \in \mathbb{R}$.

Furthermore for each $n \in \mathbb{N}$, clearly $n+1 \in \mathbb{N}$. Thus, $n+1 \leq s$ for every $n \in \mathbb{N}$ as well. However, this implies $n \leq s-1$ for every $n \in \mathbb{N}$. But this means s is not the least upper bound for the subset $\mathbb{N}$ since we have found a smaller upper bound for it, namely $s-1$. This is a contradiction. □

Similar to the rational numbers and Corollary 3.3.6, a consequence of Proposition 4.1.1 is the following:

Corollary 4.1.2 *For every positive real number $r \in \mathbb{R}_+$, there exists a natural number $n \in \mathbb{N}$ such that $0 < \frac{1}{n} < r$.*

Remark 4.1.3 Let us make some remarks regarding the Archimedean property of real numbers.

1. This fact sets the world of mathematics and the world of physics apart. In the physical world, space becomes indivisible at a scale of Planck's length, which is roughly 1.6×10^{-35}m. It is stated that the Planck's length is the shortest physically measurable distance, so any distance smaller than the Planck's length is physically meaningless.
2. In contrast, for mathematicians, we can keep going as small as we like by using the Archimedean property of the real numbers. In fact, Proposition 4.1.1 and Corollary 4.1.2 tell us that in the world of real numbers, things are allowed to be infinitely large or infinitely small, which can make some mathematical concepts go unintuitive for us in the physical world.
3. To quote Anaxagoras (c. 500B.C.–428B.C.):

> There is no smallest among the small and no largest among the large, but always something still smaller and something still larger.

Similar to Corollary 3.3.7, we can always find a real number in between any two distinct real numbers. More specifically, in between any two distinct real numbers, we can always find both a rational number and an irrational number.

Proposition 4.1.4 *Let $x, y \in \mathbb{R}$ be such that $x < y$. Then, we can find a rational number $r \in \mathbb{Q}$ and an irrational number $q \in \bar{\mathbb{Q}}$ such that $x < r < y$ and $x < q < y$.*

Proof WLOG, let us assume that $0 < x < y$. The cases for other signs of x and y can be easily adapted from the following proof. We find these numbers r and q separately.

1. We first find a rational number between x and y. Since $y - x > 0$, by the Archimedean property in Corollary 4.1.2, we can find a natural number $n \in \mathbb{N}$ such that $0 < \frac{1}{n} < y - x$. From this inequality, we obtain $nx + 1 < ny$.
By invoking the Archimedean property in Proposition 4.1.1, since $nx + 1 > 0$ there are natural numbers which are strictly bigger than $nx + 1$. Let A be the set of such integers, namely $A = \{z \in \mathbb{N} : nx + 1 < z\}$. Since $A \subseteq \mathbb{N}$, by the well-ordering principle, we can find the smallest element m in this set. Clearly $m \geq 2$. By minimality of m, the element m satisfies $m - 1 \leq nx + 1 < m$. Thus, $nx < m - 1 \leq nx + 1 < ny$ and hence we have $x < \frac{m-1}{n} < y$. So we have found a rational number $r = \frac{m-1}{n}$ between x and y.
2. To find an irrational number between x and y, note that since $0 < x < y$, we must have $0 < \frac{x}{\sqrt{2}} < \frac{y}{\sqrt{2}}$ and, by the previous part, there exists a $p \in \mathbb{Q}$ between these two numbers; that is $\frac{x}{\sqrt{2}} < p < \frac{y}{\sqrt{2}}$. Multiplying through with $\sqrt{2}$, we get $x < \sqrt{2}p < y$. We claim that $\sqrt{2}p$ is an irrational number that we are looking for. Indeed, if it is not, then $\sqrt{2}p = \frac{m}{n}$ for some natural numbers $m, n \in \mathbb{N}$. This implies $\sqrt{2} = \frac{m}{np} \in \mathbb{Q}$, which is a contradiction. Thus we have found an irrational number $q = \sqrt{2}p$ between x and y. □

Remark 4.1.5 In fact, we apply Proposition 4.1.4 inductively and conclude that in between any two distinct real numbers $x, y \in \mathbb{R}$, we can find infinitely many rational numbers and irrational numbers. These results are called the density of rational and irrational numbers in $\mathbb{R}$.

In the proof of Proposition 4.1.4, by a mix of Archimedean property and the well-ordering principle, for the real number $nx + 1 \in \mathbb{R}$, we can find an integer m such that $m - 1 \leq nx + 1 < m$. This is true in more generality; every real number must be located in between two consecutive integers. Thus we can define:

Definition 4.1.6 (Floor, Ceiling of a Number) Let $x \in \mathbb{R}$. We define the floor and ceiling of x as the integers:

$$\lfloor x \rfloor = \max\{n \in \mathbb{Z} : n \leq x\} \quad \text{and} \quad \lceil x \rceil = \min\{n \in \mathbb{Z} : n \geq x\}.$$

In words, for a real number $x \in \mathbb{R}$, the floor of x is the largest integer smaller than or equal to it and the ceiling of x is the smallest integer larger than or equal to it. Thus, we must have the inequalities $\lfloor x \rfloor \leq x < \lfloor x \rfloor + 1$ and $\lceil x \rceil - 1 < x \leq \lceil x \rceil$.

Supremum, Infimum, Minimum, Maximum

Let us look at an example on how we can utilise the Archimedean property and density of rational and irrational numbers in $\mathbb{R}$.

Example 4.1.7 We know that both of the subsets $X = \{x \in \mathbb{R} : x^2 \leq 2\}$ and $Y = \{x \in \mathbb{Q} : x^2 \leq 2\}$ of $\mathbb{R}$ are bounded from above. We also know that in the reals, by the completeness axiom, a set which is bounded from above has a supremum. Can we find the supremum of the sets X and Y?

1. For the set X, we claim that the supremum is $\sqrt{2} \in \mathbb{R}$. We check both the conditions for supremum.
   (a) First, for any $x \in X$, we must have $x^2 \leq 2$. Necessarily, $-\sqrt{2} \leq x \leq \sqrt{2}$ for all $x \in X$ and thus $\sqrt{2}$ is an upper bound for the set X.
   (b) To check that this is the least upper bound, we suppose for a contradiction that it is not. This means there must exist a smaller upper bound for the set X. We call this smaller upper bound $a \in \mathbb{R}$ so that $a < \sqrt{2}$. However, since $(\sqrt{2})^2 \leq 2$ we must have $\sqrt{2} \in X$ and so a is strictly smaller than at least one element of X (namely $\sqrt{2}$). This means a cannot be an upper bound of X, which is a contradiction.

   Thus $\sup(X) = \sqrt{2}$.
2. We can think of Y as either a subset of $\mathbb{Q}$ or $\mathbb{R}$. As we have seen in Example 3.6.9(2), the set Y does not have a supremum in $\mathbb{Q}$. However, the set Y does have a supremum in $\mathbb{R}$ by the completeness property. Moreover, the supremum is $\sqrt{2}$. Indeed:
   (a) We note that clearly $\sqrt{2}$ is an upper bound for Y by the same argument for X.
   (b) To show that this is the least upper bound, suppose for contradiction that there exists a smaller upper bound $a < \sqrt{2}$ for the set Y. By Proposition 4.1.4, there exists an $r \in \mathbb{Q}$ such that $a < r < \sqrt{2}$. Thus, by ordered field axioms, we have $a^2 < r^2 < 2$ and $r^2 \in \mathbb{Q}$, which implies $r \in Y$. Therefore, we have found an element $r \in Y$ such that $a < r$, which means a cannot be an upper bound for the set Y. This gives us the required contradiction.

   Thus, we conclude that $\sup(Y) = \sqrt{2}$.

Remark 4.1.8 If we think of the set Y as a subset of the ordered field $\mathbb{Q}$, we have seen that the supremum does not exist. We have seen a proof of this in Example 3.6.9 which involves picking a special number δ that would allow us to achieve our goal. Here is a shorter proof of this using Proposition 4.1.4. Suppose for a contradiction that $\sup(Y)$ exists in $\mathbb{Q}$ and $\sup(Y) = a$. Then, necessarily $a > \sqrt{2}$ or $a < \sqrt{2}$.

1. If $a > \sqrt{2}$, by Proposition 4.1.4, there exists a rational number $r \in \mathbb{Q}$ such that $\sqrt{2} < r < a$. Thus, for all $x \in Y$, we have $x^2 \leq 2 < r^2 < a^2$. Hence, r is a smaller upper bound for the set Y, which is a contradiction.

2. If $a < \sqrt{2}$, by Proposition 4.1.4, there exists a rational number $r \in \mathbb{Q}$ such that $a < r < \sqrt{2}$. But this implies $a^2 < r^2 < 2$. Since $r \in \mathbb{Q}$, we must have $r \in Y$ and therefore a cannot be an upper bound for the set Y, which is another contradiction.

In general, the above argument and Proposition 4.1.4 is a useful way of proving that a number is the supremum of a set: we usually claim a number $a = \sup(X)$ and show by contradiction that $\sup(X)$ cannot be smaller or greater than a. This also gives us a very important characterisation:

Proposition 4.1.9 (Characterisation of Supremum and Infimum) *Let $X \subseteq \mathbb{R}$ be a non-empty subset of the real numbers.*

1. *If X is bounded from above and $a = \sup(X)$, then for every $\varepsilon > 0$ there exists an element $x_\varepsilon \in X$ such that $a - \varepsilon < x_\varepsilon \leq a$.*
2. *If X is bounded from below and $b = \inf(X)$, then for every $\varepsilon > 0$ there exists an element $x_\varepsilon \in X$ such that $b \leq x_\varepsilon < b + \varepsilon$.*

Proof We only prove the first assertion since the second one can be proven in a similar manner.

1. Fix any $\varepsilon > 0$. Clearly $x \leq a$ for all $x \in X$. We now aim to show that there exists some $x_\varepsilon \in X$ bigger than $a - \varepsilon$ for this choice of ε.
   For a contradiction, suppose that there are no element of X bigger than $a - \varepsilon$. This means $x \leq a - \varepsilon$ for all $x \in X$. In other words, $a - \varepsilon$ is an upper bound for the set X. Hence we have found an upper bound for X strictly smaller than $a = \sup(X)$, a contradiction. Thus, we conclude that there must be some $x_\varepsilon \in X$ such that $a - \varepsilon < x_\varepsilon \leq a$. □

An intuition of the characterisation above is that: if we stand at the supremum of the set X, say $a = \sup(X)$, moving any small distance $\varepsilon > 0$ to the left (downwards along the real number line) would result in us passing by at least one element of the set X (possibly a, if $a \in X$ to begin with).

We also have a lemma that tells us how supremum and infimum behave under arithmetic operations. Extending the set operations in Definition 3.8.1 to real sets, we have:

Proposition 4.1.10 (Properties of Supremum and Infimum) *Let $X, Y \subseteq \mathbb{R}$ be non-empty bounded subsets of the real numbers.*

1. *We have* $\inf(-X) = -\sup(X)$.
2. *If $Y \subseteq X$, then* $\sup(Y) \leq \sup(X)$ *and* $\inf(Y) \geq \inf(X)$.
3. *The set $X \cap Y$ is also bounded. If $X \cap Y \neq \varnothing$, we have the ordering* $\sup(X \cap Y) \leq \min(\sup(X), \sup(Y))$ *and* $\inf(X \cap Y) \geq \max(\inf(X), \inf(Y))$.

4. *The set* $X \cup Y$ *is also bounded with* $\sup(X \cup Y) = \max(\sup(X), \sup(Y))$ *and* $\inf(X \cup Y) = \min(\inf(X), \inf(Y))$.
5. *If* $\lambda \in \mathbb{R}$ *is a real constant, then:*

$$\sup(\lambda X) = \lambda \sup(X) \quad \text{and} \quad \inf(\lambda X) = \lambda \inf(X) \quad \text{if } \lambda > 0,$$
$$\sup(\lambda X) = \lambda \inf(X) \quad \text{and} \quad \inf(\lambda X) = \lambda \sup(X) \quad \text{if } \lambda < 0.$$

6. *The set* $X + Y$ *is also bounded with* $\sup(X + Y) = \sup(X) + \sup(Y)$ *and* $\inf(X + Y) = \inf(X) + \inf(Y)$.
7. *If all the elements of* X *and* Y *are non-negative, then the set* XY *is also bounded with* $\sup(XY) = \sup(X)\sup(Y)$ *and* $\inf(XY) = \inf(X)\inf(Y)$.
8. *If all the elements of* X *are positive, we define the reciprocal set* $\frac{1}{X} = \{\frac{1}{x} : x \in X\}$. *If* $\inf(X) > 0$ *then the set* $\frac{1}{X}$ *is bounded with* $\sup\left(\frac{1}{X}\right) = \frac{1}{\inf(X)}$ *and* $\inf\left(\frac{1}{X}\right) = \frac{1}{\sup(X)}$.

Proof We shall only prove assertions 2, 3, 7, and 8. Assertion 1 was proven in Lemma 3.6.8. The rest are left as Exercise 4.11.

2. Let $\sup(X) = a$. For all $x \in X$, we have $x \leq a$. Thus, for any $y \in Y \subseteq X$, we must also have $y \leq a$. So a is an upper bound for the set Y and hence $\sup(Y) \leq a = \sup(X)$. Similar argument can be used to show $\inf(X) \leq \inf(Y)$.
3. Since X and Y are bounded, there exist some $M, N > 0$ such that each element $x \in X$ satisfies $x \leq M$ and each $y \in Y$ satisfies $y \leq N$. Thus, each element in $X \cap Y$ is bounded from above by both M and N.
   Assuming that the intersection is non-empty, by completeness of $\mathbb{R}$, the quantity $\sup(X \cap Y)$ exists. We note that $X \cap Y \subseteq X$ and $X \cap Y \subseteq Y$. Using the first assertion, we have $\sup(X \cap Y) \leq \sup(X)$ and $\sup(X \cap Y) \leq \sup(Y)$ which implies $\sup(X \cap Y) \leq \min(\sup(X), \sup(Y))$. The other inequality for infimum is similarly done.
7. Since each set is bounded and non-negative, then there exist some $M, N > 0$ such that for any $x \in X$ and $y \in Y$, we have $0 \leq x \leq M$ and $0 \leq y \leq N$. Then, $0 \leq xy \leq MN$ for all $x \in X$ and $y \in Y$ so the set XY is also bounded. This means $\sup(XY)$ exists.
   Now we show that the supremum of the product is the product of the supremum. If either X or Y is the singleton set $\{0\}$ so that its supremum is also 0, then the equalityclearly holds. Now suppose that both X and Y are not the singleton set $\{0\}$ so that they contain some positive real numbers and thus $\sup(X), \sup(Y) > 0$. Pick any element $z \in XY$. Then, $z = xy$ for some $x \in X$ and $y \in Y$. Since $0 \leq x \leq \sup(X)$ and $0 \leq y \leq \sup(Y)$, we then have $0 \leq z = xy \leq \sup(X)\sup(Y)$. Since $z \in XY$ is arbitrary, we have shown that $\sup(X)\sup(Y)$ is an upper bound for the set XY and thus $\sup(XY) \leq \sup(X)\sup(Y)$.
   For the opposite inequality, $\sup(XY)$ is an upper bound of the set XY. This means for any arbitrarily fixed nonzero $y \in Y$, for any $x \in X$ we have $xy \in XY$ and thus

$xy \leq \sup(XY)$. Therefore, $x \leq \frac{\sup(XY)}{y}$ for all $x \in X$. Thus, the number $\frac{\sup(XY)}{y}$ is an upper bound for the set X which then means $\sup(X) \leq \frac{\sup(XY)}{y}$. Likewise, since $y \in Y \setminus \{0\}$ was arbitrarily fixed, we have $y \leq \frac{\sup(XY)}{\sup(X)}$ which means the number $\frac{\sup(XY)}{\sup(X)}$ is an upper bound for the set Y. Thus, we have $\sup(Y) \leq \frac{\sup(XY)}{\sup(X)}$ which then implies $\sup(X)\sup(Y) \leq \sup(XY)$.
We conclude that $\sup(X)\sup(Y) = \sup(XY)$. The equality for infimum can also be proven in a similar way.
8. If $\inf(X) > 0$, then $x \geq \inf(X) > 0$ for all $x \in X$. Therefore, we have $0 < \frac{1}{x} \leq \frac{1}{\inf(X)}$ for each $x \in X$. This means the set $\frac{1}{X} = \{\frac{1}{x} : x \in X\}$ is bounded from above and hence its supremum exists and is non-zero.
Now let us show that $\sup\left(\frac{1}{X}\right) = \frac{1}{\inf(X)}$. Clearly, for any $y \in \frac{1}{X}$, we have $\frac{1}{y} \in X$. This means $\frac{1}{y} \geq \inf(X)$. Therefore, we must have $\frac{1}{\inf(X)} \geq y$ for every $y \in \frac{1}{X}$, implying that $\frac{1}{\inf(X)}$ is an upper bound for the set $\frac{1}{X}$. Hence, we have $\sup\left(\frac{1}{X}\right) \leq \frac{1}{\inf(X)}$.
For the reverse inequality, $\sup\left(\frac{1}{X}\right)$ is an upper bound for the set $\frac{1}{X}$. Then, for any $x \in X$ we have $\frac{1}{x} \in \frac{1}{X}$ and so $\frac{1}{x} \leq \sup\left(\frac{1}{X}\right)$. This implies $\frac{1}{\sup(\frac{1}{X})} \leq x$ for any $x \in X$. So $\frac{1}{\sup(\frac{1}{X})}$ is a lower bound for the set X which means $\frac{1}{\sup(\frac{1}{X})} \leq \inf(X)$.
By algebra, we deduce $\frac{1}{\inf(X)} \leq \sup\left(\frac{1}{X}\right)$.
Hence, $\sup\left(\frac{1}{X}\right) = \frac{1}{\inf(X)}$. The other equality can be proven in a similar way. □

In Example 4.1.7, we saw that $\sup(X) = \sqrt{2} \in X$ whereas $\sup(Y) = \sqrt{2} \notin Y$. As we have seen in Definition 3.6.1, we call the former a maximum of the set X. This happens when the supremum is attained by some element in the set.

The difference between the supremum and the maximum of a subset $X \subseteq \mathbb{R}$ is that the supremum may be an element inside or outside the set, but a maximum of X must be an element inside X. From Proposition 3.6.7, the supremum of a set X is a maximum of the set X if and only if it lies in X. In this case, we have $\sup(X) = \max(X)$. Similar holds true for infimum and minimum.

Using this correspondence, an extension of Proposition 4.1.10(1) is the following:

Lemma 4.1.11 *If* $X \subseteq \mathbb{R}$ *is a subset of the real numbers with* $\max(X) = a$, *then* $\min(-X)$ *exists and* $\min(-X) = -\max(X)$.

Another important note is that a finite collection of real numbers must attain its maximum and minimum. The proof of the following result is similar to Lemma 2.3.9 via induction.

Lemma 4.1.12 *Let* $X \subseteq \mathbb{R}$ *be a finite subset of* $\mathbb{R}$. *Then, there exist elements* $a, b \in X$ *such that* $\max(X) = a$ *and* $\min(X) = b$.

Now we state an important lemma that we need to use later. This lemma allows us to switch the order of two supremum when we are varying over two parameters:

Lemma 4.1.13 (Iterated Supremum) *Let $\{f(x, y) : x \in X, y \in Y\}$ be a bounded set of real numbers parametrised by two parameters in X and Y. Then:*

$$\begin{aligned}\sup\{f(x, y) : x \in X, y \in Y\} &= \sup\{\sup\{f(x, y) : y \in Y\} : x \in X\}\\ &= \sup\{\sup\{f(x, y) : x \in X\} : y \in Y\}.\end{aligned}$$

The proof of this is left as Exercise 4.5. We can view $\sup\{\sup\{f(x, y) : y \in Y\} : x \in X\}$ as: for each fixed $x \in X$, we find the supremum of $f(x, y)$ over the set of parameters Y. Then, for this set of supremum (which now depends only on the parameter $x \in X$), we find its supremum over all $x \in X$. Thanks to Lemma 4.1.13, because there is no ambiguity on which supremum we carry out first, we can also write this as:

$$\sup_{x\in X, y\in Y} f(x, y) = \sup_{x\in X}(\sup_{y\in Y} f(x, y)) = \sup_{y\in Y}(\sup_{x\in X} f(x, y)),$$

which is a more common notation if we view f as a real-valued function on the set $X \times Y$. A similar result for infimum can also be obtained. However, switching the order of infimum and supremum does not guarantee equality, which the readers shall prove in Exercise 4.6 later.

4.2 Exponentiation

We have seen how we can satisfactorily define addition and multiplication of real numbers, along with their inverses, using Dedekind cuts. Now we are going to define another algebraic operation on the set of real numbers, which is called exponentiation. Exponentiation involves two quantities, namely the base a and the exponent x. If $x \in \mathbb{N}$, we have seen that exponentiation with exponent x is multiplying the number a with itself x times, namely $a^x = \underbrace{a \times a \times a \times \ldots \times a}_{x \text{ times}}$.

Since x is a natural number, this operation can be defined for any $a \in \mathbb{R}$ by carrying out the multiplication operations finitely many times inductively. Exponentiation for negative integers x can also be defined as $a^x = \frac{1}{a^{-x}}$ which we know how to evaluate since $-x \in \mathbb{N}$.

Thus, for any $x, y \in \mathbb{Z} \setminus \{0\}$ we have the following rules, which are commonly known as the law of indices:

Proposition 4.2.1 (Law of Indices) *Let $a \in \mathbb{R} \setminus \{0\}$ and $x, y \in \mathbb{Z} \setminus \{0\}$.*

1. $a^{x+y} = a^x a^y$.
2. $a^{xy} = (a^x)^y$.

3. If $x < y$, then $\begin{cases} a^x < a^y & \text{if } a > 1, \\ a^x > a^y & \text{if } 0 < a < 1. \end{cases}$

To complete the definition for zero exponent, we want to ensure that the law of indices remain consistent. We note that for any $a \in \mathbb{R} \setminus \{0\}$, we have $1 = a^{-x}a^x = a^{-x+x} = a^0$. So we set $a^0 = 1$ for any $a \in \mathbb{R} \setminus \{0\}$ to be consistent with the identities above. However, the value of 0^0 is left undefined. Note that for $a > 0$, we have $a^x > 0$ for all $x \in \mathbb{N}$. On the other hand, if $a < 0$, then $a^x > 0$ if x is even and $a^x < 0$ if x is odd.

Rational Exponents

Now we want to extend this exponentiation operation to non-integer exponents. Let us first define what it means to have positive rational exponents. Let $a \in \mathbb{R}$ and $p \in \mathbb{N}$. We want to first define $a^{\frac{1}{p}}$ satisfactorily. Using Proposition 4.2.1 as a guide, if we want the law of indices to be consistent for rational exponents, we expect that $(a^{\frac{1}{p}})^p = a^{\frac{1}{p}+\frac{1}{p}+\dots+\frac{1}{p}} = a^{\frac{p}{p}} = a^1 = a$, so $a^{\frac{1}{p}}$ must be a real number such that when it is multiplied with itself p times, the result is a.

We call $a^{\frac{1}{p}}$ the p-th root of a, if there is such a number. However, we cannot exactly guarantee that such a real number always exist. For example, if a is negative and p is even, this number would not exist. Indeed, any even powered number would be non-negative and thus $0 \leq (a^{\frac{1}{p}})^p = a < 0$, a contradiction.

On the other hand, for non-negative base a, we can show that such a number exists and is unique. If $a = 0$, then clearly this number is 0. Otherwise, if $a > 0$, the existence and uniqueness of $a^{\frac{1}{p}}$ can be shown using Dedekind cuts in Exercise 3.17: if the Dedekind cut of a is $M \succ L_0$, we have shown that this unique number $a^{\frac{1}{p}}$ corresponds to the Dedekind cut $L = \{x \in \mathbb{Q} : x \geq 0, x^p \in M\} \cup \{x \in \mathbb{Q} : x < 0\}$. Moreover, this Dedekind cut satisfies $L \succ L_0$, namely $a^{\frac{1}{p}} > 0$.

Similarly, we can define the p-th root of a negative number, but only for odd integers p. This is simply $a^{\frac{1}{p}} = -(-a)^{\frac{1}{p}}$. To generalise to other positive rational exponents $x = \frac{q}{p} \in \mathbb{Q}_+$, we define:

$$a^{\frac{q}{p}} = \begin{cases} (a^{\frac{1}{p}})^q = (a^q)^{\frac{1}{p}} & \text{if } a \geq 0, \\ (-(-a)^{\frac{1}{p}})^q & \text{if } a < 0, p \text{ is odd.} \end{cases}$$

Note that in the definition, for $a \geq 0$, we have two possible definitions for $a^{\frac{q}{p}}$, namely $(a^{\frac{1}{p}})^q$ or $(a^q)^{\frac{1}{p}}$. They are actually the same regardless which exponentiation operation we carry out first. Indeed, if we set $a^{\frac{1}{p}} = b$ and use Proposition 4.2.1 as well as the definition of p-th root, we have:

$$(a^{\frac{1}{p}})^q = b^q = ((b^q)^p)^{\frac{1}{p}} = ((b^p)^q)^{\frac{1}{p}} = (((a^{\frac{1}{p}})^p)^q)^{\frac{1}{p}} = (a^q)^{\frac{1}{p}},$$

so there is no ambiguity in which exponent we apply first. Negative rational exponents $x \in \mathbb{Q}_-$ can then be treated using reciprocals via $a^x = \frac{1}{a^{-x}}$.

Because negative bases requires some restriction on the exponent in order for them to be defined, from now on let us just focus on positive bases. From the definition above, clearly we have $a^x > 0$ for any $x = \frac{p}{q} \in \mathbb{Q}$. Indeed, by Exercise 3.17, we have $a^{\frac{1}{p}} > 0$ and so $a^x = (a^{\frac{1}{p}})^q > 0$. We also have $1^x = 1$ for all $x \in \mathbb{Q}$. Moreover, we can extend the identities for integer exponents in Proposition 4.2.1 to derive following results:

Proposition 4.2.2 *Let $a, b > 0$ and $x, y \in \mathbb{Q}$.*

1. $a^{x+y} = a^x a^y$.
2. $a^{xy} = (a^x)^y$.
3. *If* $x < y$, *then* $\begin{cases} a^x < a^y & \text{if } a > 1, \\ a^x > a^y & \text{if } 0 < a < 1. \end{cases}$
4. *If* $x < 0 < y$, *then* $\begin{cases} a^x < a^0 = 1 < a^y & \text{if } a > 1, \\ a^x > a^0 = 1 > a^y & \text{if } 0 < a < 1. \end{cases}$
5. $a^x b^x = (ab)^x$.
6. *If* $0 < a < b$, *then* $\begin{cases} a^x < b^x & \text{if } x > 0, \\ a^x > b^x & \text{if } x < 0. \end{cases}$

Proof Many of these proofs hinge on the fact that they are true for integer exponents in Proposition 4.2.1.

1. Suppose that $x = \frac{p}{q}$ and $y = \frac{r}{s}$ with $q, s > 0$. Using the definition for rational exponent and Proposition 4.2.1, we have:

$$a^{x+y} = a^{\frac{p}{q}+\frac{r}{s}} = a^{\frac{ps+rq}{qs}} = (a^{\frac{1}{qs}})^{ps+rq} = (a^{\frac{1}{qs}})^{ps}(a^{\frac{1}{qs}})^{rq} = a^{\frac{ps}{qs}}a^{\frac{rq}{qs}} = a^{\frac{p}{q}}a^{\frac{r}{s}}$$
$$= a^x a^y.$$

2. Suppose that $x = \frac{p}{q}$ and $y = \frac{r}{s}$ with $q, s > 0$. First, let us show that $a^{\frac{1}{qs}} = (a^{\frac{1}{q}})^{\frac{1}{s}}$. Set $b = (a^{\frac{1}{q}})^{\frac{1}{s}}$. Then, b is the unique positive number such that $b^s = a^{\frac{1}{q}}$. Setting $c = b^s = a^{\frac{1}{q}}$, then c is the unique positive number such that $c^q = a$. This means $a = c^q = (b^s)^q = b^{qs}$ so b is the unique positive number such that $b^{qs} = a$.

On the other hand, clearly $(a^{\frac{1}{qs}})^{qs} = a$. So, by uniqueness, we must have $a^{\frac{1}{qs}} = b = (a^{\frac{1}{q}})^{\frac{1}{s}}$.

For the general case, using the fact above, the definition of rational exponents, and the law of indices, we have:

$$a^{xy} = a^{\frac{pr}{qs}} = (a^{\frac{1}{qs}})^{pr} = ((a^{\frac{1}{q}})^{\frac{1}{s}})^{pr} = ((a^{\frac{1}{q}})^{pr})^{\frac{1}{s}} = (((a^{\frac{1}{q}})^{p})^{r})^{\frac{1}{s}} = (a^{\frac{p}{q}})^{\frac{r}{s}} = (a^{x})^{y}.$$

3. For $a > 1$, suppose for contradiction that we have the opposite inequality, namely $a^x \geq a^y$. Writing $x = \frac{p}{q}$ and $y = \frac{r}{s}$ as ratios with $q, s > 0$, since $x < y$ we must have $ps < qr$ as integers. Thus, Proposition 4.2.1(3) implies $a^{ps} < a^{qr}$. However, by assumption and Proposition 4.2.1(3), we also have:
$$a^x \geq a^y \quad \Rightarrow \quad a^{\frac{p}{q}} \geq a^{\frac{r}{s}} \quad \Rightarrow \quad (a^{\frac{p}{q}})^{qs} \geq (a^{\frac{r}{s}})^{qs} \quad \Rightarrow \quad a^{ps} \geq a^{qr},$$
which is a contradiction. The proof for $0 < a < 1$ is also similarly done.
4. This is a direct consequence of the previous assertion.
5. If $x \in \mathbb{Z}$, this is true using the repeated multiplication definition for integer exponents. Now suppose that $x = \frac{p}{q}$ where $p, q \in \mathbb{Z}$ and $q > 0$. By definition, $a^{\frac{p}{q}} = (a^{\frac{1}{q}})^p$. So, by commutativity of multiplication of real numbers, it is enough to show the case for $p = 1$. By commutativity and associativity of multiplication, we have $(a^{\frac{1}{q}}b^{\frac{1}{q}})^q = \underbrace{(a^{\frac{1}{q}}b^{\frac{1}{q}}) \times (a^{\frac{1}{q}}b^{\frac{1}{q}}) \times \ldots \times (a^{\frac{1}{q}}b^{\frac{1}{q}})}_{q \text{ times}} = a^{\frac{q}{q}}b^{\frac{q}{q}} = ab$. Thus, $a^{\frac{1}{q}}b^{\frac{1}{q}}$ is the unique positive number such that its q-th power is ab.
On the other hand, we also have $((ab)^{\frac{1}{q}})^q = ab$ so $(ab)^{\frac{1}{q}}$ is also the unique positive number such that its q-th power is ab. Since this number is unique, they must be equal, namely $a^{\frac{1}{q}}b^{\frac{1}{q}} = (ab)^{\frac{1}{q}}$.
6. For the case $x > 0$, assume for contradiction that $a^x \geq b^x$. Since $a^x > 0$, we also have $a^{-x} > 0$. Thus:
$$a^x \geq b^x \quad \Rightarrow \quad a^{-x}a^x \geq a^{-x}b^x \quad \Rightarrow \quad 1 \geq \left(\frac{b}{a}\right)^x.$$
On the other hand, by assertion 4, since $\frac{b}{a} > 1$ and $x > 0$ we have $\left(\frac{b}{a}\right)^x > 1$, giving us a contradiction. So we must have $a^x < b^x$. The proof for $x < 0$ can be done similarly using reciprocals. □

Irrational Exponents

Now we arrive at a difficult question: how do we define the exponentiation for irrational exponents? The main difficulty with irrational numbers is that, unlike the rational numbers, we do not have explicit representations for them. Therefore, we either have to work using the Dedekind cuts or the completeness property of the real numbers. Again, we focus our attention only on the positive bases.

We have seen earlier in Proposition 3.8.5 that any irrational number x can be expressed as the supremum of the set of rational numbers smaller than it, namely $x = \sup\{p \in \mathbb{Q} : p \leq x\}$. Let us use this characterisation to build the exponential with irrational exponents.

1. The trivial case is when $a = 1$. Since $1^p = 1$ for all $p \in \mathbb{Q}$, for any $x \in \mathbb{R}$ we must have $1^x = \sup\{1^p : p \in \mathbb{Q}, p \leq x\} = \sup\{1 : p \in \mathbb{Q}, p \leq x\} = 1$.
2. For $a > 1$, we have shown in Proposition 4.2.2(3) that a^p is increasing with rational p, namely if $p, q \in \mathbb{Q}$ are such that $p < q$, then $a^p < a^q$. Therefore, let us define:
$$a^x = \sup\{a^p : p \in \mathbb{Q}, p \leq x\}. \tag{4.1}$$
This is a well defined number since the set $\{a^p : p \in \mathbb{Q}, p \leq x\}$ is bounded from above by the number $a^{\lceil x \rceil}$ and so its supremum must exist in $\mathbb{R}$.
3. Likewise, if $0 < a < 1$, Proposition 4.2.2(3) says that a^p is decreasing with rational p, so we define:
$$a^x = \inf\{a^p : p \in \mathbb{Q}, p \leq x\}. \tag{4.2}$$

Remark 4.2.3 In fact, the definition above coincides with the whole definitions that we have discussed for integer and rational exponents and positive base. Indeed, when x is an integer or a rational number, the supremum and infimum in the definitions (4.1) and (4.2) above are attained by a^x.

Thus, we have the following general definition:

Definition 4.2.4 (Exponentiation) Let $a, x \in \mathbb{R}$ with $a > 0$. We define:

$$a^x = \begin{cases} \sup\{a^p : p \in \mathbb{Q}, p \leq x\} & \text{if } a > 1, \\ 1 & \text{if } a = 1, \\ \inf\{a^p : p \in \mathbb{Q}, p \leq x\} & \text{if } 0 < a < 1. \end{cases}$$

Moreover, the readers shall prove the following lemma in Exercise 4.13 later:

Lemma 4.2.5 *Let $x \in \mathbb{R}$. Then:*

1. *For $a > 1$, we have $\sup\{a^p : p \in \mathbb{Q}, p \leq x\} = \inf\{a^r : r \in \mathbb{Q}, r \geq x\}$.*
2. *For $0 < a < 1$, we have $\inf\{a^p : p \in \mathbb{Q}, p \leq x\} = \sup\{a^r : r \in \mathbb{Q}, r \geq x\}$.*

As a result of Lemma 4.2.5, the exponentiation in Definition 4.2.4 can also be restated equivalently as:

$$a^x = \begin{cases} \inf\{a^r : r \in \mathbb{Q}, r \geq x\} & \text{if } a > 1, \\ 1 & \text{if } a = 1, \\ \sup\{a^r : r \in \mathbb{Q}, r \geq x\} & \text{if } 0 < a < 1. \end{cases}$$

We now show that this definition also has the same law of indices as the integer and rational exponents.

Proposition 4.2.6 *Let $a, x, y \in \mathbb{R}$ with $a > 0$.*

1. *$a^{-x} = \frac{1}{a^x} = \left(\frac{1}{a}\right)^x$.*
2. *If $x < y$, then $\begin{cases} a^x < a^y & \text{if } a > 1, \\ a^x > a^y & \text{if } 0 < a < 1. \end{cases}$*
3. *Suppose that $a \neq 1$. Then, $a^x = 1$ if and only if $x = 0$.*
4. *$a^{x+y} = a^x a^y$.*
5. *$a^{xy} = (a^x)^y$.*

Proof We prove the assertions for the case $a > 1$ only. The case of $0 < a < 1$ can be treated in the same manner using the appropriate definition of exponent in Definition 4.2.4.

1. We prove the first equality only. Using Proposition 4.1.10 and Lemma 4.2.5, we have:

$$\begin{aligned} \frac{1}{a^x} = \frac{1}{\sup\{a^p : p \in \mathbb{Q}, p \leq x\}} &= \inf\{a^{-p} : p \in \mathbb{Q}, p \leq x\} \\ &= \inf\{a^r : -r \in \mathbb{Q}, -r \leq x\} \\ &= \inf\{a^r : -r \in \mathbb{Q}, r \geq -x\} \\ &= \inf\{a^r : r \in \mathbb{Q}, r \geq -x\} = a^{-x}. \end{aligned}$$

The other equality can be proven in the same way.

2. For any $x, y \in \mathbb{R}$ with $x < y$, by Proposition 4.1.4, we can find distinct $r, s \in \mathbb{Q}$ such that $x < r < s < y$. Using the definition of exponents in Definition 4.2.4, we have:

$$a^x = \sup\{a^p : p \in \mathbb{Q}, p \leq x\} \leq a^r < a^s \leq \sup\{a^p : p \in \mathbb{Q}, p \leq x\} = a^y.$$

3. We already know that $a^0 = 1$. Moreover, by previous assertion, for $x > 0$ we have $a^x > a^0 = 1$ and for $x < 0$ we have $a^x < a^0 = 1$. Thus, there are no other $x \in \mathbb{R}$ for which $a^x = 1$ apart from $x = 0$.
4. We want to show $a^x a^y = a^{x+y}$, namely:

$$\sup\{a^p : p \in \mathbb{Q}, p \leq x\} \sup\{a^q : q \in \mathbb{Q}, q \leq y\} = \sup\{a^r : r \in \mathbb{Q}, r \leq x + y\}.$$

Luckily, this looks like Proposition 4.1.10(7), so let us work towards that. Clearly the sets $\{a^p : p \in \mathbb{Q}, p \leq x\}$ and $\{a^q : q \in \mathbb{Q}, q \leq y\}$ are bounded and consist of non-negative numbers. The product of the sets is:

$$\{a^p : p \in \mathbb{Q}, p \leq x\}\{a^q : q \in \mathbb{Q}, q \leq y\} = \{a^p a^q : p, q \in \mathbb{Q}, p \leq x, q \leq y\}.$$

We now simplify this set by introducing $r = p + q$.

$$\begin{aligned}
&\{a^p : p \in \mathbb{Q}, p \leq x\}\{a^q : q \in \mathbb{Q}, q \leq y\} \\
&= \{a^{p+q} : p, q, r \in \mathbb{Q}, p \leq x, q \leq y, r = p + q\} \\
&= \{a^r : p, r \in \mathbb{Q}, p \leq x, r - p \leq y\} \\
&= \{a^r : p, r \in \mathbb{Q}, p \leq x, r \leq y + p\} \\
&= \{a^r : r \in \mathbb{Q}, r \leq x + y\}.
\end{aligned}$$

Thus, applying Proposition 4.1.10 gives us the desired equality.

5. This identity is clearly true if at least one of x or y is 0. Let us prove the other cases:
   (a) First, assume that $x, y > 0$. We want to show $(a^x)^y = a^{xy}$. Let us start with the LHS. Since $x > 0$, we have $a^x > a^0 = 1$. Using the appropriate definition for exponentiation, we have:

$$\begin{aligned}
(a^x)^y &= \sup\{(a^x)^q : q \in \mathbb{Q}, q \leq y\} \\
&= \sup\{(a^x)^q : q \in \mathbb{Q}, 0 < q \leq y\},
\end{aligned} \tag{4.3}$$

   where the latter is obtained by removing the non-positive rational exponents q since they do not contribute to the supremum.

Moreover, since $a > 1$, by the same reasoning, we also have:

$$a^x = \sup\{a^p : p \in \mathbb{Q}, p \leq x\} = \sup\{a^p : p \in \mathbb{Q}, 0 < p \leq x\}, \tag{4.4}$$

Substituting (4.4) into (4.3) we get:

$$(a^x)^y = \sup\{\sup\{a^p : p \in \mathbb{Q}, 0 < p \leq x\}^q : q \in \mathbb{Q}, 0 < q \leq y\}.$$

In Exercise 4.9, the readers will prove that for any $q > 0$ we have $\sup\{a^p : p \in \mathbb{Q}, 0 < p \leq x\}^q = \sup\{a^{pq} : p \in \mathbb{Q}, 0 < p \leq x\}$. Using this fact and the iterated supremum in Lemma 4.1.13, we have:

$$\begin{aligned}(a^x)^y &= \sup\{\sup\{a^p : p \in \mathbb{Q}, 0 < p \leq x\}^q : q \in \mathbb{Q}, 0 < q \leq y\} \\ &= \sup\{\sup\{a^{pq} : p \in \mathbb{Q}, 0 < p \leq x\} : q \in \mathbb{Q}, 0 < q \leq y\} \\ &= \sup\{a^{pq} : p, q \in \mathbb{Q}, 0 < p \leq x, 0 < q \leq y\}.\end{aligned}$$

To simplify this, we set $r = pq$ and hence:

$$\begin{aligned}(a^x)^y &= \sup\{a^{pq} : p, q \in \mathbb{Q}, 0 < p \leq x, 0 < q \leq y\}. \\ &= \sup\{a^{pq} : p, q, r \in \mathbb{Q}, 0 < p \leq x, 0 < q \leq y, r = pq\} \\ &= \sup\left\{a^r : p, r \in \mathbb{Q}, 0 < p \leq x, 0 < \frac{r}{p} \leq y\right\}, \\ &= \sup\{a^r : p, r \in \mathbb{Q}, 0 < p \leq x, 0 < r \leq py\}, \\ &= \sup\{a^r : r \in \mathbb{Q}, 0 < r \leq xy\} \\ &= \sup\{a^r : r \in \mathbb{Q}, r \leq xy\} = a^{xy},\end{aligned}$$

where including the non-positive exponents r in the final line does not change the supremum.

(b) Now suppose that $x > 0$ and $y < 0$. Then, there exists a $z > 0$ such that $y = -z$. Using the case for positive exponents proven above, we have $(a^x)^y = (a^x)^{-z} = \frac{1}{(a^x)^z} = \frac{1}{a^{xz}} = a^{-xz} = a^{xy}$.

(c) Next, suppose that $x < 0$ and $y > 0$. Then, there exists a $w > 0$ such that $x = -w$. By the case of positive exponents above, we have $a^{xy} = a^{-wy} = \frac{1}{a^{wy}} = \frac{1}{(a^w)^y}$. Using the first assertion, this is equal to $\left(\frac{1}{a^w}\right)^y = (a^{-w})^y = (a^x)^y$.

(d) Finally, suppose that $x, y < 0$. Then, there exists a $z > 0$ such that $y = -z$. Thus, $(a^x)^y = (a^x)^{-z} = \frac{1}{(a^x)^z}$. By the previous case, this is equal to $\frac{1}{a^{xz}} = \frac{1}{a^{-xy}} = a^{xy}$. □

To conclude this section, we are going to extend Proposition 4.2.2(5) and (6) to irrational exponents. To do this, we prove a useful lemma first.

Lemma 4.2.7 *Let $a, b \geq 1$ and $x \in \mathbb{R}$.*

1. $a^x b^x = (ab)^x$.
2. $\left(\frac{a}{b}\right)^x = \frac{a^x}{b^x}$.

Proof We prove the assertions separately.

1. Denote $A = \{a^p : p \in \mathbb{Q}, p \leq x\}$, $B = \{b^q : q \in \mathbb{Q}, q \leq x\}$, and $C = \{(ab)^r : r \in \mathbb{Q}, r \leq x\}$. We want to show that:
$$\sup\{a^p : p \in \mathbb{Q}, p \leq x\} \sup\{b^q : q \in \mathbb{Q}, q \leq x\} = \sup\{(ab)^r : r \in \mathbb{Q}, r \leq x\},$$
namely $\sup(A)\sup(B) = \sup(C)$. Define the product of the sets A and B as $D = AB = \{a^p b^q : p, q \in \mathbb{Q}, p, q \leq x\}$. By Proposition 4.1.10, we have the equality $\sup(A)\sup(B) = \sup(D)$. Now we wish to show that $\sup(D) = \sup(C)$. First note that $C \subseteq D$ which means $\sup(C) \leq \sup(D)$. Next, we show that for any element in D, there exists an element in C which is bigger than or equal to it. Pick any element in D, say $a^p b^q$ for some $p, q \in \mathbb{Q}$ with $p, q \leq x$. Set $r = \max\{p, q\} \leq x$ and so, by Proposition 4.2.2, we have $a^p b^q \leq a^r b^r \in C$. Hence, $a^p b^q \leq \sup(C)$. Since $a^p b^q$ is an arbitrary element in D, this implies $\sup(D) \leq \sup(C)$. Putting the two inequalities together, we have $\sup(C) = \sup(D)$. Thus, we conclude that:
$$a^x b^x = \sup(A)\sup(B) = \sup(D) = \sup(C) = (ab)^x.$$
2. First, assume that $a \geq b \geq 1$. Then, $\frac{a}{b} \geq 1$ and, by the previous assertion, we have $\left(\frac{a}{b}\right)^x b^x = \left(\frac{a}{b} b\right)^x = a^x$. Thus, $\left(\frac{a}{b}\right)^x = \frac{a^x}{b^x}$. For the other case where $b \geq a \geq 1$, using the above, we deduce $\left(\frac{b}{a}\right)^x = \frac{b^x}{a^x}$ and then take its reciprocal. □

Proposition 4.2.8 *Let $a, b > 0$ and $x \in \mathbb{R}$.*

1. $a^x b^x = (ab)^x$.
2. *If $0 < a < b$, then* $\begin{cases} a^x < b^x & \text{if } x > 0, \\ a^x > b^x & \text{if } x < 0. \end{cases}$

Proof We prove the assertions one by one.

1. The case $a, b \geq 1$ has been shown in Lemma 4.2.7. Here we shall look at the other combinations of values for a and b.
   (a) For the first case, assume that exactly one of a or b is less than 1. WLOG, suppose that $0 < a < 1$ and $b \geq 1$. Then, there exists a $c > 1$ such that $a = \frac{1}{c}$. Using Lemma 4.2.7 we get $a^x b^x = \left(\frac{1}{c}\right)^x b^x = \frac{1}{c^x} b^x = \frac{b^x}{c^x} = \left(\frac{b}{c}\right)^x = (ab)^x$.

(b) Now assume that both a, b are less than 1. Then, there exist $c, d > 1$ such that $a = \frac{1}{c}$ and $b = \frac{1}{d}$. Using Lemma 4.2.7 again, we have $a^x b^x = \frac{1}{c^x}\frac{1}{d^x} = \frac{1}{(cd)^x} = \left(\frac{1}{cd}\right)^x = (ab)^x$.

2. Assume first that $x > 0$. By Proposition 4.2.6, for any $r < 1$ we have $r^x < r^0 = 1$. Therefore, if $a < b$, then $\frac{a}{b} < 1$ and so, by the previous assertion, we have $\frac{a^x}{b^x} = \left(\frac{a}{b}\right)^x < 1$. This implies $a^x < b^x$. The case for which $x < 0$ can be proven in a similar manner using reciprocals. □

4.3 Logarithm

In the previous section, we have seen how to define the exponential a^x for $a > 0$ and $x \in \mathbb{R}$. Suppose now we want to do the opposite, namely given numbers $a, b \in \mathbb{R}$ such that $a > 0$, can we guarantee that there exists an $x \in \mathbb{R}$ such that $a^x = b$?

In general, this is not true. First, if $b \leq 0$, then no such x exists since we have seen that a^x is necessarily positive. Moreover, if $a = 1$ and $b \neq 1$, there are no x for which $a^x = b$ either.

So let us focus our attention to $a, b > 0$ and $a \neq 1$. Since we have distinct definitions of exponentiation for the bases $0 < a < 1$ and $a > 1$ respectively, let us consider the case $a > 1$ first. In this specific case, by the property of exponentials in Proposition 4.2.6, if $b < 1$ we expect that $x < 0$ and if $b > 1$ we expect that $x > 0$. How do we actually find this x?

Using Bernoulli's inequality for rational exponents from Exercise 4.7, for any $r \in \mathbb{Q}$ and $z \in \mathbb{R}$ with $r > 1$ and $z > -1$ we have $(1+z)^r \geq 1 + rz$. Since $a > 1$, we can then set $a = 1 + z$ for some $z > 0$ so that the Bernoulli's inequality reads as $a^r \geq 1 + r(a-1)$.

Note that since $a > 1$ is a constant, the RHS can be made arbitrarily large by choosing a very large $r \in \mathbb{Q}$. Hence, for the fixed b, we can find a large r such that $a^r \geq 1 + r(a-1) \geq b$ (any positive rational $r \geq \frac{b-1}{a-1}$ would do). Thus, the set of rational exponents p for which $a^p \leq b$, namely $\{p \in \mathbb{Q} : a^p \leq b\}$, is bounded from above by this r and hence the supremum of this set exists. We call this supremum:

$$\log_a(b) = \sup\{p \in \mathbb{Q} : a^p \leq b\}.$$

We claim that this is the number x that we were looking for. Indeed, let us check that $a^x = a^{\log_a(b)} = b$.

1. We first show that $a^{\log_a(b)} \leq b$. By definition of exponentiation, since $a > 1$ we have:

$$\begin{aligned} a^{\log_a(b)} &= \sup\{a^q : q \in \mathbb{Q}, q \leq \log_a(b)\} \\ &= \sup\{a^q : q \in \mathbb{Q}, q \leq \sup\{p \in \mathbb{Q} : a^p \leq b\}\}. \end{aligned}$$

Note that the parametrising set satisfies $\{q \in \mathbb{Q} : q \le \sup\{p \in \mathbb{Q} : a^p \le b\}\} = \{q \in \mathbb{Q} : a^q \le b\}$. Indeed, by double inclusion, since $a > 1$ and hence a^p is increasing with respect to the exponent as shown in Proposition 4.2.6, we have:

$$\begin{aligned} r \in \{q \in \mathbb{Q} : q \le \sup\{p \in \mathbb{Q} : a^p \le b\}\} &\Leftrightarrow r \le \sup\{p \in \mathbb{Q} : a^p \le b\}, r \in \mathbb{Q} \\ &\Leftrightarrow a^r \le b, r \in \mathbb{Q} \\ &\Leftrightarrow r \in \{q \in \mathbb{Q} : a^q \le b\} \end{aligned}$$

Therefore, we have the inequality:

$$\begin{aligned} a^{\log_a(b)} &= \sup\{a^q : q \in \mathbb{Q}, q \le \sup\{p \in \mathbb{Q} : a^p \le b\}\} \\ &= \sup\{a^q : q \in \mathbb{Q}, a^q \le b\} \le b. \end{aligned}$$

2. Now we show the reverse inequality $a^{\log_a(b)} \ge b$ via some clever analysis. For brevity, denote $c = a^{\log_a(b)} = \sup\{a^q : q \in \mathbb{Q}, a^q \le b\}$ as from the previous part. Assume for contradiction that $c < b$. Since $a^q > 0$ for any $q \in \mathbb{Q}$, we have $c = \sup\{a^q : q \in \mathbb{Q}, a^q \le b\} > 0$. Moreover, by definition of c and assumption, there are no rational numbers $q \in \mathbb{Q}$ such that $c < a^q \le b$. Let us define two sets $A = \{q \in \mathbb{Q} : a^q \le c\}$ and $B = \{q \in \mathbb{Q} : a^q > b\}$. Clearly $A \cap B = \varnothing$.
   By Proposition 4.2.2, the set A is closed downwards and the set B is closed upwards. By the earlier argument, we then have $\mathbb{Q} = A \cup B$.
   Moreover, since A and B are bounded above and below respectively, $\sup(A)$ and $\inf(B)$ both exist with $\sup(A) \le \inf(B)$. If they are distinct, namely $\sup(A) < \inf(B)$, by Proposition 4.1.4, we can find a rational number $p \in \mathbb{Q}$ such that $\sup(A) < p < \inf(B)$. However, this means $p \notin A \cup B = \mathbb{Q}$, an absurdity. Thus, $\sup(A) = \inf(B)$.
   Now set $\varepsilon = \min\{\frac{b-c}{c(a-1)}, 1\} > 0$ and pick any rational number $\delta \in \mathbb{Q}$ with $0 < \delta < \varepsilon$. By Proposition 4.1.9, we can find a rational number $p \in A$ with $\sup(A) - \frac{\delta}{2} < p \le \sup(A)$. Hence, $\inf(B) = \sup(A) < p + \delta$ which then means the rational number $p + \delta$ is contained in B. See Fig. 4.2 for a depiction of these points.
   By definitions of the sets A and B, we have $a^p \le c$ and $b < a^{p+\delta}$. Thus:

$$b - c < a^{p+\delta} - a^p = a^p(a^\delta - 1). \tag{4.5}$$

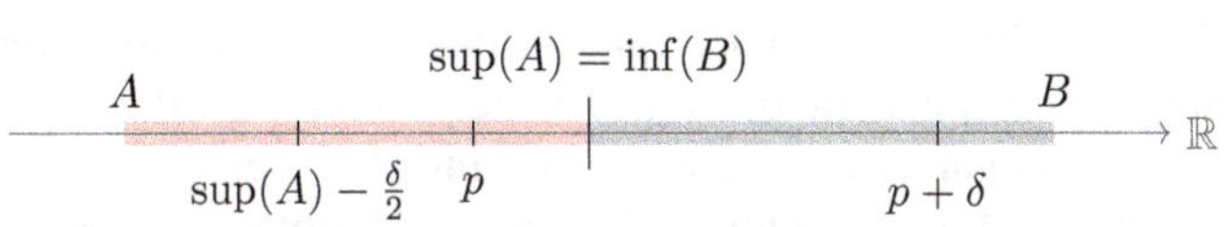

Fig. 4.2 Positions of rational numbers $p \in A$ and $p + \delta \in B$

Since $0 < \delta < 1$ and $\delta \in \mathbb{Q}$, by writing $a = 1 + z$ for some $z > 0$ and using the Bernoulli's inequality from Exercise 4.7, we get $1 < a^\delta = (1+z)^\delta \leq 1 + \delta z = 1 + \delta(a-1)$ which is equivalent to $0 < a^\delta - 1 \leq \delta(a-1)$. Putting this in Eq. (4.5) and using the facts that $0 < a^p \leq c$ and $0 < \delta < \varepsilon < \frac{b-c}{c(a-1)}$, we arrive at:

$$b - c < a^p(a^\delta - 1) \leq a^p\delta(a-1) \leq \frac{c(b-c)(a-1)}{c(a-1)} = b - c,$$

which gives us the desired contradiction. This means our initial assumption that $b > c$ must be false. So, we necessarily have $b \leq c = a^{\log_a(b)}$.

Putting the two inequalities together, we must have the equality $b = \sup\{a^q : q \in \mathbb{Q}, a^q \leq b\} = a^{\log_a(b)}$. Moreover, there is only one such x that gives us $a^x = b$. Indeed, if there are two of them, namely $x, y \in \mathbb{R}$, then we have $a^x = b = a^y$ which implies $a^{x-y} = 1$. Since $a > 1$, Proposition 4.2.6 implies $x - y = 0$ and so they are actually the same exponent.

For the case $0 < a < 1$, we note that $a^x = b$ is equivalent to $c^x = d$ where $c = \frac{1}{a} > 0$ and $d = \frac{1}{b} > 0$. Thus, we can solve this for x using the earlier case to get a unique $x = \log_c(d) = \log_{\frac{1}{a}}(\frac{1}{b})$ which we also write as $\log_a(b)$.

As we have discussed above, this modern interpretation of logarithm is the inverse procedure to the exponentiation which we have just proven to exist when $a, b > 0$ and $a \neq 1$. We define it more precisely as:

Definition 4.3.1 (Logarithm) Let $a, b > 0$ and $a \neq 1$. Then the logarithm for b with respect to the base a, written as $x = \log_a(b)$, is the unique real number x such that $a^x = b$, namely:

$$x = \log_a(b) = \begin{cases} \sup\{p \in \mathbb{Q} : a^p \leq b\} & \text{if } a > 1, \\ \sup\{p \in \mathbb{Q} : \left(\frac{1}{a}\right)^p \leq \frac{1}{b}\} = \sup\{p \in \mathbb{Q} : a^p \geq b\} & \text{if } 0 < a < 1. \end{cases}$$

The base a can be chosen to be any positive real number (except 1). Usual choices for the base a are 10, 2, and e which is the Euler-Napier constant. The final choice is the most commonly used base in mathematics and we shall see what this number is in Example 5.4.4. With the base e, the logarithm is called the natural logarithm and is written as $\log_e(b) = \ln(b)$.

Remark 4.3.2 Introduced by John Napier (1550–1617), this operation is called the logarithm, hence the notation log. The term was coined by Napier from from the Greek portmanteau *logos-arithmos*, literally meaning "ratio-number". This was published in his book entitled *Mirifici logarithmorum canonis descriptio* (Description of the Wonderful Canon of Logarithms) which includes logarithm tables to help simplify calculations and as a reference for astronomical studies and celestial navigation. This work was later refined by Henry Briggs (1561–1631).

However, in his work, Napier never thought of a logarithm via exponentials as we have just did above. One of the ways he interpreted it was in terms on a mechanics problem via the motion of particles moving on a straight line with speed proportional to its distance from a fixed point. In the language of differential equations (see Sect. 14.4), this is $\frac{dx}{dt} = \lambda x$ for some constant $\lambda \in \mathbb{R}$, which some readers might recognise for having a solution of the form $x(t) = Ae^{\lambda t}$, the inverse to logarithms. However, this was very much unknown at the time of Napier!

Using the duality between logarithms and exponentials as well as Proposition 4.2.6, we can prove the following properties of the logarithm operation, which are left as Exercise 4.14:

Proposition 4.3.3 *Let $a, b, c > 0$ with $a \neq 1$.*

1. $\log_a(1) = 0$.
2. $\log_a\left(\frac{b}{c}\right) = \log_a(b) - \log_a(c)$.
3. $\log_a(bc) = \log_a(b) + \log_a(c)$.
4. $\log_a(b^c) = c\log_a(b)$.
5. *If $b < c$, then* $\begin{cases} \log_a(b) < \log_a(c) & \text{if } a > 1, \\ \log_a(b) > \log_a(c) & \text{if } 0 < a < 1. \end{cases}$
6. *Change of base: If $d > 0$ with $d \neq 1$, then* $\log_a(b) = \frac{\log_d(b)}{\log_d(a)}$.

4.4 Decimal Representation of the Real Numbers

In the previous chapter and sections, we have constructed the real numbers and defined some algebraic operations on them. These were quite fiddly to construct, given that we do not have an explicit description of the irrational numbers and hence has to resort to the use of Dedekind cuts, rational numbers, and supremum/infimum arguments. In this section, we shall try and find an explicit description for these irrational numbers.

For each rational number, we have the (non-unique) representation of $\frac{p}{q}$ where $p, q \in \mathbb{Z}$ with $q \neq 0$. We have also seen in school that we can find its decimal representation. The word decimal comes from the Latin word *decimus* meaning "tenth". As the name suggests, a decimal representation is also known as a base-10 representation. We define this representation properly now:

Definition 4.4.1 (Decimal Representation) A decimal representation of a rational number $r \in \mathbb{Q}$ is given by the formal infinite sum:

$$r \sim \sum_{j=0}^{\infty} \frac{a_j}{10^j} = a_0 + \frac{a_1}{10} + \frac{a_2}{10^2} + \frac{a_3}{10^3} + \dots, \tag{4.6}$$

where $a_0 \in \mathbb{Z}$ and $a_j \in \{0, 1, 2, 3, ..., 9\}$ for each $j \in \mathbb{N}$.

Note that the $\sim$ symbol in (4.6) does not denote a relation. For simplicity, we may also write the infinite sum above as the string:

$$r \sim a_0.a_1a_2a_3\ldots, \tag{4.7}$$

where the dot symbol between a_0 and a_1 is called the decimal point and the number a_j for $j \in \mathbb{N}$ is called the j-th decimal place.

Remark 4.4.2 We give a few important remarks here.

1. As we have noted in Remark 2.5.2, the infinite sum and the infinite string in Definition 4.4.1 are simply symbolic, for now. At the moment, it is not a real number since the field and ring axioms only allow the sum of finitely many elements in $\mathbb{R}$. This infinite sum has infinitely many terms which is impossible to carry out or quantify yet: we would need an infinite amount of time to do this sum which is impossible to do even with a supercomputer.
2. Therefore, for the time being, the infinite sum $\sum_{j=0}^{\infty} \frac{a_j}{10^j}$ and the infinite string $a_0.a_1a_2\ldots$ are simply used as identification tags or name for the real numbers. The decimal representation is a name/symbol consisting of numerals, but itself is not a number (at least, not yet). This is analogous to my name as a string of letters forming a word and me as a person, two unequal but related objects where one is used to represent/identify the other. This is the reason why we use the symbol $\sim$ instead of $=$ in the identifications (4.6) and (4.7).
3. However, we shall see in Proposition 4.4.3 that for some rational number r, this infinite sum is actually a finite sum and hence terminates after some index $n \in \mathbb{N}_0$. In this case, the name/representation $\sum_{j=0}^{n} \frac{a_j}{10^j}$ for the number r can also be treated as a number and its value is equal to r.
4. But if the sum does not terminate, we have to think about this later in Chaps. 5 and 7 once we understood how to interpret the sum of infinitely many numbers. We shall see that this infinite sum $\sum_{j=0}^{\infty} \frac{a_j}{10^j}$ is actually equal to the number r in the sense of "limits".
   After we have established this, we can safely replace all the $\sim$ in (4.6) and (4.7) with $=$. Readers can refer to Example 5.2.8 for this. Again, this is what sets algebra and analysis apart: algebra allows us to do the sum for finitely many terms but analysis pushes this limit (pun intended) further to allow infinitely many terms by using these so called "limits".

Decimal Representation for Rational Numbers

First, how do we construct such a representation for a given rational number? Suppose that $r = \frac{p}{q}$ is a positive rational number. WLOG, suppose $p, q \in \mathbb{N}$. By long division in Exercise 2.26(a), we can find unique $a_0 \in \mathbb{N}_0$ and $b_1 \in$

$\{0, 1, 2, \ldots, q-1\}$ such that $p = a_0 q + b_1$. In $\mathbb{Q}$, this expression is equivalent to:

$$\frac{p}{q} = a_0 + \frac{b_1}{q} \quad \Leftrightarrow \quad \frac{p}{q} = a_0 + \frac{1}{10}\left(\frac{10b_1}{q}\right). \tag{4.8}$$

The second step is as follows: if $b_1 = 0$, then we are done as we have found a way of writing r in the form of (4.6) by setting $a_j = 0$ for all $j \geq 1$. Otherwise, $b_1 \in \{1, 2, \ldots, q-1\}$ and so $0 < \frac{10b_1}{q} < 10$. Thus we can carry out the long division process above again to get $10b_1 = a_1 q + b_2$ or equivalently $\frac{10b_1}{q} = a_1 + \frac{b_2}{q}$ for some $a_1 \in \{0, 1, 2, \ldots, 9\}$ and $b_2 \in \{0, 1, 2, \ldots, q-1\}$. Putting this in Eq. (4.8), we have:

$$\frac{p}{q} = a_0 + \frac{1}{10}\left(a_1 + \frac{b_2}{q}\right) = a_0 + \frac{a_1}{10} + \frac{1}{10}\left(\frac{b_2}{q}\right) = a_0 + \frac{a_1}{10} + \frac{1}{10^2}\left(\frac{10b_2}{q}\right).$$

We repeat the long division process with $10b_2$ and q to get the next terms a_2 and b_3 and so on and so forth. At the n-th step, we would have constructed a rational number:

$$r_{n-1} = a_0 + \frac{a_1}{10} + \frac{a_2}{10^2} + \ldots + \frac{a_{n-1}}{10^{n-1}} = \sum_{j=0}^{n-1} \frac{a_j}{10^j} \in \mathbb{Q},$$

where $a_j \in \{0, 1, \ldots, 9\}$ for $j = 1, 2, \ldots, n-1$ satisfying:

$$\frac{p}{q} = r_{n-1} + \frac{1}{10^{n-1}}\left(\frac{b_n}{q}\right), \tag{4.9}$$

with $b_n \in \{0, 1, 2, \ldots, q-1\}$. By construction, for every $n \in \mathbb{N}$ we have $0 \leq \frac{b_n}{q} < 1$ and so:

$$0 \leq \frac{p}{q} - r_{n-1} < \frac{1}{10^{n-1}}. \tag{4.10}$$

Now we claim that either this iterative process stops after finitely many steps or the process repeats itself periodically.

Proposition 4.4.3 *Let $r = \frac{p}{q} \in \mathbb{Q}$ be a rational number. Then, either the decimal representation of r terminates after finitely many terms or the representation becomes periodic eventually.*

Proof WLOG, let $p, q \in \mathbb{N}$ so that r is a positive rational number. At the j-th stage of the decimal construction, we will get $b_j \in \{0, 1, 2, \ldots, q-1\}$. We have two cases:

1. Suppose that during the process, we obtain $b_n = 0$ for some $n \in \mathbb{N}$. Then the process terminates here and hence the decimal representation of r would be:

$$r_{n-1} = a_0 + \frac{a_1}{10} + \ldots + \frac{a_{n-1}}{10^{n-1}},$$

or $a_0.a_1a_2\ldots a_{n-1}$. In fact, this value is exactly equal to r by the equality in (4.9), so $r = a_0 + \frac{a_1}{10} + \ldots + \frac{a_{n-1}}{10^{n-1}}$.

2. Now suppose that none of the b_j during the iterative process are 0 so that the process does not terminate. Then, $b_j \in \{1, 2, \ldots, q-1\}$ for all $j \in \mathbb{N}$. Since there are infinitely many b_j but they can only take at most $q-1$ different values, we must have some repetition in the numbers b_j. Suppose that b_m and b_n is the first time this repetition happens with $m < n$. Thus, from the process at the $(m+1)$-th step and $(n+1)$-th step, we have:

$$\frac{10b_m}{q} = a_m + \frac{b_{m+1}}{q} \quad \text{and} \quad \frac{10b_n}{q} = a_n + \frac{b_{n+1}}{q} \quad \Rightarrow \quad a_m + \frac{b_{m+1}}{q} = a_n + \frac{b_{n+1}}{q}.$$

Since $0 < \frac{b_{m+1}}{q} < 1$, $0 < \frac{b_{n+1}}{q} < 1$, and $a_m, a_n \in \{0, 1, \ldots, 9\}$, taking the floor function on both sides, we must have $a_m = a_n$ and thus $b_{m+1} = b_{n+1}$. Repeating this, we can show inductively that $a_{m+k} = a_{n+k}$ and $b_{m+k} = b_{n+k}$ for all $k \in \mathbb{N}_0$. Setting $j = m + k$, we have $a_j = a_{j+(n-m)}$ for all $j \geq m$. Hence, the decimal representation is periodic with period $n - m$ starting at a_m, namely:

$$r \sim a_0.a_1a_2\ldots \underbrace{a_ma_{m+1}\ldots a_{n-1}}\,\underbrace{a_ma_{m+1}\ldots a_{n-1}}_{\text{this string repeats}}\,\underbrace{a_ma_{m+1}\ldots a_{n-1}}\ldots.$$

We write this as $r \sim a_0.a_1a_2\ldots\dot{a}_m\dot{a}_{m+1}\ldots\dot{a}_{n-1}$ or $a_0.a_1a_2,\ldots\overline{a_m\ldots a_{n-1}}$ to indicate which string of digits repeats. □

This gives us a new way of representing rational numbers: instead of a ratio of two integers, we can write it as a string of numbers between 0 and 9 with a decimal point somewhere, knowing for a fact that the representation is finite or becomes periodic eventually. Thus, in both cases, for any rational number, we only need a finite amount data in the form of finite string of numbers (which may repeat indefinitely) to represent the number in full using the decimal notation.

However, a drawback of the decimal representation of rational numbers is that some rational numbers can be represented by two different decimal representations, so we might not have uniqueness of such representation. Let us look at an example of this.

Example 4.4.4 Let us assume for the time being that the rational number is exactly equal to their decimal representation. This fact will be proven later when we in Chap. 6 when we have an interpretation for the value of an infinite sum of numbers. Consider the rational number $\frac{1}{3}$. By carrying out the long division process as outlined above, we can find a decimal representation of this number, which is given by $\frac{1}{3} = 0.33333\ldots = 0.\dot{3}$. If we multiply the decimal representation with 3, we expect that we get a decimal representation of the number 1. However:

$$1 = 3 \times \frac{1}{3} = 3 \times 0.\dot{3} = 0.\dot{9}.$$

So the decimal representation of the number 1 is $0.\dot{9}$. But we know that a decimal representation of 1 is simply $1.\dot{0}$ by the long division argument! So we have an ambiguity in the decimal representation for the number 1.

In fact, any rational number with finite decimal representation can be represented by two different decimal representations: one by the finite representation and the other by subtracting the final non-zero term in the finite decimal representation by 1 and appending at the end of this representation with an infinite string of 9s. For example, 1.4 and $1.3\dot{9}$ are both decimal representations for the rational number $\frac{7}{5}$.

From Example 4.4.4, we have seen that some numbers may have two distinct identifying names or decimal representations. Ideally we would like every real number to have a unique name attached to it for clarity and to avoid confusion. We do not want any double agent situation where a rational number has two different names!

To deal with this ambiguity and non-uniqueness, we simply declare the equality of decimal representations $0.\dot{9} = 1.\dot{0}$ and avoid representing any decimals with an infinite repeating string of 9s. We shall justify this equality in Example 5.2.8(2) once we have properly defined limits as a way of interpreting an infinite decimal representation as a number.

The good side of decimal representation of numbers is that it allows us to also write down irrational numbers concretely. Well, almost.

We have seen that the rational numbers can be written down as a finite decimal representation or a decimal representation which becomes periodic eventually. The converse is also true: any finite decimal representation or decimal representation which become periodic eventually corresponds to a rational number. This will be proven when we have found a way to make sense of an infinite sum numerically in Chap. 7 and Exercise 7.5. As a result, we expect that the decimal representation of an irrational number, if it exists, must be infinite and non-periodic.

Decimal Representation of Irrational Numbers

But how do we construct a decimal representation of an irrational number without the division process as we did for rational numbers? For rational numbers, we have the representation $\frac{p}{q}$ for $p, q \in \mathbb{Z}$ so we can just apply long division repeatedly starting with the pair of integers p, q. For irrational numbers, we do not have these information to begin with. However, we can appeal to the floor and ceiling functions in Definition 4.1.6.

Let $r \in \bar{\mathbb{Q}}$ be an irrational number. WLOG, let us assume that $r > 0$. For the first step, since it is a real number, we choose $a_0 = \lfloor r \rfloor$ so that $a_0 \leq r < a_0 + 1$ which implies that $0 \leq r - a_0 < 1$. But since $r - a_0$ is also irrational, it cannot be equal to 0, so we must have $0 < r - a_0 < 1$.

The next step is to consider the quantity $0 < 10(r - a_0) < 10$. We can then choose $a_1 = \lfloor 10(r - a_0) \rfloor \in \{0, 1, 2, \ldots, 9\}$ so that $a_1 < 10(r - a_0) < a_1 + 1$. By algebra, this is equivalent to $0 < r - \left(a_0 + \frac{a_1}{10}\right) < \frac{1}{10}$.

We repeat the construction with the number $0 < 10^2\left(r - a_0 - \frac{a_1}{10}\right) < 10$ to find an integer $a_2 = \lfloor 10^2\left(r - a_0 - \frac{a_1}{10}\right) \rfloor \in \{0, 1, 2, \ldots, 9\}$ such that $0 < r - \left(a_0 + \frac{a_1}{10} + \frac{a_2}{10^2}\right) < \frac{1}{10^2}$. We continue with this construction ad infinitum to get the decimal representation of the irrational number:

$$r \sim \sum_{j=0}^{\infty} \frac{a_j}{10^j} = a_0.a_1a_2a_3\ldots,$$

where $a_0 \in \mathbb{N}_0$ and $a_j \in \{0, 1, 2, \ldots, 9\}$ for $j \in \mathbb{N}$. Therefore, just like the rational numbers, each irrational number also has a decimal representation. We stress here that this representation cannot be finite and cannot be eventually periodic since r is an irrational number. Therefore, unlike the rational numbers, we can never represent an irrational number in decimal notation with a finite amount of data because it is an infinite process and is non-periodic.

However, from the construction above, we have defined the sequence of rational numbers obtained during the construction as:

$$r_n = a_0 + \frac{a_1}{10} + \frac{a_2}{10^2} + \ldots + \frac{a_n}{10^n} = \sum_{j=0}^{n} \frac{a_j}{10^j} \in \mathbb{Q}.$$

Furthermore, by construction, for each $n \in \mathbb{N}$, we have the estimates for the differences between the irrational number r and its rational approximations r_n:

$$0 < r - r_n < \frac{1}{10^n},$$

which is similar to the estimate that we have for rational numbers in (4.10). Thus, the difference of the irrational number r to the rational number r_n is smaller than $\frac{1}{10^n}$. So instead of having the full decimal representation of the irrational number r, we have its rational approximations which we can take to any desired degree of accuracy. In other words, we can get to the irrational number r as close as we like with rational numbers by choosing an n as large as we wish.

As a conclusion, the full decimal representation for irrational numbers is impossible to obtain for us mere mortals. Thus, we can never call an irrational number r by its full decimal representation name. However, we can be as close as we like by giving it shorter nicknames; we can call them by the rational approximations r_n and this is close enough to the full name of r.

Example 4.4.5 Recall that the number $\sqrt{2}$ is an irrational number. If we cheat and use the calculator, we can get a decimal representation of the number $\sqrt{2}$, which is given by $\sqrt{2} = 1.41421356237\ldots$. Even a calculator or a super computer would not be able to figure out the whole decimal representation of $\sqrt{2}$ because it would take an infinitely long time to do the calculations. However, the calculators and computers can approximate the number $\sqrt{2}$ by some choices of rational numbers, with increasing degree of accuracy: $1, 1.4, 1.414, 1.4142, 1.41421, \ldots$.

Cardinality of $\mathbb{R}$

In Exercise 3.19 we have shown that the cardinality of the irrationals is at least countably infinite. Now that we know how to represent the real numbers in an explicit (sort of) way, we are finally able to show that the set of real numbers $\mathbb{R}$ is actually bigger than this.

Proposition 4.4.6 *The set of real numbers* $\mathbb{R}$ *is uncountably infinite.*

Proof In fact, we show a stronger statement, namely: the set of real numbers between 0 and 1 is uncountably infinite. Let us call this set $X = \{x \in \mathbb{R} : 0 < x < 1\}$. Clearly, it is an infinite set because this set contains all the rational numbers of the form $\frac{1}{n}$ for all $n \in \mathbb{N} \setminus \{1\}$, so the set X is at least countably infinite.

Suppose for a contradiction that it is countably infinite. Thus, we can list down all these numbers by bijectively mapping them to the natural numbers. Let us list down all these numbers $X = \{q_1, q_2, q_3, q_4, \ldots\}$ by their infinite decimal representation expression as thus:

$$
\begin{aligned}
q_1 &\sim 0.a_{1,1}a_{1,2}a_{1,3}a_{1,4}\ldots\\
q_2 &\sim 0.a_{2,1}a_{2,2}a_{2,3}a_{2,4}\ldots\\
q_3 &\sim 0.a_{3,1}a_{3,2}a_{3,3}a_{3,4}\ldots\\
q_4 &\sim 0.a_{4,1}a_{4,2}a_{4,3}a_{4,4}\ldots\\
&\vdots
\end{aligned}
$$

where each $a_{i,j} \in \{0, 1, 2, \ldots, 9\}$ for every $i, j, \in \mathbb{N}$. Of course, every finite decimal representation is appended with a string of 0s and we avoid recurring 9s to make sure that each number in X is represented only once in the list. Since we have assumed that the numbers in X is countable, this list is complete.

However, we claim that this list does not represent all the numbers in X. We do this by constructing a new number $r \in X$ with decimal representation $r \sim 0.b_1b_2b_3b_4\ldots$ which is not in the list above by looking at the digits coloured in red.

Since $a_{1,1}$ is one of the integers between and including 0 and 9, we can select another number $b_1 \in \{1, 2, \ldots 8\}$ which is not equal to $a_{1,1}$. So $r \neq q_1$ immediately. Next, we select b_2 from the set $\{1, 2, \ldots, 8\}$ which is distinct from $a_{2,2}$ so that $r \neq q_2$.

We repeat this construction for each $j \in \mathbb{N}$ by selecting b_j from the set $\{1, 2, \ldots, 8\}$ distinct from $a_{j,j}$ so that $r \neq q_j$. Thus, we have constructed a new number r represented by $r \sim 0.b_1b_2b_3b_4\ldots$ which is distinct from all the q_j and hence not in the list above. Therefore, the list above is incomplete, which is a contradiction. Thus, our assumption that X is countable is false and hence X is uncountable. □

Remark 4.4.7 Here are some remarks rgarding the proof of Proposition 4.4.6:

1. Note that we avoided the integers 0 and 9 in the selection for each b_j above to avoid constructing $r = 0$ or a decimal representation with recurring 9s.
2. The argument used in Proposition 4.4.6 is called the Cantor diagonal argument and is a very useful trick in mathematics. We shall see other applications of it later in Exercises 6.10 and 11.29.

We end this section by stating that the number base for our representation is completely arbitrary. We have used a decimal (base-10) representation simply for convenience of the Hindu-Arabic numeral system. There are also many different bases available to use. For example, computer systems use the binary (base-2) system where, for example, every number between 0 and 1 can be represented as $\sum_{j=1}^{\infty} \frac{a_j}{2^j} = 0.a_1a_2a_3\ldots$ where $a_j \in \{0, 1\}$. The construction is exactly similar to decimal representation of a real number. The only difference is we carry out the division with powers of 2 instead of powers of 10.

Example 4.4.8 Let us see how to compute binary representations with concrete examples:

1. The decimal representation of $\frac{3}{4}$ is 0.75. Let us find the binary representation of $\frac{3}{4}$. We have:

$$\frac{3}{4} = \frac{1}{2}\left(\frac{2 \cdot 3}{4}\right) = \frac{1}{2}\left(1 + \frac{2}{4}\right) = \frac{1}{2} + \frac{1}{4} = \frac{1}{2} + \frac{1}{2^2} = (0.11)_2,$$

where we enclose the string of numbers with a bracket and a subscript 2 to emphasise that the representation is in base-2.

2. On the other hand, for integers, to find its binary operation, we repeatedly divide the quotients with 2. As an example, let us find the binary representation for 23 here:

$$\begin{aligned}23 &= 2(11) + 1 = 2(2(5) + 1) + 1 = 2^2(5) + 2 + 1 = 2^2(2(2) + 1) + 2 + 1 \\ &= 2^4 + 2^2 + 2^1 + 2^0 = (10111)_2.\end{aligned}$$

Moreover, ancient civilisations such as the Sumerians and Babylonians worked in sexagesimal or base-60 representation. Other useful number bases are octal (base-8), hexadecimal (base-16), and ternary (base-3). We shall see how the ternary number base can be useful in Exercise 4.32.

4.5 Topology on $\mathbb{R}$

In this section, we are going to briefly talk about the topology of sets in $\mathbb{R}$. We have seen the elements of the set $\mathbb{R}$ and how we can put an ordered field structure on it. Now we are going to look at the subsets of real numbers and how we can represent them.

As usual, finite discrete subsets of the real numbers can be written as a list $\{a_1, a_2, \dots, a_n\}$ in the roster notation. However, for sets which are not countable, for example the set of real numbers between 0 and 1 that we have seen in Proposition 4.4.6, it is impossible to list all of the elements one by one.

Instead, we can write the set of numbers between 0 and 1 in set builder notation as $\{x \in \mathbb{R} : 0 < x < 1\}$. Now we are going to introduce some new notations to make this easier to write.

Intervals

We first define:

Definition 4.5.1 (Interval) An interval $I \subseteq \mathbb{R}$ is a subset of $\mathbb{R}$ such that if $c, d \in I$ with $c < d$, then every $x \in \mathbb{R}$ between c and d, namely $c < x < d$, is contained in I as well. In symbols, I is an interval if:

$$\forall c, d \in I, \forall x \in \mathbb{R}, (c < x < d \Rightarrow x \in I).$$

The empty set is also trivially an interval since it does satisfy the property above. An interval of real numbers can be described as one of the following:

Definition 4.5.2 (Real Number Intervals) For $a, b \in \mathbb{R}$ with $a \leq b$, we define these notations:

$$\begin{aligned}
\{x \in \mathbb{R} : a < x < b\} &= (a, b), \\
\{x \in \mathbb{R} : a \leq x < b\} &= [a, b), \\
\{x \in \mathbb{R} : a < x \leq b\} &= (a, b], \\
\{x \in \mathbb{R} : a \leq x \leq b\} &= [a, b], \\
\{x \in \mathbb{R} : a < x\} &= (a, \infty), \\
\{x \in \mathbb{R} : a \leq x\} &= [a, \infty), \\
\{x \in \mathbb{R} : x < b\} &= (-\infty, b), \\
\{x \in \mathbb{R} : x \leq b\} &= (-\infty, b].
\end{aligned}$$

Sets which look like any of these are called real number intervals.

The round brackets are called open brackets (which means the set does not include the the number/quantity next to it) and the square brackets are called closed bracket (which means the set includes the number/quantity next to it). We note that we included the symbols $\pm\infty$ in some of the intervals to denote unbounded intervals. We never put square brackets next to $\pm\infty$ here since these elements are not real numbers and thus are not contained in $\mathbb{R}$. The set of a single element $\{a\}$ is also technically an interval since:

$$\{a\} = \{x \in \mathbb{R} : a \leq x \leq a\} = [a, a].$$

This notation also gives us an alternative way to characterise intervals:

Definition 4.5.3 (Interval) An interval $I \subseteq \mathbb{R}$ is a subset of $\mathbb{R}$ such that if $c, d \in I$ with $c \leq d$, then $[c, d] \subseteq I$ as well.

Since these objects are sets, we can do all sorts of set operations on them: union, intersection, complement, difference, Cartesian product, et cetera. Note that not all subsets of $\mathbb{R}$ can be written as a single interval. For example, consider the set of real numbers bigger than 1 or smaller than -1 which is $X = \{x \in \mathbb{R} : x < -1 \text{ or } x > 1\}$. This set cannot be written as a single interval as in Definition 4.5.2. Indeed, $-2, 2 \in X$ but $[-2, 2] \nsubseteq X$. However, we can write it as a union of two intervals, namely $X = (-\infty, -1) \cup (1, \infty)$.

Moreover, we have:

Lemma 4.5.4 *Let $I, J \subseteq \mathbb{R}$ be two non-empty intervals in $\mathbb{R}$.*

1. *The intersection $I \cap J$ is an interval.*

2. *If $I \cap J \neq \varnothing$, then the union $I \cup J$ is also an interval.*

The readers will prove Lemma 4.5.4 in Exercise 4.20.

From the list of intervals in Definition 4.5.2, let us classify some of them further:

Definition 4.5.5 (Open and Closed Real Number Intervals) Let $a, b \in \mathbb{R}$ such that $a < b$. Then:

1. We call the intervals (a, b), (a, ∞), $(-\infty, a)$, $\mathbb{R}$, and $\varnothing$ open intervals.
2. We call the intervals $[a, b]$, $[a, \infty)$, $(-\infty, a]$, $\mathbb{R}$, and $\varnothing$ closed intervals.

Furthermore, the singleton set $\{a\}$ is also called a closed interval since it is of the degenerate form $[a, a]$.

Remark 4.5.6 Let us make some remarks regarding Definition 4.5.5:

1. The whole interval $(-\infty, \infty) = \mathbb{R}$ and the empty interval $\varnothing$ are called clopen intervals (a portmanteau of closed and open) because they are both open and closed as declared in Definition 4.5.5. This is a convention that comes from the study of topology. See Example 4.5.15(5) for the justification.
2. Intervals of the form $(a, b]$ and $[a, b)$ are not classified in Definition 4.5.5. So we call them half-open (or half-closed) intervals.

Open and Closed Sets

Now where do these terms "open" and "closed" intervals come from? In the study of topology, topologists are interested in collections of objects in a general set X which are called open and closed sets. These collection of sets behave in a certain nice way and adds a new structure to the set. This structure is called the topological structure.

In our universe which is $\mathbb{R}$, roughly speaking, open sets are sets which are made up from unions of disjoint open intervals. Closed sets are simply complements of open sets. This is a very informal definition for open and closed sets in $\mathbb{R}$. We shall see the proper definition of an open and a closed set later in Definition 4.5.14 once we have defined some prerequisite ideas for it.

Before we do that, in order to define an open set, we need to define an object called a ball in $\mathbb{R}$. We first define the modulus operation on the real numbers. A modulus is an operation that measures the size of the number or the distance of the number from 0, giving us a geometrical (length and distance) structure on the set $\mathbb{R}$. More concretely:

Definition 4.5.7 (Modulus) Let $a \in \mathbb{R}$. We define the modulus of a, denoted as $|a|$, to be:

$$|a| = \begin{cases} a & \text{if } a > 0, \\ 0 & \text{if } a = 0, \\ -a & \text{if } a < 0. \end{cases}$$

Clearly, from the definition above, we must have $|a| \geq 0$ for any $a \in \mathbb{R}$. Therefore, a modulus is a function sending elements from $\mathbb{R}$ to $\mathbb{R}_{\geq 0}$. Moreover, by definition, we must have $|a| = 0$ if and only if $a = 0$. Here are some properties of the modulus function:

Proposition 4.5.8 *Let $a, b \in \mathbb{R}$.*

1. $|-a| = |a|$.
2. $|a|^2 = a^2$.
3. $|ab| = |a||b|$ *and* $\left|\frac{a}{b}\right| = \frac{|a|}{|b|}$ *if* $b \neq 0$.
4. $-|a| \leq a \leq |a|$.
5. *If* $b \geq 0$*, then* $|a| \leq b$ *if and only if* $-b \leq a \leq b$.
6. *Triangle inequality:* $|a + b| \leq |a| + |b|$.
7. *Reverse triangle inequality:* $||a| - |b|| \leq |a - b|$.

Proof The first four assertions can be checked by cases. The readers will prove them in Exercise 4.18 later. We only prove assertions 5, 6, and 7 here.

5. We prove the implications one by one.

   ($\Rightarrow$): Assume that $|a| \leq b$. Then, we also have $-|a| \geq -b$. By assertion 4 and transitivity of ordering, we have:

   $$-b \leq -|a| \leq a \leq |a| \leq b,$$

   which yields the result.

   ($\Leftarrow$): Suppose that $-b \leq a \leq b$. Then, both $a \leq b$ and $-a \leq b$ are true. Since $|a|$ is equal to either $-a$ or a, we must have $|a| \leq b$.

6. We note that for any $a, b \in \mathbb{R}$, we have $-|a| \leq a \leq |a|$ and $-|b| \leq b \leq |b|$. Adding these two inequalities together, we get $-(|a| + |b|) \leq a + b \leq |a| + |b|$. By using assertion 5, we obtain the result.
7. We note that $|a| = |a + b - b|$. Using the triangle inequality from assertion 6, we get $|a| \leq |a + b| + |-b| = |a + b| + |b|$ and so $|a| - |b| \leq |a + b|$. Likewise, the fact $|b| = |a + b - a|$ yields $|b| \leq |a + b| + |a|$ and so $|b| - |a| \leq |a + b|$. Putting these two inequalities together, we have $-|a + b| \leq |a| - |b| \leq |a + b|$. Finally, applying assertion 5 gives us the result.

□

Remark 4.5.9 The final two inequalities that we have seen in Proposition 4.5.8 namely the triangle and the reverse triangle inequalities are extremely important in analysis. We shall be using them regularly in this book.

Intuitively, the modulus $|a|$ defines the distance of the number a from 0 in $\mathbb{R}$. Thus, if we have two real numbers $a, b \in \mathbb{R}$, we note that the quantity $|a - b|$ measures the length of the segment of the real number line between the numbers $a - b$ and 0. Hence, by translation, the non-negative quantity $|a - b|$ also measures the distance between the numbers a and b. See Fig. 4.3 for a visualisation.

Using the modulus as a way to measure distances, we can then define:

Definition 4.5.10 (Open Ball in $\mathbb{R}$) For a fixed $c \in \mathbb{R}$ and a non-negative number $r \geq 0$, we define the open ball of radius r centred at c as the set:

$$B_r(c) = \{x \in \mathbb{R} : |x - c| < r\}.$$

Intuitively, from its definition, $B_r(c)$ is simply the set of points $x \in \mathbb{R}$ such that their distance from the point c is strictly smaller than r. Thus, any $x \in B_r(c)$ satisfies $-r < x - c < r$ or equivalently $c - r < x < c + r$ by Proposition 4.5.8(5). This means we can also express the open ball as $B_r(c) = (c - r, c + r)$, which is an open interval according to Definition 4.5.5.

Similarly, we define:

Definition 4.5.11 (Closed Ball in $\mathbb{R}$) For a fixed $c \in \mathbb{R}$ and a non-negative number $r \geq 0$, we define the closed ball of radius r centred at c as the set:

$$\bar{B}_r(c) = \{x \in \mathbb{R} : |x - c| \leq r\}.$$

Similar to open balls, we can express the closed ball as the closed interval $\bar{B}_r(c) = [c - r, c + r]$. The difference between the closed and the open ball is the "shell" of the ball, namely the two endpoints $c - r$ and $c + r$ of the interval. We call this shell a sphere.

Definition 4.5.12 (Sphere in $\mathbb{R}$) For a fixed $c \in \mathbb{R}$ and a non-negative number $r \geq 0$, we define the sphere of radius r centred at c as the set:

$$S_r(c) = \{x \in \mathbb{R} : |x - c| = r\} = \{c - r, c + r\}.$$

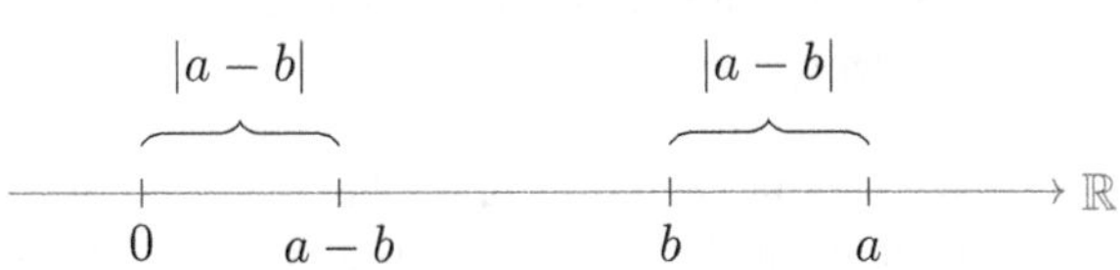

Fig. 4.3 Translating the distance $|a - b|$

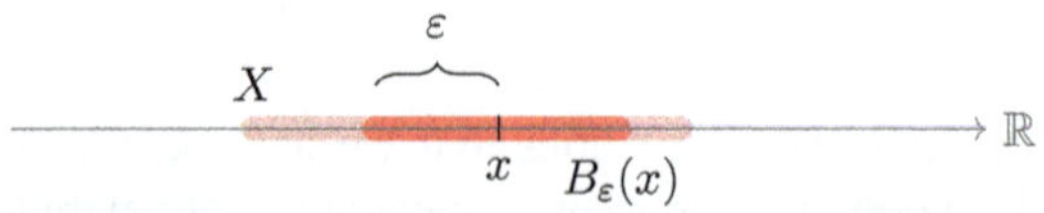

Fig. 4.4 Example of an open set X in $\mathbb{R}$. At the point $x \in X$, we have $B_\varepsilon(x) \subseteq X$. If x is closer to the edge of the set, the ε that is required to work would be smaller

Therefore, a closed ball is simply the union of an open ball and its boundary sphere, namely:

$$\bar{B}_r(c) = B_r(c) \cup S_r(c).$$

Remark 4.5.13 Why do these objects are called balls and spheres even though they do not look round like actual balls and spheres in real life such as a beach ball or an orange? They are actually mathematical 1-dimensional balls and 0-dimensional spheres respectively. Beach balls and oranges are 3-dimensional balls, whereas a disc is a 2-dimensional ball.

Using the definition of open balls, we can now define open and closed sets in $\mathbb{R}$.

Definition 4.5.14 (Open Set, Closed Set) Let $X \subseteq \mathbb{R}$.

1. The subset X is called open if for all $x \in X$, there exists an $\varepsilon > 0$ such that the ball of radius ε and centred at x is contained fully in X, namely $B_\varepsilon(x) \subseteq X$. In symbols:

$$X \text{ is an open set} \quad \text{if} \quad \forall x \in X, \exists \varepsilon > 0 : B_\varepsilon(x) \subseteq X.$$

2. The subset X is called closed if its complement is open, namely the set $X^c = \mathbb{R} \setminus X$ is open.

The intuition for open sets is that at any point within the set, we can wiggle around at the point for some small radius $\varepsilon > 0$ without leaving the set.

We note that in the definition for open set, the radius $\varepsilon > 0$ would depend on the centre of the ball x. This is because nearer to the boundary of the set X, intuitively we would need smaller radius to be able to fit the open ball $B_\varepsilon(x)$ in the set X. See Fig. 4.4 for a visualisation of this open ball.

Example 4.5.15 We can show that the open and closed intervals in $\mathbb{R}$ that we have seen earlier in Definition 4.5.5 are indeed open and closed sets respectively (and hence their names).

1. We have seen in Definition 4.5.5 that the interval $I = (a, b)$ for $a < b$ is called an open interval. To show that it is an open set in the sense of Definition 4.5.14,

we need to be able to fit in a small enough open ball around every point $x \in I$ inside the set I.
Fix $x \in I$. Thus, we must have $x - a > 0$ and $b - x > 0$. We need to find a suitable radius $\varepsilon > 0$ for the ball $B_\varepsilon(x)$ so that this ball is contained properly within I. We choose $\varepsilon = \min\left\{\frac{x-a}{2}, \frac{b-x}{2}\right\} > 0$. We now show that the open ball $B_\varepsilon(x) = (x - \varepsilon, x + \varepsilon)$ is fully contained in $I = (a, b)$, namely $(x - \varepsilon, x + \varepsilon) \subseteq (a, b)$. We do this by checking that $a < x - \varepsilon$ and $x + \varepsilon < b$. Recall that $x > a$, so:

$$x - a > \frac{x - a}{2} \geq \varepsilon \quad \Rightarrow \quad a < x - \varepsilon.$$

Similarly, we have $x < b$ and therefore:

$$b - x > \frac{b - x}{2} \geq \varepsilon \quad \Rightarrow \quad x + \varepsilon < b,$$

so we are done.
2. The interval $I = (-\infty, a)$ is also an open set. Fix a point $x \in I$. Necessarily $a - x > 0$. We need to find a radius of the ball $B_\varepsilon(x)$ centred at x so this ball is entirely within I. Pick $\varepsilon = \frac{a-x}{2} > 0$. Now we show that the ball $B_\varepsilon(x) = (x - \varepsilon, x + \varepsilon) \subseteq I = (-\infty, a)$. It is enough to show that $x + \varepsilon < a$. We know that $a - x > \frac{a-x}{2} = \varepsilon$. Then, $x + \varepsilon < a$ and we are done. Similarly, the interval $I = (b, \infty)$ is also an open set.
3. Now to show that the set $I = [a, b]$ for $a \leq b$ is closed. In order to show that this set is closed, by definition of closed sets in Definition 4.5.14, we need to show that the complement of this set is open. The complement of this set is $I^c = [a, b]^c = \mathbb{R} \setminus [a, b] = (-\infty, a) \cup (b, \infty)$.
Pick any point $x \in I^c$. Then, either $x \in (-\infty, a)$ or $x \in (b, \infty)$. From the previous example, we know that each of the sets $(-\infty, a)$ and (b, ∞) are open. Thus, in both cases, for any point $x \in I^c$, we can fit an open ball of some small enough radius $\varepsilon > 0$ which is fully contained in I^c. So the set I^c is open and we conclude that I is closed.
Here we can see that the union of two open sets is open. In fact, we shall see in Proposition 4.5.16 later that the union of any number of open sets is always open.
4. The set $I = [a, b)$ is neither closed nor open. Indeed, we can find a point in I such that any ball centred at this point would not be contained in I. If we draw a ball centred at a, then $B_\varepsilon(a)$ will not be contained in I for any radius $\varepsilon > 0$ at all, no matter how small ε is picked. For concreteness, let $\varepsilon > 0$ be any radius. Then, $a - \frac{\varepsilon}{2} \in B_\varepsilon(a) = (a - \varepsilon, a + \varepsilon)$. However, $a - \frac{\varepsilon}{2} < a$ so this point $a - \frac{\varepsilon}{2}$ is not contained in I. Thus, the ball $B_\varepsilon(a)$ is not entirely contained in the set I for any $\varepsilon > 0$ and hence I cannot be open.
Similarly to show that I is not closed, we take the complement of I, namely $I^c = [a, b)^c = (-\infty, a) \cup [b, \infty)$. By using a similar argument as above, this set is not open since any ball centred at b would not remain in this set. Since I^c is not open, I is not closed. Therefore the set I is neither open nor closed.

5. The set $\mathbb{R}$ is clearly open because for any point $x \in \mathbb{R}$, the unit radius ball $B_1(x)$ is fully contained in the set $\mathbb{R}$. This means $\mathbb{R}^c = \mathbb{R} \setminus \mathbb{R} = \varnothing$ is closed. On the other hand, the set $\varnothing$ is also open since any ball of any radius centred at a point of $\varnothing$ (which is empty), lies fully in $\varnothing$. Therefore $\varnothing^c = \mathbb{R} \setminus \varnothing = \mathbb{R}$ is closed. Thus, the sets $\mathbb{R}$ and $\varnothing$ are both open and closed. Hence, this is why they are called clopen sets, similar to the discussion in Remark 4.5.6.

We present a basic result which allows us to construct open and closed sets with unions and intersections. We have:

Proposition 4.5.16 *Let $X, Y \subseteq \mathbb{R}$ be subsets in $\mathbb{R}$.*

1. *If X and Y are both open, then $X \cup Y$ and $X \cap Y$ are also open sets.*
2. *If X and Y are both closed, then $X \cup Y$ and $X \cap Y$ are also closed sets.*

More generally:

3. *If X_n are all open sets in $\mathbb{R}$ for $n \in \mathbb{N}$, then $\bigcup_{n\in\mathbb{N}} X_n$ is open.*
4. *If Y_n are all closed sets in $\mathbb{R}$ for $n \in \mathbb{N}$, then $\bigcap_{n\in\mathbb{N}} Y_n$ is closed.*

The readers are invited to prove these results and construct some counterexamples for the infinite unions and intersections in Exercises 4.21 and 4.22. Next, reiterating Definition 3.6.3, we state:

Definition 4.5.17 (Bounded Set) A set $X \subseteq \mathbb{R}$ is called:

1. Bounded below if there exists an $a \in \mathbb{R}$ such that $a \leq x$ for all $x \in X$.
2. Bounded above if there exists a $b \in \mathbb{R}$ such that $x \leq b$ for all $x \in X$.
3. Bounded if it is both bounded above and below. In other words, there exist $a, b \in \mathbb{R}$ such that $a \leq x \leq b$ for all $x \in X$. Equivalently, there exists an $M > 0$ such that $|x| \leq M$ for all $x \in X$.

Bounded sets in the set of real numbers, by the completeness axiom on $\mathbb{R}$, have a supremum and an infimum. Moreover, if the set is an interval, the set would have a very specific form. The following result tells us the supremum and infimum of some intervals.

Proposition 4.5.18 *Let I be an interval in $\mathbb{R}$. Suppose $a, b \in \mathbb{R}$ with $a < b$.*

1. *If I is of the form $(-\infty, b]$, $[a, b]$, or $(a, b]$, then $\sup(I) = b$.*
2. *If I is of the form $(-\infty, b)$, $[a, b)$, or (a, b), then $\sup(I) = b$.*
3. *If I is of the form $[a, \infty)$, $[a, b]$, or $[a, b)$, then $\inf(I) = a$.*
4. *If I is of the form (a, ∞), $(a, b]$, or (a, b), then $\inf(I) = a$.*

Proof We prove the first two assertions since the other two assertions can be proven similarly. For both of these assertions, the interval I is bounded from above and so the supremum must exist. We find a candidate of the supremum for the set I by finding a suitable upper bound.

1. Clearly, b is an upper bound for I since for any $x \in I$, we have $x \leq b$. We claim b is the smallest upper bound of I. Indeed, if there is a smaller upper bound y of the set I, we must have $y < b$. However, y cannot be an upper bound for the set I since it is smaller than at least one element of I, namely b. This gives us a contradiction and therefore, $\sup(I) = b$.
2. Again, clearly, b is an upper bound for I since for any $x \in I$, we have $x < b$ by definition. Now we show that b is the smallest upper bound. Suppose for contradiction that there exists a smaller upper bound y for the set I. Since $y < b$, there exists at least one real number in between these two numbers, for example $y < \frac{y+b}{2} < b$. However $\frac{y+b}{2} \in I$ since it is strictly smaller than b. So there exists an element of I which is bigger than the upper bound y, a contradiction. Therefore $\sup(I) = b$. □

From Proposition 4.5.18, we can see that for a bounded interval I of $\mathbb{R}$, if I is a closed set then it contains its supremum and infimum. Therefore, it attains its maximum and minimum. However, if I is an open set, it will never attain both of its supremum and infimum.

Example 4.5.19 Consider the set $X = \{x \in \mathbb{R} : 0 < x \leq 1 \text{ or } 2 \leq x \leq 4\} = (0, 1] \cup [2, 4]$. This set is bounded from above and below, so its supremum and infimum must exist. By Proposition 4.5.18, $\sup(X) = 4$ and $\inf(X) = 0$. Furthermore, since $4 \in X$ and $0 \notin X$, we have $\max(X) = 4$ but $\min(X)$ does not exist.

We have seen in Examples 4.5.15(1) and (2) that open intervals are open sets. However, there are other open sets which are not intervals, such as the set $X = (-\infty, -1) \cup (1, \infty)$ in Example 4.5.15(4). Do we have a general expression for how open sets in $\mathbb{R}$ look like? The following theorem is a very important and useful characterisation of open sets in $\mathbb{R}$.

Theorem 4.5.20 *Any open subset $U \subseteq \mathbb{R}$ is a union of at most countably many disjoint open intervals, namely $U = \bigcup_{n \in \mathbb{N}} I_n$ where $I_n \subseteq \mathbb{R}$ are open intervals with $I_m \cap I_n = \varnothing$ for $m \neq n$.*

Proof Fix $x \in U$. Let $A = \{a \in \mathbb{R} : (a, x) \subseteq U\}$ and $B = \{b \in \mathbb{R} : (x, b) \subseteq U\}$ be subsets of $\mathbb{R}$. Since U is open, there exists an $\varepsilon > 0$ such that $(x - \varepsilon, x + \varepsilon) \subseteq U$, meaning $x - \varepsilon \in A$ and $x + \varepsilon \in B$. This means both A and B are non-empty. Define $I_x = (a_x, b_x)$ where $a_x = \inf(A)$ if A is bounded below or $a_x = -\infty$ otherwise, and $b_x = \sup(B)$ if B is bounded above or $b_x = \infty$ otherwise. We note the following facts:

1. Since $a_x < x < b_x$, we have $x \in (a_x, b_x) = I_x$.
2. We claim that $I_x \subseteq U$. Suppose for contradiction that this is not true. Then, there exists a $y \in I_x$ which is not in U. WLOG, suppose that $x < y < b_x$. We have two cases:
   (a) If b_x is finite, by characterisation of supremum, for $\varepsilon = \frac{b_x - y}{2} > 0$, we can find a $b' \in B$ such that $b_x - \varepsilon < b' \leq b_x$. This means $x < y < \frac{b_x+y}{2} = b_x - \varepsilon < b' \leq b_x$ and so $y \in (x, b') \subseteq U$, a contradiction.
   (b) If $b_x = \infty$, then the set B is not bounded above. By negation of Definition 4.5.17, this means for any $K > 0$ at all, there exists an element $z \in B$ such that $K < z$. Setting $K = y$, we can find a $z \in B$ such that $y < z$. By definition of the set B, this means $(x, z) \subseteq U$. Moreover, since $x < y < z$, we then have $y \in (x, z) \subseteq U$, a contradiction.
3. If b_x is finite, we must have $b_x \notin U$. Suppose for contradiction that $b_x \in U$. Since U is open, there exists an $\varepsilon > 0$ such that $(b_x - \varepsilon, b_x + \varepsilon) \subseteq U$. This implies $(x, b_x + \varepsilon) \subseteq U$, which means $b_x + \varepsilon \in B$ is greater than $\sup(B) = b_x$, giving us a contradiction. Similar arguments also show that if a_x is finite, then $a_x \notin U$.
4. I_x is the largest or maximal open interval in U which contains x. Indeed, suppose that (a, b) is any other interval in U which contains x. Then (a, x) and (x, b) are both in U as well. By definition of a_x and b_x, we must have $a_x \leq a$ and $b \leq b_x$ which then implies $(a, b) \subseteq (a_x, b_x) = I_x$.

We thus have built a maximal interval I_x containing the point $x \in U$ such that $I_x \subseteq U$. By repeating this construction for all points $x \in U$, we have the union $\bigcup_{x \in U} I_x$. Now we claim that this union is actually equal to U. The inclusion $\bigcup_{x \in U} I_x \subseteq U$ is clear since all the I_x are contained in U. For the reverse inclusion, pick any $x \in U$. Then, $x \in I_x$ and thus $x \in \bigcup_{x \in U} I_x$. Since $x \in U$ is arbitrary, we have $U \subseteq \bigcup_{x \in U} I_x$. Therefore, we have the equality of sets $U = \bigcup_{x \in U} I_x$.

Now for any two points $x, y \in U$, we claim that either $I_x = I_y$ or $I_x \cap I_y = \varnothing$. Indeed, supposing that the intersection $I_x \cap I_y$ is non-empty, the union $I_x \cup I_y$ is also an open interval according to Lemma 4.5.4 and Proposition 4.5.16. Also, $I_x \cup I_y$ contains x and, by maximality of I_x, this union is contained I_x. This means $I_x \cup I_y \subseteq I_x$ and so $I_y \subseteq I_x$ by Exercise 1.18. By symmetry, we can show the inclusion $I_x \subseteq I_y$ and hence conclude that $I_x = I_y$.

Thus, we can discard any repeated sets in the union $U = \bigcup_{x \in U} I_x$ to get the union $U = \bigcup_{j \in J} I_j$ where J is some indexing set (finite or infinite) and I_j are all disjoint from each other.

Finally, to show that there is at most countably many such I_j, note that since I_j is an open interval, it would contain at least one rational number. Thus, there is an injection from the set of intervals $\{I_j\}_{j \in J}$ to $\mathbb{Q}$. Since $\mathbb{Q}$ is countable, the set of intervals $\{I_j\}_{j \in J}$ is at most countable and so we can relabel them as $\{I_n : n \in \mathbb{N}\}$. Hence $U = \bigcup_{n \in \mathbb{N}} I_n$. □

So any open set in $\mathbb{R}$ is a union of countably many open intervals of the form (a, b), $(-\infty, a)$, and (b, ∞). A related question is: can we do the same for a closed set in $\mathbb{R}$? We shall get an answer to this question at the end of Exercise 4.32.

Compact Sets

We have seen finite sets which are nice since we can list down all of the elements one by one in finite time. An advantage of finite sets in $\mathbb{R}$ is that they always have a maximum and a minimum. However, the sad news is very few subsets of $\mathbb{R}$ are finite. The next best thing after finite sets are compact sets. We first define:

Definition 4.5.21 (Open Cover) Let $X \subseteq \mathbb{R}$. An open cover for the set X is a collection $\mathcal{U} = \{U_j\}_{j\in J}$ for some indexing set J such that U_j are all open sets and $X \subseteq \bigcup_{j\in J} U_j$.

Intuitively, an open cover of a set X is simply a collection of open sets in $\mathbb{R}$ whose union covers or contains the whole of X. A set X can have many possible open covers.

Example 4.5.22 Consider the set $X = (0, 1) \subseteq \mathbb{R}$. Let us look at some open covers for X.

1. The collection $\mathcal{U} = \{\mathbb{R}\}$ is trivially an open cover for X. Indeed, the sets in $\mathcal{U}$ are all open and the union is $\mathbb{R}$ which contains X.
2. The collection $\mathcal{V} = \{(\frac{1}{n}, 1) : n \in \mathbb{N}, n \geq 2\}$ is also an open cover for the set X. Indeed, all of the sets in $\mathcal{V}$ are open. Moreover, for any $x \in X$, since $x > 0$ we can find an $n \in \mathbb{N}$ such that $\frac{1}{n} < x$ by the Archimedean property and so $x \in (\frac{1}{n}, 1) \subseteq \bigcup_{n\geq 2}(\frac{1}{n}, 1)$. Since $x \in X$ is arbitrary, we have the inclusion $X \subseteq \bigcup_{n\geq 2}(\frac{1}{n}, 1)$.
3. The collection $\mathcal{W} = \{(-n, n) : n \in \mathbb{N}\}$ is also an open cover for X since all of the sets in $\mathcal{W}$ are open and the union of all the sets in $\mathcal{W}$ is the whole of $\mathbb{R}$ which contains X.

For some open cover $\mathcal{U}$ for a set X, we might not need all of the sets in $\mathcal{U}$ to still cover X. We define:

Definition 4.5.23 (Subcover) Suppose that $X \subseteq \mathbb{R}$ with an open cover $\mathcal{U} = \{U_j\}_{j\in J}$ for some indexing set J. A subcover of $\mathcal{U}$ for X is a subset $\mathcal{U}' \subseteq \mathcal{U}$ such that $\mathcal{U}'$ is also an open cover for X.

Example 4.5.24 Recall Example 4.5.22(3) where $X = (0, 1)$ and $\mathcal{W} = \{(-n, n) : n \in \mathbb{N}\}$. We have seen that $\mathcal{W}$ forms an open cover for X. However, we do not need all of the sets in $\mathcal{W}$ to cover X. Indeed, if we define $\mathcal{W}' = \{(-1, 1)\} \subseteq \mathcal{W}$, the collection $\mathcal{W}'$ still covers X. Thus $\mathcal{W}'$ is a subcover of $\mathcal{W}$ for the set X.

Now we can define compact sets:

Definition 4.5.25 (Compact Set) A set $X \subseteq \mathbb{R}$ is called compact if for any of its open cover $\mathcal{U} = \{U_j\}_{j \in J}$, there is a finite subcover $\{U_j\}_{j=1}^n \subseteq \mathcal{U}$ for some $n \in \mathbb{N}$, namely $X \subseteq \bigcup_{j=1}^n U_j$.

The following is a paraphrased analogy used by Hermann Weyl (1885–1955) to describe compactness:

> If a city is compact, it can be guarded by a finite number of arbitrarily short-sighted policemen.

A big warning here: a set is compact if given any open cover at all we can find a finite subcover from this open cover. It does not say that X has a finite subcover: this is silly because any set $X \subseteq \mathbb{R}$ always has a finite open cover, namely $\{\mathbb{R}\}$ itself!

Example 4.5.26 Let us look at some examples here:

1. Consider a finite subset of the real numbers $X = \{a_1, a_2, \ldots, a_n\}$. Let $\mathcal{U} = \{U_j\}_{j \in J}$ be any open cover of the set X. We claim that the set X is compact. Indeed, for each $k \in \{1, 2, \ldots, n\}$, there exists a covering set $U_{j(k)} \in \mathcal{U}$ that contains the point a_k. Therefore, the collection $\{U_{j(k)}\}_{k=1}^n$ forms a subcover of $\mathcal{U}$ for X since every element of X is contained in at least one of the $U_{j(k)}$. Moreover, it is a finite subcover since it contains at most n open sets. Thus, any finite set is compact.
2. Recall Example 4.5.22 where $X = (0, 1)$. We have seen three open covers of X, namely $\mathcal{U} = \{\mathbb{R}\}$, $\mathcal{V} = \{(\frac{1}{n}, 1) : n \in \mathbb{N}, n \geq 2\}$, and $\mathcal{W} = \{(-n, n) : n \in \mathbb{N}\}$. For the open covers $\mathcal{U}$ and $\mathcal{W}$, we can find finite subcovers from each of them that still covers X. Does this mean that the set X is compact?
   No! As warned above, compactness requires every open cover to have a finite subcover. In this case, the cover $\mathcal{V}$ does not have a finite subcover. Suppose for contradiction that $\mathcal{V}$ has a finite subcover $\mathcal{V}' = \{V_{j_n}\}_{n=1}^m$ where $V_{j_n} = (\frac{1}{j_n}, 1)$ for some natural numbers $j_n \geq 2$. Set $j = \max\{j_n : n = 1, \ldots, m\}$ and thus the union of all the sets in $\mathcal{V}'$ would be $\bigcup_{n=1}^m V_{j_n} = (\frac{1}{j}, 1)$. However this does not cover X since $\frac{1}{2j} \in X$ is not in the union, giving us the desired contradiction.
   Since we have found a cover $\mathcal{V}$ of X that does not have a finite subcover, we conclude that X is not a compact set.
3. Now we claim that for any $a, b \in \mathbb{R}$ with $a \leq b$, the closed interval $[a, b]$ is compact. The case for which $a = b$ is trivial since it is a finite set.
   Suppose that $a < b$. Let $\mathcal{U} = \{U_j\}_{j \in J}$ be any open cover for $[a, b]$. We define $S \subseteq [a, b]$ as the set of all $x \in [a, b]$ for which the interval $[a, x]$ has a finite

subcover in $\mathcal{U}$, namely:

$$S = \{x \in [a, b] : [a, x] \text{ has a finite subcover in } \mathcal{U}\}.$$

Clearly this set is nonempty as $a \in S$. Furthermore, the set S is bounded from above by b. Thus, $\sup(S)$ must exist and $\sup(S) \leq b$. We want to show that $\sup(S) = b$. Suppose for contradiction that $\sup(S) < b$. Denote $x_0 = \sup(S)$. Clearly $x_0 > a$ since any open set that contains a would also contain $a + \varepsilon$ for some $\varepsilon > 0$.

Now take any open set $U_0 \in \mathcal{U}$ that contains x_0. Since U_0 is open, there exists some $\varepsilon > 0$ such that $(x_0 - \varepsilon, x_0 + \varepsilon) \subseteq U_0$. WLOG, by shrinking ε as necessary, we can assume that ε is such that $a < x_0 - \varepsilon$. Note that $(x_0 - \varepsilon, x_0 + \frac{\varepsilon}{2}] \subseteq (x_0 - \varepsilon, x_0 + \varepsilon) \subseteq U_0$. Since $x_0 = \sup(S)$, by characterisation of supremum, for this ε there exists an element $x \in S$ such that $a < x_0 - \varepsilon < x \leq x_0$. See Fig. 4.5 for a visualisation.

Since $x \in S$, by definition, there exists a finite subcover $\mathcal{U}' \subseteq \mathcal{U}$ for the interval $[a, x]$. Furthermore, the subcover $\mathcal{U}' \cup \{U_0\}$ of $\mathcal{U}$ is also finite and covers the closed interval $[a, x] \cup (x_0 - \varepsilon, x_0 + \frac{\varepsilon}{2}] = [a, x_0 + \frac{\varepsilon}{2}] \subseteq [a, x_0 + \varepsilon)$. This means $x_0 + \frac{\varepsilon}{2} \in S$, so x_0 cannot be the supremum of the set S. Thus, we have arrived at a contradiction.

Therefore $\sup(S) = b$ and, by a similar argument using the characterisation of supremum, we can show that $b \in S$. This means the interval $[a, b]$ has a finite subcover from $\mathcal{U}$ and hence must be compact.

Now that we have seen some examples, how can we tell whether a set in $\mathbb{R}$ is compact? From Example 4.5.26(2), we can see that some interval in $\mathbb{R}$ may not be compact but Example 4.5.26(3) shows that the closed bounded interval is compact. Extending from this example, we can prove that any compact set is in fact closed and bounded. In fact, this is a two-way implication and is called the Heine-Borel theorem, after Eduard Heine (1821–1881) and Émile Borel (1871–1956).

Theorem 4.5.27 (Heine-Borel Theorem) *Let $X \subseteq \mathbb{R}$. The set X is compact if and only if it is closed and bounded.*

Proof We prove the implications one by one:

($\Leftarrow$): Since the set X is bounded, there is some $M > 0$ such that $X \subseteq [-M, M] = Y$. Let $U_0 = (M - 1, M + 1) \cap X^c$. This is an open set since X is closed. Let $\mathcal{U} = \{U_j\}_{j \in J}$ be an open cover of the set X. Then, the collection $\mathcal{V} =$

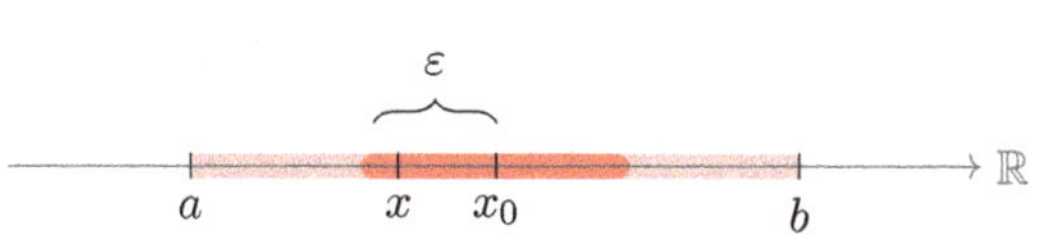

Fig. 4.5 Configuration of $x_0 = \sup(S)$ and $x \in S$

$\mathcal{U} \cup \{U_0\}$ covers Y. From Example 4.5.26(3), there is a finite subcover for Y within $\mathcal{V}$, which we call $\mathcal{V}' \subseteq \mathcal{V}$. Since $\mathcal{V}'$ covers Y, it also covers X. Since U_0 does not contain X, if $U_0 \in \mathcal{V}'$, we can remove the covering set U_0 and the remaining covering sets $\mathcal{V}' \setminus \{U_0\}$ still cover X. Then, $\mathcal{V}' \setminus \{U_0\} \subseteq \mathcal{V} \setminus \{U_0\} = \mathcal{U}$ is a finite subcover of $\mathcal{U}$ for X. Thus, X is compact.

($\Rightarrow$): Suppose that the set X is compact. We prove that it is bounded and closed separately.

1. Consider $\mathcal{U} = \{U_n\}_{n\in\mathbb{N}}$ where $U_n = (-n, n)$. Since $X \subseteq \mathbb{R} = \bigcup_{n=1}^{\infty} U_n$, necessarily $\mathcal{U}$ is a cover for the set X. Since X is compact, there is a finite subcover $\mathcal{U}' = \{U_{k_n}\}_{n=1}^{N}$ of $\mathcal{U}$ for some finite $N \in \mathbb{N}$. Let $K = \max\{k_n : n = 1, 2, \dots, N\}$ so that $X \subseteq \bigcup_{n=1}^{N} U_{k_n} = (-K, K)$. This implies the set X is bounded since every $x \in X$ satisfies $|x| \leq K$.
2. To prove that the set X is closed, we prove X^c is open. Fix an $x \in X^c$. We define a collection of open sets $\mathcal{U} = \{U_n\}_{n=1}^{\infty}$ where $U_n = \left[x - \frac{1}{n}, x + \frac{1}{n}\right]^c = \left(-\infty, x - \frac{1}{n}\right) \cup \left(x + \frac{1}{n}, \infty\right)$. Then, the union of these sets satisfy:

$$\bigcup_{n=1}^{\infty} U_n = \bigcup_{n=1}^{\infty} \left(\left(-\infty, x - \frac{1}{n}\right) \cup \left(x + \frac{1}{n}, \infty\right)\right)$$
$$= (-\infty, x) \cup (x, \infty) = \mathbb{R} \setminus \{x\},$$

which contains X. Thus, $\mathcal{U}$ forms an open cover for the set X. Since the set X is compact, we can find a finite subcover for X from $\mathcal{U}$, which we call $\mathcal{U}' = \{U_{k_n}\}_{n=1}^{N}$. Setting $K = \max\{k_n : n = 1, 2, \dots, N\}$, we have:

$$X \subseteq \bigcup_{n=1}^{N} U_{k_n} = \left(-\infty, x - \tfrac{1}{K}\right) \cup \left(x + \tfrac{1}{K}, \infty\right)$$
$$\Rightarrow \quad X^c \supseteq \left(\left(-\infty, x - \tfrac{1}{K}\right) \cup \left(x + \tfrac{1}{K}, \infty\right)\right)^c = \left[x - \tfrac{1}{K}, x + \tfrac{1}{K}\right].$$

Thus, we have the inclusion $B_{\frac{1}{K}}(x) \subseteq [x - \frac{1}{K}, x + \frac{1}{K}] \subseteq X^c$. Since $x \in X^c$ is arbitrarily fixed, this implies that X^c is open and hence the set X is closed. □

4.6 Real n-Space and Complex Numbers

Finally, we end this chapter with a glimpse of algebraic and topological structures in other spaces that we can obtain from $\mathbb{R}$.

Real n-Space

An important object in mathematics which generalises the real numbers is the real n-space and is studied extensively in linear algebra. This space is obtained by constructing the Cartesian product of finitely many real lines so that each point in this space can be described uniquely by an ordered list of real numbers, one from each of the constituent real lines. This ordered list is called the coordinates of that point.

Definition 4.6.1 (Real n-space) The real n-space for some $n \in \mathbb{N}$ is the Cartesian product of n copies of the set $\mathbb{R}$. Namely:

$$\mathbb{R}^n = \underbrace{\mathbb{R} \times \mathbb{R} \times \cdots \times \mathbb{R}}_{n \text{ times}} = \{(a_1, a_2, \ldots, a_n) : a_j \in \mathbb{R} \text{ for } j = 1, 2, \ldots, n\}.$$

However, different to the single $\mathbb{R}$, the real n-spaces $\mathbb{R}^n$ for $n \geq 2$ are not canonically ordered number fields because the elements do not even satisfy the field axioms (even though each component of the element does). The set $\mathbb{R}^n$ is not even a ring because we do not have a canonical multiplication operation on it which satisfies the ring axioms.

Instead, the space $\mathbb{R}^n$ forms an algebraic structure which is called a real vector space (or, in the language of abstract algebra, an $\mathbb{R}$-module). A real vector space is a different algebraic structure compared to rings and fields altogether. Elements in a real vector space can be added together, but they may not have a multiplication operation amongst themselves. Instead, we can multiply elements in a real vector space with elements in $\mathbb{R}$ in a process called scaling or scalar multiplication. A general real vector space is defined as:

Definition 4.6.2 (Real Vector Space) A real vector space is the triple $(V, +, \cdot)$ where V is a set with the addition and scalar multiplication operations $+ : V \times V \to V$ and $\cdot : \mathbb{R} \times V \to V$ satisfying the following axioms:

1. $\mathbf{v} + \mathbf{w} = \mathbf{w} + \mathbf{v}$ for all $\mathbf{v}, \mathbf{w} \in V$.
2. $\mathbf{u} + (\mathbf{v} + \mathbf{w}) = (\mathbf{u} + \mathbf{v}) + \mathbf{w}$ for all $\mathbf{u}, \mathbf{v}, \mathbf{w} \in V$.
3. There exists a unique identity element $\mathbf{0} \in V$ such that $\mathbf{0} + \mathbf{v} = \mathbf{v}$ for all $\mathbf{v} \in V$.
4. For every $\mathbf{v} \in V$, there exists an inverse element $-\mathbf{v} \in V$ such that $\mathbf{v} + (-\mathbf{v}) = 0$.
5. $1 \cdot \mathbf{v} = \mathbf{v}$ for any $\mathbf{v} \in V$.
6. $a \cdot (b \cdot \mathbf{v}) = (ab) \cdot \mathbf{v}$ for all $a, b \in \mathbb{R}$ and $\mathbf{v} \in V$.
7. $a \cdot (\mathbf{v} + \mathbf{w}) = a \cdot \mathbf{v} + a \cdot \mathbf{w}$ for all $a \in \mathbb{R}$ and $\mathbf{v}, \mathbf{w} \in V$.
8. $(a + b) \cdot \mathbf{v} = a \cdot \mathbf{v} + b \cdot \mathbf{v}$ for all $a, b \in \mathbb{R}$ and $\mathbf{v} \in V$.

The objects $\mathbf{v} \in V$ are called vectors and $a \in \mathbb{R}$ are called scalars. The scalar multiplication is usually suppressed as $a \cdot \mathbf{v} = a\mathbf{v}$. Due to their rather simple structure, vector spaces are studied in more detail in linear algebra and abstract algebra. Moreover, they occur naturally in various areas of mathematics such as functional analysis and functions spaces. We shall see some vector spaces in this book but we will not get into too much details with them.

Example 4.6.3 Let us look at some examples of real vector spaces.

1. Consider $\mathbb{R}^n$. For $n = 2$, we get the Cartesian plane which is a product of two copies of the real line, namely $\mathbb{R}^2 = \mathbb{R} \times \mathbb{R} = \{(x, y) : x, y \in \mathbb{R}\}$. Two numbers, which are called the coordinates, are needed to describe any point on the Cartesian plane. At the moment, $\mathbb{R}^2$ is just a set of points. Nothing more, nothing less.
   We can turn $\mathbb{R}^2$ into a vector space if we equip it with the appropriate algebraic operations. Addition and scalar multiplication on this space is done term-wise, namely if $(x, y), (w, z) \in \mathbb{R}^2$ and $\lambda \in \mathbb{R}$, we define:
   $$(x, y) + (w, z) = (x + w, y + z) \quad \text{and} \quad \lambda(x, y) = (\lambda x, \lambda y),$$
   which can be shown to satisfy the vector space axioms in Definition 4.6.2. Thus $(\mathbb{R}^2, +, \cdot)$ is a real vector space.
2. By setting $n = 1$ in the example above, the set of real numbers $\mathbb{R}$ can also be considered as a real vector space.
3. Recall the set of complex numbers from Exercise 3.24:
   $$\mathbb{C} = \{a + ib : a, b \in \mathbb{R}, i^2 = -1\}.$$
   As a set, the complex number $\mathbb{C}$ can also be thought of as the real 2-space $\mathbb{R}^2$ since $\mathbb{C} = \{a + ib : a, b \in \mathbb{R}\} \equiv \{(a, b) : a, b \in \mathbb{R}\}$ since every complex number is parametrised or described uniquely by two real numbers. For a complex number $z = a + ib$, we call a the real part of z and b the imaginary part of z, denoted by $\text{Re}(z)$ and $\text{Im}(z)$ respectively. We also define the complex conjugate of the number z as $\bar{z} = a - ib$.
   Similar to the representation of $\mathbb{R}^2$ as the Cartesian plane, complex numbers can be represented in an Argand diagram, named after Jean-Robert Argand. This diagram is similar to the Cartesian plane where the horizontal axis represents the real part of the complex number and the vertical axis represents the imaginary part of the number. Therefore, on the Argand diagram, the complex number $z = a + ib$ can be represented by the point with coordinate (a, b) as in Fig. 4.6. The set $\mathbb{C}$ can also be treated as a real vector space.

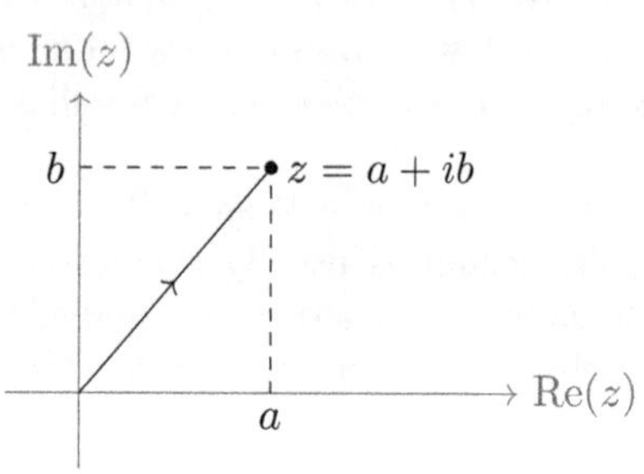

Fig. 4.6 The Argand diagram representing $\mathbb{C}$

In the space of real numbers $\mathbb{R}$, we have seen the modulus operation in Definition 4.5.7 which measures how far a number $x \in \mathbb{R}$ from the point 0. This can also be seen as the size of the number x. A generalisation of this concept to real vector spaces is called a norm.

Definition 4.6.4 (Norm) Let V be a real vector space. A norm $||\cdot|| : V \to \mathbb{R}$ is a function on V satisfying:

1. $||\mathbf{v}|| \geq 0$ for all $\mathbf{v} \in V$ with equality if and only if $\mathbf{v} = \mathbf{0}$,
2. $||\lambda\mathbf{v}|| = |\lambda|||\mathbf{v}||$ for all $\lambda \in \mathbb{R}$ and $\mathbf{v} \in V$, and
3. Triangle inequality: $||\mathbf{v} + \mathbf{w}|| \leq ||\mathbf{v}|| + ||\mathbf{w}||$ for all $\mathbf{v}, \mathbf{w} \in V$.

Sometimes the norm on a vector space is denoted as $|\cdot|$ instead of $||\cdot||$. In this case, context matters. A real vector space V equipped with a norm $||\cdot||$ is called a normed real vector space. Every real vector space can be endowed with a norm, but it may be possible that a vector space has many different norms that can be equipped to it. The choice of a norm on a vector space depends on how we want to measure the sizes of the elements in the space.

Norms are important in vector spaces because we want to be able to transfer from non-ordered vector spaces like $\mathbb{R}^n$ or $\mathbb{C}$ to a totally ordered set $\mathbb{R}_{\geq 0}$. We shall see how this is important shortly.

Example 4.6.5 Let us look at some examples:

1. Based on Proposition 4.5.8, we have seen that the modulus operation satisfies the norm axioms in Definition 4.6.4, so it is a norm on $\mathbb{R}$ if we view $\mathbb{R}$ as a real vector space as in Example 4.6.3(2).
2. Let $\mathbf{a} = (a_1, a_2, \dots a_n) \in \mathbb{R}^n$. We define the Euclidean norm of $\mathbf{a}$ to be:

$$||\mathbf{a}|| = \sqrt{a_1^2 + a_2^2 + \ldots + a_n^2}.$$

   One can easily show that this is indeed a norm by checking the norm axioms, which we leave as Exercise 4.33. The vector space $\mathbb{R}^n$ with the Euclidean norm above is called the Euclidean space.
3. We can also endow the real n-space $\mathbb{R}^n$ with a different norm. If $\mathbf{a} = (a_1, a_2, \dots a_n) \in \mathbb{R}^n$, then we can define $||\mathbf{a}|| = \max\{|a_1|, |a_2|, \dots, |a_n|\}$ which is also a norm. There are many other norms on the set $\mathbb{R}^n$, some of which we shall see in Exercise 4.33.
4. Likewise, since $\mathbb{C} = \mathbb{R}^2$ as real vector spaces, we can define a norm on the set of complex numbers as $|a + ib| = \sqrt{a^2 + b^2}$.

Complex Numbers

However, the complex number in Examples 4.6.3(3) and 4.6.5(4) is much more than just a real 2-space. The number i is called an imaginary unit which satisfies $i^2 = -1$. This structure of the imaginary unit i allows us to define a multiplication operation on $\mathbb{C}$. Using the fact that $i^2 = -1$, we can define:

$$(a + ib) + (c + id) = (a + c) + i(b + d),$$

$$(a + ib) \times (c + id) = ac + iad + ibc + i^2bd = (ac - bd) + i(ad + bc),$$

which allows us to show that the complex numbers $\mathbb{C}$ is, unlike an ordinary $\mathbb{R}^2$ space, a number field. The readers were asked to verify this in Exercise 3.24.

As a result of the multiplication definition above, the norm of a complex number $z \in \mathbb{C}$ in Example 4.6.5(4) can also be written as:

$$|z| = \sqrt{a^2 + b^2} = \sqrt{z\bar{z}},$$

where $\bar{z} = a - ib$ is the complex conjugate of z. We call this the complex norm.

If we plot the complex number z on an Argand diagram, the norm $|z| = \sqrt{a^2 + b^2}$ is geometrically the distance of the point z from the origin. This is simply the application of the Pythagorean theorem. So the norm measures how far the point z is from the origin. More generally, by translation, if we have two complex numbers $y, z \in \mathbb{C}$, the norm $|y - z|$ measures how far the two numbers are from each other on the complex plane.

The norm on complex numbers above is an important component for the study of complex numbers. Here are some properties of the norm on the complex numbers:

Proposition 4.6.6 *Let $y, z \in \mathbb{C}$. Then:*

1. $|-z| = |z|$.
2. $|z| = |\bar{z}|$.
3. $|yz| = |y||z|$.
4. $Re(z) \leq |Re(z)| \leq |z|$ *and* $Im(z) \leq |Im(z)| \leq |z|$.
5. *Triangle inequalities: We have* $|y + z| \leq |y| + |z|$ *and* $||y| - |z|| \leq |y - z|$.

Proof The first three assertions can be checked by computations. We prove assertions 4 and 5.

4. Let $z = a + ib$ where $a = \mathrm{Re}(z)$ and $b = \mathrm{Im}(z)$. Then, $|\mathrm{Re}(z)| = |a| \leq \sqrt{a^2 + b^2} = |z|$ and $|\mathrm{Im}(z)| = |b| \leq \sqrt{a^2 + b^2} = |z|$.

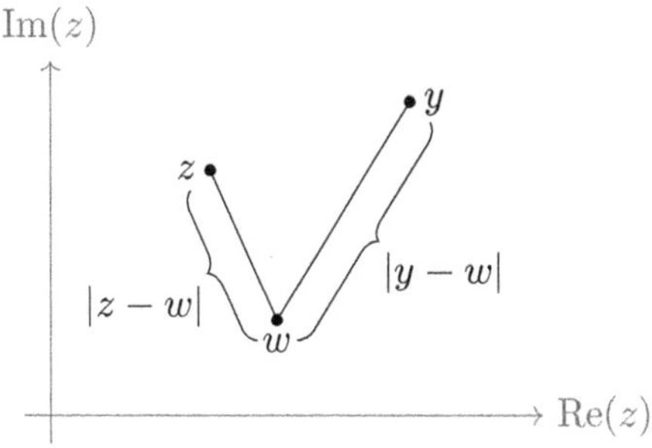

Fig. 4.7 Since $|z - w| < |y - w|$, we can say that z is closer to w than y is to w

5. We first note that $|y+z|^2 = (y+z)(\bar{y}+\bar{z}) = y\bar{y}+z\bar{z}+y\bar{z}+z\bar{y} = |y|^2+|z|^2+2\text{Re}(z\bar{y})$. Using the previous assertions, we get:

$$\begin{aligned}|y+z|^2 = |y|^2+|z|^2+2\text{Re}(z\bar{y}) &\leq |y|^2+|z|^2+2|\text{Re}(z\bar{y})| \\ &\leq |y|^2+|z|^2+2|z\bar{y}| \\ &= |y|^2+|z|^2+2|z||\bar{y}| \\ &= |y|^2+|z|^2+2|z||y| = (|z|+|y|)^2,\end{aligned}$$

and taking the square root on both sides yields the triangle inequality. The reverse triangle inequality can be deduced from the triangle inequality. □

However, as seen in Exercise 3.29, the complex number field does not have a total order which is compatible with the field structure above. Therefore, the concept of inequalities, supremum, infimum, maximum, minimum, ceiling and floor functions, and all the other field-compatible ordering concepts that we have seen for real numbers do not apply for complex numbers.

Due to this reason, norms are useful for us to transfer from the space $\mathbb{C}$ with no canonical ordering to the totally ordered set $\mathbb{R}_{\geq 0}$. We may not be able to compare (with a field-compatible ordering) the elements $z, y \in \mathbb{C}$.

But, by using norms, we can compare their distances from some other fixed point since distances, which are non-negative real numbers, can be ordered naturally. With norms, we can now determine which of the points z or y in the complex plane is closer to another fixed point $w \in \mathbb{C}$ by comparing the real numbers $|z - w|$ and $|y - w|$. To elaborate, if $|z - w| < |y - w|$, the point z is closer to the point w than the point y is to w. This can be visualised in Fig. 4.7. This notion will be important in Sect. 6.3 later.

The Argand diagram in Fig. 4.6 gives us another way to represent the complex number. If we draw a line segment connecting the non-zero point $z = a + ib$ to the origin $0 + i0$, we can compute the length of this segment using the complex norm which we denote as $r = |z| > 0$.

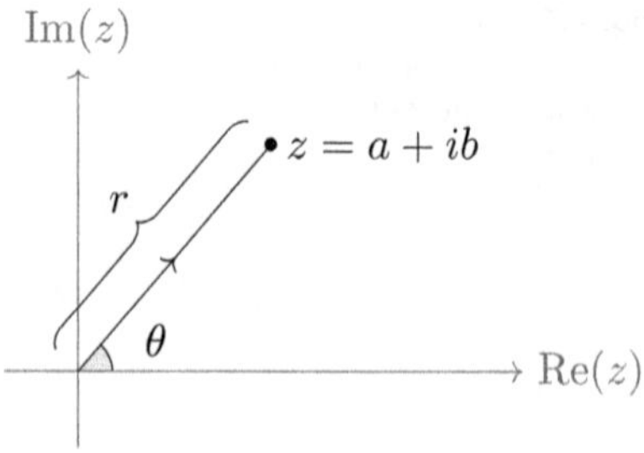

Fig. 4.8 The Argand diagram representing $\mathbb{C}$. The point $z = a + ib$ can be represented in polar form as $z = r(\cos(\theta) + i\sin(\theta))$ using its modulus r and principal argument θ

We call this quantity the modulus of z. Furthermore, we can also find the angle between the positive real axis and this line segment in a counter-clockwise orientation. This quantity is called the principal argument of the number z and is denoted as $\text{Arg}(z) = \theta \in (-180^\circ, 180^\circ]$ such that $a = r\cos(\theta)$ and $b = r\sin(\theta)$. Explicitly:

$$\theta = \begin{cases} \arctan\left(\frac{b}{a}\right)^\circ & \text{if } a > 0, \\ 90^\circ & \text{if } b > 0 \text{ and } a = 0, \\ -90^\circ & \text{if } b < 0 \text{ and } a = 0, \\ 180^\circ - \arctan\left(\frac{b}{|a|}\right)^\circ & \text{if } b > 0 \text{ and } a < 0, \\ -180^\circ + \arctan\left(\frac{b}{a}\right)^\circ & \text{if } b < 0 \text{ and } a < 0, \\ 180^\circ & \text{if } b = 0 \text{ and } a < 0. \end{cases}$$

Thus, a non-zero complex number can also be written as $z = a+ib = r(\cos(\theta)+i\sin(\theta))$ (Fig. 4.8).

In fact, we can extend the principal argument to any other arguments by adding an integer multiple of 360° to the principal argument. This form is called the polar form of the complex number. In this form, the conjugate of z can be written as $\bar{z} = a - ib = r(\cos\theta - i\sin\theta) = r(\cos(-\theta) + i\sin(-\theta))$. A very nice result that can be obtained via this representation is the De Moivre's identity which can be proven via induction:

Theorem 4.6.7 (De Moivre's Identity) *For any $z = a + ib \neq 0$ and $n \in \mathbb{Z}$, we have:*

$$z^n = (a + ib)^n = r^n(\cos\theta + i\sin\theta)^n = r^n(\cos(n\theta) + i\sin(n\theta)).$$

Moreover, for any two complex numbers $z_1 = r(\cos\theta + i\sin\theta)$ and $z_2 = s(\cos\phi + i\sin\phi)$, by using the standard trigonometric identities, we can show that their arguments are additive under multiplication, namely:

$$z_1z_2 = rs(\cos(\theta + \phi) + i\sin(\theta + \phi)).$$

As a result, we can also write the complex number in the form $z = r(\cos\theta + i\sin\theta) = re^{i\theta}$ where e is the Euler's number which we shall see in Example 5.4.4. This identity will be proven by the readers in Exercise 17.9. This profound relationship is called the Euler's formula and the notation is called the exponential form. This form is used to exploit the additivity of the exponent under multiplication and allows us to multiply and manipulate complex numbers easily. Thus the multiplication and conjugation in this form are:

$$z_1 z_2 = r(\cos(\theta) + i\sin(\theta)) \times s(\cos(\phi) + i\sin(\phi)) = re^{i\theta}se^{i\varphi} = rse^{i(\theta+\phi)},$$
$$\bar{z}_1 = r(\cos(-\theta) + i\sin(-\theta)) = re^{-i\theta}.$$

Moreover, De Moivre's identity can then be written as:

$$z_1^n = (re^{i\theta})^n = r^n e^{in\theta},$$

for $n \in \mathbb{N}$. As a result, the exponential form also allows us to clearly generalise De Moivre's identity from integer exponents $n \in \mathbb{Z}$ to any exponent in $\mathbb{R}$ (and even complex exponents!).

Topology on $\mathbb{R}^n$ and $\mathbb{C}$

So why are norms important? On a general real vector space, norms also allow us to define distances between any two points in the space. Using the Euclidean norm, if $\mathbf{x}, \mathbf{y} \in \mathbb{R}^n$ are of the form $\mathbf{x} = (x_1, x_2, \ldots, x_n)$ and $\mathbf{y} = (y_1, y_2, \ldots, y_n)$, we can define the Euclidean distance between these two points as:

$$\begin{aligned}||\mathbf{x} - \mathbf{y}|| &= ||(x_1, x_2, \ldots, x_n) - (y_1, y_2, \ldots, y_n)|| \\ &= \sqrt{(x_1 - y_1)^2 + (x_2 - y_2)^2 + \ldots + (x_n - y_n)^2}.\end{aligned}$$

Remark 4.6.8 On a more general set, we can define distances by a metric (or distance) function. We shall define what metric and metric spaces are in Chap. 6. But to give an idea, a metric is a function that assigns distances between elements in an arbitrary set.

Geometrically, this can be seen as the length of the line segment joining two points in the spaces using Pythagorean theorem. With distances between any two points in these spaces defined, we can now define other geometrical objects in $\mathbb{R}^n$.

Similar to Definitions 4.5.10, 4.5.11, and 4.5.12, in $\mathbb{R}^2$ we can define the open balls, closed balls, and sphere of radius $r \geq 0$ with centre $\mathbf{c} = (c_1, c_2) \in \mathbb{R}^2$ as the sets:

$$\begin{aligned}
B_r(\mathbf{c}) &= \{(x, y) \in \mathbb{R}^2 : ||(x, y) - (c_1, c_2)|| < r\} \\
&= \{(x, y) \in \mathbb{R}^2 : (x - c_1)^2 + (y - c_2)^2 < r^2\}, \\
\bar{B}_r(\mathbf{c}) &= \{(x, y) \in \mathbb{R}^2 : ||(x, y) - (c_1, c_2)|| \leq r\} \\
&= \{(x, y) \in \mathbb{R}^2 : (x - c_1)^2 + (y - c_2)^2 \leq r^2\}, \\
S_r(\mathbf{c}) &= \{(x, y) \in \mathbb{R}^2 : ||(x, y) - (c_1, c_2)|| = r\} \\
&= \{(x, y) \in \mathbb{R}^2 : (x - c_1)^2 + (y - c_2)^2 = r^2\}.
\end{aligned}$$

The objects above are obtained from Definitions 4.5.10, 4.5.11, and 4.5.12 by replacing the universe set $\mathbb{R}$ with $\mathbb{R}^2$ and distance measurement $|\cdot|$ with $||\cdot||$. An example of the open ball in $\mathbb{R}^2$ is given in Fig. 4.9.

This can also be extended to $\mathbb{R}^n$ for other $n \in \mathbb{N}$. In particular, for $n = 3$, the balls and spheres look like the usual beach balls and oranges in real life!

Likewise, we also have the definition for open balls, closed balls, and spheres in $\mathbb{C}$. For radius $r \geq 0$ and centre $c \in \mathbb{C}$, we have:

$$\begin{aligned}
B_r(c) &= \{z \in \mathbb{C} : |z - c| < r\}, \\
\bar{B}_r(c) &= \{z \in \mathbb{C} : |z - c| \leq r\}, \\
S_r(c) &= \{z \in \mathbb{C} : |z - c| = r\},
\end{aligned}$$

which are the open ball, closed ball, and sphere respectively. Again, these definitions are obtained from Definitions 4.5.10, 4.5.11, and 4.5.12 simply by changing the universe set from $\mathbb{R}$ to $\mathbb{C}$ and the distance measurement from modulus to complex norm. Geometrically, these objects look like a disc with no boundary, a disc with boundary, and a circle respectively in the complex plane. An open ball in the complex plane looks just like the open ball Fig. 4.9.

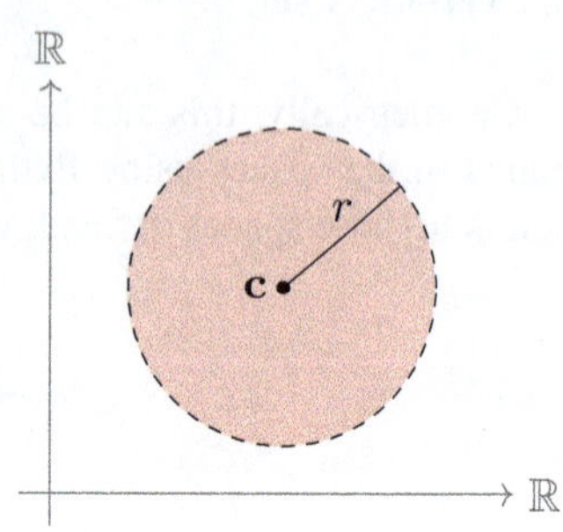

Fig. 4.9 Open ball $B_r(\mathbf{c})$ of radius r and centre $\mathbf{c}$ in $\mathbb{R}^2$ The dashed boundary line is $S_r(\mathbf{c})$ and not included in $B_r(\mathbf{c})$

With the open balls defined, we can thus define open sets in $\mathbb{R}^n$ and $\mathbb{C}$ as:

Definition 4.6.9 (Open and Closed Set) Let $X \subseteq \mathbb{R}^n$ or $\mathbb{C}$.

1. The subset X is called open if for all $x \in X$, there exists an $\varepsilon > 0$ such that the ball of radius ε and centred at x is contained fully in X. In other words, $B_\varepsilon(x) \subseteq X$.
2. The subset X is called closed if its complement is open. In other words, the set $X^c = \mathbb{R}^n \setminus X$ or $\mathbb{C} \setminus X$ is open.

Exercises

4.1 (a) Let $f : \mathbb{N} \to \mathbb{N}$ be defined as the function mapping the integer n to the smallest even integer greater than or equal to n. Write down the function f in closed form and show that it is increasing, namely $f(n+1) \geq f(n)$ for any $n \in \mathbb{N}$.

(b) Let $g : \mathbb{N} \to \mathbb{N}$ be defined as the function mapping the integer n to the smallest odd integer greater than or equal to n. Write down the function g in closed form and show that it is also increasing.

4.2 $(*)$ Let $a, b \in \mathbb{Q}$ and $c, d, \in \bar{\mathbb{Q}}$.

(a) Show that $a + b$ and $a \times b$ are in $\mathbb{Q}$.

(b) Is $c + d$ necessarily in $\bar{\mathbb{Q}}$?

(c) Is $c \times d$ necessarily in $\bar{\mathbb{Q}}$?

(d) Is c^d necessarily in $\bar{\mathbb{Q}}$?

Because irrational numbers are difficult to write down, many things about them are still unknown to us. For example, two of the most popular irrational numbers are the Euler-Napier constant e and the Archimedes constant π. In later chapters, we shall see why these numbers are fundamentally important in mathematics. However, even though they are fundamental constants, to this date, nobody knows whether the numbers $\pi \pm e$, πe, $\frac{\pi}{e}$, and π^e are irrational!

4.3 (a) Suppose that $a, b, c, d \in \mathbb{N}$ such that $\gcd(a, b) = \gcd(c, d) = 1$. Prove that if $ad = bc$, then $a = c$ and $b = d$.

(b) Let $p \in \mathbb{N}$ with $p > 1$. Show that for all $n \in \mathbb{N}$ we have $p^n > n$.

(c) Suppose that $r, s \in \mathbb{Q}_+$ are such that $r^r = s$. Prove that $r \in \mathbb{N}$. Hence, conclude that $s \in \mathbb{N}$.

4.4 Let $X, Y \subseteq \mathbb{R}$ such that X is bounded and $Y \subseteq X$. Given that for any $x \in X$ there exists a $y \in Y$ such that $x < y$, show that $\sup(X) = \sup(Y)$.

4.5 $(*)$ Prove Lemma 4.1.13, namely:

Let $\{f(x, y) : x \in X, y \in Y\}$ be a bounded set of real numbers parametrised by two parameters $x \in X$ and $y \in Y$. Prove that:

$$\sup\{\sup\{f(x, y) : y \in Y\} : x \in X\} = \sup\{\sup\{f(x, y) : x \in X\} : y \in Y\}.$$

4.6 (∗) Let $\{f(x, y) : x \in X, y \in Y\}$ be a bounded set of real numbers parametrised by two parameters $x \in X$ and $y \in Y$.
(a) Prove that:

$$\sup\{\inf\{f(x, y) : y \in Y\} : x \in X\} \leq \inf\{\sup\{f(x, y) : x \in X\} : y \in Y\}.$$

(b) Give an example for which the inequality in part (a) is strict.

4.7 (∗) In this question, we want to extend Bernoulli's inequality in Exercise 3.15 to rational exponents.
(a) Using the AM-GM inequality, prove the Bernoulli's inequality with natural number exponent in Exercise 3.15.
(b) Using the AM-GM inequality, extend the Bernoulli's inequality to rational exponent $r = \frac{p}{q} \in \mathbb{Q}$ where $0 \leq r < 1$. Namely, show that $(1 + y)^r \leq 1 + ry$ for $y > -1$.
(c) Hence, prove Bernoulli's inequality $(1 + x)^r \geq 1 + rx$ where $x > -1$ for the case of rational numbers $r > 1$.
We shall prove the case for irrational exponent r in Exercise 4.15.

4.8 (∗) Let $a, b, c > 0$ are real numbers. Suppose that there is a $d > 0$ such that for every $0 < \varepsilon < d$ we have $a \geq b - \varepsilon c$. Prove that necessarily $a \geq b$.
This is a common argument in analysis which we will employ repeatedly.

4.9 (∗) Let $A \subseteq \mathbb{R}$ be a non-empty bounded set of positive real numbers and $q \in \mathbb{Q}_+$. Define the set $A^q = \{x^q : x \in A\}$. We want to show that $\sup(A^q) = \sup(A)^q$ for $q \in \mathbb{Q}_+$.
For $q = 1$, we do not have to do anything. For $q > 1$, we follow these steps:
(a) Explain why $\sup(A)$ exists and is positive.
Show that the set A^q is also bounded and hence its supremum exists and is positive.
(b) First show that $\sup(A^q) \leq \sup(A)^q$.
(c) Show that for all $0 < \varepsilon < \sup(A)$ we have: $\sup(A^q) \geq \sup(A)^q - \varepsilon q \sup(A)^{q-1}$.
(d) Deduce that $\sup(A^q) \geq \sup(A)^q$.
(e) Conclude that $\sup(A^q) = \sup(A)^q$ for rationals $q > 1$.
Now we show that this is also true for rational exponents $0 < q < 1$:
(f) Show the equality of sets: $(A^q)^{\frac{1}{q}} = A$.
(g) By using parts (e) and (f), prove that $\sup(A^q) = \sup(A)^q$ for $0 < q < 1$.

4.10 Let $X \subseteq \mathbb{R}$ be a non-empty set. Determine all sets X for which $\sup(X) \leq \inf(X)$.
Can we find a set $X \subseteq \mathbb{R}$ such that $\sup(X) < \inf(X)$?

4.11 (∗) Prove the remaining assertions in Proposition 4.1.10, namely:
Let $X, Y \subseteq \mathbb{R}$ be bounded non-empty subsets of the real numbers and $\lambda \in \mathbb{R}$ be a real constant.
(a) Prove that the set $X \cup Y$ is also bounded with $\sup(X \cup Y) = \max(\sup(X), \sup(Y))$ and $\inf(X \cup Y) = \min(\inf(X), \inf(Y))$.

(b) Prove that:

$$\sup(\lambda X) = \lambda \sup(X) \quad \text{and} \quad \inf(\lambda X) = \lambda \inf(X) \quad \text{if } \lambda > 0,$$
$$\sup(\lambda X) = \lambda \inf(X) \quad \text{and} \quad \inf(\lambda X) = \lambda \sup(X) \quad \text{if } \lambda < 0.$$

(c) Prove that the set $X + Y$ is bounded with $\sup(X + Y) = \sup(X) + \sup(Y)$ and $\inf(X + Y) = \inf(X) + \inf(Y)$.

(d) For all sets $X, Y \subseteq \mathbb{R}$ for which $X \cap Y$ is non-empty, we have seen that $\sup(X \cap Y) \leq \min(\sup(X), \sup(Y))$. Give an example of sets X and Y for which strict inequality occurs.

4.12 Let $\{A_j\}_{j\in\mathbb{N}}$ be an infinite collection of bounded sets.

(a) Prove that for any finite $n \in \mathbb{N}$ we have:

$$\sup\left(\bigcup_{j=1}^{n} A_j\right) = \max_{1\leq j\leq n}(\sup(A_j)) \text{ and } \inf\left(\bigcup_{j=1}^{n} A_j\right) = \min_{1\leq j\leq n}(\inf(A_j)).$$

(b) Is it necessarily true that:

$$\sup\left(\bigcup_{j=1}^{\infty} A_j\right) = \sup_{j\in\mathbb{N}}(\sup(A_j)) \quad \text{and} \quad \inf\left(\bigcup_{j=1}^{\infty} A_j\right) = \inf_{j\in\mathbb{N}}(\inf(A_j)).$$

Provide a proof or counterexamples.

4.13 ($*$) Prove Lemma 4.2.5, namely:

Let $x \in \mathbb{R}$. Then:

(a) For $a > 1$, we have $\sup\{a^p : p \in \mathbb{Q}, p \leq x\} = \inf\{a^r : r \in \mathbb{Q}, r \geq x\}$.

(b) For $0 < a < 1$, we have $\inf\{a^p : p \in \mathbb{Q}, p \leq x\} = \sup\{a^r : r \in \mathbb{Q}, r \geq x\}$.

4.14 ($*$) Provide the proof for each assertion in Proposition 4.3.3 regarding the properties of the logarithm.

4.15 ($*$) Now let us complete the Bernoulli's inequality for irrational exponents. One need to recall Exercises 4.7 and 4.13 for this.

(a) For an irrational number $r > 1$ and any real number $x > -1$, by considering the cases for $-1 < x < 0$ and $x > 0$ separately, prove that $(1 + x)^r \geq 1 + rx$.

(b) Likewise, for an irrational number $0 \leq r < 1$ and any real number $x > -1$, by considering the cases for $-1 < x < 0$ and $x > 0$ separately, prove that $(1 + x)^r \leq 1 + rx$.

Hence we can encapsulate the Bernoulli's inequality for all real exponents in a compact form via the following important and useful proposition:

Proposition 4.7.10 (Bernoulli's Inequality) *For real numbers $r \geq 0$ and $x > -1$, the following inequalities hold:*

$$(1+x)^r \geq 1+rx \qquad \text{if } r \geq 1,$$

$$(1+x)^r \leq 1+rx \qquad \text{if } 0 \leq r < 1.$$

4.16 Prove that the set $\{2^x : x \in \mathbb{R}\}$ is not bounded from above.

4.17 ($*$) For each of the following sets, determine whether their infimum, minimum, supremum, and maximum exist. If they exist, find their values.
(a) $A = \mathbb{Q} \cap [0, \sqrt{2}]$.
(b) $B = \{(-1)^n + \frac{2}{n} : n \in \mathbb{N}\}$.
(c) $C = \{\frac{1}{2^m} + \frac{1}{3^n} : m, n \in \mathbb{N}\}$.
(d) $D = \bigcup_{n=1}^{\infty} [\frac{1}{2n}, \frac{1}{2n-1}]$.
(e) $E = \mathbb{N}$.
(f) $F = \left\{\frac{n^2+2n+1}{n^2+2n} : n \in \mathbb{N}\right\}$.
(g) $G = \left\{2^{-n^2} : n \in \mathbb{Z}\right\}$.
(h) $H = \left\{\frac{2^n}{n^2} : n \in \mathbb{Z} \setminus \{0\}\right\}$.

4.18 ($*$) Let $a, b \in \mathbb{R}$. Prove:
(a) $|-a| = |a|$.
(b) $|a|^2 = a^2$.
(c) $|ab| = |a||b|$ and $\left|\frac{a}{b}\right| = \frac{|a|}{|b|}$ if $b \neq 0$.
(d) $-|a| \leq a \leq |a|$.
(e) $b \leq |a|$ if and only if $a \leq -b$ or $a \geq b$.

4.19 Let $I \subseteq \mathbb{R}$ be a subset that is bounded from above. Suppose that the set I is closed downwards, namely if $x \in I$ and $y < x$, then $y \in I$ as well.
(a) Show that I must be an interval of the form $(-\infty, \sup(I))$ or $(-\infty, \sup(I)]$.
(b) Hence, prove that $\inf(I^c) = \sup(I)$.

4.20 ($*$) Prove Lemma 4.5.4:
Let $I, J \subseteq \mathbb{R}$ be two intervals in $\mathbb{R}$.
(a) The intersection $I \cap J$ is an interval.
(b) If $I \cap J \neq \varnothing$, then the union $I \cup J$ is also an interval.
Would this assertion still be true if we remove the non-empty intersection condition?

4.21 ($*$)Prove Proposition 4.5.16, namely:
Let $X, Y \subseteq \mathbb{R}$. Prove:
(a) If both X and Y are open sets, then the sets $X \cup Y$ and $X \cap Y$ are open.
(b) If both X and Y are closed sets, then the sets $X \cup Y$ and $X \cap Y$ are closed.

4.22 ($*$) Now let $\{X_n : n \in \mathbb{N}\}$ be an infinite collection of open sets and $\{Y_n : n \in \mathbb{N}\}$ be an infinite collection of closed sets in $\mathbb{R}$.
(a) Show that $\bigcup_{n \in \mathbb{N}} X_n$ is also open.
(b) By using De Morgan's law, show that $\bigcap_{n \in \mathbb{N}} Y_n$ is also closed.

(c) Find an example of a collection of open sets X_n such that $\bigcap_{n\in\mathbb{N}} X_n$ is not open.
(d) Find an example of a collection of closed sets Y_n such that $\bigcup_{n\in\mathbb{N}} Y_n$ is not closed.

4.23 ($*$) Determine whether the following subsets in $\mathbb{R}$ are open, closed, clopen, or neither:
(a) $A = (100, \infty)$.
(b) $B = [0, 2) \cup (3, 10]$.
(c) $C = [0, \infty)$.
(d) $D = \mathbb{Z}$.
(e) $E = \bigcup_{j=1}^{100} (j, j+1)$.
(f) $F = \mathbb{Q}$.
(g) $G = \bigcap_{j=1}^{\infty} [0, \frac{1}{j})$.
(h) $H = \{\frac{1}{n} : n \in \mathbb{N}\}$.

4.24 ($\diamond$) In this question, we want to prove Lemma 3.4.11.
WLOG, let $I = \mathbb{N}$ since I is bijective to $\mathbb{N}$. Let $\{A_n\}_{n\in\mathbb{N}}$ be the countable sets and $A = \bigcup_{j=1}^{\infty} A_j$
(a) Define a new collection of sets $\{B_n\}_{n\in\mathbb{N}}$ by $B_1 = A_1$ and $B_n = A_n \setminus \left(\bigcup_{j=1}^{n-1} A_j\right)$. Show that the sets B_n are countable, pairwise disjoint, and $\bigcup_{j=1}^{\infty} B_j = A$.
(b) Show that for each $n \in \mathbb{N}$, there is an injective map $\phi_n : B_n \to \mathbb{N}$.
(c) Hence, show that there is an injective map $\phi : A \to \mathbb{N} \times \mathbb{N}$ and deduce that A is countable.

4.25 Show that the set $A = \{(m, n) : m, n \in \mathbb{N}, m \leq n\}$ is countably infinite.

4.26 ($\diamond$) Let A be the set of all finite subsets of $\mathbb{N}$, namely $A = \{X \subseteq \mathbb{N} : |X| < \infty\}$.
(a) For each $n \in \mathbb{N}$, prove that the set $A_n = \{X \in A : |X| = n\} \subseteq A$ is countably infinite.
(b) Hence, deduce that A is countaby infinite.

4.27 (a) Via a suitable bijection, show that the sets $(0, 1)$ and $(0, \infty)$ have equal cardinalities.
(b) Via a suitable bijection, show that the sets $(0, 1)$ and $\mathbb{R}$ have equal cardinalities.
(c) Show that the sets $(0, 1)$ and $[0, 1)$ have equal cardinalities.

4.28 ($*$) Cantor's theorem in Theorem 3.9.6 states that for any set X, we have $|X| < |\mathcal{P}(X)|$. Since $\mathbb{N}$ is countably infinite set, we expect that the cardinality of its power set is at least uncountably infinite. In this question, we want to prove that $|\mathcal{P}(\mathbb{N})| = |\mathbb{R}|$.
(a) Explain why it is sufficient to instead show that $|[0, 1)| = |\mathcal{P}(\mathbb{N})|$.
(b) Construct an injection $f : [0, 1) \to \mathcal{P}(\mathbb{N})$.
(c) Conversely, construct an injective function $g : \mathcal{P}(\mathbb{N}) \to [0, 1)$.
(d) Conclude that $|\mathcal{P}(\mathbb{N})| = |\mathbb{R}|$.

4.29 Prove that there are no set X such that $|\mathcal{P}(X)| = |\mathbb{N}|$.

4.30 (◇) An algebraic number is a complex number that is a root of some polynomial with integer coefficients. Clearly, any rational number $\frac{p}{q} \in \mathbb{Q}$ is an algebraic number coming from a linear polynomial $P : \mathbb{R} \to \mathbb{R}$ defined as $P(x) = qx - p$. However, there are also irrational numbers which are algebraic.

(a) By constructing suitable polynomials, show that $\sqrt{2}$, $\sqrt{2}+\sqrt{3}$, $\varphi = \frac{1+\sqrt{5}}{2}$, and i are algebraic numbers.

(b) Denote the set of algebraic numbers as $\mathbb{A}$. Prove that the set $\mathbb{A}$ is countably infinite.

(c) On the other hand, numbers which are not algebraic are called transcendental. What is the cardinality of transcendental numbers?

This exercise gave us something to think about: rational numbers and algebraic numbers are (rather) easy to describe. However, they are not as abundant as the irrational numbers and transcendental numbers, both of which are more difficult to describe. So the set of numbers that we know how to write down forms just a tiny portion of all the numbers out there. One can see this in Fig. 4.10. As commented by Eli Maor (1937-):

> In 1874 ... Georg Cantor made the startling discovery that there are more irrational numbers than rational ones, and more transcendental numbers than algebraic ones. In other words, rather than being oddities, most real numbers are irrational; and among irrational numbers, most are transcendental.

As mentioned above, mathematicians do know that transcendental numbers exist and many conjectures were made. They became an interest of many mathematicians since they lie at the interface of analysis, algebra, and number theory. The first concrete example of a transcendental number was constructed by Joseph Liouville (1809–1882) which is the number with decimal representation $\sum_{j=1}^{\infty} \frac{1}{10^{j!}}$. The proof of this is given in [19].

After that, many other examples of transcendental numbers appear, such as $2^{\sqrt{2}}$, π, and e. The latter two can be proven to be transcendental by the Lindemann–Weierstrass theorem in algebra and number theory. There are many numbers out there which are still unknown whether to be algebraic or transcendental, for example the Euler-Mascheroni constant γ (see Exercise 12.27), the sum $\pi + e$, and the exponent π^e.

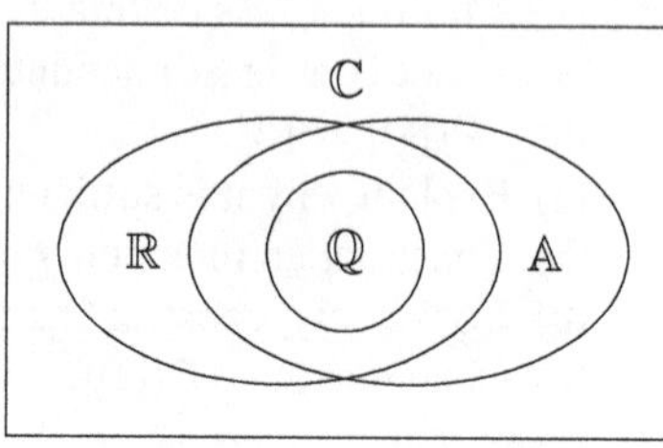

Fig. 4.10 The Venn diagram for the number sets in this question

4.31 (∗) In this question, we are going to prove the rational root theorem and use it to show irrationality of some real numbers. We state:

Theorem 4.7.11 (Rational Root Theorem) *Let $P : \mathbb{R} \to \mathbb{R}$ be a polynomial of degree $n \in \mathbb{N}$ with integer coefficients of the form $P(x) = \sum_{j=0}^{n} a_j x^j$ for $a_j \in \mathbb{Z}$ for $j = 0, 1, \ldots, n$. If $r = \frac{p}{q}$ is a rational root of P expressed in the lowest form, then $p|a_0$ and $q|a_n$.*

(a) By considering the equation $P(\frac{p}{q}) = 0$, prove the theorem above.
(b) Using the rational root theorem, deduce that any rational root of a monic polynomial with integer coefficients must be an integer.
(c) In Exercise 4.30 we have shown that $\sqrt{2}+\sqrt{3}$ is algebraic. Use the rational root theorem to prove that it is irrational.
(d) Show also that $\sqrt[3]{\sqrt{3}-1}$, $\sqrt{3-2\sqrt{5}}$, and $\sqrt{4+2\sqrt{6}}$ are irrational.

4.32 (∗) The Cantor set is a set obtained by a recursive construction. We start with the interval $C_0 = [0, 1]$ and remove the open middle third of this interval, namely $(\frac{1}{3}, \frac{2}{3})$. Thus we have $C_1 = [0, \frac{1}{3}] \cup [\frac{2}{3}, 1]$. In the next step, we remove the open middle thirds of each interval in C_1. This results in 4 disjoint closed intervals $C_2 = [0, \frac{1}{9}] \cup [\frac{2}{9}, \frac{3}{9}] \cup [\frac{6}{9}, \frac{7}{9}] \cup [\frac{8}{9}, 1]$.
Inductively, we carry out this construction to get a sequence of nested intervals $C_n \subseteq C_{n-1}$ where each C_n is made up of 2^n closed intervals of lengths $\frac{1}{3^n}$ obtained from removing the open middle third of the 2^{n-1} intervals in C_{n-1}. See Fig. 4.11 for a visualisation of this process.
The Cantor set C is then defined as the intersection $C = \bigcap_{n \in \mathbb{N}_0} C_n$.

(a) Explain how we can get the set C_n from C_{n-1} via scaling and translation. Hence, show that $C_{n+1} = \frac{1}{3}C_n \cup \left(\frac{2}{3} + \frac{1}{3}C_n\right)$ for all $n \in \mathbb{N}_0$.
(b) By induction, show that for $n \in \mathbb{N}$ we have:

$$C_n = \bigcap_{m=1}^{n} \bigcup_{j=0}^{\frac{3^m-1}{2}} \left[\frac{2j}{3^m}, \frac{2j+1}{3^m}\right].$$

(c) Hence, prove that C is a closed and non-empty set in $\mathbb{R}$.
(d) Any element in the set C can be represented in a ternary (base-3) representation of the form $\sum_{j=1}^{\infty} \frac{a_j}{3^j}$ where $a_j \in \{0, 1, 2\}$. Prove that the

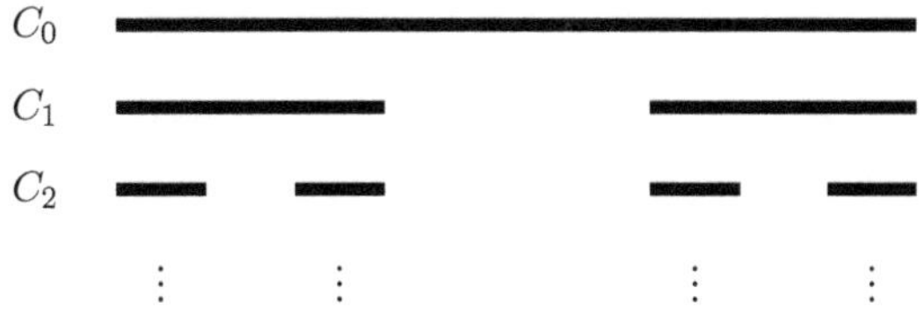

Fig. 4.11 Visualisation of the construction for C_0, C_1, and C_2

elements in C_1 are real numbers that can be represented by a unique ternary representation such that $a_1 \neq 1$.

(e) Thus, using part (a), show by induction that each number in C_n can be represented by a unique ternary representation such that $a_j \neq 1$ for $j = 1, 2, \dots, n$.

(f) Deduce that all elements in C can be represented uniquely by ternary representation consisting of the digits 0s and 2s only.

(g) By induction and part (b), show that any element of $[0, 1]$ with ternary representation consisting of only the digits 0s and 2s in the first n digits must be in C_n.

(h) Deduce that all numbers in $[0, 1]$ with only 0 and 2 in its ternary representation must be contained in C.

(i) Hence, show that C is an uncountably infinite set.

(j) Prove that C does not contain any closed interval of the form $[a, b]$ where $a \neq b$.

(k) Comment on the existence of the analogue to Theorem 4.5.20 for closed sets in $\mathbb{R}$.

4.33 ($\diamond$) Let $\mathbb{R}^n$ be the real n-space where $n \geq 5$. Show that the following definitions of $\|\cdot\| : V \to \mathbb{R}_{\geq 0}$ for $\mathbf{a} = (a_1, a_2, \dots a_n) \in \mathbb{R}^n$, satisfy the norm axioms:

(a) $\|\mathbf{a}\| = \sqrt{a_1^2 + a_2^2 + \dots + a_n^2}$.

(b) $\|\mathbf{a}\| = \max\{|a_1|, |a_2|, \dots, |a_n|\}$.

(c) $\|\mathbf{a}\| = |a_1| + |a_2| + \dots + |a_n|$.

(d) $\|\mathbf{a}\| = |a_1| + 7|a_2| + \max\{|a_3|, 5|a_4|\} + \sqrt{a_5^2 + \dots + a_n^2}$.

5 Real Sequences

A sequence works in a way a collection never can.

— George Murray, poet

In Chap. 4, during the construction of decimal representations for real numbers, we have defined an ordered infinite list of digits $a_0, a_1, a_2, \ldots$ as a name or identifier for a particular real number. For an irrational number r, we have also created a list of rational number $r_1, r_2, r_3, \ldots$ which approximate it. Thus, these infinite list of numbers is a useful concept. We would like to understand more about this infinite list of numbers.

In this chapter, we will look further into these lists and study their behaviour. We begin by defining what these lists are.

Definition 5.0.1 (Sequence of Real Numbers) A sequence of real numbers or a real sequence is a function $a : \mathbb{N} \to \mathbb{R}$. For simplicity, we also use the notation $(a_n)_{n\in\mathbb{N}}$ or simply (a_n) to denote a sequence. We write the n-th term of the sequence as $a(n)$ or a_n.

Thus, in general, sequences are just ordered infinite lists of objects. It is important to note that a sequence $(a_n)_{n\in\mathbb{N}}$ is different from a set $\{a_n\}_{n\in\mathbb{N}}$. A sequence has more information than a set. This is mainly because of two reasons:

1. The elements in a sequence must be ordered according to the indices n whereas the elements in a set can be ordered in any way we like.
2. Furthermore, for a set, each element can only appear once in the notation; if they appear multiple times in the notation, this is redundant as they are considered to be the same element. On the other hand, in a sequence, an element may appear multiple times in different places and remain distinct.

S. Johar, *The Big Book of Real Analysis*,
https://doi.org/10.1007/978-3-031-30832-1_5

Therefore, one has to be careful with the concept and notation. We have seen two specific real sequences from the previous chapter, so now let us look at more examples of real sequences.

Example 5.0.2 Writing down a sequence may be difficult or even impossible because it is a list of infinitely many things. However, if we know a pattern of a sequence, we can list the sequence completely by utilising the indices $n \in \mathbb{N}$.

1. If we have a sequence of $(a_n) = (-1, 1, -1, 1, -1, \ldots)$ which alternates between -1 and 1, we can write this sequence in full compactly by writing the n-th term as $a_n = (-1)^n$ for $n \in \mathbb{N}$.
2. If we have a sequence of even natural numbers $(b_n) = (2, 4, 6, 8, 10, \ldots)$, this sequence can be described succinctly as having the n-th term $b_n = 2n$ for $n \in \mathbb{N}$.
3. We can also describe a sequence via a recursive relation. For example, recall the Fibonacci sequence in Exercise 2.15. This sequence is given by $(f_n) = (1, 1, 2, 3, 5, 8, 13, \ldots)$. It can be described succinctly as $f_1 = f_2 = 1$ and $f_n = f_{n-1} + f_{n-2}$ for all integers $n \geq 3$. This recursive relation describes the sequence in full as the information given above is sufficient for us to determine the value of any term in the sequence.

Remark 5.0.3 In fact, the definition of real sequences generalises to any codomain.

1. We have defined a sequence of rational numbers that approximate $\sqrt{2}$ in Example 4.4.5. More concretely, this sequence is given by the list of numbers:

$$\begin{aligned} a_1 &= 1, \\ a_2 &= 1.4, \\ a_3 &= 1.41, \\ a_4 &= 1.414, \\ &\vdots \end{aligned}$$

   where each $a_n \in \mathbb{Q}$. The sequence above is called a rational sequence since each term in the sequence a_n is in $\mathbb{Q}$. We write this as $(a_n) \subseteq \mathbb{Q}$.
2. The point is that the codomain of a sequence $a : \mathbb{N} \to X$ can be any set X at all: the set of complex numbers $\mathbb{C}$, the set of integers $\mathbb{Z}$, the set of students in a class, the set of objects in a box, the set of letters in the alphabet, et cetera. We write this as $(a_n) \subseteq X$.
3. The real sequences in Example 5.0.2 are also integer sequences since each term in the sequences is an integer.

4. Another thing to note is that a sequence may also be finite if we define it as $a : \mathcal{N} \to \mathbb{R}$ where $\mathcal{N} = \{1, 2, \ldots, n\}$ for some finite number $n \in \mathbb{N}$. In this book, we distinguish these sequences by calling them finite sequences.

5.1 Algebra of Real Sequences

For the time being, let us restrict our attention to real sequences. This is because the real numbers form a field and we know how real numbers behave: we can add real numbers together, multiply with some other real number, and order them. Now we would like to see how real sequences behave.

First, using the field properties of real numbers, we can define or create new sequences from old sequences. Indeed, suppose that we have two real sequences (a_n) and (b_n). Then, we can define the sum of these sequences by taking the sum of the elements term-wise, namely we can construct a new sequence of sums $(a_n + b_n)$.

Using the same idea, we can also multiply the sequences term-wise to get a new sequence of products $(a_n b_n)$ and, if all of the b_n are non-zero, we can define the real sequence of quotients $\left(\frac{a_n}{b_n}\right)$. Also, if $\lambda \in \mathbb{R}$ is a real constant, we can define a new sequence by scaling the whole sequence (a_n) with this constant to get a new sequence (λa_n).

Example 5.1.1 Recall the two real sequences (a_n) and (b_n) in Example 5.0.2. From these two sequences, we can create many new sequences. Here are some examples:

1. $a_n b_n = (-1)^n 2n$ so $(a_n b_n) = (-2, 4, -6, 8, \ldots)$,
2. $2a_n = 2(-1)^n$ so $(2a_n) = (-2, 2, -2, 2, \ldots)$,
3. $\frac{1}{n} b_n = 2$ so $\left(\frac{1}{n} b_n\right) = (2, 2, 2, 2, \ldots)$,
4. $\frac{a_n}{b_n} = \frac{(-1)^n}{2n}$ so $\left(\frac{a_n}{b_n}\right) = \left(-\frac{1}{2}, \frac{1}{4}, -\frac{1}{6}, \frac{1}{8}, \ldots\right)$.

As the name suggests, infinite sequences do not have an end. This is another example of a potential infinity posited by Aristotle in Remark 2.3.8: we can keep going down the list indefinitely. In analysis, we are interested to see what the general behaviour of the sequence as we go far along it. In other words, what is the behaviour of the sequence towards infinity?

Example 5.1.2 Recall the sequences in Examples 5.0.2 and 5.1.1.

1. The sequence $\left(\frac{1}{n} b_n\right)$ is a constant sequence of 2s, so it just remains a fixed value, which is 2, forever.
2. For the sequence (a_n), the terms alternate between -1 and 1, so we know that it does not approach a fixed value: it keeps alternating between two values.

3. Finally, the sequence $\left(\frac{a_n}{b_n}\right)$ can be seen to get closer to the value 0. The terms alternate between being negative and positive. However, the size of the terms get smaller as we increase n. So it seems that the sequence converges to a common point 0.

5.2 Limits and Convergence

The words "closer", "converge", and "approach" that we have used above are very vague. What does it mean by real numbers being "close" to each other and what does it mean for a sequence to "converge" to a fixed value? Mathematics is a precise language, so we need to define exactly what these terms mean. The term "closer" indicates that there is an idea of distance between the numbers, so we need to first quantify distances between numbers.

We have developed the idea of distance between two real numbers in the previous chapter. Recall from Definition 4.5.7 that for real numbers $a, b \in \mathbb{R}$, the non-negative quantity $|a - b|$ gives a measure for the distance between these two numbers.

Bounded Sequences

From this definition of distance, we can first define bounded real sequences. A bounded real sequence is a sequence such that all of its terms are within a constant finite distance from a fixed reference point. A convenient reference point in $\mathbb{R}$ would be the number 0. We define:

Definition 5.2.1 (Bounded Real Sequence) Let (a_n) be a real sequence.

1. If there exists a constant $M \in \mathbb{R}$ such that $M \leq a_n$ for all $n \in \mathbb{N}$, then we call the sequence bounded from below.
2. If there exists a constant $M \in \mathbb{R}$ such that $a_n \leq M$ for all $n \in \mathbb{N}$, then we call the sequence bounded from above.
3. If there exists a constant $M > 0$ such that $|a_n| \leq M$ for all $n \in \mathbb{N}$, then we call the sequence bounded. In other words, a bounded sequence is bounded from above and below since $|a_n| \leq M$ is equivalent to saying $-M \leq a_n \leq M$ for all $n \in \mathbb{N}$.

   In terms of balls, a sequence is bounded if it stays within the closed ball $[-M, M] = \bar{B}_M(0)$. This is equivalent to saying tht the distance between any element in the sequence to 0 is at most M.

Remark 5.2.2 Since a real sequence can also be treated as a subset of the real numbers, Definition 5.2.1 is equivalent to Definition 4.5.17.

Example 5.2.3 Let us look at some examples of bounded and unbounded sequences from Example 5.1.1.

1. The sequence (a_n) defined as $a_n = (-1)^n$ is bounded from above and below. We note that $|a_n| = |(-1)^n| = 1$. This means $|a_n| \leq 1$ for any $n \in \mathbb{N}$ and hence the sequence is bounded.
2. Recall the sequence (b_n) defined as $b_n = 2n$. Clearly, $b_n = 2n > 0$ for all $n \in \mathbb{N}$ which means the sequence is bounded from below. However, the sequence is not bounded from above. Suppose for contradiction that it is bounded from above, namely there exists some $M > 0$ such that $b_n \leq M$ for all $n \in \mathbb{N}$. In particular, if $n = \lceil M \rceil \in \mathbb{N}$, we would have $b_n = 2n = 2\lceil M \rceil \geq 2M > M$ which contradicts the assumption that $b_n \leq M$ for all $n \in \mathbb{N}$. Thus, this sequence cannot be bounded from above.

Convergent Sequences

Since we have a notion of distances between two points in $\mathbb{R}$, we are now ready to define what the vague term "converge" means. Let us make some observations first to determine what we want it to mean.

Recall from Sect. 4.4 that we have constructed rational approximations of an irrational number $r \sim a_0.a_1a_2a_3a_4\ldots$. For any $n \in \mathbb{N}$, we have the sequence of rational approximations (r_n) for the number r defined as $r_n = \sum_{i=0}^{n} \frac{a_j}{10^j} \in \mathbb{Q}$.

1. The construction implies that for any $n \in \mathbb{N}$, we have the inequality $0 < r - r_n < \frac{1}{10^n}$, which means the distance between the irrational number r and its n-th rational approximation r_n is less than $\frac{1}{10^n}$. This distance can be made as small as we like since $\frac{1}{10^n}$ can be arbitrarily small if we pick a very large n.
2. Furthermore, for a fixed large index N, we know that $0 < r - r_N < \frac{1}{10^N}$, but this is also true for any larger index $n \geq N$ since $0 < r - r_n < \frac{1}{10^n} \leq \frac{1}{10^N}$ for any $n \geq N$ at all. This means the distances between r and any r_n for indices n larger than N are all also smaller than $\frac{1}{10^N}$. Thus, beyond this index N, we know that there are no outliers in the sequence (r_n) deviating a distance more than $\frac{1}{10^N}$ away from the number r.

These two observations would form a model for our definition for the convergence of a sequence: no matter how small a distance is fixed, there always exists an index $N \in \mathbb{N}$ for which the distance from the N-th term of the sequence, along with all the terms that come after it, to the convergence point would be smaller than this fixed distance.

This proper definition was first coined by Bolzano in 1816, but was very little noticed at the time. Later this idea was picked up and popularised by Cauchy in his seminal textbook *Cours d'Analyse* (Analysis Course) by a wordy definition:

> When the values successively attributed to the same variable indefinitely approach a fixed value in such a way as to end by differing from it as little as one wishes, this latter is called the limit of all others.

This qualitative description is then fine-tuned by Karl Theodor Wilhelm Weierstrass (1815–1897) into the modern definition which is more quantitative in nature. This definition is given by:

Definition 5.2.4 (Convergent Real Sequence) A real sequence (a_n) is convergent or converges if there exists a number $L \in \mathbb{R}$ such that for any $\varepsilon > 0$, there exists an index $N(\varepsilon) \in \mathbb{N}$ such that for all $n \geq N(\varepsilon)$ we have $|a_n - L| < \varepsilon$. We call the number L the limit of the sequence (a_n). We write this as:

$$\lim_{n\to\infty} a_n = L \quad \text{or} \quad a_n \xrightarrow{n\to\infty} L \quad \text{or} \quad a_n \to L.$$

The definition above is sometimes referred to the ε-N definition of convergence for obvious reasons. Symbolically, this definition is written with quantifiers as:

$$(a_n) \text{ converges} \quad \text{if} \quad \exists L \in \mathbb{R} : \forall \varepsilon > 0, \exists N(\varepsilon) \in \mathbb{N} : \forall n \geq N(\varepsilon), |a_n - L| < \varepsilon,$$

or, if the value of the limit L is known:

$$a_n \to L \quad \text{if} \quad \forall \varepsilon > 0, \exists N(\varepsilon) \in \mathbb{N} : \forall n \geq N(\varepsilon), |a_n - L| < \varepsilon.$$

If there is no such real number L, we call the sequence (a_n) divergent.

Remark 5.2.5 There is a lot of things going on in Definition 5.2.4 so let us make some remarks and clarifications on this definition:

1. The definition for convergence of real sequences went through a lot of revisions over hundreds of years until its rigour is satisfied by the mathematical community. Interested readers may consult Chap. 22 of [42] for a brief timeline of various definitions for limits ranging from Leibniz, Isaac Newton (1642–1726), Colin Maclaurin (1698–1746), Jean-Baptiste le Rond D'Alembert (1717–1783), Sylvestre François Lacroix (1765–1843), and finally the modern definition above.
2. If $|a_n - L| < \varepsilon$ for all $n \geq N(\varepsilon)$, we say that the terms a_n for $n \geq N(\varepsilon)$ are ε-close to L. In other words, they are of distance less than ε from L. See Fig. 5.1 for a graphical representation of this.
3. The requirement $|a_n - L| < \varepsilon$ for all $n \geq N(\varepsilon)$ is equivalent to:

$$|a_n - L| < \varepsilon \Leftrightarrow L - \varepsilon < a_n < L + \varepsilon \Leftrightarrow a_n \in (L - \varepsilon, L + \varepsilon) \Leftrightarrow a_n \in B_\varepsilon(L),$$

   for all $n \geq N(\varepsilon)$. Thus, the gist of convergence is: no matter how small we set the radius $\varepsilon > 0$, for the open ball $B_\varepsilon(L)$ there always exists an index $N(\varepsilon)$ such that all the terms of the sequence beginning from $a_{N(\varepsilon)}$ are contained in this open ball.
4. Another important thing to note here is that the index $N(\varepsilon)$ depends on the value $\varepsilon > 0$. As a rough guide, the smaller the value of ε, the larger the index $N(\varepsilon)$ is

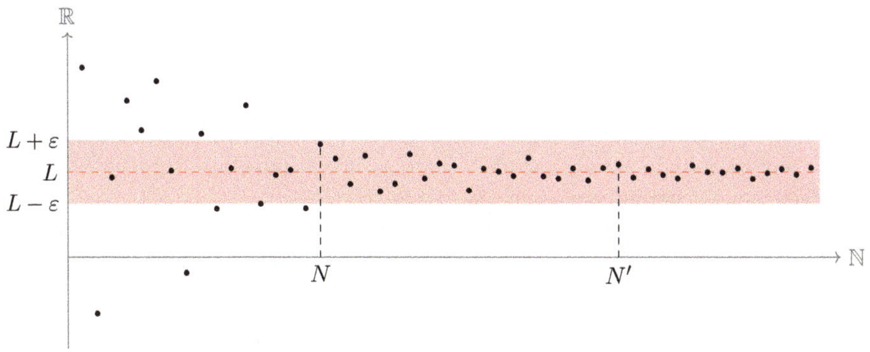

Fig. 5.1 The first 50 terms in a real sequence (a_n) with limit L. For the fixed $\varepsilon > 0$, starting from the index N, all the terms in the sequence are ε-close to L (they all lie within the red rectangle). Note also that the choice of N for this $\varepsilon > 0$ is not unique. The N in the diagram above is the smallest possible N that we can find for this fixed ε, but the definition did not require that it should be the smallest such N. We can choose any $N' \geq N$ and still all the terms in the sequence starting from the index N' lie ε-close to L

required to be. With a bit of imagination, we can see this in Fig. 5.1 if we vary the size of ε: the smaller the value of ε, the further we have to go along the sequence to find an $N(\varepsilon)$ that works.

5. We may sometimes write $N(\varepsilon)$ simply as N to declutter. However we still need to implicitly remember that N depends on ε!
6. Note also that the definition for convergence requires the existence of at least one such N for every fixed $\varepsilon > 0$. In fact, if we can find such an N, we can find infinitely many of them. Indeed, if $\forall n \geq N, |a_n - L| < \varepsilon$ is true, then for any $N' \geq N$ the statement $\forall n \geq N', |a_n - L| < \varepsilon$ is also true. See Fig. 5.1 for further explanation.

Example 5.2.6 Let us apply Definition 5.2.4 to some of the sequences we have constructed in Example 5.1.1:

1. Recall the sequence $\frac{1}{n}b_n = 2$ for all $n \in \mathbb{N}$. Let us call this constant sequence (c_n). We show that this sequence converges to some real number L. Intuitively, this number must be $L = 2$ since the sequence remains the same forever. Let us prove this properly using Definition 5.2.4.
   Fix $\varepsilon > 0$. We now aim to find a $N \in \mathbb{N}$ such that for all $n \geq N$, we have $|c_n - 2| < \varepsilon$. But since $c_n = 2$ for any $n \in \mathbb{N}$, we have $|c_n - 2| = |2 - 2| = 0$ for every $n \in \mathbb{N}$. So if we pick $N = 1$, then we are done since for every $n \geq N = 1$, we have $|c_n - 2| = 0 < \varepsilon$. Thus we can conclude that the sequence (c_n) converges to 2.

2. Let us define $d_n = \frac{a_n}{b_n} = \frac{(-1)^n}{2n}$ so that $(d_n) = \left(-\frac{1}{2}, \frac{1}{4}, -\frac{1}{6}, \frac{1}{8}, \dots\right)$. This is a bounded sequence since $|d_n| = \frac{|(-1)^n|}{|2n|} = \frac{1}{2n} \leq 1$ for all $n \in \mathbb{N}$.
We can also intuitively see that the sequence converges to $L = 0$ since the size of the terms get smaller as n gets larger. Let us prove this using Definition 5.2.4. We first need to do the dirty work of guessing for a suitable N for a given ε. Let us show two different approaches to finding this N:
   (a) We begin with the rough work. This is very important for us to guess what N is needed for a given ε. Once we have done the rough work, the final proper proof looks clean, just like magic.
   Rough work: Set $\varepsilon > 0$. We now find $N \in \mathbb{N}$ such that $|d_n - 0| = |d_n| < \varepsilon$ for every $n \geq N$. This means we want $|d_n| = \left|\frac{(-1)^n}{2n}\right| = \frac{1}{2n} < \varepsilon$ for every $n \geq N$. Thus, we need $n > \frac{1}{2\varepsilon}$ for every $n \geq N$. To this end, it is enough to just pick any integer $N > \frac{1}{2\varepsilon}$. Let us pick $N = \lceil \frac{1}{2\varepsilon} \rceil + 1 \in \mathbb{N}$. Now we check that this choice of N works for the fixed ε.
   The clean work starts here: Fix $\varepsilon > 0$. Choose $N = \lceil \frac{1}{2\varepsilon} \rceil + 1 \in \mathbb{N}$. Then, for all $n \geq N$, we have:
$$|d_n - 0| = |d_n| = \frac{1}{2n} \leq \frac{1}{2N} = \frac{1}{2\lceil \frac{1}{2\varepsilon} \rceil + 2} < \frac{1}{2\lceil \frac{1}{2\varepsilon} \rceil} \leq \frac{1}{2(\frac{1}{2\varepsilon})} = \varepsilon. \quad (5.1)$$
   Since for any $\varepsilon > 0$ at all we can find such an $N(\varepsilon)$, we conclude $d_n \to 0$.
   (b) Another way to do this is to bound $|d_n - L|$ with a simple term first before we apply the required ε bound at the end.
   Rough work: Recall that we want $|d_n - 0| = |d_n|$ to be less than ε for all $n \geq N$ where N is to be determined. So, we bound the quantity $|d_n - 0|$ from above as $|d_n - 0| = |d_n| = \frac{1}{2n} < \frac{1}{n}$ with a very simple term. Using the fact that $n \geq N$, this means $|d_n - 0| < \frac{1}{n} \leq \frac{1}{N}$. Since we want $|d_n - 0| < \varepsilon$, we can just set $\frac{1}{N} \leq \varepsilon$ which is equivalent to $\frac{1}{\varepsilon} \leq N$. Thus, $N = \lceil \frac{1}{\varepsilon} \rceil$ is a suitable candidate for N.
   Fix $\varepsilon > 0$. Choose $N = \lceil \frac{1}{\varepsilon} \rceil \in \mathbb{N}$. Then, for all $n \geq N$, we have:
$$|d_n - 0| = |d_n| = \frac{1}{2n} < \frac{1}{n} \leq \frac{1}{N} = \frac{1}{\lceil \frac{1}{\varepsilon} \rceil} \leq \frac{1}{\frac{1}{\varepsilon}} = \varepsilon.$$
   Thus, we conclude that $d_n \to 0$.

There are several things to note in Example 5.2.6(2) above.

1. For the same ε, we could get different N. In fact, as mentioned in Remark 5.2.5, there are infinitely many N that would work. For example, if $\varepsilon = 0.01$, using the first method yields $N = 51$ whereas the second method gives us $N = 100$. In fact, any index $N \geq 51$ would work for $\varepsilon = 0.01$. Picking any one of these is

correct because, from the definition, we just need to show there exists at least one N that works for each ε.

2. Note that we have to ensure at least one of the connecting equalities/inequalities in the process is a strict inequality to make sure that the requirement $|a_n - L| < \varepsilon$ in the definition is satisfied. This is the reason why we need the $+1$ term in the choice $N = \lceil \frac{1}{2\varepsilon} \rceil + 1$ in the first method: to ensure we have a strict inequality in (5.1) somewhere. If we do not have the $+1$ in the choice for N, the connecting inequalities in (5.1) are all weak inequalities. Thus, we get $|d_n| \leq \varepsilon$, which does not fulfill Definition 5.2.4.
3. Most of the time, the second approach is easier to work with since we would get a simpler inequality involving N and ε that we need to deal with in the end.

Example 5.2.7 Now let us look at some non-examples.

1. For the sequence (a_n) defined by $a_n = (-1)^n$, we have seen that this sequence is bounded in Example 5.2.3. However it does not converge to any real number. Suppose for contradiction that it converges to some number $L \in \mathbb{R}$. Then, by definition of convergence, for every $\varepsilon > 0$ we can find an index $N \in \mathbb{N}$ such that $|a_n - L| < \varepsilon$ for each $n \geq N$. Since this is true for any $\varepsilon > 0$, it must be true for $\varepsilon = \frac{1}{2}$ and so we know that there exists an $N \in \mathbb{N}$ such that $|a_n - L| < \frac{1}{2}$ for all $n \geq N$. Consider the terms a_N and a_{N+1}. Since the sequence alternates between -1 and 1, one of them is -1 and the other is 1, so we must have $|a_N - a_{N+1}| = 2$. However, by triangle inequality, we have:
$$2 = |a_N - a_{N+1}| = |a_N - L - a_{N+1} + L| \leq |a_N - L| + |a_{N+1} - L| < \frac{1}{2} + \frac{1}{2} = 1,$$
which says $2 < 1$, a contradiction! Therefore, our initial assumption must be false and thus the sequence (a_n) must be divergent.
2. The sequence (b_n) defined by $b_n = 2n$ is not bounded as seen in Example 5.2.3. To show that it is divergent, suppose for contradiction that it is convergent to some number $L \in \mathbb{R}$. Then, for $\varepsilon = \frac{1}{2}$, there exists some $N \in \mathbb{N}$ such that $|b_n - L| < \frac{1}{2}$ for all $n \geq N$. So:
$$|b_N - b_{N+1}| \leq |b_N - L| + |b_{N+1} - L| < \frac{1}{2} + \frac{1}{2} = 1.$$
On the other hand, $|b_N - b_{N+1}| = |2N - 2(N + 1)| = 2$. So, from the above we have $2 < 1$, which is a contradiction!

Example 5.2.8 In the previous chapter, we have seen that the decimal representation notation of a real number is merely symbolic. Now we are going to make sense of it via sequences and limits.

1. For any real number $r \in \mathbb{R}$, we have shown that we can find its decimal representation as the string of digits $a_0.a_1a_2a_3\ldots$ where $a_0 \in \mathbb{Z}$ and $a_j \in \{0, 1, \ldots, 9\}$ for all $j \in \mathbb{N}$. We can define this notation as the limit of a sequence of rational numbers (r_n) defined as $r_n = a_0.a_1\ldots a_n = \sum_{j=0}^{n} \frac{a_j}{10^j}$ for each $n \in \mathbb{N}$. Based on the construction the decimal representation, we know that $0 < r - r_n < \frac{1}{10^n}$ for all $n \in \mathbb{N}$. We now want to provide a proper proof that $\lim_{n\to\infty} r_n = r$.

   Rough work: We note that, by the Bernoulli's inequality, for any $n \in \mathbb{N}$ we have $10^n = (1+9)^n \geq 1 + 9n > n$ and so $\frac{1}{10^n} < \frac{1}{n}$. Thus, we have $0 < |r - r_n| < \frac{1}{10^n} < \frac{1}{n}$, which is a simple bound for $|r - r_n|$ that we can work with similar to Example 5.2.6(2)(b). We choose a similar N here as well, namely $N = \lceil \frac{1}{\varepsilon} \rceil$.

   Now for the clean work. Fix $\varepsilon > 0$. Set $N = \lceil \frac{1}{\varepsilon} \rceil \in \mathbb{N}$. For all $n \geq N$, we have:

$$0 < |r - r_n| < \frac{1}{10^n} < \frac{1}{n} \leq \frac{1}{N} = \frac{1}{\lceil \frac{1}{\varepsilon} \rceil} \leq \frac{1}{\frac{1}{\varepsilon}} = \varepsilon.$$

   In short, $|r - r_n| < \varepsilon$ for all $n \geq N$ and thus $r_n \to r$.
   As a conclusion, we can think of the decimal representation of a real number r as the limit of the sequence of truncated decimal representations (r_n), namely:

$$\lim_{n\to\infty} r_n = r \quad \Leftrightarrow \quad \lim_{n\to\infty} a_0.a_1a_2\ldots a_n = r \quad \Leftrightarrow \quad a_0.a_1a_2\ldots = r,$$

   where we removed the limit notation in the final equality but the unbounded ellipsis is understood to mean that it is taken in the infinite limiting sense. This then allows us to attach a numerical meaning to an infinite decimal representation, which we could not do using algebra alone.
2. We recall a mysterious rule that we had regarding the recurring 9s in a decimal expansion in Example 4.4.4. In particular, we declared $0.\dot{9} = 1$. Why is this true? As mentioned in the previous example, the expression $0.\dot{9}$ can be seen as the limit of the sequence of rational numbers (a_n) where $a_n = 0.\underbrace{999\ldots9}_{n \text{ times}} \in \mathbb{Q}$.

   Let us show that this sequence converges to 1 as $n \to \infty$. Similar to the argument above, for a fixed $\varepsilon > 0$, we choose $N = \lceil \frac{1}{\varepsilon} \rceil \in \mathbb{N}$. Then, for all $n \geq N$ we have:

$$0<|1 - x_n| = 1 - 0.\underbrace{999\ldots9}_{n \text{ times}} = 0.\underbrace{000\ldots0}_{n-1 \text{ times}}1 = \frac{1}{10^n} < \frac{1}{n} \leq \frac{1}{N} = \frac{1}{\lceil \frac{1}{\varepsilon} \rceil} \leq \frac{1}{\frac{1}{\varepsilon}} = \varepsilon.$$

   Thus, we conclude that:

$$\lim_{n\to\infty} x_n = 1 \quad \Leftrightarrow \quad \lim_{n\to\infty} 0.\underbrace{999\ldots9}_{n \text{ times}} = 1 \quad \Leftrightarrow \quad 0.999\ldots = 0.\dot{9} = 1.$$

3. Recall the sequence of rational approximations of $\sqrt{2}$ in Example 4.4.5. This sequence is (x_n) where $x_1 = 1, x_2 = 1.4, x_3 = 1.41, x_4 = 1.414$, and so on. Due to the reasoning above, we have $x_n \to \sqrt{2}$. In particular we have the equality $\sqrt{2} = 1.414\ldots$.

Remark 5.2.9 We make several remarks on Example 5.2.8.

1. This example gives us a new idea on how to manipulate irrational numbers. Every irrational number $r \in \bar{\mathbb{Q}}$ has a decimal representation. The construction of the representation allows us to create a sequence of increasing rational numbers (r_n) that converges to this irrational number r. In other words, we can approximate an irrational number as close as we can with sequences of rational numbers.
2. Suppose we want to find the value of $\sqrt{2} + \sqrt{3}$. We know that there exist sequences of rational approximations for these numbers, which we denote by (a_n) and (b_n) respectively. Then, a way to find the numerical value of this sum is to find limit of the sequence (c_n) where $c_n = a_n + b_n$ as $n \to \infty$. We know what the terms $c_n = a_n + b_n$ are explicitly for each $n \in \mathbb{N}$: they are simply rational numbers which we can compute by the usual arithmetic. Therefore, the sum $\sqrt{2} + \sqrt{3}$ would be the limit of this sequence (c_n).
The quick readers might have realised that there is another problem here: even though the sequences (a_n) and (b_n) converge, does this new sequence (c_n) even converge? And if it does converge, must it necessarily converge to the sum of the limits of (a_n) and (b_n)? We shall see later that it does indeed via a result called the algebra of limits in Theorem 5.9.1.
3. The algebra of limits also guarantee that more complicated operations on irrational numbers like multiplication, division, and exponentiation can also be done with this idea.

As we have seen in the numerous examples before, the idea of proving the convergence of a sequence is usually an ad hoc process. However, we can use the following steps as a guide:

1. Make a conjecture about the limit, say the limit is $L \in \mathbb{R}$. If the limit is given, then we do not need to do this.
2. For every $\varepsilon > 0$, we need to find an index $N(\varepsilon) \in \mathbb{N}$ such that all the terms a_n for $n \geq N(\varepsilon)$ stay within $B_\varepsilon(L)$.
   (a) This is done by first fixing $\varepsilon > 0$ and then, by algebraic manipulations, find a suitable N by manipulating $|a_N - L| < \varepsilon$. This N would then depend on ε, so $N : \mathbb{R}_+ \to \mathbb{N}$ is some function of ε. Make sure that the image is a natural number! This can be done by using the ceiling or floor functions suitably.
   (b) Next, we check that with this choice of $N(\varepsilon)$, all the other $n \geq N(\varepsilon)$ also satisfy the required condition $|a_n - L| < \varepsilon$.
   (c) Since N is a function of ε, we can then vary ε to get different $N(\varepsilon)$ that works for different values of $\varepsilon > 0$.

3. Finally, we write down the proper proof nicely and in order. Steps 1 and 2 are just the rough work, so once we found the $N(\varepsilon)$, we can rewrite from the beginning and by setting $n \geq N(\varepsilon)$, we would (seemingly magically) get $|a_n - L| < \varepsilon$ by a series of inequalities that we need to justify based on the choice of $N(\varepsilon)$ we made in the rough work.

Later on, we will see many automatic results that would make the process of finding and proving limits easier. However, it is still an essential skill to be able to know how to do this from first definition because the proofs of these advanced results depend on it and these automatic results can only help us in very specific and special cases.

Example 5.2.10 Let us look at another example. Consider a sequence (a_n) where $a_n = \frac{n\sin(n^2)}{3n^3+1}$. We want to show that this sequence converges to 0. We start by fixing $\varepsilon > 0$ and proceed to find an $N \in \mathbb{N}$ such that $|a_n - 0| < \varepsilon$ for all $n \geq N$.

Rough work: Using the fact that $|\sin(x)| \leq 1$ for any $x \in \mathbb{R}$, we have:

$$|a_n - 0| = \left|\frac{n\sin(n^2)}{3n^3+1}\right| \leq \frac{n}{3n^3+1} < \frac{n}{3n^3} = \frac{1}{3n^2} < \frac{1}{n^2} \leq \frac{1}{n},$$

for all $n \in \mathbb{N}$. So, if this last term is less than or equal to ε, then $|a_n - 0| < \varepsilon$ as well. In particular, we want an $N \in \mathbb{N}$ such that $|a_N - 0| < \frac{1}{N} < \varepsilon$. We can pick $N = \lceil \frac{1}{\varepsilon} \rceil$.

For a fixed $\varepsilon > 0$, set $N = \lceil \frac{1}{\varepsilon} \rceil \in \mathbb{N}$. For all $n \geq N$ we then have:

$$|a_n - 0| = \left|\frac{n\sin(n^2)}{3n^3+1}\right| \leq \frac{n}{3n^3+1} < \frac{n}{3n^3} = \frac{1}{3n^2} < \frac{1}{n^2} \leq \frac{1}{n} \leq \frac{1}{N} = \frac{1}{\lceil \frac{1}{\varepsilon} \rceil} \leq \frac{1}{\frac{1}{\varepsilon}} = \varepsilon,$$

and thus we have proven that $a_n \to 0$.

Since convergence of a sequence only depends on the terms with large indices, we can chop off finitely many terms at the beginning of the sequence. In Remark 5.2.5, for convergence, we require all the terms $a_N, a_{N+1}, a_{N+2}, \ldots$ to be in the ball $B_\varepsilon(L) = (L - \varepsilon, L + \varepsilon)$. Effectively, what we did here is we ignored the first $N - 1$ terms of the original sequence. We call this new sequence a tail of the original sequence.

Definition 5.2.11 (Tail of a Sequence) Given a real sequence (a_n). For any $N \in \mathbb{N}$ we call the sequence $(a_{N+j}) = (a_{N+1}, a_{N+2}, a_{N+3}, \ldots)$ a tail of the sequence (a_n).

A sequence has many tails, depending on the N where want want to start the tail from. It is useful to remember that the convergence of a sequence depends on the behaviour of the sequence for large indices, towards infinity. So, to study its convergence, it is enough to look at tails of the sequence. In other words, we can ignore the first finite number of terms in a sequence during the analysis since they do not contribute to the behaviour of the sequence towards infinity. Here is an example:

Example 5.2.12 Consider the real sequence (q_n) defined by $q_n = \frac{1}{\left(n-\frac{9}{2}\right)^2}$. We guess that this sequence converges to 0.

Rough work: To prove this, we first fix $\varepsilon > 0$. Now we need to find $N \in \mathbb{N}$ such that $|q_n - 0| = q_n < \varepsilon$ for all $n \geq N$. We note that:

$$0 \leq q_n = \frac{1}{n^2 - 9n + \frac{81}{4}} < \frac{1}{n^2 - 9n}, \tag{5.2}$$

which is true only for $n \geq 10$ (otherwise we would have a negative term is bigger than 0, which is clearly absurd). We now aim to get rid of the term $-9n$ in the denominator to make the bound simpler.
Note that if $n \geq 10$, we also have $n^2 - 10n \geq 0$ which is equivalent to $n^2 - 9n \geq n$. This says for any $n \geq 10$, we must have $\frac{1}{n^2-9n} \leq \frac{1}{n}$. Thus, in the tail of the sequence starting at $n = 10$, the inequality in (5.2) satisfies the following simple bound:

$$0 \leq q_n < \frac{1}{n^2 - 9n} \leq \frac{1}{n} \quad \text{for } n \geq 10. \tag{5.3}$$

To find a suitable N here, we note that for all $n \geq N$ we have $\frac{1}{n} \leq \frac{1}{N}$. Thus, setting $\frac{1}{N} \leq \varepsilon$ is enough. We can choose any $N \geq \lceil \frac{1}{\varepsilon} \rceil$ which at the same time also satisfies $N \geq 10$ since the inequalities in (5.3) are only true for indices greater than or equal to 10. An obvious suitable choice for $N(\varepsilon)$ would be the larger of the two, namely $N = \max\{\lceil \frac{1}{\varepsilon} \rceil, 10\}$.

For a fixed $\varepsilon > 0$, we pick $N = \max\{\lceil \frac{1}{\varepsilon} \rceil, 10\} \in \mathbb{N}$. For any $n \geq N$, we have:

$$\begin{aligned}
|q_n - 0| = q_n &= \frac{1}{n^2 - 9n + \frac{81}{4}} \\
&< \frac{1}{n^2 - 9n} && (\because n \geq N \geq 10 \Rightarrow n^2 - 9n + \tfrac{81}{4} > n^2 - 9n) \\
&\leq \frac{1}{n} && (\because n \geq N \geq 10 \Rightarrow n^2 - 9n \geq n) \\
&\leq \frac{1}{N} && (\because n \geq N) \\
&\leq \varepsilon && \left(\because N \geq \left\lceil \tfrac{1}{\varepsilon} \right\rceil \geq \tfrac{1}{\varepsilon}\right),
\end{aligned}$$

and so we conclude that the sequence (q_n) converges to 0.

Again, we note that if the sequence converges, for any $\varepsilon > 0$, this choice of N is not unique. In Example 5.2.12, for a fixed $\varepsilon > 0$ we picked $N = \max\{\lceil \frac{1}{\varepsilon} \rceil, 10\}$. However, we can also pick $N = \max\{\lceil \frac{1}{\varepsilon} \rceil, 20\}$ or $N = \max\{\lceil \frac{1}{\varepsilon} \rceil, 1000\}$ or $N = \max\{\lceil \frac{1}{\varepsilon} \rceil, 20\} + 10$ and the final proof would still work. By the definition of convergence of a sequence, we just need to show that one such N exists for each ε. So we have a lot of choices to pick from!

However, the limit L for a convergent real sequence must be unique.

Proposition 5.2.13 *Let (a_n) be a convergent real sequence. Suppose that $a_n \to L_1$ and $a_n \to L_2$ where $L_1, L_2 \in \mathbb{R}$. Then $L_1 = L_2$.*

Proof Suppose for contradiction that $L_1 \neq L_2$. WLOG, let $L_2 > L_1$ so that $L_2 - L_1 > 0$. Set $\varepsilon = \frac{L_2 - L_1}{2} > 0$. Since $a_n \to L_1$, there exists an $N_1 \in \mathbb{N}$ such that $|a_n - L_1| < \varepsilon = \frac{L_2 - L_1}{2}$ for all $n \geq N_1$. Furthermore, since $a_n \to L_2$, there exists an $N_2 \in \mathbb{N}$ such that $|a_n - L_2| < \varepsilon = \frac{L_2 - L_1}{2}$ for all $n \geq N_2$. Thus, for $N = \max\{N_1, N_2\}$, the two inequalities above hold. By using the triangle inequality, we have:

$$L_2 - L_1 = |L_2 - L_1| = |L_2 - a_N + a_N - L_1| \leq |L_2 - a_N| + |a_N - L_1| < \frac{L_2 - L_1}{2} + \frac{L_2 - L_1}{2} = L_2 - L_1,$$

which is a contradiction! So our initial assumption is false and we must have $L_1 = L_2$. □

To prove that a sequence is divergent, the general strategy based on Example 5.2.7 is to assume for a contradiction that the sequence is convergent to some $L \in \mathbb{R}$. By choosing an appropriate value for $\varepsilon > 0$ and algebraic manipulations, we aim to get a contradiction to disprove our initial assumption.

However, there are some sequences that can be easily seen to be divergent at first glance. A way to spot this is to know the properties of a general convergent sequence and use *modus tollens*. If a given sequence does not have one of these properties, we can safely say that it is not convergent. A useful property for convergent sequences is that they must be bounded:

Proposition 5.2.14 *Let (a_n) be a real sequence. If (a_n) is convergent, then it is bounded.*

Proof Since (a_n) is convergent, there exists some $L \in \mathbb{R}$ such that $a_n \to L$. Then, for $\varepsilon = 1 > 0$, there exists an $N \in \mathbb{N}$ such that $|a_n - L| < 1$ for all $n \geq N$. By using the triangle inequality, we can deduce that for all $n \geq N$ we have:

$$|a_n| = |a_n - L + L| \leq |a_n - L| + |L| < 1 + |L|.$$

This means all the terms including and after the N-th term are bounded by $1+|L|$. Moreover, recall from Lemma 4.1.12 that the maximum of a finite set of real numbers must exist. So the maximum of $N-1$ terms $K = \max\{|a_1|, |a_2|, \ldots, |a_{N-1}|\}$ exists and clearly $|a_n| \leq K$ for all $n = 1, 2, \ldots, N-1$. Therefore, if we define $M = \max\{K, 1+|L|\}$, we would get $|a_n| \leq M$ for all $n \in \mathbb{N}$ and thus the sequence is bounded. □

By contrapositive of Proposition 5.2.14, if a sequence is unbounded, then the sequence is not convergent. In other words, if we see an unbounded real sequence, we can immediately infer that this sequence is divergent.

Example 5.2.15 In Example 5.2.3, the sequence (b_n) defined as $b_n = 2n$ has been shown to be unbounded and therefore we can immediately deduce that it must be divergent.

However, it is important to point out that Proposition 5.2.14 is only a one way implication. The converse is not true! Not all bounded sequences are convergent. For example, recall the sequence (a_n) defined by $a_n = (-1)^n$. We have shown that this sequence is bounded in Example 5.2.3. However, in Example 5.2.7, we have proven that it diverges.

5.3 Blowing up to Infinity

The divergent sequence in Example 5.2.15 is said to "diverge to infinity" or "blow up to infinity" because it grows arbitrarily large as we go along the sequence. We define this more precisely:

Definition 5.3.1 (Blowing Up to Infinity) Let (a_n) be an unbounded divergent real sequence.

1. The sequence (a_n) is said to be blowing up to infinity or diverges to infinity if for each positive real number $K > 0$, there exists an $N(K) \in \mathbb{N}$ such that $a_n > K$ for all $n \geq N(K)$.
2. The sequence (a_n) is said to be blowing up to negative infinity or diverges to negative infinity if for each negative real number $K < 0$, there exists an $N(K) \in \mathbb{N}$ such that $a_n < K$ for all $n \geq N(K)$.

The intuition behind the first definition is that no matter how large a number $K > 0$ is chosen, we can always go far along the sequence to find an index $N(K)$ so that all the terms after this index are greater than the chosen $K > 0$. This can be visualised in Fig. 5.2. Similar idea holds for the definition of blowing up to $-\infty$.

In an abuse of notation, we often write $\lim_{n\to\infty} a_n = \infty$ or $a_n \to \infty$ if the sequence (a_n) diverges to infinity and $\lim_{n\to\infty} a_n = -\infty$ or $a_n \to -\infty$ if the sequence (a_n) diverges to negative infinity. These notation are not strictly correct

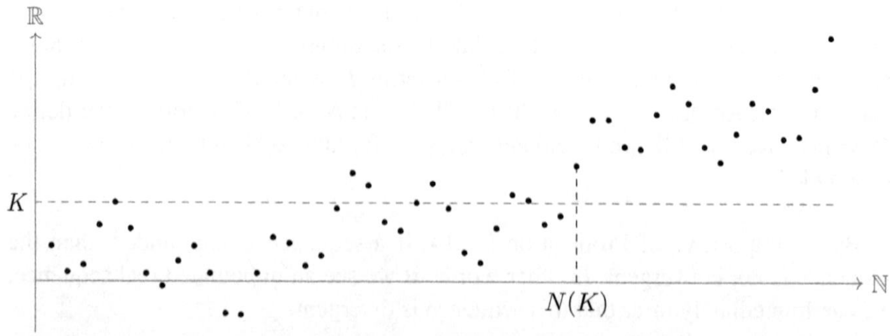

Fig. 5.2 The first 50 terms in a real sequence (a_n) which blows up to ∞. For the fixed $K > 0$, starting from the index $N(K)$, all the terms in the sequence are greater than K and thus lie above the red line. Similar to what we saw for convergent sequences, this $N(K)$ is not unique

in the sense of Definition 5.2.4 as $\pm\infty$ are not elements of the real number $\mathbb{R}$ and hence the expressions $|a_n \pm \infty|$ do not make sense in $\mathbb{R}$. But it is a widely accepted notation for sequences blowing up to infinity.

Symbolically we write the blowing-up conditions in Definition 5.3.1 using quantifiers as:

$$a_n \to \infty \quad \text{if} \quad \forall K > 0, \exists N(K) \in \mathbb{N} : \forall n \geq N(K), a_n > K,$$
$$a_n \to -\infty \quad \text{if} \quad \forall K < 0, \exists N(K) \in \mathbb{N} : \forall n \geq N(K), a_n < K.$$

For brevity, we also write $N(K)$ simply as N with the implicit knowledge that this N depends on K.

Example 5.3.2 The sequence (b_n) where $b_n = 2n$ diverges to ∞. In order to show this, we need to check that for every $K > 0$, there exists an $N \in \mathbb{N}$ such that $b_n > K$ for all $n \geq N$.

Rough work: Fix a real number $K > 0$. We want to find an $N \in \mathbb{N}$ such that $b_N = 2N > K$. In other words, we want to choose an integer N such that $N > \frac{K}{2}$. We can choose $N = \lceil \frac{K}{2} \rceil + 1 \in \mathbb{N}$. Now we check that this choice works for all $n \geq N$ as well.

Fix $K > 0$ and set $N = \lceil \frac{K}{2} \rceil + 1 \in \mathbb{N}$. For all $n \geq N$, we have:

$$b_n = 2n \geq 2N = 2\left\lceil \frac{K}{2} \right\rceil + 2 > 2\left\lceil \frac{K}{2} \right\rceil \geq K.$$

So we conclude that the sequence b_n diverges to ∞.

Remark 5.3.3 We have shown that the sequence (b_n) in Example 5.3.2 to be unbounded in Example 5.2.3. So why did we have to check that it diverges to ∞?

There are unbounded sequences that do not blow up to ∞ or $-\infty$. If we consider the sequence $(c_n) = (-2, 4, -6, 8, \dots)$ defined by $c_n = (-1)^n 2n$, even though this sequence is unbounded, it is neither tending to ∞ or $-\infty$. WLOG, assume for contradiction that it blows up to ∞. Then, for $K = 1$, there exists an index N such that for all $n \geq N$ we have $c_n > 1$. However, one of c_N or c_{N+1} must be negative, giving us a contradiction. So this sequence satisfies neither of the conditions in Definition 5.3.1.

5.4 Monotone Sequences

An important and very useful family of real sequences is the monotone sequences. Since we are working on real numbers which is an ordered field, the terms in a real sequence can be ordered or compared with one another. Therefore, we can have a sequence of terms that gets bigger or gets smaller as we go along the sequence. We define these special sequences as:

Definition 5.4.1 (Monotone Sequence) Let (a_n) be a real sequence.

1. If $a_n \leq a_{n+1}$ for all $n \in \mathbb{N}$, then we call the sequence (a_n) increasing or non-decreasing.
2. If $a_n < a_{n+1}$ for all $n \in \mathbb{N}$, then we call the sequence (a_n) strictly increasing.
3. If $a_n \geq a_{n+1}$ for all $n \in \mathbb{N}$, then we call the sequence (a_n) decreasing or non-increasing.
4. If $a_n > a_{n+1}$ for all $n \in \mathbb{N}$, then we call the sequence (a_n) strictly decreasing.

In all of the cases above, the sequences (a_n) are called monotone sequences.

Monotone real sequences have a very well understood behaviour: they either converge to some finite number or blow up to $\pm\infty$ (depending whether they are increasing or decreasing). See Fig. 5.3 for examples of the possible behaviour of increasing sequences.

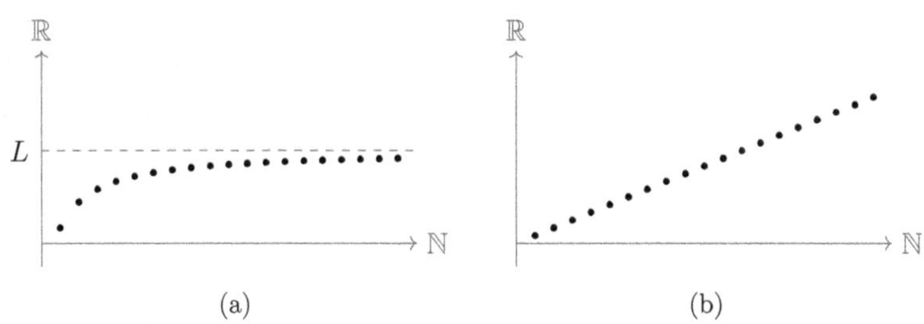

Fig. 5.3 Two possible behaviour of increasing real sequences. (**a**) Bounded increasing sequence. (**b**) Unbounded increasing sequence

Theorem 5.4.2 (Monotone Sequence Theorem) *Let (a_n) be a monotone real sequence.*

1. *Suppose that the sequence (a_n) is increasing. The sequence (a_n) converges if and only if it is bounded from above.*
2. *Suppose that the sequence (a_n) is decreasing. The sequence (a_n) converges if and only if it is bounded from below.*

Proof We are going to prove only the first assertion. The second can be proven similarly.

1. We prove the backwards implication only as the forward is true by Proposition 5.2.14.

   ($\Leftarrow$): Suppose that the increasing real sequence (a_n) is bounded from above. Therefore, if we consider the set of numbers in the sequence $A = \{a_n\}_{n=1}^{\infty}$, this set is non-empty and bounded from above. By the completeness axiom of $\mathbb{R}$, we know that the supremum of this set exists. Let $\sup(A) = L$ and we claim that $a_n \to L$.
   Fix $\varepsilon > 0$. We want to find an $N \in \mathbb{N}$ such that $|a_n - L| < \varepsilon$ for all $n \geq N$. By characterisation of supremum in Proposition 4.1.9, there exists an element $a_N \in A$ such that $L - \varepsilon < a_N \leq L$ or equivalently, $0 \leq L - a_N < \varepsilon$. We claim that this index N works.
   Since the sequence (a_n) is increasing and bounded above by L, we must have $a_N \leq a_n \leq L$ for all $n \geq N$. Thus, $0 \leq L - a_n \leq L - a_N < \varepsilon$ for all $n \geq N$. We conclude that $|a_n - L| = L - a_n < \varepsilon$ for all $n \geq N$, which proves that $a_n \to L$. □

Therefore, it is always useful to check the behaviour of the sequence before proving that it converges: it might save us a lot of time. If a sequence is monotone and bounded, then we can immediately conclude that it converges. From there, we find its supremum or infimum (depending whether the sequence is increasing or decreasing) and conclude that the sequence converges to this value.

Example 5.4.3 Consider a real sequence (a_n) defined by $a_n = 1 - \frac{1}{n}$. Let us list down the first few elements in the sequence: $(a_n) = \left(0, \frac{1}{2}, \frac{2}{3}, \frac{3}{4}, \frac{4}{5}, \frac{5}{6}, \ldots\right)$. We can clearly see that this sequence is increasing from the first few terms. To actually prove that this sequence is increasing all the time, we need to show $a_n \leq a_{n+1}$ or equivalently $a_{n+1} - a_n \geq 0$ for all $n \in \mathbb{N}$. To wit:

$$a_{n+1} - a_n = 1 - \frac{1}{n+1} - \left(1 - \frac{1}{n}\right) = \frac{1}{n} - \frac{1}{n+1} = \frac{1}{n(n+1)} \geq 0,$$

for all $n \in \mathbb{N}$. So the sequence is increasing. Therefore, by Theorem 5.4.2, it either blows up to ∞ or converges to some real number. However, we note that $a_n = 1 - \frac{1}{n} \leq 1$ for all $n \in \mathbb{N}$. So this sequence is bounded from above and hence it cannot blow up to ∞. Moreover, a simple exercise shows that the supremum of this sequence is 1. Therefore, (a_n) converges to 1.

Example 5.4.4 Consider the sequence of numbers (a_n) defined as $a_n = \left(1 + \frac{1}{n}\right)^n$. We claim that this sequence is bounded and increasing. First, we show that this sequence is increasing by showing that $a_{n+1} \geq a_n$ for all $n \in \mathbb{N}$. Since $a_n > 0$ for all $n \in \mathbb{N}$, this is equivalent to showing that $\frac{a_{n+1}}{a_n} \geq 1$ for all $n \in \mathbb{N}$. For any $n \in \mathbb{N}$, we have:

$$\begin{aligned}
\frac{a_{n+1}}{a_n} = \frac{\left(1 + \frac{1}{n+1}\right)^{n+1}}{\left(1 + \frac{1}{n}\right)^n} &= \left(\frac{n(n+2)}{(n+1)^2}\right)^{n+1} \left(1 + \frac{1}{n}\right) \\
&= \left(1 - \frac{1}{(n+1)^2}\right)^{n+1} \left(1 + \frac{1}{n}\right) \\
&\geq \left(1 - \frac{1}{n+1}\right) \left(1 + \frac{1}{n}\right) = 1,
\end{aligned}$$

where we used Bernoulli's inequality from Exercise 3.15. Thus, the sequence is increasing.

Next we show that the sequence is bounded above. Using binomial expansion in Exercise 3.14, for a fixed $n \in \mathbb{N}$ we have:

$$\begin{aligned}
a_n &= \sum_{k=0}^{n} \binom{n}{k} \frac{1}{n^k} \\
&= \sum_{k=0}^{n} \frac{n!}{k!(n-k)!} \frac{1}{n^k} \\
&= 1 + 1 + \frac{1}{2!}\left(1 - \frac{1}{n}\right) + \frac{1}{3!}\left(1 - \frac{1}{n}\right)\left(1 - \frac{2}{n}\right) + \dots \\
&\quad \dots + \frac{1}{n!}\left(1 - \frac{1}{n}\right) \dots \left(1 - \frac{n-1}{n}\right).
\end{aligned} \tag{5.4}$$

From (5.4), clearly $a_n \geq 2$ for any $n \in \mathbb{N}$. Moreover, we can bound all the (a_n) from above since each bracketed term in (5.4) are between 0 and 1. Namely:

$$\begin{aligned}
a_n \leq 1 + 1 + \frac{1}{2!} + \frac{1}{3!} + \dots + \frac{1}{n!} &\leq 2 + \frac{1}{2} + \frac{1}{2^2} + \dots + \frac{1}{2^{n-1}} \\
&= 2 + \left(1 - \frac{1}{2^{n-1}}\right) \leq 3,
\end{aligned}$$

which is true for any $n \in \mathbb{N}$ at all. Since the sequence (a_n) is increasing and bounded from above, it must converge to some number between 2 and 3. In fact, the limit of this sequence is actually a very important number:

$$\lim_{n\to\infty}\left(1+\frac{1}{n}\right)^n = 2.71828\ldots$$

Due to the importance of this constant, it is called the Euler-Napier constant or Euler's number after Leonhard Euler and John Napier. The discovery of this constant is credited to Jacob Bernoulli from his studies of compound interests in finance.

We shall see that its crops up in various mathematical facts in Example 5.7.4 and we shall discuss more about this very important number later in Chap. 8. We are also going to show that this constant is an irrational number in Exercise 12.19. Due to it being irrational (and hence does not have a convenient explicit expression) along with the fact that it is a commonly occurring constant, it is usually denoted with the symbol e for ease of notation. The first published appearance of symbol e for this constant was in Euler's *Mechanica sive motus scientia analytice exposita* (Mechanics of the Science of Motion set forth Analytically).

5.5 Subsequences

Another idea that could help us prove the divergence of bounded sequences is by looking at subsequences. Apart from multiplying, dividing, or adding sequences together, we can also create a new sequence from any given sequence by looking at only some terms in the original sequence. Since sequences must be infinitely long, we must take infinitely many terms in the original sequence as well, whilst keeping them ordered in the same way as they originally were. We define:

Definition 5.5.1 (Subsequence of a Real Sequence) Let (a_n) or $a : \mathbb{N} \to \mathbb{R}$ be a real sequence and $k : \mathbb{N} \to \mathbb{N}$ be a strictly increasing function, namely $k(n) < k(n+1)$ for all $n \in \mathbb{N}$. We define the composition $a \circ k : \mathbb{N} \to \mathbb{R}$ as a subsequence of (a_n), denoted as $(a_{k(n)})$ or (a_{k_n}) (Fig. 5.4).

We note that the function $k : \mathbb{N} \to \mathbb{N}$ is required to be strictly increasing because we want to ensure that each term in the original sequence appears at most once in the subsequence and the ordering of the terms in this new subsequence follows their ordering in the original sequence. A useful observation that we shall use repeatedly is the following:

Lemma 5.5.2 *If $k : \mathbb{N} \to \mathbb{N}$ is a strictly increasing function, then $k_n \geq n$ for all $n \in \mathbb{N}$.*

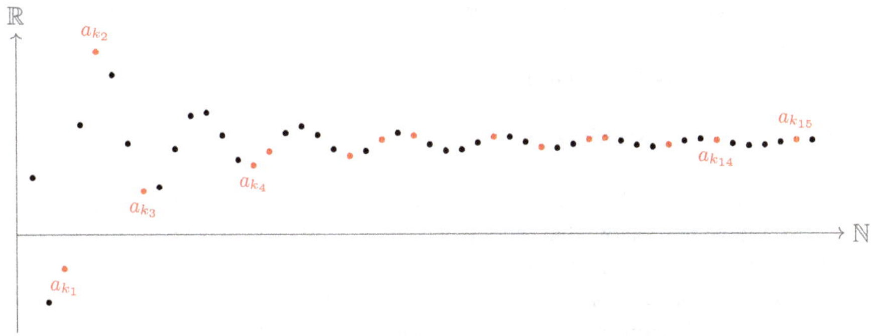

Fig. 5.4 Real sequence (a_n). A subsequence (a_{k_n}) is picked out by the red dots

Proof We prove this via induction. Since $k_1 \in \mathbb{N}$, clearly we must have $k_1 \geq 1$. Now assume that $k_m \geq m$ for some $m \in \mathbb{N}$. Since k is a strictly increasing function, we must have $k_{m+1} > k_m \geq m$. Hence, $k_{m+1} \geq m + 1$. □

Example 5.5.3 Recall the real sequences (a_n) and (b_n) in Example 5.1.1 defined by $a_n = (-1)^n$ and $b_n = 2n$.

1. If we define an increasing function $k : \mathbb{N} \to \mathbb{N}$ by $k(n) = 2n$, then we have a new sequence:

$$(a_{k_n}) = (a_{2n}) = (a_2, a_4, a_6, a_8, \ldots) = (1, 1, 1, 1, \ldots).$$

   This sequence (a_{k_n}) is a subsequence of (a_n) since we are only taking the even numbered terms in the original sequence and ignoring the rest.
2. We can also define $k(n) = n^2$ so that we can create a subsequence of (b_n) given by:

$$(b_{k_n}) = (b_{n^2}) = (b_1, b_4, b_9, b_{16}, \ldots) = (2, 8, 18, 32, \ldots),$$

   where the n-th term in this new sequence is given by $b_{k_n} = 2n^2$.
3. Another example of a subsequence that we have seen is a tail of a sequence in Definition 5.2.11. Recall that a tail of a sequence (a_n) is a sequence $(a_{N+1}, a_{N+2}, a_{N+3}, \ldots)$ for some fixed $N \in \mathbb{N}$. This tail is a subsequence obtained by precomposing the sequence (a_n) with the strictly increasing function $k(n) = N + n$. Thus, the resulting subsequence of (a_n) would be $(a_{k_n}) = (a_{N+n}) = (a_{N+1}, a_{N+2}, a_{N+3}, \ldots)$.

How can a subsequence tell us about the convergence behaviour of the original sequence?

Proposition 5.5.4 *Let (a_n) be a real sequence.*

1. *If (a_n) is a convergent sequence, then any subsequence of (a_n) converges to the same limit.*
2. *If (a_n) blows up $\pm\infty$, then any subsequence of (a_n) also blows up to $\pm\infty$.*

Proof We prove the assertions one by one.

1. Suppose that $a_n \to L$ for some $L \in \mathbb{R}$. Let (a_{k_n}) be a subsequence of (a_n). Fix $\varepsilon > 0$. We want to show that there exists an $N \in \mathbb{N}$ such that $|a_{k_n} - L| < \varepsilon$ for all $n \geq N$. For the same $\varepsilon > 0$, since $a_n \to L$, there exists an $N \in \mathbb{N}$ such that $|a_n - L| < \varepsilon$ for every $n \geq N$. Lemma 5.5.2 implies that for all $n \geq N$, we have $k_n \geq n \geq N$. Therefore, $|a_{k_n} - L| < \varepsilon$ for all $n \geq N$. Hence, we conclude that $a_{k_n} \to L$.
2. WLOG, suppose that $a_n \to \infty$ and let (a_{k_n}) be a subsequence of (a_n). Fix $K > 0$. Then there exists an index $N \in \mathbb{N}$ such that for all $n \geq N$ we have $a_n > K$. However, Lemma 5.5.2 says that $k_n \geq n$ for all $n \in \mathbb{N}$. In particular, for all $n \geq N$ we have $a_{k_n} > K$ and thus we conclude that $a_{k_n} \to \infty$. □

Thus, we can see another way of showing that a sequence is divergent by considering subsequences. From Proposition 5.5.4, if a sequence (a_n) converges, then any of its subsequence is also convergent to the same limit. Therefore, by contrapositive, if we can find either two subsequences of (a_n) which do not converge to the same limit or a divergent subsequence, then (a_n) must be divergent.

Example 5.5.5 This result can be used directly to prove the divergence of the sequence (a_n) defined by $a_n = (-1)^n$. We can find two subsequences of (a_n) that converge to different limits: the subsequence $(a_{2n}) = (1, 1, 1, \ldots)$ is a constant sequence of 1s, so it converges to 1, while the subsequence $(a_{2n-1}) = (-1, -1, -1, \ldots)$ is a constant subsequence of -1s, so it converges to -1. Hence, we have found two subsequences of (a_n) that converge to different limits and thus we conclude that the original sequence (a_n) must be divergent.

Subsequences are also useful when they are located within a monotone sequence. We have the following result which the readers will prove in Exercise 5.21.

Proposition 5.5.6 *Let (a_n) be a monotone sequence. If there exists a subsequence (a_{k_n}) of (a_n) that is convergent, then the sequence (a_n) is also convergent.*

Bolzano-Weierstrass Theorem

In Example 5.5.5, for the bounded but divergent sequence $(a_n) = (-1, 1, -1, 1, \ldots)$, we have seen that there are at least two convergent subsequences in it. This is an interesting observation, which is in fact true in greater generality, namely: every bounded sequence must have a convergent subsequence! This amazing result is called the Bolzano-Weierstrass theorem. Before we prove this theorem, we prove the following lemma.

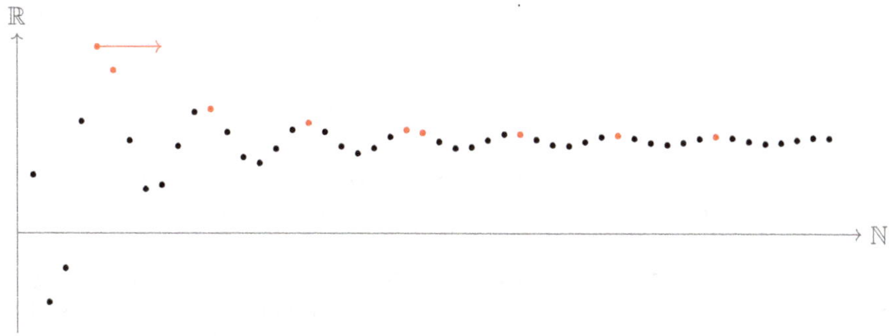

Fig. 5.5 Example of a sequence (a_n). The terms a_m where $m \in V$ are the red dots. An interpretation of the terms a_m is that, if we look to the left from these points (an arrow is shown for the first term), we will never see any other points which are greater than or equal to it. Thus, sometimes this result is also known as the scenic viewpoint lemma

Lemma 5.5.7 (Monotone Subsequence Lemma) *Let (a_n) be a real sequence. Then, (a_n) has a monotone subsequence.*

Proof We consider the subset of indices of the real sequence defined by:

$$V = \{m \in \mathbb{N} : a_j < a_m \text{ for all } j > m\}.$$

In words, the set V is the set of indices m such that all the terms in the sequence (a_n) after a_m are strictly smaller than a_m. See Fig. 5.5 for an example.

By investigating this set, we are going to construct a monotone subsequence of (a_n). There are two cases for this set: either the set V is infinite or finite.

1. If V is an infinite set, then we can order the elements of V in increasing order, say $k_1 < k_2 < k_3 < \ldots$. Consider the subsequence of (a_n) given by $(a_{k_1}, a_{k_2}, a_{k_3}, \ldots)$. We claim that this is a decreasing subsequence of (a_n). Indeed, since $k_n \in V$ for all $n \in \mathbb{N}$, by definition of the set V, we must have $a_j < a_{k_n}$ for all $j > k_n$. In particular, since $k_{n+1} > k_n$, we must have $a_{k_{n+1}} < a_{k_n}$. So for every $n \in \mathbb{N}$, we have $a_{k_{n+1}} < a_{k_n}$. Since this is true for all $n \in \mathbb{N}$, the subsequence (a_{k_n}) of (a_n) is decreasing.
2. Otherwise, if the set V is finite, then there exists a largest element of this set which we call $\max(V) = N$. Thus, every integer index greater than N is not contained in V. We construct a subsequence of (a_n) recursively using this fact. We pick the first element of the subsequence as $a_{k_1} = a_{N+1}$. Since $N + 1 \notin V$, by definition of V, there must be some index $k_2 > N + 1$ such that $a_{k_2} \geq a_{N+1} = a_{k_1}$. We set a_{k_2} as the second element of this subsequence. Next, since $k_2 > N$ and hence $k_2 \notin V$, there must exist some integer $k_3 > k_2$ such that $a_{k_3} \geq a_{k_2}$. We set a_{k_3} as the third element in the subsequence. We repeat this construction indefinitely

so that $a_{k_{n+1}} \geq a_{k_n}$ for all $n \in \mathbb{N}$. Thus, (a_{k_n}) is a subsequence of (a_n) that is increasing.

In either case, there exists a monotone subsequence (a_{k_n}) of (a_n). □

Theorem 5.5.8 (Bolzano-Weierstrass Theorem) *If (a_n) is a bounded real sequence, then there exists a subsequence of (a_n) that is convergent.*

Proof By Lemma 5.5.7, the sequence (a_n) has a monotone subsequence (a_{k_n}). Since the sequence (a_n) is bounded, this monotone subsequence (a_{k_n}) is also bounded. Thus, by monotone sequence theorem in Theorem 5.4.2, the subsequence (a_{k_n}) is convergent. □

In other words, the Bolzano-Weierstrass theorem says that if a real sequence stays in a closed ball $\bar{B}_M(0)$ for some large enough radius $M > 0$, then the sequence has a convergent subsequence.

Finally, what about unbounded sequences? Can we find a convergent subsequence in an unbounded sequence? Not necessarily, but we can definitely find a subsequence that either blows up to ∞ or $-\infty$. We prove a lemma first:

Lemma 5.5.9 *Let (a_n) be a real sequence.*

1. *If (a_n) is unbounded from above, then any tail sequence of (a_n) is unbounded from above.*
2. *If (a_n) is unbounded from below, then any tail sequence of (a_n) is unbounded from below.*

Proof We prove the first assertion only.

1. Suppose for contradiction that there is a tail sequence (a_{N+n}) of (a_n) which is bounded from above. Then, there exists a $K > 0$ such that $|a_{N+n}| \leq K$ for all $n \in \mathbb{N}$. This means $|a_n| \leq K$ for all $n \geq N+1$. Denote $M = \max\{|a_1|, \ldots, |a_N|, K\} > 0$. Then, $|a_n| \leq M$ for any $n \in \mathbb{N}$, which is a contradiction. □

Proposition 5.5.10 *Let (a_n) be a real sequence.*

1. *(a_n) is unbounded from above if and only if there exists a subsequence (a_{k_n}) of (a_n) such that $\lim_{n\to\infty} a_{k_n} = \infty$.*
2. *(a_n) is unbounded from below if and only if there exists a subsequence (a_{k_n}) of (a_n) such that $\lim_{n\to\infty} a_{k_n} = -\infty$.*

Proof We shall prove the first assertion only.

1. The converse is clearly true, so we prove the forward implication only.

($\Rightarrow$): Since the sequence (a_n) is not bounded from above, there exists an element a_{k_1} such that $a_{k_1} > 1$. Moreover, by Lemma 5.5.9, the tail sequence (a_{k_1+n}) of (a_n) is unbounded as well. So we can find an element a_{k_2} in this tail such that $a_{k_2} > \max\{a_{k_1}, 2\}$. Inductively, choose a_{k_m} in the tail sequence $(a_{k_{m-1}+n})$ such that $a_{k_m} > \max\{a_{k_{m-1}}, m\}$. By construction, the subsequence (a_{k_n}) is strictly increasing and $a_{k_n} > n$ for all $n \in \mathbb{N}$. Since the subsequence is increasing and unbounded, by Theorem 5.4.2, we must have $a_{k_n} \to \infty$. □

5.6 Comparing Sequences

Given two real sequences, we can also compare them term-wise using the ordering in $\mathbb{R}$. This comparison allows us to study the behaviour of their limits. The first result that we can prove is the preservation of weak inequalities when we apply the limits:

Proposition 5.6.1 (Preservation of Weak Inequalities) *Let (a_n) and (b_n) be two convergent real sequences such that $a_n \to L$ and $b_n \to M$. If $a_n \leq b_n$ for all $n \in \mathbb{N}$, then $L \leq M$.*

Proof Suppose for contradiction that $L > M$ so that $L - M > 0$. Setting $\varepsilon = \frac{L-M}{2} > 0$, since $a_n \to L$, there exists some $N_1 \in \mathbb{N}$ such that $|a_n - L| < \frac{L-M}{2}$ for all $n \geq N_1$. In particular, we have $\frac{M+L}{2} < a_n$ for $n \geq N_1$. Also, since $b_n \to M$, there exists some $N_2 \in \mathbb{N}$ such that $|b_n - M| < \frac{L-M}{2}$ for all $n \geq N_2$. In particular, we have $b_n < \frac{L+M}{2}$ for $n \geq N_2$. Therefore, for $N = \max\{N_1, N_2\}$, we have $\frac{M+L}{2} < a_N \leq b_N < \frac{L+M}{2}$, which is a contradiction. Thus, we conclude that $L \leq M$. □

The proposition above says that if two convergent sequences satisfy a term-wise weak ordering for each index $n \in \mathbb{N}$, then their limits also satisfy the same ordering. However, this is not true for the strict inequalities. Indeed, an example would be the two sequences (a_n) and (b_n) where $a_n = 0$ and $b_n = \frac{1}{n}$ for all $n \in \mathbb{N}$. Clearly $a_n < b_n$ for all $n \in \mathbb{N}$, but the limits are equal since both of the sequences converge to the same limit 0. However, since strict inequalities are also weak inequalities, we can immediately state:

Corollary 5.6.2 *Let (a_n) and (b_n) be two convergent real sequences such that $a_n \to L$ and $b_n \to M$. If $a_n < b_n$ for all $n \in \mathbb{N}$, then $L \leq M$.*

Proof Since $a_n < b_n$ for all $n \in \mathbb{N}$, we also have $a_n \leq b_n$ for all $n \in \mathbb{N}$. Applying Proposition 5.6.1 yields the result. □

Therefore, in the limit, any inequality becomes weak inequality. Another obvious result involving comparison of sequences is the following:

Proposition 5.6.3 *Let (a_n) and (b_n) be two real sequences such that $a_n \leq b_n$ for all $n \in \mathbb{N}$.*

1. *If $a_n \to \infty$, then $b_n \to \infty$ as well.*
2. *If $b_n \to -\infty$, then $a_n \to -\infty$ as well.*

Proof We only prove the first assertion as the second is similarly done.

1. Fix any $K > 0$. Since $a_n \to \infty$, there exists an N such that $a_n > K$ for every $n \geq N$. Since $b_n \geq a_n$ for every $n \in \mathbb{N}$, we also have $b_n > K$ for every $n \geq N$. Thus, we conclude that $b_n \to \infty$. □

Next, we have the sandwich or squeeze lemma which compares three sequences, two of which are tending to the same limit. The full statement of this result is as follows:

Proposition 5.6.4 (Sandwich Lemma) *Let (a_n), (b_n), and (c_n) be three real sequences such that $a_n \leq b_n \leq c_n$ for all $n \in \mathbb{N}$. Suppose that the sequences (a_n) and (c_n) are convergent with $a_n \to L$ and $c_n \to L$. Then, (b_n) is also a convergent sequence with $b_n \to L$.*

Proof Fix $\varepsilon > 0$. Our aim is to find an $N \in \mathbb{N}$ such that $|b_n - L| < \varepsilon$ for all $n \geq N$. Since $a_n \to L$, for this $\varepsilon > 0$, there exists an $N_1 \in \mathbb{N}$ such that $|a_n - L| < \varepsilon$ for all $n \geq N_1$. In other words, $L - \varepsilon < a_n < L + \varepsilon$ for all $n \geq N_1$. Likewise, since $c_n \to L$ as well, there exists an $N_2 \in \mathbb{N}$ such that $L - \varepsilon < c_n < L + \varepsilon$ for all $n \geq N_2$.

Pick $N = \max\{N_1, N_2\}$. For all $n \geq N$, we have both $L-\varepsilon < a_n$ and $c_n < L+\varepsilon$. Putting these two facts together with b_n gives us:

$$L - \varepsilon < a_n \leq b_n \leq c_n < L + \varepsilon \quad \Rightarrow \quad |b_n - L| < \varepsilon \quad \text{for all } n \geq N,$$

and thus the sequence (b_n) also converges with $b_n \to L$. □

Proposition 5.6.4 is also useful in proving the convergence of a sequence by comparing it with some known or simpler sequences.

Example 5.6.5 As an example, consider the sequence (b_n) defined by $b_n = \frac{n}{3n^2+1}$. We note that for each $n \in \mathbb{N}$, we have the ordering:

$$0 \leq b_n = \frac{n}{3n^2 + 1} < \frac{n}{3n^2} = \frac{1}{3n} < \frac{1}{n},$$

so let us define new sequences (a_n) and (c_n) where $a_n = 0$ and $c_n = \frac{1}{n}$ for all $n \in \mathbb{N}$. From the inequality above, we have the ordering $a_n \leq b_n \leq c_n$ for all

$n \in \mathbb{N}$. Applying the sandwich lemma, since $a_n \to 0$ and $c_n \to 0$, we conclude $b_n \to 0$ as well.

In fact, for Propositions 5.6.1, 5.6.3, and 5.6.4, we do not even require the ordering $a_n \leq b_n \leq c_n$ for all $n \in \mathbb{N}$. We can weaken this condition to only requiring that there exists some $N \in \mathbb{N}$ such that $a_n \leq b_n \leq c_n$ for all $n \geq N$. In other words, we are just interested in their tail to study their convergence behaviour. This is true since limits only depend on the tails of the sequence, so ignoring a finite number of terms at the beginning of the sequence does not affect the limit. We look at an example here.

Example 5.6.6 Consider the sequence (q_n) defined by $q_n = \frac{1}{\left(n-\frac{5}{2}\right)^2}$. Clearly all the terms are non-negative. To bound this sequence from above, we expand the bracket to get:

$$0 \leq q_n = \frac{1}{n^2 - 5n + \frac{25}{4}} < \frac{1}{n^2 - 5n}, \tag{5.5}$$

which is true for $n \geq 6$ only. We further note that for $n \geq 6$, we have $n^2 - 6n \geq 0$ and hence $n^2 - 5n \geq n$. By taking reciprocals, we have $\frac{1}{n^2-5n} \leq \frac{1}{n}$. So for $n \geq 6$, the inequality in (5.5) can be bound further as such:

$$0 \leq q_n < \frac{1}{n^2 - 5n} \leq \frac{1}{n} \quad \text{for } n \geq 6,$$

and thus by applying sandwich lemma for $n \geq 6$, we have $q_n \to 0$.

5.7 Asymptotic Notations

As seen in the previous section, comparing complicated sequences with a simpler sequence is a good idea since their convergence are easier to investigate. We expect that sequences which behave in a similar manner at infinity would have the same convergence or divergence property. But what does "behave in a similar manner at infinity" mean? We characterise them as such:

Definition 5.7.1 (Asymptotically Equivalent Sequences) Let (a_n) and (b_n) be non-zero real sequences. The sequences are called asymptotically equivalent if $\lim_{n\to\infty} \frac{a_n}{b_n} = 1$. Sequences which are asymptotically equivalent are denoted as $(a_n) \sim (b_n)$ or $a_n \sim b_n$.

Remark 5.7.2 The term asymptote comes from the Greek word *asumptōtos* which means "not falling together". This term was introduced by Apollonius of Perga (c. 240B.C.–190B.C.).

The notation $\sim$ employed in Definition 5.7.1 suggests some kind of relation between these sequences. Indeed, we can check that being asymptotically equivalent is an equivalence relation. This means the relation $\sim$ is reflexive, symmetric, and transitive. In particular, writing $(a_n) \sim (b_n)$ is similar to writing $(b_n) \sim (a_n)$. The readers will show these in Exercise 5.23 and this proof can be made simple once we cover algebra of limits in Theorem 5.9.1.

Lemma 5.7.3 *Let $\sim$ be the relation on the set of non-zero real sequences where $(a_n) \sim (b_n)$ iff $\lim_{n\to\infty} \frac{a_n}{b_n} = 1$. Then, this relation is an equivalence relation.*

Example 5.7.4 Let us look at some examples of asymptotically equivalent sequences.

1. A simple example would be the sequences (a_n) and (b_n) defined as $a_n = n^2 + n$ and $b_n = n^2$. Note that as n gets very large, if we look at (a_n), the term n^2 dominates the term n. In other words, the term n does not contribute much in comparison to the n^2 term, so the sequence eventually roughly behaves like $n^2 = b_n$. So we expect $(a_n) \sim (b_n)$. Now let us prove this rigorously using Definition 5.7.1. We have:
$$\lim_{n\to\infty} \frac{a_n}{b_n} = \lim_{n\to\infty} \frac{n^2+n}{n^2} = \lim_{n\to\infty} \left(1 + \frac{1}{n}\right) = 1,$$
which can be shown easily via the ε-N definition. So $n^2 + n$ is asymptotically equivalent to n^2.
2. Recall the sequence (b_n) given by $b_n = \frac{n}{3n^2+1}$ that we have seen in Example 5.6.5. This sequence is asymptotically equivalent to the sequence (a_n) defined as $a_n = \frac{1}{3n}$. To guess this, we algebraically manipulate b_n to get $b_n = \frac{n}{3n^2+1} = \frac{1}{3n+\frac{1}{n}}$. We note as $n \to \infty$, the term $\frac{1}{n}$ in the denominator becomes very small in comparison to the $3n$ term. Therefore we say that the $3n$ term dominates the $\frac{1}{n}$ term and thus as n gets larger, the sequence resembles $\frac{1}{3n} = a_n$ instead.
To prove this properly and concretely, we check:
$$\lim_{n\to\infty} \frac{a_n}{b_n} = \lim_{n\to\infty} \frac{\frac{1}{3n}}{\frac{n}{3n^2+1}} = \lim_{n\to\infty} \frac{3n^2+1}{3n^2} = \lim_{n\to\infty} \left(1 + \frac{1}{3n^2}\right) = 1.$$
So we conclude that $(a_n) \sim (b_n)$.

3. Let (a_n) be a real sequence defined as $a_n = \sqrt{n^2+1} - n$. We claim that towards infinity, this sequence behaves asymptotically like $\frac{1}{2n}$. Indeed, we check:

$$\lim_{n\to\infty} \frac{\sqrt{n^2+1}-n}{\frac{1}{2n}} = \lim_{n\to\infty} \frac{(\sqrt{n^2+1}-n)(\sqrt{n^2+1}+n)}{\frac{\sqrt{n^2+1}+n}{2n}}$$

$$= \lim_{n\to\infty} \frac{2}{\sqrt{1+\frac{1}{n^2}}+1} = 1,$$

via a simple exercise. Thus, $\sqrt{n^2+1} - n \sim \frac{1}{2n}$.
4. A non-trivial example is the asymptotic equivalence $n! \sim \sqrt{2\pi n}\left(\frac{n}{e}\right)^n$. The number e here is the Euler-Napier constant that we saw in Example 5.4.4 and π is the Archimedes constant that will be introduced in Chap. 6 when we look at circular measures. This asymptotic formulation of the factorial is called the Stirling's approximation and we shall prove this much later in Exercise 16.30.
5. The partition function $p : \mathbb{N} \to \mathbb{N}$ is a function for which $p(n)$ denotes the number of distinct possible ways to express the integer n as a sum of positive integers. For example, $p(4) = 5$ since we can write $4 = 1+1+1+1 = 2+1+1 = 2+2 = 3+1$ in five different ways.
   Values of $p(n)$ can be computed by a recurrence relationship or generating function which may get complicated for big values of n. The asymptotic behaviour of this function was discovered by G.H. Hardy, Srinivasa Ramanujan (1887–1920), and J.V. Uspensky (1883–1947) as $p(n) \sim \frac{1}{4n\sqrt{3}} e^{\pi\sqrt{\frac{2n}{3}}}$. Notice the constants e and π that appear in this formula.
6. Another non-trivial asymptotic equivalence involves the prime numbers. We define a function $P : \mathbb{N} \to \mathbb{N}$ so that $P(n)$ is equal to the number of prime numbers less than equal to n. The function P is called the prime counting function and behaves asymptotically as $P(n) \sim \frac{n}{\ln(n)}$.
   This amazing fact was proven independently by Jacques Hadamard (1865–1963) and Charles de la Vallée Poussin (1866–1962). A consequence of this is: if (p_n) is the sequence of integers where p_n is the n-th prime number, then we have $p_n \sim n\ln(n)$. Here $\ln(n)$ is the logarithm of the number n taken with respect to the base e. Again, another appearance of the ubiquitous constant e.

In the spirit of comparison of sequences, the asymptotic equivalence could help us express or approximate limits of complicated sequences by comparing it to some known sequences. Since the behaviour of asymptotically equivalent sequences are similar at infinity, we expect that if one of the sequences converge, the other would converge as well. Likewise, identical behaviour for divergence to $\pm\infty$ is expected. Indeed, we have:

Proposition 5.7.5 *Let (a_n) and (b_n) be non-zero real sequences such that $(a_n) \sim (b_n)$.*

1. *Suppose that $L \in \mathbb{R}$. Then, $b_n \to L$ if and only if $a_n \to L$.*
2. *$b_n \to \pm\infty$ if and only if $a_n \to \pm\infty$.*

Proof We only prove some of the assertions as they are symmetric.

1. Suppose that $(a_n) \sim (b_n)$ and $b_n \to L$. We want to show that $a_n \to L$ as well. Fix $\varepsilon > 0$. Since the sequence (b_n) converges, it must be bounded, say $|b_n| \leq M$ for some $M > 0$. We know that $b_n \to L$ so there exists an $N_1 \in \mathbb{N}$ such that $|b_n - L| < \frac{\varepsilon}{2}$ for all $n \geq N_1$. Moreover, since $\frac{a_n}{b_n} \to 1$, there exists an $N_2 \in \mathbb{N}$ such that for all $n \geq N_2$ we have:
$$\left|\frac{a_n}{b_n} - 1\right| < \frac{\varepsilon}{2M} \quad \Rightarrow \quad |a_n - b_n| < \frac{\varepsilon|b_n|}{2M} \leq \frac{\varepsilon M}{2M} = \frac{\varepsilon}{2}.$$
Thus, if we set $N = \max\{N_1, N_2\}$, for all $n \geq N$ we have:
$$|a_n - L| = |a_n - b_n + b_n - L| \leq |a_n - b_n| + |b_n - L| < \frac{\varepsilon}{2} + \frac{\varepsilon}{2} = \varepsilon.$$
Therefore, we conclude that $a_n \to L$ as well. The converse of this is similarly proven.
2. Suppose that $b_n \to \infty$. Fix $K > 0$. We want to find an $N \in \mathbb{N}$ such that $a_n > K$ for all $n \geq N$. Since $b_n \to \infty$, we can find an $N_1 \in \mathbb{N}$ such that $b_n > 2K$ for all $n \geq N_1$. Moreover, since $\frac{a_n}{b_n} \to 1$, there exists an $N_2 \in \mathbb{N}$ such that:
$$\left|\frac{a_n}{b_n} - 1\right| < \frac{1}{2} \quad \Rightarrow \quad \frac{1}{2} < \frac{a_n}{b_n}.$$
Set $N = \max\{N_1, N_2\}$. Then, for any $n \geq N$, we have $b_n > 2K$ and, in particular, must be positive. So the inequality above implies:
$$\frac{1}{2} < \frac{a_n}{b_n} \quad \Rightarrow \quad \frac{b_n}{2} < a_n \quad \Rightarrow \quad K < a_n,$$
for all $n \geq N$. We conclude that $a_n \to \infty$ as well. The converse and the case for blowing up to $-\infty$ can be proven in a similar manner. □

Proposition 5.7.5 is a valuable result to know along with some standard convergent or divergent sequences. This allows us to determine some convergence or divergence properties of complicated sequences via some simpler sequences. In fact, we shall use more of this result in Chap. 7 when we study real series.

Example 5.7.6 Let us recall some sequences from Example 5.7.4.

1. Recall the sequences (a_n) and (b_n) defined as $a_n = n^2 + n$ and $b_n = n^2$. They are asymptotically equivalent. Moreover, we know that b_n diverges to ∞, so Proposition 5.7.5 implies that $a_n \to \infty$ as well.
2. The sequences (a_n) and (b_n) given by $a_n = \frac{1}{3n}$ and $b_n = \frac{n}{3n^2+1}$ are asymptotically equivalent. Since we know that $a_n \to 0$, we can conclude that $b_n \to 0$ as well.
3. Let (a_n) and (b_n) be real sequences defined as $a_n = \sqrt{n^2+1} - n$ and $b_n = \frac{1}{2n}$. At first glance, it is not clear how the sequence (a_n) behaves like at infinity since both of the terms in it go to ∞ separately. However, since it is asymptotic to a sequence (b_n) that converges to 0, we can confidently say that (a_n) converges to 0 as well by Proposition 5.7.5. This is interesting because even though the quantities $\sqrt{n^2+1}$ and n both diverge to ∞, they get closer to each other at a rate of $\frac{1}{2n}$ as $n \to \infty$.

Asymptotic equivalence is one of the many asymptotic behaviour notation that mathematicians and scientists use. These notations are useful when one wants to encapsulate the behaviour of some complicated quantity at infinity with a simpler and more well-understood term. This also allows us to deduce convergence results more easily.

Big-O and Little-o Notations

Two other frequently used asymptotic notations are:

Definition 5.7.7 (Big-O and Little-o Notations) Let (a_n) and (b_n) be real sequences such that $b_n > 0$. We write:

1. $a_n \in O(b_n)$ if there exist a $K > 0$ and an $N \in \mathbb{N}$ such that $|a_n| \leq Kb_n$ for all $n \geq N$. We say the sequence (a_n) is (of order) big-O of (b_n). In symbols:

$$a_n \in O(b_n) \quad \text{if} \quad \exists K > 0 : \exists N \in \mathbb{N} : \forall n \geq N, |a_n| \leq Kb_n.$$

2. $a_n \in o(b_n)$ if for every $K > 0$, there exists an $N \in \mathbb{N}$ such that $|a_n| \leq Kb_n$ for all $n \geq N$. We say the sequence (a_n) is (of order) little-o of (b_n). In symbols:

$$a_n \in o(b_n) \quad \text{if} \quad \forall K > 0, \exists N \in \mathbb{N} : \forall n \geq N, |a_n| \leq Kb_n.$$

In Exercise 5.26, the readers will prove their limit definitions. These notations are commonly used in discrete mathematics, number theory, graph theory, computer science, network science, physics, and engineering when one wants to analyse the complexity of algorithms or approximate sizes and order of magnitudes of large quantities. They are part of a bigger family of asymptotic notations called the Bachmann–Landau notations, introduced by Paul Bachmann (1837–1920) and Edmund Landau (1877–1938) in the study of analytic number theory.

The intuition behind the big-O notation is that $a_n \in O(b_n)$ if the sequence (a_n) is eventually bounded from above by a constant scale of the sequence (b_n). Roughly speaking, this means that (a_n) and (b_n) are of the same order of magnitude.

The little-o notation is more restrictive: if $a_n \in o(b_n)$, then the sequence (a_n) is dominated by the sequence (b_n). In looser terms, the sequence (a_n) is eventually negligible compared to (b_n). These notations are often used to simplify the terms in a sequence by just looking at the dominating or significant term.

Example 5.7.8 If we consider the expression $(n+1)^3 = n^3 + 3n^2 + 3n + 1$ for $n \in \mathbb{N}$, we can simplify this by writing $(n+1)^3 = n^3 + O(n^2)$ because all of the other terms (which we call lower order terms) are of the same order as n^2. Moreover, for large n, the terms $3n^2 + 3n + 1$ become negligible in comparison to the leading term n^3. So we can also write $(n+1)^3 = n^3 + o(n^3)$. Notice the difference in the usage above, namely:

$$(n+1)^3 = n^3 + O(n^2) = n^3 + o(n^3).$$

Therefore, one has to be careful with the big-O and little-o notation as they are not interchangeable. We shall see in Example 5.7.10(1) that only one implies the other.

Furthermore, we can immediately see how useful these notations are in limit comparison. An immediate consequence from the definition and sandwiching argument are:

Proposition 5.7.9 *Let (a_n) and (b_n) be real sequences and $b_n > 0$ for all $n \in \mathbb{N}$.*

1. *If $a_n \in O(b_n)$ and $b_n \to 0$, then $a_n \to 0$ as well.*
2. *If $a_n \in o(b_n)$ and (b_n) converges, then $a_n \to 0$.*

Example 5.7.10 For now, let us recall some examples from Example 5.7.4.

1. Let (a_n) be a sequence defined as $a_n = n$. Then, we have $a_n \in o(n^2)$ since for any fixed $K > 0$, we can set $N = \lceil \frac{1}{K} \rceil \in \mathbb{N}$ so that for all $n \geq N$, we have: $\frac{a_n}{n^2} = \frac{n}{n^2} = \frac{1}{n} \leq \frac{1}{N} = \frac{1}{\lceil \frac{1}{K} \rceil} \leq \frac{1}{\frac{1}{K}} = K$ which says $a_n \leq Kn^2$.
   Moreover, we also have $a_n \in O(n^2)$. This is clear since $a_n = n \leq n^2$ for all $n \in \mathbb{N}$.
   In fact, this is true for any general sequences, namely: if $a_n \in o(b_n)$, then $a_n \in O(b_n)$.
2. Let (a_n) be defined as $a_n = 3n^3 + 2n^2 + n + 10$. Notice that for any $n \in \mathbb{N}$ we have $n^3 \geq n^2, n$. So $a_n = 3n^3 + 2n^2 + n + 10 \leq (3+2+1+10)n^3 = 16n^3$ for all $n \in \mathbb{N}$. This implies $a_n \in O(n^3)$.
   However, $a_n \notin o(n^3)$. Indeed, for $K = 1$, we cannot find an $N \in \mathbb{N}$ such that $a_n \leq n^3$ for all $n \geq N$ since $\frac{a_n}{n^3} = 3 + \frac{2}{n} + \frac{1}{n^2} + \frac{10}{n^3} \geq 3 > 1$ for any $n \in \mathbb{N}$.

3. Consider the sequence (a_n) defined as $a_n = \frac{(-1)^n}{n^2+n^3}$. We can show that $a_n \in O(\frac{1}{n})$, so there must exist a constant $K > 0$ and an $N \in \mathbb{N}$ such that $|a_n| \leq \frac{K}{n}$ for all $n \geq N$. Moreover, since $\frac{1}{n} \to 0$ as $n \to \infty$, by Proposition 5.7.9, we must have $a_n \to 0$.
4. The first assertion in Proposition 5.7.9 is not true when the dominating sequence (b_n) does not converge to 0. An example of this is the sequence (a_n) defined as $a_n = (-1)^n$. Clearly $a_n \in O(1)$ and the constant sequence of 1s converges to $1 \neq 0$. However, the sequence (a_n) itself does not converge.

5.8 Cauchy Sequences

Another type of sequence that we often work with in analysis is the Cauchy sequence, which is attributed to Augustin-Louis Cauchy. This type of sequence is actually more often looked at in analysis as opposed to convergent sequences. The reason why is that, whilst convergent sequences are nice, we need to actually know what the limit L is in its definition. For Cauchy sequences, we do not need to know this information and thus can be defined in more generality. We define:

Definition 5.8.1 (Cauchy Sequences) A real sequence (a_n) is called a Cauchy sequence if for every $\varepsilon > 0$, there exists an $N \in \mathbb{N}$ such that for every $m, n \geq N$ we have $|a_n - a_m| < \varepsilon$.

Symbolically, this is written with quantifiers as:

$$(a_n) \text{ is Cauchy} \quad \text{if} \quad \forall \varepsilon > 0, \exists N \in \mathbb{N} : \forall m, n \geq N, |a_m - a_n| < \varepsilon.$$

The definition says for every $\varepsilon > 0$, there exists a tail of the sequence such that all the terms in the tail are within ε of each other. So intuitively this means that the terms in the sequence are getting closer to each other as $n \to \infty$.

From this intuitive definition, we expect that the sequence converges to some limit. For real sequences, this is indeed true. Thus, the Cauchy property can be useful in certain situations when we do not actually have a numerical value for the limit L of a real sequence to write down the definition for convergence.

In order to prove the fact that real Cauchy sequences are the same as real convergent sequences, we first need to prove a lemma. This lemma states that a Cauchy sequence must be bounded, which is similar to Proposition 5.2.14 for convergent sequences.

Lemma 5.8.2 *If (a_n) is a Cauchy real sequence, then the sequence (a_n) is bounded.*

Proof Fix $\varepsilon = 1$. Then, there exists an $N \in \mathbb{N}$ such that $|a_n - a_m| < 1$ for every $m, n \geq N$. In particular, $|a_n - a_N| < 1$ for every $n \geq N$. Using triangle inequality, we have $|a_n| = |a_n - a_N + a_N| \leq |a_n - a_N| + |a_N| < 1 + |a_N|$ for all $n \geq N$. So all the terms after the N-th term are bounded by the constant $|a_N|+1$. Furthermore, if we let $M = \max\{|a_1|, |a_2|, \dots, |a_{N-1}|\}$ then clearly $|a_n| \leq M$ for all $n = 1, 2, \dots, N-1$. Therefore, if we define $K = \max\{M, |a_N| + 1\}$, we can deduce $|a_n| \leq K$ for all $n \in \mathbb{N}$ and thus the whole sequence is bounded. □

Theorem 5.8.3 *Let (a_n) be a real sequence. (a_n) is convergent if and only if it is Cauchy.*

Proof We prove the implications one by one.

($\Rightarrow$): Suppose that the real sequence (a_n) is convergent to some $L \in \mathbb{R}$. Fix $\varepsilon > 0$. Then, there exists an $N \in \mathbb{N}$ such that $|a_n - L| < \frac{\varepsilon}{2}$ for every $n \geq N$. Thus, for every $m, n \geq N$, we have $|a_n - L| < \frac{\varepsilon}{2}$ and $|a_m - L| < \frac{\varepsilon}{2}$. By triangle inequality, for every $m, n \geq N$, we have:

$$|a_n - a_m| = |a_n - L - a_m + L| \leq |a_n - L| + |a_m - L| < \frac{\varepsilon}{2} + \frac{\varepsilon}{2} = \varepsilon.$$

Thus, the sequence (a_n) is also Cauchy.

($\Leftarrow$): Suppose that the sequence (a_n) is Cauchy. To show that this sequence converges, we need to find a candidate L for its limit. By Lemma 5.8.2, the sequence (a_n) is bounded. Since this sequence is bounded, Bolzano-Weierstrass theorem says there exists a subsequence (a_{k_n}) of (a_n) that is convergent, say to $L \in \mathbb{R}$. We now want to prove that the whole sequence (a_n) converges to L.

Fix $\varepsilon > 0$. Since the sequence (a_n) is Cauchy, there exists an $N_1 \in \mathbb{N}$ such that $|a_n - a_m| < \frac{\varepsilon}{2}$ for every $m, n \geq N_1$. Furthermore, since the subsequence (a_{k_n}) is convergent to L, there exists an $N_2 \in \mathbb{N}$ such that $|a_{k_n} - L| < \frac{\varepsilon}{2}$ for every $n \geq N_2$.

Let $N = \max\{N_1, N_2\}$. Then, both of the above hold which means $|a_{k_N} - L| < \frac{\varepsilon}{2}$ and $|a_n - a_m| < \frac{\varepsilon}{2}$ for any $m, n \geq N$. In particular, since $k_N \geq N$, by setting $m = k_N$ in the latter, we have $|a_n - a_{k_N}| < \frac{\varepsilon}{2}$ for any $n \geq N$. Thus, by using the triangle inequality and these inequalities, for every $n \geq N$ we have:

$$|a_n - L| = |a_n - a_{k_N} + a_{k_N} - L| \leq |a_n - a_{k_N}| + |a_{k_N} - L| < \frac{\varepsilon}{2} + \frac{\varepsilon}{2} = \varepsilon,$$

and so $a_n \to L$. Therefore, the Cauchy sequence (a_n) is convergent. □

Another example of a situation where Cauchy sequences are used is when we want to prove the divergence of a sequence (a_n): since real convergent sequences are also Cauchy, we simply have to show that (a_n) is not Cauchy to establish that it is divergent via the negation of Theorem 5.8.3.

Example 5.8.4 Let us look at some examples where we can use the Cauchy property to our advantage.

1. Consider a real sequence (a_n) for which $|a_{n+1} - a_n| \leq \frac{1}{2^n}$ for all $n \in \mathbb{N}$. If we expand this, we have $-\frac{1}{2^n} + a_{n-1} < a_n < a_{n-1} + \frac{1}{2^n}$.
   We want to show that this sequence is convergent, but we do not have any idea what its limit is in order for us to carry out the ε-N proof. We do not have an explicit form for the terms to make a guess for its limit. Furthermore, we do not know whether it is monotone since the recursive relation only gives us a rough distance of where the n-th term are in relative to the $(n-1)$-th term, but not its exact location. So we cannot use monotone sequence theorem here.
   To get around these problems, we show that this sequence is Cauchy. First, let us estimate the difference $|a_n - a_m|$ for any $n > m$. Using the triangle inequality repeatedly, we have:
$$\begin{aligned} |a_n - a_m| &= |a_n - a_{n-1} + a_{n-1} - \ldots - a_{m+1} + a_{m+1} - a_m| \\ &\leq |a_n - a_{n-1}| + \ldots + |a_{m+1} - a_m| \\ &\leq \frac{1}{2^{n-1}} + \ldots + \frac{1}{2^m} = \frac{\frac{1}{2^m}\left(1 - \frac{1}{2^{n-m}}\right)}{1 - \frac{1}{2}} < \frac{1}{2^{m-1}} \leq \frac{1}{m}, \end{aligned} \tag{5.6}$$
   by using Bernoulli's inequality.
   To show that the sequence (a_n) is Cauchy, we fix $\varepsilon > 0$ and claim $N = \lceil \frac{1}{\varepsilon} \rceil$ is enough to ensure the Cauchy property. Indeed, for any $n > m \geq N$, using the estimate in (5.6), we have:
$$|a_n - a_m| < \frac{1}{m} \leq \frac{1}{N} = \frac{1}{\lceil \frac{1}{\varepsilon} \rceil} \leq \frac{1}{\frac{1}{\varepsilon}} = \varepsilon,$$
   from which we conclude that (a_n) is Cauchy. Thus, applying Theorem 5.8.3, we can therefore conclude that (a_n) converges to some real number $L \in \mathbb{R}$.
2. Recall the sequence (a_n) defined as $a_n = (-1)^n$. We have proven that this sequence diverges by using the ε-N definition or looking at subsequences. Let us present a different method for proving this by showing that it is not Cauchy.
   Set $\varepsilon = \frac{1}{2}$. There are no $N \in \mathbb{N}$ for which $|a_n - a_m| < \frac{1}{2}$ for all $n > m \geq N$. Indeed, if there is such an N, we can set $n = N + 1$ and $m = N$ so that $2 = |a_{N+1} - a_N| < \frac{1}{2}$, a contradiction. Hence, the sequence (a_n) is not Cauchy and thus is not convergent.

As mentioned earlier in this section, Cauchy sequences are very important in analysis. Its most important use is in the construction of the real numbers from the rationals. The idea is exactly similar to the construction of $\sqrt{2}$ in Example 4.4.5: we look at all the Cauchy sequences in $\mathbb{Q}$. Some of these sequences may converge to an element not in $\mathbb{Q}$. For example, the rational approximations (r_n) of $\sqrt{2}$ is a Cauchy sequence of rational numbers but it does not have a limit in $\mathbb{Q}$. Its limit is outside the set of rational numbers. These new limits are then added on to our number field $\mathbb{Q}$.

The resulting number field would then be the union of the old numbers and these new numbers: $\mathbb{Q} \cup \bar{\mathbb{Q}}$. Thus, any Cauchy sequence in this new set would converge to an element in this set as well. We call such spaces complete, as in Definition 6.4.11 (compare with the nomenclature for the completeness axiom for $\mathbb{R}$ in Definition 3.6.11). For details of this, readers are invited to carry out this construction rigorously in Exercise 6.27.

5.9 Algebra of Limits

One might have seen the algebra of limits during an introductory class on calculus, but these were often stated without their proofs. In these results, we usually create new sequences from old ones by addition or multiplication and we would like to know what happens to their limits.

Here, we shall state the results and prove them from definitions. The most important assumption in the next theorem is that the sequences (a_n) and (b_n) are known to converge before we can apply the results.

Theorem 5.9.1 (Algebra of Limits, AOL) *Let (a_n) and (b_n) be convergent real sequences such that $a_n \to L$ and $b_n \to M$ where $L, M \in \mathbb{R}$.*

1. *For a constant $\lambda \in \mathbb{R}$, the sequence (λa_n) converges to λL. In other words:*

$$\lim_{n\to\infty} \lambda a_n = \lambda \lim_{n\to\infty} a_n.$$

2. *The sequence $(|a_n|)$ converges to $|L|$. In other words:*

$$\lim_{n\to\infty} |a_n| = |\lim_{n\to\infty} a_n|.$$

3. *The sequence $(a_n + b_n)$ converges to $L + M$. In other words:*

$$\lim_{n\to\infty} (a_n + b_n) = \lim_{n\to\infty} a_n + \lim_{n\to\infty} b_n.$$

4. *The sequence $(a_n - b_n)$ converges to $L - M$. In other words:*

$$\lim_{n\to\infty} (a_n - b_n) = \lim_{n\to\infty} a_n - \lim_{n\to\infty} b_n.$$

5. *The sequence $(a_n b_n)$ converges to LM. In other words:*

$$\lim_{n\to\infty} a_n b_n = \left(\lim_{n\to\infty} a_n\right)\left(\lim_{n\to\infty} b_n\right).$$

6. *Assume that* $a_n \neq 0$ *for all* $n \in \mathbb{N}$. *If* $L \neq 0$, *then the sequence* $\left(\frac{1}{a_n}\right)$ *converges to* $\frac{1}{L}$. *In other words:*

$$\lim_{n\to\infty} \frac{1}{a_n} = \frac{1}{\lim_{n\to\infty} a_n}.$$

7. *Assume that* $a_n \neq 0$ *for all* $n \in \mathbb{N}$. *If* $L \neq 0$, *then the sequence* $\left(\frac{b_n}{a_n}\right)$ *converges to* $\frac{M}{L}$. *In other words:*

$$\lim_{n\to\infty} \frac{b_n}{a_n} = \frac{\lim_{n\to\infty} b_n}{\lim_{n\to\infty} a_n}.$$

Proof We prove the assertions one by one.

1. If $\lambda = 0$, then the sequence (λa_n) is a constant sequence of 0s which clearly converges to 0. Suppose that $\lambda \neq 0$. Fix $\varepsilon > 0$. Our goal is to find an $N \in \mathbb{N}$ such that $|\lambda a_n - \lambda L| < \varepsilon$ for all $n \geq N$.
   We note that $\frac{\varepsilon}{|\lambda|} > 0$ as well. Using this in the definition of convergence for the sequence (a_n), there exists an $N \in \mathbb{N}$ such that for all $n \geq N$ we have:
   $$|a_n - L| < \frac{\varepsilon}{|\lambda|} \quad \Leftrightarrow \quad |\lambda a_n - \lambda L| < \varepsilon.$$
   So for this choice of $N \in \mathbb{N}$, we must have $|\lambda a_n - \lambda L| < \varepsilon$ for every $n \geq N$. Thus, we have found our desired N and hence $\lambda a_n \to \lambda L$.
2. Fix $\varepsilon > 0$. Our goal is to find an $N \in \mathbb{R}$ such that $||a_n| - |L|| < \varepsilon$ for all $n \geq N$. Using the definition of convergence for the sequence (a_n), there exists an $N \in \mathbb{N}$ such that for all $n \geq N$, we have $|a_n - L| < \varepsilon$. So, for the same choice of $N \in \mathbb{N}$, by using reverse triangle inequality, we must have $||a_n| - |L|| < |a_n - L| < \varepsilon$ for every $n \geq N$. Thus, we have found our desired N and hence $|a_n| \to |L|$.
3. Fix $\varepsilon > 0$. We want to find an $N \in \mathbb{N}$ such that $|a_n + b_n - (L + M)| < \varepsilon$ for all $n \geq N$.
   Since $a_n \to L$, there must exist an $N_1 \in \mathbb{N}$ such that $|a_n - L| < \frac{\varepsilon}{2}$ for all $n \geq N_1$. Also, since $b_n \to M$, there must exist an $N_2 \in \mathbb{N}$ such that $|b_n - M| < \frac{\varepsilon}{2}$ for all $n \geq N_2$. So if we set $N = \max\{N_1, N_2\}$, for any $n \geq N$ both inequalities above must hold. By triangle inequality, we then have:
   $$|a_n + b_n - (L + M)| = |a_n - L + b_n - M| \leq |a_n - L| + |b_n - M| < \frac{\varepsilon}{2} + \frac{\varepsilon}{2} = \varepsilon,$$
   for all $n \geq N$. This means $a_n + b_n \to L + M$.
4. This is an application of the first and third assertions: since $a_n \to L$ and $-b_n \to -M$, we have $a_n - b_n = a_n + (-b_n) \to L - M$.

5. Since both of the sequences (a_n) and (b_n) converge, by Proposition 5.2.14 these sequences are bounded. In particular, there is a $K > 0$ such that $|a_n| \leq K$ for all $n \in \mathbb{N}$.
   Fix $\varepsilon > 0$. Our goal is to show that there exists some $N \in \mathbb{N}$ such that $|a_n b_n - LM| < \varepsilon$ for all $n \geq N$. Since $a_n \to L$, there must exist some $N_1 \in \mathbb{N}$ such that $|a_n - L| < \frac{\varepsilon}{K+|M|}$ for all $n \geq N_1$. Similarly, since $b_n \to M$, there must exist an $N_2 \in \mathbb{N}$ such that $|b_n - M| < \frac{\varepsilon}{K+|M|}$ for all $n \geq N_2$. Let us pick $N = \max\{N_1, N_2\}$ so that for any $n \geq N$, both of the inequalities above hold. Then, for all $n \geq N$, we have:

$$\begin{aligned}|a_n b_n - LM| &= |a_n b_n - a_n M + a_n M - LM| \\ &= |a_n(b_n - M) + M(a_n - L)| \\ &\leq |a_n||b_n - M| + |M||a_n - L| \\ &< K\frac{\varepsilon}{K+|M|} + |M|\frac{\varepsilon}{K+|M|} = \varepsilon,\end{aligned}$$

   which implies $a_n b_n \to LM$.
6. First, we show that for large enough n, the terms a_n can be bounded away from 0. Indeed, since $a_n \to L \neq 0$, there exists an $N_1 \in \mathbb{N}$ such that $|a_n - L| < \frac{|L|}{2}$ for all $n \geq N_1$. So, by triangle inequality, we have $|L| = |L - a_n + a_n| \leq |a_n - L| + |a_n|$. This then implies:

$$|a_n| \geq |L| - |a_n - L| > |L| - \frac{|L|}{2} = \frac{|L|}{2} > 0, \tag{5.7}$$

   for all $n \geq N_1$.
   Now we show the desired limit. Fix $\varepsilon > 0$. We want to show that there exists an $N \in \mathbb{N}$ such that $\left|\frac{1}{a_n} - \frac{1}{L}\right| < \varepsilon$. We note that for $n \geq N_1$, the inequality in (5.7) implies $\frac{1}{|a_n|} \leq \frac{2}{|L|}$. Furthermore, since $a_n \to L$, there exists an $N_2 \in \mathbb{N}$ such that $|a_n - L| < \frac{\varepsilon|L|^2}{2}$ for all $n \geq N_2$. By picking $N = \max\{N_1, N_2\}$, for all $n \geq N$ we have:

$$\left|\frac{1}{a_n} - \frac{1}{L}\right| = \frac{|a_n - L|}{|a_n L|} = \frac{1}{|a_n|}\frac{|a_n - L|}{|L|} \leq \frac{2}{|L|^2}|a_n - L| < \frac{2}{|L|^2}\frac{\varepsilon|L|^2}{2} = \varepsilon,$$

   and this implies $\frac{1}{a_n} \to \frac{1}{L}$.
7. This is an application of the fifth and sixth assertions: since $\frac{1}{a_n} \to \frac{1}{L}$ and $b_n \to M$, we have $\frac{b_n}{a_n} = b_n \times \frac{1}{a_n} \to \frac{M}{L}$. □

As a corollary of the algebra of limits, if the sequence (a_n) converges to $L \in \mathbb{R}$, we can use induction to show that the sequence (a_n^k) for any fixed $k \in \mathbb{N}$ also converges and it converges to L^k. In fact, we also have the following:

Proposition 5.9.2 *Let (a_n) be a sequence of non-negative numbers. If the sequence (a_n) converges to $L \geq 0$, then the sequence $(\sqrt{a_n})$ also converges and it converges to $\sqrt{L}$.*

Proof Fix $\varepsilon > 0$. There are two possibilities:

1. If $L = 0$, since $a_n \to 0$, there exists an $N \in \mathbb{N}$ such that when $n \geq N$, we have $|a_n - L| = |a_n| < \varepsilon^2$. Thus, for the same N, we have $|\sqrt{a_n}| < \varepsilon$ for all $n \geq N$, which means $\sqrt{a_n} \to 0$.
2. If $L > 0$, since $a_n \to L$, there exists an $N \in \mathbb{N}$ such that $|a_n - L| < \varepsilon\sqrt{L}$ when $n \geq N$. For the same N, we have:

$$|\sqrt{a_n} - \sqrt{L}| = \left|\frac{(\sqrt{a_n}-\sqrt{L})(\sqrt{a_n}+\sqrt{L})}{\sqrt{a_n}+\sqrt{L}}\right| = \frac{|a_n - L|}{\sqrt{a_n}+\sqrt{L}} \leq \frac{|a_n - L|}{\sqrt{L}} < \frac{\varepsilon\sqrt{L}}{\sqrt{L}} = \varepsilon,$$

for all $n \geq N$. This means $\sqrt{a_n} \to \sqrt{L}$. □

We would like to put great emphasis here that the algebra of limits only work for sequences which are convergent. Thus we can only use algebra of limits when we are certain that the sequences that we are working with are convergent.

Moreover, the converse of any of the statements in the algebra of limits may not be true. For example, Theorem 5.9.1(2) says that if $a_n \to L$, then we must have $|a_n| \to |L|$. However, the converse of this is not always true! Indeed, for the sequence (a_n) with $a_n = (-1)^n$, the modulus of this sequence is $|a_n| = 1$ for all $n \in \mathbb{N}$, which converges. On the other hand, we have seen that (a_n) itself does not converge in Example 5.2.7.

However, we can guarantee that this particular converse is true when $L = 0$.

Lemma 5.9.3 *Let (a_n) be a real sequence. Then, $\lim_{n\to\infty} |a_n| = 0$ if and only if $\lim_{n\to\infty} a_n = 0$.*

Proof We prove the forward implication only since the converse implication is immediately true by algebra of limits.

($\Rightarrow$): Fix $\varepsilon > 0$. Then, there exists an $N \in \mathbb{N}$ such that $||a_n| - 0| < \varepsilon$ for all $n \geq N$. Therefore, $|a_n - 0| = |a_n| = ||a_n| - 0| < \varepsilon$. Thus, we conclude that $a_n \to 0$ as well. □

Now let us look at the algebra of limits in action.

Example 5.9.4 Consider a real sequence (a_n) defined recursively as:

$$a_1 = \sqrt{2} \quad \text{and} \quad a_{n+1} = \sqrt{2 + a_n} \text{ for } n \geq 1.$$

It is very tempting to use the algebra of limits on the recursive relationship, but we have to be aware that we do not yet know whether the sequence (a_n) converges in order to do so!

Therefore, our first mission is to determine that it does converge. Clearly, the sequence is bounded below from 0 and we can inductively show that the sequence is bounded from above by 2 very easily. Furthermore, for all $n \geq 1$, using the fact that $0 \leq a_n \leq 2$, we have:

$$a_{n+1}^2 = 2 + a_n = \frac{(2+a_n)^2}{2+a_n} = \frac{4+4a_n+a_n^2}{2+a_n} \geq \frac{a_n^2+2a_n^2+a_n^2}{4} = a_n^2,$$

which means $a_{n+1} \geq a_n$ and so the sequence is increasing. Thus, by monotone sequence theorem, we conclude that (a_n) converges to some $L \in [\sqrt{2}, 2]$. Now that we know $a_n \to L$, we can apply the algebra of limits on the recursive relationship to get:

$$\lim_{n\to\infty} a_{n+1}^2 = \lim_{n\to\infty}(2+a_n) \quad \Rightarrow \quad L^2 = 2 + L \quad \Rightarrow \quad L = -1 \text{ or } 2.$$

But since $L \in [\sqrt{2}, 2]$, we must have $L = 2$.

Furthermore, notice that if we have sequences which blow up to $\pm\infty$, we cannot apply the algebra of limits in Theorem 5.9.1 since the proofs just work for the cases of the sequences having finite limits. However, we may use the following results:

Proposition 5.9.5 *Let (a_n) be a real sequence.*

1. *If $a_n > 0$ for all $n \in \mathbb{N}$, then $a_n \to \infty$ if and only if $\frac{1}{a_n} \to 0$.*
2. *If $a_n < 0$ for all $n \in \mathbb{N}$, then $a_n \to -\infty$ if and only if $\frac{1}{a_n} \to 0$.*

Proof We prove the first assertion only.

1. We prove the implications separately.

($\Rightarrow$): Fix $\varepsilon > 0$. Our goal is to find an $N \in \mathbb{N}$ such that $\left|\frac{1}{a_n} - 0\right| = \frac{1}{a_n} < \varepsilon$ for all $n \geq N$. Since $a_n \to \infty$, by using the definition of sequences blowing up to infinity, for $K = \frac{1}{\varepsilon} > 0$, there exists an $n \in \mathbb{N}$ such that $a_n > K = \frac{1}{\varepsilon}$ for every $n \geq N$. In other words, for every $n \geq N$ we have $\frac{1}{a_n} < \varepsilon$. So we have found our N.

($\Leftarrow$): Since a_n are all positive, $\frac{1}{a_n}$ are all positive as well. Fix $K > 0$. Our goal is to show that there exists an $N \in \mathbb{N}$ such that $a_n > K$ for all $n \geq N$. Since $\frac{1}{a_n} \to 0$ and $a_n > 0$, by using $\varepsilon = \frac{1}{K} > 0$ there must exist an $N \in \mathbb{N}$ such that $0 < \frac{1}{a_n} < \frac{1}{K}$ for all $n \geq N$. In other words, for all $n \geq N$ we have $a_n > K$. So we have found our desired N for the above.

□

Moreover, we have the following results:

Proposition 5.9.6 *Let (a_n) and (b_n) be real sequences such that $a_n \to L \in \mathbb{R}$ and $b_n \to \infty$.*

1. *The sequence $(a_n + b_n)$ diverges to ∞.*
2. *If $L > 0$, then $a_n b_n \to \infty$.*
3. *If $L < 0$, then $a_n b_n \to -\infty$.*

Likewise, if $b_n \to -\infty$, we have:

4. *The sequence $(a_n + b_n)$ diverges to $-\infty$.*
5. *If $L > 0$, then $a_n b_n \to -\infty$.*
6. *If $L < 0$, then $a_n b_n \to \infty$.*

Proof We prove the cases for $b_n \to \infty$ only. The cases for $b_n \to -\infty$ can be similarly done.

1. Fix $K > 0$. Since (a_n) converges, it must be bounded, so there exists an $M > 0$ such that $-M \leq a_n \leq M$. Moreover, since $b_n \to \infty$, there exists an $N \in \mathbb{N}$ such that $b_n > K + M$ for all $n \geq N$. Thus, for all $n \geq N$ we have $a_n + b_n > -M + (K + M) = K$, proving that $(a_n + b_n)$ diverges to ∞.
2. Fix $K > 0$. Since $a_n \to L > 0$, there exists an $N_1 \in \mathbb{N}$ such that $|a_n - L| < \frac{L}{2}$ for all $n \geq N_1$. This means $a_n > \frac{L}{2} > 0$ for all such n. On the other hand, $b_n \to \infty$ implies there exists an $N_2 \in \mathbb{N}$ such that $b_n > \frac{2K}{L}$ for all $n \geq N_2$. Set $N = \max\{N_1, N_2\}$. Whenever $n \geq N$ we have $a_n b_n > \frac{L}{2}\frac{2K}{L} = K$ and thus $(a_n b_n)$ diverges to ∞.
3. This is similarly done as the previous case. □

Example 5.9.7 Here are some more examples of applications for the algebra of limits:

1. Consider the sequence (a_n) defined by $a_n = \frac{n+1}{n}$ and we want to find its limit as $n \to \infty$. We can split this sequence into a quotient of two non-zero sequences (b_n) and (c_n) defined by $b_n = n + 1$ and $c_n = n$ so that $a_n = \frac{b_n}{c_n}$ for all $n \in \mathbb{N}$. However, since $b_n \to \infty$ and $c_n \to \infty$, this case is not covered in the algebra of limits.
   We can instead consider $a_n = \frac{n+1}{n} = 1 + \frac{1}{n}$ so that (a_n) is the sum of a constant sequence of 1s and a sequence made up of $\frac{1}{n}$. Since $n \to \infty$, Proposition 5.9.5 says $\frac{1}{n} \to 0$. Thus, both of these sequences converge to 1 and 0 respectively and hence we can use algebra of limits to conclude that $a_n = 1 + \frac{1}{n} \to 1 + 0 = 1$.
2. Consider the sequence (d_n) defined by $d_n = \frac{n^2+n+1}{n^2+2}$. Again, we cannot apply the algebra of limits straight away here since $\lim_{n\to\infty} n^2 + n + 1 = \infty$ and

$\lim_{n\to\infty} n^2 + 2 = \infty$. However, we can algebraically manipulate the terms slightly as thus:

$$d_n = \frac{n^2\left(1 + \frac{1}{n} + \frac{1}{n^2}\right)}{n^2\left(1 + \frac{2}{n^2}\right)} = \frac{1 + \frac{1}{n} + \frac{1}{n^2}}{1 + \frac{2}{n^2}}. \tag{5.8}$$

We now note that since $\frac{1}{n}$, $\frac{1}{n^2}$, and $\frac{2}{n^2}$ all converge to 0 using the fact that $n, n^2 \to \infty$ and Proposition 5.9.5, by the algebra of limits, both the numerator and the denominator in (5.8) converge to 1. Thus, applying the algebra of limits again, the sequence (d_n) converges with $d_n \to \frac{1}{1} = 1$.

As we noted in the examples above, we are not able to determine the limit of the quotient if both the numerator and denominator blow up to ∞. This is part of a wider scope of degenerate cases for algebra of limits. The cases where we have the limits of the form $\infty - \infty$, $0 \times \infty$, or $\frac{\pm\infty}{\pm\infty}$ are called indeterminate forms: we cannot determine the limits of the sums, products, or quotients straight away. So one has to be clever in manipulating the terms in order to apply the algebra of limits correctly as we have seen in Example 5.9.7.

Even so, in these indeterminate cases, the limits may or may not exist!

Example 5.9.8 Let us look at some examples:

1. Suppose that (a_n) and (b_n) are real sequences such that $a_n = \frac{n}{n^2}$ and $b_n = \frac{n^2}{n}$. Both of them are of the indeterminate form $\frac{\infty}{\infty}$. However, after cancellations, we can see that the sequence (a_n) is convergent to 0 whereas the sequence (b_n) diverges to ∞.
2. The indeterminate case $\infty - \infty$ could result in the limit $\pm\infty$. Consider the sequences (a_n) and (b_n) defined as $a_n = 2n$ and $b_n = n$. If we consider them separately, the sequences $(a_n - b_n)$ and $(b_n - a_n)$ both have the indeterminate form of $\infty - \infty$ since both $a_n, b_n \to \infty$. However, we note that $a_n - b_n = 2n - n = n$, so this sequence diverges to ∞. On the other hand, we have $b_n - a_n = -n \to -\infty$.
3. The indeterminate case $\infty - \infty$ may also result in a finite limit! We saw in Examples 5.7.4 and 5.7.6 the sequence (a_n) where $a_n = \sqrt{n^2 + 1} - n$. This sequence is also of the indeterminate form $\infty - \infty$, so algebra of limits is not applicable here. However, we have seen that this sequence is asymptotically equivalent to the sequence (b_n) where $b_n = \frac{1}{2n}$. So, by Proposition 5.7.5, the sequence (a_n) does converge to 0, which is finite!

This is why these forms are called indeterminate: at first glance, it may be impossible to determine whether the sequence converges or not and it could go either way. Therefore one needs to work extra hard and extra carefully to determine the conclusion. Let us look at another type of indeterminate form, which is ∞^0. This limit will come in handy later.

Example 5.9.9 We want to find the limit $\lim_{n\to\infty} n^{\frac{1}{n}}$. Since the base converges to ∞ and the exponent converges to 0, this is an indeterminate form ∞^0. However, we claim that this sequence converges.

Note that since $n \geq 1$, for each $n \in \mathbb{N}$ we must have $n^{\frac{1}{n}} \geq 1$ or otherwise $n^{\frac{1}{n}} < 1$ implies $n < 1^n = 1$, which is absurd! So we can write $n^{\frac{1}{n}} = 1 + a_n$ for some sequence of non-negative numbers (a_n). This implies $n = (1 + a_n)^n$ and by binomial expansion for $n \geq 2$, we get:

$$n = (1+a_n)^n = \sum_{k=0}^{n} \binom{n}{k} a_n^k \geq \binom{n}{2} a_n^2 = \frac{n(n-1)}{2} a_n^2 \quad \Rightarrow \quad 0 \leq a_n \leq \sqrt{\frac{2}{n-1}}.$$

Using Proposition 5.9.5 and sandwiching, by taking the limit as $n \to \infty$, we get $a_n \to 0$. Therefore, by taking the limit as $n \to \infty$ on both sides of $n^{\frac{1}{n}} = 1 + a_n$, we can use algebra of limits to conclude that $\lim_{n\to\infty} n^{\frac{1}{n}} = 1$.

Another way to show this result is to use the AM-GM inequality. For any $n \geq 3$, the AM-GM inequality from Exercise 3.27(b) with $n - 2$ copies of 1 and 2 copies of $\sqrt{n}$ implies:

$$\frac{(n-2) + 2\sqrt{n}}{n} \geq n^{\frac{1}{n}} \geq 1 \quad \Rightarrow \quad 1 - \frac{2}{n} + \frac{2}{\sqrt{n}} \geq n^{\frac{1}{n}} \geq 1.$$

By sandwiching, we then deduce $\lim_{n\to\infty} n^{\frac{1}{n}} = 1$.

5.10 Limit Superior and Limit Inferior

As we have seen earlier, not all sequences have limits. Sequences which do not have limits are called divergent sequences. Divergent sequences (a_n) may behave in various different ways.

1. If the sequence is unbounded, then the sequence may blow up to $\pm\infty$, for example when $a_n = n^2$ or $a_n = -n^2$. However, they may also not tend to either $\pm\infty$ as we can see with the sequence defined by $a_n = (-1)^n 2n$.
2. On the other hand, if the sequence is bounded, divergence means that the sequence may oscillate in a bounded region without tending to a specific point in $\mathbb{R}$. Sometimes it might be useful to know how much do they oscillate as they go to infinity.

Let us study the second case above. Suppose that we have a bounded real sequence (a_n). Since (a_n) is a bounded sequence, as a set in $\mathbb{R}$, its supremum exists.

Furthermore, this also means for any subsequence of (a_n), their supremum must also exist. Thus, we have no problem in defining a new real sequence (b_n) where:

$$b_n = \sup_{m \geq n} a_m = \sup\{a_m : m \geq n\}.$$

Clearly, for all $n \in \mathbb{N}$ we have $a_n \leq \sup_{m \geq n} a_m = b_n$. Since the sequence (a_n) is bounded from below, the sequence (b_n) is also bounded from below. Moreover, we note that the sequence (b_n) is decreasing. Indeed, for any $n \in \mathbb{N}$ we have the inclusion of sets $\{a_m : m \geq n+1\} \subseteq \{a_m : m \geq n\}$ which, by Proposition 4.1.10, implies the inequality $b_{n+1} = \sup\{a_m : m \geq n+1\} \leq \sup\{a_m : m \geq n\} = b_n$.

Thus by monotone sequence theorem, the sequence (b_n) converges and it converges to the infimum of the set $\{b_n\}_{n \in \mathbb{N}}$. In other words, $\lim_{n\to\infty} b_n = \lim_{n\to\infty}(\sup_{m \geq n} a_m)$ exists and is equal to $\inf_{n \in \mathbb{N}} b_n = \inf_{n \in \mathbb{N}}(\sup_{m \geq n} a_m)$. We can repeat the construction with the supremum replaced by infimum and obtain the following definitions:

Definition 5.10.1 (Limit Superior, Limit Inferior) Let (a_n) be a bounded real sequence. We define the limit superior of this sequence as:

$$\limsup_{n\to\infty} a_n = \lim_{n\to\infty}\left(\sup_{m \geq n} a_m\right) = \inf_{n \in \mathbb{N}}\left(\sup_{m \geq n} a_m\right).$$

Similarly, the limit inferior of this sequence is defined as:

$$\liminf_{n\to\infty} a_n = \lim_{n\to\infty}\left(\inf_{m \geq n} a_m\right) = \sup_{n \in \mathbb{N}}\left(\inf_{m \geq n} a_m\right).$$

Examples of the limit superior and limit inferior of a bounded sequence can be seen in Fig. 5.6. On the other hand, for unbounded sequences, we cannot guarantee that the limit superior and limit inferior exist from the construction above. Indeed, if a sequence (a_n) is not bounded above, then any of its tail is also not bounded

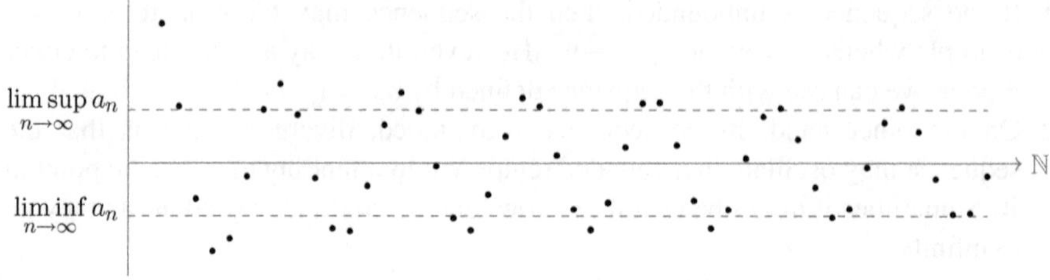

Fig. 5.6 Limit superior and limit inferior of a sequence (a_n) which is denoted by the black dots. As $n \to \infty$, the quantity $\sup_{m \geq n} a_m$ gets smaller and the quantity $\inf_{m \geq n} a_m$ gets larger. Eventually they converge to $\limsup_{n\to\infty} a_n$ and $\liminf_{n\to\infty} a_n$ respectively

above by Lemma 5.5.9. Thus, the quantities $b_n = \sup_{m\geq n} a_m$ do not exist for any $n \in \mathbb{N}$. However, for this case we can declare $\limsup_{n\to\infty} a_n = \infty$. Likewise, if the sequence is not bounded from below, we declare $\liminf_{n\to\infty} a_n = -\infty$.

On the other hand, for sequences that are not bounded above, its limit inferior may or may not exist and, similarly, the limit superior for sequences that are not bounded below may or may not exist. These situations have to be dealt with on a case-by-case basis from the first definition.

Example 5.10.2 Let us look at some examples:

1. Recall the real sequence (a_n) defined as $a_n = (-1)^n$. This sequence does not converge. But since it is bounded, its limit inferior and limit superior both exist. We find:
$$\limsup_{n\to\infty} a_n = \lim_{n\to\infty} \sup_{m\geq n}(-1)^m = \lim_{n\to\infty} 1 = 1.$$
Similarly, $\liminf_{n\to\infty} a_n = -1$.
2. Let (a_n) be the real sequence defined as $a_n = (-1)^n\left(1+\frac{1}{n}\right)$. A plot of the first few terms of this sequence can be seen in Fig. 5.7. This sequence does not converge since the subsequences (a_{2n}) and (a_{2n+1}) converge to 1 and -1 respectively. However this sequence is bounded since $|a_n| \leq 2$ for all $n \in \mathbb{N}$ and so its limit superior and limit inferior must exist. Let us compute these.
Notice that all the even-indexed terms in this sequence are positive and all the odd-indexed terms are negative. Hence, when we are looking for the supremum, we can ignore the odd-indexed terms as they are clearly smaller than any of the even-indexed terms. So:
$$\sup_{m\geq n} a_m = \sup_{\substack{m\geq n\\ m \text{ even}}} a_m = \sup_{\substack{m\geq n\\ m \text{ even}}} \left(1+\frac{1}{m}\right) = 1+\frac{1}{p(n)},$$
where $p : \mathbb{N} \to \mathbb{N}$ is the function such that $p(n)$ is the smallest even integer greater than or equal to n. As we have seen in Exercise 3.1, this function is $p(n) = 2\lceil\frac{n}{2}\rceil$ and it is increasing and unbounded. Hence, by Proposition 5.9.5, we deduce $\limsup_{n\to\infty} a_n = \lim_{n\to\infty}(\sup_{m\geq n} a_m) = \lim_{n\to\infty}\left(1+\frac{1}{p(n)}\right) = 1$. Using a similar argument, we can show that $\liminf_{n\to\infty} a_n = -1$.
3. Consider a real sequence (a_n) defined as $a_n = n^{(-1)^n}$. This sequence is not bounded from above. To see this, we note that the subsequence (a_{2n}) is given by $a_{2n} = 2n$ which blows up to infinity and hence, by Proposition 5.5.10, the whole sequence is not bounded from above. So we immediately get $\limsup_{n\to\infty} a_n = \infty$ by definition.
However, this sequence has a limit inferior. Notice that the even-indexed terms are all greater than 1 and the odd-indexed terms are all between 0 and 1, so we can ignore the even-indexed terms when we are looking for the infimum since all the odd-indexed terms are smaller than any of the even-indexed terms. Thus, for

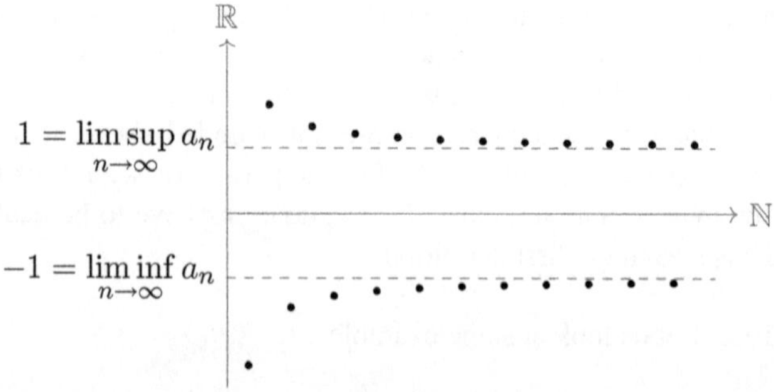

Fig. 5.7 The sequence (a_n) where $a_n = (-1)^n(1+\frac{1}{n})$

any $n \in \mathbb{N}$ we have:

$$\inf_{m\geq n} a_m = \inf_{\substack{m\geq n\\ m \text{ odd}}} a_m = \inf_{\substack{m\geq n\\ m \text{ odd}}} m^{(-1)^m} = \inf_{\substack{m\geq n\\ m \text{ odd}}} \frac{1}{m} = 0,$$

which implies $\liminf_{n\to\infty} a_n = 0$.

4. Now consider the sequence (a_n) defined as $a_n = n$. This sequence is not bounded from above, so immediately $\limsup_{n\to\infty} a_n = \infty$. What about the limit inferior? We can compute this manually using the definition $\liminf_{n\to\infty} a_n = \lim_{n\to\infty} \inf_{m\geq n} a_n = \lim_{n\to\infty} \inf_{m\geq n} n = \lim_{n\to\infty} n = \infty$. Thus $\limsup_{n\to\infty} a_n = \liminf_{n\to\infty} a_n = \infty$.

Limit superior and limit inferior are useful concepts to work with for any bounded sequence because these quantities always exist, unlike limits. So for cases when limits of a sequence do not exist, its limit superior and limit inferior can be used as good substitutes when we want to study the behaviour of a sequence towards infinity. Let us look at some of their properties.

Lemma 5.10.3 *Let (a_n) and (b_n) be bounded real sequences.*

1. $\limsup_{n\to\infty}(-a_n) = -\liminf_{n\to\infty} a_n$.
2. *Preservation of weak inequalities: If $a_n \leq b_n$ for all $n \in \mathbb{N}$, then $\limsup_{n\to\infty} a_n \leq \limsup_{n\to\infty} b_n$ and $\liminf_{n\to\infty} a_n \leq \liminf_{n\to\infty} b_n$.*
3. *For the sequence (a_n), we have $\inf_{n\in\mathbb{N}} a_n \leq \liminf_{n\to\infty} a_n \leq \limsup_{n\to\infty} a_n \leq \sup_{n\in\mathbb{N}} a_n$,*
4. *The limit superior is subadditive, namely:*

$$\limsup_{n\to\infty}(a_n+b_n) \leq \limsup_{n\to\infty} a_n + \limsup_{n\to\infty} b_n$$

5. *The limit inferior is superadditive, namely:*

$$\liminf_{n\to\infty}(a_n + b_n) \geq \liminf_{n\to\infty} a_n + \liminf_{n\to\infty} b_n$$

Proof The first assertion is an easy proof. We prove the others here.

2. For any fixed $n \in \mathbb{N}$, for any $j \geq n$ we have $a_j \leq b_j \leq \sup_{m\geq n} b_m$. Therefore, $\sup_{m\geq n} b_m$ is an upper bound for the set $\{a_j : j \geq n\}$ and thus $\sup_{j\geq n} a_j \leq \sup_{m\geq n} b_m$. Taking the limit as $n \to \infty$ and using Proposition 5.6.1 yields the result. The inequality for limit inferior is similarly deduced.
3. For the middle inequality, by definition, we have the ordering $\inf_{m\geq n} a_m \leq \sup_{m\geq n} a_m$ for all $n \in \mathbb{N}$. Taking the limit as $n \to \infty$ on both sides and using the preservation of weak inequalities in Proposition 5.6.1, we obtain the inequality $\liminf_{n\to\infty} a_n \leq \limsup_{n\to\infty} a_n$.
   For the first inequality, by Proposition 4.1.10(2), the infimum of a set is always smaller than the infimum of any of its subset. Thus, for any $n \in \mathbb{N}$ we have $\inf_{j\in\mathbb{N}} a_j \leq \inf_{m\geq n} a_m$. Taking the limit as $n \to \infty$, by preservation of weak inequality, we get $\inf_{j\in\mathbb{N}} a_j \leq \lim_{n\to\infty} \left(\inf_{m\geq n} a_m\right) = \liminf_{n\to\infty} a_n$. The final inequality is done in a similar way.
4. We first note that the sequence $(a_n + b_n)$ is also bounded, so its limit superior exists. For any fixed $n \in \mathbb{N}$, for all $j \geq n$ we have $a_j + b_j \leq \sup_{m\geq n} a_m + \sup_{m\geq n} b_m$. Taking the supremum over $j \geq n$, we have $\sup_{j\geq n}(a_j + b_j) \leq \sup_{m\geq n} a_m + \sup_{m\geq n} b_m$. Hence, taking the limit as $n \to \infty$ on both sides of the inequality and applying the algebra of limits, we obtain the desired inequality.

The proof for the final assertion is similar to the fourth assertion. □

For non-negative sequences, the limit superior and limit inferior satisfy the following:

Lemma 5.10.4 *Let (a_n) and (b_n) be non-negative bounded real sequences. Then:*

1. $\limsup_{n\to\infty}(a_n b_n) \leq (\limsup_{n\to\infty} a_n)(\limsup_{n\to\infty} b_n)$.
2. $\liminf_{n\to\infty}(a_n b_n) \geq (\liminf_{n\to\infty} a_n)(\liminf_{n\to\infty} b_n)$.

Proof We shall prove the first assertion only. The second assertion can be proven in a similar manner.

1. Since the sequences (a_n) and (b_n) are both bounded, it is easy to show that the sequence $(a_n b_n)$ is also bounded and so its limit superior exists. Note that since the sequences are non-negative, the quantities $\sup_{m\geq n} a_m$ and $\sup_{m\geq n} b_m$ are also non-negative. Hence, for any $j \geq n$ we have $a_j b_j \leq (\sup_{m\geq n} a_m)(\sup_{m\geq n} b_m)$. Taking the supremum over $j \geq n$, we get

$\sup_{j\geq n}(a_j b_j) \leq (\sup_{m\geq n} a_m)(\sup_{m\geq n} b_m)$. Finally, taking the limit as $n \to \infty$ on both sides of the inequality and applying the algebra of limits (which can be done because all the limits exist), we arrive at the conclusion. □

Another interpretation of the limit superior and limit inferior is the following:

Proposition 5.10.5 *Let (a_n) be a real bounded sequence. Then:*

$$\limsup_{n\to\infty} a_n = \sup\{r \in \mathbb{R} : a_n > r \text{ for infinitely many } n\},$$

$$\liminf_{n\to\infty} a_n = \inf\{r \in \mathbb{R} : a_n < r \text{ for infinitely many } n\}.$$

Proof We prove the first equality only. Denote the set $A = \{r \in \mathbb{R} : a_n > r \text{ for infinitely many } n\} \subseteq \mathbb{R}$.

1. First, we show that $\limsup_{n\to\infty} a_n \geq \sup(A)$. Pick any $x \in A$. By definition of the set A, there are infinitely many indices $n \in \mathbb{N}$ for which $a_n > x$. This means for any fixed $n \in \mathbb{N}$, there is an index $j \geq n$ such that $a_j > x$. Hence, we must have $\sup_{m\geq n} a_m > x$. Taking the limit as $n \to \infty$, we have $\limsup_{n\to\infty} a_n \geq x$. Since $x \in A$ is arbitrary, this means the set A is bounded from above by $\limsup_{n\to\infty} a_n$ and so we have the inequality $\limsup_{n\to\infty} a_n \geq \sup(A)$.
2. To show the opposite inequality, we show $\limsup_{p\to\infty} a_p - \varepsilon \in A$ for all $\varepsilon > 0$. Fix $\varepsilon > 0$. To fulfil the conditions on the set A, we need to find an infinite number of terms a_n that is greater than the quantity $\limsup_{p\to\infty} a_p - \varepsilon$.
   Note that for $\sup_{m\geq n} a_m$ with $n = 1$, by characterisation of supremum, we can find an index $k_1 \geq n = 1$ such that $\sup_{m\geq 1} a_m - \varepsilon < a_{k_1} \leq \sup_{m\geq 1} a_m$. This also implies:

   $$\limsup_{p\to\infty} a_p - \varepsilon = \inf_{p\in\mathbb{N}} \sup_{m\geq p} a_m - \varepsilon \leq \sup_{m\geq 1} a_m - \varepsilon < a_{k_1}.$$

   We have found our first term in the sequence that is greater than $\limsup_{p\to\infty} a_p - \varepsilon$. For the second term, for $\sup_{m\geq k_1+1} a_m$, by characterisation of supremum, we can find an index $k_2 \geq k_1+1$ such that $\sup_{m\geq k_1+1} a_m - \varepsilon < a_{k_2} \leq \sup_{m\geq k_1+1} a_m$. This then implies:

   $$\limsup_{p\to\infty} a_p - \varepsilon = \inf_{p\in\mathbb{N}} \sup_{m\geq p} a_m - \varepsilon \leq \sup_{m\geq k_1+1} a_m - \varepsilon < a_{k_2}.$$

   Inductively, for $j \in \mathbb{N}$ by using $n = k_j + 1$ in the above argument, we can find another index $k_{j+1} \geq n = k_j + 1 > k_j$ such that $\limsup_{p\to\infty} a_p - \varepsilon < a_{k_{j+1}}$. Hence, there are infinitely many terms of (a_n) that are greater than $\limsup_{p\to\infty} a_p - \varepsilon$.

Thus, by definition, $\limsup_{p\to\infty} a_p - \varepsilon \in A$. This means $\limsup_{p\to\infty} a_p - \varepsilon \leq \sup(A)$ for all $\varepsilon > 0$. Using Exercise 4.8, we conclude that $\limsup_{p\to\infty} a_p \leq \sup(A)$.

Putting the two inequalities together yields the desired equality. □

A similar result to Proposition 5.10.5, which the readers shall prove in Exercise 5.33, is:

Proposition 5.10.6 *Let (a_n) be a real bounded sequence. Then:*

$$\limsup_{n\to\infty} a_n = \inf\{r \in \mathbb{R} : a_n > r \text{ for only finitely many } n\},$$

$$\liminf_{n\to\infty} a_n = \sup\{r \in \mathbb{R} : a_n < r \text{ for only finitely many } n\}.$$

Via Propositions 5.10.5 and 5.10.6, an intuition behind these quantities is that they capture the bounding behaviour of the sequence at infinity: limit superior and limit inferior provide an upper bound and a lower bound of the sequence at infinity respectively.

More specifically, from the interpretations in Proposition 5.10.5, we can find subsequences of (a_n) that converge to the numbers $\liminf_{n\to\infty} a_n$ and $\limsup_{n\to\infty} a_n$. This is intuitively true since in Proposition 5.10.5, we have seen that there are infinitely many terms in the sequence (a_n) which are arbitrarily close to each of $\limsup_{n\to\infty} a_n$ and $\liminf_{n\to\infty} a_n$. From these terms, we can build subsequences converging to the limit superior and limit inferior.

To construct these sequences rigorously, we state a short lemma first. The readers were invited to prove this lemma in Exercise 2.33.

Lemma 5.10.7 *Let $X \subseteq \mathbb{N}$ be an infinite subset of the natural numbers and $Y = \{x_1, \ldots, x_n\} \subseteq \mathbb{N}$ be a finite subset of $\mathbb{N}$. Then, the set $X \setminus Y$ is non-empty and there exists an $x \in X \setminus Y$ such that $x > x_j$ for all $j = 1, 2, \ldots, n$.*

Proposition 5.10.8 *Let (a_n) be a real bounded sequence. Then, there exist subsequences (a_{k_n}) and (a_{j_n}) such that:*

$$\lim_{n\to\infty} a_{k_n} = \limsup_{n\to\infty} a_n \quad \textit{and} \quad \lim_{n\to\infty} a_{j_n} = \liminf_{n\to\infty} a_n.$$

Proof We construct a subsequence (a_{k_n}) that tends to $\limsup_{n\to\infty} a_n$. Denote $L = \limsup_{n\to\infty} a_n$ and:

$$A = \{r \in \mathbb{R} : a_n > r \text{ for infinitely many } n\}.$$

By Proposition 5.10.5, since $L = \sup(A)$, by characterisation of supremum, for each $j \in \mathbb{N}$ there exists $x_j \in A$ such that $L - \frac{1}{j} < x_j \leq L$. Since $x_j \in A$, by definition of the set A, there exists infinitely many indices $n \in \mathbb{N}$ such that $a_n > x_j$. We denote the set of all such indices as $N_j = \{n \in \mathbb{N} : L - \frac{1}{j} < x_j < a_n\}$ and all of these sets are infinite.

Now we construct the desired subsequence (a_{k_n}) term-by-term inductively. For the first term, pick any $k_1 \in N_1$. By this choice of index, we have $L - 1 < x_1 < a_{k_1} \leq \sup_{m \geq k_1} a_m \leq \sup_{m \geq 1} a_m$. Note that the final inequality is true since $k_1 \geq 1$.

Inductively, for $n \geq 2$, by virtue of Lemma 5.10.7, we can always find an index $k_n \in N_n \setminus \{k_1, k_2, \ldots, k_{n-1}\}$ for which $k_{n-1} < k_n$. The term with this index satisfies the inequality:

$$L - \frac{1}{n} < x_n < a_{k_n} \leq \sup_{m \geq k_n} a_m \leq \sup_{m \geq n} a_m. \tag{5.9}$$

We claim that the resulting subsequence (a_{k_n}) is the sequence that we are looking for. Indeed, by taking the limit as $n \to \infty$ in the first and final terms in inequality (5.9), we obtain:

$$\lim_{n \to \infty} \left(L - \frac{1}{n} \right) = L \quad \text{and} \quad \lim_{n \to \infty} \sup_{m \geq n} a_m = \limsup_{n \to \infty} a_n = L.$$

Thus, by sandwiching, we must also have $a_{k_n} \to L$, giving us the desired subsequence.

Similar construction can be done to find a subsequence that converges to the value $\liminf_{n \to \infty} a_n$ □

Furthermore, from Propositions 5.10.5 and 5.10.6, $\liminf_{n \to \infty} a_n$ and $\limsup_{n \to \infty} a_n$ are the smallest and largest possible limits for any convergent subsequence of (a_n). In other words:

Proposition 5.10.9 *Let (a_n) be a bounded sequence. If (a_{k_n}) is a convergent subsequence of (a_n), then:*

$$\liminf_{n \to \infty} a_n \leq \lim_{n \to \infty} a_{k_n} \leq \limsup_{n \to \infty} a_n.$$

We leave the proof of this to the readers as Exercise 5.33.

Remark 5.10.10 Note that if we have a bounded sequence (a_n) such that $\liminf_{n \to \infty} a_n < \limsup_{n \to \infty} a_n$, we can find subsequences of (a_n) converging to either of these values. However, it is not necessarily true that there is a subsequence that converge to any number L strictly in between them, namely $L \in (\limsup_{n \to \infty} a_n, \liminf_{n \to \infty} a_n)$.

An example of this is the sequence (a_n) with $a_n = (-1)^n$. We have $\liminf_{n \to \infty} a_n = -1$ and $\limsup_{n \to \infty} a_n = 1$. Clearly the subsequences (a_{2n+1})

and (a_{2n}) converge to the limit inferior and superior -1 and 1 respectively. However, there are no subsequences of (a_n) that converge to any number $L \in (-1, 1)$.

Recall from Proposition 5.5.4 that for any convergent sequences, all of its subsequences converge to the same limit. Therefore, based on Proposition 5.10.8, if (a_n) is a convergent sequence, we expect that the limit superior and limit inferior would coincide. In fact, the converse is also true.

Proposition 5.10.11 *Let (a_n) be a real sequence. The sequence (a_n) converges if and only if we have the equality* $\limsup_{n\to\infty} a_n = \liminf_{n\to\infty} a_n$.

Proof We prove the implications one by one.

($\Rightarrow$): Suppose that $a_n \to L \in \mathbb{R}$. By Proposition 5.10.8, there are subsequences of (a_n) that converge to each of $\limsup_{n\to\infty} a_n$ and $\liminf_{n\to\infty} a_n$ respectively. By Proposition 5.5.4, we know that any subsequence of a convergent sequence converges to the same limit as the original sequence. Thus, the limits of the two subsequences, namely $\limsup_{n\to\infty} a_n$ and $\liminf_{n\to\infty} a_n$, must also both be L.

($\Leftarrow$): Suppose that $\limsup_{n\to\infty} a_n = \liminf_{n\to\infty} a_n = L$. Fix $\varepsilon > 0$. Since $L = \liminf_{n\to\infty} a_n = \sup_{n\in\mathbb{N}} \inf_{m\geq n} a_n$, using the characterisation of supremum, there exists an $N_1 \in \mathbb{N}$ such that $L - \varepsilon < \inf_{m\geq N_1} a_m \leq L$. This says for all $n \geq N_1$, we have $L - \varepsilon < a_n$ which implies $L - a_n < \varepsilon$ for all $n \geq N_1$.
On the other hand, by a similar argument using the characterisation of infimum in the definition of limit superior, we can find an $N_2 \in \mathbb{N}$ such that $-\varepsilon < L - a_n$ for all $n \geq N_2$. Setting $N = \max\{N_1, N_1\}$, we have found an $N \in \mathbb{N}$ such that $|a_n - L| < \varepsilon$ for all $n \geq N$. Thus, the sequence (a_n) converges to L. □

Via Proposition 5.10.11, we have a direct corollary:

Corollary 5.10.12 *If (a_n) is a convergent real sequence, then* $\lim_{n\to\infty} a_n = \limsup_{n\to\infty} a_n = \liminf_{n\to\infty} a_n$.

Using Proposition 5.10.11 and Lemmas 5.10.3 and 5.10.4, we have the following algebra of limits results. The proof of the following result is left as Exercise 5.34 for the readers.

Proposition 5.10.13 *Suppose that (a_n) and (b_n) are real sequences such that (a_n) is a bounded sequence and (b_n) converges to some $b \in \mathbb{R}$. Then:*

1. $\limsup_{n\to\infty}(a_n + b_n) = \limsup_{n\to\infty} a_n + b$.
2. $\liminf_{n\to\infty}(a_n + b_n) = \liminf_{n\to\infty} a_n + b$.

Furthermore, if $b \geq 0$, then:

3. $\limsup_{n\to\infty}(a_n b_n) = b \limsup_{n\to\infty} a_n$.
4. $\liminf_{n\to\infty}(a_n b_n) = b \liminf_{n\to\infty} a_n$.

Exercises

5.1 (∗) For each of the real sequence (a_n) where a_n are defined as follows, show that $a_n \to 0$. In other words, for a fixed $\varepsilon > 0$, find an $N(\varepsilon) \in \mathbb{N}$ such that $|a_n| < \varepsilon$ for all $n \geq N(\varepsilon)$:
(a) $a_n = \frac{1}{n^2+3}$.
(b) $a_n = \frac{1}{n-\frac{5}{2}}$.
(c) $a_n = \frac{1}{n(n-\frac{7}{2})}$.
(d) $a_n = \frac{1}{\sqrt{5n-1}}$.
(e) $a_n = \frac{\sin(n)}{n}$.
(f) $a_n = \begin{cases} \frac{1}{2^n} & \text{if } n \text{ is prime or } 1, \\ \frac{1}{3^n} & \text{if } n \text{ is not prime.} \end{cases}$
(g) $a_n = \sqrt{n+1} - \sqrt{n}$.
(h) $a_n = n - \sqrt{n^2 + \sqrt{n}}$.

5.2 (∗) Let (a_n) be a real sequence defined by $a_n = (-1)^n + \frac{(-1)^{n+1}}{n}$ for $n \geq 1$.
(a) Show that this sequence is bounded.
(b) Consider the subsequence (a_{2n}). Show that this subsequence converges to 1 using the ε-N definition.
(c) Consider the subsequence (a_{2n+1}). Using the algebra of limits, find the limit of this subsequence.
(d) Explain why the sequence (a_n) does not converge.

5.3 Let (a_n) and (b_n) be real sequences such that $a_n \to 0$ and (b_n) is a bounded sequence. Show that the sequence $(a_n b_n)$ converges to 0.

5.4 (∗) For each of the following, give an example of a real sequence (a_n) with $a_n \to 0$ and another real sequence (b_n) such that:
(a) (b_n) is unbounded but $a_n b_n \to 0$.
(b) $a_n b_n \to \infty$.
(c) $(a_n b_n)$ converges to a nonzero limit.
(d) $(a_n b_n)$ is bounded but does not converge.

5.5 Suppose that (a_n) is a real sequence such that $a_n \geq 0$ for all $n \in \mathbb{N}$. If the sequence (a_n) converges to some real number L, prove using the ε-N definition that $L \geq 0$.

5.6 In this question, we are going to prove a result due to Harlan brothers [32]. Recall the Pascal triangle in Exercise 3.14.

(a) Prove that for $n \in \mathbb{N}_0$, the product p_n of the entries in row n of the Pascal triangle, is given by:

$$p_n = \frac{(n!)^{n+1}}{\prod_{j=0}^{n}(j!)^2}.$$

(b) Hence, prove that $\lim_{n\to\infty} \frac{p_{n+1}p_{n-1}}{p_n^2} = e$.

5.7 (∗) Let (a_n) be a real sequence.

(a) Suppose that the even-indexed subsequence (a_{2n}) and the odd-indexed subsequence (a_{2n+1}) both converge to the same limit. Prove that the whole sequence (a_n) also converges to the same limit.

(b) More generally, assume that the subsequences (a_{3n}), (a_{3n+1}), and (a_{3n+2}) all converge to the same limit. Show that the whole sequence (a_n) also converges to the same limit.

5.8 (a) Suppose that the sequences (a_n) and (b_n) both converge to the same limit. Using the ε-N definition, show that the sequence (c_n) defined as $c_n = |a_n - b_n|$ converges to 0.

(b) Find two real sequences (a_n) and (b_n) such that the sequence (c_n) defined as $c_n = |a_n - b_n|$ converges to 0 but neither (a_n) nor (b_n) converges.

5.9 (∗) Determine whether the following sequences (a_n), where a_n are defined as follows, converge. If they do, state their limits. You may use the algebra of limits, sandwiching lemma, or the comparison theorems.

(a) $a_n = \frac{n^2+n+1}{n+1}$.

(b) $a_n = \frac{n^2}{n!}$.

(c) $a_n = \frac{n^{\frac{3}{4}}}{\sqrt{5n-1}}$.

(d) $a_n = \frac{1}{n^2+1} + \frac{1}{n^2+2} + \ldots + \frac{1}{n^2+n} = \sum_{j=1}^{n} \frac{1}{n^2+j}$.

5.10 (∗) Let (a_n) be a real sequence defined as $a_n = \frac{1}{n+1} + \frac{1}{n+2} + \ldots + \frac{1}{n+n} = \sum_{j=1}^{n} \frac{1}{n+j}$. Show that this sequence converges and find positive finite lower and upper bounds for this sequence.

5.11 (a) Let $p, q \in \mathbb{R}$. Prove that the maximum $\max\{p, q\}$ and minimum $\min\{p, q\}$ can be written as $\max\{p, q\} = \frac{1}{2}(p + q + |p - q|)$ and $\min\{p, q\} = \frac{1}{2}(p + q - |p - q|)$ respectively.

(b) Suppose that (a_n) and (b_n) are two real sequences such that $a_n \to L$ and $b_n \to K$. Using part (a), show that the sequences (c_n) and (d_n) defined as $c_n = \max\{a_n, b_n\}$ and $d_n = \min\{a_n, b_n\}$ converge to $\max\{L, K\}$ and $\min\{L, K\}$ respectively.

5.12 (∗) Let (a_n) and (b_n) be two divergent real sequences. Prove that:

(a) If $a_n, b_n \to \infty$, then $a_n + b_n, a_nb_n \to \infty$.

(b) If $a_n, b_n \to -\infty$, then $a_n + b_n \to -\infty$ and $a_nb_n \to \infty$.

(c) If $a_n \to -\infty$ and $b_n \to \infty$, then $a_nb_n \to -\infty$.

5.13 Let (a_n) and (b_n) be real sequences such that $b_n \to \pm\infty$. Prove that if the sequence (a_nb_n) converges to a real number, then $a_n \to 0$.

5.14 (∗) Let (a_n) and (b_n) be two real sequences such that $0 \leq b_1 < a_1$. The terms in the sequences are defined recursively for all $n \geq 1$ as:

$$a_{n+1} = \frac{a_n + b_n}{2} \quad \text{and} \quad b_{n+1} = \sqrt{a_n b_n}.$$

(a) Show that $0 \leq b_n < a_n$ for all $n \in \mathbb{N}$.
(b) Prove that (a_n) is a decreasing sequence whereas (b_n) is an increasing sequence.
(c) Deduce that these sequences converge.
Moreover, show that their limits are the same.

5.15 (∗) We want to find the limit $\lim_{n\to\infty} r^n$ where $r \in \mathbb{R}$ is a constant. Clearly when $r = 0$ or $r = 1$ the limit is 0 or 1 respectively. For the other values of r, we split this into several cases.
(a) Let $r \in (0, 1)$. Prove that $\lim_{n\to\infty} r^n = 0$.
(b) Deduce that if $r \in (-1, 0)$, then $\lim_{n\to\infty} r^n = 0$.
(c) Show that if $r > 1$, then $\lim_{n\to\infty} r^n = \infty$.
(d) If $r \leq -1$, is it true that $\lim_{n\to\infty} r^n = -\infty$?

5.16 (a) Let (a_n) be a real sequence such that $a_n \neq 0$ for any $n \in \mathbb{N}$. Suppose that we have $\lim_{n\to\infty} \frac{a_{n+1}}{a_n} = r$ where $|r| < 1$. Show that $\lim_{n\to\infty} a_n = 0$.
For each of the real sequence (a_n) defined as follows, show that $\lim_{n\to\infty} a_n = 0$.
(b) $a_n = \frac{n^k}{r^n}$ for some $r > 1$ and $k > 0$
(c) $a_n = \frac{r^n}{n!}$ for some $r \in \mathbb{R}$
(d) $a_n = nr^n$ for some $|r| < 1$.

5.17 (∗) In this question, we want to prove that for any $r > 0$, we have $\lim_{n\to\infty} r^{\frac{1}{n}} = 1$.
(a) First, for any constant $r \geq 1$, prove that $r^{\frac{1}{n}} \to 1$ as $n \to \infty$.
(b) Hence, prove that for any constant $r \in (0, 1)$ we have $r^{\frac{1}{n}} \to 1$ as well.
(c) Now let (a_n) be a sequence that converges to a positive constant. Prove that the sequence $(a_n^{\frac{1}{n}})$ also converges and determine its limit.
(d) Suppose that $r_1, r_2, \ldots r_k \in \mathbb{R}$ are k positive constants. Let (a_n) be the sequence $a_n = (r_1^n + r_2^n + \ldots + r_k^n)^{\frac{1}{n}}$. Find the limit of this sequence.

5.18 Recall Example 5.9.9 in which we proved that $\lim_{n\to\infty} n^{\frac{1}{n}} = 1$.
(a) Using this fact and Exercise 5.17, find $\lim_{n\to\infty}(k + n)^{\frac{1}{n}}$ for any $k \in \mathbb{R}$.
(b) Find the limit $\lim_{n\to\infty} \sqrt[n]{2^n - n}$.

5.19 (∗) We now wish to generalise the results from Exercise 5.17 by considering the limit $\lim_{n\to\infty} r^{a_n}$ where $r > 0$ and (a_n) is some convergent real sequence.
(a) Let $r \geq 1$. Prove that $|r^x - 1| \leq r^{|x|} - 1$ for all $x \in \mathbb{R}$.
(b) Consider a real sequence (a_n) that converges to $a \in \mathbb{R}$. Using part (a), prove that $\lim_{n\to\infty} r^{a_n} = r^a$ for any fixed $r \geq 1$.
(c) Extend the result in part (b) to any $0 < r < 1$.

5.20 (∗) Recall that the sequence (a_n) defined as $a_n = \left(1 + \frac{1}{n}\right)^n$ in Example 5.4.4. We have shown that this sequence is increasing and converges to some real number e where $2 < e < 3$.

(a) Let (b_n) be a real sequence defined as $b_n = \left(1 + \frac{1}{n}\right)^{n+1}$. Prove that this sequence is decreasing and converges.
(b) Using the algebra of limits, show that $\lim_{n\to\infty} b_n = e$ as well.
(c) Show that for all $n \in \mathbb{N}$ we have $\ln(a_n) \leq 1 \leq \ln(b_n)$.
(d) Hence, deduce that for all $n \in \mathbb{N}$ we have:

$$\frac{1}{n+1} \leq \ln\left(1 + \frac{1}{n}\right) \leq \frac{1}{n}.$$

Using the sequence (a_n), prove the limits of the following similar-looking sequences.

(e) $\lim_{n\to\infty} \left(1 + \frac{1}{n^2}\right)^n = 1$,
(f) $\lim_{n\to\infty} \left(1 + \frac{1}{\sqrt{n}}\right)^n = \infty$.

5.21 (∗) Prove Proposition 5.5.6, namely:
Let (a_n) be a monotone sequence. If there exists a convergent subsequence (a_{k_n}), prove that the sequence (a_n) is also convergent.

5.22 (∗) Let (a_n) be a real sequence defined as $a_n = \frac{\ln(n)}{n}$.

(a) Show that this sequence is bounded below.
(b) Show that there exists some $N \in \mathbb{N}$ such that (a_n) is decreasing for all $n \geq N$.

Hence, by monotone sequence theorem, the sequence (a_n) converges. Now we would like to find its limit.

(c) Using induction, show that for all $n \geq 4$ we have $n^2 \leq 2^n$.
(d) Deduce that $n^2 < e^n$ for all $n \geq 4$.
(e) Using part (d), show that $\frac{\ln(n^2)}{n^2} \to 0$.
(f) Hence conclude that $a_n = \frac{\ln(n)}{n} \to 0$.

5.23 (∗) Prove Lemma 5.7.3, namely:
Let $\sim$ be the relation on set of real sequences where $(a_n) \sim (b_n)$ iff $\lim_{n\to\infty} \frac{a_n}{b_n} = 1$. Show that this relation is an equivalence relation.

5.24 Let (a_n), (b_n), and (c_n) be three non-zero real sequences. Suppose that $a_n \leq b_n \leq c_n$ for all $n \in \mathbb{N}$. Prove that if $(a_n) \sim (c_n)$, then $(a_n) \sim (b_n)$.

5.25 (∗) Let (a_n) and (b_n) be non-zero real sequences as follows. Show that the sequences are asymptotically equivalent and hence determine their convergence/divergence properties.

(a) $a_n = \frac{2^n+1}{3^n+1}$ and $b_n = \frac{2^n}{3^n}$.
(b) $a_n = n^2 + 2^{-n}$ and $b_n = \sin(n) + n^2$.
(c) $a_n = \sqrt{n+1} - \sqrt{n}$ and $b_n = \frac{1}{n}$.
(d) $a_n = \frac{1}{\sqrt{n}} - \frac{1}{\sqrt{n+1}}$ and $b_n = \frac{2}{n\sqrt{n}}$.

5.26 Let (a_n) and (b_n) be two real sequences such that $b_n > 0$ for all $n \in \mathbb{N}$. Prove that:
(a) $a_n \in O(b_n)$ if and only if $\limsup_{n\to\infty} \frac{|a_n|}{b_n} < \infty$.
(b) $a_n \in o(b_n)$ if and only if $\lim_{n\to\infty} \frac{|a_n|}{b_n} = 0$.

5.27 Let (a_n) be a real sequence. Write $a_n \in O(b_n)$ for some dominant term b_n in the expression for a_n.
(a) $a_n = n^3 + 1000n$.
(b) $a_n = \frac{10}{n^2} + \frac{5}{n} + \frac{1}{\sqrt{n}}$.
(c) $a_n = n^5 + 5^n$.
(d) $a_n = n^2 \ln(n) + n(\ln(n))^2$.

5.28 ($\diamond$) Let (a_n) be a real sequence such that $|a_n - a_{n-1}| < r|a_{n-1} - a_{n-2}|$ for all $n \geq N$ for some integer $N \geq 3$ and positive real number $0 < r < 1$.
(a) Show that for $n > m \geq 3$ we have $|a_n - a_m| \leq \frac{|a_2-a_1|}{1-r} r^{m-1}$.
(b) Prove that the sequence (a_n) is Cauchy and hence convergent.
(c) Give an example for which $r = 1$ and the sequence (a_n) diverges.

5.29 ($\diamond$) Consider the real sequence (a_n) defined as $a_n = \cos(n)$. This sequence is clearly bounded. Prove that this sequence diverges.

5.30 ($*$) Compute the limit superior and limit inferior of the following sequences (a_n) where:
(a) $a_n = 8 + (-1)^n \frac{n}{n+8}$.
(b) $a_n = \sin(\frac{n180^\circ}{4})$.
(c) $a_n = (1 + (-1)^n + \frac{1}{2^n})^{\frac{1}{n}}$.
(d) $a_n = n^{\sin(\frac{n180^\circ}{2})}$.
(e) $a_n = \begin{cases} \frac{n+1}{n} & \text{if } n \text{ is odd,} \\ 0 & \text{if } n \text{ is even.} \end{cases}$
(f) $a_1 = 0$, and $a_{2n} = \frac{a_{2n-1}}{2}$, $a_{2n+1} = \frac{1}{2} + a_{2n}$ for $n \in \mathbb{N}$.

5.31 ($*$) Let (a_n) be a positive real sequence. Prove the following:
(a) If $\limsup_{n\to\infty} \frac{a_{n+1}}{a_n} = L < 1$, then $\lim_{n\to\infty} a_n = 0$.
(b) If $\liminf_{n\to\infty} \frac{a_{n+1}}{a_n} = L > 1$, then $\lim_{n\to\infty} a_n = \infty$.

5.32 ($\diamond$) Construct two bounded real sequences (a_n) and (b_n) for which $a_n \geq b_n$ for all $n \in \mathbb{N}$ but $\liminf_{n\to\infty} a_n \leq \limsup_{n\to\infty} b_n$.

5.33 (a) Prove Propostion 5.10.6, namely:
Let (a_n) be a real bounded sequence. Show that:

$$\limsup_{n\to\infty} a_n = \inf\{r \in \mathbb{R} : a_n > r \text{ for only finitely many } n\},$$

$$\liminf_{n\to\infty} a_n = \sup\{r \in \mathbb{R} : a_n < r \text{ for only finitely many } n\}.$$

(b) Hence, prove Proposition 5.10.9, namely:
Let (a_n) be a bounded sequence. If (a_{k_n}) is a convergent subsequence of (a_n), then:

$$\liminf_{n\to\infty} a_n \leq \lim_{n\to\infty} a_{k_n} \leq \limsup_{n\to\infty} a_n.$$

5.34 ($*$) Prove Proposition 5.10.13, namely:
Suppose that (a_n) and (b_n) are real sequences such that (a_n) is a bounded sequence and (b_n) converges to some $b \in \mathbb{R}$. Prove:
(a) $\limsup_{n\to\infty}(a_n + b_n) = \limsup_{n\to\infty} a_n + b$.
(b) $\liminf_{n\to\infty}(a_n + b_n) = \liminf_{n\to\infty} a_n + b$.
Now suppose that (a_n) is a positive sequence. If $b > 0$, prove:
(c) $\limsup_{n\to\infty}(a_n b_n) = b \limsup_{n\to\infty} a_n$.
(d) $\liminf_{n\to\infty}(a_n b_n) = b \liminf_{n\to\infty} a_n$.
In the final two cases, what happens if $b < 0$?

5.35 ($\diamond$) Let (a_n) be the sequence of sums of squares, namely $a_n = 1^2 + 2^2 + \ldots + n^2 = \sum_{j=1}^n j^2$.
(a) Show that $a_n \in O(n^3)$.
(b) From part (a), we can see that the sequence (a_n) cannot grow faster than a cubic. Let us guess that for any $n \in \mathbb{N}$, we can write $a_n = an^3 + bn^2 + cn + d$ for some constants, $a, b, c, d \in \mathbb{R}$. Determine values of these constants.
(c) Verify the formula obtained in part (b) for all $n \in \mathbb{N}$ via induction.
(d) Hence, deduce that $a_n \sim \frac{n^3}{3}$.
The formula for sums of squares, cubes, and powers of four were computed by Abu Ali al-Hasan ibn al-Hasan ibn al-Haytham (c. 965–1040). The formula for the sum of cubes and sum of fourth powers will be derived by the readers in Exercise 7.29. These formula form a basis for al-Haytham's integration of the functions $f(x) = x^n$ for $x \in \mathbb{R}$ where $n = 1, 2, 3, 4$, which shall be discussed in Chap. 15.

Some Applications of Real Sequences 6

Arithmetic is based on counting, the epitome of a discrete process. The facts of arithmetic can be clearly understood as outcomes of certain counting processes, and one does not expect them to have any meaning beyond this. Geometry, on the other hand, involves continuous rather than discrete objects, such as lines, curves, and surfaces. Continuous objects cannot be built from simple elements by discrete processes, and one expects to see geometric facts rather than arrive at them through calculation.

— John Stillwell, mathematician

Now that we have seen real sequences and saw some of their properties, let us look at where they come in handy. We have seen that they allow us to express any irrational number as a limit of a sequence of rational numbers. Using this idea, we shall see how the circumference of a circle was calculated using sequences. This is a very important milestone in the history of mathematics as it leads to the discovery of an important mathematical constant, π.

Next, we shall return to the study of topology on the real line by defining limit points for a subset $X \subseteq \mathbb{R}$. Limit points arise as points in the superspace $\mathbb{R}$ that can be approached (in the limiting sense) by a sequence of points contained solely within X. This definition will be instrumental when we study limits of functions in Chap. 9.

Finally, we will extend the definition of convergent real sequences to the field of complex numbers, the real n-space, and general metric spaces. This will be useful in the future when we turn to more advanced topics of complex analysis, multivariable calculus, and topology. Fundamentally, they are just applications of real sequences from Chap. 5!

In the exercises at the end of this chapter, there are various other applications of real number sequences. Two notable applications are the Newton-Raphson method

S. Johar, *The Big Book of Real Analysis*,
https://doi.org/10.1007/978-3-031-30832-1_6

in Exercise 6.6 and an alternative construction of the real numbers in Exercise 6.27 using Cauchy sequences instead of Dedekind cuts that we saw in Chap. 3.

6.1 Circular Arclength

Let us look at an example which uses limits before it was even formalised in the eighteenth century. One of the major applications of sequences (which are discrete objects) is in geometry (which is a study of "continuous" objects), as pointed out by John Stillwell (1942-) [69] in the quote at the beginning of this chapter.

Using arithmetic is the primitive idea and process utilised by many in order to find the circumference of a circle. In older times, people used to approximate a circle by a discrete process of circumscribed and inscribed regular polygons. The number of sides for these polygons are increased gradually so they approximate the circumference of the circle in the limit. Let us formalise this process, but for a sector of a circle instead as a general case.

We note that in ancient times, before the constant π was discovered and linked to circular measures, there are no standard way of measuring angles. Therefore, the measurements were done in various arbitrary units such as turns, quadrants, sextants, and degrees.

Euclid, the father of geometry, used right angles or turns as a reference unit for measuring angles. This can be seen in the way the Euclid's axioms were phrased in Definition 1.0.2. The Arabs used the unit zams where 224 zams is a full circle. Similar forms of the degree unit were used in ancient civilisations such as Persian, Babylonian, and Indian as a reflection of their calendar days. Despite these various conventions, many consistent trigonometric and geometric identities were developed during this era.

In more modern times, the angular unit grad was used by the French in which a full circle is measured as 400 grad. The unit degree is used as a common convention whence 360° denotes a full circle. However, no matter which angle measurement is used, the trigonometric identities still hold because they are defined as ratios of the sidelengths in a right triangle. Here, for demonstration, we begin with the arbitrary (and more familiar) angular unit of degree.

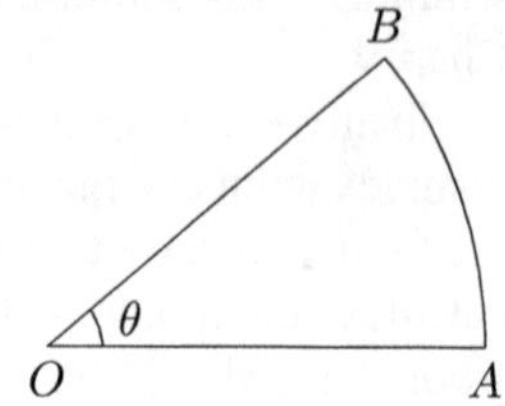

Fig. 6.1 Sector of unit circle with angle θ radians and arc AB

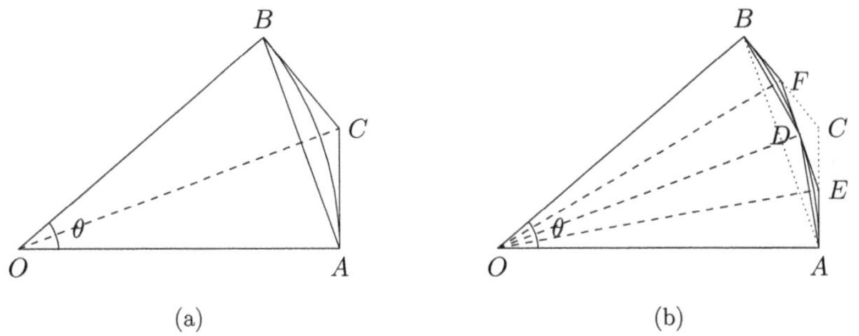

Fig. 6.2 Approximating the arclength AB with secant and tangent line segments

Approximating Arclength

Consider a sector of a unit circle AOB of acute angle $0 < \theta \leq 90°$. We have the lengths $OA = OB = 1$ and we wish to find the length of the circular arc AB as depicted in Fig. 6.1.

In Fig. 6.2a, we first approximate the arclength AB with some straight lines. The line segment AB is a secant line to the arc whereas the line segments AC and BC are tangents to the arc at A and B respectively. These are crude approximations of the arc AB, but this is a good place to start.

Using trigonometry, the length AC is given by $\tan\left(\frac{\theta}{2}\right)$ and so the total length of the tangent line segments AC and BC are given by $2\tan\left(\frac{\theta}{2}\right)$. On the other hand, the secant line AB has length $2\sin\left(\frac{\theta}{2}\right)$. We denote the length of the secant line as $a_1 = 2\sin\left(\frac{\theta}{2}\right)$ and the tangents as $b_1 = 2\tan\left(\frac{\theta}{2}\right)$. By applying triangle inequality in ΔABC, we have the inequality $a_1 = AB \leq AC + BC = b_1$.

In the second step, we add more points to the arc AB. Let D be the intersection point of the arc AB with the line OC. The point D is the midpoint of the arc AB. At D, we construct another tangent to the arc, which would intersect the existing tangents at points E and F respectively. At the same time, we construct the secants AD and BD. This construction is depicted in Fig. 6.2b.

Note that by using triangle inequality, the total length of the secants, which we are going to call a_2, is non-decreasing since $a_2 = AD + BD \geq AB = a_1$. On the other hand, the total lengths of the tangents, which we are going to call b_2, is non-increasing since $b_2 = BF+FE+EA \leq BF+(FC+CE)+EA = BC+CA = b_1$ by using triangle inequality. Their actual lengths can be obtained via trigonometry again, namely: $a_2 = 4\sin\left(\frac{\theta}{4}\right)$ and $b_2 = 4\tan\left(\frac{\theta}{4}\right)$. Also, by using triangle inequality on ΔADE and ΔBDF, we can show that $a_2 \leq b_2$.

Inductively, we continue this construction by adding more and more equispaced points to the arc, each time doubling the number of secants to the sector. We can see in Fig. 6.2b that the tangent lines and the secant lines get closer to the arc AB so we hope that both of these can approach the arclength AB in the limit.

At the n-th step, the total length of the secant line segments joining these points is $a_n = 2^n \sin\left(\frac{\theta}{2^n}\right)$, whereas the total length of tangents at these points would be $b_n = 2^n \tan\left(\frac{\theta}{2^n}\right)$. Furthermore, by the triangle inequality, we have the inequalities

$a_n \leq b_n$, $a_{n-1} \leq a_n$, and $b_n \leq b_{n-1}$. The latter two say the sequence (a_n) is increasing whereas the sequence (b_n) is decreasing.

Now since $a_1 \leq a_n \leq b_n \leq b_1$ for all $n \in \mathbb{N}$, the sequence (a_n) is bounded from above and increasing whereas the sequence (b_n) is bounded from below and decreasing. Using monotone sequence theorem, both of these sequences converge, say $a_n \to a$ and $b_n \to b$ where $a, b \in \mathbb{R}$. Furthermore, since $a_n \leq b_n$ for all $n \in \mathbb{N}$, by preservation of weak inequalities, we get $a \leq b$.

Finally, we want to show that these sequences converge to the same limit, namely $a = b$. We consider the difference $b_n - a_n$. For any $n \in \mathbb{N}$, we have:

$$\begin{aligned} 0 \leq b_n - a_n &= 2^n \tan\left(\frac{\theta}{2^n}\right) - 2^n \sin\left(\frac{\theta}{2^n}\right) \\ &= 2^n \sin\left(\frac{\theta}{2^n}\right) \frac{1 - \cos\left(\frac{\theta}{2^n}\right)}{\cos\left(\frac{\theta}{2^n}\right)} \\ &= 2^n \sin\left(\frac{\theta}{2^n}\right) \frac{1 - \cos^2\left(\frac{\theta}{2^n}\right)}{\cos\left(\frac{\theta}{2^n}\right)\left(1 + \cos\left(\frac{\theta}{2^n}\right)\right)} \\ &= 2^n \frac{\sin^3\left(\frac{\theta}{2^n}\right)}{\cos\left(\frac{\theta}{2^n}\right) + \cos^2\left(\frac{\theta}{2^n}\right)} \leq 2^n \frac{\sin^3\left(\frac{\theta}{2^n}\right)}{\cos\left(\frac{\theta}{2^n}\right)}. \end{aligned} \tag{6.1}$$

Moreover, since $0 < \theta \leq 90°$, we have $0 < \frac{\theta}{2^n} \leq 45°$. Thus $\frac{1}{\sqrt{2}} \leq \cos\left(\frac{\theta}{2^n}\right)$ which implies $\frac{1}{\cos\left(\frac{\theta}{2^n}\right)} \leq \sqrt{2} < 2$. Using this estimate in (6.1) yields:

$$0 \leq b_n - a_n \leq 2^n \frac{\sin^3\left(\frac{\theta}{2^n}\right)}{\cos\left(\frac{\theta}{2^n}\right)} < 2^{n+1} \sin^3\left(\frac{\theta}{2^n}\right) = \frac{2}{2^{2n}} a_n^3 \leq \frac{2}{2^{2n}} b_n^3 \leq \frac{2}{2^{2n}} b_1^3,$$

where we used the fact that $a_n \leq b_n \leq b_1$.

Since $\lim_{n\to\infty} \frac{2}{2^{2n}} b_1^3 = 0$, by sandwiching, we have $\lim_{n\to\infty}(b_n - a_n) = 0$ and thus $b = a$. This limit is then defined to be the arclength AB for the sector of a unit circle of angle $0 < \theta \leq 90°$. Via symmetry of the circle, we can also compute the circumference of the whole unit circle, denoted as C_1, by piecing together different sectors to make up the whole of the circle.

In fact, the above limiting construction can also be done for circles of any radius $r > 0$ other than 1 from which we can deduce that its circumference C_r is the radius r multiplied by the circumference of a unit circle, namely $C_r = rC_1$. From this, we have the equality of ratios $\frac{C_r}{2r} = \frac{C_1}{2}$ for any $r > 0$. Thus, the ratio of the circumference of any circle to its diameter is a constant. This gives us the following definition:

Definition 6.1.1 (Archimedes Constant, π) The ratio of a circumference of a circle to its diameter is a constant which we denote as π.

Using the definition above, we have $C_1 = 2\pi$ and $C_r = 2\pi r$.

Remark 6.1.2 The first instance of the symbol π used to denote this constant is by William Jones (1675–1749), most likely after the Greek word *periphery*.

From construction, it is expected that this constant appears frequently in geometry. However, this constant also crops up in other surprising places. We have seen the combinatorial results of Stirling's approximation for $n!$ and partition functions in Example 5.7.4 and these too feature the constant π. Other than that, it also appears in other branches of mathematics including topology, complex analysis, number theory, and statistics.

Value of π

At this step, we know that the constant π exists, but what is its numerical value? To determine the value of this constant, Archimedes painstakingly computed the circumference of the unit circle by approximating it with regular n-gon for some large number n. Using the polygon method, Archimedes managed to bound the value of this constant π to be between $\frac{223}{71} < \pi < \frac{22}{7}$ using a 96-gon, which has a closed form of:

$$48\sqrt{2-\sqrt{2+\sqrt{2+\sqrt{2+\sqrt{3}}}}}.$$

His important work on it results in the constant π being called the Archimedes' constant. The algorithm for his approximation is shown in Exercise 6.5.

This idea was continued by Viéte who found the value of π accurate to 9 decimal places using a $393,216$-gon and by Ludolph van Ceulen (1540–1610) who spent a huge chunk of his lifetime to compute the first 36 digits of π using a 2^{62}-gon. After Ceulen's death, the digits 3.14159265358979323846264338327950288 were engraved on his tombstone in Leiden, the Netherlands.

We shall see much later in Exercise 16.26 that this constant is actually irrational. Its value is usually cited as $3.14159\ldots$.

Remark 6.1.3 Being a very fundamental (probably the most fundamental) constant in mathematics, the constant π is often cited in pop culture. Let us make some amusing remarks regarding this constant:

1. Due to the irrationality of the number π, its decimal represenation is non-terminating and non-periodic. Thus, many people took it as a challenge to memorise as many digits of π as possible, a practice known as piphilology. The current world record is held by Rajveer Meena who memorised 70,000 digits and recited them in 9 h and 27 min in 2015.
2. Kate Bush, in her album *Aerial*, wrote a song in which she describes a man infatuated with calculating the digits of π. She then crooned the digits of π,

correct to 53 decimal places. She went slightly mysterious after that by reciting two wrong digits and omitting a huge chunk of the digits. In total, she recited 115 digits in that song.
3. More impressively, musician Weird Al Yankovic once claimed: *"I've memorized all the digits of π. Just not in the right order."* Touché.
4. However, mathematician James Grime has determined that 39 digits of π is sufficient to calculate the circumference of the known universe to the accuracy of a width of the hydrogen atom. Indeed, any greater number of digits would then wander below the Planck's length which have no physical meaning as discussed in Remark 4.1.3.
5. According to Marc Ryman, the Chief Engineer for Mission Operations and Science at the NASA's Jet Propulsion Laboratory, the approximation 3.141592653589793 is used in their highest accuracy calculations for interplanetary navigation. They seem to do very well without needing to use 70,000 digits.
6. Another amusing (and slightly alarming) anecdote regarding the value of π is that in 1897, Indiana's state representatives in the USA voted to declare 3.2 the legal value of π. The bill, under the long title *A Bill for an act introducing a new mathematical truth and offered as a contribution to education to be used only by the State of Indiana free of cost by paying any royalties whatever on the same, provided it is accepted and adopted by the official action of the Legislature of 1897* or often succinctly referred to as *Indiana Pi Bill*, was written by a medical doctor Edward Goodwin who dabbled in mathematics in his free time. This bill never became a law, largely thanks to the intervention of mathematician Clarence Waldo (1852–1926) after giving the senators a crash course on elementary geometry. Whew!

Numerical approximations to π can be obtained to any desired degree of accuracy by truncating sequences and infinite sums that are known to converge to π. Some well-known series that lead to π are the Basel problem (in Exercise 8.12), the Leibniz formula for π (in Exercise 16.23), and a family of Machin-like formula (in Exercise 16.24). Another formula that can be used for approximating the value of π that we shall see later is Wallis's formula (in Exercise 16.27).

Radians

The constant π also allows us to introduce a new unit for angular measurement. Since the circle is rotationally symmetric, we can then determine the arclength AB of the unit circle in Fig. 6.1 by using ratios. If θ° is the angle of the sector AOB in degrees, then we have the ratios:

$$\frac{AB}{C_1} = \frac{\theta^\circ}{360^\circ} \quad \Leftrightarrow \quad AB = C_1 \frac{\theta^\circ}{360^\circ} = 2\pi \frac{\theta^\circ}{360^\circ}.$$

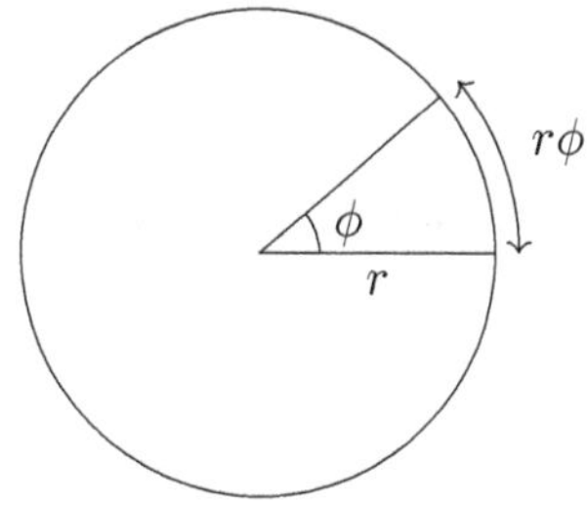

Fig. 6.3 Length of an arc of a circle with radius r subtended by angle ϕ radians

Instead of using degrees, a new angle unit can be introduced here as the number $\angle AOB = 2\pi \frac{\theta}{360}$. As a result, this angle measurement is dimensionless. However, in most literature, the dimension of the angle is given as radians and this is recognised as one of the derived International System of Units (or SI unit for short).

The advantage of using this angular unit is that the arclength of a circle can be described very easily in comparison to other angle units. Indeed, we saw that the circumference of a unit circle C_1, subtended by an angle $360° = 2\pi$ radian, is 2π units (which is the same value as the measure of the angle in radians). Therefore, by ratios, any sector of angle ϕ radians from this unit circle has arclength ϕ units. For circles of other radius $r > 0$, we can just scale the circumference and arclength of the unit circle accordingly. Namely, the length of an arc on a circle of radius r subtended by an angle of ϕ in Fig. 6.3 is $r\phi$. Very simple expression.

Furthermore, since it is now dimensionless, we can take it to any power without worrying about its meaning. We shall see later how this is useful in the Chap. 17 when we attempt to express the trigonometric functions such as sine and cosine as some "infinite degree polynomial" of their argument. This allows us to have an alternative explicit and geometry-free definition for the trigonometric functions in terms of their argument.

6.2 Limit Points and Topology

Consider a real sequence (a_n) in a bounded open interval $A = (0, 10)$ defined by $a_n = \frac{1}{n}$. Clearly, this sequence stays within the interval A. This sequence is monotone and bounded from below. Then, by monotone sequence theorem, this sequence converges to the infimum of the set $\{\frac{1}{n} : n \in \mathbb{N}\}$. Using the Archimedean property, one can show that the infimum is actually 0.

Note that this point 0 is not contained in the set A. However, we can approach or approximate the point 0 by the sequence (a_n) which is strictly contained in A. That is, we can get as close as possible to the point 0 without leaving the set A. This point 0 is called a limit point of the set A.

Informally, a limit point x for a subset $X \subseteq \mathbb{R}$ is any point in $\mathbb{R}$ that can be approximated to any degree of accuracy by points within the set X. This means we can get as close as possible (but not equal) to the point x via points in X. As a result, the point x may or may not even lie in the set X. For the latter to happen, this point x must be extremely "close" to the set X. We define this formally as:

Definition 6.2.1 (Limit Point of a Set, Definition 1) Let $X \subseteq \mathbb{R}$. A point $x \in \mathbb{R}$ is a limit point of the set X if there exists a sequence (x_n) such that $x_n \in X \setminus \{x\}$ for all $n \in \mathbb{N}$ and $\lim_{n\to\infty} x_n = x$. We denote the set of all limit points for the set X as X'.

Using quantifiers, this is written as:

$$x \in X' \quad \text{if} \quad \exists (x_n) \subseteq X \setminus \{x\} : \lim_{n\to\infty} x_n = x.$$

Sometimes limit points are also called accumulation points or cluster points. If we do not like sequences, we may use another definition for the limit points, which is given as follows:

Definition 6.2.2 (Limit Point of a Set, Definition 2) Let $X \subseteq \mathbb{R}$. A point $x \in \mathbb{R}$ is a limit point of the set X if for any $\varepsilon > 0$, the set $X \cap (B_\varepsilon(x) \setminus \{x\}) = X \cap B_\varepsilon(x) \setminus \{x\}$ is non-empty.

Using quantifiers, this is written as:

$$x \in X' \quad \text{if} \quad \forall \varepsilon > 0, X \cap B_\varepsilon(x) \setminus \{x\} \neq \varnothing.$$

Sometimes, we affectionately call the set $B_\varepsilon(x) \setminus \{x\}$ a punctured ball since it is an open ball with one point (which is the centre) removed. Now we have two definitions for limit points which could be a problem with consistency. Which one do we use?

Definitions 6.2.1 and 6.2.2 are in fact equivalent. The readers are invited to show this fact in the Exercise 6.7. Thus we may choose either definition to work with, depending on the situation. This will be demonstrated in Example 6.2.3(1).

An analogy that I like to use when thinking about limit points via Definition 6.2.2 is that no matter how bad our myopia (short-sightedness) is, when we stand at any limit point of a set and look around, we can clearly see at least one element of the set.

Example 6.2.3 Let us look at some examples:

1. Recall that $A = (0, 10) \subseteq \mathbb{R}$. We have seen that $0 \in A'$. There is another limit point of the set A which is not in A, namely 10. This is true because it can be approximated by another sequence (a_n) where $a_n = 10 - \frac{1}{n}$ are all contained in A.

Next, any point $a \in A$ is also a limit point of the set A. Let us show this in two different ways using the two different (but equivalent) definitions of limit points that we have seen.

(a) Using Definition 6.2.1: Fix $a \in A$. We construct a sequence (a_n) in $A \setminus \{a\}$ that tends to a. Since A is an open set, there exists an $\varepsilon > 0$ such that $B_\varepsilon(a) \subseteq A$. Thus, we can find a point in $a_1 \in B_\varepsilon(a) \setminus \{a\}$ to be the first element in the sequence. Next, we pick $a_2 \in B_{\frac{\varepsilon}{2}}(a) \setminus \{a\}$ to be the second element in the sequence.
Inductively, we pick $a_n \in B_{\frac{\varepsilon}{n}}(a) \setminus \{a\}$ so that, by construction, we have $0 < |a_n - a| < \frac{\varepsilon}{n}$ for all $n \in \mathbb{N}$. Clearly $a_n \in A \setminus \{a\}$ for all $n \in \mathbb{N}$ and, by sandwiching, we get $a_n \to a$. Thus $a \in A'$.

(b) Using Definition 6.2.2: Fix $a \in A$. Since A is open, there exists an $\varepsilon > 0$ such that $B_\varepsilon(a) \subseteq A$. Therefore, $B_\varepsilon(a) \setminus \{a\} \subseteq A$, namely $A \cap B_\varepsilon(a) \setminus \{a\} \neq \varnothing$. So $a \in A'$.

Either way, since $a \in A$ was arbitrary, we have the inclusion $A \subseteq A'$. Therefore, we have $A \cup \{0, 10\} = [0, 10] \subseteq A'$
Finally, we have to show that the set A has no other limit points in $\mathbb{R}$ apart from $[0, 10]$. Indeed, suppose for contradiction that there is a limit point $x \in A'$ outside of $[0, 10]$, namely $x \in [0, 10]^c$. Since $[0, 10]^c$ is open, there exists an $\varepsilon > 0$ such that $B_\varepsilon(x) \subseteq [0, 10]^c$. In other words, $A \cap B_\varepsilon(x) \setminus \{x\} \subseteq [0, 10] \cap B_\varepsilon(x) \setminus \{x\} = \varnothing$. This contradicts Definition 6.2.2 and so x cannot be a limit point of A. Thus, we conclude that $A' = [0, 10]$ only.

2. Consider the set $X = (0, 1] \cup [2, 4] \cup \{5\}$ as in Fig. 6.4. This set is a union of three intervals, so in order to find the limit points of X, we need to check each interval separately. For $(0, 1]$, the limit points would be the set $[0, 1]$. On the other hand, the limit points for the set $[2, 4]$ is $[2, 4]$ itself.
However, the set of limit points for the interval $\{5\}$ is empty. One can see this is true because there are no sequences in $X \setminus \{5\}$ that would converge to 5 since the closest point to 5 in X would be 4, which is of distance 1 away from 5.
To prove this rigorously using definitions, we note that for $\varepsilon = \frac{1}{2}$, we have $X \cap B_{\frac{1}{2}}(5) \setminus \{5\} = \{5\} \setminus \{5\} = \varnothing$, not fulfilling Definition 6.2.2. So 5 is not a limit point of the set X. Therefore, the limit points of X is the set $X' = [0, 1] \cup [2, 4]$.

As we have seen above, there may be points in the set of limit points X' that do not lie in the set X and, conversely, there may also be points in the set X that do not lie in the set of limit points X'. Here is another funny example:

Example 6.2.4 Consider the set $X = \{\frac{1}{n} : n \in \mathbb{N}\} \subseteq \mathbb{R}$. This is an infinite subset of $\mathbb{R}$. The limit point of this set is only $X' = \{0\}$. Clearly we can find a sequence in $X \setminus \{0\}$ that converges to 0, for example $\frac{1}{n} \to 0$. Thus $0 \in X'$.

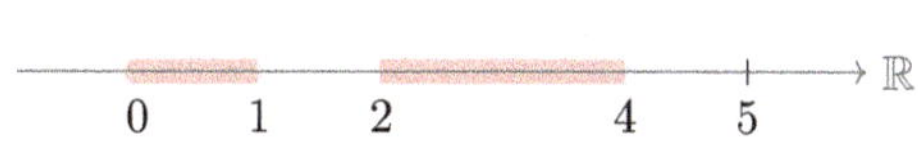

Fig. 6.4 The set $X = (0, 1] \cup [2, 4] \cup \{5\}$

Next, we claim that there are no other limit points of X outside of the set X apart from 0. We check these cases:

1. If $x \in (-\infty, 0)$, the punctured ball of radius $\varepsilon = \frac{|x|}{2} > 0$ does not intersect X. Hence $x \notin X'$.
2. If $x \in (1, \infty)$, the punctured ball of radius $\varepsilon = \frac{x-1}{2} > 0$ does contain any element of X. Hence $x \notin X'$.
3. If $x \in (0, 1)\setminus X$, then it must be in between two elements of X, say $\frac{1}{N+1} < x < \frac{1}{N}$ for some $N \in \mathbb{N}$. Thus, the punctured ball of radius $\varepsilon = \min\left\{\frac{x-\frac{1}{N+1}}{2}, \frac{\frac{1}{N}-x}{2}\right\} > 0$ does not contain any points of X. Hence, $x \notin X'$.

Finally, none of the elements in X is a limit point of the set X. Assume that there is one, say $\frac{1}{N} \in X'$ for some $N \in \mathbb{N}$. If we pick $\varepsilon = \frac{\frac{1}{N}-\frac{1}{N+1}}{2} > 0$, then we have $X \cap B_\varepsilon(\frac{1}{N}) \setminus \{\frac{1}{N}\} = \{\frac{1}{N}\} \setminus \{\frac{1}{N}\} = \varnothing$. This contradicts Definition 6.2.2.

In conclusion, $X' = \{0\}$. Thus, all of the points in X are not in X' and vice versa.

From Example 6.2.4, we have a name for such points of X which are not in X'.

Definition 6.2.5 (Isolated Points) Let $X \subseteq \mathbb{R}$. The points in $X \setminus X'$ are called isolated points of X. In other words, $x \in X$ is an isolated point if there exists an $\varepsilon > 0$ such that $B_\varepsilon(x) \cap X = \{x\}$.

The naming in Definition 6.2.5 comes from the fact that we can isolate any these point away from the rest of the set by an open ball.

One result that we should take note of and will be important later is that a set is closed if and only if it contains all of its limit points. This gives us another characterisation of closed sets in $\mathbb{R}$.

Lemma 6.2.6 *Let $X \subseteq \mathbb{R}$. Then, X is closed if and only if $X' \subseteq X$.*

Proof We prove the implications separately:

($\Rightarrow$): Suppose for a contradiction that $X' \nsubseteq X$. This means there exists a limit point $x \in X'$ outside of X, namely $x \in X^c$. Since X is closed, its complement X^c is open and so there exists an $\varepsilon > 0$ such that $B_\varepsilon(x) \subseteq X^c$. Thus, $X \cap B_\varepsilon(x) \setminus \{x\} \subseteq X \cap B_\varepsilon(x) = \varnothing$, contradicting Definition 6.2.2. Thus, we have $X' \subseteq X$.

($\Leftarrow$): If $X' \subseteq X$, then all points in X^c are not limit points for the set X. So, by Definition 6.2.2, for each $x \in X^c$ we can find an $\varepsilon > 0$ such that $B_\varepsilon(x) \setminus \{x\} \subseteq X^c$. Moreover, since $x \in X^c$ as well, we have $B_\varepsilon(x) \subseteq X^c$. Thus, we conclude that X^c is open and hence X is closed. □

As a corollary of Lemma 6.2.6, we have yet another way to characterise closed sets in $\mathbb{R}$:

Corollary 6.2.7 *Let $X \subseteq \mathbb{R}$.*

1. *Suppose that (x_n) is a sequence in X. If $x_n \to x \in \mathbb{R}$, then $x \in X \cup X'$.*
2. *If X is closed and $(x_n) \subseteq X$ converges to some element $x \in \mathbb{R}$, then $x \in X$ as well.*
3. *X is closed if and only if every Cauchy sequence in X converges to an element in X.*

Proof We prove the assertions separately.

1. Suppose for contradiction that $x \in (X \cup X')^c = X^c \cap (X')^c$. Since $x \notin X'$, there exists an $\varepsilon > 0$ such that $X \cap B_\varepsilon(x) \setminus \{x\} = (X \setminus \{x\}) \cap B_\varepsilon(x) = \varnothing$. Moreover, since $x \notin X$, we have $X \setminus \{x\} = X$ and so $X \cap B_\varepsilon(x) = \varnothing$. This implies $B_\varepsilon(x) \subseteq X^c$. Since $x_n \to x$, there exists an $N \in \mathbb{N}$ such that $|x_n - x| < \varepsilon$ for all $n \geq N$, namely $x_n \in B_\varepsilon(x) \subseteq X^c$ for all $n \geq N$. But this is a contradiction since these x_n are assumed to be contained in X.
2. Since X is closed, by Lemma 6.2.6, we have $X' \subseteq X$. By the first assertion, we have $x \in X \cup X' = X$.
3. We prove the implications separately.

   ($\Rightarrow$): Let (x_n) be a Cauchy sequence in X. Since it is also a Cauchy sequence in the superspace $\mathbb{R}$, it must converge to some $x \in \mathbb{R}$. By the second assertion, necessarily $x \in X$.

   ($\Leftarrow$): Pick any $x \in X'$. By definition, there exists a sequence (x_n) in $X \setminus \{x\}$ such that $x_n \to x$. Since (x_n) is convergent, it must also be Cauchy. From the assumption, any Cauchy sequence converges to a point in X, so $x \in X$. Since $x \in X'$ is arbitrary, we must have $X' \subseteq X$. Thus, by Lemma 6.2.6, the set X is closed. □

Example 6.2.8 Let us look at some examples here:

1. We have seen previously that the limit points of the closed interval $X = [2, 4]$ is the set $X' = [2, 4] = X$ itself. Thus, $X' \subseteq X$, satisfying Lemma 6.2.6.
2. Another closed interval that is interesting is the singleton set $\{a\}$. The limit point of this interval is empty $\varnothing$ which is also a subset of $\{a\}$, in accordance to Lemma 6.2.6.
3. Consider the subset $(0, 1] \subseteq \mathbb{R}$. The Cauchy sequence (a_n) defined as $a_n = \frac{1}{n}$ is fully contained in $(0, 1]$. As a sequence in $\mathbb{R}$, the sequence is Cauchy and hence converges to the limit 0. However, this limit is not in $(0, 1]$ and thus, by Corollary 6.2.7, we conclude that it is not a closed set.

6.3 Sequences in $\mathbb{C}$ and $\mathbb{R}^n$

So far, we have been focusing mainly on real sequences. Real numbers are nice because they form a complete ordered field. However, there are many other number fields out there.

We have seen some introduction to complex numbers in Chap. 4. Similar to the real numbers, they form a number field. However, in contrast to $\mathbb{R}$, there are no strict total order that we can define on $\mathbb{C}$ which is compatible with its field structure.

Similar to real numbers, we can define a complex sequence:

Definition 6.3.1 (Sequence of Complex Numbers) A sequence of complex numbers is a function $z : \mathbb{N} \to \mathbb{C}$ usually denoted as $(z_n)_{n\in\mathbb{N}}$ or simply (z_n).

Example 6.3.2 Here are some examples of complex sequences (z_n) in $\mathbb{C}$:

1. $z_n = \frac{1}{n} + \frac{(-1)^n}{n} i$.
2. $z_n = \left(\frac{1}{1+i}\right)^n$.
3. $z_n = 1 + (-1)^n i$.

Due to the field structure on $\mathbb{C}$, we can add, subtract, multiply, divide, and scale complex sequences, similar to what we can do with real sequences. However, how do we define bounded or convergent complex sequences?

For real sequences, we have defined the modulus operation which assigns a non-negative size to each real number or distance to each pair of real numbers. This is crucial as it allowed us to make sense of and quantify what "near" means in $\mathbb{R}$. Using this, we can define convergence of a real sequence (a_n) to a real number L by investigating the distance between the numbers a_n and the point L via Definition 5.2.4.

For complex numbers, we have seen the complex norm in Example 4.6.5(4). This can be viewed as an analogue of the modulus in $\mathbb{R}$ and thus is used to quantify the size of an element in $\mathbb{C}$. Using this norm, we can define bounded complex sequences. Geometrically, similar to the real sequences, bounded complex sequences are simply complex sequences for which each term in the sequence is within a fixed finite size. In more detail, we define:

Definition 6.3.3 (Bounded Complex Sequence) Let (z_n) be a complex sequence. The sequence (z_n) is bounded if there exists a positive real constant $M > 0$ such that $|z_n| \leq M$ for all $n \in \mathbb{N}$. In other words, the sequence stays in the closed ball $\bar{B}_M(0) \subseteq \mathbb{C}$.

Norms can also be used to measure how far apart two complex numbers $z_1, z_2 \in \mathbb{C}$ are from each other. This distance is quantified by the quantity $|z_1 - z_2|$. With this, we can define convergent complex sequences as follows:

Definition 6.3.4 (Convergent Complex Sequence) A complex sequence (z_n) is convergent or converges to a complex number $L \in \mathbb{C}$ if for any $\varepsilon > 0$, there exists an $N \in \mathbb{N}$ such that $|z_n - L| < \varepsilon$ for all $n \geq N$. We call the number L the limit of the sequence (z_n).

With quantifiers, we write this as:

$$z_n \to L \quad \text{if} \quad \forall \varepsilon > 0, \exists N \in \mathbb{N} : \forall n \geq N, |z_n - L| < \varepsilon.$$

The definition is almost exactly the same as the definition for convergent real sequences in Definition 5.2.4. The only difference is that the complex norm is used in place of the modulus (but the notation used here is still the same).

In fact, many concepts and definitions from the study of real sequences carry forward into the study of complex sequences. For example, the definitions for tail of a sequence and subsequences remain the same for complex sequences. On the other hand, there are no increasing, decreasing, or monotone sequences in $\mathbb{C}$ since we cannot order complex numbers in a natural way that fits with the field structure.

Example 6.3.5 Recall the complex sequences (z_n) in Example 6.3.2. We want to determine whether they converge using Definition 6.3.4.

1. For $z_n = \frac{1}{n} + \frac{(-1)^n}{n} i$, for any $n \in \mathbb{N}$ we compute $|z_n| = \sqrt{\frac{1}{n^2} + \frac{1}{n^2}} = \frac{\sqrt{2}}{n} \leq \sqrt{2}$. Thus the sequence (z_n) is bounded. Moreover, we claim that the sequence converges to 0.

   Rough work: Set $\varepsilon > 0$. We now find an $N \in \mathbb{N}$ such that $|z_n - 0| = |z_n| < \varepsilon$ for every $n \geq N$. This means $|z_n| = \frac{\sqrt{2}}{n} < \varepsilon$ for every $n \geq N$. In other words, we need $n > \frac{\sqrt{2}}{\varepsilon}$ for every $n \geq N$. Let us pick $N = \lceil \frac{\sqrt{2}}{\varepsilon} \rceil + 1 \in \mathbb{N}$. Now we check that this choice of N works for the fixed ε.

   Fix $\varepsilon > 0$. We choose $N = \lceil \frac{\sqrt{2}}{\varepsilon} \rceil + 1 \in \mathbb{N}$. Then, for all $n \geq N$ we have:

$$|z_n - 0| = \frac{\sqrt{2}}{n} \leq \frac{\sqrt{2}}{N} = \frac{\sqrt{2}}{\lceil \frac{\sqrt{2}}{\varepsilon} \rceil + 1} < \frac{\sqrt{2}}{\lceil \frac{\sqrt{2}}{\varepsilon} \rceil} \leq \frac{\sqrt{2}}{(\frac{\sqrt{2}}{\varepsilon})} = \varepsilon,$$

   and thus we conclude $z_n \to 0$.
2. If $z_n = \left(\frac{1}{1+i}\right)^n$, then this sequence is also bounded with $|z_n| \leq 1$. Moreover, it also converges to 0. Indeed, for a fixed $\varepsilon > 0$, we choose $N = 2\lceil \frac{1}{\varepsilon} \rceil \in \mathbb{N}$. Then, for $n \geq N$, by Bernoulli's inequality, we have $2^{\frac{n}{2}} = (1+1)^{\frac{n}{2}} \geq 1 + \frac{n}{2}$. Using

this, we have:

$$|z_n - 0| = \left|\left(\frac{1}{1+i}\right)^n\right| = \left|\frac{1-i}{2}\right|^n = \frac{2^{\frac{n}{2}}}{2^n} = \frac{1}{2^{\frac{n}{2}}}$$
$$\leq \frac{1}{1+\frac{n}{2}}$$
$$< \frac{1}{\frac{N}{2}} = \frac{1}{\lceil\frac{1}{\varepsilon}\rceil} \leq \varepsilon.$$

Therefore, $z_n \to 0$.

3. For $z_n = 1+(-1)^n i$, this sequence is bounded since $|z_n| = \sqrt{1^2+(-1)^{2n}} = \sqrt{2}$ for all $n \in \mathbb{N}$. However, it cannot converge to any complex number. Assume for contradiction that it converges to some $z = a+ib \in \mathbb{C}$. Then, for $\varepsilon = \frac{1}{2}$, there exists an index $N \in \mathbb{N}$ such that $|z_n - z| < \frac{1}{2}$ for all $n \geq N$. Pick any $m, n \geq N$ such that m is odd and n is even. Then:
$$\frac{1}{2} > |z_m - z| = \sqrt{(1-a)^2 + ((-1)^m - b)^2} \geq |-1-b| \quad \Rightarrow \quad b < -\frac{1}{2},$$
$$\frac{1}{2} > |z_n - z| = \sqrt{(1-a)^2 + ((-1)^n - b)^2} \geq |1-b| \quad \Rightarrow \quad \frac{1}{2} < b,$$
giving us a contradiction.

Note that each element in a complex sequence (z_n) can be written as $z_n = a_n + ib_n$ where (a_n) and (b_n) are two real sequences. Since complex sequences are made up of two real sequences, we can still use the theory we developed for real sequences in Chap. 5 to study complex sequences. In particular, we have the following result:

Theorem 6.3.6 *Let (z_n) be a complex sequence. Denote $z_n = a_n + ib_n$ for every $n \in \mathbb{N}$ so (a_n) and (b_n) are two real sequences induced from the complex sequence (z_n). The complex sequence (z_n) converges if and only if both of the real sequences (a_n) and (b_n) converge.*
In particular, $z_n \to L_1 + iL_2$ if and only if $a_n \to L_1$ and $b_n \to L_2$.

Proof We prove each implication separately.

($\Rightarrow$): Suppose that $z_n \to L_1+iL_2$. We want to show that $a_n \to L_1$ and $b_n \to L_2$. Fix $\varepsilon > 0$. Then, there exists an $N \in \mathbb{N}$ such that $|z_n - (L_1+iL_2)| < \varepsilon$ for every $n \geq N$. In other words, we have $|a_n - L_1 + i(b_n - L_2)| < \varepsilon$ for all $n \geq N$. Using Proposition 4.6.6(4), both the real and imaginary parts satisfy $|a_n - L_1| < \varepsilon$ and $|b_n - L_2| < \varepsilon$ for all $n \geq N$. So, $a_n \to L_1$ and $b_n \to L_2$.

($\Leftarrow$): Now we want to show that $z_n \to L_1 + iL_2$ provided that $a_n \to L_1$ and $b_n \to L_2$. Fix $\varepsilon > 0$. Since $a_n \to L_1$, there exists an $N_1 \in \mathbb{N}$ such that $|a_n - L_1| < \frac{\varepsilon}{2}$ for all $n \geq N_1$. Also, since $b_n \to L_2$, there exists an $N_2 \in \mathbb{N}$

such that $|b_n - L_2| < \frac{\varepsilon}{2}$ for all $n \geq N_2$. Our goal now is to find an $N \in \mathbb{N}$ such that $|z_n - (L_1 + iL_2)| < \varepsilon$ for all $n \geq N$. We pick $N = \max\{N_1, N_2\}$ and show that this works.
By applying the triangle inequality, for every $n \geq N$, we have:

$$\begin{aligned}|z_n - (L_1 + iL_2)| &= |a_n - L_1 + i(b_n - L_2)| \\ &\leq |a_n - L_1| + |i(b_n - L_2)| \\ &= |a_n - L_1| + |i||b_n - L_2| \\ &= |a_n - L_1| + |b_n - L_2| < \frac{\varepsilon}{2} + \frac{\varepsilon}{2} = \varepsilon,\end{aligned}$$

and so $z_n \to L_1 + iL_2$. □

Therefore, to study the convergence of a complex sequence, it is enough to study the convergence of the real and imaginary parts of the sequence. We have developed a lot of theories for real sequences in Chap. 5, so we know exactly how to do this!

Example 6.3.7 Recall the complex sequences (z_n) in Example 6.3.2. We have shown their convergence/divergence in Example 6.3.5 using the ε-N definition. Let us revisit some of them using Theorem 6.3.6.

1. For $z_n = \frac{1}{n} + \frac{(-1)^n}{n} i$, this sequence converges to $0 + i0$ since both the real and imaginary parts $(\frac{1}{n})$ and $(\frac{(-1)^n}{n})$ converge to 0.
2. For $z_n = 1 + (-1)^n i$ this sequence diverges since the imaginary part $((-1)^n)$ diverges.

As a corollary of Theorem 6.3.6, we have:

Corollary 6.3.8 *Let $(z_n) = (a_n + ib_n)$ be a complex sequence.*

1. *If (z_n) is a convergent sequence, then its limit must be unique.*
2. *If (z_n) is a convergent sequence, then this sequence must be bounded.*
3. *If (z_n) is a bounded complex sequence and the real sequences (a_n) and (b_n) are both monotone, then the sequence (z_n) converges.*
4. *The complex sequence (z_n) is convergent if and only if it is Cauchy.*

Proof We prove the assertions one by one:

1. Suppose that $z_n \to L_1 + iL_2$ and $z_n \to K_1 + iK_2$. Since the complex sequence converges, its real part converges with $a_n \to L_1$ and $a_n \to K_1$. By uniqueness of limits in $\mathbb{R}$ from Proposition 5.2.13, we have $L_1 = K_1$. Similarly, $L_2 = K_2$ and thus $L_1 + iL_2 = K_1 + iK_2$.
2. Since (z_n) is a convergent complex sequence, its real and imaginary parts also converge. By Proposition 5.2.14, the real and imaginary sequences are bounded,

say $|a_n| \leq K$ and $|b_n| \leq M$ for all $n \in \mathbb{N}$ and some constants $M, K > 0$. Then, we have $|z_n| = \sqrt{a_n^2 + b_n^2} \leq \sqrt{K^2 + M^2}$ for all $n \in \mathbb{N}$ which says that the complex sequence (z_n) is bounded.

3. Since the complex sequence (z_n) is bounded, there exists an $M > 0$ such that $|z_n| \leq M$. Moreover, since $|a_n| = |\mathrm{Re}(z_n)| \leq |z_n| \leq M$ and $|b_n| = |\mathrm{Im}(z_n)| \leq |z_n| \leq M$, the real sequences (a_n) and (b_n) are bounded too. Furthermore, by assumption, these real sequences are monotone and by monotone sequence theorem, $a_n \to L_1$ and $b_n \to L_2$ for some $L_1, L_2 \in \mathbb{R}$. Thus, the complex sequence (z_n) converges to $L_1 + iL_2$.
4. We first prove that the complex sequence (z_n) is Cauchy if and only if the real sequences (a_n) and (b_n) are Cauchy. We prove the implications one by one:

($\Rightarrow$): Suppose that the sequence (z_n) is Cauchy. Fix $\varepsilon > 0$. Then, there exists an $N \in \mathbb{N}$ such that $|z_n - z_m| < \varepsilon$ for every $m, n \geq N$. In other words, we have $|a_n - a_m + i(b_n - b_m)| < \varepsilon$ for all $m, n \geq N$. Using Proposition 4.6.6(4), both the real and imaginary parts satisfy $|a_n - a_m| < \varepsilon$ and $|b_n - b_m| < \varepsilon$ for all $m, n \geq N$, implying that both of them are also Cauchy.

($\Leftarrow$): Fix $\varepsilon > 0$. Since (a_n) is Cauchy, there exists an $N_1 \in \mathbb{N}$ such that $|a_m - a_n| < \frac{\varepsilon}{2}$ for all $m, n \geq N_1$. Also, since (b_n) is Cauchy, there exists an $N_2 \in \mathbb{N}$ such that $|b_m - b_n| < \frac{\varepsilon}{2}$ for all $m, n \geq N_2$. Pick $N = \max\{N_1, N_2\}$. By applying the triangle inequality, for every $m, n \geq N$, we have:

$$|z_m - z_n| = |a_m - a_n + i(b_m - b_n)| \leq |a_m - a_n| + |b_m - b_n| < \varepsilon,$$

and so (z_n) is Cauchy as well.

Therefore, along with Theorem 6.3.6, we have a chain of equivalences:

$$(z_n) \text{ is Cauchy} \Leftrightarrow (a_n) \text{ and } (b_n) \text{ are Cauchy} \Leftrightarrow (a_n) \text{ and } (b_n) \text{ converge}$$
$$\Leftrightarrow (z_n) \text{ converges},$$

giving us the desired conclusion. □

Furthermore, the algebra of limits also remain true for complex sequences. Even Bolzano-Weierstrass theorem remains true for complex sequences. However, since we used an ordering argument in the proof for Bolzano-Weierstrass theorem for real sequences (via the monotone subsequence lemma), we need to provide an alternative proof for Bolzano-Weierstrass theorem for complex sequences.

Theorem 6.3.9 (Bolzano-Weierstrass Theorem) *If (z_n) is a bounded complex sequence, then there exists a subsequence of (z_n) that is convergent.*

Proof For the complex sequence (z_n), suppose that $z_n = a_n + ib_n$. Consider the real sequences (a_n) and (b_n). Since the complex sequence (z_n) is bounded, there is an $M > 0$ such that $|z_n| \leq M$ for every $n \in \mathbb{N}$. This implies the real sequences (a_n) and (b_n) are bounded too. By applying the Bolzano-Weierstrass theorem to the real sequence (a_n), we can find a subsequence (a_{k_n}) which converges, namely $a_{k_n} \to L_1 \in \mathbb{R}$. We note that every subsequence of (a_{k_n}) converges to L_1 as well by Proposition 5.5.4.

Now consider the subsequence (z_{k_n}) of (z_n). Again, this sequence is bounded so, in particular, the sequence of imaginary parts (b_{k_n}) is bounded. By applying the Bolzano-Weierstrass theorem to the real sequence (b_{k_n}), we can find a convergent subsequence $b_{l_{k_n}} \to L_2$. Thus, if we consider the subsequence $(z_{l_{k_n}})$ of (z_n), we note that the imaginary part converges to L_2 and the real part converges to L_1 since it is a subsequence of a convergent sequence (a_{k_n}). Therefore, we have found a convergent subsequence of (z_n). □

In fact, since the complex numbers $\mathbb{C}$ can be seen as the vector space space $\mathbb{R}^2$, the ideas from this section carry forward to sequences in real n-spaces $\mathbb{R}^n$. The main idea is the same: we do not have an ordering in the real n-space $\mathbb{R}^n$, but we use norms to compare distances between elements in $\mathbb{R}^n$.

The space $\mathbb{R}^n$ is not a field, but it is a real vector space where we can add, subtract, and scale elements in this space by doing these operations componentwise as a vector space. Naturally, we can also define sequences in this space. Sizes and distances between two points in $\mathbb{R}^n$ can be defined by using a norm on this set. There are various norms available on the real n-space $\mathbb{R}^n$ such as the Euclidean norm, taxicab norm, maximum norm, and l^p norm, some of which we saw in Example 4.6.5. Geometrically, this is simply the distance from the point $\mathbf{a} = (a_1, a_2, \ldots, a_n)$ to the origin or the size of the vector $\mathbf{a}$.

All of these norms can be used to measure distances between two points. For any two points $\mathbf{a}, \mathbf{b} \in \mathbb{R}^n$, we can define the distance between them by using the norm as $||\mathbf{a} - \mathbf{b}||$. With distances defined in this space, convergence of sequences in $\mathbb{R}^n$ is also well defined:

Definition 6.3.10 (Convergent Sequence in $\mathbb{R}^n$) Let $(\mathbb{R}^n, ||\cdot||)$ be a normed vector space. A sequence $(\mathbf{x}_n)$ in $\mathbb{R}^n$ is convergent or converges to a point $\mathbf{a} \in \mathbb{R}^n$ if for any $\varepsilon > 0$, there exists an $N \in \mathbb{N}$ such that $||\mathbf{x}_n - \mathbf{a}|| < \varepsilon$ for all $n \geq N$. We call the point $\mathbf{a}$ the limit of the sequence $(\mathbf{x}_n)$.

With quantifiers, we write this as:

$$\mathbf{x}_n \to \mathbf{a} \quad \text{if} \quad \forall \varepsilon > 0, \exists N \in \mathbb{N} : \forall n \geq N, ||\mathbf{x}_n - \mathbf{a}|| < \varepsilon.$$

Example 6.3.11 For demonstration, we pick the Euclidean norm on $\mathbb{R}^n$ which is given by:

$$||\mathbf{a}|| = ||(a_1, a_2, \ldots, a_n)|| = \sqrt{a_1^2 + a_2^2 + \ldots + a_n^2}.$$

The resulting normed vector space $(\mathbb{R}^n, ||\cdot||)$ is called the Euclidean space. For the Euclidean norm, since the modulus of a component of an element $\mathbf{a} = (a_1, a_2, \ldots, a_n)$ is smaller than the Euclidean norm of $\mathbf{a}$, namely $|a_j| \leq ||\mathbf{a}||$ for any $j = 1, 2, \ldots, n$ and $||\mathbf{a}|| \leq n \max_j |a_j|$, the convergence of points in $\mathbb{R}^n$ is equivalent to the convergence of each of its component in $\mathbb{R}$ similar to what we did in Theorem 6.3.6.

Therefore, we can deduce convergence in the Euclidean space $\mathbb{R}^n$ via the convergences for each of the components in $\mathbb{R}$, which we know how to do and have an abundance of theory to work with from Chap. 5.

6.4 Introduction to Metric Spaces

In the previous section, we have seen how we can define convergence of sequences in the sets $\mathbb{C}$ and $\mathbb{R}^n$ for some $n \in \mathbb{N}$. In order to define convergence in these sets, we need to define distances between any two elements in the set so that we can quantify what "near" means. Since these sets could be derived from the $\mathbb{R}$ (on which we have defined distances using the modulus function), the distances on $\mathbb{C}$ and $\mathbb{R}^n$ could be derived from the distance on $\mathbb{R}$ via norms and geometry. Can we do so in a more general set?

For example, recall the set of family members $X = \{F, M, D, S\}$ that we saw in Example 2.1.4. At the moment, this is just a set of points. Nothing more, nothing less. We would like to put some geometric structure on it by defining how "near" the elements of this set are relative to each other. How can we legitimately define distances between them?

As we have seen from the real numbers, distances should satisfy some properties or axioms. For any real numbers $x, y, z \in \mathbb{R}$, the modulus function that measures the distance between any of these two points has been shown to satisfy:

1. $|x - y| \geq 0$ with equality if and only if $x = y$,
2. $|x - y| = |y - x|$,
3. $|x - y| \leq |x - z| + |z - y|$.

So, this forms a model of what distances on a more general set should be. In fact, this is how we view distances in real life: distances are non-negative, measuring the distance from x to y is the same as measuring the distance from y to x, and taking a detour via a third point z can only increase the total distance. Using this model, on a set X, we define the distance or metric function as:

Definition 6.4.1 (Metric Function) Let X be a set. The distance or metric function on X is a function $d : X \times X \to \mathbb{R}$ such that for any $x, y, z \in X$ it satisfies the metric axioms:

1. Non-negativity: $d(x, y) \geq 0$ with equality if and only if $x = y$.
2. Symmetry: $d(x, y) = d(y, x)$.
3. Triangle inequality: $d(x, y) \leq d(x, z) + d(z, y)$.

Any set X with a choice of metric d is called a metric space.

Definition 6.4.2 (Metric Space) A metric space (X, d) is a set X with a metric function $d : X \times X \to \mathbb{R}$.

Example 6.4.3 Let us look at some examples:

1. Of course, we have seen an example of a metric space which is the set of real numbers $\mathbb{R}$ endowed with a metric derived from the modulus. The metric that we have seen and used repeatedly in Chap. 5, defined as $d(x, y) = |x - y|$, is called the Euclidean metric on $\mathbb{R}$. This metric satisfies all three of the metric axioms.
2. Recall the set of family $X = \{F, M, D, S\}$. We can define a distance function on them as:

$$d : X \times X \to \mathbb{R}$$

$$(x, y) \mapsto \begin{cases} 0 & \text{if } x = y, \\ 1 & \text{if } x \neq y. \end{cases}$$

This function says that the distance between any two distinct element in the set is always 1. We can check that this is indeed a metric on X by showing that the function d satisfies the axioms in Definition 6.4.1.

(a) Clearly $d(x, y) = 0, 1$ for any $x, y \in X$ so its image is non-negative. Furthermore, by definition, $d(x, y) = 0$ if and only if $x = y$.

(b) By definition, $d(x, y) = d(y, x)$. Indeed, if $x = y$ we have $d(x, y) = 0 = d(y, x)$. Otherwise if $x \neq y$, we have $d(x, y) = 1 = d(y, x)$.

(c) To check triangle inequality, we have several cases:

i. If $x = y$, then clearly $d(x, y) = 0 \leq d(x, z) + d(z, y)$ for any z.

ii. If $x \neq y$ and z is equal to at least one of x or y, necessarily z can only be equal to one of them. WLOG, suppose that $x = z$ and $y \neq z$. Thus, $d(x, y) = 1$, $d(x, z) = 0$, and $d(z, y) = 1$ which then satisfies the triangle inequality too.

iii. Finally, if all of x, y, z are distinct, we have $d(x, y) = 1, d(x, z) = 1$, and $d(z, y) = 1$, which again satisfies the triangle inequality.

Therefore, d is a metric on the set X.

3. Let X be a set and $\mathcal{P} = \{A \subseteq X : |A| < \infty\}$. In other words, $\mathcal{P}$ is the set of all finite subsets of X. We define a function $d : \mathcal{P} \times \mathcal{P} \to \mathbb{R}$ as $d(A, B) = |A \triangle B|$. This is a well-defined function since $A \triangle B \subseteq A$ which is a finite set and thus $|A \triangle B| \in \mathbb{N}_0 \subseteq \mathbb{R}$ for any $A, B \in \mathcal{P}$. Now we check that this indeed defines a metric.
   (a) For any $A, B \in \mathcal{P}$, we have $|A \triangle B| \in \mathbb{N}_0$ and so $d(A, B) \geq 0$. Clearly if $A = B$, then $d(A, B) = |A \triangle B| = |\varnothing| = 0$. Conversely, if $d(A, B) = 0$ then $A \triangle B = \varnothing$. This is equivalent to $A \setminus B = \varnothing$ and $B \setminus A = \varnothing$ which imply $A \subseteq B$ and $B \subseteq A$ respectively. Thus, $A = B$.
   (b) For any $A, B \in \mathcal{P}$, by Exercise 1.21, we have $A \triangle B = B \triangle A$ so $d(A, B) = d(B, A)$.
   (c) For any $A, B, C \in \mathcal{P}$, from Exercise 1.21, we have $A \triangle B \subseteq (A \triangle C) \cup (C \triangle B)$. Thus, using Lemma 3.4.8, we have $|A \triangle B| \leq |(A \triangle C) \cup (C \triangle B)| = |(A \triangle C)| + |(C \triangle B)| - |(A \triangle C) \cap (C \triangle B)| \leq |(A \triangle C)| + |(C \triangle B)|$ giving us the triangle inequality $d(A, B) \leq d(A, C) + d(C, B)$.

   Therefore, d is a metric function on $\mathcal{P}$. What is the interpretation of this "distance" function? For any two sets $A, B \in \mathcal{P}$ the metric d counts how many of the elements in $A \cup B$ is not common to both A and B. In other words, it gives a measure of how far the sets A and B are from being identical sets.
4. Let $(V, ||\cdot||)$ be a normed real vector space. Then, we can define a metric from this norm (which we call the norm metric or induced metric) on the set V by $d : V \times V \to \mathbb{R}$ via $d(\mathbf{v}, \mathbf{w}) = ||\mathbf{v} - \mathbf{w}||$. Let us check that this satisfies the metric axioms. We need to recall the properties of a norm from Definition 4.6.4 for this.
   (a) For any $\mathbf{u}, \mathbf{v} \in V$, we have $||\mathbf{u} - \mathbf{v}|| \geq 0$. Moreover, we have $||\mathbf{u} - \mathbf{v}|| = 0$ if and only if $\mathbf{u} - \mathbf{v} = \mathbf{0}$, namely $\mathbf{u} = \mathbf{v}$.
   (b) For any $\mathbf{u}, \mathbf{v} \in V$, we have $||\mathbf{u} - \mathbf{v}|| = ||(-1)(\mathbf{v} - \mathbf{u})|| = |-1|||\mathbf{v} - \mathbf{u}|| = ||\mathbf{v} - \mathbf{u}||$.
   (c) For any $\mathbf{u}, \mathbf{v}, \mathbf{w} \in V$, we have $||\mathbf{u} - \mathbf{v}|| = ||\mathbf{u} - \mathbf{w} + \mathbf{w} - \mathbf{v}|| \leq ||\mathbf{u} - \mathbf{w}|| + ||\mathbf{w} - \mathbf{v}||$.

   Whilst both norm and metric are geometric structures on a vector space, we note the subtle difference between a norm and a metric on a real vector space: norms measure sizes whereas metrics measure distances.

   If we have a norm on a vector space, we can induce a metric from it. However, if we have a metric on a vector space, we may not be able to induce a norm from it (unless we have some additional special properties on the metric). Moreover, norms can only be defined on vector spaces whereas a metric can be defined on any set at all.
5. Let us look at a different set, namely the set of bounded functions on $[a, b]$ where $a, b \in \mathbb{R}$ with $a < b$. We denote this set as $B([a, b]; \mathbb{R})$ or simply $B([a, b])$ so:

$$B([a, b]) = \{f : [a, b] \to \mathbb{R} : \exists M > 0 : \forall x \in [a, b], |f(x)| \leq M\}.$$

   This set is an example of functions spaces, since the elements of this set are functions defined on the same domain and to the same codomain. The study of

such spaces are important in functional analysis. This is because the set $B([a, b])$ forms a real-vector space which has a lot of structure.
We can equip this vector space with a norm $||\cdot|| : B([a, b]) \to \mathbb{R}$ defined as $||f|| = \sup_{x\in[a,b]} |f(x)|$. By the previous example, any norm on a vector space induces a metric, so this norm induces a metric d_∞ on the set $B([a, b])$ as:

$$d_\infty(f, g) = ||f - g|| = \sup_{x\in[a,b]} |f(x) - g(x)|.$$

The readers will show that this whole construction is well-defined in Exercise 6.18. This metric d_∞ is called the uniform metric or the supremum metric on $B([a, b])$.

In fact, for any set, there may be more than one kind of metric that we can put on it, depending on how we want to measure distances:

Example 6.4.4 Consider the set $\mathbb{R}^n$. We present some metrics $d : \mathbb{R}^n \times \mathbb{R}^n \to \mathbb{R}$ on this space. If $\mathbf{x}, \mathbf{y} \in \mathbb{R}^n$ where $\mathbf{x} = (x_1, x_2, \ldots, x_n)$ and $\mathbf{y} = (y_1, y_2, \ldots, y_n)$ for $x_j, y_j \in \mathbb{R}$, we define:

1. Euclidean metric: In Example 4.6.5, we have defined the Euclidean norm $||\cdot||$. From this, we define the Euclidean metric:

$$d(\mathbf{x}, \mathbf{y}) = ||\mathbf{x} - \mathbf{y}|| = \left(\sum_{j=1}^{n}(x_j - y_j)^2\right)^{\frac{1}{2}}.$$

2. Taxicab or Manhattan metric: $d(\mathbf{x}, \mathbf{y}) = \sum_{j=1}^{n} |x_j - y_j|$. The naming of this metric comes from the grid-like stricture of Manhattan traffic network in which an imaginary taxicab can only move east-west or north-south. This means the taxicab is not allowed to drive along Broadway.
3. Maximum metric: $d(\mathbf{x}, \mathbf{y}) = \max_{1\leq j\leq n} |x_j - y_j|$.
4. Discrete metric: Similar to Example 6.4.3(2), we can also define the metric:

$$d(\mathbf{x}, \mathbf{y}) = \begin{cases} 0 & \text{if } \mathbf{x} = \mathbf{y}, \\ 1 & \text{if } \mathbf{x} \neq \mathbf{y}. \end{cases}$$

5. Railway metric:

$$d(\mathbf{x}, \mathbf{y}) = \begin{cases} ||\mathbf{x} - \mathbf{y}|| & \text{if } \mathbf{x} = k\mathbf{y} \text{ for some } k \in \mathbb{R}, \\ ||\mathbf{x}|| + ||\mathbf{y}|| & \text{otherwise,} \end{cases}$$

where $||\cdot||$ is the Euclidean norm on $\mathbb{R}^n$ from Example 4.6.5. The reason for this naming comes from a made-up railway network in a made-up country $\mathbb{R}^n$.

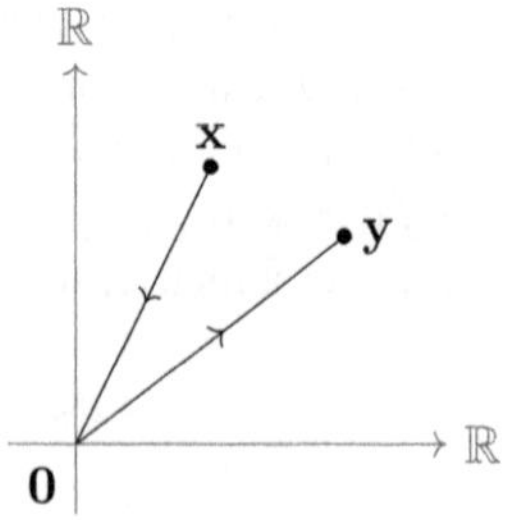

Fig. 6.5 Measuring the distance from **x** to **y** in $\mathbb{R}^2$ using the railway metric

Imagine **0** is the capital city in the country $\mathbb{R}^n$ with straight radial railway lines going in and out from **0**. To get from the city **x** to **y**, if they line on the same radial railway line, we can just take the train directly from **x** to **y**. But if they do not lie on the same radial railway line, we have to commute from **x** to the capital **0** first before moving back out to reach **y**. This can be seen for $\mathbb{R}^2$ in Fig. 6.5.

The readers are invited to prove that these are all indeed legitimate metric functions on $\mathbb{R}^n$ in Exercise 6.17.

Now that we have distances defined on an arbitrary set, we can define convergent sequences in it. As we have noted in Remark 5.0.3, a sequence in a set X is:

Definition 6.4.5 (Sequence in X) A sequence points in a set X is a function $x : \mathbb{N} \to X$ usually denoted as $(x_n)_{n\in\mathbb{N}}$ or simply (x_n).

Analogous to the definition of bounded, convergent, and Cauchy sequences in Chap. 5, once we have endowed the set X with a distance function, we can define the following:

Definition 6.4.6 (Bounded Sequence) A sequence (x_n) in a metric space (X, d) is bounded if there exists a point $x \in X$ and a real number $M > 0$ such that $d(x_n, x) \leq M$ for all $n \in \mathbb{N}$.

Remark 6.4.7 In a generic set X, unlike the sets $\mathbb{R}$, $\mathbb{C}$, or $\mathbb{R}^n$ that we have previously seen, we do not have a canonical origin to measure sizes from. Therefore, in Definition 6.4.6, we have to specify which reference point x we are measuring sizes from. However, this choice of point is irrelevant: a set is bounded no matter where we measure its size from. Indeed, if we choose another reference point $y \in X$, by triangle inequality, we have $d(x_n, y) \leq d(x_n, x) + d(x, y) \leq M + d(x, y)$ for all $n \in \mathbb{N}$ (note that $M + d(x, y) \in \mathbb{R}$ is a finite constant since x and y are fixed reference points).

Definition 6.4.8 (Convergent Sequence) A sequence (x_n) in a metric space (X, d) is convergent or converges if there exists a point $x \in X$ such that for any $\varepsilon > 0$, there exists an index $N \in \mathbb{N}$ where we have $d(x_n, x) < \varepsilon$ for all $n \geq N$. We call the point x the limit of the sequence (x_n). We write this as:

$$\lim_{n\to\infty} x_n = x \quad \text{or} \quad x_n \xrightarrow{n\to\infty} x \quad \text{or} \quad x_n \to x.$$

Symbolically, this definition is written with quantifiers as:

$$(x_n) \text{ converges} \quad \text{if} \quad \exists x \in X : \forall \varepsilon > 0, \exists N \in \mathbb{N} : \forall n \geq N, d(x_n, x) < \varepsilon,$$

or, if the value of the limit x is known:

$$x_n \to x \quad \text{if} \quad \forall \varepsilon > 0, \exists N \in \mathbb{N} : \forall n \geq N, d(x_n, x) < \varepsilon.$$

Definition 6.4.9 (Cauchy Sequences) A sequence (x_n) in a metric space (X, d) is called a Cauchy sequence if for every $\varepsilon > 0$, there exists an $N \in \mathbb{N}$ such that for every $m, n \geq N$ we have $d(x_n, x_m) < \varepsilon$.

In symbols:

$$(x_n) \text{ is Cauchy} \quad \text{if} \quad \forall \varepsilon > 0, \exists N \in \mathbb{N} : \forall m, n \geq N, d(x_m, x_n) < \varepsilon.$$

Remark 6.4.10 Notice that we obtained Definitions 6.4.8 and 6.4.9 by simply changing the $|x - y|$ in Definitions 5.2.4 and 5.8.1 for real numbers to $d(x, y)$. However, not all concepts and definitions from the previous chapter on real sequences carries through since the set X may not be an ordered field like $\mathbb{R}$. Therefore concepts that require ordering (such as monotone sequences and sandwich lemma) or the field structure (like algebra of limits) do not generalise to a general metric space since the arbitrary set X may lack these structures.

We have seen that in the real Euclidean space $\mathbb{R}^n$ and complex space $\mathbb{C}$ Cauchy sequences are equivalent to convergent sequences. However, in general metric spaces (X, d), this is not true. It is trivially true that any convergent sequence is Cauchy, but the converse may not be true. For example, if we consider the metric space of rational numbers $(\mathbb{Q}, |\cdot|)$ with the modulus metric, there are many Cauchy sequences which do not converge to an element in $\mathbb{Q}$. One such sequence is the sequence of decimal truncations of $\sqrt{2}$, which does not have a limit in $\mathbb{Q}$.

Therefore, in a general metric space, being Cauchy and convergent are not equivalent traits. We have a special name for metric spaces in which Cauchy sequences converge, namely:

Definition 6.4.11 (Complete Spaces) Let (X, d) be a metric space. This space is called a complete metric space if any Cauchy sequence (x_n) in X converges to an element in X.

We shall see later how this is important in the construction of real numbers in Exercise 6.27.

Exercises

6.1 We want to find the limit superior of the sequence (a_n) defined as $a_n = |\sin(n)|^{\frac{1}{n}}$.
(a) Show that $|\sin(x)| > \frac{1}{2}$ for $x \in \left(\frac{\pi}{6} + 2\pi m, \frac{5\pi}{6} + 2\pi m\right)$ for any $m \in \mathbb{N}$.
(b) Hence, show that there is a subsequence (a_{k_n}) of (a_n) for which $\left(\frac{1}{2}\right)^{\frac{1}{k_n}} < a_{k_n} \leq 1$.
(c) Find the limit of this subsequence and deduce that $\limsup_{n\to\infty} a_n = 1$.

6.2 $(*)$ Recall from Exercise 2.15 that the Fibbonaci sequence (f_n) is a sequence of natural numbers defined recursively via $f_1 = f_2 = 1$ and $f_n = f_{n-1} + f_{n-2}$ for all $n \geq 3$. We want to derive the Binet's formula that gives us the n-th Fibonacci number. This formula is named in honour of Jacques Binet (1786–1856). Let $P : \mathbb{R} \to \mathbb{R}$ be the polynomial $P(x) = x^2 - x - 1$.
(a) Show that the roots of this polynomial are both irrational.
(b) Let the bigger and smaller root of P be called φ and ψ respectively. Show that $\psi = 1 - \varphi = -\frac{1}{\varphi}$.
(c) Show by induction that for all $n \geq 2$ we have $\psi^{n-1} = f_n - \varphi f_{n-1}$.
(d) Hence, show that $f_n = \sum_{j=0}^{n-1} \varphi^{n-1-j}\psi^j$.
(e) Deduce Binet's formula for f_n:

$$f_n = \frac{\varphi^n - \psi^n}{\sqrt{5}} = \frac{\varphi^n - (1-\varphi)^n}{\sqrt{5}}.$$

(f) If (a_n) is a real sequence defined as $a_n = \frac{\varphi^n}{\sqrt{5}}$, show that $(f_n) \sim (a_n)$.

The irrational constant φ is called the golden ratio. The reason for this naming comes from the interpretation that this ratio is aesthetically pleasing. As a result, many work of art, design, and architecture features such proportions. Some notable examples are the painting The Sacrament of the Last Supper by Salvador Dalí, the flag of Togo by Paul Ahyi, and the designs by architect Le Corbusier.

6.3 $(*)$ Let (f_n) be the Fibonacci sequence. Define a new sequence (r_n) where $r_n = \frac{f_{n+1}}{f_n}$ is the ratio of two consecutive terms in the Fibonacci sequence.
(a) Show that the sequence (f_n) sequence is increasing and hence $r_n > 1$ for all $n \geq 2$.
(b) Prove that for all $n \geq 3$ we have $|r_n - r_{n-1}| = \frac{|r_{n-1} - r_{n-2}|}{r_{n-1}r_{n-2}} < \frac{1}{2}|r_{n-1} - r_{n-2}|$.
(c) Using induction, show that $|r_n - r_{n-1}| < \frac{1}{2^{n-2}}$ for all $n \geq 3$.
(d) Hence, show that the sequence (r_n) is Cauchy and converges to some real number.
(e) Show that $\lim_{n\to\infty} r_n = \varphi$, the golden ratio in Exercise 6.2.

6.4 The technique in Exercise 6.3 is a common technique employed in numerical analysis when we want to determine whether a problem has a solution. We first define a contractive sequence as:

Definition 6.4.12 (Contractive Sequence) A real sequence (a_n) is called a contractive sequence if there exists a constant $0 < K < 1$ and an $N \in \mathbb{N}$ such that $|a_{n+2} - a_{n+1}| \leq K|a_{n+1} - a_n|$ for all $n \geq N$.

(a) Show that the sequence (a_n) defined as $a_n = \frac{6^n}{n!}$ is contractive.
(b) Show that any contractive sequence is Cauchy and thus converges in $\mathbb{R}$.
Now let us apply this to find a positive root to the polynomial $P(x) = x^2 + 2x - 1$. In other words, we want to solve the equation $x^2 + 2x - 1 = 0$ for $x > 0$. But where are the sequences?
Let us first rewrite this equation as $x = \frac{1}{2+x}$. In order to solve for x on the LHS, we need to evaluate the RHS, but in order to evaluate the RHS, we need to know what x is, so this is a cyclical problem!
To get over this, from the equation, we define a sequence (x_n) recursively where $x_1 = c > 0$ is some constant and $x_{n+1} = \frac{1}{2+x_n}$ for all $n \in \mathbb{N}$. The number x_1 is our initial guess for the solution, so at each iteration, we are trying to improve our guess and hopefully we converge to a solution.
Note that x_n is a solution to the equation if and only if $x_{n+1} = x_n$. But in general, unless we are very lucky with the initial guess x_1, this is not true so we have to keep iterating until we get a stable numerical value. This stable value is the limit of the sequence (x_n), if it converges at all.
(c) Show that $x_n > 0$ for all $n \in \mathbb{N}$.
(d) Hence show that the sequence is contractive and converges to some $L \geq 0$.
(e) Conclude that the limit L solves the equation $x^2 + 2x - 1 = 0$ for $x \geq 0$ and deduce that $L > 0$.
For parts (f) and (g), the readers are encouraged to write a computer program to help with the computations.
(f) By starting with an initial guess $x_1 = 1$, approximate a solution to $P(x) = 0$ by computing several terms of the sequence (x_n) until it stabilises to 4 decimal places.
(g) Using the same idea as parts (c)–(f), now show that there is a solution to $P(x) = x^3 - 4x + 1 = 0$ in $[0, 1]$ with a starting guess of $x_1 = 1$. Compute the approximate solution correct to 4 decimal places.

6.5 (∗) Let (a_n) and (b_n) be two real sequences defined recursively with $0 < b_1 < a_1$ and:

$$a_{n+1} = \frac{2}{\frac{1}{a_n} + \frac{1}{b_n}} \quad \text{and} \quad b_{n+1} = \sqrt{a_{n+1} b_n}. \tag{6.2}$$

(a) Show that this recursive operation is well defined, namely $a_n, b_n > 0$ for all $n \in \mathbb{N}$.

(b) Show inductively that for all $n \in \mathbb{N}$ we have $b_n < b_{n+1} < a_{n+1} < a_n$.
(c) Hence, conclude that there are positive real numbers $a, b > 0$ such that $\lim_{n\to\infty} a_n = a$ and $\lim_{n\to\infty} b_n = b$ with $b \leq a$.
(d) Prove that for any $n \in \mathbb{N}$ we have $|a_{n+1} - b_{n+1}| < \frac{1}{2}|a_n - b_n|$.
Hence, show that $|a_{n+1} - b_{n+1}| < \frac{1}{2^n}|a_1 - b_1|$.
(e) Conclude that the limits of the two sequences are equal, namely $a = b$.

Starting with $a_1 = 2\sqrt{3}$ and $b_1 = 3$, this is the algorithm used by Archimedes to approximate the value of π. For $n \in \mathbb{N}$, Archimedes defined the angles $\theta_n = \frac{60°}{2^n}$. Then, the quantities $a_n = 3 \cdot 2^n \cdot \tan(\theta_n)$ and $b_n = 3 \cdot 2^n \cdot \sin(\theta_n)$ denote the semiperimeters (half of the perimeter) of the circumscribed and inscribed polygons of $3 \cdot 2^n$ sides respectively.
(f) Check that these quantities satisfy the recursive formulas in (6.2).

6.6 Another application of convergence of sequences is called Newton-Raphson root-finding method. This method, named after Isaac Newton and Joseph Raphson (c. 1668–1715), allows us to build a machine for finding a root of some function via sequences starting with an initial guess.

Suppose that $f : \mathbb{R} \to \mathbb{R}$ is a differentiable function (which will be discussed later in Chap. 13). The Newton-Raphson sequence for this function is given by (x_n) defined recursively as $x_{n+1} = x_n - \frac{f(x_n)}{f'(x_n)}$ where f' is the derivative of the function f (which geometrically denotes the slope of the tangent to the graph). We need to equip this with an initial guess $x_1 = c$. The resulting real sequence may or may not converge; there is no reason why it should converge! However, if it does, the limit $\lim_{n\to\infty} x_n$ would be a zero of the function f. See Fig. 6.6 for an intuitive demonstration.

Now let us look at an example. Suppose that $f : \mathbb{R} \to \mathbb{R}$ is defined as $f(x) = x^2 - 2$ so that $f'(x) = 2x$. Algebraically, we know that the roots of f are $\pm\sqrt{2}$. The Newton-Raphson sequence is then given by:

$$x_{n+1} = x_n - \frac{f(x_n)}{f'(x_n)} = x_n - \frac{x_n^2 - 2}{2x_n} = \frac{1}{2}\left(x_n + \frac{2}{x_n}\right).$$

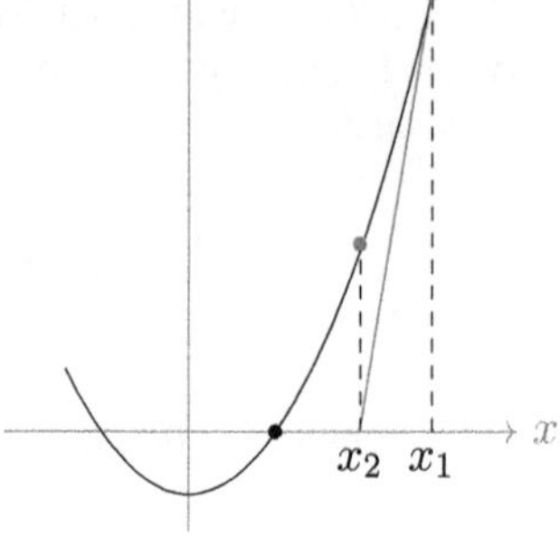

Fig. 6.6 The graph of f and the tangent line to the graph at the initial guess $x = x_1$ with slope $f'(x_1)$. This tangent line crosses the x-axis at $x = x_2$. We repeat this construction with the hope that the sequence of points (x_n) converges to the root of f denoted by the black dot

(a) Suppose that $x_1 = 2$. By induction, show that $x_n > 0$ for all $n \in \mathbb{N}$.
(b) Prove that $x_n^2 - 2 \geq 0$ for all $n \in \mathbb{N}$.
(c) Show that the sequence (x_n) is decreasing.
(d) Conclude that this sequence converges to some positive real number L.
(e) Find the limit of this sequence.
(f) Now suppose that we start with an initial guess $x_1 = 1$. Would the new sequence (x_n) converge? If it does, would it have the same limit L as above?
(g) How about if we begin with the initial guess $x_1 = -1$?

Thus, we can find a rational approximation for the value of $\sqrt{2}$ using this algorithm up to any desired degree of accuracy starting from a suitable initial point.

(h) Using the algorithm above and a computer program, find an approximation of $\sqrt{2}$ which is correct to four decimal places.
(i) Explain how we can approximate $\sqrt{k}$ for any integer $k > 2$ using this method.
(j) Using the algorithm above and a computer program, find approximations of $\sqrt{3}$, $\sqrt{5}$, and $\sqrt{7}$ which are correct to four decimal places.
(k) Now suppose that $f : \mathbb{R} \to \mathbb{R}$ is a function given by $f(x) = \sqrt[3]{x}$. Clearly there is a root of this function at $x = 0$. We can compute the derivative $f'(x) = \frac{1}{3\sqrt[3]{x^2}}$ at $x \neq 0$. Show that if we start with a non-zero guess, the Newton-Raphson sequence cannot converge to the root.

Therefore, when we are carrying out the Newton-Raphson method, we could either converge to the desired limit, converge to a different limit, or diverge. Therefore, one needs to think of a robust way of carrying out this algorithm or a good initial guess. Sometimes we may not even get the desired outcome and analysis is used to explain this.

Many computer programs and calculators use this kind of algorithm to solve our mathematical problems (which you have coded in parts (h) and (j)). Of course, some algorithms can be improved and therefore it is very important for us to understand what is going on in the background! Hence, this is one of the reasons why we study analysis.

6.7 (∗) Prove that Definitions 6.2.1 and 6.2.2 are equivalent.

6.8 (∗) Find the limit points of the following subsets of $\mathbb{R}$.
(a) $A = \mathbb{Z}$.
(b) $B = (-10, 0) \cup \{7\}$.
(c) $C = (-1, 0) \cup (0, 1)$.
(d) $D = \bigcup_{n=1}^{\infty}(2n, 2n + 1)$.
(e) $E = \mathbb{Q}$.
(f) $F = \bar{\mathbb{Q}}$.

6.9 (∗) Let $X \subseteq \mathbb{R}$ be any subset of the real numbers.
(a) If $x_0 \in X'$, prove that for every $\varepsilon > 0$, the open ball $B_\varepsilon(x_0)$ contains infinitely many points of X.
(b) Let $X \subseteq \mathbb{R}$ be a finite subset of $\mathbb{R}$. Prove that $X' = \varnothing$.

6.10 (◇) Let $X \subseteq \mathbb{R}$. Using Exercise 6.9(a), show that for any $x_0 \in X'$ there are uncountably many sequences $(x_n) \subseteq X \setminus \{x_0\}$ such that $\lim_{n\to\infty} x_n = x_0$.

6.11 (◇) Let $X \subseteq \mathbb{R}$. Prove that the set of isolated points of X is at most countable.

6.12 (∗) Let $X \subseteq \mathbb{R}$ and X' be the set of its limit points.

(a) Prove that $X'' = (X')' \subseteq X'$.

(b) Hence, prove that the set X' is closed.

(c) Let $A, B \subseteq \mathbb{R}$. Prove that $(A \cup B)' = A' \cup B'$.

(d) Define $Y = X \cup X'$. Prove that Y is closed.

The set Y in part (d) is called the closure of the set X, denoted as $\text{cl}(X)$ or $\bar{X}$. The closure of the set X is the smallest closed set that contains X. We can rewrite it as the union of disjoint sets $\text{cl}(X) = X \cup (X' \setminus X)$. In this form, we can see that the points $X' \setminus X$ is the minimal amount of points required to be added to the set X in order to turn it into a closed set.

6.13 (◇) We define:

Definition 6.4.13 (Sequentially Compact) A subset $X \subseteq \mathbb{R}$ is called sequentially compact in $\mathbb{R}$ if any sequence $(a_n) \subseteq X$ for all $n \in \mathbb{N}$ has a convergent subsequence that converges to an element in X.

(a) Prove that a set $X \subseteq \mathbb{R}$ is sequentially compact if and only if it is closed and bounded.

(b) Hence, prove $X \subseteq \mathbb{R}$ is compact if and only if it is sequentially compact.

The above correspondence is not necessarily true in other metric spaces.

6.14 (∗) Let (z_n) be a complex sequence.

(a) Prove that if (z_n) is convergent, then $(|z_n|)$ is convergent as well.

(b) Does the converse hold?

(c) Prove that $z_n \to 0$ if and only if $|z_n| \to 0$.

6.15 (∗) Determine whether the following complex sequences (z_n) converge and justify your answers. If they converge, find their limits.

(a) $z_n = \frac{1+i}{n}$.

(b) $z_n = (1 - i)^n$.

(c) $z_n = \frac{n}{(1+i)^n}$.

(d) $z_n = \frac{(a+ib)^n}{n!}$ for some real numbers $a, b \in \mathbb{R}$.

(e) $z_n = (a + ib)^n$ where $a^2 + b^2 > 1$.

(f) $z_n = (a + ib)^n$ where $0 < a^2 + b^2 < 1$.

6.16 Using the ε-N proof, show that if a complex sequence (z_n) converges to some $L \in \mathbb{C}$, then its conjugate sequence $(\bar{z}_n)$ converges to $\bar{L}$.

6.17 (◇) Prove that all the functions $d : \mathbb{R}^n \times \mathbb{R}^n \to \mathbb{R}$ in Example 6.4.4 are metrics on $\mathbb{R}^n$.

6.18 (◇) Recall the set of bounded real functions from the interval $[a, b]$, which we denote as $B([a, b])$.

(a) Show that $B([a, b])$ is a real-vector space.

(b) Show that the function $||\cdot|| : B([a, b]) \to \mathbb{R}$ defined as $||f|| = \sup_{x\in[a,b]} |f(x)|$ is a norm.

(c) Show that the function $d_\infty : B([a,b]) \times B([a,b]) \to \mathbb{R}$ defined as $d_\infty(f,g) = ||f-g|| = \sup_{x\in[a,b]} |f(x)-g(x)|$ is a well-defined function.

6.19 ($\diamond$) Let $\mathbb{N}$ be the set of natural numbers and define a function $d : \mathbb{N}\times\mathbb{N} \to \mathbb{R}$ as $d(m,n) = \left|\frac{1}{n} - \frac{1}{m}\right|$.

(a) Show that d is a metric on the set $\mathbb{N}$.

(b) If $a_n \to L \in \mathbb{N}$ with respect to the metric above, show that $|a_n| \leq M$ for some constant $M > 0$.

(c) Hence, show that any convergent sequence in this metric space is eventually constant.

(d) Show that the sequence (a_n) where $a_n = n$ is Cauchy in this space.

(e) Hence, deduce that this space is not complete.

6.20 Let $X = \{(a,b,c) : a,b,c \in \{0,1\}\} \subseteq \mathbb{Z}^3$ be the set of binary strings of length 3 and $d : X \times X \to \mathbb{R}$ be defined as $d(\mathbf{x},\mathbf{y}) =$ number of differing components for $\mathbf{x}$ and $\mathbf{y}$. For example, if $\mathbf{x} = (0,0,1)$ and $\mathbf{y} = (0,1,0)$, then $d(\mathbf{x},\mathbf{y}) = 2$ since they differ in the second and third components.

Show that (X,d) is a metric space. This metric can also be generalised to the set of longer binary strings.

This metric is called the Hamming distance, named after Richard Hamming (1915–1998) who pioneered in the study of error-correction codes. It is used in computer science and coding.

6.21 ($\diamond$) Let $\mathbb{Z}\setminus\{0\}$ be the set of non-zero integers and $\mathbb{Z}_{\geq 0}$ be the set of non-negative integers. Let μ be the 2-valuation function $\mu : \mathbb{Z}\setminus\{0\} \to \mathbb{Z}_{\geq 0}$ defined as:

$$\mu(x) = \max\{k \in \mathbb{Z}_{\geq 0} : 2^k \text{ divides } x\}.$$

For example, $\mu(56) = 3$ since $56 = 2^3 \cdot 7$ and $\mu(-45) = 0$ since $-45 = -3^2 \cdot 5$.

(a) Prove that $\mu(x+y) \geq \min\{\mu(x), \mu(y)\}$ and $\mu(xy) = \mu(x) + \mu(y)$ for any $x, y \in \mathbb{Z}\setminus\{0\}$.

(b) Define the function $d : \mathbb{Z}\times\mathbb{Z} \to \mathbb{R}$ as:

$$d(m,n) = \begin{cases} 0 & \text{if } m = n, \\ 2^{-\mu(m-n)} & \text{if } m \neq n. \end{cases}$$

Show that d is a metric on $\mathbb{Z}$.

(c) Show that the space $(\mathbb{Z}, d)$ is bounded.

(d) Find all $n \in \mathbb{Z}$ for which $d(n,1) = 1$.

6.22 ($\diamond$) We now extend the valuation function μ in Exercise 6.21 to the rationals by defining $\nu : \mathbb{Q}\setminus\{0\} \to \mathbb{R}$ as $\nu(\frac{p}{q}) = \mu(p) - \mu(q)$.

(a) Show that this extension is well-defined, namely if $\frac{p}{q} = \frac{r}{s}$ as rational numbers, then $\nu(\frac{p}{q}) = \nu(\frac{r}{s})$.

(b) Prove that $\nu(x + y) \geq \min\{\nu(x), \nu(y)\}$ for any $x, y \in \mathbb{Q} \setminus \{0\}$.
(c) Define a function $d : \mathbb{Q} \times \mathbb{Q} \to \mathbb{R}$ as:

$$d(r, s) = \begin{cases} 0 & \text{if } r = s, \\ 2^{-\nu(r-s)} & \text{if } r \neq s. \end{cases}$$

Show that d is a metric on $\mathbb{Q}$.

The metric d above is called the 2-adic metric on $\mathbb{Q}$. The number 2 can also be changed to any other prime number p to derive the p-adic metric. This metric is used in number theory and it allows us to create an alternative measure of nearness between rational numbers. In fact, it is used in the work of proving Fermat's last theorem, which eluded many mathematicians for hundreds of years, by Andrew Wiles (1953-).

6.23 Let $(\mathbb{Q}, d)$ be the set of rational numbers with the discrete metric d in Example 6.4.4(4). Show that:
(a) If (x_n) is a Cauchy sequence in $(\mathbb{Q}, d)$ then it is eventually constant.
(b) Hence, conclude that $(\mathbb{Q}, d)$ is a complete metric space.

6.24 Show that the spaces $\mathbb{C}$ and $\mathbb{R}^k$ for some $k \in \mathbb{N}$ with the usual complex and Euclidean norms are complete.

6.25 Why is the set $\mathbb{R} \setminus \{0\}$ with the usual Euclidean metric not complete?

6.26 ($\diamond$) The axioms of real number has the completeness axiom which says any bounded subset of the real numbers must have a supremum. In this question, we are going to state a condition that is equivalent to the completeness axiom on $\mathbb{R}$. We define:

Definition 6.4.14 (Nested Interval Property) The nested interval property says that if $\{I_n\}$ is any sequence of closed intervals of finite length $d_n > 0$ such that $I_{n+1} \subseteq I_n$ for all $n \in \mathbb{N}$ and $d_n \to 0$, then there is exactly one point $x \in \mathbb{R}$ such that $x \in I_n$ for all $n \in \mathbb{N}$.

Note: A closed interval $I = [a, b]$ has length $d = b - a$.
(a) Prove that the completeness axiom of the $\mathbb{R}$ implies the nested interval property.
(b) Conversely, prove that the nested interval property implies the completeness axiom.

6.27 ($\diamond$) In Chap. 3, we have seen that the real numbers can be constructed from $\mathbb{Q}$ using Dedekind cuts. In this question, we are going to construct the real numbers using Cauchy sequences in $\mathbb{Q}$ instead.

In fact, the resulting fields from these two constructions are the same (or in a more precise algebraic language, isomorphic). The proof of this requires extra knowledge from abstract algebra, so we are not going to do it here. Interested readers may consult any advanced abstract algebra textbooks for the proof.

We forget the real numbers that we have built so far and restart from the ordered field $\mathbb{Q}$. We state two definitions which are crucial for this construction:

Definition 6.4.15 (Convergent Rational Sequence) A sequence of rational numbers (a_n) converges to $L \in \mathbb{Q}$ if for every $\varepsilon \in \mathbb{Q}_+$ there exists an index $N \in \mathbb{N}$ such that for all $n \geq N$ we have $|a_n - L| < \varepsilon$. We write this as $\lim_{n\to\infty} a_n = L$.

Definition 6.4.16 (Cauchy Rational Sequence) A sequence of rational numbers (a_n) is called a Cauchy sequence if for every $\varepsilon \in \mathbb{Q}_+$ there exists an index $N \in \mathbb{N}$ such that for all $m, n \geq N$ we have $|a_m - a_n| < \varepsilon$.

Note that the definitions above and their ideas are the same as Definitions 5.2.4 and 5.8.3 except that the limit L and distance ε must be rational numbers (since we have forgotten about the real numbers and are now trying to rebuild it again). In the field $\mathbb{Q}$, convergent sequences are necessarily Cauchy, which is easy to prove using the definition above.
However, the converse is false! Indeed, the sequence of decimal truncations of $\sqrt{2}$ is Cauchy in $\mathbb{Q}$ but does not converge to any element in $\mathbb{Q}$. The following construction will extend the set of rational numbers by adding in a new set of numbers so that this converse becomes true. First, we prove two basic results:

(a) Prove that if the rational sequence (a_n) is Cauchy, then it must be bounded.
(b) Suppose that (a_n) and (b_n) are rational Cauchy sequences. Prove that the sequences (c_n) and (d_n) where $c_n = a_n + b_n$ and $d_n = a_n b_n$ are rational Cauchy sequences as well.

Let $\mathcal{C} = \{(a_n) : (a_n) \text{ is a rational Cauchy sequence}\}$ be the set of all Cauchy sequences in $\mathbb{Q}$. To identify sequences which behave in a similar way asymptotically, we define a relation $\sim$ on $\mathcal{C}$ such that $(a_n) \sim (b_n)$ iff $\lim_{n\to\infty}(a_n - b_n) = 0$.

(c) Show that the relation $\sim$ is an equivalence relation.
(d) Denote $\mathcal{C}/\!\sim\; = \{[(a_n)] : [(a_n)] \text{ is the equivalence class of } (a_n)\}$. Show that the addition $[(a_n)] \oplus [(b_n)] = [(a_n) + (b_n)]$ and multiplication $[(a_n)] \otimes [(b_n)] = [(a_n)(b_n)]$ operations are well defined in $\mathcal{C}/\!\sim$.
(e) Show that the additive and multiplicative identity in $\mathcal{C}/\!\sim$ are $[(0)]$ and $[(1)]$ respectively, where (0) and (1) are the constant sequences of 0 and 1.
Show also that the additive inverse of $[(a_n)] \in \mathcal{C}/\!\sim$ is $[(-a_n)]$.
(f) Let $[(a_n)] \neq [(0)]$. Show that there exists a $K \in \mathbb{Q}_+$ and an $N \in \mathbb{N}$ such that $|a_n| > K > 0$ for all $n \geq N$.
(g) From the sequence (a_n) in part (f), define a rational sequence (b_n) where $b_n = 0$ for $1 \leq n \leq N-1$ and $b_n = \frac{1}{a_n}$ for $n \geq N$. Prove that the sequence (b_n) is also Cauchy.
(h) Hence, show that $[(b_n)] \in \mathcal{C}/\!\sim$ is a multiplicative inverse of $[(a_n)]$.

(i) So we have defined the addition $\oplus$ and multiplication $\otimes$ operations along with their identities and inverses on $\mathcal{C}/\sim$. Prove the other field axioms in Definition 3.1.1, namely $\oplus$ and $\otimes$ are commutative and associative, and that $\otimes$ is distributive over $\oplus$.

Recall that $[(a_n)] = [(b_n)]$ if and only if $(a_n - b_n) \to 0$. We define an ordering $\prec$ on $\mathcal{C}/\sim$ with $[(a_n)] \prec [(b_n)]$ if and only if there exists a rational constant $K > 0$ and an index $N \in \mathbb{N}$ such that $b_n - a_n > K$ for all $n \geq N$. We call $[(a_n)]$ a positive element if $[(a_n)] \succ [(0)]$ and a negative element if $[(a_n)] \prec [(0)]$.

(j) Prove that $\prec$ is a strict total order on $\mathcal{C}/\sim$.

(k) Hence, prove that $\prec$ satisfies the ordered field axioms in Definition 3.3.1, namely the compatibility of $\prec$ with $\oplus$ and $\otimes$.

So $\mathcal{C}/\sim$ with the ordering $\prec$ and operations $\oplus$ and $\otimes$ forms an ordered field. In fact, it contains a copy of $\mathbb{Q}$ in the form of the equivalence classes of constant sequences $\{[(r)] : r \in \mathbb{Q}\}$. We call such elements the rational elements, denoted as $\mathcal{C}/\sim_{\mathbb{Q}}$. Note that the class $[(r)]$ consists of all the Cauchy sequences in $\mathbb{Q}$ that converge to $r \in \mathbb{Q}$. Indeed, if $(a_n) \in [(r)]$, then by definition we have $\lim_{n\to\infty}(a_n - r) = 0$ or equivalently $\lim_{n\to\infty} a_n = r$.

On the other hand, the elements of $\mathcal{C}/\sim$ which are not in $\mathcal{C}/\sim_{\mathbb{Q}}$ are exactly the classes of Cauchy sequences that do not converge in $\mathbb{Q}$. We call these the irrational elements.

Finally, we want to show that $\mathcal{C}/\sim$ satisfies the completeness axiom. Let $S \subseteq \mathcal{C}/\sim$ be a non-empty bounded set. We want to show that its least upper bound exists.

Pick a class of sequences $[(u)] \in \mathcal{C}/\sim_{\mathbb{Q}}$ which is an upper bound of S. Pick another class of sequence $[(l)] \in \mathcal{C}/\sim_{\mathbb{Q}}$ which is not an upper bound of S. Define the rational sequences (u_n) and (l_n) starting with $u_1 = u$ and $l_1 = l$ recursively as follows: for each $n \in \mathbb{N}$, define $m_n = \frac{u_n + l_n}{2} \in \mathbb{Q}$. Let $[(m_n)]$ be the class for the constant sequence (m_n) and assign:

$$\begin{cases} u_{n+1} = m_n \text{ and } l_{n+1} = l_n & \text{if } [(m_n)] \text{ is an upper bound of } S, \\ u_{n+1} = u_n \text{ and } l_{n+1} = m_n & \text{if } [(m_n)] \text{ is not an upper bound of } S. \end{cases}$$

This means for any fixed $N \in \mathbb{N}$, $[(u_N)]$ is an upper bound of S and $[(l_N)]$ is not an upper bound of S.

(l) Show that the rational sequence (u_n) is decreasing and the rational sequence (l_n) is increasing. Furthermore, show that both of the sequences are Cauchy.
Show also that $|u_n - l_n| \leq \frac{1}{2^n}|u - l|$ for all $n \in \mathbb{N}$.

(m) Show that $[(u_n)] = [(l_n)]$.

(n) Prove that the element in part (m) is an upper bound for the set S.

(o) By contradiction, show that the element in part (m) is the least upper bound for S.

Thus, since $\mathcal{C}/\sim$ equipped with $\prec$, $\oplus$, and $\otimes$ satisfies the field, ordering, and completeness axioms, it forms the real numbers as per Definition 3.6.13. Now that $\mathcal{C}/\sim$ is shown to be the real numbers, we have the converse that we wanted, namely: a Cauchy sequence in $\mathcal{C}/\sim$ converges.

This construction of the real numbers is called the Cauchy completion of $\mathbb{Q}$. The difference between this construction and the approach using Dedekind cut in Chap. 3 is the structure that was used. Here, the structure that is used is the metric on $\mathbb{Q}$ whereas the structure that was utilised in Dedekind cuts is the order on $\mathbb{Q}$.

The completion approach is more flexible as we can apply it to any other metric spaces or even on the same set $\mathbb{Q}$ endowed with other metrics. For example, if we used the p-adic metric on $\mathbb{Q}$ from Exercise 6.19, we would get a totally different number system which is called the p-adic number system $\mathbb{Q}_p$. In contrast, if we used the discrete metric on $\mathbb{Q}$ to start with, its completion is the same space because it is already a complete space as seen in Exercise 6.23!

6.28 ($\diamond$) Using Dedekind cuts, we have shown that the exponentiation operation can be defined, namely for any $a > 0$ and $r \in \mathbb{R}$, there is a number $b > 0$ such that $b = a^r$.

Now let us show a constructive approach to the definition by merely using the axioms of real numbers. For $n \in \mathbb{Z}$, we have no issues with the exponentiation a^n and their identities. Now suppose that $r = \frac{1}{p}$ for some $p \in \mathbb{N}$. We want to first define the exponentiation a^r.

(a) Show that there are non-negative real numbers $x_1 < y_1$ such that $x_1^p \leq a \leq y_1^p$.

For each $n \in \mathbb{N}$, define recursively the real numbers x_{n+1} and y_{n+1} as:

$$\begin{cases} x_{n+1} = \frac{x_n+y_n}{2} \text{ and } y_{n+1} = y_n & \text{if } \left(\frac{x_n+y_n}{2}\right)^p < a, \\ x_{n+1} = x_n \text{ and } y_{n+1} = \frac{x_n+y_n}{2} & \text{if } \left(\frac{x_n+y_n}{2}\right)^p \geq a. \end{cases}$$

(b) Prove that $x_n \leq y_n$ for all $n \in \mathbb{N}$.

(c) Show that the sequences (x_n) and (y_n) are increasing and decreasing respectively.

(d) Denote $I_n = [x_n, y_n]$ and define a sequence (d_n) where $d_n = y_n - x_n$. Show that $I_{n+1} \subseteq I_n$ for all $n \in \mathbb{N}$ and $\lim_{n\to\infty} d_n = 0$.

(e) Using Exercise 6.26, deduce that there is a unique point $b \in \mathbb{R}$ such that $b \in I_n$ for all $n \in \mathbb{N}$.

(f) Hence, show that $b^p = a$ so that $b = a^{\frac{1}{p}} = a^r$.

(g) Show that a $b \in \mathbb{R}_+$ that satisfies $b^p = a$ is unique.

Thus, we can define exponentials for base $a > 0$ with rational exponent $\frac{q}{p}$ where $q \in \mathbb{Z}$ and $p \in \mathbb{N}$ as:

$$a^{\frac{q}{p}} = (a^{\frac{1}{p}})^q = (a^q)^{\frac{1}{p}}.$$

The method above is called the bisection method due to the fact that at every step, we bisect (divide into two equal parts) the interval I_n to trap the solution in the limiting interval. This method is used in root-finding algorithms in numerical methods and the proof for Theorem 10.4.1 later.

6.29 (◇) Finally, let us extend the exponentiation to irrational exponents.
(a) Let $a \geq 1$. Prove that $|a^x - 1| \leq a^{|x|} - 1$ for all $x \in \mathbb{Q}$.
(b) Consider a rational sequence (x_n) that converges to $x \in \mathbb{R}$. Using part (a), prove that $\lim_{n\to\infty} a^{x_n}$ exists for any fixed $a > 1$.
(c) Suppose that (x_n) and (y_n) are any two rational sequences that converge to $x \in \mathbb{R}$. For a fixed $a > 1$, show that $\lim_{n\to\infty} a^{x_n} = \lim_{n\to\infty} a^{y_n}$.
(d) Deduce the same results as parts (b) and (c) for $0 < a < 1$.
Thus, we can uniquely define a^x for any irrational x as the limit $a^x = \lim_{n\to\infty} a^{x_n}$ for any rational sequence (x_n) that converges to x. This is well-defined as it is independent of the choice of sequence (x_n).

6.30 Using Exercise 6.29, we can extend the Bernoulli's inequality to irrational exponents in a different way than what we have seen in Exercise 4.15. Show that:
(a) $(1+x)^a \geq 1+ax$ where $x > -1$ for the case of irrational numbers $a > 1$.
(b) $(1+x)^a \leq 1+ax$ where $x > -1$ for the case of irrational numbers $0 < a < 1$.

7 Real Series

The petty cares, the minute anxieties, the infinite littles which go to make up the sum of human experience, like the invisible granules of powder, give the last and highest polish to a character.

— *William Matthews, poet*

In Chap. 4, we have defined the decimal representation for a real number $r \in \mathbb{R}$ as the formal infinite sum:

$$r \sim a_0 + \frac{a_1}{10} + \frac{a_2}{10^2} + \frac{a_3}{10^3} + \ldots = \sum_{j=0}^{\infty} \frac{a_j}{10^j} \quad \text{or} \quad a_0.a_1a_2a_3\ldots,$$

where $a_0 \in \mathbb{Z}$ and $a_j \in \{0, 1, 2\ldots, 9\}$ for all $j \in \mathbb{N}$ with no recurring 9s in the representation. This sum, which is an infinite sum, was stated formally (hence why the symbol $\sim$ was used).

Algebraically, the problem with the summation above is the amount of summands involved. Summing up two real numbers is well-defined so, by induction, we can sum up n real numbers for any $n \in \mathbb{N}$, no matter how large n is. But how do we sum up infinitely many quantities? Of course, this is a physically impossible task since this process will never end. This is an example of potential infinity coined by Aristotle as discussed in Remark 2.3.8. Purportedly, Aristotle remarked:

> As an example of a potentially infinite series in respect to increase, one number can always be added after another in the series that starts 1,2,3,... but the process of adding more and more numbers cannot be exhausted or completed.

Therefore we have to think of ways to make sense of this sum other than actually attempting to sum up all the numbers because this approach is futile.

S. Johar, *The Big Book of Real Analysis*,
https://doi.org/10.1007/978-3-031-30832-1_7

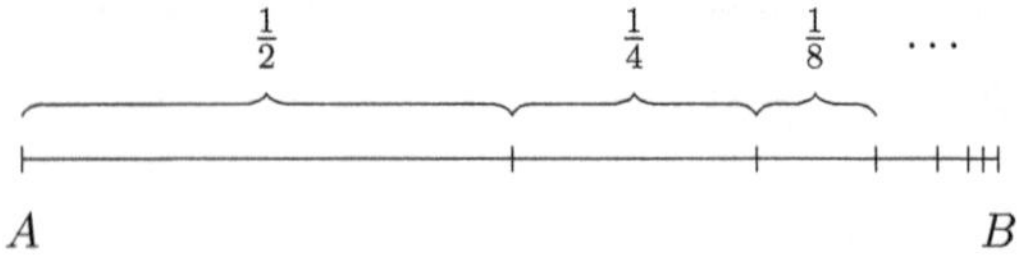

Fig. 7.1 Zeno's dichotomy paradox

This philosophical problem has been discussed since the time of Zeno of Elea (c. 490B.C.-430B.C.). The dichotomy paradox, as it is usually referred to, is proposed by Zeno as follows: if we are at point A and we want to get to the point B, we must first travel halfway across the line segment, then halfway from the point where we previously stopped, and so on and so forth as depicted in Fig. 7.1. We need to keep repeating this until we reach B. But Zeno insists that we will never actually reach the the point B! This is because with each motion, we only cover half the distance of the previous steps, so we need an infinite amount of motion to get to B. Therefore, Zeno concluded that any motion is impossible. Aristotle summarised this as:

> That which is in locomotion must arrive at the half-way stage before it arrives at the goal.

The argument by Zeno also implies that the infinite sum $\frac{1}{2}+\frac{1}{4}+\frac{1}{8}+\ldots$ cannot be finite.

This paradox comes about from the assumption that an infinite number of things cannot be performed in a finite amount of time. However, there are many ways to resolve this paradox. In particular, as argued by Aristotle and Archimedes in their own works, this is done by asserting that an infinite amount of quantities can add up to a finite number. But how?

7.1 Partial Sums

In Example 5.2.8, we have made sense of an infinite sum $\sum_{j=0}^{\infty}\frac{a_j}{10^j}$ as the limit of the rational sequence (r_n) where for each $n \in \mathbb{N}$ we defined $r_n = a_0.a_1a_2\ldots a_n = a_0 + \frac{a_1}{10} + \frac{a_2}{10^2} + \ldots + \frac{a_n}{10^n}$ which is the sum of only the first $n+1$ terms in the infinite sum. This is a perfectly reasonable definition as finite sums of real numbers, which are the r_n, can always be defined. With this definition, we have shown that this sequence (r_n) converges to r and hence:

$$r = \lim_{n\to\infty} r_n = \lim_{n\to\infty} a_0.a_1a_2\ldots a_n = a_0.a_1a_2a_3\ldots = \sum_{j=0}^{\infty}\frac{a_j}{10^j}. \tag{7.1}$$

Therefore, we shall use this idea of looking at the sequence of finite sums to generalise this to other infinite sums. Let us first define what a general infinite sum or series is.

Definition 7.1.1 (Series or Infinite Sum) Let (a_n) be a real sequence. A series for this sequence is the formal infinite sum $\sum_{j=1}^{\infty} a_j$.

Remark 7.1.2 The index in the notation for the series may begin at a different integer. In (7.1), the series began with the index 0.

For the time being, we could not attach a value to the infinite sum in Definition 7.1.1 because we do not know how to evaluate it. Furthermore, since there are infinitely many numbers, as Zeno suggested, would the sum blow up to infinity? From this infinite sum, let us define a different object that makes perfect sense algebraically, which are the partial sums of the series. As we have noted earlier, by the ring and field axioms and induction, we can always add up finitely many terms of real numbers, no matter how large the number of terms are. So we can define:

Definition 7.1.3 (Partial Sums) Let (a_n) be a real sequence and $\sum_{j=1}^{\infty} a_j$ be its series. The n-th partial sum of this series is a real number defined as the sum of the first n terms of the sequence, namely $s_n = \sum_{j=1}^{n} a_j$.

The list of partial sums form another real sequence (s_n) which may or may not have a limit. Thus, we can make sense of the value of the infinite sum in Definition 7.1.1 as the limit of the sequence of its partial sums (s_n), if the limit exists.

7.2 Convergent Series

From the discussion in the previous section, we define:

Definition 7.2.1 (Convergent Series) A real series $\sum_{j=1}^{\infty} a_j$ is called a convergent series if the sequence of its partial sums (s_n) converges. We then define the value of the series to be the limit of its partial sums. In other words, if $\lim_{n\to\infty} s_n = s$, we assign the value s to the series:

$$\sum_{j=1}^{\infty} a_j = \lim_{n\to\infty} \sum_{j=1}^{n} a_n = \lim_{n\to\infty} s_n = s.$$

Otherwise, if the sequence of the partial sums (s_n) diverges, the series $\sum_{n=1}^{\infty} a_n$ is called a divergent series.

Example 7.2.2 Let us look at some examples of a real series:

1. Recall the decimal representation of a real number $r \sim \sum_{j=0}^{\infty} \frac{a_j}{10^j}$ by a formal series. We have discussed how to make sense of this via sequences of

rational approximations in Example 5.2.8. In the construction of this decimal representation, we have obtained the rational approximation $r_n = \sum_{j=0}^{n} \frac{a_j}{10^j}$ which is the n-th partial sum of the series. This partial sum satisfies the bound:

$$0 \leq r - r_n < \frac{1}{10^n} \quad \Rightarrow \quad |r - r_n| < \frac{1}{10^n},$$

for all $n \in \mathbb{N}$. By sandwich lemma, the sequence of partial sums converges and its limit is r. So the formal series $\sum_{j=0}^{\infty} \frac{a_j}{10^j}$ is a convergent series and has the value of r in this limiting sense. We can then replace all the $\sim$ symbols in the decimal representation with $=$ where the equality $r = \sum_{j=0}^{\infty} \frac{a_j}{10^j}$ is understood to be in the limiting sense.

2. Suppose that we have an arithmetic progression or sequence of real numbers (a_n) in which we have a common difference of terms $a_{n+1} - a_n = d$. Thus, the n-th term of the sequence is given by $a_n = a_1 + (n-1)d$. Consider the series $\sum_{j=1}^{\infty} a_j$. From Exercise 3.13, the n-th partial sum of this series is given by:

$$s_n = \sum_{j=1}^{n} a_j = na_1 + \frac{n(n-1)d}{2} = \frac{n}{2}(2a_1 + (n-1)d).$$

We can show, by splitting into cases, that the sum blows up to $\pm\infty$ depending on the values of a_1 and d.
   (a) If $d = 0$, then $s_n = na_1$ which implies that the sequence (s_n) diverges to $\pm\infty$ depending on the sign of a_1.
   (b) If $d > 0$, for large enough n, the term $2a_1 + (n-1)d$ would be positive and thus the sequence of partial sums (s_n) diverges to ∞.
   (c) If $d < 0$, for large enough n, the term $2a_1 + (n-1)d$ would be negative and thus the sequence of partial sums (s_n) diverges to $-\infty$.

   In all of the cases above, the series $\sum_{j=1}^{\infty} a_j$ is divergent as the limit of the sequence of partial sums (s_n) does not exist.
3. Suppose that we have a geometric progression or sequence of real numbers (a_n) in which we have a common ratio of terms $\frac{a_{n+1}}{a_n} = r \neq 0$. The n-th term of the sequence is given by $a_n = r^{n-1}a_1$. Consider the series $\sum_{j=1}^{\infty} a_n$. To study this series, we have to consider the following cases:
   (a) If $r = 1$, then the n-th partial sum of this series is $s_n = \sum_{j=1}^{n} a_j = na_1$ which diverges to $\pm\infty$, depending on the sign on a_1.
   (b) If $r \neq 1$, from Exercise 3.13, the n-th partial sum of this series is given by:

$$s_n = \sum_{j=1}^{n} a_j = \frac{a_1(1 - r^n)}{1 - r}.$$

Using algebra of limits, we can show that s_n converges if $|r| < 1$ and diverges if $|r| > 1$ or $r = -1$. Indeed:

$$\lim_{n\to\infty} s_n = \frac{a_1}{1-r} \lim_{n\to\infty} (1 - r^{n-1}).$$

As seen in Exercise 5.15, we have:

i. If $|r| < 1$, then $\lim_{n\to\infty} r^n = 0$.
ii. If $r > 1$, then $\lim_{n\to\infty} r^n = \infty$.
iii. If $r \leq -1$, then $\lim_{n\to\infty} r^n$ does not exist.

Therefore, the geometric series is convergent if and only if $|r| < 1$. If the series is convergent, its value is:

$$\sum_{j=1}^{\infty} a_j = \lim_{n\to\infty} s_n = \frac{a_1}{1-r}.$$

Using this fact, we can finally prove that every periodic decimal representation correspond to a rational number as we have claimed in Chap. 4. This is Exercise 7.5 for the readers.

4. Consider the series $\sum_{j=1}^{\infty} \frac{1}{j(j+1)}$. We can rewrite the terms in the series as $\frac{1}{j(j+1)} = \frac{1}{j} - \frac{1}{j+1}$ for every $j \in \mathbb{N}$, so the series can be rewritten as:

$$\sum_{j=1}^{\infty} \frac{1}{j(j+1)} = \sum_{j=1}^{\infty} \left(\frac{1}{j} - \frac{1}{j+1}\right).$$

If we consider the sequence of partial sums (s_n) of this series, we note that many of the terms cancel each other since:

$$s_n = \sum_{j=1}^{n} \left(\frac{1}{j} - \frac{1}{j+1}\right) = 1 - \frac{1}{2} + \frac{1}{2} - \frac{1}{3} + \ldots + \frac{1}{n} - \frac{1}{n+1} = 1 - \frac{1}{1+n}.$$

Therefore, if we apply the algebra of limits to the partial sums, we get:

$$\lim_{n\to\infty} s_n = 1 - \lim_{n\to\infty} \frac{1}{n+1} = 1,$$

and hence the sequence of partial sums converge. Thus, we conclude that $\sum_{j=1}^{\infty} \frac{1}{j(j+1)} = \lim_{n\to\infty} s_n = 1$.

In general, a real series that can be written in the form $\sum_{j=1}^{\infty} (f(j) - f(j+1))$ for some function $f : \mathbb{N} \to \mathbb{R}$ is called a telescoping series. Due to cancellations, the partial sums of this series will then be of the form $s_n = f(1) - f(n+1)$ for all $n \in \mathbb{N}$, which is easier to analyse.

5. Consider the sequence (a_n) where $a_n = (-1)^n$ for $n \in \mathbb{N}$. We can formally construct a series $\sum_{j=1}^{\infty} a_j$ but what is its value? One can naïvely write down the terms and put the brackets strategically to cancel out some terms as such:

$$\sum_{j=1}^{\infty} a_j = (-1+1)+(-1+1)+(-1+1)+\ldots = 0+0+0+\ldots = 0,$$

or a different kind of bracketing as:

$$\sum_{j=1}^{\infty} a_j = -1+(1-1)+(1-1)+(1-1)+\ldots = -1+0+0+0+\ldots = -1.$$

Equating them, we get $0 = -1$. What is going on here? This series is called the Grandi's series after Guido Grandi (1671–1742) and caused a lot of discussion (and heated debates) on what it means to add infinitely many real numbers.
If we refer to Definition 7.2.1, both of the interpretations of the Grandi's series above are not correct. Indeed, we defined the value of a series as the limit of the sequence of its partial sums, so let us now investigate the partial sums of this series instead.
If we write down the partial sums (s_n), we see that the sequence given is by $s_n = -1$ for odd n and $s_n = 0$ for even n, so the sequence alternates between these two values. Since we can find two different subsequences of (s_n) that converge to different values, the sequence of partial sums do not converge and hence the series corresponding to it, namely $\sum_{j=1}^{\infty}(-1)^j$, is a divergent series.

Remark 7.2.3 From Example 7.2.2(5) above, we give some warning here.

1. For finite sums, manipulating or rearranging the terms is perfectly permissible. This is due to the associativity and commutativity of addition in the ring and field axioms from Definitions 2.5.1 and 3.1.1. Using induction, these properties can be shown to work for any finite (no matter how large) number of terms.
2. However, we are not able to manipulate the terms in an infinite sum by rearranging, grouping them in a certain way by adding brackets, multiplying with a scalar, or even combining two series before we know that the series converges! This is because there are infinitely many terms in the sum and the value of a series is defined in terms of a limit. Thus, manipulating the terms in a certain way may change this limit.
3. We shall see in Propositions 7.2.8 and 7.2.9 that for convergent series, it is permissible to combine them, scale them, and remove or add finitely many terms. We shall also see in Chap. 8 that rearranging and grouping the terms in a series is permissible only in some cases.
4. Therefore, if we have a series, the most important thing to investigate first is whether it converges because many series manipulations require this as a

prerequisite! One might work with a divergent series and yield contradictory results, as with the Grandi's series in Example 7.2.2(5) using which we "showed" that $0 = -1$. Niels Henrik Abel (1802–1829) wrote in a letter to his teacher Bernt Michael Holmboe (1795–1850):

> Divergent series are, in general, something terrible and it is a shame to base any proof on them. We can prove anything by using them and they have caused so much misery and created so many paradoxes.

Remark 7.2.4 If we were to look at a series made up of complex numbers in place of real numbers, we can split this series into the real and imaginary parts. Indeed, if $\sum_{j=1}^{\infty} z_j$ is a complex series, we can write each term in the series as $z_j = a_j + ib_j$ where $a_j, b_j \in \mathbb{R}$. The n-th partial sum of this series is then given by:

$$s_n = \sum_{j=1}^{n} z_j = \sum_{j=1}^{n} (a_j + ib_j) = \sum_{j=1}^{n} a_j + i \sum_{j=1}^{n} b_j.$$

By definition of convergent series, the complex series converges if the sequence of partial sums (s_n) converge. Since (s_n) a complex sequence, by Theorem 6.3.6, it is convergent if and only if both the real and imaginary parts converge. In other words:

$$\sum_{j=1}^{\infty} z_j = \sum_{j=1}^{\infty} (a_j + ib_j) \text{ converges} \quad \Leftrightarrow \quad \sum_{j=1}^{\infty} a_j \text{ and } \sum_{j=1}^{\infty} b_j \text{ converge.}$$

Therefore, studying real series is useful for us to form a basis for the concepts in complex series.

Using the definition of convergent series, in order to study real series, we need to study the real sequence of partial sums. We have seen many theories and results on real sequences in Chap. 5 which are sufficient to deduce many of the results on series. In this chapter, we shall see more specialised results for series that we can derive from the theory of sequences.

First, an obvious fact for a convergent real series $\sum_{j=1}^{\infty} a_j$ is the sequence of terms in the series (a_j) must converge to 0 as $j \to \infty$.

Proposition 7.2.5 *If the real series* $\sum_{j=1}^{\infty} a_j$ *converges, then* $\lim_{j\to\infty} a_j = 0$.

Proof Since the series converges, by definition, the sequence of partial sums $s_n = \sum_{j=1}^{n} a_j$ converges, say $s_n \to s \in \mathbb{R}$. Note that $a_n = s_n - s_{n-1}$. By taking the limit as $n \to \infty$ on both sides and applying the algebra of limits, we have:

$$\lim_{n\to\infty} a_n = \lim_{n\to\infty} (s_n - s_{n-1}) = \lim_{n\to\infty} s_n - \lim_{n\to\infty} s_{n-1} = s - s = 0,$$

and we are done. □

This provides us a quick first test to determine whether a series diverges by *modus ponens*, namely: if the terms in the series do not converge to 0 (either the sequence of terms converges to a non-zero number or diverges), we can immediately say that the series diverges.

Example 7.2.6 For example, recall Grandi's series $\sum_{j=1}^{\infty}(-1)^j$ in Example 7.2.2(5). The terms of the series form a real sequence (a_n) where $a_n = (-1)^n$. We have shown that this sequence does not converge to 0 (or even converge at all) in Example 5.2.7(1). Thus, Proposition 7.2.5 says this series is not convergent.

However, a word of caution: the converse of Proposition 7.2.5 may not be true! Not all series for which the terms $a_n \to 0$ is convergent! Here is a very important counterexample.

Example 7.2.7 We define the harmonic series as the infinite sum of the reciprocals of natural numbers. It can be written formally as:

$$\sum_{j=1}^{\infty} \frac{1}{j} = 1 + \frac{1}{2} + \frac{1}{3} + \frac{1}{4} + \dots$$

The name harmonic series is derived from the concept of overtones or harmonics in music. Clearly the terms in this series form a sequence $\left(\frac{1}{n}\right)$ which converges to 0. Seems harmless. However, Alexandre Borovik (1956-) gave a very stern caution:

> I know, you're looking at this series and you don't see what I'm warning you about. You look and it and you think, "I trust this series. I would take candy from this series. I would get in a car with this series." But I'm going to warn you, this series is out to get you. Always remember: The harmonic series diverges. Never forget it.

How do we prove this fact? There are many different ways of proving this result, here are three elementary proofs:

1. This first proof was discovered by Oresme circa 1350. Assume for contradiction that the series converges. This means the sequence of partial sums (s_n) converges. Proposition 5.5.4 then tells us that all subsequences of (s_n) also converge to the same limit. Our goal is to get a contradiction by constructing a subsequence of (s_n) that diverges. Let us look at the subsequence of the form (s_{2^n}). By bounding the terms from below, one can show that for every $n \in \mathbb{N}$, we have $s_{2^n} \geq 1 + \frac{n}{2}$. Indeed, for each $2 \leq k \leq n$ the sum of 2^{k-1} terms beginning after $\frac{1}{2^{k-1}}$ up to $\frac{1}{2^k}$ all satisfy the following inequality:

$$\sum_{j=2^{k-1}+1}^{2^k} \frac{1}{j} = \frac{1}{2^{k-1}+1} + \frac{1}{2^{k-1}+2} + \dots \frac{1}{2^k - 1} + \frac{1}{2^k} \geq \underbrace{\frac{1}{2^k} + \dots \frac{1}{2^k}}_{2^{k-1} \text{ times}} = \frac{1}{2}. \tag{7.2}$$

For example, for $k = 2, 3, 4$, inequality (7.2) says that $\frac{1}{3}+\frac{1}{4} \geq \frac{1}{2}$, $\frac{1}{5}+\ldots+\frac{1}{8} \geq \frac{1}{2}$, and $\frac{1}{9}+\ldots+\frac{1}{16} \geq \frac{1}{2}$. Therefore, if we look at the subsequence (s_{2^n}) of (s_n), we can list some of the terms in this subsequence as:

$$s_{2^1} = s_2 = 1 + \frac{1}{2},$$
$$s_{2^2} = s_4 = s_2 + \frac{1}{3} + \frac{1}{4} \geq s_2 + \frac{1}{2} = 1 + \frac{2}{2},$$
$$s_{2^3} = s_8 = s_4 + \frac{1}{5} + \ldots + \frac{1}{8} \geq s_4 + \frac{1}{2} \geq 1 + \frac{3}{2},$$

and so on. By using induction and inequality (7.2), we can show that the n-th term in this subsequence can be bounded from below as follows:

$$s_{2^n} = \sum_{j=1}^{2^n} \frac{1}{j} \geq 1 + \underbrace{\frac{1}{2} + \ldots + \frac{1}{2}}_{n \text{ times}} = 1 + \frac{n}{2}.$$

Thus, by Proposition 5.6.3, the subsequence (s_{2^n}) diverges to infinity, a contradiction.

2. An alternative way of proving this is to, again, first assume for a contradiction that the harmonic series converges. This means the sequence of partial sums (s_n) converges. In particular, by Theorem 5.8.3, the sequence (s_n) is Cauchy. Therefore, for $\varepsilon = \frac{1}{2} > 0$, there exists an $N \in \mathbb{N}$ such that $|s_n - s_m| < \frac{1}{2}$ for all $m, n \geq N$. Moreover, we can find a $k \in \mathbb{N}$ such that $2^k \geq N$. So, by setting $m = 2^k$ and $n = 2^{k+1}$, we have:

$$\frac{1}{2} > |s_{2^{k+1}} - s_{2^k}| = \underbrace{\frac{1}{2^{k+1}} + \ldots + \frac{1}{2^k+2} + \frac{1}{2^k+1}}_{2^k \text{ terms}} \geq \frac{1}{2^{k+1}}(2^k) = \frac{1}{2},$$

which is a contradiction! Thus, the harmonic series must be divergent.

3. The third proof is due to Jacob Bernoulli which uses fundamental ideas of arithmetic and geometric series that we saw in Example 7.2.2. Readers can refer to [19] for the full proof.

Thus, the harmonic series diverges. However, since the terms of the harmonic series converges to 0, the series is very slowly divergent. Indeed, it takes 12, 367 terms for the series to first exceed 10 and 272, 400, 600 terms to exceed 20.

Algebra of Series

Another direct consequence of looking at the sequence of partial sums is that convergent series behave well under scaling with a constant real number and addition. This can be obtained easily via the algebra of limits and is left for the readers to check in Exercise 7.6.

Proposition 7.2.8 *Let $\sum_{j=1}^{\infty} a_j$ and $\sum_{j=1}^{\infty} b_j$ be convergent real series. Then:*

1. *For any $\lambda \in \mathbb{R}$, the series $\sum_{j=1}^{\infty} \lambda a_j$ converges and is equal to $\lambda \sum_{j=1}^{\infty} a_j$.*
2. *The series $\sum_{j=1}^{\infty}(a_j + b_j)$ converges and is equal to $\sum_{j=1}^{\infty} a_j + \sum_{j=1}^{\infty} b_j$.*

Moreover, since the convergence of a series is a limiting behaviour, we can ignore or add any finitely many terms at the beginning of the series safely without disturbing the convergence property. We have:

Proposition 7.2.9 *Let $\sum_{j=1}^{\infty} a_j$ be a real series.*

1. *If there is an $N \in \mathbb{N}$ such that the series $\sum_{j=N}^{\infty} a_j$ converges, then the series $\sum_{j=1}^{\infty} a_j$ converges as well with $\sum_{j=1}^{\infty} a_j = \sum_{j=1}^{N-1} a_j + \sum_{j=N}^{\infty} a_j$.*
2. *If the series $\sum_{j=1}^{\infty} a_j$ converges, then for any $N \in \mathbb{N}$ the series $\sum_{j=N}^{\infty} a_j$ converges as well.*

Proof We prove the assertions one by one.

1. For $n \geq N$, consider the sequence of partial sums (t_n) where $t_n = \sum_{j=N}^{n} a_j$. Let (s_n) be the sequence of partial sums where $s_n = \sum_{j=1}^{n} a_j$. For $n \geq N$ we have $s_n = \sum_{j=1}^{N-1} a_j + t_n = K + t_n$ where $K = \sum_{j=1}^{N-1} a_j \in \mathbb{R}$ is a real constant. Since (t_n) converges as $n \to \infty$, by algebra of limits, we conclude that (s_n) converges as well. Moreover, we have $\sum_{j=1}^{\infty} a_j = \lim_{n\to\infty} s_n = \lim_{n\to\infty}(K + t_n) = K + \lim_{n\to\infty} t_n = \sum_{j=1}^{N-1} a_j + \sum_{j=N}^{\infty} a_j$.
2. Fix $N \in \mathbb{N}$ and for $n \geq N$, consider the sequence of partial sums (t_n) where $\sum_{j=N}^{n} a_j$. Let (s_n) be the sequence of partial sums for the series $\sum_{j=1}^{\infty} a_j$ where $s_n = \sum_{j=1}^{n} a_j$. Then, for any $n \geq N$ we have $t_n = s_n - \sum_{j=1}^{N-1} a_j = s_n - K$ where $K = \sum_{j=1}^{N-1} a_j \in \mathbb{R}$ is a real constant. Since (s_n) converges, by algebra of limits, we conclude that the sequence (t_n) also converges. □

Monotone Series

One special family of real series that is easy to handle is the monotone series in which the terms in the series, apart from the first one, all have the same sign. We define:

Definition 7.2.10 (Monotone Series) Let $\sum_{j=1}^{\infty} a_j$ be a real series. The series is called increasing if $a_j \geq 0$ or decreasing if $a_j \leq 0$ for all $j \geq 2$. Either way, the series is called a monotone series.

This definition, as the name suggests, ensures that the sequence of partial sums (s_n) of the series is a monotone sequence. As an example, consider an increasing series $\sum_{j=1}^{\infty} a_j$. The differences between consecutive partial sums are $s_{n+1} - s_n = a_{n+1} \geq 0$ for all $n \in \mathbb{N}$ which implies that the sequence of partial sums is increasing. Hence, by monotone sequence theorem in Theorem 5.4.2, for this special family of series, a direct consequence is:

Proposition 7.2.11 *Let $\sum_{j=1}^{\infty} a_j$ be a real series.*

1. *If the series is increasing and the sequence of partial sums is bounded from above, then the series converges.*
2. *If the series is decreasing and the sequence of partial sums is bounded from below, then the series converges.*

7.3 Absolute and Conditional Convergence

From Example 7.2.7(2), we have seen that we can use the Cauchy property of the sequence of partial sums to deduce convergence of a series. Indeed, from the correspondence of convergent and Cauchy real sequences, we have the following chain of equivalences:

$$\sum_{j=1}^{\infty} a_n \text{ converges} \quad \Leftrightarrow \quad (s_n) \text{ converges} \quad \Leftrightarrow \quad (s_n) \text{ is Cauchy},$$

which proves the following result:

Proposition 7.3.1 (Cauchy Criterion for Convergence of a Series) *The real series $\sum_{j=1}^{\infty} a_j$ converges if and only if for every $\varepsilon > 0$, there exists an $N \in \mathbb{N}$ such that for every $n > m \geq N$ we have $|s_n - s_m| = |a_{m+1} + a_{m+2} + \ldots + a_n| < \varepsilon$.*

Proposition 7.3.1 is a very useful characterisation for convergence series. To demonstrate its usefulness, we first define a stronger version of series convergence, which is called absolute convergence.

Definition 7.3.2 (Absolute Convergence) A real series $\sum_{j=1}^{\infty} a_j$ is called absolutely convergent if the corresponding absolute series $\sum_{j=1}^{\infty} |a_j|$ converges.

Why is this definition considered stronger than the usual convergence? Using Proposition 7.3.1, if a series is absolutely convergent, then the series converges. Let us prove this:

Proposition 7.3.3 *If the real series $\sum_{j=1}^{\infty} a_j$ is absolutely convergent, then the series is convergent. In other words:*

$$\sum_{j=1}^{\infty} |a_j| \text{ converges} \quad \Rightarrow \quad \sum_{j=1}^{\infty} a_j \text{ converges.}$$

Proof Define the partial sums $s_n = \sum_{j=1}^{n} a_j$ and $S_n = \sum_{j=1}^{n} |a_n|$. Fix $\varepsilon > 0$. Since the series is absolutely convergent, the sequence (S_n) is Cauchy. Thus, we can find an $N \in \mathbb{N}$ such that for all $n > m \geq N$, we have:

$$\begin{aligned} |S_n - S_m| &= ||a_{m+1}| + |a_{m+2}| + \ldots + |a_n|| \\ &= |a_{m+1}| + |a_{m+2}| + \ldots + |a_n| < \varepsilon \end{aligned} \tag{7.3}$$

We now claim that this N also works for the sequence (s_n). Indeed, for all $n > m \geq N$, by applying the triangle inequality and using (7.3), we have:

$$|s_n - s_m| = |a_{m+1} + a_{m+2} + \ldots + a_n| \leq |a_{m+1}| + |a_{m+2}| + \ldots + |a_n| < \varepsilon.$$

Thus, by Proposition 7.3.1, the series $\sum_{j=1}^{\infty} a_j$ converges. □

An important and useful thing to note here is that the absolute series $\sum_{j=1}^{\infty} |a_j|$ is made up of non-negative terms. Therefore, the absolute series is an increasing series. Hence, by virtue of Propositions 7.2.11 and 7.3.3, it is enough to show that the sequence of partial sums (S_n) where $S_n = \sum_{j=1}^{n} |a_j|$ is bounded from above to deduce that the series $\sum_{j=1}^{\infty} a_j$ is convergent since we have the following chain of implications:

$$(S_n) \text{ is bounded above} \quad \Rightarrow \quad \sum_{j=1}^{\infty} |a_j| \text{ converges} \quad \Rightarrow \quad \sum_{j=1}^{\infty} a_j \text{ converges.}$$

Remark 7.3.4 Absolute convergence is a particularly useful idea to use for normed vector spaces and complex series: it turns a series of vectors or complex series into a real series for which we have a variety of tools to study with. Recall from Remark 7.2.4 that a complex series can be written as:

$$\sum_{j=1}^{\infty} z_j = \sum_{j=1}^{\infty} (a_j + ib_j) = \sum_{j=1}^{\infty} a_j + i \sum_{j=1}^{\infty} b_j,$$

and to show convergence of the complex series, we need to show that both the real and imaginary parts are convergent series.

However, by using Proposition 7.3.3, instead of showing that the two real series coming from the real and imaginary parts converge, it is enough to show that the complex series is absolutely convergent to conclude the convergence of the complex series $\sum_{j=1}^{\infty} z_j$. Indeed, if the real series $\sum_{j=1}^{\infty} |z_j|$ converges, then its sequence of partial sums $S_n = \sum_{j=1}^{n} |z_j| = \sum_{j=1}^{n} \sqrt{a_j^2 + b_j^2}$ converges and so is bounded by some constant $M > 0$. Thus, for all $n \in \mathbb{N}$ we have $\sum_{j=1}^{n} |a_j| \leq S_n \leq M$ and $\sum_{j=1}^{n} |b_j| \leq S_n \leq M$.

Moreover, since both series $\sum_{j=1}^{\infty} |a_j|$ and $\sum_{j=1}^{\infty} |b_j|$ are increasing and bounded, by monotone sequence theorem, the series of real and imaginary parts both absolutely converge and hence converge by Proposition 7.3.3. This implies the complex series $\sum_{j=1}^{\infty} z_j$ converge. In short, we have:

$$\sum_{j=1}^{\infty} |z_j| \text{ converges} \quad \Rightarrow \quad \sum_{j=1}^{\infty} |a_j| \text{ and } \sum_{j=1}^{\infty} |b_j| \text{ converge}$$

$$\Rightarrow \quad \sum_{j=1}^{\infty} a_j \text{ and } \sum_{j=1}^{\infty} b_j \text{ converge} \quad \Rightarrow \quad \sum_{j=1}^{\infty} z_j \text{ converges.}$$

Finally, a very important thing to note is that Proposition 7.3.3 is only a one-way implication. There are many series which are not absolutely convergent but is convergent. Such series are called:

Definition 7.3.5 (Conditional Convergence) A real series $\sum_{j=1}^{\infty} a_j$ is called conditionally convergent if $\sum_{j=1}^{\infty} a_j$ converges but $\sum_{j=1}^{\infty} |a_j|$ diverges to infinity.

An important distinction between the absolutely convergent and conditionally convergent series is that the terms in an absolutely convergent series can be rearranged without changing the value of the series whereas the terms in a conditionally convergent series can be rearranged so that the the rearranged series converges to any number in $\mathbb{R}$ or even diverges to $\pm\infty$. This is called the Riemann rearrangement theorem and we shall prove it in Theorem 8.1.5.

7.4 Alternating Series

To create an example of a conditionally convergent sequence, let us define alternating series. As the name suggests, alternating series is a real series made up of terms with alternating signs.

Definition 7.4.1 (Alternating Series) A real series is called alternating if it is of the form:

$$\sum_{j=1}^{\infty} (-1)^j b_j \quad \text{or} \quad \sum_{j=1}^{\infty} (-1)^{j-1} b_j,$$

where $b_j > 0$ for all $j \in \mathbb{N}$.

As usual with real series, the series converges if its sequence of partial sums converge. But we have another method of showing convergence of alternating series, which is called the alternating series test or Leibniz test:

Theorem 7.4.2 (Alternating Series Test) *An alternating series of the form* $\sum_{j=1}^{\infty}(-1)^j b_j$ *or* $\sum_{j=1}^{\infty}(-1)^{j-1} b_j$ *with* $b_j > 0$ *converges if the terms* (b_j) *is decreasing and* $b_j \to 0$.

Proof WLOG, consider the alternating series of the form $\sum_{j=1}^{\infty}(-1)^{j-1} b_j$ where the first term in the series is positive. Let (s_n) be its sequence of partial sums. We consider the subsequence of even-indexed and odd-indexed partial sums, namely (s_{2n}) and (s_{2n-1}). We note that for the subsequence of even-indexed partial sums, by some grouping some consecutive terms together, we have:

$$\begin{aligned} s_{2n} &= b_1 - b_2 + b_3 - b_4 + \ldots - b_{2n-2} + b_{2n-1} - b_{2n} \\ &= b_1 - (b_2 - b_3) - \ldots - (b_{2n-2} - b_{2n-1}) - b_{2n} \leq b_1, \end{aligned}$$

since $b_j \geq b_{j+1}$ for all $j \in \mathbb{N}$. Furthermore, we also have:

$$s_{2(n+1)} - s_{2n} = -b_{2n+2} + b_{2n+1} \geq 0 \quad \Rightarrow \quad s_{2(n+1)} \geq s_{2n},$$

for all $n \in \mathbb{N}$. Thus, the subsequence of partial sums (s_{2j}) is bounded from above and increasing. By the monotone sequence theorem, the subsequence of even-indexed partial sums (s_{2n}) converges.

By using similar arguments, we can show that the subsequence of odd-indexed partial sums (s_{2n-1}) is bounded from below and decreasing. Therefore, applying the monotone sequence theorem, the subsequence of odd-indexed partial sums (s_{2n-1}) converges as well.

Furthermore, from the relationship $-b_{2n} = s_{2n} - s_{2n-1}$, by taking the limit on both sides and applying the algebra of limits, we obtain:

$$0 = -\lim_{n\to\infty} b_{2n} = \lim_{n\to\infty}(s_{2n} - s_{2n-1}) = \lim_{n\to\infty} s_{2n} - \lim_{n\to\infty} s_{2n-1},$$

and thus $\lim_{n\to\infty} s_{2n} = \lim_{n\to\infty} s_{2n-1}$. So the limit of the even-indexed and odd-indexed partial sums are equal, say s. Finally, by Exercise 5.7(a), the whole sequence of partial sums (s_n) converges to the same limit s, which says the series is convergent. □

Thus, for alternating series, we can deduce that the series converges simply by checking that the sequence of absolute values of the terms is decreasing and converges to 0.

Example 7.4.3 Consider the series $\sum_{j=1}^{\infty} \frac{(-1)^j}{j}$. The terms of the series are $a_1 = -1, a_2 = \frac{1}{2}, a_3 = -\frac{1}{3}, a_4 = \frac{1}{4}, \ldots$ and so this is an alternating series since the terms a_j have alternate signs. If we consider the sequence $(|a_j|)$, we get $(1, \frac{1}{2}, \frac{1}{3}, \frac{1}{4}, \ldots)$ which is a decreasing sequence that converges to 0. Then, by alternating series test, we can immediately conclude that the series $\sum_{j=1}^{\infty} \frac{(-1)^j}{j}$ converges.

However, this series is not absolutely convergent. Indeed, the absolute series $\sum_{j=1}^{\infty} \left|\frac{(-1)^j}{j}\right| = \sum_{j=1}^{\infty} \frac{1}{j}$ is the harmonic series that we have seen not to converge in Example 7.2.7. Thus the series $\sum_{j=1}^{\infty} \frac{(-1)^j}{j}$ is not absolutely convergent, giving us an example of a conditionally convergent series.

7.5 Comparison Tests

For real sequences, we have seen that limits preserve weak inequalities and the sandwich lemma. These results can be used to help us compare or bound the limits of a sequence with commonly known sequences. Now we want to be able to do the same for series. We have seen some basic examples of series which converge and diverge so we have some standard families of series to compare to.

Direct Comparison Test

The first convergence test is the direct comparison test for series. The idea is simple and intuitive: suppose that we have two series with non-negative terms such that one of the series is term-wise larger than the other. If the series with the larger terms converges, then the series with the smaller terms will also converge. Likewise, if the series with the smaller terms diverges, necessarily the series with the larger terms will diverge as well. We state:

Proposition 7.5.1 (Direct Comparison Test) *Let $\sum_{j=1}^{\infty} a_j$ and $\sum_{j=1}^{\infty} b_j$ be two real series such that $0 \leq a_j \leq b_j$ for all $j \in \mathbb{N}$.*

1. *If the series $\sum_{j=1}^{\infty} b_j$ converges, then the series $\sum_{j=1}^{\infty} a_j$ converges as well.*
2. *If the series $\sum_{j=1}^{\infty} a_j$ diverges to ∞, then the series $\sum_{j=1}^{\infty} b_j$ diverges to ∞ as well.*

Proof Let $s_n = \sum_{j=1}^{n} a_j$ and $t_n = \sum_{j=1}^{n} b_j$ so that (s_n) and (t_n) are the sequences of partial sums of the series respectively. We note that since $a_j, b_j \geq 0$, both of the sequences (s_n) and (t_n) are increasing. Moreover, the condition $0 \leq a_j \leq b_j$ implies $0 \leq s_n \leq t_n$ for all $n \in \mathbb{N}$. We now prove the assertions one by one.

1. First, note that the sequence (t_n) converges since the series $\sum_{j=1}^{\infty} b_j$ converges. Therefore, this sequence must be bounded, say $t_n \leq M$ for all $n \in \mathbb{N}$ and some $M > 0$. Thus, $s_n \leq t_n \leq M$ for every $n \in \mathbb{N}$ and so the sequence (s_n) is also

bounded from above. Applying the monotone sequence theorem, the sequence (s_n) converges and so the series $\sum_{j=1}^{\infty} a_j$ is also a convergent series.
2. Since the series $\sum_{j=1}^{\infty} a_n$ diverges, we note that $s_n \to \infty$. Since $s_n \leq t_n$ for all $n \in \mathbb{N}$, this implies the sequence (t_n) also diverges to infinity by Proposition 5.6.3. □

A corollary of this for series with mixed signs can be obtained by looking at the absolute series.

Corollary 7.5.2 *Let $\sum_{j=1}^{\infty} a_j$ and $\sum_{j=1}^{\infty} b_j$ be two real series such that $|a_j| \leq |b_j|$ for all $j \in \mathbb{N}$.*

1. *If the series $\sum_{j=1}^{\infty} |b_j|$ converges, then the series $\sum_{j=1}^{\infty} a_j$ is absolutely convergent.*
2. *If the series $\sum_{j=1}^{\infty} |a_j|$ diverges to ∞, then the series $\sum_{j=1}^{\infty} b_j$ is not absolutely convergent (but may still be conditionally convergent).*

Example 7.5.3 Let us look at some examples:

1. Consider the series $\sum_{j=1}^{\infty} \frac{1}{j^2}$. We note that $\frac{1}{j^2} \leq \frac{1}{j}$ for all $j \in \mathbb{N}$. So we can compare this series with the harmonic series $\sum_{j=1}^{\infty} \frac{1}{j}$. But this does not tell us anything because the direct comparison test in Proposition 7.5.1 requires either the smaller series to diverge to the larger series to converge in order for us to draw some conclusion. Here, the larger series, which is the harmonic series, diverges and so nothing could be concluded here. Therefore we need to consider a different series to compare $\sum_{j=1}^{\infty} \frac{1}{j^2}$ to.
Note that for all $j \in \mathbb{N}$ we have $2j^2 \geq j^2 + j = j(j+1)$ which means $0 \leq \frac{1}{j^2} \leq \frac{2}{j(j+1)}$ for all $j \in \mathbb{N}$. So let us compare the series with $\sum_{j=1}^{\infty} \frac{1}{j(j+1)}$. We have seen this new series before: it is a telescoping series in Example 7.2.2(4) which converges to 1. By Proposition 7.2.8, via scaling, the series $\sum_{j=1}^{\infty} \frac{2}{j(j+1)}$ converges to 2. Hence, we can conclude that the series $\sum_{j=1}^{\infty} \frac{1}{j^2}$ converges by direct comparison test.
Moreover, for any $p > 2$, the series $\sum_{j=1}^{\infty} \frac{1}{j^p}$ converges. This can be shown by directly comparing it to the convergent series $\sum_{j=1}^{\infty} \frac{1}{j^2}$ since $\frac{1}{j^p} \leq \frac{1}{j^2}$ for any $j \in \mathbb{N}$.
2. Suppose that (a_j) and (b_j) are sequences such that $a_j = \frac{1}{j}$ and $b_n = \frac{(-1)^n}{\sqrt{j}}$. Then, we have $|a_j| \leq |b_j|$ for all $j \in \mathbb{N}$. By Corollary 7.5.2, since $\sum_{j=1}^{\infty} |a_j| = \sum_{j=1}^{\infty} \frac{1}{j}$ diverges, we conclude that the series $\sum_{j=1}^{\infty} b_j$ is not absolutely convergent.
However, clearly the series $\sum_{j=1}^{\infty} b_j$ converges since it satisfies the alternating series test. So this series is only conditionally convergent.

Limit Comparison Test

The second comparison test that we are going to show is called limit comparison test. The crucial factor that determines whether a series converge or not is how the terms behave as we go towards infinity. We know that a necessary condition for the series to converge is that the terms must decay to 0. However, this alone is not sufficient to imply convergence of the series as we have seen with the harmonic series. Thus, the terms in the series must decay fast enough for series convergence to happen.

The rough idea of the limit comparison test is to compare the decay rates of the terms in two series. If the asymptotic behaviour of the terms in two series are similar (up to a finite scale), namely $(a_n) \sim L(b_n)$ for some constant $L \in (0, \infty)$, then both of the series must have the same convergence/divergence property. We state and prove this rigorously.

Proposition 7.5.4 (Limit Comparison Test) *Let $\sum_{j=1}^{\infty} a_j$ and $\sum_{j=1}^{\infty} b_j$ be two real series such that $a_j \geq 0$ and $b_j > 0$ for all $j \in \mathbb{N}$. Suppose that $\lim_{j\to\infty} \frac{a_j}{b_j} = L$ for some $0 < L < \infty$. Then, either both series converge or both series diverge. In other words:*

$$\sum_{j=1}^{\infty} a_j \text{ converges} \quad \Leftrightarrow \quad \sum_{j=1}^{\infty} b_j \text{ converges.}$$

Proof Since $\lim_{j\to\infty} \frac{a_j}{b_j} = L$, for $\varepsilon = \frac{L}{2} > 0$, there exists an $N \in \mathbb{N}$ such that $\left|\frac{a_j}{b_j} - L\right| < \frac{L}{2}$ for all $j \geq N$. Equivalently, for any $j \geq N$ we have:

$$\frac{L}{2} b_j < a_j < \frac{3L}{2} b_j.$$

We shall now prove the implications one by one:

($\Rightarrow$): By assumption, the series $\sum_{j=1}^{\infty} a_j$ converges. Therefore, the series $\frac{2}{L}\sum_{j=1}^{\infty} a_j = \sum_{j=1}^{\infty} \frac{2}{L} a_j$ also converges by Proposition 7.2.8. Since we have $0 < b_j < \frac{2}{L} a_j$ for all $j \geq N$, by direct comparison test, the series $\sum_{j=N}^{\infty} b_j$ also converges. Finally by Proposition 7.2.9, the full series $\sum_{j=1}^{\infty} b_j$ converges.

($\Leftarrow$): Similar to the opposite implication, since $0 \leq a_j < \frac{3L}{2} b_j$ for all $j \geq N$ and the series $\sum_{j=N}^{\infty} \frac{3L}{2} b_j$ converges, the series $\sum_{j=N}^{\infty} a_j$ also converges by direct comparison test. Thus, we conclude that the full series $\sum_{j=1}^{\infty} a_j$ converges. □

We note that the limit comparison test above only works when $0 < L < \infty$ since in the proof, we have to explicitly use the quantity $\frac{L}{2} > 0$. However, for the degenerate cases $L = 0$ or $L = \infty$, we have partial results. The idea of the proof for the following is similar to the proof of Proposition 7.5.4 and we leave it as an exercise for the readers in Exercise 7.22.

Proposition 7.5.5 (Limit Comparison Test—Degenerate Case) *Let $\sum_{j=1}^{\infty} a_j$ and $\sum_{j=1}^{\infty} b_j$ be two real series such that $a_j \geq 0$ and $b_j > 0$ for all $j \in \mathbb{N}$.*

1. *If $\lim_{j\to\infty} \frac{a_j}{b_j} = 0$ and the series $\sum_{j=1}^{\infty} b_j$ converges, then the series $\sum_{j=1}^{\infty} a_j$ converges.*
2. *If $\lim_{j\to\infty} \frac{a_j}{b_j} = \infty$ and the series $\sum_{j=1}^{\infty} b_j$ diverges, then the series $\sum_{j=1}^{\infty} a_j$ diverges.*

Example 7.5.6 Now let us look at some examples.

1. We want to determine whether the series $\sum_{j=1}^{\infty} \frac{2^{\frac{1}{j}}}{j^2}$ converges. In order to use the limit comparison test, we need to first guess the behaviour of the terms in the series as j becomes large. This would enable us to guess which standard series we can compare this series with.
   Note that, by Exercise 5.17, the numerator $2^{\frac{1}{j}}$ tends to 1 as $j \to \infty$. Therefore, the series terms behave like $\frac{1}{j^2}$ when j gets really large, so let us do a limit comparison test with the convergent series $\sum_{j=1}^{\infty} \frac{1}{j^2}$. All of the terms in these series are positive so, applying the limit comparison test, we get:
$$\lim_{j\to\infty} \frac{\frac{2^{\frac{1}{j}}}{j^2}}{\frac{1}{j^2}} = \lim_{j\to\infty} \frac{2^{\frac{1}{j}}}{j^2} j^2 = \lim_{j\to\infty} 2^{\frac{1}{j}} = 1 \in (0, \infty).$$
   Therefore, this implies the series $\sum_{j=1}^{\infty} \frac{2^{\frac{1}{j}}}{j^2}$ converges.
2. Let us determine the convergence of the series $\sum_{j=2}^{\infty} \frac{1}{j^2-1}$. As j gets very large, the denominator behaves like j^2 so we expect that this series to also behave like $\sum_{j=1}^{\infty} \frac{1}{j^2}$ far enough down the series.
   We know that the series $\sum_{j=2}^{\infty} \frac{1}{j^2}$ converges. We cannot apply the direct comparison test immediately since $\frac{1}{j^2} \leq \frac{1}{j^2-1}$ for all $j \geq 2$, where the inequality is the wrong way round for us to apply the direct comparison. Let us try the limit comparison test here. We note that:
$$\lim_{j\to\infty} \frac{\frac{1}{j^2-1}}{\frac{1}{j^2}} = \lim_{j\to\infty} \frac{j^2}{j^2-1} = \lim_{j\to\infty} \frac{1}{1-\frac{1}{j^2}} = 1 \in (0, \infty),$$

by using the algebra of limits. Thus, we deduce that the series $\sum_{j=2}^{\infty} \frac{1}{j^2-1}$ converges.

3. How about the series $\sum_{j=0}^{\infty} \frac{1}{2j+10}$? If we go far along the series, the 10 in the denominator will be minuscule compared to the $2j$ term, so we expect that the series to behave like $\sum_{j=1}^{\infty} \frac{1}{2j}$. This is true because:

$$\lim_{j\to\infty} \frac{\frac{1}{2j+10}}{\frac{1}{2j}} = \lim_{j\to\infty} \frac{2j}{2j+10} = \lim_{j\to\infty} \frac{2}{2+\frac{10}{j}} = 1 \in (0,\infty).$$

Thus, since the series $\sum_{j=1}^{\infty} \frac{1}{2j}$ diverges, the series $\sum_{j=1}^{\infty} \frac{1}{2j+10}$ is also divergent.

4. We have proven that $\sum_{j=1}^{\infty} \frac{1}{j}$ diverges and $\sum_{j=1}^{\infty} \frac{1}{j^2}$ converges. So by comparison, the series $\sum_{j=1}^{\infty} \frac{1}{j^p}$ diverges for any $p \leq 1$ and converges for any $p \geq 2$. How about for any index $p \in (1,2)$? Let us try to prove their convergence using the limit comparison test.
For example, suppose that $p = \frac{3}{2}$. We consider the series $\sum_{j=1}^{\infty} \frac{1}{j^{\frac{3}{2}}}$ and compare it with $\sum_{j=1}^{\infty} \frac{1}{j^2}$ using the limit comparison test. The ratio of the terms satisfy the following limit:

$$\lim_{j\to\infty} \frac{\frac{1}{j^{\frac{3}{2}}}}{\frac{1}{j^2}} = \lim_{j\to\infty} \frac{j^2}{j^{\frac{3}{2}}} = \lim_{j\to\infty} j^{\frac{1}{2}} = \infty.$$

This limit blows up to ∞, so we have to use the degenerate case of the limit comparison test in Proposition 7.5.5. However, the series $\sum_{j=1}^{\infty} \frac{1}{j^2}$ converges, so (sadly) the test does not tell us anything about the convergence or divergence of $\sum_{j=1}^{\infty} \frac{1}{j^{\frac{3}{2}}}$.
Similarly, if we compare it with the series $\sum_{j=1}^{\infty} \frac{1}{j}$, we would not get any result from the limit comparison test. We can also try the same with any other $p \in (1,2)$ but we still could not conclude whether these series converge using the limit comparison test. We shall revisit this problem later in Exercise 7.12 and Example 16.4.15.

7.6 Ratio and Root Tests

Two of the most important series convergence tests are the root and ratio tests. These are usually the first go-to tests that one uses to determine whether a series converges because they are very easy to use.

Ratio Test

The ratio test was first published by D'Alembert which uses the idea of comparing the series that we are interested in with a geometric series. We know from Example 7.2.2 that a geometric series $\sum_{j=1}^{\infty} a_j$ converges if the common ratio of the terms r is such that $\left|\frac{a_{n+1}}{a_n}\right| = |r| < 1$ and the geometric series diverges if $|r| \geq 1$.

Of course, any general series $\sum_{j=1}^{\infty} a_j$ that we are interested in is not necessarily a geometric series and thus there is no common ratio for the terms: the ratios of consecutive terms $\frac{a_{j+1}}{a_j}$ for $j \in \mathbb{N}$ may not be constant over the sequence. However, we can consider the limiting behaviour of the ratios of the terms and see if it goes above or below 1. This would then tell us the long-term behaviour of the terms in the series. This analysis is called the ratio test.

Theorem 7.6.1 (Ratio Test) *Let $\sum_{j=1}^{\infty} a_j$ be a real series such that $a_j \neq 0$ for all $j \in \mathbb{N}$. Let $L = \lim_{j\to\infty} \left|\frac{a_{j+1}}{a_j}\right| \geq 0$.*

1. *If $L < 1$, then the series converges absolutely.*
2. *If $L > 1$, then the series diverges.*

Proof We prove the assertions separately.

1. Suppose that $\lim_{j\to\infty} \left|\frac{a_{j+1}}{a_j}\right| = L < 1$. Then, for $\varepsilon = \frac{1-L}{2} > 0$, there exists an $N \in \mathbb{N}$ such that $\left|\left|\frac{a_{n+1}}{a_n}\right| - L\right| < \frac{1-L}{2}$ for all $n \geq N$. This implies $\frac{|a_{n+1}|}{|a_n|} < \frac{1+L}{2}$ for all $n \geq N$. Denote $r = \frac{1+L}{2} < 1$ so that $|a_{n+1}| < r|a_n|$ for all $n \geq N$. By induction, we can show that $|a_{k+N}| < r^k |a_N|$ for all $k \in \mathbb{N}$.
   Let us compare the tail of the series $\sum_{j=N+1}^{\infty} |a_j| = \sum_{k=1}^{\infty} |a_{k+N}|$ with the geometric series $\sum_{k=1}^{\infty} r^k |a_N|$. Clearly the geometric series converges since $r < 1$. By direct comparison test, since $|a_{k+N}| < r^k |a_N|$ for all $k \in \mathbb{N}$, the tail of this series $\sum_{k=1}^{\infty} |a_{k+N}| = \sum_{j=N+1}^{\infty} |a_j|$ also converges. Proposition 7.2.9 then implies that the whole series $\sum_{j=1}^{\infty} |a_j|$ converges.
2. Suppose that $\lim_{j\to\infty} \left|\frac{a_{j+1}}{a_j}\right| = L > 1$. By similar argument as the previous part, if we choose $\varepsilon = \frac{L-1}{2} > 0$, we can show that there exists an $N \in \mathbb{N}$ such that $\frac{1+L}{2} < \frac{|a_{n+1}|}{|a_n|}$ for all $n \geq N$. Denote $r = \frac{1+L}{2} > 1$, so that $0 < |a_N| < r^k |a_N| < |a_{k+N}|$ for all $k \in \mathbb{N}$. Since $r^k |a_N| \to \infty$, we have $|a_{k+N}| \to \infty$ as well. This means $\lim_{j\to\infty} |a_j| \neq 0$ and so $\lim_{j\to\infty} a_j \neq 0$ by Lemma 5.9.3. Thus, the series $\sum_{j=1}^{\infty} a_j$ cannot converge by Proposition 7.2.5. □

We note that the ratio test applies only when $L > 1$ or $L < 1$. The test is inconclusive if $L = 1$ because we can find series which satisfies $L = 1$ but can be either convergent or divergent. Indeed, consider the two series $\sum_{j=1}^{\infty} \frac{1}{j^2}$ and

$\sum_{j=1}^{\infty} \frac{1}{j}$. We know that the former converges whereas the latter diverges. However, for both series, we can compute that the limit of the ratios of consecutive terms are both $L = 1$.

Example 7.6.2 Now let us look at some examples on how to use the ratio test:

1. Consider the series $\sum_{j=1}^{\infty} \frac{3^j}{(-2)^j j^2}$. Let us apply the ratio test to this series to check its convergence. We have:

$$\lim_{j\to\infty} \frac{\left|\frac{3^{j+1}}{(-2)^{j+1}(j+1)^2}\right|}{\left|\frac{3^j}{(-2)^j j^2}\right|} = \lim_{j\to\infty} \frac{3^{j+1}}{2^{j+1}(j+1)^2}\frac{2^j j^2}{3^j} = \frac{3}{2}\lim_{j\to\infty}\frac{1}{(1+\frac{1}{j})^2} = \frac{3}{2} > 1,$$

   which implies that this series is divergent.
2. Let us consider the series $\sum_{j=1}^{\infty} \frac{j!}{j^j}$. We consider the following limit of ratios of consecutive terms:

$$\lim_{j\to\infty} \frac{\left|\frac{(j+1)!}{(j+1)^{j+1}}\right|}{\left|\frac{j!}{j^j}\right|} = \lim_{j\to\infty} \frac{(j+1)!}{(j+1)^{j+1}}\frac{j^j}{j!} = \lim_{j\to\infty}\frac{j^j}{(j+1)^j} = \lim_{j\to\infty}\frac{1}{(1+\frac{1}{j})^j} = \frac{1}{e},$$

   by using the limit that we have seen in Example 5.4.4. Furthermore, since $2 \leq e \leq 3$, this limit is strictly less than 1, which implies that the series is convergent. As an exercise, try to prove that this series converges using direct comparison test.

Root Test

A similar idea of comparing a series to a geometric series was used by Cauchy to prove the following series convergence test:

Theorem 7.6.3 (Root Test) *Let $\sum_{j=1}^{\infty} a_j$ be a real series such that $a_j \neq 0$ for all $j \in \mathbb{N}$. Let $L = \lim_{j\to\infty} \sqrt[j]{|a_j|} \geq 0$.*

1. *If $L < 1$, then the series converges absolutely.*
2. *If $L > 1$, then the series diverges.*

Proof We prove the assertions separately.

1. Suppose that $\lim_{j\to\infty} \sqrt[j]{|a_j|} = L < 1$. Then, for $\varepsilon = \frac{1-L}{2} > 0$, there exists an $N \in \mathbb{N}$ such that $|\sqrt[n]{|a_n|} - L| < \frac{1-L}{2}$ for all $n \geq N$. This implies $|a_n| < \left(\frac{1+L}{2}\right)^n$ for all $n \geq N$. Denote $r = \frac{1+L}{2} < 1$.

Clearly the geometric series $\sum_{j=N}^{\infty} r^j$ converges since $r < 1$ and by direct comparison test, since $|a_n| < r^n$ for all $n \geq N$, the tail $\sum_{j=N}^{\infty} |a_j|$ also converges. Proposition 7.2.9 then tells us that the whole series $\sum_{j=1}^{\infty} |a_j|$ converges.

2. Suppose that $\lim_{j\to\infty} \sqrt[j]{|a_j|} = L > 1$. By similar argument as the previous part, if we choose $\varepsilon = \frac{L-1}{2} > 0$, we can show that there exists an $N \in \mathbb{N}$ such that $\left(\frac{1+L}{2}\right)^n < |a_n|$ for all $n \geq N$. Denote $r = \frac{1+L}{2} > 1$, so that $r^n < |a_n|$ for all $n \geq N$. Since $r^n \to \infty$, we must have $|a_n| \to \infty$ as well. This means $\lim_{j\to\infty} |a_j| \neq 0$ and so $\lim_{j\to\infty} a_j \neq 0$ by Lemma 5.9.3. Thus, the series $\sum_{j=1}^{\infty} a_j$ does not converge by Proposition 7.2.5. □

Similar to the ratio test, the root test is only applicable when $L > 1$ or $L < 1$. When $L = 1$, the test is inconclusive. As an example, again, consider the two series $\sum_{j=1}^{\infty} \frac{1}{j^2}$ and $\sum_{j=1}^{\infty} \frac{1}{j}$ that we have checked before. We know that the former converges whereas the latter diverges. However, we can show that $L = 1$ for both series by using the fact that $\lim_{j\to\infty} j^{\frac{1}{j}} = 1$ from Example 5.9.9.

Example 7.6.4 Let us look at some examples of an application of the root test.

1. Consider the series $\sum_{j=1}^{\infty} \frac{j^2}{2^j}$. To determine whether this series converges or diverges, we find the limit of the j-th root of the terms:

$$\lim_{j\to\infty} \sqrt[j]{\frac{|j^2|}{|2^j|}} = \lim_{j\to\infty} \frac{j^{\frac{2}{j}}}{2} = \frac{1}{2} < 1,$$

and hence we conclude that the series converges.

2. Let $x \in \mathbb{R}$ be some real number. Consider the series $\sum_{j=0}^{\infty} x^j$. We would like to know for which real number x does this series converge. A direct application of the root test tells us that the series converges if the following limit is strictly smaller than 1 and diverges if it is strictly greater than 1:

$$\lim_{j\to\infty} \sqrt[j]{|x^j|} = \lim_{j\to\infty} |x| = |x|.$$

So the series converges exactly when $|x| < 1$ and diverges when $|x| > 1$. Does it converge anywhere else? We check the remaining cases $x = \pm 1$ separately:

(a) If we substitute $x = 1$ in the series, the n-th partial sum is $s_n = n$ which diverges to ∞.

(b) If we substitute $x = -1$ in the series, we get the Grandi's series which does not converge.

In both cases, the series does not converge and so the series converges only for $x \in (-1, 1)$.

This is an example of a variable series or functions series which we call a power series. We shall see more of these series in Chap. 12.

The ratio and root tests are the first simple tests that we can use to check the convergence of a series. The ratio test is usually easier to use due to its simple form. However, the root test is a stronger test. This is due to the following proposition:

Proposition 7.6.5 *Suppose that (a_n) is a sequence of positive real numbers. If the limit $\lim_{n\to\infty} \frac{a_{n+1}}{a_n}$ exists, then the limit $\lim_{n\to\infty} \sqrt[n]{a_n}$ also exists. Moreover, the two limits are equal.*

Proof Assume first that $\lim_{n\to\infty} \frac{a_{n+1}}{a_n} = L > 0$. We want to show that $\lim_{n\to\infty} \sqrt[n]{a_n} = L$ as well. Fix an arbitrary $0 < \varepsilon \leq L$. Then, there exists an $N \in \mathbb{N}$ such that for all $n \geq N$ we have $|\frac{a_{n+1}}{a_n} - L| < \varepsilon$. In other words, for all $n \geq N$ we have $0 \leq (L - \varepsilon)a_n < a_{n+1} < (L + \varepsilon)a_n$. Using this, inductively, we can show that for any $k \in \mathbb{N}$ we have $(L - \varepsilon)^k a_N < a_{N+k} < (L + \varepsilon)^k a_N$. By relabelling the indices with $N + k = n$, for all $n \geq N + 1$ we have:

$$(L - \varepsilon)^{1-\frac{N}{n}} a_N^{\frac{1}{n}} < \sqrt[n]{a_n} < (L + \varepsilon)^{1-\frac{N}{n}} a_N^{\frac{1}{n}}. \tag{7.4}$$

At the moment, we do not know whether the limit of the sequence $(\sqrt[n]{a_n})$ exists. However, inequality (7.4) implies that this sequence is bounded since the lower and upper terms converge (to $L - \varepsilon$ and $L + \varepsilon$ respectively) and hence are bounded. So the limit superior and limit inferior for $(\sqrt[n]{a_n})$ exist.

Thus, by Lemma 5.10.3 and Exercise 5.17, from the final inequality in (7.4) we have:

$$\limsup_{n\to\infty} \sqrt[n]{a_n} \leq \limsup_{n\to\infty} (L + \varepsilon)^{1-\frac{N}{n}} a_N^{\frac{1}{n}} = \lim_{n\to\infty} (L + \varepsilon)^{1-\frac{N}{n}} a_N^{\frac{1}{n}} = L + \varepsilon.$$

Likewise, by looking at the first inequality in (7.4), we can deduce $L - \varepsilon \leq \liminf_{n\to\infty} \sqrt[n]{a_n}$. Putting these two inequalities together, we have:

$$0 \leq \limsup_{n\to\infty} \sqrt[n]{a_n} - \liminf_{n\to\infty} \sqrt[n]{a_n} \leq L + \varepsilon - (L - \varepsilon) = 2\varepsilon.$$

Since $\varepsilon > 0$ can be chosen to be arbitrarily small, the limit superior and limit inferior are equal. Thus, by Proposition 5.10.11, $\lim_{n\to\infty} \sqrt[n]{a_n}$ exists. Using this fact in (7.4), by taking the limit as $n \to \infty$ on all sides, we deduce $L - \varepsilon \leq \lim_{n\to\infty} \sqrt[n]{a_n} \leq L + \varepsilon$ or equivalently $|\lim_{n\to\infty} \sqrt[n]{a_n} - L| \leq \varepsilon$. Since $\varepsilon > 0$ is arbitrary, we conclude that $\lim_{n\to\infty} \sqrt[n]{a_n} = L$. The case for $L = 0$ can also be similarly proven. □

Due to Proposition 7.6.5, we conclude that if the limit of the ratio of terms in a series exist, the limit of the roots of the terms also exist and they are equal. However, the converse may be false. Let us look at an example here.

Example 7.6.6 Consider the series $\sum_{j=1}^{\infty} a_j$ where $a_j = \frac{1}{2^j}$ if j is odd and $a_j = \frac{1}{2^{j+1}}$ if j is even. This series converges by direct comparison with the series $\sum_{j=1}^{\infty} \frac{1}{2^j}$, but here we want to demonstrate that the root test works better than the ratio test. Using ratio test, we can check that the ratios of consecutive terms are:

$$\left|\frac{a_{j+1}}{a_j}\right| = \begin{cases} 1 & \text{if } j \text{ is even,} \\ \frac{1}{4} & \text{if } j \text{ is odd,} \end{cases}$$

so $\lim_{n\to\infty} |\frac{a_{n+1}}{a_n}|$ does not exist.

However, we may try the root test by considering the sequence $(\sqrt[n]{|a_n|})$. To study its convergence, we look at the even-indexed and odd-indexed subsequences. For the odd-indexed subsequence, we have $\sqrt[2n+1]{a_{2n+1}} = \sqrt[2n+1]{\frac{1}{2^{2n+1}}} = \frac{1}{2}$. For the even-indexed subsequence, we have $\sqrt[2n]{a_{2n}} = \sqrt[2n]{\frac{1}{2^{2n+1}}} = \frac{1}{2}(\frac{1}{2})^{\frac{1}{2n}} \to \frac{1}{2}$. By Exercise 5.7(a), we conclude the limit of the whole sequence is $\lim_{n\to\infty} \sqrt[n]{a_n} = \frac{1}{2}$ and thus, by the root test, the series converges.

Therefore, we can see that the root test is a stronger test as it can determine the convergence behaviour of some series for which the root test cannot achieve.

Generalised Ratio and Root Tests

The ratio and root tests above require that the limit of the ratios to exist first and foremost before we can obtain a conclusion. However, sometimes these limits may not exist. For example, consider the real series given by $\sum_{j=1}^{\infty} a_j = \sum_{j=1}^{\infty} 2^{(-1)^j - 3j}$. The terms in the series are given by $a_j = 2^{-1-3j} = \frac{1}{2 \cdot 8^j}$ when j is odd and $a_j = 2^{1-3j} = \frac{2}{8^j}$ when j is even. Then, the ratios of consecutive terms are:

$$\left|\frac{a_{j+1}}{a_j}\right| = \begin{cases} \frac{1}{32} & \text{if } j \text{ is even,} \\ \frac{1}{2} & \text{if } j \text{ is odd,} \end{cases}$$

and we can clearly see that this sequence of ratios does not converge and so the usual ratio test does not apply here.

There exists a more general version of the ratio and root tests using limit superior and limit inferior which can cover more cases. The proof is similar to the usual ratio and root test except that one needs to use the definition of limit superior and limit inferior instead.

Theorem 7.6.7 (Generalised Ratio Test) *Let* $\sum_{j=1}^{\infty} a_j$ *be a real series such that* $a_j \neq 0$ *for all* $j \in \mathbb{N}$.

1. *If* $\limsup_{j\to\infty} \left|\frac{a_{j+1}}{a_j}\right| < 1$, *then the series converges absolutely.*
2. *If* $\liminf_{j\to\infty} \left|\frac{a_{j+1}}{a_j}\right| > 1$, *then the series diverges.*

The proof of the above result is in Exercise 7.24 by using Proposition 7.6.10. Here, we shall prove the generalised root test.

Theorem 7.6.8 (Generalised Root Test) *Let* $\sum_{j=1}^{\infty} a_j$ *be a real series such that* $a_j \neq 0$ *for all* $j \in \mathbb{N}$.

1. *If* $\limsup_{j\to\infty} \sqrt[j]{|a_j|} < 1$, *then the series converges absolutely.*
2. *If* $\limsup_{j\to\infty} \sqrt[j]{|a_j|} > 1$, *then the series diverges.*

Proof We prove the assertions separately.

1. Suppose that $L = \limsup_{j\to\infty} \sqrt[j]{|a_j|} < 1$. Recall the definition $\limsup_{j\to\infty} \sqrt[j]{|a_j|} = \inf_{n\geq 1} \left(\sup_{j\geq n} \sqrt[j]{|a_j|}\right)$. By setting $\varepsilon = \frac{1-L}{2} > 0$ and using the characterisation of infimum, we can find an $N \in \mathbb{N}$ such that $L \leq \sup_{j\geq N} \sqrt[j]{|a_j|} < L + \varepsilon = L + \frac{1-L}{2} = \frac{L+1}{2}$. If we set $r = \frac{1+L}{2} < 1$, this means for all $j \geq N$ we have $\sqrt[j]{|a_j|} < r$ which then implies $|a_j| < r^j$. By comparing the series $\sum_{j=N}^{\infty} |a_j|$ with the geometric series $\sum_{j=N}^{\infty} r^j$, we can see that the series $\sum_{j=N}^{\infty} |a_j|$ converges. Finally, by adding the first $N-1$ terms of the series, we conclude that the full series $\sum_{j=1}^{\infty} |a_j|$ converges.
2. Now suppose that $L = \limsup_{j\to\infty} \sqrt[j]{|a_j|} > 1$. Proposition 5.10.8 says there must exist some subsequence of $(\sqrt[j]{|a_j|})$ that converges to L. Let us call this subsequence $(\sqrt[k_j]{|a_{k_j}|})$. For $\varepsilon = \frac{L-1}{2} > 0$, there exists an $N \in \mathbb{N}$ such that $|\sqrt[k_j]{|a_{k_j}|} - L| < \frac{L-1}{2}$ for all $j \geq N$. This means $\sqrt[k_j]{|a_{k_j}|} > \frac{L+1}{2} > 1$ and thus $|a_{k_j}| > 1$ for every $j \geq N$. This implies that the whole sequence a_j cannot be converging to 0 and hence the series diverges. □

Remark 7.6.9 An important detail to take note of is that for the generalised root test, the convergent case utilises limit superior while the divergent case uses limit inferior. However, for the generalised root test, both of the cases use limit superior.

Using the generalised root test, one can then prove the generalised ratio test using the following result. This is also an exercise left to the readers in Exercise 7.24.

Lemma 7.6.10 *Let (a_n) be a sequence of non-zero terms. We have:*

$$\liminf_{j\to\infty}\left|\frac{a_{j+1}}{a_j}\right| \leq \liminf_{j\to\infty}\sqrt[j]{|a_j|} \leq \limsup_{j\to\infty}\sqrt[j]{|a_j|} \leq \limsup_{j\to\infty}\left|\frac{a_{j+1}}{a_j}\right|.$$

Remark 7.6.11 Lemma 7.6.10 can also be used to prove Proposition 7.6.5.

Note that the generalised ratio and root tests also include the standard ratio and root tests since if the limits of the ratios and roots exist, they coincide with the limit superior and limit inferior as shown in Proposition 5.10.11.

Example 7.6.12 Let us look at some examples:

1. Recall the real series given by $\sum_{j=1}^{\infty} a_j = \sum_{j=1}^{\infty} 2^{(-1)^j-3j}$. The terms in the series are given by $a_j = 2^{-1-3j} = \frac{1}{2\cdot 8^j}$ when j is odd and $a_j = 2^{1-3j} = \frac{2}{8^j}$ when j is even. The ratios of consecutive terms are:

$$\left|\frac{a_{j+1}}{a_j}\right| = \begin{cases} \frac{1}{32} & \text{if } j \text{ is even,} \\ \frac{1}{2} & \text{if } j \text{ is odd.} \end{cases}$$

   Let us apply the generalised ratio test here. We note that for each $n \in \mathbb{N}$, we have $\sup_{j\geq n}\left|\frac{a_{j+1}}{a_j}\right| = \frac{1}{2}$. Taking the limit as $n \to \infty$, we have $\limsup_{n\to\infty}\left|\frac{a_{n+1}}{a_n}\right| = \frac{1}{2} < 1$ and thus, by the generalised ratio test, we conclude that this series converges.
2. Let the terms in a real series $\sum_{j=1}^{\infty} a_j$ be described as $a_j = 2^{-j}$ if j is odd and $a_j = 2^{j+1}$ if j is even. Clearly, this series diverges as the terms in the series do not converge to 0. But here we are going to demonstrate an important fact about the generalised root test. Finding the roots of these terms, we get:

$$\sqrt[j]{|a_j|} = \begin{cases} \frac{1}{2} & \text{if } j \text{ is even,} \\ 2^{1+\frac{1}{j}} & \text{if } j \text{ is odd.} \end{cases}$$

   This sequence does not converge as the even-indexed and odd-indexed subsequences do not converge to the same limit and so the usual root test does not apply. Thus, we aim to use the generalised root test. Since the even-indexed terms are strictly smaller than all the odd-indexed terms and the sequence of odd-indexed terms $2^{1+\frac{1}{j}}$ is decreasing, the supremum of these terms starting from the n-th term is given by:

$$\sup_{j\geq n}\sqrt[j]{|a_j|} = \begin{cases} 2^{1+\frac{1}{n+1}} & \text{if } n \text{ is even,} \\ 2^{1+\frac{1}{n}} & \text{if } n \text{ is odd.} \end{cases}$$

Taking the limit as $n \to \infty$, we obtain $\limsup_{n\to\infty} \sqrt[n]{|a_n|} = 2 > 1$. Thus, by the generalised root test, we conclude that the series diverges.
We note that this is an example of why the negative result of the root test uses the limit superior instead of the limit inferior. Notice that in this example $\inf_{j\geq n} \sqrt[j]{|a_j|} = \frac{1}{2}$ for all $n \in \mathbb{N}$. Thus, we have $\liminf_{n\to\infty} \sqrt[j]{|a_j|} = \frac{1}{2} < 1$. However, as mentioned at the beginning of this example, this series is obviously diverging by simply looking at the limit of the terms alone.

7.7 Raabe's Test

In the proof for the ratio test, we have seen that the limit for the ratio of the terms must be equal to some $L < 1$ to guarantee series convergence. In particular, far enough into the sequence, the ratios $\frac{a_{j+1}}{a_j}$ must be uniformly bounded from above by a constant strictly smaller than 1 (say $\frac{L+1}{2} < 1$) for us to be able to deduce convergence of the series.

However, what happens if the limit of the sequence of ratios cannot be uniformly bounded away from 1? In other words, the ratios may be smaller than 1, but can get arbitrarily close to the knife edge of 1. As an example, for the series $\sum_{j=1}^{\infty} \frac{1}{j^2}$ and $\sum_{j=1}^{\infty} \frac{1}{j}$ the ratios of consecutive terms are both strictly smaller than 1 but can get arbitrarily close to 1. Due to this, they both yield inconclusive results for the ratio test. But we have seen that the former series converges whereas the latter series diverges. So is there a better test to distinguish these cases?

The Raabe's test addresses this and is a good follow-up attempt if we reach the inconclusive case for the ratio test. Credited to Joseph Ludwig Raabe (1801–1859), the test is formulated as follows:

Theorem 7.7.1 (Raabe's Test) *Let $\sum_{j=1}^{\infty} a_j$ be a real series such that $a_j \neq 0$ for all $j \in \mathbb{N}$.*

1. *Suppose that there exists a constant $C > 1$ and an index $N \in \mathbb{N}$ such that $\left|\frac{a_{n+1}}{a_n}\right| \leq 1 - \frac{C}{n}$ for all $n \geq N$. Then, the series $\sum_{j=1}^{\infty} a_j$ absolutely converges.*
2. *Suppose that there exists an index $N \in \mathbb{N}$ such that $\left|\frac{a_{n+1}}{a_n}\right| \geq 1 - \frac{1}{n}$ for all $n \geq N$. Then, the series $\sum_{j=1}^{\infty} |a_j|$ diverges.*

Proof We shall prove the first assertion only.

1. From the assumption, for all $n \geq N$ we can rewrite the inequality as $(C-1)|a_n| \leq (n-1)|a_n| - n|a_{n+1}|$. Since $C > 1$, for all $n \geq N$ we have:

$$n|a_{n+1}| \leq (n-1)|a_n|. \tag{7.5}$$

Define a new sequence (b_n) as $b_n = (n-1)|a_n| - n|a_{n+1}|$ and consider the series $\sum_{j=1}^{\infty} b_j$. The partial sums of this series is $s_n = \sum_{j=1}^{n} b_j = \sum_{j=1}^{n}((j-1)|a_j| - j|a_{j+1}|) = -n|a_{n+1}| < 0$ for all $n \in \mathbb{N}$. Notice also that for $n \geq N$, by using (7.5), we have $s_n = -n|a_{n+1}| \geq -(n-1)|a_n| = s_{n-1}$. Thus, the partial sums (s_n) is increasing beginning from index N. Combined with the fact that the partial sums are all negative and hence bounded from above by 0, by monotone sequence theorem, the series $\sum_{j=1}^{\infty} b_j$ converges.
Finally, note that since $b_n \geq (C-1)|a_n|$ for all $n \geq N$, the series $\sum_{j=N}^{\infty} |a_j|$ converges by direct comparison with $\sum_{j=N}^{\infty} b_j$. Hence, the whole series $\sum_{j=1}^{\infty} |a_j| = \sum_{j=1}^{N-1} |a_j| + \sum_{j=N}^{\infty} |a_j|$ is convergent.

The second assertion is left for the readers to prove in Exercise 7.25. □

The Raabe's test addresses some of the cases for which the ratios get close to the inconclusive case for ratio test. However, as we have seen in the conditions of the test, for the Raabe's test to work, these ratios are allowed to get arbitrarily close to 1 as $n \to \infty$ but not too fast: this convergence cannot be faster than $\frac{1}{n}$.

Example 7.7.2 Let us look at some examples:

1. Recall the series $\sum_{j=1}^{\infty} \frac{1}{j^2}$ which we have seen before. We can compute:

$$\frac{a_{n+1}}{a_n} = \frac{\frac{1}{(n+1)^2}}{\frac{1}{n^2}} = \frac{n^2}{n^2+2n+1} = 1 - \frac{2n+1}{n^2+2n+1}.$$

Note that for $n \geq 5$ we have $2 + \frac{1}{n} \leq \frac{n}{2}$. This means:

$$\frac{2n+1}{n^2+2n+1} \geq \frac{2n}{n^2+2n+1} = \frac{2}{n+2+\frac{1}{n}} \geq \frac{2}{n+\frac{n}{2}} = \frac{4}{3n},$$

for $n \geq 5$. Using this estimate we have:

$$\frac{a_{n+1}}{a_n} = 1 - \frac{2n+1}{n^2+2n+1} \leq 1 - \frac{4}{3n},$$

for $n \geq 5$. Thus, by Raabe's test with $C = \frac{4}{3} > 1$ and $N = 5$, this series converges.

2. Consider the series $\sum_{j=1}^{\infty} \frac{(2j)!}{(j!)^2 4^n}$. If we compute the ratio of consecutive terms, we get:

$$\frac{a_{n+1}}{a_n} = \frac{\frac{(2n+2)!}{((n+1)!)^2 4^{n+1}}}{\frac{(2n)!}{(n!)^2 4^n}} = \frac{2n+1}{2n+2},$$

whose limit as $n \to \infty$ is 1. Therefore the ratio test is inconclusive here. Let us try Raabe's test. From the above, we have:

$$\frac{a_{n+1}}{a_n} = \frac{2n+1}{2n+2} = 1 - \frac{1}{2n+2} \geq 1 - \frac{1}{n},$$

since for all $n \geq 1$ we have $n \leq 2n+2 \Leftrightarrow -\frac{1}{2n+2} \geq -\frac{1}{n}$. Therefore, Raabe's test concludes that the series diverges.

There also exists a Raabe's test in limit form. Despite the simple form of Raabe's test in Theorem 7.7.1, it may be tricky to use as we need to work with many different estimates to get it in the desired form and find a pair N and C that works as we have seen in Example 7.7.2(1). As a result, the limit form is usually easier to use.

Theorem 7.7.3 (Raabe's Test—Limit Form) *Let $\sum_{j=1}^{\infty} a_j$ be a real series such that $a_j \neq 0$ for all $j \in \mathbb{N}$. Suppose further that there exists $L \in \mathbb{R}$ such that:*

$$L = \lim_{n\to\infty} n\left(\left|\frac{a_n}{a_{n+1}}\right| - 1\right). \tag{7.6}$$

Then:

1. *If $L > 1$, then the series converges absolutely.*
2. *If $L < 1$, then the series $\sum_{j=1}^{\infty} |a_j|$ diverges.*

Proof We prove the assertions one by one.

1. From the given limit, for $\varepsilon = \frac{L-1}{2} > 0$, there exists an $N_1 \in \mathbb{N}$ such that for all $n \geq N_1$ we have:

$$\left|n\left(\left|\frac{a_n}{a_{n+1}}\right| - 1\right) - L\right| < \varepsilon = \frac{L-1}{2} \quad \Rightarrow \quad 1 + \frac{L+1}{2n} < \left|\frac{a_n}{a_{n+1}}\right|$$
$$\Rightarrow \quad \left|\frac{a_{n+1}}{a_n}\right| < 1 - \frac{L+1}{2n+L+1}. \tag{7.7}$$

Set $N_2 = \left\lceil \frac{(L+3)(L+1)}{2(L-1)} \right\rceil \in \mathbb{N}$. For any $n \geq N_2$ we then have $2n(L-1) \geq (L+3)(L+1)$ which is equivalent to $\frac{L+1}{2n+L+1} \geq \frac{L+3}{4n}$ after some algebraic manipulation. Thus, for all $n \geq \max\{N_1, N_2\}$, substituting this in (7.7) we get:

$$\left|\frac{a_{n+1}}{a_n}\right| < 1 - \frac{L+1}{2n+L+1} \leq 1 - \frac{L+3}{4n}.$$

Thus, applying Theorem 7.7.1 with $N = \max\{N_1, N_2\}$ and $C = \frac{L+3}{4} > 1$, we conclude that the series converges absolutely.

2. From the given limit, for $\varepsilon = 1 - L > 0$ there exists an $N \in \mathbb{N}$ such that for all $n \geq N$ we have:

$$\left| n\left(\left|\frac{a_n}{a_{n+1}}\right| - 1\right) - L\right| < \varepsilon = 1 - L \quad \Rightarrow \quad \left|\frac{a_n}{a_{n+1}}\right| < \frac{1+n}{n}.$$

By algebra, we get $\left|\frac{a_{n+1}}{a_n}\right| > \frac{n}{1+n} = 1 - \frac{1}{1+n} > 1 - \frac{1}{n}$. Applying Theorem 7.7.1 for the divergent case gives us the result. □

Remark 7.7.4 We have some remarks regarding the test above:

1. Similar to the ratio and root tests, if the limit in (7.6) is $L = 1$, then the test is inconclusive.
2. Raabe's test also works if $\lim_{n\to\infty} n\left(\left|\frac{a_n}{a_{n+1}}\right| - 1\right) = \infty$ for which we conclude that the series converges absolutely. The readers are invited to prove this in Exercise 7.27.
3. We can also prove the limit form of Raabe's test in Theorem 7.7.3 independently of Theorem 7.7.1 using another series convergence test called Kummer's test. This will be done in Exercise 7.34.

Similar to the ratio and root tests, Raabe's test can also be carried out with limit superior and limit inferior if the limit (7.6) does not exist. We state the following result and the proof is left as Exercise 7.27.

Theorem 7.7.5 (Generalised Raabe's Test—Limit Form) *Let $\sum_{j=1}^{\infty} a_j$ be a real series such that $a_j \neq 0$ for all $j \in \mathbb{N}$. Define the sequence (b_n) where $b_n = n\left(\left|\frac{a_n}{a_{n+1}}\right| - 1\right)$.*

1. *If $\liminf_{n\to\infty} b_n > 1$, then the series converges absolutely.*
2. *If $\limsup_{n\to\infty} b_n < 1$, then the series $\sum_{j=1}^{\infty} |a_j|$ diverges.*

Example 7.7.6 Let us look at the following examples:

1. Consider the series $\sum_{j=1}^{\infty} \frac{1}{j^2}$. We have seen that the ratio test fails to conclude anything with regards to this series. Let us see if the Raabe's test could tell us something. We have:

$$\lim_{j\to\infty} j\left(\frac{a_j}{a_{j+1}} - 1\right) = \lim_{j\to\infty} j\left(\frac{(j+1)^2}{j^2} - 1\right) = \lim_{j\to\infty} \frac{2j+1}{j} = 2 > 1,$$

from which we conclude that the series converges absolutely. Hence Raabe's test is a success here.

2. Suppose that $\sum_{j=1}^{\infty} \frac{(2j)!}{(j!)^2 4^j}$ is a real series of positive terms which we want to investigate. The usual first plan of action is to apply the ratio test to get:

$$\lim_{j\to\infty} \frac{a_{j+1}}{a_j} = \lim_{j\to\infty} \frac{(2j+2)!}{((j+1)!)^2 4^{j+1}} \frac{(j!)^2 4^j}{(2j)!} = \lim_{j\to\infty} \frac{(2j+2)(2j+1)}{4(j+1)^2} = 1.$$

Therefore the ratio test does not tell us anything here. Let us try Raabe's test instead:

$$\lim_{j\to\infty} j\left(\frac{a_j}{a_{j+1}} - 1\right) = \lim_{j\to\infty} j\left(\frac{4(j+1)^2}{(2j+2)(2j+1)} - 1\right) = \lim_{j\to\infty} \frac{j}{2j+1} = \frac{1}{2} < 1,$$

and thus we conclude that the series diverges.

3. Consider the harmonic series $\sum_{j=1}^{\infty} \frac{1}{j}$. The ratio test could not conclude anything for it. Let us try Raabe's test next. We compute:

$$\lim_{j\to\infty} j\left(\frac{\frac{1}{j}}{\frac{1}{j+1}} - 1\right) = \lim_{j\to\infty} j\left(\frac{j+1}{j} - 1\right) = \lim_{j\to\infty} 1 = 1.$$

So the Raabe's test in limit form is also inconclusive here.

7.8 Dirichlet's and Abel's Tests

Next, we are going to look at Dirichlet's and Abel's tests for series convergence. Before we do that, we would like to introduce summation by parts. This trick is also called Abel's lemma or Abel transformation and allows us to rewrite a sum products in a different way.

Lemma 7.8.1 (Summation by Parts) *Given two series $\sum_{j=1}^{\infty} a_j$ and $\sum_{j=1}^{\infty} b_j$. Suppose that the former has partial sums $s_n = \sum_{j=1}^{n} a_j$ for $n \in \mathbb{N}$ and $s_0 = 0$. Then, for any $m, n \in \mathbb{N}$ with $m < n$ we have the following identity:*

$$\sum_{j=m}^{n} a_j b_j = (s_n b_n - s_{m-1} b_m) - \sum_{j=m}^{n-1} s_j (b_{j+1} - b_j).$$

The reason why this is called summation by parts is because it is an analogue of integration by parts in Proposition 16.1.7 for summation: the summation can be seen as the integral and the terms in the bracket outside the summation are the boundary terms. The proof of this is entirely computational and we leave this to the readers to verify in Exercise 7.28. It is a very useful identity to have, which we demonstrate in the following example:

Example 7.8.2 Consider the sum of squares $\sum_{j=1}^{n} j^2$. We have found an expression for this in Exercise 5.35. Let us work through a different proof by using the summation by parts identity with $a_j = b_j = j$. Thus, $\sum_{j=1}^{n} a_j = \sum_{j=1}^{n} b_j = \sum_{j=1}^{n} j$. This means $s_n = \frac{n(n+1)}{2}$ and $s_0 = 0$. With $m = 1$, summation by parts says:

$$\sum_{j=1}^{n} j^2 = \frac{n(n+1)}{2}n - \sum_{j=1}^{n-1} \frac{j(j+1)}{2}(j+1-j) = \frac{n^2(n+1)}{2} - \sum_{j=1}^{n-1} \frac{j^2}{2} - \sum_{j=1}^{n-1} \frac{j}{2}$$

Using algebraic manipulations, we get:

$$\frac{3}{2}\sum_{j=1}^{n-1} j^2 + n^2 = \frac{n^2(n+1)}{2} - \frac{(n-1)n}{4}$$

$$\Rightarrow \quad \sum_{j=1}^{n} j^2 = \frac{n^2}{3} + \frac{n^2(n+1)}{3} - \frac{(n-1)n}{6} = \frac{n(2n+1)(n+1)}{6}.$$

The following series convergence test is attributed to Dirichlet and was published posthumously in 1862.

Theorem 7.8.3 (Dirichlet's Test) *Suppose that:*

1. *(a_n) is a real sequence with bounded partial sums, and*
2. *(b_n) is a monotone sequence of real numbers such that* $\lim_{n\to\infty} b_n = 0$.

Then, the series $\sum_{j=1}^{\infty} a_j b_j$ converges.

Proof WLOG, suppose that (b_n) is a decreasing sequence of non-negative numbers with $b_n \to 0$. Let us denote the n-th partial sum of the series $\sum_{j=1}^{\infty} a_j$ and $\sum_{j=1}^{\infty} a_j b_j$ as (s_n) and (t_n) respectively and introduce $s_0 = 0$. From the assumption, since the partial sum (s_n) is bounded, there exists some $M > 0$ such that $|s_n| \leq M$ for all $n \in \mathbb{N}$. We can rewrite the partial sum t_n using summation by parts as:

$$t_n = \sum_{j=1}^{n} a_j b_j = s_n b_n - \sum_{j=1}^{n-1} s_j (b_{j+1} - b_j). \tag{7.8}$$

Now we show that both $s_n b_n$ and $\sum_{j=1}^{n-1} s_j (b_{j+1} - b_j)$ converge as $n \to \infty$.

1. For the first one, note that (s_n) is bounded from above by M so we have the bounds $0 \leq |s_n b_n| \leq M b_n$. Since $b_n \to 0$, the sandwich lemma implies $|s_n b_n| \to 0$ and Lemma 5.9.3 says $s_n b_n \to 0$ as well.
2. For the second one, we prove that the series converges absolutely. Using the fact that (b_n) is decreasing, we first note that the series $\sum_{j=1}^{\infty} M(b_j - b_{j+1})$ is an increasing series since all the terms are non-negative. Moreover, it is a telescoping sum. Thus:

$$\sum_{j=1}^{n-1} M(b_j - b_{j+1}) = M(b_1 - b_n)$$

$$\Rightarrow \quad \sum_{j=1}^{\infty} M(b_j - b_{j+1}) = \lim_{n\to\infty} M(b_1 - b_n) = M b_1.$$

Using this fact, for any $n \in \mathbb{N}$ we have:

$$\sum_{j=1}^{n-1} |s_j(b_{j+1} - b_j)| = \sum_{j=1}^{n-1} |s_j||b_{j+1} - b_j| \leq \sum_{j=1}^{n-1} M(b_j - b_{j+1})$$
$$\leq \sum_{j=1}^{\infty} M(b_j - b_{j+1}) = M b_1.$$

Applying Proposition 7.2.11, the series $\sum_{j=1}^{\infty} s_j(b_{j+1} - b_j)$ is absolutely convergent and hence convergent.

Thus, taking the limit as $n \to \infty$ in (7.8), we conclude that (t_n) is convergent and this proves the theorem. □

Remark 7.8.4 We note that the Dirichlet's test is a generalisation of the alternating series test in Theorem 7.4.2. Indeed, for an alternating series $\sum_{j=1}^{\infty}(-1)^j c_j$ where $c_j > 0$ and decreasing to 0, if we choose $a_n = (-1)^n$ and $b_n = c_n$, these two choices fulfil the conditions in Dirichlet's test and hence the series converges.

A convergence test that is related to Dirichlet's test, credited to Niels Henrik Abel, is the following:

Theorem 7.8.5 (Abel's Test) *Suppose that:*

1. *(a_n) is a real sequence such that the series $\sum_{j=1}^{\infty} a_j$ is convergent, and*
2. *(b_n) is a monotone and bounded sequence.*

Then, the series $\sum_{j=1}^{\infty} a_j b_j$ is convergent.

Proof WLOG, suppose that the sequence (b_n) is a decreasing and bounded sequence. By Theorem 5.4.2, it converges to some $L \in \mathbb{R}$. Define a new sequence (c_n) where $c_n = b_n - L$. The sequence (c_n) is also decreasing and it converges to 0.

Furthermore, since the series $\sum_{j=1}^{\infty} a_j$ converges, its sequence of partial sums is bounded. Therefore, we can apply Dirichlet's test to conclude that the series $\sum_{j=1}^{\infty} a_j c_j$ converges. Next, we note that:

$$\sum_{j=1}^{n} a_j b_j = \sum_{j=1}^{n} a_j c_j + \sum_{j=1}^{n} a_j L = \sum_{j=1}^{n} a_j c_j + L \sum_{j=1}^{n} a_j,$$

for any $n \in \mathbb{N}$. By taking the limit as $n \to \infty$ on both sides, since both series on the RHS are convergent, we conclude that $\sum_{j=1}^{\infty} a_j b_j$ is also convergent. □

Example 7.8.6 Consider the series $\sum_{j=1}^{\infty} \frac{\cos(j)}{j}$. We want to show that this series converges. Direct comparison test does not work here since $|\cos(j)| \leq 1$ and we would have to compare this series with the divergent harmonic series. Thus we need to look at more sophisticated comparison tests.

We shall attempt to use Dirichlet's test here. First, we need to decide which term will be (a_n) and (b_n) in the conditions of the test. Since we require (b_n) to be monotone and convergent to 0, we pick $b_n = \frac{1}{n}$ and that leaves $a_n = \cos(n)$. Now we check if this choice works: whether the partial sums $s_n = \sum_{j=1}^{n} \cos(j)$ are bounded.

To show this, we could use complex numbers and geometric series, but we are going to leave this routine computational method as an exercise to the readers. Here we are going to use a clever trick to turn this sum into a telescoping sum. What we do is multiply this partial sum s_n with $2\sin(\frac{1}{2})$ and use the product-to-sum formula $2\sin(x)\cos(y) = \sin(y+x) - \sin(y-x)$. We get:

$$\begin{aligned} 2\sin\left(\frac{1}{2}\right) s_n = \sum_{j=1}^{n} 2\sin\left(\frac{1}{2}\right)\cos(j) &= \sum_{j=1}^{n} \left(\sin\left(j+\frac{1}{2}\right) - \sin\left(j-\frac{1}{2}\right)\right) \\ &= \sin\left(n+\frac{1}{2}\right) - \sin\left(\frac{1}{2}\right), \end{aligned}$$

where the consecutive terms cancel each other. Thus, using triangle inequality, we get:

$$\left|2\sin\left(\frac{1}{2}\right) s_n\right| = \left|\sin\left(n+\frac{1}{2}\right) - \sin\left(\frac{1}{2}\right)\right| \leq \left|\sin\left(n+\frac{1}{2}\right)\right| + \left|\sin\left(\frac{1}{2}\right)\right| \leq 2,$$

which implies $|s_n| \leq \frac{1}{\sin(\frac{1}{2})}$ for all $n \in \mathbb{N}$. So all the requirements for the Dirichlet's test are satisfied and thus we can conclude that the series $\sum_{j=1}^{\infty} \frac{\cos(j)}{j}$ converges.

In fact, there are many other series convergence tests out there. The readers will prove the p-series test in Exercise 7.12, the Cauchy condensation test in Exercise 7.32, Kummer's test in Exercise 7.34, and Bertrand's test in Exercise 7.35. Later on, we shall see integral test in Theorem 16.4.14 once we have defined improper Riemann integrals. There is also the Gauss's test which pushes the results for Raabe's test further. For more information, readers are directed to the comprehensive book on infinite series [10].

We close this chapter with a sombre remark that there is no universal test that would allow us to determine whether any given real series converges. In fact, there are many series out there for which their convergence is still unknown. Here are some notable examples:

1. Consider the series $\sum_{j=1}^{\infty} \frac{\csc^2(j)}{j^3}$. This innocent looking series is called the Flint Hills series and was introduced by Clifford A. Pickover (1957-). To this day, its convergence/divergence behaviour is still unknown.
2. A related series to the previous example is the Cookson Hills series $\sum_{j=1}^{\infty} \frac{\sec^2(j)}{j^3}$, whose convergence is also still unknown.
3. Yet another series whose convergence behaviour is still an open problem is the series $\sum_{j=1}^{\infty}(-1)^j \frac{j}{p_j}$ where p_j is the j-th prime number seen in Example 5.7.4(6). From the asymptotics in the example, it is tempting to claim that we can deduce its convergence via the alternating series test since $\frac{n}{p_n} \to 0$ by Exercise 5.13. However, we have one setback: it is unknown whether the sequence $(\frac{n}{p_n})$ is decreasing which is another important prerequisite for using the alternating series test.

Exercises

7.1 Consider the real sequence (a_n) defined recursively as $a_1 = \sqrt{2}$ and $a_n = \sqrt{2a_{n-1}}$ for all $n \geq 2$. Find the limit of a_n as $n \to \infty$.

7.2 $(*)$ Let (a_n) be a sequence non-negative real numbers. Suppose that the series $\sum_{j=1}^{\infty} a_j$ converges. Define a new sequence (b_n) as $b_n = \frac{a_n + a_{n+1}}{2}$. Prove that the series $\sum_{j=1}^{\infty} b_j$ also converges.

7.3 $(*)$ Prove that the series $\sum_{j=1}^{\infty}(j!)^{\frac{1}{j}}$ does not converge.

7.4 $(\diamond)$ Let (a_n) be a real sequence. Define a real sequence (c_j) where $c_j = \frac{1}{j}\sum_{k=1}^{j} a_k$ is the arithmetic mean of the first j terms in the sequence (a_n). Show that if $a_n \to a \in \mathbb{R}$, then $c_n \to a$ as well.

7.5 $(*)$ Prove that any real number with a periodic decimal representation is a rational number.

7.6 $(*)$ Prove Proposition 7.2.8, namely:
Let $\sum_{j=1}^{\infty} a_j$ and $\sum_{j=1}^{\infty} b_j$ be convergent real series.
(a) For any $\lambda \in \mathbb{R}$, the series $\sum_{j=1}^{\infty} \lambda a_j$ converges and is equal to $\lambda \sum_{j=1}^{\infty} a_j$.
(b) The series $\sum_{j=1}^{\infty}(a_j + b_j)$ converges and is equal to $\sum_{j=1}^{\infty} a_j + \sum_{j=1}^{\infty} b_j$.

7.7 Let (a_n) be a sequence of real numbers. Prove or provide a counterexample to the following statements:
(a) If the sequence $(n^2 a_n)$ converges to 0, then the series $\sum_{k=1}^{\infty} a_j$ converges.
(b) If the series $\sum_{j=1}^{\infty} a_j$ converges, then the series $\sum_{j=1}^{\infty} a_j^2$ converges.
(c) If the series $\sum_{j=1}^{\infty} a_j$ converges absolutely, then the series $\sum_{j=1}^{\infty} a_j^2$ converges.
(d) If the series $\sum_{j=1}^{\infty} a_j^2$ converges, then the series $\sum_{j=1}^{\infty} \frac{a_j}{j}$ converges absolutely.

7.8 $(*)$ Suppose that the real series $\sum_{j=1}^{\infty} a_j$ converges.
(a) Prove that $\lim_{n\to\infty} \sum_{j=n}^{\infty} a_j = 0$.
(b) Hence show that for any $\varepsilon > 0$ there exists an $N \in \mathbb{N}$ such that $|\sum_{j=n}^{\infty} a_j| < \varepsilon$ for all $n \geq N$. In other words, the tail of a convergent series can become arbitrarily small.

7.9 $(*)$ Using direct or limit comparison test, determine whether the following series converges:
(a) $\sum_{j=1}^{\infty} \frac{2j+1}{(j+1)(j+2)^2}$.
(b) $\sum_{j=1}^{\infty} \frac{10^j}{4^{2j+1}(j+1)}$.
(c) $\sum_{j=1}^{\infty} \frac{j+2^j}{j2^j}$.
(d) $\sum_{j=1}^{\infty} (\sqrt{j+1} - \sqrt{j})$.
(e) $\sum_{j=1}^{\infty} \frac{\sqrt{j+2}-\sqrt{j+1}}{j}$.
(f) $\sum_{j=1}^{\infty} \frac{1}{3^j+j}$.
(g) $\sum_{j=1}^{\infty} \frac{(2j+1)(3j-1)}{(j+1)(j+2)^2}$.
(h) $\sum_{j=1}^{\infty} \frac{1}{j \cdot j^{\frac{1}{j}}}$.

7.10 $(*)$ Using ratio or root test, determine whether the following series converges:
(a) $\sum_{j=1}^{\infty} \frac{10^j}{4^{2j+1}(j+1)}$.
(b) $\sum_{j=1}^{\infty} \frac{9^j}{(-1)^{j+1}j}$.
(c) $\sum_{j=1}^{\infty} \frac{j^2+2j+1}{3^j+2}$.
(d) $\sum_{j=1}^{\infty} \frac{5^j}{j^j}$.
(e) $\sum_{j=1}^{\infty} \frac{j^3}{j!}$.
(f) $\sum_{j=1}^{\infty} \frac{j^j}{j!}$.
(g) $\sum_{j=1}^{\infty} \frac{(-2)^j}{j}$.
(h) $\sum_{j=1}^{\infty} \left(1+\frac{1}{j}\right)^{j^2}$.

7.11 $(*)$ Show that for the following series the ratio test is inconclusive. Thus use a different test to determine whether the following series converges:
(a) $\sum_{j=1}^{\infty} \frac{(-1)^j}{j^2+1}$.

(b) $\sum_{j=1}^{\infty} \frac{j+1}{2j+7}$.
(c) $\sum_{j=1}^{\infty} \frac{1}{(2j-1)(2j)}$.
(d) $\sum_{j=1}^{\infty} \frac{\sqrt{j+1}}{(2j^2-3j+1)(\ln(j)+(\ln(j))^2)}$
(e) $\sum_{j=1}^{\infty} \binom{r}{j}$ for $r \in \mathbb{R}_{>0} \setminus \mathbb{N}$ where $\binom{r}{j}$ are the generalised binomial coefficients given by:

$$\binom{r}{j} = \prod_{k=1}^{j} \frac{r-k+1}{k} = \frac{r}{1} \cdot \frac{r-1}{2} \cdot \dots \cdot \frac{r-j+1}{j}$$
$$= \frac{r(r-1)\dots(r-j+1)}{j!}.$$

(f) $\sum_{j=1}^{\infty} \binom{r}{j}$ for $r \leq -1$ where $\binom{r}{j}$ are the generalised binomial coefficients.
(g) $\sum_{j=1}^{\infty} \binom{r}{j}$ for $r \in (-1, 0)$ where $\binom{r}{j}$ are the generalised binomial coefficients.
(h) $\sum_{j=1}^{\infty} \frac{(2j-1)!!}{(2j)!!}$ where the double factorial on the natural number $n \in \mathbb{N}$ is defined as:

$$n!! = \begin{cases} \prod_{j=1}^{\frac{n}{2}} 2j & \text{if } n \text{ is even,} \\ \prod_{j=1}^{\frac{n+1}{2}} (2j-1) & \text{if } n \text{ is odd.} \end{cases}$$

In other words, the double factorial is similar to the factorial, but we only multiply out the positive integers smaller than or equal to n that have the same parity as n.

7.12 (∗) In this question, we are going to prove the p-series test.

Theorem 7.8.7 (**p-series Test**) *Let $\sum_{j=1}^{\infty} \frac{1}{j^p}$ be a real series for some $p \in \mathbb{R}$. The series diverges for $p \leq 1$ and converges for $p > 1$.*

Most of the work for the proof has been done throughout the chapter. We have seen the following results:

1. For $p \leq 0$, the series diverges since the terms in the series does not converge to 0.
2. For $p = 1$, the series is a harmonic series which diverges as in Example 7.2.7.
3. For $0 < p < 1$, the series diverges by direct comparison test with the harmonic series.
4. For $p = 2$, the series converges as seen in Example 7.5.3.
5. For $p > 2$, the series converges by direct comparison test with the case of $p = 2$.

Now we want to complete the list above by looking at the case $1 < p < 2$. This proof is from [30]. Fix any such p and let (s_n) be the partial sum $s_n = \sum_{j=1}^{n} \frac{1}{j^p}$.

(a) Prove that for all $n \in \mathbb{N}$ we have $s_n < s_{2n} < 1 + \frac{2}{2^p} s_n$.
(b) Deduce that $0 < (1 - \frac{2}{2^p})s_n < 1$.
(c) Show that (s_n) is an increasing and bounded sequence.
Hence, deduce that it converges.

7.13 (a) Let (a_n) and (b_n) be two non-negative real sequences such that (b_n) is a bounded sequence. Prove that if $\sum_{j=1}^{\infty} a_j$ converges, then the series $\sum_{j=1}^{\infty} a_j b_j$ is also convergent.
(b) With counterexamples, show that the above result is not true if we remove the non-negativity condition on the sequences (a_n) and (b_n).

7.14 Let $\sum_{j=1}^{\infty} a_j$ and $\sum_{j=1}^{\infty} b_j$ be two real series such that for all $n \in \mathbb{N}$ we have $0 \leq a_n \leq b_n \leq a_{n+1}$. Show that the series $\sum_{j=1}^{\infty} a_j$ converges if and only if the series $\sum_{j=1}^{\infty} b_j$ converges.

7.15 Let (a_n) and (b_n) be real sequences such that $\sum_{j=1}^{\infty} a_j^2$ and $\sum_{j=1}^{\infty} b_j^2$ both converge. Deduce that the series $\sum_{j=1}^{\infty} a_j b_j$ converges absolutely.

7.16 (∗) Let (a_n) and (b_n) be two real sequences such that the series $\sum_{j=1}^{\infty} a_j$ and $\sum_{j=1}^{\infty} b_j$ both converge absolutely. Let (c_j) and (d_j) be sequences defined as $c_n = \min\{a_n, b_n\}$ and $d_n = \max\{a_n, b_n\}$ for all $n \in \mathbb{N}$. Prove that the series $\sum_{j=1}^{\infty} c_j$ and $\sum_{j=1}^{\infty} d_j$ both also converge absolutely.

7.17 Construct two positive sequences (a_n) and (b_n) for which $\sum_{j=1}^{\infty} a_j$ and $\sum_{j=1}^{\infty} b_j$ diverge but $\sum_{j=1}^{\infty} c_j$ where $c_n = \min\{a_n, b_n\}$ converges.

7.18 (∗) Let $p, q > 0$ such that some real numbers such that $q < p$. Discuss the convergence or divergence of the series $\sum_{j=2}^{\infty} \frac{1}{j^p - j^q}$ for all the possible values of p and q.

7.19 (∗) Let $p, q > 0$ be positive real numbers and define the sequence (a_n) as $a_n = \frac{p(p+1)\ldots(p+n-1)}{q(q+1)\ldots(q+n-1)}$. Determine when series $\sum_{j=1}^{\infty} a_n$ converges and when it diverges.

7.20 Show that the series $\sum_{j=1}^{\infty} \frac{\sin(\frac{j\pi}{3})}{\sqrt{j+1}}$ does not converge absolutely.

7.21 (◊) Suppose that (a_n) is a decreasing sequence such that the series $\sum_{j=1}^{\infty} a_j$ converges.
(a) Prove that $\lim_{n\to\infty} n a_n = 0$.
(b) Thus, show that the series $\sum_{j=1}^{\infty} j(a_j - a_{j+1})$ also converges and is equal to $\sum_{j=1}^{\infty} a_j$.

7.22 (∗) Prove Proposition 7.5.5, namely:
Let $\sum_{j=1}^{\infty} a_j$ and $\sum_{j=1}^{\infty} b_j$ be two real series such that $a_j \geq 0$ and $b_j > 0$ for all $j \in \mathbb{N}$.
(a) If $\lim_{j\to\infty} \frac{a_j}{b_j} = 0$ and the series $\sum_{j=1}^{\infty} b_j$ converges, then the series $\sum_{j=1}^{\infty} a_j$ converges.

(b) If $\lim_{j\to\infty} \frac{a_j}{b_j} = \infty$ and the series $\sum_{j=1}^{\infty} b_j$ diverges, then the series $\sum_{j=1}^{\infty} a_j$ diverges.

7.23 Suppose that the real series $\sum_{j=1}^{\infty} j a_j$ converges. Prove that the series $\sum_{j=1}^{\infty} j a_{j+1}$ also converges.

7.24 (∗) We are going to prove the generalised ratio test in Theorem 7.6.7. We follow the following steps:

(a) Let $a, b \in \mathbb{R}$ be two fixed real numbers. Suppose that for each $r \geq b$ we have $r \geq a$ as well. Prove that $a \leq b$.

(b) Using part (a), prove the inequality in Lemma 7.6.10, namely:
Let (a_n) be a sequence of non-zero terms. Then:

$$\limsup_{j\to\infty} \sqrt[j]{|a_j|} \leq \limsup_{j\to\infty} \left|\frac{a_{j+1}}{a_j}\right|.$$

(c) Deduce the positive case of the generalised ratio test, namely:
Let $\sum_{j=1}^{\infty} a_j$ be a real series such that $a_j \neq 0$ for all $j \in \mathbb{N}$. If $\limsup_{j\to\infty} \left|\frac{a_{j+1}}{a_j}\right| < 1$, then the series converges absolutely.

(d) Using analogous arguments from parts (a)-(b), prove the following inequality in Lemma 7.6.10:

$$\liminf_{j\to\infty} \left|\frac{a_{j+1}}{a_j}\right| \leq \liminf_{j\to\infty} \sqrt[j]{|a_j|} \leq \limsup_{j\to\infty} \sqrt[j]{|a_j|}.$$

Thus, deduce the negative case for the generalised ratio test, namely:
Let $\sum_{j=1}^{\infty} a_j$ be a real series such that $a_j \neq 0$ for all $j \in \mathbb{N}$. If $\liminf_{j\to\infty} \left|\frac{a_{j+1}}{a_j}\right| > 1$, then the series diverges.

(e) Can you provide a direct proof for the generalised ratio test using ε-N definition?

7.25 Using Exercise 7.24, provide another proof for the limit $\lim_{n\to\infty} n^{\frac{1}{n}} = 1$.

7.26 (∗) Prove the second case of Raabe's test in Theorem 7.7.1, namely:
Let $\sum_{j=1}^{\infty} a_j$ be a real series such that $a_j \neq 0$ for all $j \in \mathbb{N}$. Suppose further that there exists an index $N \in \mathbb{N}$ such that $\left|\frac{a_{n+1}}{a_n}\right| \geq 1 - \frac{1}{n}$ for all $n \geq N$. Show that the series $\sum_{j=1}^{\infty} |a_j|$ diverges.

7.27 In this question, we are going to prove the degenerate case for Raabe's theorem and Theorem 7.7.5. Let $\sum_{j=1}^{\infty} a_j$ be a real series such that $a_j \neq 0$ for all $j \in \mathbb{N}$. Define the sequence (b_n) where $b_n = n\left(\left|\frac{a_n}{a_{n+1}}\right| - 1\right)$. Prove that:

(a) If $\lim_{n\to\infty} b_n = \infty$, then the series converges absolutely.

(b) If $\liminf_{n\to\infty} b_n > 1$, then the series converges absolutely.

(c) If $\limsup_{n\to\infty} b_n < 1$, then the series $\sum_{j=1}^{\infty} |a_j|$ diverges.

7.28 (∗) Prove the summation by parts formula in Lemma 7.8.1, namely:
Given two series $\sum_{j=1}^{\infty} a_j$ and $\sum_{j=1}^{\infty} b_j$. Suppose that the former has partial sums $s_n = \sum_{j=0}^{n} a_j$ for $n \in \mathbb{N}$ and $s_0 = 0$. Then, for any $m, n \in \mathbb{N}$ with $m < n$ show that:

$$\sum_{j=m}^{n} a_j b_j = (s_n b_n - s_{m-1} b_m) - \sum_{j=m}^{n-1} s_j (b_{j+1} - b_j).$$

7.29 Using the summation by parts formula, find the value of the following sums in terms of n:
(a) $\sum_{j=1}^{n} j 2^j$.
(b) $\sum_{j=1}^{n} j^3$.
Using the sum of integers, sum of squares (see Example 7.8.2), and sum of cubes in part (b), deduce the formula for the sum of fourth powers $\sum_{j=1}^{n} j^4$.

7.30 Let (a_j) be a sequence of distinct positive integers. Prove that $\sum_{j=1}^{n} \frac{a_j}{j^2} \geq \sum_{j=1}^{n} \frac{1}{j}$ for any $j \in \mathbb{N}$.
This question appeared in the 1978 International Mathematical Olympiad which is the most prestigious annual mathematical competition for high school students.

7.31 (◊) Recall Example 7.8.6 for which the series $\sum_{j=1}^{\infty} \frac{\cos(j)}{j}$ converges. Does this series converge absolutely?

7.32 In this question, we are going to prove another series convergence test which is called the Cauchy condensation test, which is a generalisation of the proof for the divergence of the harmonic series by Oresme in Example 7.2.7(1). Prove the following result.

Theorem 7.8.8 (Cauchy Condensation Test) *Let (a_n) be a non-negative real sequence that is monotone decreasing. Then:*

$$\sum_{j=1}^{\infty} a_j \text{ converges} \quad \Leftrightarrow \quad \sum_{j=1}^{\infty} 2^j a_{2^j} \text{ converges.}$$

7.33 Using the Cauchy condensation test from Exercise 7.32, determine whether the following series converge:
(a) $\sum_{j=1}^{\infty} \frac{1}{j}$.
(b) $\sum_{j=2}^{\infty} \frac{1}{j \ln(j)}$.
(c) $\sum_{j=2}^{\infty} \frac{1}{(\ln(j))^2}$.
(d) $\sum_{j=2}^{\infty} \frac{1}{j (\ln(j))^2}$.

7.34 In this question, we are going to prove and apply Kummer's test due to Ernst Kummer (1810–1893).

Theorem 7.8.9 (Kummer's Test) *Let $\sum_{j=1}^{\infty} a_j$ be a real series such that $a_j \neq 0$ for all $j \in \mathbb{N}$. Let (ζ_n) be a positive real sequence. Define the sequence (b_n) where $b_n = \zeta_n \left|\frac{a_n}{a_{n+1}}\right| - \zeta_{n+1}$.*

1. If $\lim_{n\to\infty} b_n > 0$, then the series converges absolutely.

2. If $\lim_{n\to\infty} b_n < 0$ and the series $\sum_{j=1}^{\infty} \frac{1}{\zeta_j}$ diverges, then the series $\sum_{j=1}^{\infty} |a_j|$ diverges.

We prove the first assertion in parts (a)-(d) and the second assertion in part (e).

(a) Assuming $\lim_{n\to\infty} b_n > 0$, show that there exists an $N \in \mathbb{N}$ and a $k > 0$ such that $k|a_{n+1}| \leq \zeta_n|a_n| - \zeta_{n+1}|a_{n+1}|$ for all $n \geq N$.

(b) Hence, show that the sequence $(\zeta_n|a_n|)$ is decreasing after the index N and thus converges

(c) Prove that the series $\sum_{j=1}^{\infty}(\zeta_j|a_j| - \zeta_{j+1}|j_{n+1}|)$ converges.

(d) Conclude that the series $\sum_{j=1}^{\infty} |a_j|$ converges.

(e) Now assume that $\lim_{n\to\infty} b_n < 0$ and the series $\sum_{j=1}^{\infty} \frac{1}{\zeta_j}$ diverges. By using similar ideas from parts (a)-(b), prove that the series $\sum_{j=1}^{\infty} |a_n|$ diverges.

Kummer's test is a very general test since we get to choose what the positive sequence (ζ_n) to work with. In fact, the ratio test and Raabe's test can be obtained from Kummer's test.

(f) By choosing a suitable sequence (ζ_n) in Kummer's test, deduce the ratio test.

(g) By choosing a suitable sequence (ζ_n) in Kummer's test, deduce Raabe's test (limit form).

7.35 Now we want to prove Bertrand's test using Kummer's test. This test is credited to Joseph Bertrand (1822–1900) and De Morgan. We state:

Theorem 7.8.10 (Bertrand's Test) *Let $\sum_{j=1}^{\infty} a_j$ be a real series such that $a_j \neq 0$ for all $j \in \mathbb{N}$. Let $b_n = n\ln(n)\left(\left|\frac{a_n}{a_{n+1}}\right| - 1\right) - \ln(n)$. Then:*

1. If $\lim_{n\to\infty} b_n > 1$, then the series converges absolutely.

2. If $\lim_{n\to\infty} b_n < 1$, then the series $\sum_{j=1}^{\infty} |a_j|$ diverges.

To prove Bertrand's test, we follow part (a)-(c):

(a) Prove the asymptotics $(n+1)\ln\left(1+\frac{1}{n}\right) \sim 1+\frac{1}{n}$.

(b) Thus, show that $\lim_{n\to\infty}(n+1)\ln\left(1+\frac{1}{n}\right) = 1$.

(c) Hence, by choosing a suitable sequence (ζ_n) in Kummer's test, deduce Bertrand's test.

Consider the series $\sum_{j=1}^{\infty}\left(\frac{(2j-1)!!}{(2j)!!}\right)^2$ where the double factorial is defined in Exercise 7.11(h).

(d) Show that the ratio and Raabe's tests are both inconclusive for this series. Hence, use Bertrand's test to deduce its convergence.

Additional Topics in Real Series

8

The fact is that the same sequence of days can arrange themselves into a number of different stories.

—Jane Smiley, novelist and Pulitzer prize winner

This short chapter will be devoted to three questions on real series, namely:

1. Can we rearrange the order of addition for the terms in a series?
2. Can we group together terms in a series?
3. We have seen how to add, subtract, and scale series. Can we multiply series too?

The first two question questions are very important. We have seen in Example 7.2.2(5) if we group the terms in a series together in some way, the series could have different convergence behaviour. Likewise, if we shuffle the terms around, something bad might happen. On the other hand, according to the ring and field axioms in Definitions 2.5.1 and 3.1.1, grouping and shuffling terms around in a finite sum is perfectly permissible. Namely, addition of real numbers is associative and commutative. So why are we not allowed to do it for series?

The subtle difference here is that in the ring and field axioms, associativity and commutativity are allowed for addition of finitely many terms via induction. In a series, we have infinitely many terms. Moreover, because the value of a series is defined as a limit, the order of the sequence of addition matters. Therefore, rearranging the series or grouping them in some way could potentially cause some trouble. Except for some cases. We shall see when can we do them in the first two sections of this chapter.

For the third question, multiplication can be tricky to define for a series. Again, we have the distributivity axiom of multiplication over addition of real numbers in Definitions 2.5.1 and 3.1.1, but this only holds for finite sums. Therefore, for infinite sums, we need to think of another way to carry out this multiplication.

S. Johar, *The Big Book of Real Analysis*,
https://doi.org/10.1007/978-3-031-30832-1_8

Moreover, even if we can define a nice multiplication operation on two series, we may have two issues: the resulting series might not converge even if the two original series converge and if it does converge, it might not converge to the value that we expect (namely the numerical product of the original series). We shall explore these questions in the third section.

8.1 Rearrangement of Series

Before we proceed with our rearrangement theorems, we first define some terminologies. For a real series $\sum_{j=1}^{\infty} a_j$ we define two non-negative real sequences (a_n^+) and (a_n^-) where:

$$a_n^+ = \frac{|a_n| + a_n}{2} = \begin{cases} a_n & \text{if } a_n \geq 0, \\ 0 & \text{if } a_n < 0, \end{cases}$$

and

$$a_n^- = \frac{|a_n| - a_n}{2} = \begin{cases} -a_n & \text{if } a_n \leq 0, \\ 0 & \text{if } a_n > 0. \end{cases}$$

This can also be written using maximum and minimum as $a_j^+ = \max\{a_j, 0\}$ and $a_j^- = -\min\{a_j, 0\}$. We call these terms the positive and negative parts of the sequence. We note that we can recover the terms of the original sequence and and its absolute value via $a_n = a_n^+ - a_n^-$ and $|a_n| = a_n^+ + a_n^-$.

Let us consider the positive and negative parts of the series, namely: $\sum_{j=1}^{\infty} a_j^+$ and $\sum_{j=1}^{\infty} a_j^-$. The former only picks up the positive terms in the original series whereas the latter is just the sum of all the negative terms but each turned positive (from the definition) of the original series. For an absolutely convergent series, we have the following lemma:

Lemma 8.1.1 *Let $\sum_{j=1}^{\infty} a_j$ be a real series.*

1. *The series is absolutely convergent if and only if both the positive and negative parts of the series are convergent.*
2. *If the series is absolutely convergent, we have the equality $\sum_{j=1}^{\infty} a_j = \sum_{j=1}^{\infty} a_j^+ - \sum_{j=1}^{\infty} a_j^-$.*

Proof We prove the assertions one by one.

1. We prove the implications separately.

(⇒): By triangle inequality, we note that we have the ordering $0 \leq a_n^+ \leq |a_n|$ for all $n \in \mathbb{N}$. Since the series $\sum_{j=1}^{\infty} |a_j|$ converges, by comparison, the series $\sum_{j=1}^{\infty} a_j^+$ converges as well. By a similar argument, we can show that the series $\sum_{j=1}^{\infty} a_j^-$ converges too.

(⇐): For the converse, we note that $|a_n| = a_n^+ + a_n^-$ for all $n \in \mathbb{N}$. Thus, we have the partial sums $\sum_{j=1}^{n} |a_j| = \sum_{j=1}^{n} a_j^+ + \sum_{j=1}^{n} a_j^-$. Taking the limits on both sides and using the fact that the negative and positive parts of the series both converge, we get:

$$\begin{aligned}\sum_{j=1}^{\infty} |a_j| = \lim_{n\to\infty} \sum_{j=1}^{n} |a_j| &= \lim_{n\to\infty} \left(\sum_{j=1}^{n} a_j^+ + \sum_{j=1}^{n} a_j^- \right) \\ &= \lim_{n\to\infty} \sum_{j=1}^{n} a_j^+ + \lim_{n\to\infty} \sum_{j=1}^{n} a_j^- \\ &= \sum_{j=1}^{\infty} a_j^+ + \sum_{j=1}^{\infty} a_j^- < \infty.\end{aligned}$$

This means the series $\sum_{j=1}^{\infty} a_j$ is absolutely convergent.

2. We note that $a_n = a_n^+ - a_n^-$ for all $n \in \mathbb{N}$ so the partial sums satisfy $\sum_{j=1}^{n} a_j = \sum_{j=1}^{n} a_j^+ - \sum_{j=1}^{n} a_j^-$. Taking the limits on both sides of the equation and applying the algebra of limits, since each of the positive and negative parts of the series are convergent by the previous assertion, we obtain:

$$\begin{aligned}\sum_{j=1}^{\infty} a_j = \lim_{n\to\infty} \sum_{j=1}^{n} a_j &= \lim_{n\to\infty} \left(\sum_{j=1}^{n} a_j^+ - \sum_{j=1}^{n} a_j^- \right) \\ &= \lim_{n\to\infty} \sum_{j=1}^{n} a_j^+ - \lim_{n\to\infty} \sum_{j=1}^{n} a_j^- \\ &= \sum_{j=1}^{\infty} a_j^+ - \sum_{j=1}^{\infty} a_j^-,\end{aligned}$$

giving us the desired equality. □

Now we shall prove that an absolutely convergent series remains the same after rearrangement of the terms, a result due to Dirichlet. We first properly define what a rearrangement of a series means:

Definition 8.1.2 (Rearrangement of a Series) Let $\sum_{j=1}^{\infty} a_j$ be a series. A rearrangement of this series is a series $\sum_{j=1}^{\infty} a_{\sigma(j)}$ where $\sigma : \mathbb{N} \to \mathbb{N}$ is some bijection of the indices.

Theorem 8.1.3 *Let $\sum_{j=1}^{\infty} a_j$ be a real series and $\sum_{j=1}^{\infty} a_{\sigma(j)}$ be its rearrangement. If the series $\sum_{j=1}^{\infty} a_j$ is absolutely convergent with $\sum_{j=1}^{\infty} a_j = L$, then the rearrangement $\sum_{j=1}^{\infty} a_{\sigma(j)}$ is also absolutely convergent with $\sum_{j=1}^{\infty} a_{\sigma(j)} = L$ as well.*

Proof By Lemma 8.1.1, since the series $\sum_{j=1}^{\infty} a_j$ is absolutely convergent, its positive part $\sum_{j=1}^{\infty} a_j^+$ with partial sums (s_n) converges to some $s \in \mathbb{R}$. Note also that since the terms (a_n^+) are non-negative, the sequence of partial sums (s_n) is increasing, so we have $s_m \leq s$ for all $m \in \mathbb{N}$.

Now if we consider the rearranged positive series $\sum_{j=1}^{\infty} a_{\sigma(j)}^+$, this series is also increasing since all the terms are non-negative. Let us denote the partial sums of this series as (t_n) and we want to show that it converges. For any fixed $n \in \mathbb{N}$, the set of indices $\{\sigma(j)\}_{j=1}^n$ is a subset of of the indices $\{1, 2, \ldots, m\}$ for $m = \max\{\sigma(j) : j = 1, \ldots, n\} \in \mathbb{N}$. Thus, the partial sums t_n satisfy:

$$t_n = \sum_{j=1}^{n} a_{\sigma(j)}^+ \leq \sum_{j=1}^{m} a_j^+ = s_m \leq s,$$

for all $n \in \mathbb{N}$.

By monotone sequence theorem, since the sequence (t_n) is increasing and bounded from above, it must converge to some $t \leq s$. Thus, we have the ordering $\sum_{j=1}^{\infty} a_{\sigma(j)}^+ \leq \sum_{j=1}^{\infty} a_j^+$. Using a similar argument, we can show that $\sum_{j=1}^{\infty} a_{\sigma(j)}^- \leq \sum_{j=1}^{\infty} a_j^-$. In particular, since both the positive and negative parts of the rearranged series $\sum_{j=1}^{\infty} a_{\sigma(j)}$ converge, Lemma 8.1.1 implies that the rearranged series is also absolutely convergent.

By symmetry, since the series $\sum_{j=1}^{\infty} a_j$ is a rearrangement of the absoutely convergent series $\sum_{j=1}^{\infty} a_{\sigma(j)}$, using similar arguments as the above, we have $\sum_{j=1}^{\infty} a_j^+ \leq \sum_{j=1}^{\infty} a_{\sigma(j)}^+$ and $\sum_{j=1}^{\infty} a_j^- \leq \sum_{j=1}^{\infty} a_{\sigma(j)}^-$. Therefore, putting all these inequalities together, we obtain:

$$\sum_{j=1}^{\infty} a_j^+ = \sum_{j=1}^{\infty} a_{\sigma(j)}^+ \quad \text{and} \quad \sum_{j=1}^{\infty} a_j^- = \sum_{j=1}^{\infty} a_{\sigma(j)}^-. \tag{8.1}$$

Finally, by applying Lemma 8.1.1 and using the equalities in (8.1) we deduce:

$$\sum_{j=1}^{\infty} a_{\sigma(j)} = \sum_{j=1}^{\infty} a_{\sigma(j)}^{+} - \sum_{j=1}^{\infty} a_{\sigma(j)}^{-} = \sum_{j=1}^{\infty} a_j^{+} - \sum_{j=1}^{\infty} a_j^{-} = \sum_{j=1}^{\infty} a_j = L,$$

which is what we wanted to prove. □

Thus rearranging an absolutely convergent series does not change its convergence property and its value. In contrast, conditionally convergent series have a different behaviour. For starters, their positive and negative parts both diverge.

Lemma 8.1.4 *Let $\sum_{j=1}^{\infty} a_j$ be a conditionally convergent real series. Then, both the positive and negative parts of this series diverge to ∞.*

Proof We first prove that if one of the positive or negative parts converge, the other must converge too. We prove that if the positive parts converge, then the negative parts converge too. Note that $a_j = a_j^+ - a_j^-$ implies $a_j^- = a_j^+ - a_j$. Thus, by taking the limits on the partial sums and using the algebra of limits, since the full series and its positive part both converge, we have:

$$\sum_{j=1}^{\infty} a_j^- = \lim_{n\to\infty} \sum_{j=1}^{n} a_j^- = \lim_{n\to\infty} \left(\sum_{j=1}^{n} a_j^+ - \sum_{j=1}^{n} a_j \right) = \sum_{j=1}^{\infty} a_j^+ - \sum_{j=1}^{\infty} a_j,$$

which is finite. Therefore, the negative part of the series converges as well. In a similar manner, the positive part of the series can be proven to converge if the negative part converges. So for a convergent series, if one of the positive or negative parts converge, then the other would converge too.

However, Lemma 8.1.1 says that if both the positive and negative parts of the series are convergent, then the series must be absolutely convergent. But based on our assumption, our series is only conditionally convergent. So this implies that both of the positive and negative parts of the series are divergent.

Finally, these series of positive and negative parts are monotone series since all the terms are non-negative. By Proposition 7.2.11, since these monotone series diverge, they must be unbounded and hence diverge to ∞. □

Lemma 8.1.4 shows that the positive and negative parts of a conditionally convergent series behave differently from absolutely convergent series. Therefore, we can expect that a rearrangement of a conditionally convergent series would have a different behaviour as well. Indeed it does. More surprisingly, given any real number at all we can rearrange the terms of a conditionally convergent series so that this rearranged series converges to the chosen real number. This theorem is called the Riemann rearrangement theorem or Riemann series theorem due to Bernhard Riemann (1826–1866).

Theorem 8.1.5 (Riemann Rearrangement Theorem) *Let $\sum_{j=1}^{\infty} a_j$ be a conditionally convergent real series and $L \in \mathbb{R}$ be any real number. Then, there is a rearrangement of the series $\sum_{j=1}^{\infty} a_j$ such that it converges to L, namely $\sum_{j=1}^{\infty} a_{\sigma(j)} = L$ for some bijection $\sigma : \mathbb{N} \to \mathbb{N}$.*

Proof WLOG, assume that $a_j \neq 0$ for any $j \in \mathbb{N}$ and let $L > 0$. We denote the non-negative and non-positive terms of the series as (p_j) and (q_j) respectively, namely $p_j = \max\{a_j, 0\} = a_j^+$ and $q_j = \min\{a_j, 0\} = -a_j^-$. We note that both of the sequences have infinitely many non-zero terms, precisely because the positive and negative parts of the series diverge as shown in Lemma 8.1.4.

From the same lemma, we have two monotone series (one increasing and the other decreasing) with $\sum_{j=1}^{n} p_j \to \infty$ and $\sum_{j=1}^{n} q_j \to -\infty$. Moreover, since the original series $\sum_{j=1}^{\infty} a_j$ converges, Proposition 7.2.5 tells us that $\lim_{j\to\infty} a_j = 0$ which then implies that both of the sequences (p_j) and (q_j) also converge to 0 by Exercise 5.11. Now we are ready to rearrange the terms to form a new series that converges to L.

Since $\sum_{j=1}^{n} p_j \to \infty$ is a monotone series, we can find an index $j_1 \in \mathbb{N}$ such that the sum of the first j_1 terms in this series equals or just exceeds L but any fewer terms would not reach L. We denote the sum of this first j_1 terms as s and, by this choice, we have:

$$s = p_1 + p_2 + \ldots + p_{j_1-1} + p_{j_1} \geq L,$$
$$s - p_{j_1} = p_1 + p_2 + \ldots + p_{j_1-1} < L.$$

Now let us add some of the negative terms to this sum. Since $\sum_{j=1}^{n} q_j \to -\infty$ is also a monotone series, we can add as big of a negative quantity from this series as we wish to s. However the aim is to bring the sum s down to equal or just below L which would require the first few, say i_1, terms of this negative sequence and no fewer. We call this new sum s_{k_1} which satisfies:

$$s_{k_1} = s + q_1 + q_2 + \ldots + q_{i_1-1} + q_{i_1} \leq L,$$
$$s_{k_1} - q_{i_1} = s + q_1 + q_2 + \ldots + q_{i_1-1} > L.$$

As a result, we have $|L - s_{k_1}| < -q_{i_1}$. We repeat this construction indefinitely, each time taking just enough of the positive terms to go slightly above the number L and then taking just enough negative terms to go slightly below the number L. At the n-th stage, we would have the partial sum:

$$s_{k_n} = \underbrace{p_1 + \ldots + p_{j_1}}_{\text{positive terms}} + \underbrace{q_1 + \ldots + q_{i_1}}_{\text{negative terms}} + \underbrace{p_{j_1+1} + \ldots + p_{j_2}}_{\text{positive terms}} + \ldots + \underbrace{q_{i_{n-1}+1} + \ldots + q_{i_n}}_{\text{negative terms}},$$

which is the partial sum of some rearrangement of the terms a_j. Along with it, this partial sums satisfie $s_{k_n} \leq L$ and $|L - s_{k_n}| < -q_{i_n}$ for all $n \in \mathbb{N}$. Since $q_n \to 0$, the subsequence (q_{i_n}) also converges to 0. By applying the sandwich lemma, we have:

$$0 \leq \lim_{n\to\infty} |L - s_{k_n}| \leq \lim_{n\to\infty} (-q_{i_n}) \quad \Rightarrow \quad \lim_{n\to\infty} |s_{k_n} - L| = 0,$$

and, by using Lemma 5.9.3, we have the convergence $\lim_{n\to\infty} s_{k_n} = L$.

Now we need to show that the whole rearranged series converges to L. Extending the notation we used above, we denote the n-th partial sum of this rearranged series as s_n and fix $\varepsilon > 0$. Since $s_{k_n} \to L$, we can find an $N_1 \in \mathbb{N}$ such that $|s_{k_n} - L| < \varepsilon$ for all $n \geq N_1$. Furthermore, since $p_n, q_n \to 0$, there exist $N_2, N_3 \in \mathbb{N}$ such that $|p_n| < \varepsilon$ for all $n \geq N_2$ and $|q_n| < \varepsilon$ for all $n \geq N_3$. Pick $N = \max\{N_1, N_2, N_3\}$. Thus, for any $n \geq N$, we have $L - \varepsilon < s_{k_n} \leq L$.

Fix any $n \geq N$. Then, we have:

$$s_{k_{n+1}} = s_{k_n} + \underbrace{p_{j_n+1} + \ldots + p_{j_{n+1}}}_{\text{positive terms}} + \underbrace{q_{i_n+1} + \ldots + q_{i_{n+1}}}_{\text{negative terms}}. \tag{8.2}$$

We note that since the positive terms in (8.2) are all smaller than ε and the index j_{n+1} is the first time the sum $s_{k_n} + p_{j_n+1} + \ldots + p_{j_{n+1}}$ exceeds L, all the partial sums after adding the positive terms (namely the partial sums s_m for $m = k_n + 1, k_n + 2, \ldots, k_n + (j_{n+1} - j_n)$) must be contained in $[s_{k_n}, L + \varepsilon) \subseteq (L - \varepsilon, L + \varepsilon)$. By the same argument, the partial sums after adding up the negative terms, namely the partial sums s_m where $m = k_n + (j_{n+1} - j_n) + 1, \ldots, k_{n+1} - 1$, are all between $(L - \varepsilon, L + \varepsilon)$.

This implies the partial sums s_m for all $m \in \{k_n, k_n + 1, k_n + 2, \ldots, k_{n+1}\}$ satisfy $L - \varepsilon < s_m < L + \varepsilon$, namely $|s_m - L| < \varepsilon$. Since $n \geq N$ was arbitrarily fixed, we have shown that $|s_m - L| < \varepsilon$ for all $m \geq k_N$. Therefore, we conclude that the full rearranged series $\sum_{j=1}^{\infty} a_{\sigma(j)}$ converges to L. □

The Riemann rearrangement theorem above is the reason why we should be wary about rearranging the terms in a series unless it is absolutely convergent for which Theorem 8.1.3 says it is safe to do so. If we rearrange the terms of a conditionally convergent series, we might unwittingly change its value! In fact, one can also prove that there are rearrangements of a conditionally convergent series that result with series that blow up to $\pm\infty$. This is a problem that the readers can think about in Exercise 8.4.

However, as we have did many times before, it is perfectly safe to withhold or add finitely many terms at the beginning of a series as we have seen in Proposition 7.2.9. This is true because these sums are finite and they would not affect the convergence and limit behaviour of the sequence of partial sums (s_n). Due to the same reason, we are also allowed to rearrange finitely many terms in a convergent series without changing its value.

Proposition 8.1.6 *Let $\sum_{j=1}^{\infty} a_j$ be a convergent real series. Suppose that $\sigma : \mathbb{N} \to \mathbb{N}$ is a bijection such that there exists an index $N \in \mathbb{N}$ where $\sigma(n) = n$ for all*

$n \geq N$. Then, the rearranged series $\sum_{j=1}^{\infty} a_{\sigma(j)}$ also converges and $\sum_{j=1}^{\infty} a_j = \sum_{j=1}^{\infty} a_{\sigma(j)}$.

Proof Write (s_n) and (t_n) as the sequences of partial sums of the series $\sum_{j=1}^{\infty} a_j$ and $\sum_{j=1}^{\infty} a_{\sigma(j)}$ respectively. By assumption, since $\sigma(n) = n$ for all $n \geq N$ and σ is an injection, for all $1 \leq j \leq N-1$, the image $\sigma(j)$ cannot be any integer greater than or equal to N. So $\sigma(j) \in \{1, 2, \ldots, N-1\}$ for all $j \in \{1, 2, \ldots, N-1\}$. Define $X = \{1, 2 \ldots, N-1\}$ and $Y = \{\sigma(1), \sigma(2), \ldots, \sigma(N-1)\}$. The above observation tells us $Y \subseteq X$. Moreover, since σ is an injection, we have $|Y| = N - 1 = |X|$. Note that $X = Y \cup (X \setminus Y)$ is a disjoint union so, by Lemma 3.4.8 we have $|X| = |X \setminus Y| + |Y|$. Thus, $|X \setminus Y| = 0$ and so $X \setminus Y = \varnothing$. Exercise 1.18 then says $X \subseteq Y$. Therefore, $X = Y$.

The above says each of the first $N-1$ terms in the series $\sum_{j=1}^{\infty} a_j$ is also in the first $N-1$ terms of the rearranged series $\sum_{j=1}^{\infty} a_{\sigma(j)}$ and vice versa. Hence $s_{N-1} = t_{N-1}$. Moreover, for any $n \geq N$, we have $s_n = s_{N-1} + a_N + \ldots + a_n = t_{N-1} + a_{\sigma(N)} + \ldots + a_{\sigma(n)} = t_n$ since $a_n = a_{\sigma(n)}$ for all $n \geq N$. Taking the limit as $n \to \infty$, we conclude that the rearranged series converges to the same limit as the original series. □

8.2 Bracketing of Series

There is another issue that we have to address regarding series, namely bracketing or grouping terms in a series together. But first, we need to define what is bracketing.

Simply put, bracketing is the operation of grouping consecutive terms together with a bracket $(\ldots)$ so that we carry out the operations within the bracket first. For example, we can bracket the sum $1+2+3+4$ in five different ways as:

$$\begin{aligned}
(1+2)+(3+4) &= 3+7 = 10, \\
(1+(2+3))+4 &= (1+5)+4 = 10, \\
((1+2)+3)+4 &= (3+3)+4 = 6+4 = 10, \\
1+(2+(3+4)) &= 1+(2+7) = 1+9 = 10, \\
1+((2+3)+4) &= 1+(2+7) = 1+9 = 10.
\end{aligned}$$

By associativity of addition axiom on real numbers, these are all the same number. So the value of a finite sum is independent of its bracketing.

For a real series, which is an infinite sum, there are many more ways to place the brackets. There are two kinds of bracket we can distinguish, namely infinite brackets and finite brackets. Consider the real series $\sum_{j=1}^{\infty} a_j$.

1. Infinite bracket is a bracket where there exists an $n \in \mathbb{N}$ such that the bracket (is placed before the term a_n and the other bracket is placed at infinity. This means

before we evaluate the full series, we have to evaluate the subseries $a_n + a_{n+1} + \ldots = \sum_{j=n}^{\infty} a_j$ first.

2. Finite brackets are brackets placed around finitely many terms, namely there are $n \geq m$ such that the left bracket (is placed before the term a_m and the right bracket) is placed after the a_n term so we have $(a_m + a_{m+1} + \ldots + a_n)$ grouped in the series. This means before we evaluate the whole series as a limit, we have to carry out the finite sums in the brackets first.

Remark 8.2.1 Note that these brackets may be nested, namely a pair of brackets is contained within another pair bracket. However, by associativity of addition for real numbers, we can remove any nested brackets from within finite brackets since finite brackets correspond to a sum of finitely many real numbers, which is independent of bracketing. Therefore, WLOG, we can assume that any finite brackets have no nested brackets within them.

Infinite brackets are easier to handle. This is because if we have the bracket before the term a_n in the series $\sum_{j=1}^{\infty} a_j$, the resulting series is simply $a_1 + a_2 + \ldots + a_{n-1} + (a_n + a_{n+1} + \ldots) = \sum_{j=1}^{n-1} a_j + (\sum_{j=n}^{\infty} a_j)$. Now this becomes a finite sum since there are a total of n terms with the final term being the value (if it exists) of the series $\sum_{j=n}^{\infty} a_j$. We saw from Proposition 7.2.9 that the bracketed term $\sum_{j=n}^{\infty} a_j$ converges if and only if the full series $\sum_{j=1}^{\infty} a_j$ converges. Moreover, if convergence does happen, then their values are the same, namely $\sum_{j=1}^{n-1} a_j + (\sum_{j=n}^{\infty} a_j) = \sum_{j=1}^{\infty} a_j$.

For finite brackets, if there are finitely many pairs of such brackets, then we have no problem with the bracketed series. Indeed, suppose that the final) bracket is placed right after the n-th term in the original series. Therefore, this bracketed series is simply $F(a_1, \ldots, a_n) + a_{n+1} + a_{n+2} + \ldots = F(a_1, \ldots, a_n) + \sum_{j=n+1}^{\infty} a_j$ where $F(a_1, \ldots, a_n)$ is the sum of the first n terms with arbitrary bracketing inserted. However, note that the quantity $F(a_1, \ldots, a_n)$ is well defined since it is a sum of finitely many real numbers and, by the associativity axiom, the value of $F(a_1, \ldots, a_n)$ is independent of the bracketing. Therefore, by Proposition 7.2.9, $F(a_1, \ldots, a_n) + \sum_{j=n+1}^{\infty} a_j = (a_1 + a_2 + \ldots + a_n) + \sum_{j=n+1}^{\infty} a_j = \sum_{j=1}^{n} a_j + \sum_{j=n+1}^{\infty} a_j$ converges if and only if the series $\sum_{j=n+1}^{\infty} a_j$ converges and its value is $\sum_{j=1}^{\infty} a_j$.

So far we have seen that infinite brackets and finitely many finite brackets pose no real threat to the value of a convergent real series. Therefore, the case that we have to worry about is when we have infinitely many finite brackets. If we recall Example 7.2.2(5), we saw that the infinite bracketing of the divergent Grandi's series

$\sum_{j=1}^{\infty}(-1)^j$ gave two conflicting convergence results, namely:

$$\sum_{j=1}^{\infty} a_j = (-1+1)+(-1+1)+(-1+1)+\ldots = 0+0+0+\ldots = 0,$$

$$\sum_{j=1}^{\infty} a_j = -1+(1-1)+(1-1)+(1-1)+\ldots = -1+0+0+0+\ldots = -1,$$

Using Definition 7.2.1, let us investigate what went wrong with the result after bracketing. Suppose that we have a real series $\sum_{j=1}^{\infty} a_j$ and define its sequence of partial sums as (s_n) where $s_n = \sum_{j=1}^{n} a_j$.

Now let us place infinitely many finite brackets in the series. By Remark 8.2.1, we can assume that there are no nested brackets. Moreover, for terms which are not contained in any brackets, we can place them in a single bracket, namely contain them in a bracket by itself. Now every term in the sequence is contained within exactly one bracket.

For every $j \in \mathbb{N}$, we define k_j as the index of the final term in the j-th bracket. As a convention, denote $k_0 = 0$. For $j \in \mathbb{N}$, the sum of the terms in the j-th bracket is a real number as it is a finite sum and we denote it as $b_j = a_{k_{j-1}+1}+a_{k_{j-1}+2}+\ldots+a_{k_j}$. Therefore, the bracketed series can be written as the real series $\sum_{j=1}^{\infty} b_j$.

Let us now denote the partial sums of the bracketed series as (t_n) where $t_n = \sum_{j=1}^{n} b_j = \sum_{j=1}^{n}(a_{k_{j-1}+1} + a_{k_{j-1}+2} + \ldots + a_{k_j})$. In this form, we can see that t_n is simply the sum of the first k_n terms in the original series, namely $t_n = s_{k_n}$. So the sequence of partial sums in the bracketed series (t_n) is a subsequence of the partial sums (s_n). With this observation, we can state the following result:

Proposition 8.2.2 *Let $\sum_{j=1}^{\infty} a_j$ be a real series and $\sum_{j=1}^{\infty} b_j$ be any of its bracketed series with infinitely many terms.*

1. *If the series $\sum_{j=1}^{\infty} a_j$ converges, then the series $\sum_{j=1}^{\infty} b_j$ also converges.*
2. *If the series $\sum_{j=1}^{\infty} a_j$ diverges to $\pm\infty$, then the series $\sum_{j=1}^{\infty} b_j$ also diverges to $\pm\infty$.*

Proof Denote (s_n) and (t_n) as the partial sums of the series $\sum_{j=1}^{\infty} a_j$ and $\sum_{j=1}^{\infty} b_j$ respectively. Recall that (t_n) is a subsequence of (s_n). Applying Proposition 5.5.4, we now note that if (s_n) converges, then (t_n) converges to the same number. Likewise, by the same proposition, if (s_n) blows up to $\pm\infty$, (t_n) also blows up to $\pm\infty$. □

On the other hand, if we have a series that diverges but does not blow up to $\pm\infty$, bracketing the series may not give the same convergence property of the original series. Indeed, by Bolzano-Weierstrass theorem, we have seen that for any bounded sequence, even if the sequence does not converge, we may still find a convergent subsequence within it. This is what we saw with Grandi's series $\sum_{j=1}^{\infty}(-1)^j$, namely: its sequence of partial sums (s_j) is bounded but does not converge, but the subsequences (s_{2n}) and (s_{2n+1}) both converge to 0 and -1 respectively.

As a conclusion, regrouping the terms in a series does not change its value if the series converges or diverges to $\pm\infty$. But if a series does neither, this may not be a good idea!

8.3 Cauchy Product

Now we are going to answer the final question posed at the beginning of this chapter: how can we multiply two series together? First, we need to figure out a candidate for the product of two series. Let us guess how we can do this by looking at products of finite sums first.

For two real finite sums $\sum_{j=1}^{n} a_j$ and $\sum_{j=1}^{n} b_j$, their product can be calculated by using the distributivity of multiplication over addition as:

$$\begin{aligned}\left(\sum_{j=1}^{n} a_j\right)\left(\sum_{j=1}^{n} b_j\right) =& a_1b_1 + a_1b_2 + a_1b_3 + \ldots + a_1a_n\\ &+ a_2b_1 + a_2b_2 + a_2b_3 + \ldots + a_2b_n\\ &+ \vdots + \vdots + \vdots + \vdots + \vdots\\ &+ a_nb_1 + a_nb_2 + a_nb_3 + \ldots + a_nb_n.\end{aligned}$$

How can we encapsulate this sum into one compact summation notation? One way to do this is to group together all the products a_kb_i such that their indices k and i add up to the same integer. Namely, we define $c_j = \sum_{k=1}^{j} a_kb_{j-k+1}$ to be the sum of all the product of terms where the indices add up to $j+1$. From the product above, since the possible sum of the indices run from 2 to $2n$, we therefore have to sum up c_j from $j = 1$ to $j = 2n-1$. In order to do so, we need to define what a_j and b_j are for indices larger than n. For these terms, because they do not contribute anything to the series, we just define them all to be 0. Namely $a_j = b_j = 0$ for all $j > n$. Hence, the sum above can be written compactly as:

$$\left(\sum_{j=1}^{n} a_j\right)\left(\sum_{j=1}^{n} b_j\right) = a_1b_1 + (a_1b_2 + a_2b_1) + \ldots + a_nb_n = \sum_{j=1}^{2n-1} c_j,$$

where $c_j = \sum\limits_{k=1}^{j} a_kb_{j-k+1}$. Now let us add more and more terms in the finite sums $\sum_{j=1}^{n} a_j$ and $\sum_{j=1}^{n} b_j$, namely we consider as n goes to infinity. We define:

Definition 8.3.1 (Cauchy Product) Let $\sum_{j=1}^{\infty} a_j$ and $\sum_{j=1}^{\infty} b_j$ be two real series. Their Cauchy product is defined to be the series:

$$\sum_{j=1}^{\infty} c_j \quad \text{where} \quad c_j = \sum_{k=1}^{j} a_k b_{j-k+1}.$$

Remark 8.3.2 In many literature, the Cauchy product is defined on the series $\sum_{j=0}^{\infty} a_j$ and $\sum_{j=0}^{\infty} b_j$ as the series $\sum_{j=0}^{\infty} c_j$ where $c_j = \sum_{k=0}^{j} a_k b_{j-k}$ (note the difference between the indices and their range in the three series here and Definition 8.3.1). This definition is equivalent to our definition above, up to a relabeling of indices. This convention/notation is more useful when we deal with power series later in Chap. 12.

We note now that for finite sums, thanks to the ring axioms, rearranging the terms $a_k b_{j-k+1}$ so that the terms with indices adding up to the same integer are grouped together is perfectly allowed without changing the value of the product.

However, for infinite series, in general, we are not allowed to change the ordering of the terms as we have seen in Riemann rearrangement theorem. Therefore, in general, even if the series $\sum_{j=1}^{\infty} a_j$ and $\sum_{j=1}^{\infty} b_j$ both converge, there is no guarantee that the product of these two sums is equal to their Cauchy product $\sum_{j=1}^{\infty} c_j$ or even if the Cauchy product converges at all!

However, the good news is that if at least one of $\sum_{j=1}^{\infty} a_j$ or $\sum_{j=1}^{\infty} b_j$ converges absolutely, then we can guarantee that their product is equal to their Cauchy product. This result was proven by Franz Mertens (1840–1927).

Theorem 8.3.3 (Mertens' Theorem) *Let $\sum_{j=1}^{\infty} a_j$ and $\sum_{j=1}^{\infty} b_j$ be two convergent real series such that one of them converges absolutely. Then, their Cauchy product converges to the product $\left(\sum_{j=1}^{\infty} a_j\right)\left(\sum_{j=1}^{\infty} b_j\right)$, namely:*

$$\sum_{j=1}^{\infty} c_j = \left(\sum_{j=1}^{\infty} a_j\right)\left(\sum_{j=1}^{\infty} b_j\right) \quad \text{where} \quad c_j = \sum_{k=1}^{j} a_k b_{j-k+1}.$$

Proof WLOG, suppose that the series $\sum_{j=1}^{\infty} a_j$ converges absolutely. Let $\sum_{j=1}^{\infty} c_j$ where $c_j = \sum_{k=1}^{j} a_k b_{j-k+1}$ be the Cauchy product of the two series. Define the partial sums $s_n = \sum_{j=1}^{n} a_j$, $S_n = \sum_{j=1}^{n} |a_j|$, $t_n = \sum_{j=1}^{n} b_j$, and $u_n = \sum_{j=1}^{n} c_j$. Suppose further that $s_n \to s$, $S_n \to S \geq 0$, and $t_n \to t$. We aim to show that $\lim_{n\to\infty} u_n = st$.

Fix $\varepsilon > 0$. Note that we can rewrite st as $st = (s - s_n)t + \sum_{j=1}^{n} a_j t$. Furthermore, $u_n = \sum_{j=1}^{n} c_j$ is, by definition, the sum of all the products $a_k b_i$ such that $k + i \in$

$\{2, 3, \ldots, n+1\}$. Thus, we can rewrite it using the partial sums (t_n) as:

$$u_n = \sum_{j=1}^{n} a_j \left(\sum_{k=1}^{n-j+1} b_k \right) = \sum_{j=1}^{n} a_j t_{n-j+1}.$$

Taking their difference and using triangle inequality, we have:

$$\begin{aligned} |u_n - st| = \left| \sum_{j=1}^{n} a_j t_{n-j+1} - (s - s_n)t - \sum_{j=1}^{n} a_j t \right| &= \left| \sum_{j=1}^{n} a_j (t_{n-j+1} - t) - (s - s_n)t \right| \\ &\leq \sum_{j=1}^{n} |a_j||t_{n-j+1} - t| + |s - s_n||t|. \end{aligned} \tag{8.3}$$

Now we list down some estimates that will help us bound this inequality.

1. Since $s_n \to s$, there exists an $N_1 \in \mathbb{N}$ such that $|s_n - s| < \frac{\varepsilon}{3(|t|+1)}$ for all $n \geq N_1$.
2. Since $t_n \to t$, there exists an $N_2 \in \mathbb{N}$ such that $|t_n - t| < \frac{\varepsilon}{3(S+1)}$ for all $n \geq N_2$.
3. Since $|t_n - t| \to 0$, this sequence is bounded. Namely, there exists a $K > 0$ such that $|t_n - t| \leq K$ for all $n \in \mathbb{N}$.
4. Since the series $\sum_{j=1}^{\infty} |a_j|$ converges, we have $|a_n| \to 0$. So there exists an $N_3 \in \mathbb{N}$ such that for all $n \geq N_3$ we have $|a_n| < \frac{\varepsilon}{3KN_2}$ where $N_2 \in \mathbb{N}$ is the index in the second estimate above.

Set $N = \max\{N_1, N_2 + N_3\}$. We claim that for all $n \geq N$, we have $|u_n - st| < \varepsilon$. Indeed, let us look at the terms in (8.3) one by one. The second term is simpler to deal with, namely for all $n \geq N \geq N_1$, we have $|s_n - s||t| < \frac{\varepsilon|t|}{3(|t|+1)} < \frac{\varepsilon}{3}$ via the first estimate.

For the first term in (8.3), which is a summation up to $n \geq N > N_2$, we split the sum into two parts:

$$\sum_{j=1}^{n} |a_j||t_{n-j+1} - t| = \underbrace{\sum_{j=1}^{n-N_2} |a_j||t_{n-j+1} - t|}_{(1)} + \underbrace{\sum_{j=n-N_2+1}^{n} |a_j||t_{n-j+1} - t|}_{(2)}.$$

We note that the indices for t_k in summation (1) run from $N_2 + 1$ to n and so the second estimate above holds here. Using this estimate, we obtain:

$$(1) = \sum_{j=1}^{n-N_2} |a_j||t_{n-j+1} - t| \leq \sum_{j=1}^{n-N_2} |a_j|\frac{\varepsilon}{3(S+1)} = \frac{\varepsilon}{3(S+1)} \sum_{j=1}^{n-N_2} |a_j|$$
$$\leq \frac{\varepsilon S}{3(S+1)} < \frac{\varepsilon}{3},$$

Moreover, the indices for a_j in summation (2) run from $n - N_2 + 1$ to n. Since the lowest index satisfies $n - N_2 + 1 \geq N - N_2 + 1 \geq N_2 + N_3 - N_2 + 1 > N_3$, we can then apply the third and fourth estimates to get:

$$(2) = \sum_{j=n-N_2+1}^{n} |a_j||t_{n-j+1} - t| \leq \sum_{j=n-N_2+1}^{n} |a_j|K = K \sum_{j=n-N_2+1}^{n} |a_j|$$
$$< K\frac{N_2\varepsilon}{3KN_2} = \frac{\varepsilon}{3},$$

since all the indices for $|a_j|$ are bigger than N_3.

Putting both of these estimates together in (8.3), for all $n \geq N$ we have $|u_n - st| < \frac{\varepsilon}{3} + \frac{\varepsilon}{3} + \frac{\varepsilon}{3} = \varepsilon$ and hence we conclude that $\lim_{n\to\infty} u_n = st$. □

Note that Mertens' theorem requires at least one of the series to be absolutely convergent for us to guarantee that their Cauchy product converges. However, this does not tell us anything about the case for which both of the series are only conditionally convergent. For this case, the Cauchy product could either converge or diverge!

Example 8.3.4 Let us look at some examples investigating the Cauchy product of two conditionally convergent series.

1. Let $\sum_{j=1}^{\infty} a_j$ and $\sum_{j=1}^{\infty} b_j$ be two real series with $a_j = b_j = \frac{(-1)^j}{j}$. Both of these series only converge conditionally. We claim that their Cauchy product converges as well. Indeed, for each $j \in \mathbb{N}$ we compute:

$$c_j = \sum_{k=1}^{j} a_k b_{j-k+1} = \sum_{k=1}^{j} \frac{(-1)^k}{k}\frac{(-1)^{j-k+1}}{j-k+1} = (-1)^{j+1}\sum_{k=1}^{j} \frac{1}{k(j-k+1)}. \tag{8.4}$$

Let us call the final sum in (8.4) d_j. Notice that:

$$d_j = \sum_{k=1}^{j} \frac{1}{k(j-k+1)} = \sum_{k=1}^{j} \frac{1}{j+1}\left(\frac{1}{k} + \frac{1}{j-k+1}\right)$$
$$= \frac{1}{j+1}\left(\sum_{k=1}^{j} \frac{1}{k} + \sum_{k=1}^{j} \frac{1}{j-k+1}\right)$$
$$= \frac{2}{j+1}\sum_{k=1}^{j} \frac{1}{k},$$

after relabelling the indices in the second sum. By straightforward algebra, one can show that this sequence (d_j) is strictly decreasing. Furthermore, since the sequence is strictly positive, it is bounded from below and hence, by monotone sequence theorem, it converges to some $d \geq 0$.

To show that $d = 0$, we can directly apply a result from Exercise 8.4. Here we shall give a different proof for a bit of variety. We note that for all $j \geq 2$ we can bound the harmonic sum as follows:

$$\sum_{k=1}^{j} \frac{1}{k} \leq \sum_{k=1}^{\lfloor \sqrt{j} \rfloor} \frac{1}{k} + \sum_{k=\lceil \sqrt{j} \rceil}^{j} \frac{1}{k} \leq \sum_{k=1}^{\lfloor \sqrt{j} \rfloor} 1 + \sum_{k=\lceil \sqrt{j} \rceil}^{j} \frac{1}{\lceil \sqrt{j} \rceil} \leq \lfloor \sqrt{j} \rfloor + \frac{j}{\lceil \sqrt{j} \rceil}$$
$$\leq \sqrt{j} + \frac{j}{\sqrt{j}} = 2\sqrt{j}.$$

This means $0 \leq d_j = \frac{2}{j+1}\sum_{k=1}^{j} \frac{1}{k} \leq \frac{4\sqrt{j}}{j+1} \leq \frac{4}{\sqrt{j}}$ for all $j \geq 2$. Thus, (d_j) converges to 0 by sandwiching. Therefore, the Cauchy product $\sum_{j=1}^{\infty} c_j = \sum_{j=1}^{\infty} (-1)^{j+1} d_j$ converges by the alternating series theorem. In Exercise 8.1, the readers will show that this Cauchy product only converges conditionally.

2. Now let $\sum_{j=1}^{\infty} a_j$ and $\sum_{j=1}^{\infty} b_j$ be two real series with $a_j = b_j = \frac{(-1)^j}{\sqrt[4]{j}}$. Both of these series converge conditionally by the alternating series test. Let us now compute its Cauchy product $\sum_{j=1}^{\infty} c_j$. For each $j \in \mathbb{N}$ we compute:

$$c_j = \sum_{k=1}^{j} a_k b_{j-k+1} = \sum_{k=1}^{j} \frac{(-1)^k}{\sqrt[4]{k}} \frac{(-1)^{j-k+1}}{\sqrt[4]{j-k+1}} = (-1)^{j+1} \sum_{k=1}^{j} \frac{1}{\sqrt[4]{k(j-k+1)}}.$$

Notice that $k(j-k+1) \leq j^2 - 1 + j < (j+1)^2$ for all $1 \leq k \leq j$. Thus, we have:

$$|c_j| = \sum_{k=1}^{j} \frac{1}{\sqrt[4]{k(j-k+1)}} > \sum_{k=1}^{j} \frac{1}{\sqrt[4]{(j+1)^2}} = \frac{j}{\sqrt{j+1}},$$

which diverges to ∞ as $j \to \infty$. Thus, $|c_j| \not\to 0$ and therefore $c_j \not\to 0$. This means the Cauchy product $\sum_{j=1}^{\infty} c_j$ of the series $\sum_{j=1}^{\infty} a_j$ and $\sum_{j=1}^{\infty} b_j$ does not converge even though both of the series converge.

Exercises

8.1 Recall the Cauchy product in Example 8.3.4 given by $\sum_{j=1}^{\infty} c_j$ where $c_j = \frac{2(-1)^{j+1}}{j+1} \sum_{k=1}^{j} \frac{1}{k}$. Show that this series only converges conditionally.

8.2 Let $\sum_{j=1}^{\infty} a_j$ and $\sum_{j=1}^{\infty} b_j$ be two real series with $a_j = b_j = \frac{(-1)^j}{\sqrt{j}}$.
(a) Show that these two series converge.
(b) Let c_j for $j \in \mathbb{N}$ be defined as the sum $c_j = \sum_{k=1}^{j} a_k b_{j-k+1}$. Prove that for all $j \in \mathbb{N}$ we have $|c_j| > \alpha$ where $\alpha > 0$ is some positive constant.
(c) Hence, deduce that the Cauchy product of the two series $\sum_{j=1}^{\infty} a_j$ and $\sum_{j=1}^{\infty} b_j$ does not converge.
Explain why this does not contradict Mertens' theorem.

8.3 (∗) Mertens' theorem require at least one of the series to converge absolutely for their Cauchy product to converge. Proving the theorem where both series converge absolutely is much easier. Let $\sum_{j=1}^{\infty} a_j$ and $\sum_{j=1}^{\infty} b_j$ be two absolutely convergent real series which converge to A and B respectively.
(a) Prove that their Cauchy product $\sum_{j=1}^{\infty} c_j$ where $c_j = \sum_{k=1}^{j} a_k b_{j-k+1}$ also converges absolutely to some $C \in \mathbb{R}$.
(b) Hence, by using Riemann rearrangement theorem, show that $C = AB$.

8.4 (◇) Recall Riemann rearrangement theorem in Theorem 8.1.5 which states that if $\sum_{j=1}^{\infty} a_j$ is a conditionally convergent series and $L \in \mathbb{R}$ is any real number, we can rearrange this series such that it converges to L. Using the same idea, prove that there is a rearrangement of the series that diverges to $\pm\infty$.

8.5 (∗) By alternating series test, the series $s = \sum_{j=1}^{\infty} \frac{(-1)^{j+1}}{j}$ converges conditionally. Denote $a_j = \frac{(-1)^{j+1}}{j}$. Now we shall construct a rearrangement of this series that converges to another number. Define a new rearranged series:

$$\begin{aligned} t &= a_1 + a_2 + a_4 + a_3 + a_6 + a_8 + a_5 + a_{10} + a_{12} + \dots \\ &= 1 - \frac{1}{2} - \frac{1}{4} + \frac{1}{3} - \frac{1}{6} - \frac{1}{8} + \frac{1}{5} - \frac{1}{10} - \frac{1}{12} + \dots, \end{aligned}$$

where we alternate the terms with odd denominators with pairs of consecutive terms with even denominators. Denote the n-th partial sums of s and t as s_n and t_n respectively.

(a) Let $h_n = \sum_{j=1}^{n} \frac{1}{j}$. Show that for all $n \in \mathbb{N}$, $h_{2n} - h_n = s_{2n}$.
(b) Using part (a), show that $t_{3n} = \frac{s_{2n}}{2}$.
(c) Deduce that $\lim_{n\to\infty} t_{3n} = \frac{s}{2}$.
(d) Show that $t_{3n+1} = t_{3n} + \frac{1}{2n+1}$ and $t_{3n+2} = t_{3n} + \frac{1}{2(2n+1)}$. Find their limits.
(e) Using Exercise 4.7, deduce that $\lim_{n\to\infty} t_n = \frac{s}{2}$.

Define a new rearranged series:

$$\begin{aligned} u &= a_1 + a_3 + a_2 + a_5 + a_7 + a_4 + a_9 + a_{11} + a_6 + \dots \\ &= 1 + \frac{1}{3} - \frac{1}{2} + \frac{1}{5} + \frac{1}{7} - \frac{1}{4} + \frac{1}{9} + \frac{1}{11} - \frac{1}{6} + \dots, \end{aligned}$$

where we pair the terms with odd denominators and alternate these pairs with terms containing even denominators. We denote the n-th partial sum of this rearranged series as u_n.

(f) Using the same idea as parts (b)-(e), show that $\lim_{n\to\infty} u_n = \frac{3s}{2}$.

The conclusion that we can draw from this exercise is when we rearrange the original series s, the new series could converge to $\frac{s}{2}$ or $\frac{3s}{2}$ (or to other values in $\mathbb{R}$ or diverge to $\pm\infty$ according to Riemann rearrangement theorem).

8.6 ($\diamond$) Now let us prove that any positive rational number can be written as the sum of reciprocals of distinct natural numbers. In other words, for any $r \in \mathbb{Q}_+$ there are distinct positive integers $n_1, n_2, n_2, \dots, n_k \in \mathbb{N}$ such that $r = \sum_{j=1}^{k} \frac{1}{n_j}$. Such an expression is called an Egyptian fraction for r. Fix $r \in \mathbb{Q}_+$.

(a) Let $h_n = \sum_{j=1}^{n} \frac{1}{j}$ and $h_0 = 0$. Prove that there is an $n \in \mathbb{N}_0$ such that $h_n \leq r < h_{n+1}$.
(b) Define $x_1 = r - h_n \geq 0$. If $x_1 = 0$, then we are done. Otherwise, we can write $x_1 = \frac{y_1}{z_1}$ for some coprime $y_1, z_1 \in \mathbb{N}$. Prove that there exists a natural number $m_1 > n$ such that $\frac{1}{m_1+1} \leq x_1 < \frac{1}{m_1}$.
(c) Define $x_2 = x_1 - \frac{1}{m_1+1} \geq 0$. If $x_2 = 0$, then we are done. Otherwise, we can write $x_2 = \frac{y_2}{z_2}$ for some coprime $y_2, z_2 \in \mathbb{N}$. Show that:

$$x_2 = \frac{y_2}{z_2} = \frac{y_1(m_1+1) - z_1}{z_1(m_1+1)}.$$

Prove that $y_1(m_1+1) - z_1 < y_1$ and hence deduce that, in lowest forms, the numerator of x_2 is strictly smaller than the numerator for x_1.
(d) Continue the construction of $x_3, x_4, \dots$ using the same argument. Explain why $x_k = 0$ for some finite $k \in \mathbb{N}$.
(e) Deduce that r is a finite sum of reciprocals of distinct natural numbers.

(f) Write a computer program that takes a positive rational number r as an input and generates the Egyptian fraction expression for r.

One of the open problems in number theory is the Erdős-Straus conjecture which states that any rational number of the form $\frac{4}{n}$ for $n \geq 2$ can be written as an Egyptian fraction with exactly three terms. In other words, for any integer $n \geq 2$, there are $x, y, z \in \mathbb{N}$ such that $\frac{4}{n} = \frac{1}{x} + \frac{1}{y} + \frac{1}{z}$. This problem may look deceptively simple, but it remains unproven since 1948!

8.7 (◇) Let (p_j) be a real sequence where p_j is the j-th prime number. We want to investigate the series $\sum_{j=1}^{\infty} \frac{1}{p_j}$. This seems smaller than the harmonic series since the j-th term of this series is strictly smaller than the j-th term in the harmonic series. We shall show that this series diverges via contradiction. This proof is due to Paul Erdős (1913–1996) a mathematician known for his eccentricity and ingenuity.

Assume for contradiction that the series converges.

(a) Show that there exists a $k \in \mathbb{N}$ such that $\sum_{j=k}^{\infty} \frac{1}{p_j} < \frac{1}{2}$.

Fix any $x \in \mathbb{N}$ and let $X = \{1, 2, \dots, x\}$. Define the following subset:

$$M(x) = \{n \in X : p_j \text{ does not divide } n \text{ for all } j \geq k\} \subseteq X.$$

(b) By the fundamental theorem of arithmetic in Exercise 2.30, it is easy to see that we can express any $m \in M(x)$ as $m = yz^2$ where y is square-free (not divisible by any integer squared) and $z \in \mathbb{N}$. Using this expression, show that $|M(x)| \leq 2^{k-1}\sqrt{x}$.

(c) The set $X \setminus M(x)$ is the set of all positive integers smaller than or equal to x which is divisible by some p_j where $j \geq k$. For any $j \geq k$ denote the set $N^j(x) = \{n \in X : p_j \text{ divides } x\} \subseteq X \setminus M(x)$. Explain why $\bigcup_{j=k}^{\infty} N^j(x) = X \setminus M(x)$.

(d) Prove that for any $j \geq k$ we have $|N^j(x)| \leq \frac{x}{p_j}$.

(e) Prove that $\frac{x}{2} < |M(x)|$.

(f) Finally, derive a contradiction.

8.8 In this question, we are going to prove a funny result due to Aubrey Kempner (1880–1973). The result says: if we remove from the harmonic series all the terms $\frac{1}{n}$ where the number n contains a digit 9 in its decimal representation, the series would then converge.

Let (a_j) be an integer sequence such that a_j is the j-th natural number with no digit 9 in its decimal representation. So the first twenty elements in this sequence are:

$$1, 2, 3, 4, 5, 6, 7, 8, \cancel{9}, 10, 11, 12, 13, 14, 15, 16, 17, 18, \cancel{19}, 20, 21, 22, \dots.$$

We wish to prove that the series $\sum_{j=1}^{\infty} \frac{1}{a_j}$, which is known as Kempner series, converges.

(a) Let $k : \mathbb{N}_0 \to \mathbb{N}_0$ be the function $k_0 = 0$ and $k_j = \max\{m : a_m \text{ has at most } j \text{ digits}\}$ for $j > 0$. Determine the value of k_j for any $j \in \mathbb{N}$.

(b) For any $j \in \mathbb{N}$, denote the sum $c_j = \sum_{m=k_{j-1}+1}^{k_j} \frac{1}{a_m}$ which is the sum of the reciprocals of terms in the sequence with exactly j digits. Explain why the sum $\sum_{j=1}^{\infty} c_j$ is a bracketing of the series $\sum_{j=1}^{\infty} \frac{1}{a_j}$.

(c) Show that the series $\sum_{j=1}^{\infty} c_j$ converges.

(d) Deduce that the series $\sum_{j=1}^{\infty} \frac{1}{a_j}$ also converges.

In fact, one can also show that the harmonic series with the reciprocals of integers not containing any fixed finite string of digits also converge and their sums can be computed by an algorithm in [67]. Kempner estimated his series in part (d) to have value less than 90, but based on the algorithm, this sum is even smaller: it is roughly 22.92068.

The intuitive reasoning for these convergence is that for larger numbers, the occurrences of integers containing a specific string of digits get more common. For example, amongst 10-digits positive integers, fewer than 35% of them do not contain the digit 9. Moreover, for 100-digits positive integers, fewer than 0.003% of them are left after we remove all the numbers with 9 in its decimal representation. Thus, removing these terms prevents the series from diverging.

8.9 (∗) We are now going to prove Tannery's theorem (or dominated convergence theorem for series) which allows us to switch the order of limit and infinite sums. In general, this process is forbidden because we would be intermixing two limits and limits do not generally commute. For example, if we consider the sequence $(a_{m,n})$ doubly-indexed by $m, n \in \mathbb{N}$ where $a_{m,n} = \frac{m^2}{m^2+n^2}$, we would have:

$$\lim_{m\to\infty}\left(\lim_{n\to\infty}\frac{m^2}{m^2+n^2}\right) = \lim_{m\to\infty} 0 = 0, \qquad \text{and}$$

$$\lim_{n\to\infty}\left(\lim_{m\to\infty}\frac{m^2}{m^2+n^2}\right) = \lim_{n\to\infty} 1 = 1.$$

Tannery's theorem, named after Jules Tannery (1848–1910), states that:

Theorem 8.4.5 (Tannery's Theorem) *Consider a real sequence $(a_{m,n})$ doubly-indexed by $m, n \in \mathbb{N}$. Suppose that:*

1. for all $n \in \mathbb{N}$ the series $s_n = \sum_{j=1}^{\infty} a_{j,n}$ are all convergent,

2. for all m we have $\lim_{n\to\infty} a_{m,n} = a_m$, and

3. for each $m \in \mathbb{N}$ there exists an $M_m > 0$ such that $|a_{m,n}| \leq M_m$ for all $n \in \mathbb{N}$ and the series $\sum_{m=1}^{\infty} M_m$ converges.

Then, we can switch the following limits:

$$\lim_{n\to\infty} s_n = \lim_{n\to\infty} \sum_{j=1}^{\infty} a_{j,n} = \sum_{j=1}^{\infty} \lim_{n\to\infty} a_{j,n} = \sum_{j=1}^{\infty} a_j.$$

We shall prove this theorem step by step.

(a) Show that for all $m \in \mathbb{N}$ we have $|a_m| \leq M_m$ as well. Hence, deduce that the series $\sum_{j=1}^{\infty} a_j$ converges absolutely.

(b) Show that for each $n \in \mathbb{N}$ the series $\sum_{j=1}^{\infty}(a_{j,n}-a_j)$ converges absolutely.

(c) Fix $\varepsilon > 0$. Show that there exists an index $K \in \mathbb{N}$ such that $\sum_{k=K+1}^{\infty} M_k < \frac{\varepsilon}{4}$.

(d) Hence, derive the following estimate:

$$\left|s_n - \sum_{j=1}^{\infty} a_j\right| \leq \sum_{j=1}^{K} |a_{j,n} - a_j| + \left|\sum_{j=K+1}^{\infty} (a_{j,n}-a_j)\right| < \sum_{j=1}^{K} |a_{j,n}-a_j| + \frac{\varepsilon}{2}. \tag{8.5}$$

(e) Next, show that for each $j = 1, 2, \ldots, K$, we can find an index $N_j \in \mathbb{N}$ such that for all $n \geq N_j$ we have $|a_{j,n} - a_j| < \frac{\varepsilon}{2K}$.

(f) Finally, complete the proof by finding a suitable N for this whole quantity in (8.5) to be smaller than ε for all $n \geq N$.

8.10 (∗) Using Tannery's theorem, show that for any $x \in \mathbb{R}$ we have:

$$\lim_{n\to\infty} \left(1 + \frac{x}{n}\right)^n = \sum_{j=0}^{\infty} \frac{x^j}{j!}.$$

Carefully specify the double sequence $(a_{m,n})$ in the theorem for this problem.

8.11 (a) Let $(a_{m,n})$ be a real sequence indexed by $m, n \in \mathbb{N}$ such that:

$$a_{m,n} = \begin{cases} 1 & \text{if } m = n, \\ -1 & \text{if } m = n+1, \\ 0 & \text{otherwise.} \end{cases}$$

Show that:

$$\sum_{n=1}^{\infty} \sum_{m=1}^{\infty} a_{m,n} \neq \sum_{m=1}^{\infty} \sum_{n=1}^{\infty} a_{m,n}.$$

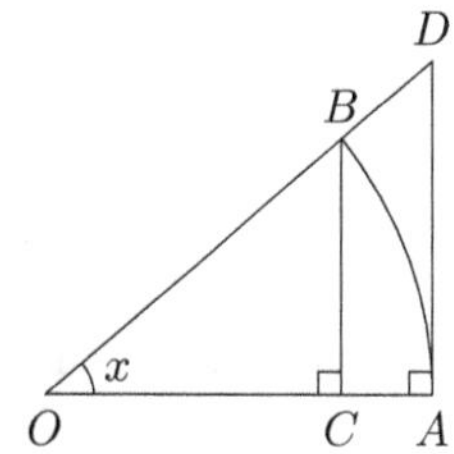

Fig. 8.1 Geometric configuration for Exercise 8.12(a)

(b) Let $(b_{m,n})$ be a real sequence doubly indexed by $m, n \in \mathbb{N}$. Suppose that:
   i. for each $n \in \mathbb{N}$ the series $s_n = \sum_{j=1}^{\infty} b_{j,n}$ are all convergent, and
   ii. for each $m \in \mathbb{N}$ the series $\sum_{k=1}^{\infty} b_{m,k}$ is absolutely convergent to some M_m such that the series $\sum_{m=1}^{\infty} M_m$ converges.

   Using Tannery's theorem, show that:

$$\sum_{m=1}^{\infty}\sum_{n=1}^{\infty} b_{m,n} = \sum_{n=1}^{\infty}\sum_{m=1}^{\infty} b_{m,n}.$$

8.12 (∗) We have seen that the series $\sum_{j=1}^{\infty} \frac{1}{j^2}$ converges, but we do not know its value. The problem of finding the numerical value of this series is called the Basel problem after the hometown of Leonhard Euler who first solved it in 1734. To date, there are various different ways to do it, including infinite products, Fourier series, and Parseval's identity. We are going to find its value in this question via a proof by Cauchy.

(a) Consider a sector of a unit circle AOB with angle $0 < x < \frac{\pi}{2}$ radians and $OA = OB = 1$. Suppose further that C is on the line OA and D is on the line OB such that BC and DA are both perpendicular to OA. Refer to Fig. 8.1.
   Explain why $BC = \sin(x)$ and $DA = \tan(x)$.
   Prove that $0 \leq \sin(x) \leq x \leq \tan(x)$.
(b) Using part (a), show that $\cot^2(x) < \frac{1}{x^2} < \csc^2(x)$ for all $x \in (0, \frac{\pi}{2})$.
(c) Fix $n \in \mathbb{N}$. For all $j = 1, 2, \ldots, n$ we have $0 < \frac{j\pi}{2n+1} < \frac{\pi}{2}$. Show that:

$$\frac{\pi^2}{(2n+1)^2}\sum_{j=1}^{n} \cot^2\left(\frac{j\pi}{2n+1}\right) < \sum_{j=1}^{n}\frac{1}{j^2} < \frac{\pi^2}{(2n+1)^2}\sum_{j=1}^{n} \csc^2\left(\frac{j\pi}{2n+1}\right). \tag{8.6}$$

We now want to express the sums of the trigonometric expressions in (8.6) in a closed form.

(d) Using complex numbers, De Moivre's theorem, and binomial theorem, find an expression for $\sin(mx)$ for $m \in \mathbb{N}$ in terms of sines and cosines.

Hence, show that for $x \in (0, \frac{\pi}{2})$ we have:

$$\sin((2n+1)x) = \sin^{2n+1}(x) \sum_{k=1}^{n} (-1)^k \binom{2n+1}{2k+1} (\cot^2(x))^{n-k}.$$

(e) Hence, show that for each $j = 1, 2, \dots, n$:

$$0 = \sum_{k=0}^{n} (-1)^k \binom{2n+1}{2k+1} \left(\cot^2\left(\frac{j\pi}{2n+1}\right)\right)^{n-k}.$$

(f) List down all the n roots of the n-th degree real polynomial $P : \mathbb{R} \to \mathbb{R}$ given by:

$$P(y) = \sum_{k=0}^{n} (-1)^k \binom{2n+1}{2k+1} y^{n-k},$$

and show that they are all distinct.

(g) Show that the coefficient of y^{n-1} in the polynomial P is the sum of the roots of the polynomial P multiplied with the negative of the leading coefficient. Hence, show that:

$$\sum_{j=1}^{n} \cot^2\left(\frac{j\pi}{2n+1}\right) = \frac{n(2n-1)}{3}.$$

Using this, deduce also that:

$$\sum_{j=1}^{n} \csc^2\left(\frac{j\pi}{2n+1}\right) = \frac{2n(n+1)}{3}.$$

(h) Finally, show that for any $n \in \mathbb{N}$ we have:

$$\frac{\pi^2}{3}\frac{n(2n-1)}{(2n+1)^2} < \sum_{j=1}^{n} \frac{1}{j^2} < \frac{\pi^2}{3}\frac{n(2n+2)}{(2n+1)^2},$$

and determine the value of $\sum_{j=1}^{\infty} \frac{1}{j^2}$.

We shall see another way to prove this identity in Exercise 20.22.

8.13 We have shown the set $\mathbb{R}$ is uncountable using Cantor's diagonal argument in Proposition 4.4.6. Let us provide another proof to show that the set $\mathbb{R}$ is uncountable using series. Define a function $f : \mathcal{P}(\mathbb{N}) \to \mathbb{R}$ as the infinite sum:

$$f(X) = \sum_{n \in X} \frac{1}{3^n},$$

for every $X \neq \varnothing$ and $f(\varnothing) = 0$.

(a) Show that the function f is well-defined. Explain also why the value of $f(X)$ is independent of how we carry out the sum over the elements in X.

We now aim to prove that this function is injective. Suppose for contradiction that $f(X) = f(Y)$ for some sets $X \neq Y$ in $\mathcal{P}(\mathbb{N})$. Then, $X \Delta Y \neq \varnothing$. Since it is a non-empty subset of $\mathbb{N}$, by well-ordering principle, there is a minimal element $m \in X \Delta Y$.

(b) Show that for all $n \in \mathbb{N}$ with $n < m$, either $n \in X \cap Y$ or $n \in X^c \cap Y^c$.
(c) WLOG, suppose that $m \in X$. Show that if $Y \setminus X$ is non-empty, then every element in $Y \setminus X$ is greater than m.
(d) Hence, show that $f(X) - f(Y) > 0$ and deduce that f is injective.
(e) Conclude the proof.

Functions and Limits

9

If the limit never approaches anything... the limit does not exist. The limit does not exist!

—Cady Heron, mean girl and mathlete

In Chap. 5, we have seen how sequences of real numbers and their limits behave. Now we are going to look at how limits of sequences behave under mappings. These mappings are called functions. We have seen in Chap. 1 the general idea of functions and some terminologies.

In this chapter, instead of general sets X and Y, we are going to focus our attention to $X \subseteq \mathbb{R}$ and $Y = \mathbb{R}$ since we already know how to study sequences and convergence in these sets from Chap. 5. The resulting function $f : X \to \mathbb{R}$ is then called a real-valued function. So, given a sequence (x_n) of elements in $X \subseteq \mathbb{R}$, the function f would map this sequence to another sequence $(f(x_n))$ in the codomain $Y = \mathbb{R}$. We call the sequence $(f(x_n))$ the image sequence of (x_n) under f.

If we know certain information about the real sequence (x_n) such as convergence, monotonicity, or any other sequential properties, what can we say about its image sequence $(f(x_n))$ in $Y = \mathbb{R}$? In truth, not a lot. This is because the function $f : X \to \mathbb{R}$ can be as wild and arbitrary as we can define it. Therefore, we need to distinguish some family of nice functions to work with. First, let us lay out the algebra of real-valued functions.

9.1 Algebra of Real-Valued Functions

Suppose that we have two real functions $f : X \to \mathbb{R}$ and $g : Y \to \mathbb{R}$ where $X, Y \subseteq \mathbb{R}$. We want to be able to define algebraic operations such as modulus, reciprocal, sum, difference, product, and quotient on these functions. Since the

S. Johar, *The Big Book of Real Analysis*,
https://doi.org/10.1007/978-3-031-30832-1_9

images are contained in the ordered field $\mathbb{R}$, this could be done easily by utilising the algebraic structure on $\mathbb{R}$.

1. Similar to real numbers, we can scale functions by a real number. More precisely, for a constant $\lambda \in \mathbb{R}$, we define the function $\lambda f : X \to \mathbb{R}$ by $(\lambda f)(x) = \lambda f(x)$ for each $x \in X$. Clearly, this is a well-defined function on X and scales the images of f by the constant λ.
2. Another function that we can create from the original function $f : X \to \mathbb{R}$ is the modulus function $|f| : X \to \mathbb{R}$. This function has the same domain as f, which is X, but defined as $|f|(x) = |f(x)|$ for every $x \in X$.
3. We can also define the reciprocal for a function $f : X \to \mathbb{R}$ by taking the reciprocal of the value $f(x)$ for each $x \in X$. Of course, this operation is allowed as long as the value of $f(x)$ is non-zero. So, the domain of the reciprocal function must be restricted to a subset $Z \subseteq X$ where f does not vanish, namely $Z = \{x \in X : f(x) \neq 0\}$. Thus, the reciprocal function $\frac{1}{f} : Z \to \mathbb{R}$ defined as $\frac{1}{f}(x) = \frac{1}{f(x)}$ makes sense.
4. For two functions f and g, the sum $f+g$ may be a function. Clearly the codomain is $\mathbb{R}$ but what is the domain of this function? For any x, if we want to declare $(f + g)(x) = f(x) + g(x)$, we need to ensure that both $f(x)$ and $g(x)$ are defined (and have values in $\mathbb{R}$).
Thus, the element x must be in both the domain of f and g, namely $x \in X \cap Y$. Otherwise, one of $f(x)$ or $g(x)$ does not have a value and hence the sum is non-existent. So this new function can be defined as $f + g : X \cap Y \to \mathbb{R}$ via $(f+g)(x) = f(x)+g(x)$ for all $x \in X\cap Y$. Similar result holds for the functions $f - g$ and $f \times g$.
5. For quotients, similar to addition, subtraction, and multiplication, if we want to define $\frac{f}{g}(x)$ as $\frac{f(x)}{g(x)}$, the domain of the function $\frac{f}{g}$ would be limited to the x in the domain of f and g, namely $x \in X \cap Y$. Moreover, since we cannot divide real numbers with 0, g must also not vanish for this x. Thus, the domain is given by $Z = X \cap Y \cap \{y \in Y : g(y) \neq 0\}$, where we can meaningfully define $\frac{f}{g} : Z \to \mathbb{R}$ as $\frac{f}{g}(x) = \frac{f(x)}{g(x)}$.
6. Recall also that for two real numbers $a, b \in \mathbb{R}$, since we have the ordering axiom on $\mathbb{R}$ which allows us to compare either $a < b$, $a = b$, or $a > b$, we can define the maximum and minimum of these two numbers by $\max\{a, b\}$ and $\min\{a, b\}$ respectively. From Exercise 5.11, we can also write these quantities down algebraically as:

$$\max\{a, b\} = \frac{a + b + |b - a|}{2} \quad \text{and} \quad \min\{a, b\} = \frac{a + b - |b - a|}{2}.$$

By comparing the values of the functions f and g pointwise at each x, we can also define this operation on functions. The resulting function would be the $\max(f, g)$ and $\min(f, g)$ functions. The domain of these functions would then be the set of x on which both $f(x)$ and $g(x)$ are well-defined real numbers, which is again $X \cap Y$. For each $x \in X \cap Y$, we can thus define the functions $\max(f, g), \min(f, g) : X \cap Y \to \mathbb{R}$ as:

$$\max(f, g)(x) = \max\{f(x), g(x)\} = \frac{f(x) + g(x) + |f(x) - g(x)|}{2},$$

$$\min(f, g)(x) = \min\{f(x), g(x)\} = \frac{f(x) + g(x) - |f(x) - g(x)|}{2}.$$

Another operation that we can carry out on real-valued functions is called composition. We have seen this in Definition 1.5.14 with more generality, but let us reiterate this definition for real-valued functions:

Definition 9.1.1 (Composition of Functions) Let $f : X \to \mathbb{R}$ and $g : Y \to \mathbb{R}$ be functions defined on $X \subseteq \mathbb{R}$ and $Y \subseteq \mathbb{R}$ respectively. Suppose further that $f(X) \subseteq \text{Dom}(g) = Y$. Then, we define the composition $g \circ f$ of the functions f and g as the function:

$$\begin{aligned} g \circ f : X &\to \mathbb{R} \\ x &\mapsto g(f(x)). \end{aligned}$$

Clearly, the condition $f(X) \subseteq \text{Dom}(g)$ is necessary because in order for the quantity $g(f(x))$ to make sense, for every $x \in X$ the number $f(x)$ must be in the domain for the function g. Otherwise $g(f(x))$ does not have a value. If we insist on composing two functions where $f(X) \not\subseteq \text{Dom}(g)$, we have to restrict the domain of the first function f so that the new image lies within Y. This restriction has been introduced in Definition 1.5.20 and an example for this process was given in Example 1.5.21.

Example 9.1.2 Let us look at some more examples:

1. Suppose that $f, g : \mathbb{R} \to \mathbb{R}$ are two functions defined by $f(x) = x^2$ and $g(x) = x + 1$.
   (a) Let us find the composition $g \circ f : \mathbb{R} \to \mathbb{R}$. We first check that the image of the function f lies in the domain of the function g, namely: $f(\mathbb{R}) \subseteq \mathbb{R}$. This is true by definition, so the composition makes sense. Therefore, we have:

   $$(g \circ f)(x) = g(f(x)) = g(x^2) = x^2 + 1.$$

   (b) Likewise, the image of the function g lies in the domain of the function f and so the composition $f \circ g$ exists. This composition is given by:

   $$(f \circ g)(x) = f(g(x)) = f(x + 1) = (x + 1)^2 = x^2 + 2x + 1.$$

2. Suppose that $f : \mathbb{R} \to \mathbb{R}$ is a function defined as $f(x) = 2x + 1$ and $g : \mathbb{R}_{\geq 0} \to \mathbb{R}$ is a function defined as $g(x) = \sqrt{x}$. We would like to compose these functions together.
   (a) Note that the image of the function g lies in the domain of the function f since $g(\mathbb{R}_+) = \mathbb{R}_+ \subseteq \mathbb{R}$ Thus, the composition $g \circ f$ exists with:

$$(f \circ g)(x) = f(g(x)) = f(\sqrt{x}) = 2\sqrt{x} + 1.$$

   (b) On the other hand, we note that $f(\mathbb{R}) = \mathbb{R}$ since it is a surjective function. Therefore, we cannot compose it with the function g as $f(\mathbb{R}) = \mathbb{R} \not\subseteq \text{Dom}(g) = \mathbb{R}_{\geq 0}$.
   If we still want to compose these functions meaningfully, we need to restrict the domain of the function f so that its image lies in the domain of g. In other words, we want to limit our attention to the points $x \in \text{Dom}(f) = \mathbb{R}$ for which $f(x) \in \text{Dom}(g) = \mathbb{R}_{\geq 0}$. Hence, these points must satisfy $f(x) = 2x + 1 \geq 0$ or equivalently $x \geq -\frac{1}{2}$.
   Thus, if we restrict the function f to the set $X = \{x \in \mathbb{R} : x \geq -\frac{1}{2}\}$, we can make sense of the composition. Indeed, on this set we have $f(X) = \{f(x) : x \in X\} = \{f(x) : x \geq -\frac{1}{2}\} = \mathbb{R}_{\geq 0} = \text{Dom}(g)$.
   The restricted function is denoted as $f|_X : X \to \mathbb{R}$. The composition is then given by the function $g \circ f|_X : X \to \mathbb{R}$ which is:

$$g \circ f|_X(x) = g(f|_X(x)) = g(2x + 1) = \sqrt{2x + 1}.$$

   Clearly, the restriction process is necessary here since the quantity $\sqrt{2x - 1}$ does not exist in $\mathbb{R}$ for $x < -\frac{1}{2}$.

We now want to define some new terminologies for real-valued functions. If we have a function $f : X \to \mathbb{R}$ on a subset $X \subseteq \mathbb{R}$, then the image set $f(X)$ is a subset of the real numbers, which is an ordered field. Hence, we can define:

Definition 9.1.3 (Bounded Functions) Let $f : X \to \mathbb{R}$ where $X \subseteq \mathbb{R}$.

1. The function f is called bounded from above if there exists an $M \in \mathbb{R}$ such that for all $x \in X$ we have $f(x) \leq M$. In other words, the image set $f(X) \subseteq \mathbb{R}$ is bounded from above.
2. The function f is called bounded from below if there exists an $M \in \mathbb{R}$ such that for all $x \in X$ we have $M \leq f(x)$. In other words, the image set $f(X) \subseteq \mathbb{R}$ is bounded from below.
3. The function f is called bounded if there exists an $M > 0$ such that for all $x \in X$ we have $|f(x)| \leq M$. In other words, the image set $f(X) \subseteq \mathbb{R}$ is a bounded set.

From the above, we have the following special family of functions:

Definition 9.1.4 (Positive, Negative Functions) Let $f : X \to \mathbb{R}$.

1. f is called a positive function if $f(x) > 0$ for all $x \in X$. We write this as $f > 0$.
2. f is called a non-negative function if $f(x) \geq 0$ for all $x \in X$. We write this as $f \geq 0$.
3. f is called a negative function if $f(x) < 0$ for all $x \in X$. We write this as $f < 0$.
4. f is called a non-positive function if $f(x) \leq 0$ for all $x \in X$. We write this as $f < 0$.
5. f is called a zero function if $f(x) = 0$ for all $x \in X$. We write this as $f = 0$ or $f \equiv 0$.

If we have two functions $f, g : X \to \mathbb{R}$, for each $x \in X$ we can compare the two values $f(x)$ and $g(x)$.

1. If $f(x) \geq g(x)$ for all $x \in X$, we write this as $f \geq g$.
2. If $f(x) > g(x)$ for all $x \in X$, we write this as $f > g$.

Moreover, if the image set $f(X) \subseteq \mathbb{R}$ is bounded from above, then by the completeness axiom of the real numbers, the supremum of the set $f(X)$ exists. Similarly, if the image set $f(X)$ is bounded from below, then its infimum exists. We denote these as:

Definition 9.1.5 (Infimum, Supremum of a Function) Let $f : X \to \mathbb{R}$ be a function on $X \subseteq \mathbb{R}$. We define:

$$\inf f(X) = \inf_{x \in X} f(x) = \inf\{f(x) : x \in X\},$$

$$\sup f(X) = \sup_{x \in X} f(x) = \sup\{f(x) : x \in X\},$$

whenever these exist.

We note that the infimum and the supremum value of a function may not be mapped from any point in the domain. In other words, even though they may exist in $\mathbb{R}$, it is not guaranteed that $\inf f(X), \sup f(X) \in f(X)$.

Example 9.1.6 The function $f : \mathbb{R}_+ \to \mathbb{R}$ defined by $f(x) = \frac{1}{x}$ is bounded from below and hence has an infimum. The infimum of this function is $\inf_{x \in \mathbb{R}_+} f(x) = \inf\{\frac{1}{x}, x > 0\} = 0$. However, this value is not assumed by any value $x \in \mathbb{R}$ since the equation $f(x) = 0$ has no solution. So $\inf_{x \in \mathbb{R}_+} f(X) \notin f(\mathbb{R}_+)$.

However, by characterisation of infimum and supremum, we can always find some sequence of points in X for which their image sequences converge to these values.

Example 9.1.7 The function $f : \mathbb{R}_+ \to \mathbb{R}$ in Example 9.1.6 defined by $f(x) = \frac{1}{x}$ has $\inf_{x\in\mathbb{R}_+} f(x) = 0$ which is not assumed by any value $x \in \mathbb{R}$. However, the sequence $(x_n) \subseteq \mathbb{R}_+$ where $x_n = n$ has image $(f(x_n)) = (\frac{1}{n})$ which converges to $\inf_{x\in\mathbb{R}_+} f(x)$.

In fact, such a sequence may not even be unique. These sequences are called the minimising and maximising sequences respectively.

Definition 9.1.8 (Minimising, Maximising Sequences) Let $f : X \to \mathbb{R}$ be a function on $X \subseteq \mathbb{R}$ such that $\inf f(X)$ and $\sup f(X)$ exist. Then, we call a sequence (x_n) in X a

1. minimising sequence if $\lim_{n\to\infty} f(x_n) = \inf f(X)$,
2. maximising sequence if $\lim_{n\to\infty} f(x_n) = \sup f(X)$.

In some cases, the infimum and supremum may be attained by some point in the domain. For these cases, we refer to the attained infimum or supremum as the global minimum or maximum of the function respectively.

Definition 9.1.9 (Global Minimum, Global Maximum of a Function) Let $f : X \to \mathbb{R}$ be a function on $X \subseteq \mathbb{R}$. If the infimum of the function is attained by some $\xi \in X$, namely $f(\xi) = \inf f(X)$, we define:

$$\min_{x\in X} f(x) = \min f(X) = f(\xi).$$

Similarly, if the supremum of the function is attained by some $\zeta \in X$, namely $f(\zeta) = \sup f(X)$, we define:

$$\max_{x\in X} f(x) = \max f(X) = f(\zeta).$$

These points $\xi, \zeta \in X$ are called the global minimum and global maximum of the function f respectively while the values $f(\xi), f(\zeta) \in \mathbb{R}$ are called the global minimum value and global maximum value respectively.

Note that a global maximum or a global minimum, if they exist, may not be unique. For example, the function $f : \mathbb{R} \to \mathbb{R}$ defined as $f(x) = \sin(x)$ has global maximum value of 1 at $x = 2\pi n + \frac{\pi}{2}$ for any $n \in \mathbb{Z}$ and has global minimum value of -1 at $x = 2\pi n + \frac{3\pi}{2}$ for any $n \in \mathbb{Z}$.

A related concept to global minimum and global maximum is:

Definition 9.1.10 (Local Minimum, Local Maximum of a Function) Let $f : X \to \mathbb{R}$ be a function on $X \subseteq \mathbb{R}$ and $x_0 \in X$.

1. If there exists a $\delta > 0$ such that $f(x_0) \geq f(x)$ for every $x \in (x_0 - \delta, x_0 + \delta) \cap X \subseteq X$, then the point x_0 is called a local maximum point of the function f. In symbols:

$$x_0 \text{ is a local maximum for } f \quad \text{if} \quad \exists\delta{>}0 : \forall x{\in}X, |x - x_0|{<}\delta \Rightarrow f(x){\leq}f(x_0).$$

2. If there exists a $\delta > 0$ such that $f(x_0) \leq f(x)$ for every $x \in (x_0{-}\delta, x_0{+}\delta){\cap}X \subseteq X$, then the point x_0 is called a local minimum point of the function f. In symbols:

$$x_0 \text{ is a local minimum for } f \quad \text{if} \quad \exists\delta{>}0 : \forall x{\in}X, |x - x_0|{<}\delta \Rightarrow f(x){\geq}f(x_0).$$

The intuition for Definition 9.1.10 is that the local maximum $x_0 \in X$ might not be mapped to the greatest value attained by the function over its whole domain, but $f(x_0)$ is the greatest value attained by the function over some small region around it, hence the adjective "local" in the terminology.

Next, if the function preserves or reverses ordering of the domain in the image, this function is called monotone. More specifically:

Definition 9.1.11 (Monotone Functions) Let $f : X \to \mathbb{R}$ where $X \subseteq \mathbb{R}$.

1. If $f(x) \leq f(y)$ whenever $x \leq y$ in X, then we call the function f increasing or non-decreasing.
2. If $f(x) < f(y)$ whenever $x < y$ in X, then we call the function f strictly increasing.
3. If $f(x) \geq f(y)$ whenever $x \leq y$ in X, then we call the function f decreasing or non-increasing.
4. If $f(x) > f(y)$ whenever $x < y$ in X, then we call the function f strictly decreasing.

In all of the cases above, the functions f are called monotone functions.

9.2 Limit of a Function

Consider a real-valued function $f : X \to \mathbb{R}$ where $X \subseteq \mathbb{R}$ is a subset of $\mathbb{R}$. Then, for any limit point of the set X, say $x_0 \in X'$, this point may or may not lie in X. However, by Definition 6.2.1 of limit points, there exists at least one sequence (x_n) in $X \setminus \{x_0\}$ that converges to x_0.

Now let us look at the image of that sequence, namely $(f(x_n))$, in the range $f(X)$. Clearly, this new sequence makes sense because $x_n \in X$ for all $n \in \mathbb{N}$ so $f(x_n)$ are all well-defined real numbers. The sequence (x_n) converges in $\mathbb{R}$ to the point x_0, but there is no guarantee that the image sequence $(f(x_n))$ converges in the codomain $\mathbb{R}$. This depends on the behaviour of function the f itself.

Example 9.2.1 Let us look at the set $X = (0, 10)$ and define the following function on X:

$$f : X \to \mathbb{R},$$
$$x \mapsto \begin{cases} (-1)^n & \text{if } x \in (\frac{1}{n+1}, \frac{1}{n}] \quad \text{for } n \in \mathbb{N}, \\ 0 & \text{otherwise.} \end{cases}$$

As we saw in Example 6.2.3(1), the domain of this function has limit points $X' = [0, 10]$. Consider first the limit point $0 \in X'$.

1. This limit point can be approached via a sequence (x_n) in X where $x_n = \frac{1}{n}$. Using the definition of the the function f, we have $f(x_n) = f(\frac{1}{n}) = (-1)^n$. So this sequence induces the sequence $(f(x_n))$ in the codomain which alternates between -1 and 1. Even though the sequence in the domain (x_n) converges, this image sequence $(f(x_n))$ does not converge.
2. On the other hand, we can find another sequence (y_n) in X defined as $y_n = \frac{1}{2n}$ which also converges to the limit point 0. The images of the terms in this sequence are then given by $f(y_n) = (-1)^{2n} = 1$. So $(f(y_n))$ is a constant sequence which converges to 1.

From the above, we have found two sequences in X that converge to the limit point 0 such that their image sequence diverges and converges respectively.

Thus, even if the sequence in the domain converges, the corresponding image sequence in the codomain is not guaranteed to converge. In fact, even if we have two sequences in the domain that converge to the same limit point, the convergence behaviour of their image sequences might be different as we have seen in the above.

Let us look at the other end of the domain, namely at the limit point $10 \in X$.

3. There are many sequences in X that converge to 10 and one of them is (x_n) where $x_n = 10 - \frac{1}{n}$. For this sequence, we can see that $f(x_n) = f(10 - \frac{1}{n}) = 0$ for all $n \in \mathbb{N}$. So the image of this sequence converges, namely $f(x_n) \to 0$.
4. Now consider the sequence (y_n) in X defined as $y_n = 10 - \frac{1}{n^2}$. This sequence also converges to the limit point 10. The image sequence $(f(y_n))$ is also given by the constant sequence of 0s, which clearly converges to 0.

In Example 9.2.1(3) and (4) above, we only considered only two particular sequences that approach the limit point 10 and both of their image sequences converge. Moreover, these image sequences converge to the same value. However, according to Exercise 6.10, there are infinitely many other sequences in X that approach the same limit point 10. If the limits of the image sequences $(f(x_n))$ for every sequence $(x_n) \subseteq X \setminus \{10\}$ tending to 10 are the same, we call this common value the limit of the function f as x approaches 10.

More concretely, we define:

Definition 9.2.2 (Limit of a Function at x_0, Definition 1) Let $f : X \to \mathbb{R}$ where $X \subseteq \mathbb{R}$ and $x_0 \in X'$. We say that the function f has the limit $L \in \mathbb{R}$ as $x \to x_0$ if for every sequence $(x_n) \subseteq X \setminus \{x_0\}$ with $x_n \to x_0$, the image sequence $(f(x_n))$ converges to $L \in \mathbb{R}$. We write this as:

$$\lim_{x \to x_0} f(x) = L \quad \text{or} \quad f(x) \xrightarrow{x \to x_0} L.$$

In symbols, this definition is written as:

$$f(x) \xrightarrow{x \to x_0} L \quad \text{if} \quad \forall (x_n) \subseteq X \setminus \{x_0\}, x_n \to x_0 \Rightarrow f(x_n) \to L.$$

Remark 9.2.3 Let us make some important remarks regarding Definition 9.2.2.

1. The intuition behind this definition is that the function f would have a well-defined limit at some limit point x_0 if the behaviour of the values $f(x)$ for x close enough to the limit point x_0 are roughly the same no matter how we approach the limit point.
   The various ways one can approach the limit point depends on the direction we approach it from and also the rate at we approach it. So we have to check for all the ways we can approach the limit point x_0 via sequences in $X \setminus \{x_0\}$.
2. We also note that this is the importance of limit points of the domain set: they are the points at which we can ask whether a limit of a function exist. The limit of the function itself may not exist here as we have seen for the limit point 0 in Example 9.2.1, but it is still legitimate to ask whether it exists beforehand.
3. If a point x_0 is not a limit point of the domain X for the function f, then asking what is $\lim_{x \to x_0} f(x)$ is not even allowed!
   Indeed, if $x_0 \notin X'$, there exists an $\varepsilon > 0$ such that $B_\varepsilon(x_0) \setminus \{x_0\} \subseteq X^c$. Therefore, given any sequence $(x_n) \subseteq X \setminus \{x_0\}$ converging to the point x_0, there exists an index $N \in \mathbb{N}$ such that for all $n \geq N$ we have $0 < |x_n - x_0| < \varepsilon$ or equivalently $x_n \in B_\varepsilon(x_0) \setminus \{x_0\} \subseteq X^c$. This means the sequence (x_n) would eventually leave the set X after the index $N \in \mathbb{N}$. As a result, the image sequence $(f(x_n))$ would not have values for any $n \geq N$ since the function f is only defined on the domain X.

Example 9.2.4 Let us look at some examples here:

1. Recall Example 9.2.1. For the function $f : X \to \mathbb{R}$, if we look at the limit point $x = 0$ there is at least one sequence (x_n) in $X \setminus \{0\}$ such that $x_n \to 0$ but $(f(x_n))$ does not converge. Therefore, Definition 9.2.2 is not fulfilled at the point 0 and so $\lim_{x\to 0} f(x)$ does not exist.
2. On the other hand, at the limit point 10, the limit $\lim_{x\to 10} f(x)$ does exist and is equal to 0. We have shown in Example 9.2.1(3) and (4) that $f(x_n) \to 0$ for two specific sequences $(x_n) \subseteq X \setminus \{10\}$ that both converge to 10. However, this is not enough. We need to show this is true for all such sequences in order to fulfill the requirement in Definition 9.2.2. It is difficult to check one by one since there are so many of them! Instead, let us do that checking cleverly here.
   Pick any arbitrary sequence $(x_n) \subseteq X \setminus \{10\}$ such that $x_n \to 10$. Then, for $\varepsilon = 1$, there exists an $N \in \mathbb{N}$ such that for all $n \geq N$, we have $0 < |x_n - 10| < 1$ (or equivalently $9 < x_n < 10$). Hence, for all $n \geq N$, we have $f(x_n) = 0$ since $x_n > 9$ and f is identically zero on the interval $(9, 10)$. This means the sequence $(f(x_n))$ is eventually constant 0 after the index N and so it converges to 0. Since the sequence (x_n) was arbitrarily chosen, we can conclude that $\lim_{x\to 10} f(x) = 0$.

Example 9.2.4(2) above is an easy exception, but Definition 9.2.2 is a tricky definition to work with. This is because we need to check that for all sequences $(x_n) \subseteq X \setminus \{x_0\}$ with $x_n \to x_0$, the image sequences $(f(x_n))$ converge to the same number L.

However, this definition is useful if we want to show that the function does not have a limit as $x \to x_0$ by contrapositive as we did in Example 9.2.4(1) at $x = 0$. We simply have to either:

1. find a sequence (x_n) in $X \setminus \{x_0\}$ with $x_n \to x_0$ such that the image sequence $(f(x_n))$ does not converge, or
2. find two sequences (x_n) and (y_n) in $X \setminus \{x_0\}$ with $x_n \to x_0$ and $y_n \to x_0$ such that the image sequences $(f(x_n))$ and $(f(y_n))$ converge to two different limits.

Example 9.2.5 Consider the function:

$$f : (0, \infty) \to \mathbb{R}$$
$$x \mapsto \sin(\tfrac{1}{x}),$$

where its graph is given in Fig. 9.1.

The point 0 is a limit point of the domain $(0, \infty)$. So we can ask: what is $\lim_{x\to 0} f(x)$?

The answer is: the limit of this function does not exist as $x \to 0$. To see this, we find two different sequences of real numbers (a_n) and (b_n) in $(0, \infty)$ both of which tending to 0 but their images have different limits, namely: $\lim_{n\to\infty} f(a_n) \neq$

Fig. 9.1 Graph of $y = f(x)$

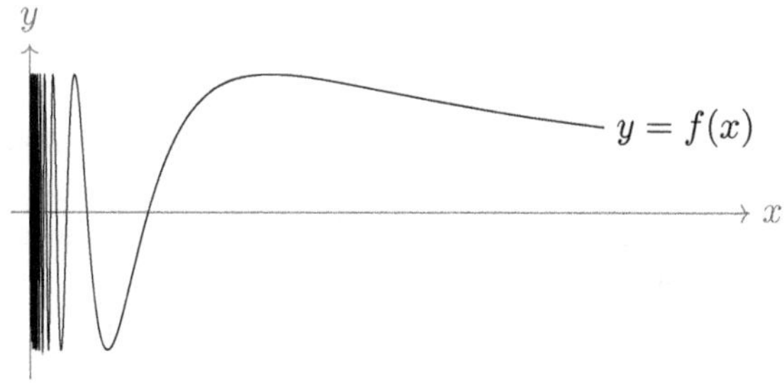

$\lim_{n\to\infty} f(b_n)$. To wit, let us define two sequences:

$$a_n = \frac{1}{n\pi} \quad \text{and} \quad b_n = \frac{1}{(2n+\frac{1}{2})\pi},$$

in $(0, \infty)$. Both of these sequences tend to 0 as $n \to \infty$. However, we have:

$$\lim_{n\to\infty} f(a_n) = \lim_{n\to\infty} \sin(n\pi) = \lim_{n\to\infty} 0 = 0,$$

$$\lim_{n\to\infty} f(b_n) = \lim_{n\to\infty} \sin((2n+\tfrac{1}{2})\pi) = \lim_{n\to\infty} 1 = 1,$$

so the function f does not have a limit as $x \to 0$.

Easier still, we can just find a sequence (c_n) in $(0, \infty)$ for which $c_n \to 0$ but the image sequence $(f(c_n))$ diverges. We pick $c_n = \frac{1}{\left(n+\frac{1}{2}\right)\pi}$ so that $c_n \to 0$. Using this sequence, we have $f(c_n) = (-1)^n$ and hence the image sequence $(f(c_n))$ does not converge. Thus, we conclude that $\lim_{x\to 0} f(x)$ does not exist.

Either way, we have showed that the function f does not have a limit as $x \to 0$.

As we have mentioned earlier, by using Definition 9.2.2, it may be impossible to check that $f(x_n) \to L$ for each and every sequences $x_n \to x_0$ in $X \setminus \{x_0\}$ since there are so many such sequences (x_n). In fact, there are uncountably many such sequences as we have seen in Exercise 6.10. Therefore, a slick way of checking for all these sequences at once is via the following definition by Bolzano, Cauchy, and Weierstrass:

Definition 9.2.6 (Limit of a Function at x_0, Definition 2) Let $f : X \to \mathbb{R}$ where $X \subseteq \mathbb{R}$ and $x_0 \in X'$. The function f has the limit $L \in \mathbb{R}$ as $x \to x_0$ if for every $\varepsilon > 0$ there exists a $\delta(\varepsilon) > 0$ such that for all $x \in X$ with $0 < |x - x_0| < \delta(\varepsilon)$ we have $|f(x) - L| < \varepsilon$.

Symbolically, this can be written with quantifiers as:

$$f(x) \xrightarrow{x\to x_0} L \quad \text{if}$$

$$\forall \varepsilon>0, \exists \delta(\varepsilon)>0 : \forall x\in X, 0 < |x - x_0|<\delta(\varepsilon) \Rightarrow |f(x) - L| < \varepsilon.$$

If there is no such real number L, then the function f does not have a limit as $x \to x_0$.

Remark 9.2.7 Let us make some remarks on Definition 9.2.6 .

1. For obvious reasons, we refer to this definition as the ε-δ definition for limits of functions.
2. We can also write the definition above in terms of open balls, namely:

$$f(x) \xrightarrow{x \to x_0} L \quad \text{if}$$

$$\forall \varepsilon > 0, \exists \delta(\varepsilon) > 0 : \forall x \in X, x \in B_{\delta(\varepsilon)}(x_0) \setminus \{x_0\} \Rightarrow f(x) \in B_\varepsilon(L).$$

3. Note that for the definition to make sense, again, we must require that $x_0 \in X'$. Otherwise, if $x_0 \notin X'$, for any small enough $\delta > 0$ we have $\{x \in X : 0 < |x - x_0| < \delta\} = X \cap B_\delta(x_0) \setminus \{x_0\} = \varnothing$ and hence the expression $|f(x) - L|$ in Definition 9.2.6 above does not have a value to make sense.
4. It does not matter what the value of f at x_0 or whether f is even defined at x_0 since we are just looking at all the images of the points x around x_0. More specifically, from the definition, we are only considering the values $f(x)$ on the set $\{x \in X : 0 < |x - x_0| < \delta(\varepsilon)\}$ for some small $\delta(\varepsilon) > 0$. Note that this set does not include the point $x = x_0$, so the value of $f(x_0)$ (or the lack thereof) is irrelevant to the definition.
5. The radius $\delta(\varepsilon)$ depends on the value ε. Roughly speaking, the smaller the value of ε, the smaller the radius $\delta(\varepsilon)$ is required to be.
6. Furthermore for each ε, there may not be a unique value of $\delta(\varepsilon)$ that would fulfil this definition. However, the existence of one such $\delta(\varepsilon)$ for every ε is enough for our purposes. This is similar to the definition of convergence of real sequences in which we require the existence one $N(\varepsilon)$ for every ε for Definition 5.2.4 to hold. Most of the time, we write $\delta(\varepsilon)$ simply as δ to declutter: however we still need to implicitly remember that δ depends on ε.
7. Using Example 1.4.8, we can find the negation of this definition. In other words, if $\lim_{x \to x_0} f(x) \neq L$, it must fulfil:

$$\exists \varepsilon > 0 : \forall \delta > 0, \exists x \in X : 0 < |x - x_0| < \delta \wedge |f(x) - L| \geq \varepsilon.$$

We now have two definitions of limits, so which one do we use? The two definitions in Definitions 9.2.2 and 9.2.6 are in fact equivalent. So one may use either definition, whichever more convenient, when dealing with limits. We show their equivalence here:

Lemma 9.2.8 (Equivalence of Limit Definitions) *Definitions 9.2.2 and 9.2.6 are equivalent.*

Proof We prove Definition 9.2.2 $\Leftrightarrow$ Definition 9.2.6.

($\Leftarrow$): We want to show that for any sequence $(x_n) \subseteq X \setminus \{x_0\}$ such that $x_n \to x_0$, we have $f(x_n) \to L$. In other words, for every $\varepsilon > 0$, there exists an $N \in \mathbb{N}$ such that $|f(x_n) - L| < \varepsilon$ for all $n \geq N$.
Pick any such sequence (x_n) and fix $\varepsilon > 0$. Assuming Definition 9.2.6 is true, there exists a $\delta > 0$ such that for all $x \in X$ with $0 < |x - x_0| < \delta$ we must have $|f(x) - L| < \varepsilon$. Since $x_n \to x_0$, for the same $\delta > 0$ as above, there exists an $N \in \mathbb{N}$ such that $0 < |x_n - x_0| < \delta$ for all $n \geq N$. By our assumption, this means $|f(x_n) - L| < \varepsilon$ for all $n \geq N$. Thus, we have Definition 9.2.2.

($\Rightarrow$): We prove this via contradiction by constructing a suitable sequence of points in X. Suppose that Definition 9.2.2 is true and Definition 9.2.6 is not true. By negation of the latter in Remark 9.2.7(7), there exists an $\varepsilon > 0$ such that for all $\delta > 0$, there exists an $x \in X$ with $0 < |x - x_0| < \delta$ and $|f(x) - L| \geq \varepsilon$.
With this $\varepsilon > 0$, by setting $\delta = 1$, there exists an element $x \in X$ such that $0 < |x - x_0| < 1$ and $|f(x_n) - L| \geq \varepsilon$. Set this x as the first element of the sequence. Inductively, for $\delta = \frac{1}{n}$ where $n \in \mathbb{N}$, we can find an element $x_n \in X$ with $0 < |x_n - x_0| < \frac{1}{n}$ and $|f(x_n) - L| \geq \varepsilon$. From this, we have constructed a sequence $(x_n) \subseteq X \setminus \{x_0\}$ such that $x_n \to x_0$ and $|f(x_n) - L| \geq \varepsilon$ for all $n \in \mathbb{N}$.
The latter means the sequence $(f(x_n))$ is not converging to L, contradicting the assumption that Definition 9.2.2 is true. Thus, we conclude that Definition 9.2.6 must be true. $\square$

Example 9.2.9 Let us look at some examples for finding limits of a function:

1. Consider the function $f : (0, \infty) \to \mathbb{R}$ defined as $f(x) = x \sin(\frac{1}{x})$. The point 0 is a limit point of the domain $(0, \infty)$ and so we can ask whether $\lim_{x\to 0} f(x)$ exists. If we plot the graph of this function, we see that function oscillates up and down but the amplitude of the oscillation gets smaller as we approach 0. We claim that $\lim_{x\to 0} f(x) = 0$.

   Rough work: Fix $\varepsilon > 0$. We want to find a $\delta > 0$ such that if $x \in (0, \infty)$ with $0 < |x - 0| = |x| < \delta$, then $|f(x) - 0| = |f(x)| < \varepsilon$. From the definition of the function f, we note that $|f(x)| = |x \sin(\frac{1}{x})| \leq |x| < \delta$ since the sine function is bounded between -1 and 1. So, we can pick $\delta = \varepsilon$.

   Fix $\varepsilon > 0$. Set $\delta = \varepsilon > 0$. Then, whenever $0 < |x| < \delta = \varepsilon$, we would get $|f(x)| = |x \sin(\frac{1}{x})| \leq |x| < \varepsilon$. Thus, we have $f(x) \xrightarrow{x\to 0} 0$.
2. Consider the function $f : \mathbb{R} \to \mathbb{R}$ defined as $f(x) = 2x + 1$. The point 2 lies in the limit point of the domain so we can try and find the limit of this function as $x \to 2$. If we plot the graph of this function, we see that as x gets closer to 2, the value $f(x)$ approaches 5. Let us try and prove this analytically.

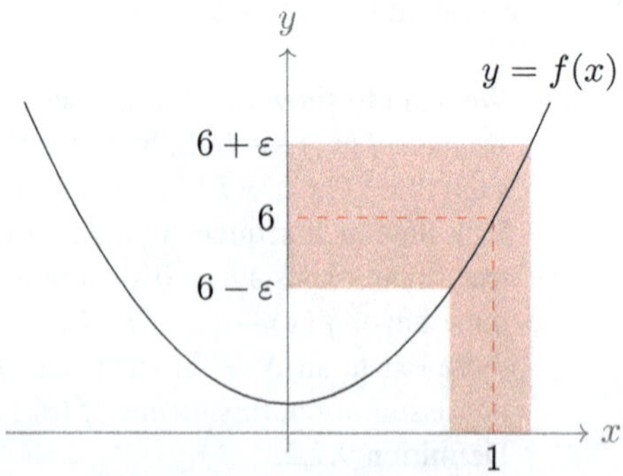

Fig. 9.2 Diagram for finding $\lim_{x\to 1} f(x)$

Rough work: Fix $\varepsilon > 0$. We want to find a $\delta > 0$ such that if $0 < |x-2| < \delta$ then $|f(x)-5| < \varepsilon$. The latter can be rewritten as $|f(x)-5| = |2x+1-5| = |2x-4| = 2|x-2| < 2\delta$. We want this quantity to be smaller than ε, so a suitable choice for δ would be $\delta = \frac{\varepsilon}{2}$.

Fix $\varepsilon > 0$. Set $\delta = \frac{\varepsilon}{2} > 0$. If $0 < |x-2| < \delta = \frac{\varepsilon}{2}$, then $|f(x)-5| = 2|x-2| < 2\frac{\varepsilon}{2} = \varepsilon$. Thus, we conclude that $f(x) \xrightarrow{x\to 2} 5$.

3. Consider the function $f : \mathbb{R} \to \mathbb{R}$ defined as $f(x) = 5x^2 + 1$. The point 1 is a limit point of the domain so we can ask: what is $\lim_{x\to 1} f(x)$? We claim that the limit is 6.

Rough work: Fix $\varepsilon > 0$. We want to find a $\delta > 0$ such that if $0 < |x-1| < \delta$, then $|f(x)-6| < \varepsilon$. A diagram for this can be seen in Fig. 9.2.
Working from the latter, we have:

$$|f(x)-6| = |5x^2+1-6| = |5x^2-5| = 5|x-1||x+1|. \tag{9.1}$$

Now we have reached a tricky point because we have two expressions that contain x here, namely $|x-1|$ and $|x+1|$. The term $|x-1|$ appears in the condition for δ, so we try to keep this term and get rid of the other term by bounding it with some constant. We do this by putting an initial assumption on the condition for $|x-1|$ which we will combine the choice for δ later.
Say we place an assumption that $|x-1| < 1$. With this condition, we have $1 < x+1 < 3$, which means we have the bound $|x+1| < 3$. Hence, if we put in the assumption $|x-1| < 1$, we can continue from the Eq. (9.1) with:

$$|f(x)-6| = 5|x-1||x+1| < 15|x-1| \quad \text{for} \quad |x-1| < 1.$$

We have got rid of the $|x+1|$ term. Now we can set another condition on $|x-1|$ so that this quantity is less than ε. We can thus set $|x-1| < \frac{\varepsilon}{15}$ which then implies $|f(x)-6| < 15\frac{\varepsilon}{15} = \varepsilon$, our desired goal.
To recap, we have placed two conditions to reach the desired goal, namely: $|x-1| < 1$ and $|x-1| < \frac{\varepsilon}{15}$. Thus, an appropriate choice of $\delta > 0$ that satisfy both of these would be $\delta = \min\{1, \frac{\varepsilon}{15}\} > 0$. Now we write this down properly.

Fix $\varepsilon > 0$. Set $\delta = \min\{1, \frac{\varepsilon}{15}\} > 0$. Suppose that $0 < |x-1| < \delta = \min\{1, \frac{\varepsilon}{15}\}$. Then, we have:

$$\begin{aligned}|f(x)-6| = 5|x-1||x+1| &< 15|x-1| \quad (\because |x-1| < 1 \Rightarrow |x+1| < 3)\\ &< 15\frac{\varepsilon}{15} = \varepsilon \quad (\because |x-1| < \tfrac{\varepsilon}{15}).\end{aligned}$$

Therefore, we conclude that $f(x) \xrightarrow{x\to 1} 6$.

4. We could also try a different initial bound on the term $|x-1|$ in the previous example. Say we set $|x-1| < 3$ instead of $|x-1| < 1$. With this initial bound, we have $-1 < x+1 < 5$ which implies $|x+1| < 5$. We can then choose $\delta = \min\{3, \frac{\varepsilon}{25}\} > 0$. Hence, for any $|x-1| < \delta$, we have:

$$\begin{aligned}|f(x)-6| = 5|x-1||x+1| &< 25|x-1| \quad (\because |x-1| < 3 \Rightarrow |x+1| < 5)\\ &< 25\frac{\varepsilon}{25} = \varepsilon \quad (\because |x-1| < \tfrac{\varepsilon}{25}),\end{aligned}$$

showing that $f(x) \xrightarrow{x\to 1} 6$. The lesson here is there might be more than one possible initial estimate that we could make, but one has to be careful since not all initial guess estimates work! We shall see this in the next example.

5. Consider the function $f : \mathbb{R}\backslash\{-2, 1\} \to \mathbb{R}$ defined as $f(x) = \frac{x^2-1}{x^2+x-2}$. The point 1 is a limit point of the domain, so we can ask what is the limit of the function as $x \to 1$. We claim that this limit is actually $\frac{2}{3}$. How do we prove it?

Rough work: Fix $\varepsilon > 0$. We want to find $\delta > 0$ such that if $x \in \mathbb{R} \setminus \{-2, 1\}$ with $0 < |x-1| < \delta$, then $|f(x) - \frac{2}{3}| < \varepsilon$. Let us simplify the end goal a little bit so we know what to work towards. We write:

$$|f(x) - \tfrac{2}{3}| = \left|\frac{x^2-1}{x^2+x-2} - \frac{2}{3}\right| = \left|\frac{(x-1)(x+1)}{(x+2)(x-1)} - \frac{2}{3}\right| = \frac{|x-1|}{3|x+2|}, \tag{9.2}$$

and we want to bound this with the chosen ε.

First, we get rid of the term $\frac{1}{|x+2|}$ by bounding it from above with a positive constant. Let us place an initial restriction of $|x-1| < 3$. This estimate means $0 < x+2 < 6$. But this is not good enough since we cannot bound $\frac{1}{|x+2|}$ from above by a positive constant as this term can get arbitrarily large!

Now let us try a smaller initial estimate of $|x-1| < 1$ instead so that $2 < x+2 < 4$. This implies $|x+2| > 2$ and hence $\frac{1}{|x+2|} < \frac{1}{2}$. This is much better! Using this assumption, we can bound (9.2) by:

$$|f(x) - \tfrac{2}{3}| = \frac{|x-1|}{3|x+2|} < \frac{|x-1|}{6} \quad \text{for} \quad |x-1| < 1.$$

Now we put the condition $|x-1| < \varepsilon$, so that the term above can be bounded as:

$$|f(x) - \tfrac{2}{3}| = \frac{|x-1|}{3|x+2|} < \frac{|x-1|}{6} < \frac{\varepsilon}{6} < \varepsilon \quad \text{for} \quad |x-1| < 1 \text{ and } |x-1| < \varepsilon,$$

to reach our desired goal. So there are two conditions $|x-1| < 1$ and $|x-1| < \varepsilon$ that need to be satisfied. Thus, a reasonable choice for δ would be $\delta = \min\{1, \varepsilon\} > 0$.

For a fixed $\varepsilon > 0$, choose $\delta = \min\{1, \varepsilon\} > 0$. If $x \in \mathbb{R} \setminus \{-2, 1\}$ with $0 < |x-1| < \delta$, then:

$$\begin{aligned} |f(x) - \tfrac{2}{3}| = \frac{|x-1|}{3|x+2|} &< \frac{|x-1|}{6} && (\because |x-1| < 1 \Rightarrow \tfrac{1}{|x+2|} < \tfrac{1}{2}) \\ &< \frac{\varepsilon}{6} < \varepsilon && (\because |x-1| < \varepsilon). \end{aligned}$$

This implies $f(x) \xrightarrow{x \to 1} \frac{2}{3}$.

6. Consider the function $f : \mathbb{R} \to \mathbb{R}$ defined as:

$$f(x) = \begin{cases} 2x+1 & \text{if } x \in \mathbb{R} \setminus \{2\}, \\ 7 & \text{if } x = 2. \end{cases}$$

We note that this is the same function as we have seen in the second example above, but its value at $x = 2$ is different. If we plot the graph of this function, we see that as x gets closer to 2, the value $f(x)$ approaches 5. However, $f(2) = 7$, so there is a hole in the graph at $x = 2$. The graph of the function f is given in Fig. 9.3.

Again, the point 2 lies in the limit point of the domain, so we can ask: what is $\lim_{x \to 2} f(x)$? Let us prove that the limit of this function as x approaches 2 is 5 (and not $f(2) = 7$).

Fig. 9.3 Graph of $y = f(x)$

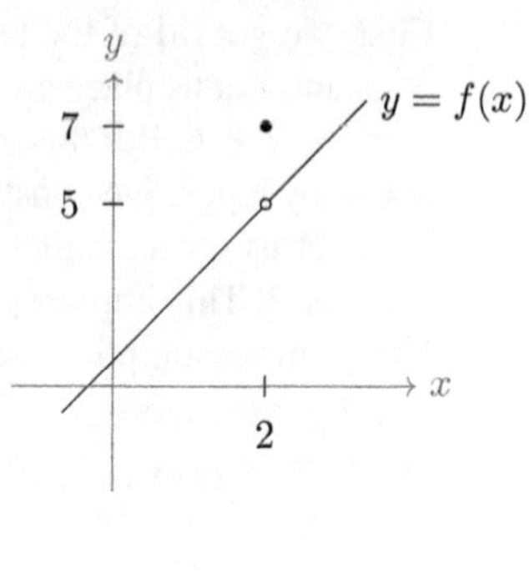

Fig. 9.4 Graph of $y = f(x)$

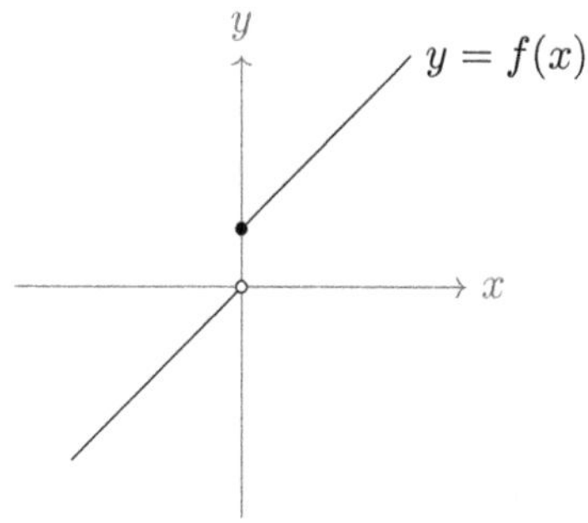

Fix $\varepsilon > 0$. Choose $\delta = \frac{\varepsilon}{2}$. Assume that $0 < |x-2| < \delta = \frac{\varepsilon}{2}$. Notice that since $0 < |x-2|$, we have $x \neq 2$ and hence $f(x) = 2x+1$ for these x. So:

$$|f(x)-5| = |(2x+1)-5| = |2x-4| = 2|x-2| < 2\frac{\varepsilon}{2} = \varepsilon \quad (\because |x-2|<\tfrac{\varepsilon}{2}),$$

and hence we conclude $f(x) \xrightarrow{x\to 2} 5$.

7. Consider the function $f : \mathbb{R} \to \mathbb{R}$ defined as:

$$f(x) = \begin{cases} x & \text{if } x < 0, \\ x+1 & \text{if } x \geq 0. \end{cases}$$

The graph of the function f is given in Fig. 9.4.
The point 0 lies in the limit point of the domain, so we can try and find the limit of this function as $x \to 0$. However, the limit of this function f as $x \to 0$ does not exist since we can find two different sequences (x_n) and (y_n) in $\mathbb{R} \setminus \{0\}$ such that $x_n \to 0$ and $y_n \to 0$ but $\lim_{n\to\infty} f(x_n) \neq \lim_{n\to\infty} f(y_n)$.
For example, consider $x_n = -\frac{1}{n}$ and $y_n = \frac{1}{n}$. Clearly these two sequences converge to the limit point 0. However, we have $f(x_n) = f(-\frac{1}{n}) = -\frac{1}{n} \to 0$ and $f(y_n) = f(\frac{1}{n}) = \frac{1}{n} + 1 \to 1$, so the limits are different. Hence, the function f does not have a limit as $x \to 0$.

8. Consider the function $f : (0, \infty) \to \mathbb{R}$ defined as $f(x) = \frac{1}{x}$. The point 0 is a limit point of the domain, so we can try and find a limit of this function as $x \to 0$. However, the limit of this function as $x \to 0$ does not exist.
To prove this, suppose for contradiction that the limit exists and is equal to $L \in \mathbb{R}$, namely $\lim_{x\to 0} f(x) = L$. By Definition 9.2.2, for any sequence of points (x_n) in $(0, \infty)$ such that $x_n \to 0$, we must have $f(x_n) \to L$.
Pick one such sequence, say $x_n = \frac{1}{n}$. Since this sequence converges to 0, the sequence $(f(x_n)) = (f(\frac{1}{n}))$ must converge to L. In particular, by Proposition 5.2.14, the sequence $(f(\frac{1}{n}))$ is bounded. However, this sequence is actually $(f(\frac{1}{n})) = (1, 2, 3, 4, 5, \ldots)$ which is an unbounded sequence, giving us the desired contradiction. Thus, $\lim_{x\to 0} f(x)$ does not exist.

Remark 9.2.10 Let us comment some of the examples above:

1. From Example 9.2.9(1) and (5), we have seen that the limit of function f as $x \to x_0$ is defined even if the limit point x_0 does not lie in the domain X of the functions.
2. Furthermore, even if x_0 lies in X so that $f(x_0)$ has a value in $\mathbb{R}$, the limit of the function as $x \to x_0$ does not depend on the value of $f(x_0)$ as we have seen in Example 9.2.9(2) and (5). In fact, we do not care at all about the value $f(x_0)$ at this stage (we will later in Chap. 10). The limit just depends on the values of the function f around x_0, namely on $B_\delta(x_0) \setminus \{x_0\}$ for some $\delta > 0$.
3. As a result of the observations above, the limit captures how the function behaves near (but not at) the point x_0. Hence, the limit is often referred to as a local behaviour or local property of a function.

The idea of proving the limit of a function f as $x \to x_0$ is usually similar. We follow the following steps as a rough guide:

1. Make a conjecture about the limit, say the limit is L. If the limit is given, then we do not need to do this.
2. For every $\varepsilon > 0$, we need to find a $\delta > 0$ such that for any $x \in X$ with $0 < |x - x_0| < \delta$, we have $|f(x) - L| < \varepsilon$.
   (a) This is done by first fixing $\varepsilon > 0$ and by algebraic manipulations, find a suitable δ by trying to get a bound of the form $|f(x) - L| \leq K|x - x_0|^p < K\delta^p$ for some constant $K > 0$ and $p > 0$. One may need to put some additional assumptions or estimates on $|x - x_0|$ to get to this form.
   (b) Since we want $|f(x) - L| < \varepsilon$, it is enough to set $K\delta^p = \varepsilon$ and from here we can manipulate the inequality to get δ as a function of ε. It is very important to remember that we also need to take into account of all the assumptions we made for $|x - x_0|$ to reach $|f(x) - L| \leq K|x - x_0|^p$. One then construct a function $\delta : \mathbb{R}_+ \to \mathbb{R}_+$ of ε.
   (c) Since δ is a function of ε, we can then vary ε to get different $\delta(\varepsilon)$ that works for different $\varepsilon > 0$.
3. Write down the proof nicely and in order. Steps 1 and 2 are just the rough work. Once we found the $\delta(\varepsilon)$, we can rewrite it from the beginning and by setting $0 < |x - x_0| < \delta(\varepsilon)$, we would magically get $|f(x) - L| < \varepsilon$ by a series of inequalities that we need to justify based on the choices we made for $\delta(\varepsilon)$.

For each $\varepsilon > 0$, the choice for $\delta(\varepsilon) > 0$ is not unique! Luckily, we just need one for each $\varepsilon > 0$ for the proof to work. So there are lots of functions $\delta(\varepsilon)$ that we could make. However, similar to the limits of real sequences, if the limit of a real-valued function f as $x \to x_0$ exists, then this limit must be unique.

Proposition 9.2.11 *Let $f : X \to \mathbb{R}$ where $X \subseteq \mathbb{R}$ and $x_0 \in X'$. Suppose that the limit of the function $f(x)$ as $x \to x_0$ exists, then $\lim_{x \to x_0} f(x)$ is unique.*

Proof Suppose for contradiction that there are two different limits $L_1 < L_2$ of f as $x \to x_0$. Using the definition of limits with $\varepsilon = \frac{L_2-L_1}{2} > 0$, there exists a $\delta_1 > 0$ such that for all $x \in X$ with $0 < |x - x_0| < \delta_1$ we have $|f(x) - L_1| < \varepsilon = \frac{L_2-L_1}{2}$ and also a $\delta_2 > 0$ such that for all $x \in X$ with $0 < |x - x_0| < \delta_2$ we have $|f(x) - L_2| < \varepsilon = \frac{L_2-L_1}{2}$. Setting $\delta = \min\{\delta_1, \delta_2\} > 0$, for all $x \in X$ such that $0 < |x - x_0| < \delta$, both of the inequalities above are true and we have:

$$L_2-L_1=|L_2-L_1|=|L_2-f(x)+f(x)-L_1| \leq |L_1-f(x)|+|f(x)-L_2|<L_2-L_1,$$

which is a contradiction. Thus, we conclude that the limit must be unique. □

9.3 One-Sided Limits

When we defined limits of a function in Definition 9.2.2, we took into account every sequence $(x_n) \subseteq X \setminus \{x_0\}$ such that $x_n \to x_0$. However, since the domain X is a subset of an ordered set $\mathbb{R}$ we can restrict these sequences by requiring that either $x_n > x_0$ for all $n \in \mathbb{N}$ or $x_n < x_0$ for all $n \in \mathbb{N}$.

For the former, geometrically, we are approaching the limit point x_0 via a sequence of numbers strictly bigger than it or we say we approach x_0 on the real number line from the right/above. Analogously, for the latter, we are approaching the limit point x_0 via a sequence of numbers strictly smaller than it or we say we approach x_0 on the real number line from the left/below.

This might be useful in some cases, as we have seen in Example 9.2.9(7): we can approach the limit point from left and from the right and the limit of the image sequences are different. These restrictions motivate the idea of one-sided limits. Restricting Definitions 9.2.2 and 9.2.6, we have:

Definition 9.3.1 (One-Sided Limits) Let $f : X \to \mathbb{R}$ where $X \subseteq \mathbb{R}$ and $x_0 \in X'$.

1. We say that the function f has the right-limit $L \in \mathbb{R}$ as $x \to x_0$ if for every sequence (x_n) in X with $x_n > x_0$ such that $x_n \to x_0$, the image sequence $(f(x_n))$ converges to L. We write this as:

$$\lim_{x \downarrow x_0} f(x) = L \quad \text{or} \quad f(x) \xrightarrow{x \downarrow x_0} L.$$

   Equivalently, for any $\varepsilon > 0$ there exists a $\delta > 0$ such that for all $x \in X$ with $0 < x - x_0 < \delta$ we must have $|f(x) - L| < \varepsilon$. In symbols:

$$f(x) \xrightarrow{x \downarrow x_0} L \quad \text{if} \quad \forall \varepsilon > 0, \exists \delta > 0 : \forall x \in X, 0 < x-x_0 < \delta \Rightarrow |f(x)-L| < \varepsilon.$$

2. We say that the function f has the left-limit $L \in \mathbb{R}$ as $x \to x_0$ if for every sequence (x_n) in X with $x_n < x_0$ such that $x_n \to x_0$, the image sequence $(f(x_n))$ converges to L. We write this as:

$$\lim_{x \uparrow x_0} f(x) = L \quad \text{or} \quad f(x) \xrightarrow{x \uparrow x_0} L.$$

Equivalently, for any $\varepsilon > 0$ there exists a $\delta > 0$ such that for all $x \in X$ with $0 < x_0 - x < \delta$ we must have $|f(x) - L| < \varepsilon$. In symbols:

$$f(x) \xrightarrow{x \uparrow x_0} L \quad \text{if} \quad \forall \varepsilon > 0, \exists \delta > 0 : \forall x \in X, 0 < x_0 - x < \delta \Rightarrow |f(x) - L| < \varepsilon.$$

Remark 9.3.2 Sometimes, the right and left limits are also denoted as $\lim_{x \to x_0^+} f(x)$ and $\lim_{x \to x_0^-} f(x)$ respectively. This tells us whether we are approaching the limit point x_0 from the positive (right/above) side or the negative (left/below) side.

Even when a real-valued function f does not have a limit as $x \to x_0$, the one sided limits may still exist. We have seen an example of this in Example 9.2.9(7).

Example 9.3.3 Recall the function $f : \mathbb{R} \to \mathbb{R}$ defined as:

$$f(x) = \begin{cases} x & \text{if } x < 0, \\ x + 1 & \text{if } x \geq 0. \end{cases}$$

The limit of this function as $x \to 0$ does not exist since we can find two sequences of points converging to 0 such that their image sequences converge to two different numbers. However, the left- and right-limits exist and they are given by:

$$\lim_{x \uparrow 0} f(x) = 0 \quad \text{and} \quad \lim_{x \downarrow 0} f(x) = 1.$$

We prove that this is true for the left-limit and the right-limit can be proven analogously.

Rough work: Fix $\varepsilon > 0$. We want to find a $\delta >$ such that if $0 < 0 - x < \delta$, we have $|f(x) - 0| < \varepsilon$. This is equivalent to finding a $\delta > 0$ such that $-\delta < x < 0$ implies $|f(x)| < \varepsilon$. Since $x < 0$, we must have $f(x) = x$ and hence the latter inequality is simply $|f(x)| = |x| < \varepsilon$. A suitable choice would then be $\delta = \varepsilon$.

Fix $\varepsilon > 0$. Set $\delta = \varepsilon$. Then, $-\delta = -\varepsilon < x < 0$. In particular, $|x| < \varepsilon$. Moreover, for any such x, since $x < 0$, we have $f(x) = x$ here and thus $|f(x) - 0| = |x - 0| = |x| < \varepsilon$. Therefore, we have proven that the left-limit of the function f at $x = 0$ is indeed 0.

Clearly, if the full limit of a function as $x \to x_0$ exists, then the left- and right-limits at x_0 exist as well because $0 < |x - x_0| < \delta$ implies both $0 < x_0 - x < \delta$ and $0 < x - x_0 < \delta$. The converse might not be true, as we saw in Example 9.3.3. However, if the left- and the right-limits of a function f as $x \to x_0$ both exist and are equal to each other, then the full limit exists.

Proposition 9.3.4 *Let $f : X \to \mathbb{R}$ where $X \subseteq \mathbb{R}$ and $x_0 \in X'$. Suppose that the left- and the right-limits of the function as $x \to x_0$ exist and are equal, namely $\lim_{x\uparrow x_0} f(x) = \lim_{x\downarrow x_0} f(x) = L \in \mathbb{R}$. Then, the limit $\lim_{x\to x_0} f(x)$ exists and is equal to L as well.*

Proof We want to show the limit of f equals to L as $x \to x_0$. Namely, for each $\varepsilon > 0$ we want to find a $\delta > 0$ such that for any $x \in X$ with $0 < |x - x_0| < \delta$, we have $|f(x) - L| < \varepsilon$.

Fix $\varepsilon > 0$. Since $\lim_{x\uparrow x_0} f(x) = L$, there exists a $\delta_1 > 0$ such that if $x \in X$ with $-\delta_1 < x - x_0 < 0$, then $|f(x) - L| < \varepsilon$. Similarly, since $\lim_{x\downarrow x_0} f(x) = L$, there exists a $\delta_2 > 0$ such that if $x \in X$ with $0 < x - x_0 < \delta_2$, then $|f(x) - L| < \varepsilon$.

Pick $\delta = \min\{\delta_1, \delta_2\} > 0$. If $x \in X$ is such that $0 < |x - x_0| < \delta$, then either:

$$-\delta_1 \leq -\delta < x - x_0 < 0 \quad \text{or} \quad 0 < x - x_0 < \delta \leq \delta_2,$$

since δ is chosen such that $\delta \leq \delta_1$ and $\delta \leq \delta_2$. In both cases, from the existence of both the left- and right-limits above, for such x we must have $|f(x) - L| < \varepsilon$, which is what we wanted to show. □

9.4 Blowing Up and Limits at Infinity

Blowing Up to $\pm\infty$

We have seen in Example 9.2.9(8) that the function $f(x) = \frac{1}{x}$ defined for $x > 0$ grows in an unbounded manner as we approach the limit point $x \to 0$ via the sequence (x_n) where $x_n = \frac{1}{n}$. If this also holds true for any sequence (x_n) with $x_n \to 0$, the function is said to diverge to infinity as we approach the limit point 0. We define the general scenario as:

Definition 9.4.1 (Blowing Up to Infinity) Let $f : X \to \mathbb{R}$ where $X \subseteq \mathbb{R}$ and $x_0 \in X'$.

1. The function f is said to be blowing up to infinity or diverges to infinity as $x \to x_0$ if for any sequence $(x_n) \subseteq X \setminus \{x_0\}$ such that $x_n \to x_0$, we have $f(x_n) \to \infty$. We write this as:

$$f(x) \xrightarrow{x\to x_0} \infty.$$

An equivalent formulation is: for all $K > 0$, there exists a $\delta > 0$ such that for all $x \in X$ with $0 < |x - x_0| < \delta$ we have $f(x) > K$. In symbols:

$$f(x) \xrightarrow{x \to x_0} \infty \quad \text{if} \quad \forall K > 0, \exists \delta > 0 : \forall x \in X, 0 < |x - x_0| < \delta \Rightarrow f(x) > K.$$

2. The function f is said to be blowing up to negative infinity or diverges to negative infinity as $x \to x_0$ if for any sequence $(x_n) \subseteq X \setminus \{x_0\}$ such that $x_n \to x_0$, we have $f(x_n) \to -\infty$. We write this as:

$$f(x) \xrightarrow{x \to x_0} -\infty.$$

An equivalent formulation is: for all $K < 0$, there exists a $\delta > 0$ such that for all $x \in X$ with $0 < |x - x_0| < \delta$ we have $f(x) < K$. In symbols:

$$f(x) \xrightarrow{x \to x_0} -\infty \quad \text{if} \quad \forall K < 0, \exists \delta > 0 : \forall x \in X, 0 < |x - x_0| < \delta \Rightarrow f(x) < K.$$

In an abuse of notation, we often write $\lim_{x \to x_0} f(x) = \infty$ if the function f blows up to infinity as $x \to x_0$ and $\lim_{x \to x_0} f(x) = -\infty$ if the function f blows up to negative infinity as $x \to x_0$. This notation is not strictly correct in the sense of Definitions 9.2.2 and 9.2.6 as $\pm\infty$ are not elements of the real number set $\mathbb{R}$ and hence the expressions $|f(x) \pm \infty|$ do not make sense. However, this is a widely accepted notation for functions blowing up to infinity.

Example 9.4.2 Define a function $f : \mathbb{R} \to \mathbb{R}$ as:

$$f(x) = \begin{cases} \frac{1}{x^2} & \text{if } x \neq 0, \\ 1 & \text{if } x = 0. \end{cases}$$

We now show that this function blows up to ∞ as we approach $x = 0$.

Rough work: Fix $K > 0$. We want to find a $\delta > 0$ such that whenever $x \in X$ with $0 < |x - 0| < \delta$ we have $f(x) > K$. In other words, whenever $0 < |x| < \delta$ we want $\frac{1}{x^2} > K$. After some algebra, we can claim that $\delta = \frac{1}{\sqrt{K}} > 0$ is enough.

Fix $K > 0$. Set $\delta = \frac{1}{\sqrt{K}} > 0$. Thus, if $x \in \mathbb{R}$ is such that $0 < |x| < \delta = \frac{1}{\sqrt{K}}$, we have:

$$f(x) = \frac{1}{x^2} > \frac{1}{\delta^2} = \frac{1}{\frac{1}{\sqrt{K}^2}} = K,$$

which is what we wanted.

Note in Example 9.4.2 that the value of the function at $x = 0$, namely $f(0) = 1$, does not play a part at all in the limit. This is because for limits, we only care about the behaviour of the function around (but not at) the limit point $x = 0$. This is similar to the discussion in Remark 9.2.10(2) for finite limits.

Remark 9.4.3 The definition and results for one-sided limits also hold true in Definition 9.4.1 analogously. For example, for the function $f : \mathbb{R}\setminus\{0\} \to \mathbb{R}$ defined by $f(x) = \frac{1}{x}$, we have the left-limit of $f(x)$ as $x \uparrow 0$ is negative infinity whereas the right-limit of $f(x)$ as $x \downarrow 0$ is positive infinity, namely:

$$\lim_{x\uparrow 0} f(x) = -\infty \quad \text{and} \quad \lim_{x\downarrow 0} f(x) = \infty.$$

However, this function does not blow up to either $\pm\infty$ as $x \to 0$ since the left- and right-limits at $x = 0$ do not coincide.

Limits at $\pm\infty$

Another variation of limits for real-valued functions with real domain is when the domain of the function X is unbounded, say $X = (a, \infty)$. Even though ∞ is not a limit point of the domain X, we can still pick a sequence of points (x_n) in the domain such that $x_n \to \infty$ in the sense of Definition 5.3.1. This would also give rise to another sequence $(f(x_n))$ in the image which may or may not converge.

For the case that the image sequence $(f(x_n))$ converges to a real number $L \in \mathbb{R}$ for any sequence $x_n \to \infty$, we call this number L the limit of the function f at infinity. More concretely, we define:

Definition 9.4.4 (Limit at Infinity) Let $f : X \to \mathbb{R}$ where $X \subseteq \mathbb{R}$.

1. If the domain X is unbounded above, we say that the function f has the limit $L \in \mathbb{R}$ as $x \to \infty$ if for any sequence $(x_n) \subseteq X$ such that $x_n \to \infty$, we have $f(x_n) \to L$. We write this as:

$$f(x) \xrightarrow{x\to\infty} L \quad \text{or} \quad \lim_{x\to\infty} f(x) = L.$$

   Equivalently, for any $\varepsilon > 0$ there exists a $K > 0$ such that for all $x \in X$ with $x > K$ we have $|f(x) - L| < \varepsilon$. In symbols:

$$f(x) \xrightarrow{x\to\infty} L \quad \text{if} \quad \forall\varepsilon > 0, \exists K > 0 : \forall x \in X, x > K \Rightarrow |f(x) - L| < \varepsilon.$$

2. If the domain X is unbounded below, we say that the function f has the limit $L \in \mathbb{R}$ as $x \to -\infty$ if for any sequence $(x_n) \subseteq X$ such that $x_n \to -\infty$, we have $f(x_n) \to L$. We write this as:

$$f(x) \xrightarrow{x\to-\infty} L \quad \text{or} \quad \lim_{x\to-\infty} f(x) = L.$$

Equivalently, for any $\varepsilon > 0$ there exists a $K < 0$ such that for all $x \in X$ with $x < K$ we have $|f(x) - L| < \varepsilon$. In symbols:

$$f(x) \xrightarrow{x \to -\infty} L \quad \text{if} \quad \forall \varepsilon > 0, \exists K < 0 : \forall x \in X, x < K \Rightarrow |f(x) - L| < \varepsilon.$$

Remark 9.4.5 The limits at infinity are always one-sided: we can only approach ∞ from below and $-\infty$ from above. So there is no ambiguity when we write $\lim_{x\to\infty} f(x)$ and $\lim_{x\to-\infty} f(x)$ which can also be written as $\lim_{x\uparrow\infty} f(x)$ and $\lim_{x\downarrow-\infty} f(x)$ respectively.

Example 9.4.6 Let $f : (0, \infty) \to \mathbb{R}$ be defined as $f(x) = \frac{1}{x}\sin(x)$. If we plot the graph of this function, we can see that the amplitude of this function gets smaller as $x \to \infty$. We want to show that the limit of this function at infinity is 0.

Rough work: Fix $\varepsilon > 0$. We want to find a $K > 0$ such that $|f(x) - 0| = |f(x)| < \varepsilon$ for all $x > K$. Simplifying, we have:

$$|f(x) - 0| = |f(x)| = \left|\frac{1}{x}\sin(x)\right| = \left|\frac{1}{x}\right| |\sin(x)| \leq \left|\frac{1}{x}\right| = \frac{1}{x},$$

which we want to be smaller than ε whenever $x > K$. So, we can pick $K = \frac{1}{\varepsilon} > 0$.

Fix $\varepsilon > 0$. Set $K = \frac{1}{\varepsilon} > 0$. For all $x > K$ we have $|f(x) - 0| = \left|\frac{1}{x}\sin(x)\right| \leq \frac{1}{x} < \frac{1}{K} = \varepsilon$. Thus, we can conclude that $f(x) \to 0$ as $x \to \infty$.

We may also have the following blowing up at infinity behaviour:

Definition 9.4.7 (Blowing Up at Infinity) Let $f : X \to \mathbb{R}$ where $X \subseteq \mathbb{R}$.

1. If the domain X is unbounded above, we say that the function f blows up to ∞ as $x \to \infty$ if for any $M > 0$, there exists a $K > 0$ such that for all $x \in X$ with $x > K$ we have $f(x) > M$. We often write $\lim_{x\to\infty} f(x) = \infty$ or symbolically:

$$f(x) \xrightarrow{x \to \infty} \infty \quad \text{if} \quad \forall M > 0, \exists K > 0 : \forall x \in X, x > K \Rightarrow f(x) > M.$$

2. If the domain X is unbounded above, we say that the function f blows up to $-\infty$ as $x \to \infty$ if for any $M < 0$, there exists a $K > 0$ such that for all $x \in X$ with $x > K$ we have $f(x) < M$. We often write $\lim_{x\to\infty} f(x) = -\infty$ or symbolically:

$$f(x) \xrightarrow{x \to \infty} -\infty \quad \text{if} \quad \forall M < 0, \exists K > 0 : \forall x \in X, x > K \Rightarrow f(x) < M.$$

3. If the domain X is unbounded below, we say that the function f blows up to ∞ as $x \to -\infty$ if for any $M > 0$, there exists a $K < 0$ such that for all $x \in X$ with $x < K$ we have $f(x) > M$. We often write $\lim_{x\to-\infty} f(x) = \infty$ or symbolically:

$$f(x) \xrightarrow{x\to-\infty} \infty \quad \text{if} \quad \forall M > 0, \exists K < 0 : \forall x \in X, x < K \Rightarrow f(x) > M.$$

4. If the domain X is unbounded below, we say that the function f blows up to $-\infty$ as $x \to -\infty$ if for any $M < 0$, there exists a $K < 0$ such that for all $x \in X$ with $x < K$ we have $f(x) < M$. We often write $\lim_{x\to-\infty} f(x) = -\infty$ or symbolically:

$$f(x) \xrightarrow{x\to-\infty} -\infty \quad \text{if} \quad \forall M < 0, \exists K < 0 : \forall x \in X, x < K \Rightarrow f(x) < M.$$

9.5 Algebra of Limits

We have seen algebra of limits for real sequences in which we allow the limit operation to be switched with some algebraic operations in the field $\mathbb{R}$ under mild assumptions. Since the limit of functions is essentially a limit of sequences $(f(x_n))$ in the codomain $\mathbb{R}$, we also have analogous algebra of limits for functions. We first show the following useful result.

Proposition 9.5.1 *Let $f : X \to \mathbb{R}$ where $X \subseteq \mathbb{R}$ and $x_0 \in X'$. Suppose that $\lim_{x\to x_0} f(x)$ exists. Then, there exists a $\delta > 0$ such that for all $x \in X$ with $0 < |x - x_0| < \delta$ we have $|f(x)| \leq M$ for some $M > 0$.*
In other words, there exists a punctured ball $X \cap B_\delta(x_0) \setminus \{x_0\}$ over which the function f is bounded.

Proof Suppose that $\lim_{x\to x_0} f(x) = L \in \mathbb{R}$. Fix $\varepsilon = 1$. Then, there exists a $\delta > 0$ such that for all $x \in X$ with $0 < |x - x_0| < \delta$ we must have $|f(x) - L| < 1$. Using triangle inequality, we get:

$$|f(x)| = |f(x) - L + L| \leq |f(x) - L| + |L| < 1 + |L|,$$

so we can define $M = 1 + |L|$ and hence $|f(x)| \leq M$ for all $x \in X$ with $0 < |x - x_0| < \delta$. □

Now we state and prove the algebra of limits.

Theorem 9.5.2 (Algebra of Limits, AOL) *Let $f, g : X \to \mathbb{R}$ where $X \subseteq \mathbb{R}$ and $x_0 \in X'$. Suppose that $\lim_{x\to x_0} f(x) = L$ and $\lim_{x\to x_0} g(x) = M$ where $L, M \in \mathbb{R}$. Then:*

1. *For a constant* $\lambda \in \mathbb{R}$, *function* λf *converges to* λL *as* $x \to x_0$. *In other words:*

$$\lim_{x \to x_0} \lambda f(x) = \lambda \lim_{x \to x_0} f(x).$$

2. *The function* $|f|$ *converges to* $|L|$ *as* $x \to x_0$. *In other words:*

$$\lim_{x \to x_0} |f|(x) = | \lim_{x \to x_0} f(x)|.$$

3. *The function* $f + g$ *converges to* $L + M$ *as* $x \to x_0$. *In other words:*

$$\lim_{x \to x_0} (f + g)(x) = \lim_{x \to x_0} f(x) + \lim_{x \to x_0} g(x).$$

4. *The function* $f - g$ *converges to* $L - M$ *as* $x \to x_0$. *In other words:*

$$\lim_{x \to x_0} (f - g)(x) = \lim_{x \to x_0} f(x) - \lim_{x \to x_0} g(x).$$

5. *The function* $f \times g$ *converges to* LM *as* $x \to x_0$. *In other words:*

$$\lim_{x \to x_0} (f \times g)(x) = \left(\lim_{x \to x_0} f(x) \right) \left(\lim_{x \to x_0} g(x) \right).$$

6. *If* $L \neq 0$, *then the function* $\frac{1}{f}$ *converges to* $\frac{1}{L}$ *as* $x \to x_0$. *In other words:*

$$\lim_{x \to x_0} \frac{1}{f}(x) = \frac{1}{\lim_{x \to x_0} f(x)}.$$

7. *If* $L \neq 0$, *then the function* $\frac{g}{f}$ *converges to* $\frac{M}{L}$ *as* $x \to x_0$. *In other words:*

$$\lim_{x \to x_0} \frac{g}{f}(x) = \frac{\lim_{x \to x_0} g(x)}{\lim_{x \to x_0} f(x)}.$$

Proof We prove the assertions one by one.

1. If $\lambda = 0$, then the function λf is a constant 0 function, so the limit as $x \to x_0$ would be the limit of a constant zero sequence which is 0. Hence the limit of $\lambda f(x)$ as $x \to x_0$ exists and is equal to 0.
   Suppose now that $\lambda \neq 0$. Fix $\varepsilon > 0$. Our goal is to find a $\delta > 0$ such that for all $x \in X$ with $0 < |x - x_0| < \delta$ we must have $|\lambda f(x) - \lambda L| < \varepsilon$.

We note that $\frac{\varepsilon}{|\lambda|} > 0$ as well. Using this in the definition of $\lim_{x\to x_0} f(x) = L$, there exists a $\delta > 0$ such that for all $x \in X$ with $0 < |x - x_0| < \delta$, we have:

$$|f(x) - L| < \frac{\varepsilon}{|\lambda|} \quad \Leftrightarrow \quad |\lambda f(x) - \lambda L| < \varepsilon.$$

So, for this same choice of $\delta > 0$, for all $x \in X$ with $0 < |x - x_0| < \delta$ we must have $|\lambda f(x) - \lambda L| < \varepsilon$. Thus, $\lambda f(x) \to \lambda L$ as $x \to x_0$.

2. Fix $\varepsilon > 0$. Our goal is to find a $\delta > 0$ such that for any $x \in X$ with $0 < |x - x_0| < \delta$ we must have $||f(x)| - |L|| < \varepsilon$. Via definition of the limits, there exists a $\delta > 0$ such that for all $x \in X$ with $0 < |x - x_0| < \delta$, we have $|f(x) - L| < \varepsilon$. Using reverse triangle inequality, for the latter, we also have $||f(x)| - |L|| < |f(x) - L| < \varepsilon$. So $|f|(x) \to |L|$ as $x \to x_0$.
3. Fix $\varepsilon > 0$, we want to find a $\delta > 0$ such that for $x \in X$, if $0 < |x - x_0| < \delta$ then $|f(x)+g(x)-(L+M)| < \varepsilon$. From the definition of limits, there exists a $\delta_1 > 0$ such that for any $x \in X$ with $0 < |x - x_0| < \delta_1$ we have $|f(x) - L| < \frac{\varepsilon}{2}$. Also, there exists a $\delta_2 > 0$ such that if $0 < |x - x_0| < \delta_2$ then $|g(x) - M| < \frac{\varepsilon}{2}$. Therefore, if we pick $\delta = \min\{\delta_1, \delta_2\} > 0$, both of the conditions above hold. As a result, for every $x \in X$ with $0 < |x - x_0| < \delta$, by triangle inequality, we have:

$$\begin{aligned}|f(x)+g(x) - (L+M)| = |f(x)-L+g(x)-M| &\leq |f(x) - L| + |g(x) - M| \\ &< \tfrac{\varepsilon}{2} + \tfrac{\varepsilon}{2} = \varepsilon,\end{aligned}$$

and so we have found our desired $\delta > 0$. Thus, $f(x) + g(x) \xrightarrow{x\to x_0} L + M$.
4. This is an application of the first and third assertions, namely: $f(x) \to L$ and $-g(x) \to -M$ as $x \to x_0$, so $(f - g)(x) = f(x) - g(x) \to L - M$ as $x \to x_0$.
5. Fix $\varepsilon > 0$. Our goal is to find a $\delta > 0$ such that for any $x \in X$ with $0 < |x-x_0| < \delta$ we have $|f(x)g(x) - LM| < \varepsilon$. We note that the function f has a finite limit at x_0 so, by Proposition 9.5.1, there exists a $\delta_1 > 0$ such that for $x \in X$ with $0 < |x - x_0| < \delta_1$ we must have $|f(x)| \leq K$ for some $K > 0$.

   Next, since $g(x) \to M$ as $x \to x_0$, there exists a $\delta_2 > 0$ such that if $x \in X$ and $0 < |x - x_0| < \delta_2$, then $|g(x) - M| \leq \frac{\varepsilon}{K+|M|}$. Finally, since $f(x) \to L$ as $x \to x_0$, there exists a $\delta_3 > 0$ such that if $x \in X$ and $0 < |x - x_0| < \delta_3$, then $|f(x) - L| \leq \frac{\varepsilon}{K+|M|}$.

   So if we pick $\delta = \min\{\delta_1, \delta_2, \delta_3\} > 0$, for any $x \in X$ with $0 < |x - x_0| < \delta$, all three of the conditions above hold and therefore:

$$\begin{aligned}|f(x)g(x) - LM| &= |f(x)(g(x) - M) + M(f(x) - L)| \\ &\leq |f(x)||g(x) - M| + |M||f(x) - L| \\ &\leq K|g(x) - M| + |M||f(x) - L| \\ &< K\frac{\varepsilon}{K + |M|} + |M|\frac{\varepsilon}{K + |M|} = \varepsilon.\end{aligned}$$

   Thus, we conclude $(f \times g)(x) = f(x)g(x) \to LM$ as $x \to x_0$.

6. First we note that there exists a $\delta_1 > 0$ such that for which $f(x)$ is bounded away from 0 for any $x \in X$ such that $0 < |x - x_0| < \delta_1$. Indeed since $f(x) \to L$ as $x \to x_0$, if we pick $\varepsilon = \frac{|L|}{2} > 0$, there exists a $\delta_1 > 0$ such that for all $x \in X$ with $0 < |x - x_0| < \delta_1$ we have $|f(x) - L| < \frac{|L|}{2}$. By triangle inequality, we have:
$$|f(x)| \geq |L| - |f(x) - L| > |L| - \frac{|L|}{2} = \frac{|L|}{2} > 0,$$
which says $|f(x)| \geq \frac{|L|}{2}$ for all $x \in X$ with $0 < |x - x_0| < \delta_1$.
Fix $\varepsilon > 0$. We want to find $\delta > 0$ such that for all $x \in X$ with $0 < |x - x_0| < \delta$ we must have $\left|\frac{1}{f(x)} - \frac{1}{L}\right| < \varepsilon$. We know $f(x) \to L$ as $x \to x_0$, so there exists a $\delta_2 > 0$ such that for all $x \in X$ with $0 < |x - x_0| < \delta_2$ we have $|f(x) - L| < \frac{\varepsilon|L|^2}{2}$. Therefore, let us pick $\delta = \min\{\delta_1, \delta_2\} > 0$. For $x \in X$ with $0 < |x - x_0| < \delta$, we must have both $\frac{|L|}{2} \leq |f(x)|$ and $|f(x) - L| < \frac{\varepsilon|L|^2}{2}$. Using both of these estimates, we have:
$$\left|\frac{1}{f(x)} - \frac{1}{L}\right| = \frac{|f(x) - L|}{|f(x)||L|} \leq \frac{2|f(x) - L|}{|L|^2} < \frac{2}{|L|^2}\frac{\varepsilon|L|^2}{2} = \varepsilon,$$
and so we have found our required $\delta > 0$. Thus, $\frac{1}{f(x)} \to \frac{1}{L}$ as $x \to x_0$.
7. This is an application of the fifth and sixth assertions, namely: $\frac{1}{f(x)} \to \frac{1}{L}$ and $g(x) \to M$ as $x \to x_0$, so $\frac{g(x)}{f(x)} \to \frac{M}{L}$ as $x \to x_0$. □

The algebra of limits simplifies the process of proving limits analytically since we may split a complicated function into smaller, more manageable pieces.

Example 9.5.3 Here are some examples of the application of algebra of limits.

1. We have seen in Example 9.2.9(3) in which we found the limit of the function $f : \mathbb{R} \to \mathbb{R}$ defined as $f(x) = 5x^2 + 1$ as $x \to 1$. We have found the limit to be $\lim_{x\to 1} f(x) = 6$. However, we can find the limit using Theorem 9.5.2 if the limit of the function $g : \mathbb{R} \to \mathbb{R}$ defined as $g(x) = x^2$ as $x \to 1$ exists. It does and the limit is 1.
We can check this easily. For a given $\varepsilon > 0$, if we choose $\delta = \min\{1, \frac{\varepsilon}{3}\} > 0$, then for any $x \in \mathbb{R}$ with $0 < |x - 1| < \delta = \min\{1, \frac{\varepsilon}{3}\}$ we must have $1 < x + 1 < 3$ and hence:
$$|g(x) - 1| = |x^2 - 1| = |x - 1||x + 1| < 3|x - 1| < \frac{3\varepsilon}{3} = \varepsilon.$$

Thus, $\lim_{x\to 1} x^2 = 1$. Applying the algebra of limits on the function f, we have:

$$\lim_{x\to 1} f(x) = \lim_{x\to 1}(5x^2+1) = \lim_{x\to 1}(5x^2) + \lim_{x\to 1} 1 = 5\lim_{x\to 1} x^2 + 1 = 5+1 = 6.$$

2. Another example that we have seen is the function $f : \mathbb{R} \setminus \{-2, 1\} \to \mathbb{R}$ defined as $f(x) = \frac{x^2-1}{x^2+x-2}$ and we wanted to find the limit of this function as $x \to 1$ in Example 9.2.9(5). Let us first find the limits of the numerator and the denominator which are given by the functions $g, h : \mathbb{R} \setminus \{-2, 1\} \to \mathbb{R}$ where $g(x) = x^2 - 1$ and $h(x) = x^2 + x - 2$ respectively. Using algebra of limits and the facts that $\lim_{x\to 1} x^2 = 1$ and $\lim_{x\to 1} x = 1$, we can immediately see that:

$$\lim_{x\to 1} g(x) = \lim_{x\to 1} x^2 - 1 = 1 - 1 = 0,$$

$$\lim_{x\to 1} h(x) = \lim_{x\to 1} x^2 + \lim_{x\to 1} x - 2 = 1 + 1 - 2 = 0.$$

Hence, the limits of the numerator and denominator separately exist as $x \to 1$. However, since the limit of the denominator is 0, we are not able to apply the algebra of limit for quotients in Theorem 9.5.2(7) here since it specifically requires the limit of the denominator to be non-zero.
Nonetheless, note that we can rewrite the function $f(x)$ as:

$$f(x) = \frac{x^2-1}{x^2+x-2} = \frac{(x-1)(x+1)}{(x-1)(x+2)},$$

and since the domain does not include the point 1, we have $x - 1 \neq 0$ everywhere and so we can cancel out the $(x - 1)$ factors in the numerator and the denominator by using the field axioms to get $f(x) = \frac{x+1}{x+2}$. Again, we can check that the limits of the numerator and the denominator exist as $x \to 1$ which are equal to 2 and 3 respectively. Thus, we can now apply the algebra of limits to get:

$$\lim_{x\to 1} f(x) = \lim_{x\to 1} \frac{x+1}{x+2} = \frac{\lim_{x\to 1}(x+1)}{\lim_{x\to 1}(x+2)} = \frac{2}{3}.$$

Remark 9.5.4 In Example 9.5.3(2), we have encountered a limit of the form $\frac{0}{0}$. This limit is called the indeterminate form because we could not determine the limit of the original function straight away without manipulating the functions slightly to get a form on which we are allowed to apply the algebra of limits. We shall see other indeterminate forms of limits and how we can deal with them more systematically later in Chap. 14.

We note that the algebra of limits for functions (and non-zero denominator for the fraction case) as $x \to x_0$ only work for functions which have finite limits as

$x \to x_0$. So, if we have functions which blow up to ∞ or $-\infty$, we cannot apply the algebra of limits from Theorem 9.5.2. However, there are variants of results for functions which blow up to $\pm\infty$ and these need to be investigated on a case-by-case basis. Hence, this is why we should be fluent with the definitions and the tricks of manipulating ε and δ.

Example 9.5.5 Suppose that we have two functions $f, g : X \to \mathbb{R}$ where $X \subseteq \mathbb{R}$ and $x_0 \in X'$. Suppose further that $\lim_{x\to x_0} f(x) = M > 0$ and $\lim_{x\to x_0} g(x) = \infty$. We claim that the limit $\lim_{x\to x_0} \frac{f(x)}{g(x)}$ exists and is equal to 0. To prove this, for every $\varepsilon > 0$ we need to find a $\delta > 0$ such that for $x \in X$ with $0 < |x - x_0| < \delta$ we must have $\left|\frac{f(x)}{g(x)} - 0\right| = \left|\frac{f(x)}{g(x)}\right| < \varepsilon$.

Fix $\varepsilon > 0$. Since $\lim_{x\to x_0} f(x)$ exists, by Proposition 9.5.1, there exists a $\delta_1 > 0$ such that $|f(x)| \leq K$ for some number $K > 0$ for any $x \in X$ with $0 < |x - x_0| < \delta_1$. Furthermore, since $g(x) \to \infty$ as $x \to x_0$, there exists a $\delta_2 > 0$ such that $g(x) > \frac{K}{\varepsilon} > 0$ for all $x \in X$ with $0 < |x - x_0| < \delta_2$. Therefore, if we pick $\delta = \min\{\delta_1, \delta_2\} > 0$, for all $x \in X$ with $0 < |x - x_0| < \delta$, both of the conditions above apply. Thus, for any such x we have:

$$\frac{|f(x)|}{|g(x)|} \leq \frac{K}{|g(x)|} < K\frac{\varepsilon}{K} = \varepsilon.$$

Thus, we conclude that $\lim_{x\to x_0} \frac{f(x)}{g(x)} = 0$.

More similar examples can be seen in Exercises 9.8 and 9.14.

Likewise, Theorem 9.5.2(7) specifies that in order to apply the algebra of limits for quotients, we require the limit of the denominator to be non-vanishing. For the case where the limit of the denominator vanishes, we have the following result:

Proposition 9.5.6 *Let $f : X \to \mathbb{R}$ where $X \subseteq \mathbb{R}$ and $x_0 \in X'$.*

1. *If there exists a $\delta > 0$ such that $f(x) > 0$ for all $x \in (x_0, x_0 + \delta) \cap X$ and $\lim_{x\downarrow x_0} f(x) = 0$, then $\lim_{x\downarrow x_0} \frac{1}{f(x)} = \infty$.*
2. *If there exists a $\delta > 0$ such that $f(x) < 0$ for all $x \in (x_0, x_0 + \delta) \cap X$ and $\lim_{x\downarrow x_0} f(x) = 0$, then $\lim_{x\downarrow x_0} \frac{1}{f(x)} = -\infty$.*
3. *If there exists a $\delta > 0$ such that $f(x) > 0$ for all $x \in (x_0 - \delta, x_0) \cap X$ and $\lim_{x\uparrow x_0} f(x) = 0$, then $\lim_{x\uparrow x_0} \frac{1}{f(x)} = \infty$.*
4. *If there exists a $\delta > 0$ such that $f(x) < 0$ for all $x \in (x_0 - \delta, x_0) \cap X$ and $\lim_{x\uparrow x_0} f(x) = 0$, then $\lim_{x\uparrow x_0} \frac{1}{f(x)} = -\infty$.*

Proof We prove the first assertion only as the others are similarly done.

1. Fix $K > 0$. Since $\lim_{x\downarrow x_0} f(x) = 0$, there exists a $\delta_1 > 0$ such that for all $x \in X$ with $0 < x - x_0 < \delta_1$ we have $|f(x)| < \frac{1}{K}$. Let $\delta_2 = \min\{\delta, \delta_1\}$. Thus, for any

$x \in X$ with $0 < x - x_0 < \delta_2$ we have $0 < f(x) < \frac{1}{K}$. This means $\frac{1}{f(x)} > K$ for all such x. Therefore, we conclude that $\lim_{x \downarrow x_0} \frac{1}{f(x)} = \infty$. □

As a corollary:

Corollary 9.5.7 *Let $f : X \to \mathbb{R}$ where $X \subseteq \mathbb{R}$ and $x_0 \in X'$.*

1. *If there exists a $\delta > 0$ such that $f(x) > 0$ for all $x \in (x_0 - \delta, x_0 + \delta) \setminus \{x_0\} \cap X$ and $\lim_{x \to x_0} f(x) = 0$, then $\lim_{x \to x_0} \frac{1}{f(x)} = \infty$.*
2. *If there exists a $\delta > 0$ such that $f(x) < 0$ for all $x \in (x_0 - \delta, x_0 + \delta) \setminus \{x_0\} \cap X$ and $\lim_{x \to x_0} f(x) = 0$, then $\lim_{x \to x_0} \frac{1}{f(x)} = -\infty$.*

In fact, the converses to the results in Corollary 9.5.7 are also true.

Example 9.5.8 Let us look at an example of this.

1. Consider the function $g : \mathbb{R} \to \mathbb{R}$ given by $g(x) = x^2$. Note that $g(x) > 0$ for any $x \neq 0$. Thus, Corollary 9.5.7 says $\lim_{x \to 0} \frac{1}{g(x)} = \lim_{x \to 0} \frac{1}{x^2} = \infty$.
2. On the other hand, consider the function $f : \mathbb{R} \to \mathbb{R}$ given by $f(x) = x$. Since $f(x) > 0$ for $x > 0$ and $f(x) < 0$ for $x < 0$, by using Proposition 9.5.6, we have the one-sided limits:

$$\lim_{x \uparrow 0} \frac{1}{f(x)} = -\infty \quad \text{and} \quad \lim_{x \downarrow 0} \frac{1}{f(x)} = \infty.$$

However, as the sign of f is not constant over any neighbourhood of the point $x = 0$, Corollary 9.5.7 does not hold for the function $\frac{1}{f(x)}$. In fact, according to the one-sided limits above, we can see that the limit $\lim_{x \to 0} \frac{1}{f(x)}$ does not exist as the one-sided limits do not coincide.

Another result that we can prove is for limits at infinity. The limits at infinity is handled easier than functions blowing up to $\pm\infty$ and we have the exact same results as Theorem 9.5.2. The proof is almost similar to Theorem 9.5.2, but one need to adapt Definition 9.4.4 for the proofs of limits of a function at infinity and work with K instead of δ. The proof of the following is left for the readers to try in Exercise 9.22.

Theorem 9.5.9 (Algebra of Limits at Infinity, AOL at Infinity) *Let $f, g : X \to \mathbb{R}$ be functions on a domain X that is unbounded from above. Suppose that $\lim_{x \to \infty} f(x) = L$ and $\lim_{x \to \infty} g(x) = M$. Then:*

1. *For a constant $\lambda \in \mathbb{R}$, function λf converges to λL as $x \to \infty$. In other words:*

$$\lim_{x \to \infty} \lambda f(x) = \lambda \lim_{x \to \infty} f(x).$$

2. *The function $|f|$ converges to $|L|$ as $x \to x_0$. In other words:*

$$\lim_{x\to\infty} |f|(x) = |\lim_{x\to\infty} f(x)|.$$

3. *The function $f + g$ converges to $L + M$ as $x \to \infty$. In other words:*

$$\lim_{x\to\infty} (f+g)(x) = \lim_{x\to\infty} f(x) + \lim_{x\to\infty} g(x).$$

4. *The function $f - g$ converges to $L - M$ as $x \to \infty$. In other words:*

$$\lim_{x\to\infty} (f-g)(x) = \lim_{x\to\infty} f(x) - \lim_{x\to\infty} g(x).$$

5. *The function $f \times g$ converges to LM as $x \to \infty$. In other words:*

$$\lim_{x\to\infty} (f\times g)(x) = \left(\lim_{x\to\infty} f(x)\right)\left(\lim_{x\to\infty} g(x)\right).$$

6. *If $L \neq 0$, then the function $\frac{1}{f}$ converges to $\frac{1}{L}$ as $x \to \infty$. In other words:*

$$\lim_{x\to\infty} \frac{1}{f}(x) = \frac{1}{\lim_{x\to\infty} f(x)}.$$

7. *If $L \neq 0$, then the function $\frac{g}{f}$ converges to $\frac{M}{L}$ as $x \to \infty$. In other words:*

$$\lim_{x\to\infty} \frac{g}{f}(x) = \frac{\lim_{x\to\infty} g(x)}{\lim_{x\to\infty} f(x)}.$$

Another remark is that Theorem 9.5.9 also holds for limit of functions at $-\infty$. The end note is that as long as the limits exist (and is non-zero for denominators in the fraction cases), one can apply the algebra of limits!

9.6 Asymptotic Notations

To finish up this chapter, we are going to introduce the asymptotic notations for functions. We have seen in Chap. 5 the asymptotic equivalence, big-O, and little-o notations for sequences. In a similar vein, we can define the same notations here:

Definition 9.6.1 (Asymptotic Notations) Let $f, g : [a, \infty) \to \mathbb{R}$.

1. We say that the functions f and g are asymptotically equivalent as $x \to \infty$ if we have $\lim_{x\to\infty} \frac{f(x)}{g(x)} = 1$. We write this as $f \sim g$ as $x \to \infty$.

2. If $g > 0$, we say that f is (of order) big-O of g if there exists an $M > 0$ and a $K > a$ such that for every $x \geq K$ we have $|f(x)| \leq Mg(x)$. We write this as $f \in O(g)$ as $x \to \infty$. In symbols:

$$f \in O(g) \text{ as } x \to \infty \quad \text{if} \quad \exists M > 0 : \exists K > a : x \geq K \Rightarrow |f(x)| \leq Mg(x).$$

3. If $g > 0$, we say that f is (of order) little-o of g if for every $\varepsilon > 0$, there exists a $K > a$ such that for every $x \geq K$ we have $|f(x)| \leq \varepsilon g(x)$. We write this as $f \in o(g)$ as $x \to \infty$. In symbols:

$$f \in o(g) \text{ as } x \to \infty \quad \text{if} \quad \forall \varepsilon > 0 : \exists K > a : x \geq K \Rightarrow |f(x)| \leq \varepsilon g(x).$$

Similar to their sequential counterparts, they are usually used as a simplification when discussing growth rates of functions as $x \to \infty$. In words:

1. $f \sim g$ means these functions approach one another at infinity and so they behave in the same way towards infinity.
2. $f \in O(g)$ means the function f is eventually bounded above by a constant multiple of g.
3. $f \in o(g)$ means the function f is eventually negligible compared to the function g.

There are also limit definitions for the big-O and little-o notations, which we shall leave for the readers to verify in Exercise 9.33.

In addition to Definition 9.6.1, we can also define the asymptotic notations as the limit when we approach a finite limit point of the domain instead of infinity. More specifically:

Definition 9.6.2 (Asymptotic Notations at Limit Points) Let $f, g : X \to \mathbb{R}$ where $X \subseteq \mathbb{R}$. Let $x_0 \in X'$.

1. We say that the functions f and g are asymptotically equivalent as $x \to x_0$ if we have $\lim_{x \to x_0} \frac{f(x)}{g(x)} = 1$. We write this as $f \sim g$ as $x \to x_0$.
2. If $g > 0$, we say that f is (of order) big-O of g if there exists an $M > 0$ and a $\delta > 0$ such that for every $x \in X$ with $0 < |x - x_0| < \delta$ we have $|f(x)| \leq Mg(x)$. We write this as $f \in O(g)$ as $x \to x_0$. In symbols:

$$f \in O(g) \text{ as } x \to x_0 \quad \text{if}$$
$$\exists M > 0 : \exists \delta > 0 : \forall x \in X, 0 < |x - x_0| < \delta \Rightarrow |f(x)| \leq Mg(x).$$

3. If $g > 0$, we say that f is (of order) little-o of g if for every $\varepsilon > 0$, there exists a $\delta > 0$ such that for every $x \in X$ with $0 < |x - x_0| < \delta$ we have $|f(x)| \leq \varepsilon g(x)$. We write this as $f \in o(g)$ as $x \to x_0$. In symbols:

$$f \in o(g) \text{ as } x \to x_0 \quad \text{if}$$

$$\forall \varepsilon > 0 : \exists \delta > 0 : \forall x \in X, 0 < |x - x_0| < \delta \Rightarrow |f(x)| \leq \varepsilon g(x).$$

Therefore, unlike for sequences, when we use the asymptotic notations for functions, we need to specify the point at which we are comparing the two functions, whether a finite limit point or $\pm\infty$. These notations will be very useful and convenient when we talk about differentiation and convergence of power series in later chapters.

Exercises

9.1 Suppose that $f, g : X \to \mathbb{R}$ are two bounded functions on $X \subseteq \mathbb{R}$. Prove that:

$$\sup_{x,y\in X} (f(x) - g(y)) = \sup_{x\in X} f(x) - \inf_{y\in X} g(y).$$

9.2 (∗) A function $f : I \to \mathbb{R}$ over an interval I is called a convex function if the secant line segment joining two points $(x_1, f(x_1))$ and $(x_2, f(x_2))$ for any two points $x_1 < x_2$ in I does not lie below the graph of f over $[x_1, x_2]$. This can be seen graphically in Fig. 9.5.
Mathematically, the function f is convex if for any $x_1, x_2 \in I$ and all $t \in [0, 1]$ we have:

$$f(tx_1 + (1 - t)x_2) \leq tf(x_1) + (1 - t)f(x_2).$$

Show that the following functions $f : \mathbb{R} \to \mathbb{R}$ are convex.
(a) $f(x) = ax + b$ for some $a, b \in \mathbb{R}$.
(b) $f(x) = |x|$.

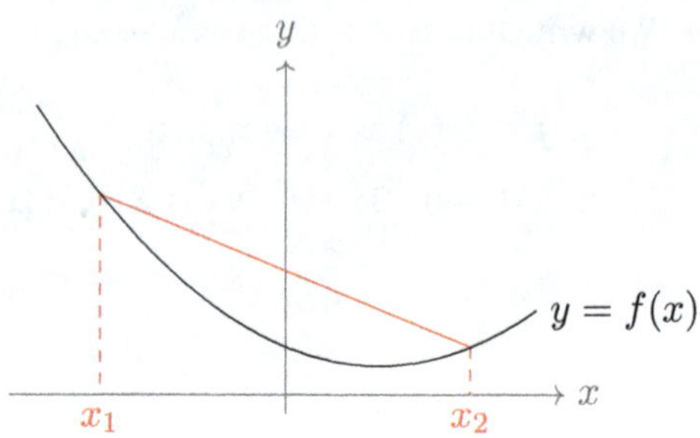

Fig. 9.5 Example of a strictly convex function. The red secant line segment joining the points $(x_1, f(x_1))$ and $(x_2, f(x_2))$ lies completely above the graph of f

A convex function is called strictly convex if for any $x_1, x_2 \in I$ and all $t \in (0, 1)$ we have:

$$f(tx_1 + (1-t)x_2) < tf(x_1) + (1-t)f(x_2),$$

or, in other words, the graph and the secant line segment intersect only at the endpoints. Show that the following functions $f : \mathbb{R} \to \mathbb{R}$ are all strictly convex.

(c) $f(x) = x^2$.
(d) $f(x) = \frac{1}{x}$ for $x > 0$.
(e) $f(x) = a^x$ for some $a > 1$.

9.3 (∗) Following Exercise 9.2, a function $f : I \to \mathbb{R}$ over an interval I is called a concave function if for any $x_1, x_2 \in I$ and all $t \in [0, 1]$ we have:

$$f(tx_1 + (1-t)x_2) \geq tf(x_1) + (1-t)f(x_2),$$

and is called strictly concave if for any $x_1, x_2 \in I$ and all $t \in (0, 1)$ we have:

$$f(tx_1 + (1-t)x_2) > tf(x_1) + (1-t)f(x_2).$$

(a) Prove that if $f : I \to \mathbb{R}$ is a convex function, then the function $-f$ is concave.
(b) Suppose that $f : I \to f(I)$ is a strictly monotone and strictly convex function with inverse $f^{-1} : f(I) \to I$. Show that f^{-1} is a strictly monotone and strictly concave.

9.4 (∗) We are now going to use the idea of convexity in Exercises 9.2 and 9.3 to prove some well-known inequalities.

(a) Prove by induction that if $f : I \to \mathbb{R}$ is a convex function, then it satisfies the Jensen's inequality:

$$f\left(\sum_{j=1}^{n} t_j a_j\right) \leq \sum_{j=1}^{n} t_j f(a_j), \tag{9.3}$$

where $a_j \in I$ and $t_j \geq 0$ are non-negative real numbers for $j = 1, 2, \ldots, n$ such that $t_1 + t_2 + \ldots + t_n = 1$.
If f is a concave function instead, then the inequality in (9.3) reverses. This inequality is named after Johan Jensen (1859–1925).

(b) Using part (a), provide a different proof of the AM-GM inequality that we have proven inductively in Exercise 3.27, namely: for any n non-negative real numbers $a_1, a_2, \ldots, a_n \geq 0$ we have:

$$\frac{a_1 + a_2 + \ldots + a_n}{n} \geq \sqrt[n]{a_1 a_2 \ldots a_n}.$$

(c) Prove Young's inequality which states that for any $p, q > 0$ with $\frac{1}{p} + \frac{1}{q} = 1$ and any $a, b \geq 0$ we have:

$$ab \leq \frac{a^p}{p} + \frac{b^q}{q}.$$

This inequality is named after William Henry Young (1863–1942).

(d) Using Young's inequality, prove that for real numbers $a_1, a_2, \ldots, a_n, b_1, b_2, \ldots, b_n \in \mathbb{R}$ and $p, q > 0$ such that $\frac{1}{p} + \frac{1}{q} = 1$ we have:

$$|a_1 b_1 + \ldots + a_n b_n| \leq \sqrt[p]{|a_1|^p + \ldots + |a_n|^p} \sqrt[q]{|b_1|^q + \ldots + |b_n|^q}.$$

This is called Hölder's inequality, which is named after Otto Hölder (1859–1937). The special case for which $p = q = 2$ is called the Cauchy-Schwarz inequality that we saw in Exercise 3.25.

9.5 (∗) Using the ε-δ definition of limits, prove the following limits:
(a) $\lim_{x \to -1} 2 - 7x = 9$.
(b) $\lim_{x \to 1} x^2 - 2x = -1$.
(c) $\lim_{x \to 0} x(x^2 + 1) \cos(x) = 0$.
(d) $\lim_{x \to 2} x^2 = 4$.
(e) $\lim_{x \to 5} x^2 - 2x - 14 = 1$.
(f) $\lim_{x \to 1} \frac{x+1}{x+2} = \frac{2}{3}$.
(g) $\lim_{x \to 1} \frac{x^4 - 1}{x - 1} = 4$.
(h) $\lim_{x \to 1} \frac{x+3}{1+\sqrt{x}} = 2$.

9.6 (∗) Using the appropriate definitions of limits, prove that:
(a) $\lim_{x \to \infty} x^2 = \infty$.
(b) $\lim_{x \to 2} \frac{1}{(x-2)^2} = \infty$.
(c) $\lim_{x \to \infty} \frac{6x+1}{2x+1} = 3$.
(d) $\lim_{x \downarrow 1} \frac{1}{x-1} = \infty$.
(e) $\lim_{x \uparrow 1} \frac{1}{x-1} = -\infty$.
(f) $\lim_{x \uparrow 4} \frac{\sqrt{x}}{(x-4)^3} = -\infty$.

9.7 (a) Let $f : \mathbb{R} \to \mathbb{R}$ be the Dirichlet function defined as $f(x) = \begin{cases} 1 \text{ if } x \in \mathbb{Q}, \\ 0 \text{ if } x \in \bar{\mathbb{Q}}. \end{cases}$

Show that this function does not have a limit as $x \to x_0$ for any $x_0 \in \mathbb{R}$.

(b) Now let $g : \mathbb{R} \to \mathbb{R}$ be defined as $g(x) = \begin{cases} x \text{ if } x \in \mathbb{Q}, \\ 0 \text{ if } x \in \bar{\mathbb{Q}}. \end{cases}$

Show that $\lim_{x \to 0} g(x) = 0$ but $\lim_{x \to x_0} g(x)$ does not exist for any $x_0 \neq 0$.

9.8 ($*$) Suppose that $f : \mathbb{R} \setminus \{0\} \to \mathbb{R}$ and $g : \mathbb{R} \to \mathbb{R}$. Which of the following statements are true? Give a proof or a counterexample:
(a) If $\lim_{x\to 0} f(x) = \infty$ and $\lim_{x\to 0} g(x) = 0$, then $\lim_{x\to 0} f(x)g(x) = 1$.
(b) If $\lim_{x\to 0} f(x) = \infty$ and $\lim_{x\to 0} g(x) = 1$, then $\lim_{x\to 0} f(x)g(x) = \infty$.
(c) If $\lim_{x\to 0} f(x) = \infty$ and $\lim_{x\to 0} f(x)g(x) = 0$, then $\lim_{x\to 0} g(x) = 0$.

9.9 Let $f : (0, \infty) \to \mathbb{R}$ be defined as $f(x) = -\frac{3\sin(3x)}{x} + 4$. Find $\lim_{x\to\infty} f(x)$ and $\lim_{x\downarrow 0} f(x)$.

9.10 ($*$) In this question, we are going to prove some comparison results for limits of functions akin to Propositions 5.6.1 and 5.6.4 for sequences.
Suppose $f, g : X \to \mathbb{R}$ and $x_0 \in X'$ are such that $\lim_{x\to x_0} f(x) = L$ and $\lim_{x\to x_0} g(x) = M$ with $L, M \in \mathbb{R}$.
(a) Suppose further that there exists some $\delta > 0$ such that $f(x) \leq g(x)$ for all $x \in B_\delta(x_0) \setminus \{x_0\}$. Prove that $L \leq M$.
(b) Suppose now there exists another function $h : X \to \mathbb{R}$ with $f(x) \leq h(x) \leq g(x)$ whenever $x \in B_\delta(x_0) \setminus \{x_0\}$ for some $\delta > 0$. Prove the sandwich lemma where if $L = M$, then $\lim_{x\to x_0} h(x)$ also exists and is equal to L.
The results above are also true if the limit point x_0 is replaced with $\pm\infty$.

9.11 ($*$) We have a partial converse to the first assertion in Exercise 9.10(a). Suppose that $f, g : X \to \mathbb{R}$ and $x_0 \in X'$ are such that $\lim_{x\to x_0} f(x) = L$ and $\lim_{x\to x_0} g(x) = M$ with $L, M \in \mathbb{R}$.
(a) Prove that if $L < M$, then there is a $\delta > 0$ such that $f(x) < g(x)$ for all $x \in B_\delta(x_0) \setminus \{x_0\}$.
(b) Is the result in part (a) still true if $L = M$?

9.12 ($\diamond$) Let $f : X \to \mathbb{R}$ be a real-valued function with $X \subseteq \mathbb{R}$ and $x_0 \in X'$.
(a) Suppose that $\lim_{x\to x_0} f(x) = L \in \mathbb{R}$. Prove that for all $\varepsilon > 0$, there exists a $\delta > 0$ such that for any $x, y \in B_\delta(x_0) \setminus \{x_0\}$ we have $|f(x) - f(y)| < \varepsilon$.
(b) The converse to part (a) is also true: Suppose that for all $\varepsilon > 0$, there exists a $\delta > 0$ such that for any $x, y \in B_\delta(x_0) \setminus \{x_0\}$ we have $|f(x) - f(y)| < \varepsilon$. Prove that $\lim_{x\to x_0} f(x)$ exists.

9.13 ($*$) Let $f, g : \mathbb{R} \to \mathbb{R}$. Suppose that $\lim_{x\to\infty} f(x) = \infty$ and $\lim_{x\to\infty} g(x) = K \in \mathbb{R}$.
(a) If $K > 0$, prove that $\lim_{x\to\infty} f(x)g(x) = \infty$.
(b) If $K < 0$, prove that $\lim_{x\to\infty} f(x)g(x) = -\infty$.
(c) What can we say if $K = 0$?

9.14 ($*$) In this question, we want to study the limits of polynomials at $\pm\infty$.
(a) Let $f : (0, \infty) \to \mathbb{R}$ be a function defined as $f(x) = \frac{1}{x^n}$ for some $n \in \mathbb{N}$. Prove that $\lim_{x\to\infty} f(x) = 0$.
Let $P : \mathbb{R} \to \mathbb{R}$ be an n-th degree polynomial $P(x) = \sum_{j=0}^{n} a_j x^j$ with $n \in \mathbb{N}$ and leading coefficient $a_n > 0$.
(b) Using part (a) and Exercise 9.13, prove that $\lim_{x\to\infty} P(x) = \infty$.
(c) Find the limit $\lim_{x\to-\infty} P(x)$.
(d) What happens if the leading coefficient of the polynomial P satisfies $a_n < 0$ instead?

9.15 (∗) Let $f : X \to \mathbb{R}$ be defined as $f(x) = \sqrt{x + \sqrt{x}} - \sqrt{x - \sqrt{x}}$.
(a) Find the maximal domain $X \subseteq \mathbb{R}$ for this function. In other words, determine the largest subset $X \subseteq \mathbb{R}$ for which the mapping above has a real value.
(b) Explain why $\lim_{x\to 0} f(x)$ does not exist.
(c) Find $\lim_{x\to\infty} f(x)$.
(d) Find $\inf_{x\in X} f(x)$, $\sup_{x\in X} f(x)$, $\min_{x\in X} f(x)$, and $\max_{x\in X} f(x)$.

9.16 Suppose that $f, g : X \to \mathbb{R}$ and $x_0 \in X'$ are such that $\lim_{x\to x_0} f(x) = L$ and $\lim_{x\to x_0} g(x) = M$ with $L, M \in \mathbb{R}$. Show that:
(a) $\lim_{x\to x_0} \max\{f(x), g(x)\} = \max\{L, M\}$,
(b) $\lim_{x\to x_0} \min\{f(x), g(x)\} = \min\{L, M\}$.

9.17 (∗) Consider the subset $A = (0, 1) \cup (1, 2) \subseteq \mathbb{R}$.
(a) Prove that the set A is open.
(b) Find (with proof) all the limit points of the set A.
(c) Let $f : A \to \mathbb{R}$ be a function defined as $f(x) = \frac{x^2-1}{x^2+2x-3}$. Using the ε-δ definition of limits, show that $\lim_{x\to 1} f(x) = \frac{1}{2}$.

9.18 Define a function $f : \mathbb{R} \setminus \{0\} \to \mathbb{R}$ as $f(x) = \frac{x}{|x|}$. Show that $\lim_{x\to 0} f(x)$ does not exist.

9.19 Redo Exercise 9.5 by carefully using the algebra of limits instead of the ε-δ definition.

9.20 In Theorem 9.5.2, we have seen that for a function $f : X \to \mathbb{R}$ with $X \subseteq \mathbb{R}$ and $x_0 \in X'$, if $\lim_{x\to x_0} f(x) = L$ for some $L \in \mathbb{R}$, then we have $\lim_{x\to x_0} |f|(x) = |L|$.
(a) Is the converse true? Namely if we know that $\lim_{x\to x_0} |f|(x) = |L|$, is it necessarily true that $\lim_{x\to x_0} f(x) = L$?
(b) Now suppose that $L = 0$. Is the converse now true?

9.21 Let $f : \mathbb{R} \to \mathbb{R}$ be a function such that:

$$f(x) = \begin{cases} \frac{1}{x}\lfloor x \rfloor & \text{for} \quad x \neq 0, \\ 0 & \text{for} \quad x = 0. \end{cases}$$

(a) Write down the function f as a piecewise reciprocal function without the floor function in the numerator.
(b) Find the limits $\lim_{x\to\infty} f(x)$, $\lim_{x\to-\infty} f(x)$, $\lim_{x\uparrow 0} f(x)$, and $\lim_{x\downarrow 0} f(x)$.

9.22 (∗) Prove Theorem 9.5.9, namely the algebra of limits at infinity.

9.23 (∗) Let $f : (0, \infty) \to \mathbb{R}$ be a function defined as $f(x) = x^{\frac{1}{x}}$.
(a) Show that for all $x \geq 1$ we have $f(x) \geq 1$.
(b) For a fixed $x \geq 3$, define $n = \lfloor x \rfloor \leq x$. Show that:

$$\frac{(n-2) + 2\sqrt{x}}{n} \geq f(x).$$

(c) Hence, by using the sandwiching result in Exercise 9.10(b) and the algebra of limits, conclude that $\lim_{x\to\infty} x^{\frac{1}{x}} = 1$.

9.24 (◇) Fix $k > 0$. Let $g : (0, \infty) \to \mathbb{R}$ be a function defined as $f(x) = x^{\frac{1}{x^k}}$. Using the same method as in Exercise 9.23, prove that $\lim_{x\to\infty} x^{\frac{1}{x^k}} = 1$.

9.25 ($*$) We first define:

Definition 9.7.3 (Periodic Function) Let $f : \mathbb{R} \to \mathbb{R}$ be a real-valued function. The function f is called periodic if there exists a $P > 0$ such that $f(x + P) = f(x)$ for all $x \in \mathbb{R}$.

If f is a periodic function and $\lim_{x\to\infty} f(x) = L$ for some $L \in \mathbb{R}$, prove that f is a constant function.

9.26 ($*$) Let $f : [a, \infty) \to \mathbb{R}$ be an increasing function. Prove the following results:

(a) If f is a bounded function, then $\lim_{x\to\infty} f(x)$ exists.

(b) If there exists an increasing sequence $(x_n) \subseteq [a, \infty)$ such that $x_n \to \infty$ and $\lim_{n\to\infty} f(x_n)$ exists, then $\lim_{x\to\infty} f(x)$ exists and is equal to $\lim_{n\to\infty} f(x_n)$.

Analogous results can also be deduced if f is a decreasing function.

9.27 Let $f, g : \mathbb{R} \to \mathbb{R}$ be real functions defined as:

$$f(x) = \begin{cases} 1 & \text{if } x \neq 0, \\ 0 & \text{if } x = 0, \end{cases} \quad \text{and} \quad g(x) = \begin{cases} x & \text{if } x \in \mathbb{Q}, \\ 0 & \text{if } x \in \bar{\mathbb{Q}}. \end{cases}$$

(a) Find the composite function $f \circ g : \mathbb{R} \to \mathbb{R}$.

(b) Find the limits $\lim_{x\to 0} f(g(x))$ and $f(\lim_{x\to 0} g(x))$.

(c) Deduce that limits and composition of functions do not generally commute.

9.28 ($*$) As we have seen in Exercise 9.27, limits and composition in general do not commute. However, we can place some extra conditions for this result to be true.

Let $f, g : \mathbb{R} \to \mathbb{R}$ be real functions such that $\lim_{x\to M} f(x) = L$ and $\lim_{x\to x_0} g(x) = M$ where $L, M \in \mathbb{R}$.

(a) Prove that if there exists a $\delta > 0$ such that $g(x) \neq M$ for all $x \in B_\delta(x_0) \setminus \{x_0\}$, then $\lim_{x\to x_0} f(g(x)) = \lim_{y\to M} f(y)$.

(b) Prove that if $L = f(M)$, then $\lim_{x\to x_0} f(g(x)) = f(M) = f(\lim_{x\to x_0} g(x))$.

The second condition above is called the continuity condition at the point M. This conditions says that $\lim_{x\to M} f(x) = f(M)$ so the limit of the function f as it approaches the point M is exactly the value of the function at the point M. Continuity is a very desirable condition to have on a function. We shall devote Chap. 10 to look at this condition in more detail.

9.29 ($*$) Let $g : (a, b) \to \mathbb{R}$ be a bounded and increasing function.

(a) Show that $\lim_{x\uparrow b} g(x) = \sup_{x\in(a,b)} g(x)$ and $\lim_{x\downarrow a} g(x) = \inf_{x\in(a,b)} g(x)$.
(b) For any $c \in (a, b)$, show that $\lim_{x\uparrow c} g(x) = \sup\{g(x) : a < x < c\}$ and $\lim_{x\downarrow c} g(x) = \inf\{g(x) : c < x < b\}$.

Analogously, if $g : (a, b) \to \mathbb{R}$ is a bounded and decreasing function, then we have $\lim_{x\uparrow b} g(x) = \inf_{x\in(a,b)} g(x)$ and $\lim_{x\downarrow a} g(x) = \sup_{x\in(a,b)} g(x)$. For any $c \in (a, b)$ we also have $\lim_{x\uparrow c} = \inf\{g(x) : a < x < c\}$ and $\lim_{x\downarrow c} = \sup\{g(x) : c < x < b\}$.

9.30 (◇) Let $f : X \to \mathbb{R}$ be a bounded function and $x_0 \in X'$.
(a) Define a function $S : (0, \infty) \to \mathbb{R}$ as $S(h) = \sup\{f(x) : x \in B_h(x_0) \cap X \setminus \{x_0\}\}$. Show that S is also a bounded function.
(b) Using Exercise 9.29, explain why $\lim_{h\downarrow 0} S(h)$ exists.

This value is called the limit superior of the function f at x_0 and is denoted as:

$$\limsup_{x\to x_0} f(x) = \lim_{h\downarrow 0} \sup\{f(x) : x \in B_h(x_0) \cap X \setminus \{x_0\}\}.$$

Similarly, the limit inferior can be also defined at x_0 as:

$$\liminf_{x\to x_0} f(x) = \lim_{h\downarrow 0} \inf\{f(x) : x \in B_h(x_0) \cap X \setminus \{x_0\}\}.$$

Analogous to the limit superior and inferior in real sequences, these limits can be used as a substitute when the limit of a function at x_0 does not exist because these quantities always exist for bounded functions.
(c) Prove that $\liminf_{x\to x_0} f(x) \leq \limsup_{x\to x_0} f(x)$.
(d) Prove that for any sequence $(x_n) \subseteq X \setminus \{x_0\}$ with $x_n \to x_0$, we have:

$$\liminf_{x\to x_0} f(x) \leq \liminf_{n\to\infty} f(x_n) \leq \limsup_{n\to\infty} f(x_n) \leq \limsup_{x\to x_0} f(x).$$

(e) By sandwiching, construct a sequence $(x_n) \subseteq X \setminus \{x_0\}$ such that $x_n \to x_0$ and $\lim_{n\to\infty} f(x_n) = \limsup_{x\to x_0} f(x)$.
Similarly, we can find a sequence $(y_n) \subseteq X \setminus \{x_0\}$ such that $y_n \to x_0$ and $\lim_{n\to\infty} f(y_n) = \liminf_{x\to x_0} f(x)$.
(f) Explain why the limit superior and inferior of f at x_0 is respectively the largest and smallest possible limit for any image sequence $(f(x_n))$ with $x_n \to x_0$.
(g) Prove that $\lim_{x\to x_0} f(x)$ exists if and only if $\liminf_{x\to x_0} f(x) = \limsup_{x\to x_0} f(x)$.
The limit is then the common value:

$$\lim_{x\to x_0} f(x) = \liminf_{x\to x_0} f(x) = \limsup_{x\to x_0} f(x).$$

(h) Define the functions $f, g : (0, 1) \to \mathbb{R}$ as $f(x) = \sin(\frac{1}{x})$ and $g(x) = \sin(x)$. Find:
   i. $\limsup_{x\to 0} f(x)$ and $\liminf_{x\to 0} f(x)$,
   ii. $\limsup_{x\to 0} g(x)$ and $\liminf_{x\to 0} g(x)$.

9.31 (◇) In fact, we can also extend the limit inferior and limit superior to points at $\pm\infty$. Suppose that $f : [a, \infty) \to \mathbb{R}$ is a bounded function. We define the limit superior of f at ∞ as:

$$\limsup_{x\to\infty} f(x) = \lim_{K\to\infty} \sup\{f(x) : x \in [K, \infty)\}.$$

Similarly, the limit inferior of f at ∞ is defined at x_0 as:

$$\liminf_{x\to\infty} f(x) = \lim_{K\to\infty} \inf\{f(x) : x \in [K, \infty)\}.$$

(a) Explain why the quantities $\liminf_{x\to\infty} f(x)$ and $\limsup_{x\to\infty} f(x)$ both exist and $\liminf_{x\to\infty} f(x) \leq \limsup_{x\to\infty} f(x)$.

(b) Prove that $\lim_{x\to\infty} f(x)$ exists if and only if $\liminf_{x\to\infty} f(x) = \limsup_{x\to\infty} f(x)$. The limit is then the common value:

$$\lim_{x\to\infty} f(x) = \liminf_{x\to\infty} f(x) = \limsup_{x\to\infty} f(x).$$

(c) Define the functions $f, g : (0, \infty) \to \mathbb{R}$ as:

$$f(x) = \begin{cases} 1 & \text{if } x \in (2n, 2n+1], n \in \mathbb{N}, \\ -\frac{1}{x} & \text{if } x \in (2n+1, 2(n+1)], n \in \mathbb{N}, \end{cases}$$

$$g(x) = \left(\frac{1}{x} + \frac{1}{2}\right) \sin(x).$$

Find:
   i. $\limsup_{x\to 0} f(x)$ and $\liminf_{x\to 0} f(x)$,
   ii. $\limsup_{x\to 0} g(x)$ and $\liminf_{x\to 0} g(x)$.

9.32 Suppose that $f, g : [a, \infty) \to \mathbb{R}$ with $g > 0$. Show that:
(a) $f \in O(g)$ as $x \to \infty$ if and only if $\limsup_{x\to\infty} \frac{|f(x)|}{g(x)} < \infty$.
(b) $f \in o(g)$ as $x \to \infty$ if and only if $\lim_{x\to\infty} \frac{|f(x)|}{g(x)} = 0$.
Next, suppose that $f, g : X \to \mathbb{R}$ with $g > 0$. If $x_0 \in X'$, show that:
(c) $f \in O(g)$ as $x \to x_0$ if and only if $\limsup_{x\to x_0} \frac{|f(x)|}{g(x)} < \infty$.
(d) $f \in o(g)$ as $x \to x_0$ if and only if $\lim_{x\to x_0} \frac{|f(x)|}{g(x)} = 0$.

9.33 Suppose that $f_1, f_2, g_1, g_2 : X \to \mathbb{R}$ with $g_1, g_2 > 0$ and $x_0 \in X$. If $f_1 \in O(g_1)$ and $f_2 \in O(g_2)$ as $x \to x_0$ and $\lambda, \mu \in \mathbb{R}$ are real constants, show that $\lambda f_1 + \mu f_2 \in O(\max\{g_1, g_2\})$ as $x \to x_0$.

Continuity 10

What is life? A continuous praise and blame.

—Friedrich Nietzsche, philosopher

In Chap. 9, we have seen many examples of limits for real functions. For a function $f : X \to \mathbb{R}$ where $X \subseteq \mathbb{R}$, its limits $\lim_{x\to x_0} f(x)$ can only be asked for at the limit points of the domain, namely $x_0 \in X'$. As a result, these limit points may not even lie in the domain X where the function is defined.

Furthermore, even if the limit point x_0 that we are looking at lies within the domain X, as we have discussed in Remark 9.2.10(2), the limit $\lim_{x\to x_0} f(x)$ does not depend on the value of the function at this point and its value may be different than the value of $f(x_0)$. We have seen this happens for the function $f : \mathbb{R} \to \mathbb{R}$ in Example 9.2.9(6) defined as:

$$f(x) = \begin{cases} 2x+1 & \text{if } x \in \mathbb{R} \setminus \{2\}, \\ 7 & \text{if } x = 2. \end{cases}$$

at $x = 2$. This point is a limit point of the domain, so we can ask what is the limit of the function here. Furthermore, this point also lies in the domain of the function with value $f(2) = 7$, so is the limit of the function here equal to this value? We have seen that:

$$\lim_{x\to 2} f(x) = 5 \neq 7 = f(2).$$

If we were refer to the graph of this function in Fig. 9.3, at this point, the graph of the function jumps to a different value. So the graph is broken by the jump or hole.

S. Johar, *The Big Book of Real Analysis*,
https://doi.org/10.1007/978-3-031-30832-1_10

10.1 Continuous Functions

However, if this hole in the graph is filled, in the sense that $\lim_{x\to 2} f(x) = f(2)$, then we get a "continuous" graph. What we mean by "continuous" here is that the graph does not stop and appear elsewhere on the Cartesian plane. Thus, we can draw the graph of this function across the point $x = 2$ without lifting our pencil from the paper. This is true for the function $f : \mathbb{R} \to \mathbb{R}$ defined as $f(x) = 2x + 1$ without the jump at $x = 2$. We call this function continuous at $x = 2$. In general, we define:

Definition 10.1.1 (Continuity at x_0, Definition 1) Let $f : X \to \mathbb{R}$ where $X \subseteq \mathbb{R}$ and $x_0 \in X$. We say that the function f is continuous at x_0 if for any $\varepsilon > 0$, there exists a $\delta(\varepsilon) > 0$ such that for all $x \in X$ with $|x - x_0| < \delta$ we have $|f(x) - f(x_0)| < \varepsilon$. Symbolically, this is written with quantifiers as:

$$\forall \varepsilon > 0, \exists \delta(\varepsilon) > 0 : \forall x \in X, |x - x_0| < \delta(\varepsilon) \Rightarrow |f(x) - f(x_0)| < \varepsilon.$$

Remark 10.1.2 We note that the ε-δ definition in Definition 10.1.1 above is almost similar to Definition 9.2.6 for limits at the limit point x_0 of the domain. The two important differences that set them apart are:

1. The limit L in Definition 9.2.6 is set to be $L = f(x_0)$ in Definition 10.1.1. As a result, continuity of a function can only be asked for at points within the domain, namely at $x_0 \in X$ where $f(x_0)$ has a value.
   This contrasts with limits since we may legitimately ask for the value of limits of a function at points not within the domain (as long as it is a limit point of the domain) at which f is not even defined on. As a result, unlike limits, we actually care about the value of the function f at the point x_0 for continuity.
2. Because $f(x_0)$ must be defined as mentioned above, we can remove the requirement $0 < |x - x_0|$ from Definition 9.2.6. This is true because for x which satisfies $|x - x_0| = 0$ (namely $x = x_0$), trivially we have $|f(x) - f(x_0)| = 0 < \varepsilon$ for any $\varepsilon > 0$ at all.

Note also that, similar to limits, the quantity $\delta(\varepsilon) >$ depends on the chosen $\varepsilon > 0$. Usually we write this quantity simply as δ to declutter. However, we must always remember its dependence on ε.

From the symbolic notation for continuity, we can write the definition of continuity in terms of open balls, namely:

$$\forall \varepsilon > 0, \exists \delta > 0 : \forall x \in X, x \in B_\delta(x_0) \Rightarrow f(x) \in B_\varepsilon(f(x_0)).$$

This can be rewritten more succinctly as:

$$\forall \varepsilon > 0, \exists \delta > 0 : f(B_\delta(x_0) \cap X) \subseteq B_\varepsilon(f(x_0)).$$

The gist of the definition is no matter how small an open ball $B_\varepsilon(f(x_0))$ centred at $f(x_0)$ in the codomain is, we can always find an open ball $B_\delta(x_0)$ centred at x_0 in the domain that is mapped into $B_\varepsilon(f(x_0))$. In even looser words, numbers close to x_0 in the domain are mapped to numbers close to $f(x_0)$ in the codomain.

Example 10.1.3 Recall the function $f : \mathbb{R} \to \mathbb{R}$ in Example 9.2.9(6) defined as:

$$f(x) = \begin{cases} 2x+1 & \text{if } x \in \mathbb{R} \setminus \{2\}, \\ 7 & \text{if } x \in \mathbb{R}. \end{cases}$$

We have shown that $\lim_{x\to 2} f(x) = 5 \neq 7 = f(2)$, so the function is not continuous at $x = 2$. If we refer to the plot of the graph of f in Fig. 9.3, we see that numbers close to $x = 2$ in the domain are not mapped from numbers close to $f(2) = 7$ in the domain since all of them have images around the point $y = 5$ (which are "far" from 7).

Now let us prove this rigorously. Suppose for contradiction that f is continuous at $x = 2$. Then, in particular, for $\varepsilon = 1$ we can find a $\delta > 0$ such that whenever $|x - 2| < \delta$, we must have $|f(x) - f(2)| = |f(x) - 7| < 1$. Equivalently, this says $f(B_\delta(2)) \subseteq B_1(7) = (6, 8)$. However, using the definition of the function f we must have:

$$f(B_\delta(2)) = (5 - 2\delta, 5) \cup \{7\} \cup (5, 5 + 2\delta).$$

Thus, for any $\delta > 0$ at all, $f(B_\delta(2))$ contains an element smaller than 5, which contradicts the fact that $f(B_\delta(2)) \subseteq (6, 8)$. Therefore, f cannot be continuous at $x = 2$.

On top of the Definition 10.1.1 above, similar to limits, we also have a second equivalent definition for continuity in terms of sequences:

Definition 10.1.4 (Continuity at x_0, Definition 2) Let $f : X \to \mathbb{R}$ where $X \subseteq \mathbb{R}$ and $x_0 \in X$. We say that the function f is continuous at x_0 if for any sequence (x_n) in X such that $x_n \to x_0$, we have:

$$\lim_{n\to\infty} f(x_n) = f(\lim_{n\to\infty} x_n) = f(x_0).$$

Remark 10.1.5 Let us make some remarks regarding this definition.

1. From Definition 10.1.4, note that $x_0 \in X$ and the sequence (x_n) in the definition are allowed to take the value x_0 as well. Therefore, $x_0 \in X$ need not be a limit point of X.

2. For a point $x_0 \in X$, we either have $x_0 \in X \cap X'$ or $x_0 \in X \setminus X'$, where the former means x_0 is a limit point of X whereas the latter means x_0 is an isolated point of X.
   (a) If $x_0 \in X \cap X'$, continuity at x_0 in Definition 10.1.4 is the same as:

$$\lim_{x \to x_0} f(x) = f(\lim_{x \to x_0} x) = f(x_0).$$

   This notation also implies the limit of f as $x \to x_0$ exists and is equal to $f(x_0)$.
   (b) If $x_0 \in X \setminus X'$, continuity at the isolated points are always guaranteed by default. Indeed, fix $\varepsilon > 0$. By definition of isolated points in Definition 6.2.5, we can always find a $\delta > 0$ small enough such that $B_\delta(x_0) \cap X = \{x_0\}$. In other words, the only $x \in X$ that satisfies $|x - x_0| < \delta$ is $x = x_0$. Then, for all $x \in X$ with $|x - x_0| < \delta$ (namely $x = x_0$ only), necessarily we have $|f(x) - f(x_0)| = |f(x_0) - f(x_0)| = 0 < \varepsilon$, fulfilling Definition 10.1.1.
3. Therefore, since continuity at isolated points of the domain is trivially guaranteed, by an abuse/simplification of notation introduced in Definition 9.2.2 (where x_0 is only allowed to be a limit point), sometimes we write continuity as $\lim_{x \to x_0} f(x) = f(\lim_{x \to x_0} x) = f(x_0)$ regardless of whether x_0 is a limit point or an isolated point in the domain.

The proof to show that that Definitions 10.1.1 and 10.1.4 are equivalent is in the same vein as the proof for equivalence of limit definitions in Lemma 9.2.8. We leave this for the readers to prove in Exercise 10.3.

Lemma 10.1.6 (Equivalence of Continuity Definitions) *Definitions 10.1.1 and 10.1.4 are equivalent.*

If the function $f : X \to \mathbb{R}$ is continuous at each and every $x_0 \in X$, we call the function a continuous function.

Definition 10.1.7 (Continuous Functions) Let $f : X \to \mathbb{R}$ where $X \subseteq \mathbb{R}$. We say that the function f is continuous on X if the function is continuous at all $x_0 \in X$.

Symbolically, this is written with quantifiers as:

$$\forall x_0 \in X, \forall \varepsilon > 0, \exists \delta(\varepsilon, x_0) > 0 : \forall x \in X, |x - x_0| < \delta(\varepsilon, x_0) \Rightarrow |f(x) - f(x_0)| < \varepsilon.$$

We note now the quantity $\delta > 0$ depends on both ε and x_0. This is true since for different points $x_0 \in X$ but the same $\varepsilon > 0$, we might need different values of δ to ensure that the definition of continuity in Definition 10.1.1 holds. Let us look at an example of this:

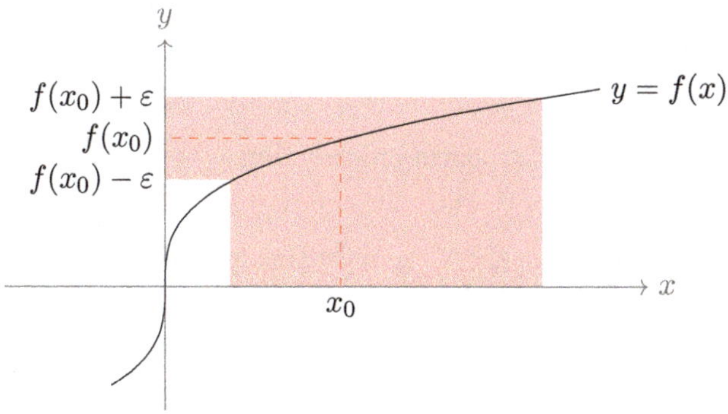

Fig. 10.1 Graph of $y = f(x) = \sqrt[3]{x}$

Example 10.1.8 Consider the function $f : \mathbb{R} \to \mathbb{R}$ defined as $f(x) = \sqrt[3]{x}$. We want to show that it is continuous everywhere. First, we fix the point $x_0 \in \mathbb{R}$ where we want to check the continuity at and $\varepsilon > 0$. Our goal is to find a $\delta > 0$ such that for any $|x - x_0| < \delta$ we have $|\sqrt[3]{x} - \sqrt[3]{x_0}| < \varepsilon$. A diagram for this can be seen in Fig. 10.1.

1. The case $x_0 = 0$ is easy to deal with since we want to find a δ for which $|x - 0| = |x| < \delta$ implies $|\sqrt[3]{x} - \sqrt[3]{0}| = |\sqrt[3]{x}| < \varepsilon$. An obvious choice for δ for this case would be $\delta = \varepsilon^3$.
2. For the other case, namely $x_0 \neq 0$, let us lay out the rough work first.

 Rough work: We work from $|\sqrt[3]{x} - \sqrt[3]{x_0}|$ and try to bound it by a term involving $|x - x_0|$. Notice that $|\sqrt[3]{x} - \sqrt[3]{x_0}||\sqrt[3]{x}^2 + \sqrt[3]{xx_0} + \sqrt[3]{x_0}^2| = |x - x_0|$ where we have a variable term $|\sqrt[3]{x}^2 + \sqrt[3]{xx_0} + \sqrt[3]{x_0}^2|$. If we can bound this term from below by some positive constant, then we can control $|\sqrt[3]{x} - \sqrt[3]{x_0}|$ using $|x - x_0|$.
 Note first that if we put a restriction on x such that $|x - x_0| < \frac{|x_0|}{2}$, then the quantities x and x_0 are both of the same sign. Thus, $xx_0 > 0$ and hence:

$$|\sqrt[3]{x}^2 + \sqrt[3]{xx_0} + \sqrt[3]{x_0}^2| = \sqrt[3]{x}^2 + \sqrt[3]{xx_0} + \sqrt[3]{x_0}^2 > \sqrt[3]{x_0}^2.$$

 Using this estimate in $|\sqrt[3]{x} - \sqrt[3]{x_0}||\sqrt[3]{x}^2 + \sqrt[3]{xx_0} + \sqrt[3]{x_0}^2| = |x - x_0|$, we get:

$$\sqrt[3]{x_0}^2|\sqrt[3]{x} - \sqrt[3]{x_0}| < |x - x_0| \quad \Rightarrow \quad |\sqrt[3]{x} - \sqrt[3]{x_0}| < \frac{|x - x_0|}{\sqrt[3]{x_0}^2},$$

and we want this whole RHS to be smaller than ε. Therefore, we can make an appropriate choice for δ that takes into account the previous restriction that we set, namely $\delta = \min\left\{\frac{|x_0|}{2}, \varepsilon\sqrt[3]{x_0}^2\right\} > 0$.

Now we write the argument down properly. Fix $x_0 \in \mathbb{R} \setminus \{0\}$ and $\varepsilon > 0$. Set $\delta = \min\left\{\frac{|x_0|}{2}, \varepsilon\sqrt[3]{x_0}^2\right\} > 0$. Then, for any $|x - x_0| < \delta$, since $|x - x_0| < \delta \leq \frac{|x_0|}{2}$, the numbers x and x_0 are of the same sign and thus $xx_0 > 0$. This implies $\sqrt[3]{x}^2 + \sqrt[3]{xx_0} + \sqrt[3]{x_0}^2 > \sqrt[3]{x_0}^2 > 0$ and hence:

$$|\sqrt[3]{x} - \sqrt[3]{x_0}| = \frac{|x - x_0|}{\sqrt[3]{x}^2 + \sqrt[3]{xx_0} + \sqrt[3]{x_0}^2} < \frac{|x - x_0|}{\sqrt[3]{x_0}^2} < \frac{\varepsilon\sqrt[3]{x_0}^2}{\sqrt[3]{x_0}^2} = \varepsilon,$$

where we used the fact $|x - x_0| < \delta \leq \varepsilon\sqrt[3]{x_0}^2$ for the final inequality.

Example 10.1.8 shows that at different points x_0 in the domain, we may need different values of $\delta(x_0, \varepsilon)$ for the definition of continuity to be fulfilled.

10.2 Algebra of Continuous Functions

In order to show whether a function is continuous, we need to check continuity at all points on the domain. This can be a difficult task indeed in some cases. For instance, in the example above, we have seen that we have to treat the cases for $x_0 = 0$ and $x_0 \neq 0$ separately. For even more complicated functions, things could get really involved.

Luckily, by the same proof of algebra of limits in Theorem 9.5.2, we can split a complicated function into smaller more elementary functions and investigate the continuity of these elementary functions separately instead. If all these separate parts are continuous and behave nicely, we can combine them back together to conclude the continuity of the original function.

Using the exact same proof for the algebra of limits in Theorem 9.5.2, by replacing L with $f(x_0)$, M with $g(x_0)$, and $0 < |x - x_0| < \delta$ with $|x - x_0| < \delta$ (all of these replacements do not change any argument in the proof of Theorem 9.5.2 at all), we have:

Theorem 10.2.1 (Algebra of Continuous Functions) *Let $f, g : X \to \mathbb{R}$ where $X \subseteq \mathbb{R}$ and $x_0 \in X$. Suppose that the functions f and g are continuous at x_0. Then:*

1. *For a constant $\lambda \in \mathbb{R}$, the function λf is continuous at x_0.*
2. *The function $|f|$ is continuous at x_0.*
3. *The function $f \pm g$ is continuous at x_0.*
4. *The function $f \times g$ is continuous at x_0.*

5. *If* $f(x_0) \neq 0$, *then the function* $\frac{1}{f}$ *is continuous at* x_0.
6. *If* $f(x_0) \neq 0$, *then the function* $\frac{g}{f}$ *is continuous at* x_0.

Furthermore, continuity is also preserved by composition of functions. This is also a useful fact to know on top of those results in Theorem 10.2.1.

Theorem 10.2.2 *Let* $f : X \to \mathbb{R}$ *and* $g : Y \to \mathbb{R}$ *with* $f(X) \subseteq Y$. *If the function* f *is continuous at* $x_0 \in X$ *and the function* g *is continuous at* $f(x_0) \in Y$, *then the composition function* $h = g \circ f : X \to \mathbb{R}$ *is continuous at* x_0.

Proof Fix $\varepsilon > 0$. We want to find a $\delta > 0$ such that for all $x \in X$ with $|x - x_0| < \delta$ we must have $|h(x) - h(x_0)| = |g(f(x)) - g(f(x_0))| < \varepsilon$.

Since the function g is continuous at $f(x_0)$, there exists a $\delta_1 > 0$ such that if $y \in Y$ with $|y - f(x_0)| < \delta_1$ then $|g(y) - g(f(x_0))| < \varepsilon$. Moreover, since f is continuous at x_0, there exists a $\delta > 0$ such that any $x \in X$ with $|x - x_0| < \delta$ would imply $|f(x) - f(x_0)| < \delta_1$. This is the $\delta > 0$ that we require. Indeed, by setting $y = f(x)$, we have a chain of implications:

$$|x-x_0| < \delta \Rightarrow |f(x)-f(x_0)| < \delta_1 \Rightarrow |g(f(x))-g(f(x_0))| = |h(x)-h(x_0)| < \varepsilon,$$

and thus the composition $h = g \circ f$ is continuous at x_0. □

Using a combination Theorems 10.2.1 and 10.2.2, we can prove the continuity of many functions. We are going to show that every polynomial is continuous everywhere. We first prove the following lemma, which is for monomials instead:

Lemma 10.2.3 *Let* $f : \mathbb{R} \to \mathbb{R}$ *be defined as* $f(x) = x^n$ *for some* $n \in \mathbb{N}$. *Then,* f *is a continuous function.*

Proof We need to show that for any $x_0 \in \mathbb{R}$, the function f is continuous at x_0. In other words, we need to show that $\lim_{x \to x_0} x^n = x_0^n$. This is clearly true for $n = 1$, so let us prove it for the cases $n \neq 1$.

Fix $x_0 \in \mathbb{R}$ and $\varepsilon > 0$. We need to find a $\delta > 0$ such that for all $x \in \mathbb{R}$ with $|x - x_0| < \delta$ we have $|x^n - x_0^n| < \varepsilon$. Working from the latter, via factorisation and triangle inequality, we have:

$$\begin{aligned}
|x^n - x_0^n| &= |(x - x_0)(x^{n-1} + x^{n-2}x_0 + \ldots + x_0^{n-1})| \\
&= |x - x_0||x^{n-1} + x^{n-2}x_0 + \ldots + x_0^{n-1}| \\
&\leq |x - x_0| \left(|x|^{n-1} + |x|^{n-2}|x_0| + \ldots + |x_0|^{n-1}\right). \qquad (10.1)
\end{aligned}$$

Let us put a condition on $|x - x_0|$ to simplify the bracketed terms in (10.1). If we set the restriction $|x - x_0| < 1$, by triangle inequality, we would get $|x| \leq |x - x_0| + |x_0| < 1 + |x_0|$. Hence, inequality (10.1) can be bound further as:

$$\begin{aligned}|x^n - x_0^n| &\leq |x - x_0| \left(|x|^{n-1} + |x|^{n-2}|x_0| + \ldots + |x_0|^{n-1}\right)\\ &< |x - x_0| \left((|x_0| + 1)^{n-1} + (|x_0| + 1)^{n-2}|x_0| + \ldots + |x_0|^{n-1}\right)\\ &= |x - x_0| F(x_0),\end{aligned}$$

where we denoted $F(x_0) = (|x_0| + 1)^{n-1} + (|x_0| + 1)^{n-2}|x_0| + \ldots + |x_0|^{n-1}$ for brevity. Notice that $F(x_0)$ is a positive constant greater than or equal to 1. So, if we pick $\delta = \min\left\{1, \frac{\varepsilon}{F(x_0)}\right\} > 0$, whenever $|x - x_0| < \delta$ we would have:

$$\begin{aligned}|x^n - x_0^n| &\leq |x - x_0| \left((|x_0| + 1)^{n-1} + (|x_0| + 1)^{n-2}|x_0| + \ldots + |x_0|^{n-1}\right)\\ &= |x - x_0| F(x_0) < \frac{\varepsilon}{F(x_0)} F(x_0) = \varepsilon \qquad (\because |x - x_0| < \tfrac{\varepsilon}{F(x_0)}),\end{aligned}$$

thus the monomial function is continuous at x_0. Since $x_0 \in \mathbb{R}$ is arbitrary, we conclude that f is continuous everywhere. □

Remark 10.2.4 The above proof is a classical long-winded proof for continuity that is done for the sake of exercise. One can be a bit cleverer and define a function $g(x) = x$ on $\mathbb{R}$ so that $f(x) = x^n = g(x) \times g(x) \times \ldots \times g(x)$ is a product of the function g n times. The function $g(x) = x$ is continuous everywhere: simply pick $\delta = \varepsilon$ in the proof and we are done. Thus, at any $x_0 \in \mathbb{R}$, by using Theorem 10.2.1 for products n times, f is continuous here.

Theorem 10.2.5 *Any real polynomial on the real numbers is a continuous function.*

Proof Let $P(x) = a_0 + a_1x + a_2x^2 + \ldots + a_nx^n$ be our polynomial. We want to show that $\lim_{x\to x_0} P(x) = P(x_0)$ for all $x_0 \in \mathbb{R}$. Fix $x_0 \in \mathbb{R}$. We know from Lemma 10.2.3 that each term x^n are continuous and hence has the limit x_0^n as $x \to x_0$. Therefore, we can apply Theorem 10.2.1 onto the function P, namely:

$$\begin{aligned}\lim_{x\to x_0} P(x) &= \lim_{x\to x_0} (a_0 + a_1x + a_2x^2 + \ldots + a_nx^n)\\ &= a_0 \lim_{x\to x_0} 1 + a_1 \lim_{x\to x_0} x + a_2 \lim_{x\to x_0} x^2 + \ldots + a_n \lim_{x\to x_0} x^n\\ &= a_0 + a_1x_0 + a_2x_0^2 + \ldots + a_nx_0^n = P(x_0),\end{aligned}$$

and so the polynomial is a continuous function. □

Another useful construction is the extension of a continuous function to its limit point. The readers will prove the following result in Exercise 10.5.

Proposition 10.2.6 *Let* $f : (a, b] \to \mathbb{R}$ *be a continuous function such that* $\lim_{x\to a} f(x)$ *exists. Then, the extension* $\tilde{f} : [a, b] \to \mathbb{R}$ *defined as:*

$$\tilde{f}(x) = \begin{cases} \lim_{x\to a} f(x) & \text{if } x = a, \\ f(x) & \text{if } x \in (a, b], \end{cases}$$

is also a continuous function.

Finally, for ease of notation, let us give a name to the collection of all continuous real-valued functions defined on a fixed set X.

Definition 10.2.7 (Space of Continuous Functions) The space of all continuous functions with domain X and codomain $\mathbb{R}$ is denoted as $C^0(X; \mathbb{R})$. In other words, $C^0(X; \mathbb{R})$ is the set:

$$C^0(X; \mathbb{R}) = \{f : X \to \mathbb{R} : f \text{ is continuous on } X\}.$$

Sometimes, if the codomain is clear, this space is simply denoted as $C^0(X)$ for brevity.

Remark 10.2.8 For those of us who are inclined to linear algebra, note that from Theorem 10.2.1(1) and (3), the set of continuous functions on X is preserved under scalar multiplication and addition. Thus, the set $C^0(X; \mathbb{R})$ is a real vector space (see Definition 4.6.2) with the zero element $f = 0$.

10.3 One-Sided Continuity

Similar to limits, we also have the property of one-sided continuity. By simply amending the definition of one-sided limits in Definition 9.3.1 to ensure that x_0 is in the domain of the function and $L = f(x_0)$, we have:

Definition 10.3.1 (One-Sided Continuity) Let $f : X \to \mathbb{R}$ where $X \subseteq \mathbb{R}$ and $x_0 \in X$.

1. We say that the function f is right-continuous at x_0 if for any $\varepsilon > 0$, there exists a $\delta > 0$ such that for all $x \in X$ with $0 \leq x - x_0 < \delta$ we have $|f(x) - f(x_0)| < \varepsilon$. Symbolically:

$$\forall \varepsilon > 0, \exists \delta > 0 : \forall x \in X, 0 \leq x - x_0 < \delta \Rightarrow |f(x) - f(x_0)| < \varepsilon.$$

If x_0 is a limit point of the domain X, this is the same as:

$$\lim_{x \downarrow x_0} f(x) = f(x_0).$$

2. We say that the function f is left-continuous at x_0 if for any $\varepsilon > 0$, there exists a $\delta > 0$ such that for all $x \in X$ with $0 \leq x_0 - x < \delta$ we have $|f(x) - f(x_0)| < \varepsilon$. Symbolically:

$$\forall \varepsilon > 0, \exists \delta > 0 : \forall x \in X, 0 \leq x_0 - x < \delta \Rightarrow |f(x) - f(x_0)| < \varepsilon.$$

If x_0 is a limit point of the domain X, this is the same as:

$$\lim_{x \uparrow x_0} f(x) = f(x_0).$$

Example 10.3.2 Recall the function in Examples 9.2.9(7) and 9.3.3 which was given by:

$$f(x) = \begin{cases} x & \text{if } x < 0, \\ x + 1 & \text{if } x \geq 0. \end{cases}$$

The left- and right-limits at $x = 0$ exist and they are:

$$\lim_{x \uparrow 0} f(x) = 0 \neq f(0) \quad \text{and} \quad \lim_{x \downarrow 0} f(x) = 1 = f(0).$$

Thus, this function is right-continuous but not left-continuous at $x = 0$.

Clearly, if a function f is continuous at x_0, then it is both left- and right-continuous at x_0 This is just obtained by restricting our attention from $|x - x_0| < \delta$ to each half of the interval, namely to $0 \leq x - x_0 < \delta$ and $0 \leq x_0 - x < \delta$. The converse is also true, namely: if a function is both left- and right-continuous at x_0, then the function f is continuous at x_0 as well. The proof of this result is similar to the proof of Proposition 9.3.4.

Proposition 10.3.3 *Let $f : X \to \mathbb{R}$ where $X \subseteq \mathbb{R}$ and $x_0 \in X$. If the function f is both left- and right-continuous at x_0, then it is continuous at x_0.*

Proof Fix any $x_0 \in X$ and $\varepsilon > 0$. Since f is right-continuous at x_0, there exists a $\delta_1 > 0$ such that for any $x \in X$ with $0 \leq x - x_0 < \delta_1$ we have $|f(x) - f(x_0)| < \varepsilon$. Likewise, since it is left-continuous at x_0, there exists a $\delta_2 > 0$ such that for any $x \in X$ with $0 \leq x_0 - x < \delta_2$ we have $|f(x) - f(x_0)| < \varepsilon$.

Set $\delta = \min\{\delta_1, \delta_2\} > 0$. Then, for any $x \in X$ with $|x - x_0| < \delta$, either one of the inequalities $0 \leq x - x_0 < \delta_1$ and $0 \leq x_0 - x < \delta_2$ hold, from which we conclude $|f(x) - f(x_0)| < \varepsilon$. Thus, the function f is continuous at x_0. □

Now suppose that the above does not hold, namely: the function is not both left- and right-continuous at a point $x_0 \in X$. We call such function discontinuous. We have three different cases for which a discontinuity can happen.

Definition 10.3.4 (Types of Discontinuities) Let $f : X \to \mathbb{R}$ and $x_0 \in X$.

1. If $\lim_{x \uparrow x_0} f(x)$ and $\lim_{x \downarrow x_0} f(x)$ both exist and are equal (or in other words, the limit $\lim_{x \to x_0} f(x)$ exists) but not equal to $f(x_0)$, then we say there is a removable discontinuity at the point x_0.
2. If $\lim_{x \uparrow x_0} f(x)$ and $\lim_{x \downarrow x_0} f(x)$ both exist but are not equal to each other, then we say there is a jump discontinuity at the point x_0.
3. If either one of $\lim_{x \uparrow x_0} f(x)$ and $\lim_{x \downarrow x_0} f(x)$ does not exist, then we say there is an essential discontinuity at the point x_0.

Example 10.3.5 Let us look at examples of these discontinuities and comment on them.

1. Recall the function $f : \mathbb{R} \to \mathbb{R}$ in Example 9.2.9(6) defined as:

$$f(x) = \begin{cases} 2x + 1 & \text{if } x \in \mathbb{R} \setminus \{2\}, \\ 7 & \text{if } x \in \mathbb{R}. \end{cases}$$

   We have shown that $\lim_{x \to 2} f(x) = 5 \neq 7 = f(2)$, so the function is not continuous at $x = 2$. Hence, we have a removable discontinuity here.
   The discontinuity is called removable because we can remove it by simply redefining the function at that point. Indeed, we can modify the function f by redefining its value at $x = 2$ to be $\lim_{x \to 2} f(x) = 5$. The resulting function $\tilde{f} : \mathbb{R} \to \mathbb{R}$ given as $\tilde{f}(x) = 2x + 1$ is now continuous at $x = 2$. Thus the original discontinuity at $x = 2$ is "removed".
2. We have seen an example of a jump discontiuity before. Recall the function:

$$\begin{aligned} f : \mathbb{R} &\to \mathbb{R} \\ x &\mapsto \begin{cases} x & \text{if } x < 0, \\ x + 1 & \text{if } x \geq 0, \end{cases} \end{aligned}$$

   from Example 10.3.2. Both the left- and right-limits at $x = 0$ exist with $\lim_{x \uparrow 0} f(x) = 0$ and $\lim_{x \downarrow 0} f(x) = 1 = f(1)$. However, since they are not equal, we have a jump discontinuity here.

3. Finally, consider the function:

$$f : \mathbb{R} \to \mathbb{R}$$

$$x \mapsto \begin{cases} 0 & \text{if } x \leq 0, \\ \sin(\frac{1}{x}) & \text{if } x > 0. \end{cases}$$

We can show that the left limit at $x = 0$ exists and is equal to $\lim_{x\uparrow 0} f(x) = 0 = f(0)$. On the other hand, the limit $\lim_{x\downarrow 0} f(x) = \lim_{x\downarrow 0} \sin(\frac{1}{x})$ does not exist. Indeed, as we have seen in Example 9.2.5, we can find a sequence of points in $(0, \infty) \subseteq \mathbb{R} \setminus \{0\}$ which converges to 0 but its image sequence is divergent. Thus, the discontinuity here is essential.

Therefore, discontinuities can range in behaviour from mild as in the removable discontinuity, to pathological as in the essential discontinuity. On the other hand, continuous functions are nice because their behaviour is more predictable. We shall now discuss some properties and results that can be obtained from continuous functions.

10.4 Intermediate Value Theorem

In this section, we are going to look at continuous functions defined on intervals of $\mathbb{R}$ and their nice properties. The first result is called the intermediate value theorem (or the IVT for short).

Suppose that $f : I \to \mathbb{R}$ is a continuous function defined on a compact interval $I = [a, b]$. We can intuitively plot its graph without lifting our pencil from the paper. This means if we start drawing the graph from the point $(a, f(a))$ in the Cartesian xy-plane, the graph will connect this point to the point $(b, f(b))$.

Therefore, for each real number c between $f(a)$ and $f(b)$, we will have to pass our pencil across the line $y = c$ somewhere (may even be at more than one point) when we are drawing the graph. In other words, the function assumes the value c somewhere in the domain. A visualisation of this is given in Fig. 10.2.

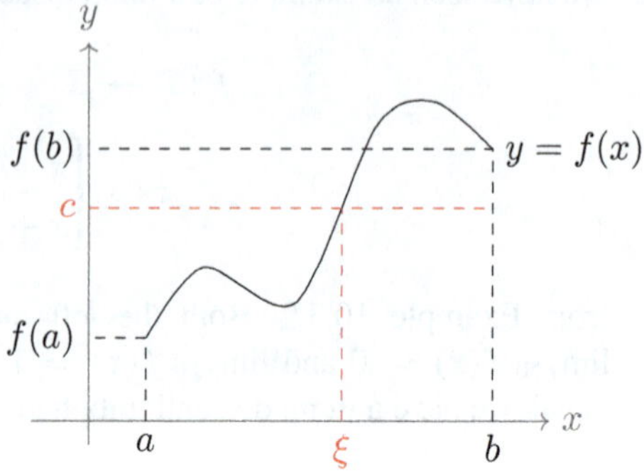

Fig. 10.2 The continuous function f attains the value $c \in [f(a), f(b)]$ at $x = \xi$

As described above, this is a very intuitive result and was treated as obvious pre-19th century. It was only proven rigorously in 1817 by Bolzano and brushed up by Cauchy in 1821.

Theorem 10.4.1 (Intermediate Value Theorem, IVT) *Let $f \in C^0(I)$ be a continuous real-valued function defined on the compact interval $I = [a, b]$. If c is a number between $f(a)$ and $f(b)$, then there exists a point $\xi \in [a, b]$ such that $f(\xi) = c$.*

Proof If $c = f(a)$ or $c = f(b)$, then the result is trivial. WLOG, assume that $f(a) < f(b)$ and c is strictly between these numbers, namely $f(a) < c < f(b)$. Define a new function $g : I \to \mathbb{R}$ as $g(x) = f(x) - c$ which is also a continuous function by virtue of Theorem 10.2.1. Thus, finding a point $\xi \in (a, b)$ such that $f(\xi) = c$ is the same as finding a zero of the function g.

We do this by constructing two sequences (x_n) and (y_n) as follows. We first define $x_1 = a$ and $y_1 = b$. We note that $g(a) < 0$ and $g(b) > 0$ and so $g(a)g(b) < 0$. Now we look at the midpoint $z_1 = \frac{x_1+y_1}{2}$. At this point, by trichotomy of strict ordering in $\mathbb{R}$, we either have $g(z_1) = 0$, $g(z_1) < 0$, or $g(z_1) > 0$. From these cases, we proceed as follows:

1. If $g(z_1) = 0$, then we then we have found our ξ.
2. If $g(z_1) < 0$, we define $x_2 = z_1$ and $y_2 = y_1$. Note that $x_2 \geq x_1$ and also $g(x_2)g(y_2) = g(z_1)g(y_1) = g(z_1)g(b) < 0$.
3. If $g(z_1) > 0$, we define $x_2 = x_1$ and $y_2 = z_1$. Note that $y_2 \leq y_1$ and also $g(x_2)g(y_2) = g(x_1)g(z_1) = g(a)g(z_2) < 0$.

If $g(z_1) \neq 0$, either way the distance between the pair of points x_2 and y_2 would be halved from the previous pair, namely $|y_2 - x_2| = \frac{1}{2}|y_1 - x_1| = \frac{1}{2}|b - a|$ with $x_2 \geq x_1$, $y_2 \leq y_1$, and $g(x_2)g(y_2) < 0$.

We repeat the construction above by looking at the subinterval $[x_2, y_2] \subseteq [a, b]$. Eventually, we will either find the required ξ or we continue the construction indefinitely and create two infinite sequences (x_n) and (y_n) such that $|y_n - x_n| = \frac{1}{2^{n-1}}|b - a|$, (x_n) increasing, (y_n) decreasing, and $g(x_n)g(y_n) < 0$ for each $n \in \mathbb{N}$.

For the latter case, both of the sequences (x_n) and (y_n) are bounded. So, by the monotone sequence theorem and Corollary 6.2.7, they converge to some elements in I, namely: $x_n \to x$ and $y_n \to y$ where $x, y \in I$. By using the algebra of limits, we have:

$$|x - y| = |\lim_{n\to\infty} x_n - \lim_{n\to\infty} y_n| = \lim_{n\to\infty} |x_n - y_n| = \lim_{n\to\infty} \frac{1}{2^{n-1}}|b - a| = 0,$$

which means $x = y$. We denote this common limit as $z \in [a, b]$ now. Finally, using the facts that limits preserve weak inequalities and the function g is continuous, we have:

$$g(x_n)g(y_n) < 0 \Rightarrow \lim_{n\to\infty} (g(x_n)g(y_n)) \leq 0 \Rightarrow g(z)^2 \leq 0 \Rightarrow g(z)^2 = 0,$$

so that $g(z) = 0$. Thus, we have found our desired $\xi = z$. □

The proof of the IVT above is constructive, meaning that we can actually find what the value of ξ is. This method is called the bisection method and is used in root-finding algorithms in numerical methods. We also have seen a form of this argument when we were trying to define a root function in Exercise 6.28.

Remark 10.4.2 There is another proof for the IVT which is done by considering a special subset of the domain $I = [a, b]$ and proving that the supremum of this set is the solution $f(\xi) = c$. The proof is as follows:

WLOG, assume that $f(a) < f(b)$ and c is strictly between these numbers, namely $f(a) < c < f(b)$. Consider the set $H = \{x \in I : f(x) < c\} \subseteq I$. This set is non-empty because $a \in H$ and this set is bounded from above by b. By the completeness axiom of real numbers, this set has a supremum $\sup(H) = \xi \in I$. We claim that $f(\xi) = c$. Suppose for contradiction that this is not true. Then, either $f(\xi) < c$ or $f(\xi) > c$ is true. We prove that both of these cases cannot hold.

1. Assume that $f(\xi) < c$, which is equivalent to $0 < c - f(\xi)$. By continuity of the function f at ξ, for $\varepsilon = c - f(\xi) > 0$ there exist a $\delta > 0$ such that for all $x \in I$ with $|x - \xi| < \delta$, we have $|f(x) - f(\xi)| < \varepsilon = c - f(\xi)$.
   So, in particular, for the point $x_0 = \xi + \frac{\delta}{2} > \xi$ we have $|x_0 - \xi| = |\xi + \frac{\delta}{2} - \xi| = \frac{\delta}{2} < \delta$ and therefore $|f(x_0) - f(\xi)| < c - f(\xi)$ which implies $f(x_0) < c$. This means $x_0 \in H$. However, $x_0 = \xi + \frac{\delta}{2} > \xi = \sup(H)$, a contradiction. So this case cannot happen.
2. For the second case, if $f(\xi) > c$, for $\varepsilon = f(\xi) - c > 0$ there exists a $\delta > 0$ such that for all $x \in I$ with $|x - \xi| < \delta$, we have $|f(x) - f(\xi)| < f(\xi) - c$. In particular, $c < f(x)$ here. This means for all $x \in (\xi - \delta, \xi)$ we have $f(x) > c$, which then implies ξ is not the supremum for the set H as we can find a strictly smaller upper bound for the set H, for example $\xi - \frac{\delta}{2}$. This gives us another contradiction and therefore this case cannot happen either.

Hence, we conclude that $f(\xi) = c$.

We note that from both of the proofs of the IVT presented above, we have found one such ξ where $f(\xi) = c$. However, there may be more than one such ξ.

Example 10.4.3 The IVT is very useful for to determine whether a complicated equation has any solutions. Consider the polynomial $f(x) = x^5 - 2x^3 - 2$ for $x \in \mathbb{R}$. We want to show that this polynomial has a real root. If we set $x = 0$, we would get $f(0) = -2$ and if we set $x = 2$, we would get $f(2) = 2^5 - 2(2^3) - 2 = 14$. Since $f(0) < 0 < f(2)$ and f is continuous, by the IVT, we conclude that there is a $\xi \in (0, 2)$ such that $f(\xi) = 0$.

Here, we do not know what the value of the root ξ is and how many roots are there in the interval $[0, 2]$. However, we do know that there is at least one solution to the equation $f(x) = 0$ here! An exact solution may be approximated by the constructive bisection method as we have seen in the proof of the IVT. Even so, by construction, the bisection method only gives us one root to the polynomial.

Example 10.4.4 Here are some amusing applications of the IVT.

1. Suppose on Saturday, we climb up a mountain from the base camp starting at 1pm and arriving at the top at 8pm. Our altitude at $t$pm is given by the function $f : [1, 8] \to \mathbb{R}$. Tired, we decided to stay overnight at the summit. On Sunday, we depart for the base camp at 1pm and by 8pm we are already at the base camp (we could have doubled back if we dropped something or we could have stopped to enjoy the scenery or we could have arrived at the base camp earlier than 8pm). Our altitude at $t$pm on Sunday is given by the function $g : [1, 8] \to \mathbb{R}$.
   Note that the functions f and g are continuous since motion is continuous (no teleportation is allowed). As we can see in Fig. 10.3, there is a time $T \in (1, 8)$ on both of the days at which our altitude is the same. How do we show this T exists rigorously?
   Define a new function $h : [1, 8] \to \mathbb{R}$ as $h(t) = f(t) - g(t)$. Note that $h(1) = f(1) - g(1) < 0$ and $h(8) = f(8) - g(8) > 0$. So, by the IVT, since h is continuous, there must exist a time $T \in (1, 8)$ at which $h(T) = 0$, namely $f(T) = g(T)$. In other words, at time Tpm on Sunday, we are at the same altitude as we were exactly 24 hours ago.

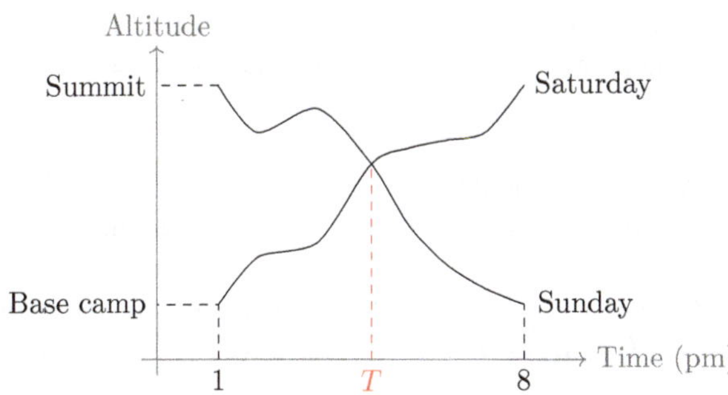

Fig. 10.3 The altitude of ascent and descent

2. Another example which is closer to everyday life is pointed out by Eugenia Cheng (1976-) in her article *The Calculus of a Shower That's Either Too Hot or Too Cold*:

> I used a shower at the gym for the first time in a year and a half and couldn't get it to the right temperature. First it was too cold, but if I turned the handle ever so slightly it immediately became too hot. I turned it the tiniest amount back, and it became too cold again. No position seemed to exist where it was at the right temperature for me, which means it violated a theorem from calculus called the Intermediate Value Theorem ... I resigned myself to the fact that this particular one's heat function was not continuous, and I moved to a different shower.

3. Yet another application of the IVT in daily life is how to deal with annoying wobbly tables. In their paper [6], using the IVT and geometrical arguments, Bill Baritompa, Rainer Löwen, Burkrard Polster, and Marty Ross demonstrated that one can always stabilise a table placed on an idealised continuous floor by rotating it in its place. One simply has to find the right angle to rotate the table.

Let us leave the real world problems behind and return to mathematics. A direct consequence of the IVT is that the image of any interval is also an interval.

Proposition 10.4.5 *Let $f \in C^0(I)$ be a continuous real-valued function and I is an interval in $\mathbb{R}$. Then, the image $f(I) \subseteq \mathbb{R}$ is an interval as well.*

Proof From Definition 4.5.1, we are required to show that for any two points $p, q \in f(I)$ with $p < q$, all elements $c \in (p, q)$ are contained in $f(I)$. Let $p, q \in f(I)$ be arbitrary. Since they lie in the image of the function f, there must be points $a, b \in I$ such that $f(a) = p$ and $f(b) = q$. Then, we either have $a < b$ or $a > b$. WLOG, assume that $a < b$

Consider the restriction of the function f to the interval $[a, b]$, namely $f|_{[a,b]} : [a, b] \to \mathbb{R}$. Since f is continuous, the restriction $f|_{[a,b]}$ must be continuous as well. Furthermore, $[a, b]$ is a compact interval, so we may apply the IVT to this restricted function. The IVT says that for every c in between $f(a) = p$ and $f(b) = q$ there exists a $\xi \in [a, b]$ with $f(\xi) = c$. Thus, every point c between p and q must be an image of some point $\xi \in [a, b] \subseteq I$ and therefore $c = f(\xi) \in f([a, b]) \subseteq f(I)$. □

Proposition 10.4.5 works for any interval I, so the result holds for unbounded or open intervals I too. If the interval I is compact, we can get more information:

Proposition 10.4.6 *Let $f \in C^0(I)$ be a continuous real-valued function defined on the compact interval $I = [a, b]$. Then, the image $f(I) \subseteq \mathbb{R}$ is a bounded interval.*

Proof Proposition 10.4.5 says that the image $f(I)$ is an interval. Now we have to show that $f(I)$ is bounded.

Suppose for contradiction that it is not bounded. Then, for each $n \in \mathbb{N}$, there exists an element $x_n \in I$ such that $|f(x_n)| > n$. From this, we can construct a sequence (x_n) in I with a corresponding image sequence $(f(x_n))$ in $f(I)$ which satisfies $|f(x_n)| > n$ for all $n \in \mathbb{N}$. The interval $I = [a, b]$ is a bounded set so the sequence (x_n) is a bounded sequence. Applying Bolzano-Weierstrass theorem, we can extract a convergent subsequence of (x_n), say (x_{k_n}). Moreover, since I is a closed interval, by Corollary 6.2.7, the sequence (x_{k_n}) converges to a point $x \in I$. Since the function f is continuous, the corresponding sequence $(f(x_{k_n}))$ must converge to $f(x) \in \mathbb{R}$ and hence is necessarily bounded. However, by construction, we have $|f(x_{k_n})| > k_n \geq n$ for all $n \in \mathbb{N}$ and so the sequence $(f(x_{k_n}))$ is unbounded, giving us the required contradiction. □

Example 10.4.7 We note that we require the function f to be continuous and defined on a compact (closed and bounded) interval for Proposition 10.4.6 to hold. If we do not have either one of these conditions, then the proposition may not hold. Consider the following counterexamples:

1. Suppose that $f : (0, 1] \to \mathbb{R}$ is defined as $f(x) = \frac{1}{x}$. This function is clearly continuous by algebra of limits and the domain is bounded. However, the image is not a bounded interval since $f(x) \to \infty$ as $x \to 0$. This is because the domain is not a closed set.
2. Suppose that $f : [0, \infty) \to \mathbb{R}$ is defined as $f(x) = x$. This function is clearly continuous and the domain is closed. However the image is unbounded as $f(x) \to \infty$ as $x \to \infty$. This is because the domain is an unbounded set.
3. Suppose that $f : [0, 1] \to \mathbb{R}$ is defined as:

$$f(x) = \begin{cases} \frac{1}{x} & \text{if } x \neq 0, \\ 0 & \text{if } x = 0. \end{cases}$$

   This function is defined on a compact interval. However, the image is not a bounded interval since $f(x) \to \infty$ as $x \to 0$. This is because the function f is not continuous over its domain.

Therefore, all three conditions are necessary to ensure that the image is a bounded interval.

10.5 Extreme Value Theorem

From Propositions 10.4.5 and 10.4.6, we know that the image of a compact interval is a bounded interval and so the image set has an infimum and a supremum. Is this image interval closed?

Remarkably, yes and this result is called the extreme value theorem (or the EVT for short) which says if the function is continuous and defined on a compact interval, these infimum and supremum are attained by some point in the domain. In other words, this function has a global minimum and a global maximum.

Theorem 10.5.1 (Extreme Value Theorem, EVT) *Let $f \in C^0(I)$ be a continuous real-valued function defined on a compact interval $I = [a, b]$. Then, there exist points $\xi, \zeta \in I$ such that:*

$$f(\xi) = \min f(I) \quad \textit{and} \quad f(\zeta) = \max f(I).$$

In other words, the function f attains its infimum and supremum somewhere and hence has a global minimum and global maximum.

Proof We prove the existence of minimum only. If $a = b$, then the statement is trivial. Assume now that $a < b$. By Proposition 10.4.6, the image set $f(I)$ is bounded and therefore the infimum $m = \inf f(I)$ of the image set exist.

We now find an $\xi \in I$ whose image $f(\xi)$ is equal to the infimum. By the characterisation of infimum in Proposition 4.1.9, for every $\varepsilon > 0$ there exists a point $y \in f(I)$ such that $m \leq y < m + \varepsilon$. Since $y \in f(I)$, we must have $y = f(x)$ for some $x \in I$, so there exists a point $x \in I$ such that $m \leq f(x) < m + \varepsilon$. In particular, for every $n \in \mathbb{N}$, by setting $\varepsilon = \frac{1}{n}$, there exists an $x_n \in I$ such that $m \leq f(x_n) < m + \frac{1}{n}$. This gives rise to a sequence (x_n) in I.

The sequence (x_n) is contained in I and hence it is a bounded sequence. So, by the Bolzano-Weierstrass theorem, we can extract a convergent subsequence (x_{k_n}) which converges to some $\xi \in I$ by Corollary 6.2.7. Since $x_{k_n} \to \xi$ and the function f is continuous, we must have $f(x_{k_n}) \to f(\xi)$. Furthermore, by construction, we have:

$$m \leq f(x_{k_n}) < m + \frac{1}{k_n} \leq m + \frac{1}{n},$$

and so, by taking the limit as $n \to \infty$ and sandwiching, we have:

$$\begin{aligned} m \leq \lim_{n\to\infty} f(x_{k_n}) \leq m + \lim_{n\to\infty} \frac{1}{n} \quad &\Rightarrow \quad m \leq f(\xi) \leq m \\ &\Rightarrow \quad f(\xi) = m = \inf f(I). \end{aligned}$$

Since the infimum is attained by the point ξ, this point is a global minimum for the function f and hence $f(\xi) = \min f(I)$. □

Remark 10.5.2 A shorter proof of the EVT can be obtained by contradiction as follows:

Suppose for contradiction that $f(x) > m = \inf f(I)$ for all $x \in I$. In other words, none of the points in the domain is mapped to m. Define a function $g : I \to \mathbb{R}$ as $g(x) = \frac{1}{f(x)-m}$. This function is continuous by Theorem 10.2.1 since $f(x) - m \neq 0$ anywhere. Furthermore, by Proposition 10.4.6, the image is bounded and so there exists some $K > 0$ such that $g(x) = \frac{1}{f(x)-m} \leq K$ for all $x \in I$. This is equivalent to $f(x) \geq m + \frac{1}{K} > m$ for all $x \in I$. Hence, m is not the greatest lower bound for the image set $f(I)$, giving us the desired contradiction.

Similar to the IVT, the EVT simply tells us that the infimum and supremum are attained somewhere, but it does not tell us how many times and where exactly. The EVT along with Proposition 10.4.6 directly proves the following result.

Corollary 10.5.3 *Let $f \in C^0(I)$ be a continuous real-valued function on a compact interval $I = [a, b]$. Then, $f(I) = [m, M]$ where $m = \min f(I)$ and $M = \max f(I)$.*
In other words, the image of a compact interval under a continuous function is also a compact interval.

Next, if a continuous function f is strictly monotone (either strictly decreasing or strictly increasing) over its domain $[a, b]$, we can show that the function $f : [a, b] \to [m, M]$ is a bijection.

WLOG, let us assume that the function f is strictly increasing. Clearly, the function $f : [a, b] \to [m, M]$ is surjective by Corollary 10.5.3. To show that it is injective, suppose that $f(x) = f(y)$ for some $x, y \in [a, b]$. Assume for contradiction that $x \neq y$ so that we either have $x < y$ or $x > y$. Since f is strictly increasing, for the former we must have $f(x) < f(y)$ and for the latter we must have $f(y) < f(x)$. Either way, both cases contradict the assumption that $f(x) = f(y)$. Thus, $x = y$ and we conclude that the function f is injective.

Therefore, the bijective function $f : [a, b] \to [m, M]$ above has an inverse function $f^{-1} : [m, M] \to [a, b]$. If the original function f is continuous and strictly monotone, what can we say about the inverse function f^{-1}? We have the following result:

Theorem 10.5.4 *Let $f : [a, b] \to [c, d]$ be a strictly monotone bijective real-valued function with inverse $f^{-1} : [c, d] \to [a, b]$. If $f \in C^0([a, b])$, then f^{-1} is also strictly monotone and continuous.*

Proof WLOG, assume that the function $f : [a, b] \to [c, d]$ is strictly increasing. We show that f^{-1} is strictly increasing and continuous separately.

1. We first show that the inverse function $f^{-1} : [c, d] \to [a, b]$ is also strictly increasing. Pick any $y, z \in [c, d]$ with $y < z$. Then, there exist some $w, x \in [a, b]$ such that $f^{-1}(y) = x$ and $f^{-1}(z) = w$. We want to show that $x = f^{-1}(y) < f^{-1}(z) = w$. Assume for a contradiction that $x \geq w$. Since f is strictly increasing, we must have $f(x) \geq f(w)$, namely $y \geq z$. But this contradicts the choice $y < z$ and thus we must have $x < w$ which is $f^{-1}(y) < f^{-1}(z)$. Thus, the inverse function f^{-1} is strictly increasing as well.
2. To show that the inverse function f^{-1} is continuous on $f([a, b]) = [c, d]$, we first show continuity inside the interval (c, d). Pick a point $y_0 \in (c, d)$. Since

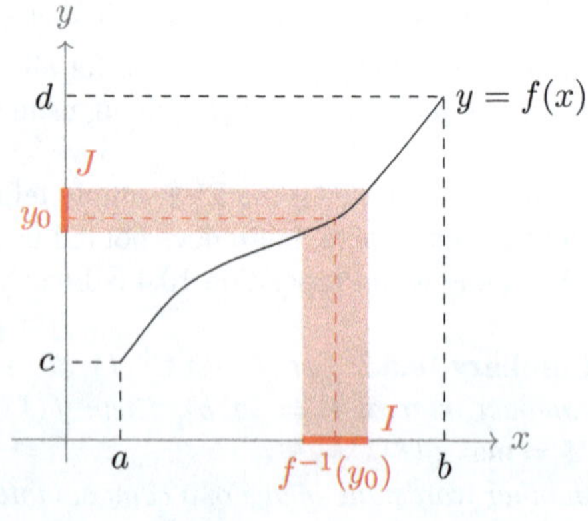

Fig. 10.4 The graph of a strictly increasing function f. The interval highlighted in red on the x-axis is I whereas the interval highlighted in red on the y-axis is J

f^{-1} is also strictly increasing, we have $a = f^{-1}(c) < f^{-1}(y_0) < f^{-1}(d) = b$. Fix $\varepsilon > 0$. We want to show that there exists a $\delta > 0$ such that for all $y \in [c, d]$ with $|y - y_0| < \delta$, we must have $|f^{-1}(y) - f^{-1}(y_0)| < \varepsilon$.

The set $I = (f^{-1}(y_0) - \varepsilon, f^{-1}(y_0) + \varepsilon) \cap [a, b]$ is an interval in $[a, b]$ and, since the function f is continuous, the image $f(I)$ is also an interval in $[c, d]$ by Proposition 10.4.5. We call $f(I) = J$. Refer to Fig. 10.4 for a visualisation.

Note that the interval J contains y_0 since $f^{-1}(y_0) \in I$. Moreover, y_0 cannot be either of the endpoints of J. Indeed, the interval I contains $\min\{f^{-1}(y_0) + \frac{\varepsilon}{2}, b\} > f^{-1}(y_0)$ and hence, by using the strict monotonicity of the function f, there always exists an element larger than y_0 in J and thus $\sup(J) - y_0 > 0$. Similar argument shows that $y_0 - \inf(J) > 0$. To find a suitable $\delta > 0$, we choose the smallest distance from y_0 to either endpoints of the interval J, namely $\delta = \min\{\sup J - y_0, y_0 - \inf J\} > 0$.

To check that this choice of $\delta > 0$ works, pick any $y \in [c, d]$ such that $|y - y_0| < \delta$. This means $y_0 - \delta < y < y_0 + \delta$ and, since $\delta \leq \sup J - y_0$ and $\delta \leq y_0 - \inf J$, we have:

$$y_0 - \delta < y < y_0 + \delta \quad \Rightarrow \quad \inf(J) < y < \sup(J),$$

which implies $y \in J$. Thus:

$$f^{-1}(y) \in f^{-1}(J) = I \subseteq (f^{-1}(y_0) - \varepsilon, f^{-1}(y_0) + \varepsilon),$$

which means $|f^{-1}(y) - f^{-1}(y_0)| < \varepsilon$ and we are done. To show that the function f^{-1} is left-continuous at d and right-continuous at c, similar argument can be employed. □

Remark 10.5.5 Theorem 10.5.3 is also true if we were to replace the intervals $[a, b]$ and $[c, d]$ with any other intervals of $\mathbb{R}$ as long as the function f is a strictly monotone bijection between these intervals.

10.6 Uniform and Lipschitz Continuity

Recall the definition of a continuous real-valued function $f : X \to \mathbb{R}$, namely:

$$\forall x_0 \in X, \forall \varepsilon>0, \exists \delta(\varepsilon, x_0)>0 : \forall x \in X, |x - x_0| < \delta(\varepsilon, x_0) \Rightarrow |f(x) - f(x_0)| < \varepsilon.$$

We note that the choice of $\delta > 0$ for continuous functions, on top of being dependent on the ε, may also depend on the point x_0 where we are looking at. We have seen an example of this in Lemma 10.2.3 where we chose $\delta(\varepsilon, x_0) = \min\left\{1, \frac{\varepsilon}{F(x_0)}\right\} > 0$ to make the analysis work.

Uniform Continuity

However, if this δ can be chosen independent of x_0, we would have:

$$\forall x_0 \in X, \forall \varepsilon > 0, \exists \delta(\varepsilon) > 0 : \forall x \in X, |x - x_0| < \delta(\varepsilon) \Rightarrow |f(x) - f(x_0)| < \varepsilon.$$

This kind of continuity was pointed out by Heine and his mentor Weierstrass as they observed that it contributes to some interesting mathematical phenomena. Due to this, we have a special name for this kind of continuity:

Definition 10.6.1 (Uniformly Continuous Functions, Definition 1) Let $f : X \to \mathbb{R}$ where $X \subseteq \mathbb{R}$. We say that the function f is uniformly continuous on X if for any $x_0 \in X$ and $\varepsilon > 0$, there exists a $\delta(\varepsilon) > 0$ such that for all $x \in X$ with $|x - x_0| < \delta(\varepsilon)$ we must have $|f(x) - f(x_0)| < \varepsilon$.

Example 10.6.2 Consider the function $f : \mathbb{R} \to \mathbb{R}$ defined as $f(x) = x$. For any fixed $\varepsilon > 0$ and point $x_0 \in \mathbb{R}$, the proof for continuity can be obtained by choosing $\delta = \varepsilon$ which is independent of the point x_0. Thus, the function f is uniformly continuous over $\mathbb{R}$.

Due to the independence of the point of interest, we can rewrite Definition 10.6.1 as follows. Using the implication $|x - x_0| < \delta(\varepsilon) \Rightarrow |f(x) - f(x_0)| < \varepsilon$, by choosing any two $x, y \in X$ that satisfies the assumption, we have $|x - y| = |x - x_0 + x_0 - y| \leq |x - x_0| + |y - x_0| < 2\delta$ implies $|f(x) - f(y)| = |f(x) - f(x_0) + f(x_0) - f(y)| \leq |f(x) - f(x_0)| + |f(y) - f(x_0)| < 2\varepsilon$. So we also have the the independence of the whole definition on the point x_0 too. Thus, we have the following equivalent formulation for uniform continuity:

Definition 10.6.3 (Uniformly Continuous Functions, Definition 2) Let $f : X \to \mathbb{R}$ where $X \subseteq \mathbb{R}$. We say that the function f is uniformly continuous on X if for any $\varepsilon > 0$, there exists a $\delta(\varepsilon) > 0$ such that for all $x, y \in X$ with $|x - y| < \delta(\varepsilon)$ we must have $|f(x) - f(y)| < \varepsilon$.

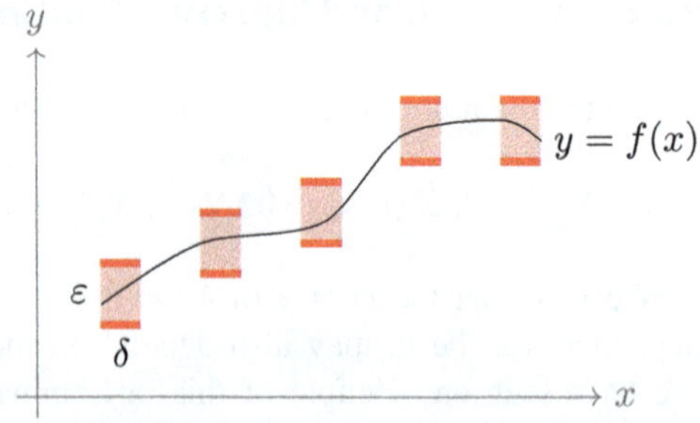

Fig. 10.5 For each $\varepsilon > 0$, we can find a $\delta > 0$ such that the graph of a uniformly continuous function can be threaded all the way by a rectangle of height ε and width δ

Symbolically, this is written with quantifiers as:

$$\forall \varepsilon > 0, \exists \delta(\varepsilon) > 0 : \forall x, y \in X, |x - y| < \delta(\varepsilon) \Rightarrow |f(x) - f(y)| < \varepsilon.$$

Intuitively, this means for any $\varepsilon > 0$, there exists a $\delta > 0$ such that any two elements in X which are of distance at most δ to each other are mapped to elements which are at most ε away from each other in the codomain. The visual representation is as follows: for a fixed $\varepsilon > 0$, we can find a constant $\delta > 0$ such that if we "thread" the graph with a rectangle of width δ and height ε, the graph would never leave the top and the bottom of the rectangle. See Fig. 10.5 for a demonstration of this.

Clearly, all uniformly continuous functions are continuous. However, not all continuous functions are uniformly continuous over its domain as we have seen for the monomials $f(x) = x^n$ for $n \in \mathbb{N} \setminus \{1\}$ on $\mathbb{R}$ in Lemma 10.2.3 because the value $\delta > 0$ chosen in that proof has to depend on the value x_0.

In Example 10.6.2, we saw that the function $f(x) = x$ on $\mathbb{R}$ is uniformly continuous over $\mathbb{R}$. Here is another example of a uniformly continuous function.

Example 10.6.4 Let us show that the trigonometric function $\sin(x)$ defined on the whole of $\mathbb{R}$ is continuous. We first show the estimate $|\sin(x)| \leq |x|$ for all $x \in \mathbb{R}$.

1. This is clearly true for $x = 0$ since $\sin(0) = 0$.
2. From Exercise 8.12(a), we have seen that $0 \leq \sin(x) \leq x$ for any $x \in (0, \frac{\pi}{2})$.
3. For $x \in (-\frac{\pi}{2}, 0)$, since $-x \in (0, \frac{\pi}{2})$, we then have $0 \leq \sin(-x) \leq -x$. This implies $0 \leq |\sin(-x)| \leq |-x|$ and thus $0 \leq |\sin(x)| \leq |x|$ since $\sin(-x) = -\sin(x)$.
4. For other values of $x \in (-\infty, -\frac{\pi}{2}] \cup [\frac{\pi}{2}, \infty)$, we note that $|\sin(x)| \leq 1$ but $|x| \geq \frac{\pi}{2} > \frac{2}{2} = 1$. Hence, the estimate $0 \leq |\sin(x)| \leq |x|$ also holds here.

Now we are ready to show uniform continuity of $\sin(x)$.

Rough work: Fix $\varepsilon > 0$. We want to find a $\delta > 0$ such that for any $x, y \in \mathbb{R}$ such that $|x - y| < \delta$, we get $|\sin(x) - \sin(y)| < \varepsilon$. This can be achieved via the sum-to-product trigonometric formula, namely: $\sin(a) - \sin(b) = 2\cos\left(\frac{a+b}{2}\right)\sin\left(\frac{a-b}{2}\right)$. Putting x and y in this formula, we get:

$$\begin{aligned}|\sin(x) - \sin(y)| &= \left|2\cos\left(\frac{x+y}{2}\right)\sin\left(\frac{x-y}{2}\right)\right| \\ &= 2\left|\cos\left(\frac{x+y}{2}\right)\right|\left|\sin\left(\frac{x-y}{2}\right)\right| \\ &\leq 2\left|\sin\left(\frac{x-y}{2}\right)\right| \leq 2\left|\frac{x-y}{2}\right| = |x-y|,\end{aligned}$$

by the estimate that we have obtained earlier. Hence, a suitable choice for δ that would yield $|\sin(x) - \sin(y)| < \varepsilon$ would be $\delta = \varepsilon$.

Fix $\varepsilon > 0$. Set $\delta = \varepsilon > 0$. Whenever $|x - y| < \delta = \varepsilon$, we have:

$$\begin{aligned}|\sin(x) - \sin(y)| &= \left|2\cos\left(\frac{x+y}{2}\right)\sin\left(\frac{x-y}{2}\right)\right| \\ &\leq 2\left|\sin\left(\frac{x-y}{2}\right)\right| \leq 2\left|\frac{x-y}{2}\right| = |x-y| < \varepsilon \quad (\because |\sin(h)| \leq h).\end{aligned}$$

Thus, we conclude that the sine function is uniformly continuous over $\mathbb{R}$.

Uniform continuity can also be viewed in terms of sequences, similar to the definition of limits and continuity. We have the following sequential characterisation of uniform continuity:

Proposition 10.6.5 *Let $f : X \to \mathbb{R}$ be a function on $X \subseteq \mathbb{R}$. The function f is uniformly continuous on X if and only if for any sequences (x_n) and (y_n) in X such that $x_n - y_n \to 0$, we have $f(x_n) - f(y_n) \to 0$.*

Proof We prove the implications one by one.

($\Rightarrow$): Fix $\varepsilon > 0$. By definition of uniform continuity, there is a constant $\delta > 0$ such that for any $x, y \in X$ with $|x - y| < \delta$ we have $|f(x) - f(y)| < \varepsilon$. Pick any two sequences (x_n) and (y_n) in X such that $x_n - y_n \to 0$. Since $x_n - y_n \to 0$, there exists an $N \in \mathbb{N}$ such that $|x_n - y_n| < \delta$ for all $n \geq N$. By uniform continuity, this means for all $n \geq N$ we also have $|f(x_n) - f(y_n)| < \varepsilon$. Hence, we conclude that $f(x_n) - f(y_n) \to 0$.

($\Leftarrow$): Suppose for contradiction that f is not uniformly continuous over X. By negation, there exists an $\varepsilon > 0$ such that for all $\delta > 0$, there are $x, y \in X$ such that $|x - y| < \delta$ and $|f(x) - f(y)| \geq \varepsilon$.
Now we construct the sequences (x_n) and (y_n) in X as follows: for every $n \in \mathbb{N}$, pick $x_n, y_n \in X$ such that $|x_n - y_n| < \frac{1}{n}$ with $|f(x_n) - f(y_n)| \geq \varepsilon$.

By construction, this means $x_n - y_n \to 0$ and, by assumption, we must have $f(x_n) - f(y_n) \to 0$. However, by the fact that limits preserve weak inequalities and since $|f(x_n) - f(y_n)| \geq \varepsilon$ for all $n \in \mathbb{N}$, we have:

$$0 < \varepsilon \leq \lim_{n\to\infty} |f(x_n) - f(y_n)| = |\lim_{n\to\infty} (f(x_n) - f(y_n))| = 0,$$

which is a contradiction. Thus, f is necessarily uniformly continuous. □

Since we have three equivalent definitions of uniform continuity, we can use either one of them as we wish. Similar as before, in certain cases, one definition may be easier to use than the others.

Example 10.6.6 Let $f : (0, \infty) \to \mathbb{R}$ be defined as $f(x) = \frac{1}{x}$. This function is clearly continuous by Theorem 10.2.1. However, it is not uniformly continuous over $(0, \infty)$. We can show this in various ways:

1. We show this using the characterisation of uniform continuity in Proposition 10.6.5. We need to pick two sequences (x_n) and (y_n) in $(0, \infty)$ so that they get closer to each other but their images do not get closer to each other. Pick $x_n = \frac{1}{n}$ and $y_n = \frac{1}{n+1}$. Then, we have $|x_n - y_n| = |\frac{1}{n} - \frac{1}{n+1}| = \frac{1}{n(n+1)} \to 0$ but $|f(x_n) - f(y_n)| = |\frac{1}{\frac{1}{n}} - \frac{1}{\frac{1}{n+1}}| = |n - n - 1| = 1$. The latter implies that $f(x_n) - f(y_n)$ does not converge to 0. Hence, f cannot be uniformly continuous over $(0, \infty)$.
2. We can also show this using Definition 10.6.3 and contradiction. Suppose for contradiction that $f(x) = \frac{1}{x}$ is uniformly continuous on $(0, \infty)$. For a fixed $\varepsilon = 1$, there is a $\delta > 0$ such that for any $x, y \in (0, \infty)$ satisfying $|x - y| < \delta$, we must have $|f(x) - f(y)| < 1$. This means if we fix $x = \delta \in (0, \infty)$, for all the points $y \in (0, 2\delta)$ we would have $|x - y| < \delta$ and thus $|f(x) - f(y)| < 1$. However, this means:

$$\left|\frac{1}{\delta} - \frac{1}{y}\right| < 1 \quad \Rightarrow \quad -1 < \frac{1}{\delta} - \frac{1}{y} < 1 \quad \Rightarrow \quad \frac{1}{y} < \frac{1}{\delta} + 1,$$

for all $y \in (0, 2\delta)$. This is a contradiction since $\frac{1}{y}$ cannot be bounded by a constant here since it gets arbitrarily large as $y \to 0$.
The geometric interpretation of this argument is as follows: as we reach the lower end of the interval $(0, \infty)$ with the uniform rectangle of dimensions $\delta \times 1$, by the interpretation in Fig. 10.5, this traps the lower end of the function to be within this rectangle and hence the graph cannot escape to $\pm\infty$. This would then contradict the fact that $\lim_{x\downarrow 0} \frac{1}{x} = \infty$.

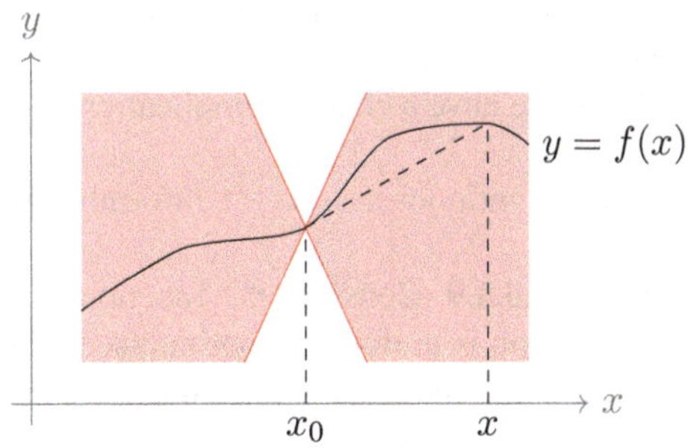

Fig. 10.6 The double cone of slope K at $(x_0, f(x_0))$ is denoted by the red lines. All the other points on the graph lie outside the double cone in the red region. The slope of the secant line connecting the points $(x_0, f(x_0))$ and any $(x, f(x))$ is between $-K$ and K

Lipschitz Continuity

Another special type of continuity for functions is Lipschitz continuity. Named after Rudolf Lipschitz (1832–1903), this continuity is defined as:

Definition 10.6.7 (Lipschitz Continuous Functions) Let $f : X \to \mathbb{R}$ where $X \subseteq \mathbb{R}$. We say that the function f is Lipschitz continuous on X if there exists a $K > 0$ such that for all $x, y \in X$, we have $|f(x) - f(y)| \leq K|x - y|$.

The positive constant K is called a Lipschiz constant for f. Visually, we can think of this condition as finding a constant $K > 0$ such that if we fix a double cone of slope K at any point $(x_0, f(x_0))$ on the graph of f, all the other points on the graph are outside this cone. See Fig. 10.6 for the visualisation.

This fact is evident because for any fixed $x_0 \in X$, the Lipschitz condition is equivalent to:

$$\left|\frac{f(x_0) - f(x)}{x_0 - x}\right| \leq K \quad \Leftrightarrow \quad -K \leq \frac{f(x_0) - f(x)}{x_0 - x} \leq K,$$

for all $x \neq x_0$. This means the slopes of the secant lines connecting $(x_0, f(x_0))$ and any other point $(x, f(x))$ on the graph are all between $-K$ and K.

Example 10.6.8 We have seen an example of a Lipschitz continuous function on $\mathbb{R}$, namely the sine function. This is true because we have proven that $|\sin(x) - \sin(y)| \leq |x - y|$ for all $x, y \in \mathbb{R}$ in Example 10.6.4, so the Lipschitz constant here is $K = 1$.

Relationship Between Different Types of Continuities

We note that all Lipschitz continuous functions are also uniformly continuous. Clearly, this is true because we can just pick $\delta = \frac{\varepsilon}{K} > 0$ independent of x_0 to establish the continuity of Lipschitz continuous functions from first principles. However, not all uniformly continuous functions are Lipschitz continuous as the dependence of δ on ε in uniform continuity may not be linear as in Lipschitz continuity.

Thus, we have a chain of implications for functions $f : X \to \mathbb{R}$:

$$\text{Lipschitz continuous} \quad \Rightarrow \quad \text{Uniformly continuous} \quad \Rightarrow \quad \text{Continuous functions.}$$

Example 10.6.9 Consider the function $f : [0, 1] \to \mathbb{R}$ defined by $f(x) = \sqrt{x}$. This function is uniformly continuous, and hence continuous, in the interval $[0, 1]$, which is a task the readers are required to show in Exercise 10.13. However, it is not Lipschitz continuous.

Assume for contradiction that it is, namely there is a $K > 0$ such that for all $x, y \in [0, 1]$, we have $|f(x) - f(y)| \leq K|x - y|$. In particular, if we set $y = 0$, we have $\sqrt{x} \leq Kx$ for any $x \in (0, 1]$. This is equivalent to $\frac{1}{\sqrt{x}} \leq K$ for all $x \in (0, 1]$, which says that the quantity $\frac{1}{\sqrt{x}}$ is bounded for all $x \in (0, 1]$. However, this is obviously a contradiction since this quantity gets arbitrarily large as $x \to 0$.

The problem here lies at the point $y = 0$. Indeed, for any point $x > 0$, the slope of the secant line joining the points $(0, f(0))$ and $(x, f(x))$ is given by $\frac{f(x)-f(0)}{x-0} = \frac{1}{\sqrt{x}}$ which gets arbitrarily large as $x \to 0$. This implies that there cannot be a uniform bounding cone as in Fig. 10.6 at $y = 0$.

However, if we restrict our function f to a subset $[a, 1]$ for some $0 < a < 1$, the function immediately becomes Lipschitz continuous over this new restricted domain with $K = \frac{1}{2\sqrt{a}}$. To show this, for any $x, y \in [a, 1]$ we have:

$$|\sqrt{x} - \sqrt{y}||\sqrt{x} + \sqrt{y}| = |x - y| \quad \Rightarrow \quad |\sqrt{x} - \sqrt{y}| = \frac{|x - y|}{\sqrt{x} + \sqrt{y}} \leq \frac{1}{2\sqrt{a}}|x - y|.$$

Note that the converses of the implications for the three types of continuities that we saw above may or may not be true and this depends on various factors. One of the important factors is the domain of the function, as we have seen in Example 10.6.9 in which restricting the domain of a uniformly continuous function to a smaller set could upgrade the type of continuity of the function to Lipschitz continuity.

A general useful result is: for functions defined on compact interval $I = [a, b]$, any continuous function is also uniformly continuous.

Theorem 10.6.10 (Heine-Cantor Theorem) *Let $f \in C^0(I)$ be a continuous real-valued function on a compact interval $I = [a, b]$. Then, the function f is also uniformly continuous on I.*

Proof Suppose for a contradiction that the function $f : I \to \mathbb{R}$ is not uniformly continuous. So, there exists an $\varepsilon > 0$ such that for any $\delta > 0$ there are points $x, y \in [a, b]$ with $|x - y| < \delta$ but $|f(x) - f(y)| \geq \varepsilon$. Since this is true for all $\delta > 0$, for each $n \in \mathbb{N}$ we can set $\delta = \frac{1}{n}$ and find $x_n, y_n \in [a, b]$ with $|x_n - y_n| < \frac{1}{n}$ but $|f(x_n) - f(y_n)| \geq \varepsilon$.

Thus, we have two sequences of points (x_n) and (y_n) in $[a, b]$. Since the sequence (x_n) is bounded, by Bolzano-Weierstrass theorem, it has a convergent subsequence (x_{k_n}) which converges to some $z \in I$. Furthermore, the corresponding subsequence (y_{k_n}) of (y_n) also converges to the same point z since:

$$0 \leq |y_{k_n} - z| = |y_{k_n} - x_{k_n} + x_{k_n} - z| \leq |y_{k_n} - x_{k_n}| + |x_{k_n} - z| < \frac{1}{k_n} + |x_{k_n} - z| \leq \frac{1}{n} + |x_{k_n} - z| \to 0,$$

as $n \to \infty$ and hence, by sandwich lemma, we have $|y_{k_n} - z| \to 0$. This then implies $y_{k_n} \to z$.

Furthermore, from the assumption and choice of the sequences (x_n) and (y_n), we have $0 < \varepsilon \leq |f(x_{k_n}) - f(y_{k_n})|$ for all $n \in \mathbb{N}$. Thus, when we take the limit as $n \to \infty$, by using the fact that limits preserve weak inequalities and the continuity of f, we have:

$$0 < \varepsilon \leq \lim_{n\to\infty} |f(x_{k_n}) - f(y_{k_n})| = |\lim_{n\to\infty} f(x_{k_n}) - \lim_{n\to\infty} f(y_{k_n})| = |f(z) - f(z)| = 0,$$

which is a contradiction. Therefore, we conclude that f must be uniformly continuous on I. □

Example 10.6.11 The conditions that the domain is closed and bounded (hence compact) in Theorem 10.6.10 are both necessary. We can find counterexamples for domains which are neither closed nor bounded, which are given below:

1. Suppose that $f : [0, \infty) \to \mathbb{R}$ is defined as $f(x) = x^2$. We have seen that this function is continuous everywhere. However, it is not uniformly continuous. The observation comes from the proof of Lemma 10.2.3. For a fixed $\varepsilon > 0$ and $x_0 \in [0, \infty)$, we have set $|x_0 - x| < \delta(\varepsilon, x_0) = \min\{1, \frac{\varepsilon}{2|x_0|+1}\}$ to achieve $|f(x_0) - f(x)| < \varepsilon$. From this, we can see that regardless of the choice of ε, the larger x_0 gets, the smaller the required δ is. Since the domain is unbounded, as $x_0 \to \infty$, we would require δ to be arbitrarily small. As a result, we may not be able to find a uniform constant $\delta > 0$ that works for all points $x_0 \in [0, \infty)$ at the same time. Let us now prove this rigorously.
   Assume for contradiction that f is uniformly continuous over $[0, \infty)$. Then, for $\varepsilon = 1$, there exists a $\delta > 0$ such that if $|x - y| < \delta$ then $|x^2 - y^2| < 1$. However, we can pick $x = \frac{1}{\delta} + \frac{\delta}{2}$ and $y = \frac{1}{\delta} + \delta$ so that $|x - y| = \frac{\delta}{2} < \delta$ but $|x^2 - y^2| = |x - y||x + y| = \frac{\delta}{2}(\frac{2}{\delta} + \frac{3\delta}{2}) > 1$, a contradiction.
   On the other hand, if the domain is bounded, say we restrict the function $f(x) = x^2$ defined on $[0, \infty)$ to $f|_{[0,a]} : [0, a] \to \mathbb{R}$ for some finite $a > 0$, we can find a new $\tilde{\delta}(\varepsilon)$ independent of x_0 that would work for all $x_0 \in [0, a]$. To do this, we take the infimum of the $\delta(\varepsilon, x_0)$ above over x_0. Namely, we choose $\tilde{\delta}(\varepsilon) = \inf_{x_0 \in [0,a]} \left(\min\{1, \frac{\varepsilon}{2|x_0|+1}\}\right) = \min\{1, \frac{\varepsilon}{2a+1}\} > 0$. This infimum exists

and is non-zero. Furthermore, this new $\tilde{\delta}$ is independent of x_0 and so we can conclude that $f|_{[0,a]}$ is uniformly continuous over the compact interval $[0, a]$, just as Theorem 10.6.10 claims.

2. We have seen in Example 10.6.6 that the function $f : (0, \infty) \to \mathbb{R}$ on an interval that is not closed defined as $f(x) = \frac{1}{x}$ is not uniformly continuous.
3. Consider the function $f : (0, \infty) \to \mathbb{R}$ defined as $f(x) = \sin(\frac{1}{x})$. This function is also not uniformly continuous. Clearly, it is continuous since it is a composition of two continuous functions. However, we can see from Fig. 9.1 that as $x \to 0$, the function oscillates more rapidly with a constant amplitude. Therefore, if we were to think geometrically via the $\delta \times \varepsilon$ rectangle visualisation as in Fig. 10.5, the rapid oscillation may cause a problem here for uniform continuity.
Using the same argument as above, suppose for contradiction that the function is uniformly continuous. Then, for $\varepsilon = \frac{1}{2}$ there exists a $\delta > 0$ such that for any $x, y \in (0, \infty)$ with $|x - y| < \delta$, we would have $|f(x) - f(y)| < \frac{1}{2}$. By the Archimedean property, there exists an $n \in \mathbb{N}$ such that $\frac{1}{n} < \delta$. Thus, we can pick $x = \frac{1}{n\pi}$ and $y = \frac{1}{n\pi + \frac{\pi}{2}}$ so that $|x - y| = \frac{1}{2n(n\pi + \frac{\pi}{2})} < \frac{1}{n} < \delta$ and hence $|f(x) - f(y)| < \frac{1}{2}$. However:
$$|f(x)-f(y)| = \left| f\left(\frac{1}{n\pi}\right) - f\left(\frac{1}{n\pi + \frac{\pi}{2}}\right) \right| = |\sin(n\pi) - \sin(n\pi + \tfrac{\pi}{2})| = 1 > \frac{1}{2},$$
which gives us a contradiction.

Remark 10.6.12 As we have seen in Example 10.6.6 with the function $f(x) = \frac{1}{x}$ on $(0, \infty)$, the problem with continuous functions on an open interval is that it might blow up as it gets closer to the boundary and this prevents uniform continuity. Moreover, uniform continuity might also be prevented by continuous functions which oscillates a lot on $(0, \infty)$ as it approaches the boundary, for example $f(x) = \sin(\frac{1}{x})$ in Example 10.6.11(3). In both of these cases, the limits of the functions as we approach the boundaries do not exist.

What happens if the limits of the function at the open boundary do exist? For a function $f : (a, b) \to \mathbb{R}$, if both $\lim_{x \downarrow a} f(x)$ and $\lim_{x \uparrow b} f(x)$ exist, we can extend this function to an extended function on $[a, b]$ as:

$$\tilde{f} : [a, b] \to \mathbb{R},$$
$$x \mapsto \begin{cases} \lim_{x \downarrow a} f(x) & \text{if } x = a, \\ f(x) & \text{if } x \in (a, b), \\ \lim_{x \uparrow b} f(x) & \text{if } x = b. \end{cases}$$

This is a continuous function on $[a, b]$ by Proposition 10.2.6 and hence must be uniformly continuous as well by Theorem 10.6.10. Thus, by restricting x and y in the definition of uniform continuity to (a, b) and using the same $\delta(\varepsilon)$ as $\tilde{f}$, we have proven that the original function f is uniformly continuous on (a, b).

The examples that we have seen so far gave us a general guideline on how to show uniform continuity. We follow the following steps:

1. Show that the function $f : X \to \mathbb{R}$ is continuous at every point $x_0 \in X$.
   (a) Fix $x_0 \in X$. Then for every $\varepsilon > 0$, we need to find a $\delta(\varepsilon, x_0) > 0$ such that for any $x \in X$ with $|x - x_0| < \delta$, we have $|f(x) - f(x_0)| < \varepsilon$
   (b) This is done by first fixing $\varepsilon > 0$ and by algebraic manipulations, find a suitable δ by trying to obtain a bound of the form $|f(x) - f(x_0)| \leq K|x - x_0|^p < K\delta^p$ for some constant $K(x_0) > 0$ and $p > 0$. One may need to put some additional assumptions on $|x - x_0|$ to get to this form. Note that the constant $K(x_0)$ may also depend on x_0.
   (c) Since we want $|f(x) - f(x_0)| < \varepsilon$, it is enough to set $K\delta^p = \varepsilon$ and from here we can manipulate the inequality to get δ as a function of ε and x_0. It is very important to remember that we also need to take into account of all the assumptions we made on $|x - x_0|$ to get $|f(x) - f(x_0)| \leq K|x - x_0|^p$. One then construct the function $\delta(\varepsilon, x_0)$.
2. We now vary the x_0 and try to find a uniform $\delta(\varepsilon)$ that would work for all $x_0 \in X$. This can be done by taking the smallest such $\delta(\varepsilon, x_0)$ over all x_0, namely find $\delta(\varepsilon) = \inf_{x_0 \in X} \delta(\varepsilon, x_0)$. This $\delta(\varepsilon)$ must be chosen to be strictly positive for all $\varepsilon > 0$.

A word of caution here is that the final step in the guide above may not work all the time. It worked in Example 10.6.11(1) but we might not be so lucky all the time.

Therefore, one has to be a bit creative in trying to figure out the final step. We might have to use some non-constructive methods using the established propositions that we have seen regarding uniformly continuous functions.

Example 10.6.13 An example that we can look at is the cube root function given by $f : \mathbb{R} \to \mathbb{R}$ as $f(x) = \sqrt[3]{x}$. We have shown that this function is continuous everywhere in Example 10.1.8. In that example, for a fixed $\varepsilon > 0$, we have chosen the $\delta > 0$ to be $\delta = \varepsilon^3$ at $x_0 = 0$ and $\delta = \min\left\{\frac{|x_0|}{2}, \varepsilon\sqrt[3]{\frac{|x_0|^2}{2}}\right\}$ at $x_0 \neq 0$. If we were to follow the guide above, we would have to take the infimum of δ over all $x_0 \in \mathbb{R}$, which would be 0 here!

To get around this issue, we approach this problem non-constructively. We split the domain into three overlapping regions, namely $(-\infty, -2] \cup [-3, 3] \cup [2, \infty)$. Notice that these regions are chosen to overlap on intervals of length 1. This is necessary because when we have $x, x_0 \in \mathbb{R}$ such that $|x - x_0| < \delta$, both of them could lie in different intervals and this would make the analysis difficult.

Thus to make sure that both of them always lie in a common interval, we overlap the intervals by length 1 and we set $\delta \leq 1$ as one of the conditions later. Now we analyse the continuity of the function f at different x_0 in these regions separately. Fix $\varepsilon > 0$.

1. Suppose $x_0 \in (-\infty, -2] \cup [2, \infty)$. Clearly, there exists a uniform $\delta_1 > 0$ in the regions $(-\infty, -2]$ and $[2, \infty)$ that ensure uniform continuity over these regions: using the analysis in Example 10.1.8, we can take the infimum of $\delta = \min\left\{\frac{|x_0|}{2}, \varepsilon\sqrt[3]{\frac{|x_0|^2}{2}}\right\}$ over $x_0 \in (-\infty, -2] \cup [2, \infty)$ which gives us $\delta_1 = \min\{1, \varepsilon\sqrt[3]{2}\} > 0$. Thus, if $x \in (-\infty, -2] \cup [2, \infty)$ is such that $|x - x_0| < \delta_1$, we have $|f(x) - f(x_0)| < \varepsilon$.
2. Now suppose $x_0 \in [-3, 3]$. Since f is continuous in the region $[-3, 3]$ and this domain is compact, f is uniformly continuous here by Theorem 10.6.10. Thus, we can find a $\delta_2 > 0$ such that for any $x \in [-3, 3]$ satisfying $|x - x_0| < \delta_2$, we have $|f(x) - f(x_0)| < \varepsilon$.

Combining all the domains back together, we can choose the constant $\delta = \min\{\delta_1, \delta_2, 1\} > 0$ which is independent of x_0. At any $x_0 \in \mathbb{R}$, for any $x \in \mathbb{R}$ such that $|x - x_0| < \delta \leq 1$, both x and x_0 are either in $(-\infty, -2] \cup [2, \infty)$ or $[-3, 3]$. For the former, since $|x - x_0| < \delta \leq \delta_1$, we have $|f(x) - f(x_0)| < \varepsilon$. For the latter, since $|x - x_0| < \delta \leq \delta_2$, we also have $|f(x) - f(x_0)| < \varepsilon$. Thus, we can conclude that the cube root function is uniformly continuous over the whole of $\mathbb{R}$.

As we have seen in Example 10.6.6(2), by geometric argument, if a continuous function blows up to ∞ at a finite point in $\mathbb{R}$, we would lose uniform continuity. Furthermore, we have also seen that if the function blows up to $\pm\infty$ as $x \to \infty$, uniform continuity may not be achieved if the function grows very quickly as we have seen for the function $f(x) = x^2$ in Example 10.6.11(1).

However, we may still have uniform continuity if a continuous function remains bounded at infinity. This is an extension of Theorem 10.6.10 onto an unbounded domain and the discussion about limits at the endpoints of the domain in Remark 10.6.12.

Proposition 10.6.14 *Let $f \in C^0([0, \infty))$ be a continuous real valued function. If the function f is continuous over $[0, \infty)$ and $\lim_{x\to\infty} f(x)$ exists, then f is uniformly continuous on $[0, \infty)$.*

Proof Suppose that $\lim_{x\to\infty} f(x) = L \in \mathbb{R}$. Fix $\varepsilon > 0$. Then, there exists a $K > 0$ such that $|f(x) - L| < \frac{\varepsilon}{2}$ for all $x > K$. Note that the interval $[0, K + 1]$ is compact. By Theorem 10.6.10, f is uniformly continuous here. So there exists a $\delta_1 > 0$ such that whenever $x, y \in [0, K + 1]$ with $|x - y| < \delta_1$ we have $|f(x) - f(y)| < \varepsilon$. Furthermore, for all $x, y > K$ with $|x - y| < \delta_1$ we also have $|f(x) - f(y)| = |f(x) - L + L - f(y)| \leq |f(x) - L| + |f(y) - L| < \frac{\varepsilon}{2} + \frac{\varepsilon}{2} = \varepsilon$.

Now we choose $\delta = \min\{\delta_1, 1\} > 0$. Thus, any $x, y \in [0, \infty)$ such that $|x - y| < \delta \leq 1$ can be both contained in the same interval $[0, K + 1]$ or (K, ∞).

Therefore, within either of the intervals, since $|x - y| < \delta \leq \delta_1$, we then have $|f(x) - f(y)| < \varepsilon$. □

Remark 10.6.15 In Example 10.6.13 and the proof of Proposition 10.6.14, we used an overlapping trick in which we overlap the intervals over some fixed length and take into account of overlap in the end choice of δ. This is to ensure that any two points $x, y \in X$ we chose with $|x - y| < \delta$ can always be contained in the same constituent interval, which makes the analysis easier since we know the behaviour of the function in each of these constituent intervals. This is a very useful trick to know.

Example 10.6.16 Consider the function $f : [0, \infty) \to \mathbb{R}$ defined as $f(x) = \frac{2x^2}{x^2+1}$. Clearly this function is continuous everywhere on its domain as the numerator and denominator are both polynomials with the numerator never being 0. In fact, we can instantly deduce that this function is uniformly continuous everywhere using Proposition 10.6.14 since $\lim_{x\to\infty} f(x) = 2$.

The converse of Proposition 10.6.14 does not hold necessarily. As we have noted earlier, the linear function $f(x) = x$ and the sine function $f(x) = \sin(x)$ are both uniformly continuous on $\mathbb{R}$, but these functions diverge as $x \to \infty$. The following result tells us that any uniformly continuous function must be finite everywhere and cannot grow faster than the linear function at $\pm\infty$.

Proposition 10.6.17 *Let $f : \mathbb{R} \to \mathbb{R}$ be a uniformly continuous function. Then, there are constants $a, b \geq 0$ such that $|f(x)| < a|x| + b$.*

Proof Fix $\varepsilon = 1$. Then, there exists a $\delta > 0$ such that whenever $|x - y| < \delta$, we have $|f(x) - f(y)| < 1$. Fix $x \in \mathbb{R}$. There must exist a $k \in \mathbb{N}$ such that $|x| < k\delta$, so the set $\{k \in \mathbb{N} : |x| < k\delta\}$ is non-empty. By the well-ordering principle for subsets of $\mathbb{N}$, let m (which depends on x) be the minimum of this set so that $(m-1)\delta \leq |x|$ and thus $m \leq \frac{|x|}{\delta} + 1$.

On the other hand, by triangle inequality, we have:

$$\begin{aligned} |f(0) - f(x)| &= \left| f(0) - \sum_{j=1}^{m-1} f\left(\frac{jx}{m}\right) + \sum_{j=1}^{m-1} f\left(\frac{jx}{m}\right) - f(x) \right| \\ &\leq \sum_{j=1}^{m} \left| f\left(\frac{(j-1)x}{m}\right) - f\left(\frac{jx}{m}\right) \right| < m, \end{aligned} \tag{10.2}$$

since for all $j = 1, \ldots, m$ we have $\left|\frac{(j-1)x}{m} - \frac{jx}{m}\right| = \frac{|x|}{m} < \delta$ and so $\left| f\left(\frac{(j-1)x}{m}\right) - f\left(\frac{jx}{m}\right) \right| < 1$.

Hence, by using the estimate (10.2), we have $|f(x)| = |f(x) - f(0) + f(0)| \leq |f(x) - f(0)| + |f(0)| < m + f(0) \leq \frac{|x|}{\delta} + 1 + |f(0)|$. Since $x \in \mathbb{R}$ was arbitrarily chosen, we obtain the desired result with constants $a = \frac{1}{\delta}$ and $b = 1 + |f(0)|$. □

A consequence of this is that any uniformly continuous function on a bounded domain must be bounded since the above result says it must bounded from above and below by linear functions which cannot blow up over a bounded domain. We state this:

Corollary 10.6.18 *Let $f : X \to \mathbb{R}$ be a uniformly continuous function. If $X \subseteq \mathbb{R}$ is a bounded set, then there must exist an $M > 0$ such that $|f(x)| \leq M$ on X.*

Corollary 10.6.18 gives us an even quicker way to prove that the function $f(x) = \frac{1}{x}$ on $(0, 1]$ is not uniformly continuous over $(0, 1]$. However, we have to remember that both Proposition 10.6.17 and Corollary 10.6.18 are one-way implications. The converses may not be true! Let us look at the following example to see this:

Example 10.6.19 Consider the function $f : \mathbb{R} \to \mathbb{R}$ defined as $f(x) = \sin(x^2)$. This function satisfies $|f(x)| \leq 1$ and hence can be bounded by a linear function of the form $a|x| + b$ with $a = 0$ and $b = 1$. It is also continuous everywhere since it is a composition of two continuous functions by Proposition 10.2.2.

However, it is not uniformly continuous on $\mathbb{R}$. This can be seen roughly as follows: when $x \to \infty$, the function $\sin(x^2)$ oscillates more rapidly, so if we have a uniform rectangle of dimensions $\delta \times \frac{1}{2}$ threading the graph of the function, eventually the graph would leave the rectangle through the top and bottom.

Now we prove this rigorously. Suppose for contradiction that the function f is uniformly continuous on $\mathbb{R}$. Then, for $\varepsilon = \frac{1}{2}$, there exists a uniform $\delta > 0$ such that whenever $|x - y| < \delta$, we would have $|f(x) - f(y)| < \frac{1}{2}$. Pick $x = \sqrt{n\pi}$ and $y = \sqrt{n\pi + \frac{\pi}{2}}$ for some $n \in \mathbb{N}$ that we will determine later. Then:

$$|x - y| = \sqrt{n\pi + \frac{\pi}{2}} - \sqrt{n\pi} = \frac{\frac{\pi}{2}}{\sqrt{n\pi + \frac{\pi}{2}} + \sqrt{n\pi}} < \frac{\frac{\pi}{2}}{2\sqrt{n\pi}} = \frac{\sqrt{\pi}}{4\sqrt{n}} < \frac{1}{\sqrt{n}}.$$

So, if we pick $n = \lceil \frac{1}{\delta^2} \rceil$, we would get $|x - y| < \delta$. This implies $|f(x) - f(y)| < \frac{1}{2}$. However, we can calculate:

$$|f(x) - f(y)| = |\sin(x^2) - \sin(y^2)| = \left|\sin(n\pi) - \sin\left(n\pi + \frac{\pi}{2}\right)\right| = 1 > \frac{1}{2},$$

which is a contradiction. Hence, this function cannot be uniformly continuous over $\mathbb{R}$.

Exercises

10.1 (∗) Let $f : \mathbb{R} \to \mathbb{R}$ be defined as

$$f(x) = \begin{cases} 8x - 3 \text{ if } x < 0, \\ 4x - 2 \text{ if } x \geq 0. \end{cases}$$

Find $\lim_{x \downarrow 0} f(x)$ and $\lim_{x \uparrow 0} f(x)$.
Hence, show that this function is discontinuous at $x = 0$ but continuous everywhere else.

10.2 Let $f : \mathbb{R} \to \mathbb{R}$ be the floor function $f(x) = \lfloor x \rfloor$. Determine the subsets of $\mathbb{R}$ for which the function is continuous, only left continuous, and only right continuous.

10.3 (∗) Prove Lemma 10.1.6 in which we claimed the two definitions of continuity, namely Definitions 10.1.1 and 10.1.4, are equivalent.

10.4 (∗) If $f, g : \mathbb{R} \to \mathbb{R}$ are continuous functions, prove that the functions $\max(f, g), \min(f, g) : \mathbb{R} \to \mathbb{R}$ are also continuous.

10.5 (∗) Using the ε-δ definition for continuity, prove Proposition 10.2.6.

10.6 (∗) The topologists' sine function $f : \mathbb{R} \to \mathbb{R}$ is defined as

$$f(x) = \begin{cases} \sin(\frac{1}{x}) & \text{if } x \neq 0, \\ 0 & \text{if } x = 0. \end{cases}$$

A graph for this function is given in Fig. 10.7. Show that this function is continuous at any $x \neq 0$ but discontinuous at $x = 0$.

10.7 Let $f : \mathbb{R} \to \mathbb{R}$ be defined as

$$f(x) = \begin{cases} x & \text{if } x \in \mathbb{Q}, \\ x^3 & \text{if } x \in \bar{\mathbb{Q}}. \end{cases}$$

(a) Prove that f is continuous at $x = -1, 0, 1$.
(b) Show that f cannot be continuous everywhere else.

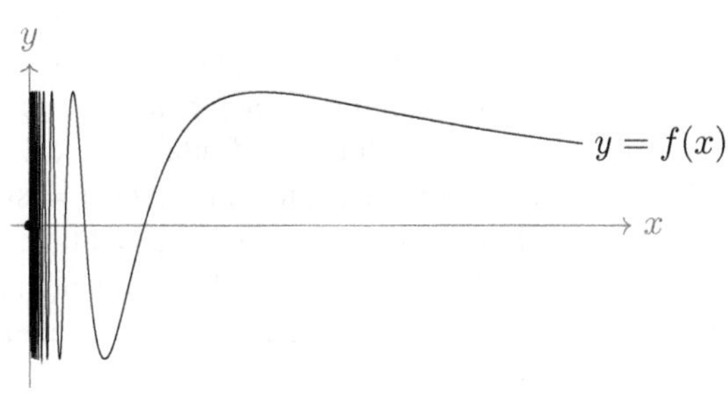

Fig. 10.7 Topologists' sine function. It rapidly oscillates as x gets closer to 0

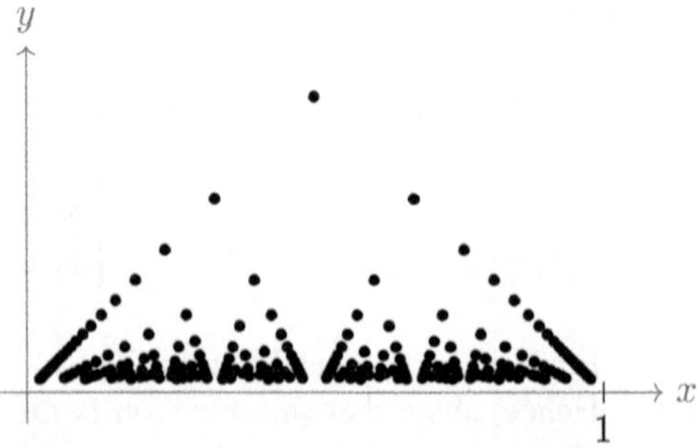

Fig. 10.8 Thomae's function. John Horton Conway (1937–2020) poetically called it the Stars over Babylon

10.8 (◇) Let $f : \mathbb{R} \to \mathbb{R}$ be defined as:

$$f(x) = \begin{cases} \frac{1}{q} & \text{if } x \in \mathbb{Q} \text{ with } x = \frac{p}{q} \text{ where } p, q \text{ are coprime,} \\ 1 & \text{if } x = 0, \\ 0 & \text{if } x \in \bar{\mathbb{Q}}. \end{cases}$$

This function is called Thomae's function after Carl Johannes Thomae (1840–1921). Some also called it the raindrop function because it looked like water droplets falling down instead. Other names include: popcorn function, ruler function, countable cloud function, and modified Dirichlet function (see Exercise 9.7 for the original Dirichlet function). See Fig. 10.8 for a visualisation of this function and decide for yourself what you want to call it.

(a) Show that f is a periodic function with period 1.

(b) For any $n \in \mathbb{N}$, define the set:

$$Q_n = \left\{ \frac{p}{q} \in [0, 1] \cap \mathbb{Q} : p \text{ and } q \text{ are coprime, } q \leq n \right\}.$$

Show that the set Q_n is finite.

(c) Hence, for any fixed $x_0 \in \bar{\mathbb{Q}} \cap [0, 1]$, show that the set $\{|x_0 - r| : r \in Q_n\}$ has a positive minimum. Describe this set and the minimum in words.

(d) Deduce that the function f is continuous at any $x_0 \in \bar{\mathbb{Q}} \cap [0, 1]$ and hence at any $x_0 \in \bar{\mathbb{Q}}$.

(e) By constructing a suitable sequence in the domain, show that the function f is not continuous at any point $x_0 \in \mathbb{Q}$.

(f) Show that the function f attains a local maximum at every $x_0 \in \mathbb{Q}$.

10.9 (a) Prove that the equation $\cos(\sin(x)) = \sin(\cos(x)) + 1$ where $x \in \mathbb{R}$ has a solution within the interval $[\frac{\pi}{2}, \pi]$.

(b) Show that the equation $x^3 + 2 = \sin(x)$ has a solution for x in $\mathbb{R}$.

10.10 (∗) Let $f : X \to \mathbb{R}$, where X is an interval in $\mathbb{R}$, be a continuous and injective function.

(a) Prove that f must be strictly monotone.
Now let $X = [0, 1]$ and $Y = (0, 1)$.
(b) Prove that there are no continuous bijections $f : X \to Y$.
(c) Likewise, prove that there are no continuous bijections $f : Y \to X$.
However, if we remove the continuous requirement, there are bijections between the two sets X and Y. This is clearly true since the sets X and Y have the same cardinality. As an explicit example of such a function, define:

$$f : X \to Y$$
$$x \mapsto \begin{cases} \frac{1}{2} - \frac{1}{2(n+1)} & \text{if } x = \frac{1}{2} - \frac{1}{2n} \text{ for } n \in \mathbb{N}, \\ \frac{1}{2} + \frac{1}{2(n+1)} & \text{if } x = \frac{1}{2} + \frac{1}{2n} \text{ for } n \in \mathbb{N}, \\ x & \text{otherwise,} \end{cases}$$

where we shifted some rational points in each of the interval $[0, \frac{1}{2})$ and $(\frac{1}{2}, 1]$ inwards along the interval.
(d) Prove that the function f is a bijection and is not continuous at $x = 1$.
(e) Construct a bijection $g : Y \to X$ and show that it is not continuous.

10.11 Let $f : \mathbb{R} \to \mathbb{R}$ be a monotone function.
(a) Prove that at every $x_0 \in \mathbb{R}$, the limits $\lim_{x \uparrow x_0} f(x)$ and $\lim_{x \downarrow x_0} f(x)$ both exist.
(b) Thus deduce that the only type of discontinuity that a monotone function can have is jump discontinuity.
(c) Furthermore, prove that the total number of jump discontinuities for a monotone function is at most countably infinite.

10.12 (∗) Suppose that $f : [0, 1] \to \mathbb{R}$ is a continuous function such that $f(0) = f(1)$.
(a) Show that there is a point $\xi \in [0, \frac{1}{2}]$ such that $f(\xi) = f(\xi + \frac{1}{2})$.
(b) Now fix any $n \in \mathbb{N}$. Show that there is a point $\xi \in [0, 1 - \frac{1}{n}]$ such that $f(\xi) = f(\xi + \frac{1}{n})$.

10.13 (∗) Let $f : [0, \infty) \to \mathbb{R}$ be a real function defined as $f(x) = \sqrt{x}$.
(a) Prove that this function is continuous everywhere.
(b) Prove that this function is also uniformly continuous on $[0, \infty)$.

10.14 Suppose that $f : X \to \mathbb{R}$ is a real-valued function and (x_n) is a Cauchy sequence in X.
(a) Prove that if $X = (0, 1)$ and f is uniformly continuous, then the image sequence $(f(x_n))$ is also Cauchy.
(b) Give an example of a continuous (but not uniformly continuous) function $f : (0, 1) \to \mathbb{R}$ and a Cauchy sequence (x_n) where the image sequence $(f(x_n))$ may not be Cauchy.
(c) Show that if $X = \mathbb{R}$ and f is continuous (but not necessarily uniformly continuous), then the image sequence $(f(x_n))$ is Cauchy.

10.15 Let $f, g : \mathbb{R} \to \mathbb{R}$ be real functions such that f is continuous at M and $\lim_{x\to\infty} g(x) = M$ where $M \in \mathbb{R}$. Prove that $\lim_{x\to\infty} f(g(x)) = f(\lim_{x\to\infty} g(x)) = f(M)$.

10.16 $(*)$ Let $f : \mathbb{R} \to \mathbb{R}$ be a continuous function. For any open set U in the codomain, show that its preimage $f^{-1}(U)$ must be open in the domain. This property of continuous function is used as the prime definition for continuous functions in the study of topology where the domain and codomain of the function f could be a different set to $\mathbb{R}$.

10.17 $(*)$ Let $f : I \to \mathbb{R}$ be a continuous function on an interval $I \subseteq \mathbb{R}$. Prove that for any $n \in \mathbb{N}$ and any $x_1, x_2, \ldots, x_n \in I$ we can find a $\xi \in I$ such that $\frac{1}{n}\sum_{j=1}^{n} f(x_j) = f(\xi)$.

10.18 $(*)$ Show that for a fixed constant $k > 0$, the equation $\tan(x) = kx$ for $x \in \mathbb{R} \setminus \{m\pi + \frac{\pi}{2} : m \in \mathbb{Z}\}$ has a solution in the interval $(n\pi, n\pi + \frac{\pi}{2})$ for any $n \in \mathbb{Z}_{\geq 0}$ and a solution in the interval $(n\pi - \frac{\pi}{2}, n\pi)$ for any $n \in \mathbb{Z}_{\leq 0}$.

10.19 $(*)$ Let $f : \mathbb{R} \to \mathbb{R}$ be a continuous function.
(a) Suppose that $\lim_{x\to\infty} f(x) = \lim_{x\to-\infty} f(x) = \infty$. Show that the function attains its global minimum somewhere in $\mathbb{R}$.
(b) Likewise, suppose that $\lim_{x\to\infty} f(x) = \lim_{x\to-\infty} f(x) = -\infty$. Show that the function attains its global maximum somewhere in $\mathbb{R}$.
(c) Demonstrate with examples that parts (a) and (b) cannot be true if f is not continuous.
(d) Suppose that $\lim_{x\to\infty} f(x) = \lim_{x\to-\infty} f(x) = K$ for some constant $K \in \mathbb{R}$. Show that the function attains a global minimum or a global maximum somewhere in $\mathbb{R}$.
(e) Does the result in part (d) still hold true if the limits of the function f at $\pm\infty$ exist but $\lim_{x\to\infty} f(x) \neq \lim_{x\to-\infty} f(x)$?

10.20 Let $f : \mathbb{R} \to \mathbb{R}$ be a continuous function. Show that f is a bijection if and only if f is strictly monotone and unbounded.

10.21 Let $f, g : X \to \mathbb{R}$ for some $X \subseteq \mathbb{R}$ be two uniformly continuous functions and $\lambda \in \mathbb{R}$ be a constant.
(a) Prove that the functions $f \pm g, \lambda f : X \to \mathbb{R}$ are also uniformly continuous.
(b) Give a counterexample of domain X and functions f and g over this domain for which the functions $\frac{f}{g}, f \times g : X \to \mathbb{R}$ are not uniformly continuous respectively.

10.22 $(*)$ Define a function $f : (0, \infty) \to \mathbb{R}$ as $f(x) = x\sin(\frac{1}{x})$.
(a) Show that this function is uniformly continuous on $[1, \infty)$.
(b) By extending the function appropriately to $x = 0$, show that this function is uniformly continuous on $(0, \infty)$.

10.23 Let $f : \mathbb{R} \to \mathbb{R}$ be a continuous function.
(a) If f is periodic with some period $P > 0$, prove that f is uniformly continuous.
(b) If $\lim_{x\to\infty} f(x)$ and $\lim_{x\to-\infty} f(x)$ are both finite, show that the function f is uniformly continuous over $\mathbb{R}$.

10.24 Let $S \subseteq \mathbb{R}$ be a subset of the real numbers. Define the function $f : \mathbb{R} \to \mathbb{R}$ as the distance function to the set S, namely $f(x) = \inf\{|x - s| : s \in S\}$.
(a) Show that $f(x) \geq 0$ with equality if and only if $x \in S \cup S'$.
(b) Prove that the function f is uniformly continuous over $\mathbb{R}$.

10.25 (a) Prove that for any $n \in \mathbb{N}$, the monomial $f : [0, 1] \to \mathbb{R}$ defined as $f(x) = x^n$ is Lipschitz continuous.
(b) Show that the monomial $f : \mathbb{R} \to \mathbb{R}$ defined as $f(x) = x^n$ for any integer $n \geq 2$ is not Lipschitz continuous.

10.26 ($*$) Let $P : \mathbb{R} \to \mathbb{R}$ be a polynomial $P(x) = \sum_{j=0}^{n} a_j x^j$ of degree $n \in \mathbb{N}_0$ with leading coefficient $a_n > 0$.
(a) Prove that if n is even, then there exists a point where the global minimum of the polynomial occurs.
(b) Hence, show that the polynomial function is not surjective for even degrees.
(c) Prove that the degree n is odd if and only if $P(\mathbb{R}) = \mathbb{R}$.
(d) Deduce that any polynomial with an odd degree has at least one real root.
(e) Show that if P is bounded, then P must be of degree 0.

10.27 Let $f : \mathbb{R} \to \mathbb{R}$ be a continuous function.
(a) If there exists a $c \in \mathbb{R}$ such that $f(c)f(-c) < 0$, prove that the equation $f(x) = 0$ has a real solution.
(b) Give an example of a surjective continuous function f such that $f(x)f(-x) \geq 0$ for all $x \in \mathbb{R}$.

10.28 ($\diamond$) For a constant $\alpha > 0$, a function $f : (a, b) \to \mathbb{R}$ is said to have the α-Hölder condition if there exists a constant $C > 0$ such that for all $x, y \in (a, b)$ we have $|f(x) - f(y)| < C|x - y|^\alpha$.
(a) Prove that for any $a, b \geq 0$ and $p \in (0, 1]$ we have $(a + b)^p \leq a^p + b^p$.
(b) Deduce that for $x, y \geq 0$ and $p \in (0, 1]$ we have $|x^p - y^p| \leq |x - y|^p$.
(c) Consider a function $g : [0, 1] \to \mathbb{R}$ where $g(x) = x^\beta$ with $0 < \beta \leq 1$. Show that g is α-Hölder if and only if $0 < \alpha \leq \beta$.
(d) Show that if $h : \mathbb{R} \to \mathbb{R}$ is α-Hölder for $\alpha > 1$, then h must be a constant.
(e) Prove that if f satisfies the α-Holder condition, then it is continuous on its domain.

Hence, any function satisfying the α-Hölder condition is also called α-Hölder continuous. If $\alpha = 1$, then α-Hölder condition is the same as Lipschitz continuity condition.

10.29 ($*$) Let $f : \mathbb{R} \to \mathbb{R}$ be a function such that $\lim_{x\to 0} f(x) = 1$ and $f(x+y) = f(x)f(y)$ for all $x, y \in \mathbb{R}$.
(a) Show that there exists a $\delta > 0$ such that $f(x) > 0$ on $B_\delta(0) \setminus \{0\}$.
(b) Show that $f(0) = 0$ or 1. Using part (a), deduce that $f(0) = 1$.
(c) For an integer $n > 0$, show that $f(nx) = (f(x))^n$ for all $x \in \mathbb{R}$.
(d) Using parts (a) and (c), show that $f(x) > 0$ for all $x \in \mathbb{R}$.
(e) For any $x_0 \in \mathbb{R}$, show that f is continuous at x_0.
(f) For a rational number $r > 0$, show that $f(rx) = (f(x))^r$ for all $x \in \mathbb{R}$.
(g) If $f(1) = a > 0$, find $f(x)$ for $x \in \mathbb{Q}$.
(h) Via continuity obtained in part (e), determine the whole function f.

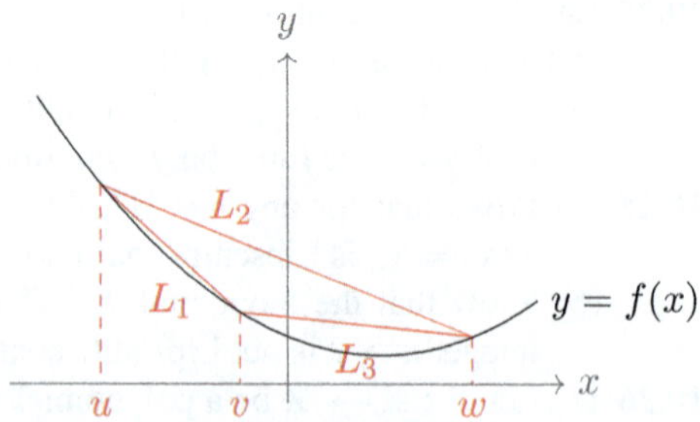

Fig. 10.9 The interpretation of the inequalities in part (a) is that the slopes of the secant line segments on a convex function above satisfy the ordering Slope(L_1) $\leq$ Slope(L_2) $\leq$ Slope(L_3)

(i) Using part (h), determine all functions $g : \mathbb{R} \to \mathbb{R}$ for which $\lim_{x\to 0} g(x) = 0$ and $g(x + y) = g(x) + g(y)$ for any $x, y \in \mathbb{R}$. This equation is called the Cauchy functional equation.

10.30 (∗) Suppose that $I = (a, b)$ is an open bounded interval and $f : I \to \mathbb{R}$ is a convex function.

(a) Prove that for any $u, v, w \in I$ such that $a < u < v < w < b$, we have:

$$\frac{f(v) - f(u)}{v - u} \leq \frac{f(w) - f(u)}{w - u} \leq \frac{f(w) - f(v)}{w - v}.$$

An interpretation of these inequalities is given geometrically as comparing the slopes of the secant lines for the graph of f as in Fig. 10.9.

(b) Show that f is bounded from above over any closed interval $[c, d] \subseteq I$.

(c) Show that f is bounded from below over any closed interval $[c, d] \subseteq I$.

(d) Hence, show that f is Lipschitz continuous over the closed interval $[c, d] \subseteq I$.

(e) Deduce that f is continuous at any point $x_0 \in I$.

(f) Does continuity over I still hold true if I is a closed interval instead, say $I = [a, b]$?

As we can see, convexity is a very strong condition which can be used to prove many things including continuity, analytical results, inequalities, and optimisation problems. In fact, there is a whole subarea of mathematics called convex analysis which explores the properties and applications of convex functions. We shall see more of convex functions in Chap. 14.

10.31 (∗) Let $f : \mathbb{R} \to \mathbb{R}$ be the exponential function $f(x) = e^x$. The graph of the exponential function and a linear function are plotted in Fig. 10.10. We are going to prove that the exponential function always lies above the given linear function.

(a) Using convexity, show that the function f is continuous everywhere on its domain.

(b) Hence, conclude that its inverse function, namely the natural logarithm $\ln : (0, \infty) \to \mathbb{R}$, is also continuous everywhere.

(c) Show that for any $x \geq 1$, $e^x \geq x + 1$.

(d) Now show that $e^x \geq 1 + x$ for any $x < 1$.

Fig. 10.10 The graphs of $y = e^x$ and $y = x + 1$. They coincide only at $x = 0$

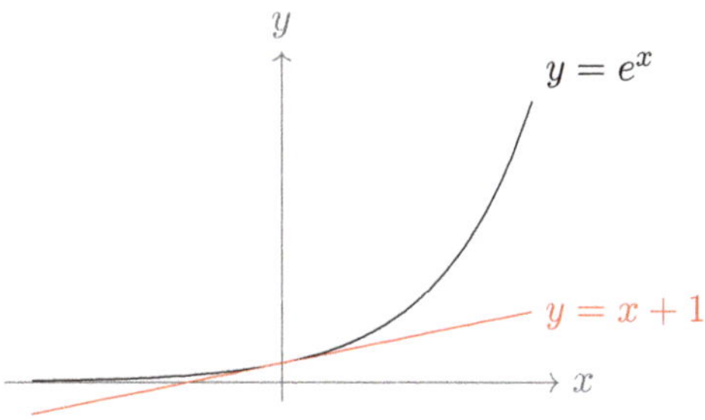

(e) Show that $\ln(x) \leq x - 1$ for all $x > 0$.
(f) Prove that the natural logarithm function is Lipschitz continuous on $[1, \infty)$ but cannot be uniformly continuous on $(0, 1)$.
(g) For any fixed $k > 0$, show that $\lim_{x\to\infty} \frac{\ln(x)}{x^k} = 0$ and thus $\ln(x) \in o\left(x^k\right)$ as $x \to \infty$.

10.32 ($\diamond$) Let us now prove a result called the Banach fixed point theorem, proven originally by Stefan Banach (1892–1945) in 1922. This result is a continuous analogue to the contractive sequence that we have seen in Exercise 6.4.

Theorem 10.7.20 (Banach Fixed Point Theorem) *Let $f : \mathbb{R} \to \mathbb{R}$ be a Lipschitz continuous function with Lipschitz constant $0 < K < 1$. Then, there exists a unique point $x \in \mathbb{R}$ such that $f(x) = x$.*

In this case, the function f is called a contraction map since the distance between the images of any two points are strictly smaller than their original distances.

(a) Pick any point $x_1 \in \mathbb{R}$ and define a sequence recursively as $x_{n+1} = f(x_n)$ for all $n \in \mathbb{N}$. For any $n \in \mathbb{N}$, show inductively that $|x_{n+2} - x_{n+1}| \leq K^n|x_2 - x_1|$.
(b) Hence, show that (x_n) is a Cauchy sequence and hence converges in $\mathbb{R}$.
(c) Let the limit of this sequence (x_n) be $x \in \mathbb{R}$. Show that it is a fixed point, namely $f(x) = x$.
(d) Show that this fixed point is unique.
(e) Hence, conclude that if we begin with any other initial point $y_1 \in \mathbb{R}$ and define the recursive sequence (y_n) in the same manner as previously, its limit is also x as in part (c).

One of the most important application of the Banach fixed point theorem is the Picard-Lindelöf theorem in the study of differential equations. This theorem provides a sufficient condition for an ordinary differential equation (ODE) problem to have a unique solution. We shall prove this theorem in Exercise 16.34.

Functions Sequence and Series 11

After all, our lives are but a sequence of accidents - a clanking chain of chance events. A string of choices, casual or deliberate, which add up to that one big calamity we call life.

—Rohinton Mistry, writer

In Chap. 5, we have defined sequences of real numbers and studied their properties and convergence. But we noted in Remark 5.0.3 that a sequence can be made up of any object at all. As a generalisation of the real sequence, we have briefly seen sequences of complex numbers, sequences of points in $\mathbb{R}^n$, and sequences in metric spaces in Sects. 6.3 and 6.4. In this chapter, we are going to define sequences of real-valued functions and study some of their properties. We first define what a sequence of functions is.

Definition 11.0.21 (Sequence of Functions) Let $X \subseteq \mathbb{R}$ be a subset of the real numbers. A sequence of real-valued functions or functions sequence is a list of functions (f_n) where the element f_n is a function $f_n : X \to \mathbb{R}$ for each $n \in \mathbb{N}$.

We would like to define what convergence of a functions sequence defined on $X \subseteq \mathbb{R}$ means. Let us call the set of real functions defined on $X \subseteq \mathbb{R}$ as $F(X; \mathbb{R}) = \{f : X \to \mathbb{R}\}$ and this sequence $(f_n) \subseteq F(X; \mathbb{R})$.

In order to define convergence of (f_n) in $F(X; \mathbb{R})$, we need to have a metric on the set of functions $F(X; \mathbb{R})$ as per Definition 6.4.8. At the moment, we do not have a metric or distance function defined on this set. However, even in the absence of a metric on $F(X; \mathbb{R})$, we can still define some notion of convergence for the sequence (f_n) using the metric on the codomain space $\mathbb{R}$, namely the modulus.

S. Johar, *The Big Book of Real Analysis*,
https://doi.org/10.1007/978-3-031-30832-1_11

11.1 Pointwise Convergence

Recall in Example 7.6.4 that we have seen a series of the form $\sum_{j=0}^{\infty} x^j$ for real numbers $x \in X = (-1, 1)$. This series was thought of as the limit of partial sums $s_n(x) = \sum_{j=0}^{n} x^j$ defined on X. Hence, the partial sums (s_n) is a sequence in $F(X; \mathbb{R})$.

For any fixed $x_0 \in X$, when we take the limit as $n \to \infty$, the sequence of real numbers $(s_n(x_0))$ converges to the infinite sum $s(x_0) = \sum_{j=0}^{\infty} x_0^j = \frac{1}{1-x_0}$. Since x_0 is arbitrary, we have the convergence $s_n(x) \to \frac{1}{1-x} = s(x)$ for every $x \in X$. We say that s_n converges pointwise to s since at every point $x \in X$ we have $s_n(x) \to s(x)$ as real numbers.

Let us lay this out in a more general setting. For a sequence of real-valued functions (f_n) all defined on a fixed domain $X \subseteq \mathbb{R}$, for each fixed point $x_0 \in X$ we have a sequence of real numbers $(f_n(x_0))$. This real sequence may or may not converge as $n \to \infty$. If it does, let us call the limiting number $f(x_0)$. Now we vary x_0 over X.

If for each $x \in X$ the sequence of real numbers $(f_n(x))$ converge to some $f(x) \in \mathbb{R}$, since limits in $\mathbb{R}$ are unique, we can treat this assignment as a function $f : X \to \mathbb{R}$ where $f(x) = \lim_{n\to\infty} f_n(x)$ for each $x \in X$. We call this resulting function the pointwise limit of the sequence (f_n). More specifically:

Definition 11.1.1 (Pointwise Convergence and Limit) Let $X \subseteq \mathbb{R}$ and (f_n) be a sequence of functions $f_n : X \to \mathbb{R}$. The sequence (f_n) converges pointwise to a function $f : X \to \mathbb{R}$ if for every $x_0 \in X$ and $\varepsilon > 0$, there exists an $N(x_0, \varepsilon) \in \mathbb{N}$ such that $|f_n(x_0) - f(x_0)| < \varepsilon$ for all $n \geq N(x_0, \varepsilon)$. The function $f : X \to \mathbb{R}$ is called the pointwise limit of the sequence (f_n). We write this as:

$$\lim_{n\to\infty} f_n(x) = f(x) \text{ for all } x \in X \quad \text{or} \quad f_n \xrightarrow{pw} f \text{ on } X.$$

Symbolically, this is written as:

$$f_n \xrightarrow{pw} f \text{ on X} \quad \text{if}$$
$$\forall x_0 \in X, \forall \varepsilon > 0, \exists N(\varepsilon, x_0) \in \mathbb{N} : \forall n \geq N(\varepsilon, x_0), |f_n(x_0) - f(x_0)| < \varepsilon.$$

Remark 11.1.2 Let us comment on Definition 11.1.1 and conventions above.

1. We have to note a very important technical distinction here. The objects f and $f(x)$ are two different objects. The former is a function, so it is a mapping from X to $\mathbb{R}$. The latter is the image of x under f, so it is a real number.
2. Thus, if we write $f_n \xrightarrow{pw} f$, we mean $f_n(x) \to f(x)$ (as real numbers) for each $x \in X$.

3. Using the definition above and the fact that convergent and Cauchy sequences are equivalent in $\mathbb{R}$, we can also deduce that the sequence of real functions (f_n) converges pointwise to some function f if and only if for each $x_0 \in X$, the sequence of real numbers $(f_n(x_0))$ is a Cauchy sequence. We will use this fact in some of our proofs later.

Pointwise convergence allows us to use the convergence results in real numbers, a topic which we have studied in the previous chapters. However, instead of having just one sequence of real numbers as we have studied in Chap. 5, from a functions sequence (f_n) we have a family of real sequences $\{(f_n(x)) : x \in X\}$ parametrised by the points in the domain X.

Thus, instead of bounded sequence and monotone sequence, we have the following definitions of uniform boundedness and uniform monotone sequence of functions which take into account all the sequences $(f_n(x))$ for all $x \in X$ in the family at once.

Definition 11.1.3 (Uniformly Bounded Functions Sequence) Let (f_n) be a sequence of functions $f_n : X \to \mathbb{R}$. The sequence is called:

1. Uniformly bounded from above if there exists a constant $M > 0$ such that $f_n \leq M$ on X for all $n \in \mathbb{N}$. In other words, for every $x \in X$ and $n \in \mathbb{N}$, we have $f_n(x) \leq M$.
2. Uniformly bounded from below if there exists a constant $M < 0$ such that $f_n \geq M$ on X for all $n \in \mathbb{N}$. In other words, for every $x \in X$ and $n \in \mathbb{N}$, we have $f_n(x) \geq M$.
3. Uniformly bounded if there exists a constant $M > 0$ such that $|f_n| \leq M$ on X for all $n \in \mathbb{N}$. In other words, for every $x \in X$ and $n \in \mathbb{N}$, we have $|f_n(x)| \leq M$.

Definition 11.1.4 (Pointwise Monotone Functions Sequence) Let (f_n) be a sequence of functions $f_n : X \to \mathbb{R}$. The sequence is called:

1. Pointwise increasing if for every $n \in \mathbb{N}$ we have $f_n \leq f_{n+1}$ on X. In other words, for every $x \in X$ and $n \in \mathbb{N}$, we have $f_n(x) \leq f_{n+1}(x)$.
2. Pointwise decreasing if for every $n \in \mathbb{N}$ we have $f_n \geq f_{n+1}$ on X. In other words, for every $x \in X$ and $n \in \mathbb{N}$, we have $f_n(x) \leq f_{n+1}(x)$.
3. Pointwise monotone if for any fixed $x \in X$ either $f_n(x) \leq f_{n+1}(x)$ for every $n \in \mathbb{N}$ or $f_n(x) \geq f_{n+1}(x)$ for all $n \in \mathbb{N}$. In other words, at any fixed point $x \in X$ the real sequence $(f_n(x))$ is either increasing or decreasing.

From the definitions above, we have the following result, which we leave as Exercise 11.4.

Proposition 11.1.5 *Let (f_n) be a sequence of functions $f_n : X \to \mathbb{R}$.*

1. *If (f_n) is pointwise increasing and uniformly bounded from above, then $f_n \xrightarrow{pw} f$ for some function $f : X \to \mathbb{R}$. We write this as $f_n \uparrow f$.*
2. *If (f_n) is pointwise decreasing and uniformly bounded from below, then $f_n \xrightarrow{pw} f$ for some function $f : X \to \mathbb{R}$. We write this as $f_n \downarrow f$.*
3. *If (f_n) is pointwise monotone and uniformly bounded, then $f_n \xrightarrow{pw} f$ for some function $f : X \to \mathbb{R}$.*

In all of the cases above, the limiting function f must also be bounded.

The pointwise limit f of a sequence of functions (f_n), if it exists, may be a very wild function because for each $x_0 \in X$ we are only looking at the values of the sequence $(f_n(x_0))$ without taking into account what is going on around the point x_0 in each function f_n. In other words, any local properties of the functions f_n may be diminished in the pointwise limit. As a result, continuity of the pointwise limit function f is not clear even if we know that all of the functions f_n in the sequence are nice and continuous. Let us look at this closely.

Suppose that we have a sequence of functions (f_n) where $f_n : X \to \mathbb{R}$ are all continuous. Thus for any $x_0 \in X$ and $n \in \mathbb{N}$, we have $\lim_{x\to x_0} f_n(x) = f_n(x_0)$. To check whether a limiting function f of this sequence is continuous at some $x_0 \in X$, by definition, we need to show that $\lim_{x\to x_0} f(x) = f(x_0)$. However, f is also defined as a limit of a sequence of functions, namely $f(x) = \lim_{n\to\infty} f_n(x)$ for each $x \in X$. This means if we want to show continuity of the limiting function f at x_0, we need to show that the following two quantities are equal:

$$\lim_{x\to x_0} f(x) = \lim_{x\to x_0} (\lim_{n\to\infty} f_n(x)), \text{ and}$$

$$f(x_0) = \lim_{n\to\infty} f_n(x_0) = \lim_{n\to\infty} (\lim_{x\to x_0} f_n(x)).$$

Notice that if (and this is a very big if) we can switch the order of the limits $\lim_{n\to\infty}$ and $\lim_{x\to x_0}$, the equality would be immediate because the RHS of both equations above are then equal. However, the sad news is that we cannot switch the order of the limits all the time! Here is an example where switching the order of limits is forbidden:

Example 11.1.6 Consider the sequence of functions (f_n) where $f_n : [0, 1] \to \mathbb{R}$ are defined as:

$$f_n(x) = \begin{cases} 1 - nx & \text{if } 0 \leq x \leq \frac{1}{n}, \\ 0 & \text{if } \frac{1}{n} \leq x \leq 1. \end{cases}$$

It is easy to check that all of the functions f_n are continuous. Now we want to find the pointwise limit of the sequence (f_n). We claim that the pointwise limit is the function:

$$f(x) = \begin{cases} 1 & \text{if } x = 0, \\ 0 & \text{if } 0 < x \leq 1. \end{cases}$$

Intuitively, this is true since the region where $f_n(x) > 0$ is exactly $[0, \frac{1}{n})$ which gets smaller as $n \to \infty$. In the limiting case, the only point for which the function remains non-zero is at $x = 0$. We now show this rigorously. Fix $x \in [0, 1]$. We have two cases:

1. If $x = 0$, the sequence $(f_n(0)) = (1, 1, 1, \ldots)$ converges to $f(0) = 1$.
2. If $x \neq 0$, by the Archimedean property, there exists an $N \in \mathbb{N}$ such that $\frac{1}{N} \leq x$. Thus, $f_N(x) = 0$. Moreover, for all $n \geq N$, we have $\frac{1}{n} \leq \frac{1}{N} \leq x$ and so $f_n(x) = 0$ for all $n \geq N$ as well. So the terms in the sequence $(f_n(x))$ will all be constantly 0 after the N-th term and hence $f_n(x) \to 0 = f(x)$.

Therefore we conclude that the function f is the pointwise limit of the sequence (f_n). Now, note that:

$$\lim_{x\to 0}(\lim_{n\to\infty} f_n(x)) = \lim_{x\to 0} f(x) = 0, \qquad \text{and}$$

$$\lim_{n\to\infty}(\lim_{x\to 0} f_n(x)) = \lim_{n\to\infty} 1 = 1,$$

and so changing the order of the limits here give very different results. Note also that the functions f_n are all continuous at $x = 0$ but the limiting function f is not continuous here. Thus, the continuity of each of the function in the sequence (f_n) at $x = 0$ is diminished in the limit.

The example above shows that pointwise convergence is quite weak as we may lose some local properties of the functions in the sequence. Therefore, we need to strengthen it into something more substantial. We note that in the definition of pointwise convergence, the index $N(\varepsilon, x_0)$ depends on the point x_0 we are looking at, namely: at some points x_0 in the domain we may need a large N to ensure that $f_n(x_0)$ are ε-close to $f(x_0)$ for all $n \geq N$, whereas a smaller N might be sufficient at other points.

This results in the possibility of different rates of convergence at different points in the domain: slow convergence if we require a large N, and fast convergence if we require a small N. We saw this in Example 11.1.6. At points $x \in (0, 1]$ which are closer to 1, the sequence $(f_n(x))$ converges quickly to $f(x)$. On the other hand, at points $x \in (0, 1]$ closer to 0, the sequence (f_n) converges pointwise slowly to f.

Remark 11.1.7 Mathematicians understood the concept of sequences and series of constant terms very well. However, sequences and series of variable terms might even confuddle some great minds. A notable example of this is the great Cauchy.

In his seminal analysis textbook *Cours d'Analyse*, he provided an erroneous proof claiming that the pointwise limit $f : X \to \mathbb{R}$ for the sequence of functions (f_n) which are all continuous at a point $x_0 \in X$ is also continuous at x_0. This came to be known as "Cauchy's famous wrong proof".

However, scientific and mathematical discoveries go onwards and upwards: Abel spotted this error several years after the publication of this book and gave a counterexample. Abel noted that Cauchy's proof may work for some special cases. Later on, Weierstrass realised what these special cases are and managed to fix Cauchy's wrong proof. His argument utilises the concept of uniform convergence which is a stronger condition than the pointwise convergence assumed by Cauchy.

11.2 Uniform Convergence

For Definition 11.1.1, if for each $\varepsilon > 0$ we can find an N that is independent of the point x_0, the rate of convergence for $f_n \xrightarrow{pw} f$ would then be the same for all points in the domain. This would be more desirable as we have more predictability and control over the convergence. With this idea, we define a new kind of convergence for functions sequence:

Definition 11.2.1 (Uniform Convergence and Limit) Let $X \subseteq \mathbb{R}$ and (f_n) be a sequence of functions $f_n : X \to \mathbb{R}$. The sequence (f_n) converges uniformly to a function $f : X \to \mathbb{R}$ over the domain X if for every $\varepsilon > 0$, there exists an $N(\varepsilon) \in \mathbb{N}$ such that for all $x \in X$ we have $|f_n(x) - f(x)| < \varepsilon$ for all $n \geq N(\varepsilon)$. The function $f : X \to \mathbb{R}$ is called the uniform limit of the sequence (f_n). We write this as:

$$f_n \xrightarrow{u} f \text{ over } X \quad \text{or} \quad f_n \rightrightarrows f \text{ over } X.$$

Symbolically, we have:

$$f_n \xrightarrow{u} f \text{ on } X \quad \text{if} \quad \forall \varepsilon > 0, \exists N(\varepsilon) \in \mathbb{N} : \forall x \in X, \forall n \geq N(\varepsilon), |f_n(x) - f(x)| < \varepsilon.$$

Remark 11.2.2 Note the subtle but important difference between pointwise convergence and uniform convergence: refer carefully to the symbolic notations of these two convergence modes to see the differences.

Let us think what uniform convergence mean geometrically. Note first that $|f_n(x) - f(x)| < \varepsilon$ is equivalent to $f(x) - \varepsilon < f_n(x) < f(x) + \varepsilon$ for all $x \in X$. So, for a fixed $\varepsilon > 0$, if we draw a "ribbon" of width 2ε centred at the graph of the limiting function f, then all of the graphs of the functions f_n for $n \geq N(\varepsilon)$ would have to lie within this ribbon to satisfy uniform convergence. This can be seen in Fig. 11.1.

By decreasing the value of ε towards 0, for large enough index, the functions in the sequence look very much similar to the limiting function globally since the width of the ribbon gets smaller uniformly. Therefore, we might be able to pass some

Fig. 11.1 If $f_n \xrightarrow{u} f$, for each $\varepsilon > 0$ we can find an index $N \in \mathbb{N}$ such that the graph of f_n for $n \geq N$ all lie within the red ribbon bounded between $f - \varepsilon$ and $f + \varepsilon$. This ribbon is also referred to as a "belt" by William Fogg Osgood (1864–1943)

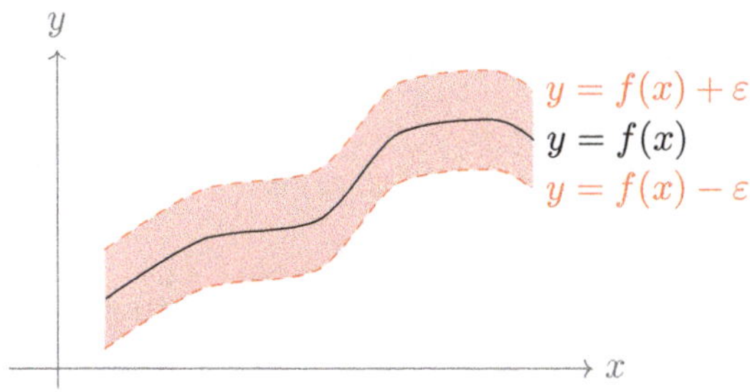

Fig. 11.2 First four terms of the function sequence (f_n) where $f_n(x) = \frac{x^n}{n}$

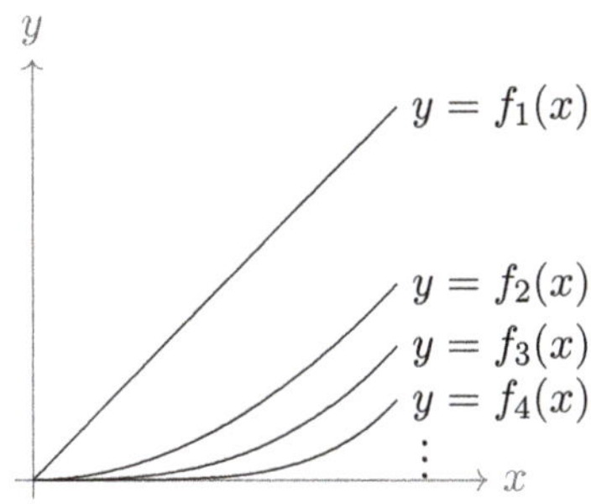

local properties of the sequence (f_n) to its limiting function f when we squeeze the sequence to get closer to f.

Example 11.2.3 Let us look at an example and a non-example of uniform convergence:

1. Consider the sequence of functions (f_n) where $f_n : [0, 1] \to \mathbb{R}$ is defined as $f_n(x) = \frac{x^n}{n}$. The graphs of the first for terms in this functions sequence is given in Fig. 11.2.
   This sequence of functions converges pointwise to the constant 0 function since for any fixed $x_0 \in [0, 1]$ we have $|f_n(x_0) - 0| = \left|\frac{x_0^n}{n}\right| \leq \frac{1}{n} \to 0$ as $n \to \infty$.
   Moreover, notice that in the argument above, the bound at the end does not depend explicitly on the point x_0. Indeed, if we switch to check convergence at a different point $y_0 \in [0, 1]$, the argument would be identical. This implies that this convergence is stronger than pointwise: it is uniform and the rate of convergence of the sequence $(f_n(x))$ to 0 at every point $x \in [0, 1]$ is faster than the rate at which $\frac{1}{n} \to 0$ (which is independent of x).
   Let us prove this rigorously according to Definition 11.2.1. Fix $\varepsilon > 0$. We claim that $N = \lceil \frac{1}{\varepsilon} \rceil + 1 \in \mathbb{N}$ can be used to show the uniform convergence over [0, 1]. Indeed, for any $n \geq N$, at any $x \in [0, 1]$ we have:

$$|f_n(x) - 0| = \left|\frac{x^n}{n}\right| \leq \frac{1}{n} \leq \frac{1}{N} = \frac{1}{\lceil \frac{1}{\varepsilon} \rceil + 1} < \frac{1}{\lceil \frac{1}{\varepsilon} \rceil} \leq \frac{1}{\frac{1}{\varepsilon}} = \varepsilon.$$

   Thus, we conclude that $f_n \xrightarrow{u} 0$ on [0, 1].

2. In Example 11.1.6, we have seen a sequence of continuous functions (f_n) defined as $f_n : [0, 1] \to \mathbb{R}$ by:

$$f_n(x) = \begin{cases} 1 - nx & \text{if } 0 \leq x \leq \frac{1}{n}, \\ 0 & \text{if } \frac{1}{n} \leq x \leq 1, \end{cases}$$

which converges pointwise to the function:

$$f(x) = \begin{cases} 1 & \text{if } x = 0, \\ 0 & \text{if } 0 < x \leq 1. \end{cases}$$

However, this convergence is not uniform. One can see that the ribbon of a small width $\varepsilon > 0$ around the function f is a discontinuous ribbon. For example, if we set $\varepsilon = \frac{1}{3}$, the ribbon is given by the set $\{(0, y) : \frac{2}{3} < y < \frac{4}{3}\} \cup \{(x, y) : 0 < x \leq 1, -\frac{1}{3} < y < \frac{1}{3}\}$. Since this ribbon is not "connected", we cannot have any continuous function contained wholly within this ribbon. Hence, none of the f_n, which are all continuous, can fully stay inside this disconnected ribbon. Mathematically, we show the negation of uniform convergence is true, namely:

$$\exists \varepsilon > 0 : \forall N \in \mathbb{N}, \exists x \in X : \exists n \geq N : |f_n(x) - f(x)| \geq \varepsilon.$$

For $\varepsilon = \frac{1}{3}$ and any $n \in \mathbb{N}$, there are always points $x \in (0, 1]$ such that:

$$|f_n(x) - f(x)| = |f_n(x) - 0| = |f_n(x)| \geq \varepsilon = \frac{1}{3},$$

which can be found by solving $f_n(x) = \frac{1}{3}$, namely $x = \frac{2}{3n} > 0$. Thus, we conclude that the convergence $f_n \to f$ is pointwise, but not uniform.

Clearly, if a sequence of functions (f_n) converges uniformly to the function f, then the sequence of functions (f_n) converges pointwise to f, but not vice versa! This is why we say that uniform convergence is a stronger version of pointwise convergence.

However, some pointwise convergence with additional conditions can lead to uniform convergence. An example of this is via Dini's theorem, in which we require the sequence of functions to be pointwise increasing or decreasing and the limiting function to be continuous. This theorem, named after Ulisse Dini (1845–1918), will be proven by the readers in Exercise 11.14 later.

Theorem 11.2.4 (Dini's Theorem) *Let (f_n) be a sequence of functions where $f_n : [a, b] \to \mathbb{R}$ is such that $f_n \downarrow f$ to some continuous function $f : [a, b] \to \mathbb{R}$. Then, this convergence is uniform over $[a, b]$.*

We can simplify the idea in Definition 11.2.1 slightly by thinking of uniform convergence in terms of the ribbon description that we have seen in Fig. 11.1. Instead of looking at all the x as in the definition, we could consider the greatest (via supremum) difference between $f(x)$ and $f_n(x)$ over all the $x \in X$. This is intuitively true in one direction: if the supremum of $|f_n(x) - f(x)|$ is smaller than ε, then $|f_n(x) - f(x)|$ for any x must also be smaller than ε. In fact, the converse is also true. We prove the following characterisation of uniform convergence:

Proposition 11.2.5 *Let $X \subseteq \mathbb{R}$ and (f_n) be a sequence of functions $f_n : X \to \mathbb{R}$. Then, $f_n \xrightarrow{u} f$ over X if and only if $\sup_{x\in X} |f_n(x) - f(x)| \to 0$.*

Proof We prove the implications separately.

($\Leftarrow$): Suppose that $\sup_{x\in X} |f_n(x) - f(x)| \to 0$. By definition of limits, for every $\varepsilon > 0$ there exists an $N \in \mathbb{N}$ such that $\sup_{x\in X} |f_n(x) - f(x)| < \varepsilon$ for all $n \geq N$. This means for all $n \geq N$, for all $x \in X$ we have:

$$|f_n(x) - f(x)| \leq \sup_{x\in X} |f_n(x) - f(x)| < \varepsilon,$$

which is the definition of uniform convergence.

($\Rightarrow$): Fix $\varepsilon > 0$. Our goal is to find an $N \in \mathbb{N}$ such that $\sup_{x\in X} |f_n(x) - f(x)| < \varepsilon$ for all $n \geq N$. Since f_n uniformly converges to f, there exists an $N \in \mathbb{N}$ such that for all $x \in X$ and $n \geq N$ we have $|f_n(x) - f(x)| < \frac{\varepsilon}{2}$. This means for each $n \geq N$, the set $A_n = \{|f_n(x) - f(x)| : x \in X\} \subseteq \mathbb{R}$ is bounded above by $\frac{\varepsilon}{2}$ and hence its supremum is also bounded from above by $\frac{\varepsilon}{2}$. Thus, for all $n \geq N$ we have:

$$\sup_{x\in X} |f_n(x) - f(x)| \leq \frac{\varepsilon}{2} < \varepsilon.$$

So this N works. □

Therefore, symbolically, uniform convergence $f_n \xrightarrow{u} f$ over X can also be written as:

$$f_n \xrightarrow{u} f \quad \text{if} \quad \forall \varepsilon > 0, \exists N(\varepsilon) \in \mathbb{N} : \forall n \geq N(\varepsilon), \sup_{x\in X} |f_n(x) - f(x)| < \varepsilon.$$

Thus using the supremum might be easier for us because we would have one thing less to worry about, which is varying x.

Remark 11.2.6 From Proposition 11.2.5, we can see what a suitable metric on the space of functions is. Notice that by using the supremum, instead of having to consider the value of the functions at each point in X as we did with pointwise convergence, we are attaching a single value which covers the of the whole X.

However, instead of the function set $F(X; \mathbb{R})$, we restrict our attention to the set of bounded functions over X, which we label as:

$$B(X; \mathbb{R}) = \{f \in F(X; \mathbb{R}) : \exists M > 0 \text{ such that } |f(x)| \leq M \text{ for all } x \in X\}.$$

This set is a real vector space and we can equip this vector space with a norm $||\cdot|| : B(X; \mathbb{R}) \to \mathbb{R}$ defined as $||f|| = \sup_{x\in X} |f(x)|$. Using this norm, we can then equip this set with a metric $d_\infty : B(X; \mathbb{R}) \times B(X; \mathbb{R}) \to \mathbb{R}$ induced from this norm defined as $d_\infty(f, g) = ||f - g|| = \sup_{x\in X} |f(x) - g(x)|$. We have shown all this in Example 6.4.3 and Exercise 6.18 for the domain $X = [a, b]$. The readers are invited to show why this metric is not well-defined on the whole of $F(X; \mathbb{R})$ in Exercise 11.5.

Another way to express the idea of uniform convergence of a sequence of functions is via the Cauchy criterion:

Proposition 11.2.7 (Cauchy Criterion for Uniform Convergence) *Let $X \subseteq \mathbb{R}$ and (f_n) be a sequence of functions $f_n : X \to \mathbb{R}$. The sequence (f_n) converges uniformly over X to a function $f : X \to \mathbb{R}$ if and only if for every $\varepsilon > 0$ there exists an $N \in \mathbb{R}$ such that for all $m, n \geq N$ we have:*

$$\sup_{x\in X} |f_n(x) - f_m(x)| < \varepsilon.$$

Proof We prove the implications separately.

($\Rightarrow$): Fix $\varepsilon > 0$. Since $f_n \xrightarrow{u} f$ on X, there exists an $N \in \mathbb{N}$ such that for all $n \geq N$ and $x \in X$, we have $|f_n(x) - f(x)| < \frac{\varepsilon}{4}$. Thus, if $m, n \geq N$, for all $x \in X$, by triangle inequality, we have:

$$\begin{aligned}|f_m(x) - f_n(x)| &= |f_n(x) - f(x) + f(x) - f_n(x)| \\ &\leq |f_m(x) - f(x)| + |f_n(x) - f(x)| \leq \frac{\varepsilon}{4} + \frac{\varepsilon}{4} = \frac{\varepsilon}{2},\end{aligned}$$

and thus taking the supremum over all $x \in X$, we get:

$$\sup_{x\in X} |f_m(x) - f_n(x)| \leq \frac{\varepsilon}{2} < \varepsilon,$$

for any $m, n \geq N$, which is what we wanted.

($\Leftarrow$): From our assumption, for any $\tilde{\varepsilon} > 0$, there exists an $\tilde{N} \in \mathbb{N}$ such that $\sup_{x\in X} |f_n(x) - f_m(x)| < \tilde{\varepsilon}$ for all $m, n \geq \tilde{N}$. This means for every $x \in X$, we have:

$$|f_n(x) - f_m(x)| \leq \sup_{x\in X} |f_n(x) - f_m(x)| < \tilde{\varepsilon}.$$

Thus, for any $x \in X$, the real sequence $(f_n(x))$ is Cauchy and hence convergent. By Remark 11.1.2(3), the functions sequence (f_n) hence converges pointwise to some function $f : X \to \mathbb{R}$.
We now show the convergence $f \xrightarrow{pw} f$ is in fact uniform. Fix $\varepsilon > 0$. By using our assumption, there exists an $N \in \mathbb{N}$ such that for all $m, n \geq N$ and $x \in X$ we have:

$$|f_n(x) - f_m(x)| \leq \sup_{x \in X} |f_n(x) - f_m(x)| < \frac{\varepsilon}{2}.$$

By fixing n, we take the limit as $m \to \infty$. Since the sequence of functions (f_m) converges pointwise to f, for all $x \in X$ we have $f_m(x) \to f(x)$ as $m \to \infty$. Since limits preserve weak inequalities, we then have:

$$\frac{\varepsilon}{2} \leq \lim_{m \to \infty} |f_n(x) - f_m(x)| = |f_n(x) - \lim_{m \to \infty} f_m(x)| = |f_n(x) - f(x)|,$$

for all $x \in X$. Taking the supremum over $x \in X$ yields:

$$\sup_{x \in X} |f_n(x) - f(x)| \leq \frac{\varepsilon}{2} < \varepsilon,$$

for all $n \geq N$. So, by Proposition 11.2.5, we conclude that $f_n \xrightarrow{u} f$ on X. □

Example 11.2.8 We have seen examples and non-examples of uniformly convergent sequence of functions. Here are some more of them:

1. Consider a sequence of real functions (f_n) defined on the whole of $\mathbb{R}$ as $f_n(x) = \frac{x}{n(1+x^2)}$. This sequence of functions converges pointwise to the zero function. Indeed, if we fix $x \in \mathbb{R}$, then the sequence $(f_n(x))$ is simply a constant multiple of the sequence $\left(\frac{1}{n}\right)$ which we know converges to 0 as $n \to \infty$. Hence, we have $f_n(x) \to 0$ as $n \to \infty$ for all $x \in \mathbb{R}$, namely $f_n \xrightarrow{pw} 0$.
Is this convergence uniform over $\mathbb{R}$? The answer is yes. We recall first the useful AM-GM inequality from Exercise 3.27 that says $\frac{a+b}{2} \geq \sqrt{ab}$ for all $a, b \geq 0$. Armed with this inequality, let us proceed to showing the uniform convergence.

   Rough work: Fix $\varepsilon > 0$. We need to find an $N \in \mathbb{N}$ such that $\sup_{x \in \mathbb{R}} |f_n(x) - 0| = \sup_{x \in \mathbb{R}} |f_n(x)| < \varepsilon$ for all $n \geq N$. At $x = 0$, we have $f_n(0) = 0$ for any $n \in \mathbb{N}$. Otherwise, we note that for all $x \in \mathbb{R} \setminus \{0\}$ we have:

$$|f_n(x)| = \left|\frac{x}{n(1+x^2)}\right| = \frac{1}{n}\frac{|x|}{1+x^2} = \frac{1}{n}\frac{1}{\frac{1}{|x|} + |x|} \leq \frac{1}{2n},$$

where we used the AM-GM inequality $\frac{1}{|x|} + |x| \geq 2\sqrt{\frac{|x|}{|x|}} = 2$ in the denominator. Taking the supremum over all $x \in \mathbb{R}$ (including $x = 0$), we then obtain:

$$\sup_{x\in\mathbb{R}} |f_n(x)| \leq \frac{1}{2n} < \frac{1}{n},$$

which we want to be smaller than ε. Therefore we can pick $N = \lceil \frac{1}{\varepsilon} \rceil$.

Now we write this properly. Fix $\varepsilon > 0$. Set $N = \lceil \frac{1}{\varepsilon} \rceil > 0$. For all $n \geq N$, if $x = 0$ then $|f_n(0)| = 0 < \varepsilon$. Otherwise, for any $x \neq 0$, via the AM-GM inequality, we have:

$$|f_n(x)| \leq \left| \frac{x}{n(1+x^2)} \right| = \frac{1}{n} \frac{|x|}{1+x^2} = \frac{1}{n} \frac{1}{\frac{1}{|x|} + |x|} \leq \frac{1}{2n}.$$

Taking the supremum over $x \in \mathbb{R}$, we obtain:

$$\sup_{x\in\mathbb{R}} |f_n(x)| \leq \frac{1}{2n} < \frac{1}{n} \leq \frac{1}{N} = \frac{1}{\lceil \frac{1}{\varepsilon} \rceil} \leq \frac{1}{\frac{1}{\varepsilon}} = \varepsilon,$$

for all $n \geq N$. Hence, we conclude that $f_n \xrightarrow{u} 0$ on $\mathbb{R}$.

2. Consider the sequence of real functions (f_n) defined on $\mathbb{R}$ as $f_n(x) = \frac{nx}{1+n^2x^2}$. This sequence of functions also converge pointwise to the zero function. This is clearly true at $x = 0$. For any other fixed $x \in \mathbb{R} \setminus \{0\}$ the absolute value of the sequence $(f_n(x))$ is given by the sequence of non-negative numbers:

$$0 \leq |f_n(x)| = \left| \frac{nx}{1+n^2x^2} \right| \leq \frac{n|x|}{n^2x^2} = \frac{1}{n|x|},$$

so, by sandwich lemma, we have $|f_n(x)| \to 0$ and hence $f_n(x) \to 0$ as $n \to \infty$. Thus, $f_n \xrightarrow{pw} 0$.

However, this sequence of functions (f_n) does not converge uniformly to 0. This can be seen as follows: for all $x \in \mathbb{R} \setminus \{0\}$, by using the AM-GM inequality, we have:

$$|f_n(x)| = \frac{n|x|}{1+n^2x^2} = \frac{1}{\frac{1}{n|x|} + n|x|} \leq \frac{1}{2}.$$

So for every $n \in \mathbb{N}$, the function f_n is bounded between $-\frac{1}{2}$ and $\frac{1}{2}$. In fact, for each $n \in \mathbb{N}$, this upper bound is attained by some $x \in \mathbb{R}$. This can be seen in Fig. 11.3. To find the point at which is value is attained, we solve the equation $\frac{1}{n|x|} + n|x| = 2$ to get $x = \pm\frac{1}{n}$. By checking signs, we conclude that the maximum

Fig. 11.3 First four terms of the function sequence (f_n) where $f_n(x) = \frac{nx}{1+n^2x^2}$

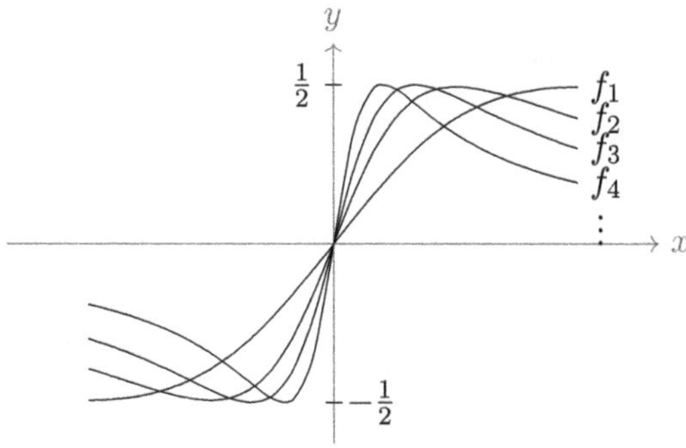

of the function f_n is $\frac{1}{2}$ which occurs at $x = \frac{1}{n}$. Therefore, for each $n \in \mathbb{N}$, we have:

$$\sup_{x\in\mathbb{R}} |f_n(x) - 0| = \sup_{x\in\mathbb{R}} |f_n(x)| = f_n(\tfrac{1}{n}) = \frac{1}{2},$$

which does not approach 0 as n goes to infinity. So this convergence is not uniform.

However, if we restrict the domain of the sequence of functions (f_n) above to $x \in [1, \infty)$, this would then remove the problem points $x = \frac{1}{n}$ at which f_n is equal to $\frac{1}{2}$. Hence, we expect that we could get uniform convergence. Indeed, for all $x \in [1, \infty)$ we have:

$$|f_n(x) - 0| = \frac{nx}{1+n^2x^2} \leq \frac{nx}{n^2x^2} = \frac{1}{nx} \leq \frac{1}{n}$$
$$\Rightarrow \quad \sup_{x\in[1,\infty)} |f_n(x) - 0| \leq \frac{1}{n} \to 0,$$

and therefore we have uniform convergence $f_n \xrightarrow{u} 0$ over the restricted domain $[1, \infty)$.

3. Consider the sequence of real functions (f_n) where $f_n : [0, 1] \to \mathbb{R}$ is defined as $f_n(x) = x^2 + \sin(\frac{x}{n})$. We want to find the pointwise limit of this function as $n \to \infty$. Clearly the x^2 term is not affected by the limit as $n \to \infty$ but the sine term will be. In fact, for any $x \in [0, 1]$, as $n \to \infty$, the sine term will vanish since $\frac{x}{n} \to 0$. Thus we guess that the pointwise limit is the function $f(x) = x^2$. Indeed, rigorously we have:

$$|f_n(x) - f(x)| = \left|x^2 + \sin\left(\frac{x}{n}\right) - x^2\right| = \left|\sin\left(\frac{x}{n}\right)\right| \leq \frac{x}{n} \leq \frac{1}{n}, \tag{11.1}$$

where we used the estimate $|\sin(h)| \leq |h|$ from Example 10.6.4. So, at any $x \in [0, 1]$ we have $f_n(x) \to f(x) = x^2$.
In fact, this convergence is uniform. Indeed, from (11.1), we have seen that for all $n \in \mathbb{N}$ we have:

$$|f_n(x) - f(x)| \leq \frac{1}{n} \quad \Rightarrow \quad \sup_{x\in[0,1]} |f_n(x) - f(x)| \leq \frac{1}{n}.$$

Taking the limit as $n \to \infty$ and sandwiching, we get $\sup_{x\in[0,1]} |f_n(x) - f(x)| \to 0$. Thus, we conclude that $f_n \xrightarrow{u} f$ on $[0, 1]$.

Example 11.2.8 tells us how to prove the uniform convergence of a sequence of function. We follow the following steps:

1. Find the pointwise limit $f : X \to \mathbb{R}$ for the sequence of functions (f_n). This step may require a bit of guesswork to decide what the limiting function is.
2. For every $\varepsilon > 0$, we need to find an $N(\varepsilon) > 0$ such that for any $n \geq N(\varepsilon)$, we have $\sup_{x\in X} |f_n(x) - f(x)| < \varepsilon$
   (a) This is done by first fixing $\varepsilon > 0$ and by algebraic manipulations, find a suitable N by trying to obtain a bound of the form $|f_N(x) - f(x)| < F(N)$ where $F(N)$ is a non-constant function of N which is decreasing to 0 and is independent of x. It is then enough to solve $F(N) = \frac{\varepsilon}{2}$ to get the desired N. We have to ensure that the choice of N is a natural number by using ceiling and floor functions, if needed.
   (b) Taking the supremum over $x \in X$ yields the required inequality $\sup_{x\in X} |f_N(x) - f(x)| \leq \frac{\varepsilon}{2} < \varepsilon$.
   (c) Finally, we check that with this choice of $N(\varepsilon)$, we have $\sup_{x\in X} |f_n(x) - f(x)| < \varepsilon$ for all $n \geq N(\varepsilon)$ as well.

Another useful trick is to consider the uniform convergence of the functions series in finitely many separate regions of the domain and then combine them in the end. To wit:

Proposition 11.2.9 *Let (f_n) be a sequence of functions $f : X \to \mathbb{R}$ and $Y, Z \subseteq X$ such that $Y \cup Z = X$. If $f_n \xrightarrow{u} f$ on Y and Z separately, then $f \xrightarrow{u} f$ on $Y \cup Z = X$.*

Proof Fix $\varepsilon > 0$. Via the assumption, there exist $N_1, N_2 \in \mathbb{N}$ such that for all $n \geq N_1$, we have $\sup_{x\in Z} |f_n(x) - f(x)| < \varepsilon$ and for all $n \geq N_2$, we have $\sup_{x\in Y} |f_n(x) - f(x)| < \varepsilon$. Setting $N = \max\{N_1, N_2\}$, for all $n \geq N$ we have $\sup_{x\in Y\cup Z} |f_n(x) - f(x)| < \varepsilon$. Therefore, we conclude that $f_n \xrightarrow{u} f$ on $Y \cup Z = X$. □

11.3 Consequences of Uniform Convergence

Now we ask ourselves: why is uniform convergence more desirable than pointwise convergence? As we have mentioned when discussing the geometric interpretation of uniform convergence, the limiting function can be better controlled by the functions in the sequence and so some of the properties of the functions in the sequence may be passed onto the limiting function. First, we can prove a boundedness result.

Proposition 11.3.1 *Let $X \subseteq \mathbb{R}$ and (f_n) be a sequence of bounded functions $f_n : X \to \mathbb{R}$. If (f_n) converges uniformly to a function $f : X \to \mathbb{R}$, then the uniform limit f must be bounded as well.*

Proof Since $f_n \xrightarrow{u} f$, for $\varepsilon = 1$, we can find an $N \in \mathbb{N}$ such that for all $n \geq N$ and $x \in X$, we have $|f_n(x) - f(x)| < 1$. In particular, $|f(x)| = |f(x) - f_N(x) + f_N(x)| \leq |f(x) - f_N(x)| + |f_N(x)| < 1 + |f_N(x)|$ for all $x \in X$. Since the function f_N is bounded, there exists an $M > 0$ such that $|f_N(x)| \leq M$ for all $x \in X$. Thus, $|f(x)| \leq M + 1$ for all $x \in X$ and hence the result. □

Note that the functions sequence (f_n) is not required to be uniformly bounded. We simply require each f_n to be a bounded function. In other words, the bounds can still depend on the index n. With this in mind, let us look at an example on how we can use the result above.

Example 11.3.2 An example where we can apply this result is to determine whether a sequence of functions converge uniformly. Define $f_n : (0, 1) \to \mathbb{R}$ as $f_n(x) = \frac{n}{nx+1}$ for each $n \in \mathbb{N}$. For each n, the function f_n is bounded since $|f_n(x)| = \frac{n}{nx+1} \leq n$. Moreover, this sequence converges pointwise to the function $f(x) = \frac{1}{x}$.

However, since all the functions f_n are bounded and its pointwise limit f is an unbounded function, by the contrapositive of Proposition 11.3.1, we can conclude that the convergence $f_n \to f$ cannot be uniform over $(0, 1)$.

The next important local property that remains preserved under uniform limit is continuity. We have seen in Example 11.1.6 a sequence of continuous functions (f_n) that converges pointwise (but not uniformly) to a non-continuous function. In this example, we have shown that different orders of the limits $x \to x_0$ and $n \to \infty$ give different values, namely:

$$\lim_{x \to x_0} (\lim_{n \to \infty} f_n(x)) \neq \lim_{n \to \infty} (\lim_{x \to x_0} f_n(x)).$$

However, if we are working on a sequence of functions that converge uniformly, we are allowed to switch these limits. The following result is named after Eliakim Hastings Moore (1862–1932) and Osgood.

Theorem 11.3.3 (Moore-Osgood Theorem) *Let $X \subseteq \mathbb{R}$ and (f_n) be a sequence of functions $f_n : X \to \mathbb{R}$ that converges uniformly over X to a function $f : X \to \mathbb{R}$. Assume that for $x_0 \in X'$, both $\lim_{x\to x_0} f(x)$ and $\lim_{x\to x_0} f_n(x)$ for all $n \in \mathbb{N}$ exist. Then:*

$$\lim_{x\to x_0}(\lim_{n\to\infty} f_n(x)) = \lim_{n\to\infty}(\lim_{x\to x_0} f_n(x)).$$

Proof First note that since the sequence (f_n) converges uniformly to f, this convergence is also pointwise namely $f(x) = \lim_{n\to\infty} f_n(x)$ for all $x \in X$. So we want to prove the equality:

$$\lim_{x\to x_0} f(x) = \lim_{n\to\infty}(\lim_{x\to x_0} f_n(x)). \tag{11.2}$$

If we let $p_n = \lim_{x\to x_0} f_n(x)$ for each $n \in \mathbb{N}$ and $p = \lim_{x\to x_0} f(x)$, proving equation (11.2) is equivalent to showing the convergence of real sequence $p_n \to p$.

Fix $\varepsilon > 0$. Since (f_n) converges uniformly to f, there exists an $N \in \mathbb{N}$ such that for all $n \geq N$, we have $|f_n(x) - f(x)| < \frac{\varepsilon}{2}$ for any $x \in X$. Thus, we can take the limit as $x \to x_0$ on both sides. Since limits preserve weak inequalities as we have seen in Exercise 9.10, we get $\lim_{x\to x_0} |f_n(x) - f(x)| \leq \frac{\varepsilon}{2} < \varepsilon$. By applying the algebra of limits, we then have:

$$\begin{aligned} |p_n - p| &= |\lim_{x\to x_0} f_n(x) - \lim_{x\to x_0} f(x)| = |\lim_{x\to x_0}(f_n(x) - f(x))| \\ &= \lim_{x\to x_0} |f_n(x) - f(x)| < \varepsilon, \end{aligned}$$

and thus for all $n \geq N$ we have $|p_n - p| < \varepsilon$ which is what we wanted to prove. □

As a consequence, knowing that the functions in (f_n) are continuous everywhere is enough to guarantee that its uniform limit is also continuous everywhere.

Theorem 11.3.4 (Uniform Limit Theorem) *Let $X \subseteq \mathbb{R}$ and (f_n) be a sequence of functions $f_n : X \to \mathbb{R}$ that converges uniformly over X to a function $f : X \to \mathbb{R}$. If $f_n \in C^0(X)$ for all $n \in \mathbb{N}$, then $f \in C^0(X)$ as well.*

Proof We show that f is continuous at an arbitrary point. Fix $x_0 \in X$ and $\varepsilon > 0$. We need to find a $\delta > 0$ such that if $x \in X$ with $|x - x_0| < \delta$ then $|f(x) - f(x_0)| < \varepsilon$. Since $f_n \xrightarrow{u} f$, there exists an $N \in \mathbb{N}$ such that $\sup_{x\in X} |f_n(x) - f(x)| < \frac{\varepsilon}{3}$ for all $n \geq N$. In particular, for any $x \in X$ and $n = N$, we have:

$$|f_N(x) - f(x)| \leq \sup_{x\in X} |f_N(x) - f(x)| < \frac{\varepsilon}{3}.$$

Moreover, the function $f_N : X \to \mathbb{R}$ is continuous by assumption, so it is continuous at x_0 and hence there exists a $\delta > 0$ such that for any $x \in X$ with

$|x - x_0| < \delta$ we have $|f_N(x) - f_N(x_0)| < \frac{\varepsilon}{3}$. Therefore, for any $x \in X$ with $|x - x_0| < \delta$, by using triangle inequality and the inequalities above, we have:

$$\begin{aligned}|f(x) - f(x_0)| &= |f(x) - f_N(x_0) + f_N(x_0) - f_N(x) + f_N(x) - f(x_0)| \\ &\leq |f(x) - f_N(x)| + |f_N(x) - f_N(x_0)| + |f_N(x_0) - f(x_0)| \\ &< \frac{\varepsilon}{3} + \frac{\varepsilon}{3} + \frac{\varepsilon}{3} = \varepsilon,\end{aligned}$$

and so we are done. □

Example 11.3.5 In Example 11.1.6, we have seen a sequence of continuous functions (f_n) defined as $f_n : [0, 1] \to \mathbb{R}$ by:

$$f_n(x) = \begin{cases} 1 - nx & \text{if } 0 \leq x \leq \frac{1}{n}, \\ 0 & \text{if } \frac{1}{n} \leq x \leq 1, \end{cases}$$

which converges pointwise to the function:

$$f(x) = \begin{cases} 1 & \text{if } x = 0, \\ 0 & \text{if } 0 < x \leq 1. \end{cases}$$

We note that all of the f_n are continuous, but the limiting function f is not continuous. Hence, by contrapositive of Theorem 11.3.4, we can immediately deduce here that the convergence cannot be uniform.

11.4 Functions Series

In Chap. 7, we have seen that a real series can be assigned a value if we treat it as a limit of the sequence of partial sums. Let us now generalise real series and define a series of functions from an infinite collection of functions:

Definition 11.4.1 (Functions Series) Let (f_n) be a sequence of real-valued functions $f_n : X \to \mathbb{R}$. A functions series is the formal infinite sum $\sum_{j=1}^{\infty} f_j$ of functions.

Example 11.4.2 There are various examples of functions series.

1. We have seen an example of one at the beginning of this chapter, namely the series $\sum_{j=0}^{\infty} x^j$ where each term is a monomial $f_j : \mathbb{R} \to \mathbb{R}$ defined as $f_j(x) = x^j$.
2. To generalise, we have the power series where each term in the series has the form $f_j(x) = a_j(x - c)^j$ for some constant real numbers $a_j, c \in \mathbb{R}$.
3. In the future, one might see Fourier series, asymptotic series, and Laurent series in further analysis courses.

For the time being, the infinite sum in Definition 11.4.1 is just a formal sum as we have not defined what its values are or whether it is even a function. However, we do have some ideas on how to define them based on what we have seen in the previous section.

These infinite sum of functions, similar to real series, cannot be calculated by adding the functions one by one to infinity. However, we can always add up finitely many functions defined on a common domain X. Hence, the functions series can be defined as the limit of these finite sums. These finite sums are called the partial sums.

Definition 11.4.3 (Partial Sums) Let $\sum_{j=1}^{\infty} f_j$ be a functions series where $f_j : X \to \mathbb{R}$ for all $j \in \mathbb{N}$. The n-th partial sum of this series is a function $s_n : X \to \mathbb{R}$ defined as the sum of the first n terms of the series, namely $s_n = \sum_{j=1}^{n} f_j$.

Pointwise Convergence of Functions Series

Thus, at each $x \in X$ and $n \in \mathbb{N}$, the quantity $s_n(x) = \sum_{j=1}^{n} f_j(x)$ are all well-defined real numbers. Moreover, since the domains of the functions f_j for all $j \in \mathbb{N}$ are all X, the domain of the partial sum s_n for any $n \in \mathbb{N}$ is also X. We aim to define the functions series as the pointwise limit of the partial sums, namely for all $x \in X$:

$$\sum_{j=1}^{\infty} f_j(x) = \lim_{n\to\infty} s_n(x).$$

However, this may not be well-defined at every point in X. When we try to sum up infinitely many functions pointwise, on some parts of the domain X this sum may not converge or blows up. However, on other parts of the domain, the functions series may behave well such that the sequence of partial sums evaluated at these points converge.

We have seen this before: the partial sums $s_n(x) = \sum_{j=0}^{n} x^j$ are polynomials defined on the whole of $\mathbb{R}$ but in the limit, the series $\sum_{j=0}^{\infty} x^j$ can only be defined for $x \in (-1, 1)$. Therefore, we need to isolate the parts where the series converges from the original domain X. We define the points in the domain over which the series makes sense as:

Definition 11.4.4 (Domain of Convergence) Let (f_n) be a sequence of real-valued functions $f_n : X \to \mathbb{R}$ and $\sum_{j=1}^{\infty} f_j$ be its functions series. The domain of convergence for this series is the subset $D \subseteq X$ such that:

$$D = \left\{ x \in X : \sum_{j=1}^{\infty} f_j(x) \text{ converges} \right\}.$$

Example 11.4.5 Let us look at some examples.

1. As seen earlier, the series $\sum_{j=1}^{\infty} x^j$ only converges for $x \in (-1, 1)$. Thus, for this functions series, we have $D = (-1, 1)$.
2. The functions series $\sum_{j=0}^{\infty} j!x^j$ only converges for $x = 0$. Indeed, for any fixed $x \neq 0$, by using the ratio test, we have $\left|\frac{(j+1)!x^{j+1}}{j!x^j}\right| = (j+1)|x| \to \infty$ as $j \to \infty$. So the series diverges here. Hence, $D = \{0\}$ only.
3. Consider the functions series $\sum_{j=1}^{\infty} j^x$ for $x \in \mathbb{R}$. We can rewrite this in a more familiar form by setting $x = -p$ so that the series becomes $\sum_{j=1}^{\infty} \frac{1}{j^p}$. Now we can see that this is a p-series as seen in Exercise 7.12 which converges if and only if $p > 1$. Namely, the series converges if and only if $x < -1$ and so $D = (-\infty, -1)$.

Uniform Convergence of Functions Series

Since the functions series is the limiting function of its partial sums, the definition of uniform convergence and Cauchy criterion for uniform convergence can be immediately adapted from Definition 11.2.1 and Propositions 11.2.7 and 11.2.5 as follows:

Definition 11.4.6 (Uniform Convergence of Functions Series) Let (f_n) be a sequence of real-valued functions $f_n : X \to \mathbb{R}$ and $\sum_{j=1}^{\infty} f_j$ be its functions series with n-th partial sum s_n. The functions series $\sum_{j=1}^{\infty} f_j$ converges uniformly over X to a function $s : X \to \mathbb{R}$ if for every $\varepsilon > 0$, there exists an $N \in \mathbb{N}$ such that for all $x \in X$ we have $|s_n(x) - s(x)| < \varepsilon$ for all $n \geq N$. In other words, the functions series $\sum_{j=1}^{\infty} f_j$ converges uniformly to s if $s_n \xrightarrow{u} s$.

Proposition 11.4.7 *Suppose that (f_n) is a sequence of real-valued functions $f_n : X \to \mathbb{R}$ and $\sum_{j=1}^{\infty} f_j$ is its functions series with n-th partial sum s_n. The functions series $\sum_{j=1}^{\infty} f_j$ converges uniformly over X to a function $s : X \to \mathbb{R}$ if and only if $\sup_{x\in X} |s_n(x) - s(x)| \to 0$.*

Proposition 11.4.8 (Cauchy Criterion for Uniform Convergence of Functions Series) *Let (f_n) be a sequence of real-valued functions $f_n : X \to \mathbb{R}$ and $\sum_{j=1}^{\infty} f_j$ be its functions series with n-th partial sum s_n. The series $\sum_{j=1}^{\infty} f_j$ converges uniformly over X to a function $s : X \to \mathbb{R}$ if and only if for every $\varepsilon > 0$ there exists an $N \in \mathbb{N}$ such that for all $m, n \geq N$ with $n > m$ we have:*

$$\sup_{x\in X} |s_n(x) - s_m(x)| = \sup_{x\in X} \left| \sum_{j=m+1}^{n} f_j(x) \right| < \varepsilon.$$

From the Cauchy criterion above, we have a simple test to check whether a functions series converges uniformly. The idea is, instead of checking a functions series, we check a real constant series consisting of bounds of the terms in the functions series instead. This test is called the Weierstrass M-test:

Theorem 11.4.9 (Weierstrass M-Test) *Let (f_n) be a sequence of real-valued functions $f_n : X \to \mathbb{R}$ and $\sum_{j=1}^{\infty} f_j$ be its functions series. Suppose that:*

1. *for each $n \in \mathbb{N}$ there exists an $M_n \geq 0$ such that $\sup_{x\in X} |f_n(x)| \leq M_n$, and*
2. *the series of real numbers $\sum_{j=1}^{\infty} M_j$ converges.*

Then, the functions series $\sum_{j=1}^{\infty} f_j$ converges uniformly on X. Moreover, we have the bound:

$$\sup_{x\in X} \left| \sum_{j=1}^{\infty} f_j(x) \right| \leq \sum_{j=1}^{\infty} M_j.$$

Proof To show uniform convergence of the functions series, we simply show that it satisfies the Cauchy criterion in Proposition 11.4.8. Fix $\varepsilon > 0$. We want to find an $N \in \mathbb{N}$ such that for all $m, n \geq N$ we have $\sup_{x\in X} |s_n(x) - s_m(x)| < \varepsilon$ where s_n is the n-th partial sum for the functions series.

Since the series $\sum_{j=1}^{\infty} M_j$ converges, by Cauchy criterion for real series in Proposition 7.3.1, there exists an $N \in \mathbb{N}$ such that for all $m, n \geq N$ with $n > m$ we have $|\sum_{j=1}^{n} M_j - \sum_{j=1}^{m} M_j| = |\sum_{j=m+1}^{n} M_j| = \sum_{j=m+1}^{n} M_j < \frac{\varepsilon}{2}$. With this same N, for all $m, n \geq N$ with $n > m$ we have:

$$|s_n(x) - s_m(x)| = \left| \sum_{j=m+1}^{n} f_j(x) \right| \leq \sum_{j=m+1}^{n} |f_j(x)| \leq \sum_{j=m+1}^{n} M_j < \frac{\varepsilon}{2}, \tag{11.3}$$

where we used the triangle inequality and the fact that $|f_j(x)| \leq M_j$ for all $j \in \mathbb{N}$ and $x \in X$. Taking the supremum over $x \in X$ in the inequality (11.3), we have $\sup_{x\in X} |s_n(x) - s_m(x)| \leq \frac{\varepsilon}{2} < \varepsilon$ and so the functions series converges uniformly by Proposition 11.4.8.

The final assertion is simply obtained by taking limits. Indeed, for any $n \in \mathbb{N}$, by using the triangle inequality, we have:

$$|s_n(x)| = \left| \sum_{j=1}^{n} f_j(x) \right| \leq \sum_{j=1}^{n} |f_j(x)| \leq \sum_{j=1}^{n} M_j.$$

Because limits preserve weak inequalities, by taking the limit as $n \to \infty$, we get:

$$\lim_{n\to\infty} |s_n(x)| \le \lim_{n\to\infty} \sum_{j=1}^{n} M_j \quad \Rightarrow \quad |\lim_{n\to\infty} s_n(x)| \le \sum_{j=1}^{\infty} M_j$$

$$\Rightarrow \quad \left|\sum_{j=1}^{\infty} f_j(x)\right| \le \sum_{j=1}^{\infty} M_j,$$

and, finally, taking the supremum over all $x \in X$ yields the result. □

Example 11.4.10 Consider the functions series $\sum_{j=1}^{\infty} \frac{j\sin(jx)}{e^j}$ where $x \in \mathbb{R}$. We note that $\sup_{x\in\mathbb{R}} |\frac{j\sin(jx)}{e^j}| \le \frac{j}{e^j}$ for each $j \in \mathbb{N}$ and the series $\sum_{j=1}^{\infty} \frac{j}{e^j}$ converges by the ratio test. Thus, by Weierstrass M-test, we can conclude that the functions series $\sum_{j=1}^{\infty} \frac{j\sin(jx)}{e^j}$ converges uniformly over $\mathbb{R}$.

Remark 11.4.11 We would like to stress here that the Weierstrass M-test is only a one-way statement. The converse is not true! We can find a functions series that converges uniformly over its domain but without the constant bounds $M_j > 0$ whose infinite sum converges. Consider the sequence of functions (f_j) where $f_j : (0, \infty) \to \mathbb{R}$ is defined as:

$$f_j(x) = \begin{cases} \frac{1}{j} & \text{if } x \in (j-1, j), \\ 0 & \text{otherwise.} \end{cases}$$

The series $\sum_{j=1}^{\infty} f_j$ converges uniformly everywhere by the Cauchy criterion. Indeed, for a fixed $\varepsilon > 0$, choose $N = \lceil \frac{1}{\varepsilon} \rceil \in \mathbb{N}$ so that for all $n > m \ge N$, we have:

$$\sup_{x\in X} |s_n(x) - s_m(x)| = \sup_{x\in(0,\infty)} \left| \sum_{j=m+1}^{n} f_j(x) \right|$$

$$= \frac{1}{m+1} \le \frac{1}{N+1} < \frac{1}{N} = \frac{1}{\lceil \frac{1}{\varepsilon} \rceil} \le \frac{1}{\frac{1}{\varepsilon}} = \varepsilon.$$

Next, notice that $\sup_{x\in(0,\infty)} |f_j(x)| = \frac{1}{j}$ and, by definition, this is the lowest upper bound that we can get for $|f_j(x)|$ over $x \in (0, \infty)$. So we can set $M_j = \frac{1}{j}$ for all $j \in \mathbb{N}$. However, the series $\sum_{j=1}^{\infty} M_j = \sum_{j=1}^{\infty} \frac{1}{j}$ does not converge since it is the harmonic series. Therefore, the Weierstrass M-test fails for this series even though we have proven that it converges uniformly.

Dirichlet's and Abel's Tests for Functions Series

Two other tests that we can use to check whether a functions series is uniformly convergent are the Dirichlet's and Abel's tests. We have seen the constant version of these tests in Theorems 7.8.3 and 7.8.5, so now we present the functions series versions of them. We state:

Theorem 11.4.12 (Dirichlet's Test for Uniform Convergence) *Let (f_n) and (g_n) be two sequences of real-valued functions $f_n, g_n : X \to \mathbb{R}$. Suppose that:*

1. *the sequence (s_n) of partial sums $s_n = \sum_{j=1}^{n} f_j$ is uniformly bounded over X,*
2. *(g_n) is pointwise monotone, and*
3. *$g_n \xrightarrow{u} 0$ over X.*

Then, the series $\sum_{j=1}^{\infty} f_j g_j$ is uniformly convergent over X.

The proof of Theorem 11.4.12 is left as an exercise for the readers in Exercise 11.23.

Example 11.4.13 Consider the functions series $\sum_{j=0}^{\infty} \frac{(-1)^j x^{3j+1}}{3j+1}$ where $x \in [0, 1]$. This series converges pointwise on $[0, 1]$ by the alternating series test. Now we want to find out whether this convergence is uniform on $[0, 1]$.

The Weierstrass M-test could not help us here. Indeed, we have $\sup_{x\in[0,1]} \left|\frac{(-1)^j x^{3j+1}}{3j+1}\right| = \sup_{x\in[0,1]} \left|\frac{x^{3j+1}}{3j+1}\right| = \frac{1}{3j+1}$ for each $j \in \mathbb{N}$ so these numbers are the smallest possible candidates for each $M_j > 0$. However, the series $\sum_{j=1}^{\infty} M_j$ does not converge and therefore the M-test does not apply.

But this does not mean that the series does not converge uniformly. This just means the Weierstrass M-test could not conclude anything here. Let us try the Dirichlet's test next. We can pick $f_j, g_j : [0, 1] \to \mathbb{R}$ as $f_j(x) = (-1)^j$ and $g_j(x) = \frac{x^{3j+1}}{3j+1}$. With these choices, we check:

1. The partial sums $s_n = \sum_{j=0}^{n} f_j$ are uniformly bounded on $[0, 1]$ since their value is either 0 or 1.
2. For any $x \in [0, 1]$ and $j \in \mathbb{N}$, we have $g_j(x) = \frac{x^{3j+1}}{3j+1} \geq \frac{x^{3(j+1)+1}}{3(j+1)+1} = g_{j+1}(x)$, so the functions sequence (g_j) is pointwise decreasing.
3. We have $g_j \xrightarrow{u} 0$ on $[0, 1]$ because $\sup_{x\in[0,1]} |g_j(x) - 0| = \sup_{x\in[0,1]} \left|\frac{x^{3j+1}}{3j+1}\right| = \frac{1}{3j+1} \to 0$.

Thus, since all three of its conditions are fulfilled, we can apply the Dirichlet's test to conclude that the series $\sum_{j=0}^{\infty} f_j g_j = \sum_{j=0}^{\infty} \frac{(-1)^j x^{3j+1}}{3j+1}$ converges uniformly on $[0, 1]$.

Now let us look at another uniform convergence test:

Theorem 11.4.14 (Abel's Test for Uniform Convergence) *Let (f_n) and (g_n) be two sequences of real-valued functions $f_n, g_n : X \to \mathbb{R}$. Suppose that:*

1. *the functions series $\sum_{j=1}^{\infty} f_j$ converges uniformly over X, and*
2. *(g_n) is pointwise monotone and uniformly bounded over X.*

Then, the series $\sum_{j=1}^{\infty} f_j g_j$ is uniformly convergent over X.

Proof Suppose that $M > 0$ is the uniform bound for the functions sequence (g_n), namely $|g_n| \leq M$ for all $n \in \mathbb{N}$. Fix $\varepsilon > 0$. Since $\sum_{j=1}^{\infty} f_j$ is uniformly convergent, by denoting its partial sums as s_n and applying the Cauchy criterion, there exists an $N \in \mathbb{N}$ such that for all $n > m \geq N$ we have $|s_n(x) - s_m(x)| < \frac{\varepsilon}{6M}$. In particular, $|s_n(x) - s_N(x)| < \frac{\varepsilon}{6M}$ for all $n \geq N$. For any $n \in \mathbb{N}$, let us write $t_n(x) = s_n(x) - s_N(x)$ so that $|t_n(x)| < \frac{\varepsilon}{6M}$ for all $n \geq N$.

Now we show that the partial sums of the series $\sum_{j=1}^{\infty} f_j g_j$ also satisfies the Cauchy criterion. For the same N as above, for any $n > m \geq N + 1$, summation by parts gives us:

$$\sum_{j=m+1}^{n} f_j g_j = s_n g_n - s_{m-1} g_m - \sum_{j=m}^{n-1} s_j (g_{j+1} - g_j).$$

Substituting $t_n = s_n - s_N$ in the summation by parts above, we can compute easily that all the terms with s_N cancel each other, leaving us with:

$$\sum_{j=m+1}^{n} f_j g_j = t_n g_n - t_{m-1} g_m - \sum_{j=m}^{n-1} t_j (g_{j+1} - g_j), \tag{11.4}$$

upon which we can apply the estimate on t_n that we have obtained earlier. Taking the absolute value of (11.4), by using the triangle inequality, telescoping sum, and the assumption that the functions sequence (g_n) is pointwise monotone, for all $x \in X$ we have:

$$\begin{aligned}\left|\sum_{j=m+1}^{n} f_j(x) g_j(x)\right| &\leq |t_n(x) g_n(x)| + |t_{m-1}(x) g_m(x)| \\ &\quad + \sum_{j=m}^{n-1} |t_j(x)(g_{j+1}(x) - g_j(x))|\end{aligned}$$

$$< \frac{\varepsilon}{6M}M + \frac{\varepsilon}{6M}M + \frac{\varepsilon}{6M}\sum_{j=m}^{n-1}|g_{j+1}(x) - g_j(x)|$$
$$= \frac{\varepsilon}{3} + \frac{\varepsilon}{6M}|g_n(x) - g_m(x)|$$
$$\leq \frac{\varepsilon}{3} + \frac{\varepsilon}{6M}(|g_n(x)| + |g_m(x)|)$$
$$\leq \frac{\varepsilon}{3} + \frac{\varepsilon}{6M}2M = \frac{\varepsilon}{3} + \frac{\varepsilon}{3} = \frac{2\varepsilon}{3}.$$

Taking the supremum over all X, for all $n > m \geq N + 1$ we have:

$$\sup_{x\in X}\left|\sum_{j=m+1}^{n} f_j(x)g_j(x)\right| \leq \frac{2\varepsilon}{3} < \varepsilon,$$

from which we recognise the Cauchy criterion on the series $\sum_{j=1}^{\infty} f_j g_j$. Hence, the series converges uniformly over X. □

Example 11.4.15 Let us look at some examples and non-examples of uniform convergence of functions series:

1. Recall that the series $\sum_{j=0}^{\infty} x^j$ converges pointwise for all $|x| < 1$. Using the formula for geometric series, it converges to the function $f(x) = \frac{1}{1-x}$ on $|x| < 1$. However, this series does not converge uniformly to f on the interval $(-1, 1)$. We prove this via the Cauchy criterion.
   Suppose for a contradiction that the series converges uniformly to f on $(-1, 1)$. Then, for $\varepsilon = \frac{1}{2}$, there exists an $N \in \mathbb{N}$ such that for all $m, n \geq N$, we have $\sup_{x\in(-1,1)} |s_n(x) - s_m(x)| < \frac{1}{2}$. So if we set $m = N$ and $n = N + 1$, we would have:
   $$\sup_{x\in(-1,1)} |s_{N+1}(x) - s_N(x)| < \frac{1}{2} \quad \Rightarrow \quad \sup_{x\in(-1,1)} |x|^{N+1} = 1 < \frac{1}{2},$$
   which is a contradiction. Thus, the series only converges pointwise to f over the interval $(-1, 1)$.
   An alternative way to prove this is by using Proposition 11.3.1. Note that for any $n \in \mathbb{N}$ the partial sums $s_n(x) = \sum_{j=0}^{n} x^j$ are all bounded on $B_1(0) = (-1, 1)$ with $|s_n(x)| \leq n$. Thus, by Proposition 11.3.1, if the convergence is uniform, we deduce that the limit must be bounded too. However, clearly the limit function $\frac{1}{1-x}$ on $(-1, 1)$ is not bounded as it blows up to ∞ as $x \to 1$. This gives us the desired contradiction.
2. Consider the functions series $\sum_{j=1}^{\infty} \frac{1}{x^2+j^2}$ on the whole of $\mathbb{R}$. To find out where the series converges, we can try the ratio test. However, we would get an

inconclusive result. This is true because any fixed $x \in \mathbb{R}$, we have the limit:

$$\lim_{n\to\infty} \frac{\frac{1}{x^2+(n+1)^2}}{\frac{1}{x^2+n^2}} = \lim_{n\to\infty} \frac{x^2+n^2}{x^2+(n+1)^2} = 1.$$

Whenever the ratio test is inconclusive, we usually try Raabe's test next. For any fixed $x \in \mathbb{R}$, we have:

$$\lim_{n\to\infty} n\left(\frac{\frac{1}{x^2+n^2}}{\frac{1}{x^2+(n+1)^2}} - 1\right) = \lim_{n\to\infty} n\left(\frac{x^2+n^2+2n+1}{x^2+n^2} - 1\right)$$
$$= \lim_{n\to\infty} \frac{2n^2+n}{x^2+n^2} = 2 > 1,$$

which means the series converges at this fixed x. Since $x \in X$ was arbitrary, we conclude that the series converges pointwise everywhere on $\mathbb{R}$.
However, we can prove something stronger, namely this functions series converges uniformly over $\mathbb{R}$. This can be achieved via the Weierstrass M-test. Note first that each term of the functions series can be bounded independent of x as $\frac{1}{x^2+n^2} \leq \frac{1}{n^2}$ for all $n \in \mathbb{N}$. Furthermore, we have seen in Example 7.5.3 that the series $\sum_{j=1}^{\infty} \frac{1}{j^2}$ is convergent. Thus, by applying the Weierstrass M-test with $M_j = \frac{1}{j^2}$, we conclude that the functions series $\sum_{j=1}^{\infty} \frac{1}{x^2+j^2}$ converges uniformly on the whole of $\mathbb{R}$.

Now that we know the various ways on how one can check for uniform convergence of functions series, applying the Moore-Osgood theorem in Theorem 11.3.3 to uniformly convergent functions series, we can switch the order of limits and infinite sums.

Theorem 11.4.16 *Let (f_n) be a sequence of real-valued functions $f_n : X \to \mathbb{R}$ and $\sum_{j=1}^{\infty} f_j$ be its functions series with n-th partial sum s_n. Assume that the series converges uniformly over X to a function $s : X \to \mathbb{R}$. If for $x_0 \in X'$ the limits $\lim_{x\to x_0} s(x)$ and $\lim_{x\to x_0} s_n(x)$ for all $n \in \mathbb{N}$ exist, then:*

$$\lim_{x\to x_0} \sum_{j=1}^{\infty} f_j(x) = \lim_{x\to x_0} s(x) = \lim_{x\to x_0} (\lim_{n\to\infty} s_n(x))$$
$$= \lim_{n\to\infty} (\lim_{x\to x_0} s_n(x)) = \sum_{j=1}^{\infty} \lim_{x\to x_0} f_j(x).$$

Similarly, we would have the continuity property of the uniform limiting function if the functions series is made up of continuous functions. This follows directly from

Theorem 11.3.4 since a finite sum of continuous function is always continuous by Theorem 10.2.1.

Theorem 11.4.17 (Uniform Limit Theorem for Functions Series) *Let (f_n) be a sequence of continuous real-valued functions in $C^0(X)$ and $\sum_{j=1}^{\infty} f_j$ be its functions series with n-th partial sum s_n. If the series converges uniformly over X to a function $s : X \to \mathbb{R}$, then the limit function s is also continuous.*

Example 11.4.18 Let $\sum_{j=0}^{\infty} a^j \cos(b^j \pi x)$ where $a \in (0, 1)$ is a constant and b is a positive odd integer such that $ab > 1 + \frac{3\pi}{2}$. This series is called a Weierstrass function and was a debacle at the time of its conception due to its unexpected behaviour. We shall see why later in Exercise 14.33. For now, notice that we have $\left|a^j \cos(b^j \pi x)\right| \leq a^j$. So, by using the Weierstrass M-test with $M_j = a^j$, since $\sum_{j=0}^{\infty} M_j = \frac{1}{1-a}$, the series converges uniformly and hence pointwise on $\mathbb{R}$. Moreover, since the partial sums of the series are all continuous, we conclude that the limit itself is also continuous.

One final result that we can discuss is a necessary condition on the series terms (f_n) for the functions series to converge uniformly. This is very useful as a test in order to determine whether a functions series converges uniformly.

Proposition 11.4.19 *Let (f_n) be a sequence of real-valued functions $f_n : X \to \mathbb{R}$ and $\sum_{j=1}^{\infty} f_j$ be its functions series with n-th partial sum s_n. If the series converges uniformly on X to a function $s : X \to \mathbb{R}$, then $\sup_{x \in X} |f_n(x)| \to 0$.*

Proof Fix $\varepsilon > 0$. Since (s_n) converges uniformly to s, there exists an $N \in \mathbb{N}$ such that $\sup_{x \in X} |s_n(x) - s(x)| < \frac{\varepsilon}{3}$ for all $n \geq N$. By triangle inequality, for all $x \in X$ and $n \geq N + 1$ we have:

$$|f_n(x)| = |s_n(x) - s_{n-1}(x)| \leq |s_n(x) - s(x)| + |s(x) - s_{n-1}(x)| < \frac{\varepsilon}{3} + \frac{\varepsilon}{3} = \frac{2\varepsilon}{3},$$

and thus $\sup_{x \in X} |f_n(x)| \leq \frac{2\varepsilon}{3} < \varepsilon$ for all $n \geq N + 1$ which proves the result. □

Remark 11.4.20 We make some remarks here:

1. One can notice that the test above is similar to the necessary condition for a real series to converge which we saw in Proposition 7.2.5, namely: if $\sum_{j=1}^{\infty} a_j$ converges, then $a_j \to 0$.
2. Note that Proposition 11.4.19 only goes one way! If we have $\sup_{x \in X} |f_n(x)| \to 0$, that does not mean that the functions series $\sum_{j=1}^{\infty} f_j$ is uniformly convergent over X. Again, this is similar to real series: if the terms in the series converge 0, it does not imply that the series is convergent (for example, the harmonic series).

3. Moreover, Proposition 11.4.19 may not be true for pointwise convergence. Indeed, let (f_n) be a sequence of functions where $f_n : [0, \infty) \to \mathbb{R}$ are defined as:

$$f_n(x) = \begin{cases} 1 & \text{if } x \in (n-1, n), \\ -1 & \text{if } x \in (n, n+1), \\ 0 & \text{otherwise.} \end{cases}$$

Then, the functions series $\sum_{j=1}^{\infty} f_j$ converges pointwise to the function:

$$f(x) = \begin{cases} 1 & \text{if } x \in (0, 1), \\ 0 & \text{otherwise.} \end{cases}$$

However, we have $\sup_{x \in [0,\infty)} |f_n(x)| = 1$ for all $n \in \mathbb{N}$.

Example 11.4.21 Here are some examples:

1. Recall Example 11.4.15(1) in which we show the pointwise convergence of the series $\sum_{j=0}^{\infty} x^j$ for $|x| < 1$ to $f(x) = \frac{1}{1-x}$. We have also shown that this convergence is not uniform using two different approaches, namely the Cauchy criterion and uniform limit theorem. We can also prove this by using Proposition 11.4.19. Indeed, we have $\lim_{n\to\infty}(\sup_{x\in(-1,1)} |x^n|) = \lim_{n\to\infty}(1) = 1 \neq 0$ which means the convergence cannot be uniform.
2. Let $\sum_{j=1}^{\infty} j \sin\left(\frac{x}{j^3}\right)$. Notice that this series converges pointwise for all $x \in \mathbb{R}$ by comparison. Indeed, for any fixed $x \in \mathbb{R}$, using the estimate $|\sin(h)| \leq |h|$ from Exercise 10.6.4, we have $\left|j \sin\left(\frac{x}{j^3}\right)\right| \leq j\frac{|x|}{j^3} = \frac{|x|}{j^2}$. Since the series $\sum_{j=1}^{\infty} \frac{|x|}{j^2}$ converges for any $x \in \mathbb{R}$, by direct comparison, the series $\sum_{j=1}^{\infty} j \sin\left(\frac{x}{j^3}\right)$ converges (absolutely) pointwise.
However, this series cannot converge uniformly over $\mathbb{R}$. Indeed, for any $n \in \mathbb{N}$ we have $\sup_{x\in\mathbb{R}} \left|n \sin\left(\frac{x}{n^3}\right)\right| = n$ which is attained at $x = \frac{n^3\pi}{2}$. Therefore:

$$\lim_{n\to\infty}\left(\sup_{x\in\mathbb{R}}\left|n \sin\left(\frac{x}{n^3}\right)\right|\right) = \lim_{n\to\infty} n = \infty,$$

which violates Proposition 11.4.19.

We close this chapter by noting that a functions series can come in many different form since the functions f_n can be any function at all. As a result, it can be difficult to develop a general theory for the various kinds of functions. Therefore, many subclasses of functions series are studied individually instead. Here are some examples:

1. One example of this is the Fourier series which has the form:

$$a_0 + \sum_{j=1}^{\infty}(a_j \cos(2\pi jx) + b_j \sin(2\pi jx)), \tag{11.5}$$

for some constants $a_j, b_j \in \mathbb{R}$. This series appeared in the study of heat and differential equations by Joseph Fourier (1768-1830) and first published in *Mémoire sur la Propagation de la Chaleur dans les Corps Solides* (Treatise on the Propagation of Heat in Solid Bodies).
One thing of note is that this series was used by Abel as a counterexample to Cauchy's wrong proof that we have discussed in Remark 11.1.7. By some choice of constants a_j and b_j, Abel constructed a series made up of continuous functions (the trigonometric functions are all continuous) with discontinuous infinite sum. We shall see the series constructed by Abel in Exercises 11.22 and 16.25.
2. If all of the a_j in the Fourier series (11.5) are zero, then the series is called a sine series. Likewise, if all the b_j in (11.5) vanish, the series is then called a cosine series.
3. Another simple class of functions series is the power series. The general form for these functions series is:

$$\sum_{j=0}^{\infty} a_j(x-c)^j,$$

where $c, a_j \in \mathbb{R}$ are some constants. We have seen an example of this series in Examples 7.6.4 and 11.4.15(1). This series is easier to study due to the standard and simple form of the terms in the series. Therefore, we shall devote the next chapter to study this family of functions series.

Exercises

11.1 $(*)$ Let (f_n) be a sequence of functions $f_n : \mathbb{R} \to \mathbb{R}$ defined as follows. Show that each of the following functions sequence converges pointwise to some function $f : \mathbb{R} \to \mathbb{R}$.
(a) $f_n(x) = (x + \frac{1}{n})^2$.
(b) $f_n(x) = \frac{\sin(nx)}{1+n^2x^2}$.

11.2 $(*)$ Consider the sequence of functions (f_n) defined as $f_n : [0, 1] \to \mathbb{R}$ where $f_n(x) = x^n$. Find the pointwise limit of this function.

11.3 $(*)$ For each of the following functions sequence (f_n), find its pointwise limit and show that the convergence is not uniform over the domain.
(a) $f_n : [0, 1] \to \mathbb{R}$ defined as $f_n(x) = x^2 + x^n$ for all $n \in \mathbb{N}$.
(b) $f_n : [0, \infty) \to \mathbb{R}$ defined as $f_n(x) = 2^{\frac{x}{n}}$ for all $n \in \mathbb{N}$.
(c) $f_n : \mathbb{R} \to \mathbb{R}$ defined as $f_n(x) = \sin^n(x)$ for all $n \in \mathbb{N}$.

11.4 (∗) Prove Proposition 11.1.5.

11.5 Let $F(X;\mathbb{R})$ be the set of real-valued functions on X and define $d_\infty : F(X;\mathbb{R}) \times F(X;\mathbb{R}) \to \mathbb{R}$ as $d_\infty(f,g) = \sup_{x\in X}|f(x)-g(x)|$. Explain why this is not a well-defined function.

11.6 (∗) Let (f_n) be a sequence of functions $f_n : \mathbb{R} \to \mathbb{R}$ defined as follows. Show that each of the following functions sequence converges uniformly to some function $f : \mathbb{R} \to \mathbb{R}$.

(a) $f_n(x) = \frac{nx^3}{nx^2+1}$.
(b) $f_n(x) = \frac{\sin(nx)}{1+n^2|x|}$.
(c) $f_n(x) = \frac{x^2}{\sqrt{x^2+\frac{1}{n}}}$.

11.7 (∗) For the following sequence of functions (f_n) where $f_n : [0,1] \to \mathbb{R}$, find the pointwise limits of the functions and determine whether the convergence is uniform.

(a) $f_n(x) = \begin{cases} \frac{1}{nx} & \text{if } x \neq 0, \\ \frac{1}{n} & \text{if } x = 0. \end{cases}$

(b) $f_n(x) = \begin{cases} 1 & \text{if } x = 1, \frac{1}{2}, \dots, \frac{1}{n}, \\ 0 & \text{otherwise.} \end{cases}$

(c) $f_n(x) = \begin{cases} x & \text{if } x = 1, \frac{1}{2}, \dots, \frac{1}{n}, \\ 0 & \text{otherwise.} \end{cases}$

11.8 (∗) Let (f_n) and (g_n) be sequences of functions $f_n, g_n : X \to \mathbb{R}$ where $X \subseteq \mathbb{R}$. Suppose that $f_n \xrightarrow{u} f$ and $g_n \xrightarrow{u} g$ on X for some functions $f, g : X \to \mathbb{R}$. Show that:

(a) For any constant $\lambda \in \mathbb{R}$, we have $\lambda f_n \xrightarrow{u} \lambda f$ on X.
(b) $f_n + g_n \xrightarrow{u} f + g$ on X.
(c) $f_n g_n \xrightarrow{pw} fg$.
Is it necessarily true that this convergence is uniform on X?
(d) Suppose further that the limit functions f and g are bounded. Show that $f_n g_n \xrightarrow{u} fg$ on X.

11.9 (◇) Suppose that (f_n) is a sequence of functions $f_n : X \to \mathbb{R}$ where $X \subseteq \mathbb{R}$ is an interval in $\mathbb{R}$. Assume that $f_n \xrightarrow{u} f$ on X to some function $f : X \to \mathbb{R}$.

(a) If there exists a $K \in \mathbb{N}$ such that each f_n has at most K points of discontinuities, prove that the uniform limit function f also has at most K points of discontinuities.
(b) If each f_n has finitely many discontinuities on X, is it necessarily true that f has finitely many discontinuities as well?

11.10 (∗) Suppose that (f_n) is a sequence of functions $f_n : X \to \mathbb{R}$ where $X \subseteq \mathbb{R}$ is an interval in $\mathbb{R}$. Assume that $f_n \xrightarrow{pw} f$ to some function $f : X \to \mathbb{R}$. Show that if each f_n is increasing, the limiting function f is also increasing.

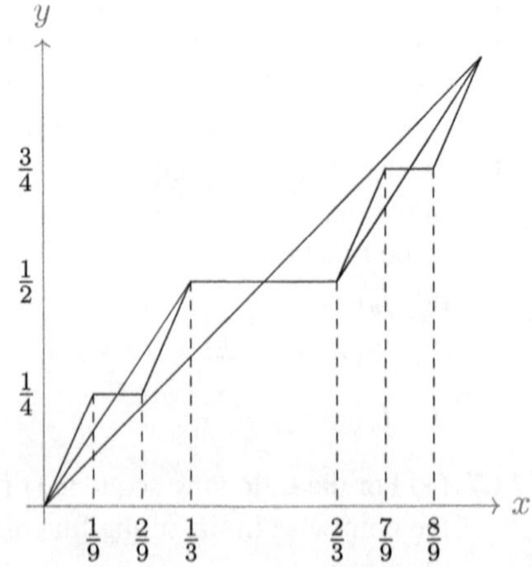

Fig. 11.4 Plots of the graphs of f_0, f_1, and f_2 in the same diagram

11.11 (∗) The Cantor's staircase is a function $f : [0, 1] \to [0, 1]$ defined as the pointwise limit of the sequence of functions $f_n : [0, 1] \to [0, 1]$ defined iteratively as:

$$f_0(x) = x \quad \text{and} \quad f_n(x) = \begin{cases} \frac{f_{n-1}(3x)}{2} & \text{if } x \in \left[0, \frac{1}{3}\right], \\ \frac{1}{2} & \text{if } x \in \left[\frac{1}{3}, \frac{2}{3}\right], \\ \frac{1}{2} + \frac{f_{n-1}(3x-2)}{2} & \text{if } x \in \left[\frac{2}{3}, 1\right], \end{cases} \quad \text{for all } n \in \mathbb{N}.$$

The first three iterations of this construction is given in Fig. 11.4.

(a) By induction, show that for each $n \in \mathbb{N}$:

$$\sup_{x\in[0,1]} |f_{n+1}(x) - f_n(x)| \le \frac{1}{2} \sup_{x\in[0,1]} |f_n(x) - f_{n-1}(x)|.$$

(b) Prove that the sequence of functions (f_n) converges uniformly over $[0, 1]$.
(c) Hence, deduce that limiting function f, which we call Cantor's staircase, is continuous.
(d) Hence, prove that the function f is surjective.
(e) Prove that each of the functions f_n are increasing.
Hence, deduce that the function f is also increasing.

This function is also called the Cantor ternary function, the Lebesgue function, the Cantor–Vitali function, and (more ominously) the Devil's staircase. It has many interesting and surprising properties that we shall see later in Exercises 13.11, 18.27, and 19.9.

11.12 Let (f_n) be a sequence of functions $f_n : X \to \mathbb{R}$ all of which are uniformly continuous. Suppose that $f_n \xrightarrow{u} f$ on X where $f : X \to \mathbb{R}$. We know by Proposition 11.3.4 that this limiting function f is also continuous. Prove that it is uniformly continuous on X.

11.13 Find an example of a sequence of functions (f_n) where $f_n : [0, 1] \to \mathbb{R}$ are all Lipschitz continuous but converges uniformly over $[0, 1]$ to a non-Lipschitz continuous function.

11.14 $(*)$ In this question, we shall prove Dini's theorem for uniform convergence.

Theorem 11.5.22 (Dini's Theorem) *Let (f_n) be a sequence of continuous functions where $f_n : [a, b] \to \mathbb{R}$ is such that $f_n \downarrow f$ to some continuous function $f : [a, b] \to \mathbb{R}$. Then, this convergence is uniform over $[a, b]$.*

Clearly, $f_n \geq f$ for all $n \in \mathbb{N}$. Suppose for contradiction that this convergence is not uniform.

(a) Show that there exists an $\varepsilon > 0$ such that there is a subsequence of the functions (which we call (f_n) again for simplicity and to declutter) which satisfies $\sup_{x\in[a,b]} |f_n(x) - f(x)| \geq 2\varepsilon > 0$.

(b) For each $n \in \mathbb{N}$, show that there exists a point $x_n \in [a, b]$ such that $|f_n(x_n) - f(x_n)| \geq \varepsilon > 0$.
Hence, deduce that there is a subsequence (x_{k_n}) of (x_n) which converges to some $x_0 \in [a, b]$.

(c) Show that for all index k_n we have $f_{k_n}(x_0) - f(x_0) \geq \varepsilon > 0$ and deduce the contradiction.

This theorem also holds if $f_n \uparrow f$ to some continuous function f and if the domain $[a, b]$ is replaced with any compact subset of $\mathbb{R}$.

11.15 $(*)$ Let (f_n) be a sequence of continuous functions $f_n : [a, b] \to \mathbb{R}$ that converges pointwise to a continuous function f. If the sequence (f_n) is pointwise monotone, prove that $f_n \xrightarrow{u} f$ on $[a, b]$.

11.16 $(*)$ Define a sequence of functions (f_n) where $f_n : [-1, 1] \to \mathbb{R}$ with $f_1(x) = 0$ and the others defined recursively as follows:

$$f_{n+1}(x) = f_n(x) + \frac{x^2 - f_n(x)^2}{2} \quad \text{for } x \in [-1, 1] \text{ and } n \geq 1.$$

(a) Prove inductively that $0 \leq f_n(x) \leq |x|$ for all $n \in \mathbb{N}$ and $x \in [-1, 1]$.

(b) Show that for each $x \in [-1, 1]$ and $n \in \mathbb{N}$ we have $f_{n+1}(x) \geq f_n(x)$.
Hence, deduce that the functions sequence converges pointwise and determine its limiting function.

(c) Show that this convergence is uniform over $[-1, 1]$.

11.17 Consider the real functions series $\sum_{j=1}^{\infty} je^{-jx}$. Determine its domain of convergence and whether it uniformly converges here.

11.18 (a) Recall that we have shown $\sum_{j=1}^{\infty} \frac{\cos(j)}{j}$ converges in Example 7.8.6. Using the same method, show that $\sum_{j=1}^{\infty} \frac{\sin(j)}{j}$ converges.
(b) Using the Cauchy criterion and part (a), show that the series $\sum_{j=1}^{\infty} \frac{\cos(j+x)}{j}$ converges uniformly on $\mathbb{R}$.

11.19 (a) Prove that $1 - \frac{1}{x} \leq \ln(x) \leq x - 1$ for all $x > 0$.
(b) Hence, show that the series $\sum_{j=1}^{\infty} \frac{\ln(j+x)-\ln(j)}{j}$ converges uniformly on $[0, 1]$.

11.20 (a) Let (f_n) be a sequence of functions $f : \mathbb{R} \to \mathbb{R}$ that converges pointwise to a function $f : \mathbb{R} \to \mathbb{R}$. If all the functions f_n are periodic with period $T > 0$, prove that f is also periodic with period T.
(b) Hence, for a convergent functions series $\sum_{j=1}^{\infty} f_j(x)$, if each of the terms f_j are periodic with period $T > 0$, prove that its limit is also periodic with period T.

11.21 (a) Show that:

$$\frac{\cos(x)}{1 \pm \sin(x)} = \sin\left(\frac{\pi \pm 2x}{4}\right) \csc\left(\frac{\pi \mp 2x}{4}\right),$$

whenever they are defined.
(b) Consider the series $\sum_{j=1}^{\infty} \sin(x)^j \cos(x)$ and $\sum_{j=1}^{\infty} (-1)^j \sin(x)^j \cos(x)$ for $x \in \mathbb{R}$. Investigate their pointwise and uniform convergence.

11.22 (◇) Consider the real functions series $\sum_{j=1}^{\infty} (-1)^{j+1} \frac{2\sin(jx)}{j}$. Denote its sequence of partial sums by (s_n). The plots for the partial sums s_5, s_{15}, s_{25}, and s_{35} are given in Fig. 11.5.
(a) By the same method as Exercise 11.18(a), show that for any fixed $x \in [-\pi, \pi]$, if $t_n = \sum_{j=1}^{n} (-1)^{j+1} \sin(jx)$, then there exists a constant $K(x) > 0$ (depending only on x) such that $|t_n| \leq K(x)$ for all $n \in \mathbb{N}$.
(b) Hence, by investigating (s_n), show that the functions series $\sum_{j=1}^{\infty} (-1)^{j+1} \frac{2\sin(jx)}{j}$ converges pointwise on $[-\pi, \pi]$. Deduce that this series converges pointwise for all $x \in \mathbb{R}$.
(c) Using Cauchy criterion for uniform convergence of functions series, show that the series does not converge uniformly over $[-\pi, \pi]$.

This is an example of a Fourier or a sine series. We shall see this series again later in Exercise 16.25 to investigate its limiting function. Can you guess what function will this functions series converge to based on Fig. 11.5?

11.23 (∗) Prove the Dirichlet's test for uniform convergence in Theorem 11.4.12, namely:
Let (f_n) and (g_n) be two sequences of real-valued functions $f_n, g_n : X \to \mathbb{R}$. Suppose that:
1. the sequence (s_n) of partial sums $s_n = \sum_{j=1}^{n} f_j$ is uniformly bounded over X,
2. (g_n) is pointwise monotone, and
3. $g_n \xrightarrow{u} 0$ over X.

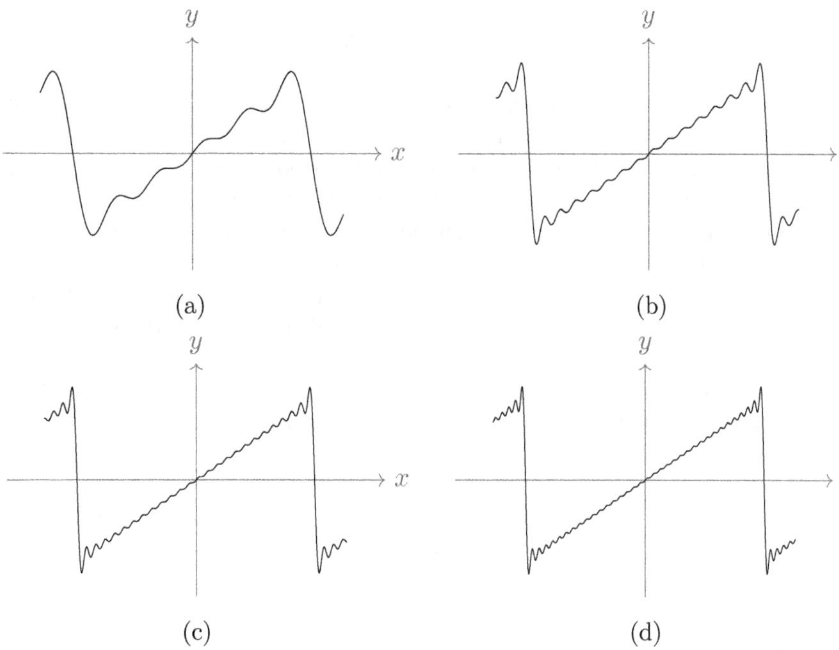

Fig. 11.5 Four examples of partial sums for the functions series. (**a**) s_5. (**b**) s_{15}. (**c**) s_{25}. (**d**) s_{35}

Prove that the series $\sum_{j=1}^{\infty} f_j g_j$ is uniformly convergent over X.

11.24 Suppose that $f_n, g_n : X \to \mathbb{R}$ are functions such that the series $\sum_{j=1}^{\infty} f_j$ and $\sum_{j=1}^{\infty} g_j$ converge uniformly on X.
(a) Show that the series $\sum_{j=1}^{\infty}(f_j + g_j)$ also converges uniformly over X.
(b) Let $h : X \to \mathbb{R}$ be a bounded function on X. Prove that the series $\sum_{j=1}^{\infty} hf_j$ also converges uniformly over X.

11.25 (a) Show that the functions series $\sum_{j=0}^{\infty} \frac{x^j}{1+x^{2j}}$ converges pointwise on $\mathbb{R}$ except at two points.
(b) Find all possible intervals in $\mathbb{R}$ over which the series above converges uniformly.

11.26 (∗) Show that the series $\sum_{j=0}^{\infty} \frac{x^2}{(1+x^2)^j}$ converges pointwise on $\mathbb{R}$ and find the function that it converges to.
Hence, deduce that the series does not converge uniformly on $\mathbb{R}$.

11.27 Consider the functions series $\sum_{j=0}^{\infty} \frac{\cos(jx)}{e^{jx}}$.
(a) Determine the subset of $[0, \infty)$ on which the series converges pointwise.
(b) For every $k > 0$, show that the series converges uniformly on $[k, \infty)$. Can the series converge uniformly on $(0, \infty)$?

11.28 Show that the functions series $\sum_{j=1}^{\infty} \frac{(-1)^j}{j+x^2}$ converges uniformly on $\mathbb{R}$.

11.29 (◇) Recall the Bolzano-Weierstrass theorem for real sequences that says any bounded real sequence (a_n) has a convergent subsequence. For functions sequences and pointwise convergence, this may not be true. As an example, the sequence (f_n) where $f_n : [0, \pi] \to \mathbb{R}$ defined as $f_n(x) = \sin(nx)$ is bounded since $|f_n(x)| = |\sin(nx)| \leq 1$ for all $n \in \mathbb{N}$ and $x \in [0, \pi]$ but there are no subsequence (f_{k_n}) that converges pointwise. A proof of this is deferred to Example 16.5.14 once we have enough tools to approach it.

However, if the domain of the functions sequence is countable, the result is true. Suppose that $X = \{x_j : j \in \mathbb{N}\}$ and $F_0 = (f_n)$ is a sequence of functions $f_n : X \to \mathbb{R}$. Suppose further that this sequence is uniformly bounded, namely there exists an $M > 0$ such that $|f_n(x)| \leq M$ for all $n \in \mathbb{N}$ and $x \in X$.

(a) Show that the real sequence $(f_n(x_1))$ has a convergent subsequence. Call the corresponding subsequence of functions $F_1 = (f_{k_{1,n}}) \subseteq F_0$.

(b) Next, show that there exists a subsequence $F_2 = (f_{k_{2,n}}) \subseteq F_1$ such that the real sequence $(f_{k_{2,n}}(x_2))$ converges.

(c) Iteratively, we can find a subsequence $F_n \subseteq F_{n-1}$ such that the real sequence of the functions in F_n evaluated at x_n converges. From this, construct a subsequence of F which converges pointwise on X.

11.30 (◇) As a follow up to Exercise 11.29, we are going to prove the Arzelà-Ascoli theorem originally by Cesare Arzelà (1847–1912) and Giulio Ascoli (1843–1896):

Theorem 11.5.23 (Arzelà-Ascoli Theorem) *Let $I = [a, b] \subseteq \mathbb{R}$ be a compact interval for some $a < b$. Suppose that (f_n) is a sequence of continuous functions $f_n : I \to \mathbb{R}$ which are uniformly bounded and are uniformly equicontinuous. Then, there exists a subsequence (f_{k_n}) which converges uniformly over I.*

There is a new term in the theorem above that we have not seen before, which is "uniform equicontinuous". We define it here:

Definition 11.5.24 (Uniform Equicontinuity) A sequence of functions (f_n) where $f_n : X \to \mathbb{R}$ for some $X \subseteq \mathbb{R}$ is called uniformly equicontinuous if for every $\varepsilon > 0$, there exists a $\delta(\varepsilon) > 0$ such that whenever $x, y \in X$ with $|x - y| < \delta(\varepsilon)$ we have $|f_n(x) - f_n(y)| < \varepsilon$ for all $n \in \mathbb{N}$.

Note that in the definition above, the choice δ is only allowed to depend on ε. In other words, the same δ works for any pairs $x, y \in X$ (uniform continuity over X) and any index n (equally for all f_n), hence the name uniform-equicontinuity. Now we prove the theorem as follows:

(a) Using Exercise 11.29, show that there exists a subsequence (f_{k_n}) of (f_n) which converges pointwise at every rational point in I.

We rename the functions in this subsequence as $g_n = f_{k_n}$ to declutter. At the moment we only know that (g_n) converges pointwise only at the rational points, but not the whole of I. We claim that it does converge pointwise everywhere. Since we do not have a candidate for the limiting function to prove this pointwise convergence, let us use the Cauchy criterion instead.

(b) Fix $\varepsilon > 0$. By equicontinuity, there is a $\delta > 0$ such that whenever $|x - y| < \delta$ we have $|g_n(x) - g_n(y)| < \frac{\varepsilon}{3}$ for all $n \in \mathbb{N}$. For the same $\delta > 0$, show that there are finitely many rational numbers $\{q_j\}_{j=1}^k \subseteq I \cap \mathbb{Q}$ such that $I \subseteq \bigcup_{j=1}^k B_{\frac{\delta}{2}}(q_j)$.

(c) Since we know that the real sequences $(g_n(q_j))$ for $j = 1, 2, \ldots, k$ converge, each of them must be Cauchy as well. Prove that there exists an $N \in \mathbb{N}$ independent of j such that whenever $m, n \geq N$ we have $|g_n(q_j) - g_m(q_j)| < \frac{\varepsilon}{3}$ for all $j = 1, 2, \ldots, k$.

(d) Now fix any $x \in I$. By some suitable triangle inequality, show that $|g_n(x) - g_m(x)| < \varepsilon$ for all $m, n \geq N$ in part (c).

Hence, conclude the proof using Proposition 11.2.7.

11.31 ($*$) In this question, we are going to prove the Stone-Weierstrass theorem. The original version of this result was established by Weierstrass in 1885 and Marshall Stone (1903–1989) extended it further onto more general domains. We are going to state and prove the first version of the result here.

Theorem 11.5.25 (Stone-Weierstrass Theorem) *Let $f : X \to \mathbb{R}$ be a real function on a compact interval $X \subseteq \mathbb{R}$. Then, there exists a sequence of polynomials (P_n), where $P_n : X \to \mathbb{R}$ is a degree n polynomial, for which $P_n \xrightarrow{u} f$.*

There are many ways to prove this. Here we are going to construct these approximating polynomials explicitly. WLOG, let $X = [0, 1]$.

(a) For any $j, n \in \mathbb{N}$ with $0 \leq j \leq n$, define the Bernstein basis polynomials $b_{j,n} : X \to \mathbb{R}$ defined as $b_{j,n}(x) = \binom{n}{j} x^j (1-x)^{n-j}$. Show that $b_{j,n}(x) \geq 0$ for all $x \in X$.

(b) Prove that for any $n \in \mathbb{N}$ and $x \in X$ we have $\sum_{j=0}^n b_{j,n}(x) = 1$.

(c) Show that for any $n \in \mathbb{N}$ we have:

i. $j\binom{n}{j} = n\binom{n-1}{j-1}$ for $1 \leq j \leq n$, and

ii. $j(j-1)\binom{n}{j} = n(n-1)\binom{n-2}{j-2}$ for $2 \leq j \leq n$.

(d) Using part (c), show $\sum_{j=0}^n j b_{j,n}(x) = nx$ and $\sum_{j=0}^n j^2 b_{j,n}(x) = n^2x^2 + nx(1-x)$.

(e) Hence, show that $\sum_{j=0}^n (nx - j)^2 b_{j,n}(x) = nx(1-x) \leq \frac{n}{4}$.

Define the n-th Bernstein polynomial for the function f as:

$$B_n^f(x) = \sum_{j=0}^n f\left(\frac{j}{n}\right) b_{j,n}(x).$$

Fix $\varepsilon > 0$. Since f is continuous on $X = [0, 1]$, it is uniformly continuous by Heine-Cantor theorem in Theorem 10.6.10. So there exists a $\delta > 0$ such that for any $x, y \in X$ with $|x - y| < \delta$, we have $|f(x) - f(y)| < \frac{\varepsilon}{2}$. Furthermore, by the EVT, for all $x \in X$ we have $|f(x)| \leq M$ for some $M > 0$. Our aim is to find an $N \in \mathbb{N}$ such that for all $n \geq N$ we have:

$$\sup_{x \in X} |B_n^f(x) - f(x)| < \varepsilon.$$

(f) Show that for a fixed $x \in X$ we have $|B_n^f(x) - f(x)| \leq \sum_{j=0}^{n} \left| f\left(\frac{j}{n}\right) - f(x) \right| b_{j,n}(x)$.

(g) Split the indices $J = \{0, 1, 2, \ldots, n\}$ into two disjoint sets $H = \{j \in J : |\frac{j}{n} - x| < \delta\}$ and $I = \{j \in J : |\frac{j}{n} - x| \geq \delta\}$. Derive the following estimates:
   i. $\sum_{j \in H} \left| f\left(\frac{j}{n}\right) - f(x) \right| b_{j,n}(x) < \frac{\varepsilon}{2}$.
   ii. $\sum_{j \in I} \left| f\left(\frac{j}{n}\right) - f(x) \right| b_{j,n}(x) \leq \frac{M}{2n\delta^2}$.

(h) Hence, find a suitable N that would ensure $\sup_{x \in X} |B_n^f(x) - f(x)| < \varepsilon$ for all $n \geq N$. Conclude the proof.

Polynomials are easier to manage since any polynomial requires only finitely many data in the form of the polynomial coefficients. Therefore, it is easier for us to approximate any continuous functions to a certain degree of accuracy using polynomials, as shown above. As a result, the Bernstein polynomials (named after Sergei Natanovich Bernstein (1880–1968)) are used widely in numerical analysis as well as computer graphics and design in the form of Bézier curves. The latter application was particularly useful when Pierre Bézier (1910–1999) patented and used it to design fancy cars!

12 Power Series

There is no good and evil, there is only power and those too weak to seek it.

—Lord Voldemort, very evil wizard

In this chapter, we are going to look at a special family of functions series, namely the power series. Due to the concrete form of these series, we could derive many more specialised results than the ones we saw in Chap. 11 for a general functions series. We first define what power series are:

Definition 12.0.1 (Power Series) A real power series is a functions series of the standard form $\sum_{j=0}^{\infty} a_j(x-c)^j$ where $c, a_j \in \mathbb{R}$ are constants for all $j \in \mathbb{N}_0$ and $x \in \mathbb{R}$ is a real variable. The series is said to be centred at c or expanded about c.

Generalisations of this family of series are the Laurent series (named after Pierre Laurent (1813–1854)) where the indices j are also allowed to be negative integers and Puiseux series (named after Victor Puiseux (1820–1883)) where the indices j are allowed to be rational numbers. The Laurent series are used widely in the study of complex analysis but we shall leave the exploration of these series for another time.

12.1 Convergence of Power Series

For the power series, each of the terms in the series, namely $a_j(x-c)^j$, is a function defined on the whole of the real line. Again, the summation notation is purely formal: we do not know beforehand where the series converges and where it does not. If we set $x = c$, clearly we get the zero series and hence the series trivially

S. Johar, *The Big Book of Real Analysis*,
https://doi.org/10.1007/978-3-031-30832-1_12

converges here. We are now interested to know for which other $x \in \mathbb{R}$ does the power series converge. We first show a very important result.

Proposition 12.1.1 *Let $\sum_{j=0}^{\infty} a_j(x-c)^j$ be a real power series where $c, a_j \in \mathbb{R}$ are constants. If $x_0 \in \mathbb{R}\backslash\{c\}$ is such that the real series $\sum_{j=0}^{\infty} a_j(x_0-c)^j$ converges, then the power series also converges for any $x \in \mathbb{R}$ such that $|x-c| < |x_0-c|$.*

Proof WLOG, let us assume that $c = 0$. According to the assumption, the series $\sum_{j=0}^{\infty} a_j x_0^j$ converges for $x_0 \neq 0$. This means the sequence $(a_n x_0^n)$ converges to 0 and hence must be bounded. Thus, there exists some real number $M > 0$ such that $|a_n x_0^n| \leq M$ for all $n \in \mathbb{N}_0$.

Now we want to check that the power series converges for all $x \in \mathbb{R}$ such that $|x| < |x_0|$, namely the series converges in the open ball $B_{|x_0|}(0)$. Note that for all such x, we have:

$$|a_n x^n| = |a_n x_0^n| \left|\frac{x^n}{x_0^n}\right| \leq M \left|\frac{x}{x_0}\right|^n.$$

Since $\left|\frac{x}{x_0}\right| < 1$, the geometric series $\sum_{j=1}^{\infty} M \left|\frac{x}{x_0}\right|^j$ converges and therefore, by direct comparison test, the series $\sum_{j=1}^{\infty} |a_j x^j|$ and hence $\sum_{j=1}^{\infty} a_j x^j$ converge as well. □

A visualisation for the convergence region of the power series as discussed in Proposition 12.1.1 is given in Fig. 12.1.

Example 12.1.2 Let us look at an example. Consider the power series $\sum_{j=1}^{\infty} \frac{(-1)^j}{j} x^j$ centred at $x = 0$. We can easily check that the series converges at $x = 1$ by the alternating series test. Therefore, Proposition 12.1.1 immediately implies that the power series converges for all $x \in \mathbb{R}$ such that $|x-0| < |1-0|$, namely for all $x \in (-1, 1)$ as well.

However, the convergence at the other end of the interval, namely at $x = -1$, is false. Indeed, if we substitute $x = -1$ in the series, we get $\sum_{j=1}^{\infty} \frac{(-1)^j}{j}(-1)^j = \sum_{j=1}^{\infty} \frac{1}{j}$ which is the divergent harmonic series.

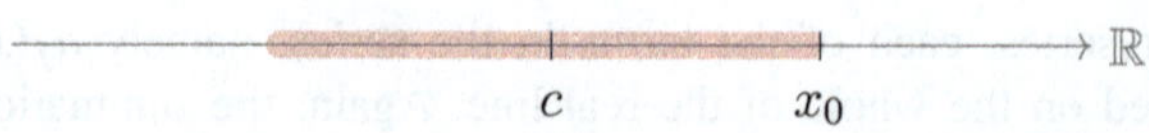

Fig. 12.1 Proposition 12.1.1 says that if a power series centred at c converges at $x_0 \in \mathbb{R}$, it must also converge at all $x \in \mathbb{R}$ such that $|x-c| < |x_0-c|$, namely all the x on the red line. It may or may not converge at the other end of the red line

From this observation, we prove:

Corollary 12.1.3 *Let $\sum_{j=0}^{\infty} a_j(x-c)^j$ be a real power series where $c, a_j \in \mathbb{R}$ are constants. There exists an $0 \leq R \leq \infty$ such that the series converges if $|x-c| < R$ and diverges if $|x-c| > R$.*

Proof WLOG, assume that $c = 0$. Let $X = \{|x| \geq 0 : \sum_{j=0}^{\infty} a_j x^j \text{ converges}\} \subseteq \mathbb{R}_{\geq 0}$. We then define $R = \sup(X)$ if X is a bounded set and $R = \infty$ if it is unbounded. By definition, the series diverges for all $x \in \mathbb{R}$ with $|x| > R$. Now we have three cases:

1. If $R = 0$, the series converges only for $x = 0$.
2. If $R > 0$, by characterisation of supremum, for any $\varepsilon > 0$ there exists an $r \in X$ such that $R - \varepsilon < r \leq R$ for which the series $\sum_{j=0}^{\infty} a_j x^j$ converges for any x with $|x| = r$. Hence, by Proposition 12.1.1, the power series also converges for all x where $|x| < r$. In particular, it converges for all x with $|x| < R - \varepsilon$. Since ε can be picked to be arbitrarily small, the series converges for all x where $|x| < R$.
3. If $R = \infty$, then the series converges for all $x \in \mathbb{R}$. Indeed, since X is unbounded from above, for any $x \in \mathbb{R}$ we can find an $r \in X$ such that $|x| < r$. Since the series converges at r, by Proposition 12.1.1, it must converge at x. □

Radius of Convergence

The quantity R from Corollary 12.1.3 has a name.

Definition 12.1.4 (Radius of Convergence, ROC) Let $\sum_{j=0}^{\infty} a_j(x-c)^j$ be a real power series where $c, a_j \in \mathbb{R}$ are constants. The radius of convergence for this series is the quantity $0 \leq R \leq \infty$ such that:

1. $R = 0$ if the series converges at $x = c$ only.
2. $0 < R < \infty$ if the series converges at any $x \in \{y \in \mathbb{R} : |y-c| < R\} = B_R(c)$ and diverges at any $x \in \{y \in \mathbb{R} : |y-c| > R\} = \mathbb{R} \setminus \bar{B}_R(c)$.
3. $R = \infty$ if the series converges everywhere on $\mathbb{R}$.

Remark 12.1.5 We make several remarks here:

1. As an added bonus, the proof of Proposition 12.1.1 actually shows a stronger convergence for the series, namely: if the radius of convergence of a series centred at c is R, then the power series converges absolutely in $B_R(c)$. This is an important fact that we are going to use repeatedly.

2. Another important note here is that Corollary 12.1.3 and Definition 12.1.4 did not specify any convergence or divergence behaviour on the sphere $S_R(c)$, namely for those $x \in \mathbb{R}$ that satisfies $|x-c| = R$. The power series may or may not converge here and to determine this, we have to check it manually just like what we did in Example 12.1.2.

Example 12.1.6 Here are some examples of radius of convergence:

1. In Example 11.4.15, we have seen a power series centred at $c = 0$ given by $\sum_{j=1}^{\infty} x^j$ which converges for $|x| < 1$. Its radius of convergence is $R = 1$. We have also seen its behaviour at $x = \pm 1$, namely: at $x = 1$ the series diverges to ∞ whereas at $x = -1$ the series alternates between -1 and 0. So the series does not converge at any point on $S_1(0) = \{\pm 1\}$. Thus, this power series converges only on $(-1, 1)$.
2. The series $\sum_{j=1}^{\infty} \frac{(-1)^j}{j} x^j$ in Example 12.1.2 has radius of convergence $R = 1$ as well. This is the largest possible R that we can find. Indeed, supposing for contradiction that $R > 1$, then $-1 \in B_R(0)$. However, as we saw in Example 12.1.2, the series diverges at this point, giving us the desired contradiction!
Since $R = 1$, this guarantees that it converges in $B_1(0)$. Additionally, we have seen that it also converges at $x = 1$. Thus, this power series converges only on $(-1, 1]$.

As we saw in Example 12.1.6(2), these power series may converge only for some points on the sphere $S_R(c)$. We will see more examples of such power series in Example 12.1.9.

Domain of Convergence

Similar to the definition we have seen for functions series, we call all the points on which the series converges the domain of convergence for the series. Elaborating from the domain of convergence of a functions series in Definition 11.4.4, we define:

Definition 12.1.7 (Domain of Convergence, DOC) Let $\sum_{j=0}^{\infty} a_j(x-c)^j$ be a real power series where $c, a_j \in \mathbb{R}$ are constants. Let $0 \leq R \leq \infty$ be the radius of convergence of this power series. The domain of convergence is the subset $D \subseteq \mathbb{R}$ on which the power series converges. Namely:

$$D = \begin{cases} \mathbb{R} & \text{if } R = \infty, \\ \left\{x \in \mathbb{R} : \sum_{j=0}^{\infty} a_j(x-c)^j \text{ converges}\right\} \subseteq \bar{B}_R(c) & \text{if } R < \infty. \end{cases}$$

Remark 12.1.8 We make some remarks regarding Definition 12.1.7:

1. Clearly, if R is a finite radius of convergence for some power series, the open ball $B_R(c)$ is necessarily contained in the domain of convergence D by definition of radius of convergence. However, we may find other points on the boundary of the ball $S_R(c) = \{c - R, c + R\}$ for which the series converge. Together, they form the domain of convergence.
2. In short, a real power series converges only on an interval of $\mathbb{R}$ with centre c. This interval must be one of the following forms:
   (a) If $R = \infty$, it must be $\mathbb{R}$.
   (b) If $R = 0$, it must be $\{c\}$.
   (c) If $0 < R < \infty$, it must either be a closed bounded interval $[c - R, c + R]$, an open bounded interval $(c - R, c + R)$, or a half-closed-half-open bounded interval $(c - R, c + R]$ or $[c - R, c + R)$.

We have talked about where a series converges and its radius of convergence, but how do we actually find this radius of convergence? So far, in Example 12.1.6, the radius of convergence of the power series were found by guessing. We want a more systematic way of finding R.

Finding Radius of Convergence

Since power series are just lots of different real series parametrised by x, a natural way to find this R is to appeal to the convergence and comparison tests that we have seen in Chap. 7. The simplest way to do this is via the root or ratio test in Theorem 7.6.3. Consider the power series $\sum_{j=0}^{\infty} a_j (x - c)^j$.

1. The root test says that if $\lim_{j\to\infty} \sqrt[j]{|a_j||x - c|^j} < 1$ then the series converges and if $\lim_{j\to\infty} \sqrt[j]{|a_j||x - c|^j} > 1$ then the series diverges. Hence, from here, provided that the limit $\lim_{j\to\infty} \sqrt{|a_j|}$ exists and is nonzero, we can conclude that:
   (a) if $|x - c| < \frac{1}{\lim_{j\to\infty} \sqrt[j]{|a_j|}}$, then the series converges, and
   (b) if $|x - c| > \frac{1}{\lim_{j\to\infty} \sqrt[j]{|a_j|}}$, then the series diverges.
   Using this, we can deduce that the radius of convergence for this power series is $R = \frac{1}{\lim_{j\to\infty} \sqrt[j]{|a_j|}}$. But what if $\lim_{j\to\infty} \sqrt{|a_j|} = 0$ or ∞? The former case is even nicer because that would mean:
$$\lim_{j\to\infty} \sqrt[j]{|a_j||x - c|^j} = \lim_{j\to\infty} \sqrt[j]{|a_j|}|x - c| = 0 < 1,$$

for any $x \in \mathbb{R}$ at all. This means the series converges for any x. In this case, we would have $R = \infty$. Otherwise, if $\lim_{j\to\infty} \sqrt[j]{|a_j|} = \infty$, the power series converges only for $x = c$ and hence have zero radius of convergence.

2. Of course, the limit involving the j-th root can be tricky to evaluate, so we can also use the ratio test to find the radius of convergence. The ratio test says that the power series converges if $\lim_{j\to\infty} \left|\frac{a_{j+1}(x-c)^{j+1}}{a_j(x-c)^j}\right| < 1$ and diverges if $\lim_{j\to\infty} \left|\frac{a_{j+1}(x-c)^{j+1}}{a_j(x-c)^j}\right| > 1$. Provided that the limit $\lim_{j\to\infty} \left|\frac{a_{j+1}}{a_j}\right|$ exists and is nonzero, this is equivalent to:
(a) if $|x-c| < \frac{1}{\lim_{j\to\infty}\left|\frac{a_{j+1}}{a_j}\right|}$, then the series converges, and
(b) if $|x-c| > \frac{1}{\lim_{j\to\infty}\left|\frac{a_{j+1}}{a_j}\right|}$, then the series diverges.
Thus, we can deduce that the radius of convergence for this power series is $R = \frac{1}{\lim_{j\to\infty}\left|\frac{a_{j+1}}{a_j}\right|}$. The cases for $\lim_{j\to\infty} \left|\frac{a_{j+1}}{a_j}\right| = 0$ and ∞ are the same as the root test in the previous part.

Example 12.1.9 Let us determine the radius and domain of convergence of some power series using the method of ratio or root test for power series that we have developed above.

1. Consider the real power series $\sum_{j=0}^{\infty} 2^j x^j$. To find its radius of convergence, we can use either the ratio test or the root test. The root test is easier to implement here. We find the limit:

$$\lim_{j\to\infty} \sqrt[j]{|2|^j} = \lim_{j\to\infty} 2 = 2.$$

Hence, the radius of convergence is $R = \frac{1}{2}$. This means any point $x \in B_{\frac{1}{2}}(0)$ ensures that the series converges and any point $x \in \mathbb{R} \setminus \bar{B}_{\frac{1}{2}}(0)$ results in the series being divergent.
What about the points on the boundary of this open ball $S_{\frac{1}{2}}(0)$, namely $x = \pm\frac{1}{2}$? Can we deduce their convergence/divergence property? We can check these cases separately:
(a) For $x = \frac{1}{2}$, the series becomes $\sum_{j=0}^{\infty} 1$ which diverges.
(b) For $x = -\frac{1}{2}$, the series becomes $\sum_{j=0}^{\infty}(-1)^j$ which also diverges.
Therefore, the series does not converge for any points on $S_{\frac{1}{2}}(0)$ and thus the domain of convergence of this series is simply $D = B_{\frac{1}{2}}(0) = (-\frac{1}{2}, \frac{1}{2})$.

2. Consider the real power series defined by $\sum_{j=0}^{\infty} \frac{x^j}{j!}$. To find its radius of convergence, we use the ratio test. We compute the limit for the ratios of the coefficients to obtain:

$$\lim_{j\to\infty}\left|\frac{\frac{1}{(j+1)!}}{\frac{1}{j!}}\right| = \lim_{j\to\infty}\left|\frac{1}{j+1}\right| = \lim_{j\to\infty}\frac{1}{j+1} = 0.$$

This says the radius of convergence is $R = \infty$ and the domain of convergence is $D = \mathbb{R}$.

3. Consider the real power series defined by $\sum_{j=0}^{\infty}\frac{j^{2j}(2x+1)^j}{4^j}$. Since all the terms in the series is raised to some power of j, we appeal to the root test here. Looking at the coefficients of the power series, we find:

$$\lim_{j\to\infty}\sqrt[j]{\left|\frac{j^{2j}}{4^j}\right|} = \lim_{j\to\infty}\frac{j^2}{4} = \infty.$$

Thus, the radius of convergence of this series is $R = 0$ and so the domain of convergence is just the singleton set $D = \{-\frac{1}{2}\}$.

4. Consider the real power series defined by $\sum_{j=1}^{\infty}\frac{x^j}{j^j}$. We could use the ratio or root test to find the radius of convergence for this series, but let us try something else from our toolbox of series convergence tests in Chap. 7.
Let us aim to use the comparison test with a known power series. We note that for all $j \in \mathbb{N}$ we have $j^j \geq j!$ and thus $\frac{|x|^j}{j^j} \leq \frac{|x|^j}{j!}$ for any $x \in \mathbb{R}$. However, we have seen that the series $\sum_{j=0}^{\infty}\frac{x^j}{j!}$ converges absolutely for all $x \in \mathbb{R}$. So, by direct comparison test, the series $\sum_{j=0}^{\infty}\frac{x^j}{j^j}$ converges absolutely and hence converges for all $x \in \mathbb{R}$. This means its radius of convergence is ∞.

5. Consider the real power series defined by $\sum_{j=1}^{\infty}\frac{x^j}{2^j j^4}$. Let us apply the ratio test here to find its radius of convergence. By looking at the coefficients of the power series, we have:

$$\lim_{j\to\infty}\left|\frac{\frac{1}{2^{j+1}(j+1)^4}}{\frac{1}{2^j j^4}}\right| = \lim_{j\to\infty}\left|\frac{1}{2\left(1+\frac{1}{j}\right)^4}\right| = \lim_{j\to\infty}\frac{1}{2\left(1+\frac{1}{j}\right)^4} = 2.$$

Hence, the radius of convergence is $R = 2$.
To find the domain of convergence, we now check the convergence on the boundary of the open ball $B_2(0)$, namely at the points $x = \pm 2$. For these values of x, the power series is either $\sum_{j=1}^{\infty}\frac{1}{j^4}$ or $\sum_{j=1}^{\infty}\frac{(-1)^j}{j^4}$. Using our knowledge on real series, we can easily see that both of these series converge.
So, in addition to the open ball $B_2(0)$, the power series $\sum_{j=1}^{\infty}\frac{x^j}{2^j j^4}$ also converges for $x = \pm 2$. Thus, the domain of convergence for this power series is the set $D = \bar{B}_2(0) = [-2, 2]$.

6. Consider the power series $\sum_{j=0}^{\infty}\frac{x^{2j}}{2^j}$. If we write this series in its standard form as in Definition 12.0.1, it would look like $1 + 0x + \frac{x^2}{2} + 0x^3 + \frac{x^4}{4} + 0x^5 + \frac{x^6}{8} + \ldots$.

We note that this power series only has non-zero coefficient for even-powered monomials. Therefore, some of the ratios of the coefficients $\left|\frac{a_{j+1}}{a_j}\right|$ would be undefined and thus the ratio test for power series fails here. The root test also fails here since the sequence $(\sqrt[j]{|a_j|})$ alternates between $\frac{1}{2}$ and 0 and thus does not converge. Therefore, we need to think of a way to get around these problems. Let us define $y = x^2$ so that the series becomes $\sum_{j=0}^{\infty} \frac{y^j}{2^j}$ which is now in the standard form with non-zero coefficients and thus we can apply the ratio test. We then deduce that the series (in the variable y) converges for $|y| < 2$ and diverges everywhere else. Translating back to the x variable, we get convergence if and only if $|x^2| < 2$ or simply $|x| < \sqrt{2}$. Thus, its radius of convergence is $\sqrt{2}$.

However, the ratio and root tests may not always work when we are finding the radius of convergence for a power series. We note for both the root and ratio tests above, we can find the radius of convergence provided that the limits $\lim_{j\to\infty}\left|\frac{a_{j+1}}{a_j}\right|$ and $\lim_{j\to\infty}\sqrt[j]{|a_j|}$ exist respectively, even after a clever substitution as in Example 12.1.9(6).

If the limits of these quantities do not exist, we cannot apply algebraic manipulations on the inequalities to get R since the limits are not real numbers (or even exist!). Furthermore, there are power series which has a positive radius of convergence but for which both of these tests fail. Here is an example:

Example 12.1.10 Let $\sum_{j=1}^{\infty} a_j x^j$ be a power series where $a_j = 1$ when j is odd and $a_j = \frac{1}{2^j}$ when j is even. We note that this series diverges for all $x \geq 1$ since $a_j x^j \not\to 0$ for any such x.

However, it converges for any $0 < x < 1$. Indeed, pick an arbitrary $x_0 \in (0, 1)$ and denote $s_n(x_0)$ as the n-th partial sum of this series at $x = x_0$. Note first that $(s_n(x_0))$ is a monotone increasing sequence. Furthermore, the subsequence $(s_{2n}(x_0))$ is convergent. Indeed, we have:

$$\begin{aligned}
s_{2n}(x_0) &= x_0 + \frac{x_0^2}{2^2} + x_0^3 + \frac{x_0^4}{2^4} + \ldots + x_0^{2n-1} + \frac{x_0^{2n}}{2^{2n}}\\
&= x_0\left(1 + \frac{x_0}{2^2}\right) + x_0^3\left(1 + \frac{x_0}{2^4}\right) + \ldots + x_0^{2n-1}\left(1 + \frac{x_0}{2^{2n}}\right)\\
&\leq x_0\left(1 + \frac{x_0}{2}\right) + x_0^3\left(1 + \frac{x_0}{2}\right) + \ldots + x_0^{2n-1}\left(1 + \frac{x_0}{2}\right)\\
&= \left(1 + \frac{x_0}{2}\right)\sum_{j=1}^{n} x_0^{2j-1} < \left(1 + \frac{x_0}{2}\right)\sum_{j=1}^{\infty} x_0^{2j-1} = \left(1 + \frac{x_0}{2}\right)\frac{x_0}{1 - x_0^2},
\end{aligned}$$

for any $n \in \mathbb{N}$. Since this subsequence is bounded and monotone, it converges. Thus, by Proposition 5.5.6, the whole sequence $(s_n(x_0))$ itself is convergent. Since x_0 is arbitrary, by Corollary 12.1.3, this means $(s_n(x))$ converges for any $|x| < 1$ and hence its radius of convergence is 1.

Now let us see what happens if we try to use the root and ratio tests:

1. If we were to apply the ratio test, the limit $\lim_{j\to\infty}\left|\frac{a_{j+1}}{a_j}\right|$ does not exist since:

$$\left|\frac{a_{j+1}}{a_j}\right| = \begin{cases} \frac{1}{2^{j+1}} & \text{for odd } j, \\ 2^j & \text{for even } j, \end{cases}$$

and so there are subsequences of these ratios that go to ∞ and 0.
2. On the other hand, when we apply the root test, the sequence $(\sqrt[j]{|a_j|})$ alternates between 1 and $\frac{1}{2}$. Hence, this sequence also does not converge for one to apply the root test successfully.

Thus, as seen in this cautionary example, we have to be aware that these techniques derived from automation of ratio and root tests may not always give us the full picture of the series since they might fail.

Therefore, we need a more solid way of determining the radius of convergence for a power series. We recall that, apart from the ratio and root tests, we have also developed the generalised ratio and root tests in Theorems 7.6.7 and 7.6.8 that utilise the more dependable limit superior and limit inferior instead.

These tests are more reliable since the limit superior and limit inferior still exist (if we include the unbounded limits $\pm\infty$) for sequences that do not converge. Furthermore, if the limits do exist, it would coincide with both of these quantities. Therefore, limit superior and inferior allow us to extend the root and ratio tests technique for determining the radius of convergence that we saw earlier. This result is called the Cauchy-Hadamard theorem:

Theorem 12.1.11 (Cauchy-Hadamard Theorem) *Let $\sum_{j=0}^{\infty} a_j(x-c)^j$ be a real power series where $c, a_j \in \mathbb{R}$ are constants. Then, the radius of convergence of this series is given by:*

$$R = \begin{cases} \frac{1}{\limsup_{j\to\infty}\sqrt[j]{|a_j|}} & \text{if } \limsup_{j\to\infty}\sqrt[j]{|a_j|} \text{ is finite}, \\ 0 & \text{if } \limsup_{j\to\infty}\sqrt[j]{|a_j|} = \infty, \\ \infty & \text{if } \limsup_{j\to\infty}\sqrt[j]{|a_j|} = 0. \end{cases}$$

Proof WLOG, let $c = 0$ and we aim to show that the series converges for $|x| < R$ and diverges for $|x| > R$. By Theorem 7.6.8, we know that the series converges absolutely at any $x \in \mathbb{R}$ such that $\limsup_{j\to\infty}\sqrt[j]{|a_j x^j|} = \limsup_{j\to\infty}\sqrt[j]{|a_j|}|x| < 1$ and diverges at $x \in \mathbb{R}$ for which $\limsup_{j\to\infty}\sqrt[j]{|a_j|}|x| > 1$. Then, we have three cases:

1. If $\limsup_{j\to\infty} \sqrt[j]{|a_j|}$ is finite and non-zero, we get the first R.
2. If $\limsup_{j\to\infty} \sqrt[j]{|a_j|} = \infty$, then for any $x \neq 0$ we would have $\limsup_{j\to\infty} \sqrt[j]{|a_j|}|x| = \infty > 1$, which means the series does not converge for any $x \neq 0$. Hence, $R = 0$.
3. If $\limsup_{j\to\infty} \sqrt[j]{|a_j|} = 0$, then for any $x \in \mathbb{R}$ we would have $\limsup_{j\to\infty} \sqrt[j]{|a_j|}|x| = 0 < 1$, which means the series converges for any $x \in \mathbb{R}$. Hence, $R = \infty$. □

Therefore, the Cauchy-Hadamard theorem gives us a sure-fire way of determining the radius of convergence of a power series.

Example 12.1.12 Let us look at some examples:

1. Referring back to Example 12.1.10, we have seen that for the power series $x + \frac{x^2}{2^2} + x^3 + \frac{x^4}{2^4} + x^5 + \frac{x^6}{2^6} + \ldots$, the sequence of roots $(\sqrt[j]{|a_j|})$ alternates between 1 and $\frac{1}{2}$. This implies that $\limsup_{j\to\infty} \sqrt[j]{|a_j|} = 1$ and thus, by applying the Cauchy-Hadamard theorem, the radius of convergence for this series is indeed $R = 1$.
2. Consider the power series $\sum_{j=0}^{\infty} a_j x^j$ where:

$$a_j = \begin{cases} 2^{\frac{j}{2}} & \text{if } j \text{ is even,} \\ 3^{\frac{j+1}{2}} & \text{if } j \text{ is odd.} \end{cases}$$

To find the radius of convergence of this series, aiming to use the Cauchy-Hadamard theorem, we consider the j-th root of the coefficient of x^j in the series, namely:

$$\sqrt[j]{|a_j|} = \begin{cases} \sqrt{2} & \text{if } j \text{ is even,} \\ \sqrt{3}\,\sqrt[2j]{3} & \text{if } j \text{ is odd.} \end{cases}$$

Since $\sqrt{3}\,\sqrt[2j]{3} > \sqrt{2}$ for all $j \in \mathbb{N}$, we can then ignore the terms with even indices when we are looking for the supremum of the roots of the coefficients. Thus:

$$\sup_{m\geq j} \sqrt[j]{|a_j|} = \sup_{\substack{m\geq j \\ m \text{ odd}}} \sqrt[j]{|a_j|} = \sqrt{3}3^{\frac{1}{2q(j)}},$$

where $q(j)$ is the smallest odd integer greater than or equal to j. Since $q(j) \geq j$ for any $j \in \mathbb{N}$, by sandwiching, we have $1 \leq \lim_{j\to\infty} 3^{\frac{1}{2q(j)}} \leq \lim_{j\to\infty} 3^{\frac{1}{2j}} = 1$. Therefore:

$$\limsup_{j\to\infty} \sqrt[j]{|a_j|} = \sqrt{3} \lim_{j\to\infty} 3^{\frac{1}{2q(j)}} = \sqrt{3}.$$

Thus, by the Cauchy-Hadamard theorem, the radius of convergence for this series is $\frac{1}{\sqrt{3}}$.

3. Consider the power series $\sum_{k=0}^{\infty}(-1)^k x^{2^k}$. We note that the coefficients of x^j are only non-zero when j are powers of 2. Therefore, we cannot apply the standard ratio test here since most of the ratios do not make sense. The root test does not hold here too since:

$$\sqrt[j]{|a_j|} = \begin{cases} 1 & \text{if } j = 2^k \text{ for } k \in \mathbb{N}, \\ 0 & \text{otherwise.} \end{cases}$$

Hence, the limit of these roots does not exist. Therefore, to find the radius of convergence, we use the Cauchy-Hadamard theorem. From above, we note that for any $j \in \mathbb{N}$, we have $\sup_{m \geq j} \sqrt[m]{|a_m|} = 1$ and hence $\limsup_{j\to\infty} \sqrt[j]{|a_j|} = 1$. Thus, the radius of convergence of this series is the reciprocal of this value, which is 1.

12.2 Continuity of Power Series

From Remark 12.1.5(1), we have noted that a power series is absolutely convergent within $B_R(c)$ where R is its radius of convergence. This is a very nice bonus as it implies that we can rearrange the terms series without changing its value, as per Theorem 8.1.3. In addition to this, over any compact subset of $B_R(c)$, this convergence is uniform to its limit. The proof of this result extends from the proof for Corollary 12.1.3.

Proposition 12.2.1 *Let $\sum_{j=0}^{\infty} a_j(x-c)^j$ be a real power series where $c, a_j \in \mathbb{R}$ are constants with radius of convergence $R > 0$.*

1. *For any r such that $0 \leq r < R$, the power series converges uniformly in the set $\bar{B}_r(c)$.*
2. *More generally, the power series converges uniformly on any compact interval $K \subseteq B_R(c)$ if R is finite and any compact interval $K \subseteq \mathbb{R}$ if R is infinite.*

Proof WLOG, let us assume that $c = 0$. We prove the assertions one by one.

1. Pick any x_0 such that $r < |x_0| < R$. Since the series $\sum_{j=0}^{\infty} a_j x_0^j$ converges, the sequence $(a_n x_0^n)$ converges to 0 and hence are bounded by some $M > 0$, namely $|a_n x_0^n| \leq M$ for all $n \in \mathbb{N}_0$. For all $x \in \bar{B}_r(0)$, we have the bound:

$$|a_n x^n| \leq |a_n| r^n = |a_n x_0^n| \left|\frac{r^n}{x_0^n}\right| \leq M \left|\frac{r}{x_0}\right|^n \quad \Rightarrow \quad \sup_{x \in \bar{B}_r(0)} |a_n x^n| \leq M \left|\frac{r}{x_0}\right|^n,$$

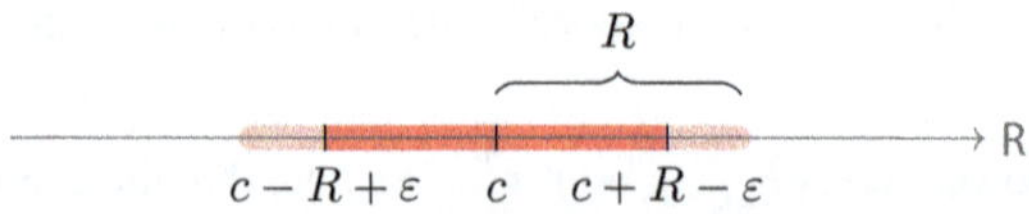

Fig. 12.2 Suppose that a series centred at c has radius of convergence $R > 0$. It converges for any x the red open interval. Moreover, Proposition 12.2.1 says that it converges uniformly on any compact interval $K \subseteq B_R(c)$. An example is the compact interval $[c - R + \varepsilon, c + R - \varepsilon]$ where $\varepsilon > 0$ in darker red above

for all $n \in \mathbb{N}_0$. Furthermore, we note that the series $\sum_{j=1}^{\infty} M \left|\frac{r}{x_0}\right|^j$ converges since it is a geometric series with common ratio $\left|\frac{r}{x_0}\right| < 1$. We can thus apply the Weierstrass M-test and conclude that the series $\sum_{j=0}^{\infty} a_j x^j$ converges uniformly on $\bar{B}_r(0)$.

2. For a general compact interval $K = [a, b] \subseteq B_R(0)$, if we set $r = \max\{|b|, |a|\}$, then $0 \leq r < R$ and thus $K \subseteq \bar{B}_r(0) \subseteq B_R(0)$. From the previous assertion, the power series converges uniformly on $\bar{B}_r(0)$ and hence on the interval K. □

A visualisation for the region where the power series converges uniformly as discussed in Proposition 12.2.1 is given in Fig. 12.2. However, the uniform convergence may not be necessarily true on the whole open ball $B_R(c)$. Indeed, Proposition 11.3.1 says that the uniform limit of a sequence of bounded functions must also be bounded. But there are many power series which does not converge on the boundary and hence becomes unbounded near the boundaries of the ball $B_R(c)$.

Example 12.2.2 An example of this can be obtained from Example 11.4.15 where we saw $\sum_{j=0}^{\infty} x^j$ converges to $\frac{1}{1-x}$ pointwise, but not uniformly, in $B_1(0)$. However, for any small $\varepsilon > 0$ at all, we have $B_{1-\varepsilon}(0) \subseteq \bar{B}_{1-\varepsilon}(0) \subseteq B_1(0)$. As guaranteed by Proposition 12.2.1, this convergence is uniform on $\bar{B}_{1-\varepsilon}(0)$.

Regardless, Proposition 12.2.1 is already a very strong result since uniform convergence is very desirable when it comes to functions sequences as it preserves local properties of the function sequences. We shall see that this result will be useful in Chaps. 13 and 14 when we look at another local property of functions which is differentiability.

A direct corollary that we can present now is the local property of continuity:

Corollary 12.2.3 *Let $\sum_{j=0}^{\infty} a_j (x - c)^j$ be a real power series where $c, a_j \in \mathbb{R}$ are constants with radius of convergence $R > 0$. Then, it is continuous in $B_R(c)$ if R is finite and everywhere on $\mathbb{R}$ if R is infinite.*

Proof WLOG, let $c = 0$ and assume that R is finite. We show that the series is continuous at any arbitrary $x_0 \in B_R(0)$. Since $B_R(0)$ is open, there exists an $\varepsilon > 0$

such that $B_\varepsilon(x_0) \subseteq B_R(0)$. Consider the compact interval $[x_0 - \frac{\varepsilon}{2}, x_0 + \frac{\varepsilon}{2}] = \bar{B}_{\frac{\varepsilon}{2}}(x_0) \subseteq B_\varepsilon(x_0) \subseteq B_R(0)$ which contains x_0. By Proposition 12.2.1, the power series converges uniformly here. Since all the partial sums are continuous at x_0, by Theorem 11.4.16, the power series must also be continuous at x_0. Finally, since $x_0 \in B_R(0)$ is arbitrary, we conclude that the series is continuous everywhere in $B_R(0)$. The case for infinite R can also be proven in the same manner. □

We have seen in Proposition 12.2.1 that we can at best guarantee uniform convergence of a power series in a compact subset bounded away from the boundaries of the domain of convergence. This is necessary to address the cases for which the power series blows up at the boundary of its domain of convergence such as in Example 12.2.2.

But what if the power series does converge at either of the boundaries? We have seen several examples of such series in Examples 12.1.6(2) and 12.1.9(5). Can we extend the uniform convergence of the series to include these boundary points? In this case, the answer is yes and this result is called Abel's theorem.

Theorem 12.2.4 (Abel's Theorem) *Let $\sum_{j=0}^\infty a_j(x-c)^j$ be a real power series where $c, a_j \in \mathbb{R}$ are constants with finite radius of convergence $R > 0$. If the power series converges at $x = c + R$, then it converges uniformly on $[c, c+R]$.*

Proof WLOG, suppose that $c = 0$. Let us rewrite the power series as $\sum_{j=0}^\infty a_j R^j \frac{x^j}{R^j}$. By assumption, the constant real series $\sum_{j=0}^\infty a_j R^j$ converges and, trivially, this convergence is uniform on $[0, R]$ as the series is independent of x.

Moreover, the sequence of functions (g_j) where $g_j : [0, R] \to \mathbb{R}$ is defined as $g_j(x) = \frac{x^j}{R^j}$ is uniformly bounded since $|g_j(x)| \leq 1$ for all $j \in \mathbb{N}_0$ and $x \in [0, R]$. In addition, the sequence is pointwise decreasing on $[0, R]$ since for any $x \in [0, R]$ and $j \in \mathbb{N}_0$ we have $g_j(x) = \frac{x^j}{R^j} \geq \frac{x^j}{R^j}\frac{x}{R} = g_{j+1}(x)$.

Thus, we can apply Abel's test for uniform convergence from Theorem 11.4.14 to conclude that $\sum_{j=0}^\infty a_j R^j \frac{x^j}{R^j} = \sum_{j=0}^\infty a_j x^j$ converges uniformly on $[0, R]$. □

By the same argument, if we have convergence of the series $\sum_{j=0}^\infty a_j(x-c)^j$ at the other end of the domain of convergence, namely at $x = c - R$, we can also show that the series converges uniformly on the closed interval $[c - R, c]$.

The following consequence of Abel's theorem extends Corollary 12.2.3. It states that the continuity of the power series can also be extended to the boundary of the domain of convergence if the power series converges at the boundary. Namely:

Corollary 12.2.5 *Let $\sum_{j=0}^{\infty} a_j(x-c)^j$ be a real power series where $c, a_j \in \mathbb{R}$ are constants with finite radius of convergence $R > 0$. If the power series converges at $x = c + R$, then the it is continuous on $[c, c + R]$ and the left-limit at $x = c + R$ is given as:*

$$\lim_{x \uparrow c+R} \sum_{j=0}^{\infty} a_j(x-c)^j = \sum_{j=0}^{\infty} a_j R^j.$$

Proof WLOG, suppose that $c = 0$. Let $s_n(x) = \sum_{j=0}^{n} a_j x^j$ for all $n \in \mathbb{N}$. The partial sums are all continuous and converges uniformly on $[0, R]$. By Theorem 11.4.17, the limit $s(x) = \sum_{j=0}^{\infty} a_j x^j$ is also continuous on $[0, R]$. Moreover, since $\lim_{x \uparrow R} s(x)$ and $\lim_{x \uparrow R} s_n(x)$ for any $n \in \mathbb{N}$ all exist, by Theorem 11.4.16, we can commute the infinite summation and limit as:

$$\lim_{x \uparrow R} \sum_{j=0}^{\infty} a_j x^j = \sum_{j=0}^{\infty} \lim_{x \uparrow R} a_j x^j = \sum_{j=0}^{\infty} a_j R^j,$$

and we are done. □

An analogous result can be obtained if the power series converges at the other endpoint of the domain of convergence $x = c - R$.

12.3 Algebra of Power Series

Since we know how to reliably find the radius of convergence of a series using Cauchy-Hadamard theorem, let us define addition and scalar multiplication of power series (similar to what we have done in Proposition 7.2.8) and determine where they converge.

Proposition 12.3.1 *Let $\sum_{j=0}^{\infty} a_j(x-c)^j$ and $\sum_{j=0}^{\infty} b_j(x-c)^j$ be two power series for some $c \in \mathbb{R}$ with radius of convergence R_1 and R_2 respectively. Then:*

1. *For any $\lambda \in \mathbb{R}$, the power series $\sum_{j=0}^{\infty} \lambda a_j(x-c)^j$ converges as well with radius of convergence R_1 and is equal to $\lambda \sum_{j=0}^{\infty} a_j(x-c)^j$.*
2. *For any $k \in \mathbb{N}$, the power series $\sum_{j=0}^{\infty} a_j(x-c)^{k+j}$ converges as well with radius of convergence R_1 and is equal to $(x-c)^k \sum_{j=0}^{\infty} a_j(x-c)^j$.*
3. *For any $x \in B_{\min\{R_1,R_2\}}(c)$ we have:*

$$\sum_{j=0}^{\infty} a_j(x-c)^j + \sum_{j=0}^{\infty} b_j(x-c)^j = \sum_{j=0}^{\infty} (a_j + b_j)(x-c)^j. \tag{12.1}$$

Furthermore, the radius of convergence R of the series on the RHS of (12.1) satisfies $R \geq \min\{R_1, R_2\}$.

Proof We shall prove the third assertion only because the first two are straightforward.

3. WLOG, let $c = 0$. Pick any $x \in \mathbb{R}$ such that $|x| < \min\{R_1, R_2\}$. Then, we know both $\sum_{j=0}^{\infty} a_j x^j$ and $\sum_{j=0}^{\infty} b_j x^j$ converge. Thus, the sequence of partial sums $s_n = \sum_{j=0}^{n}(a_j + b_j)x^j = \sum_{j=0}^{n} a_j x^j + \sum_{j=0}^{n} b_j x^j$ converges by the algebra of limits. Furthermore, the algebra of limits also gives us the equality:
$$\sum_{j=0}^{\infty}(a_j + b_j)x^j = \sum_{j=0}^{\infty} a_j x^j + \sum_{j=0}^{\infty} b_j x^j \quad \text{for all } |x| < \min\{R_1, R_2\}.$$
This means the radius of convergence R for the sum is at least $\min\{R_1, R_2\}$. □

Remark 12.3.2 We make several remarks regarding Proposition 12.3.1:

1. The equality (12.1) is intuitively true because within the smaller radius of convergence, we know that each of the series converge and has a value in $\mathbb{R}$. Hence, they can be added pointwise here. Therefore, the convergence of the series $\sum_{j=0}^{\infty}(a_j + b_j)(x - c)^j$ within the open ball $B_{\min\{R_1, R_2\}}(c)$ is guaranteed.
2. However, the resulting series might converge in an even bigger set. This might happen because summing up the two original series could cancel out their diverging parts. Thus, the resulting power series $\sum_{j=0}^{\infty}(a_j + b_j)(x - c)^j$ could actually converge in a bigger set $B_R(c)$ for some $R \geq \min\{R_1, R_2\}$ even though the two original series converge in a strictly smaller set.

Example 12.3.3 Let us look at some examples to explain Remark 12.3.2 better.

1. Consider two power series $\sum_{j=0}^{\infty} x^j$ and $\sum_{j=0}^{\infty}(2x)^j$ which has radius of convergence $R_1 = 1$ and $R_2 = \frac{1}{2}$ respectively. From Proposition 12.3.1, the sum of these two series is given by $\sum_{j=0}^{\infty}(1 + 2^j)x^j$ which converges for $|x| < \min\{R_1, R_2\} = \frac{1}{2}$. Now we check that the radius of convergence for this series is exactly $\frac{1}{2}$. To do this, we use the ratio test to deduce:
$$\lim_{j\to\infty}\left|\frac{a_{j+1}}{a_j}\right| = \lim_{j\to\infty}\frac{1 + 2^{j+1}}{1 + 2^j} = 2,$$
and hence the series has radius of convergence $R = \frac{1}{2}$.
2. As we have mentioned in Remark 12.3.2(2), the radius of convergence for the sum might be strictly greater than the minimum radius of convergence of the two power series.

On the extreme case, it might even be bigger than the larger radius. A simple example would be the power series $\sum_{j=0}^{\infty} x^j$ and $\sum_{j=0}^{\infty} -x^j$, both of which having the radius of convergence 1. Proposition 12.3.1 says that the radius of convergence R for the sum of these two series is at least 1. However, we know that the sum of the power series on their domain of convergence is identically 0, which does not only converge in $|x| < 1$, but on the whole of $\mathbb{R}$. So we have the strict inequality $R = \infty > \min\{R_1, R_2\} = 1$.

3. Here is a non-trivial example for the above. Consider two power series $\sum_{j=0}^{\infty}(\frac{1}{2^j} - 1)x^j$ and $\sum_{j=0}^{\infty} x^j$. One can check that the radius of these power series are both equal to 1. So we expect from Proposition 12.3.1 that the radius of convergence for their sum is at least 1. When we attempt to add up the two power series, they do not make sense for any point $|x| \geq 1$ because both of them are not real numbers here. In short, we have the equality:

$$\sum_{j=0}^{\infty}\left(\frac{1}{2^j} - 1\right)x^j + \sum_{j=0}^{\infty} x^j = \sum_{j=0}^{\infty} \frac{x^j}{2^j} \quad \text{only for } |x| < 1,$$

where the first two series make sense. But the sum of the series, which is $\sum_{j=0}^{\infty} \frac{x^j}{2^j}$ actually converges in a larger set. Indeed, by using the ratio test, we can easily show that the RHS power series has radius of convergence $R = 2 > \min\{R_1, R_2\} = 1$.

We have seen how we can scale and add power series to create a new power series in Proposition 12.3.1. Now we want to look how we can multiply power series together. If the radius of convergence for these series are non-zero, then their product can be expressed as their Cauchy product.

The convergence and equality of the product to the Cauchy product in Mertens' theorem (Theorem 8.3.3) hinge on the condition that at least one of the series converges absolutely within its domain of convergence. However, since the convergence of a power series is also absolute as have noted in Remark 12.1.5(1), we get the equality of the product to their Cauchy product for free!

Note that since the power series are defined with indices starting from 0, we use the convention in Remark 8.3.2 for the next result.

Proposition 12.3.4 *Let $\sum_{j=0}^{\infty} a_j(x-c)^j$ and $\sum_{j=0}^{\infty} b_j(x-c)^j$ be real power series with $c, a_j, b_j \in \mathbb{R}$. Suppose that the radius of convergence of these series are R_1 and R_2 respectively. If their Cauchy product is $\sum_{j=0}^{\infty} c_j(x-c)^j$ where $c_j = \sum_{i=0}^{j} a_i b_{j-i}$, then for any $x \in B_{\min\{R_1,R_2\}}(c)$ we have the equality:*

$$\left(\sum_{j=0}^{\infty} a_j(x-c)^j\right)\left(\sum_{j=0}^{\infty} b_j(x-c)^j\right) = \sum_{j=0}^{\infty} c_j(x-c)^j.$$

Furthermore, the radius of convergence R of the Cauchy product satisfies $R \geq \min\{R_1, R_2\}$.

Proof WLOG, let us assume that $c = 0$. Pick any $x \in \mathbb{R}$ such that $|x| < \min\{R_1, R_2\}$. Since x is within the radius of convergence of both of the series, both of the series converge at x. In fact, both of these series converge absolutely at x by the proof of Proposition 12.1.1. Thus, we can apply Mertens' theorem that says their Cauchy product is equal to their product at x, namely:

$$\sum_{j=0}^{\infty} c_j x^j = \left(\sum_{j=0}^{\infty} a_j x^j\right)\left(\sum_{j=0}^{\infty} b_j x^j\right).$$

This equality is true for any x such that $x \in B_{\min\{R_1,R_2\}}(0)$, giving us the desired equality. As a result, the Cauchy product converges for $|x| < \min\{R_1, R_2\}$ and hence its radius of convergence R is at least $\min\{R_1, R_2\}$. □

Example 12.3.5 Let us look at some examples:

1. We note that, similar to the sum of power series, the radius of convergence for a Cauchy product might be strictly bigger than the minimum of the two radius of convergence for the constituent power series.
   Consider the two power series $1 - x$ and $\sum_{j=0}^{\infty} x^j$. Note that the former is also a power series of the form $\sum_{j=0}^{\infty} a_j x^j$ where $a_j = 0$ for $j \geq 2$. The radius of convergence for these power series are ∞ and 1 respectively. By Proposition 12.3.4, we expect that their Cauchy product would have radius of convergence $R \geq 1$.
   If we set (a_j) and (b_j) for $j \in \mathbb{N}_0$ as the coefficients of the power series $1-x$ and $\sum_{j=0}^{\infty} x^j$ respectively, we can compute the Cauchy product as $\sum_{j=0}^{\infty} c_j x^j$ where $c_j = \sum_{i=0}^{j} a_i b_{j-i} = \sum_{i=0}^{j} a_i$ so that we get $c_0 = 1$ and $c_j = 0$ for all $j \geq 1$. This means the Cauchy product is simply 1 and we have the equality:

$$(1 - x)\sum_{j=0}^{\infty} x^j = 1 \quad \text{only for } |x| < 1.$$

   However, we notice that the Cauchy product, which is simply 1, converges not only for $|x| < 1$, but the whole of $\mathbb{R}$ since it is just a constant. Thus, $R = \infty > \min\{R_1, R_2\} = 1$.
   An alternative way to get this Cauchy product is to note that $\sum_{j=0}^{\infty} x^j = \frac{1}{1-x}$ for $|x| < 1$ and proceed algebraically to get $(1-x)\sum_{j=0}^{\infty} x^j = \frac{1-x}{1-x} = 1$ on $|x| < 1$.

2. Consider two power series $\sum_{j=0}^{\infty} x^j$ and $\sum_{j=0}^{\infty}(2x)^j$. We have seen that the radius of convergence for these series are 1 and $\frac{1}{2}$ respectively. Their Cauchy product is given by:

$$\left(\sum_{j=0}^{\infty} x^j\right)\left(\sum_{j=0}^{\infty}(2x)^j\right) = \sum_{j=0}^{\infty} c_j x^j, \tag{12.2}$$

where $c_j = \sum_{i=0}^{j} 2^{j-i} = 2^{j+1} - 1$. According to Proposition 12.3.4, the radius of convergence of this Cauchy product is at least $\frac{1}{2}$. To find the exact radius of convergence, we use the ratio test on the Cauchy product. We find:

$$\lim_{j\to\infty}\left|\frac{c_{j+1}}{c_j}\right| = \lim_{j\to\infty}\left|\frac{2^{j+2}-1}{2^{j+1}-1}\right| = \lim_{j\to\infty}\frac{2^{j+2}-1}{2^{j+1}-1} = 2,$$

and hence the radius of convergence is exactly $\frac{1}{2}$. We can easily check that the power series does not converge at $x = \pm\frac{1}{2}$.
Furthermore, we know what the closed form of each of the power series on the LHS since they are both geometric series. Namely, we have:

$$\sum_{j=0}^{\infty} x^j = \frac{1}{1-x} \quad \text{for } |x| < 1 \quad \text{and} \quad \sum_{j=0}^{\infty}(2x)^j = \frac{1}{1-2x} \quad \text{for } |x| < \frac{1}{2}.$$

Thus, the equality (12.2) can be rewritten for $|x| < \frac{1}{2}$ where both of the above hold as:

$$\frac{1}{1-x}\frac{1}{1-2x} = \frac{1}{1-3x+2x^2} = \sum_{j=1}^{\infty}(2^{j+1}-1)x^j.$$

If we denote $f : \mathbb{R}\backslash\{1, \frac{1}{2}\} \to \mathbb{R}$ as $f(x) = \frac{1}{1-3x+2x^2}$, this function has a domain of all real numbers except two points. However, the series $\sum_{j=1}^{\infty}(2^{j+1}-1)x^j$ above converges only on $(-\frac{1}{2}, \frac{1}{2})$. Therefore, this equality only holds within a subset of the domain of f. We call this power series as the power series for the function f expanded at or centred at $x = 0$.
3. Conversely, we can also find the power series for the function f above by using partial fractions. Indeed, we have:

$$f(x) = \frac{1}{1-3x+2x^2} = \frac{1}{1-x}\frac{1}{1-2x} = -\frac{1}{1-x} + \frac{2}{1-2x},$$

from which we can then find the respective power series and radius of convergence. As a result, we can sum them up over a suitable domain of convergence using Proposition 12.3.1. We should also get the series $\sum_{j=1}^{\infty}(2^{j+1}-1)x^j$ which converges only for $|x| < \frac{1}{2}$.

4. To find the power series expansion of the same function f at other points, say $x = 3$, we expand the terms $\frac{1}{1-x}$ and $\frac{1}{1-2x}$ using geometric series. To do this, we rewrite the fractions in a suitable form and apply our knowledge on geometric series, namely:

$$\frac{1}{1-x} = \frac{1}{-2-(x-3)} = \frac{-\frac{1}{2}}{1+\frac{(x-3)}{2}} = -\frac{1}{2}\sum_{j=0}^{\infty}\left(-\frac{(x-3)}{2}\right)^j$$
$$= \sum_{j=0}^{\infty}\frac{1}{(-2)^{j+1}}(x-3)^j,$$

$$\frac{1}{1-2x} = \frac{1}{-5-2(x-3)} = \frac{-\frac{1}{5}}{1+\frac{2(x-3)}{5}} = -\frac{1}{5}\sum_{j=0}^{\infty}\left(-\frac{2(x-3)}{5}\right)^j$$
$$= \sum_{j=0}^{\infty}\frac{-(-2)^j}{5^{j+1}}(x-3)^j,$$

where the former is true if $|x-3| < 2$ and the latter is true if $|x-3| < \frac{5}{2}$ by ratio test. The Cauchy product of these series then has radius of convergence $R \geq \min\{2, \frac{5}{2}\} = 2$ and is given by $\sum_{j=0}^{\infty} c_j(x-3)^j$ where:

$$c_j = \sum_{i=0}^{j}\frac{1}{(-2)^{i+1}}\frac{-(-2)^{j-i}}{5^{j-i+1}} = \frac{1}{10}\left(\frac{-2}{5}\right)^j\sum_{i=0}^{j}\left(\frac{5}{4}\right)^i = \frac{(-2)^{j+1}}{5^{j+1}} + \frac{(-1)^j}{2^{j+1}}.$$

Thus, we also have the power series expansion:

$$\frac{1}{1-x}\frac{1}{1-2x} = \frac{1}{1-3x+2x^2} = \sum_{j=1}^{\infty}\left(\frac{(-2)^{j+1}}{5^{j+1}} + \frac{(-1)^j}{2^{j+1}}\right)(x-3)^j,$$

which is valid only for $|x-3| < 2$ or equivalently for $1 < x < 5$.

This is in contrast to the series that we have found earlier which equals to f at $-\frac{1}{2} < x < \frac{1}{2}$.

In Examples 12.3.5(3) and (4), we have seen that the function $f(x) = \frac{1}{1-3x+2x^2}$ can be expressed as two different power series with different centres, namely:

$$f(x) = \sum_{j=1}^{\infty}(2^{j+1} - 1)x^j \qquad \text{only for } |x| < \frac{1}{2},$$

$$f(x) = \sum_{j=1}^{\infty}\left(\frac{(-2)^{j+1}}{5^{j+1}} + \frac{(-1)^j}{2^{j+1}}\right)(x-3)^j \quad \text{only for } |x-3| < 2.$$

As demonstrated above, depending on where we chose to expand it at, the power series might look different and converge for different radius of convergence. We shall see more of this later when we try to expand more general functions as power series in Chap. 16. A useful result to prepare us in that direction is the following:

Proposition 12.3.6 *Consider the power series $\sum_{j=0}^{\infty} a_j(x-c)^j$ and $\sum_{j=1}^{\infty} ja_j(x-c)^{j-1}$ for some constants $c, a_j \in \mathbb{R}$. If the radius of convergence for both of these series are non-zero, then their radius of convergence are equal.*

Proof WLOG, we assume that $c = 0$. Suppose that the series $\sum_{j=0}^{\infty} a_j x^j$ and $\sum_{j=1}^{\infty} ja_j x^{j-1}$ have radius of convergence R_1 and R_2 respectively (which could also be ∞). We aim to show that $R_2 = R_1$.

1. First, we show that $R_2 \leq R_1$. Pick any $x \in \mathbb{R}$ such that $x \in B_{R_2}(0)$ so that the series $\sum_{j=1}^{\infty} ja_j x^{j-1}$ converges absolutely. For any $j \geq 1$, we have:

$$|a_j x^j| = |x||a_j x^{j-1}| \leq |x||ja_j x^{j-1}|.$$

   Since the series $\sum_{j=1}^{\infty} |x||ja_j x^{j-1}| = |x|\sum_{j=1}^{\infty}|ja_j x^{j-1}|$ converges, by direct comparison test, the series $\sum_{j=1}^{\infty} |a_j x^j|$ converges and thus the series $\sum_{j=0}^{\infty} |a_j x^j|$ also converges by adding in the $|a_0|$ term. Hence, the point x also lies in the domain of convergence of the series $\sum_{j=0}^{\infty} a_j x^j$, namely $x \in D \subseteq \bar{B}_{R_1}(0)$. Therefore $B_{R_2}(0) \subseteq \bar{B}_{R_1}(0)$. In other words, $(-R_2, R_2) \subseteq [R_1, R_1]$ which proves $R_2 \leq R_1$.
2. To show that $R_1 \leq R_2$, first notice that for $x = 0$, both of the series converge. So let us pick any $x \in \mathbb{R}$ such that $0 < |x| < R_1$. Then, there exists an $r \in \mathbb{R}$ such that $|x| < r < R_1$. We know that r lies within the domain of convergence of the series $\sum_{j=0}^{\infty} a_j x^j$, so the series $\sum_{j=1}^{\infty} a_j r^j$ converges. This means the terms of this series converge to 0 and hence the sequence $(a_j r^j)$ is bounded, say $|a_j r^j| \leq M$ for all $j \in \mathbb{N}$ for some $M > 0$. Thus, for all such j, we have:

$$|ja_j x^{j-1}| = \frac{j|a_j r^j|}{|x|}\left|\frac{x}{r}\right|^j \leq \frac{Mj}{|x|}\left|\frac{x}{r}\right|^j. \tag{12.3}$$

Consider the series $\sum_{j=1}^{\infty} \frac{Mj}{|x|} \left|\frac{x}{r}\right|^j$. By the ratio test, we find:

$$\lim_{j\to\infty} \left| \frac{\frac{M(j+1)}{|x|} \frac{x^{j+1}}{r^{j+1}}}{\frac{Mj}{|x|} \frac{x^j}{r^j}} \right| = \left|\frac{x}{r}\right| \lim_{j\to\infty} \frac{j+1}{j} = \left|\frac{x}{r}\right| < 1,$$

and so this series converges. Thus, by comparison test using the inequality (12.3), the series $\sum_{j=1}^{\infty} |ja_j x^{j-1}|$ also converges. Therefore, we have shown that any $x \in B_{R_1}(0)$ also lies in the domain of convergence of the series $\sum_{j=1} ja_j x^{j-1}$ and so $x \in \bar{B}_{R_2}(0)$. This proves $R_1 \leq R_2$.

Thus, we conclude that $R_1 = R_2$. □

Remark 12.3.7 An alternative proof of Proposition 12.3.6 is as follows. WLOG, we assume that $c = 0$. Suppose that the series $\sum_{j=0}^{\infty} a_j x^j$ and $\sum_{j=1}^{\infty} ja_j x^{j-1}$ have nonzero radius of convergence R_1 and R_2 respectively.

Consider the series $x\sum_{j=1}^{\infty} ja_j x^{j-1} = \sum_{j=1}^{\infty} ja_j x^j$ with radius of convergence R. We can view this as the Cauchy product of two series x and $\sum_{j=1}^{\infty} ja_j x^{j-1}$ and so, by Proposition 12.3.4, we have $R \geq R_2$. However, we also have the opposite inequality. Indeed, pick any $x_0 \in B_R(0)$. If $x_0 = 0$, then both series clearly converge. Otherwise, we have:

$$\frac{1}{x_0} \sum_{j=1}^{\infty} ja_j x_0^j < \infty \quad \Rightarrow \quad \sum_{j=1}^{\infty} ja_j x_0^{j-1} < \infty,$$

which implies $x_0 \in B_{R_2}(0)$ as well. Thus, we have $R \leq R_2$ and so $R = R_2$.

On the other hand, by using Proposition 5.10.13, we have:

$$\limsup_{n\to\infty} \sqrt[n]{n|a_n|} = \lim_{n\to\infty} n^{\frac{1}{n}} \limsup_{n\to\infty} \sqrt[n]{|a_n|} = \limsup_{n\to\infty} \sqrt[n]{|a_n|},$$

and hence the Cauchy-Hadamard theorem implies $R = R_1$. Putting everything together we have $R_1 = R = R_2$, thus completing the proof.

By change of index labelling in Proposition 12.3.6, we also have the following result:

Proposition 12.3.8 *Consider the power series $\sum_{j=0}^{\infty} a_j(x-c)^j$ and $\sum_{j=0}^{\infty} \frac{a_j}{j+1}(x-c)^{j+1}$ for some constants $c, a_j \in \mathbb{R}$. If the radius of convergence for both of these series are non-zero, then their radius of convergence are equal.*

Based on the above propositions, the power series $\sum_{j=0}^{\infty} a_j(x - c)^j$, $\sum_{j=1}^{\infty} ja_j(x - c)^{j-1}$, and $\sum_{j=0}^{\infty} \frac{a_j}{j+1}(x - c)^{j+1}$ all have the same radius of convergence, if they are non-zero. However, this does not mean that their domain of convergence are also all the same. Let us look at an example when this happens.

Example 12.3.9 Let $\sum_{j=0}^{\infty}(-1)^j x^j$, $\sum_{j=1}^{\infty} j(-1)^j x^{j-1}$, and $\sum_{j=0}^{\infty} \frac{(-1)^j x^{j+1}}{j+1}$ be three power series. The radius of convergence for these series are all $R = 1$. Let us investigate the convergence of these series at ± 1. Both the first and second series do not converge at both $x = \pm 1$. On the other hand, the third series converges at $x = 1$ by alternating series test, but diverges at $x = -1$. Therefore, the domain of convergence for the first and second series are identical, but the third series has a bigger domain of convergence.

12.4 Exponentiation and Logarithm Revisited

Let us now look at a very important power series, namely $\sum_{j=0}^{\infty} \frac{x^j}{j!}$. We have seen that this series converges for all $x \in \mathbb{R}$ in Example 12.1.9. Since this series has a real value for any $x \in \mathbb{R}$, it defines a real-valued function on the set of real numbers. Let us call this series $E(x)$ where $E : \mathbb{R} \to \mathbb{R}$.

Now let us find $E(x+y)$ for two real numbers $x, y \in \mathbb{R}$. Putting this in the power series and using the binomial expansion, we get:

$$\begin{aligned} E(x+y) = \sum_{j=0}^{\infty} \frac{(x+y)^j}{j!} &= \sum_{j=0}^{\infty} \frac{1}{j!} \sum_{k=0}^{j} \binom{j}{k} x^k y^{j-k} \\ &= \sum_{j=0}^{\infty} \frac{1}{j!} \sum_{k=0}^{j} \frac{j!}{k!(j-k)!} x^j y^{j-k} \\ &= \sum_{j=0}^{\infty} \sum_{k=0}^{j} \frac{x^k}{k!} \frac{y^{j-k}}{(j-k)!}, \end{aligned}$$

and now we note that this is a Cauchy product (using the convention in Remark 8.3.2, with the indices in the series starting from 0) of two series $\sum_{j=0}^{\infty} \frac{x^j}{j!}$ and $\sum_{j=0}^{\infty} \frac{y^j}{j!}$. Furthermore, since both of these power series converge absolutely as noted in Remark 12.1.5, we apply Mertens' theorem from Theorem 8.3.3 to write it as:

$$\begin{aligned} E(x+y) = \sum_{j=0}^{\infty} \frac{(x+y)^j}{j!} &= \sum_{j=0}^{\infty} \sum_{k=0}^{j} \frac{x^k}{k!} \frac{y^{j-k}}{(j-k)!} \\ &= \left(\sum_{j=0}^{\infty} \frac{x^j}{j!} \right) \left(\sum_{j=0}^{\infty} \frac{y^j}{j!} \right) = E(x)E(y), \end{aligned}$$

and so the series satisfies the equation $E(x + y) = E(x)E(y)$ for any x, $y \in \mathbb{R}$.

Moreover, since this series converges uniformly in $[-1, 1]$, it is continuous at $x = 0$. Thus, by using the functional equation in Exercise 10.29, we can conclude that $E(x) = E(1)^x$ for any $x \in \mathbb{R}$.

Clearly $E(1) = \sum_{j=0}^{\infty} \frac{1}{j!} > 0$ and so $E(x) = E(1)^x > 0$ for any $x \in \mathbb{R}$.

But what is this number $E(1)$? We have actually seen this number before under a different guise in Example 5.4.4. It is the Euler-Napier constant e. Let us show this. In fact, we shall prove a more general equality first.

Proposition 12.4.1 *For all $x \in \mathbb{R}$, we have $E(x) = \sum_{j=0}^{\infty} \frac{x^j}{j!} = \lim_{n\to\infty} \left(1 + \frac{x}{n}\right)^n$.*

Proof For $x = 0$, this is clearly true. We inspect positive and negative x separately.

1. Fix any $x > 0$.
   (a) We show $\lim_{n\to\infty} \left(1 + \frac{x}{n}\right)^n \leq \sum_{j=0}^{\infty} \frac{x^j}{j!}$ first. This can be obtained by using the binomial expansion. For a fixed $n \in \mathbb{N}_0$ we have:

$$
\begin{aligned}
\left(1+\frac{x}{n}\right)^n &= \sum_{j=0}^{n} \binom{n}{j} \frac{x^j}{n^j} \\
&= \sum_{j=0}^{n} \frac{n!}{j!(n-j)!} \frac{x^j}{n^j} \\
&= \sum_{j=0}^{n} \frac{x^j}{j!} \frac{n}{n} \frac{n-1}{n} \frac{n-2}{n} \cdots \frac{n-(j-1)}{n} \\
&= \sum_{j=0}^{n} \frac{x^j}{j!} \left(1-\frac{1}{n}\right)\left(1-\frac{2}{n}\right)\cdots\left(1-\frac{j-1}{n}\right) \qquad (12.4) \\
&< \sum_{j=0}^{n} \frac{x^j}{j!} \leq \sum_{j=0}^{\infty} \frac{x^j}{j!} = E(x),
\end{aligned}
$$

   since all of the bracketed terms in (12.4) are all strictly smaller than 1. Therefore, the sequence (a_n) where $a_n = \left(1 + \frac{x}{n}\right)^n$ is bounded from above. Moreover, we can show that this sequence is increasing in the same way as we did in Example 5.4.4. This means the limit $\lim_{n\to\infty} \left(1 + \frac{x}{n}\right)^n$ exists. Thus, by taking the limit as $n \to \infty$ on both sides, since limits preserve weak inequalities, we get the desired inequality.

   (a) To get the opposite inequality, from the above binomial expansion, note that all of the terms in the summation (12.4) are positive. So, for any $0 \leq m \leq n$ we have the inequality:

$$
\left(1+\frac{x}{n}\right)^n \geq \sum_{j=0}^{m} \frac{x^j}{j!} \left(1-\frac{1}{n}\right)\left(1-\frac{2}{n}\right)\cdots\left(1-\frac{j-1}{n}\right).
$$

By keeping m fixed, let us take the limit as $n \to \infty$ on both sides of the inequality. We can then apply the algebra of limits on the RHS since it is a finite summation of products of finitely many terms to get:

$$\lim_{n\to\infty}\left(1+\frac{x}{n}\right)^n \geq \lim_{n\to\infty}\sum_{j=0}^{m}\frac{x^j}{j!}\left(1-\frac{1}{n}\right)\left(1-\frac{2}{n}\right)\dots\left(1-\frac{j-1}{n}\right)$$
$$=\sum_{j=0}^{m}\frac{x^j}{j!}\lim_{n\to\infty}\left(1-\frac{1}{n}\right)\dots\lim_{n\to\infty}\left(1-\frac{j-1}{n}\right)=\sum_{j=0}^{m}\frac{x^j}{j!}.$$

We then get the inequality $\lim_{n\to\infty}\left(1+\frac{x}{n}\right)^n \geq \sum_{j=1}^{m}\frac{x^j}{j!}$ which holds for any $m \in \mathbb{N}$. Finally, taking the limit as m goes to infinity then gives us the opposite inequality $\lim_{n\to\infty}\left(1+\frac{x}{n}\right)^n \geq \sum_{j=1}^{\infty}\frac{x^j}{j!}$ which is what we wanted.
Thus, for any positive $x > 0$, we have the desired equality.

2. Now fix $x < 0$. Denote $x = -y$ for some $y > 0$. Then, for any $\mathbb{N}_0$ we have:

$$\left(1+\frac{x}{n}\right)^n = \left(1-\frac{y}{n}\right)^n = \frac{\left(1-\frac{y}{n}\right)^n\left(1+\frac{y}{n}\right)^n}{\left(1+\frac{y}{n}\right)^n} = \frac{\left(1-\frac{y^2}{n^2}\right)^n}{\left(1+\frac{y}{n}\right)^n}. \tag{12.5}$$

For the numerator, we have:

$$1 \geq \left(1-\frac{y^2}{n^2}\right)^n = \sum_{j=0}^{n}\frac{n!}{(n-j)!j!}\frac{(-1)^j y^{2j}}{n^{2j}} > 1-\sum_{j=1}^{n}\frac{n!}{(n-j)!j!}\frac{y^{2j}}{n^{2j}}. \tag{12.6}$$

However, note that for all $j = 1, 2, \dots, n$ we have:

$$\frac{n!}{(n-j)!j!}\frac{y^{2j}}{n^{2j}} \leq \frac{1}{j!}\frac{y^{2j}}{n^j} \leq \frac{y^{2j}}{n^j}.$$

So, for all $n > y^2$ we have the bound:

$$\sum_{j=1}^{n}\frac{n!}{(n-j)!j!}\frac{y^{2j}}{n^{2j}} \leq \sum_{j=1}^{n}\frac{y^{2j}}{n^j} < \sum_{j=1}^{\infty}\frac{y^{2j}}{n^j} = \frac{\frac{y^2}{n}}{1-\frac{y^2}{n}}.$$

Putting this estimate in (12.6), for all $n > y^2$ we have:

$$1 \geq \left(1-\frac{y^2}{n^2}\right)^n > 1-\frac{y^2}{n-y^2} = \frac{n}{n-y^2}.$$

Thus by sandwiching, we get $\lim_{n\to\infty}\left(1-\frac{y^2}{n^2}\right)^n = 1$. This means that, by taking the limit as $n \to \infty$ in (12.5), we have:

$$\lim_{n\to\infty}\left(1+\frac{x}{n}\right)^n = \lim_{n\to\infty}\frac{\left(1-\frac{y^2}{n^2}\right)^n}{\left(1+\frac{y}{n}\right)^n} = \frac{\lim_{n\to\infty}\left(1-\frac{y^2}{n^2}\right)^n}{\lim_{n\to\infty}\left(1+\frac{y}{n}\right)^n}$$
$$= \frac{1}{E(y)} = \frac{1}{E(-x)} = E(x),$$

where the final equality is obtained by using the additive property of E seen earlier, namely $1 = E(0) = E(x)E(-x)$. □

Remark 12.4.2 In fact, a more straightforward proof for Proposition 12.4.1 is to use Tannery's theorem as we have seen in Exercise 8.10.

Thus, putting $x = 1$, we can then determine what $E(1)$ is:

Corollary 12.4.3 *We have:*

$$E(1) = \sum_{j=0}^{\infty}\frac{1}{j!} = \lim_{n\to\infty}\left(1+\frac{1}{n}\right)^n = e.$$

Therefore, $E(x) = e^x > 0$ for all $x \in \mathbb{R}$. We have seen the exponentiation function e^x for $x \in \mathbb{R}$ in Chap. 4 in which we defined it using supremum and infimum. However, it is delightful to see a different way of expressing it in terms of a limit and a power series. Now, just for fun, we prove some of the (already proven) properties of the exponential function using the power series approach.

Proposition 12.4.4 *The exponential function $e^x : \mathbb{R} \to \mathbb{R}$ satisfies the following:*

1. *e^x is a continuous function.*
2. *e^x is a bijection between $\mathbb{R}$ and $\mathbb{R}_+$.*

Proof We prove the assertions one by one.

1. For any $r > 0$, the power series $e^x = \sum_{j=0}^{\infty}\frac{x^j}{j!}$ converges uniformly in the interval $[-r, r]$ by Proposition 12.2.1. Since the partial sums $s_n(x) = \sum_{j=0}^{n}\frac{x^j}{j!}$ are all continuous and they converge uniformly to $e^x = \sum_{j=0}^{\infty}\frac{x^j}{j!}$ in the interval $[-r, r]$, Theorem 11.4.17 tells us that the limit e^x is also continuous on $[-r, r]$. Finally, since r is arbitrary, we can conclude that e^x is continuous everywhere in $\mathbb{R}$.
2. To show that e^x is a bijection between $\mathbb{R}$ and $\mathbb{R}_+$, we have to show that it is both an injection and a surjection.

To show that it is injective, we show that it is strictly increasing. We split this into cases:

(a) If $x > y \geq 0$, we have $E(x) - E(y) = \sum_{j=0}^{\infty} \frac{x^j - y^j}{j!} = \sum_{j=0}^{\infty} \frac{(x-y)(x^{j-1}+\ldots+y^j)}{j!} > 0$ since all of the terms in the series are positive. Thus $E(x) > E(y)$. Moreover, note that $E(y) = \sum_{j=0}^{\infty} \frac{y^j}{j!} \geq 1$ with equality if and only if $y = 0$.

(b) If $y < x \leq 0$, we have $-y > -x \geq 0$ and so $E(-y) > E(-x)$ by the first case. Multiplying with $E(x)E(y) > 0$ then yields $E(x) > E(y)$. Moreover, since $E(x)E(-x) = 1$ and $E(-x) \geq 1$ from the first case, we necessarily have $E(x) \leq 1$ with equality for $x = 0$.

(c) If $x < 0 < y$, from the previous cases, we then have $E(x) < 1 < E(y)$.

Thus, $E(x)$ is strictly increasing with respect to x and hence is injective.

Next we show that it is surjective in two steps:

(a) We first show that the exponential function attains any value greater than or equal to 1. Clearly it attains the value 1 at $x = 0$. Fix a value $y > 1$. At the point $x = y$, we have $e^y = \sum_{j=0}^{\infty} \frac{y^j}{j!} > y$. By using the fact that the exponential function is continuous, we can apply the IVT on the interval $[0, y]$ and since $1 < y < e^y$, there exists a point $\xi \in [0, y]$ such that $e^\xi = y$. Since $y > 1$ is arbitrary, we conclude that e^x attains any value in $[1, \infty)$.

(b) To show that it attains any value in $(0, 1)$, we first fix a value $y \in (0, 1)$. Then, $\frac{1}{y} > 1$ and hence there exists an $x > 0$ such that $e^x = \frac{1}{y}$. This means $y = e^{-x}$. Since y is arbitrary, the exponential function also attains any value in $(0, 1)$.

Thus, the exponential function attains any value in $(0, \infty)$ and hence surjects onto $\mathbb{R}_+$. Hence, e^x is a bijection between $\mathbb{R}$ and $\mathbb{R}_+$. □

From Proposition 12.4.4, we now note that the function $e^x : \mathbb{R} \to \mathbb{R}_+$ is a bijection. Hence, there exists an inverse to this function. We call this function the natural logarithm or simply logarithm, denoted as $\ln : \mathbb{R}_+ \to \mathbb{R}$.

When we do exponents, we put in a number x and ask what is the answer if we multiply the number e with itself x times. The logarithm is the reverse: we put in a positive number x and ask how many times do we have to multiply e with itself to get x. Again, we have seen this function in Chap. 4.

Since inverse functions reverses the domain and codomain, we can deduce that $\ln(x)$ is negative for $x \in (0, 1)$, vanishes at $x = 1$, and is positive for $x > 1$. Furthermore, from Theorem 10.5.4, since the exponential function is continuous and strictly increasing, the logarithm function also has the same properties. Another useful properties of the logarithm function are the following:

Proposition 12.4.5 *Let $x, y \in \mathbb{R}_+$. Then:*

1. $\ln(xy) = \ln(x) + \ln(y)$.
2. *For any* $p \in \mathbb{R}$, *we have* $\ln(x^p) = p\ln(x)$.

Proof We prove the assertions one by one.

1. Note that by a property of the exponential function, we have $e^{\ln(x)+\ln(y)} = e^{\ln(x)}e^{\ln(y)} = xy = e^{\ln(xy)}$ since the composition of the exponential and logarithm is the identity. Because the exponential function is injective, we then conclude that $\ln(x) + \ln(y) = \ln(xy)$.
2. By similar arguments to the above, we have $e^{\ln(x^p)} = x^p = \left(e^{\ln(x)}\right)^p = e^{p\ln(x)}$ and thus we conclude that $\ln(x^p) = p\ln(x)$. □

This logarithm function is useful to generalise the exponentiation function to any number other than e. Notice that since they are inverses of each other, their composition is the identity map. In particular, for any positive real number $a > 0$, we have $e^{\ln(a)} = a$. Thus, for any $x \in \mathbb{R}$ we have $e^{x\ln(a)} = a^x$ and this defines the base-a exponentiation map $a^x : \mathbb{R} \to \mathbb{R}_+$. Therefore, this base-$a$ exponentiation also has a power series given by:

$$a^x = \sum_{j=0}^{\infty} \frac{(x\ln(a))^j}{j!},$$

which, again, converges for all $x \in \mathbb{R}$.

Furthermore, if $a > 1$, since a^x is simply the exponential map e^x with the argument scaled by a positive constant $\ln(a)$, the function a^x shares the same properties as e^x as in Proposition 12.4.4. However, if $a \in (0, 1)$, then $\ln(a)$ is negative and so $a^x = e^{x\ln(a)}$ reverses the monotonicity of e^x, namely the function a^x is strictly decreasing instead. Either way, the function a^x is also a bijection from $\mathbb{R}$ to $\mathbb{R}_+$, so it has an inverse which we call the base-a logarithm function $\log_a : \mathbb{R}_+ \to \mathbb{R}$.

Since we can write a base-a exponential function in terms of e^x, we must be able to write the base-a logarithm in terms of $\ln(x)$. Indeed we can. Note that $x = a^{\log_a x}$ since the base-a logarithm and the base-a exponential are inverses of each other. We can apply the natural logarithm on both sides of the equation and use Proposition 12.4.5 to get:

$$\ln(x) = \ln\left(a^{\log_a x}\right) \quad \Rightarrow \quad \ln(x) = \log_a x \cdot \ln a \quad \Rightarrow \quad \log_a x = \frac{\ln x}{\ln a},$$

if $a \neq 1$. Thus, since the base-a logarithm is a constant scale of the natural logarithm, it is also continuous, strictly monotone (increasing or decreasing, depending on the value of a here), and satisfies Proposition 12.4.5.

Exercises

12.1 ($*$) Determine the radius of convergence and the domain of convergence in $\mathbb{R}$ for the following power series:

(a) $\sum_{j=0}^{\infty} j^{2015} x^j$.
(b) $\sum_{j=1}^{\infty} \frac{x^j}{2^j j^4}$.
(c) $\sum_{j=1}^{\infty} j^j x^j$.
(d) $\sum_{j=1}^{\infty} \frac{(-1)^j j}{4^j} x^j$.
(e) $\sum_{j=0}^{\infty} (-1)^j j^2 x^j$.
(f) $\sum_{j=1}^{\infty} j^{\frac{1}{j}} x^j$.
(g) $\sum_{j=0}^{\infty} \frac{2^j (2j)! x^j}{(j!)^2}$.
(h) $\sum_{j=1}^{\infty} \ln(j) x^j$.

12.2 ($*$) Find the radius of convergence for the real series $\sum_{j=0}^{\infty} j! x^{j!}$. Justify your answer.

12.3 ($*$) Let $p \in \mathbb{R}$ be a constant. Find the radius of convergence for the real power series $\sum_{j=1}^{\infty} \frac{(-1)^j}{4^j j^p} x^{2j}$ and hence deduce the domain of convergence for the series.

12.4 ($\diamond$) For a positive integer k, the rising Pochhammer symbol $(k)_n$ for $n \in \mathbb{N}_0$ was introduced by Leo August Pochhammer (1841–1920) and defined as:

$$(k)_n = \begin{cases} 1 & \text{if } n = 0, \\ k(k+1)(k+2)\dots(k+n-1) & \text{if } n > 0. \end{cases}$$

(a) Find the radius of convergence of the hypergeometric series:

$$F(a, b, c; x) = \sum_{j=0}^{\infty} \frac{(a)_j (b)_j}{(c)_j} \frac{x^j}{j!},$$

for positive integers $a, b, c \in \mathbb{N}$.

(b) Show that $F(a, b, c; 1)$ is convergent if $c > a + b$ and divergent if $c < a + b$.

12.5 Find the radius and domain of convergence for the real power series $\sum_{j=1}^{\infty} \frac{\sin(\frac{j\pi}{6})}{2^j} (x-2)^j$.

12.6 ($*$) Let (a_n) be a real sequence defined recursively as $a_0 = a_1 = 5$ and $a_{n+1} = a_n - 6a_{n-1}$ for all $n \geq 1$.

(a) Supposing that the series $S = \sum_{j=0} a_j x^j$ has radius of convergence $R \geq 0$, show that within the radius of convergence, $S = \frac{5}{1-x-6x^2}$.
(b) Write $\frac{5}{1-x-6x^2}$ as a sum of two geometric series and determine their radius of convergence.
(c) Write down the sum of the series in part (b) and determine its minimum radius of convergence.
(d) Check that the coefficients of the series in part (c) satisfies the recurrence relation for the sequence (a_n) and thus is exactly equal to (a_n).
Hence, determine the radius of convergence for the series $\sum_{j=0}^{\infty} a_j x^j$.

12.7 (a) Suppose that the coefficients of the power series $\sum_{j=0}^{\infty} a_j x^j$ are non-zero integers. Prove that the radius of convergence is at most 1.
(b) Give two examples of power series with non-zero integer coefficients having radius of convergence exactly 1 and exactly 0 respectively.

12.8 (∗) Let $\sum_{j=0}^{\infty} a_j(x-c)^j$ and $\sum_{j=0}^{\infty} b_j(x-c)^j$ be two real power series centred at $c \in \mathbb{R}$ that converge pointwise to the same values. Suppose that the both power series have positive radius of convergence $R > 0$. Prove that $a_j = b_j$ for all $j \in \mathbb{N}_0$.

12.9 (∗) The Bessel function of order $p \in \mathbb{N}_0$ is defined as the power series:

$$J_p(x) = \sum_{j=0}^{\infty} \frac{(-1)^j}{j!(j+p)!} \left(\frac{x}{2}\right)^{2j+p}.$$

Prove that this power series converges everywhere on $\mathbb{R}$.
These functions were first defined by Daniel Bernoulli (1700–1782) (one of the mathematicians in the Bernoulli family) and generalised by Friedrich Bessel (1784–1846). These functions appear in many applied mathematics problems such as heat conduction, vibrations, and electromagnetism. We shall see more of this function in Exercise 14.12.

12.10 (∗) Recall the binomial expansion $(1+x)^n$ for $n \in \mathbb{N}$ which can be stated as $(1+x)^n = \sum_{j=0}^{n} \binom{n}{j} x^j$ where $\binom{n}{j} = \frac{n!}{j!(n-j)!}$ are the binomial coefficients. We can extend the definition of binomial coefficients to non-integer $r \in \mathbb{R}$. We have seen the generalised binomial coefficients in Exercise 7.11(e), namely:

$$\binom{r}{j} = \frac{r(r-1)(r-2)\dots(r-j+1)}{j!} \text{ for } j \in \mathbb{N} \quad \text{and} \quad \binom{r}{0} = 1.$$

Newton's binomial theorem states the following extension to the usual binomial theorem:

Theorem 12.5.6 (Newton's Binomial Theorem) *For $r \in \mathbb{R}$, consider the real series:*

$$\sum_{j=0}^{\infty} \binom{r}{j} x^j.$$

1. *If $r \in \mathbb{N}_0$, then the series converges for all $x \in \mathbb{R}$.*
2. *If $r > 0$ is non-integer, then the series converges only for $|x| \leq 1$.*
3. *If $-1 < r < 0$, then the series converges only for $|x| < 1$ and $x = 1$.*
4. *If $r \leq -1$, then the series converges only for $|x| < 1$.*

We are now going to prove this theorem via the following steps:
(a) Explain why for any $r \in \mathbb{N}_0$ this series converges everywhere and agrees with the binomial expansion.
(b) Show that for any $r \in \mathbb{R} \setminus \mathbb{N}_0$, the radius of convergence for this series is 1.
(c) If $r \geq 0$, prove that the series converges absolutely at both $x = \pm 1$.
(d) If $-1 < r < 0$, prove that the series converges at $x = 1$ and diverges at $x = -1$.
(e) If $r \leq -1$, prove that the series diverges at both $x = \pm 1$.

12.11 (∗) Consider the following real power series:

$$C(x) = \sum_{j=0}^{\infty} \frac{(-1)^j}{(2j)!} x^{2j} \quad \text{and} \quad S(x) = \sum_{j=0}^{\infty} \frac{(-1)^j}{(2j+1)!} x^{2j+1}.$$

(a) Find the radius of convergence for each series.
(b) Prove that for any $r > 0$ each of the series converges uniformly on $[-r, r]$.
(c) Deduce that the series are both continuous at any point in $\mathbb{R}$.
(d) Find the power series for $C(x)^2$ and $S(x)^2$ and express the coefficients in terms of binomial coefficients. State their radius of convergence.
(e) Hence, show that that $C(x)^2 + S(x)^2 = 1$ everywhere where they are both defined.
(f) Conclude that $|C(x)| \leq 1$ and $|S(x)| \leq 1$ for all $x \in \mathbb{R}$.
(g) Explain why these series cannot converge uniformly on $\mathbb{R}$.

12.12 (∗) We state a definition first:

Definition 12.5.7 (Odd, Even Functions) Let $f : \mathbb{R} \to \mathbb{R}$.
1. The function f is called an odd function if $f(-x) = -f(x)$ for all $x \in \mathbb{R}$.
2. The function f is called an even function if $f(-x) = f(x)$ for all $x \in \mathbb{R}$.

Recall the power series $C(x)$ and $S(x)$ from Exercise 12.11.

(a) Show that the series $C(x)$ and $S(x)$ define an even and an odd function on $\mathbb{R}$ respectively.
(b) Show that for any $x, y \in \mathbb{R}$, we have $C(x+y) = C(x)C(y) - S(x)S(y)$ and $S(x+y) = S(x)C(y) + S(y)C(x)$.
(c) Hence, show that $S(2x) = 2S(x)C(x)$ and $C(2x) = 2C(x)^2 - 1 = 1 - 2S(x)^2$ for all $x \in \mathbb{R}$.

So these series $C(x)$ and $S(x)$ behave suspiciously like the cosine and sine functions that we know from geometry...

12.13 (∗) For more evidence, we now show that the power series $C(x)$ and $S(x)$ are periodic.

(a) Show that:

$$C(x) = \sum_{j=0}^{\infty} \frac{x^{4j}}{(4j)!}\left(1 - \frac{x^2}{(4j+1)(4j+2)}\right).$$

Hence, deduce that for $0 < x < \sqrt{2}$ we have $C(x) > 0$.

(b) Show that:

$$C(x) = \frac{1}{4!}(x^4 - 12x^2 + 24) - x^6 \sum_{j=0}^{\infty} \frac{x^{4j}}{(4j+6)!}\left(1 - \frac{x^2}{(4j+7)(4j+8)}\right).$$

Hence, deduce that for $6 - \sqrt{12} < x^2 < 6 + \sqrt{12}$ we have $C(x) < 0$.

(c) Using parts (a) and (b), prove that there is a solution to the equation $C(x) = 0$ within the interval $(\sqrt{2}, \sqrt{6 - \sqrt{12}})$.
(d) Show that $S(0) = 0$ and $C(0) = 1$.
Let the smallest positive solution of $C(x) = 0$ be called τ. Note that $\sqrt{2} < \tau < \sqrt{6 - \sqrt{12}}$. Prove that $S(4\tau) = 0$ and $C(4\tau) = 1$.
(e) Hence, show that $C(x)$ and $S(x)$ has a period of 4τ.
(f) Using similar arguments as in part (a), show that $S(x) > 0$ for $0 < x < \sqrt{6}$.
(g) Hence, determine the values of $S(\tau)$, $C(2\tau)$, $S(2\tau)$, $C(3\tau)$, and $S(3\tau)$.

12.14 (∗) We now show that the smallest period for the power series $C(x)$ and $S(x)$ is 4τ. Suppose that y is a solution for $C(y) = 1$ where $0 < y < 4\tau$.

(a) If $0 < y < 2\tau$, show that $S(\frac{y}{2}) = 0$ and hence get a contradiction.
(b) Otherwise, if $2\tau < y < 4\tau$, write $-\tau < \frac{y-4\tau}{2} < 0$ and show that $S(\frac{y-4\tau}{2}) = 0$.
Hence, deduce another contradiction.
(c) Thus, conclude that 4τ is the smallest period for $C(x)$.
(d) Now show that if $S(x+z) = S(x)$ for all $x \in \mathbb{R}$, then $C(z) = 1$.
Hence, conclude that 4τ is also the smallest period for $S(x)$.

So it looks like these two power series behave very much like the cosine and sine function if $2\tau = \pi$. How can we prove this? We shall gather more evidence in Exercise 14.14 and look at this in more detail in Chap. 16.

12.15 Consider the function $f : \mathbb{R} \setminus \{1\} \to \mathbb{R}$ defined as $f(x) = \frac{1}{1-x}$. For any $a \neq 1$, find the power series expansion of f centred at a and determine its radius of convergence.

12.16 (∗) Using Mertens' theorem, find a power series centred at $x = 0$ for the following expressions where $x \in \mathbb{R}$ and determine their radius of convergence.
(a) $\frac{3}{(x^2+4)(x^2-1)}$.
(b) $\frac{1}{1-2x-x^2+2x^3}$.

12.17 Determine the coefficient of:
(a) x^6 in the power series expansion of $\frac{3}{1-2x^2}$ centred at 0.
(b) x^3 in the power series expansion of $\frac{1}{(1-x)^4}$ centred at 0.
(c) x^3 in the power series expansion of $\frac{1}{(1-x)^2(1-2x)^2}$ centred at 0.

12.18 (a) Show that the series $\sum_{j=0}^{\infty} \frac{j+1}{3^j}$ converges.
(b) Find the power series of $\frac{1}{(1-x)^2}$ centred at 0 and determine its radius of convergence.
(c) Hence, determine the value of $\sum_{j=0}^{\infty} \frac{j+1}{3^j}$.

12.19 (∗) We are going to prove that the number e is irrational in this question.
(a) Prove that for any $n \in \mathbb{N}$, we have $0 < e - \sum_{j=0}^{n} \frac{1}{j!} < \frac{1}{n!}\frac{1}{n}$.
(b) Hence, via contradiction, show that the number e is irrational.

12.20 (∗) Let $f : \mathbb{R} \setminus \{0\} \to \mathbb{R}$ be defined as $f(x) = \frac{e^x-1}{x}$. Find the limit $\lim_{x\to 0} f(x)$.

12.21 (a) Show that for all $n \in \mathbb{N}$ we have $n! > \left(\frac{n}{e}\right)^n$.
(b) Deduce the following inequality:

$$n\ln(n) - n \leq \ln(n!) \leq n\ln(n).$$

(c) Hence, show that $\ln(n!) \sim n\ln(n)$.

12.22 (∗) We define the hyperbolic trigonometry functions sinh, cosh : $\mathbb{R} \to \mathbb{R}$ as:

$$\sinh(x) = \frac{e^x - e^{-x}}{2} \quad \text{and} \quad \cosh(x) = \frac{e^x + e^{-x}}{2}.$$

(a) Show $\cosh^2(x) - \sinh^2(x) = 1$ for all $x \in \mathbb{R}$.
(b) Explain why cosh is greater than or equal to 1 everywhere.
(c) Show that cosh is an even function and sinh is an odd function.
(d) Prove that cosh and sinh are both strictly increasing for $x \geq 0$. Hence, comment on the the monotonic behaviour of these functions over the whole domain $\mathbb{R}$.
(e) Prove that for all $x, y \in \mathbb{R}$ we have $\sinh(x + y) = \sinh(x)\cosh(y) + \cosh(x)\sinh(y)$.

(f) Prove that for all $x, y \in \mathbb{R}$ we have $\cosh(x+y) = \cosh(x)\cosh(y) + \sinh(x)\sinh(y)$.

(g) Find the power series of $\sinh(x)$ and $\cosh(x)$ and state their radius of convergence.

12.23 (a) Using Exercise 12.22, determine the domain of the inverse functions for cosh and sinh. We denote the inverse functions as arccosh and arcsinh respectively.

(b) Hence, find the closed form of arccosh(x) and arcsinh(x) over their domains.

(c) Show that arcsinh(x) is an odd function.

12.24 Define a new function $\tanh : \mathbb{R} \to \mathbb{R}$ as $\tanh(x) = \frac{\sinh(x)}{\cosh(x)} = \frac{e^x - e^{-x}}{e^x + e^{-x}}$.

(a) Show that tanh is an odd function.

(b) Show that $\lim_{x\to\infty} \tanh(x) = 1$ and hence deduce $\lim_{x\to-\infty} \tanh(x) = -1$.

(c) Show that tanh is strictly increasing over the whole domain $\mathbb{R}$.

(d) Conclude that tanh is invertible and determine the domain of the inverse function.

(e) We denote the inverse function of tanh as arctanh. Find the closed form for arctanh(x) on its domain.

12.25 Extending the domain of sinh and cosh to complex numbers, show that for $x \in \mathbb{R}$ and imaginary unit i we have:

(a) $\cosh(ix) = C(x)$.

(b) $\sinh(ix) = iS(x)$.

12.26 ($*$) Using Tannery's theorem from Exercise 8.9, prove that:

$$\lim_{n\to\infty} \left(\left(\frac{1}{n}\right)^n + \left(\frac{2}{n}\right)^n + \ldots + \left(\frac{n-1}{n}\right)^n + \left(\frac{n}{n}\right)^n \right) = \frac{e}{e-1}.$$

12.27 ($*$) Consider the real sequence (a_n) defined as $a_n = \sum_{j=1}^n \frac{1}{j} - \ln(n)$.

(a) Show that (a_n) is decreasing and $a_n \leq 1$ for all $n \in \mathbb{N}$.

(b) Define a sequence (b_n) as $b_n = \sum_{j=1}^n \frac{1}{j} - \ln(n+1)$. Show that (b_n) is increasing and $b_n > 0$ for all $n \in \mathbb{N}$.

(c) Deduce that $0 < a_n \leq 1$.

Conclude that the sequence (a_n) converges.

The limit of the sequence (a_n) is called the Euler-Mascheroni constant, denoted as γ and named after Leonhard Euler and Lorenzo Mascheroni (1750–1800). Its approximate value is $0.57721\ldots$. Similar to the Euler-Napier constant e, this number appears in many different areas of mathematics. However, unlike e which has been proven to be irrational in Exercise 12.19, to this day, nobody knows whether γ is a rational or an irrational number!

12.28 For a fixed non-integer constant $r > 0$, define the series:

$$\sum_{j=1}^{\infty} \ln\left|1 - \frac{r}{j}\right|,$$

where each term in the series exists.

(a) Show that the series diverges to $-\infty$.

(b) Hence, show that:

$$\lim_{n\to\infty}\left(1-\frac{r}{1}\right)\left(1-\frac{r}{2}\right)\dots\left(1-\frac{r}{n}\right) = 0.$$

(c) Deduce that for $r > -1$ we have $\lim_{n\to\infty}\binom{r}{n} = 0$ where $\binom{r}{n}$ is the generalised binomial coefficient in Exercise 12.10.

12.29 (∗) Recall the sequence (f_n) of integers defined recursively as $f_1 = f_2 = 1$ and $f_n = f_{n-1} + f_{n-2}$ for $n \geq 3$ is called the Fibonacci sequence. We have seen in Exercise 6.2 that there is a closed form for the terms in the sequence, namely: the n-th Fibonacci number is given by the Binet formula $f_n = \frac{\varphi^n-(1-\varphi)^n}{\sqrt{5}} = \frac{\varphi^n-\psi^n}{\sqrt{5}}$ where $\varphi = \frac{1+\sqrt{5}}{2}$ is the golden ratio and $\psi = 1 - \varphi = -\frac{1}{\varphi}$. This formula was derived by Binet using recurrence relationships. In fact, this formula was known to De Moivre and Daniel Bernoulli a century earlier. The method used by them utilises techniques involving power series instead. We shall demonstrate their method in this question.

(a) Define a power series $F(x) = \sum_{j=1}^{\infty} f_j x^j$. Show that this series has a strictly positive radius of convergence $R > 0$.

(b) Show that for any $x \in B_R(0)$, we have $F(x) - xF(x) - x^2F(x) = x$.

(c) Deduce that $F(x) = \frac{x}{1-x-x^2} = -\frac{x}{(x+\varphi)(x+\psi)}$ on $B_R(0)$ where φ and ψ are as in the above.

(d) Using partial fractions, show that $F(x) = \frac{x}{\sqrt{5}}\left(\frac{1}{x+\psi} - \frac{1}{x+\varphi}\right)$.

(e) By expressing F as power series and equating coefficients, show that $f_n = \frac{\varphi^n-\psi^n}{\sqrt{5}}$ for all $n \in \mathbb{N}$.

The technique above is called generating functions. A generating function is a powerful tool used to encapsulate an infinite sequence compactly in a single function. This infinite sequence can then be recovered by expanding the function as a power series and extracting the coefficients. It is described by George Pólya (1887–1985) as:

> A generating function is a device somewhat similar to a bag. Instead of carrying many little objects detachedly, which could be embarrassing, we put them all in a bag, and then we have only one object to carry, the bag.

In this question, the infinite sequence (the little objects) is the Fibonacci sequence (f_n) and its generating function (the bag) is the function $F(x) = \frac{x}{1-x-x^2}$.

Generating functions are also analogised by Herbert Wilf (1931–2012) as:

> A generating function is a clothesline on which we hang up a sequence of numbers for display.

Due to the convenience of condensing a lot of information in a compact form which can then be manipulated more easily, generating functions are used frequently in combinatorial and recurrence problems. In probability theory, they are commonly used to express the n-th moments of a probability distribution. We shall see more generating functions in Exercise 14.34.

Differentiation

13

With an absurd oversimplification, the 'invention' of the calculus is sometimes ascribed to two men, Newton and Leibniz. In reality, the calculus is the product of a long evolution that was neither initiated nor terminated by Newton and Leibniz, but in which both played a decisive part.

— Richard Courant, mathematician

In this chapter, we are going to look at the concept of differentiation that we have seen and used in any class on introductory calculus.

The concept of differentiability that we are going to discuss here is for functions defined on subsets of $\mathbb{R}$. Though the definitions also carry forward to complex-valued functions, calculus on complex domains contains very different theories thanks to the additional complex structure in the domain. This makes things more rigid and well-behaved there, resulting in a rich area of study called complex analysis. But for now, we focus our attention to real-valued functions.

13.1 Derivatives

Let $f : X \to \mathbb{R}$ be a real-valued function and fix a point $x_0 \in X' \cap X$. We can define a new function on the set $X \setminus \{x_0\}$, which is called a difference quotient, as follows:

Definition 13.1.1 (Difference Quotient at x_0) Let $f : X \to \mathbb{R}$ be a real-valued function and $x_0 \in X' \cap X$. The difference quotient of the function f at the point x_0 is the function $\Delta_{x_0} f : X \setminus \{x_0\} \to \mathbb{R}$ defined as $\Delta_{x_0} f(x) = \frac{f(x)-f(x_0)}{x-x_0}$.

S. Johar, *The Big Book of Real Analysis*,
https://doi.org/10.1007/978-3-031-30832-1_13

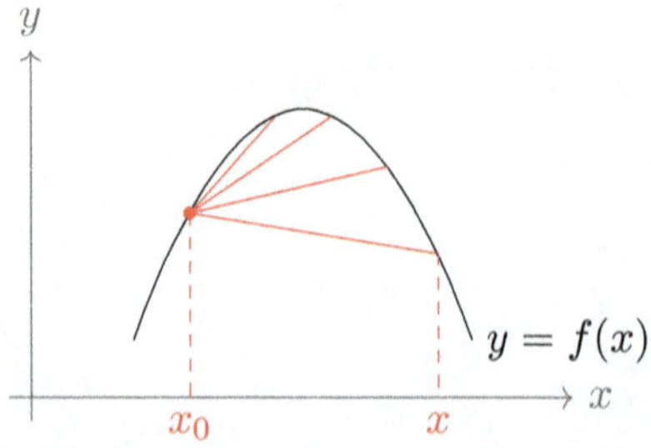

Fig. 13.1 The secant lines joining $(x_0, f(x_0))$ and $(x, f(x))$ for various x. Their slopes are the values of the difference quotients $\Delta_{x_0} f(x)$

For any $x \neq x_0$, the difference quotient $\Delta_{x_0} f(x)$ can be seen as the slope of the secant line to the graph of the function f passing through the two points $(x_0, f(x_0))$ and $(x, f(x))$ as depicted in Fig. 13.1.

The early concept of calculus was proposed by Gottfried Wilhelm von Leibniz and Isaac Newton in the seventeenth century by considering the difference quotient with very small denominator. This leads to a massive rivalry between the two mathematicians, one accusing the other of plagiarism. However, today, both are credited to this discovery. As reconciled by Alfred Rupert Hall (1920–2009) in his book *Philosophers at War*:

> It was certainly Isaac Newton who first devised a new infinitesimal calculus and elaborated it into a widely extensible algorithm, whose potentialities he fully understood; of equal certainty, the differential and integral calculus, the fount of great developments flowing continuously from 1684 to the present day, was created independently by Gottfried Wilhelm Leibniz. Whatever we may feel of the relations between these two men, we cannot but admire their analogous creative achievements with as much impartiality as our emotions will admit.

In addition, the two came up with the idea of calculus from two separate directions; Leibniz was interested in the philosophical content whereas Newton was motivated by problems in applied mathematics and physics. However, what they do have in common is the language of limits were not yet established during their time, so they resorted to the construction using "infinitesimals".

Infinitesimals, also referred to indivisibles, are infinitely small insignificant quantities which can be treated essentially as 0 in calculations. For example, in contrast to the symbol ∞ to denote infinitely large quantities, Newton denoted o (not to be confused with zero 0) as an infinitesimal. For the function $f(x) = x^2$ defined on $\mathbb{R}$, at a point $x_0 \in \mathbb{R}$ he considered the difference quotient:

$$\frac{f(x_0 + o) - f(x_0)}{o} = \frac{(x_0 + o)^2 - x^2}{o} = \frac{2x_0 o + o^2}{o} = 2x_0 + o, \tag{13.1}$$

and declared that the slope of the tangent for the graph of f at x_0 is $2x_0$ by setting o to be zero.

Quick readers might have noticed that if we set o to be zero, then the original difference quotient would have not made any algebraic sense in $\mathbb{R}$. Thus, the introduction of these infinitesimals was faced with the problem of division by zero. Indeed, even Leibniz used the same argument by declaring that infinitesimals are "smaller than any arbitrary assignable quantity" or "smaller than any prescribed positive number".

Of course, due to the lack of rigour as well as these hand-wavy description and treatment of the infinitesimals, the original ideas of calculus were greatly criticised at the time. Bishop George Berkeley famously derided the infinitesimals which are something yet nothing at the same time. Despite not disputing the validity of the end result, in his book *The Analyst: A Discourse Addressed to an Infidel Mathematician* [27] Berkeley ridiculed the leap in logical reasoning employed by Newton:

> And what are these same evanescent increments? They are neither finite quantities nor quantities infinitely small, nor yet nothing. May we not call them the ghosts of departed quantities?

Others, such as the great Euler himself, are more open to it. Published in *Institutiones calculi differentialis*, Euler interpreted the infinitesimal described by Leibniz above as simply zero [19]. He wrote:

> There is no doubt that every quantity can be diminished until it vanishes completely and is reduced to nothing. But an infinitely small quantity is simply an evanescent quantity and therefore actually equal to 0 ...

It is not until the nineteenth century when the language of limits were rigorously established by Bolzano, Cauchy, and Weierstrass to complete the arguments by Newton and Leibniz and remove the ghostly infinitesimals. Thus, the birth of calculus is complete. Now, we can fully appreciate the quote by Courant at the beginning of this chapter: the theory of calculus went through a lot of evolution via contributions from many mathematicians to get to its current rigorous state.

Remark 13.1.2 We make some remarks regarding the term calculus.

1. Newton called his invention "the method of fluxions". It was Leibniz who coined the term calculus for this branch of mathematics.
2. The word *calculus* is Latin for "small pebble" which was used for basic counting, arithmetic, and computations in the olden days. It is also the root word for other counting-related words like calculator and calculation. Its usage in mathematics can be traced back to as early as Marcus Tullius Cicero (106B.C.–43B.C.).
3. Outside of mathematics, this word is also used in medicine and physiology closer to its original meaning. Imagine my bewilderment as I sat there in a chair and blinded by a white light, when my dentist told me that I have calculus on my teeth. I was hoping for a more divine mathematical experience similar to the Coen brothers film *A Serious Man*, but all I got was a bill of $60.

4. In a tragic irony, Steven Strogatz pointed out that both Newton and Leibniz died in pain while suffering from some kind of medical calculus: a bladder stone for Newton and a kidney stone for Leibniz [72].

To utilise the concept of limits in the arguments devised by Newton and Leibniz, we first note in Definition 13.1.1 that since $x_0 \in X' \cap X$, this point is a limit point of the domain for the difference quotient $\Delta_{x_0} f$, namely $x_0 \in (X \setminus \{x_0\})'$. Therefore we can ask: what is the limit of the difference quotient as we approach the point x_0? This limit, if it exists, must be unique and is called the derivative of a function at x_0:

Definition 13.1.3 (Derivative at x_0**)** Let $f : X \to \mathbb{R}$ be a real-valued function and $x_0 \in X' \cap X$. The derivative of the function f at the point x_0 is defined as the limit of the difference quotient:

$$\lim_{x \to x_0} \Delta_{x_0} f(x) = \lim_{x \to x_0} \frac{f(x) - f(x_0)}{x - x_0},$$

if it exists. If this limit exists, we call the function f differentiable at x_0. We denote the value of this limit as $f'(x_0)$.

In symbols, f is differentiable at x_0 with derivative $f'(x_0)$ if:

$$\forall \varepsilon > 0, \exists \delta > 0 : \forall x \in X, 0 < |x - x_0| < \delta \Rightarrow \left| \frac{f(x) - f(x_0)}{x - x_0} - f'(x_0) \right| < \varepsilon.$$

Thus, using Fig. 13.1, we can deduce a geometric interpretation for the derivative of f at x_0. The quantity $f'(x_0)$ is the limit of the slopes of secant lines of the function f passing through the points $(x_0, f(x_0))$ and $(x, f(x))$ as $x \to x_0$. Hence, the quantity $f'(x_0)$, if it exists, can be thought of as the slope of the tangent line to the graph of the function f at the point x_0.

The operation of finding the derivative of a function f at a point x_0 is called differentiation with respect to the variable x at x_0. This differentiation operation is usually written as the mapping $\left.\frac{d}{dx}\right|_{x_0}$ from the space of functions defined on a set $X \subseteq \mathbb{R}$ such that $x_0 \in X \cap X'$ to the set of real numbers, given as:

$$\left.\frac{d}{dx}\right|_{x_0} f(x) = f'(x_0).$$

As a result, we can also write $f'(x_0) = \left.\frac{df}{dx}\right|_{x_0} = \frac{df}{dx}(x_0)$.

Remark 13.1.4 We make some remarks regarding this definition:

1. There are various notations used for derivatives. Joseph Louis Lagrange (1736–1813) introduced the notation $f'(x_0)$, Newton used the notation $\dot{f}(x_0)$, and Euler used the notation $D_{x_0} f$.
2. Probably the most common notation for derivatives, which was introduced by Leinbiz, is $\frac{df}{dx}\big|_{x_0}$. His notation was literally taken to be a fraction of the difference in f over the infinitesimal dx, which made Berkeley very upset. Nowadays, after the advent of limits, one has to be aware that this notation does not represent a quotient or a fraction as the notation suggests. It is simply symbolic.
3. Note that due to the presence of limits, derivatives of a function can only be defined at points of the domain X which are also limits points of X, namely at $X' \cap X$. Hence, we cannot find the derivative of a function at isolated points of its domain.

From the definition of derivatives using limits, we can see that the derivative of a function f at the point x_0 depends not only on the value of $f(x_0)$, but also on the values of $f(x)$ for x such that $0 < |x - x_0| < \delta$ for some $\delta > 0$. In other words, it also depends on how f behaves at points x near x_0. Thus the derivative of a function is called a local behaviour of the function, similar to continuity and limits.

Remark 13.1.5 A different, yet useful, way to write derivatives is to localise our attention to the point x_0. Since we are taking the limit as $x \to x_0$, we can instead define the difference $h = x - x_0 \neq 0$ and consider the limit as $h \to 0$ instead. Thus, we sometimes use the following equivalent local definition of derivatives:

$$f'(x_0) = \lim_{x \to x_0} \frac{f(x) - f(x_0)}{x - x_0} = \lim_{h \to 0} \frac{f(x_0 + h) - f(x_0)}{h}. \tag{13.2}$$

Similar to limits and continuity, we also have the concept of one-sided differentiability:

Definition 13.1.6 (One-Sided Differentiability) Let $f : I \to \mathbb{R}$.

1. If $I = (a, b]$ and $x_0 \in (a, b]$, we define the left-derivative of the function f at x_0 as:

$$\lim_{x \uparrow x_0} \frac{f(x) - f(x_0)}{x - x_0},$$

if the limit exists. In symbols:

$$\forall \varepsilon > 0, \exists \delta > 0 : \forall x \in (a, b], 0 < x_0 - x < \delta \Rightarrow \left| \frac{f(x) - f(x_0)}{x - x_0} - f'(x_0) \right| < \varepsilon.$$

2. If $I = [a, b)$ and $x_0 \in [a, b)$, we define the right-derivative of the function f at x_0 as:

$$\lim_{x \downarrow x_0} \frac{f(x) - f(x_0)}{x - x_0},$$

if the limit exists. In symbols:

$$\forall \varepsilon > 0, \exists \delta > 0 : \forall x \in [a, b), 0 < x - x_0 < \delta \Rightarrow \left| \frac{f(x) - f(x_0)}{x - x_0} - f'(x_0) \right| < \varepsilon.$$

Remark 13.1.7 Here are some remarks on Definition 13.1.6:

1. A function $f : [a, b] \to \mathbb{R}$ defined on a closed interval is called a differentiable function if it is differentiable in (a, b), left-differentiable at b, and right-differentiable at a.
2. By uniqueness of limits, the value of $f'(x_0)$, if it exists, is unique. Moreover, if the function $f : [a, b] \to \mathbb{R}$ is differentiable at $x_0 \in (a, b)$, then it must be both left- and right-differentiable at x_0 and these one-sided derivatives are equal. Conversely, if the left- and right-derivatives of the function f exists at a point x_0 and are equal, then the function is differentiable at x_0. These facts are due to the properties of limits in Proposition 9.3.4.
3. On the other hand, if the function is continuous at x_0 but its left- and right-derivatives at x_0 do not coincide and have opposite signs, we call this point a cusp. This term also covers the case for which the left- and right-derivatives blow up to $\pm\infty$ respectively or $\mp\infty$ respectively. Pictorially, cusps appear as sharp corners on the graph of functions. We shall see an example of this in Example 13.1.9(4).

Remark 13.1.8 Let us make some remarks for generalisations.

1. Of course, in the definitions above, we can define f to be a complex-valued function with domain $X \subseteq \mathbb{R}$ since the definition still would make sense.
2. However, if we change the domain to be the set of complex numbers $\mathbb{C}$ or the Euclidean space $\mathbb{R}^n$ for some $n > 1$, we would have more constraints since the derivative would depend on the direction one approaches the point where we want to find the derivative.
3. On $\mathbb{R}$, based on Remark 13.1.7, we need to ensure that the left- and right-derivatives at x_0 are the same to deduce that the function is differentiable at x_0. However on $\mathbb{C}$ and $\mathbb{R}^n$, for differentiability at a point x_0, we need to ensure that the derivatives are the same for all directions we are approaching x_0. In these domains, there are way more than two general directions to approach a particular point, in contrast to what we have in $\mathbb{R}$.
4. Furthermore, there is an additional complex structure which exists in $\mathbb{C}$ in the form of multiplication by the imaginary unit i. This additional structure gives

more rigidity to complex derivatives which produces many interesting outcomes. Therefore, complex analysis has a somewhat different feel to it and has more restrictions when compared to real analysis and analysis on $\mathbb{R}^n$.

Example 13.1.9 Let us compute the derivatives of some elementary functions. We use the definition of derivatives using the localised convention as (13.2).

1. Let $f : \mathbb{R} \to \mathbb{R}$ be defined as $f(x) = x^2$ and $x_0 \in \mathbb{R}$. We have seen that, using Newton's argument via the ghostly infinitesimals in (13.1), the slope of the curve for this function at $x_0 \in \mathbb{R}$ is $2x_0$. Now we are going to prove this is indeed true rigorously by using limits.

   The difference quotient of this function is given by $\frac{f(x_0+h)-f(x_0)}{h} = \frac{(x_0+h)^2-x_0^2}{h}$. Hence, the derivative at the point x_0 is:

$$\lim_{h\to 0} \frac{(x_0+h)^2 - x_0^2}{h} = \lim_{h\to 0} \frac{x_0^2 + 2x_0h + h^2 - x_0^2}{h} = \lim_{h\to 0}(2x_0 + h) = 2x_0.$$

   Since this limit exists for every $x_0 \in \mathbb{R}$, we have the derivative $\frac{d}{dx}x^2 = 2x$ for all $x \in \mathbb{R}$. In general, by using binomial expansion, the readers will show that $\frac{d}{dx}x^n = nx^{n-1}$ for any $n \in \mathbb{N}$ in Exercise 13.1.
2. Let $f : \mathbb{R} \setminus \{0\} \to \mathbb{R}$ be defined as $f(x) = \frac{1}{x^2}$ and $x_0 \in \mathbb{R} \setminus \{0\}$. The difference quotient of this function is $\frac{1}{h}\left(\frac{1}{(x_0+h)^2} - \frac{1}{x_0^2}\right) = \frac{x_0^2-(x_0+h)^2}{h(x_0+h)^2x_0^2}$ for $0 < |h| < |x_0|$. Hence, the derivative can be obtained by taking the limit as $h \to 0$, namely:

$$\lim_{h\to 0} \frac{x_0^2 - (x_0+h)^2}{h(x_0+h)^2x_0^2} = \lim_{h\to 0} \frac{-h^2 - 2x_0h}{h(x_0+h)^2x_0^2} = -\lim_{h\to 0} \frac{h + 2x_0}{(x_0+h)^2x_0^2} = -\frac{2}{x_0^3},$$

   where we used the algebra of limits in the last equality. Since this limit makes sense for every $x_0 \in \mathbb{R} \setminus \{0\}$, we have the derivative $\frac{d}{dx}\frac{1}{x^2} = -\frac{2}{x^3}$ for all $x \in \mathbb{R} \setminus \{0\}$. In general, by using binomial expansion again, we can deduce $\frac{d}{dx}\frac{1}{x^n} = -\frac{n}{x^{n+1}}$ for any $n \in \mathbb{N}$ and $x \in \mathbb{R} \setminus \{0\}$.
3. So far we have seen that $\frac{d}{dx}x^n = nx^{n-1}$ for any non-zero integers. This in fact holds for other exponents as well. Let $f : \mathbb{R}_{\geq 0} \to \mathbb{R}$ be defined as $f(x) = \sqrt{x}$ and $x_0 \in \mathbb{R}_{\geq 0}$. We have two cases:
   (a) If $x_0 > 0$, then the derivative at this point is given by:

$$\begin{aligned}\lim_{h\to 0} \frac{\sqrt{x_0+h} - \sqrt{x_0}}{h} &= \lim_{h\to 0} \frac{(\sqrt{x_0+h} - \sqrt{x_0})(\sqrt{x_0+h} + \sqrt{x_0})}{h(\sqrt{x_0+h} + \sqrt{x_0})} \\ &= \lim_{h\to 0} \frac{h}{h(\sqrt{x_0+h} + \sqrt{x_0})} = \frac{1}{2\sqrt{x_0}}.\end{aligned}$$

(b) If $x_0 = 0$, we can compute the right-derivative as:

$$\lim_{h \downarrow 0} \frac{\sqrt{h} - 0}{h} = \lim_{h \downarrow 0} \frac{1}{\sqrt{h}},$$

which does not exist.

Hence, the function f is only differentiable at any $x > 0$ with derivative $\frac{d}{dx}\sqrt{x} = \frac{1}{2\sqrt{x}}$.

More generally, we shall show later in Propositions 13.2.6 and 13.7.5 that $\frac{d}{dx}x^r = rx^{r-1}$ for any $r \in \mathbb{R}$ and $x \in \mathbb{R}_+$. This is called the power rule for derivatives.

4. Consider the function $f : \mathbb{R} \to \mathbb{R}$ defined as $f(x) = |x|$. We wish to find its derivative at $x = 0$. We compute its left- and right-derivatives here, which are given by:

$$\lim_{h \uparrow 0} \frac{f(h) - f(0)}{h} = \lim_{h \uparrow 0} \frac{-h - 0}{h} = \lim_{h \uparrow 0} -1 = -1,$$

$$\lim_{h \downarrow 0} \frac{f(h) - f(0)}{h} = \lim_{h \downarrow 0} \frac{h - 0}{h} = \lim_{h \uparrow 0} 1 = 1.$$

Note that the values of $f(h)$ used above are different in both cases because in the left-derivative, h is taken to be to the left of 0 and for the right-derivative, only h to the right of 0 are considered. The left- and right-derivatives calculated above do not agree. Hence the function is not differentiable at $x = 0$. The point at $x = 0$ is an example of a cusp which is a sharp corner as depicted in Fig. 13.2.

However, for any other x in the domain of the function, the limit of the difference quotients all exist. For $x > 0$, the values are $f'(x) = 1$ and for $x < 0$, the values are $f'(x) = -1$.

5. Let $f : (a, b) \to \mathbb{R}$ be a constant function $f(x) = C$ for some $C \in \mathbb{R}$. This function is differentiable everywhere. Indeed, for a fixed point $x_0 \in (a, b)$, we have:

$$\lim_{h \to 0} \frac{f(x_0 + h) - f(x_0)}{h} = \lim_{h \to 0} \frac{C - C}{h} = \lim_{h \to 0} 0 = 0,$$

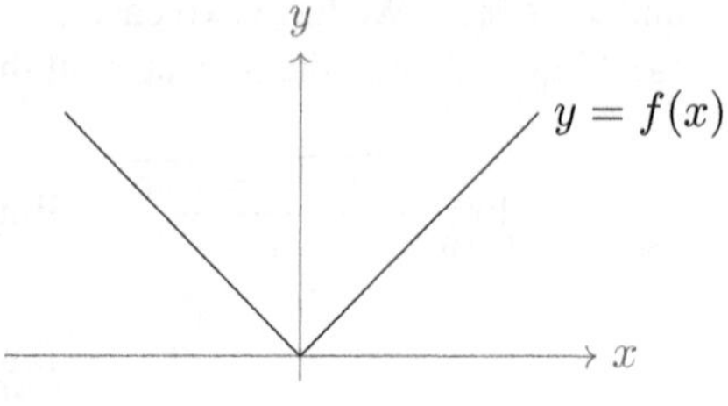

Fig. 13.2 Graph of $f(x) = |x|$. The point $x = 0$ is a cusp

and so $f'(x_0) = 0$. Since we can vary x_0 over the domain, we conclude that $f'(x) = 0$ for all $x \in (a, b)$. So, constant functions have vanishing derivative.

6. Consider the sine function $\sin : \mathbb{R} \to \mathbb{R}$. Using trigonometric identities, the limit of the difference quotient for this function at $x_0 \in \mathbb{R}$ is given by:

$$\begin{aligned}\lim_{h\to 0} \frac{\sin(x_0 + h) - \sin(x_0)}{h} &= \lim_{h\to 0} \frac{\sin(x_0)\cos(h) + \sin(h)\cos(x_0) - \sin(x_0)}{h} \\ &= \lim_{h\to 0} \left(\sin(x_0)\frac{(\cos(h) - 1)}{h} + \cos(x_0)\frac{\sin(h)}{h} \right).\end{aligned} \tag{13.3}$$

We now need to evaluate the limits of the terms with h in (13.3) so that we know they exist and hence can apply the algebra of limits accordingly. We know from Exercise 8.12(a) that for $h \in (0, \frac{\pi}{2})$, we have $0 \leq \sin(h) \leq h \leq \tan(h)$. This implies $0 \leq \frac{\sin(h)}{h} \leq 1$ and $\cos(h) \leq \frac{\sin(h)}{h}$. We can then combine these inequalities as $\cos(h) \leq \frac{\sin(h)}{h} \leq 1$.

(a) If we take the limit as $h \downarrow 0$ for this inequality, since $\cos(h) \downarrow 1$, by sandwiching, we obtain $\frac{\sin(h)}{h} \downarrow 1$ as well. Replacing h with $-h$ gives us the left limit $\frac{\sin(h)}{h} \uparrow 1$ and thus we can conclude that $\lim_{h\to 0} \frac{\sin(h)}{h} = 1$.

(b) Next, we want to find the limit $\lim_{h\to 0} \frac{\cos(h)-1}{h}$. By using trigonometric identities, for small $h \in (-\frac{\pi}{2}, \frac{\pi}{2})$ we have:

$$\frac{\cos(h)-1}{h} = \frac{(\cos(h)-1)(\cos(h)+1)}{h(\cos(h)+1)} = \frac{\cos^2(h) - 1}{h(\cos(h)+1)} = \frac{\sin^2(h)}{h(\cos(h)+1)}.$$

Using part (a) and the algebra of limits, we then have:

$$\lim_{h\to 0} \frac{\cos(h) - 1}{h} = \lim_{h\to 0} \frac{\sin(h)}{h} \frac{\sin(h)}{\cos(h)+1} = 1 \cdot 0 = 0.$$

Thus, since both of these limits exist, we can apply the algebra of limits on the expression (13.3) to get that the derivative of $\sin(x)$ at the point x_0 is:

$$\left.\frac{d}{dx}\right|_{x_0} \sin(x) = \sin(x_0) \lim_{h\to 0} \frac{(\cos(h) - 1)}{h} + \cos(x_0) \lim_{h\to 0} \frac{\sin(h)}{h} = \cos(x_0).$$

In a similar manner, we can prove:

$$\left.\frac{d}{dx}\right|_{x_0} \cos(x) = -\sin(x_0).$$

7. We can also compute the derivative of other more complicated functions from first principles. Let us compute the derivative of the function $f : \mathbb{R} \setminus \{-2\} \to \mathbb{R}$

defined as $f(x) = \frac{1-x^2}{2+x}$ at some point $x_0 \neq -2$. The difference quotient at the point x_0 for $h \neq -2 - x_0$ is given as:

$$\frac{f(x_0+h) - f(x_0)}{h} = \frac{\frac{1-(x_0+h)^2}{2+x_0+h} - \frac{1-x_0^2}{2+x_0}}{h} = -\frac{h^2 + hx_0 + x_0^2 + 4x_0 + 1}{(x_0+2)(h+x_0+2)},$$

and thus taking the limit as $h \to 0$, by the algebra of limits, we have:

$$\begin{aligned} f'(x_0) = \lim_{h\to 0} \frac{f(x_0+h) - f(x_0)}{h} &= \lim_{h\to 0} -\frac{h^2 + hx_0 + x_0^2 + 4x_0 + 1}{(x_0+2)(h+x_0+2)} \\ &= -\frac{x_0^2 + 4x_0 + 1}{(x_0+2)^2}, \end{aligned}$$

which exists for any $x_0 \neq -2$.

8. Let us find the derivative of the exponential function $f : \mathbb{R} \to \mathbb{R}$ defined as $f(x) = e^x$. Fix $x_0 \in \mathbb{R}$. For $h \neq 0$, we compute:

$$\lim_{h\to 0} \frac{f(x_0+h) - f(x_0)}{h} = \lim_{h\to 0} \frac{e^{x_0}(e^h - 1)}{h} = e^{x_0} \lim_{h\to 0} \frac{e^h - 1}{h} = e^{x_0} \cdot 1 = e^{x_0},$$

by Exercise 12.20. So, we have:

$$\left.\frac{d}{dx}\right|_{x_0} e^x = e^{x_0}.$$

A direct corollary that we can deduce from differentiable functions is the following:

Proposition 13.1.10 *Let $f : X \to \mathbb{R}$ for some $X \subseteq \mathbb{R}$ be differentiable at $x_0 \in X$. Then, the function f is continuous at x_0.*

Proof To show that f is continuous at x_0, we have to show $\lim_{x\to x_0} f(x) = f(x_0)$ or equivalently $\lim_{x\to x_0}(f(x) - f(x_0)) = 0$. Since $f'(x_0) = \lim_{x\to x_0} \frac{f(x)-f(x_0)}{x-x_0}$ exists, by the algebra of limits, we have:

$$\begin{aligned} \lim_{x\to x_0} (f(x) - f(x_0)) &= \lim_{x\to x_0} \frac{(f(x) - f(x_0))(x - x_0)}{x - x_0} \\ &= \lim_{x\to x_0} \frac{f(x) - f(x_0)}{x - x_0} \lim_{x\to x_0} (x - x_0) \\ &= f'(x_0) \lim_{x\to x_0} (x - x_0) = 0, \end{aligned}$$

which is what we wanted to prove. □

However, the converse of Proposition 13.1.10 does not hold, namely: not all functions which are continuous at $x = x_0$ is differentiable at x_0. Indeed, we have seen in Example 13.1.9(4) that the function $f(x) = |x|$ defined on $\mathbb{R}$ is continuous at $x = 0$ but not differentiable here.

13.2 Algebra of Derivatives

We have seen many examples on how to compute derivatives from first principles in Example 13.1.9. For more complicated functions, things might not be as easy. Therefore, we would like to know some more properties of derivatives in order for us to manipulate them and differentiate more complicated functions.

This can be done by breaking the complicated functions into smaller manageable pieces that are easy to differentiate and combining them back together via some rules. Using the definition of derivatives and properties of limits, we can directly prove the following results:

Proposition 13.2.1 (Algebra of Derivatives) *Suppose that $f, g : X \to \mathbb{R}$ for some $X \subseteq \mathbb{R}$ are differentiable at $x_0 \in X$ and $\lambda \in \mathbb{R}$ is a constant. Then:*

1. *The function λf is differentiable at x_0 with:*
$$(\lambda f)'(x_0) = \lambda f'(x_0).$$
2. *The function $f \pm g$ is differentiable at x_0 with:*
$$(f \pm g)'(x_0) = f'(x_0) \pm g'(x_0).$$
3. *Product rule: The function $f \times g$ is differentiable at x_0 with:*
$$(f \times g)'(x_0) = f'(x_0)g(x_0) + f(x_0)g'(x_0).$$
4. *If $g(x_0) \neq 0$, then the function $\frac{1}{g}$ is differentiable at x_0 with:*
$$\left(\frac{1}{g}\right)'(x_0) = -\frac{g'(x_0)}{g(x_0)^2}.$$
5. *Quotient rule: If $g(x_0) \neq 0$, then the function $\frac{f}{g}$ is differentiable at x_0 with:*
$$\left(\frac{f}{g}\right)'(x_0) = \frac{f'(x_0)g(x_0) - f(x_0)g'(x_0)}{g(x_0)^2}.$$

Proof The proof of the first two assertions are straightforward via the algebra of limits. Let us prove the other assertions.

3. By adding and subtracting the term $f(x_0)g(x_0 + h)$ in the numerator, the difference quotient of the product $f \times g$ can be rewritten as:

$$\begin{aligned}
&\lim_{h\to 0} \frac{f(x_0+h)g(x_0+h) - f(x_0)g(x_0)}{h} \\
&= \lim_{h\to 0} \frac{f(x_0+h)g(x_0+h) - f(x_0)g(x_0+h)+f(x_0)g(x_0+h)-f(x_0)g(x_0)}{h} \\
&= \lim_{h\to 0} \left(g(x_0+h)\frac{f(x_0+h)-f(x_0)}{h} + f(x_0)\frac{g(x_0+h)-g(x_0)}{h} \right)
\end{aligned} \tag{13.4}$$

In order to split the sum within the limit, we need to use the algebra of limits. But in order to do so, we need to check first whether each separate limits exist. Indeed, the second term is clear as the function g is differentiable, namely:

$$\lim_{h\to 0} f(x_0)\frac{g(x_0+h)-g(x_0)}{h} = f(x_0)g'(x_0).$$

For the first term, we note that since g is differentiable at x_0, it is also continuous here and so $\lim_{h\to 0} g(x_0+h) = g(x_0)$. Applying the algebra of limits, we have:

$$\begin{aligned}
&\lim_{h\to 0} \left(g(x_0+h)\frac{f(x_0+h)-f(x_0)}{h} \right) \\
&= \lim_{h\to 0} g(x_0+h) \lim_{h\to 0} \frac{f(x_0+h)-f(x_0)}{h} = g(x_0)f'(x_0).
\end{aligned}$$

Therefore, the limit in (13.4) becomes:

$$\lim_{h\to 0} \frac{f(x_0+h)g(x_0+h) - f(x_0)g(x_0)}{h} = g(x_0)f'(x_0) + f(x_0)g'(x_0),$$

which gives us the product rule.

4. We first show that the quantity $\frac{1}{g(x)}$ is defined for all x near x_0. In other words, we want to show that $g(x) \neq 0$ where $|x - x_0| < \delta$ for some $\delta > 0$. Indeed, since g is differentiable at x_0, it must be continuous at x_0. Suppose that $g(x_0) = L \neq 0$. For $\varepsilon = \frac{|L|}{2} > 0$, there exists some $\delta > 0$ such that for all $x \in X$ with $|x - x_0| < \delta$ we have $|g(x) - g(x_0)| = |g(x) - L| < \varepsilon = \frac{|L|}{2}$. By using triangle inequality, we then have:

$$|g(x)| \geq |g(x_0)| - |g(x) - g(x_0)| > |L| - \frac{|L|}{2} = \frac{|L|}{2} > 0,$$

for all $x \in (x_0 - \delta, x_0 + \delta) \cap X$, which says g is bounded away from 0 here. Now we consider the difference quotient of $\frac{1}{g}$. If $0 < |h| < \delta$, we can then write it as:

$$\lim_{h \to 0} \frac{\frac{1}{g(x_0+h)} - \frac{1}{g(x_0)}}{h} = \lim_{h \to 0} \frac{1}{g(x_0)g(x_0+h)} \frac{g(x_0) - g(x_0+h)}{h}$$
$$= \frac{1}{g(x_0)^2}(-g'(x_0)) = -\frac{g'(x_0)}{g(x_0)^2},$$

where we used the algebra of limits and the fact that g is continuous at x_0.

5. This is an application of the third and fourth assertions, namely: the derivative of f and $\frac{1}{g}$ at x_0 are $f'(x_0)$ and $-\frac{g'(x_0)}{g(x_0)^2}$ respectively. So, by applying the product rule, we have:

$$\left(\frac{f}{g}\right)'(x_0) = \frac{f'(x_0)}{g(x_0)} - \frac{f(x_0)g'(x_0)}{g(x_0)^2} = \frac{g(x_0)f'(x_0) - g'(x_0)f(x_0)}{g(x_0)^2},$$

which gives us the quotient rule. □

The results above help us fill in the gaps when we want to find the derivatives of more complicated functions for which the first principles would be too messy.

Example 13.2.2 Let us look at some examples:

1. We have seen how we can find the derivatives of the monomial x^2 which we can extend to the case x^n for any $n \in \mathbb{N}$ using the binomial expansion and first principles. This will be done by the readers in Exercise 13.1 later.
   Using the product rule in Theorem 13.2.1, we can also find the derivative of $f(x) = x^n$ by induction. Our claim is $f'(x_0) = nx_0^{n-1}$ for any $n \in \mathbb{N}$ and $x_0 \in \mathbb{R}$. This is clearly true for the case $n = 1$. Suppose that it is also true for the case $n = k$, namely $\frac{d}{dx}\Big|_{x_0} x^k = kx_0^{k-1}$. Now we show that this is also true for the case $n = k + 1$. Indeed, by using product rule and the inductive hypothesis, we have:

$$\frac{d}{dx}\Big|_{x_0} x^{k+1} = \frac{d}{dx}\Big|_{x_0} (x \cdot x^k) = 1 \cdot x_0^k + x_0 \cdot \frac{d}{dx}\Big|_{x_0} x^k = x_0^k + x_0 \cdot kx_0^{k-1}$$
$$= x_0^k + kx_0^k = (k+1)x_0^k,$$

   completing the inductive step.
2. Furthermore, using the first example above and the fact that derivatives can be scaled and added, we can also find the derivative for any polynomial $P : \mathbb{R} \to \mathbb{R}$

given by $P(x) = \sum_{j=0}^{n} a_j x^j$ where $a_j \in \mathbb{R}$ are some constants. For any $x_0 \in \mathbb{R}$, we can distribute the derivative over this finite sum as thus:

$$\left.\frac{d}{dx}\right|_{x_0} P(x) = \left.\frac{d}{dx}\right|_{x_0} \sum_{j=0}^{n} a_j x^j = \sum_{j=0}^{n} \left.\frac{d}{dx}\right|_{x_0} (a_j x^j)$$
$$= \sum_{j=0}^{n} a_j \left.\frac{d}{dx}\right|_{x_0} x^j = \sum_{j=1}^{n} j a_j x_0^{j-1}.$$

Therefore, the derivative of the polynomial $P(x) = 3x^2 + 2x + 1$ at x_0 is simply obtained by differentiating term-wise to yield $P'(x_0) = 6x_0 + 2$.
Notice that this is only true for finite sums and we shall see later that it might not be the case for functions series. This phenomenon occurs because in order to switch a derivative and an infinite sum, we need to switch the order of two different limits: one coming from the derivative and the other from the series. We have seen many examples for which limits cannot be swapped unless we have some kind of uniform convergence condition. In Chap. 14, we shall see when we can switch the order of limits and derivatives.

3. Consider the cosecant function which is given as $\csc(x) = \frac{1}{\sin(x)}$. This function is only defined on $\mathbb{R} \setminus \{n\pi : n \in \mathbb{Z}\}$ where the denominator does not vanish. Pick any x_0 in this set. We can compute the derivative of this function by using Theorem 13.2.1(4) as follows:

$$\left.\frac{d}{dx}\right|_{x_0} \csc(x) = \left.\frac{d}{dx}\right|_{x_0} \frac{1}{\sin(x)} = -\frac{\left.\frac{d}{dx}\right|_{x_0} \sin(x)}{\sin^2(x_0)} = -\frac{\cos(x_0)}{\sin^2(x_0)} = -\cot(x_0)\csc(x_0).$$

4. Consider the function $f : \mathbb{R} \setminus \{-\frac{1}{2}\} \to \mathbb{R}$ defined as $f(x) = \frac{\sqrt{x}}{2x+1}$. We know how to find the derivatives of the numerator and the denominator from the first principles separately as we have seen in Example 13.1.9.
But together as a fraction, this can be complicated. Luckily for us, we have the quotient rule to help us out here. If we let $g(x) = \sqrt{x}$ and $h(x) = 2x + 1$ both defined on $\mathbb{R} \setminus \{-\frac{1}{2}\}$, we can compute $g'(x_0) = \frac{1}{2\sqrt{x_0}}$ which is only defined for $x_0 \neq 0$ and $h'(x_0) = 2$. Thus, the quotient rule says:

$$f'(x_0) = \frac{g'(x_0)h(x_0) - g(x_0)h'(x_0)}{h(x_0)^2} = \frac{\frac{2x_0+1}{2\sqrt{x_0}} - 2\sqrt{x_0}}{(2x_0+1)^2} = \frac{1 - 2x_0}{2\sqrt{x_0}(2x_0+1)^2},$$

for $x_0 \neq -\frac{1}{2}, 0$.

Another very useful result for differentiation is the chain rule.

Theorem 13.2.3 (Chain Rule) *Let $X, Y \subseteq \mathbb{R}$. Suppose that $f : X \to \mathbb{R}$ and $g : Y \to \mathbb{R}$ are such that $f(X) \subseteq Y$ with:*

1. *f is differentiable at $x_0 \in X$, and*
2. *g is differentiable at $y_0 = f(x_0) \in Y$.*

Then, the composite function $g \circ f : X \to \mathbb{R}$ is differentiable at x_0 with $(g \circ f)'(x_0) = g'(f(x_0))f'(x_0)$.

Proof Consider the following function $h : Y \to \mathbb{R}$ defined as:

$$h(y) = \begin{cases} \frac{g(y)-g(y_0)}{y-y_0} - g'(y_0) & \text{if } y \neq y_0, \\ 0 & \text{if } y = y_0. \end{cases}$$

As $y \to y_0$, we have $h(y) \to 0 = h(y_0)$ and so this function is continuous at y_0. If we precompose this function with f, we then have $h \circ f : X \to \mathbb{R}$ defined as:

$$(h \circ f)(x) = \begin{cases} \frac{g(f(x))-g(y_0)}{f(x)-y_0} - g'(y_0) & \text{if } f(x) \neq y_0, \\ 0 & \text{if } f(x) = y_0. \end{cases} \tag{13.5}$$

Furthermore, since f is continuous at x_0 and h is continuous at $y_0 = f(x_0)$, this composite function is continuous at $x_0 \in X$ with $\lim_{x \to x_0}(h \circ f)(x) = 0$. We can then rewrite (13.5) as the equality:

$$(g(f(x)) - g(y_0)) = g'(y_0) \cdot (f(x) - y_0) + (h \circ f)(x) \cdot (f(x) - y_0), \tag{13.6}$$

which is valid for all $x \in X$.

We wish to find the limit of the difference quotient $\Delta_{x_0}(g \circ f)(x) = \frac{g(f(x))-g(f(x_0))}{x-x_0}$ as $x \to x_0$, so we may divide both sides of Eq. (13.6) with $x - x_0$ since the limit will be taken for $x \neq x_0$. Thus:

$$\frac{g(f(x))-g(f(x_0))}{x - x_0} = \frac{g'(f(x_0)) \cdot (f(x)-f(x_0))}{x - x_0} + \frac{(h \circ f)(x) \cdot (f(x)-f(x_0))}{x - x_0}. \tag{13.7}$$

Taking the limit as $x \to x_0$ on both sides of (13.7) and using the algebra of limits, by noting that $f'(x_0)$ exists and $\lim_{x \to x_0}(h \circ f)(x) = 0$, we have:

$$\begin{aligned} &\lim_{x \to x_0} \frac{g(f(x)) - g(f(x_0))}{x - x_0} \\ &= g'(f(x_0)) \cdot \lim_{x \to x_0} \frac{f(x) - f(x_0)}{x - x_0} + \lim_{x \to x_0} (h \circ f)(x) \cdot \lim_{x \to x_0} \frac{f(x) - f(x_0)}{x - x_0} \\ &= g'(f(x_0)) \cdot f'(x_0) + 0 \cdot f'(x_0) = g'(f(x_0)) \cdot f'(x_0) \end{aligned}$$

which gives us the chain rule at $x = x_0$. ◻

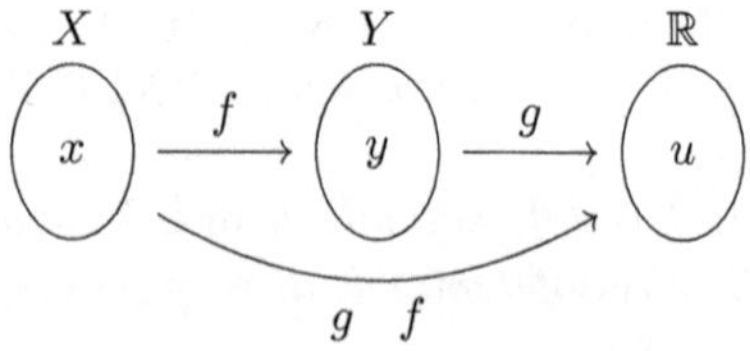

Fig. 13.3 The composition of functions g and f

In the chain rule, it may be a bit confusing to see which variables the derivatives are done with respect to. A good way to remember this is to write the functions in terms of explicit dummy variables, say $f(x) = y$ and $g(y) = u$.

Based on Fig. 13.3, the chain rule then says:

$$\left.\frac{d}{dx}\right|_{x_0} (g \circ f)(x) = \frac{dg}{dy}(y = f(x_0)) \cdot \frac{df}{dx}(x_0).$$

Example 13.2.4 Consider the function $f : \mathbb{R} \to \mathbb{R}$ defined as $f(x) = (x^2 + 3)^6$ and we wish to find its derivative at some $x_0 \in \mathbb{R}$. We can choose to expand this using the binomial theorem and apply the derivative term-wise.

However, we can also use the chain rule to do this. Let $g, h : \mathbb{R} \to \mathbb{R}$ be defined as $h(x) = x^6$ and $g(x) = x^2 + 3$. Thus, we have $f = h \circ g$. Moreover, at any $x_0 \in \mathbb{R}$, g is differentiable at x_0 with $g'(x_0) = 2x_0$ and h is differentiable at $g(x_0)$ with $h'(g(x_0)) = 6g(x_0)^5$. Therefore, we can apply the chain rule to get $h'(x_0) = h'(g(x_0))g'(x_0) = 6g(x_0)^5 \cdot 2x_0 = 12x_0(x_0^2 + 3)^5$.

In order to use the chain rule in Theorem 13.2.3, we need to first ensure that f is differentiable at x_0 and g is differentiable at $f(x_0)$ to deduce that $(g \circ f)'(x_0)$ exists. However, in some cases, we might not know the differentiablity of f but we know the differentiability of the composite function $g \circ f$ instead. With this information, we can deduce the differentiability of f under some mild conditions. This is given by the following converse to the chain rule:

Proposition 13.2.5 (Converse of Chain Rule) *Let $X, Y \subseteq \mathbb{R}$. Suppose that $f : X \to \mathbb{R}$ and $g : Y \to \mathbb{R}$ are such that $f(X) \subseteq Y$ with:*

1. *f is continuous at $x_0 \in X$,*
2. *g is differentiable at $y_0 = f(x_0) \in Y$ with $g'(y_0) \neq 0$, and*
3. *$g \circ f : X \to \mathbb{R}$ is differentiable at x_0.*

Then, the function f must be differentiable at x_0 with $f'(x_0) = \frac{(g \circ f)'(x_0)}{g'(f(x_0))}$.

Proof WLOG, assume that $g'(y_0) = g'(f(x_0)) > 0$. In the proof of Theorem 13.2.3, we have derived the equation:

$$\begin{aligned}\frac{g(f(x)) - g(f(x_0))}{x - x_0} &= \frac{g'(f(x_0)) \cdot (f(x) - f(x_0))}{x - x_0} + \frac{(h \circ f)(x) \cdot (f(x) - f(x_0))}{x - x_0}\\ &= (g'(f(x_0)) + (h \circ f)(x)) \frac{f(x) - f(x_0)}{x - x_0},\end{aligned} \tag{13.8}$$

for all $x \in X \setminus \{x_0\}$. Since $\lim_{x \to x_0} (h \circ f)(x) = 0$, there exists a $\delta > 0$ such that for all $x \in X \setminus \{x_0\}$ satisfying $0 < |x - x_0| < \delta$ we have $|(h \circ f)(x) - 0| \leq \frac{g'(f(x_0))}{2}$. This then implies $g'(f(x_0)) + (h \circ f)(x) > \frac{g'(f(x_0))}{2} > 0$ for all such x. Therefore, for all such x, from the equality (13.8), we have:

$$\frac{f(x) - f(x_0)}{x - x_0} = \frac{1}{g'(f(x_0)) + (h \circ f)(x)} \frac{g(f(x)) - g(f(x_0))}{x - x_0}. \tag{13.9}$$

Taking the limit as $x \to x_0$ in (13.9), by using the algebra of limits, we have:

$$\begin{aligned}\lim_{x \to x_0} \frac{f(x) - f(x_0)}{x - x_0} &= \lim_{x \to x_0} \left(\frac{1}{g'(f(x_0)) + (h \circ f)(x)} \frac{g(f(x)) - g(f(x_0))}{x - x_0} \right)\\ &= \lim_{x \to x_0} \frac{1}{g'(f(x_0)) + (h \circ f)(x)} \lim_{x \to x_0} \frac{g(f(x)) - g(f(x_0))}{x - x_0}\\ &= \frac{(g \circ f)'(x_0)}{g'(f(x_0))},\end{aligned}$$

which is what we wanted to prove. □

Let us look at a very important example on an application of Proposition 13.2.5. We saw in Example 13.1.9 that $\frac{d}{dx}\big|_{x_0} x^n = n x_0^{n-1}$ for any $n \in \mathbb{Z} \setminus \{0\}$ and $\frac{d}{dx}\big|_{x_0} x^{\frac{1}{2}} = \frac{1}{2} x^{-\frac{1}{2}}$. Now we generalise this to the expression x^r for any other exponent $r \in \mathbb{Q}$.

Proposition 13.2.6 *Let $f : \mathbb{R}_+ \to \mathbb{R}$ be defined as $f(x) = x^r$ for some $r \in \mathbb{Q}$. Then, $\frac{d}{dx}\big|_{x_0} f(x) = r x_0^{r-1}$ for $x_0 \neq 0$.*

Proof Let us write $r = \frac{p}{q}$ where $p \in \mathbb{Z}$ and $q \in \mathbb{N}$. Define two functions $g, h : \mathbb{R}_+ \to \mathbb{R}$ as $g(x) = x^p$ and $h(x) = x^q$. Then, we have the following composition:

$$h(f(x)) = (f(x))^q = (x^{\frac{p}{q}})^q = x^p = g(x).$$

Since $p, q \in \mathbb{Z}$, we know how to differentiate the functions g and h at any $x_0 \neq 0$, namely $g'(x_0) = p x_0^{p-1}$ and $h'(y_0) = q y_0^{q-1} \neq 0$. Moreover, f is continuous everywhere. Thus, Proposition 13.2.5 says the function f is differentiable at x_0 with:

$$f'(x_0) = \frac{(h \circ f)'(x_0)}{h'(f(x_0))} = \frac{g'(x_0)}{h'(f(x_0))} = \frac{px_0^{p-1}}{q(f(x_0))^{q-1}} = \frac{p}{q}x_0^{\frac{p}{q}-1} = rx_0^{r-1},$$

which is what we wanted to prove. □

Later, once we have the right tools for it, we shall generalise Proposition 13.2.6 to include the irrational exponents in Proposition 13.7.5.

Before we move on to the next section, whilst the definition for derivative that we have been using is very convenient to use for real functions defined on subsets of the real line, we would like to express it in a more general, but equivalent, way to cater for generalisation to other domains. Indeed, in the future, we might be interested to work with functions with domain of $\mathbb{C}$ or $\mathbb{R}^n$. We have the following characterisation of the derivative at point x_0 in the domain.

Theorem 13.2.7 *A function $f : (a, b) \to \mathbb{R}$ is differentiable at $x_0 \in (a, b)$ if and only if there exists a number $L \in \mathbb{R}$ and a continuous function $\varepsilon : (a, b) \to \mathbb{R}$ with $\lim_{x \to x_0} \varepsilon(x) = 0$ such that $f(x) = f(x_0) + L \cdot (x - x_0) + \varepsilon(x)(x - x_0)$.*

The readers will prove this theorem in Exercise 13.12. The number L is called the derivative of f at x_0 where, by the notation used earlier, $L = f'(x_0)$. Using the asymptotic notations, this says f is differentiable at x_0 if and only if there is an $L \in \mathbb{R}$ such that:

$$f(x) = f(x_0) + L \cdot (x - x_0) + o(|x - x_0|).$$

Thus, this gives us $f(x) \approx f(x_0) + L(x - x_0)$ near x_0 which says that f can be approximated by the linear function $g(x) = f(x_0) + L(x - x_0)$ near x_0. The quantity $\varepsilon(x)(x - x_0)$ in Theorem 13.2.7 is the error in the approximation which, via the little-o notation, is insignificant when compared to $|x - x_0|$ near x_0.

Due to this, the number L gives us the "best linear approximation" of the function f at x_0 via the function g. Indeed, if we pick a different number $K \neq L$, from Theorem 13.2.7, we would then have the following approximation:

$$\begin{aligned} f(x) &= f(x_0) + L \cdot (x - x_0) + \varepsilon(x)(x - x_0), \\ &= f(x_0) + K \cdot (x - x_0) + ((L - K) \cdot (x - x_0) + \varepsilon(x)(x - x_0)). \end{aligned}$$

This means the error of this approximation of $f(x)$ with the linear function $f(x_0)+K(x-x_0)$ near the point x_0 is $(L-K)\cdot(x-x_0)+\varepsilon(x)(x-x_0) \notin o(|x-x_0|)$ as $x \to x_0$. In other words, this error is of the same order as $|x - x_0|$. This tells us that this linear approximation with a line of slope K is not as good as the best one with slope L.

In more general settings, namely in higher dimensions or in general normed vector spaces, this is the main interpretation for derivatives, namely: it is the linear operator L that best approximates the function at the point of interest. Therefore, the derivative of a function $f : \mathbb{R}^m \to \mathbb{R}^n$ would be a linear map from the space $\mathbb{R}^m$ to $\mathbb{R}^n$ (which is an $m \times n$ matrix if we have chosen a basis for each of the domain and codomain).

13.3 Differentiable Functions

If the function $f : X \to \mathbb{R}$ for $X \subseteq \mathbb{R}$ is differentiable at all $x_0 \in X$, we have a name for it.

Definition 13.3.1 (Differentiable Function) Let $f : X \to \mathbb{R}$ be a real-valued function. Then we say the function f is differentiable if it is differentiable at every $x_0 \in X$.

Remark 13.3.2 Proposition 13.1.10 implies that a differentiable function f is necessarily continuous everywhere.

Thus, for a differentiable function $f : X \to \mathbb{R}$, if we vary the point x_0 in the definition for the differentiability at x_0, we get an assignment $f' : X \to \mathbb{R}$ which is called the derivative or first derivative of f, denoted as f', $D_x f$, or $\frac{df}{dx}$. Since the derivative exists at all $x_0 \in X$, the derivative assignment is a bona fide function due to uniqueness of limits. The operation sending the function $f : X \to \mathbb{R}$ to its derivative $f' : X \to \mathbb{R}$ is usually denoted as the operator $\frac{d}{dx}$, which is called the differential operator. In other words, we have $\frac{d}{dx} f = f'$.

Example 13.3.3 Let us look at some examples:

1. Recall the sine and cosine functions defined on the whole of $\mathbb{R}$ in Example 13.1.9(6). For an $x_0 \in \mathbb{R}$ we have computed:

$$\left.\frac{d}{dx}\right|_{x_0} \sin(x) = \cos(x_0) \quad \text{and} \quad \left.\frac{d}{dx}\right|_{x_0} \cos(x) = -\sin(x_0).$$

However, the point x_0 was chosen arbitrarily in $\mathbb{R}$, so we may vary it and still get the same result. Thus, we conclude that the sine and cosine functions have derivatives:

$$\frac{d}{dx} \sin(x) = \cos(x) \quad \text{and} \quad \frac{d}{dx} \cos(x) = -\sin(x),$$

for all $x \in \mathbb{R}$.

2. Similarly, we have seen in Example 13.1.9(8) that the exponential function e^x defined on the whole of $\mathbb{R}$ has a derivative e^{x_0} at any arbitrary $x_0 \in \mathbb{R}$. Thus, it is differentiable everywhere with $\frac{d}{dx}e^x = e^x$.
3. Consider the function $f : \mathbb{R} \setminus \{-\frac{1}{2}\}$ defined as $f(x) = \frac{\sqrt{x}}{2x+1}$. We saw in Example 13.2.2(4) that for $x_0 \neq -\frac{1}{2}, 0$ we have:

$$f'(x_0) = \frac{\frac{2x_0+1}{2\sqrt{x_0}} - 2\sqrt{x_0}}{(2x_0+1)^2} = \frac{1-2x_0}{2\sqrt{x_0}(2x_0+1)^2}.$$

Therefore, the function is not differentiable everywhere on its domain $\mathbb{R} \setminus \{-\frac{1}{2}\}$ because the derivative does not exist at $x_0 = 0$. Indeed, at this point the limit $\lim_{h\to 0} \frac{f(0+h)-f(0)}{h} = \lim_{h\to 0} \frac{1}{\sqrt{h}(2h+1)}$ blows up to infinity. However, we can define its derivative when we restrict the domain to not include the problem point. The derivative is then denoted as $f' : \mathbb{R} \setminus \{-\frac{1}{2}, 0\} \to \mathbb{R}$ where $f'(x) = \frac{1-2x}{2\sqrt{x}(2x+1)^2}$.

As we have noted in Example 13.3.3, differentiability of a function over a domain is determined by checking the differentiability of the function over every point of its domain. By varying the point of differentiation x_0 in the algebra of derivatives in Proposition 13.2.1 and the chain rule in Proposition 13.2.3, these results can be extended to the whole domain X. We state:

Proposition 13.3.4 (Algebra of Derivatives) *Suppose that $f, g : X \to \mathbb{R}$ for some $X \subseteq \mathbb{R}$ are differentiable on X and $\lambda \in \mathbb{R}$ is a constant. Then:*

1. *The function λf is differentiable on X with:*

$$(\lambda f)' = \lambda f'.$$

2. *The function $f \pm g$ is differentiable on X with:*

$$(f \pm g)' = f' \pm g'.$$

3. *Product rule: The function $f \times g$ is differentiable on X with:*

$$(f \times g)' = f'g + fg'.$$

4. *If $g(x) \neq 0$ for any $x \in X$, then the function $\frac{1}{g}$ is differentiable on X with:*

$$\left(\frac{1}{g}\right)' = -\frac{g'}{g^2}.$$

5. *Quotient rule: If* $g(x) \neq 0$ *for any* $x \in X$*, then the function* $\frac{f}{g}$ *is differentiable over* X *with:*

$$\left(\frac{f}{g}\right)' = \frac{f'g - fg'}{g^2}.$$

Theorem 13.3.5 (Chain Rule) *Let* $X, Y \subseteq \mathbb{R}$*. Suppose that* $f : X \to \mathbb{R}$ *is differentiable on* X*. Suppose further that* $f(X) \subseteq Y$ *so that the composition* $g \circ f : X \to \mathbb{R}$ *is well-defined and* $g : Y \to \mathbb{R}$ *is differentiable on* $f(X) \subseteq Y$*. Then the composition function* $g \circ f$ *is differentiable over* X *with* $(g \circ f)'(x) = g'(f(x))f'(x)$ *for all* $x \in X$*.*

Notice that the differentiation process can be seen as a map or operator transforming a function f to its derivative f'. This observation is important later when we talk about implicit differentiation. Moreover, Proposition 13.3.4(1) and (2) say that the differential operator is an $\mathbb{R}$-linear operation and the space of real-valued differentiable functions form an $\mathbb{R}$-vector space. Indeed, for differentiable functions $f, g : (a, b) \to \mathbb{R}$ and $\lambda \in \mathbb{R}$ we have:

$$\frac{d}{dx}(f + \lambda g) = (f + \lambda g)' = f' + \lambda g' = \frac{d}{dx} f + \lambda \frac{d}{dx} g.$$

We have seen the space of continuous functions on X, denoted as $C^0(X)$, in Definition 10.2.7. The space $C^0(X)$ forms a real vector space. Based on the observation above, we can define the a new vector space of functions:

Definition 13.3.6 (Space of Continuously Differentiable Functions) The space of continuously differentiable real-valued functions on domain $X \subseteq \mathbb{R}$ with continuous derivative is denoted as $C^1(X; \mathbb{R})$ or $C^1(X)$. In other words, $C^1(X)$ is the set:

$$C^1(X; \mathbb{R}) = \{f : X \to \mathbb{R} : f \text{ is differentiable on } X \text{ and } f' \in C^0(X)\}.$$

Note that since any differentiable function is also continuous everywhere in its domain, we must have the inclusion $C^1(X) \subsetneq C^0(X)$. This inclusion is strict since there are many continuous functions which are not differentiable everywhere.

We have seen previously that if a function $f : X \to \mathbb{R}$ is differentiable everywhere, then it would define the derivative function $f' : X \to \mathbb{R}$. By a similar process, we could attempt to differentiate the derivative f' at a point x_0 in its domain as well. In particular, for any $x_0 \in X$, we can define:

$$f''(x_0) = \lim_{h \to 0} \frac{f'(x_0 + h) - f'(x_0)}{h}, \tag{13.10}$$

if this limit exists.

So, if this new function f' is also differentiable everywhere (meaning that we can define the limit in (13.10) for all $x_0 \in X$), it then defines a new function $f'' : X \to \mathbb{R}$. This new function is called the second order derivative or second derivative of the function f and is denoted as f'', $f^{(2)}$, $D_x^2 f$, or $\frac{d^2 f}{dx^2}$. The final notation comes from the fact that the second derivative of f is obtained by applying the differential operator $\frac{d}{dx}$ twice to f, namely:

$$f'' = \frac{d}{dx}\left(\frac{d}{dx}f\right) = \left(\frac{d}{dx}\right)^2 f = \frac{d^2}{dx^2}f.$$

Inductively, as long as the n-th derivative of the function f is differentiable everywhere, we can define the $(n+1)$-th derivative of the function f, which we denote as the function $f^{(n+1)} : X \to \mathbb{R}$. Again, various notations are available for n times differentiable functions, namely $f^{(n)}$, $D_x^n f$, or $\frac{d^n f}{dx^n}$. Generalising Definitions 10.2.7 and 13.3.6, we define:

Definition 13.3.7 (Space of n-times Continuously Differentiable Functions) Let $n \in \mathbb{N}_0$. The space of n times differentiable real-valued functions on domain $X = (a, b)$ with continuous n-th derivative is denoted as $C^n(X; \mathbb{R})$ or $C^n(X)$. In other words, $C^n(X)$ is the set:

$$C^n(X) = \{f : X \to \mathbb{R} : f \text{ is } n \text{ times differentiable on } X \text{ and } f^{(n)} \in C^0(X)\}.$$

Since the differential operator $\frac{d}{dx}$ is an $\mathbb{R}$-linear operator, we can repeatedly apply the differential operator for as long as the operation makes sense. So for any n-th differentiable functions $f, g : X \to \mathbb{R}$ defined on $X \subseteq \mathbb{R}$ and $\lambda \in \mathbb{R}$, we have:

1. $(\lambda f)^{(n)} = \lambda f^{(n)}$,
2. $(f \pm g)^{(n)} = f^{(n)} \pm g^{(n)}$,

and hence the set $C^n(X)$ forms a real vector space.

Furthermore, the product rule in Proposition 13.3.4 generalises to higher order derivatives, which is called the Leibniz rule. The proof of this result is left as Exercise 13.23.

Proposition 13.3.8 (Leibniz Rule) *Let $n \in \mathbb{N}_0$. Suppose that $f, g : X \to \mathbb{R}$ for $X \subseteq \mathbb{R}$ are n times differentiable functions. Then, the n-th derivative of the product $f \times g$ is given by the sum:*

$$(f \times g)^{(n)}(x) = \sum_{j=0}^{n} \binom{n}{j} f^{(k)}(x) g^{(n-j)}(x).$$

Remark 13.3.9 The form of Leibniz rule is reminiscent of the binomial expansion formula for $(a + b)^n$ in Exercise 3.14, which is a convenient way to remember it.

Finally, if a function is said to be differentiable to any degree at all, we call it a smooth function. The class of smooth functions is defined as:

Definition 13.3.10 (Space of Smooth Functions) Let $X \subseteq \mathbb{R}$. A function $f : X \to \mathbb{R}$ is called a smooth function if the n-th derivative $f^{(n)}$ exists for any $n \in \mathbb{N}$. The space of smooth functions on domain X is denoted as $C^\infty(X; \mathbb{R})$ or $C^\infty(X)$. In other words, $C^\infty(X)$ is the set:

$$C^\infty(X)=\{f : X\to\mathbb{R} : f \text{ is } n \text{ times differentiable on } X \text{ for all } n\in\mathbb{N}\}=\bigcap_{n\in\mathbb{N}} C^n(X).$$

From definitions above, we note that $C^\infty(X) \subsetneq C^n(X)$ and $C^{n+1}(X) \subsetneq C^n(X)$ for any $n \in \mathbb{N}_0$. Similar to the space $C^n(X)$, the space of smooth functions forms a real vector space. In addition, we have the following result.

Proposition 13.3.11 *If $f, g \in C^n(X; \mathbb{R})$ for some $X \subseteq \mathbb{R}$ and $n \in \mathbb{N}$, then $fg \in C^n(X; \mathbb{R})$. In particular, the product of two smooth functions is also smooth.*

The above result can be proven using induction and Leibniz rule, which we leave to the readers as Exercise 13.26.

13.4 Implicit Differentiation

In the previous section, we have talked about the derivative of a function $f : X \to \mathbb{R}$ defined explicitly in the form of $y = f(x)$. The derivative of this is then given by $\frac{dy}{dx} = f'(x)$. However, there are many functions that may not be described explicitly.

For example, consider the function $y : \mathbb{R} \to \mathbb{R}$ of x described as $y^5 + y = x$. The quantity y is a well-defined function of the variable x since for every $x \in \mathbb{R}$, there exists one and only one value of y corresponding to it. This can be seen from the fact that $y^5 + y$ is strictly increasing and the IVT. However, we cannot (at least not easily) write y explicitly in terms of x in the form of $y = f(x)$. Therefore, the equation $y^5 + y = x$ describes the function y in terms of x implicitly.

Since we do not have an explicit representation of y in terms of x, we cannot find the derivative of y with respect to x via first principles using difference quotients easily. Due to this difficulty in writing y explicitly, how do we find the derivative of the function y with respect to its implicit variable x? We use implicit differentiation.

As remarked in the previous section, the derivative can be seen as an operation on functions. Similar to algebraic operations on equations, we can apply the derivative operation on both sides of an equation (with respect to the same variable, of course) provided that the operation is well-defined on the equation.

Referring to the example of $y^5 + y = x$, we know that y is a function of x, so the operation of differentiation (with respect to the variable x) can be applied here. For clarity, we write y as $y(x)$ so that the equation becomes $y(x)^5 + y(x) = x$. All we have to do is apply the differentiation operation on both sides and use the algebra of derivatives and chain rule to get:

$$\begin{aligned}\frac{d}{dx}(y(x)^5 + y(x)) = \frac{d}{dx}(x) \quad &\Rightarrow \quad \frac{d}{dx}y(x)^5 + \frac{d}{dx}y(x) = 1\\ &\Rightarrow \quad 5y(x)^4 y'(x) + y'(x) = 1\\ &\Rightarrow \quad y'(x) = \frac{1}{5y(x)^4 + 1},\end{aligned}$$

for any $x \in \mathbb{R}$. This is again an implicit expression for the derivative, but at least we have something tangible that we can work with!

In general, implicit differentiation allows us to differentiate an equation of the form $F(x, y) = 0$ for some function $F : \mathbb{R}^2 \to \mathbb{R}$. The most important thing that we have to check before using the technique of implicit differentiation with respect to x is that the quantity y is a genuine function of x.

There are many advanced tools that we can use to help us do this checking. In particular, there is a result called implicit function theorem which tells us where and when can y be treated as a function of x. This theorem is usually covered in a multivariable course after the students have seen partial derivatives and Jacobians, so we shall not mention it any further here. Interested readers may refer to [18] for this theorem.

Example 13.4.1 Consider the circle $C = \{(x, y) \in \mathbb{R}^2 : x^2 + y^2 = 1\}$. Necessarily if $(x, y) \in C$, we must have $x \in [-1, 1]$. Note that the quantity y cannot be written as a function of $x \in [-1, 1]$. Indeed, from the equation of the circle, we have $y = \pm\sqrt{1 - x^2}$ so every $x \in (-1, 1)$ corresponds to two values of y.

To ensure y is a genuine function over $x \in [-1, 1]$, we restrict our attention to the upper half of the circle, namely $C' = \{(x, y) \in \mathbb{R}^2 : x^2 + y^2 = 1, y \geq 0\}$. As a result, y can be written as a function of x, namely $y : [-1, 1] \to \mathbb{R}$ is $y(x) = \sqrt{1 - x^2}$. Suppose that we want to find the derivative y', there are two ways to do this:

1. Since we can write y explicitly in terms of x, we can compute the derivative using power rule and chain rule to get:

$$y'(x) = \frac{-x}{\sqrt{1 - x^2}} = -\frac{x}{y},$$

for $x \neq \pm 1$.

2. Alternatively, we can use implicit differentiation. We know that $y^2 + x^2 = 1$. Applying the derivative with respect to x on both sides and using the chain rule, we have:

$$\frac{d}{dx}(y^2 + x^2) = \frac{d}{dx}(1) \quad \Rightarrow \quad 2yy' + 2x = 0 \quad \Rightarrow \quad yy' = -x.$$

Note that $y \neq 0$ for $x \in (-1, 1)$, so the derivative of y at $x \in (-1, 1)$ is $y' = -\frac{x}{y}$. For $x = \pm 1$, the quantity y' is undefined. This is the exact same result as the above.

13.5 Extremum and Critical Points

Differentiable functions, or functions that are differentiable everywhere in its domain, are nice because we can obtain a lot of information from their derivatives. Since they are also continuous, they have all the properties of continuous functions from Chap. 10 for free.

We can get additional information about the function by recalling that the derivative of a function at a point x_0 represents the slope of the tangent line to the graph of f at $(x_0, f(x_0))$. We shall see how it can be used to deduce the behaviour of the function later, but let us first define some new terms to study these functions more carefully. We have seen these definitions in Definition 9.1.10 earlier. To reiterate:

Definition 13.5.1 (Local Extremum Points) Let $f : X \to \mathbb{R}$ where $X \subseteq \mathbb{R}$ and $x_0 \in X$.

1. If there exists a $\delta > 0$ such that $f(x_0) \geq f(x)$ for every $x \in (x_0-\delta, x_0+\delta) \cap X \subseteq X$, then the point x_0 is called a local maximum point of the function f.
2. If there exists a $\delta > 0$ such that $f(x_0) \leq f(x)$ for every $x \in (x_0-\delta, x_0+\delta) \cap X \subseteq X$, then the point x_0 is called a local minimum point of the function f.

In either of the cases, the point x_0 is called a local extremum point of the function f.

The gist of the definitions above is clear: if we compare the value of $f(x_0)$ with other $f(x)$ for any x close to x_0, a local maximum means $f(x_0)$ is greater than these other values and a local minimum means $f(x_0)$ is smaller than these other values. If we choose to compare with all the x in the domain instead, we have the following definitions which we saw in Definition 9.1.9 earlier:

Definition 13.5.2 (Global Extremum Point) Let $f : X \to \mathbb{R}$ where $X \subseteq \mathbb{R}$ and $x_0 \in X$.

1. If $f(x_0) \geq f(x)$ for every $x \in X$, then the point x_0 is called a global maximum point of the function f and $f(x_0)$ is the global maximum value of the function f over X.
2. If $f(x_0) \leq f(x)$ for every $x \in X$, then the point x_0 is called a global minimum point of the function f and $f(x_0)$ is the global minimum value of the function f over X.

In either of the cases, the point x_0 is called a global extremum point of the function f.

A global extremum point is necessarily a local extremum point, but not the other way round. Local and global extremum points are referred collectively as extremum points.

If the function f is differentiable at its extremum point x_0, then we can deduce an information about its derivative here. This result is due to Pierre de Fermat (1607–1665).

Theorem 13.5.3 (Fermat's Theorem) *Suppose that $f : X \to \mathbb{R}$ where $X \subseteq \mathbb{R}$ and $x_0 \in X$ is an extremum point of the function f. If the function f is differentiable at x_0, then $f'(x_0) = 0$.*

Proof WLOG, assume that the extremum point x_0 is a local maximum. Thus there exists a $\delta > 0$ such that $f(x) \leq f(x_0)$ for $x \in (x_0 - \delta, x_0 + \delta) \cap X$. Since the function is differentiable at x_0, the left- and right-derivatives of f at x_0 exist and are equal to each other. In other words, $\lim_{h\uparrow 0} \frac{f(x_0+h)-f(x_0)}{h} = \lim_{h\downarrow 0} \frac{f(x_0+h)-f(x_0)}{h}$. Since we are taking the limit as $h \to 0$, we can assume that $|h| < \delta$ and thus $f(x_0 + h) - f(x_0) \leq 0$ for all such h.

For the right-limit, we consider only $h > 0$. Since limits preserve weak inequalities, we have:

$$\frac{f(x_0 + h) - f(x_0)}{h} \leq 0 \quad \Rightarrow \quad \lim_{h\downarrow 0} \frac{f(x_0 + h) - f(x_0)}{h} \leq 0.$$

Using a similar argument, we show that the left limit satisfies:

$$\lim_{h\uparrow 0} \frac{f(x_0 + h) - f(x_0)}{h} \geq 0.$$

However, since these two quantities are equal, both of them must be equal to 0. □

Remark 13.5.4 We make some important remarks regarding Theorem 13.5.3.

1. Notice that the theorem only specified points at which f is differentiable but not points at which f is not differentiable. Therefore, there are probably extremum points within the set of points at which f is not differentiable. For example, the function $f : \mathbb{R} \to \mathbb{R}$ defined as $f(x) = |x|$ that we saw in Example 13.1.9(4) has a global minimum at $x = 0$, which is a point where f is not differentiable.
2. Another important thing to notice is that the implication in Theorem 13.5.3 goes only one way. The points $x_0 \in X$ at which $f'(x_0) = 0$ are not necessarily extremum points. Consider the function $f : \mathbb{R} \to \mathbb{R}$ defined as $f(x) = x^3$.

Its derivative exists everywhere and is given by the function $f'(x) = 3x^2$ which vanishes at $x = 0$.
However, it is not a local extremum point because $f(0) = 0$ and for any $\delta > 0$ at all, $f(x)$ takes both positive and negative values in the interval $(-\delta, \delta)$ since $f(-\frac{\delta}{2}) < 0 < f(\frac{\delta}{2})$. As a result, the point $x = 0$ is not an extremum point even though $f'(0) = 0$.

Following the remark above, let us consider the following subset of X:

Definition 13.5.5 (Critical Points) Let $f : X \to \mathbb{R}$ be a real-valued function. If $x_0 \in X$ is such that f is not differentiable at x_0 or $f'(x_0) = 0$, the point x_0 is called a critical point of f.

The local extremum points of a function must be contained among the set of critical points. Indeed, if x_0 is a local extremum point of f, then either $f'(x_0)$ does not exist or $f'(x_0)$ exists. For the latter, necessarily $f'(x_0) = 0$ by Theorem 13.5.3. Either way, x_0 must be a critical point of the function f.

However, Remark 13.5.4(2) suggest that not all of these points could be extremum points. There are also critical points which are not extremum points. As a result, we can find all the extremum points of a function by checking the critical points one by one.

Example 13.5.6 Let us look at some examples on how we can do this:

1. Consider the function $f : \mathbb{R} \to \mathbb{R}$ defined as $f(x) = \sqrt[3]{x}$. Its derivative is given by $f'(x) = \frac{1}{3\sqrt[3]{x^2}}$. The derivative exist everywhere except at the point $x = 0$. The derivative does not vanish anywhere, thus the point $x = 0$ is the only critical point of the function f. Hence, it is the only candidate for the extremum point of f.
   However, $x = 0$ is not a local extremum point for the function f because for any $\delta > 0$ at all, we have $f(-\delta) < f(0) < f(\delta)$. Thus, the function f does not have any extremum points.
2. Consider the polynomial $f : \mathbb{R} \to \mathbb{R}$ defined as $f(x) = x^4 - x^3$. Its derivative exists everywhere and is given by $f'(x) = 4x^3 - 3x^2 = x^2(4x - 3)$. So, its critical points are the solutions to the equation $f'(x) = 0$, namely $x = 0, \frac{3}{4}$. Let us investigate these points separately.
   (a) Near $x = 0$, we can check that for a small δ (say $\delta = \frac{1}{2}$), for $-\delta < x < 0$ we have $f(x) = x^3(x - 1) > 0$ and for $0 < x < \delta$ we have $f(x) < 0$. Thus, $x = 0$ is not an extremum point.
   (b) On the other hand, at $x = \frac{3}{4}$ we have a local minimum. To show this, let us consider a new function on $\mathbb{R}$ given as:

$$g(x) = f(x) - f\left(\frac{3}{4}\right) = x^4 - x^3 - \frac{3^4}{4^4} + \frac{3^3}{4^3} = \frac{1}{256}(4x-3)^2(16x^2+8x+3).$$

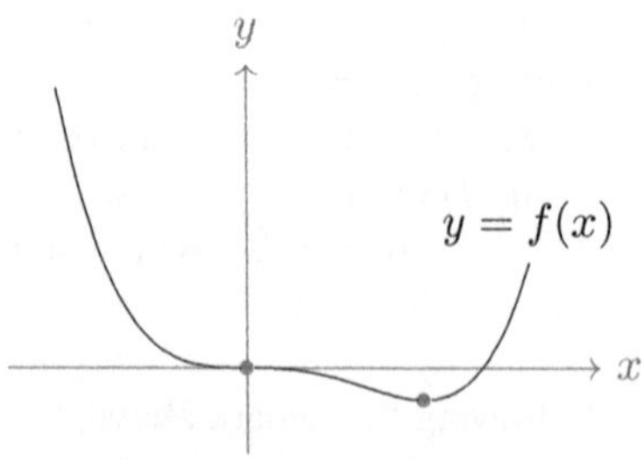

Fig. 13.4 Graph of $f(x) = x^4 - x^3$ with its critical points

Notice that the quantity $16x^2 + 8x + 3$ is strictly positive since it has negative discriminant and is positive somewhere. Thus, we have $g(x) \geq 0$ for all $x \in \mathbb{R}$ which implies $f(x) \geq f(\frac{3}{4})$ everywhere. This means that not only $x = \frac{3}{4}$ is a local minimum point, it is also a global minimum for the function f.

The conclusion here is that we have two critical points of f, but only one is an extremum point. The graph for this function is given in Fig. 13.4.

3. Consider the function $f : \mathbb{R} \to \mathbb{R}$ defined as:

$$f(x) = \begin{cases} \sqrt{-1-x} & \text{if } x \leq -1, \\ \sqrt{1-x^2} & \text{if } -1 < x < 1, \\ \sqrt{x-1} & \text{if } x \geq 1. \end{cases}$$

This function is continuous and non-negative. The derivative of this function can be computed easily in the regions $x < -1$, $-1 < x < 1$, and $x > 1$ as:

$$f'(x) = \begin{cases} -\frac{1}{2(-1-x)^{\frac{3}{2}}} & \text{if } x < -1, \\ -\frac{x}{(1-x^2)^{\frac{3}{2}}} & \text{if } -1 < x < 1, \\ \frac{1}{2(x-1)^{\frac{3}{2}}} & \text{if } x > 1. \end{cases}$$

However, the derivatives at $x = \pm 1$ do not exist since the limit of the difference quotients diverge either to $\pm\infty$, depending on how we approach the point ± 1. Namely, these points are cusps. Therefore, the three critical points for this function are $x = \pm 1$ (where the derivatives do not exist) and also $x = 0$ which we get when we solve the equation $f'(x) = 0$.

Upon checking, we obtain $f(\pm 1) = 0 \leq f(x)$ for all $x \in \mathbb{R}$ which says that these points are the local (and also global) minimum. On the other hand, the point $x = 0$ is a local maximum since for all $|x| < 1$, we have $f(0) = 1 \geq \sqrt{1-x^2}$. The global maximum does not exist since the function is unbounded from above with $\lim_{x \to \pm\infty} f(x) = \infty$. The graph of this function is depicted in Fig. 13.5.

In this example, all of the critical points of f are extremum points of f.

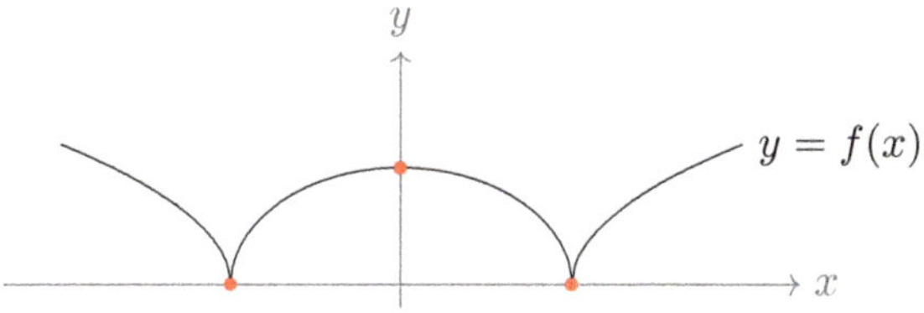

Fig. 13.5 Graph of f with its critical points

Recall the EVT in Theorem 10.5.1 in which we saw that a continuous function over a compact interval attains its global extremum somewhere. However, the theorem only states that these points exist, but does not give us any way of locating them.

As we know, any differentiable function f is also continuous, so the EVT also holds for it, as long as the domain of f is compact. The added bonus for differentiable functions is that we can locate these global extremum points by narrowing down our search just to the critical points where the extremum points are located in.

Proposition 13.5.7 (Extreme Value Theorem II, EVT II) *Let $f \in C^0(I)$ be a continuous real-valued function on a compact interval $I = [a, b]$. Then, a global extremum (maximum or minimum) point $\xi \in I$ of the function f is either:*

1. *on the boundary of $[a, b]$, namely $\xi \in \{a, b\}$, or*
2. *a critical point of the function f in (a, b), namely $\xi \in (a, b)$ with $f'(\xi) = 0$ or $f'(\xi)$ is undefined.*

Proof We prove that the global maximum satisfies one of the above. The proof for global minimum is similarly obtained.

Since f is a continuous function on a closed interval $[a, b]$, by the EVT in Theorem 10.5.1, there exists a point $\xi \in [a, b]$ such that $f(\xi) = \max_{x \in [a,b]} f(x)$. Suppose that the global maximum ξ does not occur on the boundary, namely $\xi \neq a, b$. Then, it must be inside the interval (a, b). If the function is differentiable at this point, by Theorem 13.5.3, we have $f'(\xi) = 0$. Otherwise, $f'(\xi)$ is undefined. Either way, the global maximum point ξ is a critical point of f in (a, b). □

Example 13.5.8 Let us look at some examples:

1. Let $f : [-1, 2] \to \mathbb{R}$ be defined as $f(x) = x^2$. This function is continuous in $[-1, 2]$, differentiable in $(-1, 2)$, and its graph is given in Fig. 13.6. We wish to find the global extremum points of this function. We first differentiate it and equate its derivative with 0 to locate any interior critical points. This yields $f'(x) = 2x = 0$ and so $x = 0$.
   The derivatives are defined everywhere in $(-1, 2)$, so there are no other critical points here. Thus, the only critical point of this function is then $x = 0$ with value $f(0) = 0$. To find the global extremum points, by virtue of EVT II, we also need

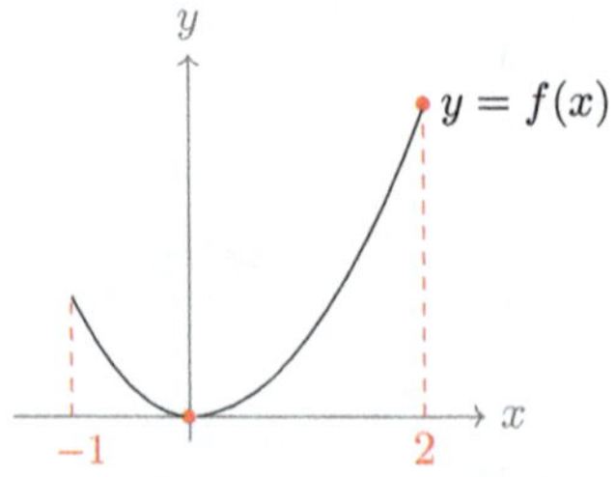

Fig. 13.6 The graph of $f(x) = x^2$ for $x \in [-1, 2]$ and its global extremum points

to check the values of the function f at the boundaries. At the boundaries, we have $f(-1) = 1$ and $f(2) = 4$.
Therefore, the global extremum points are among the points $x = -1, 0, 2$ which have values $f(-1) = 1$, $f(0) = 0$, and $f(2) = 4$. Hence, we deduce that the global minimum is at $x = 0$ with value 0 and the global maximum is at $x = 2$ with value 4. These points are labeled red in Fig. 13.6.

2. Consider the function $f : [-2, 2] \to \mathbb{R}$ defined as $f(x) = |x|$. We have seen that f is continuous everywhere and differentiable on $(-2, 2) \setminus \{0\}$ with derivative:

$$f' : (-2, 2) \setminus \{0\} \to \mathbb{R}$$

$$x \mapsto \begin{cases} -1 & \text{if } x \in (-2, 0), \\ 1 & \text{if } x \in (0, 2). \end{cases}$$

We wish to find the global extremum points, which exists in $[-2, 2]$ by the EVT. The critical points of this function is $x = 0$ only as the derivative of the function f exists and is non-zero everywhere else. The value of this function at the critical point is $f(0) = |0| = 0$. The boundary points are ± 2, which give the same value $f(\pm 2) = |\pm 2| = 2$. Therefore, the global maximum of this function is 2 at the boundaries and its global minimum is 0 at the critical point.

13.6 Rolle's Theorem and Mean Value Theorems

A consequence of EVT II is called the Rolle's theorem, proven for the polynomial cases only by Michel Rolle (1652–1719). This was later extended by Cauchy to any differentiable function on a compact domain. Graphically, this is a very intuitive result for differentiable functions. See Fig. 13.7 for a visualisation of this phenomenon.

Theorem 13.6.1 (Rolle's Theorem) *Let $f \in C^0(I)$ be a continuous real-valued function on a compact interval $I = [a, b]$. Suppose that f is differentiable in (a, b). If $f(a) = f(b)$, then there exists a point $c \in (a, b)$ such that $f'(c) = 0$.*

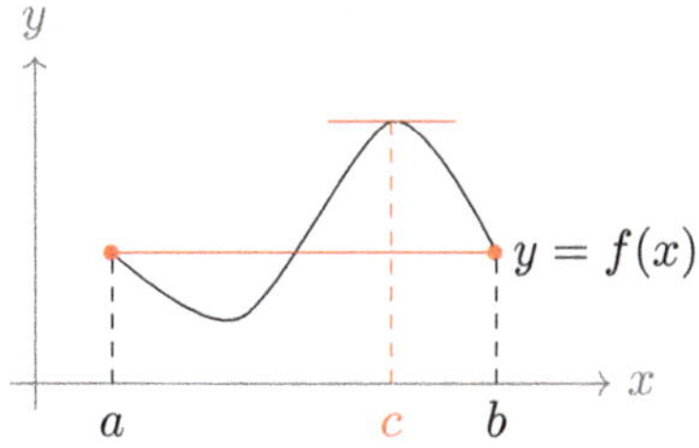

Fig. 13.7 At the point $c \in (a, b)$ we have $f'(c) = 0$. There is also another point between a and c where the derivative of f vanish

Proof Since $f \in C^0(I)$ is differentiable in (a, b), by EVT II, this function attains its global maximum/minimum either at the boundary points $\{a, b\}$ or at a critical point $c \in (a, b)$. Label the global minimum point $\xi \in I$ and the global maximum point $\zeta \in I$.

1. If either ξ or ζ lies in (a, b), then by EVT II, we have either $f'(\xi) = 0$ or $f'(\zeta) = 0$ and we have found our required point c.
2. Otherwise, if both ξ and ζ lie on the boundary of the interval I, by the assumption, we have $f(\xi) = f(\zeta)$. This means the global maximum value of the function is also its global minimum value. Thus, f is a constant function and constant functions have vanishing derivatives everywhere as we have seen in Example 13.1.9(5). Therefore, any $c \in (a, b)$ satisfies $f'(c) = 0$. □

Remark 13.6.2 We note that differentiability in the domain of f is an essential condition for Rolle's theorem. Indeed, if we consider the function $f : [-2, 2] \to \mathbb{R}$ defined by $f(x) = |x|$, we have seen in Example 13.1.9(4) that this function is differentiable on $(-2, 2) \setminus \{0\}$ with derivatives -1 for $x < 0$ and 1 for $x > 0$. Also, $f(-2) = f(2) = 2$, but there are no points in $(-2, 2)$ for which the derivative vanishes!

A direct consequence of Rolle's theorem is the mean value theorem (or the MVT for short). Geometrically, this theorem says for a differentiable function f defined over a compact interval $I = [a, b]$, if we join the endpoints of its graph with a straight line, we can always find a point within I such that the tangent line to the graph of f at this point has the same slope as the line joining the endpoints. See Fig. 13.8.

More precisely:

Theorem 13.6.3 (Mean Value Theorem, MVT) *Let $f \in C^0(I)$ be a continuous real-valued function defined on a compact interval $I = [a, b]$. Supppose that f is differentiable over (a, b). Then, there exists some $c \in (a, b)$ such that:*

$$f'(c) = \frac{f(b) - f(a)}{b - a}.$$

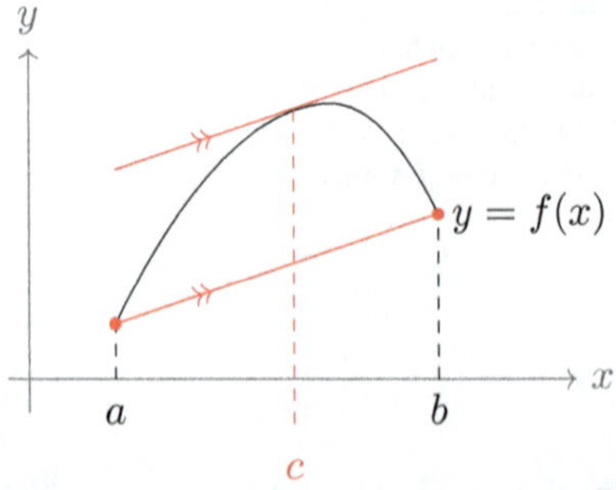

Fig. 13.8 The secant line joining $(a, f(a))$ and $(b, f(b))$ with a parallel tangent line

Proof Define $m = \frac{f(b)-f(a)}{b-a}$ and a function $g : I \to \mathbb{R}$ as $g(x) = f(x) - m(x - a)$. Clearly g is continuous on I and differentiable in (a, b) since it is a sum of differentiable functions. We then check:

$$g(a) = f(a) \quad \text{and} \quad g(b) = f(b) - \frac{f(b) - f(a)}{b - a}(b - a) = f(a).$$

Thus, we can apply Rolle's theorem on the function g which says that there exists some $c \in (a, b)$ such that $g'(c) = 0$. Unpacking definitions, we have:

$$0 = g'(c) = f'(c) - m = f'(c) - \frac{f(b) - f(a)}{b - a},$$

which gives us the theorem. □

Another variant of the MVT above is the following which is also known as Cauchy's mean value theorem:

Theorem 13.6.4 (Mean Value Theorem II, MVT II) *Let $f, g \in C^0(I)$ be continuous real-valued functions on a compact interval $I = [a, b]$. Supppose that f and g are differentiable in (a, b). Assume further that $g'(x) \neq 0$ for all $x \in (a, b)$. Then, there exists some $c \in (a, b)$ such that:*

$$\frac{f'(c)}{g'(c)} = \frac{f(b) - f(a)}{g(b) - g(a)}.$$

Proof We first note that $g(b) \neq g(a)$ so the above quotient does make sense. Indeed, if $g(b) = g(a)$, by Rolle's theorem, the derivative g' would vanish somewhere in (a, b), thus contradicting the assumption in the theorem.

Let $m = \frac{f(b)-f(a)}{g(b)-g(a)}$ and define the function $h = f - mg$ on I. Clearly, the function h is continuous in I and differentiable in (a, b). We check that:

$$h(a) = f(a) - \frac{f(b) - f(a)}{g(b) - g(a)} g(a) = \frac{f(a)g(b) - f(b)g(a)}{g(b) - g(a)},$$
$$h(b) = f(b) - \frac{f(b) - f(a)}{g(b) - g(a)} g(b) = \frac{f(a)g(b) - f(b)g(a)}{g(b) - g(a)},$$

so that $h(a) = h(b)$. Thus, we can apply Rolle's theorem, which says there exists a $c \in (a, b)$ such that $h'(c) = 0$. In other words:

$$0 = h'(c) = f'(c) - mg'(c) \quad \Rightarrow \quad \frac{f'(c)}{g'(c)} = m = \frac{f(b) - f(a)}{g(b) - g(a)},$$

which is what we wanted to prove. □

Remark 13.6.5 Let us make some remarks.

1. We note that the MVT is a special case of MVT II with the choice $g(x) = x$.
2. One thing to remember is that for EVT II, Rolle's Theorem, the MVT, and MVT II, the points of interest c may not be unique! For example, in the EVT there may be two distinct global maximum of the function f in $[a, b]$, both of which have the same value. This was seen in Example 13.5.8(2). However, the conclusion that can be obtained from the theorems are there must exist at least one of these points of interest.
3. The MVT, despite its simplicity and unassuming statement, is a very important result in differential calculus. It is widely utilised in the proofs of other results as we shall see repeatedly. As remarked by Edward Mills Purcell (1912–1997):

> The mean value theorem is the midwife of calculus – not very important or glamorous by itself, but often helping to deliver other theorems that are of major significance.

We have seen in Example 13.1.9 that for constant functions, their derivative must vanish everywhere. However, does the converse hold true? Yes, but only for a connected domain in $\mathbb{R}$, namely an interval. This can be proven by using the MVT.

Corollary 13.6.6 *Let $f : (a, b) \to \mathbb{R}$ be a differentiable real-valued function. If $f'(x) = 0$ for all $x \in (a, b)$, then $f(x) \equiv C$ for some constant $C \in \mathbb{R}$.*

Proof Pick any two arbitrary distinct numbers $\xi, \zeta \in (a, b)$. WLOG, suppose that $\xi < \zeta$. By applying the MVT over the domain $[\xi, \zeta]$, we get $f(\xi) - f(\zeta) = f'(c)(\xi - \zeta)$ for some $c \in (\xi, \zeta)$. However, since the derivative of f vanishes everywhere, we have $f(\xi) - f(\zeta) = 0$ which implies the value of f at these two points are the same. Since these two numbers were arbitrarily chosen, we arrive at the conclusion. □

Remark 13.6.7 We note that from the proof above, we can see that the result only holds in each connected interval in the domain. Suppose that we have a differentiable function $f : (a, b) \cup (c, d) \to \mathbb{R}$ such that $(a, b) \cap (c, d) = \varnothing$. If the function f has vanishing derivative everywhere, it must be constant in each of the disjoint intervals (a, b) and (c, d). Namely, f would be of the form:

$$f(x) = \begin{cases} C_1 & \text{if } x \in (a, b), \\ C_2 & \text{if } x \in (c, d), \end{cases}$$

where C_1 and C_2 are (not necessarily equal) real constants.

A direct corollary is the following:

Corollary 13.6.8 *Let $f, g : (a, b) \to \mathbb{R}$ be differentiable real-valued functions. If $f'(x) = g'(x)$ for all $x \in (a, b)$, then $f(x) = g(x) + C$ for some constant $C \in \mathbb{R}$. In other words, the functions f and g differ by an additive constant.*

13.7 Inverse Function Theorem

Recall in Theorem 10.5.4 that for a strictly increasing function $f : [a, b] \to \mathbb{R}$, if the function f is continuous, then the image of this function is also a compact interval, say $[c, d]$, and this gives rise to a continuous inverse function $f^{-1} : [c, d] \to [a, b]$.

Suppose now that the function f is also differentiable in $[a, b]$. What can we say about the derivative of its inverse? Is the inverse function f^{-1} differentiable everywhere on $[c, d]$ as well? Not necessarily.

Example 13.7.1 Let us look at a non-example. The function $f : \mathbb{R} \to \mathbb{R}$ defined as $f(x) = x^3$ is strictly increasing and differentiable everywhere with its derivative given by $f'(x) = 3x^2$. On the other hand, its inverse function $f^{-1} : \mathbb{R} \to \mathbb{R}$ is $f^{-1}(x) = \sqrt[3]{x}$. We have seen in Example 13.5.6 that f^{-1} is not differentiable at $x = 0$. So this inverse is not differentiable everywhere on its domain.

Geometrically, the graph of the inverse function f^{-1} of f is the reflection of the graph of f about the line $y = x$. Therefore, tangent line to the graph of f at the point $x = x_0$ with slope m would be reflected to tangent line of the function f^{-1} at $x = f(x_0)$ with slope $\frac{1}{m}$. Figure 13.9 depicts this phenomenon.

Thus, the tangent lines of f with vanishing slope would be mapped to vertical tangent lines of f^{-1} and hence the derivative of f^{-1} cannot be defined here. At the other points, this interpretation of reciprocal slopes does not have any issues since reciprocals can be defined for non-zero real numbers. Therefore, if we restrict our function f to have non-vanishing derivative, we might be able to deduce something. We prove:

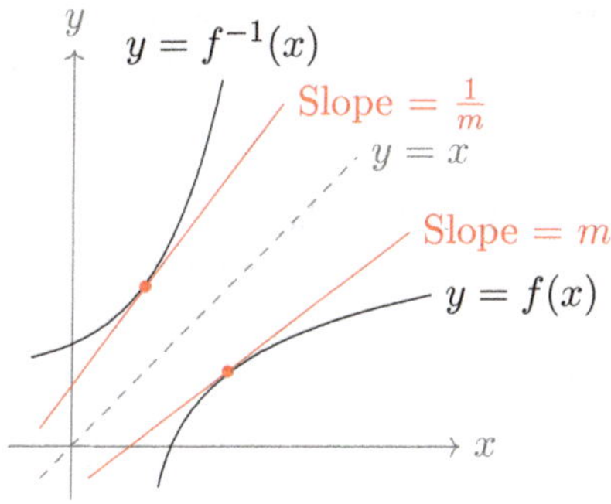

Fig. 13.9 The graphs of f and f^{-1} are reflections of each other across the line $y = x$. As a result, the slope of the tangent lines at $(x, f(x))$ on the graph of f and $(f(x), x)$ on the graph of f^{-1} are reciprocals to each other

Theorem 13.7.2 (Inverse Function Theorem) *Let $f : [a, b] \to [c, d]$ be a strictly monotone bijective continuous function. Suppose that f is differentiable in (a, b) and $f'(x) \neq 0$ for any $x \in (a, b)$. Then, the inverse function $f^{-1} : [c, d] \to [a, b]$ is also continuous in $[c, d]$, strictly monotone, and differentiable in (c, d) with:*

$$(f^{-1})'(y) = \frac{1}{f'(f^{-1}(y))},$$

for all $y \in (c, d)$.

Proof Continuity and strict monotonicity of f^{-1} have been proven in Theorem 10.5.4. Let $y_0 \in (c, d)$ so that there is some $x_0 \in (a, b)$ with $f(x_0) = y_0$. We wish to show the limit of the following quotient exists and find its value:

$$\lim_{y \to y_0} \frac{f^{-1}(y) - f^{-1}(y_0)}{y - y_0}.$$

We first rewrite this in terms of x and x_0. Since f^{-1} is also a bijection, for every $y \in (c, d)$ we have a unique $x \in (a, b)$ such that $f^{-1}(y) = x$. Furthermore, as $y \to y_0$, since f^{-1} is continuous, we have $x = f^{-1}(y) \to f^{-1}(y_0) = x_0$. Since $f'(x_0) \neq 0$, by using the algebra of limits, we then have:

$$\begin{aligned}\lim_{y \to y_0} \frac{f^{-1}(y) - f^{-1}(y_0)}{y - y_0} &= \lim_{x \to x_0} \frac{x - x_0}{f(x) - f(x_0)} \\ &= \lim_{x \to x_0} \frac{1}{\frac{f(x) - f(x_0)}{x - x_0}} \\ &= \frac{1}{\lim_{x \to x_0} \frac{f(x) - f(x_0)}{x - x_0}} = \frac{1}{f'(x_0)} = \frac{1}{f'(f^{-1}(y_0))},\end{aligned}$$

which exists and hence the function f^{-1} is differentiable at $y_0 \in (c, d)$ with value $(f^{-1})'(y_0) = \frac{1}{f'(f^{-1}(y_0))}$. Since y_0 is arbitrary, we can then vary y_0 in (c, d) to obtain the desired result. □

Remark 13.7.3 Theorem 13.7.2 is also true if we were to replace the intervals $[a, b]$ and $[c, d]$ with any other intervals of $\mathbb{R}$ as long as the function f is a strictly monotone bijection between these intervals and is differentiable everywhere with non-vanishing derivative.

Once we have known that the inverse function is differentiable, finding its derivative is straightforward. Indeed, since f and f^{-1} are inverses of each other, their composition is the identity function, namely $(f \circ f^{-1})(y) = y$ for all $y \in [c, d]$. Thus, by differentiating implicitly with respect to y, if we know that both the functions f and f^{-1} are differentiable, we can use the chain rule to obtain:

$$\begin{aligned}(f \circ f^{-1})(y) = y \quad &\Rightarrow \quad f'(f^{-1}(y)) \cdot (f^{-1})'(y) = 1 \\ &\Rightarrow \quad (f^{-1})'(y) = \frac{1}{f'(f^{-1}(y))},\end{aligned}$$

since $f'(x) \neq 0$ for any $x \in (a, b)$. However, to carry out the above analysis, it is necessary to first show that the inverse function f^{-1} is differentiable, which is a required condition for the chain rule. This is where the inverse function theorem comes in handy.

Example 13.7.4 Let us look at an application of the inverse function theorem. We have seen earlier that the derivative of the exponential function on the real line is itself, namely $\frac{d}{dx}e^x = e^x$. Recall that the inverse of the exponential function is the natural logarithm $\ln(x)$ defined on positive real numbers, so let us see if we can find the derivative of $\ln(x)$.

In order to apply the inverse function theorem, we need to check that the function e^x is strictly monotone and differentiable everywhere with nowhere vanishing derivative. We have shown these to be true in Proposition 12.4.4 and so we can apply the inverse function theorem to deduce that $\ln(x)$ is also differentiable. However, what is its derivative? We use the chain rule as suggested prior to this example.

Note that since e^x and $\ln(x)$ are inverses of each other, we have $(\exp \circ \ln)(x) = x$ on $x > 0$. Furthermore, as both the exponential and the logarithm functions are differentiable as we have shown above, we can differentiate both sides implicitly and apply the chain rule to get:

$$\begin{aligned}\frac{d}{dx}(\exp \circ \ln)(x) = \frac{d}{dx}x \quad &\Rightarrow \quad \left.\frac{d}{dy}\right|_{\ln(x)} e^y \cdot \frac{d}{dx}\ln(x) = 1 \\ &\Rightarrow \quad e^{\ln(x)}\frac{d}{dx}\ln(x) = 1,\end{aligned}$$

which implies that $\frac{d}{dx}\ln(x) = \frac{1}{x}$ for all $x > 0$.

Another important application of the inverse function theorem is completing the proof for the power rule. We have seen how to compute the derivative of $f(x) = x^r$ for non-zero rational exponents $r \neq 0$ and $x > 0$ to get $f'(x) = rx^{r-1}$ in Proposition 13.2.6. How can we extend this power rule to include the irrational exponents? Suppose that r is the irrational exponent and (r_n) is a sequence of rational numbers converging to r so that $x^{r_n} \to x^r$. To use this in the definition of derivatives via first principle might be too messy because we need to justify switching the order of the following limits:

$$\lim_{h\to 0} \frac{(x+h)^r - x^r}{h} = \lim_{h\to 0}\lim_{n\to\infty} \frac{(x+h)^{r_n} - x^{r_n}}{h}.$$

However, by using logarithms and Proposition 13.2.5, we can avoid analysing this iterated limits. We have the following result:

Proposition 13.7.5 (Power Rule) *Let $f : \mathbb{R}_+ \to \mathbb{R}$ be defined as $f(x) = x^r$ for some $r \in \mathbb{R}$. Then $\frac{d}{dx} f(x) = rx^{r-1}$.*

Proof Clearly this is true for $r \in \mathbb{Q}$ as we have seen in Proposition 13.2.6. Now assume that r is irrational and we want to differentiate $f(x) = x^r$. Since $f(x) > 0$ for all $x \in \mathbb{R}_+$, we apply logarithms on both sides to get $\ln(f(x)) = \ln(x^r) = r\ln(x)$.

If we write $h, g : \mathbb{R}_+ \to \mathbb{R}$ as $h(x) = \ln(x)$ and $g(x) = r\ln(x)$, we have the composition $h \circ f = g$. We also know that f is continuous everywhere and h and g are both differentiable everywhere with $h'(x) = \frac{1}{x} \neq 0$ and $g'(x) = \frac{r}{x}$. Using Proposition 13.2.5, we conclude that the function f is differentiable everywhere with:

$$f'(x) = \frac{(h \circ f)'(x)}{h'(f(x))} = \frac{g'(x)}{h'(f(x))} = \frac{rf(x)}{x} = rx^{r-1},$$

which finishes the proof. □

Exercises

13.1 (∗) Compute the derivative of the following functions from first principles:

(a) $f(x) = x^n$ on $\mathbb{R}$ for $n \in \mathbb{N}$.
(b) $f(x) = 2x^2 + 3x$ on $\mathbb{R}$.
(c) $f(x) = \frac{2x+3}{x-2}$ on $\mathbb{R} \setminus \{2\}$.
(d) $f(x) = \frac{1}{x} + x^2$ on $\mathbb{R} \setminus \{0\}$.
(e) $f(x) = x^{\frac{p}{q}}$ on $\mathbb{R}_+$ where $p, q \in \mathbb{N}$ are coprime natural numbers.

Hence, determine each of their critical points.

13.2 Suppose that $f : \mathbb{R} \to \mathbb{R}$ is differentiable. Show that for any $x \in \mathbb{R}$ we have $f'(x) = \lim_{h\to 0} \frac{f(x+h)-f(x-h)}{2h}$.

13.3 (∗) Find the derivatives of the tangent and cotangent functions.

13.4 (∗) Using the inverse function theorem, find the derivatives of the functions:
(a) $f : [-1, 1] \to [-\frac{\pi}{2}, \frac{\pi}{2}]$ defined as $f(x) = \arcsin(x)$.
(b) $f : [-1, 1] \to [0, \pi]$ defined as $f(x) = \arccos(x)$.
(c) $f : \mathbb{R} \to (-\frac{\pi}{2}, \frac{\pi}{2})$ defined as $f(x) = \arctan(x)$.

13.5 (∗) Let $f : \mathbb{R} \to \mathbb{R}$ be a differentiable function.
(a) If f is an even function, show that f' is an odd function.
(b) If f is an odd function, show that f' is an even function.

13.6 (∗) Let $f : (1, \infty) \to \mathbb{R}$ be defined as $f(x) = \frac{1}{1-x}$. Show that for all $n \in \mathbb{N}_0$ we have:

$$f^{(n+1)}(x) = (n+1)f^{(n)}(x)f(x).$$

13.7 Let $f : \mathbb{R} \to \mathbb{R}$ be a differentiable function. Suppose that $g, h : \mathbb{R} \to \mathbb{R}$ are defined as $g(x) = f(x-a)$ and $h(x) = f(ax)$ for some $a \in \mathbb{R} \setminus \{0\}$. Show that the functions g and h are also differentiable everywhere.

13.8 (∗) Find the derivative of the function $f : \mathbb{R} \to \mathbb{R}$ defined as $f(x) = |x(x-1)(x+1)|$.

13.9 (∗) Let $f : \mathbb{R} \to \mathbb{R}$ be a a function defined as:

$$f(x) = \begin{cases} x^3 & \text{if } x \leq 0, \\ x^2 & \text{if } x > 0. \end{cases}$$

(a) Show that this function is continuous everywhere.
(b) Show that this function is differentiable at $x = 0$ and hence deduce that $f \in C^1(\mathbb{R})$.
(c) Is $f \in C^2(\mathbb{R})$? Explain your answer.

13.10 (∗) Recall Thomae's function from Exercise 10.8 which is $f : \mathbb{R} \to \mathbb{R}$ defined as:

$$f(x) = \begin{cases} \frac{1}{q} & \text{if } x \in \mathbb{Q} \text{ with } x = \frac{p}{q} \text{ where } p, q \text{ are coprime}, \\ 1 & \text{if } x = 0, \\ 0 & \text{if } x \in \bar{\mathbb{Q}}. \end{cases}$$

(a) Explain why the function f is not differentiable at any $x_0 \in \mathbb{Q} \cap (0, 1)$.
(b) Prove that this function is not differentiable at any $x_0 \in \bar{\mathbb{Q}} \cap (0, 1)$ either.
(c) Hence, conclude that Thomae's function is differentiable nowhere.

13.11 (∗) Recall from Exercise 4.32 the Cantor set $C \subseteq [0, 1]$ which was defined recursively by removing the open middle thirds of the intervals starting from $[0, 1]$. Starting with $C_0 = [0, 1]$, we have also shown that the resulting set after the n-th step of this construction as:

$$C_n = \bigcap_{m=1}^{n} \bigcup_{j=0}^{\frac{3^m-1}{2}} \left[\frac{2j}{3^m}, \frac{2j+1}{3^m}\right],$$

which satisfies the recursive relation $C_{n+1} = \frac{1}{3}C_n \cup \left(\frac{2}{3} + \frac{1}{3}C_n\right)$ for $n \in \mathbb{N}_0$. The Cantor set is defined as $C = \bigcap_{n\in\mathbb{N}_0} C_n$. In the first part of this question, we shall prove that C can be written in a different way.

(a) For any $m \in \mathbb{N}$, define $D_m = \bigcup_{j=0}^{3^{m-1}-1} \left(\frac{3j+1}{3^m}, \frac{3j+2}{3^m}\right)$. This set is obtained by dividing the interval $[0, 1]$ into 3^{m-1} intervals of lengths $\frac{1}{3^{m-1}}$ and taking the union of the open middle third interval of length $\frac{1}{3^m}$ from each of them. Show that:

$$D_m^c = \bigcup_{j=0}^{3^{m-1}-1} \left(\left[\frac{3j}{3^m}, \frac{3j+1}{3^m}\right] \cup \left[\frac{3j+2}{3^m}, \frac{3j+3}{3^m}\right]\right).$$

(b) Define $E_0 = \varnothing$ and $E_n = \bigcup_{m=1}^{n} D_m$ for any $n \in \mathbb{N}$. Using induction, show that E_n^c satisfies the recursive relation $E_{n+1}^c = \frac{1}{3}E_n^c \cup \left(\frac{2}{3} + \frac{1}{3}E_n^c\right)$ for all $n \in \mathbb{N}_0$.

(c) Deduce that $E_n^c = C_n$ for all $n \in \mathbb{N}$ and $C = [0, 1] \setminus \left(\bigcup_{m=1}^{\infty} \bigcup_{j=0}^{3^{m-1}-1} \left(\frac{3j+1}{3^m}, \frac{3j+2}{3^m}\right)\right)$.

Now we are going to study the Cantor staircase. Recall from Exercise 11.11 that the Cantor staircase function $f : [0, 1] \to [0, 1]$ was defined as the pointwise limit of the sequence of functions (f_n) where $f_n : [0, 1] \to [0, 1]$ are defined iteratively as:

$$f_0(x) = x \quad \text{and} \quad f_n(x) = \begin{cases} \frac{f_{n-1}(3x)}{2} & \text{if } x \in \left[0, \frac{1}{3}\right], \\ \frac{1}{2} & \text{if } x \in \left[\frac{1}{3}, \frac{2}{3}\right], \\ \frac{1}{2} + \frac{f_{n-1}(3x-2)}{2} & \text{if } x \in \left[\frac{2}{3}, 1\right], \end{cases} \quad \text{for all } n \in \mathbb{N}.$$

(d) Prove by induction that for each $n \in \mathbb{N}_0$, the function f_n is strictly increasing on C_n and constant on each of the intervals of E_n.

(e) Show that for every $m \in \mathbb{N}$ and $j \in \{0, 1, \ldots, 3^{m-1} - 1\}$, there exists an $N \in \mathbb{N}$ such that f_k on the interval $I = \left(\frac{3j+1}{3^m}, \frac{3j+2}{3^m}\right)$ is the same constant for all $k \geq N$.

(f) Hence, prove that the function f is differentiable at every $x \in C^c$ with $f'(x) = 0$.

13.12 Prove Theorem 13.2.7, namely:
A function $f : (a, b) \to \mathbb{R}$ is differentiable at $x_0 \in (a, b)$ if and only if there exists a number $L \in \mathbb{R}$ and a continuous function $\varepsilon : (a, b) \to \mathbb{R}$ with $\lim_{x\to x_0} \varepsilon(x) = 0$ such that $f(x) = f(x_0) + L \cdot (x - x_0) + \varepsilon(x)(x - x_0)$.

13.13 (∗) Find and classify all critical points of the following functions:
(a) $f(x) = \frac{1}{x^2+1}$ on $\mathbb{R}$.
(b) $f(x) = \frac{x^2}{x+1}$ on $\mathbb{R} \setminus \{-1\}$.
(c) $f(x) = x^4 + \frac{1}{x^4}$ on $\mathbb{R} \setminus \{0\}$.
(d) $f(x) = x \sinh(x)$ on $\mathbb{R}$.
(e) $f(x) = \begin{cases} -2x - 2 & \text{if } x < -1, \\ x^2 - 1 & \text{if } -1 \leq x \leq 1, \\ \ln(x) & \text{if } x > 1. \end{cases}$

13.14 Let $f : (0, \infty) \to \mathbb{R}$ be a function defined as $f(x) = x^x$.
(a) Prove that $f(x) > 0$ for all $x > 0$.
(b) Find the derivative f'.
(c) Deduce all of its critical points and values.

13.15 (∗) Suppose that $f : [a, b] \to \mathbb{R}$ is a differentiable function. Show that if $f'(x) \neq 0$ for any $x \in [a, b]$, then f must be injective.

13.16 Let $f, g : [a, b] \to \mathbb{R}$ be functions continuous on $[a, b]$ and differentiable on (a, b). Suppose that $f(a) = f(b) = 0$. Prove that there is a point $c \in (a, b)$ such that $g'(c)f(c) + f'(c) = 0$.

13.17 (∗) Let $R > 0$ and $f : [-R, R] \to \mathbb{R}$ be a function continuous on $[-R, R]$ and differentiable on $(-R, R)$. Suppose that $f'(x) \leq 1$ for all $x \in (-R, R)$, $f(R) = R$, and $f(-R) = -R$. Show that f is the identity function.

13.18 (∗) Let $f : \mathbb{R} \to \mathbb{R}$ be the function $f(x) = x^5 + x + e^x + \sin(x)$. Find the number of solutions to the equation $f(x) = 0$.

13.19 (∗) Suppose that $f, g, h : (a, b) \to \mathbb{R}$ such that f and g are both twice-differentiable and satisfy the following equations on (a, b):

$$f''(x) + h(x)f(x) = 0,$$

$$g''(x) + h(x)g(x) = 0.$$

(a) Define the function $W : (a, b) \to \mathbb{R}$ as $W = f'g - fg'$. Show that W is a constant function.
(b) If $W \neq 0$ and $f(c) = f(d) = 0$ for some $a < c < d < b$, prove that there exists a $\xi \in (c, d)$ such that $g(\xi) = 0$.

The function W is called the Wronskian and we shall see more of it in Definition 14.4.19.

13.20 (∗) Prove Darboux's theorem:

> **Theorem 13.8.6 (Darboux's Theorem)** *Let $f : \mathbb{R} \to \mathbb{R}$ be a differentiable function and $a, b \in \mathbb{R}$ with $a < b$. If $k \in \mathbb{R}$ satisfies $f'(a) < k < f'(b)$, then there exists a point $c \in (a, b)$ such that $f'(c) = k$.*

13.21 (∗) Let $f : [a, b] \to \mathbb{R}$ be a differentiable function. Recall that $x \in [a, b]$ is called a fixed point of f if $f(x) = x$. Prove that if $f'(x) \neq 1$, then there is at most one fixed point of f in $[a, b]$.

13.22 Using Cauchy's MVT, prove that for any $k > 0$ and $x \geq 1$ we have $\ln(x) \leq \frac{x^k - 1}{k}$.

13.23 Let $f : \mathbb{R} \to \mathbb{R}$ be a real function that satisfies $|f(x) - f(y)| \leq (x - y)^2$ for all $x, y \in \mathbb{R}$. Prove that f is a constant function.

13.24 (∗) Using induction, prove the Leibniz rule in Proposition 13.3.8.

13.25 For each $n \in \mathbb{N}$, determine $\frac{d^n}{dx^n}(x^3 \sin(2x))$.

13.26 Prove Proposition 13.3.11, namely:
If $f, g \in C^n(X)$ for some $X \subseteq \mathbb{R}$ and $n \in \mathbb{N}$, then $fg \in C^n(X)$. Conclude that the product of any two smooth functions is also smooth.

13.27 (∗) Let $f : \mathbb{R} \to \mathbb{R}$ be a smooth function. We have seen in Corollary 13.6.6 that if $f' = 0$, then f must be a constant. We would like to extend this fact to higher derivatives.

(a) Prove that if $f'' \equiv 0$, then $f(x) = a_1 x + a_0$ for some constants $a_0, a_1 \in \mathbb{R}$.

(b) Hence, show that if $f^{(n)} \equiv 0$ for some $n \in \mathbb{N}$, then f is a real polynomial of degree $n - 1$.

13.28 (∗) Let $P : \mathbb{R} \to \mathbb{R}$ be a polynomial of degree n.

(a) Show that if P has n distinct real roots, then P' has $n - 1$ distinct real roots.

(b) Is the converse true? Provide a proof or a counterexample.

13.29 (∗) Let $P : \mathbb{R} \to \mathbb{R}$ be a polynomial of degree n.

(a) Show that if x_0 is a root of P with multiplicity m, then x_0 is also a root of P' with multiplicity $m - 1$.

(b) Hence, prove that $x_0 \in \mathbb{R}$ is a root of P with multiplicity $0 < m \leq n$ if and only if $P(x_0) = P'(x_0) = \ldots = P^{(m-1)}(x_0) = 0$ and $P^{(m)}(x_0) \neq 0$.

(c) For a polynomial of degree n given by $P(x) = \sum_{j=0}^{n} a_j x^j$ with $a_n \neq 0$, show that if it has n real roots (counted with multiplicities), then $(n - 1)a_{n-1}^2 \geq 2n a_n a_{n-2}$.

13.30 (∗) Let $P : \mathbb{R} \to \mathbb{R}$ be a cubic polynomial $P(x) = a_0 + a_1 x + a_2 x^2 + a_3 x^3$.

(a) Show that the polynomial can only have exactly 1 real root or 3 real roots (counted with multiplicity).

(b) Show that the cubic polynomial $f(x) = x^3 - 7x^2 + 25x + 8$ has a root in $(-1, 0)$.
By considering f', show that this is the only root for the polynomial f.

(c) Using Exercise 13.29(c), show that if $a_2^2 < 3a_3a_1$, then the cubic polynomial P has exactly one real root.
Verify this with the polynomial f.

(d) Is it necessarily true that if $a_2^2 \geq 3a_3a_1$, then P would have exactly 3 real roots? Provide a proof or a counterexample.

13.31 Let $P_n : \mathbb{R} \to \mathbb{R}$ be a real polynomial of degree $n \in \mathbb{N}_0$. Define a real function $Q_n : \mathbb{R} \to \mathbb{R}$ as $Q_n(x) = e^x - P_n(x)$.

(a) Show that there are at most $n + 1$ distinct solutions to the equation $Q_n(x) = 0$.

(b) For $n = 2$, give examples of P_2 such that $Q_2(x) = 0$ has no solution, has exactly 1 solution, has exactly 2 solutions, and has exactly 3 solutions respectively.

13.32 (◇) For each $n \in \mathbb{N}$ let $f_n : \mathbb{R} \to \mathbb{R}$ be the polynomial $f_n(x) = \frac{1}{2^n n!}\frac{d^n}{dx^n}(x^2 - 1)^n$.

(a) Show that $f_n(1)$ and $f_n(-1)$ are non-zero.

(b) For a fixed $n \in \mathbb{N}$ with $n \geq 2$, via induction on m, show that for any integer $1 \leq m \leq n - 1$, the polynomial $g_m(x) = \frac{d^m}{dx^m}(x^2 - 1)^n$ has at least m distinct roots within $(-1, 1)$.

(c) Hence, show that the polynomial f_n has n distinct roots which are all within the interval $(-1, 1)$.

Some Applications of Differentiation 14

In the fall of 1972 President Nixon announced that the rate of increase of inflation was decreasing. This was the first time a sitting president used the third derivative to advance his case for reelection.

— Hugo Rossi, mathematician

In Chap. 13, we have defined what differentiation means and have build up some theories as well as useful implications that can be gleaned from derivatives. We have also seen some derivatives of elementary functions, which can then be generalised to more complicated functions using the algebra of derivatives and chain rule.

The development of differential calculus opened up many doors to applications in science and technology since the real world deals more with continuous phenomenon and changes. Even the US President Nixon (implicitly) used it as a tiny nudge in his career in politics, as remarked above! In this chapter, we are going to see what information we can get from derivatives and how to apply these knowledge on some mathematical problems.

Towards the end of this chapter, we are going to introduce equations that involve derivatives of an unknown function. This will be a short introduction on the theory of differential equations which is a very big field of study in terms of theories and applications.

14.1 Graph Sketching

First, let us look at how derivatives help us understand the general behaviour or trend of a function. This information then allows us to roughly sketch the graph of the function. We have seen in Theorem 13.5.3 that points at which we have vanishing or non-existence of derivatives are possible extremum points for the function. Now

S. Johar, *The Big Book of Real Analysis*,
https://doi.org/10.1007/978-3-031-30832-1_14

we are going to look at how we can use derivatives to tell whether a function is increasing or decreasing over a certain interval.

Monotonicity of Functions

Since the derivative of a function a tells us about the slope of the tangent line to its graph, this could help us deduce the local monotonicity behaviour of a function near any point. By recalling Definition 9.1.11 of increasing and decreasing functions, we have:

Proposition 14.1.1 *Let $f : I \to \mathbb{R}$ be a differentiable real-valued function defined on an interval $I = (a, b)$.*

1. *$f'(x) \geq 0$ for all $x \in I$ if and only if f is increasing over I.*
2. *$f'(x) \leq 0$ for all $x \in I$ if and only if f is decreasing over I.*
3. *If $f'(x) > 0$ for all $x \in I$, then f is strictly increasing over I.*
4. *If $f'(x) < 0$ for all $x \in I$, then f is strictly decreasing over I.*

Proof We prove the first assertion only. The other assertions can be proven similarly.

1. We prove the implications separately.

($\Rightarrow$): Pick any $x, y \in I$ with $x < y$ so that $0 < y - x$. By the MVT, there exists some $c \in [x, y] \subseteq I$ such that $f(y) - f(x) = f'(c) \cdot (y - x)$. However, since $c \in I$, we must have $f'(c) \geq 0$ and thus $f(y) - f(x) = f'(c) \cdot (y - x) \geq 0$. This implies $f(y) \geq f(x)$. Since $x, y \in I$ are arbitrary, the function f is increasing over I.

($\Leftarrow$): Pick an arbitrary $x \in I$. Since the function is increasing, for any $0 \leq h < b - x$, we have $f(x + h) - f(x) \geq 0$ and hence $\frac{f(x+h)-f(x)}{h} \geq 0$. Note that since the function f is differentiable at x, its right-derivative at x is equal to $f'(x)$. Thus, taking the limit as h goes to 0 we have:

$$0 \leq \lim_{h \downarrow 0} \frac{f(x + h) - f(x)}{h} = f'(x),$$

which is what we wanted to prove. □

Remark 14.1.2 Note that for the strict monotonicity cases in Proposition 14.1.1(3) and (4), we only have one-way implications. The converses are false since there are functions which are strictly monotone but their derivatives might vanish somewhere. An example of this is the cubic monomial $f(x) = x^3$ for $x \in \mathbb{R}$ with derivative

$f'(x) = 3x^2$. This function is strictly increasing everywhere but its derivative vanishes at $x = 0$.

Recall that Theorem 13.5.3 narrows down the possible candidates of local extremum points in the domain of a function to the set of critical points. To check for genuine extremum points among these critical points, we can use Proposition 14.1.1. This can be done by analysing whether the function is decreasing or increasing just before and just after each of the critical points.

Proposition 14.1.3 (Test for Extremum Points I) *Let $f \in C^0(I)$ be a real-valued continuous function where $I = (a, b)$. Suppose that $x_0 \in I$ is a critical point for f.*

1. *If there exists a $\delta > 0$ such that $f'(x) \leq 0$ for all $x \in (x_0 - \delta, x_0)$ and $f'(x) \geq 0$ for all $x \in (x_0, x_0 + \delta)$, then x_0 is a local minimum of the function f.*
2. *If there exists a $\delta > 0$ such that $f'(x) \geq 0$ for all $x \in (x_0 - \delta, x_0)$ and $f'(x) \leq 0$ for all $x \in (x_0, x_0 + \delta)$, then x_0 is a local maximum of the function f.*

Proof We prove only the first assertion.

1. Suppose that f is differentiable on $(x_0 - \delta, x_0 + \delta) \setminus \{x_0\}$. By assumption and Proposition 14.1.1, f is decreasing in $(x_0 - \delta, x_0)$ and increasing over $(x_0, x_0 + \delta)$. So, for any $x \in (x_0 - \delta, x_0)$, by Exercise 9.29 and continuity of the function, we then have $f(x) \geq \inf_{x \in (x_0 - \delta, x_0)} f(x) = \lim_{x \uparrow x_0} f(x) = f(x_0)$. Likewise, for any $x \in (x_0, x_0 + \delta)$ we have $f(x) \geq \inf_{x \in (x_0, x_0 + \delta)} f(x) = \lim_{x \downarrow x_0} f(x) = f(x_0)$. Therefore, for all $x \in (x_0 - \delta, x_0 + \delta)$ we have $f(x) \geq f(x_0)$ and so x_0 is a local minimum point of f. □

Example 14.1.4 Recall the function $f : \mathbb{R} \to \mathbb{R}$ in Example 13.5.6 defined as $f(x) = x^4 - x^3$. It is differentiable everywhere with derivative $f'(x) = 4x^3 - 3x^2 = x^2(4x - 3)$ from which we can deduce the critical points $x = 0, \frac{3}{4}$. We can test these points to check whether they are extremum points using Proposition 14.1.3.

1. At $x = 0$, we can see that immediately to its left and right, we have $f'(x) < 0$. Thus, it cannot be an extremum point as f is strictly decreasing on either side of this point.
2. At $x = \frac{3}{4}$, to its left (all the way to $-\infty$) we have $f'(x) < 0$ and to its right (all the way to ∞) we have $f(x) > 0$. Thus, by Proposition 14.1.3, we conclude that the point $x = \frac{3}{4}$ is a local minimum. In fact, this point is also a global minimum.

Another way of classifying the critical points of a function f is via the second derivatives.

Proposition 14.1.5 (Test for Extremum Points II) *Let $f : I \to \mathbb{R}$ be a twice-differentiable real-valued function where $I = (a, b)$. Suppose that there exists a point $x_0 \in I$ with $f'(x_0) = 0$.*

1. *If $f''(x_0) > 0$, then x_0 is a local minimum of the function f.*
2. *If $f''(x_0) < 0$, then x_0 is a local maximum of the function f.*

Proof This is done by analysing the behaviour of the first derivative near the point x_0. We prove the first assertion only as the second one is similarly done.

1. By definition of derivatives, we have:

$$0 < f''(x_0) = \lim_{x \to x_0} \frac{f'(x) - f'(x_0)}{x - x_0} = \lim_{x \to x_0} \frac{f'(x)}{x - x_0}.$$

By choosing $\varepsilon = \frac{f''(x_0)}{2} > 0$ in the definition of limits, there exists a $\delta > 0$ such that for all $x \in (a, b)$ with $0 < |x - x_0| < \delta$, we have $\left| \frac{f'(x)}{x - x_0} - f''(x_0) \right| < \frac{f''(x_0)}{2}$, which then implies $\frac{f'(x)}{x - x_0} > \frac{f''(x_0)}{2} > 0$.

We now break into two cases:

(a) First, if $-\delta < x - x_0 < 0$ (namely for points just to the left of x_0), because $\frac{f'(x)}{x - x_0} > 0$, we must have $f'(x) < 0$.

(b) On the other hand, if $0 < x - x_0 < \delta$ (namely for points just to the right of x_0), because $\frac{f'(x)}{x - x_0} > 0$, we must have $f'(x) > 0$.

Using Proposition 14.1.3, we conclude that x_0 is a local minimum of the function f. □

Note that Proposition 14.1.5 is only applicable when $f'(x_0) = 0$ and $f''(x_0) \neq 0$. If the second derivative at x_0 vanishes, then we could not definitively conclude anything. For this case, the point x_0 may either be a local minimum, a local maximum, or neither! Let us look at an example of this.

Example 14.1.6 Consider real-valued functions $f_1, f_2, f_3 : \mathbb{R} \to \mathbb{R}$ defined respectively as:

$$f_1(x) = x^4, \quad f_2(x) = -x^4, \quad \text{and} \quad f_3(x) = x^3.$$

The graphs of these functions are given in Fig. 14.1. If we differentiate these functions, we get:

$$f_1'(x) = 4x^3, \quad f_2'(x) = -4x^3, \quad \text{and} \quad f_3'(x) = 3x^2.$$

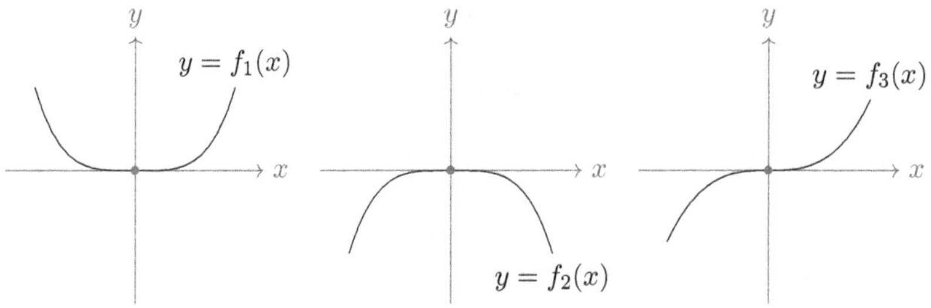

Fig. 14.1 The functions f_1, f_2, and f_3 with their critical point at $x = 0$. It is a local minimum for f_1, a local maximum for f_2, and neither for f_3

In all of the cases above, the point $x = 0$ is a critical point where the derivatives of all these functions vanish. We now try to classify the point $x = 0$ using their second derivatives. We compute:

$$f_1''(x) = 12x^2, \quad f_2''(x) = -12x^2, \quad \text{and} \quad f_3''(x) = 6x,$$

and so all their second derivatives also vanish at $x = 0$. Hence, we cannot apply Proposition 14.1.5 to determine whether the point $x = 0$ is a local minimum or a local maximum for these functions.

However, one can still classify these critical points using Proposition 14.1.3. By checking the behaviour of the function to the left and to the right of the critical point $x = 0$, we can conclude that for these example functions, the point $x = 0$ is a local minimum, local maximum, and neither for the functions f_1, f_2, and f_3 respectively. This can be seen in Fig. 14.1.

Convexity of Functions

The first and second derivatives of a function are also related to convexity of said function. Recall from Exercise 9.2 that a function $f : I \to \mathbb{R}$ over an interval I is called a convex function if for any $x_1, x_2 \in I$ and all $t \in [0, 1]$ we have:

$$f(tx_1 + (1-t)x_2) \leq tf(x_1) + (1-t)f(x_2), \tag{14.1}$$

and is called strictly convex if for any $x_1, x_2 \in I$ and all $t \in (0, 1)$ we have:

$$f(tx_1 + (1-t)x_2) < tf(x_1) + (1-t)f(x_2).$$

On the other hand, the function f is called concave or strictly concave if the inequalities above are flipped to $\geq$ and $>$ respectively. In other words, f is called concave or strictly concave if its negative $-f$ is convex or strictly convex respectively.

The geometric interpretation for a convex function is whenever we draw a secant line segment connecting any two points on its graph, this line segment stays above the graph whereas for a concave function, the secant line segment stays below the graph. Therefore, graphically, a convex function looks like a bowl (refer to graph of f_1 in Fig. 14.1) while a concave function looks like an upside-down bowl (refer to graph of f_2 in Fig. 14.1).

We shall present another convex/concave characterisation for differentiable functions. Before we do so, we prove the following lemma:

Lemma 14.1.7 *Let $f : I \to \mathbb{R}$ be a convex function over an interval $I = (a, b)$. Fix $c \in I$.*

1. *The function $F : (c, b) \to \mathbb{R}$ defined as $F(x) = \frac{f(x)-f(c)}{x-c}$ is an increasing function.*
2. *The function $G : (a, c) \to \mathbb{R}$ defined as $G(x) = \frac{f(c)-f(x)}{c-x}$ is an increasing function.*

Proof Recall from Exercise 10.30(a) that for any $u, v, w \in (a, b)$ such that $a < u < v < w < b$, we have the inequalities:

$$\frac{f(v) - f(u)}{v - u} \leq \frac{f(w) - f(u)}{w - u} \leq \frac{f(w) - f(v)}{w - v}. \tag{14.2}$$

We shall use these inequalities to prove this lemma. We prove the first assertion only as the second assertion is similarly proven.

1. Fix any two points $x_1, x_2 \in (c, b)$ with $x_1 < x_2$. Using the first inequality in (14.2) with $u = c$, $v = x_1$, and $w = x_2$, we have:

$$\frac{f(x_1) - f(c)}{x_1 - c} \leq \frac{f(x_2) - f(c)}{x_2 - c} \quad \Rightarrow \quad F(x_1) \leq F(x_2).$$

Since $x_1 < x_2$ are arbitrary in (c, b), we conclude that F is increasing on (c, b). □

Remark 14.1.8 We can make some modifications to Lemma 14.1.7:

1. If we replace the convex condition in the lemma above with strict convexity, the functions F and G are strictly increasing instead.
2. If we replace the convex condition in the lemma above with concavity, the functions F and G are decreasing instead.
3. If we replace the convex condition in the lemma above with strict concavity, the functions F and G are strictly decreasing instead.

Now we prove the following characterisation of differentiable convex and concave functions.

Proposition 14.1.9 *Let $f : I \to \mathbb{R}$ be a differentiable function where $I = (a, b)$. Then:*

1. *The function f is convex if and only if f' is increasing over I.*
2. *The function f is concave if and only if f' is decreasing over I.*
3. *The function f is strictly convex if and only if f' is strictly increasing over I.*
4. *The function f is strictly concave if and only if f' is strictly decreasing over I.*

Proof We shall prove the first assertion only. The other assertions can be proven in a similar manner.

1. We prove the implications separately.

($\Leftarrow$): Fix any two points $x_1, x_2 \in I$ such that $a < x_1 < x_2 < b$. Pick any $x \in [x_1, x_2]$. By the MVT, there are points $c \in (x_1, x)$ and $d \in (x, x_2)$ such that:

$$f'(c) = \frac{f(x) - f(x_1)}{x - x_1} \quad \text{and} \quad f'(d) = \frac{f(x_2) - f(x)}{x_2 - x}.$$

Since f' is increasing and $c < x < d$, we then have:

$$\frac{f(x) - f(x_1)}{x - x_1} = f'(c) \leq f'(d) = \frac{f(x_2) - f(x)}{x_2 - x}.$$

This can then be rewritten as:

$$f(x) \leq \frac{x_2 - x}{x_2 - x_1} f(x_1) + \frac{x - x_1}{x_2 - x_1} f(x_2),$$

which is the inequality for convexity as per (14.1) with $t = \frac{x_2 - x}{x_2 - x_1} \in [0, 1]$. Since $x_1, x_2, x \in I$ are arbitrary, we conclude that f is convex.

($\Rightarrow$): Suppose that f is differentiable over I and is convex. We want to show that f' is increasing over I. Pick any two $x_1, x_2 \in I$ such that $x_1 < x_2$. By Lemma 14.1.7, the function $F : (x_1, x_2) \to \mathbb{R}$ defined as $F(x) = \frac{f(x) - f(x_1)}{x - x_1}$ is an increasing function.
Since F is increasing, by Exercise 9.29 and continuity of the function f, for any $x \in (x_1, x_2)$, we have the upper bound $F(x) \leq \sup_{x \in (x_1, x_2)} F(x) = \lim_{x \uparrow x_2} F(x) = \frac{f(x_2) - f(x_1)}{x_2 - x_1}$. Moreover, since f is differentiable at x_1, the one-sided limit of F as $x \downarrow x_1$ is exactly $f'(x_1)$ and so we have $f'(x_1) = \lim_{x \downarrow x_1} F(x) \leq \frac{f(x_2) - f(x_1)}{x_2 - x_1}$.

On the other hand, if $G : (x_1, x_2) \to \mathbb{R}$ is a function defined as $G(x) = \frac{f(x_2)-f(x)}{x_2-x}$, this function is also increasing over (x_1, x_2) according to Lemma 14.1.7. By Exercise 9.29 and continuity of the function f, the function G is then bounded from below by $G(x) \geq \inf_{x\in(x_1,x_2)} G(x) = \lim_{x\downarrow x_1} G(x) = \frac{f(x_2)-f(x_1)}{x_2-x_1}$ for any $x \in (x_1, x_2)$. By the same argument as before, we have $f'(x_2) = \lim_{x\uparrow x_2} G(x) \geq \frac{f(x_2)-f(x_1)}{x_2-x_1}$.
Combining these estimates, we get:

$$f'(x_1) \leq \frac{f(x_2) - f(x_1)}{x_2 - x_1} \leq f'(x_2),$$

for arbitrary $x_1 < x_2$. This shows that f' is increasing over I. □

Putting Propositions 14.1.1 and 14.1.9 together, for twice-differentiable functions we have:

Proposition 14.1.10 *Let $f : I \to \mathbb{R}$ be a twice-differentiable function where $I = (a, b)$.*

1. *The function f is convex if and only if $f''(x) \geq 0$ for all $x \in I$.*
2. *The function f is concave if and only if $f''(x) \leq 0$ for all $x \in I$.*

The above results motivate the following definition:

Definition 14.1.11 (Inflexion Point) Let $f : I \to \mathbb{R}$ be a continuous function on $I = (a, b)$. An inflexion point or point of inflexion is a point $x_0 \in I$ such that there exists a $\delta > 0$ for which the function is convex in $(x_0 - \delta, x_0]$ and concave on $[x_0, x_0 + \delta)$ or vice versa. See Fig. 14.2 for an example of an inflexion point.
In other words, it is a point $x_0 \in I$ across which the function f switches its convexity.

Proposition 14.1.9 tells us how we can find inflexion points of twice-differentiable functions.

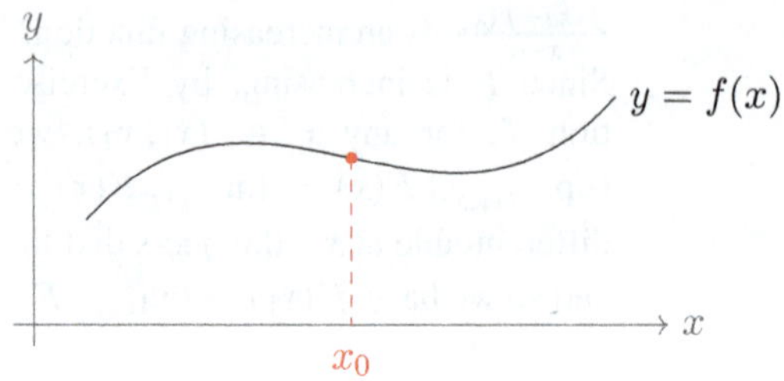

Fig. 14.2 Inflexion point x_0 of the function f. The graph of f is concave to the left of x_0 and convex to the right of x_0

Proposition 14.1.12 *Let $f : I \to \mathbb{R}$ where $I = (a, b)$ be a twice-differentiable function. If f has a point of inflexion at $x_0 \in I$, then $f''(x_0) = 0$.*

Proof WLOG, let f be convex just before x_0 and concave just after x_0. Proposition 14.1.9 says f' is increasing just before x_0 and f' is decreasing just after x_0. Therefore, the function f' has a local maximum at x_0. Since the derivative f' is differentiable, Theorem 13.5.3 says that $f''(x_0) = 0$. □

However, the converse is not true, namely if the second derivative of a function at some point vanishes, it may not be an inflexion point. The point x_0 for which $f''(x_0) = 0$ but the convexity are the same at either side of it is called an undulation point.

Example 14.1.13 Consider again the function $f : \mathbb{R} \to \mathbb{R}$ defined as $f(x) = x^4$ as seen in Fig. 14.1. Since $f''(x) = 12x^2 \geq 0$, this function is convex everywhere. In fact, it is strictly convex everywhere by Proposition 14.1.9 since $f'(x) = 4x^3$ is a strictly increasing function. Its second derivative is $f''(x) = 12x^2$ which vanishes at $x = 0$. However, there is no change in convexity across the point $x = 0$ since the function is strictly convex everywhere. Thus, x_0 is an undulation point of f.

So, for us to find the inflexion points of a twice-differentiable function, using Proposition 14.1.12, we find all the points at which the second derivative vanishes and check them one by one using Definition 14.1.11 to pick out the inflexion points. This can be done by checking the sign of f'' on either side of these points x_0 and see whether it changes (so x_0 is an inflexion point) or remains the same (so x_0 is an undulation point).

Graph Sketching

So far we have seen that the first and second derivatives of a twice-differentiable function can be used to tell us where the extremum points are, where a function is increasing or decreasing, and its convexity. This is plenty of geometric information for the function f. Using all the information that we have seen, we can sketch the graph of any twice-differentiable real-valued functions on $\mathbb{R}$. A rough guide on how to do this is:

- Determine the range and symmetries of the function.
- Find axes intercepts.
- Determine where the function is positive and negative.
- Determine the behaviour of functions at $\pm\infty$, if applicable.
- Compute first derivative and determine critical points.
- Compute second derivative and classify the critical points.
- Determine concavity of functions and points of inflexion.

Note that this is simply a rough list on what we can do to get a sketch of the graph. There is no strict order to this list, but of course, there is a natural order for some of the steps: for example, in order to compute the second derivative we need to compute the first derivative beforehand!

Example 14.1.14 Here are some worked examples:

1. Consider $f : \mathbb{R} \to \mathbb{R}$ defined as $f(x) = e^{-x^2}$.
   - Since $-x^2 \leq 0$ for all $x \in \mathbb{R}$ we have $0 < e^{-x^2} \leq 1$ with equality to 1 at $x = 0$. Thus the range of this function is $(0, 1]$. Notice that $f(-x) = e^{-(-x)^2} = e^{-x^2} = f(x)$ so this function is even and the graph is symmetric about the y-axis.
   - The y-intercept is at $f(0) = 1$. From the previous step, since the function is strictly positive everywhere, there are no x-intercepts.
   - We have $\lim_{x\to\pm\infty} e^{-x^2} = 0$ so the function is asymptotically constant 0.
   - $f'(x) = -2xe^{-x^2}$ for all $x \in \mathbb{R}$. Thus $f'(x) = 0$ only at $x = 0$. Therefore, the only critical point is at $x = 0$. By looking at the sign of f and using Proposition 14.1.1, we can see that f is strictly increasing for $x < 0$, strictly decreasing for $x > 0$, and $\lim_{x\to\pm\infty} f'(x) = 0$. See Fig. 14.3 for details. This tells us that the point $x = 0$ is a local maximum with value $f(0) = 1$. In fact, it is a global maximum since the function cannot get any bigger than 1.
   - $f''(x) = (4x^2 - 2)e^{-x^2}$ for all $x \in \mathbb{R}$. We note that $f''(0) = -2 < 0$. So, by Proposition 14.1.5, we conclude that $x = 0$ is local maximum. Note that this has been established in the previous step. There are no other local extremum points.
   - By solving $f''(x) = 0$, the inflexion points might be among $x = \pm\frac{1}{\sqrt{2}}$. We can check that for $x < -\frac{1}{\sqrt{2}}$ and $x > \frac{1}{\sqrt{2}}$ we have $f''(x) > 0$ while for $-\frac{1}{\sqrt{2}} < x < \frac{1}{\sqrt{2}}$ we have $f''(x) < 0$. So, the function f is convex for $x < -\frac{1}{\sqrt{2}}$ and $x > \frac{1}{\sqrt{2}}$ and concave for $-\frac{1}{\sqrt{2}} < x < \frac{1}{\sqrt{2}}$ according to Proposition 14.1.10. Since the convexity changes across the points $\pm\frac{1}{\sqrt{2}}$, we conclude that both of these points are inflexion points. See Fig. 14.4 for details.

   With all these information, we can then produce the sketch of the graph for f as in Fig. 14.5.

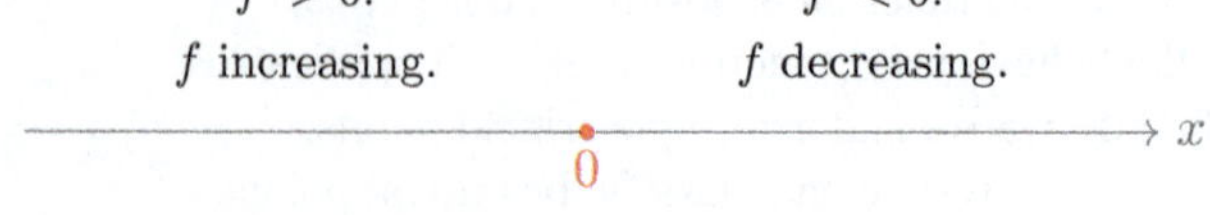

Fig. 14.3 Analysis of first derivative. There is a critical point at $x = 0$

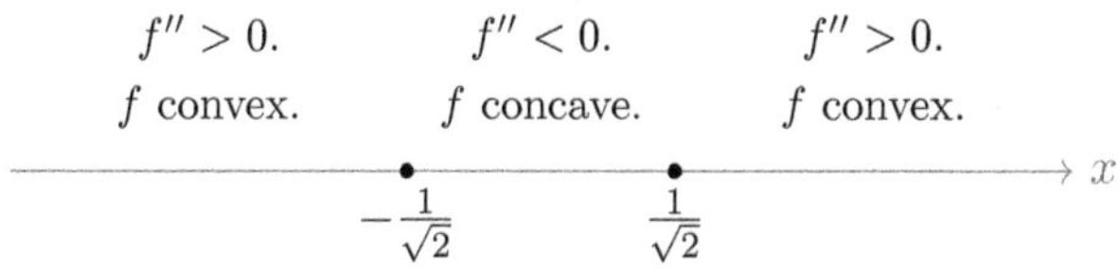

Fig. 14.4 Analysis of second derivative. There are inflexion points at $x = \pm\frac{1}{\sqrt{2}}$

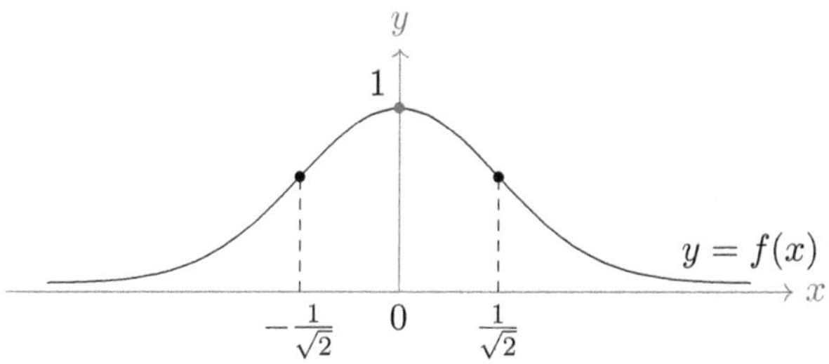

Fig. 14.5 Sketch for $f(x) = e^{-x^2}$ with its inflexion points in black and critical point in red

2. Let us analyse the function $f(x) = 3x^4 - 4x^3 + 3$ for $x \in \mathbb{R}$ and sketch its graph.

- The range is not clear to determine at this stage, so let us skip this part first.
- f is a real polynomial with even degree and positive leading coefficient. By Exercise 9.14, this function has no global maximum.
- The y-intercept is $y = 3$. The x-intercept can be obtained by solving $f(x) = 3x^4 - 4x^3 + 3 = 0$. This is a quartic equation which is not very easy to solve, but we can rewrite it as $x^3\left(x - \frac{4}{3}\right) = -1$. So, for a solution x (if any) to this equation, the quantities x and $x - \frac{4}{3}$ must have opposite signs. Since $x - \frac{4}{3} < x$, necessarily $x > 0$ and $x - \frac{4}{3} < 0$. Thus, the x-intercept can only occur within the interval $(0, \frac{4}{3})$. This is all we can deduce for now, but we will come back to this later.
- We compute $f'(x) = 12x^3 - 12x^2 = 12x^2(x - 1)$ which exists everywhere. So, the critical points of f are only $x = 0, 1$ which can obtained when we solve the equation $f'(x) = 0$.
  (a) By investigating the sign of the derivative function f', we can see that the function is strictly decreasing for $0 < x < 1$ and strictly increasing for $x > 1$. Therefore, according to Proposition 14.1.3, the point $x = 1$ is a local minimum.
  (b) By similar reasoning as above, we can see that f' is strictly decreasing immediately on either sides of the point $x = 0$. Thus, this is not an extremum point.

  See Fig. 14.6 for the information obtained from the analysis of the first derivative.

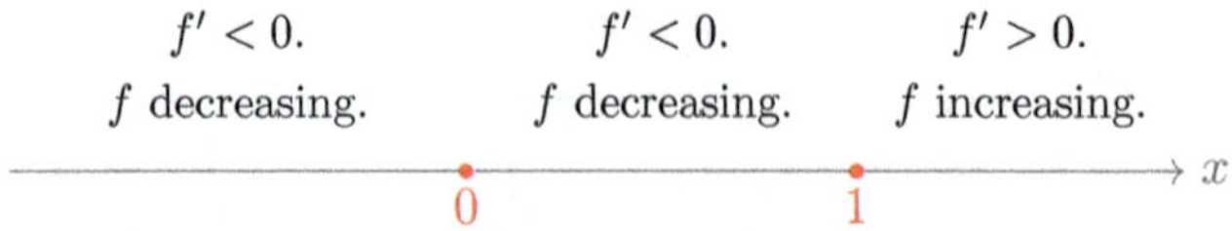

Fig. 14.6 Analysis of first derivative. There are critical points at $x = 0, 1$

With this information, we now know that there is no x-intercept between $x = 1$ and $x = \frac{4}{3}$ because the function is increasing from the value $f(1) = 2$ which means $f(x) \geq 2$ for any $x \in [1, \frac{4}{3}]$. Furthermore, there are no x-intercept in $[0, 1]$ because we know that f is only strictly decreasing here and so $f(x) \geq f(1) = 2$ for any $x \in [0, 1]$. Hence, there are no x-intercepts at all. This answers earlier question regarding x-intercepts.
Furthermore, we can conclude that $(1, 2)$ is a global minimum and the range of the function is then $f(x) \geq 2$.

- We compute $f''(x) = 36x^2 - 24x = 12x(3x - 2)$. Thus, $f''(x) = 0$ precisely when $x = 0$ and $x = \frac{2}{3}$. Furthermore, $f''(x) > 0$ when $x < 0$ and $x > \frac{2}{3}$ and hence the graph of f is convex in these regions. On the other hand $f''(x) < 0$ when $0 < x < \frac{2}{3}$ and thus the graph is concave here. Thus, the points $x = 0$ and $x = \frac{2}{3}$ are points of inflexion. See Fig. 14.7 for details.

With all these information, we can then produce the sketch of the graph for f as in Fig. 14.8.

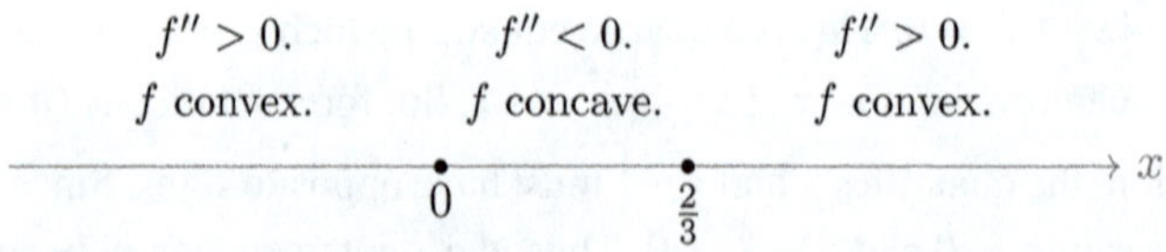

Fig. 14.7 Analysis of second derivative. There are inflexion points at $x = 0, \frac{2}{3}$

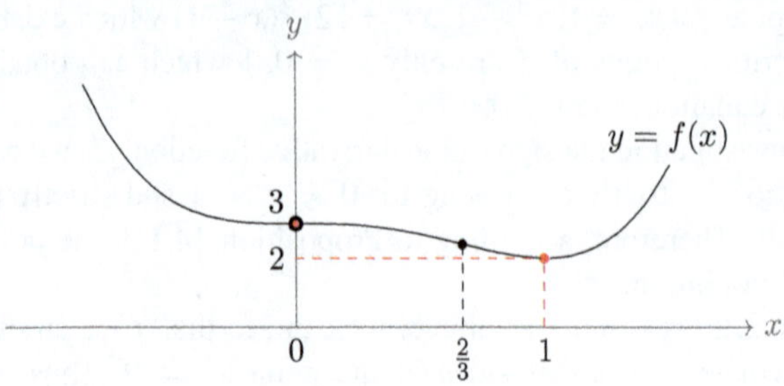

Fig. 14.8 Sketch for $f(x) = 3x^4 - 4x^3 + 3$ with its inflexion points in black and critical points in red. Notice that $x = 0$ is both an inflexion point and a critical point

14.2 Differentiation and Limits

In Chap. 11, we have seen sequences of functions and two kinds of convergence modes for these sequences, namely pointwise and uniform. Pointwise convergence is rather weak as it only takes into account the behaviour of the functions at each point separately. Thus, we saw that continuity could be diminished under pointwise convergence. On the other hand, uniform convergence is stronger and it preserves continuity under the limit.

Since differentiability, like continuity, is a local property, we expect that this property can also be passed from the sequence of functions to the limit if we have some kind of uniform convergence property (pointwise convergence alone is not enough). This is true since differentiation is essentially a limiting process, so we require some kind of uniform convergence to justify swapping the order of the limits for differentiation and limits, akin to Theorem 11.3.3.

However, uniform convergence of the function sequence by itself is not enough to guarantee the convergence of the derivatives.

Example 14.2.1 Let us look at some examples where things may go wrong:

1. Let (f_n) be a sequence of functions $f_n : \mathbb{R} \to \mathbb{R}$ defined as $f_n(x) = \sqrt{\frac{1}{n} + x^2}$. All of f_n are continuous and differentiable everywhere. Furthermore, the sequence converges uniformly to $f(x) = |x|$ over $\mathbb{R}$. Indeed, for any $x \in \mathbb{R}$, we have:

$$|f_n(x) - |x|| = \sqrt{\frac{1}{n} + x^2} - \sqrt{x^2} = \frac{\frac{1}{n}}{\sqrt{\frac{1}{n} + x^2} + \sqrt{x^2}} \leq \frac{1}{\sqrt{n}} \to 0.$$

   However, as we have seen earlier, the limiting function $f(x) = |x|$ is not differentiable at 0 even though the functions f_n are all differentiable here. Thus, we note here that even though uniform convergence preserves continuity in the limit, differentiability may be lost.
2. Consider the sequence of functions (f_n) where $f_n : \mathbb{R} \to \mathbb{R}$ is defined as $f_n(x) = \frac{\sin(nx)}{\sqrt{n}}$. This sequence converges uniformly over $\mathbb{R}$ to the 0 function which has vanishing derivative everywhere. However, if we consider the sequence (f_n') where $f_n'(x) = \sqrt{n}\cos(nx)$, we can see that this sequence cannot converge uniformly or even pointwise to any function defined on the whole of $\mathbb{R}$ since $f_n'(2\pi k) = \sqrt{n} \to \infty \neq 0$ for any $k \in \mathbb{Z}$. In other words, in this case we have:

$$0 = \frac{d}{dx} \lim_{n\to\infty} f_n(x) \neq \lim_{n\to\infty} f_n'(x).$$

Differentiation of Function Sequence

Therefore, we need a stronger condition to ensure that the limit and derivative commute. The key is we need the sequence of derivatives itself to converge uniformly. This is an an extension of the result in Theorem 11.3.3.

Theorem 14.2.2 (Differentiable Limit Theorem) *Let $X \subseteq \mathbb{R}$ be an open subset of the real numbers and (f_n) be a sequence of differentiable functions $f_n : X \to \mathbb{R}$. If:*

1. *(f_n) converges pointwise to a function $f : X \to \mathbb{R}$, and*
2. *the sequence of derivatives (f_n') converges uniformly on X to some function $g : X \to \mathbb{R}$,*

then the limit function f is differentiable with $f' = g$.
In other words, if $f_n \xrightarrow{pw} f$ and $f_n' \xrightarrow{u} g$ on X, then:

$$\frac{d}{dx} f(x) = \frac{d}{dx} \lim_{n\to\infty} f_n(x) = \lim_{n\to\infty} f_n'(x) = g(x).$$

Proof Fix $x_0 \in X$ and $\varepsilon > 0$. We want to show that $f'(x_0) = g(x_0)$ or in other words, there exists a $\delta > 0$ such that if $x \in X$ with $0 < |x - x_0| < \delta$, then $\left|\frac{f(x)-f(x_0)}{x-x_0} - g(x_0)\right| < \varepsilon$. Let us rewrite this last term as a sum of terms which we can estimate separately via triangle inequalities:

$$\begin{aligned}
&\left|\frac{f(x) - f(x_0)}{x - x_0} - g(x_0)\right| \\
&= \left|\frac{f(x) - f(x_0)}{x - x_0} - \frac{f_N(x) - f_N(x_0)}{x - x_0} + \frac{f_N(x) - f_N(x_0)}{x - x_0} - f_N'(x_0) + f_N'(x_0) - g(x_0)\right| \\
&\leq \left|\frac{f(x) - f(x_0)}{x - x_0} - \frac{f_N(x) - f_N(x_0)}{x - x_0}\right| + \left|\frac{f_N(x) - f_N(x_0)}{x - x_0} - f_N'(x_0)\right| + |f_N'(x_0) - g(x_0)|,
\end{aligned} \tag{14.3}$$

where $N \in \mathbb{N}$ is some index which we need to choose appropriately later. Let us deal with the terms in (14.3) separately.

1. We first consider the function $f_m - f_n$ for some indices $m, n \in \mathbb{N}$. This function is differentiable in X and hence in the closed interval with x_0 and x as endpoints. By the MVT, there exists a c strictly in between x_0 and x such that:

$$\begin{aligned} f_m'(c) - f_n'(c) &= \frac{(f_m(x) - f_n(x)) - (f_m(x_0) - f_n(x_0))}{x - x_0} \\ &= \frac{f_m(x) - f_m(x_0)}{x - x_0} - \frac{f_n(x) - f_n(x_0)}{x - x_0}, \end{aligned}$$

and thus we have:

$$\left| \frac{f_m(x) - f_m(x_0)}{x - x_0} - \frac{f_n(x) - f_n(x_0)}{x - x_0} \right| = |f_m'(c) - f_n'(c)| \leq \sup_{x \in X} |f_m'(x) - f_n'(x)|.$$

However, since $f_n' \xrightarrow{u} g$ on X, by the Cauchy criterion of uniform convergence, there exists an $N_1 \in \mathbb{N}$ such that for all $m, n \geq N_1$ we have $\sup_{x \in X} |f_m'(x) - f_n'(x)| < \frac{\varepsilon}{3}$. Hence:

$$\left| \frac{f_m(x) - f_m(x_0)}{x - x_0} - \frac{f_n(x) - f_n(x_0)}{x - x_0} \right| < \frac{\varepsilon}{3},$$

for all $m, n \geq N_1$. Furthermore, by taking the limit as $m \to \infty$ and using the fact that limits preserve weak inequalities and the algebra of limits, since $f_m \xrightarrow{pw} f$, we have:

$$\begin{aligned} &\lim_{m \to \infty} \left| \frac{f_m(x) - f_m(x_0)}{x - x_0} - \frac{f_n(x) - f_n(x_0)}{x - x_0} \right| \\ &= \left| \frac{f(x) - f(x_0)}{x - x_0} - \frac{f_n(x) - f_n(x_0)}{x - x_0} \right| \leq \frac{\varepsilon}{3}, \end{aligned} \tag{14.4}$$

for all $n \geq N_1$. This gives us an estimate for the first term in the RHS of inequality (14.3).

2. Next, we deal with the third term in (14.3). Since $f_n' \xrightarrow{u} g$ on X and hence $f_n' \xrightarrow{pw} g$, there exists an $N_2 \in \mathbb{N}$ such that for any $n \geq N_2$, we have $|f_n'(x_0) - g(x_0)| < \frac{\varepsilon}{3}$. Therefore, if we set $N = \max\{N_1, N_2\}$ in the inequality (14.3) and use the estimate (14.4), we then have:

$$\left| \frac{f(x) - f(x_0)}{x - x_0} - g(x_0) \right| < \frac{\varepsilon}{3} + \left| \frac{f_N(x) - f_N(x_0)}{x - x_0} - f_N'(x_0) \right| + \frac{\varepsilon}{3}. \tag{14.5}$$

3. Finally, we deal with the middle term in (14.5). Since the function f_N is differentiable at x_0, there exists a $\delta > 0$ such that for any $x \in X$ with $0 < |x - x_0| < \delta$ we would have $\left| \frac{f_N(x) - f_N(x_0)}{x - x_0} - f_N'(x_0) \right| < \frac{\varepsilon}{3}$. So, with this

$\delta > 0$, putting this estimate in (14.5) we obtain:

$$\left|\frac{f(x) - f(x_0)}{x - x_0} - g(x_0)\right| < \frac{\varepsilon}{3} + \frac{\varepsilon}{3} + \frac{\varepsilon}{3} = \varepsilon,$$

if $x \in X$ with $0 < |x - x_0| < \delta$.

Thus, we have $f'(x_0) = g(x_0)$. Since x_0 is arbitrarily chosen in X, we have shown that f is also differentiable everywhere with $f' = g$. □

We note that the pointwise convergence condition $f_n \xrightarrow{pw} f$ in Theorem 14.2.2 could be weakened further. Instead of requiring the sequence of functions (f_n) to converge pointwise on X, we only need to check that the sequence converges at a single point $x_0 \in X$ in every connected interval of the domain X to deduce Theorem 14.2.2. This is due to the following result:

Proposition 14.2.3 *Let $I \subseteq \mathbb{R}$ be an open interval of the real numbers and (f_n) is a sequence of differentiable functions $f_n : I \to \mathbb{R}$. If:*

1. *there exists an $x_0 \in X$ such that $\lim_{n\to\infty} f_n(x_0)$ exists, and*
2. *the sequence (f_n') converges uniformly on I to some function $g : I \to \mathbb{R}$,*

then the functions sequence (f_n) converges pointwise to some function $f : I \to \mathbb{R}$.

Proof Assume first that I is bounded, namely $I = (a, b)$. Fix $\varepsilon > 0$. Since $(f_n(x_0))$ converges, it must be a Cauchy sequence and hence there exists an $N_1 \in \mathbb{N}$ such that for all $m, n \geq N_1$ we have $|f_n(x_0) - f_m(x_0)| < \frac{\varepsilon}{2}$. Furthermore, since $f_n' \xrightarrow{u} g$ on I, by Cauchy criterion in Proposition 11.2.7, there exists an $N_2 \in \mathbb{N}$ such that for any $m, n \geq N_2$, we have $\sup_{x\in I} |f_n'(x) - f_m'(x)| < \frac{\varepsilon}{2(b-a)}$.

Set $N = \max\{N_1, N_2\}$. Let $m, n \in \mathbb{N}$ be such that $m, n \geq N$ and define a function $F_{m,n} : I \to \mathbb{R}$ as the difference $F_{m,n} = f_n - f_m$. The function $F_{m,n}$ is also differentiable everywhere on I.

Pick any $x \in I \setminus \{x_0\}$ and we want to show that the sequence of function (f_n) converges here as well. Consider the closed interval with endpoints x_0 and x in I. By the MVT, there exists a ξ between x and x_0 such that:

$$F_{m,n}'(\xi) = \frac{F_{m,n}(x) - F_{m,n}(x_0)}{x - x_0}.$$

This implies $f_n(x) - f_m(x) = (f_n(x_0) - f_m(x_0)) + (f_n'(\xi) - f_m'(\xi))(x_0 - x)$. By taking absolute value and applying the triangle inequality, we get:

$$\begin{aligned}|f_n(x) - f_m(x)| &\leq |f_n(x_0) - f_m(x_0)| + |f_n'(\xi) - f_m'(\xi)||x_0 - x| \\ &\leq |f_n(x_0) - f_m(x_0)| + \sup_{x\in I} |f_n'(x) - f_m'(x)|(b - a) \\ &< \frac{\varepsilon}{2} + \frac{\varepsilon}{2} = \varepsilon. \end{aligned} \tag{14.6}$$

This means the sequence $(f_n(x))$ is Cauchy and hence must converge. Since $x \in I$ was arbitrary, we can conclude that (f_n) converges pointwise to some function $f : I \to \mathbb{R}$.

Now if I is an unbounded interval, say $I = (a, \infty)$, we proceed by writing the interval I as unions of overlapping bounded intervals. Define $I_m = (m-1, m+1) \cap I$ for all $m \in \mathbb{Z}$ so that $I = \bigcup_{m \in \mathbb{Z}} I_m$. Then, $x_0 \in I_{m_0}$ for some $m_0 \in \mathbb{Z}$. By applying the result for bounded intervals, we can ensure that (f_n) converges pointwise on I_{m_0}. Since the interval I_{m_0} overlaps with the intervals I_{m_0-1} and I_{m_0+1}, we can proceed to show that (f_n) converges pointwise on I_{m_0-1} and I_{m_0+1} as well. Thus, by induction, we deduce that (f_n) converges pointwise on every I_m. This implies that $f_n \xrightarrow{pw} f$ where $f : I \to \mathbb{R}$ is some limiting function. □

Remark 14.2.4 In fact, for a bounded interval I, it is easy to see using the inequality (14.6) and Proposition 11.2.7 that the convergence $f_n \xrightarrow{pw} f$ we obtained in Proposition 14.2.3 is uniform on I.

Differentiation of Functions Series

Since a functions series is the limit of its partial sums, the results above can be applied to functions series. The analogue of Theorem 14.2.2 for functions series is:

Theorem 14.2.5 (Differentiable Limit Theorem for Functions Series) *Let (f_n) be a sequence of differentiable real-valued functions $f_n : X \to \mathbb{R}$. Suppose that $\sum_{j=1}^{\infty} f_j$ is its functions series with partial sums $s_n : X \to \mathbb{R}$. If:*

1. *the series converges pointwise to a function $s : X \to \mathbb{R}$, and*
2. *the sequence of the derivatives (s_n') converges uniformly on X to some function $t : X \to \mathbb{R}$,*

then the limit function s is also differentiable with $s' = t$. In particular:

$$\frac{d}{dx} s(x) = \frac{d}{dx} \lim_{n\to\infty} s_n(x) = \lim_{n\to\infty} s_n'(x) = t(x).$$

Written in functions series form, this says:

$$\frac{d}{dx} \sum_{j=1}^{\infty} f_j = \frac{d}{dx} \lim_{n\to\infty} \sum_{j=1}^{n} f_j = \lim_{n\to\infty} \frac{d}{dx} \sum_{j=1}^{n} f_j = \lim_{n\to\infty} \sum_{j=1}^{n} \frac{d}{dx} f_j = \sum_{j=1}^{\infty} \frac{d}{dx} f_j.$$

The analogue of Proposition 14.2.3 for functions series is:

Proposition 14.2.6 *Let $I \subseteq \mathbb{R}$ be an open interval of the real numbers and (f_n) is a sequence of differentiable functions $f_n : I \to \mathbb{R}$. Suppose that $\sum_{j=1}^{\infty} f_j$ is its functions series with partial sums $s_n : I \to \mathbb{R}$. If:*

1. *there exists an $x_0 \in X$ such that $\sum_{j=1}^{\infty} f_n(x_0)$ exists, and*
2. *the sequence (s_n') converges uniformly on I to some function $t : I \to \mathbb{R}$,*

then (s_n) converges pointwise on I to some function $s : I \to \mathbb{R}$.

Moreover, for a power series, we have the following result which allows us to differentiate a power series term-by-term within its radius of convergence.

Proposition 14.2.7 (Term-Wise Differentiation of Power Series) *Let $\sum_{j=0}^{\infty} a_j (x-c)^j$ be a power series for some constants $c, a_j \in \mathbb{R}$ with radius of convergence $R > 0$. We have:*

$$\frac{d}{dx}\sum_{j=0}^{\infty} a_j(x-c)^j = \sum_{j=0}^{\infty} \frac{d}{dx} a_j(x-c)^j = \sum_{j=1}^{\infty} j a_j (x-c)^{j-1}.$$

for any $x \in (c-R, c+R)$ if R is finite and any $x \in \mathbb{R}$ if R is infinite.

Proof WLOG, assume $c = 0$ and R is finite. Suppose that $s_n(x) = \sum_{j=0}^{n} a_j x^j$ is a sequence of partial sums of the power series. Consider the functions sequence (t_n) where $t_n(x) = \frac{d}{dx} s_n(x) = \sum_{j=1}^{n} j a_j x^{j-1}$. We have seen in Proposition 12.3.6 that the limit of this series as $n \to \infty$ given by $t(x) = \sum_{j=1}^{\infty} j a_j x^{j-1}$ also has the same radius of convergence R. We note that this power series converges uniformly over some closed interval $|x| \leq R - \delta$ for any small $\delta > 0$ by Proposition 12.2.1. Thus, by Theorem 14.2.5, we can conclude that:

$$\frac{d}{dx}\sum_{j=0}^{\infty} a_j x^j = \frac{d}{dx}\lim_{n\to\infty} s_n(x) = \lim_{n\to\infty}\frac{d}{dx} s_n(x) = \lim_{n\to\infty} t_n(x) = t(x) = \sum_{j=1}^{\infty} j a_j x^{j-1},$$

for $x \in [-R+\delta, R-\delta]$ for any small $\delta > 0$. Since $\delta > 0$ can be chosen arbitrarily small, we have the equality $\frac{d}{dx}\sum_{j=0}^{\infty} a_j x^j = \sum_{j=1}^{\infty} j a_j x^{j-1}$ for all $x \in (-R, R)$. The case for infinite R can also be proven in the same manner. □

Example 14.2.8 Now let us look at some applications of Proposition 14.2.7.

1. In Example 13.3.3(2), we saw that the derivative of the exponential function e^x is itself by using first principles. Let us show an alternative way of deriving this using its power series definition.
   Recall that the exponential function on $\mathbb{R}$ can be expressed as the power series $e^x = \sum_{j=0}^{\infty} \frac{x^j}{j!}$. This power series has infinite radius of convergence. Thus, by

Proposition 14.2.7, we can apply term-wise differentiation to the power series at any $x \in \mathbb{R}$ to get:

$$\frac{d}{dx}e^x = \frac{d}{dx}\sum_{j=0}^{\infty}\frac{x^j}{j!} = \sum_{j=0}^{\infty}\frac{d}{dx}\frac{x^j}{j!} = \sum_{j=1}^{\infty}\frac{jx^{j-1}}{j!} = \sum_{j=1}^{\infty}\frac{x^{j-1}}{(j-1)!} = \sum_{j=0}^{\infty}\frac{x^j}{j!} = e^x.$$

2. In Exercise 12.18, the readers were asked to find the power series of the function $f : \mathbb{R} \setminus \{1\} \to \mathbb{R}$ defined as $f(x) = \frac{1}{(1-x)^2}$. One can use Mertens' theorem to do this.
   Another way (a very slick one too) to do this is via differentiation. Notice that $\sum_{j=0}^{\infty} x^j = \frac{1}{1-x}$ for $|x| < 1$ by Proposition 12.3.6. Using Proposition 14.2.7, within this region we can switch the infinite series with derivative as follows:

$$\frac{1}{(1-x)^2} = \frac{d}{dx}\frac{1}{1-x} = \frac{d}{dx}\sum_{j=0}^{\infty}x^j = \sum_{j=0}^{\infty}\frac{d}{dx}x^j = \sum_{j=1}^{\infty}jx^{j-1}.$$

   Thus, the power series of $\frac{1}{(1-x)^2}$ over $|x| < 1$ is then given by $\sum_{j=1}^{\infty} jx^{j-1}$.

14.3 L'Hôpital's Rule

Another application of differentiation is for evaluating limits. Recall Theorem 9.5.2 (7), namely: if $f, g : X \to \mathbb{R}$ where $X \subseteq \mathbb{R}$ and $x_0 \in X'$ are such that $\lim_{x\to x_0} f(x) = L$ and $\lim_{x\to x_0} g(x) = M \neq 0$, then we we can evaluate the limit of the following quotient as the quotient of limits:

$$\lim_{x\to x_0}\frac{f}{g}(x) = \frac{\lim_{x\to x_0} f(x)}{\lim_{x\to x_0} g(x)} = \frac{L}{M}.$$

What what about the case where $M = 0$? If the functions f and g are continuous and $L \neq 0$, by Corollary 9.5.7, we may show that the limit of the quotient diverges to ∞ or $-\infty$, depending on the sign of L and how the function g behave around the point x_0. However, if $L = 0$ as well, what can we deduce about the limit?

This limit of the form $\frac{0}{0}$ is called an indeterminate form. We have seen an example of this in Example 9.5.3(2) where the limit of the numerator and denominator are both 0, but the limit of the quotient is finite which is $\frac{2}{3}$. In fact, there are many other indeterminate forms: the limits would have one of the following forms $\frac{0}{0}$, $\frac{\pm\infty}{\pm\infty}$, $0 \times \infty$, $\infty - \infty$, 1^{∞}, ∞^0, or 0^0.

As we have seen in Example 9.5.3(2), one may be able to evaluate the limit by first principle or by some manual analytic argument. Here we are going to show a method that uses differentiation for the indeterminate forms of the type $\frac{0}{0}$, $\frac{\pm\infty}{\pm\infty}$, and $0 \times \infty$. We first state several lemmas:

Lemma 14.3.1 (L'Hôpital's Theorem I) *Let $f, g : X \to \mathbb{R}$ where $X \subseteq \mathbb{R}$ and $x_0 \in X'$. Suppose that there exists a $\delta > 0$ such that:*

1. *the functions f and g are differentiable in $(x_0, x_0 + \delta)$,*
2. *$g'(x) \neq 0$ for any $x \in (x_0, x_0 + \delta)$,*
3. *$\lim_{x \downarrow x_0} f(x) = \lim_{x \downarrow x_0} g(x) = 0$, and*
4. *$\lim_{x \downarrow x_0} \frac{f'(x)}{g'(x)}$ exists and is equal to $L \in \mathbb{R}$.*

Then, we have:

$$\lim_{x \downarrow x_0} \frac{f(x)}{g(x)} = \lim_{x \downarrow x_0} \frac{f'(x)}{g'(x)} = L.$$

Proof WLOG, we can assume that $x_0 = 0$. We aim to prove that $\lim_{x \downarrow 0} \frac{f(x)}{g(x)} = L$. Fix $\varepsilon > 0$. We want to find a $\tilde{\delta} > 0$ such that whenever $0 < x < \tilde{\delta}$ we have $\left| \frac{f(x)}{g(x)} - L \right| < \varepsilon$.

First we note that, by the final assumption, there exists a $\delta_1 > 0$ such that for any $0 < c < \delta_1$ we have:

$$\left| \frac{f'(c)}{g'(c)} - L \right| < \varepsilon. \tag{14.7}$$

We want to use MVT II on the functions f and g. But since f and g may not even be defined at the point 0 or $f(0) \neq 0$ and $g(0) \neq 0$ (namely f and g may not be continuous at 0), we have to define new functions that include the point 0 in its domain which would then allow us to use MVT II. Define $\bar{f}, \bar{g} : X \cup \{0\} \to \mathbb{R}$ as:

$$\bar{f}(x) = \begin{cases} f(x) & \text{if } x \in X \setminus \{0\}, \\ 0 & \text{if } x = 0, \end{cases} \quad \text{and} \quad \bar{g}(x) = \begin{cases} g(x) & \text{if } x \in X \setminus \{0\}, \\ 0 & \text{if } x = 0, \end{cases}$$

so that $\bar{f}$ and $\bar{g}$ are both right-continuous at 0 and hence continuous in any interval of the form $[0, x]$ as well as differentiable in any interval $(0, x)$ for $0 < x < \delta$. Furthermore, for any $x \in (0, \delta)$, we must have $\bar{g}(x) = g(x) \neq 0$ or otherwise, by Rolle's theorem, there would be a point in $(0, \delta)$ for which g' vanishes, contradicting the second assumption.

Set $\tilde{\delta} = \min\{\delta_1, \delta\} > 0$. Since $\bar{f}$ and $\bar{g}$ satisfy the required conditions in MVT II, for any $x \in (0, \tilde{\delta})$ we can apply MVT II to the functions $\bar{f}$ and $\bar{g}$ on the interval $[0, x] \subseteq [0, \tilde{\delta})$. Thus, there exists a $0 < c < x < \tilde{\delta}$ such that:

$$\frac{\bar{f}'(c)}{\bar{g}'(c)} = \frac{\bar{f}(x) - \bar{f}(0)}{\bar{g}(x) - \bar{g}(0)} \quad \Rightarrow \quad \frac{f'(c)}{g'(c)} = \frac{f(x)}{g(x)}, \tag{14.8}$$

since $\bar{f}(x) = f(x)$ and $\bar{g}(x) = g(x)$ for $x \neq 0$ and $\bar{f}(0) = \bar{g}(0) = 0$. Thus, for any $0 < x < \tilde{\delta}$ we have:

$$\left|\frac{f(x)}{g(x)} - L\right| = \left|\frac{f'(c)}{g'(c)} - L\right| < \varepsilon,$$

by first using the equality (14.8) and then the estimate (14.7) since $0 < c < \tilde{\delta} \leq \delta_1$. This proves that $\lim_{x\downarrow 0} \frac{f(x)}{g(x)} = L$. □

The above theorem also holds for left-limits using a similar proof. We state:

Lemma 14.3.2 (L'Hôpital's Theorem II) *Let $f, g : X \to \mathbb{R}$ where $X \subseteq \mathbb{R}$ and $x_0 \in X'$. Suppose that there exists a $\delta > 0$ such that:*

1. *the functions f and g are differentiable in $(x_0 - \delta, x_0)$,*
2. *$g'(x) \neq 0$ for any $x \in (x_0 - \delta, x_0)$,*
3. *$\lim_{x\uparrow x_0} f(x) = \lim_{x\uparrow x_0} g(x) = 0$, and*
4. *$\lim_{x\uparrow x_0} \frac{f'(x)}{g'(x)}$ exists and is equal to $L \in \mathbb{R}$.*

Then, we have:

$$\lim_{x\uparrow x_0} \frac{f(x)}{g(x)} = \lim_{x\uparrow x_0} \frac{f'(x)}{g'(x)} = L.$$

In fact, the third condition in Lemma 14.3.1, namely $\lim_{x\downarrow x_0} f(x) = \lim_{x\downarrow x_0} g(x) = 0$, can be swapped out to cover the case for which $\lim_{x\downarrow x_0} f(x) = \lim_{x\downarrow x_0} g(x) = \pm\infty$ as well. This is given as follows:

Lemma 14.3.3 (L'Hôpital's Theorem III) *Let $f, g : X \to \mathbb{R}$ where $X \subseteq \mathbb{R}$ and $x_0 \in X'$. Suppose that there exists a $\delta > 0$ such that:*

1. *the functions f and g are differentiable in $(x_0, x_0 + \delta)$,*
2. *$g'(x) \neq 0$ for any $x \in (x_0, x_0 + \delta)$,*
3. *$\lim_{x\downarrow x_0} f(x) = \lim_{x\downarrow x_0} g(x) = \pm\infty$, and*
4. *$\lim_{x\downarrow x_0} \frac{f'(x)}{g'(x)}$ exists and is equal to $L \in \mathbb{R}$.*

Then, we have:

$$\lim_{x\downarrow x_0} \frac{f(x)}{g(x)} = \lim_{x\downarrow x_0} \frac{f'(x)}{g'(x)} = L.$$

Proof The idea is the same as Lemma 14.3.1: we use MVT II on some compact interval contained in $(x_0, x_0 + \delta)$. In contrast to the proof for Lemma 14.3.1, we are not able to extend the functions f and g onto the point x_0 to create a continuous function since f and g both blow up to $\pm\infty$ here. Instead, we have to apply MVT II on a different compact interval. The argument here is a bit more involved.

WLOG, suppose that $x_0 = 0$ and $\lim_{x\downarrow 0} f(x) = \lim_{x\downarrow 0} g(x) = \infty$. Fix $\varepsilon > 0$. We want to find a $\tilde{\delta} > 0$ such that whenever $0 < x < \tilde{\delta}$ we have $\left|\frac{f(x)}{g(x)} - L\right| < \varepsilon$.

By the third assumption, since $\lim_{x\downarrow 0} g(x) = \infty$, there exists a $\delta_1 > 0$ such that for all $0 < x < \delta_1$, we have $g(x) > 0$. Moreover, by the final assumption, there exists a $\delta_2 > 0$ such that for any $0 < c < \delta_2$ we have:

$$\left|\frac{f'(c)}{g'(c)} - L\right| < \frac{\varepsilon}{2}. \tag{14.9}$$

Set $\delta_3 = \min\{\delta_1, \delta_2, \delta\} > 0$. Fix any point $y \in (0, \delta_3)$. For any $0 < x < y$, since $[x, y] \subseteq (0, \delta_3) \subseteq (0, \delta)$, the derivative g' is nowhere vanishing here. Thus, we can apply MVT II on the compact interval $[x, y]$ to get:

$$\frac{f(x) - f(y)}{g(x) - g(y)} = \frac{f'(c)}{g'(c)},$$

where $c \in (x, y) \subseteq (0, \delta_3) \subseteq (0, \delta_2)$. Using this equality and the estimate (14.9), for any $0 < x < y$ we have:

$$\left|\frac{f(x) - f(y)}{g(x) - g(y)} - L\right| = \left|\frac{f'(c)}{g'(c)} - L\right| < \frac{\varepsilon}{2}, \tag{14.10}$$

since $0 < c < \delta_2$. Note also this implies the following bound, which will be useful later:

$$\begin{aligned}\left|\frac{f(x) - f(y)}{g(x) - g(y)}\right| &= \left|\frac{f(x) - f(y)}{g(x) - g(y)} - L + L\right| \\ &\leq \left|\frac{f(x) - f(y)}{g(x) - g(y)} - L\right| + |L| < \frac{\varepsilon}{2} + |L| = M, \end{aligned} \tag{14.11}$$

for any $0 < x < y$.

On the other hand, since $x, y \in (0, \delta_3) \subseteq (0, \delta_1)$, we have $g(x), g(y) > 0$ and so we can write:

$$\begin{aligned} \frac{f(x) - f(y)}{g(x) - g(y)} &= \frac{\frac{f(x)}{g(x)} - \frac{f(y)}{g(x)}}{1 - \frac{g(y)}{g(x)}} \\ \Rightarrow \quad \frac{f(x)}{g(x)} - \frac{f(x) - f(y)}{g(x) - g(y)} &= \frac{f(y)}{g(x)} - \frac{f(x) - f(y)}{g(x) - g(y)} \frac{g(y)}{g(x)}. \end{aligned}$$

Thus, by using the triangle inequality and the estimate (14.11) on this, we have:

$$\begin{aligned}\left|\frac{f(x)}{g(x)}-\frac{f(x)-f(y)}{g(x)-g(y)}\right| &\leq \left|\frac{f(y)}{g(x)}\right|+\left|\frac{f(x)-f(y)}{g(x)-g(y)}\right|\left|\frac{g(y)}{g(x)}\right| \\ &< \left|\frac{f(y)}{g(x)}\right|+M\left|\frac{g(y)}{g(x)}\right|. \end{aligned} \tag{14.12}$$

Since $f(y)$ and $g(y)$ are constants and $\lim_{x\downarrow 0} g(x) = \infty$, the limit on the RHS of the inequality (14.12) as $x \downarrow 0$ is 0. By sandwiching, we have the right-limit $\lim_{x\downarrow 0}\left|\frac{f(x)}{g(x)} - \frac{f(x)-f(y)}{g(x)-g(y)}\right| = 0$. Thus, there exists a $\delta_4 > 0$ such that for any $0 < x < \delta_4$ we have:

$$\left|\frac{f(x)}{g(x)}-\frac{f(x)-f(y)}{g(x)-g(y)}\right| < \frac{\varepsilon}{2}. \tag{14.13}$$

Finally, we combine the estimates (14.10) and (14.13) together with triangle inequality. By setting $\tilde{\delta} = \min\{\delta_4, y\} > 0$, whenever $0 < x < \tilde{\delta}$ we then have:

$$\left|\frac{f(x)}{g(x)}-L\right| \leq \left|\frac{f(x)}{g(x)}-\frac{f(x)-f(y)}{g(x)-g(y)}\right|+\left|\frac{f(x)-f(y)}{g(x)-g(y)}-L\right| < \frac{\varepsilon}{2}+\frac{\varepsilon}{2}=\varepsilon,$$

thus proving that $\lim_{x\downarrow 0}\frac{f(x)}{g(x)} = L$. □

Likewise, we also have an analogous result for the left-limit:

Lemma 14.3.4 (L'Hôpital's Theorem IV) *Let $f, g : X \to \mathbb{R}$ where $X \subseteq \mathbb{R}$ and $x_0 \in X'$. Suppose that there exists a $\delta > 0$ such that:*

1. *the functions f and g are differentiable in $(x_0, x_0+\delta)$,*
2. *$g'(x) \neq 0$ for any $x \in (x_0, x_0+\delta)$,*
3. *$\lim_{x\uparrow x_0} f(x) = \lim_{x\uparrow x_0} g(x) = \pm\infty$, and*
4. *$\lim_{x\uparrow x_0}\frac{f'(x)}{g'(x)}$ exists and is equal to $L \in \mathbb{R}$.*

Then, we have:

$$\lim_{x\uparrow x_0}\frac{f(x)}{g(x)} = \lim_{x\uparrow x_0}\frac{f'(x)}{g'(x)} = L.$$

Thus, putting all the L'Hôpital's theorems above together using Proposition 9.3.4, we get the L'Hôpital's rule:

Theorem 14.3.5 (L'Hôpital's Rule) *Let $f, g : X \to \mathbb{R}$ where $X \subseteq \mathbb{R}$ and $x_0 \in X'$. Suppose that there exists a $\delta > 0$ such that:*

1. *the functions* f *and* g *are differentiable in* $(x_0 - \delta, x_0 + \delta)$,
2. $g'(x) \neq 0$ *for any* $x \in (x_0 - \delta, x_0 + \delta) \setminus \{x_0\}$,
3. $\lim_{x \to x_0} f(x) = \lim_{x \to x_0} g(x) = 0$ *or* $\lim_{x \to x_0} f(x) = \lim_{x \to x_0} g(x) = \pm\infty$, *and*
4. $\lim_{x \to x_0} \frac{f'(x)}{g'(x)}$ *exists and is equal to* $L \in \mathbb{R}$.

Then, we have:

$$\lim_{x \to x_0} \frac{f(x)}{g(x)} = \lim_{x \to x_0} \frac{f'(x)}{g'(x)} = L.$$

This rule is named after Guillaume de L'Hôpital (1661–1704), who published it in his book *Analyse des Infiniment Petits pour l'Intelligence des Lignes Courbes* (Analysis of the Infinitely Small for the Understanding of Curved Lines). However, this result was actually introduced to L'Hôpital by one of the mathematicians in the Bernoulli family, Johann Bernoulli.

Bernoulli was hired by L'Hôpital as his tutor and as part of the contract, along with monetary payment, L'Hôpital has the right to use Bernoulli's discoveries as he wish. L'Hôpital also requested Bernoulli to inform him and him alone of any mathematical discoveries he made during this period. This is how the result ended up in L'Hôpital's book and named after him. However, in the preface of his book, L'Hôpital credited Bernoulli to the results.

Despite the rewards and this minor credit, this arrangement made Bernoulli deeply unhappy and he lamented to his close friends that he had not received enough recognition for his work. After L'Hôpital's death, Bernoulli tried to reclaim his work by publishing their letters and correspondence as proof of him as the rightful owner of these mathematical discoveries. Alas, due to his attempted theft of his brother Jacob Bernoulli's ideas in the past, this was largely dismissed by the mathematical community at the time.

Remark 14.3.6 The conclusion of L'Hôpital's rule is very nice and convenient but we need to make sure that the four conditions for applying it must hold before going straight to the conclusion. Before applying L'Hôpitals's rule, always check:

1. f and g are differentiable functions near x_0.
2. $g'(x) \neq 0$ in some open interval containing x_0.
3. It is of the indeterminate form $\frac{0}{0}$ or $\frac{\pm\infty}{\pm\infty}$.
4. The limit $\frac{f'(x)}{g'(x)}$ as $x \to x_0$ exists.

Example 14.3.7 Let us revisit the limit that we have seen earlier in Example 9.5.3(2). Recall that $h : \mathbb{R} \setminus \{-2, 1\} \to \mathbb{R}$ was defined as $h(x) = \frac{x^2-1}{x^2+x-2}$ and we wanted to find the limit of this function as $x \to 1$. We have seen that this limit is an indeterminate form since the limits of the numerator and the denominator are both 0. So, we might be able to use L'Hôpital's rule here.

Before doing so, we need to check that the numerator and denominator, which we denote as functions $f(x) = x^2 - 1$ and $g(x) = x^2 + x - 2$ both defined on $\mathbb{R} \setminus \{-2, 1\}$, satisfy the required conditions.

1. Since f and g are polynomials, they are differentiable everywhere.
2. $g'(x) = 2x + 1 \neq 0$ in some open interval containing 1 since it only vanishes at $x = -\frac{1}{2}$,
3. h is of the indeterminate form $\frac{0}{0}$.
4. The limit of $\frac{f'(x)}{g'(x)} = \frac{2x}{2x+1}$ as $x \to 1$ exists since this is simply equal to $\frac{2}{3}$.

Thus, since all the conditions for L'Hôpital's rule are satisfied, we can conclude that:

$$\lim_{x\to 1} h(x) = \lim_{x\to 1} \frac{f'(x)}{g'(x)} = \frac{2}{3}.$$

We may also apply the L'Hôpital's rule for evaluating limits as $x \to \pm\infty$. This is done by precomposing the functions f and g with the reciprocal function to bring the "limit point" at $\pm\infty$ to 0 and then apply the L'Hôpital's rule at 0. More concretely:

Theorem 14.3.8 (L'Hôpital's Rule at Infinity) *Let $f, g : (a, \infty) \to \mathbb{R}$ be real-valued functions. Suppose that:*

1. *the functions f and g are differentiable in (a, ∞),*
2. *there exists a $K > a$ such that $g'(x) \neq 0$ for any $x > K$,*
3. *$\lim_{x\to\infty} f(x) = \lim_{x\to\infty} g(x) = 0$ or $\lim_{x\to\infty} f(x) = \lim_{x\to\infty} g(x) = \pm\infty$, and*
4. *$\lim_{x\to\infty} \frac{f'(x)}{g'(x)}$ exists and is equal to $L \in \mathbb{R}$.*

Then, we have:

$$\lim_{x\to\infty} \frac{f(x)}{g(x)} = \lim_{x\to\infty} \frac{f'(x)}{g'(x)} = L.$$

Sometimes, we may need to apply the L'Hôpital's rule twice and this must be done carefully. Let us look at the following example:

Example 14.3.9 Let the function $f(x) = \frac{x^2}{e^x}$ be defined on $\mathbb{R}$. We wish to evaluate the limit of $\lim_{x\to\infty} f(x)$. First we check the conditions in Theorem 14.3.8.

1. The two expressions x^2 and e^x are differentiable everywhere.
2. The derivative of e^x does not vanish anywhere.
3. It is an indeterminate form $\frac{\infty}{\infty}$.

4. We also need to check the following limit:

$$\lim_{x\to\infty} \frac{2x}{e^x}, \tag{14.14}$$

exists. However, this is also of the indeterminate form $\frac{\infty}{\infty}$ so the algebra of limits does not work here. In order to apply L'Hôpital's rule for $\frac{x^2}{e^x}$, we need to show this limit exists.

As a side quest, we now investigate the limit (14.14). In order to use the L'Hôpital's rule for this limit, we check the following four conditions:
(a) The two expressions $2x$ and e^x are differentiable everywhere.
(b) The derivative of e^x does not vanish anywhere.
(c) It is an indeterminate form $\frac{\infty}{\infty}$.
(d) The limit $\lim_{x\to\infty} \frac{2}{e^x} = 0$ exists.
Therefore, the limit in (14.14) exists by the L'Hôpital's rule and its value is:

$$\lim_{x\to\infty} \frac{2x}{e^x} = \lim_{x\to\infty} \frac{2}{e^x} = 0.$$

Since the four conditions of the L'Hôpital's rule are satisfied, we thus have:

$$\lim_{x\to\infty} \frac{x^2}{e^x} = \lim_{x\to\infty} \frac{2x}{e^x} = \lim_{x\to\infty} \frac{2}{e^x} = 0.$$

As we have demonstrated, for this problem we need two nested layers of L'Hôpital's rule to find the desired limit.

Remark 14.3.10 Sometimes it is tempting to use the L'Hôpital's rule in some cases, for example when evaluating the limit $\frac{\sin(x)}{x}$ as $x \to 0$. However, strictly speaking, the L'Hôpital's rule cannot be used here. This is because in order to find the derivative of the numerator $\sin(x)$, as we have seen earlier in Example 13.1.9, we need to find the limit $\frac{\sin(x)}{x}$ as $x \to 0$. So if we attempt to use L'Hôpital's rule here, we will be stuck in a circular argument!

The moral of the story here is that the L'Hôpital's rule is extremely convenient, but we have to always be very careful with it since it is an extremely delicate result that requires a lot of assumptions in order to work!

So far we have seen L'Hôpital's rule for the indeterminate forms $\frac{0}{0}$ and $\frac{\pm\infty}{\pm\infty}$. What about the other indeterminate forms? For the indeterminate form $0 \cdot \pm\infty$, we can turn this into one of the indeterminate forms $\frac{0}{0}$ and $\frac{\pm\infty}{\pm\infty}$ by taking the reciprocal of one of the terms.

Example 14.3.11 Let $f : \mathbb{R} \to \mathbb{R}$ be defined as $f(x) = xe^x$. We want to evaluate the limit $\lim_{x\to-\infty} f(x)$. Note that this is an indeterminate form $-\infty \cdot 0$. Let us then write the product as a quotient:

$$\lim_{x\to-\infty} xe^x = \lim_{x\to-\infty} \frac{x}{e^{-x}}.$$

Now this limit is an indeterminate form $\frac{-\infty}{\infty}$ which we may apply the L'Hôpital's rule on. By first positively checking all the conditions for L'Hôpital's rule (which we leave for the readers to verify), we can show that:

$$\lim_{x\to-\infty} xe^x = \lim_{x\to-\infty} \frac{x}{e^{-x}} = \lim_{x\to-\infty} \frac{1}{-e^{-x}} = 0.$$

Finally, for indeterminate forms of the form 0^0, $(\pm\infty)^0$, and $1^{\pm\infty}$, we can turn this form into the $\frac{0}{0}$ or $\frac{\pm\infty}{\pm\infty}$ forms by taking logarithms.

Example 14.3.12 Let $f : \mathbb{R}_+ \to \mathbb{R}$ be defined as $f(x) = x^{\frac{1}{x}}$. We want to evaluate the limit $\lim_{x\to\infty} f(x)$ which is of the ∞^0 form. By taking logarithms on both side, we get:

$$\ln(f(x)) = \frac{1}{x}\ln(x) = \frac{\ln(x)}{x},$$

which is now of the $\frac{\infty}{\infty}$ form. We can evaluate the limit of $\ln(f(x))$ by using L'Hôpital's rule (check the conditions first!) to get:

$$\lim_{x\to\infty} \ln(f(x)) = \lim_{x\to\infty} \frac{\ln(x)}{x} = \lim_{x\to\infty} \frac{\frac{1}{x}}{1} = 0.$$

However, our original goal was to evaluate the limit of $f(x) = x^{\frac{1}{x}}$ as $x \to \infty$. Therefore, using the continuity of the exponential function, we note that:

$$x^{\frac{1}{x}} = f(x) = e^{\ln(f(x))} \quad \Rightarrow \quad \lim_{x\to\infty} x^{\frac{1}{x}} = \lim_{x\to\infty} e^{\ln(f(x))} = e^{\lim_{x\to\infty}\ln(f(x))} = e^0 = 1,$$

which is the same limit that we have obtained in Exercise 9.23.

Remark 14.3.13 The conditions in L'Hôpital's rule cannot be weakened further. Here are some remarks regarding them:

1. In some literature, the condition that $g'(x) \neq 0$ in a neighbourhood of x_0 in the L'Hôpital's rule is usually (incorrectly) omitted. This has caused many pitfalls when one tries to apply the rule.
   A counterexample by Otto Stolz (1842–1905) [70] would be the following: suppose that $f, g : \mathbb{R}_+ \to \mathbb{R}$ are defined as $f(x) = x + \sin(x)\cos(x)$ and $g(x) = (x + \sin(x)\cos(x))e^{\sin(x)}$. These two functions tend to ∞ as $x \to \infty$ and

are differentiable for every $x \in \mathbb{R}_+$. We note that the quotient:

$$\frac{f(x)}{g(x)} = \frac{x + \sin(x)\cos(x)}{(x + \sin(x)\cos(x))e^{\sin(x)}},$$

(which is of the indeterminate $\frac{\infty}{\infty}$ form) does not have a limit as $x \to \infty$ since it is actually just equal to $\frac{f(x)}{g(x)} = e^{-\sin(x)}$ via cancellations. However, the quotient of the derivatives is given by:

$$\frac{f'(x)}{g'(x)} = \frac{2\cos(x)e^{-\sin(x)}}{2\cos(x) + \sin(x)\cos(x) + x},$$

which converges to 0 as $x \to \infty$. So, the L'Hôpital's rule does not work here! This is because the derivative of the denominator, given by:

$$g'(x) = e^{\sin(x)}\cos(x)(x + \sin(x)\cos(x) + 2\cos(x)),$$

vanishes at $x = \frac{(2n-1)\pi}{2}$ for every $n \in \mathbb{Z}$ which can be arbitrarily large. So the third condition in Theorem 14.3.8 was not satisfied.

2. The condition that the limit $\lim_{x \to x_0} \frac{f'(x)}{g'(x)}$ exists also cannot be removed. Consider the quotient $\frac{f(x)}{g(x)} = \frac{x+\sin(x)}{x}$ for $x > 0$. If we were to find the limit of this quotient as $x \to \infty$, it is of the indeterminate form $\frac{\infty}{\infty}$. We can check the conditions of Theorem 14.3.8 are satisfied, except for the final one. Indeed, if we compute the limit of the quotient of derivatives, we would get:

$$\lim_{x\to\infty} \frac{f'(x)}{g'(x)} = \lim_{x\to\infty} \frac{1 + \cos(x)}{1} = \lim_{x\to\infty} 1 + \cos(x),$$

which does not exist.
However, we know that the limit of the original quotient does exist because, via simplification, we have:

$$\lim_{x\to\infty} \frac{f(x)}{g(x)} = \lim_{x\to\infty} \left(1 + \frac{\sin(x)}{x}\right) = 1,$$

since $\left|\frac{\sin(x)}{x}\right| = \frac{|\sin(x)|}{x} \leq \frac{1}{x} \to 0$ as $x \to \infty$.

14.4 Introduction to Differential Equations

This section is a short introduction to the study of differential equations. Before we define what differential equations are, we would like to pose an important question which forms the simplest kind differential equations.

Antiderivatives

Suppose that we have a function $f : X \to \mathbb{R}$ defined on some subset $X \subseteq \mathbb{R}$. Is this function a derivative of some other function? In other words, is there a function $F : X \to \mathbb{R}$ such that $\frac{d}{dx}F = f$? If there is such a function, we call it a primitive or an antiderivative of f.

Simply put, antiderivatives are just what we expect to get if we reverse the operation of taking the derivative of a function, hence the name. More specifically:

Definition 14.4.1 (Antiderivative) Let $X \subseteq \mathbb{R}$. A function $F : X \to \mathbb{R}$ is called an antiderivative or primitive of the function $f : X \to \mathbb{R}$ if $\frac{d}{dx}F(x) = f(x)$ for all $x \in X$.

We note that the antiderivative of a given function f, if it does exist, is not unique. Indeed, suppose that we have two antiderivatives F and G of f, then:

$$F' - G' = f - f \equiv 0 \quad \Rightarrow \quad \frac{d}{dx}(F - G) \equiv 0,$$

and thus from Corollary 13.6.6, if X is a connected interval we know that $F - G = C$ for some constant $C \in \mathbb{R}$. Therefore, any two antiderivatives of a given function on a connected interval differ by some additive constant. As a result, it is enough to find one antiderivative of a function f to list out all of its antiderivatives by simply shifting the one antiderivative with additive constants.

At this stage of our study, we can only guess what antiderivatives of some functions are based on our knowledge on derivatives.

Example 14.4.2 Let us look at some examples:

1. Suppose that $f : \mathbb{R} \to \mathbb{R}$ is $f(x) = x^2$. We want to find an antiderivative for f. Since this is a monomial, we expect that its antiderivative is also a monomial. Moreover, since a derivative reduces the power of a monomial by 1, we expect an antiderivative of f is of the form $F(x) = Bx^3 + C$ for some constants $B, C \in \mathbb{R}$. What are these constants? We know that $f(x) = \frac{d}{dx}F(x)$ for all $x \in \mathbb{R}$, namely $x^2 = 3Bx^2$. Putting $x = 1$, we get $B = \frac{1}{3}$. Thus, the antiderivatives of f are $F(x) = \frac{1}{3}x^3 + C$ where $C \in \mathbb{R}$ is some unknown constant. In general, for a monomial $f(x) = x^n$ where $n \geq 0$ defined on the whole of $\mathbb{R}$, any of its antiderivatives is of the form $F(x) = \frac{x^{n+1}}{n+1} + C$ for some constant $C \in \mathbb{R}$.
2. Now suppose that $n \in \mathbb{Z}_- \setminus \{-1\}$. The domain for the monomial $f(x) = x^n$ is disconnected as it is $\mathbb{R} \setminus \{0\}$. Therefore, in each connected subintervals, namely $(-\infty, 0)$ and $(0, \infty)$, we could have different choices of additive constants in the

antiderivative as we have mentioned in Remark 13.6.7. So a general antiderivative of f is:

$$F(x) = \begin{cases} \frac{x^{n+1}}{n+1} + C & \text{if } x < 0, \\ \frac{x^{n+1}}{n+1} + D & \text{if } x > 0, \end{cases}$$

for some constants $C, D \in \mathbb{R}$. Indeed, it is straightforward to check that $F' = f$ on $\mathbb{R} \setminus \{0\}$.

3. How about the case for $n = -1$? In other words, what are the antiderivatives of the function $f : \mathbb{R} \setminus \{0\} \to \mathbb{R}$ where $f(x) = \frac{1}{x}$? We know that $\frac{d}{dx} \ln(x) = \frac{1}{x}$, so $\ln(x) + C$ is a candidate for this antiderivative.
But the natural logarithm $\ln(x)$ is only defined for $x > 0$, so we still need to find an antiderivative for $x < 0$. For $x < 0$ we try $\ln(-x) + D$. This is a perfectly well defined function for $x < 0$ and, by chain rule, we have $\frac{d}{dx}(\ln(-x) + D) = \frac{1}{-x} \cdot (-1) = \frac{1}{x}$ thus satisfying the required relationship. In short, for some constants $C, D \in \mathbb{R}$, the antiderivatives of the function f are:

$$F(x) = \begin{cases} \ln(-x) + D & \text{if } x < 0, \\ \ln(x) + C & \text{if } x > 0. \end{cases}$$

4. Clearly, the exponential function $f(x) = e^x$ on $\mathbb{R}$ satisfies $\frac{d}{dx} f(x) = e^x = f(x)$. Therefore, the antiderivatives of the exponential function are simply shifts of itself, namely $F(x) = e^x + C$ for some $C \in \mathbb{R}$.
How about $f(x) = e^{-x}$? Following the above, we can check that $\frac{d}{dx} f(x) = -e^{-x} = -f(x)$. By multiplying both sides with -1 and using the linearity of derivatives, we get $\frac{d}{dx}(-f(x)) = f(x)$. This shows that the antiderivatives of f are $-f(x) + C = -e^{-x} + C$ where C is any real constant.
5. Suppose that $f : \mathbb{R} \setminus \{-\frac{1}{2}, 0\} \to \mathbb{R}$ is defined as $f(x) = \frac{1-2x}{2\sqrt{x}(2x+1)^2}$. We know from Example 13.2.2 that an antiderivative of f is given by the function $F : \mathbb{R} \setminus \{-\frac{1}{2}, 0\} \to \mathbb{R}$ described as:

$$F(x) = \begin{cases} \frac{\sqrt{x}}{2x+1} + B & \text{if } x < -\frac{1}{2}, \\ \frac{\sqrt{x}}{2x+1} + C & \text{if } -\frac{1}{2} < x < 0, \\ \frac{\sqrt{x}}{2x+1} + D & \text{if } x > 0, \end{cases}$$

for some constants $B, C, D \in \mathbb{R}$. But if we did not do Example 13.2.2 beforehand, how can we even come up with this complicated antiderivative? This is a difficult task indeed. Johann Bernoulli remarked:

> But just as much as it is easy to find the differential of a given quantity, so it is difficult to find the [antiderivative] of a given differential. Moreover, sometimes we cannot say with certainty whether the [antiderivative] of a given quantity can be found or not.

I usually think of it this way: if differentiating is like breaking an egg, finding antiderivatives is like putting the broken eggshells and its contents back together.

At this stage, finding antiderivatives relies a lot on luck and guessing, but later in Chap. 16 we shall see some advanced results that could help us with this task. In the meantime, we also have some crude tools that could help us with the guessing game. These are just the reverse of the properties of differentiation that we saw in Proposition 13.2.1.

Proposition 14.4.3 *Let $f, g : \mathbb{R} \to \mathbb{R}$. Suppose that $F, G : \mathbb{R} \to \mathbb{R}$ are their antiderivatives respectively. For some real constant $C \in \mathbb{R}$, we have:*

1. *The antiderivatives of $f \pm g$ are $F \pm G + C$.*
2. *Let $\lambda \in \mathbb{R}$. The antiderivatives of λf are $\lambda F + C$.*
3. *Reverse product rule: The antiderivatives of $fG + Fg$ are $FG + C$.*
4. *Reverse chain rule: The antiderivatives of $(f \circ G)g'$ are $F \circ G + C$.*
5. *In particular, if the function G in assertion 4 is $G(x) = \lambda x$ for some constant $\lambda \neq 0$, the antiderivatives of $f(\lambda x)$ are $\frac{F(\lambda x)}{\lambda} + C$.*
6. *If F is nowhere vanishing, the antiderivatives of $\frac{f}{F}$ are $\ln(|F|) + C$.*

Proposition 14.4.3(1) and (2) say that the process of finding an antiderivative is linear, similar to the process of differentiation.

Example 14.4.4 Let us look at some examples:

1. The antiderivative of $f(x) = x^2 + 5e^x$ can be found by the linearity of antiderivatives. We note that an antiderivative of x^2 and e^x are $\frac{x^3}{3}$ and e^x respectively. Thus, the antiderivatives of f are $\frac{x^3}{3} + 5e^x + C$ for any $C \in \mathbb{R}$.
2. Suppose that we want to find the antiderivatives of the function $h(x) = x(x^2 + 3)^5$. We could expand this and find the antiderivatives of each of the constituent monomial, but here we are going to use the reverse chain rule.
   We note that we can choose $G(x) = x^2 + 3$ and $f(x) = x^5$. From these, we can compute $g(x) = 2x$ and $F(x) = \frac{x^6}{6}$. Thus, we can then write $h(x) = \frac{1}{2} f(G(x))g'(x)$ which is similar to Proposition 14.4.3(4). So, for some $C \in \mathbb{R}$, the antiderivatives of h are:
$$\frac{1}{2}(F(G(x)) + C) = \frac{1}{2}\frac{G(x)^6}{6} + \tilde{C} = \frac{1}{2}\frac{(x^2+3)^6}{6} + \tilde{C} = \frac{(x^2+3)^6}{12} + \tilde{C},$$
   where $\tilde{C} \in \mathbb{R}$ is any real constant.
3. If we revisit the function $h(x) = e^{-x}$, we can write $G(x) = -x$ and $f(x) = e^x$. Thus, $F(x) = e^x$. By Proposition 14.4.3(5), the antiderivatives of h are $\frac{e^{-x}}{-1} + C = -e^{-x} + C$.

Ordinary Differential Equations

The problem of finding an antiderivative of a function $f : X \to \mathbb{R}$ forms a very simple type of differential equation. In the problem of finding antiderivatives, we want to solve the equation $\frac{dy}{dx} = f(x)$ for some unknown function y (which we call antiderivatives). This equation involves the first derivative of the unknown function y and a function f of x.

In general, a differential equation may involve many other terms. More generally, we define:

Definition 14.4.5 (Ordinary Differential Equations, ODEs) Let $F : \mathbb{R}^{n+2} \to \mathbb{R}$ be a function on $n+2$-variables. An ordinary differential equation or ODE is an equation of the form:

$$F(x, y(x), y'(x), y''(x), \dots, y^{(n)}(x)) = 0.$$

The highest derivative of the unknown function y that appears in the equation is called the order of the ODE.

These equations are called the ordinary differential equations in contrast to other kinds of differential equations such as partial differential equations and stochastic differential equations in which more advanced types of derivatives appear.

In ODEs, the unknown function is a function of one variable x so, for brevity, instead of writing $y(x), y'(x), \dots, y^{(n)}(x)$ to denote the dependency of these function on the variable x, usually these functions are written as $y, y', \dots, y^{(n)}$ by suppressing the dependency on x.

When we have such an equation, our aim is to find whether there is a solution $y : X \to \mathbb{R}$ for it, what is the maximal domain of existence X for such solutions, how many possible solutions are there, and how can we find them. The most regular kind of solution to such ODEs are called classical solutions:

Definition 14.4.6 (Classical Solution of an ODE) Given an ODE $F(x, y, y', y'', \dots, y^{(n)}) = 0$. A classical solution to this ODE is an n times differentiable function $u : I \to \mathbb{R}$ defined on some interval $I \subseteq \mathbb{R}$ such that for all $x \in I$ we have:

$$F(x, u, u', u'', \dots, u^{(n)}) = 0.$$

Because the ODEs in Definitions 14.4.5 and 14.4.6 are very general, we can narrow down the family of ODEs that we study. Here, we shall focus only on linear ODEs. We distinguish them as follows:

1. The ODE is linear if we can write F as a linear combination of y and its derivatives with coefficients only depending on x, namely:

$$F(x, y, y', y'', \ldots, y^{(n)}) = \sum_{j=0}^{n} a_j(x)y^{(j)} + a(x) = 0, \tag{14.15}$$

with the convention that $y^{(0)} = y$.
Furthermore, linear ODEs can be categorised further into linear homogeneous ODE when $a(x) \equiv 0$ and linear inhomogeneous ODE otherwise.
A way to determine whether a linear ODE is homogeneous or inhomogeneous is to check whether the trivial function $u \equiv 0$ is a solution to Eq. (14.15); it is a solution to the homogeneous problem, but not for the inhomogeneous problem.
2. The ODE is non-linear if it cannot be written in the form as (14.15).

Remark 14.4.7 The problem of finding an antiderivative of a function $f : \mathbb{R} \to \mathbb{R}$ that we saw earlier, which is a problem of the form $\frac{dy}{dx} = f(x)$, is a first order linear inhomogeneous ODE.

Let us now focus our attention to some low order (namely first and second order) ODEs.

First Order ODEs

From Eq. (14.15), a general form of first order linear ODE looks like:

$$\frac{dy}{dx} + a(x)y = b(x), \tag{14.16}$$

for some functions $a, b : \mathbb{R} \to \mathbb{R}$. Assuming that $a(x)$ has an antiderivative $A(x)$, this problem can be equivalently restated by multiplying through with a non-zero quantity $e^{A(x)}$.

$$y'e^{A(x)} + a(x)e^{A(x)}y = b(x)e^{A(x)}. \tag{14.17}$$

Moreover, by observing that we can use the reverse product rule on the LHS, we can write (14.17) as:

$$\frac{d}{dx}(ye^{A(x)}) = b(x)e^{A(x)}. \tag{14.18}$$

Next, assume that we know an antiderivative of $b(x)e^{A(x)}$ which we call $F(x)$. Since Eq. (14.18) implies that the derivative of $ye^{A(x)}$ is $b(x)e^{A(x)}$, necessarily $ye^{A(x)}$ is an antiderivative of $b(x)e^{A(x)}$. Thus, we must have $ye^{A(x)} = F(x)+C$ for some constant $C \in \mathbb{R}$ which then implies the solution $y = (F(x) + C)e^{-A(x)}$. We can simply differentiate and substitute y and y' in the ODE (14.16) to check that y is indeed its classical solution.

Remark 14.4.8 We make several remarks regarding the procedure outlined above and the assumptions that we made.

1. The quantity $e^{A(x)}$ in the above is called an integrating factor. It is a very important quantity here because it allows us to write the terms on the LHS of Eq. (14.17) as a derivative via reversing the product rule.
2. Note that it does not matter which antiderivative $A(x)$ of $a(x)$ is chosen in the integrating factor: if we choose the antiderivative $A(x) + C$ instead of $A(x)$ in the integrating factor, Eq. (14.17) becomes:

$$y'e^{A(x)+C} + a(x)e^{A(x)+C}y = b(x)e^{A(x)+C},$$

   which is equivalent to (14.17) since the terms $e^C \neq 0$ can be divided through the equation.
3. We assumed the existence of antiderivatives twice for the whole theoretical argument to work. Namely, we assumed that $a(x)$ and $b(x)e^{A(x)}$ both have antiderivatives. But in general, we do not know whether these are true! Thus, usually we have to work on a case-by-case basis.

Let us look at a concrete example to clarify the outlined method.

Example 14.4.9 Consider the first order linear inhomogeneous ODE $y' - y = 1$. Following the above guidelines, since $a(x) = -1$ here has an antiderivative $A(x) = -x$ (we can drop the constant as per Remark 14.4.8(2)), we multiply the whole ODE with e^{-x} to get:

$$y'e^{-x} - ye^{-x} = e^{-x} \quad \Leftrightarrow \quad \frac{d}{dx}(e^{-x}y) = e^{-x}.$$

Now the term in the bracket is an antiderivative of e^{-x}, which we have seen to be $-e^{-x} + C$ for some real constant $C \in \mathbb{R}$ in Example 14.4.2(4). Thus, we have:

$$\frac{d}{dx}(e^{-x}y) = e^{-x} \quad \Leftrightarrow \quad e^{-x}y = -e^{-x} + C \quad \Leftrightarrow \quad y = -1 + Ce^{x}.$$

Remark 14.4.10 The ODE in Example 14.4.9 can be solved according to the method that we outlined earlier because the relevant quantities do have a explicit and easy-to-find antiderivatives. At the moment, it seems like we can only work with very elementary functions.

However, in Theorem 16.1.3 in Chap. 16 we shall see that the condition of $a(x)$ and $b(x)$ being continuous is enough to guarantee that the relevant antiderivatives exist and hence the ODE can be solved. We shall use this fact for the rest of this chapter.

Initial/Boundary Value Problem

Notice that the solutions of an ODE comes with some constants which are free. In general, for an n-th order ODE, the solutions would come with n free constants. These solutions are called general solutions since the constants are not specified. Therefore, to determine what the values of these constants are, we need to augment the ODE with some restrictions.

The ODE, when supplied with some restriction, would then give us a specific solution to the ODE. In other words, these restrictions allow us to then determine what the actual values of the free constants in the general solution are. Usually, for an n-th order ODE, since there are n free constants in the general solution, we need to specify n restrictions in order to get a specific solution.

These restrictions are called initial or boundary conditions for the ODE, depending on whether the variable x in the ODE represents a temporal quantity or a spatial quantity (but, honestly, unless we are working on a physical problem, as mathematicians we do not care what it is called). The problem of an ODE coupled with these restrictions is called the initial value problem (IVP for short) or boundary value problem (BVP for short).

Example 14.4.11 Recall the solution $y = -1 + Ce^x$ to the ODE $y' - y = 1$ in Example 14.4.9. Note that this is a family of general solutions parametrised by the real constant $C \in \mathbb{R}$. To get a specific solution, we need to furnish the ODE with some restriction on the solution that we want. Suppose that we want a solution of the ODE to satisfy $y(0) = 1$. Substituting this requirement in the general solution, we get:

$$1 = y(0) = -1 + Ce^0 = -1 + C \quad \Rightarrow \quad C = 2.$$

So the solution the ODE IVP $y' - y = 1$ subject to the condition $y(0) = 1$ is $y(x) = -1 + 2e^x$. Refer to Fig. 14.9 to see all the possible solutions to the ODE and the specific solution to the ODE subject to the constraint $y(0) = 1$.

Remark 14.4.12 From Fig. 14.9, we can see that the general solutions to a first order ODE form a family of paths. As described by Pólya:

> The differential equation of the first order $\frac{dy}{dx} = f(x, y)\ldots$ prescribes the slope $\frac{dy}{dx}$ at each point of the plane ... can be conceived intuitively as a problem about the steady flow of a river: Being given the direction of the flow at each point, find the streamlines ...

where the streamlines are the paths in the figure. We would like to add to Pólya's quote that finding a particular solution to an ODE IVP is like dropping an apple (the

Fig. 14.9 There are infinitely many solutions to the ODE $y' - y = 1$ but only one that satisfies the constraint $y(0) = 1$ (namely, passing through the red dot)

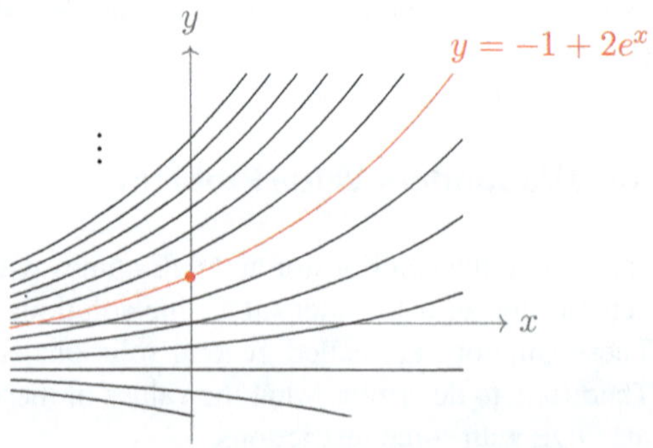

red dot in Fig. 14.9) in the steadily flowing river and determining what is the path of its flow as it floats down the river.

However, now we ask ourselves: is this the only specific solution that satisfies the ODE and the initial condition? How can we ensure that there are no other solutions to this problem? In other words, is it possible that some two streamlines in Fig. 14.9 cross somewhere so if we drop an apple at this crossing point, the apple might follow two different trajectories? This question is called the uniqueness question and can be answered, provided that the coefficients in the ODE are nice enough, by the following theorem:

Theorem 14.4.13 (Uniqueness of IVP for Linear First Order ODE) *Let*

$$\frac{dy}{dx} + a(x)y = b(x) \quad \text{and} \quad y(x_0) = \alpha,$$

be a linear first order ODE for some continuous functions $a, b : \mathbb{R} \to \mathbb{R}$ with initial condition at some point $x_0 \in \mathbb{R}$. Suppose that there is an interval $I \subseteq \mathbb{R}$ with $x_0 \in I$ such that there is a classical solution $y : I \to \mathbb{R}$. Then, this solution is unique.

Proof Suppose that we have two solutions to this ODE IVP, namely $y_1, y_2 : I \to \mathbb{R}$ so that $y_1(x_0) = y_2(x_0) = \alpha$. Now consider the function $y : I \to \mathbb{R}$ where $y = y_1 - y_2$. This new quantity satisfies the following IVP:

$$\begin{aligned}\frac{dy}{dx} + a(x)y &= \frac{dy_1}{dx} - \frac{dy_2}{dx} + a(x)(y_1 - y_2)\\ &= \frac{dy_1}{dx} + a(x)y_1 - \frac{dy_2}{dx} - a(x)y_2 = b(x) - b(x) = 0,\end{aligned}$$

and initial condition $y(x_0) = y_1(x_0) - y_2(x_0) = \alpha - \alpha = 0$.

By Remark 14.4.10, since $a(x)$ is continuous, it has an antiderivative $A(x)$ and hence we can solve this ODE by using the integrating factor to get:

$$y'e^{A(x)} + a(x)e^{A(x)}y = 0 \quad \Leftrightarrow \quad \frac{d}{dx}(ye^{A(x)}) = 0 \quad \Leftrightarrow \quad y = Ce^{-A(x)}.$$

Finally, putting in the condition $y(x_0) = 0$, we conclude that $C = 0$. Namely, y is identically 0. Hence, $y_1 - y_2 = y \equiv 0$ which implies that the two solutions are in fact the same solution. So, the solution to the IVP of the first order linear ODE, if it exists, is unique. □

So once we have found a solution to a first order ODE IVP, we do not have to worry about missing any other potential solutions since there is only one of them!

Second Order ODEs

Now let us explore further by looking at linear second order ODEs. A general linear second order ODE for a quantity $y : \mathbb{R} \to \mathbb{R}$ has the general form:

$$\frac{d^2y}{dx^2} + a(x)\frac{dy}{dx} + b(x)y(x) = c(x),$$

for some functions $a, b, c : \mathbb{R} \to \mathbb{R}$.

Similar to the IVP of the first order linear ODE, we have the existence and uniqueness results for the solutions of the ODE IVP. We shall only state this result here as the proof is beyond the scope of what we have seen so far.

Theorem 14.4.14 (Existence and Uniqueness of IVP for Second Order Linear ODE) *Let:*

$$\frac{d^2y}{dx^2} + a(x)\frac{dy}{dx} + b(x)y = c(x) \quad \text{and} \quad y(x_0) = \alpha, y'(x_0) = \beta,$$

be a linear second order ODE for some continuous functions $a, b, c : \mathbb{R} \to \mathbb{R}$ with initial conditions at some point $x_0 \in \mathbb{R}$. Then, there exists an interval $I \subseteq \mathbb{R}$ with $x_0 \in I$ where there is a unique classical solution $y : I \to \mathbb{R}$ to the ODE IVP above.

Remark 14.4.15 We shall prove an existence and uniqueness result for general first order ODEs, namely the Picard-Lindelöf theorem, in Exercise 16.33. This result can be modified to cater to higher order ODEs, in particular the ODE IVP in Theorem 14.4.14, as well. For the elaboration, readers can consult [15].

Let us look at the homogeneous case where $c(x) = 0$. For the homogeneous case, the solutions of this ODE satisfy the linear condition, namely:

Lemma 14.4.16 *Suppose that for some functions $a, b : \mathbb{R} \to \mathbb{R}$ we have a linear homogeneous second order ODE:*

$$\frac{d^2y}{dx^2} + a(x)\frac{dy}{dx} + b(x)y = 0.$$

If $y_1, y_2 : I \to \mathbb{R}$ are classical solutions of this ODE defined on some interval $I \subseteq \mathbb{R}$, then the function $y = Cy_1 + Dy_2$ for any constants $C, D \in \mathbb{R}$ also satisfies the ODE.

Proof This is a very easy check. Since we know that $y_1'' + a(x)y_1' + b(x)y_1 = 0$ and $y_2'' + a(x)y_2' + b(x)y_2 = 0$, by linearity of differentiation, we have:

$$\begin{aligned} y'' + a(x)y' + b(x)y &= (Cy_1'' + Dy_2'') + a(x)(Cy_1 + Dy_2) + b(x)(Cy_1 + Dy_2) \\ &= C(y_1''+a(x)y_1'+b(x)y_1)+D(y_2''+a(x)y_2'+b(x)y_2)=0, \end{aligned}$$

and so y also satisfies the ODE. □

Therefore, any linear combination of solutions for the ODE is also a solution to the ODE. In general, this is true for linear homogeneous ODEs of any order. In fact, as we have noted earlier, $y(x) = 0$ is also a solution. So the space of all solutions of linear homogeneous ODE forms a real vector space as defined in Definition 4.6.2.

The readers might have seen some theory of vector spaces in any basic linear algebra courses. Lemma 14.4.16 suggests that we can build a general solution by taking all the possible linear combinations of known solutions. But then we have two issues here:

1. How many of these solutions do we need to create the most general solution? In the language of linear algebra, how large is a spanning set for this vector space?
2. Do we need to still add up solutions that "look like" each other? This is a question related to linear independence of the vectors in a spanning set.

Let us address the second question first. What do we mean by solutions that "look like" each other? If we have two solutions y_1 and y_2 of the ODE and one is a scale of the other, namely $y_1(x) = \lambda y_2(x)$ for all $x \in \mathbb{R}$ and some constant $\lambda \in \mathbb{R}$, then any linear combination of these two solutions, say $Cy_1(x) + Dy_2(x)$ for come constants $C, D \in \mathbb{R}$, is merely some scale of the first solution since $Cy_1(x) + Dy_2(x) = Cy_1(x) + D\lambda y_1(x) = (C + \lambda D)y_1(x)$. So the second solution here does not really contribute to the linear combination for the general solution as it can be combined together with the first one. As a result, we can simply ignore the second solution because it "looks like" the first solution.

In the spirit of linear algebra, we call these solutions linearly dependent. In general, for a collection of several real-valued functions defined on a common domain, we define:

Definition 14.4.17 (Linear Dependence, Independence) Let $\{y_j\}_{j=1}^k$ where $y_j : X \to \mathbb{R}$ be a set of k functions.

1. This set of functions is called linearly dependent if there exists a set of real constants $\{C_j\}_{j=1}^k$ not all equal to 0 such that:

$$\sum_{j=1}^{k} C_j y_j(x) = 0 \quad \text{for all } x \in X.$$

2. This set of functions is called linearly independent if the only set of real constants $\{C_j\}_{j=1}^k$ for which:

$$\sum_{j=1}^{k} C_j y_j(x) = 0 \quad \text{for all } x \in X,$$

is $C_j = 0$ for all $j = 1, 2, \ldots, k$.

For the case of linear dependent functions, using the linear dependency, we can write one of the functions with non-zero coefficient C_j (WLOG, suppose that $C_1 \neq 0$) as a linear combination of the others, namely $y_1(x) = \frac{1}{C_1}\sum_{j=2}^k C_j y_j(x)$ for all $x \in X$. Therefore, this function y_1 can be safely discarded from the collection since it can be recovered back as a linear combination of the other functions. In particular, if $k = 2$, linear dependence simply means that y_1 and y_2 are scales of each other, as we have seen before.

Note also that for linear dependence, the same set of constants $\{C_j\}_{j=1}^k$ must work for all $x \in X$. This fact is very useful in determining linear independence of some set of functions.

Example 14.4.18 Consider the functions $y_1(x) = \sin(x)$ and $y_2(x) = x\sin(x)$ defined on $\mathbb{R}$. We claim that this set of functions is linearly independent. Assume for contradiction that they are linearly dependent. Then, there are constants $C_1, C_2 \in \mathbb{R}$ not both equal to 0 such that $C_1 y_1 + C_2 y_2 = C_1 \sin(x) + C_2 x \sin(x) = 0$ for all $x \in \mathbb{R}$. Since this is true for all $x \in \mathbb{R}$, let us fix some numbers into this equation:

$$\begin{aligned} x &= -\frac{\pi}{2} &\Rightarrow\quad & C_1 - C_2\frac{\pi}{2} = 0, \\ x &= \frac{\pi}{2} &\Rightarrow\quad & C_1 + C_2\frac{\pi}{2} = 0. \end{aligned}$$

Solving these equations simultaneously, we obtain $C_1 = C_2 = 0$, which is a contradiction. Thus, the two functions must be linearly independent.

If $y_1, y_2 : X \to \mathbb{R}$ are linearly dependent differentiable functions, then there are constants $C_1, C_2 \in \mathbb{R}$ not both equal to 0 such that $C_1 y_1(x) + C_2 y_2(x) = 0$ for all $x \in X$. Differentiating this, we get a second equation $C_1 y_1'(x) + C_2 y_2'(x) = 0$ for all $x \in X$. This means the derivatives of the solutions are also linearly dependent.

We can then put these two equations in a matrix form as:

$$\begin{pmatrix} y_1(x) & y_2(x) \\ y_1'(x) & y_2'(x) \end{pmatrix} \begin{pmatrix} C_1 \\ C_2 \end{pmatrix} = \begin{pmatrix} 0 \\ 0 \end{pmatrix} \quad \text{for all } x \in X.$$

Since C_1, C_2 are not both zero, this says the matrix equation:

$$\begin{pmatrix} y_1(x) & y_2(x) \\ y_1'(x) & y_2'(x) \end{pmatrix} \begin{pmatrix} u \\ v \end{pmatrix} = \begin{pmatrix} 0 \\ 0 \end{pmatrix} \quad \text{for all } x \in X, \tag{14.19}$$

has at least two distinct solutions, namely the trivial solution $\binom{u}{v} = \binom{0}{0}$ and a non-trivial solution $\binom{u}{v} = \binom{C_1}{C_2} \neq \binom{0}{0}$. From linear algebra, this means the 2×2 matrix in Eq. (14.19) is not invertible for every $x \in X$ and hence its determinant is always 0. We have a special name for this determinant:

Definition 14.4.19 (Wronskian) Let $y_1, y_2 : X \to \mathbb{R}$ be two differentiable functions defined on $X \subseteq \mathbb{R}$. The Wronskian $W_{y_1,y_2} : I \to \mathbb{R}$ for these functions is defined as the determinant:

$$W_{y_1,y_2}(x) = \det \begin{pmatrix} y_1(x) & y_2(x) \\ y_1'(x) & y_2'(x) \end{pmatrix} = y_1(x)y_2'(x) - y_2(x)y_1'(x).$$

The Wronskian was introduced by Józef Hoene-Wroński (1776–1853) and given that name by Thomas Muir (1844–1934). From the discussions above, the following results are immediate:

Lemma 14.4.20 *Let $y_1, y_2 : X \to \mathbb{R}$ be two differentiable functions.*

1. *If y_1 and y_2 are linearly dependent, then the Wronskian vanishes for all $x \in I$.*
2. *If the Wronskian does not vanish identically on X, then y_1 and y_2 are linearly independent.*

Remark 14.4.21 We make some remarks regarding Lemma 14.4.20:

1. The second statement in the lemma is simply the contrapositive of the first.
2. It is very important to notice that that the first assertion Lemma 14.4.20 is strictly only a one-way implication. If the Wronskian vanishes everywhere, that does not mean that the functions are linearly dependent. In other words, if the two functions are linearly independent, it may still be plausible that their Wronskian vanishes identically. We shall see examples of this in Exercises 14.25 and 14.26.
3. We can also extend the definition of Wronskian and Lemma 14.4.20 to more collection of functions. If we have k functions which are all $k-1$ times differentiable, we differentiate the functions $k-1$ times and insert the j-th derivatives of the functions as the $(j+1)$-th row of the matrix for $j =$

$0, 1, \ldots, k - 1$. This is to ensure that we have a square matrix for which a determinant can be computed. However, here we shall be interested with the case of two functions only.

Example 14.4.22 Recall the functions $y_1(x) = \sin(x)$ and $y_2(x) = x\sin(x)$ defined on $\mathbb{R}$ that we saw in Example 14.4.18. We have shown that these functions are linearly independent to each other manually. Another way to show this is by using the Wronskian and Lemma 14.4.20. We compute:

$$W_{y_1,y_2}(x) = \det\begin{pmatrix} \sin(x) & x\sin(x) \\ \cos(x) & \sin(x) + x\cos(x) \end{pmatrix} = \sin^2(x),$$

which does not vanish identically on $\mathbb{R}$. Thus, Lemma 14.4.20 says that y_1 and y_2 must be linearly independent.

The Wronskian is very important in the study of ODEs because they could also help us build independent solutions. We can show that the Wronskian of solutions to a linear homogeneous ODE either vanishes identically or is non-zero everywhere. This is called Abel's identity and the readers shall prove this result for the second order linear homogeneous ODEs in Exercise 14.28.

Theorem 14.4.23 (Abel's Identity) *Let $a, b : \mathbb{R} \to \mathbb{R}$ be some functions. Suppose that we have a linear homogeneous second order ODE:*

$$\frac{d^2y}{dx^2} + a(x)\frac{dy}{dx} + b(x)y = 0.$$

If $y_1, y_2 : I \to \mathbb{R}$ are any classical solutions of this ODE defined on some interval $I \subseteq \mathbb{R}$, then the Wronskian $W : I \to \mathbb{R}$ for these solutions satisfies the first order linear ODE:

$$\frac{dW}{dx} + a(x)W = 0.$$

Moreover, if $a(x)$ has an antiderivative, then W is either identically 0 or nowhere zero on I.

Now we would like to build up a general solution to a second order linear homogeneous ODE with linearly independent solutions only. So we address a question that we posed earlier: how many linearly independent solutions do we need to express the general solution of the ODE? For second order ODEs with nice enough coefficients, we just need two! Before we prove the main result, let us prove an interesting and useful lemma.

In general, as we have discussed in Remark 14.4.21, vanishing Wronskian of two functions does not imply linear dependence. However, if these functions are solutions to the same ODE with well-behaved coefficients (for example, continuous coefficients), then vanishing Wronskian is equivalent to linear dependence of the solutions. This is due to the following lemma:

Lemma 14.4.24 *Let $a, b : \mathbb{R} \to \mathbb{R}$ be continuous functions. Suppose that we have a linear homogeneous second order ODE:*

$$\frac{d^2y}{dx^2} + a(x)\frac{dy}{dx} + b(x)y = 0, \tag{14.20}$$

and suppose that $y_1, y_2 : I \to \mathbb{R}$ are two non-trivial classical solutions to the ODE. The solutions y_1 and y_2 are linearly dependent if and only if their Wronskian W_{y_1,y_2} is zero somewhere.

Proof We prove only the backward implication here since the forward implication is true by Lemma 14.4.20.

($\Leftarrow$) : Suppose that $W_{y_1,y_2}(x) = 0$ at some point $x = x_0$. Then, by linear algebra, the matrix equation:

$$\begin{pmatrix} y_1(x_0) & y_2(x_0) \\ y_1'(x_0) & y_2'(x_0) \end{pmatrix} \begin{pmatrix} u \\ v \end{pmatrix} = \begin{pmatrix} 0 \\ 0 \end{pmatrix},$$

has a non-trivial solution $\binom{u}{v} = \binom{C_1}{C_2} \neq \binom{0}{0}$. Set $y(x) = C_1 y_1(x) + C_2 y_2(x)$. Thus, the function y satisfies the ODE (14.20), $y(x_0) = 0$, and $y'(x_0) = 0$. Now notice that the ODE IVP:

$$y'' + a(x)y' + b(x)y = 0 \quad \text{with} \quad y(x_0) = 0,\ y'(x_0) = 0,$$

has an obvious solution, namely the zero solution $y(x) = 0$. Since $y(x) = C_1 y_1(x) + C_2 y_2(x)$ is also such a solution, by uniqueness in Theorem 14.4.14, these two solutions coincide, namely $0 = y(x) = C_1 y_1(x) + C_2 y_2(x)$. This implies the solutions y_1 and y_2 are linearly dependent since not both C_1 and C_2 are zero. □

Now we prove:

Theorem 14.4.25 *Let $a, b : \mathbb{R} \to \mathbb{R}$ be continuous functions. Suppose that we have a linear homogeneous second order ODE:*

$$\frac{d^2y}{dx^2} + a(x)\frac{dy}{dx} + b(x)y = 0, \tag{14.21}$$

If $y_1, y_2 : I \to \mathbb{R}$ are two linearly independent non-trivial classical solutions of this ODE defined on some interval $I \subseteq \mathbb{R}$, then any other solution of this ODE must be of the form $y = C_1 y_1 + C_2 y_2$ for some constants $C_1, C_2 \in \mathbb{R}$.

Proof Let $A(x)$ be an antiderivative of $a(x)$. For any two solutions u and v to the ODE (14.21), Abel's identity in Theorem 14.4.23 implies that $W_{u,v}(x)e^{A(x)} = C$ for some constant $C \in \mathbb{R}$. By Lemma 14.4.24, $C = 0$ if the two solutions are linearly dependent and $C \neq 0$ if they are linearly independent.

Now pick any solution y of the ODE. We can thus generate three equations:

$$(y_1'y - y_1y')e^{A(x)} = W_{y,y_1}(x)e^{A(x)} = D_1, \tag{14.22}$$

$$(y_2'y - y_2y')e^{A(x)} = W_{y,y_2}(x)e^{A(x)} = D_2, \tag{14.23}$$

$$(y_2'y_1 - y_2y_1')e^{A(x)} = W_{y_1,y_2}(x)e^{A(x)} = D_3 \neq 0, \tag{14.24}$$

for some constants $D_1, D_2, D_3 \in \mathbb{R}$. Note that D_3 is necessary non-zero since y_1 and y_2 are linearly independent. Multiply Eqs. (14.22) and (14.23) with y_2 and y_1 respectively, then take their difference to get $y(y_1'y_2 - y_2'y_1)e^{A(x)} = D_1y_2 - D_2y_1$. Finally, substituting (14.24) in this equation, we obtain:

$$yD_3 = D_1y_2 - D_2y_1 \quad \Rightarrow \quad y = \frac{D_1}{D_3}y_2 - \frac{D_2}{D_3}y_1 = C_1y_1 + C_2y_2,$$

for some constants $C_1, C_2 \in \mathbb{R}$. □

Therefore, it is enough to find two linearly independent solutions to determine the general solution to a second order linear homogeneous ODE. This is an amazing fact which simplifies our lives considerably!

For higher order ODEs, we have analogous results which says if the linear ODE is of order n, then n linearly independent solutions determine the general solution of the ODE. These collections of linearly independent solutions are called fundamental solutions to the ODE. In the language of linear algebra, they form a basis for the solution space of the ODE.

The next question is: how can we find these fundamental solutions? For a second order linear ODE, even though we just need to find two linearly independent solutions to generate the general solution, this may still be a difficult task! At this stage, to find the fundamental solutions, one has to make some lucky guesses.

Example 14.4.26 Let us look at some examples:

1. Suppose that we have a second order linear homogeneous ODE with constant coefficients $y''(x) - 2y'(x) - 3y(x) = 0$ where $x \in \mathbb{R}$. We want to find fundamental solutions to this equation.
   We make a guess that a fundamental solution has the form $y(x) = e^{rx}$ for some constant $r \in \mathbb{R}$ which needs to be determined. Differentiating, we get $y' = re^{rx}$ and $y'' = r^2e^{rx}$. Substituting this in the ODE, we get $r^2e^{rx} - 2re^{rx} - 3e^{rx} = 0$, namely $e^{rx}(r^2 - 2r - 3) = 0$.

Since the exponential term is never 0, we must have $r^2 - 2r - 3 = 0$ which implies $r = -3$ or $r = 1$. So we have two candidate solutions for the ODE, namely e^{-3x} or e^x, both defined on the whole of $\mathbb{R}$. We can verify that they are indeed solutions to the ODE by substituting them in the ODE.
Now we hope that these solutions are linearly independent so that they would form the set of fundamental solutions. We compute their Wronskian:

$$W(x) = \det\begin{pmatrix} e^{-3x} & e^x \\ -3e^{-3x} & e^x \end{pmatrix} = e^{-3x}e^x + 3e^{-3x}e^x = 4e^{-2x},$$

which does not vanish identically on $\mathbb{R}$. Hence, these solutions are linearly independent and thus, by Theorem 14.4.25, a general solution to the ODE is $y(x) = C_1e^{-3x} + C_2e^x$ for some constants $C_1, C_2 \in \mathbb{R}$.
The quadratic polynomial $r^2 - 2r + 3 = 0$ above is called the auxiliary or characteristic equation for the ODE $y'' - 2y' - 3y = 0$. To get the auxiliary equation, we simply replace y'', y', and y with the numbers r^2, r, and 1. As we can infer from the above, the roots of this auxiliary equation can provide us with solutions to the ODE.
Indeed, if there are two distinct solutions to the quadratic polynomial, either complex or reals, we can immediately construct two linearly independent solutions to the ODE. Supposing $r_1 \neq r_2$ are solutions to the auxiliary equation, then the general solution to the ODE is $C_1e^{r_1x} + C_2e^{r_2x}$.
However, if it has a repeated root $r_1 = r_2$, this method only gives us one solution to the ODE, namely $C_1e^{r_1x}$. As we have seen in Theorem 14.4.25, we need two linearly independent solutions to create the general solution for a second order ODE. How can we find the second solution? We shall see how to work with the case of repeated roots in Exercise 14.29 by creating another solution which is linearly independent to this first solution.

2. Let $y''(x) + 9y(x) = 0$ be an ODE. The auxiliary equation for this ODE is $r^2 + 9 = 0$ which has complex roots $r = \pm 3i$. Therefore, a general solution to this ODE is $y(x) = C_1e^{ix} + C_2e^{-ix}$ for some constants $C_1, C_2 \in \mathbb{C}$. Since $e^{i3x} = \cos(3x) + i\sin(3x)$, we can rewrite this solution as:

$$\begin{aligned} y(x) = C_1e^{3ix} + C_2e^{-3ix} &= C_1(\cos(3x) + i\sin(3x)) + C_2(\cos(3x) - i\sin(3x)) \\ &= A_1\cos(3x) + A_2\sin(3x), \end{aligned}$$

for some constants $A_1, A_2 \in \mathbb{R}$. We can also check directly that $\sin(3x)$ and $\cos(3x)$ solve the equation and are linearly independent. This can be seen by guessing a function that when differentiated twice would yield some multiple of the original function, as the ODE suggests. These functions are, of course, the sine and cosine functions which are linearly independent.

Theorem 14.4.25 only tells us how to solve a linear homogeneous second order ODE. Of course, there are other different types and higher order ODEs out there that we could not cover in this introductory section. We did not even cover the topic of inhomogeneous second order ODE in this section! Hopefully the readers shall see more of them in a proper course on differential equations in the future.

Later on, in Chap. 16, we shall see more powerful tools that could help us find antiderivatives of some functions in systematic ways. This will then enable us to solve more complicated ODEs than the ones that we have seen here without relying too much on lucky guesswork.

Exercises

14.1 ($*$) Let $f : [0, \infty) \to \mathbb{R}$ be a function twice differentiable on $(0, \infty)$ with $f(0) = 0$ and $f''(x) \geq 0$ for $x > 0$. Show that the function $g : [0, \infty) \to \mathbb{R}$ defined as $g(x) = \frac{f(x)}{x}$ is an increasing function.

14.2 ($*$) Prove the following result:

Proposition 14.5.27 *Let $f, g : [0, \infty) \to \mathbb{R}$ be continuous functions on $[0, \infty)$ and differentiable on $(0, \infty)$ with $f'(x) \leq g'(x)$ for all $x > 0$ and $f(0) \leq g(0)$. Then, $f(x) \leq g(x)$ for all $x \geq 0$.*

Hence, prove the following inequalities:

(a) $x \leq \tan(x)$ on $[0, \frac{\pi}{2})$.
(b) $1 - \frac{x^2}{2} \leq \cos(x)$ on $\mathbb{R}$.
(c) $x - \frac{x^3}{6} \leq \sin(x)$ on $\mathbb{R}$.
(d) $x - \frac{x^2}{2} \leq \ln(1 + x) \leq x - \frac{x^2}{2} + \frac{x^3}{3}$ on $[0, \infty)$.
(e) $1 - x^2 \leq e^{-x^2} \leq \frac{1}{1+x^2}$ on $[0, \infty)$.

14.3 Let $f : (0, \infty) \to \mathbb{R}$ be a function defined as $f(x) = \frac{\ln(x)}{x}$.

(a) Show that f is decreasing for $x \geq e$.
(b) Show that for any $a, b \in \mathbb{R}$ with $e \leq a < b$, we have $b^a \leq a^b$.
Is this true if $a < e$?

14.4 ($*$) Let $f : I \to \mathbb{R}$ be a differentiable function on $I = (a, b)$.

(a) Write down the equation $y = g_{x_0}(x)$ for the tangent line to the graph at x_0.
(b) Prove that f is convex on I if and only if for any $x_0 \in I$, the graph of the function f lies above the tangent line to the graph at x_0 (in other words, for any $x_0 \in I$, we have $f(x) \geq g_{x_0}(x)$ for all $x \in I$).

14.5 ($*$) Analyse and sketch the graph of the following functions on $\mathbb{R}$:

(a) $f(x) = x(x^2 - 1)$.
(b) $f(x) = (x^2 - 1)^3$.
(c) $f(x) = x(x - 2)^{\frac{2}{3}}$.
(d) $f(x) = x + e^x$.
(e) $f(x) = \cos(2x) - x$.
(f) $f(x) = \frac{8a^3}{x^2+4a^2}$ for some $a > 0$.

The graph of this function is called the "witch of Agnesi", named after Maria Gaetana Agnesi (1718–1799) who studied it extensively in her book *Instituzioni Analitiche ad uso della Gioventù Italiana* (Analytical Institutions for the Use of Italian Youth). The name of this curve might have been a mistranslation since the original word used to describe it by Grandi was *versiera* derived from the Latin *vertere* which has the meaning "to turn". This word was used to describe this curve due to its construction via a rotating circle. On the other hand, a similarly sounding Italian word *versiero* means "female devil" or "witch". Whilst the curve itself is not particularly evil, the name "witch of Agnessi" probably carries more street cred and swagger. So, the mathematical community decided to affectionately keep this amusing mistranslation.

14.6 (◇) Let $x, y : \mathbb{R} \to \mathbb{R}$ be two functions defined via a parameter $t \in \mathbb{R}$ as:

$$x(t) = t - \sin(t),$$
$$y(t) = 1 - \cos(t).$$

(a) Show that the function $x(t)$ is strictly increasing over $\mathbb{R}$. Hence, deduce that $x(t)$ is bijective.
(b) Using part (a), explain why the function y can be expressed implicitly as a continuous function of x. Denote this function $y : \mathbb{R} \to \mathbb{R}$ as $y(x)$.
(c) Prove that $y(x)$ is 2π-periodic, namely $y(x + 2\pi) = y(x)$ for all $x \in \mathbb{R}$.
(d) Explain why $y(x) \in [0, 2]$ for all $x \in \mathbb{R}$. Show that $\cos(x + \sqrt{y(x)(2 - y(x))}) + y(x) = 1$ for all $x \in \mathbb{R}$.

Since $y(x)$ is 2π-periodic, let us focus our attention to the x in the interval $[0, 2\pi]$.

(e) Determine all the points $x \in [0, 2\pi]$ for which $y(x) = 0, 2$.
(f) Using implicit differentiation with respect to x, show that for $y \neq 0, 2$ we have:

$$y'(x) = \sin(x + \sqrt{y(2 - y)}) \left(1 + \frac{(1 - y)y'}{\sqrt{y(2 - y)}}\right).$$

(g) Hence, prove that $y'(x)$ exists whenever $y \neq 0, 2$ with:

$$y'(x) = \frac{\sqrt{y(2 - y)} \sin(x + \sqrt{y(2 - y)})}{\sqrt{y(2 - y)} - (1 - y) \sin(x + \sqrt{y(2 - y)})} \in \mathbb{R}.$$

(h) Using part (f) show that $y'(x) \neq 0$ for $x \in (0, \pi) \cup (\pi, 2\pi)$. Deduce that $y(x)$ is strictly increasing over $(0, \pi)$ and strictly decreasing over $(\pi, 2\pi)$.
(i) Using part (g) show that $(y'(x))^2 = \frac{2}{y} - 1$ for $y \neq 0, 2$.
(j) Determine the limits $\lim_{x \downarrow 0} y'(x)$, $\lim_{x \uparrow \pi} y'(x)$, $\lim_{x \downarrow \pi} y'(x)$, and $\lim_{x \uparrow 2\pi} y'(x)$.

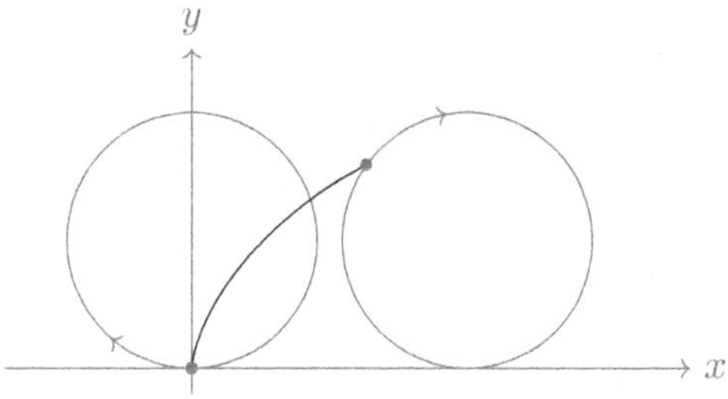

Fig. 14.10 A cycloid is a curve traced by the red point on a circle as the circle rolls on the x-axis. The readers were asked to complete the diagram in Exercise 14.6(l)

Deduce that $y'(\pi) = 0$ and the graph of $y(x)$ has cusps at the points for which $y(x) = 0$.

(k) By investigating the second derivative of $y(x)$ using part (i), show that y is concave whenever $y \neq 0, 2$.

(l) Sketch the graph of $y(x)$.

The curve determined by the parametrisation $(x(t), y(t))$ in this question is called a cycloid. It is a path traced by a point on a circle of radius 1 as the circle rolls along a flat surface. See Fig. 14.10 for reference.

14.7 (a) Let $f : \mathbb{R} \to \mathbb{R}$ be defined as $f(x) = x \sin(\frac{1}{x})$ for $x \neq 0$ and $f(0) = 0$. Prove that this function is continuous but not differentiable at $x = 0$.

(b) Now let $f : \mathbb{R} \to \mathbb{R}$ be defined as $f(x) = x^2 \sin(\frac{1}{x})$ for $x \neq 0$ and $f(0) = 0$. Prove that this function is differentiable everywhere but $f \notin C^1(\mathbb{R})$.

(c) Now let $f : \mathbb{R} \to \mathbb{R}$ be defined as $f(x) = x^3 \sin(\frac{1}{x})$ for $x \neq 0$ and $f(0) = 0$. Prove that this function is differentiable at $x = 0$ and determine the value of $f'(0)$.
Thus, show that $f \in C^1(\mathbb{R})$.

(d) Using Exercise 10.18, show that there is a sequence of positive critical points (x_n) for the function f in part (c) such that (x_{2n-1}) are nonzero local minimum points, (x_{2n}) are nonzero local maximum points of f, and $x_n \to 0$.

(e) Conclude that $x = 0$ cannot be a local extremum point for the function f in part (c).

14.8 $(*)$ Define a sequence of functions (f_n) where $f_n : \mathbb{R} \to \mathbb{R}$ as $f_n(x) = \frac{x}{1+nx^2}$.

(a) Find the pointwise limiting function $f : \mathbb{R} \to \mathbb{R}$.

(b) Find the pointwise limit $\lim_{n\to\infty} f_n'(x)$.
Is this convergence uniform on $\mathbb{R}$?

14.9 (∗) Prove that the series $\sum_{j=1}^{\infty} \frac{\cos(2^j x)}{3^j}$ converges uniformly on $\mathbb{R}$. Show further that the limit is differentiable and find its derivative.

14.10 Consider the functions series $\sum_{j=1}^{\infty} \frac{j}{j^4+x^4}$.
(a) Show that this functions series converges uniformly on $\mathbb{R}$.
(b) Find the derivative of this functions series. Explain your reasoning.

14.11 (∗) Find a closed form for the following power series:
(a) $\sum_{j=0}^{\infty} jx^j$ for $|x| < 1$.
(b) $\sum_{j=0}^{\infty} j^2 x^j$ for $|x| < 1$.
(c) $\sum_{j=0}^{\infty} \frac{(j+1)x^j}{j!}$ for $x \in \mathbb{R}$.
Hence, determine the values of the following series: $\sum_{j=0}^{\infty} \frac{j}{2^j}$, $\sum_{j=0}^{\infty} \frac{j^2}{2^j}$, and $\sum_{j=0}^{\infty} \frac{j+1}{j!}$.

14.12 (∗) Recall from Exercise 12.9 the Bessel function of order $p \in \mathbb{N}_0$ which was defined as:

$$J_p(x) = \sum_{j=0}^{\infty} \frac{(-1)^j}{j!(j+p)!} \left(\frac{x}{2}\right)^{2j+p}.$$

We have seen that this series converges pointwise everywhere on $\mathbb{R}$. Show that for all $x \in \mathbb{R}$, the following identities hold:
(a) $J_0'(x) = -J_1(x)$.
(b) For $p \in \mathbb{N}$ we have $J_p'(x) = \frac{J_{p-1}(x) - J_{p+1}(x)}{2}$.
(c) $x^2 J_p''(x) + xJ_p'(x) + (x^2 - p^2)J_p(x) = 0$.
The final identity tells us that the Bessel function satisfies a certain second order ODE with an initial condition $J_p(0) = 0$. The integer parameter p can also be made more general using the gamma function in place of the factorials, which we shall see in Exercise 16.20.

14.13 Is there a real power series centred at $x = 0$ with non-zero radius of convergence that converges to the function $f : [-1, 1] \to \mathbb{R}$ defined as $f(x) = |x|$?

14.14 (∗) In Exercise 12.11, we have seen that the following power series:

$$C(x) = \sum_{j=0}^{\infty} \frac{(-1)^j}{(2j)!} x^{2j} \quad \text{and} \quad S(x) = \sum_{j=0}^{\infty} \frac{(-1)^j}{(2j+1)!} x^{2j+1},$$

converge pointwise on $\mathbb{R}$ and uniformly on any subset $[-r, r] \subseteq \mathbb{R}$ where $r > 0$. We suspect that they are related to the cosine and sine functions (in fact, they could even be equal). Let us gather more evidence that could point us towards this.
(a) Find the term-wise differentiated power series for $C(x)$ and $S(x)$.
(b) Deduce that $C'(x) = -S(x)$ and $S'(x) = C(x)$ everywhere on $\mathbb{R}$.
We shall the the answers in Example 17.2.7(1) and Exercise 17.7.

Fig. 14.11 The function F

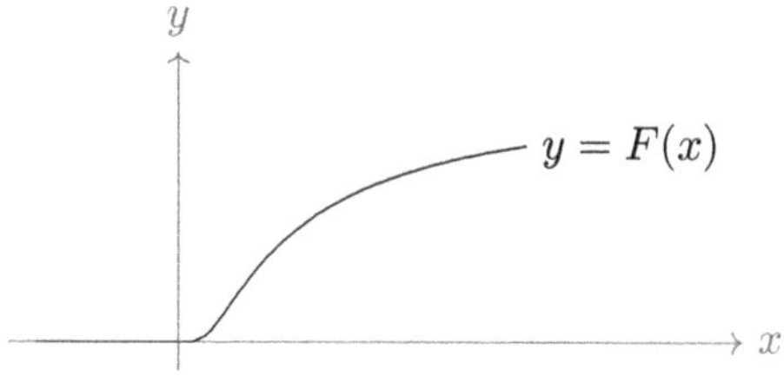

14.15 ($*$) Classify the following indeterminate forms. Hence find the following limits:

(a) $\lim_{x\to 0}\frac{1-\cos(x)}{e^x-1}$.

(b) $\lim_{x\to 0}(1+x)^{\frac{1}{x}}$.

(c) $\lim_{x\to e}\frac{\ln(\ln(x))}{\sin(x-e)}$

(d) $\lim_{x\to 1}\frac{\sin(\pi x)}{\ln(x)}$.

(e) $\lim_{x\to\infty}\left(\frac{x+1}{x-1}\right)^{\sqrt{x^2-1}}$.

(f) $\lim_{x\to-\infty} e^{x^2}\sin(e^x)$.

14.16 In Exercise 9.24, we have proven that for any $k>0$, we have $\lim_{x\to\infty} x^{\frac{1}{x^k}} = 1$ and $\lim_{x\to\infty}\frac{\ln(x)}{x^k}=0$. Prove these limits using L'Hôpital's rule.

14.17 ($*$) Let $P:\mathbb{R}\to\mathbb{R}$ be any polynomial with real coefficients. By using induction on the degree of P, prove that $\lim_{x\to\infty}P(x)e^{-x}=0$.

14.18 ($*$) Define a function $F:\mathbb{R}\to\mathbb{R}$ as $F(x)=e^{-\frac{1}{x}}$ for $x>0$ and 0 for $x\le 0$. The graph of this function is given in Fig. 14.11.

(a) Prove that $F\in C^1(\mathbb{R})$ and find its derivative.

(b) By induction, show that the n-th derivative of the function F at $x>0$ is given by $F^{(n)}(x)=F(x)P_n(\frac{1}{x})$ where $P_n:\mathbb{R}\to\mathbb{R}$ is some real polynomial of degree n.

(c) Hence, show that for all $n\in\mathbb{N}$ the function F is also n-times differentiable at $x=0$ with $F^{(n)}(0)=0$.

(d) Deduce that the function F is smooth everywhere.

14.19 ($*$) Show that the function $\Psi:\mathbb{R}\to\mathbb{R}$ defined as:

$$\Psi(x)=\begin{cases} e^{-\frac{1}{1-x^2}} & \text{for } -1<x<1,\\ 0 & \text{otherwise,}\end{cases}$$

is also smooth. The graph of this function is depicted in Fig. 14.12.

This function is called the bump function and is used in many different areas of mathematics. We shall see an application of the bump function in Exercise 20.25.

14.20 ($*$) Let $f:\mathbb{R}\to\mathbb{R}$ be defined as $f(x)=xe^x$.

(a) Find the limits $\lim_{x\to\infty}f(x)$ and $\lim_{x\to-\infty}f(x)$.

Fig. 14.12 Bump function Ψ

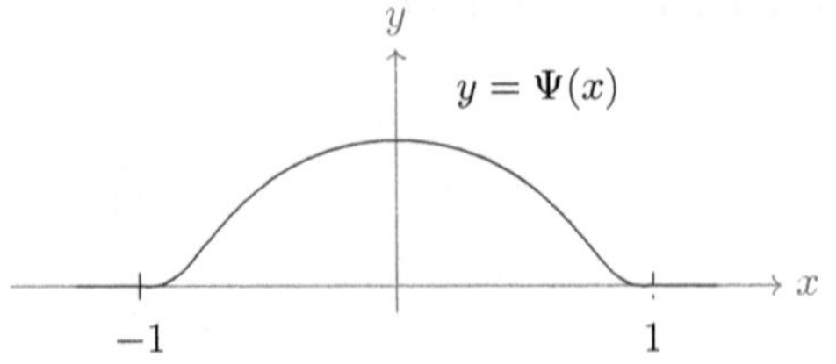

(b) Show that there is only one critical point x_0 for the function f and classify it.
(c) Determine the monotonicity behaviour of the function f in the regions $(-\infty, x_0)$ and (x_0, ∞) of the domain.
(d) Hence, for each $k \in \mathbb{R}$ determine the number of solutions to the equation $xe^x = k$.

The inverse for the function f above is then a multivalued function $w : [f(x_0), \infty) \to \mathbb{R}$. This multivalued function can be made into a genuine function by a process called branch-cutting. This function is called the Lambert W function (named after Johann Heinrich Lambert (1728–1777)) or product logarithm.

14.21 Let $f, g : (-1, 1)\backslash\{0\} \to \mathbb{R}$ be functions defined as $f(x) = \ln(1-x)-\sin(x)$ and $g(x) = 1 - \cos^2(x)$.
(a) Show that the limit $\lim_{x\to 0} \frac{f(x)}{g(x)}$ is of the indeterminate form $\frac{0}{0}$.
(b) Show that the limit $\lim_{x\to 0} \frac{f'(x)}{g'(x)}$ does not exist in $\mathbb{R}$.
Hence, we cannot use the L'Hôpital's rule in Theorem 14.3.5 to determine the limit in (a). Instead, we consider the limit of the reciprocal, namely $\lim_{x\to 0} \frac{g(x)}{f(x)}$.
(c) Show carefully that $\lim_{x\to 0} \frac{g(x)}{f(x)} = 0$.
(d) Thus, deduce that $\lim_{x\uparrow 0} \frac{f(x)}{g(x)} = \infty$ and $\lim_{x\uparrow 0} \frac{f(x)}{g(x)} = -\infty$.
Hence, the limit $\lim_{x\to 0} \frac{f(x)}{g(x)}$ does not exist. This question appeared in the penultimate showdown in the film *Mean Girls*.

14.22 (∗) Find the antiderivatives of the following functions:
(a) $f(x) = x^2 e^{x^3}$ for $x \in \mathbb{R}$.
(b) $f(x) = (x^3 + x)(x^4 + 2x^2 + 5)^3$ for $x \in \mathbb{R}$.
(c) $f(x) = \ln(x)$ for $x > 0$.
(d) $f(x) = \sinh(x)$ for $x \in \mathbb{R}$.
(e) $f(x) = x^2 \sin(x) - 2x\cos(x)$ for $x \in \mathbb{R}$.
(f) $f(x) = \sin(x)e^{\cos(x)} + x^5$ for $x \in \mathbb{R}$.
(g) $f(x) = \tan(x)$ for $|x| < \frac{\pi}{2}$.
(h) $f(x) = x\tan(x^2)$ for $|x| < \sqrt{\frac{\pi}{2}}$.
(i) $f(x) = \sec(x)$ for $|x| < \frac{\pi}{2}$.
(j) $f(x) = |x|$ for $x \in \mathbb{R}$.

14.23 (∗) Find the general solution to the following first order linear ODEs:

(a) $\frac{dy}{dx} + y = 1 + e^x$.

(b) $x\frac{dy}{dx} + y = \sqrt{x}$ for $x > 0$.

(c) $\frac{dy}{dx} + \cos(x)y = 0$.

(d) $\frac{dy}{dx} + 3x^2y = x^2$.

(e) $\frac{dy}{dx} + y = 2\cosh(x)$.

(f) $\frac{dy}{dx} + \frac{3}{x}y = x^2$ for $x > 0$.

14.24 Existence and uniqueness of some ODE problems can be a very big issue, even for a very simple ODE! Consider the following non-linear first order ODE IVP:

$$\frac{dy}{dx} = y^{\frac{1}{3}} \quad \text{with} \quad y(0) = 0. \tag{14.25}$$

Show that $y(x) = 0$, $y(x) = \left(\frac{2x}{3}\right)^{\frac{3}{2}}$, and $y(x) = -\left(\frac{2x}{3}\right)^{\frac{3}{2}}$ are all solutions to the ODE IVP (14.25) on $[0, \infty)$.

Thus, we have existence but not uniqueness of the solution for the ODE IVP (14.25).

14.25 (∗) Lemma 14.4.20 says that if two functions are linearly dependent, then their Wronskian vanishes everywhere. By considering the functions $y_1(x) = x^2$ and $y_2(x) = |x|x$ defined on $\mathbb{R}$, show that the converse is not true.

14.26 (a) Construct two smooth functions defined on $\mathbb{R}$ which are linearly independent but their Wronskian vanishes identically.

(b) Construct two smooth linearly independent functions on $\mathbb{R}$ which are zero nowhere but there exists a subset $I \subseteq \mathbb{R}$ for which the restriction of these functions to I are linearly dependent.

14.27 Let $y_1, y_2 : I \to \mathbb{R}$ where $I \subseteq \mathbb{R}$ be linearly independent solutions to the second order linear ODE $y'' + a(x)y' + b(x)y = 0$. Show that the functions $y_3, y_4 : I \to \mathbb{R}$ defined as $y_3 = y_1 + y_2$ and $y_4 = y_1 - y_2$ respectively are also linearly independent.

14.28 (a) Prove Abel's identity in Theorem 14.4.23.

(b) Deduce that the Wronskian W_{y_1,y_2} is independent of the functions y_1 and y_2.

14.29 (∗) Consider the second order linear homogeneous ODE $y'' - 4y' + 4y = 0$.

(a) Find a solution for this ODE which is of the form $y_1 = e^{rx}$ for some constant r.

Now we want to find another solution linearly independent to the first solution by using Abel's identity.

(b) Using Abel's identity, show that a Wronskian for this ODE is $W(x) = Ce^{4x}$ for some constant $C \in \mathbb{R}$.

(c) Assume y_2 is another solution which is linearly independent from y_1, show that $y_1y_2' - y_1'y_2 = Ce^{4x}$ for some $C \in \mathbb{R} \setminus \{0\}$.

(d) Derive a first order linear ODE for y_2 and solve it.
(e) Hence, deduce that a general solution for the ODE is $y = C_1e^{2x} + C_2xe^{2x}$ for constants $C_1, C_2 \in \mathbb{R}$.

14.30 ($\diamond$) In fact, the construction in Exercise 14.29 is also true in more generality, namely: for any second order homogeneous linear ODE, if we know one solution for the ODE, we can always generate another linearly independent solution to it using Abel's identity.
Consider the following linear second order homogeneous ODE:

$$y'' - \frac{(x+2)}{x}y' + \frac{(x+2)}{x^2} = 0 \quad \text{for } x > 0. \tag{14.26}$$

(a) Show that $y_1(x) = x$ is a solution to the ODE in (14.26).
(b) Using Abel's identity, show that the Wronskian for this ODE is $W(x) = Cx^2e^x$ for $x > 0$ and some constant $C \in \mathbb{R}$.
(c) Suppose that $y_2 : (0, \infty) \to \mathbb{R}$ is another solution to the ODE (14.26) which is linearly independent to y_1. Using part (b), show that $y_2' - \frac{y_2}{x} = Cxe^x$ for some constant $C \in \mathbb{R} \setminus \{0\}$.
(d) Hence, show that $y_2(x) = Cxe^x + Dx$ for some constants $C, D \in \mathbb{R}$ with $C \neq 0$.
(e) Deduce that the general solution to the ODE (14.26) is given by $y(x) = C_1x + C_2xe^x$ for constants $C_1, C_2 \in \mathbb{R}$.

14.31 ($*$) An Euler equation is a second order linear homogeneous ODE of the form $x^2y'' + axy' + by = 0$ on $x > 0$ for some constants $a, b \in \mathbb{R}$. Usually, a guess for the solution of this ODE is of the form $y = x^n$ for some constant $n \in \mathbb{R}$. Find a set of fundamental solutions to the following ODEs.
(a) $x^2y'' + 2xy' - 12y = 0$.
(b) $2x^2y'' + 3xy' - y = 0$.
(c) $x^2y'' - 3xy' + 4y = 0$.

14.32 Can $y(x) = x\sin(x)$ for $x \in \mathbb{R}$ be a solution of a second order linear homogeneous ODE with constant coefficients on $\mathbb{R}$?

14.33 ($\diamond$) In Example 11.4.18, we have seen the Weierstrass function $f : \mathbb{R} \to \mathbb{R}$ defined as the functions series $f(x) = \sum_{j=0}^{\infty} a^j \cos(b^j\pi x)$ where $a \in (0, 1)$ and b is a positive odd integer such that $ab > 1 + \frac{3\pi}{2}$. We have seen that the series converges uniformly over $\mathbb{R}$ and hence is continuous over $\mathbb{R}$.
Now we are going to prove that the function is differentiable nowhere. Fix a point $x_0 \in \mathbb{R}$. For each $m \in \mathbb{N}$, let $\alpha_m \in \mathbb{Z}$ be the integer obtained from rounding $b^m x_0$ to the nearest integer so that $b^m x_0 - \alpha_m \in [-\frac{1}{2}, \frac{1}{2})$. Define $x_m = b^m x_0 - \alpha_m$ and $y_m = \frac{\alpha_m - 1}{b^m}$.

(a) Show that $y_m < x_0$ for all $m \in \mathbb{N}$ and deduce that $y_m \to x_0$.
(b) Show that for any $m \in \mathbb{N}$ we have:

$$\frac{f(y_m) - f(x_0)}{y_m - x_0} = \sum_{j=0}^{m-1} a^j \frac{\cos(b^j \pi y_m) - \cos(b^j \pi x_0)}{y_m - x_0} + \sum_{j=m}^{\infty} a^j \frac{\cos(b^j \pi y_m) - \cos(b^j \pi x_0)}{y_m - x_0}.$$

Denote:

$$A_m = \sum_{j=0}^{m-1} a^j \frac{\cos(b^j \pi y_m) - \cos(b^j \pi x_0)}{y_m - x_0},$$

$$B_m = \sum_{j=m}^{\infty} a^j \frac{\cos(b^j \pi y_m) - \cos(b^j \pi x_0)}{y_m - x_0}.$$

(c) Prove that $|A_m| \leq \frac{\pi (ab)^m}{ab-1}$ for any $m \in \mathbb{N}$.
(d) Show that $\cos(b^j \pi x_0) = (-1)^{\alpha_m} \cos(b^{j-m} \pi x_m)$ and $\cos(b^j \pi y_m) = (-1)^{\alpha_m - 1}$ for all $j \geq m$.
Hence, prove that $|B_m| \geq \frac{2(ab)^m}{3}$ for any $m \in \mathbb{N}$.
(e) Deduce that $\left| \frac{f(y_m)-f(x_0)}{y_m - x_0} \right|$ diverges to ∞ as $m \to \infty$.
(f) Thus, explain why the derivative of the Weierstrass function f at x_0 cannot exist.

Since $x_0 \in \mathbb{R}$ is arbitrary, we can conclude that the Weierstrass function is differentiable nowhere.

This function was presented by Weierstrass to Berlin Academy in 1872. The reaction was utter chaos. It has been always assumed implicitly that any continuous function is at least differentiable somewhere. But the Weierstrass function is differentiable nowhere even though it is continuous everywhere! Here are some reactions from contemporary mathematicians.

> I turn with terror and horror from this lamentable scourge of functions with no derivatives. - Charles Hermite (1822–1901).
> These functions are an outrage against common sense, an arrogant distraction. - Henri Poincaré (1854–1912).
> If Newton had known about such functions he would have never created calculus. - Émile Picard (1856–1941).

I personally find it very edgy (literally and figuratively). Have a look at Fig. 14.13.

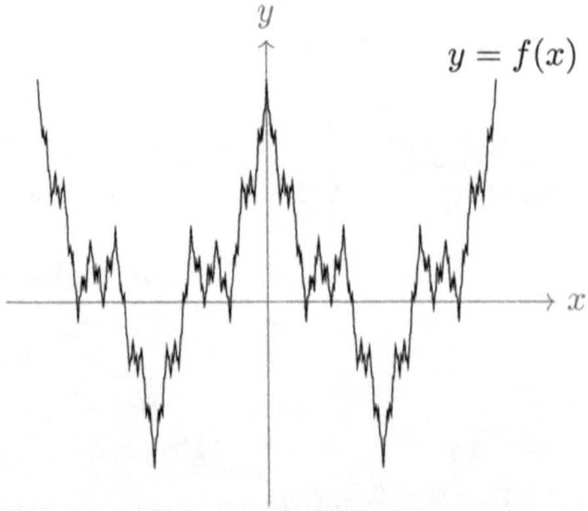

Fig. 14.13 W is for Weierstrass! This is an example of a Weierstrass function. The first term $a\cos(b\pi x)$ in the series provides the general shape of the graph. The terms $a^j\cos(b^j\pi x)$ for larger j (which are cosines with smaller amplitudes but higher frequencies) contribute to the jagged shape of the graph. For carefully selected a and b the function becomes so jagged that it is differentiable nowhere!

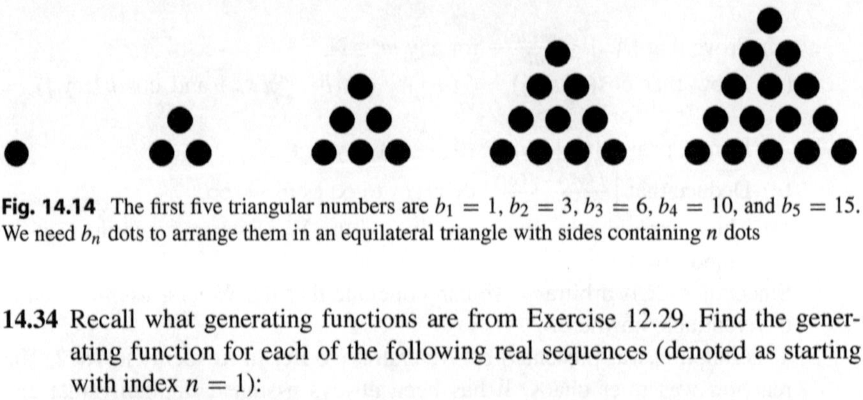

Fig. 14.14 The first five triangular numbers are $b_1 = 1$, $b_2 = 3$, $b_3 = 6$, $b_4 = 10$, and $b_5 = 15$. We need b_n dots to arrange them in an equilateral triangle with sides containing n dots

14.34 Recall what generating functions are from Exercise 12.29. Find the generating function for each of the following real sequences (denoted as starting with index $n = 1$):

(a) (a_n) where $a_n = n$.

(b) (b_n) where $b_n = \binom{n+1}{2} = \frac{n(n+1)}{2}$. These numbers are called the triangular numbers. The reason for this naming can be explained by Fig. 14.14.

(c) (c_n) where $b_n = n^2$.

15 Riemann and Darboux Integration

> *No matter how significant or life-changing your greatest hit or miss might be, neither even begins to define who you are. Each of us is a product of all our experiences and all our interactions with other people. To cite calculus, we are the area under the curve.*
>
> — *Colin Powell, politician and army officer*

Let us move on to a new topic which seems very unrelated to what we have done so far, but we shall see that it does tie in very closely to the previous two chapters. In fact, this connection is extremely important in analysis and mathematics in general to the point that it is referred to as the fundamental theorem of calculus.

Integration of a real-valued function $f : \mathbb{R} \to \mathbb{R}$ is the process of finding the area in the Cartesian plane $\mathbb{R}^2$ bounded between the graph over some compact interval $[a, b] \subseteq \mathbb{R}$ and the x-axis. This process is also called quadratures before the introduction of the modern terminology "integration" by Jacob Bernoulli in 1690. The region over which we want to find the area is called a subgraph and an example is shown in Fig. 15.1.

Definition 15.0.1 (Subgraph) Let $f : X \to \mathbb{R}$ be a function on $X \subseteq \mathbb{R}$. A subgraph $\text{Sub}(f)$ of the function f is the set of points between the graph of f and the x-axis, namely:

$$\begin{aligned}\text{Sub}(f) =& \{(x, y) \in X \times \mathbb{R} : x \in X, 0 \leq y \leq f(x) \text{ if } f(x) \geq 0 \\ &\text{and } 0 \geq y \geq f(x) \text{ if } f(x) < 0\}.\end{aligned}$$

S. Johar, *The Big Book of Real Analysis*,
https://doi.org/10.1007/978-3-031-30832-1_15

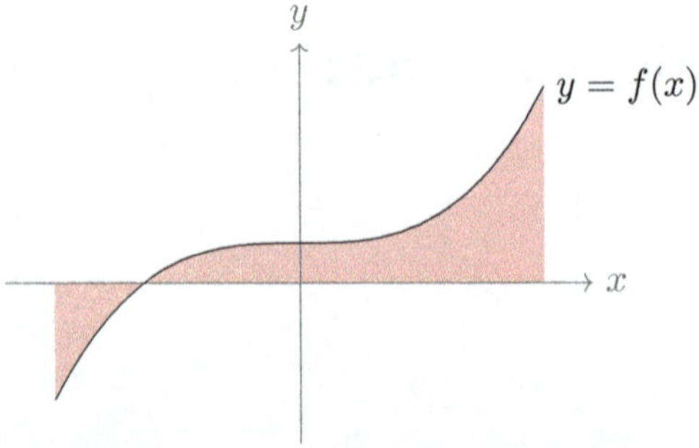

Fig. 15.1 Subgraph of a function f is the shaded region. How can we determine its area?

If the shape of the graph is regular enough, then finding the area under it can be straightforward. For example, the area under the graph of a piecewise constant function or linear function over some finite region $[a, b]$ can be easily found via elementary geometry. All we have to do is use the formula for areas of rectangles, triangles, or trapezium.

However, things get more difficult for more complicated functions. As pointed out by Stillwell in a quote at the beginning of Chap. 6, elementary arithmetic and discrete processes are not sufficient for a complete study of geometry. We have seen the difficulty in determining the arclength of a circle in Section 6.1 where we have to use limits to achieve this.

It is not much different for areas, which is another geometrical concept. The early attempts at integration by Eudoxus (c. 408B.C.–355B.C.) and Archimedes were precisely done in the same way, namely by approximating the area of the subgraph with regular enough shapes of known areas and making them as fine as possible.

For the monomial function $f(x) = x^n$ over an interval $[0, 1]$ where $n = 1, 2, 3, 4$, the area was approximated by al-Haytham using his formula for sum of powers. This was then extended all the way to $n = 9$ by Bonaventura Cavalieri (1598–1647) using Cavalieri's quadrature formula and the method of indivisibles in his work *Geometria indivisibilibus continuorum nova quadam ratione promota* (A Certain Method for the Development of a New Geometry of Continuous Indivisibles). The integration of these functions were also attempted by Fermat using rectangles.

Later, John Wallis extended Cavalieri's formula for negative and rational powers in his book *Arithmetica infinitorum* (The Arithmetic of Infinitesimals) in 1655. However, much like the work of Newton on infinitesimals, this method was non-rigorous and, as a result, was criticised. Wallis's long-time nemesis Thomas Hobbes (1588–1679) wrote a very scathing comment:

> Your scurvy book of *Arithmetica infinitorum*; where your indivisibles have nothing to do, but as they are supposed to have quantity, that is to say, to be divisibles.

which is the exact same criticism that Berkeley threw at Newton.

The major advancement and interest in the study of integrals were achieved by James Gregory (1638–1675), Isaac Barrow (1630–1677), Newton, and Leibniz via the fundamental theorem of calculus, which we shall see in Chap. 16. However, similar to the problem of infinitesimals for differentiation, their ideas lacked rigour since the language of limits were not yet laid out formally at the time.

Finally, the first rigorous treatment of integration using the idea of limits by Bernhard Riemann was published posthumously in 1868. Since then, many other types of integrals based on the various interpretations of area, domain, and type of function that is to be integrated were introduced. Contributors include Gustave Choquet (1915–2006), Jean-Gaston Darboux (1842–1917), Percy J. Daniell (1889–1946), Arnaud Denjoy (1884–1974), Jaroslav Kurzweil (1926–2022), Henri Lebesgue (1875–1941), Alfréd Haar (1885–1933), Ralph Henstock (1923–2007), Kiyosi Itô (1915–2008), Thomas Joannes Stieltjes (1856–1894), Ruslan Stratonovich (1930–1997), and Laurence Chisholm Young (1905–2000).

In this chapter, we are going to properly lay down the modern treatment for the most basic form of integration, namely the Riemann or Darboux integration process, and discuss some theories behind them. In Exercises 15.24–15.32, we are going to investigate an extension of these integrals due to Stieltjes.

15.1 Step Functions

Similar to the idea by Fermat, the intuition behind Riemann or Darboux integral is to slice the subgraph of a function defined on a compact interval into strips and approximate the area of the subgraph by using these rectangular strips. To do this, within each of these strips, the function is then approximated by a suitable constant value. Therefore, with just basic geometrical knowledge of rectangular areas, we can approximate the area under the graph using these rectangular strips.

If the function is bounded, each rectangle has a well-defined area which is simply the product of their side lengths. The total areas of the strips are finite because for bounded functions defined on a closed bounded domain, the rectangles do not get arbitrarily tall or wide. In fact, for functions bounded on $[a, b]$, say $m \leq f(x) \leq M$ for all $x \in [a, b]$ where $m, M > 0$, the total areas of these rectangles are always bounded from above and below by the quantities $M(b-a)$ and $m(b-a)$ respectively. The integral is then defined to be the "limiting" area of these rectangles if we take finer and finer slices.

Now let us properly define this process. The first step in defining an integral from scratch is to define the partition of the domain and step functions which would be our approximating rectangles.

Definition 15.1.1 (Partition of an Interval) Let $[a, b]$ be a compact interval in $\mathbb{R}$. A partition $\mathcal{P}$ of the interval $[a, b]$ is a finite collection of points $\{x_j\}_{j=0}^{n}$ in $[a, b]$ such that $a = x_0 < x_1 < x_2 < \cdots < x_n = b$.

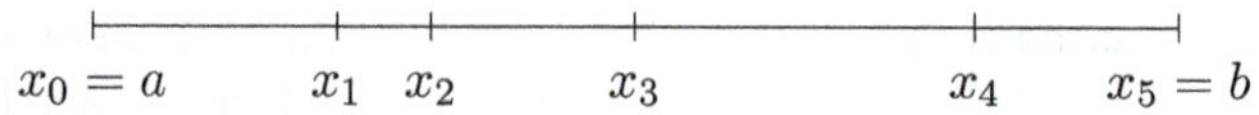

Fig. 15.2 Partition $\mathcal{P}$ of $[a, b]$ with 6 points. In this case, $||\mathcal{P}|| = x_4 - x_3$

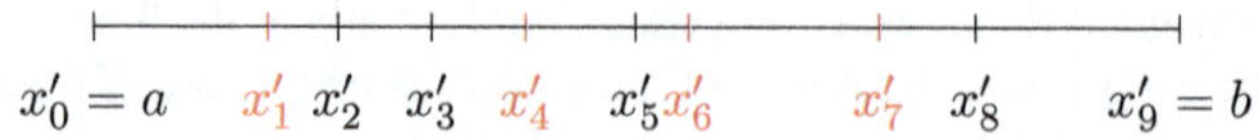

Fig. 15.3 Refinement $\mathcal{P}'$ of partition $\mathcal{P}$ in Fig. 15.2. The newly added points are in red

Definition 15.1.2 (Subintervals of a Partition) Given a compact interval $[a, b] \subseteq \mathbb{R}$ and a partition $\mathcal{P} = \{x_0, x_1, \dots, n\}$ of it, we define the subintervals of $\mathcal{P}$ as the set of intervals $\{I_j = [x_{j-1}, x_j] : j = 1, 2, \dots, n\}$.

Essentially, the partition points are the positions at which we want to slice the subgraphs. For each partition $\mathcal{P}$ of $[a, b]$, we denote $||\mathcal{P}||$ as the size or mesh of the partition, defined as $||\mathcal{P}|| = \max\{|x_j - x_{j-1}| : j = 1, 2, \dots, n\}$. This is the length of the largest subinterval in the partition $\mathcal{P}$. An example of this can be seen in Fig. 15.2.

We make finer slices by adding more points to the partition. The resulting partition after adding these new points is called a refinement of the old partition.

Definition 15.1.3 (Refinement of Partition) A partition $\mathcal{P}' = \{x'_0, x'_1, x'_2, \dots, x'_m\}$ is called a refinement of the partition $\mathcal{P} = \{x_0, x_1, x_2, \dots, x_n\}$ if $m > n$ and each x_j for $j = 0, 1, 2, \dots, n$ is equal to an x'_k for some $k \in \{0, 1, 2, \dots, m\}$.

An example of a refinement for the partition in Fig. 15.2 is given in Fig. 15.3. An important remark here is that if $\mathcal{P}_1$ and $\mathcal{P}_2$ are two different partitions of $[a, b]$, then there always exists a common refinement $\mathcal{P}_3$ of both $\mathcal{P}_1$ and $\mathcal{P}_2$. This is simply done by taking the union of the partition points from each partition and list these points in an increasing order. In other words, $\mathcal{P}_3 = \mathcal{P}_1 \cup \mathcal{P}_2$ is a refinement for both the partitions $\mathcal{P}_1$ and $\mathcal{P}_2$.

Next, using the partition points, we slice the subgraphs into vertical strips. Over each subinterval, we want to approximate the original function with a very rudimentary function, which is the constant function. Let us define these rudimentary functions first.

Definition 15.1.4 (Step Function Adapted to $\mathcal{P}$) Let $\mathcal{P} = \{x_0, x_1, \dots, x_n\}$ be a partition of $[a, b]$ with subintervals I_j. Denote $I'_j = (x_{j-1}, x_j] \subseteq I_j$ for $j = 1, 2, 3, \dots, n$. A function $\phi_\mathcal{P} : [a, b] \to \mathbb{R}$ is called a step function adapted to $\mathcal{P}$ if $\phi_\mathcal{P}$ is constant on each interval I'_j.

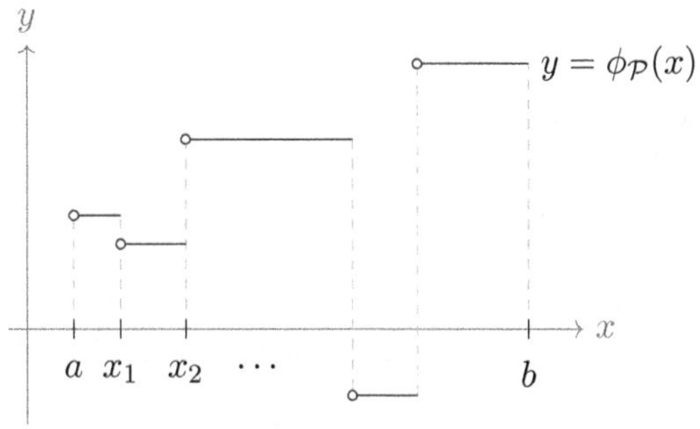

Fig. 15.4 Example of a step function adapted to a partition of $[a, b]$

An example of a step function adapted to a partition of $[a, b]$ is given in Fig. 15.4. Suppose $\phi_{\mathcal{P}} : [a, b] \to \mathbb{R}$ is a step function adapted to the partition $\mathcal{P} = \{x_0, x_1, x_2, \dots, x_n\}$ and for each $j = 1, 2, \dots, n$ we have $\phi_{\mathcal{P}}(x) = c_j \in \mathbb{R}$ when $x \in I'_j$. We can then express the step function $\phi_{\mathcal{P}}$ explicitly as the finite sum:

$$\phi_{\mathcal{P}}(x) = \sum_{j=1}^{n} c_j \mathbf{1}_{I'_j}(x),$$

where $\mathbf{1}_{I'_j} : [a, b] \to \mathbb{R}$ is the indicator function for the interval I'_j defined as:

$$\mathbf{1}_{I'_j}(x) = \begin{cases} 1 & \text{if } x \in I'_j, \\ 0 & \text{otherwise.} \end{cases}$$

The indicator function above satisfies the following lemma, which is very easy to verify.

Lemma 15.1.5 *Let $E, F \subseteq \mathbb{R}$ and $\mathbf{1}_E, \mathbf{1}_F : \mathbb{R} \to \mathbb{R}$ be indicator functions on these sets. Then:*

1. $\mathbf{1}_E \cdot \mathbf{1}_F = \mathbf{1}_{E \cap F}$.
2. $\mathbf{1}_E + \mathbf{1}_F = \mathbf{1}_{E \cup F} + \mathbf{1}_{E \cap F}$.
3. $|\mathbf{1}_E - \mathbf{1}_F| = \mathbf{1}_{E \triangle F}$.

Using Lemma 15.1.5, we can show that the set of step functions is closed under addition, namely: the sum of two step functions on $[a, b]$ is also a step function.

Proposition 15.1.6 *Let $\mathcal{P} = \{x_0, x_1, \dots, x_n\}$ and $\mathcal{Q} = \{y_0, y_1, \dots, y_m\}$ be two partitions of the interval $[a, b]$. If $\phi_{\mathcal{P}}, \phi_{\mathcal{Q}} : [a, b] \to \mathbb{R}$ are step functions adapted to the partitions $\mathcal{P}$ and $\mathcal{Q}$, then $\phi_{\mathcal{P}} + \phi_{\mathcal{Q}}$ is a step function adapted to $\mathcal{P} \cup \mathcal{Q}$.*

Proof Denote $I'_j = (x_{j-1}, x_j]$ for $j = 1, 2, \ldots, n$ and $H'_k = (y_{k-1}, y_k]$ for $k = 1, 2, \ldots, m$. Then, we can write $\phi_{\mathcal{P}} = \sum_{j=1}^{n} a_j \mathbf{1}_{I'_j}$ and $\phi_{\mathcal{Q}} = \sum_{k=1}^{m} b_k \mathbf{1}_{H'_k}$ for some constants $a_j, b_k \in \mathbb{R}$. Denote $G'_{jk} = I'_j \cap H'_k$ for $j = 1, 2, \ldots, n$ and $k = 1, 2, \ldots, m$. Then, each G'_{jk} is either empty or a half-closed interval with endpoints from the set $\mathcal{P} \cup \mathcal{Q}$. Moreover, we can easily check that the intervals G'_{jk} are pairwise disjoint.

For each $j = 1, 2, \ldots, n$ we can write $I'_j = \bigcup_{k=1}^{m} (I'_j \cap H'_k) = \bigcup_{k=1}^{m} G'_{jk}$. Moreover, since all the G'_{jk} are pairwise disjoint, by Lemma 15.1.5, we have $\mathbf{1}_{I'_j} = \sum_{k=1}^{m} \mathbf{1}_{G'_{jk}}$. Likewise, $H'_k = \bigcup_{j=1}^{n} G'_{jk}$ and $\mathbf{1}_{H'_k} = \sum_{j=1}^{n} \mathbf{1}_{G'_{jk}}$. Thus, we have:

$$\begin{aligned}\phi_{\mathcal{P}} + \phi_{\mathcal{Q}} = \sum_{j=1}^{n} a_j \mathbf{1}_{I'_j} + \sum_{k=1}^{n} b_k \mathbf{1}_{H'_k} &= \sum_{j=1}^{n} \sum_{k=1}^{m} a_j \mathbf{1}_{G'_{ij}} + \sum_{k=1}^{m} \sum_{j=1}^{n} b_k \mathbf{1}_{G'_{ij}} \\ &= \sum_{j=1}^{n} \sum_{k=1}^{m} (a_j + b_k) \mathbf{1}_{G'_{jk}},\end{aligned}$$

which is a step function adapted to the partition $\mathcal{P} \cup \mathcal{Q}$. □

Inductively, we can thus show that the sum of finitely many step functions on $[a, b]$ is also a step function.

We know exactly how to find the area for the subgraph of these step functions because the region under the graph consists of finitely many rectangles. Each subrectangle has area given by the width of the subinterval multiplied by its height which is the value of $\phi_{\mathcal{P}}$ on this interval. Adding all of them together, we get the total area under the graph for this step function. We define:

Definition 15.1.7 (Integral of Step Function) Let $\phi_{\mathcal{P}} : [a, b] \to \mathbb{R}$ be a step function with respect to some partition $\mathcal{P}$ of $[a, b]$ such that for each $j = 1, 2, \ldots, n$ we have $\phi_{\mathcal{P}}(x) = c_j \in \mathbb{R}$ for $x \in I'_j$. The integral of the function $\phi_{\mathcal{P}}$ over the interval $[a, b]$ is defined as:

$$I(\phi_{\mathcal{P}}) = \sum_{j=1}^{n} c_j |x_j - x_{j-1}|.$$

Remark 15.1.8 An example of the integral of a step function is given in Fig. 15.5. We note that this quantity measures the "signed area", in the sense that if the graph of the function is above the x-axis, the area is positive and if the graph is below the x-axis, the area would be treated as negative. This is true because the numbers c_j, which are the height of the rectangles, could either be negative or positive.

How do we adapt this procedure to general functions $f : [a, b] \to \mathbb{R}$? These functions could be very complicated, but we can still approximate them (probably

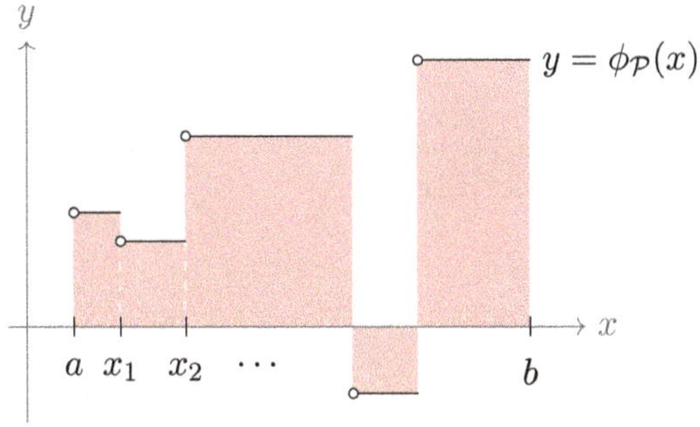

Fig. 15.5 Integral of the step function in Fig. 15.4 is the sum of the signed areas of the rectangles

crudely) by a step function adapted to some partition $\mathcal{P}$ of $[a, b]$. To get a good approximation to the original function f via a step function $\phi_{\mathcal{P}}$, how do we choose the values of the constants c_j over each subinterval of $\mathcal{P}$?

15.2 Riemann Integrals

The construction of integrals by Riemann was obtained by considering a "tagged partition" in which for each subinterval in a partition $\mathcal{P}$ a point $p_j \in I_j = [x_{j-1}, x_j]$, which we call a tag, is chosen. We denote the tagged partition as the pair $\mathcal{P}_\tau = (\mathcal{P}, \tau)$ where $\tau = \{p_1, p_2, \ldots, p_n\}$ is the set of chosen tag points from each subinterval.

Remark 15.2.1 Of course, there are countlessly many different partitions and tags that one could choose on $[a, b]$. A special partition that is useful for guessing and calculation purposes is the equispaced partition in which the subintervals in the partition all have the same size, namely $\mathcal{P} = \{x_j\}_{j=0}^n$ where $x_j = a + \frac{j(b-a)}{n}$. Along with this partition, we could then have the rightpoint tags $p_j = x_j$, leftpoint tags $p_j = x_{j-1}$, or the midpoint tags $p_j = \frac{x_{j-1}+x_j}{2}$ for all $j = 1, 2, \ldots, n$. However, the main point of this construction is that these partitions and tags are arbitrary.

In the construction by Riemann, the values c_j in the approximating step function are obtained by evaluating the function at the tag points p_j, namely we set $c_j = f(p_j)$ for each $j = 1, 2, \ldots, n$. Therefore, from the integral of the step function in Definition 15.1.7, we have a number which we call the Riemann sum with respect to the tagged partition $\mathcal{P}_\tau$. This Riemann sum is denoted as:

$$R_{f,\mathcal{P}_\tau} = \sum_{j=1}^{n} f(p_j)|x_j - x_{j-1}|.$$

The Riemann sum approximates the area under the graph of f by some rectangles. However, this approximation might be very crude if the partition sizes

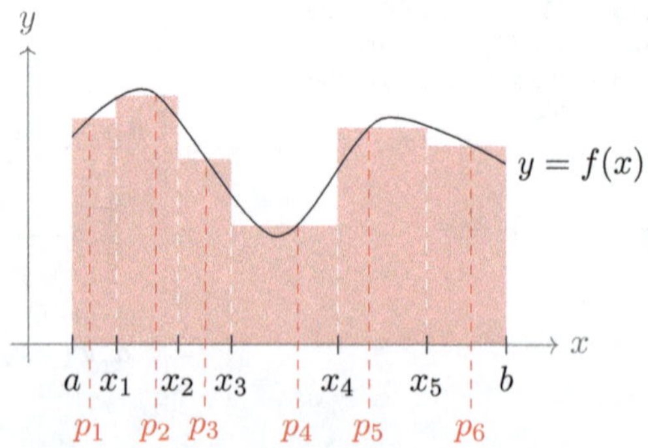

Fig. 15.6 The Riemann sum $R_{f,\mathcal{P}_\tau}$ with respect to the partition $\mathcal{P} = \{x_0, x_1, \ldots, x_6\}$ and tags $\tau = \{p_1, \ldots, p_6\}$ is the total area of the shaded region

are big as we can see in Fig. 15.6. Therefore, the Riemann integral is obtained by taking the limit as $||\mathcal{P}|| \to 0$. This can be achieved by taking refinements of the partitions, hoping that the Riemann sum converges to some fixed value. We thus define:

Definition 15.2.2 (Riemann Integral) A function $f : [a, b] \to \mathbb{R}$ is called Riemann integrable if there exists a real number $R \in \mathbb{R}$ such that for every $\varepsilon > 0$ there exists a $\delta > 0$ such that for any tagged partition $\mathcal{P}_\tau$ of $[a, b]$ with $||\mathcal{P}_\tau|| < \delta$ we have $|R_{f,\mathcal{P}_\tau} - R| < \varepsilon$. The number R is then called the Riemann integral of the function f over the interval $[a, b]$, also denoted as:

$$R = \int_a^b f(x)\,dx.$$

Symbolically, this is written with quantifiers as:

$$R = \int_a^b f(x)\,dx \quad \text{if}$$

$$\forall \varepsilon > 0, \exists \delta > 0 : \forall \text{ tagged partitions } \mathcal{P}_\tau, ||\mathcal{P}_\tau|| < \delta \Rightarrow |R_{f,\mathcal{P}_\tau} - R| < \varepsilon.$$

The set of functions for which this value R exists is called the set of Riemann integrable functions over $[a, b]$.

Definition 15.2.3 (Riemann Integrable Functions) Let $X = [a, b] \subseteq \mathbb{R}$ be a compact interval. The set of Riemann integrable functions over X is denoted as $\mathcal{R}(X)$. Namely:

$$\mathcal{R}(X) = \left\{ f : X \to \mathbb{R} : \int_a^b f(x)\,dx \text{ exists} \right\}.$$

Remark 15.2.4 We make several remarks regarding the definition and notation used.

1. The notation $\int$ was introduced by Leibniz in *De geometria recondita et analysi indivisibilium atque infinitorum* (On a Hidden Geometry and Analysis of Indivisibles and Infinites). It represents a swishy letter S which stands for *summa*, Latin for "sum". This refers to the fact that the process of integration is derived from summations of areas of regular shapes.
2. We refer to the function f under the integral sign as the integrand and the numbers a and b as the lower and upper limits of the integral respectively. The symbol dx at the end of the notation is called the differential with respect to the variable x and it refers to the variable we are integrating with respect to.
   When Leibniz came up with the notation for calculus, as mentioned in Remark 13.1.4, the term dx refers to the ghostly infinitesimals/indivisibles. So his notation $\int_a^b f(x)\,dx$ was literally meant to be taken as the *summa* of the areas of rectangles with height $f(x)$ and width dx (what is this width though? Nothing and something at the same time, as Berkeley and Hobbes said) from $x = a$ to $x = b$. Nowadays we simply use this notation symbolically rather than take it literally, just like the other Leibniz "fraction" notation $\frac{df}{dx}$ for derivatives.
3. In advanced analysis, multivariable calculus, and differential geometry, the whole expression $f(x)\,dx$ is referred to as a differential form, which are objects that we can integrate. Differential forms have many amazing properties, but we have to save that for future studies and carry on with our basic study of integrals here first.
4. In the integral notation, the variable x is called a dummy variable since the final value of the integral is a real number, which is independent of the variable x. This is similar to the dummy variable in the summation or product notation we discussed in Remark 2.5.2(3). Therefore, we can use any symbol for this dummy variable x without changing the value of the integral, namely:

$$R = \int_a^b f(x)\,dx = \int_a^b f(y)\,dy = \int_a^b f(t)\,dt = \int_a^b f(\spadesuit)\,d\spadesuit.$$

From its definition, we can guarantee that the Riemann integral of a function, if it exists, would be a unique value.

Proposition 15.2.5 *If $f \in \mathcal{R}([a, b])$, then its Riemann integral value is unique.*

Proof Suppose that we have two values for $\int_a^b f(x)\,dx$, namely R_1 and R_2. Fix $\varepsilon > 0$. By definition, we can find $\delta_1, \delta_2 > 0$ such that whenever $||\mathcal{P}_\tau|| < \delta_1$ we have $|R_{f,\mathcal{P}_\tau} - R_1| < \frac{\varepsilon}{2}$ and whenever $||\mathcal{P}_\tau|| < \delta_2$ we have $|R_{f,\mathcal{P}_\tau} - R_2| < \frac{\varepsilon}{2}$. Set $\delta = \min\{\delta_1, \delta_2\} > 0$ and pick a tagged partition $\mathcal{S}_\sigma$ of $[a, b]$ such that $||\mathcal{S}_\sigma|| < \delta$. Then, $\mathcal{S}_\sigma$ satisfies both of the estimates above which means $|R_{f,\mathcal{S}_\sigma} - R_1| < \frac{\varepsilon}{2}$ and $|R_{f,\mathcal{S}_\sigma} - R_2| < \frac{\varepsilon}{2}$ at the same time. Therefore:

$$|R_1 - R_2| = |R_1 - R_{f,\mathcal{S}_\sigma} + R_{f,\mathcal{S}_\sigma} - R_2| \leq |R_1 - R_{f,\mathcal{S}_\sigma}| + |R_{f,\mathcal{S}_\sigma} - R_2| < \frac{\varepsilon}{2} + \frac{\varepsilon}{2} = \varepsilon.$$

Thus, we have $|R_1 - R_2| < \varepsilon$ for any $\varepsilon > 0$ at all. This means $|R_1 - R_2| = 0$ which says that the two values of the integral must be equal to each other. □

Example 15.2.6 Let us compute the Riemann integral of some functions.

1. Let $f : [a, b] \to \mathbb{R}$ be the constant function $f(x) = c$ for all $x \in [a, b]$. We can guess the area under the graph of this function as $R = c(b - a)$ since it is graphically a rectangle. To show that this agrees with the definition of Riemann integral above, we pick any tagged partition $\mathcal{P}_\tau$ of $[a, b]$ where $\mathcal{P} = \{x_0, x_1, \dots, x_n\}$ and $\tau = \{p_1, p_2, \dots, p_n\}$. The Riemann sum of this function with respect to this tagged partition is:
$$R_{f,\mathcal{P}_\tau} = \sum_{j=1}^{n} f(p_j)|x_j - x_{j-1}| = \sum_{j=1}^{n} c(x_j - x_{j-1}) = c\sum_{j=1}^{n}(x_j - x_{j-1}) = c(b-a),$$
by telescoping sum. Thus, no matter which tagged partition is chosen, we would have $|R_{f,\mathcal{P}_\tau} - c(b - a)| = 0 < \varepsilon$ for any $\varepsilon > 0$ at all. Hence, we have $\int_a^b f(x)\,dx = c(b - a)$.
2. Let $f : [0, 2] \to \mathbb{R}$ be a piecewise constant function $f(x) = 1$ for all $x \in [0, 1]$ and $f(x) = 2$ for $x \in (1, 2]$. This function is left-continous at $x = 1$. Again, we can guess the area under the graph of this function as 3 since it is made up of two rectangles of areas 1 and 2 respectively. We need to be careful when computing the Riemann sum for this function since it has a jump at $x = 1$.
Fix $\varepsilon > 0$. We claim that $\delta = \varepsilon > 0$ is enough to ensure that Definition 15.2.2 holds. Pick any tagged partition $\mathcal{P}_\tau$ of $[a, b]$ such that $||\mathcal{P}_\tau|| < \delta$. We can split the set of partition points into two non-empty subsets $\mathcal{P}_1 = \{x_0, x_1, \dots, x_{m-1}\}$ and $\mathcal{P}_2 = \{x_m, x_{m+1}, \dots, x_n\}$ for some $m \in \mathbb{N}$ so that $\mathcal{P}_1 \subseteq [0, 1]$ and $\mathcal{P}_2 \subseteq (1, 2]$. Necessarily, we have $p_j \in [0, 1]$ for $j = 1, 2, \dots, m - 1$ and $p_j \in (1, 2]$ for $j = m + 1, m + 2, \dots, n$. Hence $f(p_j) = 1$ for $j \leq m - 1$ and $f(p_j) = 2$ for $j \geq m + 1$. However, the tag $p_m \in I_m = [x_{m-1}, x_m]$ could either be in $[x_{m-1}, 1]$ or in $(1, x_m]$. Therefore, the value of $f(p_m)$ could either be 1 or 2. See Fig. 15.7 for a visualisation.

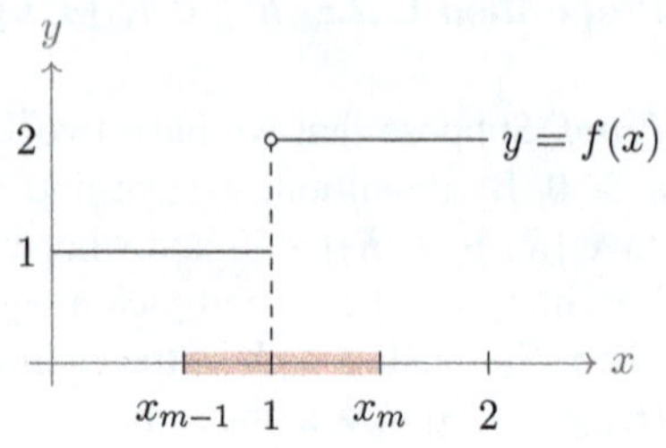

Fig. 15.7 Figure for the partition points. The tag p_m is somewhere in the red interval $[x_{m-1}, x_m]$ and so the value of $f(p_m)$ could either be 1 or 2

With this in mind, we can write down the Riemann sum as:

$$
\begin{aligned}
R_{f,\mathcal{P}_\tau} &= \sum_{j=1}^{n} f(p_j)|x_j - x_{j-1}| \\
&= \sum_{j=1}^{m-1}(x_j - x_{j-1}) + f(p_m)(x_m - x_{m-1}) + 2\sum_{j=m+1}^{n}(x_j - x_{j-1}) \\
&= x_{m-1} + f(p_m)(x_m - x_{m-1}) + 4 - 2x_m.
\end{aligned}
$$

Thus we have $R_{f,\mathcal{P}_\tau} - 3 = (f(p_m) - 1)(x_m - x_{m-1}) + (1 - x_m)$. Setting either $f(p_m) = 1$ or 2, we either have $R_{f,\mathcal{P}_\tau} - 3 = 1 - x_m$ or $R_{f,\mathcal{P}_\tau} - 3 = 1 - x_{m-1}$. Either way, since $x_{m-1} \leq 1 < x_m$ and $|x_m - x_{m-1}| \leq ||\mathcal{P}_\tau|| < \delta$, we have:

$$
-\delta < x_{m-1} - x_m \leq 1 - x_m \leq R_{f,\mathcal{P}_\tau} - 3 \leq 1 - x_{m-1} \leq x_m - x_{m-1} < \delta,
$$

which means $|R_{f,\mathcal{P}_\tau} - 3| < \delta = \varepsilon$. Hence, the function f is Riemann integrable over $[0, 1]$ with integral $R = 3$, namely $\int_0^2 f(x)\,dx = 3$.

3. We can repeat the construction above with the function $g : [0, 2] \to \mathbb{R}$ which is a piecewise constant function $g(x) = 1$ for all $x \in [0, 1)$ and $g(x) = 2$ for $x \in [1, 2]$. The only difference between the function g with the function f in the previous example is it is right-continuous at $x = 1$ instead. By the exact same argument, we can show that the integral of this function is also $R = 3$.
4. Consider the function $f : [0, 1] \to \mathbb{R}$ defined as $f(x) = x^2$. We first need to guess what the integral value R of this function is. Let us choose a special partition and tag to do this. Choose an equispaced partition $\mathcal{P} = \{x_j = \frac{j}{n}\}_{j=0}^n$ of $n + 1$ points with the n tags being the midpoint of the partition intervals, namely $\tau = \{p_j = \frac{2j-1}{2n}\}_{j=1}^n$.
We calculate the Riemann sum of this function with respect to this tagged partition $\mathcal{P}_\tau$ as:

$$
\begin{aligned}
R_{f,\mathcal{P}_\tau} = \sum_{j=1}^{n} f(p_j)|x_j - x_{j-1}| &= \sum_{j=1}^{n}\left(\frac{2j-1}{2n}\right)^2 \frac{1}{n} \\
&= \frac{1}{4n^3}\left(\sum_{j=1}^{n} 4j^2 + \sum_{j=1}^{n} 4j + \sum_{j=1}^{n} 1\right) \\
&= \frac{1}{4n^3}\left(\frac{4n^3}{3} + 4n^2 + \frac{11n}{3}\right) \\
&= \frac{1}{3} + \frac{1}{n} + \frac{11}{12n^2}.
\end{aligned}
$$

In the above, we used the formula for the sum of squares that we saw in Example 7.8.2. Therefore, with an equispaced partition and midpoint tags, as we take larger number of partition points, we can see that the limit of the Riemann sum approaches $\frac{1}{3}$.

The above computation is true for one specific choice of sequence of tagged partitions. Nevertheless, it gives us a good guess of what the value of the Riemann integral R, if it exists, should be.

Now to show that the Riemann integral is indeed $\frac{1}{3}$. Fix $\varepsilon > 0$. Choose $\delta = \varepsilon > 0$. Pick any tagged partition $\mathcal{P}_\tau$ of $[0, 1]$ such that $||\mathcal{P}_\tau|| < \delta$. Assume that it has $n+1$ partition points $\{x_j\}_{j=0}^n$ and n tags $\{p_j\}_{j=1}^n$. Thus, the Riemann sum of f with respect to this tagged partition is $R_{f,\mathcal{P}_\tau} = \sum_{j=1}^n f(p_j)|x_j - x_{j-1}|$.

Since the function f is increasing over its domain, necessarily $x_{j-1}^2 = f(x_{j-1}) \leq f(p_j) \leq f(x_j) = x_j^2$ for each $j = 1, 2, \ldots, n$. Thus, we have the bounds:

$$\sum_{j=1}^n x_{j-1}^2|x_j - x_{j-1}| - \frac{1}{3} \leq R_{f,\mathcal{P}_\tau} - \frac{1}{3} \leq \sum_{j=1}^n x_j^2|x_j - x_{j-1}| - \frac{1}{3}. \tag{15.1}$$

Next notice that $\frac{1}{3} = \frac{1}{3}\sum_{j=1}^n (x_j^3 - x_{j-1}^3)$ by telescopic sum. This is a very important observation as we can then rewrite (15.1) as:

$$R_{f,\mathcal{P}_\tau} - \frac{1}{3} \leq \sum_{j=1}^n x_j^2|x_j - x_{j-1}| - \frac{1}{3}\sum_{j=1}^n (x_j^3 - x_{j-1}^3), \text{ and}$$

$$\sum_{j=1}^n x_{j-1}^2|x_j - x_{j-1}| - \frac{1}{3}\sum_{j=1}^n (x_j^3 - x_{j-1}^3) \leq R_{f,\mathcal{P}_\tau} - \frac{1}{3}$$

The upper bound above can be rewritten and bound further by:

$$\begin{aligned}\sum_{j=1}^n x_j^2|x_j - x_{j-1}| - \frac{1}{3}\sum_{j=1}^n (x_j^3 - x_{j-1}^3) &= \sum_{j=1}^n \frac{(x_j - x_{j-1})^2(2x_j + x_{j-1})}{3} \\ &< \sum_{j=1}^n (x_j - x_{j-1})^2,\end{aligned}$$

where we used the fact that $2x_j + x_{j-1} < 3$ for all $j = 1, 2, \ldots, n$ to get the final inequality. In a similar manner, the lower bound can be simplified to:

$$\sum_{j=1}^n x_{j-1}^2|x_j - x_{j-1}| - \frac{1}{3}\sum_{j=1}^n (x_j^3 - x_{j-1}^3) > -\sum_{j=1}^n (x_j - x_{j-1})^2.$$

Hence, by using the assumption that $||\mathcal{P}_\tau|| = \max\{|x_j - x_{j-1}| : j = 1, 2, \dots, n\} < \delta$ and telescoping sum, we have:

$$\left| R_{f,\mathcal{P}_\tau} - \frac{1}{3} \right| < \sum_{j=1}^{n} (x_j - x_{j-1})^2 < \sum_{j=1}^{n} \delta(x_j - x_{j-1}) = \delta \sum_{j=1}^{n} (x_j - x_{j-1}) = \delta = \varepsilon.$$

Since $\mathcal{P}_\tau$ is an arbitrary tagged partition of size less than δ, we conclude that the function f is Riemann integrable over $[0, 1]$ with Riemann integral value of $\frac{1}{3}$, namely $\int_0^1 f(x)\,dx = \frac{1}{3}$.

The Riemann integral is a rigorous and perfectly reasonable definition for an integral for theoretical purposes. However, this definition can be quite difficult to deal with for computations. The computation can be quite complicated even with a very elementary function as we have seen in Example 15.2.6(4).

Indeed, we may not have any idea what the number R could be or whether it even exists for the function f! Furthermore, there are many parameters that we can vary here, namely the partition and the tags, so we have many things to control in this construction.

There are various special kind of partitions and tags to help us with guessing the value of the integral, which includes the equispaced partition with leftpoint tags, rightpoint tags, or midpoint tags (which we have used in Example 15.2.6(4)) that could be used in the construction. But in the end, we still have to show that the definition works for any tagged partition at all, not just these special tagged partitions!

15.3 Darboux Integrals

We now turn to a more explicit type of integral construction due to Jean-Gaston Darboux. This work was published in 1875 as a reinterpretation of the Riemann integral. We first state the choice of values for the constants c_j for the approximating step functions.

Lower and Upper Sums

For a bounded function $f : [a, b] \to \mathbb{R}$, given a partition $\mathcal{P} = \{x_0, x_1, \dots, x_n\}$ of $[a, b]$, we define a collection of numbers:

$$m_j = \inf_{x \in I_j} f(x) \quad \text{and} \quad M_j = \sup_{x \in I_j} f(x),$$

for $j = 1, 2, \ldots, n$ and $I_j = [x_{j-1}, x_j]$, which are the smallest upper bound and largest lower bound of the function f in each partition interval I_j. Note that all of these quantities exist and are finite because the function f is bounded on the whole domain. These will be used as the values of c_j for the approximating step function that we saw earlier.

Here we have two sets of these numbers, namely $\{m_j\}_{j=1}^n$ and $\{M_j\}_{j=1}^n$. So let us construct two approximations by using these sets of numbers separately. We would obtain two different step function approximations $\underline{f}_{\mathcal{P}}, \overline{f}_{\mathcal{P}} : [a, b] \to \mathbb{R}$, which are called the lower and upper approximations with respect to the partition $\mathcal{P}$ respectively. They are given as:

$$\underline{f}_{\mathcal{P}}(x) = \sum_{j=1}^{n} m_j \mathbf{1}_{I_j'}(x) \quad \text{and} \quad \overline{f}_{\mathcal{P}}(x) = \sum_{j=1}^{n} M_i \mathbf{1}_{I_j'}(x).$$

Remark 15.3.1 We make some remarks here:

1. Given a partition $\mathcal{P}$, clearly we have the pointwise ordering $\underline{f}_{\mathcal{P}}(x) \leq f(x) \leq \overline{f}_{\mathcal{P}}(x)$ for every $x \in (a, b]$. Indeed, for a fixed $x_0 \in (a, b]$, we have $x_0 \in I_j' \subseteq I_j$ for some $j \in 1, 2, \ldots, n$. Hence, $f(x_0) \leq \sup_{x \in I_j} f(x) = M_j = \overline{f}_{\mathcal{P}}(x_0)$. Since $x_0 \in (a, b]$ was chosen arbitrarily, we have the inequality $f(x) \leq \overline{f}_{\mathcal{P}}(x)$ for every $x \in (a, b]$. The other inequality can also be proven in a similar manner.
2. We do not have this inequality at the lower endpoint $x = a$ since $\underline{f}_{\mathcal{P}}(a) = \overline{f}_{\mathcal{P}}(a) = 0$ but $f(a)$ might not be 0. But this will not be a problem to our construction as it does not affect the total area of the approximating rectangles.

If we refine the partition $\mathcal{P}$ by adding an extra point somewhere in the partition, we have:

Lemma 15.3.2 *Let $f : [a, b] \to \mathbb{R}$ be a bounded function. If $\mathcal{P}$ is a partition of $[a, b]$ and $\mathcal{P}' = \mathcal{P} \cup \{c\}$ for some $c \in (a, b)$, then for every $x \in (a, b]$ we have:*

$$\underline{f}_{\mathcal{P}}(x) \leq \underline{f}_{\mathcal{P}'}(x) \leq \overline{f}_{\mathcal{P}'}(x) \leq \overline{f}_{\mathcal{P}}(x).$$

Proof The middle inequality is clearly true from Remark 15.3.1(1). Let us now prove the first inequality. WLOG, we assume that the newly added point is within the first subinterval, namely $c \in (x_0, x_1) = (a, x_1)$. The upper and lower approximations do not change outside this subinterval when we add the point c to the partition $\mathcal{P}$. Therefore, $\underline{f}_{\mathcal{P}}(x) = \underline{f}_{\mathcal{P}'}(x)$ and $\overline{f}_{\mathcal{P}'}(x) = \overline{f}_{\mathcal{P}}(x)$ for all $x \in (x_1, b]$. In other words, the orderings do not change outside the partition subinterval that includes c.

We now check the ordering in the subinterval that contains c, namely $[a, x_1]$. By Proposition 4.1.10, since $[a, c], [c, x_1] \subseteq [a, x_1]$, we have $n_1 = \inf_{x\in[a,c]} f(x) \geq \inf_{x\in[a,x_1]} f(x) = m_1$ and $n_2 = \inf_{x\in[c,x_1]} f(x) \geq \inf_{x\in[a,x_1]} f(x) = m_1$. So, if $x \in (a, x_1]$, we have:

$$\begin{aligned}\underline{f}_{\mathcal{P}}(x) - \underline{f}_{\mathcal{P}'}(x) &= m_1 \mathbf{1}_{[a,x_1]}(x) - (n_1 \mathbf{1}_{[a,c]}(x) + n_2 \mathbf{1}_{(c,x_1]}(x)) \\ &= (m_1 - n_1)\mathbf{1}_{[a,c]}(x) + (m_1 - n_2)\mathbf{1}_{(c,x_1]}(x) \leq 0,\end{aligned}$$

which gives us the first inequality. The final inequality can be obtained in the same manner. □

Thus, by finite repeated application of the above lemma, we have:

Proposition 15.3.3 *Let $f : [a, b] \to \mathbb{R}$ be a bounded function and $\mathcal{P}$ and $\mathcal{P}'$ are partitions of $[a, b]$. If $\mathcal{P}'$ is a refinement of $\mathcal{P}$, then:*

$$\underline{f}_{\mathcal{P}}(x) \leq \underline{f}_{\mathcal{P}'}(x) \leq \overline{f}_{\mathcal{P}'}(x) \leq \overline{f}_{\mathcal{P}}(x),$$

for all $x \in (a, b]$. In other words, if we refine the partition, the lower approximations get bigger pointwise and the upper approximations get smaller pointwise.

From the definition of lower and upper approximation functions, since these approximations are step functions, we can compute the areas under their graphs by using Definition 15.1.7. These areas are called the lower and upper Darboux sum with respect to a partition $\mathcal{P}$ and they are given respectively as:

$$L_{f,\mathcal{P}} = I(\underline{f}_{\mathcal{P}}) = \sum_{j=1}^{n} m_j |x_j - x_{j-1}| \quad \text{and} \quad U_{f,\mathcal{P}} = I(\overline{f}_{\mathcal{P}}) = \sum_{j=1}^{n} M_j |x_j - x_{j-1}|.$$

Since the functions $\underline{f}_{\mathcal{P}}$ and $\overline{f}_{\mathcal{P}}$ are pointwise smaller than or bigger than the function f over $(a, b]$ respectively, the Darboux sums $L_{f,\mathcal{P}}$ and $U_{f,\mathcal{P}}$ would underestimate or overestimate the area under the graph of f respectively. This can be seen clearly in Fig. 15.8.

Note also that for a given partition $\mathcal{P}$, it is clearly true that we have $L_{f,\mathcal{P}} \leq U_{f,\mathcal{P}}$ since $m_j \leq M_j$ for every $j = 1, 2, 3 \ldots, n$. This will be important later in Proposition 15.3.5. By using a similar argument as the pointwise approximation, for finer partitions, the lower Darboux sums get bigger and the upper Darboux sums get smaller. This proves:

Proposition 15.3.4 *Let $f : [a, b] \to \mathbb{R}$ be a bounded function and $\mathcal{P}$ and $\mathcal{P}'$ are partitions of $[a, b]$. If $\mathcal{P}'$ is a refinement of $\mathcal{P}$, then:*

$$L_{f,\mathcal{P}} \leq L_{f,\mathcal{P}'} \leq U_{f,\mathcal{P}'} \leq U_{f,\mathcal{P}}.$$

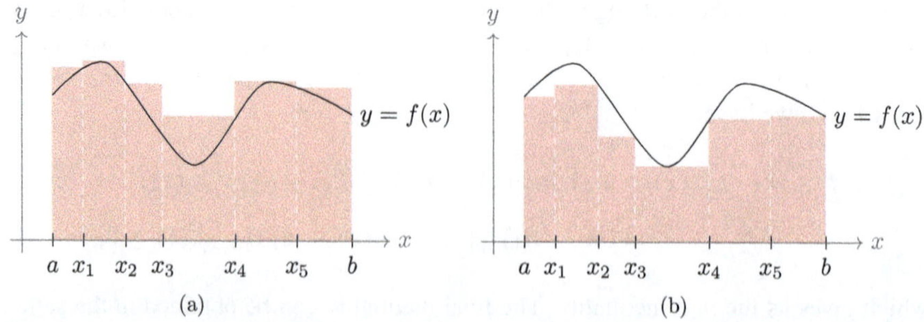

Fig. 15.8 The upper and lower Darboux sums $U_{f,\mathcal{P}}$ and $L_{f,\mathcal{P}}$ with respect to the partition $\mathcal{P} = \{x_0, x_1, \ldots, x_6\}$ are the area of the shaded region. Compare there approximations with the Riemann sum for the same function in Fig. 15.6. **(a)** $U_{f,\mathcal{P}}$. **(b)** $L_{f,\mathcal{P}}$

In other words, if we refine the partition, the lower Darboux sum increases and the upper Darboux sum decreases.

For any partition $\mathcal{P}$ of $[a, b]$, these Darboux sums are finite because each m_i and M_j are finite as f is a bounded function and $|x_j - x_{j-1}| < |b - a|$ for $j = 1, 2, \ldots, n$. In fact, for any partition $\mathcal{P}$, we have the following bounds for the Darboux sums:

$$m|b-a| \leq L_{f,\mathcal{P}} \leq U_{f,\mathcal{P}} \leq M|b-a|, \tag{15.2}$$

where $m = \inf_{x\in[a,b]} f(x)$ and $M = \sup_{x\in[a,b]} f(x)$ which are finite.

Now, consider the following subsets of $\mathbb{R}$ which are the collection of all lower and upper Darboux sums of the function f taken over all partitions of $[a, b]$:

$$\mathcal{L}=\{L_{f,\mathcal{P}} : \mathcal{P} \text{ is a partition of } [a,b]\} \quad \text{and} \quad \mathcal{U}=\{U_{f,\mathcal{P}} : \mathcal{P} \text{ is a partition of } [a,b]\}.$$

From the inequalities in (15.2), both of these sets are bounded from above and below by $M|b - a|$ and $m|b - a|$ respectively. The completeness axiom of $\mathbb{R}$ says that the infimum and supremum of the sets $\mathcal{U}$ and $\mathcal{L}$ exist and are finite. We can then define the supremum and infimum of the lower and upper Darboux sums over all possible partitions of $[a, b]$ as:

$$L_f = \sup \mathcal{L} = \sup_{\mathcal{P}} L_{f,\mathcal{P}} \quad \text{and} \quad U_f = \inf \mathcal{U} = \inf_{\mathcal{P}} U_{f,\mathcal{P}},$$

which are called the lower Darboux integral and the upper Darboux integral of f respectively. These integrals are ordered as follows:

Proposition 15.3.5 *Let $f : [a, b] \to \mathbb{R}$ be a bounded function. Then, $L_f \leq U_f$.*

Proof First we note that if $\mathcal{P}_1$ and $\mathcal{P}_2$ are two different partitions of $[a, b]$, by letting $\mathcal{P}_3$ be their common refinement and an application of Proposition 15.3.4, we have:

$$L_{f,\mathcal{P}_1} \leq L_{f,\mathcal{P}_3} \leq U_{f,\mathcal{P}_3} \leq U_{f,\mathcal{P}_2}, \tag{15.3}$$

which implies the lower Darboux sum for any partition $\mathcal{P}_1$ of $[a, b]$ is always bounded from above by the upper Darboux sum of any other partition $\mathcal{P}_2$ of $[a, b]$.

Fix an arbitrary partition $\mathcal{P}$ of $[a, b]$. By the inequality (15.3), we have $L_{f,\mathcal{P}'} \leq U_{f,\mathcal{P}}$ for any partition $\mathcal{P}'$ of $[a, b]$. Thus, the quantity $U_{f,\mathcal{P}}$ is an upper bound for the set $\mathcal{L} = \{L_{f,\mathcal{P}'} : \mathcal{P}' \text{ a partition of } [a, b]\}$. By taking the supremum over the various partitions $\mathcal{P}'$, we obtain a lower bound:

$$L_f = \sup_{\mathcal{P}'} L_{f,\mathcal{P}'} \leq U_{f,\mathcal{P}},$$

for the partition $\mathcal{P}$. Since this partition $\mathcal{P}$ is arbitrary, this inequality must hold for any $\mathcal{P}$ and hence we can take the infimum over all partitions $\mathcal{P}$ to get:

$$L_f \leq \inf_{\mathcal{P}} U_{f,\mathcal{P}} = U_f,$$

which is what we wanted to prove. □

Darboux Integral

We have seen in Proposition 15.3.5 that the lower Darboux integral of a function is always smaller than or equal to its upper Darboux integral. When these quantities coincide, we call this quantity the Darboux integral of the function.

Definition 15.3.6 (Darboux Integral) We call a bounded function $f : [a, b] \to \mathbb{R}$ Darboux integrable if the upper and lower Darboux integrals coincide, namely $L_f = U_f = D$ for some number D. This common value is called the Darboux integral of the function f and written as:

$$D = \int_a^b f(x)\, dx.$$

Definition 15.3.7 (Darboux Integrable Functions) Let $X = [a, b] \subseteq \mathbb{R}$ be a compact interval. The set of all Darboux integrable functions over X is denoted as $\mathcal{D}(X)$. Namely:

$$\mathcal{D}(X) = \left\{f : X \to \mathbb{R} : U_f = L_f < \infty\right\}.$$

What does this quantity has to do with the area under the graph of the function $f : [a, b] \to \mathbb{R}$? For any partition $\mathcal{P}$ at all, by the pointwise ordering $\underline{f}_{\mathcal{P}}(x) \leq f(x) \leq \overline{f}_{\mathcal{P}}(x)$ for all $x \in (a, b]$, the graph for the function f is sandwiched between the graphs of $\overline{f}_{\mathcal{P}}$ and $\underline{f}_{\mathcal{P}}$. So, if the area under the graph $I(f)$ for f exists, we expect that this area is bounded from above and below by the areas under the graphs of the step functions $\overline{f}_{\mathcal{P}}$ and $\underline{f}_{\mathcal{P}}$, both of which we know to exist. In other words, for any partitions $\mathcal{P}$ and $\mathcal{P}'$ of $[a, b]$, we would expect $L_{f,\mathcal{P}'} = I(\underline{f}_{\mathcal{P}'}) \leq I(f) \leq I(\overline{f}_{\mathcal{P}}) = U_{f,\mathcal{P}}$.

Now we want to find the value of $I(f)$, if it exists. As in the proof for Proposition 15.3.5, we take the supremum over all the partitions $\mathcal{P}'$ and infimum over all the partitions $\mathcal{P}$ to get $L_f \leq I(f) \leq U_f$. If the function f is Darboux integrable in $[a, b]$, Definition 15.3.6 tells us we must have $L_f = U_f = D$ which forces $I(f) = D$. Thus, the area under the graph of the function f is equal to the Darboux integral. However, if it is not Darboux integrable, we would have $L_f < U_f$ with $I(f)$ somewhere in between: its value unknown or might not even exist!

A much shorter and useful way of characterising the Darboux integrability of the function $f : [a, b] \to \mathbb{R}$ is by the ε-criterion, which comes from the characterisation of the infimum and supremum for L_f and U_f.

Theorem 15.3.8 (ε-criterion for Darboux Integration) *Let $f : [a, b] \to \mathbb{R}$ be a bounded function. Then, $f \in \mathcal{D}([a, b])$ if and only if for any $\varepsilon > 0$ there exists a partition $\mathcal{P}$ of $[a, b]$ such that $U_{f,\mathcal{P}} - L_{f,\mathcal{P}} < \varepsilon$.*

Proof We shall prove only one of the implications. The forward implication is left to the readers as Exercise 15.16.

($\Leftarrow$): For any partition $\mathcal{P}$ of $[a, b]$, we have seen the ordering $L_{f,\mathcal{P}} \leq L_f \leq U_f \leq U_{f,\mathcal{P}}$ in Proposition 15.3.5. This implies $0 \leq U_f - L_f \leq U_{f,\mathcal{P}} - L_{f,\mathcal{P}}$. If the ε-criterion holds, then for all $\varepsilon > 0$ there is a partition $\mathcal{P}'$ such that $0 \leq U_f - L_f \leq U_{f,\mathcal{P}'} - L_{f,\mathcal{P}'} < \varepsilon$. Namely, $0 \leq U_f - L_f < \varepsilon$ for all $\varepsilon > 0$. This implies $U_f - L_f = 0$ and hence $f \in \mathcal{D}([a, b])$. □

Similar to Riemann integrals, even though we do not need to worry about the tags for Darboux integral, finding a Darboux integral can still be tricky to carry out concretely because there are so many different partitions for the interval $[a, b]$ that needs to be considered.

The good news is if we can find just one sequence of partitions such that the corresponding sequence of lower Darboux sums and the upper Darboux sums approach the same limit, we can conclude that the Darboux integral does exist and is equal to this common value. This is due to the following corollary of Theorem 15.3.8.

Corollary 15.3.9 *Let $f : [a, b] \to \mathbb{R}$ be a bounded function. Then, $f \in \mathcal{D}([a, b])$ if and only if there exists a sequence of partitions $(\mathcal{P}_n)$ of $[a, b]$ such that $\lim_{n\to\infty} L_{f,\mathcal{P}_n} = \lim_{n\to\infty} U_{f,\mathcal{P}_n}$. Moreover, we have:*

$$\int_a^b f(x)\,dx = \lim_{n\to\infty} L_{f,\mathcal{P}_n} = \lim_{n\to\infty} U_{f,\mathcal{P}_n}.$$

Proof We prove the implications separately.

($\Rightarrow$): Since $f \in \mathcal{D}([a,b])$, by the ε-criterion, for each $n \in \mathbb{N}$ we can find a partition $\mathcal{Q}_n$ of $[a,b]$ such that $U_{f,\mathcal{Q}_n} - L_{f,\mathcal{Q}_n} < \frac{1}{n}$. Define a sequence of partitions $(\mathcal{P}_n)$ so that $\mathcal{P}_1 = \mathcal{Q}_1$ and $\mathcal{P}_n = \mathcal{P}_{n-1} \cup \mathcal{Q}_n$ so that $\mathcal{P}_n \subseteq \mathcal{P}_{n+1}$ for all $n \in \mathbb{N}$. Hence $(\mathcal{P}_n)$ is a sequence of refined partitions. As a result, by Proposition 15.3.4, both of the real sequences $(U_{f,\mathcal{P}_n})$ and $(L_{f,\mathcal{P}_n})$ are monotone and bounded. Thus, both of them must converge. Since $U_{f,\mathcal{P}_n} \leq U_{f,\mathcal{Q}_n}$ and $L_{f,\mathcal{P}_n} \geq L_{f,\mathcal{Q}_n}$, we have:

$$0 \leq U_{f,\mathcal{P}_n} - L_{f,\mathcal{P}_n} \leq U_{f,\mathcal{Q}_n} - L_{f,\mathcal{Q}_n} < \frac{1}{n}.$$

Taking the limit on both sides, by sandwiching, we can deduce that $\lim_{n\to\infty}(U_{f,\mathcal{P}_n} - L_{f,\mathcal{P}_n}) = 0$. The algebra of limits then implies $\lim_{n\to\infty} U_{f,\mathcal{P}_n} = \lim_{n\to\infty} L_{f,\mathcal{P}_n}$.

($\Leftarrow$): Suppose that there exists a sequence of partitions $(\mathcal{P}_n)$ of $[a,b]$ such that $\lim_{n\to\infty} L_{f,\mathcal{P}_n} = \lim_{n\to\infty} U_{f,\mathcal{P}_n}$. By algebra of limits, we have $\lim_{n\to\infty}(U_{f,\mathcal{P}_n} - L_{f,\mathcal{P}_n}) = 0$ which means for any $\varepsilon > 0$ there exists an $N \in \mathbb{N}$ such that $|U_{f,\mathcal{P}_n} - L_{f,\mathcal{P}_n} - 0| = U_{f,\mathcal{P}_n} - L_{f,\mathcal{P}_n} < \varepsilon$ for all $n \geq N$. This gives us the ε-criterion for Darboux integral and so $f \in \mathcal{D}([a,b])$.

Finally, by Proposition 15.3.5, for any $n \in \mathbb{N}$ we have $L_{f,\mathcal{P}_n} \leq L_f \leq U_f \leq U_{f,\mathcal{P}_n}$. Taking the limit as $n \to \infty$ and using the assumption, we then have $\lim_{n\to\infty} L_{f,\mathcal{P}_n} \leq L_f \leq U_f \leq \lim_{n\to\infty} U_{f,\mathcal{P}_n} = \lim_{n\to\infty} L_{f,\mathcal{P}_n}$. Thus, each inequality here is an equality and so:

$$\int_a^b f(x)\,dx = L_f = U_f = \lim_{n\to\infty} L_{f,\mathcal{P}_n} = \lim_{n\to\infty} U_{f,\mathcal{P}_n},$$

which is what we claimed. □

From its proof, Corollary 15.3.9 is also true if the sequence of partitions $(\mathcal{P}_n)$ is a sequence of refined partition. Now let us apply Corollary 15.3.9 to compute some Darboux integrals.

Example 15.3.10 The usual approach here is that we start with an equispaced partition $(\mathcal{P}_n)$ of $n+1$ points for $n \in \mathbb{N}$. Hopefully, this would give us the sequence of partitions that would lead to the desired conclusion.

1. Let $f : \mathbb{R} \to \mathbb{R}$ be defined as $f(x) = x^2$. We aim to find its Darboux integral in the domain $[0, 1]$. Choose an equispaced partition $\mathcal{P}_n = \{x_0, x_1, \ldots, x_n\}$ where $x_j = \frac{j}{n}$. For each $j = 1, 2, \ldots, n$, let I_j be the subintervals of the partition. Since

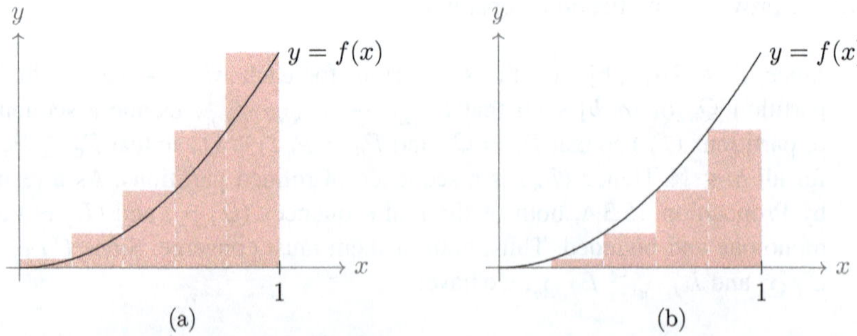

Fig. 15.9 The upper and lower Darboux sums $U_{f,\mathcal{P}_5}$ and $L_{f,\mathcal{P}_5}$ with respect to the equispaced partition $\mathcal{P}_5$ with 6 points are the area of the shaded region. **(a)** $U_{f,\mathcal{P}_5}$. **(b)** $L_{f,\mathcal{P}_5}$.

the function f is increasing over $[0, 1]$, we have:

$$m_j = \inf_{x\in I_j} f(x) = f(x_{j-1}) = \frac{(j-1)^2}{n^2},$$

$$M_j = \sup_{x\in I_j} f(x) = f(x_j) = \frac{j^2}{n^2}.$$

Using $|x_j - x_{j-1}| = \frac{1}{n}$ for all j, the upper and lower Darboux sums of with respect to this partition $\mathcal{P}_n$ are given by:

$$L_{f,\mathcal{P}_n} = \sum_{j=1}^{n} m_j|x_j - x_{j-1}| = \sum_{j=1}^{n} \frac{(j-1)^2}{n^3} = \frac{(n-1)(2n-1)}{6n^2},$$

$$U_{f,\mathcal{P}_n} = \sum_{j=1}^{n} M_j|x_j - x_{j-1}| = \sum_{j=1}^{n} \frac{j^2}{n^3} = \frac{(n+1)(2n+1)}{6n^2},$$

by using the sum of squares in Example 7.8.2.

An example of these upper and lower sums for $n = 5$ are given in Fig. 15.9. If we use greater number of partition points by letting $n \to \infty$, we have the limits:

$$\lim_{n\to\infty} L_{f,\mathcal{P}_n} = \frac{1}{3} \quad \text{and} \quad \lim_{n\to\infty} U_{f,\mathcal{P}_n} = \frac{1}{3}.$$

So, by Corollary 15.3.9, $f \in \mathcal{D}([0, 1])$ and its Darboux integral is $\frac{1}{3}$, namely $\int_0^1 f(x)\,dx = \frac{1}{3}$.

2. Now consider the function $f : \mathbb{R} \to \mathbb{R}$ defined as $f(x) = x(x-1)$. We want to find its Darboux integral over $[0, 1]$, namely the area of the shaded region in Fig. 15.10.

   For any $n \in \mathbb{N}$, pick a special partition with an odd number of points $\mathcal{P}_n = \{x_0, x_1, \ldots, x_{2n}\}$ distributed in an equispaced manner, namely $x_j = \frac{j}{2n}$.

Fig. 15.10 The graph of the function $f(x) = x(x-1)$ and the area that we want to compute is shaded in red. Since the region is fully below the x-axis, we expect that its value is negative

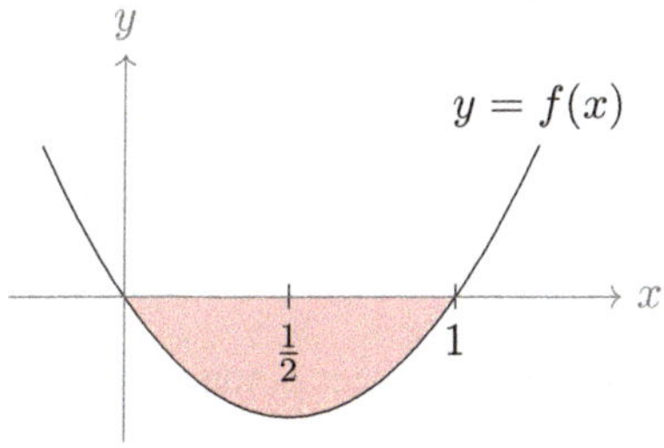

Necessarily, $x_n = \frac{1}{2}$. In the region $[0, \frac{1}{2}]$, the function is decreasing. So, for each $j = 1, 2, \ldots, n$, on the subintervals I_j we have:

$$m_j = \inf_{x \in I_j} f(x) = f(x_j) = x_j(x_j - 1) = \frac{j}{2n}\left(\frac{j}{2n} - 1\right),$$

$$M_j = \sup_{x \in I_j} f(x) = f(x_{j-1}) = x_{j-1}(x_{j-1} - 1) = \frac{j-1}{2n}\left(\frac{j-1}{2n} - 1\right).$$

On the other hand, in the interval $[\frac{1}{2}, 1]$, the function is increasing. So over this interval, for each $j = n+1, n+1, \ldots, 2n$, we have:

$$m_j = \inf_{x \in I_j} f(x) = f(x_{j-1}) = x_{j-1}(x_{j-1} - 1) = \frac{j-1}{2n}\left(\frac{j-1}{2n} - 1\right),$$

$$M_j = \sup_{x \in I_j} f(x) = f(x_j) = x_j(x_j - 1) = \frac{j}{2n}\left(\frac{j}{2n} - 1\right).$$

Since $|x_j - x_{j-1}| = \frac{1}{2n}$ for all j, the lower Darboux sums for f with respect to the partition $\mathcal{P}_n$ is:

$$\begin{aligned}
L_{f,\mathcal{P}_n} &= \sum_{j=1}^{2n} m_j |x_j - x_{j-1}| \\
&= \frac{1}{2n}\sum_{j=1}^{n} \frac{j}{2n}\left(\frac{j}{2n} - 1\right) + \frac{1}{2n}\sum_{j=n+1}^{2n} \frac{j-1}{2n}\left(\frac{j-1}{2n} - 1\right) \\
&= -\frac{(n+1)(4n-1)}{48n^2} - \frac{(n+1)(4n-1)}{48n^2} \\
&= \frac{1 - 3n - 4n^2}{24n^2},
\end{aligned}$$

where we used the sum of squares formula. On the other hand, by similar computation, the upper Darboux sum for this partition is given by:

$$U_{f,\mathcal{P}_n} = \sum_{j=1}^{2n} M_j|x_j - x_{j-1}| = \frac{1+3n-4n^2}{24n^2}.$$

If we use greater number of partition points by letting $n \to \infty$, we have:

$$\lim_{n\to\infty} L_{f,\mathcal{P}_n} = -\frac{1}{6} \quad \text{and} \quad \lim_{n\to\infty} U_{f,\mathcal{P}_n} = -\frac{1}{6},$$

and so, by Corollary 15.3.9, $f \in \mathcal{D}([0,1])$ and its Darboux integral is equal to $-\frac{1}{6}$.

3. From Exercise 9.7, recall the Dirichlet function defined as $f : \mathbb{R} \to \mathbb{R}$ with $f(x) = 1$ if $x \in \mathbb{Q}$ and 0 if $x \in \bar{\mathbb{Q}}$. We now prove that it is not Darboux integrable over any compact interval $[a,b]$ with $a < b$. Pick any partition $\mathcal{P} = \{x_0, x_1, \ldots, x_n\}$ of $[a,b]$. Note that in between any two partition points x_{j-1} and x_j, we can always find a rational number and an irrational number.
Thus, for each $j = 1, 2, \ldots, n$, necessarily $m_j = \inf_{x\in I_j} f(x) = 0$ and $M_j = \sup_{x\in I_j} f(x) = 1$. Therefore, the lower and upper Darboux sums are $L_{f,\mathcal{P}} = 0$ and $U_{f,\mathcal{P}} = 1$ for any partition $\mathcal{P}$. This means $L_f = \sup_{\mathcal{P}} L_{f,\mathcal{P}} = 0$ and $U_f = \inf_{\mathcal{P}} U_{f,\mathcal{P}} = 1$. Since they do not agree, we conclude that the Dirichlet function is not Darboux integrable on $[a,b]$.

15.4 Correspondence between Riemann and Darboux Integrals

From the previous two sections, we saw that the notations used for the Riemann and Darboux integrals in Definitions 15.2.2 and 15.3.6 are exactly the same, namely both are labelled as $\int_a^b f(x)\,dx$. Would this cause an ambiguity since we labelled them with the same notation even though they are totally different constructions?

Actually, for bounded functions there would be no ambiguity. This is because the construction by Riemann can be shown to be equivalent to the construction by Darboux. Intuitively this can be seen to be true because as the partitions get finer, the value at the tag in the Riemann sum are trapped between the value of the infimum and supremum of the function over the shrinking partitions intervals.

Furthermore, if these limits of the Riemann sum and the Darboux sums exist, we can show that they would have the same value. We have seen that this is true for the function $f : [0,1] \to \mathbb{R}$ defined as $f(x) = x^2$ in Examples 15.2.6(4) and 15.3.10(1) for which both of these functions have the same Riemann and Darboux integral value of $\frac{1}{3}$. With that, we prove:

Theorem 15.4.1 *Let $f : [a, b] \to \mathbb{R}$ be a bounded function. Then $f \in \mathcal{R}([a, b])$ if and only if $f \in \mathcal{D}([a, b])$. Moreover, the value of these Riemann and Darboux integrals coincide and this is denoted as:*

$$\int_a^b f(x)\,dx.$$

Proof We prove the implication separately.

($\Rightarrow$): Fix $\varepsilon > 0$ and suppose that $f \in \mathcal{R}([a, b])$ with Riemann integral R. Then, there exists a $\delta > 0$ such that for any tagged partition $\mathcal{P}_\tau$ of $[a, b]$ with $||\mathcal{P}_\tau|| < \delta$, we must have $|R_{f,\mathcal{P}_\tau} - R| < \frac{\varepsilon}{4}$.
Pick any such partition $\mathcal{P}$, say $\mathcal{P} = \{x_0, x_1, x_2, \dots, x_n\}$ so that $|x_j - x_{j-1}| < \delta$ for all $j = 1, 2, \dots, n$. Any tag on this partition would give us the inequality $|R_{f,\mathcal{P}_\tau} - R| < \frac{\varepsilon}{4}$ or equivalently $R - \frac{\varepsilon}{4} < R_{f,\mathcal{P}_\tau} < R + \frac{\varepsilon}{4}$. Now we would like to find a set of tags on $\mathcal{P}$ such that the Riemann sum with respect to this tagged partition is close to the upper and lower Darboux sum on $\mathcal{P}$.
On each subinterval $I_j = [x_{j-1}, x_j]$ of the partition $\mathcal{P}$, define the supremum $M_j = \sup_{x \in I_j} f(x)$ for $j = 1, 2, \dots, n$ from which we will build our upper Darboux sum from. By the characterisation of supremum, for each j there exists a point $s_j \in I_j$ such that $M_j - \frac{\varepsilon}{4(b-a)} < f(s_j)$. The collection of points $\sigma = \{s_1, s_2, \dots, s_n\}$ is the first tag on $\mathcal{P}$ that we are going to use. Multiplying each of the inequality above with the corresponding $|x_j - x_{j-1}|$ and summing up over all the indices, via telescoping sum, we get:

$$\sum_{j=1}^{n} M_j |x_j - x_{j-1}| - \sum_{j=1}^{n} \frac{\varepsilon |x_j - x_{j-1}|}{4(b-a)} < \sum_{j=1}^{n} f(s_j)|x_j - x_{j-1}|$$
$$\Rightarrow \quad U_{f,\mathcal{P}} - \frac{\varepsilon}{4} < R_{f,\mathcal{P}_\sigma}.$$

Since $||\mathcal{P}_\sigma|| < \delta$, by the earlier estimate, we must have $R_{f,\mathcal{P}_\sigma} < \frac{\varepsilon}{4} + R$ and thus the above tells us:

$$U_{f,\mathcal{P}} - \frac{\varepsilon}{4} < R_{f,\mathcal{P}_\sigma} < \frac{\varepsilon}{4} + R \quad \Rightarrow \quad U_{f,\mathcal{P}} < \frac{\varepsilon}{2} + R. \tag{15.4}$$

Now we repeat the above for the infimum $m_j = \inf_{x \in I_j} f(x)$ for $j = 1, 2, \dots, n$ instead. From the characterisation of the infimum, we can find tags $r_j \in I_j$ such that $f(r_j) < m_j + \frac{\varepsilon}{4(b-a)}$. Call these tags $\rho = \{r_1, r_2, \dots, r_n\}$. By a similar argument for the supremum, we get:

$$\sum_{j=1}^{n} f(r_j)|x_j - x_{j-1}| < \sum_{j=1}^{n} m|x_j - x_{j-1}| + \sum_{j=1}^{n} \frac{\varepsilon|x_j - x_{j-1}|}{4(b-a)}$$

$$\Rightarrow \quad R_{f,\mathcal{P}_\rho} < L_{f,\mathcal{P}} + \frac{\varepsilon}{4}.$$

Applying the assumed inequality $R - \frac{\varepsilon}{4} < R_{f,\mathcal{P}_\rho}$ we then get:

$$R - \frac{\varepsilon}{4} < R_{f,\mathcal{P}_\rho} < L_{f,\mathcal{P}} + \frac{\varepsilon}{4} \quad \Rightarrow \quad R - \frac{\varepsilon}{2} < L_{f,\mathcal{P}}. \tag{15.5}$$

Combining the inequalities (15.4) and (15.5) gives us $U_{f,\mathcal{P}} - L_{f,\mathcal{P}} < \varepsilon$ which is the ε-criterion for Darboux integration. Hence $f \in \mathcal{D}([a, b])$ with $L_f = U_f = D$. Finally, since we know that $D \leq U_{f,\mathcal{P}}$ and $D \geq L_{f,\mathcal{P}}$, by putting these in inequalities (15.4) and (15.5), we have:

$$R - \frac{\varepsilon}{2} < L_{f,\mathcal{P}} \leq D \leq U_{f,\mathcal{P}} < \frac{\varepsilon}{2} + R \quad \Rightarrow \quad |D - R| < \frac{\varepsilon}{2},$$

for any $\varepsilon > 0$. This implies $|D - R| = 0$ and hence $D = R$.

($\Leftarrow$): Assume that $f \in \mathcal{D}([a, b])$ with Darboux integral D. Fix $\varepsilon > 0$. We want to find a $\delta > 0$ such that for any tagged partition $\mathcal{P}_\tau$ with $||\mathcal{P}_\tau|| < \delta$, we have $|R_{f,\mathcal{P}_\tau} - D| < \varepsilon$.

By the ε-criterion for Darboux integration, there exists a partition $\mathcal{Q}$ of $[a, b]$ such that $U_{f,\mathcal{Q}} - L_{f,\mathcal{Q}} < \frac{\varepsilon}{2}$. Suppose that this partition contains $q + 2$ points where $q \in \mathbb{N}$, including the endpoints a and b. The function f is bounded on $[a, b]$, so there is some $M > 0$ such that $|f(x)| \leq M$ for all $x \in [a, b]$. We claim that $\delta = \frac{\varepsilon}{8Mq}$ works. We prove this in two steps.

1. The first step is to show that for any partition $\mathcal{P}$ of $[a, b]$ such that $||\mathcal{P}|| < \delta$, we have the bound $U_{f,\mathcal{P}} - L_{f,\mathcal{P}} < \varepsilon$. Pick any such partition $\mathcal{P}$ and consider the refinement $\mathcal{S} = \mathcal{P} \cup \mathcal{Q}$ where $\mathcal{Q}$ is the partition we have found above. Since $\mathcal{Q} \subseteq \mathcal{S}$, we have $L_{f,\mathcal{Q}} \leq L_{f,\mathcal{S}}$ and $U_{f,\mathcal{S}} \leq U_{f,\mathcal{Q}}$. Combining these inequalities, we get:

$$U_{f,\mathcal{S}} - L_{f,\mathcal{S}} \leq U_{f,\mathcal{Q}} - L_{f,\mathcal{Q}} < \frac{\varepsilon}{2}. \tag{15.6}$$

Now note that $\mathcal{P} \subseteq \mathcal{S}$, so we have $\mathcal{P} = \{x_0, x_1, x_2, \ldots, x_m\}$ and $\mathcal{S} = \{y_0, y_1, y_2, \ldots, y_n\}$ where $m \leq n$ and for each $j = 0, 1, 2, \ldots, m$, we have $x_j = y_k$ for some $k = 0, 1, \ldots, n$. Consider the following quantity:

$$U_{f,\mathcal{P}} - U_{f,\mathcal{S}} = \sum_{j=1}^{m} M_j|x_j - x_{j-1}| - \sum_{k=1}^{n} M'_k|y_k - y_{k-1}|, \tag{15.7}$$

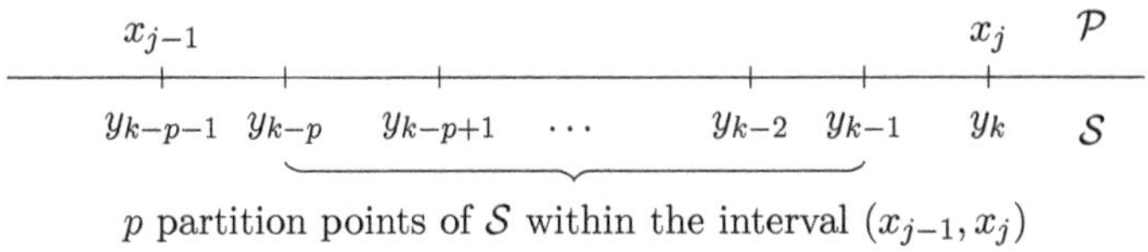

Fig. 15.11 The points $x_{j-1}, x_j \in \mathcal{P}$ and the $p+2$ points $y_{k-p-1}, \ldots, y_k \in \mathcal{S}$

where $M_j = \sup_{x \in I_j} f(x)$ and $M'_k = \sup_{y \in H_k} f(y)$ for $I_j = [x_{j-1}, x_j]$ and $H_j = [y_{k-1}, y_k]$. We would like to find an upper bound for this difference. To do this, we have to match each term in the first summation to the corresponding terms in the second summation. There would be two cases, namely: for each $j \in \{1, 2, 3 \ldots, m\}$, either there are no points of $\mathcal{S}$ in the interval (x_{j-1}, x_j) or there is at least one point of $\mathcal{S}$ in the interval (x_{j-1}, x_j).

(a) For the first case, if there are no points of $\mathcal{S}$ in the interval (x_{j-1}, x_j), then there exists a $k \in \{1, 2, \ldots, n\}$ such that $y_{k-1} = x_{j-1}$ and $y_k = x_j$ since $\mathcal{S}$ is a refinement of $\mathcal{P}$. This also implies that $M_j = M'_k$ and hence $M_j|x_j - x_{j-1}| - M'_k|y_k - y_{k-1}| = 0$.

(b) For the second case, suppose that there are $p \geq 1$ points of $\mathcal{S}$ within the interval (x_{j-1}, x_j). Then, there exists a $k \in \{p+1, \ldots, n\}$ such that $y_{k-p-1} = x_{j-1}$ and $y_k = x_j$ are the endpoints of this interval. Thus, we have p points $y_{k-p}, y_{k-p+1}, \ldots, y_{k-1} \in (x_{j-1}, x_j)$. See Fig. 15.11 for a visualisation of this case.
For this case, we can compare the one subinterval (x_{j-1}, x_j) of $\mathcal{P}$ with $p+1$ subintervals $(y_{k-p-1}, y_{k-p}), \ldots, (y_{k-1}, y_k)$ of $\mathcal{S}$ and bound the difference via telescoping as follows:

$$\begin{aligned}
&M_j|x_j - x_{j-1}| - \sum_{i=k-p}^{k} M'_i|y_i - y_{i-1}| \\
&\leq M|x_j - x_{j-1}| + \sum_{i=k-p}^{k} M|y_i - y_{i-1}| \\
&= M|x_j - x_{j-1}| + M|y_k - y_{k-p-1}| \\
&= M|x_j - x_{j-1}| + M|x_j - x_{j-1}| < 2M\delta,
\end{aligned}$$

since $-M \leq M'_i$ for all $i = k-p, \ldots, k$, $M_j \leq M$, and $||\mathcal{P}|| < \delta$.

Now we combine both of these cases to get the full bound on $U_{f,\mathcal{P}} - U_{f,\mathcal{S}}$ in (15.7). A subinterval $I_j = [x_{j-1}, x_j]$ of $\mathcal{P}$ contributes 0 to the bound on (15.7) if there are no point of $\mathcal{S}$ in (x_{j-1}, x_j) and contributes $2M\delta$ to the bound on (15.7) if there is at least one point of $\mathcal{S}$ in (x_{j-1}, x_j). For the latter, since $\mathcal{S} = \mathcal{P} \cup \mathcal{Q}$, these points (if there are any) must come from $\mathcal{Q}$.

Since there are at most q points in $\mathcal{Q}$ which are not in $\mathcal{P}$ (the endpoints a and b are in both), there would be at most q subintervals of $\mathcal{P}$ that would contribute the bound $2M\delta$ and the others would contribute nothing to the difference in (15.7). Thus, we have $U_{f,\mathcal{P}} - U_{f,\mathcal{S}} < 2M\delta q$. Using a similar argument, one can prove that $L_{f,\mathcal{S}} - L_{f,\mathcal{P}} < 2M\delta q$.
To get the desired bound, we combine these two inequalities with (15.6) to get:

$$U_{f,\mathcal{P}} - U_{f,\mathcal{S}} + L_{f,\mathcal{S}} - L_{f,\mathcal{P}} < 4M\delta q$$

$$\Rightarrow \quad U_{f,\mathcal{P}} - L_{f,\mathcal{P}} < U_{f,\mathcal{S}} - L_{f,\mathcal{S}} + 4M\delta q < \frac{\varepsilon}{2} + \frac{4Mq\varepsilon}{8Mq} = \varepsilon,$$

which is true for any partition $\mathcal{P}$ with size $||\mathcal{P}|| < \delta = \frac{\varepsilon}{8Mq}$.

2. Moving on to the second step, we now show that for any tagged partition $\mathcal{P}_\tau$ with size $||\mathcal{P}_\tau|| < \delta$, we have $|R_{f,\mathcal{P}_\tau} - D| < \varepsilon$. Pick any such tagged partition $\mathcal{P}_\tau$, say $\mathcal{P} = \{x_0, x_1, x_2, \dots, x_n\}$ and $\tau = \{p_1, p_2, \dots, p_n\}$ with $p_j \in I_j = [x_{j-1}, x_j]$. Since the function f has Darboux integral D, we have $L_{f,\mathcal{P}} \leq L_f = D = U_f \leq U_{f,\mathcal{P}}$.
On the other hand, from the definition of Riemann and Darboux sums, since in each subinterval of the partition we have $\inf_{x\in I_j} f(x) = m_j \leq f(p_j) \leq M_j = \sup_{x\in I_j} f(x)$, by multiplying with $|x_j - x_{j-1}|$ and summing up over all $j = 1, 2, \dots, n$, we have:

$$\sum_{j=1}^{n} m_j|x_j - x_{j-1}| \leq \sum_{j=1}^{n} f(p_j)|x_j - x_{j-1}| \leq \sum_{j=1}^{n} M_j|x_{j-1} - x_j|,$$

which implies $L_{f,\mathcal{P}} \leq R_{f,\mathcal{P}_\tau} \leq U_{f,\mathcal{P}}$. Combining this with the inequality $L_{f,\mathcal{P}} \leq D \leq U_{f,\mathcal{P}}$, we obtain $|R_{f,\mathcal{P}_\tau} - D| \leq U_{f,\mathcal{P}} - L_{f,\mathcal{P}}$. From the first step, we know that $U_{f,\mathcal{P}} - L_{f,\mathcal{P}} < \varepsilon$ and so we obtain $|R_{f,\mathcal{P}_\tau} - D| < \varepsilon$.

Thus, $f \in \mathcal{R}([a, b])$ with Riemann integral equal to D. □

Therefore, the construction by Riemann and Darboux give identical values for bounded functions on a closed bounded domain. How about for unbounded functions on $[a, b]$? The definition of Darboux integral specifically requires the function f to be bounded. This is necessary to ensure that the supremum and infimum of f in each of the subintervals for any partition $\mathcal{P}$ of $[a, b]$, and hence the upper and lower Darboux sums, all exist.

On the other hand, the Riemann integral definition and construction did not specify that the function f should be bounded. So there might still be an ambiguity for unbounded functions. Luckily for us, this ambiguity does not exist due to the following result:

Proposition 15.4.2 *If $f : [a, b] \to \mathbb{R}$ is a Riemann integrable function, then f is bounded.*

Proof Suppose for contradiction that the function $f : [a, b] \to \mathbb{R}$ is unbounded and is Riemann integrable with Riemann integral R. Therefore, using the definition for Riemann integrability, for $\varepsilon = 1$ there exists a $\delta > 0$ such that for any tagged partition $\mathcal{P}_\tau$ with $||\mathcal{P}_\tau|| < \delta$, we have $|R_{f,\mathcal{P}_\tau} - R| < 1$. This means:

$$|R_{f,\mathcal{P}_\tau}| = |R_{f,\mathcal{P}_\tau} - R + R| \leq |R_{f,\mathcal{P}_\tau} - R| + |R| < 1 + |R|, \tag{15.8}$$

for any such tagged partition. Pick a partition $\mathcal{P} = \{x_0, x_1, x_2, \dots, x_n\}$ such that $||\mathcal{P}|| < \delta$. We want to find a tag on this partition that would give us a contradiction.

Since the function f is unbounded in $[a, b]$, there must exist at least one of the subintervals of the partition $\mathcal{P}$ where the function is unbounded. WLOG, assume that it is $I_1 = [x_0, x_1]$. Pick any tags $p_j \in I_j = [x_{j-1}, x_j]$ for $j = 2, 3, 4, \dots, n$ and we want to choose a special tag p_1 in the first subinterval that would give us a contradiction. Since f is unbounded in I_1, for any $K > 0$ there exists an $x \in I_1$ such that $|f(x)| > K$. In particular, there exists a point $p_1 \in I_1$ for which:

$$|f(p_1)| > \frac{|R| + 1 + \sum_{j=2}^n |f(p_j)||x_j - x_{j-1}|}{|x_1 - x_0|}.$$

So, we choose the tags $\tau = \{p_1, p_2, \dots, p_n\}$ for the partition $\mathcal{P}$ and compute the Riemann sum for this tagged partition:

$$R_{f,\mathcal{P}_\tau} = \sum_{j=1}^n f(p_j)|x_j - x_{j-1}| = f(p_1)|x_1 - x_0| + \sum_{j=2}^n f(p_j)|x_j - x_{j-1}|.$$

Applying the reverse triangle inequality, we get:

$$\begin{aligned}
|R_{f,\mathcal{P}_\tau}| &= \left| f(p_1)|x_1 - x_0| + \sum_{j=2}^n f(p_j)|x_j - x_{j-1}| \right| \\
&\geq |f(p_1)||x_1 - x_0| - \sum_{j=2}^n |f(p_j)||x_j - x_{j-1}| \\
&> |R| + 1 + \sum_{j=2}^n |f(p_j)||x_j - x_{j-1}| - \sum_{j=2}^n |f(p_j)||x_j - x_{j-1}| = |R| + 1,
\end{aligned}$$

where the final inequality is obtained based on the choice for p_1. However, this contradicts inequality (15.8). Thus, we conclude that f cannot be Riemann integrable over $[a, b]$. □

So we have no worries at all now since both of these integrals are only defined for bounded functions and they are the same for these functions. Hence, there is no ambiguity with the notation we used for Darboux and Riemann integrals and we have:

Corollary 15.4.3 *Let $X \subseteq \mathbb{R}$ be a compact interval. Then, $\mathcal{D}(X) = \mathcal{R}(X)$.*

So proving any facts for one also proves for the the other, giving us the flexibility while working with them.

The definition of Darboux integral is easier to implement and more constructive as opposed to the definition of the Riemann integral. This is because for Darboux integrals we do not have extra variables in the form of the tags for the partitions that we have to deal with. However, despite the difficulty dealing with them, the definition of the Riemann integral can be very useful in some scenarios. We shall see later that its formulation is useful in the proof for Theorem 16.1.4 and in Sect. 16.2.

We end this section by noting that in many introductory real analysis literature, the Darboux integral is usually presented as the definition of the Riemann integral even though they are, strictly speaking, two different constructions. However, since they are equivalent, from now on, we are going to refer to both of them as Riemann integrals and any Riemann or Darboux integrable functions $f : [a, b] \to \mathbb{R}$ are thus called Riemann integrable functions with $R = D = \int_a^b f(x)\,dx$. The set of Darboux and Riemann integrable functions over a compact interval X is denoted as $\mathcal{R}(X)$

15.5 Properties of Riemann Integrals

Now that we have clarified the definitions of Riemann and Darboux integrals, we move on to study their properties. Since the construction for Darboux integral is more instructive, we shall prove these results using the Darboux definition. This would not cause a problem as we have seen that both of the integrals are equivalent as we have seen in Theorem 15.4.1. We prove the sublinearity of the upper Darboux integral and the superlinearity of the lower Darboux integral first.

Lemma 15.5.1 *For any bounded functions $f, g : [a, b] \to \mathbb{R}$, we have $U_{f+g} \leq U_f + U_g$ and $L_f + L_g \leq L_{f+g}$.*

Proof We prove the first inequality only. The second inequality can be proven in a similar manner. Fix an arbitrary partition $\mathcal{P} = \{x_0, x_1, \ldots, x_n\}$ of $[a, b]$. For $j = 1, 2, \ldots, n$, let $I_j = [x_{j-1}, x_j]$ be the subintervals of $\mathcal{P}$ and define:

$$M_j^f = \sup_{x\in I_j} f(x), \quad M_j^g = \sup_{x\in I_j} g(x), \quad \text{and} \quad M_j^{f+g} = \sup_{x\in I_j}(f(x) + g(x)).$$

For any $j \in \{1, 2\ldots, n\}$ and any $x \in I_j$, we clearly have the upper bound $f(x) + g(x) \leq M_j^f + M_j^g$. Thus, by taking the supremum over $x \in I_j$, we have $M_j^{f+g} \leq M_j^f + M_j^g$ for all $j \in \{1, 2, \ldots, n\}$. Hence, by definition of the upper Darboux sum, we have $U_{f+g,\mathcal{P}} \leq U_{f,\mathcal{P}} + U_{g,\mathcal{P}}$.

Since $U_{f+g} = \inf_{\mathcal{P}} U_{f+g,\mathcal{P}} \leq U_{f+g,\mathcal{P}}$ for any partition $\mathcal{P}$, we then have:

$$U_{f+g} \leq U_{f,\mathcal{P}} + U_{g,\mathcal{P}}, \tag{15.9}$$

for any partition $\mathcal{P}$ of $[a, b]$. We want to relate the RHS of (15.9) with U_f and U_g.

Fix $\varepsilon > 0$. By the characterisation of infimum, there exist partitions $\mathcal{P}_1$ and $\mathcal{P}_2$ of $[a, b]$ such that $U_{f,\mathcal{P}_1} < U_f + \frac{\varepsilon}{2}$ and $U_{g,\mathcal{P}_2} < U_g + \frac{\varepsilon}{2}$. By defining the refinement $\mathcal{P}_3 = \mathcal{P}_1 \cup \mathcal{P}_2$, using inequality (15.3), we get:

$$U_{f,\mathcal{P}_3} \leq U_{f,\mathcal{P}_1} < U_f + \frac{\varepsilon}{2} \quad \text{and} \quad U_{g,\mathcal{P}_3} \leq U_{g,\mathcal{P}_2} < U_g + \frac{\varepsilon}{2}. \tag{15.10}$$

Setting $\mathcal{P} = \mathcal{P}_3$ in the inequality (15.9) and applying the inequalities (15.10), we have:

$$U_{f+g} \leq U_{f,\mathcal{P}_3} + U_{g,\mathcal{P}_3} < U_f + U_g + \varepsilon.$$

Since $\varepsilon > 0$ is arbitrary, we obtain $U_{f+g} \leq U_f + U_g$, which is what we wanted to prove. The other inequality is also obtained in a similar way. □

We are now going to list some properties of the Riemann integral.

Proposition 15.5.2 *Suppose that $f, g \in \mathcal{R}([a, b])$. Then:*

1. *The Riemann integral is linear over $\mathbb{R}$. Namely, for constants $\lambda, \kappa \in \mathbb{R}$ we have $\lambda f + \kappa g \in \mathcal{R}([a, b])$ with:*

$$\int_a^b \lambda f(x) + \kappa g(x)\, dx = \lambda \int_a^b f(x)\, dx + \kappa \int_a^b g(x)\, dx.$$

2. *We have $fg \in \mathcal{R}([a, b])$.*

Proof We prove the assertions one by one:

1. We prove first for the case $\lambda = \kappa = 1$. Since f and g are Riemann integrable in $[a, b]$, we have $U_f = L_f$ and $U_g = L_g$. Furthermore, the lower and upper Darboux integrals must satisfy $L_{f+g} \leq U_{f+g}$. Putting these facts together with Lemma 15.5.1 we get:

$$L_f + L_g \leq L_{f+g} \leq U_{f+g} \leq U_f + U_g = L_f + L_g.$$

Thus, all the inequalities above are in fact equalities and therefore $U_{f+g} = L_{f+g}$ which says that $f + g$ is integrable. Moreover, since the above are all equalities, we have the identity $\int_a^b f(x) + g(x)\,dx = U_{f+g} = U_f + U_g = \int_a^b f(x)\,dx + \int_a^b g(x)\,dx$.

Next, we shall prove that λf is Riemann integrable for any $\lambda \in \mathbb{R}$. Clearly if $\lambda = 0$, the function $\lambda f = 0$ is Riemann integrable. For any $\lambda > 0$ and any partition $\mathcal{P}$ of $[a, b]$, if $I_j = [x_{j-1}, x_j]$ is a subinterval of the partition $\mathcal{P}$ and $m_j = \inf_{x\in I_j} f(x)$ we have:

$$\lambda m_j = \lambda \inf_{x\in I_j} f(x) = \inf_{x\in I_j} \lambda f(x),$$

which implies:

$$\lambda L_{f,\mathcal{P}} = \lambda \sum_{j=1}^{n} m_j |x_j - x_{j-1}| = \sum_{j=1}^{n} \lambda m_j |x_j - x_{j-1}| = L_{\lambda f,\mathcal{P}},$$

and thus:

$$L_{\lambda f} = \sup_{\mathcal{P}} L_{\lambda f,\mathcal{P}} = \sup_{\mathcal{P}} \lambda L_{f,\mathcal{P}} = \lambda \sup_{\mathcal{P}} L_{f,\mathcal{P}} = \lambda L_f.$$

Similarly, we can show that $U_{\lambda f} = \lambda U_f$. Thus, since $f \in \mathcal{R}([a, b])$, we have $U_{\lambda f} = \lambda U_f = \lambda L_f = L_{\lambda f}$, which implies $\lambda f \in \mathcal{R}([a, b])$ for $\lambda > 0$.

On the other hand, for $\lambda < 0$, using the same partition as above and via Lemma 3.6.8, we have:

$$\lambda m_j = \lambda \inf_{x\in I_j} f(x) = \sup_{x\in I_j} \lambda f(x),$$

and so:

$$\lambda L_{f,\mathcal{P}} = \lambda \sum_{j=1}^{n} m_j |x_j - x_{j-1}| = \sum_{j=1}^{n} \lambda m_j |x_j - x_{j-1}| = U_{\lambda f,\mathcal{P}},$$

which implies:

$$U_{\lambda f} = \inf_{\mathcal{P}} U_{\lambda f,\mathcal{P}} = \inf_{\mathcal{P}} \lambda L_{f,\mathcal{P}} = \lambda \sup_{\mathcal{P}} L_{f,\mathcal{P}} = \lambda L_f.$$

Similarly, we can show that $L_{\lambda f} = \lambda U_f$. And so, since $f \in \mathcal{R}([a, b])$, we have $U_{\lambda f} = \lambda L_f = \lambda U_f = L_{\lambda f}$ implying that $\lambda f \in \mathcal{R}([a, b])$ for $\lambda < 0$. Thus, λf is Riemann integrable for any $\lambda \in \mathbb{R}$. Furthermore, in all of the cases above, we have $\int_a^b \lambda f(x)\,dx = \lambda \int_a^b f(x)\,dx$.

Putting the above facts together, the combination $\lambda f + \kappa g$ is also Riemann integrable for any $\lambda, \kappa \in \mathbb{R}$ with:

$$\int_a^b \lambda f(x) + \kappa g(x)\, dx = \int_a^b \lambda f(x)\, dx + \int_a^b \kappa g(x)\, dx$$
$$= \lambda \int_a^b f(x)\, dx + \kappa \int_a^b g(x)\, dx.$$

2. We note that $fg = \frac{1}{4}((f+g)^2 - (f-g)^2)$. Since $f \pm g$ is Riemann integrable and scales of Riemann integrable functions are Riemann integrable, to show that fg is Riemann integrable, it is enough to show that if $h \in \mathcal{R}([a,b])$, then $h^2 \in \mathcal{R}([a,b])$.
We shall show this using the ε-criterion of integrals. Since h is Riemann integrable in $[a,b]$, it is necessarily bounded, namely there exists a $K > 0$ such that $|h(x)| \leq K$ for all $x \in [a,b]$.
Fix $\varepsilon > 0$. We aim to show that there exists a partition $\mathcal{P}$ of $[a,b]$ such that $U_{h^2,\mathcal{P}} - L_{h^2,\mathcal{P}} < \varepsilon$. With this $\varepsilon > 0$, since h is Riemann integrable, there exists a partition $\mathcal{P}$ of $[a,b]$ such that:

$$U_{h,\mathcal{P}} - L_{h,\mathcal{P}} < \frac{\varepsilon}{2K}.$$

We claim that this is the partition that we are looking for. Let $I_j = [x_{j-1}, x_j]$ be a subinterval of the partition $\mathcal{P}$. We first note that for any two points $x, y \in I_j$ we have:

$$h(x)^2 - h(y)^2 = (h(x)+h(y))(h(x)-h(y)) \leq 2K|h(x)-h(y)|$$
$$\leq 2K(M_j^h - m_j^h),$$

and since x and y are arbitrary, $2K(M_j^h - m_j^h)$ is an upper bound for the difference $h(x)^2 - h(y)^2$.
On the other hand, the supremum of this quantity over all pairs of points $x, y \in I_j$ is:

$$\sup_{x,y \in I_j} (h(x)^2 - h(y)^2) = \sup_{x \in I_j} h(x)^2 + \sup_{y \in I_j} (-h(y)^2)$$
$$= \sup_{x \in I_j} h(x)^2 - \inf_{y \in I_j} h(y)^2 = M_j^{h^2} - m_j^{h^2},$$

and so we have the important inequality $M_j^{h^2} - m_j^{h^2} \leq 2K(M_j^h - m_j^h)$ for each $j = 1, 2, \ldots, n$. Thus, by constructing the Darboux sums for the function h^2 with respect to this partition $\mathcal{P}$, we have:

$$\begin{aligned}U_{h^2,\mathcal{P}} - L_{h^2,\mathcal{P}} &= \sum_{j=1}^{n}(M_j^{h^2} - m_j^{h^2})|x_j - x_{j-1}| \\ &\leq \sum_{j=1}^{n} 2K(M_j^h - m_j^h)|x_j - x_{j-1}| \\ &= 2K(U_{h,\mathcal{P}} - L_{h,\mathcal{P}}) < 2K\frac{\varepsilon}{2K} = \varepsilon,\end{aligned}$$

which is the ε-criterion for Darboux integration. This proves $h^2 \in \mathcal{R}([a,b])$. □

Proposition 15.5.2(1) tells us that the Riemann integral is additive with respect to its integrand. The next result tells us that it is also additive over disjoint domains of integration.

Proposition 15.5.3 *Suppose that $f \in \mathcal{R}([a,b])$. If $c \in [a,b]$, then $f \in \mathcal{R}([a,c])$ and $f \in \mathcal{R}([c,b])$. Furthermore, we have the equality:*

$$\int_a^b f(x)\,dx = \int_a^c f(x)\,dx + \int_c^b f(x)\,dx. \tag{15.11}$$

Proof We first need to ensure that both the integrals $\int_a^c f(x)\,dx$ and $\int_c^b f(x)\,dx$ exist. We use the ε-criterion for Darboux integrability. Fix $\varepsilon > 0$. Since f is Riemann integrable, there exists a partition $\mathcal{P}$ of $[a,b]$ such that:

$$U_{f,\mathcal{P}} - L_{f,\mathcal{P}} < \varepsilon. \tag{15.12}$$

Define $\mathcal{P}' = \mathcal{P} \cup \{c\}$ and let $\mathcal{P}_1$ and $\mathcal{P}_2$ be a partition of $[a,c]$ and $[c,b]$ obtained from $\mathcal{P}'$ respectively. In other words, $\mathcal{P}_1 = \mathcal{P}' \cap [a,c]$ and $\mathcal{P}_2 = \mathcal{P}' \cap [c,b]$. Thus, by definition, we have:

$$\begin{aligned}U_{f,\mathcal{P}_1} + U_{f,\mathcal{P}_2} &= U_{f,\mathcal{P}'} \leq U_{f,\mathcal{P}}, \\ L_{f,\mathcal{P}_1} + L_{f,\mathcal{P}_2} &= L_{f,\mathcal{P}'} \geq L_{f,\mathcal{P}}.\end{aligned}$$

Hence, inequality (15.12) reads off as:

$$\begin{aligned}\varepsilon > U_{f,\mathcal{P}} - L_{f,\mathcal{P}} &\geq U_{f,\mathcal{P}_1} + U_{f,\mathcal{P}_2} - L_{f,\mathcal{P}_1} - L_{f,\mathcal{P}_2} \\ &= (U_{f,\mathcal{P}_1} - L_{f,\mathcal{P}_1}) + (U_{f,\mathcal{P}_2} - L_{f,\mathcal{P}_2}),\end{aligned}$$

and since both of the bracketed terms are non-negative and they add up to less than ε, each of them must be smaller than ε. To recap, for this fixed $\varepsilon > 0$, we have found two partitions $\mathcal{P}_1$ and $\mathcal{P}_2$ of $[a, c]$ and $[c, b]$ respectively such that:

$$U_{f,\mathcal{P}_1} - L_{f,\mathcal{P}_1} < \varepsilon \quad \text{and} \quad U_{f,\mathcal{P}_2} - L_{f,\mathcal{P}_2} < \varepsilon,$$

which, by the ε-criterion, implies that f is Riemann integrable in $[a, c]$ and $[c, b]$.

To show the equality (15.11), let $\mathcal{P} = \{x_0, x_1, \ldots, x_n\}$ be an arbitrary partition of $[a, b]$. Define $\mathcal{P}' = \mathcal{P} \cup \{c\}$, $\mathcal{P}_1$, and $\mathcal{P}_2$ as above. Thus, we have $U_{f,\mathcal{P}'} = U_{f,\mathcal{P}_1} + U_{f,\mathcal{P}_2}$. By definition of Darboux integrals, we have:

$$\int_a^c f(x)\,dx + \int_c^b f(x)\,dx \le U_{f,\mathcal{P}_1} + U_{f,\mathcal{P}_2} = U_{f,\mathcal{P}'} \le U_{f,\mathcal{P}},$$

and since $\mathcal{P}$ is arbitrary, by taking the infimum over $\mathcal{P}$, we have:

$$\int_a^c f(x)\,dx + \int_c^b f(x)\,dx \le \inf_{\mathcal{P}} U_{f,\mathcal{P}} = \int_a^b f(x)\,dx.$$

By reversing the argument, for the quantity $L_{f,\mathcal{P}}$ we get:

$$\int_a^c f(x)\,dx + \int_c^b f(x)\,dx \ge \sup_{\mathcal{P}} L_{f,\mathcal{P}} = \int_a^b f(x)\,dx,$$

and hence, by sandwiching, we get the desired equality. □

This result also gives us an important definition. The definition of the integral $\int_a^b f(x)\,dx$ that we have seen so far was, by construction, only defined for the limits $a < b$. Here, we would like to extend the notation of the integral where the limits are reversed.

Definition 15.5.4 For $a < b$ and Riemann integrable function $f : [a, b] \to \mathbb{R}$, we define:

$$\int_b^a f(x)\,dx = -\int_a^b f(x)\,dx.$$

where the limits of the integral are reversed.

This is motivated by Proposition 15.5.3. Namely, if we want to extend the proposition to work for any $c, d, e \in [a, b]$ (in any order) as well, we would have:

$$\int_c^e f(x)\,dx = \int_c^d f(x)\,dx + \int_d^e f(x)\,dx.$$

Since this is extended to be defined for c, d, e in any order, we can also set $e = c$ to get:

$$0 = \int_c^c f(x)\,dx = \int_c^d f(x)\,dx + \int_d^c f(x)\,dx \quad \Rightarrow \quad \int_c^d f(x)\,dx = -\int_d^c f(x)\,dx,$$

agreeing with Definition 15.5.4.

Riemann integrals also respect ordering of functions. We have the following result:

Proposition 15.5.5 *Suppose that* $f, g \in \mathcal{R}([a, b])$.

1. *If* $0 \leq f$, *then* $0 \leq \int_a^b f(x)\,dx$.
2. *If* $f \leq g$, *then* $\int_a^b f(x)\,dx \leq \int_a^b g(x)\,dx$.
3. *The function* $|f| : [a, b] \to \mathbb{R}$ *is also Riemann integrable. Moreover, we have the "triangle inequality" for integrals, namely* $|\int_a^b f(x)\,dx| \leq \int_a^b |f(x)|\,dx$.
4. *If there are constants* $m, M \in \mathbb{R}$ *such that* $m \leq f(x) \leq M$ *for all* $x \in [a, b]$, *then* $m(b-a) \leq \int_a^b f(x)\,dx \leq M(b-a)$.

Proof We prove only the first three assertions. The final assertion is simply a corollary of the second assertion.

1. For any partition $\mathcal{P}$ of $[a, b]$, the infimum of f in any of the subintervals of $\mathcal{P}$ is non-negative and hence $L_{f,\mathcal{P}} \geq 0$. Since the supremum of non-negative quantities must also be non-negative, we must have $L_f \geq 0$. Finally, since f is Riemann integrable, we have $\int_a^b f(x)\,dx = L_f \geq 0$.
2. Since $f \leq g$, the quantity $g - f$ is non-negative and therefore, by the first assertion, we must have $\int_a^b g(x) - f(x)\,dx \geq 0$. By linearity of integrals from Proposition 15.5.2 and algebraic manipulations, we get the result.
3. We shall only prove the fact that the function $|f|$ is Riemann integrable here. Fix $\varepsilon > 0$. Since f is Riemann integrable, there exists a partition $\mathcal{P} = \{x_0, x_1, \ldots, x_n\}$ of $[a, b]$ such that $U_{f,\mathcal{P}} - L_{f,\mathcal{P}} < \frac{\varepsilon}{2}$. If we denote $I_j = [x_{j-1}, x_j]$, $m_j = \inf_{x \in I_j} f(x)$, and $M_j = \sup_{x \in I_j} f(x)$ for every $j = 1, 2, \ldots, n$, for any $x, y \in I_j$ we have $m_j \leq f(x) \leq M_j$ and $-m_j \geq -f(y) \geq -M_j$. Combining these two inequalities, for any $x, y \in I_j$ we have:

$$-(M_j - m_j) \leq f(x) - f(y) \leq M_j - m_j \quad \Rightarrow \quad |f(x) - f(y)| \leq M_j - m_j.$$

Applying reverse triangle inequality to the above, we obtain:

$$|f(x)| - |f(y)| \leq |f(x) - f(y)| \leq M_j - m_j, \tag{15.13}$$

for any $x, y \in I_j$. Thus, $M_j - m_j$ is an upper bound for this quantity.

Now denote $m'_j = \inf_{x\in I_j}|f(x)|$ and $M'_j = \sup_{x\in I_j}|f(x)|$. Taking the supremum of the LHS of inequality (15.13) over all pairs of points $x, y \in I_j$, we obtain:

$$\sup_{x,y\in I_j}(|f(x)| - |f(y)|) = \sup_{x\in I_j}|f(x)| - \inf_{y\in I_j}|f(y)| = M'_j - m'_j.$$

Hence, we have the inequality $M'_j - m'_j \leq M_j - m_j$ for any $j = 1, 2, \dots, n$. So, if we compute the Darboux upper and lower sums for $|f|$ on the partition $\mathcal{P}$ and take their difference, we get:

$$\begin{aligned} U_{|f|,\mathcal{P}} - L_{|f|,\mathcal{P}} &= \sum_{j=1}^{n}(M'_j - m'_j)|x_j - x_{j-1}| \\ &\leq \sum_{j=1}^{n}(M_j - m_j)|x_j - x_{j-1}| = U_{f,\mathcal{P}} - L_{f,\mathcal{P}} < \varepsilon. \end{aligned}$$

Thus, by the ε-criterion for Darboux integrability, the function $|f|$ is also Riemann integrable over $[a, b]$.
The "triangle inequality" is a consequence of the second assertion since $-|f| \leq f \leq |f|$. □

Remark 15.5.6 We called the third assertion of the Proposition 15.5.5 the "triangle inequality" because, roughly, the integral is just a summation (which we then apply a limiting process to), reminiscent of the original triangle inequality which generalises to finite sums as $|\sum_{j=1}^{n} a_j| \leq \sum_{j=1}^{n} |a_j|$ for $a_j \in \mathbb{R}$.

15.6 Some Sufficient Conditions for Riemann Integrability

Finally, let us look at some examples of functions which are guaranteed to be integrable over a compact interval. Before we do so, we prove a useful lemma:

Lemma 15.6.1 *Let $f : [a, b] \to \mathbb{R}$ be a function which is non-zero at only finitely many points. Then, $f \in \mathcal{R}([a, b])$ with $\int_a^b f(x)\,dx = 0$.*

Proof First, we assume that the function is non-negative. The function f is strictly positive at finitely many points, so suppose that these points are $\{p_1, p_2, \dots, p_k\} \subseteq [a, b]$. Define $M = \max\{f(p_j) : j = 1, 2, \dots, k\} > 0$. Consider a family of equispaced partitions $\mathcal{P}_n = \{x_0, x_1, \dots, x_n\}$ of $[a, b]$ that consists of $n + 1$ points so that the sizes of each subinterval in the partition $\mathcal{P}_n$ is $\frac{b-a}{n}$. Clearly the lower Darboux sum for this partition is $L_{f,\mathcal{P}_n} = 0$ as the infimum of the function over any subinterval of the partition is always 0. Thus, $L_f = \sup_{\mathcal{P}} L_{f,\mathcal{P}} \geq L_{f,\mathcal{P}_n} = 0$.

If $M_j = \sup_{x \in I_j} f(x)$ where $I_j = [x_{j-1}, x_j]$, the upper Darboux sum with respect to $\mathcal{P}_n$ is:

$$U_{f,\mathcal{P}_n} = \sum_{j=1}^{n} M_j |x_j - x_{j-1}| = \sum_{j=1}^{n} M_j \frac{b-a}{n}.$$

However, only at most k subintervals of $\mathcal{P}_n$ contain any of the points $\{p_1, p_2, \ldots, p_k\}$ so only at most k of these M_j are non-zero. Therefore, at most, only k of the terms in the sum contribute to $U_{f,\mathcal{P}_n}$. Furthermore, $M_j \leq M$ for all $j \in \{1, 2, \ldots, n\}$ which then gives us:

$$U_{f,\mathcal{P}_n} = \sum_{j=1}^{n} M_j \frac{b-a}{n} \leq \sum_{j=1}^{k} M \frac{b-a}{n} = \frac{kM(b-a)}{n}.$$

So, if we take the limit as $n \to \infty$, we get $U_f = \inf_{\mathcal{P}} U_{f,\mathcal{P}} \leq U_{f,\mathcal{P}_n} \to 0$. Combining this with the bound on L_f, we get $0 \leq L_f \leq U_f \leq 0$ and thus $U_f = L_f = 0$ which says that the function f is Riemann integrable with $\int_a^b f(x)\,dx = 0$.

The general case where f has mixed signs can be obtained by considering the positive and negative parts of the function separately. Namely, define $f^+ = \max(f, 0)$ and $f^- = -\min(f, 0)$ so that $f = f^+ - f^-$. Since f^+ and f^- are non-negative functions which are non-zero at only finitely many points, they are both Riemann integrable. Hence, by Proposition 15.5.2, the sum $f = f^+ - f^-$ is also Riemann integrable. □

As a result, we have:

Proposition 15.6.2 *Let $f : [a, b] \to \mathbb{R}$ be a bounded function. If f differs from a Riemann integrable function at finitely many points, then $f \in \mathcal{R}([a, b])$.*

Proof Suppose that f differs from a Riemann integrable function $g : [a, b] \to \mathbb{R}$ at finitely many points. Then, their difference $h = f - g$ is a function on $[a, b]$ that is non-zero at finitely many points and is thus Riemann integrable by Lemma 15.6.1. Thus, the function $f = h + g$ is also integrable on $[a, b]$ since it is a sum of two Riemann integrable functions on $[a, b]$. □

Next, we have a very useful criterion that can be used to check for Riemann integrability. Any monotone function is guaranteed to be Riemann integrable, thanks to the following result.

Proposition 15.6.3 *Let $f : [a, b] \to \mathbb{R}$ be a bounded function. If f is monotone on $[a, b]$, then $f \in \mathcal{R}([a, b])$.*

Proof WLOG, assume that f is increasing. Fix $\varepsilon > 0$. Let $\mathcal{P}_n$ be a partition of $[a, b]$ with $n+1$ equispaced points. In each partition subinterval $I_j = [x_{j-1}, x_j]$, since f is increasing, we necessarily have $\sup_{x \in I_j} f(x) = f(x_j)$ and $\inf_{x \in I_j} f(x) = f(x_{j-1})$. Then:

$$\begin{aligned} U_{f,\mathcal{P}_n} - L_{f,\mathcal{P}_n} &= \sum_{j=1}^{n} (\sup_{x \in I_j} f(x) - \inf_{x \in I_j} f(x))|x_j - x_{j-1}| \\ &= \sum_{j=1}^{n} (f(x_j) - f(x_{j-1}))\frac{b-a}{n} \\ &= \frac{b-a}{n}(f(b) - f(a)), \end{aligned}$$

by telescoping summation. We can then pick any $N > \frac{(b-a)(f(b)-f(a))}{\varepsilon}$ so that $U_{f,\mathcal{P}_N} - L_{f,\mathcal{P}_N} < \varepsilon$. Hence, this gives us the ε-criterion for Darboux integrability. □

Finally, continuous functions on compact intervals are also guaranteed to be Riemann integrable.

Proposition 15.6.4 *If $f : [a, b] \to \mathbb{R}$ is continuous on $[a, b]$, then $f \in \mathcal{R}([a, b])$.*

Proof We aim to show Riemann integrability of the function f using the ε-criterion. Fix $\varepsilon > 0$. We recall that a continuous function on a compact interval $[a, b]$ is also bounded (via Proposition 10.4.6) and uniformly continuous (via Theorem 10.6.10).

From the latter, there exists a $\delta > 0$ such that for any $x, y \in [a, b]$ with $|x - y| < \delta$ we would have $|f(x) - f(y)| < \frac{\varepsilon}{b-a}$. Choose a partition $\mathcal{P}$ of $[a, b]$ such that $||\mathcal{P}|| < \delta$. Since the function is bounded, the Darboux sums for this partition exist and the difference of the Darboux sums for this partition would then be:

$$U_{f,\mathcal{P}} - L_{f,\mathcal{P}} = \sum_{j=1}^{n} (M_j - m_j)|x_j - x_{j-1}|,$$

where $M_j = \sup_{x \in I_j} f(x)$ and $m_j = \inf_{x \in I_j} f(x)$ for each subinterval $I_j = [x_{j-1}, x_j]$. Since f is continuous, by the EVT, these supremum and infimum are attained at some $\xi_j, \zeta_j \in [x_{j-1}, x_j]$ respectively. Thus, $|\xi_j - \zeta_j| \leq ||\mathcal{P}|| < \delta$ and so we have $|M_j - m_j| = |f(\xi_j) - f(\zeta_j)| < \frac{\varepsilon}{b-a}$ for all $j = 1, 2, \ldots, n$ by uniform continuity of the function f. Therefore:

$$U_{f,\mathcal{P}} - L_{f,\mathcal{P}} = \sum_{j=1}^{n} (M_j - m_j)|x_j - x_{j-1}| < \frac{\varepsilon}{b-a} \sum_{j=1}^{n} |x_j - x_{j-1}| = \varepsilon,$$

by telescoping. Hence, this gives us the ε-criterion for Darboux integrability. □

Via Proposition 15.6.4, we can see that many functions are Riemann integrable. For example, the functions defined on $[0, \infty)$ as $f(x) = \ln(\ln(x+2))$, $g(x) = e^{-x^2}$, and $h(x) = \frac{\sin(x)}{x}$ extended continuously to $x = 0$ are all integrable over any compact intervals in $[0, \infty)$. In fact, as we shall see later in Theorem 19.7.7, we can give a full characterisation of any Riemann integrable functions using (almost everywhere) continuity.

We end this chapter by noting that there are many Riemann integrable functions. However, finding what their exact values from first principles by Riemann/Darboux sums can be tricky. Are there any easier ways to do it?

Exercises

15.1 (∗) Compute the Riemann integral of the following functions defined on $\mathbb{R}$ over the stated domain using first principles.

(a) $f(x) = x + 3$ on $[5, 10]$.

(b) $f(x) = |x|$ on $[-2, 2]$.

(c) $f(x) = \begin{cases} x & \text{if } x \in [0,1], \\ 1 & \text{if } x \in (1,2], \end{cases}$ on $[0, 2]$.

(d) The tent function $f(x) = \begin{cases} x+1 & \text{if } x \in [-1,0], \\ -x+1 & \text{if } x \in (0,1], \end{cases}$ on $[-1, 1]$.

(e) $f(x) = e^x$ on $[0, y]$ for an arbitrary $y \geq 0$.

(f) $f(x) = \begin{cases} 1 & \text{if } x \in [0,1], \\ 2 & \text{if } x \in (1,2], \end{cases}$ on $[0, y]$ for an arbitrary $y \in [0, 2]$.

15.2 (∗) Let $f : \mathbb{R} \to \mathbb{R}$ be defined as:

$$f(x) = \begin{cases} 1 & \text{if } x = \frac{1}{k} \text{ for } k \in \mathbb{N}, \\ 0 & \text{otherwise.} \end{cases}$$

(a) Show that for any partition $\mathcal{P}$ of $[0, 1]$, we have $L_{f,\mathcal{P}} = 0$.

(b) For all $n \geq 3$, construct a partition $\mathcal{P}_n$ with $2n - 1$ points such that the points $x = \frac{1}{2}, \frac{1}{3}, \ldots, \frac{1}{n-1}$ all lie in separate intervals of lengths $\frac{1}{n^2}$ and the points $x = \frac{1}{n}, \frac{1}{n+1}, \ldots$ all lie in one interval.

(c) Using the partition $\mathcal{P}_n$ from part (b), compute $U_{f,\mathcal{P}_n}$.

(d) Deduce that $U_f = 0$ and hence the function f is Riemann integrable on $[0, 1]$.

15.3 (∗) Recall Thomae's function from Exercise 10.8 which is $f : \mathbb{R} \to \mathbb{R}$ defined as:

$$f(x) = \begin{cases} \frac{1}{q} & \text{if } x \in \mathbb{Q} \text{ with } x = \frac{p}{q} \text{ where } p, q \text{ are coprime}, \\ 1 & \text{if } x = 0, \\ 0 & \text{if } x \in \bar{\mathbb{Q}}. \end{cases}$$

We have seen in Exercise 13.10 that this function is not differentiable anywhere.

(a) For any $n \geq 2$, consider a the set $A_n = \{x \in [0, 1] : f(x) \geq \frac{1}{n}\}$. Show that this set is finite with cardinality at most n^2.

(b) Let $\mathcal{P}_n$ be an equispaced partition of $[0, 1]$ with $n^3 + 1$ points. Show that $U_{f,\mathcal{P}_n} < \frac{2}{n}$.

(c) Using the ε-criterion of Darboux integration, show that $f \in \mathcal{R}([0, 1])$ and state its value.

15.4 Let $f : [a, b] \to \mathbb{R}$ be an integrable function over $[a, b]$. Using the ε-criterion for Darboux integrability, prove that if $a \leq c < d \leq b$, then $f \in \mathcal{R}([c, d])$.

15.5 Let $F : \mathbb{R} \to \mathbb{R}$ be the integral function defined as $F(x) = \int_0^x |t|\, dt$.

(a) For any $x \in \mathbb{R}$, prove that $F(x) = -F(-x)$.

(b) Find the closed form of the function F using first principles.

(c) Hence, show that the function F is differentiable everywhere and find its derivative.

15.6 Suppose that $f, g \in \mathcal{R}([a, b])$. Show the Cauchy-Bunyakovsky-Schwarz inequality for integrals, namely:

$$\left(\int_a^b f(x)g(x)\, dx\right)^2 \leq \left(\int_a^b f(x)^2\, dx\right)\left(\int_a^b g(x)^2\, dx\right).$$

15.7 Suppose that $f, g \in \mathcal{R}([a, b])$. Prove that the functions $\max(f, g), \min(f, g) : [a, b] \to \mathbb{R}$ are also in $\mathcal{R}([a, b])$.

15.8 (a) Suppose that $f : [a, b] \to \mathbb{R}$ is a real-valued function and $K \in \mathbb{R}$ is a constant. Show that for any partition $\mathcal{P}$ of $[a, b]$ we have $U_{f+K,\mathcal{P}} = U_{f,\mathcal{P}} + K(b - a)$ and $L_{f+K,\mathcal{P}} = L_{f,\mathcal{P}} + K(b - a)$.

(b) Let (f_n) be a sequence of functions $f_n : [a, b] \to \mathbb{R}$ which converges uniformly on $[a, b]$ to a function $f : [a, b] \to \mathbb{R}$. Show that for any partition $\mathcal{P}$ of $[a, b]$ we have $\lim_{n\to\infty} U_{f_n,\mathcal{P}} = U_{f,\mathcal{P}}$ and $\lim_{n\to\infty} L_{f_n,\mathcal{P}} = L_{f,\mathcal{P}}$.

15.9 Let $f : [a, b] \to \mathbb{R}$ be a continuous and non-negative function.

(a) Suppose that there exists a point $c \in [a, b]$ such that $f(c) > 0$. Show that there exists a $\delta > 0$ such that $f(x) > \frac{f(c)}{2} > 0$ on $x \in (c - \delta, c + \delta) \cap [a, b]$.

(b) Hence, if there exists a $c \in [a, b]$ such that $f(c) > 0$, show that $\int_a^b f(x)\, dx > 0$.

(c) Suppose that $g : [a, b] \to \mathbb{R}$ is a continuous function. Prove that if $\int_a^b |g(x)|\, dx = 0$, then $g(x) = 0$ for all $x \in [a, b]$.

15.10 (∗) First, let us look at two short definitions:

Definition 15.7.5 (Support of a Function) The support of a real function $g : X \to \mathbb{R}$ is a subset $\text{supp}(g) \subseteq X$ in the domain defined as the closure $\text{supp}(g) = \text{cl}\{x \in X : g(x) \neq 0\}$. In other words, the function g is identically zero on $\text{supp}(g)^c$.

In Exercise 6.12, we have defined the closure of a set $X \subseteq \mathbb{R}$ as the smallest closed set in $\mathbb{R}$ that contains X. More specifically, $\text{cl}(X) = X \cup X'$.
Next we define:

Definition 15.7.6 (Compactly Supported Function) A real function $g : \mathbb{R} \to \mathbb{R}$ is called compactly supported if there exists a compact interval $[a, b] \subseteq \mathbb{R}$ such that $\text{supp}(g) \subseteq [a, b]$.

Namely, the support of the function g lies within a closed bounded (and hence compact) subset of the domain. In other words, the function g must vanish on $[a, b]^c$.
Some examples of a compactly supported function with domain $\mathbb{R}$ are the indicator function $\mathbf{1}_{[0,1]}$, the tent function in Exercise 15.1(d), and the bump function Ψ in Exercise 14.19. The first one is discontinuous, the second is continuous but not differentiable at three points, and the final one is smooth.
Now prove the following theorem:

Theorem 15.7.7 (Fundamental Theorem of Calculus of Variations) *Let $f : [a, b] \to \mathbb{R}$ be a continuous function. Suppose that for any smooth compactly supported function $g : [a, b] \to \mathbb{R}$ we have $\int_a^b f(x)g(x)\,dx = 0$. Then, $f \equiv 0$.*

As the name suggests, this is a very important result in the study of calculus of variations as well as advanced topics in analysis such as distribution theory and Fourier analysis.

15.11 (∗) Let $f : [0, 1] \to \mathbb{R}$ be a continuous function. Suppose that $\int_0^1 f(x)x^n\,dx = 0$ for all $n \in \mathbb{N}_0$. Prove that $f \equiv 0$.

15.12 (a) Let $f : [a, b] \to \mathbb{R}$ be a continuous function. Prove that if $\int_a^b f(x) = 0$, then there exists a $c \in [a, b]$ for which $f(c) = 0$.
(b) Now let $g : [a, b] \to \mathbb{R}$ be another continuous functions such that $\int_a^b f(x)\,dx = \int_a^b g(x)\,dx$. Prove that there exists a $c \in (a, b)$ such that $f(c) = g(c)$.

15.13 (∗) Let $f \in \mathcal{R}([-a, a])$ for some $a > 0$.
(a) If f is an even function, prove that for any $b \in [0, a]$ we have $\int_{-b}^0 f(x)\,dx = \int_0^b f(x)\,dx$.
Hence, deduce that $\int_{-b}^b f(x)\,dx = 2\int_0^b f(x)\,dx$.
(b) If f is an odd function, prove that for any $b \in [0, a]$ we have $\int_{-b}^b f(x)\,dx = 0$.

15.14 (∗) Let $f \in \mathcal{R}([-1, 1])$. Define a new function $g : [-1, 1] \to \mathbb{R}$ as $g(x) = f(x^2)$.
(a) Show that $\int_0^1 g(x)\,dx$ exists.
(b) Hence, conclude that $\int_{-1}^1 g(x)\,dx = 2\int_0^1 f(x^2)\,dx$.

15.15 Let $f : [0, 1] \to \mathbb{R}$. For each $n \in \mathbb{N}$, define the sum $S_n = \frac{1}{n}\sum_{j=0}^n f\left(\frac{j}{n}\right)$.
(a) Show that if $f \in \mathcal{R}([a, b])$, then $\int_0^1 f(x)\,dx = \lim_{n\to\infty} S_n$.

(b) Find an example of f for which $\lim_{n\to\infty} S_n$ exists but f is not Riemann integrable.

15.16 Prove the other implication of Theorem 15.3.8 that we left out, namely: Show that if f is Darboux integrable over $[a,b]$, then for any $\varepsilon > 0$, there exists a partition $\mathcal{P}$ of $[a,b]$ such that $U_{f,\mathcal{P}} - L_{f,\mathcal{P}} < \varepsilon$.

15.17 Let $f, g : [a,b] \to \mathbb{R}$ be two real-valued functions such that f is continuous and g is non-negative and Riemann integrable.

(a) Show that there exists a point $c \in (a,b)$ such that $\int_a^b f(x)g(x)\,dx = f(c)\int_a^b g(x)\,dx$.

(b) Would the result in part (a) hold if we take away the non-negativity condition on the function g?

15.18 (a) Let $f, g : [0,1] \to \mathbb{R}$ be defined as $f(x) = 1$ if $x \neq 0$ and $f(x) = 0$ if $x = 0$ while g is the Thomae's function in Exercise 15.3. Show that $f, g \in \mathcal{R}([0,1])$ but $f \circ g \notin \mathcal{R}([0,1])$.

Part (a) tells us that the composition of two Riemann integrable function is not necessarily Riemann integrable. Now we want to prove the following theorem that ensures Riemann integrability of composite functions under an additional condition. This generalises what we have seen in Exercise 15.14(a). Our goal now is to prove the following theorem:

Theorem 15.7.8 *If $g : [a,b] \to [c,d]$ is Riemann integrable and $f : [c,d] \to \mathbb{R}$ is a continuous function, then $f \circ g \in \mathcal{R}([a,b])$.*

By the EVT, there must be a $K > 0$ such that $|f(y)| \leq K$ for all $y \in [c,d]$.

(b) Fix $\varepsilon > 0$. Explain why there is a $\delta > 0$ such that $\delta < \frac{\varepsilon}{4K}$ and for all $x, y \in [c,d]$ with $|x-y| < \delta$ we have $|f(x) - f(y)| < \frac{\varepsilon}{2(b-a)}$.

(c) Explain why there is a partition $\mathcal{P} = \{x_0, \ldots, x_n\}$ such that $U_{g,\mathcal{P}} - L_{g,\mathcal{P}} < \delta^2$.

For each $j \in J = \{1, \ldots, n\}$, let $I_j = [x_{j-1}, x_j]$ be a subinterval of the partition $\mathcal{P}$. Define $M_j = \sup_{x\in I_j} g(x)$, $m_j = \inf_{x\in I_j} g(x)$, $M'_j = \sup_{x\in I_j}(f \circ g)(x)$, and $m'_j = \inf_{x\in I_j}(f \circ g)(x)$. We split the set of indices J into two disjoint subsets, namely $I = \{j \in J : M_j - m_j < \delta\}$ and $H = \{j \in J : M_j - m_j \geq \delta\}$ where δ is in part (b).

(d) Show that:

i. For all $j \in I$, $M'_j - m'_j < \frac{\varepsilon}{2(b-a)}$.

ii. For all $j \in H$, $\sum_{j\in H} |x_j - x_{j-1}| < \frac{\varepsilon}{4K}$.

(e) Deduce that $U_{f\circ g,\mathcal{P}} - L_{f\circ g,\mathcal{P}} < \varepsilon$ and hence $f \circ g \in \mathcal{R}([a,b])$.

15.19 (∗) Let $f, g \in \mathcal{R}([a,b])$.

(a) Let $p, q \geq 1$ such that $\frac{1}{p} + \frac{1}{q} = 1$. Prove Hölder's inequality that says:

$$\int_a^b |f(x)g(x)|\,dx \leq \left(\int_a^b |f(x)|^p\,dx\right)^{\frac{1}{p}} \left(\int_a^b |g(x)|^q\,dx\right)^{\frac{1}{q}}.$$

(b) Hence, prove Minkowski's inequality which says for any $p \geq 1$ we have:

$$\left(\int_a^b |f(x)+g(x)|^p\,dx\right)^{\frac{1}{p}} \leq \left(\int_a^b |f(x)|^p\,dx\right)^{\frac{1}{p}} + \left(\int_a^b |g(x)|^p\,dx\right)^{\frac{1}{p}}.$$

This inequality is named after Hermann Minkowski (1864–1909).

15.20 (∗) We now prove the reverse Minkowski's inequality. Suppose that $f, g \in \mathcal{R}([a,b])$ are two non-negative functions and $0 < p < 1$.

(a) Prove that:

$$\left(\int_a^b |f(x)+g(x)|^p\,dx\right)^{\frac{1}{p}} \geq \left(\int_a^b |f(x)|^p\,dx\right)^{\frac{1}{p}} + \left(\int_a^b |g(x)|^p\,dx\right)^{\frac{1}{p}}.$$

(b) Would the result still hold if we take away the non-negativity conditions on f and g?

15.21 (◇) Prove the Cauchy criterion for Riemann integrability, namely:

Proposition 15.7.9 *Let $f : [a,b] \to \mathbb{R}$ be a bounded function. Then, f is Riemann integrable with integral R if and only if for every $\varepsilon > 0$ there exists a $\delta > 0$ such that if $\mathcal{P}_\tau, \mathcal{Q}_\sigma$ are tagged partitions with $\|\mathcal{P}_\tau\|, \|\mathcal{Q}_\sigma\| < \delta$, then $|R_{f,\mathcal{P}_\sigma} - R_{f,\mathcal{Q}_\sigma}| < \varepsilon$.*

15.22 Let $f : [a,b] \to \mathbb{R}$. Prove that $f \in \mathcal{R}([a,b])$ if and only if for every $\varepsilon > 0$ there exist functions $g, h \in \mathcal{R}([a,b])$ with $g \leq f \leq h$ such that $\int_a^b h(x) - g(x)\,dx < \varepsilon$.

15.23 (∗) Let $f : [a,b] \to \mathbb{R}$ be a bounded non-negative function. Prove that for any $\varepsilon > 0$ there exists a continuous function $g : [a,b] \to \mathbb{R}$ such that $0 \leq g \leq f$ and $L_f \leq \int_a^b g(x)\,dx + \varepsilon$.

15.24 (◇) In this chapter, we have constructed the integral by defining the area of a rectangle to its length multiplied by its height. However, this is merely a convention that we have learnt in school (what is the definition of an "area" anyway?). We could define the "area" of a rectangle in a different way.

In this question, we shall define a new kind of integral called the Riemann-Stieltjes integral. This integral is a generalisation of the Riemann integral by Thomas Joannes Stieltjes which determines the area of the approximating rectangles with an integrator function g.

Let $f : [a,b] \to \mathbb{R}$ be a bounded function and $g : [a,b] \to \mathbb{R}$ be a monotone function. Suppose that $\mathcal{P} = \{x_0, x_1, \ldots, x_n\}$ is a partition of $[a,b]$ with subintervals $I_j = [x_{j-1}, x_j]$. Denote $m_j = \inf_{x\in I_j} f(x)$ and $M_j = \sup_{x\in I_j} f(x)$. Define:

$$L_{f,\mathcal{P},g} = \sum_{j=1}^n m_j |g(x_j) - g(x_{j-1})|, \quad \text{and}$$

$$U_{f,\mathcal{P},g} = \sum_{j=1}^n M_j |g(x_j) - g(x_{j-1})|.$$

(a) Show that for any partition $\mathcal{P}$ of $[a, b]$, we have:

$$m|g(b) - g(a)| \leq L_{f,\mathcal{P},g} \leq U_{f,\mathcal{P},g} \leq M|g(b) - g(a)|,$$

where $m = \inf_{x\in[a,b]} f(x)$ and $M = \sup_{x\in[a,b]} f(x)$.

(b) Suppose that $\mathcal{P}'$ is a refinement of the partition $\mathcal{P}$. Prove that we have the ordering:

$$L_{f,\mathcal{P},g} \leq L_{f,\mathcal{P}',g} \leq U_{f,\mathcal{P}',g} \leq U_{f,\mathcal{P},g}.$$

(c) Define the sets:

$$\mathcal{L}_g = \{L_{f,\mathcal{P},g} : \mathcal{P} \text{ is a partition of } [a, b]\},$$
$$\mathcal{U}_g = \{U_{f,\mathcal{P},g} : \mathcal{P} \text{ is a partition of } [a, b]\}.$$

Deduce that $\sup(\mathcal{L}_g)$ and $\inf(\mathcal{U}_g)$ both exist.

(d) Denote $L_{f,g} = \sup(\mathcal{L}_g)$ and $U_{f,g} = \inf(\mathcal{U}_g)$ as the lower and upper Riemann-Stieltjes integrals. Show that $L_{f,g} \leq U_{f,g}$.

15.25 (◇) If the lower and upper Riemann-Stieltjes integrals in Exercise 15.24(d) coincide, namely $L_{f,g} = U_{f,g}$, then the common value is called the Riemann-Stieltjes integral of f with respect to g, denoted as:

$$S = \int_a^b f(x)\,dg(x) \quad \text{or simply} \quad \int_a^b f(x)\,dg.$$

The function f is called the integrand and the function g is called the integrator. The integrator can be seen as weights on how we measure the width of the rectangles in the construction of the integral. The function f is called Riemann-Stieltjes integrable with respect to g and the class of functions which has a Riemann-Stieltjes integral with respect to the integrator g over a compact interval $X \subseteq \mathbb{R}$ is denoted as $\mathcal{RS}_g(X)$.

(a) Show that $f \in \mathcal{RS}_g([a, b])$ if and only if for every $\varepsilon > 0$ there exists a partition such that $U_{f,\mathcal{P},g} - L_{f,\mathcal{P},g} < \varepsilon$.

(b) Prove that $f \in \mathcal{RS}_g([a, b])$ if and only if there exists a sequence of partitions $(\mathcal{P}_n)$ of $[a, b]$ such that $\lim_{n\to\infty} L_{f,\mathcal{P}_n,g} = \lim_{n\to\infty} U_{f,\mathcal{P}_n,g}$.

15.26 (◇) The Riemann-Stieltjes integral is a generalisation of the Riemann integral that we saw earlier. If we pick $g(x) = x$, then the Riemann-Stieltjes integral coincides with the Riemann integral, namely $\mathcal{RS}_x(X) = \mathcal{R}(X)$. Now let us look at examples of Riemann-Stieltjes integrals with other integrator functions g.

(a) Let the integrator $g : [-1, 1] \to \mathbb{R}$ be defined as the monotone function $g(x) = \mathbf{1}_{[0,1]}$. Show that for any continuous function $f : [-1, 1] \to \mathbb{R}$ we have $\int_{-1}^{1} f(x)\,dg = f(0)$.

(b) Let the integrator $g : [0, 1] \to \mathbb{R}$ be defined as the monotone function $g(x) = x^2$. Show that if $f : [0, 1] \to \mathbb{R}$ is $f(x) = 1$, then $\int_0^1 f(x)\,dg = 1$.

The first example above formalises the idea of the Dirac delta "function" where an integral of any function f over a domain X only picks up the value of the function f at a single specific point. This is a very useful concept in the study of differential equations and physics to denote a point mass or a point charge in space. This Dirac delta "function" is not really a function, but can be interpreted as a measure, a distribution (generalised functions), or a Riemann-Stieltjes integral operator defined as above. We shall look at what measures are in Chap. 18.

15.27 (◇) The Riemann-Stieltjes can also be defined using tagged partitions. Let $f, g : [a, b] \to \mathbb{R}$ be the integrand and the integrator respectively. For a partition $\mathcal{P} = \{x_0, x_1, \ldots, x_n\}$ of $[a, b]$ with tags $\tau = \{p_1, p_2, \ldots, p_n\}$, we define the Riemann-Stieltjes sum of f with integrator g with respect to the tagged partition $\mathcal{P}_\tau$ as:

$$S_{f,\mathcal{P}_\tau,g} = \sum_{j=1}^{n} f(p_j)|g(x_j) - g(x_{j-1})|.$$

The Riemann-Stieltjes integral of f with integrator g is the number S where for every $\varepsilon > 0$ there exists a $\delta > 0$ such that for any tagged partition $\mathcal{P}_\tau$ of $[a, b]$ with $||\mathcal{P}_\tau|| < \delta$ we have $|S_{f,\mathcal{P}_\tau,g} - S| < \varepsilon$.

The equivalence of the two definitions can be proven in a similar way to the way we proved that the Riemann and Darboux integrals coincide. Moreover, all the Propositions 15.5.2, 15.5.3, 15.5.5, and 15.6.4 can be extended to Riemann-Stieltjes integral.

What we want to look at in this question are some interesting properties of the Riemann-Stieltjes integral. Assume that the integrator function g is increasing and $f \in \mathcal{RS}_g([a, b])$.

(a) Let $\lambda > 0$ be a constant. Show that the function λg is also increasing and:

$$\int_a^b f(x)\,d(\lambda g) = \int_a^b \lambda f(x)\,dg = \lambda \int_a^b f(x)\,dg.$$

(b) Let $h : [a, b] \to \mathbb{R}$ be another increasing integrator function and and $f \in \mathcal{RS}_h([a, b])$. Show that the function $g + h$ is also increasing and:

$$\int_a^b f(x)\,d(g + h) = \int_a^b f(x)\,dg + \int_a^b f(x)\,dh.$$

(c) Assume that the integrator g is continuously differentiable. Show that:

$$\int_a^b f(x)\,dg = \int_a^b f(x)g'(x)\,dx.$$

15.28 Suppose that $f, g : [a, b] \to \mathbb{R}$ are increasing functions.

(a) Let $\mathcal{P}$ be any partition of $[a, b]$. Prove that:

$$U_{f,\mathcal{P},g} + L_{g,\mathcal{P},f} = U_{g,\mathcal{P},f} + L_{f,\mathcal{P},g} = f(b)g(b) - f(a)g(a).$$

Hence, prove that $f \in \mathcal{RS}_g([a, b])$ if and only if $g \in \mathcal{RS}_f([a, b])$.

(b) Assuming that $f \in \mathcal{RS}_g([a, b])$ and $g \in \mathcal{RS}_f([a, b])$, prove that:

$$\int_a^b f(x)\,dg + \int_a^b g(x)\,df = f(b)g(b) - f(a)g(a).$$

(c) If f and g are continuously differentiable over $[a, b]$, deduce that:

$$\int_a^b f(x)g'(x)\,dx + \int_a^b g(x)f'(x)\,dx = f(b)g(b) - f(a)g(a).$$

This is called integration by parts, which we shall generalise to other classes of functions (not just monotone functions) in Theorem 16.1.7.

15.29 Let $I : \mathbb{R} \to \mathbb{R}$ be a function defined as the function $I(x) = \mathbf{1}_{\mathbb{R}_{\geq 0}}(x)$. Suppose that $c_j, d_j \in \mathbb{R}$ are positive constants for $j \in \{1, 2, \ldots, n\}$.

(a) Show that the function $g : \mathbb{R} \to \mathbb{R}$ defined as $g(x) = \sum_{j=1}^n c_j I(x - d_j)$ is increasing.

(b) Let $M = \max\{d_1, d_2, \ldots, d_n\}$. For a continuous function $f : [0, z] \to \mathbb{R}$ where $z \geq M$, show that:

$$\int_0^z f(x)\,dg = \sum_{j=1}^n c_j f(d_j).$$

15.30 The floor function is increasing on $[0, \infty)$. For any fixed $n \in \mathbb{N}$, evaluate the Riemann-Stieltjes integral:

$$\int_0^n x\,d\lfloor x \rfloor.$$

15.31 ($\diamond$) Recall in Exercise 15.9(c) where we proved that if $f : [a, b] \to \mathbb{R}$ is a continuous function such that its Riemann integral $\int_a^b |f(x)|\,dx = 0$, then $f \equiv 0$ on $[a, b]$. This is not true for Riemann-Stieltjes integral in general. Find an example of functions $f, g : [a, b] \to \mathbb{R}$ where f (not identically zero) is continuous and g is increasing such that the Riemann-Stieltjes integral $\int_a^b |f(x)|\,dg$ vanishes.

15.32 ($\diamond$) Let $I = [-1, 1]$ and $f, g : I \to \mathbb{R}$ with g increasing. Provide examples of pairs of functions f, g such that:

(a) $f \in \mathcal{R}([-1, 1])$ but not in $\mathcal{RS}_g([-1, 1])$.

(b) $f \in \mathcal{RS}_g([-1, 1])$ but not in $\mathcal{R}([-1, 1])$.

Fundamental Theorem of Calculus 16

Science is the differential calculus of the mind, art the integral calculus; they may be beautiful when apart, but are greatest only when combined.

— Ronald Ross, medical doctor and Nobel laureate

In Chap. 15, we have seen the construction and properties of Riemann integrals. We have also seen some sufficient conditions that allow us to conclude whether a function on $[a, b]$ is Riemann integrable in Sect. 15.6. However, they do not tell us anything about the value of the integral and in order to exactly know this value, we need to guess and carry out the Riemann integration construction from scratch.

The easier way to do this is to construct the Darboux integrals and use Corollary 15.3.9. Even so, this is a fiddly process since we have to start with a partition of the domain set, find the upper and lower approximations, upper and lower Darboux sums, upper and lower Darboux integrals, and finally show that these two quantities are the same.

Even though we can describe the construction step-by-step, there is a lot of computations, analysis, and numerical approximations involved. For some functions, as we have seen in Example 15.3.10, the construction can be straightforward. But other times, we may not be as lucky and carrying out the whole routine above can be messy.

However, a miraculous and surprising result due to Gregory, Barrow, Newton, and Leibniz shows that the area under the graph of some bounded function f defined on a compact interval $[a, b]$ is related to its antiderivative! If we stop and think about it, both the operations of integration and differentiation involve infinitesimals in their constructions, so this correspondence may not come as too much of a surprise after all.

S. Johar, *The Big Book of Real Analysis*,
https://doi.org/10.1007/978-3-031-30832-1_16

In this chapter, we shall present some results which relate antiderivatives and integration or area under the graph of a function and their consequences. These are called the fundamental theorem of calculus.

In order to prove the fundamental theorem of calculus, we first prove the mean value theorem for integrals. This is an application of Proposition 15.5.5 to continuous functions.

Proposition 16.0.1 (Mean Value Theorem for Riemann Integrals, MVT for Riemann Integrals) *Let $f \in \mathcal{R}([a, b])$ be a continuous function. Then, there exists a $c \in [a, b]$ such that:*

$$\int_a^b f(x)\,dx = (b-a)f(c).$$

Proof Since the function f is continuous over the interval $[a, b]$, by the EVT, there exist $\xi, \zeta \in [a, b]$ such that $f(\xi) = \min f([a, b]) = m$ and $f(\zeta) = \max f([a, b]) = M$. Due to this, we have $m \leq f(x) \leq M$ for all $x \in [a, b]$ and hence the Riemann integral is bounded as $m(b-a) \leq \int_a^b f(x)\,dx \leq M(b-a)$.

Consider the function $g : [a, b] \to \mathbb{R}$ defined as $g(x) = (b-a)f(x)$. By definition, we have the bounds $m(b-a) \leq g(x) \leq M(b-a)$ with both of the equalities attained also at $\xi, \zeta \in [a, b]$ respectively. Since f is continuous, g is also continuous and therefore takes all the values between the bounds. Since the Riemann integral $\int_a^b f(x)\,dx$ is a fixed number which is also contained between these bounds, there must exist some $c \in [a, b]$ such that:

$$\int_a^b f(x)\,dx = g(c) = (b-a)f(c),$$

which concludes the proof. □

Now we define the Riemann integral functions. These are simply Riemann integrals for which the upper limit is a variable instead of a fixed number. We have seen examples of it in Exercises 15.1(f) and 15.5.

Definition 16.0.2 (Riemann Integral Function) Let $f \in \mathcal{R}([a, b])$. We define the integral function of f as the function $I : [a, b] \to \mathbb{R}$ where:

$$I(x) = \int_a^x f(t)\,dt.$$

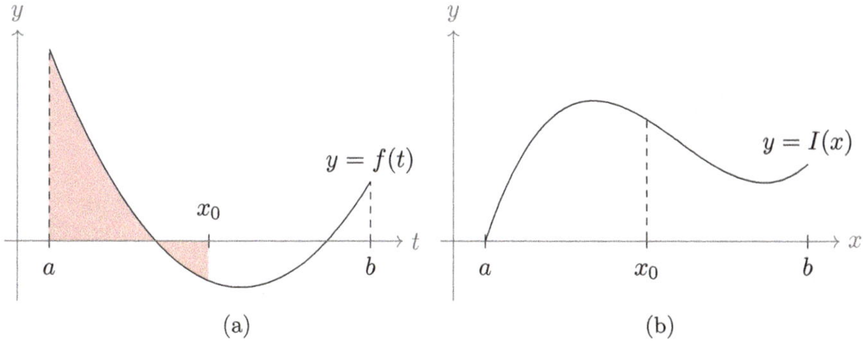

Fig. 16.1 The graphs of the function $f : [a, b] \to \mathbb{R}$ and its integral function $I(x) = \int_a^x f(t)\,dt$. The value $I(x_0) = \int_a^{x_0} f(t)\,dt$ is the (signed) area of the region shaded in red. Note that the points at which the function f vanish are critical points of the integral function I. (**a**) The graph of $f(t)$. (**b**) The graph of $I(x)$

We note that this is a perfectly well-defined function since for any $x \in [a, b]$ we can always define the Riemann integral of f over the compact subinterval $[a, x]$ by Proposition 15.5.3.

See Fig. 16.1 for an illustration of this function. One useful observation is the following:

Proposition 16.0.3 *Let $f \in \mathcal{R}([a, b])$. If f does not change sign over $[a, b]$, then its integral function I is monotone over $[a, b]$.*

Proof WLOG, suppose that $f \geq 0$ over $[a, b]$. Pick any $x, y \in [a, b]$ with $x < y$. We have $I(y) - I(x) = \int_a^y f(t)\,dt - \int_a^x f(t)\,dt = \int_x^y f(t)\,dt$. Since $f \geq 0$ over the interval $[x, y]$, by Proposition 15.5.5, this integral is non-negative and so $I(y) \geq I(x)$. Thus, I is an increasing function. □

16.1 Fundamental Theorem of Calculus

We now state and prove the first version of the fundamental theorem of calculus:

Theorem 16.1.1 (Fundamental Theorem of Calculus I, FTC I) *If $f \in \mathcal{R}([a, b])$, then its Riemann integral function $I : [a, b] \to \mathbb{R}$ is continuous on $[a, b]$.*

Proof Fix $x_0 \in (a, b)$. To show that it is continuous at x_0, we show that $\lim_{h \to 0} I(x_0 + h) = I(x_0)$. First note that since the function f is Riemann integrable, then it must be bounded, namely there exists a $K > 0$ such that

$|f(x)| \leq K$ for all $x \in [a, b]$. Furthermore, by Propositions 15.5.3 and 15.5.5, if $h > 0$, we then have:

$$\begin{aligned} |I(x_0 + h) - I(x_0)| = \left| \int_a^{x_0+h} f(t)\,dt - \int_a^{x_0} f(t)\,dt \right| &= \left| \int_{x_0}^{x_0+h} f(t)\,dt \right| \\ &\leq \int_{x_0}^{x_0+h} |f(t)|\,dt \\ &\leq \int_{x_0}^{x_0+h} K\,dt = K|h|, \end{aligned}$$

and similar inequality holds for $h < 0$. Thus, by sandwiching, as $h \to 0$ we have $|I(x_0+h) - I(x_0)| \to 0$, namely $I(x_0+h) \to I(x_0)$. To prove right-continuity at a and left-continuity at b, we use the same technique using right- and left-limits. □

At the beginning of this chapter, we have mentioned that the Riemann integral and derivatives are related to each other. However, FTC I does not say anything at all about differentiation. This is because the condition that we have placed on the function f here is too weak, which is just merely Riemann integrable. With this condition, the best that we can get is continuity of the Riemann integral function, which we shall demonstrate in the following example.

Example 16.1.2 Consider the function $f : [0, 2] \to \mathbb{R}$ which is a piecewise function defined as $f(x) = 1$ for $x \in [0, 1]$ and 2 for $x \in (1, 2]$. In Exercise 15.1(f), the readers have computed the Riemann integral function $I : [0, 2] \to \mathbb{R}$ of f, which is given by:

$$I(x) = \int_0^x f(t)\,dt = \begin{cases} x & \text{if } x \in [0, 1], \\ 1 + 2x & \text{if } x \in (1, 2]. \end{cases}$$

We can easily see that the function I is not differentiable at $x = 1$ since the left- and right-derivatives here do not agree. However, notice that this Riemann integral function is continuous, as guaranteed by FTC I. So, even if the function f is not continuous to begin with, its integral function is always a continuous function.

In order to explore the connection between Riemann integrals and derivatives, let us place a stronger condition on the function f by requiring it to be continuous on the domain $[a, b]$ (and hence must be Riemann integrable here as well by Proposition 15.6.4). With this strengthened condition, we can prove:

Theorem 16.1.3 (Fundamental Theorem of Calculus II, FTC II—Continuous Functions) *Let $f : [a, b] \to \mathbb{R}$ be a continuous function on a compact interval.*

1. *If I is the Riemann integral function of f, for any $x \in (a, b)$ we have:*

$$f(x) = \frac{d}{dx} I(x) = \frac{d}{dx} \int_a^x f(t)\, dt.$$

2. *If F is any antiderivative of the function f, then:*

$$\int_a^b f(x)\, dx = F(b) - F(a) = [F(x)]_a^b.$$

Proof We prove the assertions one by one:

1. For any $x \in (a, b)$ and small enough h, say $0 < h < \min\{|b - x|, |a - x|\}$, we have:

$$\int_x^{x+h} f(t)\, dt = \int_a^{x+h} f(t)\, dt - \int_a^x f(t)\, dt = I(x + h) - I(x).$$

By Proposition 16.0.1, there exists some $\xi \in [x, x + h]$ such that:

$$I(x + h) - I(x) = \int_x^{x+h} f(t)\, dt = hf(\xi) \quad \Rightarrow \quad f(\xi) = \frac{I(x + h) - I(x)}{h}. \tag{16.1}$$

Let us now take the limit of this equation as $h \to 0$. As h goes to 0, since $x \le \xi \le x + h$, we must also have $\xi \to x$ by sandwiching. By continuity of the function f, we have $\lim_{h \to 0} f(\xi) = f(\lim_{h \to 0} \xi) = f(x)$. On the other hand, the limit on the RHS of the equation in (16.1) is simply the right-derivative of the integral function I at x. The same can be shown for the left-derivative. Thus, we have:

$$\lim_{h \to 0} f(\xi) = \lim_{h \to 0} \frac{I(x + h) - I(x)}{h} \quad \Rightarrow \quad f(x) = \frac{d}{dx} I(x) = \frac{d}{dx} \int_a^x f(t)\, dt.$$

2. Recall that an antiderivative of a continuous function f is any differentiable function F that satisfies $\frac{d}{dx} F = f$. Pick any such antiderivative F. From the first assertion above, we also know that $\frac{d}{dx} I = f$, which implies that I is also an antiderivative of the function f. As we have seen in Corollary 13.6.8, different antiderivatives of the same function are related by an additive constant. Therefore, I and F are related as follows:

$$\int_a^x f(t)\, dt = I(x) = F(x) + C,$$

for some constant $C \in \mathbb{R}$. We now wish to find the value of this constant. By substituting $x = a$, we get:

$$0 = \int_a^a f(t)\,dt = I(a) = F(a) + C,$$

which implies $C = -F(a)$ and thus:

$$\int_a^x f(t)\,dt = I(x) = F(x) - F(a).$$

In particular, putting $x = b$ gives the desired result. □

Therefore, if we have continuity on the function f, the integral function is differentiable and is also an antiderivative of the function f. So this says any continuous function, no matter how wild and pathological it looks like, always has an antiderivative!

However, there are also discontinuous functions that have antiderivatives. Recall Exercise 14.7(b) in which the readers were asked to show that a function $g : \mathbb{R} \to \mathbb{R}$ defined as $g(x) = x^2 \sin(\frac{1}{x})$ for $x \neq 0$ and $g(0) = 0$ is differentiable everywhere but g' is discontinuous. Thus, the function $f = g'$ has an antiderivative g. However, Theorem 16.1.3 cannot be applied to f since it is discontinuous.

The good news is we do not even require the function f to be continuous for the second part of Theorem 16.1.3 to hold! Continuity allows us to use the MVT for Riemann integrals to make the proof above easier. The same result also holds if f is merely Riemann integrable and has an antiderivative. To prove this, we need to use both the Darboux and Riemann definitions for integrals and the MVT on the antiderivative F of f.

Theorem 16.1.4 (Fundamental Theorem of Calculus II, FTC II-non-continuous Functions) *Let $f \in \mathcal{R}([a, b])$. If F is any antiderivative the function f, then:*

$$\int_a^b f(x)\,dx = F(b) - F(a) = [F(x)]_a^b.$$

Proof Fix $\varepsilon > 0$. Since $f \in \mathcal{R}([a, b])$, the ε-criterion for the Darboux integral says there exists a partition $\mathcal{P}$ of $[a, b]$ such that $U_{f,\mathcal{P}} - L_{f,\mathcal{P}} < \varepsilon$. Suppose that the partition is given by $\mathcal{P} = \{x_0, x_1, x_2, \ldots, x_n\}$ and its subintervals are $I_j = [x_{j-1}, x_j]$ for $j = 1, 2, \ldots, n$. Now we want to build a Riemann sum over this partition, so we have to select some suitable tags.

Consider an antiderivative F of f, which is differentiable and hence continuous. Over each subinterval I_j of the partition, the MVT says there exists a $p_j \in (x_{j-1}, x_j)$ such that:

$$\frac{F(x_j) - F(x_{j-1})}{x_j - x_{j-1}} = F'(p_j) = f(p_j)$$
$$\Rightarrow \quad F(x_j) - F(x_{j-1}) = f(p_j)(x_j - x_{j-1}) = f(p_j)|x_j - x_{j-1}|.$$

We add up these equalities for $j = 1, 2, \dots, n$ to get:

$$\sum_{j=1}^{n}(F(x_j) - F(x_{j-1})) = \sum_{j=1}^{n} f(p_j)|x_j - x_{j-1}|. \tag{16.2}$$

Thus, if we pick the tags $\tau = \{p_1, p_2, \dots, p_n\}$ for the partition $\mathcal{P}$, the RHS of (16.2) is just the Riemann sum of the function f over the tagged partition $\mathcal{P}_\tau$. On the other hand, the LHS of (16.2) is simply $F(b) - F(a)$ by telescoping. Thus, we have $F(b) - F(a) = R_{f,\mathcal{P}_\tau}$.

Now recall that $L_{f,\mathcal{P}} \leq R_{f,\mathcal{P}_\tau} \leq U_{f,\mathcal{P}}$ since $\inf_{x\in I_j} f(x) \leq f(p_j) \leq \sup_{x\in I_j} f(x)$ for all $j = 1, 2, \dots, n$. Moreover, by Riemann integrability, we have $L_{f,\mathcal{P}} \leq \int_a^b f(t)\,dt \leq U_{f,\mathcal{P}}$. Combining these two inequalities, we get:

$$\left|\int_a^b f(t)\,dt - R_{f,\mathcal{P}_\tau}\right| \leq U_{f,\mathcal{P}} - L_{f,\mathcal{P}} \quad \Rightarrow \quad \left|\int_a^b f(t)\,dt - (F(b) - F(a))\right| < \varepsilon,$$

because $R_{f,\mathcal{P}_\tau} = F(b) - F(a)$ and $U_{f,\mathcal{P}} - L_{f,\mathcal{P}} < \varepsilon$. Finally, since $\varepsilon > 0$ is arbitrary, we conclude that $\int_a^b f(t)\,dt - (F(b) - F(a)) = 0$, which is what we wanted to prove. □

The fundamental theorem of calculus (or FTC for short) are important results in classical calculus as they tell us how to find the area under the graph of a function by using just antiderivatives. Conversely, it also tells us how to find an antiderivative of a function by looking at the area under the graph of the function.

Example 16.1.5 Let us find the area under the graph for the function $f(x) = x^2$ between $x = 0$ and $x = 1$. We have seen how we could compute the Riemann integral in Examples 15.2.6 and 15.3.10 from first principles.

Now we want to do this by using the FTC. We know that finding the Riemann integral is the same as finding the antiderivative. The monomial function f is simple enough for us to find an antiderivative; its antiderivatives are given by $G(x) = \frac{x^3}{3} + C$ for any $C \in \mathbb{R}$. Pick any of its antiderivative, say $F(x) = \frac{x^3}{3}$. By the FTC, we have:

$$\int_0^1 x^2\,dx = F(1) - F(0) = \frac{1^3}{3} - \frac{0^3}{3} = \frac{1}{3},$$

which is the same as we have calculated earlier using the Riemann and Darboux sums in Examples 15.2.6 and 15.3.10. Amazing!

We can also pick any other antiderivative of f, say $\tilde{F}(x) = \frac{x^3}{3} + 2000$, to use in the FTC. If we chose this antiderivative instead of F above, we get:

$$\int_0^1 x^2\,dx = \tilde{F}(1) - \tilde{F}(0) = \left(\frac{1^3}{3} + 2000\right) - \left(\frac{0^3}{3} + 2000\right) = \frac{1}{3},$$

which gives us the same answer. This is due to the fact that antiderivatives differ by additive constants and when we put the same choice of antiderivative in the formulation for the FTC, these additive constants will cancel each other out.

Let us look at another example on how we can use the FTC.

Example 16.1.6 Consider a function $I : [0, \infty) \to \mathbb{R}$ defined as the integral $I(x) = \int_x^{x^2} \ln(\ln(t+2))\,dt$. Suppose that we want to find the derivative of this integral function.

One way to do this is to find an antiderivative of $\ln(\ln(t+2))$, put in the limits to get a function of x, and differentiate with respect to x as usual. But finding the exact form of the antiderivative is unnecessary, could be lengthy, or even impossible if the antiderivative does not have a closed form in terms of elementary functions. So we have to be smart with this.

Since $\ln(\ln(t+2))$ is continuous, by Theorem 16.1.3(1), it has an antiderivative. Thus, we can write the above integral function implicitly as:

$$I(x) = \int_x^{x^2} \ln(\ln(t+2))\,dt = F(x^2) - F(x),$$

where F is any antiderivative of $\ln(\ln(t+2))$ (so that $F'(t) = \ln(\ln(t+2))$). Differentiating this and using the chain rule, we get:

$$\begin{aligned}\frac{d}{dx}I(x) = \frac{d}{dx}F(x^2) - \frac{d}{dx}F(x) &= F'(x^2)\cdot 2x - F'(x)\\ &= 2x\ln(\ln(x^2+2)) - \ln(\ln(x+2)).\end{aligned}$$

So we have found the answer without even knowing what the antiderivative for the integrand is. This is useful when we have to deal with complicated integrands which do not have an antiderivative in an explicit closed form.

Integration by Parts and by Change of Variable

As mentioned before, finding Riemann integrals via construction can be difficult and is nearly an impossible task for some functions. However, by the duality of antidifferentiation and Riemann integrals via the FTC, one can find the integrals by knowing antiderivatives of functions and using various tricks such as integration by parts and change of variables.

We have seen a special case of the integration by parts in Exercise 15.28(c). Let us present a more general form here:

Theorem 16.1.7 (Integration by Parts) *Suppose that $f, g : [a, b] \to \mathbb{R}$ are continuous and differentiable functions. Suppose further that $f', g' \in \mathcal{R}([a, b])$. Then, $f'g$ and fg' are also Riemann integrable and we have the equality:*

$$\int_a^b f'(x)g(x)\,dx = f(b)g(b) - f(a)g(a) - \int_a^b f(x)g'(x)\,dx$$
$$= [f(x)g(x)]_a^b - \int_a^b f(x)g'(x)\,dx.$$

Proof Define a new function $F : [a, b] \to \mathbb{R}$ as $F = fg$. By product rule, we have: $F' = f'g + fg'$. Since $f, f', g, g' \in \mathcal{R}([a, b])$, by Proposition 15.5.2, we also have $F \in \mathcal{R}([a, b])$. By using the FTC, we then have:

$$\int_a^b f'(x)g(x) + f(x)g'(x)\,dx = \int_a^b F'(x)\,dx = F(b) - F(a)$$
$$= f(b)g(b) - f(a)g(a),$$

which then gives us the result. □

Remark 16.1.8 Compare this result with the summation by parts formula of finite sums in Lemma 7.8.1.

The second trick for integration that we would like to present is the change of variable. The change of variable formula allows us to pull back the variable that we are integrating with respect to in the Riemann integral to a more convenient variable. This could be helpful since it may difficult to find an antiderivative of the integrand in one variable but easier to do so in another. We state and prove:

Theorem 16.1.9 (Change of Variable) *Suppose that $f : [a, b] \to \mathbb{R}$ is a continuous function and $\phi : [c, d] \to [a, b]$ is another continuous function with $\phi(c) = a$ and $\phi(d) = b$. Suppose further that ϕ is differentiable and $\phi' \in \mathcal{R}([c, d])$. Then:*

$$\int_a^b f(x)\,dx = \int_c^d f(\phi(t))\phi'(t)\,dt.$$

Proof Since f is continuous, by the FTC, its Riemann integral function $I : [a, b] \to \mathbb{R}$, defined as $I(x) = \int_a^x f(s)\,ds$, is a continuous and differentiable function. It is also an antiderivative of the function f. The composition $I \circ \phi : [c, d] \to \mathbb{R}$ is then also continuous and differentiable. Moreover, the chain rule gives us:

$$\frac{d}{dt}(I \circ \phi)(t) = I'(\phi(t))\phi'(t) = f(\phi(t))\phi'(t). \tag{16.3}$$

By assumption, the function $\phi'(t)$ is Riemann integrable and the composition $f(\phi(t))$ is continuous and hence Riemann integrable over $[c, d]$. Thus, by Proposition 15.5.2, the quantity in (16.3) is also Riemann integrable. By using the FTC again, we have:

$$\begin{aligned}\int_c^d f(\phi(t))\phi'(t)\,dt &= \int_c^d \frac{d}{dt}(I \circ \phi)(t)\,dt = (I \circ \phi)(d) - (I \circ \phi)(c) \\ &= I(b) - I(a) = \int_a^b f(x)\,dx,\end{aligned}$$

and we are done. □

Example 16.1.10 Let us look at some examples on how to use these results.

1. Suppose that we want to find the Riemann integral of the function $h : \mathbb{R} \to \mathbb{R}$ defined as $h(x) = xe^x$ on the compact interval $[0, 2]$. In other words, we want to find the value of $\int_0^2 xe^x\,dx$. If we know an antiderivative of the function h, then we can use the FTC to find this value.
   We know the antiderivatives of e^x and x, which are $e^x + C$ and $\frac{x^2}{2} + D$ respectively. However, this does not tell us about the antiderivative of the product xe^x. Not a problem! We can use the integration by parts formula. If we write $f'(x) = e^x$ and $g(x) = x$, integration by parts says:

$$\int_0^2 h(x)\,dx = \int_0^2 f'(x)g(x)\,dx = [f(x)g(x)]_0^2 - \int_0^2 f(x)g'(x)\,dx. \tag{16.4}$$

   Note that an antiderivative of $f'(x) = e^x$ is $f(x) = e^x + C$ for some constant $C \in \mathbb{R}$ and $g'(x) = 1$. Thus, substituting these in (16.4), we get:

$$\int_0^2 xe^x\,dx = [x(e^x + C)]_0^2 - \int_0^2 e^x + C\,dx = 2e^2 + 2C - \int_0^2 e^x + C\,dx.$$

We know the antiderivatives of the integrand on the RHS, which is $F(x) = e^x + Cx + D$ for any constant $D \in \mathbb{R}$. So, by using the FTC here, we can choose the antiderivative F with $D = 0$ to deduce:

$$\begin{aligned}\int_0^2 xe^x\,dx &= 2e^2 + 2C - \int_0^2 e^x + C\,dx\\ &= 2e^2 + 2C - (F(2) - F(0))\\ &= 2e^2 + 2C - \left(e^2 + 2C - e^0 - 0\right) = e^2 + 1.\end{aligned}$$

We note from the above that when using the integration by parts, we have to find an antiderivative for f' and there are many of them which all differ to one another by additive constants. However, we saw that this additive constant will cancel each other out in the final expression, so the best antiderivative to choose for f' to simplify the computations is the one with $C = 0$.

2. Let $f : \mathbb{R} \to \mathbb{R}$ be defined as $f(x) = (3x + 4)^5$. Clearly, this function is continuous and hence Riemann integrable over $[0, 1]$. We wish to find its value, namely:

$$\int_0^1 f(x)\,dx = \int_0^1 (3x+4)^5\,dx.$$

Of course, we could expand the integrand and get a polynomial which we could then integrate term by term by using the FTC as the antiderivative of each monomial can be obtained easily.
However, we could also change the variable to make things easier. Indeed, define a function $x : [4, 7] \to \mathbb{R}$ as $x(t) = \frac{t-4}{3}$. This is a continuous and differentiable function with $x(4) = 0$ and $x(7) = 1$. Moreover, its derivative $x'(t) = \frac{1}{3}$ is Riemann integrable over $[4, 7]$. So, by pulling back the variable of the function f from x to this new variable t, we have:

$$\int_0^1 (3x+4)^5\,dx = \int_4^7 (3x(t)+4)^5 x'(t)\,dt = \int_4^7 \frac{t^5}{3}\,dt = \frac{1}{3}\int_4^7 t^5\,dt,$$

which we know how to evaluate using the FTC since we know that an antiderivative of the integrand is $F(t) = \frac{t^6}{6}$. Thus:

$$\int_0^1 (3x+4)^5\,dx = \frac{1}{3}\int_4^7 t^5\,dt = \frac{1}{3}(F(7) - F(4)) = \frac{1}{3}\left(\frac{7^6}{6} - \frac{4^6}{6}\right) = \frac{12617}{2}.$$

3. Sometimes we might need to apply these techniques more than once to get the desired result. Consider the function $f : [0, 1] \to \mathbb{R}$ defined as $f(x) = \sqrt{1 - x^2}$ which is continuous and hence Riemann integrable. We would like to find the area under the graph of f between 0 and 1. This means we want to find the value

$\int_0^1 \sqrt{1-x^2}\,dx$. It is not clear what the antiderivative of f is, so we cannot apply the FTC straight away.

However, we can pull back our function f to a more useful coordinate by considering the function $x : [0, \frac{\pi}{2}] \to \mathbb{R}$ defined as $x(t) = \cos(t)$, which is continuous, differentiable, and has a Riemann integrable derivative over the domain. In this new variable, we have $f(x(t)) = \sqrt{1-\cos^2(t)} = \sin(t)$. Furthermore, the limits of the integral are now $t = \frac{\pi}{2}$ and $t = 0$ since $x(\frac{\pi}{2}) = 0$ and $x(0) = 1$ are the limits of the original integral with respect to x. Thus:

$$\begin{aligned}\int_0^1 \sqrt{1-x^2}\,dx = \int_{\frac{\pi}{2}}^0 \sqrt{1-x(t)^2}x'(t)\,dt &= \int_{\frac{\pi}{2}}^0 \sqrt{1-\cos^2(t)}(-\sin(t))\,dt \\ &= \int_{\frac{\pi}{2}}^0 -\sin^2(t)\,dt \\ &= \int_0^{\frac{\pi}{2}} \sin^2(t)\,dt, \end{aligned} \tag{16.5}$$

where we used Proposition 15.5.3 to switch the limits of the integral. We still do not know an antiderivative of the integrand in (16.5) to apply the FTC, but we can simplify it further using the double angle formula as follows:

$$\int_0^{\frac{\pi}{2}} \sin^2(t)\,dt = \int_0^{\frac{\pi}{2}} \frac{1}{2} - \frac{\cos(2t)}{2}\,dt = \frac{1}{2}\int_0^{\frac{\pi}{2}} 1\,dt - \frac{1}{2}\int_0^{\frac{\pi}{2}} \cos(2t)\,dt. \tag{16.6}$$

We know an antiderivative of the integrand in the first term so we can apply the FTC here. However, we do not know what the antiderivative of the integrand in the second term. We can again apply the change of variable by defining $t(u) = \frac{u}{2}$ for $u \in [0, \pi]$. So, the second integral in (16.6) becomes:

$$\int_0^{\frac{\pi}{2}} \cos(2t)\,dt = \int_0^{\pi} \cos(2t(u))t'(u)\,du = \frac{1}{2}\int_0^{\pi} \cos(u)\,du, \tag{16.7}$$

and since we know the antiderivative of $\cos(u)$, which is $-\sin(u) + C$, we can now apply the FTC here. Therefore, putting (16.7) in (16.6) and (16.5), we have:

$$\begin{aligned}\int_0^1 \sqrt{1-x^2}\,dx &= \frac{1}{2}\int_0^{\frac{\pi}{2}} 1\,dt - \frac{1}{4}\int_0^{\pi} \cos(u)\,du \\ &= \frac{1}{2}\left(\frac{\pi}{2} - 0\right) - \frac{1}{4}(-\sin(\pi) + \sin(0)) = \frac{\pi}{4}.\end{aligned}$$

This tells us that a quarter of the unit disc has area $\frac{\pi}{4}$. Therefore, a full unit disc would have an area of π. More generally, a disc of radius $r > 0$ would have an area of πr^2.

Fig. 16.2 A cycloid

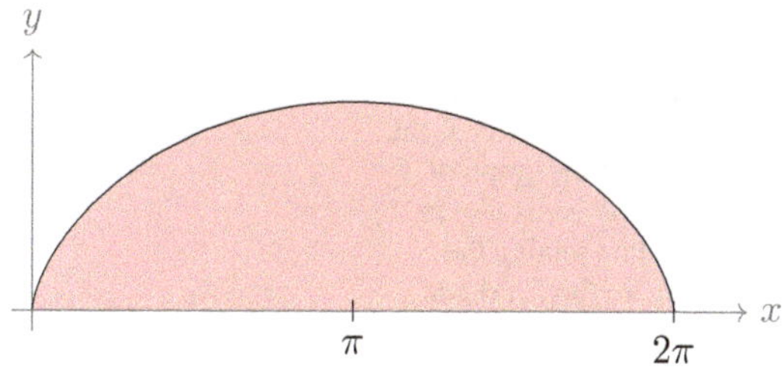

4. Recall the cycloid in Exercise 14.6. This is a curve in the Cartesian plane represented parametrically by $x(t) = t - \sin(t)$ and $y(t) = 1 - \cos(t)$ with the parameter $t \in \mathbb{R}$. The curve for the implicit function $y(x)$ where $x \in [0, 2\pi]$ is depicted in Fig. 16.2.

   We want to compute the area of the red region which is given by $\int_0^{2\pi} y(x)\,dx$. However, as we saw in Exercise 14.6, the explicit expression for y in terms of x is very difficult to obtain. So, let us move to a more convenient variable, namely we pull back the variable x to the variable t. In this variable, for $x = 0$ and $x = 2\pi$ we have $t = 0$ and $t = 2\pi$ respectively. Moreover, $x'(t) = 1 - \cos(t)$. Thus, we can compute:

$$\begin{aligned}\int_0^{2\pi} y(x)\,dx &= \int_0^{2\pi} y(x(t))x'(t)\,dt \\ &= \int_0^{2\pi} (1-\cos(t))^2\,dt \\ &= \int_0^{2\pi} 1\,dt - \int_0^{2\pi} 2\cos(t)\,dt + \int_0^{2\pi} \cos^2(t)\,dt \\ &= 2\pi + \int_0^{2\pi} \frac{\cos(2t)}{2} + \frac{1}{2}\,dt = 2\pi + \pi = 3\pi,\end{aligned}$$

   where we used the double angle formula followed by a change of variable to compute the integral of $\cos^2(t)$.

16.2 Lengths and Volumes

We have seen that the Riemann integral is, by definition and construction, the area under the graph for the function $f : [a, b] \to \mathbb{R}$. So Riemann integrals, like derivatives, also gives us some geometric information for the function f. Here we shall derive other geometric properties that can be obtained via the Riemann integration process.

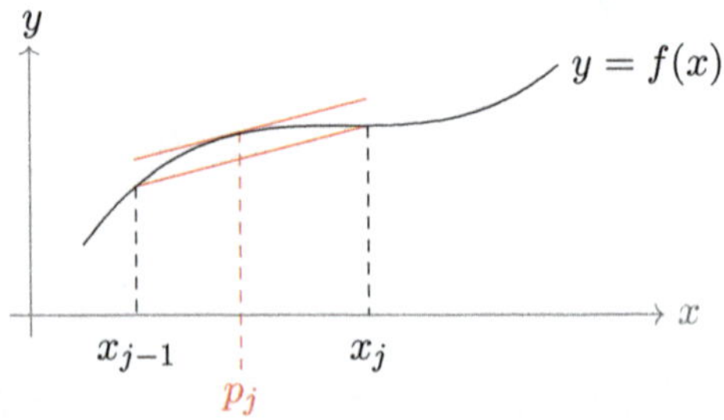

Fig. 16.3 Approximating the graph of f over the interval $[x_{j-1}, x_j]$ with a straight line. By the MVT, the tangent line to the graph of f at the point p_j is parallel to the secant line joining the points $(x_{j-1}, f(x_{j-1}))$ and $(x_j, f(x_j))$

Arclength

For a well-behaved function, the integration process can also be used to define the length of the curve described by the graph of the function.

Suppose that f is a continuously differentiable function defined over $[a, b]$. Therefore, for a partition $\mathcal{P} = \{x_0, x_1, \ldots, x_n\}$ of $[a, b]$, we can approximate the length of the curve over each subinterval $I_j = [x_{j-1}, x_j]$ with a straight line segment joining the points $(x_{j-1}, f(x_{j-1}))$ and $(x_j, f(x_j))$ as in Fig. 16.3. The length of this line segment can be computed using the Pythagorean theorem and is given by:

$$\sqrt{(x_j - x_{j-1})^2 + (f(x_j) - f(x_{j-1}))^2} = |x_j - x_{j-1}|\sqrt{1 + \frac{(f(x_j) - f(x_{j-1}))^2}{(x_j - x_{j-1})^2}}. \tag{16.8}$$

Since the function f is assumed to be differentiable, by the MVT, there exists a point $p_j \in (x_{j-1}, x_j)$ such that $\frac{f(x_j)-f(x_{j-1})}{x_j - x_{j-1}} = f'(p_j)$. Therefore, the approximate length of the graph over the subinterval I_j in (16.8) can be written as $|x_j - x_{j-1}|\sqrt{1 + f'(p_j)^2}$. See Fig. 16.3 for the diagram of this approximation.

Denoting the sum over all the subintervals of $\mathcal{P}$ as $C_{\mathcal{P}}$, we have the length approximation:

$$\begin{aligned} C_{\mathcal{P}} &= \sum_{j=1}^{n} |x_j - x_{j-1}|\sqrt{1 + \frac{(f(x_j) - f(x_{j-1}))^2}{(x_j - x_{j-1})^2}} \\ &= \sum_{j=1}^{n} |x_j - x_{j-1}|\sqrt{1+f'(p_j)^2}, \end{aligned} \tag{16.9}$$

for the curve from $x = x_0 = a$ to $x = x_n = b$.

Notice that (16.9) is simply the Riemann sum for the function $g : [a, b] \to \mathbb{R}$ defined as $g(x) = \sqrt{1 + f'(x)^2}$ with respect to the tagged partition $\mathcal{P}_\tau$ with tags $\tau = \{p_1, p_2, \ldots, p_n\}$.

Since the function g is continuous and hence Riemann integrable over $[a, b]$ with value $R = \int_a^b g(x)\,dx = \int_a^b \sqrt{1 + f'(x)^2}\,dx$, for any $\varepsilon > 0$ there is a $\delta > 0$ such that for any tagged partition $\mathcal{P}_\tau$ with $||\mathcal{P}_\tau|| < \delta$, we have $|R_{g,\mathcal{P}_\tau} - R| < \varepsilon$. Thus, if we have chosen the partition $\mathcal{P} = \{x_0, x_1, \ldots, x_n\}$ such that $||\mathcal{P}_\tau|| < \delta$ with the tags $\tau = \{p_1, p_2, \ldots, p_n\}$, Eq. (16.9) implies:

$$|C_{\mathcal{P}} - R| = \left| \sum_{j=1}^{n} |x_j - x_{j-1}| \sqrt{1 + f'(p_j)^2} - R \right| = |R_{g,\mathcal{P}_\tau} - R| < \varepsilon.$$

Moreover, since $\varepsilon > 0$ is arbitrarily set, we can make the length approximation $C_{\mathcal{P}}$ as close as we like to the value R by taking finer partitions of $[a, b]$. Thus, we can define:

Definition 16.2.1 (Length of Curve) Let $f \in C^1([a, b])$ be a continuously differentiable real-valued function. The length of the graph for this function over $[a, b]$ is defined as the Riemann integral:

$$C(f) = \int_a^b \sqrt{1 + f'(x)^2}\,dx.$$

Since Definition 16.2.1 only works for continuously differentiable functions, we need a more general definition that would cover the case of non-differentiable curves as well. We shall see this later on in Exercise 16.17, where we give another characterisation that defines the length of a curve using the supremum of approximations of the length over all possible partitions of $[a, b]$, namely:

$$C(f) = \sup_{\mathcal{P}} \left\{ C_{\mathcal{P}} = \mathcal{P} \text{ is a partition of } [a, b] \right\},$$

where $C_{\mathcal{P}} = \sum_{j=1}^{n} \sqrt{(x_j - x_{j-1})^2 + (f(x_j) - f(x_{j-1}))^2}$ for partition $\mathcal{P} = \{x_0, x_1, \ldots, x_n\}$ of $[a, b]$. Via the definition above, we can also measure the lengths of curves of functions which are not differentiable. Moreover, this supremum characterisation coincides with Definition 16.2.1 for $C^1([a, b])$ functions. Curves for which the supremum above exist (and hence finite) are called rectifiable curves.

Example 16.2.2 Let us look at some examples:

1. We can calculate the circumference of a circle using Definition 16.2.1. Recall in Sect. 6.1 that the circumference of a circle of radius 1 is defined to be 2π. Now we show that Definition 16.2.1 also agrees with this.

Consider the quarter circle $f : [0, 1] \to \mathbb{R}$ defined as $f(x) = \sqrt{1-x^2}$. Quadrupling the length of this graph over $[0, 1]$ would then give us the circumference. Thus, the circumference according to Definition 16.2.1 is:

$$4\int_0^1 \sqrt{1+f'(x)^2}\,dx = 4\int_0^1 \sqrt{1+\frac{x^2}{1-x^2}}\,dx = 4\int_0^1 \frac{1}{\sqrt{1-x^2}}\,dx. \tag{16.10}$$

Now we change the variable to a more useful coordinate via the function $x : [0, \frac{\pi}{2}] \to \mathbb{R}$ defined as $x(t) = \sin(t)$. This function is continuous, differentiable, and has a Riemann integrable derivative over the domain. In this new variable, the limits of the integral are now $t = 0$ and $t = \frac{\pi}{2}$ since $x(0) = 0$ and $x(\frac{\pi}{2}) = 1$. Thus, the integral (16.10) becomes:

$$4\int_0^1 \frac{1}{\sqrt{1-x^2}}\,dx = 4\int_0^{\frac{\pi}{2}} \frac{\cos(t)}{\sqrt{1-\sin^2(t)}}\,dt = 4\int_0^{\frac{\pi}{2}} \frac{\cos(t)}{\cos(t)}\,dt = 4\int_0^{\frac{\pi}{2}} 1\,dt = 2\pi,$$

which agrees with the known fact from Sect. 6.1.

2. Let us now look at a problem in mechanics. A projectile motion is the motion of a particle initially at $(x, y) = (0, 0)$ when it is projected at an angle θ measured from the ground and falls freely under the influence of gravity with negligible air resistance. One can imagine this as a cannonball being shot at an angle of θ from the ground.
Suppose that the gravitational acceleration is a negative constant $-g$ and the initial velocity of the particle is v. By resolving the motion into the horizontal x-direction and vertical y-direction components, we obtain a set of two differential equations for $x(t)$ and $y(t)$ parametrised by the time variable t where:

$$\frac{d^2x}{dt^2} = 0 \quad \text{with} \quad \frac{dx}{dt}(0) = v\cos\theta \quad \text{and} \quad x(0) = 0,$$
$$\frac{d^2y}{dt^2} = -g \quad \text{with} \quad \frac{dy}{dt}(0) = v\sin\theta \quad \text{and} \quad y(0) = 0.$$

Using the FTC twice for each equation (or our knowledge on ODE IVP from Sect. 14.4), we can solve these equations to get $x(t) = vt\cos(\theta)$ and $y(t) = -\frac{gt^2}{2} + vt\sin(\theta)$ for all $t \geq 0$. We can then combine them to write y in terms of x as:

$$y(x) = \tan(\theta)x - \frac{g}{2v^2\cos^2(\theta)}x^2. \tag{16.11}$$

A diagram for the path of the particle in motion is depicted in Fig. 16.4.

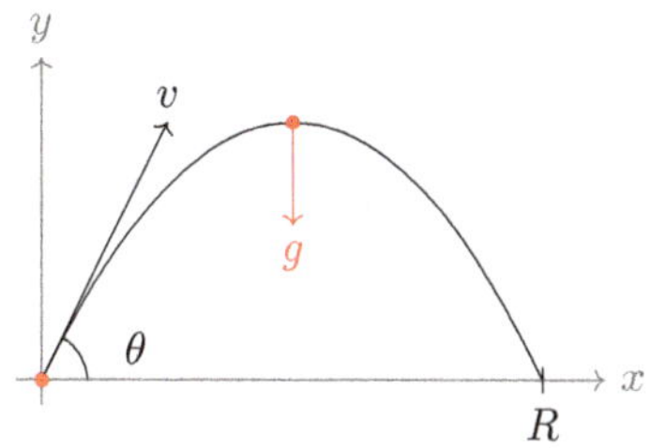

Fig. 16.4 Projectile motion of the red particle initially at the origin. The horizontal range R is the point at which the particle reaches the ground again. By setting $y(R) = 0$, Eq. (16.11) implies that the horizontal range is $R = \frac{v^2 \sin(2\theta)}{g}$

Now we wish to find the total distance travelled by the particle during its flight. To simplify the notations, we set the constants $a = \tan(\theta)$ and $b = -\frac{g}{2v^2\cos^2(\theta)}$ so the quantity y in (16.11) can be written succinctly as $y(x) = ax + bx^2$ with $b \neq 0$. Therefore, we wish to find the value $\int_0^R \sqrt{1 + y'(x)^2}\,dx = \int_0^R \sqrt{1 + (a + 2bx)^2}\,dx$. By change of variables $a + bx = u$, we get:

$$\int_0^R \sqrt{1 + (a + 2bx)^2}\,dx = \frac{1}{2b}\int_a^{a+2bR} \sqrt{1 + u^2}\,du.$$

Applying another change of variables $u = \sinh(v)$ and the double angle formula for hyperbolic cosine in Exercise 12.22(e) and (f), we get:

$$\begin{aligned}\frac{1}{2b}\int_a^{a+2bR} \sqrt{1 + u^2}\,du &= \frac{1}{2b}\int_{\operatorname{arcsinh}(a)}^{\operatorname{arcsinh}(a+bR)} \sqrt{1 + \sinh^2(v)}\cosh(v)\,dv \\ &= \frac{1}{2b}\int_{\operatorname{arcsinh}(a)}^{\operatorname{arcsinh}(a+bR)} \cosh^2(v)\,dv \\ &= \frac{1}{4b}\int_{\operatorname{arcsinh}(a)}^{\operatorname{arcsinh}(a+bR)} \cosh(2v) + 1\,dv \\ &= \frac{1}{4b}\left[\frac{\sinh(2v)}{2} + v\right]_{\operatorname{arcsinh}(a)}^{\operatorname{arcsinh}(a+bR)} \qquad (16.12)\end{aligned}$$

Therefore, by substituting the limits in (16.12), the total distance travelled by the particle is $\frac{1}{4b}((2Rb + a)\sqrt{1 + (2Rb + a)^2} - a\sqrt{1 + a^2} + \operatorname{arcsinh}(2Rb + a) - \operatorname{arcsinh}(a))$. If we substitute the actual values of the constant a, b, and R back into the expression above, the total distance travelled by the particle is then given by $\frac{v^2}{g}(\sin(\theta) + \cos^2(\theta)\operatorname{arctanh}(\sin(\theta)))$.

Solids of Revolution

We can also use the Riemann integral to define another geometric quantity, namely: volume and lateral surface area for solids of revolution. A solid of revolution is

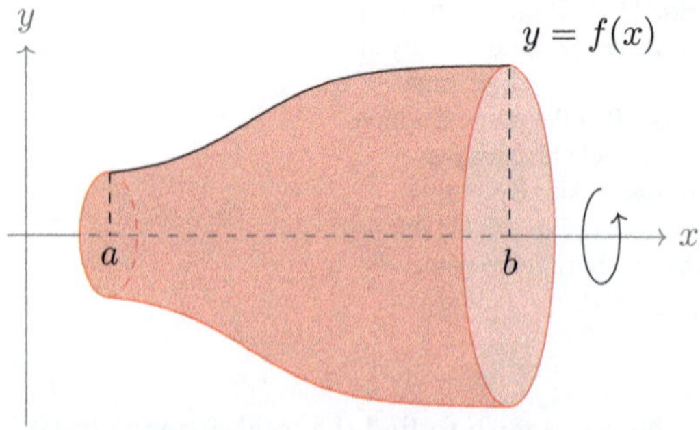

Fig. 16.5 Solid of revolution for the function f over the interval $[a, b]$. Its volume and lateral surface area are given in Definition 16.2.3

a solid generated by a positive function $f : [a, b] \to \mathbb{R}$ obtained by revolving the subgraph of f around the x-axis in one full revolution. See Fig. 16.5 for a demonstration.

So, for example, a cylinder of radius $r > 0$ and height $b - a$ can be obtained by revolving a constant function $f(x) = r$ on $[a, b]$. For more general functions, we define:

Definition 16.2.3 (Volume and Lateral Surface Area for Solid of Revolution) Let $f \in C^1([a, b])$ be a continuously differentiable positive function.

1. The volume of the solid of revolution generated by f is given by:

$$V = \pi \int_a^b f(x)^2 \, dx.$$

2. The lateral surface area of the solid of revolution generated by f is given by:

$$A = 2\pi \int_a^b f(x)\sqrt{1 + f'(x)^2} \, dx.$$

Now, we shall justify the second definition in Definition 16.2.3 only and the first one is left as Exercise 16.14 for the readers to verify. Fix $\varepsilon > 0$. First note that the functions $g, h : [a, b] \to \mathbb{R}$ defined as:

$$g(x) = 2\pi f(x)\sqrt{1 + f'(x)^2},$$
$$h(x) = 2\pi\sqrt{1 + f'(x)^2},$$

are both continuous over $[a, b]$. Then:

1. The function g is Riemann integrable with value $R = \int_a^b g(x)\,dx = \int_a^b 2\pi f(x)\sqrt{1+f'(x)^2}\,dx$. This means we can find a $\delta_1 > 0$ such that for any tagged partition $\mathcal{P}_\tau$ of $[a, b]$ with $||\mathcal{P}_\tau|| < \delta$ we have $|R_{g,\mathcal{P}_\tau} - R| < \frac{\varepsilon}{2}$. This implies:
$$R_{g,\mathcal{P}_\tau} < R + \frac{\varepsilon}{2}. \tag{16.13}$$

2. The function h is bounded by the EVT. In other words, there is an $M > 0$ such that $|h(x)| \leq M$ for all $x \in [a, b]$.

Since f is continuous over $[a, b]$ and hence uniformly continuous over $[a, b]$ by Theorem 10.6.10, we can find a $\delta_2 > 0$ such that whenever $|x - y| < \delta_2$ we have $|f(x) - f(y)| < \frac{\varepsilon}{4\pi M(b-a)}$. Set $\delta = \min\{\delta_1, \delta_2\} > 0$.

Pick any partition $\mathcal{P} = \{x_0, x_1, \ldots, x_n\}$ of $[a, b]$ with $||\mathcal{P}|| < \delta$. On each subinterval $I_j = [x_{j-1}, x_j]$ of the partition, we approximate the surface of the solid of revolution using a frustum (truncated cone or cylinder). This is obtained by joining the points $(x_{j-1}, f(x_{j-1}))$ and $(x_j, f(x_j))$ on the graph of f in $\mathbb{R}^2$ with a straight line segment and revolving the subgraph of this line segment around the x-axis to form the frustum. See Fig. 16.6 for an example of a frustum.

The surface area for the lateral side of this frustum, denoted by A_j, can then be calculated using elementary geometry of cones thus:

$$\begin{aligned} A_j &= \pi(f(x_j) + f(x_{j-1}))\sqrt{(x_j - x_{j-1})^2 + (f(x_j) - f(x_{j-1}))^2} \\ &= \pi|x_j - x_{j-1}|(f(x_j) + f(x_{j-1}))\sqrt{1 + \frac{(f(x_j) - f(x_{j-1}))^2}{(x_j - x_{j-1})^2}}. \end{aligned} \tag{16.14}$$

The sum $\sum_{j=1}^n A_j$, which we denote as $A_{\mathcal{P}}$, is thus the approximate surface area for the solid of revolution computed with respect to the partition $\mathcal{P}$. We

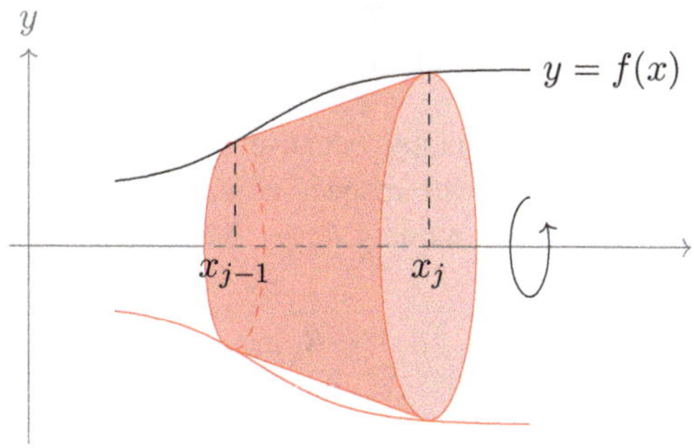

Fig. 16.6 Frustum with lateral surface of area A_j approximating the lateral surface area for the solid of revolution over the interval $[x_{j-1}, x_j]$

now aim to simplify the terms in A_j. Since the function f is differentiable and continuous in $[x_{j-1}, x_j]$, by the MVT and IVT respectively, we can find points $p_j, q_j \in (x_{j-1}, x_j)$ for which $\frac{f(x_j)-f(x_{j-1})}{x_j-x_{j-1}} = f'(p_j)$ and $\frac{f(x_{j-1})+f(x_j)}{2} = f(q_j)$. Substituting these in the expression for A_j in (16.14), we get:

$$A_j = 2\pi|x_j - x_{j-1}|f(q_j)\sqrt{1 + f'(p_j)^2}. \tag{16.15}$$

This expression looks like a term in the Riemann sum for the function g. However, it might not be so since the points p_j and q_j might be distinct points for them to be tags in the partition subinterval I_j. This seems like a dead end for us, but we are in luck: note that these two points p_j and q_j are very close to each other since they lie in the same subinterval (x_{j-1}, x_j). This means we can approximate one with the other via continuity.

Since $p_j, q_j \in (x_{j-1}, x_j)$, we have $|p_j - q_j| < |x_j - x_{j-1}| < \delta \leq \delta_2$. So, by uniform continuity of f, we have $|f(p_j) - f(q_j)| < \frac{\varepsilon}{4\pi M(b-a)}$. This implies $f(q_j) < f(p_j) + \frac{\varepsilon}{4\pi M(b-a)}$. Substituting this along with the bound $h(x) = \sqrt{1 + f'(x)^2} \leq M$ into (16.15), we have:

$$\begin{aligned}
A_j &< 2\pi|x_j - x_{j-1}|\left(f(p_j) + \frac{\varepsilon}{4\pi M(b-a)}\right)\sqrt{1 + f'(p_j)^2} \\
&\leq 2\pi|x_j - x_{j-1}|f(p_j)\sqrt{1 + f'(p_j)^2} + \frac{2\pi\varepsilon}{4\pi M(b-a)}|x_j - x_{j-1}|M \\
&= 2\pi|x_j - x_{j-1}|f(p_j)\sqrt{1 + f'(p_j)^2} + \frac{\varepsilon}{2(b-a)}|x_j - x_{j-1}|.
\end{aligned}$$

Thus, summing up the above for $j = 1, 2, \ldots, n$ and using telescoping sum, the approximate surface area $A_{\mathcal{P}}$ can then be bound as:

$$\begin{aligned}
A_{\mathcal{P}} = \sum_{j=1}^{n} A_j &< \sum_{j=1}^{n} 2\pi|x_j - x_{j-1}|f(p_j)\sqrt{1+f'(p_j)^2} + \sum_{j=1}^{n} \frac{\varepsilon}{2(b-a)}|x_j - x_{j-1}| \\
&= \sum_{j=1}^{n} 2\pi|x_j - x_{j-1}|f(p_j)\sqrt{1 + f'(p_j)^2} + \frac{\varepsilon}{2}.
\end{aligned} \tag{16.16}$$

Now we are happy because the first sum in (16.16) is a Riemann sum for the function g with respect to a tagged partition $\mathcal{P}_\tau$ of $[a, b]$ with tags $\tau = \{p_1, p_2, \ldots, p_n\}$, namely $R_{g,\mathcal{P}_\tau}$. Thus, by using the approximate in (16.13), we get:

$$A_{\mathcal{P}} < R_{g,\mathcal{P}_\tau} + \frac{\varepsilon}{2} < R + \frac{\varepsilon}{2} + \frac{\varepsilon}{2} = R + \varepsilon.$$

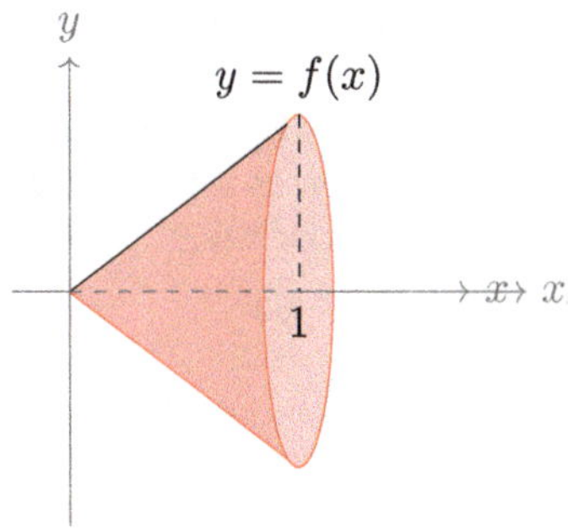

Fig. 16.7 The cone is a surface of revolution of the function $f(x) = rx$ for some $r > 0$

Repeating the above argument to get the lower bound $A_{\mathcal{P}} > R - \varepsilon$, we can conclude that $|A_{\mathcal{P}} - R| < \varepsilon$. Finally, since $\varepsilon > 0$ was set arbitrarily, no matter how small $\varepsilon > 0$ is, we can always find a surface area approximation $A_{\mathcal{P}}$ which is ε-close to the quantity R. We can thus define the lateral surface area for the solid of revolution as $A = R = 2\pi \int_a^b f(x)\sqrt{1 + f'(x)^2}\,dx$.

Example 16.2.4 Let us show that Definition 16.2.3 agrees with the volume and lateral surface area of a cone. A cone of base radius $r > 0$ and height 1 can be obtained as a solid of revolution for the function $f : [0, 1] \to \mathbb{R}$ defined as $f(x) = rx$. This can be seen in in Fig. 16.7.

1. The volume of this cone is given by

$$V = \pi \int_0^1 f(x)^2\,dx = \pi \int_0^1 r^2x^2\,dx = \pi r^2 \int_0^1 x^2\,dx = \frac{\pi r^2}{3},$$

   where we used the fact that the integral of x^2 over $[0, 1]$ is $\frac{1}{3}$ which we saw in Examples 15.2.6 and 15.3.10.
2. Now we compute the lateral surface are for the cone. We know that this area is $\pi r\sqrt{1 + r^2}$ from basic planar geometry. Now let us check that Definition 16.2.3 agrees with this known fact. Since $f'(x) = r$, this definition says:

$$A = 2\pi \int_0^1 f(x)\sqrt{1 + f'(x)^2}\,dx = 2\pi \int_0^1 rx\sqrt{1 + r^2}\,dx = \pi r\sqrt{1 + r^2},$$

   where we used the fact that $\int_0^1 x\,dx = \frac{1}{2}$. This is indeed in agreement with the known lateral surface area for a cone!

16.3 Antiderivatives and Indefinite Integrals

The operation of finding a derivative of a function $f : X \to \mathbb{R}$ is denoted as $\frac{d}{dx} f$ or f'. However, as we have seen in Chap. 14, we did not have any succinct notation to denote the operation of finding its antiderivative, so we are going to introduce it here as:

$$\int f(x)\,dx = F(x) + C \quad \text{for} \quad C \in \mathbb{R},$$

where F is any antiderivative of f.

Due to the notation, the process of antidifferentiation is also called indefinite integral (meaning that the upper and lower bounds of the integral are not defined). The terms antidifferentiation and indefinite integration are interchangeable and antiderivatives are also called indefinite integrals of f.

In contrast, the original integral that we have seen in the previous chapter (the Darboux and Riemann integrals) are referred to as definite integrals. Definite integral, if it exists, produces a definite real number: it is the area under the graph of a function between a and b. Indefinite integral, on the other hand, gives us a function instead. So they are two very different concepts, connected to each other by the FTC.

Knowing how to find antiderivatives is very useful when evaluating definite integrals, as we have seen in the FTC. Some antiderivatives can be computed easily, as we have seen in Chap. 14. But often times, finding an antiderivative is very difficult.

Luckily, the tricks that we have seen in Sect. 16.1 for the definite integral also work for indefinite integral. These tricks are helpful when we want to find antiderivatives of a function. In particular, Theorems 16.1.7 and 16.1.9 for indefinite integrals are given as follows:

Theorem 16.3.1 (Integration by Parts) *Suppose that $f, g : X \to \mathbb{R}$ are differentiable functions where $X \subseteq \mathbb{R}$. Then, we have the equality:*

$$\int f'(x)g(x)\,dx = f(x)g(x) - \int f(x)g'(x)\,dx.$$

Proof The proof is almost tautological. Let $F : X \to \mathbb{R}$ be defined as $F = fg$. The product rule says that $F' = f'g + fg'$ or equivalently $f'g = F' - fg'$. We want to find the antiderivatives (or indefinite integral) of the terms in this equality, namely:

$$\int f'(x)g(x)\,dx = \int F'(x) - f(x)g'(x)\,dx.$$

Since derivatives are linear over addition, antiderivatives are also linear over addition and so we can split the indefinite integral in the RHS to get:

$$\begin{aligned}\int f'(x)g(x)\,dx &= \int F'(x)\,dx - \int f(x)g'(x)\,dx \\ &= F(x) + C - \int f(x)g'(x)\,dx \\ &= f(x)g(x) - \int f(x)g'(x)\,dx,\end{aligned}$$

since we know the antiderivative of F'. Furthermore, the constant C from the antiderivative of F' is absorbed in the second integral to simplify the notation. □

Theorem 16.3.2 (Change of Variable) *Suppose further that $f : X \to \mathbb{R}$ where $X \subseteq \mathbb{R}$ and $\phi : Y \to X$ where $Y \subseteq \mathbb{R}$. Suppose that ϕ is differentiable and f has an antiderivative $F : X \to \mathbb{R}$. Then:*

$$\int f(\phi(t))\phi'(t)\,dt = F(\phi(t)).$$

Proof The composition $F \circ \phi : Y \to \mathbb{R}$ can be differentiated by using the chain rule as:

$$\frac{d}{dt}F(\phi(t)) = F'(\phi(t))\phi'(t) = f(\phi(t))\phi'(t).$$

Thus, by definition of antiderivatives, the antiderivative of $f(\phi(t))\phi'(t)$ is equal to the antiderivative of $\frac{d}{dt}F(\phi(t))$, which is $F(\phi(t))$. So we have the equality:

$$\int f(\phi(t))\phi'(t)\,dt = F(\phi(t)) + C,$$

for some constant $C \in \mathbb{R}$ which we can absorb in the indefinite integral on the LHS to give us the desired identity. □

Remark 16.3.3 These integration by parts and change of variables formula for indefinite integrals do not require the extra conditions as in Theorems 16.1.7 and 16.1.9. This is because we are simply finding the antiderivatives of functions which do not require the FTC in the proofs. So, in particular, the domain for the functions are not required to be compact.

Example 16.3.4 Let us revisit some of the examples in Example 16.1.10 and redo them in a different way.

1. Recall the function $h : \mathbb{R} \to \mathbb{R}$ defined as $h(x) = xe^x$. In Example 16.1.10(1), we wanted to find the Riemann integral $\int_0^2 h(x)\,dx$. If we know an antiderivative

of h, we can just simply use the FTC. At the moment, it is not clear what the antiderivative is, so let us do some work in finding it first.
Using the integration by parts, let us write $f'(x) = e^x$ and $g(x) = x$. We know an antiderivative of f' which is e^x and the derivative of g which is 1. So, by Theorem 16.3.1 we have:

$$\int h(x)\,dx = \int f'(x)g(x)\,dx = f(x)g(x) - \int f(x)g'(x)\,dx$$
$$= xe^x - \int e^x\,dx = xe^x - e^x + C.$$

Therefore, the antiderivatives of h are the functions $H : \mathbb{R} \to \mathbb{R}$ given by $H(x) = (x-1)e^x + C$ for constants $C \in \mathbb{R}$. Thus, we can put this in the FTC to solve the original problem as:

$$\int_0^2 h(x)\,dx = H(2) - H(0) = (2-1)e^2 - (-e^0) = e^2 + 1,$$

which is the same value as the one obtained in Example 16.1.10(1).

2. Let $f : [0,1] \to \mathbb{R}$ be the function $f(x) = \sqrt{1-x^2}$. In Example 16.1.10(3), we wanted to find the area under the graph of f over $[0,1]$. We can do this by finding an antiderivative of f, which we are going to call F, and applying the FTC. To find the antiderivative $\int f(x)\,dx$, we set $x(t) = \cos(t)$ for $t \in [0, \frac{\pi}{2}]$ and thus the change of variables formula in Theorem 16.3.2 says:

$$F(x(t)) = \int f(x(t))x'(t)\,dt = \int \sqrt{1-\cos^2(t)}(-\sin(t))\,dt$$
$$= \int -\sin^2(t)\,dt$$
$$= \int \frac{\cos(2t)-1}{2}\,dt$$
$$= -\frac{1}{2}\int 1\,dt + \frac{1}{2}\int \cos(2t)\,dt$$
$$= -\frac{t}{2} + \frac{1}{2}\int \cos(2t)\,dt.$$

Again, we need to apply another change of variable to find the second antiderivative. Let us set $t(u) = \frac{u}{2}$ for $u \in [0, \pi]$ to get:

$$F(x(t(u))) = -\frac{t}{2} + \frac{1}{2}\int \frac{\cos(u)}{2}\,du = -\frac{t}{2} + \frac{\sin(u)}{4} + C,$$

which is a mess of a function since all the variables are different! However, they implicitly related to one another. First, the function $t(u) = \frac{u}{2}$ is invertible, so we have $u = 2t$ and hence $F(x(t)) = -\frac{t}{2} + \frac{\sin(2t)}{4} + C$. We cannot put this in the FTC yet since the FTC says:

$$\int_0^1 f(x)\,dx = F(1) - F(0),$$

where we substitute $x = 0$ and $x = 1$ in the RHS. But so far, our function $F(x(t))$ is a function of the variable t. So to get the antiderivative in terms of x, we invert $x(t) = \cos(t)$ (which can be done, since it is a bijective function over the domain) to get $t = \arccos(x)$ and hence:

$$F(x) = -\frac{\arccos(x)}{2} + \frac{\sin(2\arccos(x))}{4} + C.$$

Putting this antiderivative in the FTC, we then get:

$$\int_0^1 f(x)\,dx = \left[-\frac{\arccos(x)}{2} + \frac{\sin(2\arccos(x))}{4}\right]_0^1 = \frac{\pi}{4},$$

which is the same answer as before.

There are many other tricks for finding antiderivatives such as using trigonometric functions, partial fractions decomposition, and using inverse functions. However, even with all the tricks, finding antiderivatives cannot be done all the time. As we have mentioned in Example 14.4.2, if finding derivatives is like breaking an egg, finding antiderivatives is like gluing the broken eggshells back together; most of the time we are not able to do it and if we can, it might be difficult.

Nevertheless, we know from FTC that all continuous functions have antiderivatives. It is just a matter of whether we have an explicit representation for them.

Example 16.3.5 Here are some examples of this:

1. Recall the function $f : [0, \infty) \to \mathbb{R}$ defined as $f(x) = \ln(\ln(x + 2))$ which we saw in Example 16.1.6. In the example, we simply wrote that its antiderivative is F. We can write:

$$F(x) = \int_0^x \ln(\ln(t + 2))\,dt,$$

as its antiderivate. In fact, we do not have a closed form for the antiderivative for this function. An antiderivative of this is usually written as $(x+2)\ln(\ln(x+2)) - \mathrm{li}(x + 2)$ which is explicit, except for the final term. This final term is called the logarithmic integral function, which is a special function.

Special functions are functions that have established names and notations due to their importance and regular occurrences, just like how some important or special rational numbers such as π or e which have specific symbols attached to them.

2. Consider the function $f(x) = e^{-x^2}$ defined for $x \in \mathbb{R}$. This function does not have an explicit antiderivative. However, since this function is continuous and bounded, it is Riemann integrable over any compact interval $[a, b]$. So, an antiderivative of this function is given by the Riemann integral function $F(x) = \int_0^x e^{-t^2}\, dt$ which exists thanks to Proposition 15.6.4.
   As mentioned earlier, a closed explicit form for this Riemann integral (which is the antiderivative of f) does not exist. This integral appears very frequently in mathematics and statistics. Since it is very common, it is another example of a special function and has a name: it is called the Gaussian error function and denoted as $\frac{\sqrt{\pi}}{2}\mathrm{erf}(x)$.
3. Another important example would be the elliptic integral, which are integrals of the form:

$$\int_0^x R(t, \sqrt{P(t)})\, dt,$$

   where P is a cubic or quartic (degree 3 or 4) polynomial with no repeated roots and R is a rational function of its two arguments. These integrals appeared in the study of arclength of ellipses and later crops up in problems in mechanics. We shall see an example of where it occurs in Exercise 16.13 later when we derive the circumference of an ellipse.

16.4 Improper Integrals

Even though we have had considerable success with them, there are some minor problems with the Darboux and Riemann integral. First, by construction, the integral can only be constructed for bounded functions defined on compact intervals $[a, b] \subseteq \mathbb{R}$. For functions defined on unbounded domains, domains which are not closed, or unbounded functions, the construction may fail. This might not always be the case if we impose some conditions on the integrand.

Example 16.4.1 Let us look first at how some of the construction for Riemann integrals might fail for these cases.

1. For the first scenario, if consider the function $f : [0, \infty) \to \mathbb{R}$, we can never find finitely many partition points $\mathcal{P}$ to partition the domain $[0, \infty)$ since the domain is unbounded. Thus, the subgraph of this function cannot be approximated by finitely many rectangles of finite widths. Thus, the first ever step in the construction of the Riemann and Darboux integral cannot be carried out for this case.

2. The construction also fails for functions which blows up to infinity at some finite point. An example of this is the function $f : [0, 1] \to \mathbb{R}$ defined as:

$$f(x) = \begin{cases} 0 & \text{for } x = 0, \\ \frac{1}{\sqrt{x}} & \text{for } x \in (0, 1]. \end{cases}$$

Note that this function is unbounded, which does not fulfil the requirement for the Darboux integration. Furthermore, we have also seen in Proposition 15.4.2 that a Riemann integrable function is necessarily bounded.
The reason why the construction fails for this particular example is that we could not even construct an upper approximating function $\bar{f}$ for any partition $\mathcal{P} = \{x_0, x_1, x_2, \ldots, x_n\}$ of $[0, 1]$ since the supremum near 0 does not exist. Namely, the quantity $M_1 = \sup_{x\in[x_0,x_1]} f(x)$ does not exist for any partition $\mathcal{P}$ of $[0, 1]$ and thus its upper Darboux sum $U_{f,\mathcal{P}}$ cannot be defined in $\mathbb{R}$.
3. Due to a similar reason as the above, the function $g : (0, 1] \to \mathbb{R}$ defined on a domain that is not closed as $g(x) = \frac{1}{\sqrt{x}}$ is also not Riemann integrable via the definitions that we have seen. So, the construction also might fail when we try to define a Riemann integral over a domain that is not closed.

In order to remedy some of these problems, we do what analysts do best: appeal to limits! We know exactly how integrals over a compact interval can be constructed, so in order to attach a value to an integral over a domain which is not closed or an unbounded domain, we integrate over a sequence of compact intervals that eventually approximate the domain that we were interested in originally. We call this operation improper Riemann integrals, in contrast to the integrals that we have seen earlier which are the proper Riemann integrals.

Definition 16.4.2 (Improper Riemann Integrals) There are two kinds of improper integrals.

1. Improper Riemann integral of the first kind: If $f : (a, b] \to \mathbb{R}$ is a real function, then we define the improper Riemann integral of f over the domain $(a, b]$ as:

$$\int_a^b f(x)\, dx = \lim_{t\downarrow a} \int_t^b f(x)\, dx,$$

if this limit exists. This can be seen in Fig. 16.8a.
Similarly, if $f : [a, b) \to \mathbb{R}$ is a real function, then we define the improper Riemann integral of f over the domain $[a, b)$ as:

$$\int_a^b f(x)\, dx = \lim_{t\uparrow b} \int_a^t f(x)\, dx,$$

if this limit exists.

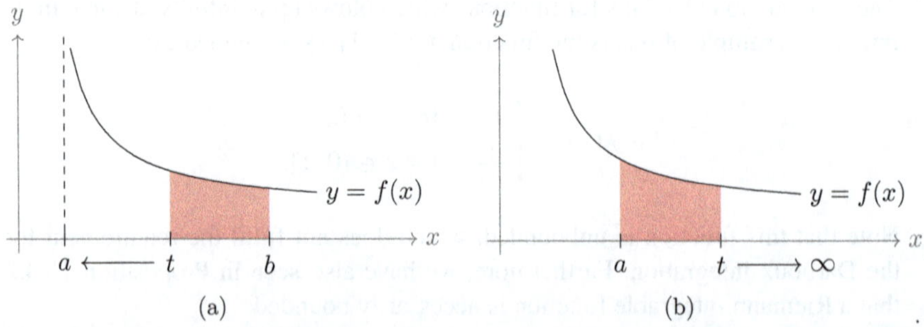

Fig. 16.8 Two kinds of improper Riemann integrals. **(a)** First kind. Take the limit as $t \downarrow a$. **(b)** Second kind. Take the limit as $t \uparrow \infty$

2. Improper Riemann integral of the second kind: If $f : [a, \infty) \to \mathbb{R}$ is a real function, then we define the improper Riemann integral of f over the domain $[a, \infty)$ as:

$$\int_a^\infty f(x)\,dx = \lim_{t\to\infty} \int_a^t f(x)\,dx,$$

if this limit exists. This can be seen in Fig. 16.8b.
Similarly, if $f : (-\infty, b] \to \mathbb{R}$ is a real function, then we define the improper Riemann integral of f over the domain $(-\infty, b]$ as:

$$\int_{-\infty}^b f(x)\,dx = \lim_{t\to-\infty} \int_t^b f(x)\,dx,$$

if this limit exists.

Remark 16.4.3 The improper Riemann integrals are akin to real series that we have seen earlier: we cannot find an actual value of the series by actual summation. So we resorted to finding the value of the infinite series via limits of its partial sums. Improper Riemann integral does exactly that: instead of finding the actual area under the graph of a function defined over a non-compact interval, we find its numerical value as a limit of sequence of integrals over compact subintervals under the graph which all exist. However, similar to real series, this improper Riemann integral may or may not exist.

Example 16.4.4 Let us look at some examples of improper Riemann integrals.

1. Let us consider the function $f : (0, 1] \to \mathbb{R}$ defined as $f(x) = \frac{1}{\sqrt{x}}$. This function blows up as $x \to 0$. As we have seen in Example 16.4.1(3), the proper Riemann

integral of this function on $(0, 1]$ cannot be defined. However, its improper Riemann integral may still exist. Since the function blows up near $x = 0$, we evaluate the following Riemann integral instead:

$$\int_t^1 \frac{1}{\sqrt{x}}\,dx \quad \text{for some small } t > 0.$$

The integrand is continuous in $[t, 1]$ and hence is Riemann integrable over this interval. By inspection, an antiderivative of f is $F(x) = 2\sqrt{x}$. Using the FTC, we have:

$$\int_t^1 \frac{1}{\sqrt{x}}\,dx = F(1) - F(t) = 2 - 2\sqrt{t}.$$

Thus, by taking the limit as $t \downarrow 0$, we can define the improper Riemann integral of f over the interval $(0, 1]$ as:

$$\int_0^1 f(x)\,dx = \lim_{t\downarrow 0} \int_t^1 f(x)\,dx = \lim_{t\downarrow 0}(2 - 2\sqrt{t}) = 2.$$

2. Consider the function $f : (0, 1] \to \mathbb{R}$ defined as $f(x) = \frac{1}{x^2}$. This function blows up as $x \to 0$. We wish to find the improper Riemann integral of this function in $(0, 1]$, if it exists. For some fixed t where $0 < t < 1$, we can evaluate the Riemann integral of this function over the compact interval $[t, 1]$ instead, namely:

$$\int_t^1 \frac{1}{x^2}\,dx = \left[-\frac{1}{x}\right]_t^1 = \frac{1}{t} - 1.$$

However, if we take the limit as $t \downarrow 0$ we have:

$$\lim_{t\downarrow 0} \int_t^1 f(x)\,dx = \lim_{t\downarrow 0} \left(\frac{1}{t} - 1\right) = \infty,$$

so the limit does not exist. Therefore, the function $f(x) = \frac{1}{x^2}$ is not improperly Riemann integrable over $(0, 1]$.

3. Let us consider the function $f : [1, \infty) \to \mathbb{R}$ defined as $f(x) = \frac{1}{\sqrt{x}}$. We wish to find the improper Riemann integral of this function over the interval $[1, \infty)$. We compute its Riemann integral over the interval $[1, t]$ for some finite $t > 1$ first, namely:

$$\int_1^t \frac{1}{\sqrt{x}} = 2\sqrt{t} - 2.$$

Thus, by taking the limit as $t \to \infty$, the limit of this integral blows up to infinity. Therefore the function is not improperly Riemann integrable over the unbounded domain $[1, \infty)$.

On the other hand, one can check that the improper Riemann integral of the function $f(x) = \frac{1}{x^2}$ over the domain $[1, \infty)$ is finite and equal to 1.

4. Recall the cycloid in Exercises 14.6 and 16.1.10(4). This is a curve represented parametrically by $x(t) = t - \sin(t)$ and $y(t) = 1 - \cos(t)$ for $t \in \mathbb{R}$ which describes the path of a point on a circle as the circle rolls along a flat surface.

Now we want to compute the total distance the point travelled as the circle completes one full rotation. In other words, we want to compute the arclength of the curve described by the implicit function $y(x)$ from $x = 0$ to $x = 2\pi$ by Definition 16.2.3. Namely, we want to evaluate the integral $\int_0^{2\pi} \sqrt{1 + y'(x)^2}\, dx$. In Exercise 14.6(i), we have seen that $y'(x)^2 = \frac{2}{y} - 1$ for all $x \in (0, 2\pi)$, so the integral simplifies to $\int_0^{2\pi} \sqrt{\frac{2}{y(x)}}\, dx$. We have two issues regarding this integral:

(a) The first issue is that the integrand blows up to ∞ as $x \downarrow 0$ and as $x \uparrow 2\pi$ since $y(x) \to 0$ at these points. Therefore, this integral has to be evaluated improperly at both ends. As we have seen in Fig. 16.2, the arclength of the cycloid seems to be finite as the curve does not exhibit any blowing up behaviour on the boundaries. Thus, we expect this improper integral to exist.

(b) The second issue here is that, similar to the conundrum in Example 16.1.10(4), the explicit expression for $y(x)$ is difficult to obtain. So, we use the variable t via the change of variable $x(t) = t - \sin(t)$ for $t \in [0, 2\pi]$ instead. This change of variable is a continuous bijection by Exercise 14.6(a) and (b). We denote its continuous inverse as $t(x)$.

Thus, for small enough $h > 0$, we compute the integral over the compact interval $[h, 2\pi - h] \subsetneq [0, 2\pi]$ as follows:

$$\begin{aligned}\int_h^{2\pi-h} \sqrt{\frac{2}{y(x)}}\, dx = \sqrt{2}\int_{t(h)}^{t(2\pi-h)} \frac{x'(t)}{\sqrt{y(t)}}\, dt &= \sqrt{2}\int_{t(h)}^{t(2\pi-h)} \sqrt{1-\cos(t)}\, dt \\ &= 2\int_{t(h)}^{t(2\pi-h)} \sin\left(\frac{t}{2}\right) dt \\ &= -4\left[\cos\left(\frac{t}{2}\right)\right]_{t(h)}^{t(2\pi-h)}.\end{aligned}$$

Thus, the improper integral is given by the following limit:

$$\int_0^{2\pi} \sqrt{\frac{2}{y(x)}}\, dx = \lim_{h\downarrow 0} -4\left[\cos\left(\frac{t}{2}\right)\right]_{t(h)}^{t(2\pi-h)}.$$

Since $t(x)$ is continuous, we have the limits $\lim_{h\downarrow 0} t(h) = t(0) = 0$ and $\lim_{h\downarrow 0} t(2\pi - h) = t(2\pi) = 2\pi$. Moreover, since cosine is also continuous everywhere, we can swap the order of the limit and cosine function to get:

$$\begin{aligned}\int_0^{2\pi}\sqrt{\frac{2}{y(x)}}\,dx &= -4\lim_{h\downarrow 0}\left[\cos\left(\frac{t(2\pi-h)}{2}\right)-\cos\left(\frac{t(h)}{2}\right)\right]\\ &= -4\left[\cos\left(\lim_{h\downarrow 0}\frac{t(2\pi-h)}{2}\right)-\cos\left(\lim_{h\downarrow 0}\frac{t(h)}{2}\right)\right]\\ &= -4\left[\cos(\pi)-\cos(0)\right] = -4(-1-1) = 8.\end{aligned}$$

Thus, the point travelled a total of 8 distance units as the circle completes one full rotation.

From Example 16.4.4(1), we have seen an example of a function that is not improperly Riemann integrable over a domain that is not closed. However, if we place some conditions on the function (such as continuity or boundedness) near the open boundary of the domain, we may get integrability on a domain that is not compact. An easy case is the following:

Proposition 16.4.5 *If $f : [a, b) \to \mathbb{R}$ is continuous on $[a, b) \subseteq \mathbb{R}$ such that $\lim_{x\uparrow b} f(x)$ exists, then f is improperly Riemann integrable on $[a, b)$.*

Proposition 16.4.5 can be easily proven by extending the function f to a continuous function $\tilde{f} : [a, b] \to \mathbb{R}$ and use the facts that we know about Riemann integral of continuous function. We leave this as an Exercise 16.5 for the readers to prove. Here, we shall prove a more general result than Proposition 16.4.5.

Proposition 16.4.6 *If $f : [a, b) \to \mathbb{R}$ is continuous and bounded on $[a, b) \subseteq \mathbb{R}$, then it is improperly Riemann integrable on $[a, b)$.*

Proof Since f is a bounded function in $[a, b)$, there exists an $M > 0$ such that $|f| \leq M$. Define an extension $\tilde{f} : [a, b] \to \mathbb{R}$ of the function f to the whole of $[a, b]$ as:

$$\tilde{f}(x) = \begin{cases} f(x) & \text{if } x \in [a, b),\\ 0 & \text{if } x = b,\end{cases}$$

so that the function $\tilde{f}$ is also continuous over $[a, b)$ and bounded with $|\tilde{f}(x)| \leq M$ for all $x \in [a, b]$.

Fix $\varepsilon > 0$. Consider the subinterval $[a, b - \frac{\varepsilon}{4M}]$ of $[a, b)$. Since this subinterval is closed and the function $\tilde{f} = f$ is continuous in this interval, it is also Riemann integrable here. Thus, there exists a partition $\mathcal{P}$ of this $[a, b - \frac{\varepsilon}{4M}]$ such that $U_{f,\mathcal{P}} - L_{f,\mathcal{P}} < \frac{\varepsilon}{2}$.

We extend this partition to the whole of $[a, b]$ by adding an extra point $\{b\}$ to obtain a partition $\mathcal{P}' = \mathcal{P} \cup \{b\}$. Furthermore, within the newly added subinterval, we know that $\tilde{f}$ is bounded from above by M and from below by $-M$. Then, the difference of the Darboux sums for this new partition can be bounded as such:

$$U_{\tilde{f},\mathcal{P}'} - L_{\tilde{f},\mathcal{P}'} \leq \left(U_{f,\mathcal{P}} + M\left(\frac{\varepsilon}{4M}\right)\right) - \left(L_{f,\mathcal{P}} - M\left(\frac{\varepsilon}{4M}\right)\right) < \frac{\varepsilon}{2} + \frac{\varepsilon}{2} = \varepsilon,$$

and hence, by the ε-criterion of Darboux integrability, the function $\tilde{f}$ is Riemann integrable over $[a, b]$. By definition, the improper Riemann integral of f over $[a, b)$ is:

$$\int_a^b f(x)\,dx = \lim_{t \uparrow b} \int_a^t f(x)\,dx = \lim_{t \uparrow b} \int_a^t \tilde{f}(x)\,dx = \int_a^b \tilde{f}(x)\,dx,$$

where we used the continuity of the integral function of $\tilde{f}$ from Theorem 16.1.1. Since $\int_a^b \tilde{f}(x)\,dx$ is a finite value, the improper Riemann integral of f over $[a, b)$ also exists and it is equal to this number. □

Of course, the proof above can be adapted to show the same also holds true for any bounded and continuous functions defined on $(a, b]$ and (a, b). Let us look at an example of this.

Example 16.4.7 Consider the function $f : (0, 1] \to \mathbb{R}$ defined as $f(x) = \sin(\frac{1}{x})$. Clearly this function is continuous on $(0, 1]$ as it is a composition of two continuous functions. Furthermore, the sine function is always bounded between -1 and 1.

Thus, by Proposition 16.4.6, we can conclude that the improper Riemann integral $\int_0^1 \sin(\frac{1}{x})\,dx$ exists. Even though we know that this Riemann integral exists, it is not expressible fully in terms of elementary functions and is given by:

$$\int_0^1 \sin\left(\frac{1}{x}\right)\,dx = \sin(1) - \mathrm{Ci}(1),$$

where $\mathrm{Ci} : (0, \infty) \to \mathbb{R}$ is the cosine integral function defined as $\mathrm{Ci}(x) = -\int_x^\infty \frac{\cos(t)}{t}\,dt$. This is another example of a special function.

Comparison Tests for Improper Riemann Integrals

Moreover, since a Riemann integral can also be thought of as an infinite sum, we also have the comparison test to determine whether an improper Riemann integral exists, similar to what we had for real series in Chap. 7. The following result is an analogue of Proposition 7.5.1.

Proposition 16.4.8 (Direct Comparison Test for Improper Riemann Integrals of the Second Kind) *Let $I = [a, \infty)$. Suppose that $f, g : I \to \mathbb{R}$ are continuous non-negative functions such that $0 \leq f(x) \leq g(x)$ for all $x \in I$.*

1. *If $\int_a^\infty g(x)\,dx$ exists, then $\int_a^\infty f(x)\,dx$ also exists.*
2. *If $\int_a^\infty f(x)\,dx$ diverges, then $\int_a^\infty g(x)\,dx$ also diverges.*

A similar result can be proven for $I = (-\infty, a]$ and improper integrals over this domain.

Proof We prove the assertions one by one. Since the functions f and g are continuous over $[a, \infty)$, these functions are Riemann integrable over the interval $[a, t]$ for any finite $t > a$.

1. Since the functions are non-negative, by ordering and additivity of domain, for any $t \geq a$ we have the bound:
$$\int_a^t f(x)\,dx \leq \int_a^t g(x)\,dx \leq \lim_{t\uparrow\infty} \int_a^t g(x)\,dx = \int_a^\infty g(x)\,dx.$$
Moreover, the integral function $F(t) = \int_a^t f(x)\,dx$ on $[a, \infty)$ is an increasing function. Thus, by Exercise 9.26, the limit of $F(t)$ as $t \to \infty$ exists since $F(t)$ is bounded by the finite number $\int_a^\infty g(x)\,dx$.
2. For any $t \geq a$ we have the ordering:
$$\int_a^t f(x)\,dx \leq \int_a^t g(x)\,dx.$$
Note that the integral function $F(t) = \int_a^t f(x)\,dx$ is increasing. Taking the limit as $t \to \infty$ on both sides, since $\lim_{t\to\infty} \int_a^t f(x)\,dx$ diverges, it must blow up to ∞. Thus, we conclude that $\lim_{t\to\infty} \int_a^t g(x)\,dx$ also diverges. □

Next, similar to real series as in Proposition 7.5.4, we have a limit form for the comparison test. The idea is still the same, namely: if the functions f and g behave in the same way asymptotically (up to some scale L) towards infinity, then their Riemann integrability at infinity are the same.

Proposition 16.4.9 (Limit Comparison Test for Improper Riemann Integrals of the Second Kind) *Let $I = [a, \infty)$. Suppose that $f, g : I \to \mathbb{R}$ are continuous positive functions. Suppose further that $\lim_{x\to\infty} \frac{f(x)}{g(x)} = L$ for some $0 < L < \infty$. Then, either both improper Riemann integrals $\int_a^\infty f(x)\,dx$ and $\int_a^\infty g(x)\,dx$ exist or both diverge. In other words:*
$$\int_a^\infty f(x)\,dx \text{ exists} \quad \Leftrightarrow \quad \int_a^\infty g(x)\,dx \text{ exists.}$$

A similar result can be proven for $I = (-\infty, a]$ with $\lim_{x\to-\infty} \frac{f(x)}{g(x)} = L$ for some $0 < L < \infty$.

Proof Since the functions f and g are continuous over $[a, \infty)$, these functions are Riemann integrable over the interval $[a, t]$ for any finite $t > a$. Moreover, since $\lim_{x\to\infty} \frac{f(x)}{g(x)} = L$, for $\varepsilon = \frac{L}{2} > 0$ there exists a $K \in [a, \infty)$ such that $\left|\frac{f(x)}{g(x)} - L\right| < \frac{L}{2}$ for all $x > K$. In other words, for any $x > K$ we have:

$$\frac{L}{2}g(x) < f(x) < \frac{3L}{2}g(x).$$

We shall now prove the implications one by one:

($\Rightarrow$): By assumption, the improper Riemann integral $\int_a^\infty f(x)\,dx = \int_a^K f(x)\,dx + \int_K^\infty f(x)\,dx$ exists and therefore the improper Riemann integral $\frac{2}{L}\int_K^\infty f(x)\,dx = \int_K^\infty \frac{2}{L} f(x)\,dx$ also exists. Since $0 < g(x) < \frac{2}{L}f(x)$ for all $x > K$, Proposition 16.4.8 says the Riemann integral $\int_K^\infty g(x)\,dx$ also exist. Furthermore, by additivity of the Riemann integral over its domain, we can conclude that the improper integral $\int_a^\infty g(x)\,dx = \int_a^K g(x)\,dx + \int_K^\infty g(x)\,dx$ also exists.

($\Leftarrow$): Similar to the above, since $0 < f(x) < \frac{3L}{2}g(x)$ for all $x > K$ and the improper Riemann integral $\frac{3L}{2}\int_K^\infty g(x)\,dx$ exists, the improper Riemann integral $\int_K^\infty f(x)\,dx$ and hence $\int_a^\infty f(x)\,dx$ also exist by Proposition 16.4.8. □

We also have analogous results of Propositions 16.4.8 and 16.4.9 for the first kind of improper Riemann integrals, which can be proven in a similar manner. We leave the proofs of the following two results as Exercise 16.7.

Proposition 16.4.10 (Direct Comparison Test for Improper Riemann Integrals of the First Kind) *Let $I = [a, b)$ or $(a, b]$. Suppose that $f, g : I \to \mathbb{R}$ are continuous non-negative functions such that $0 \leq f(x) \leq g(x)$ for all $x \in I$.*

1. *If $\int_a^b g(x)\,dx$ exists, then $\int_a^b f(x)\,dx$ also exists.*
2. *If $\int_a^b f(x)\,dx$ diverges, then $\int_a^b g(x)\,dx$ also diverges.*

Proposition 16.4.11 (Limit Comparison Test for Improper Riemann Integrals of the First Kind) *Let $I = [a, b)$. Suppose that $f, g : I \to \mathbb{R}$ are continuous positive functions. Suppose further that $\lim_{x\uparrow b} \frac{f(x)}{g(x)} = L$ for some $0 < L < \infty$. Then, either both improper Riemann integrals $\int_a^b f(x)\,dx$ and $\int_a^b g(x)\,dx$ exist or both diverge. In other words:*

$$\int_a^b f(x)\,dx \text{ exists} \quad \Leftrightarrow \quad \int_a^b g(x)\,dx \text{ exists.}$$

A similar result can be proven for $I = (a, b]$ *with* $\lim_{x\downarrow a} \frac{f(x)}{g(x)} = L$ *for some* $0 < L < \infty$.

Remark 16.4.12 In Propositions 16.4.8, 16.4.9, 16.4.10, and 16.4.11 we can also weaken the condition on the functions f and g above from being continuous to simply being Riemann integrable over any compact intervals in their domains.

Example 16.4.13 Let us look at some examples for which the comparison tests come in handy:

1. Consider the function $f(x) = \frac{\sin^2(x)}{x^3}$ on $[1, \infty)$. Over any compact interval $[1, t]$, this function is Riemann integrable as it is continuous. To show that it is improperly Riemann integrable over $[1, \infty)$, we note that for any $x \geq 1$, we have the ordering $0 \leq \frac{\sin^2(x)}{x^3} \leq \frac{1}{x^3} \leq \frac{1}{x^2}$ and we know from Example 16.4.4(3) that the improper Riemann integral $\int_1^\infty \frac{1}{x^2}\,dx$ exists. Therefore, by the comparison test, we conclude that $\int_1^\infty f(x)\,dx$ also exists.
2. Consider the function $f(x) = \frac{1}{\ln(x)}$ on $[1, \infty)$. Note that in the domain, since $x \leq e^x$ for $x \geq 1$, we have $\frac{1}{x} \leq \frac{1}{\ln(x)}$. Let us determine whether the improper Riemann integral $\int_1^\infty \frac{1}{x}\,dx$ exists. For any finite $t > 1$, by the FTC we have $\int_1^t \frac{1}{x}\,dx = \ln(t) - \ln(1) = \ln(t)$. So the improper Riemann integral of $\frac{1}{x}$ does not exist because $\lim_{t\to\infty} \int_1^t \frac{1}{x}\,dx = \lim_{t\to\infty} \ln(t) = \infty$. From this, via the comparison test, we conclude that f is also not improperly Riemann integrable over $[1, \infty)$.
3. Consider the function $f(x) = \frac{1}{\sqrt{x^2+x}}$ on $[1, \infty)$. Towards infinity, the term x in the denominator becomes very small compared to the term x^2, so we expect that the function behaves like $\frac{1}{\sqrt{x^2}} = \frac{1}{x}$ asymptotically. Indeed, we have:

$$\lim_{x\to\infty} \frac{\frac{1}{\sqrt{x^2+x}}}{\frac{1}{x}} = \lim_{x\to\infty} \sqrt{\frac{x}{x+1}} = \lim_{x\to\infty} \sqrt{\frac{1}{1+\frac{1}{x}}} = 1.$$

   Therefore, by Proposition 16.4.9, the Riemann integrability of the function f towards infinity is similar to $\frac{1}{x}$. However, we have seen in the previous example that the $\frac{1}{x}$ is not improperly Riemann integrable over $[1, \infty)$. Therefore, we conclude that f is also not improperly Riemann integrable over $[1, \infty)$.
   Another way of figuring this out is via the usual direct comparison test. We note that for $x \geq 1$ we have the inequality $x^2+x \leq 2x^2$ which implies $\frac{1}{\sqrt{2}}\frac{1}{x} \leq \frac{1}{\sqrt{x^2+x}}$. Then, the direct comparison test follows.
4. Consider the function $f : (0, 1] \to \mathbb{R}$ defined as $f(x) = \csc(x)$. In any closed interval $[t, 1]$ where $0 < t < 1$, this function is Riemann integrable. However, $\csc(x)$ blows up to ∞ as $x \downarrow 0$, so we want to know whether its

improper Riemann integral over $(0, 1]$ can be defined. So, let us first determine the asymptotic behaviour of f as x approaches 0. Note that:

$$\frac{\frac{1}{x}}{\csc x} = \frac{\frac{1}{x}}{\frac{1}{\sin(x)}} = \frac{\sin(x)}{x} \quad \Rightarrow \quad \lim_{x \downarrow 0} \frac{\frac{1}{x}}{\csc x} = \lim_{x \downarrow 0} \frac{\sin(x)}{x} = 1,$$

which we have proven in Example 13.1.9(6). Therefore $\csc(x)$ behaves asymptotically like $\frac{1}{x}$ as x approaches 0. However, $\frac{1}{x}$ is not improperly Riemann integrable over $(0, 1]$ by direct comparison with the function $\frac{1}{\sqrt{x}}$ in Example 16.4.4(1) since $\frac{1}{x} \geq \frac{1}{\sqrt{x}}$ here. So, by Proposition 16.4.11, we conclude that the same is true for $\csc(x)$.

5. Let us check the improper Riemann integrability of the function $f : [1, \infty) \to \mathbb{R}$ defined as $f(x) = \frac{\cos(x)}{x^2}$ over $[1, \infty)$. We note that this function has mixed signs, so we cannot apply the comparison test immediately. To do this, we split the function f into its positive and negative parts. Namely, we define $f^+, f^- : [1, \infty) \to \mathbb{R}$ as $f^+ = \max(f, 0)$ and $f^- = -\min(f, 0)$ so that $f = f^+ - f^-$. We note that both of the functions f^+ and f^- are non-negative and continuous. We can thus apply the direct comparison test on f^+ since $0 \leq f^+(x) \leq \frac{|\cos(x)|}{x^2} \leq \frac{1}{x^2}$. Since the latter is improperly Riemann integrable over $[1, \infty)$ as we saw in Example 16.4.4(3), by direct comparison, the improper integral $\int_1^\infty f^+(x)\,dx$ exists. By the same argument, the improper Riemann integral $\int_1^\infty f^-(x)\,dx$ also exists.

   For any finite $t > 1$, we have $\int_1^t f(x)\,dx = \int_1^t f^+(x) - f^-(x)\,dx = \int_1^t f^+(x)\,dx - \int_1^t f^-(x)\,dx$. Hence, by the algebra of limits, we have:

$$\begin{aligned}\int_1^\infty f(x)\,dx = \lim_{t \uparrow \infty} \int_1^t f(x)\,dx &= \lim_{t \uparrow \infty} \left(\int_1^t f^+(x) - f^-(x)\,dx \right) \\ &= \lim_{t \uparrow \infty} \left(\int_1^t f^+(x)\,dx - \int_1^t f^-(x)\,dx \right) \\ &= \lim_{t \uparrow \infty} \int_1^t f^+(x)\,dx - \lim_{t \uparrow \infty} \int_1^t f^-(x)\,dx,\end{aligned}$$

   which exists. Thus, we conclude that the function f is improperly Riemann integrable over $[1, \infty)$.

Integral Test for Real Series

Let us look at an application of improper Riemann integral. Recall in Chap. 7 that given a real series $\sum_{j=1}^\infty a_j$, we have an important question of determining whether it converges. We have developed many different tests to answer this question. Using improper Riemann integrals, we have yet another test for this.

Theorem 16.4.14 (Integral Test) *Let $f : [1, \infty) \to \mathbb{R}$ be a non-negative function. Suppose that f is non-increasing and $\int_n^{n+1} f(x)\,dx$ exists for all $n \in \mathbb{N}$. Then:*

$$\sum_{j=1}^{\infty} f(j) \text{ converges} \quad \Leftrightarrow \quad \int_1^{\infty} f(x)\,dx \text{ exists.}$$

Proof Let (s_n) be the sequence of partial sums of the series $\sum_{j=1}^{\infty} f(j)$, defined as $s_n = \sum_{j=1}^{n} f(j)$. We also denote the sequence (I_n) as the sequence of Riemann integrals of f over $[1, n]$, namely $I_n = \int_1^n f(x)\,dx$. Since f is non-negative, both of the sequences (s_n) and (I_n) are increasing. Moreover, by Proposition 16.0.3, the integral function $I(x) = \int_0^x f(t)\,dt$ is increasing. Let us now find the relationship between these sequences.

Since f is non-increasing, we note first that for every $j \in \mathbb{N}$, we have the ordering $f(j+1) \leq f(x) \leq f(j)$ for every $x \in [j, j+1]$. Thus, we can integrate this over the region $[j, j+1]$ to get the inequality:

$$f(j+1) = \int_j^{j+1} f(j+1)\,dx \leq \int_j^{j+1} f(x)\,dx \leq \int_j^{j+1} f(j)\,dx = f(j).$$

This inequality simply says that the area under the graph of f over $[j, j+1]$ is bounded from below by the area of rectangle of height $f(j+1)$ and width 1 and bounded from above by the area of the rectangle of height $f(j)$ and width 1. This is true for every $j \in \mathbb{N}$. So, if we sum up these inequalities from $j = 1$ to $j = n$, by domain additivity, we get:

$$\sum_{j=2}^{n+1} f(j) \leq \int_1^{n+1} f(x)\,dx \leq \sum_{j=1}^{n} f(j) \quad \Leftrightarrow \quad s_{n+1} - f(1) \leq I_{n+1} \leq s_n. \tag{16.17}$$

An example diagram for (16.17) with $n = 3$ is given in Fig. 16.9.

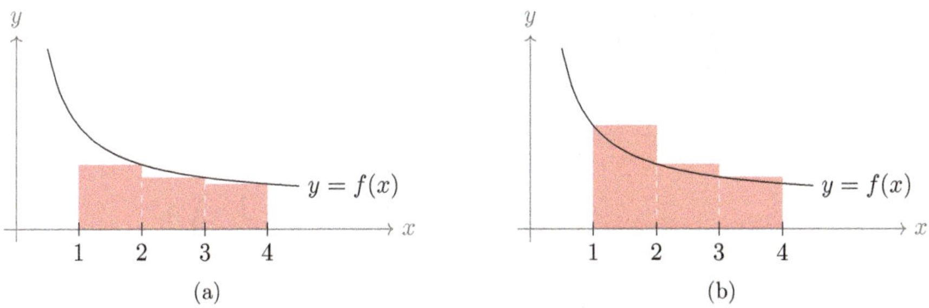

Fig. 16.9 Diagram for integral test. Areas of the shaded areas correspond to the finite sums. (**a**) $\sum_{j=2}^{4} f(j) = s_4 - f(1)$. (**b**) $\sum_{j=1}^{3} f(j) = s_3$

We now prove the implications one by one, but their ideas are the same.

($\Rightarrow$): If the series $\sum_{j=1}^{\infty} f(j)$ converges, the sequence (s_n) must be bounded. So, there exists an $M > 0$ such that $s_n \leq M$ for all $n \in \mathbb{N}$. Furthermore, from the inequality (16.17) we have $I_{n+1} \leq s_n \leq M$. Thus, the sequence (I_n) is also bounded from above. Since the sequence (I_n) is increasing and bounded, by monotone sequence theorem, it must be convergent. By Exercise 9.26(b), since I is an increasing function and $(I(n))$ converges, we conclude that the limit $\lim_{t\to\infty} I(t) = \lim_{t\to\infty} \int_0^t f(x)\,dx = \int_1^{\infty} f(x)\,dx$ exists.

($\Leftarrow$): If the improper integral $\int_1^{\infty} f(x)\,dx$ exists, then the sequence (I_n) must be convergent and hence bounded. So, there exists an $M > 0$ such that $I_n \leq M$ for all $n \in \mathbb{N}$. Therefore, from the inequality (16.17) we must have $s_n - f(1) \leq I_n \leq M$ which implies $s_n \leq f(1) + M$ for all $n \in \mathbb{N}$. Thus, the sequence (s_n) is also bounded from above. Since this sequence is also increasing, it must be convergent. Thus, the series $\sum_{j=1}^{\infty} f(j)$ is convergent. □

Example 16.4.15 Let us look at some examples on how to use the integral test:

1. Recall the harmonic series $\sum_{j=1}^{\infty} \frac{1}{j}$. We have shown that this series diverges using the definition of partial sums in Example 7.2.7. Let us try and prove that this series diverges using the integral test. The non-increasing function $f : [1, \infty) \to \mathbb{R}$ that we are interested in is $f(x) = \frac{1}{x}$. By the integral test in Theorem 16.4.14, the harmonic series $\sum_{j=1}^{\infty} \frac{1}{j}$ converges if and only if the improper Riemann integral $\int_1^{\infty} f(x)\,dx$ exists. However, we have seen in Example 16.4.13(2) that this improper integral does not exist. Hence, we conclude that the harmonic series is not convergent.
2. In Exercise 7.12, we have seen the p-series test to determine the convergence of the real series of the form $\sum_{j=1}^{\infty} \frac{1}{j^p}$ for $p \in \mathbb{R}$. We can prove this test using the integral test using the non-increasing function $f : [1, \infty) \to \mathbb{R}$ defined as $f(x) = \frac{1}{x^p}$. By the integral test, this series converges if and only if the integral $\int_1^{\infty} \frac{1}{x^p}\,dx$ exists. We can compute this improper Riemann integral:

$$\begin{aligned}\int_1^{\infty} \frac{1}{x^p}\,dx &= \lim_{t\to\infty} \int_1^t \frac{1}{x^p}\,dx \\ &= \lim_{t\to\infty} \left[-\frac{1}{(p-1)x^{p-1}}\right]_1^t \\ &= \lim_{t\to\infty} \left(-\frac{1}{(p-1)t^{p-1}} + \frac{1}{p-1}\right),\end{aligned}$$

which converges if and only if $p - 1 > 0$. Hence, the series $\sum_{j=1}^{\infty} \frac{1}{j^p}$ is convergent if and only if $p > 1$, which is exactly what the p-series test says.

16.5 Integration and Limits

The next problem that we may encounter with the Riemann integral is that even if we are integrating a function over a compact interval, not all bounded functions defined here are Riemann integrable. We noted that the Riemann integral exists if and only if the lower and upper Darboux integrals exist and are equal to each other. For some functions, they might both exist, but are not equal to each other.

Example 16.5.1 Recall the Dirichlet function $f : [0, 1] \to \mathbb{R}$ defined as:

$$f(x) = \begin{cases} 1 & \text{if } x \in \mathbb{Q}, \\ 0 & \text{if } x \in \bar{\mathbb{Q}}. \end{cases}$$

This function is not Riemann integrable as we have seen in Example 15.3.10. However, since $\mathbb{Q} \cap [0, 1]$ is countable, we can enumerate the rational numbers as $r_1, r_2, \ldots$ and thus define a sequence of functions (f_n) where $f_n : [0, 1] \to \mathbb{R}$ is given by:

$$f_n(x) = \begin{cases} 1 & \text{if } x = r_1, r_2, \ldots, r_n, \\ 0 & \text{otherwise,} \end{cases}$$

so that $f_n \xrightarrow{pw} f$. Moreover, for all $n \in \mathbb{N}$, since f_n differs to the constant zero function at finitely many points, they are all Riemann integrable by Proposition 15.6.2 with Riemann integral 0. However, we have:

$$0 = \lim_{n\to\infty} \int_0^1 f_n(x)\,dx \neq \int_0^1 \lim_{n\to\infty} f_n(x)\,dx = \int_0^1 f(x)\,dx,$$

since the latter does not even exist.

As we have seen in Example 16.5.1, limits and Riemann integrals may not behave very well together. The Dirichlet function in the example can be seen as a pointwise limit of a sequence of Riemann integrable functions, but is itself not Riemann integrable.

Moreover, even if the limiting function of a sequence of Riemann integrable functions is itself Riemann integrable, there is no guarantee that the Riemann integral of the limiting function is equal to the limit of the Riemann integrals. Here is an example of such phenomenon:

Example 16.5.2 Let (f_n) be a sequence of functions $f_n : [0, 1] \to \mathbb{R}$ defined as:

$$f_n(x) = \begin{cases} n & \text{if } x \in (0, \frac{1}{n}], \\ 0 & \text{otherwise.} \end{cases}$$

The pointwise limit of this sequence of functions is the constant function $f(x) = 0$, which is clearly Riemann integrable on $[0, 1]$. However, we have:

$$\lim_{n\to\infty} \int_0^1 f_n(x)\,dx = \lim_{n\to\infty} 1 = 1 \quad \text{and} \quad \int_0^1 \lim_{n\to\infty} f_n(x)\,dx = \int_0^1 0\,dx = 0,$$

and thus the limit of the Riemann integrals is not equal to the Riemann integral of the limit.

Integrable Limit Theorem

From Examples 16.5.1 and 16.5.2, we have evidence that limits and integrals do not commute in general, similar to what we have seen for sequence of functions, functions series, and differentiation. However, if we place a stronger condition, we might be able to get some convergence results for Riemann integrals.

Indeed, Riemann integration, like differentiation, is a limiting process. We have seen several times before that in order to be able to switch the order of limits, we require some uniform convergence property. If we place this condition on our functions, we have the following result:

Theorem 16.5.3 (Integrable Limit Theorem) *Let (f_n) be a sequence of Riemann integrable functions $f_n : [a, b] \to \mathbb{R}$. If (f_n) converges uniformly to a function $f : [a, b] \to \mathbb{R}$, then the limit function f is Riemann integrable as well. Furthermore, the limit of the Riemann integrals of f_n is equal to the Riemann integral of the limit f. In other words, if $f_n \xrightarrow{u} f$ on $[a, b]$, then:*

$$\lim_{n\to\infty} \int_a^b f_n(x)\,dx = \int_a^b \lim_{n\to\infty} f_n(x)\,dx = \int_a^b f(x)\,dx.$$

Proof First we need to show that $f \in \mathcal{R}([a, b])$. We show this via the ε-criterion of Darboux integrability. Since $f_n \xrightarrow{u} f$ on $[a, b]$, there exists an $N \in \mathbb{N}$ such that $\sup_{x\in[a,b]} |f_N(x) - f(x)| < \frac{\varepsilon}{3(b-a)}$. This means for any $x \in [a, b]$ we have:

$$|f_N(x) - f(x)| < \frac{\varepsilon}{3(b-a)} \quad \Rightarrow \quad f_N(x) - \frac{\varepsilon}{3(b-a)} < f(x) < f_N(x) + \frac{\varepsilon}{3(b-a)}. \tag{16.18}$$

Moreover, by assumption that the function f_N is Riemann integrable in $[a, b]$, there exists a partition $\mathcal{P} = \{x_0, x_1, \ldots, x_n\}$ of $[a, b]$ such that $U_{f_N,\mathcal{P}} - L_{f_N,\mathcal{P}} < \frac{\varepsilon}{3}$. For this partition $\mathcal{P}$, let us denote $M_j^f = \sup_{x\in[x_{j-1},x_j]} f(x)$ and $M_j^{f_N} = \sup_{x\in[x_{j-1},x_j]} f_N(x)$ so that, via the inequality in (16.18), we have $M_j^f \leq M_j^{f_N} + \frac{\varepsilon}{3(b-a)}$ for $j = 1, 2, \ldots, n$. Thus, the upper Darboux sums for f and f_N satisfy:

$$\begin{aligned} U_{f,\mathcal{P}} = \sum_{j=1}^{n} M_j^f |x_{j-1} - x_j| &\leq \sum_{j=1}^{n} \left(M_j^{f_N} + \frac{\varepsilon}{3(b-a)} \right) |x_{j-1} - x_j| \\ &= \sum_{j=1}^{n} M_j^{f_N} |x_{j-1} - x_j| + \sum_{j=1}^{n} \frac{\varepsilon}{3(b-a)} |x_{j-1} - x_j| \\ &= U_{f_N,\mathcal{P}} + \frac{\varepsilon}{3(b-a)}(b-a) = U_{f_N,\mathcal{P}} + \frac{\varepsilon}{3}. \end{aligned}$$

Similarly, by using the other inequality in (16.18), we can prove that $L_{f_N,\mathcal{P}} - \frac{\varepsilon}{3} \leq L_{f,P}$. Combining the three inequalities together, we get:

$$U_{f,\mathcal{P}} - L_{f,\mathcal{P}} \leq \frac{\varepsilon}{3} + U_{f_N,\mathcal{P}} - L_{f_N,\mathcal{P}} + \frac{\varepsilon}{3} < \frac{\varepsilon}{3} + \frac{\varepsilon}{3} + \frac{\varepsilon}{3} = \varepsilon.$$

Thus, the function f satisfies the ε-criterion for Darboux integrability and hence is Riemann integrable.

Now we want to show that the limit of the Riemann integrals is actually equal to the Riemann integral of the limit. If we denote the Riemann integral of f_n over $[a, b]$ as I_n and the Riemann integral of f over $[a, b]$ as I, we simply need to show that $\lim_{n\to\infty} I_n = I$. In other words, we want to show that for every $\varepsilon > 0$, there exists an $N \in \mathbb{N}$ such that for all $n \geq N$ we have:

$$|I_n - I| = \left| \int_a^b f_n(x)\,dx - \int_a^b f(x)\,dx \right| < \varepsilon.$$

Fix $\varepsilon > 0$. Since $f_n \xrightarrow{u} f$ on $[a, b]$, there is an $N \in \mathbb{N}$ such that $\sup_{x\in[a,b]} |f_n(x) - f(x)| < \frac{\varepsilon}{b-a}$ for all $n \geq N$. By applying Propositions 15.5.2 and 15.5.5, for all $n \geq N$ we have:

$$\begin{aligned} |I_n - I| = \left| \int_a^b f_n(x)\,dx - \int_a^b f(x)\,dx \right| &\leq \int_a^b |f_n(x) - f(x)|\,dx \\ &\leq \int_a^b \sup_{x\in[a,b]} |f_n(x) - f(x)|\,dx \\ &< \frac{\varepsilon}{b-a} \int_a^b dx = \varepsilon, \end{aligned}$$

and hence we are done. □

Example 16.5.4 Recall Example 14.2.1 where (f_n) is defined as a sequence of functions $f_n : \mathbb{R} \to \mathbb{R}$ given as $f_n(x) = \sqrt{\frac{1}{n} + x^2}$. We have proven that this sequence of functions converges uniformly to $f(x) = |x|$ everywhere. All of these functions are continuous over $[-1, 1]$, so we can define their Riemann integrals here. We can compute the antiderivatives of (f_n) by a change of variable of $x = \frac{\sinh(y)}{\sqrt{n}}$ so that:

$$\begin{aligned}\int f_n(x)\,dx = \int \sqrt{\frac{1}{n} + x^2}\,dx &= \frac{1}{n}\int \sqrt{1+\sinh^2(y)}\cosh(y)\,dy \\ &= \frac{1}{2n}(y + \sinh(y)\cosh(y)) + C \\ &= \frac{\operatorname{arcsinh}(\sqrt{n}x)}{2n} + \frac{x\sqrt{1+nx^2}}{2\sqrt{n}} + C,\end{aligned}$$

for some constant $C \in \mathbb{R}$. Thus, by the FTC, the Riemann integral of f_n over $[-1, 1]$ is:

$$\begin{aligned}\int_{-1}^{1} f_n(x)\,dx &= \left[\frac{\operatorname{arcsinh}(\sqrt{n}x)}{2n} + \frac{x\sqrt{1+nx^2}}{2\sqrt{n}}\right]_{-1}^{1} \\ &= \frac{\operatorname{arcsinh}(\sqrt{n})}{n} + \frac{\sqrt{1+n}}{\sqrt{n}} \\ &= \frac{\ln(\sqrt{n} + \sqrt{n+1})}{n} + \frac{\sqrt{1+n}}{\sqrt{n}}.\end{aligned}$$

By the algebra of limits and sandwiching, we can find the limit:

$$1 \leq \frac{\ln(\sqrt{n} + \sqrt{n+1})}{n} + \frac{\sqrt{1+n}}{\sqrt{n}} < \frac{\ln(2n)}{n} + \frac{\sqrt{1+n}}{\sqrt{n}} \to 0 + 1 = 1,$$

as $n \to \infty$. This limit agrees with the value of $\int_{-1}^{1} |x|\,dx = 1$. So, here we do have the equality $\lim_{n\to\infty} \int_{-1}^{1} f_n(x)\,dx = \int_{-1}^{1} \lim_{n\to\infty} f_n(x)\,dx = \int_{-1}^{1} f(x)\,dx$ as guaranteed by Theorem 16.5.3.

Furthermore, by applying Theorem 16.5.3 to functions series, we have an analogous theorem for functions series, namely:

Theorem 16.5.5 (Integrable Limit Theorem for Functions Series) *Let (f_n) be a sequence of Riemann integrable functions $f_n : [a, b] \to \mathbb{R}$ and $s = \sum_{j=1}^{\infty} f_j$ be its functions series. If the series converges uniformly on $[a, b]$, then $s \in \mathcal{R}([a, b])$.*

In other words, if the sequence of partial sums (s_n) *where* $s_n = \sum_{j=1}^{n} f_j$ *converges uniformly on* $[a, b]$*, then we can apply term-by-term Riemann integration to the series, namely:*

$$\int_a^b \sum_{j=1}^{\infty} f_j(x)\, dx = \sum_{j=1}^{\infty} \int_a^b f_j(x)\, dx.$$

Example 16.5.6 Let us look at some examples on how we can use Theorem 16.5.5.

1. Recall the series $\sum_{j=1}^{\infty} \frac{j\sin(jx)}{e^j}$ in Example 11.4.10. We have seen that this series converges uniformly over $\mathbb{R}$. So if we want to integrate it over $[0, \pi]$ we can do it term-wise as:

$$\begin{aligned}\int_0^{\pi} \sum_{j=1}^{\infty} \frac{j\sin(jx)}{e^j}\, dx = \sum_{j=1}^{\infty} \int_0^{\pi} \frac{j\sin(jx)}{e^j}\, dx &= \sum_{j=1}^{\infty} \frac{1-\cos(j\pi)}{e^j} \\ &= \sum_{j=1}^{\infty} \frac{1-(-1)^j}{e^j} \\ &= \sum_{\substack{j=1 \\ j \text{ odd}}}^{\infty} \frac{2}{e^j} \\ &= \sum_{j=1}^{\infty} \frac{2}{e^{2j-1}} = \frac{2e}{e^2-1}.\end{aligned}$$

2. By Proposition 12.2.1, the power series $\sum_{j=0}^{\infty} x^j$ has radius of convergence 1 and so it converges uniformly over any interval $[-1+\delta, 1-\delta]$ for any small $\delta > 0$. Furthermore, over this interval we have the equality:

$$\frac{1}{1-x} = \sum_{j=0}^{\infty} x^j.$$

Since this convergence is uniform over $[-1+\delta, 1-\delta]$, we can integrate this series term-wise from 0 to any $x \in [-1+\delta, 1-\delta]$ to get:

$$\begin{aligned}
\int_0^x \frac{1}{1-y}\,dy &= \int_0^x \sum_{j=0}^{\infty} y^j\,dy = \sum_{j=0}^{\infty} \int_0^x y^j\,dy \\
\Rightarrow \quad [-\ln(1-y)]_0^x &= \sum_{j=0}^{\infty} \left[\frac{y^{j+1}}{j+1}\right]_0^x \\
\Rightarrow \quad -\ln(1-x) &= \sum_{j=0}^{\infty} \frac{x^{j+1}}{j+1} = \sum_{j=1}^{\infty} \frac{x^j}{j}.
\end{aligned}$$

We note that $\delta > 0$ was an arbitrarily small number, thus this equality holds for any $x \in (-1, 1)$ at all. Therefore, the above gives us a power series expression for $-\ln(1-x)$ centred at $x = 0$.

More generally, based on the observation in Example 16.5.6(2), we have the following result for power series.

Proposition 16.5.7 (Term-Wise Integration of Power Series) *Let $\sum_{j=0}^{\infty} a_j(x-c)^j$ be a power series for some constants $c, a_j \in \mathbb{R}$ with radius of convergence $R > 0$. We have:*

$$\int_c^x \sum_{j=0}^{\infty} a_j(y-c)^j\,dy = \sum_{j=0}^{\infty} \int_c^x a_j(y-c)^j\,dy = \sum_{j=0}^{\infty} \frac{a_j(x-c)^{j+1}}{j+1},$$

for any $x \in (c-R, c+R)$ if R is finite and any $x \in \mathbb{R}$ if R is infinite.

Proof WLOG, assume $c = 0$ and R is finite. Suppose that $s_n(x) = \sum_{j=0}^{n} a_j x^j$ is the sequence of partial sums for the power series. Consider the sequence of functions (t_n) where $t_n : (-R, R) \to \mathbb{R}$ is the integral function of the n-th partial sum s_n, namely:

$$t_n(x) = \int_0^x s_n(y)\,dy = \sum_{j=0}^{n} \int_0^x a_j y^j\,dy = \sum_{j=0}^{n} \frac{a_j x^{j+1}}{j+1}.$$

We have seen in Proposition 12.3.8 that the limit of this series as $n \to \infty$ given by $t(x) = \sum_{j=0}^{\infty} \frac{a_j x^{j+1}}{j+1}$ also has the same radius of convergence R.

We note that this power series converges uniformly over the closed interval $|x| \leq R - \delta$ for any small $\delta > 0$ by Proposition 12.2.1. Thus, by Theorem 16.5.5, for $x \in [-R+\delta, R-\delta]$ where $\delta > 0$ is any small number, we can conclude that:

$$\int_0^x s(y)\,dy = \int_0^x \lim_{n\to\infty} s_n(y)\,dy = \lim_{n\to\infty} \int_0^x s_n(y)\,dy = \lim_{n\to\infty} t_n(x) = t(x). \tag{16.19}$$

Since $\delta > 0$ can be chosen arbitrarily small, the equality (16.19) implies:

$$\int_0^x \sum_{j=0}^{\infty} a_j y^j \, dy = \int_c^x s(y)\, dy = t(x) = \sum_{j=0}^{\infty} \frac{a_j x^{j+1}}{j+1},$$

for any $x \in (-R, R)$. □

As a corollary of Proposition 16.5.7, we have the following result:

Corollary 16.5.8 *Let $\sum_{j=0}^{\infty} a_j (x-c)^j$ be a power series for some constants $c, a_j \in \mathbb{R}$ with radius of convergence $R > 0$. We have:*

$$\int_a^b \sum_{j=0}^{\infty} a_j (y-c)^j \, dy = \sum_{j=0}^{\infty} \int_a^b a_j (y-c)^j \, dy,$$

for any compact subinterval $[a, b] \subseteq (c - R, c + R)$ if R is finite and any compact subinterval $[a, b] \subseteq \mathbb{R}$ if R is infinite.

The results above tell us that we can apply the Riemann integration term-wise for a power series over any compact subinterval within its radius of convergence, which is a very handy result!

Example 16.5.9 Let us look at how we can utilise Proposition 16.5.7 and Corollary 16.5.8.

1. Recall the power series:

$$\frac{1}{1-x}\frac{1}{1-2x} = \frac{1}{1-3x+2x^2} = \sum_{j=1}^{\infty} (2^{j+1} - 1) x^j,$$

for $|x| < \frac{1}{2}$ from Example 12.3.5. Applying Proposition 16.5.7, for any x such that $|x| < \frac{1}{2}$, we have the equality:

$$\int_0^x \frac{1}{1-3y+2y^2} \, dy = \int_0^x \sum_{j=1}^{\infty} (2^{j+1} - 1) y^j \, dy = \sum_{j=1}^{\infty} \frac{(2^{j+1} - 1)}{j+1} x^{j+1}. \tag{16.20}$$

But what is the term on the LHS of (16.20)? We can either find its value by completing the square and apply a trigonometric substitution, or we can split the term into two fractions using partial fraction decomposition. Using the latter, we have:

$$\int_0^x \frac{1}{1-3y+2y^2}\,dy = \int_0^x -\frac{1}{1-y} + \frac{2}{1-2y}\,dy = \ln(1-x) - \ln(1-2x)$$
$$= \ln\left(\frac{1-x}{1-2x}\right).$$

In short, from (16.20) we have the equality:

$$\ln\left(\frac{1-x}{1-2x}\right) = \sum_{j=1}^{\infty} \frac{(2^{j+1}-1)}{j+1} x^{j+1} \quad \text{for } |x| < \frac{1}{2}. \tag{16.21}$$

In fact, we can get more than this. Notice that the series in (16.21) also converges at $x = -\frac{1}{2}$ by the alternating series test. Therefore, Abel's theorem in Theorem 12.2.4 says the series converges uniformly on $[-\frac{1}{2}, 0]$. Consequently, by Corollary 12.2.5, we also have:

$$\lim_{x\downarrow -\frac{1}{2}} \ln\left(\frac{1-x}{1-2x}\right) = \lim_{x\downarrow -\frac{1}{2}} \sum_{j=1}^{\infty} \frac{(2^{j+1}-1)}{j+1} x^{j+1} = \sum_{j=1}^{\infty} \lim_{x\downarrow -\frac{1}{2}} \frac{(2^{j+1}-1)}{j+1} x^{j+1}.$$

Hence, the equality (16.21) is true for $x \in [-\frac{1}{2}, \frac{1}{2})$. In particular, at $x = -\frac{1}{2}$ we have the equality:

$$\ln\left(\frac{3}{4}\right) = \sum_{j=1}^{\infty} \frac{(2^{j+1}-1)}{j+1} \left(-\frac{1}{2}\right)^{j+1}.$$

2. The Riemann integral $\int_{-\frac{1}{2}}^{\frac{1}{2}} \frac{1}{1+x^2}\,dx$ can be easily obtained since we note that the integrand is simply the derivative of $\arctan(x)$ and we can proceed with the FTC. However, suppose that we want to evaluate the Riemann integral $\int_{-\frac{1}{2}}^{\frac{1}{2}} \frac{1}{1+x^3}\,dx$. We know this integral exists on any compact intervals in $\mathbb{R} \setminus \{-1\}$ since the integrand is continuous here. However, it is not clear what the antiderivative is and thus we cannot use the FTC directly to get an explicit integral value. See Remark 16.5.10 for the antiderivative of the integrand (warning: it is a very messy function).
An alternative way to evaluate it is to use power series. We can express the function as a power series, namely $\frac{1}{1+x^3} = \sum_{j=0}^{\infty}(-1)^j x^{3j}$ which converges only for $|x| < 1$. The domain of convergence and equality is small. However, it is enough for our purposes here since we only want to evaluate the Riemann integral over the compact interval $[-\frac{1}{2}, \frac{1}{2}]$, which is well within the domain of

convergence. Applying Corollary 16.5.8, we get:

$$\begin{aligned}\int_{-\frac{1}{2}}^{\frac{1}{2}} \frac{1}{1+x^3}\,dx = \int_{-\frac{1}{2}}^{\frac{1}{2}} \sum_{j=0}^{\infty}(-1)^j x^{3j}\,dx &= \sum_{j=0}^{\infty}\int_{-\frac{1}{2}}^{\frac{1}{2}}(-1)^j x^{3j}\,dx \\ &= \sum_{j=0}^{\infty}\left[\frac{(-1)^j x^{3j+1}}{3j+1}\right]_{-\frac{1}{2}}^{\frac{1}{2}} \\ &= \sum_{j=0}^{\infty}\frac{(-1)^j}{8^j(3j+1)}\frac{(1+(-1)^j)}{2} \\ &= \sum_{\substack{j=0\\ j \text{ even}}}^{\infty}\frac{(-1)^j}{8^j(3j+1)} \\ &= \sum_{j=0}^{\infty}\frac{1}{8^{2j}(6j+1)}.\end{aligned}$$

3. Now suppose that we want to find the value of the Riemann integral $\int_0^1 \frac{1}{1+x^3}\,dx$. We know this integral exists since the integrand is continuous over $[0, 1]$. Again, let us use the power series method to determine its value. The convergence and equality of the power series $\frac{1}{1+x^3} = \sum_{j=0}^{\infty}(-1)^j x^{3j}$ is only true on the interval $(-1, 1)$, so we might have to use some kind of limiting argument to get the upper limit of the integral to reach 1.
Define $I(t) = \int_0^t \frac{1}{1+x^3}\,dx$ for $t \in [0, 1]$. Using Proposition 16.5.7, we can integrate the power series $\sum_{j=0}^{\infty}(-1)^j x^{3j}$ over the interval $[0, t]$ where $0 \leq t < 1$ to get:

$$\begin{aligned}I(t) = \int_0^t \frac{1}{1+x^3}\,dx = \int_0^t \sum_{j=0}^{\infty}(-1)^j x^{3j}\,dx &= \sum_{j=0}^{\infty}\left[\frac{(-1)^j x^{3j+1}}{3j+1}\right]_0^t \\ &= \sum_{j=0}^{\infty}\frac{(-1)^j t^{3j+1}}{3j+1}.\end{aligned}$$

Note that the equality above is valid only any t with $0 \leq t < 1$. To find the integral over the interval $[0, 1]$, by continuity of Riemann integrals from Theorem 16.1.1, we have $\lim_{t\uparrow 1} I(t) = I(1)$. Thus:

$$\int_0^1 \frac{1}{1+x^3}\,dx = I(1) = \lim_{t\uparrow 1} I(t) = \lim_{t\uparrow 1}\int_0^t \frac{1}{1+x^3}\,dx = \lim_{t\uparrow 1}\sum_{j=0}^{\infty}\frac{(-1)^j t^{3j+1}}{3j+1}. \tag{16.22}$$

Now we need to justify switching the limit and the final expression of (16.22). We have proven that this series converges uniformly on [0, 1] in Example 11.4.13 using the Dirichlet's test. This allows us to switch the limit and the infinite sum to get:

$$\int_0^1 \frac{1}{1+x^3}\,dx = \lim_{t\uparrow 1}\sum_{j=0}^{\infty}\frac{(-1)^j t^{3j+1}}{3j+1} = \sum_{j=0}^{\infty}\lim_{t\uparrow 1}\frac{(-1)^j t^{3j+1}}{3j+1} = \sum_{j=0}^{\infty}\frac{(-1)^j}{3j+1}.$$

An alternative way to do this is to apply Abel's theorem in Theorem 12.2.4 and Corollary 12.2.5 to show that the power series $\sum_{j=0}^{\infty}\frac{(-1)^j t^{3j+1}}{3j+1}$ converges uniformly over [0, 1] and thus switching the order of the limit and the infinite sum is permitted.

Remark 16.5.10 The antiderivative of the function $f:(-1,\infty)\to\mathbb{R}$ defined as $f(x)=\frac{1}{1+x^3}$ in Examples 16.5.9(2) and (3) is $F(x)=\frac{1}{\sqrt{3}}\arctan\left(\frac{2x-1}{\sqrt{3}}\right)+\frac{1}{6}\ln\left(\frac{x^2+2x+1}{x^2-x+1}\right)+C$ which can be obtained by writing the function f in terms of partial fractions and carrying out some change of variables. We leave this for the readers to check in Exercise 16.4!

Monotone and Dominated Convergence Theorems

Two other important theorems that allow us to switch limit and integral are the monotone and dominated convergence theorems.

Theorem 16.5.11 (Monotone Convergence Theorem for Riemann Integrals, MCT for Riemann Integrals) *Let (f_n) be a sequence of Riemann integrable functions $f_n:[a,b]\to\mathbb{R}$ such that $f_n\downarrow f$ where $f:[a,b]\to\mathbb{R}$ is a continuous function. Then:*

$$\lim_{n\to\infty}\int_a^b f_n(x)\,dx = \int_a^b \lim_{n\to\infty} f_n(x)\,dx = \int_a^b f(x)\,dx.$$

Proof By Dini's theorem in Theorem 11.2.4, since (f_n) is pointwise decreasing and the limiting function is continuous, we have uniform convergence $f_n \xrightarrow{u} f$ on $[a,b]$. Thus, by the integrable limit theorem, we can conclude the desired result. □

Remark 16.5.12 The result above holds if $f_n\uparrow f$ where f is a continuous function as well.

Theorem 16.5.13 (Dominated Convergence Theorem for Riemann Integrals, DCT for Riemann Integrals) *Let (f_n) be a sequence of Riemann integrable* *functions* $f_n : [a, b] \to \mathbb{R}$ *such that* $f_n \xrightarrow{pw} f$ *where* $f : [a, b] \to \mathbb{R}$ *is a continuous function. Suppose that the sequence* (f_n) *is uniformly bounded. Then:*

$$\lim_{n\to\infty} \int_a^b |f_n(x) - f(x)|\, dx = 0.$$

In particular, this implies:

$$\lim_{n\to\infty} \int_a^b f_n(x)\, dx = \int_a^b \lim_{n\to\infty} f_n(x)\, dx = \int_a^b f(x)\, dx.$$

We shall prove this result in Exercise 16.32 following the proof by W.A.J. Luxemburg (1929–2018). We finish this chapter with a proof that we have delayed in Exercise 11.29.

Example 16.5.14 Consider the sequence of functions (f_n) where $f_n : [0, \pi] \to \mathbb{R}$ is defined as $f_n(x) = \sin(nx)$. This sequence of functions is uniformly bounded but we claim that it does not have a pointwise convergent subsequence.

Suppose for contradiction that it does, namely (f_{k_n}) is a pointwise convergent subsequence. Define (g_n) as the functions $g_n : [0, \pi] \to \mathbb{R}$ with $g_n = (f_{k_{n+1}} - f_{k_n})^2$. Since the sequence (f_{k_n}) converges pointwise, by the algebra of limits, we have $g_n \xrightarrow{pw} 0$. Moreover, by using triangle inequality, we have the uniform bound $|g_n(x)| \leq 4$ for all $n \in \mathbb{N}$ and $x \in [0, \pi]$. Thus, by Theorem 16.5.13, we have:

$$\lim_{n\to\infty} \int_0^\pi g_n(x)\, dx = \int_a^b \lim_{n\to\infty} g_n(x)\, dx = 0.$$

On the other hand, we can compute the integral of g_n by using the FTC, the double angle formula, and the product-to-sum formula of sines to get:

$$\int_0^\pi g_n(x)\, dx = \int_0^\pi (f_{k_{n+1}}(x) - f_{k_n}(x))^2\, dx = \int_0^\pi (\sin(k_{n+1}x) - \sin(k_n x))^2\, dx = \pi,$$

for any $n \in \mathbb{N}$. So, we have $\pi = \int_0^\pi g_n(x)\, dx \to 0$, which gives us a contradiction.

Exercises

16.1 Let $f \in \mathcal{R}([a, b])$. Recall that its Riemann integral function $I : [a, b] \to \mathbb{R}$ is given by:

$$I(x) = \int_a^x f(t)\, dt.$$

(a) Prove that the function I is uniformly and Lipschitz continuous on $[a, b]$.
Suppose that f is monotone and $g \in \mathcal{R}([a, b])$ is a non-negative function.
(b) Prove that there exists a $c \in [a, b]$ such that:

$$\int_a^b f(x)g(x)\,dx = f(a)\int_a^c g(x)\,dx + f(b)\int_c^b g(x)\,dx.$$

(c) Suppose now that f is decreasing and non-negative. Show that there exists a $c \in [a, b]$ such that:

$$\int_a^b f(x)g(x)\,dx = f(a)\int_a^c g(x)\,dx.$$

16.2 Using the FTC, find the derivatives of the following functions defined on $(0, \infty)$:
(a) $I(x) = \int_1^{\sqrt{x}} (3t^3 + 2)^4\,dt$.
(b) $I(x) = \int_0^{x^2} \sqrt{1 + t^2}\,dt$.
(c) $I(x) = \int_0^{\ln(x)+4} \sin(t)\,dt$.
(d) $I(x) = \int_{\frac{1}{x}}^{x^3} \ln(t)\,dt$.

16.3 Fix $\lambda > 0$. Let $F : (0, \infty) \to \mathbb{R}$ be defined as the function $F(x) = \int_x^{\lambda x} \frac{1}{t}\,dt$. Without using logarithms, show that F is a constant function.

16.4 $(*)$ Find all the antiderivatives of the following functions:
(a) $f(x) = x^2 \sin(x)$ on $\mathbb{R}$.
(b) $f(x) = x^2 e^{x^3}$ on $\mathbb{R}$.
(c) $f(x) = 2^x \sin(x)$ on $\mathbb{R}$.
(d) $f(x) = \frac{1}{x \ln(x)}$ on $(0, \infty)$.
(e) $f(x) = x^2 \ln(x)$ on $(0, \infty)$.
(f) $f(x) = \sqrt{1 - x^2}$ on $[0, 1]$.
(g) $f(x) = \arcsin(x)$ on $[-1, 1]$.
(h) $f(x) = \frac{\arcsin(x)}{\sqrt{1-x^2}}$ on $(-1, 1)$.
(i) $f(x) = \cos(\ln(x))$ on $(0, \infty)$.
(j) $f(x) = \frac{1}{1+x^3}$ on $(-1, \infty)$.

16.5 Prove Proposition 16.4.5, namely:
If $f : [a, b) \to \mathbb{R}$ is continuous on $[a, b) \subseteq \mathbb{R}$ such that $\lim_{x\to b} f(x)$ exists, then f is improperly Riemann integrable on $[a, b)$.

16.6 $(*)$ Find the following improper Riemann integrals of the function f over the specified domain:
(a) $f(x) = \frac{1}{(x+1)\sqrt{x}}$ on $(0, \infty)$.
(b) $f(x) = xe^{-x^2}$ on $\mathbb{R}$.
(c) $f(x) = \frac{1}{1+x^2}$ on $\mathbb{R}$.
(d) $f(x) = \frac{1}{\sqrt{5-x}}$ on $[0, 5)$.
(e) $f(x) = x \ln(x)$ on $(0, 1]$.

16.7 Prove Propositions 16.4.10 and 16.4.11 which are the comparison and limit comparison tests for improper Riemann integrals of the first kind.

16.8 (*) Define a function $\zeta : (1, \infty) \to \mathbb{R}$ as $\zeta(s) = \sum_{n=1}^{\infty} \frac{1}{n^s}$. This defines a bona fide function on the domain since the series converges for any such s. Show that:

$$\zeta(s) = s \int_0^\infty \frac{\lfloor x \rfloor}{x^{s+1}}\, dx \quad \text{and} \quad \zeta(s) = \frac{s}{s-1} - s \int_1^\infty \frac{x - \lfloor x \rfloor}{x^{s+1}}\, dx.$$

We can extend the domain of this function to a complex subset $S = \{s \in \mathbb{C} : \text{Re}(s) > 1\} \subseteq \mathbb{C}$. Furthermore, from the theory of complex analysis, this extended function is analytic (complex differentiable) on S which allows us to extend it further to the whole of $\mathbb{C}$ uniquely via a method called analytic continuation. Of course, this extended function would not have a series expression equal to $\sum_{n=1}^{\infty} \frac{1}{n^s}$ elsewhere since this series only converges in S.

This resulting extended function is called the Riemann zeta function, which is very important in the study of prime numbers. One of the Millennium prize problems worth US$1,000,000 is the Riemann hypothesis which involves studying the roots of the zeta function

16.9 (a) Let $f : (0, \infty) \to \mathbb{R}$ be a function defined by $f(x) = \frac{1}{x^p}$ where $p > 0$. Prove that:

i. f is improperly Riemann integrable over $(0, 1]$ if and only if $0 < p < 1$.

ii. f is improperly Riemann integrable over $[1, \infty)$ if and only if $p > 1$.

(b) Consider the following improper Riemann integrals:

$$I_1 = \int_0^1 \frac{\sin(x)}{x^p}\, dx \quad \text{and} \quad I_2 = \int_0^1 \frac{\sin(x)^p}{x}\, dx.$$

Determine the range of values of real values p for which I_2 and I_2 are defined respectively.

16.10 Let $f : [a, b] \to [f(a), f(b)]$ be a strictly increasing and differentiable function.

(a) Deduce that the inverse function $f^{-1} : [f(a), f(b)] \to [a, b]$ exists, is strictly increasing, and is continuous.

(b) Explain why the functions f and its inverse f^{-1} are Riemann integrable over their domains.

(c) Hence, show that:

$$\int_a^b f(x)\, dx + \int_{f(a)}^{f(b)} f^{-1}(x)\, dx = bf(b) - af(a).$$

(d) Explain this result geometrically.

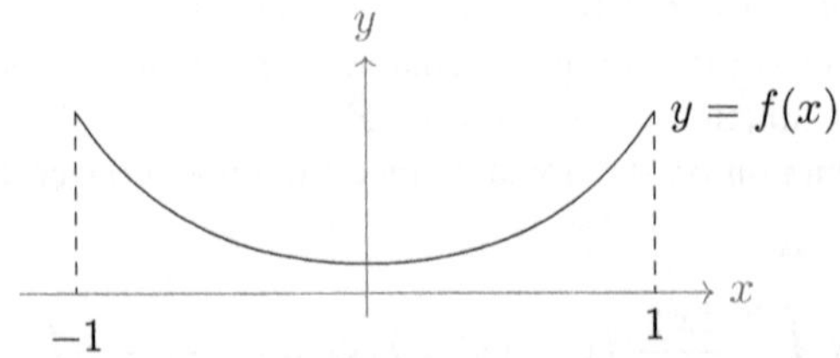

Fig. 16.10 A catenary models the curve that a chain, cable, or rope makes under the influence of its own weight when supported at the ends $x = \pm 1$. The term catenary comes from Latin word *catena*, which means "chain". It was a popular belief that the chain would form a parabola under its own weight. However, Johann Bernoulli, Leibniz, and Christiaan Huygens (1629–1695) proved independently that it forms a catenary instead

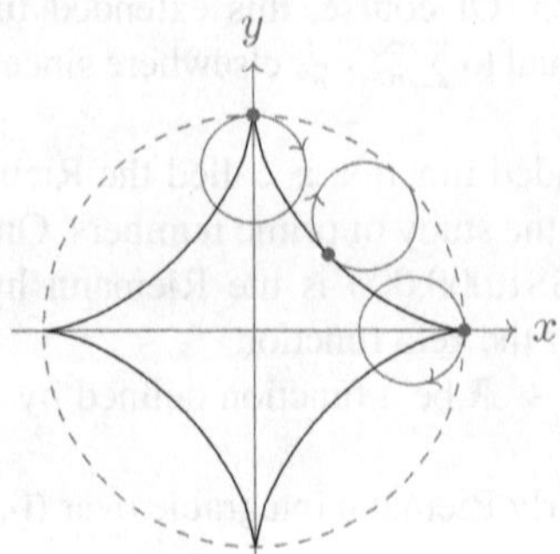

Fig. 16.11 An astroid is a curve traced by a point on a circle of radius $\frac{a}{4}$ (labelled red) as the circle rolls along inside a larger circle of radius a

16.11 ($*$) Find the length of the curves described by the following functions:

(a) The catenary $f(x) = a\cosh(\frac{x}{a})$ for $x \in [-1, 1]$ (Fig. 16.10).

(b) An astroid curve described by $x^{\frac{2}{3}} + y^{\frac{2}{3}} = a$ for $x \in [-a, a]$ where $a > 0$ is a constant (Fig. 16.11).

(c) The square root function $f(x) = \sqrt{x}$ for $x \in [0, 4]$

16.12 Recall that the area of a unit disc is given by π. This was obtained in Example 16.1.10 by evaluating the integral $\int_0^1 \sqrt{1 - x^2}\, dx$ and multiplying it by 4. By using Darboux integral from first principle, show that:

$$\pi = \lim_{n\to\infty} \frac{4}{n^2} \sum_{j=1}^{n} \sqrt{n^2 - j^2}.$$

16.13 ($\diamond$) An ellipse in $\mathbb{R}^2$ of semi-major radius $b > 0$ and semi-minor radius $a > 0$ where $a \leq b$ is described by the equation $\frac{y^2}{a^2} + \frac{x^2}{b^2} = 1$. See Fig. 16.12 for its plot.

Fig. 16.12 An ellipse

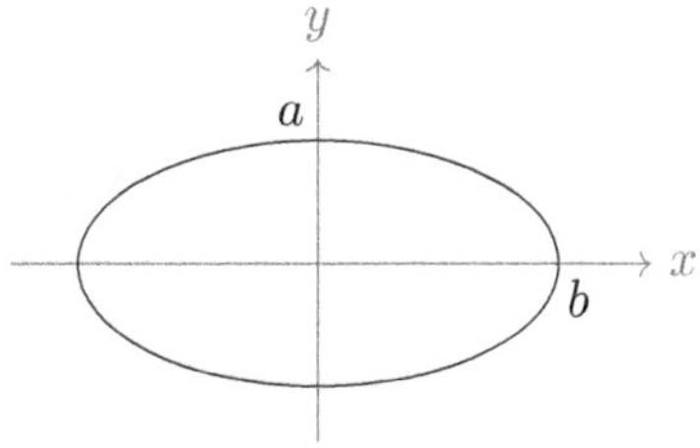

(a) Find the area enclosed by the ellipse.
(b) Show that the circumference of the ellipse is given by the Riemann integral:

$$C = 4b \int_0^1 \sqrt{\frac{1-\epsilon^2 t^2}{1-t^2}}\, dt = 4b \int_0^{\frac{\pi}{2}} \sqrt{1-\epsilon^2 \sin^2(\theta)}\, d\theta,$$

where $\epsilon = \sqrt{1-\frac{a^2}{b^2}} < 1$ is a constant called the eccentricity of the ellipse.
The ellipse is a circle if and only if $\epsilon = 0$. The integral $\int_0^1 \sqrt{\frac{1-\epsilon^2 t^2}{1-t^2}}\, dt$ is called a complete elliptic integral of the second kind and cannot be expressed exactly. This is an example of the integral in Example 16.3.5(3). We shall derive a power series representation for it in Exercise 17.21.

16.14 (∗) In this question, we are going to derive the formula for the volume of solid of revolution in Definition 16.2.3. Let $f : [a, b] \to \mathbb{R}_{\geq 0}$ be a continuously differentiable function.
(a) For a tagged partition $\mathcal{P}_\tau$ of $[a, b]$, find the approximate the volume of the solid of revolution $V_{\mathcal{P}_\tau}$ of the function f obtained by cylindrical approximations.
(b) Thus, show that for any $\varepsilon > 0$ there exists a $\delta > 0$ such that for any partition $\mathcal{P}_\tau$ of $[a, b]$ which satisfies $||\mathcal{P}_\tau|| < \delta$, we have $|V_{\mathcal{P}_\tau} - R| < \varepsilon$ where $R = \pi \int_a^b f(x)^2\, dx$.
(c) By using an appropriate solid of revolution, find the volume and surface area of a 3-dimensional ball with radius $r > 0$.

16.15 Let $f : [1, \infty) \to \mathbb{R}$ be a function $f(x) = \frac{1}{x}$. By using improper Riemann integrals, show that the volume for the solid of revolution of this function over the region $[1, \infty)$ is finite, but its surface area is infinite.
This object is called Gabriel's horn or Torricelli's trumpet, named after Evangelista Torricelli (1608–1647) who first studied it. This object gives rise to the painter's paradox: we cannot paint the trumpet with any finite amount of paint, but can actually fill the trumpet with a finite amount of paint! Regarding this object, Hobbes remarked:

> To understand this for sense, it is not required that a man should be a geometrician or a logician, but that he should be mad.

However, there are many ways to resolve this paradox. One of them is to realise that these objects live in abstract and idealised mathematical world where a paint coat on a surface has zero thickness. Thus, basing its interpretation on the physical world could potentially be misleading!

16.16 A torus (or a doughnut) of radius $R, r > 0$ with $R > r$ is a geometrical object obtained by revolving a circle $(y-R)^2+x^2 = r^2$ defined on $\mathbb{R}^2$ about the x-axis. One can think of it as the union of two surfaces of revolution defined via the functions $f_1, f_2 : [-r, r] \to \mathbb{R}$ as $f_1(x) = R + \sqrt{r^2 - x^2}$ and $f_2(x) = R - \sqrt{r^2 - x^2}$.
(a) Show that the surface area of the torus is $A = 4\pi^2 Rr$.
(b) Show that the volume enclosed by the torus is $V = 2\pi^2 Rr^2$.

16.17 Recall that for a continuously differentiable function $f : [a, b] \to \mathbb{R}$, the length of this curve is defined as $C(f) = \int_a^b \sqrt{1 + f'(x)^2}\, dx$ in Definition 16.2.1. For a partition $\mathcal{P} = \{x_0, x_1, \dots, x_n\}$ of $[a, b]$, we denoted the length approximation obtained from $\mathcal{P}$ as:

$$C_{\mathcal{P}} = \sum_{j=1}^{n} \sqrt{(x_j - x_{j-1})^2 + (f(x_j) - f(x_{j-1}))^2}.$$

Show that if f is continuously differentiable, then:

$$C(f) = \sup_{\mathcal{P}}\{C_{\mathcal{P}} : \mathcal{P} \text{ is a partition of } [a, b]\}.$$

16.18 ($\diamond$) In this question, we are going to prove that the length operation on curves does not behave very well under limits. Let (f_n) be a sequence of functions $f_n : [0, 1] \to \mathbb{R}$ defined as $f_n(x) = \frac{\sin(2n^2\pi x)}{n}$.
(a) Show that f_n converges uniformly to the zero function $f(x) \equiv 0$.
(b) For any fixed $n \in \mathbb{N}$, show that $C(f_n) = \int_0^1 \sqrt{1 + (2\pi n^2 \cos(2n^2\pi x))^2}\, dx > 4n$.
(c) Hence, deduce that $C(f_n) \not\to C(f)$ despite $f_n \xrightarrow{u} f$.
So we can see that even if $f_n \xrightarrow{u} f$, the lengths of the curves (f_n) do not necessarily converge to the length of the curve of f. Instead, the length operation satisfies the following limit:

Proposition 16.5.15 *Let $f : [a, b] \to \mathbb{R}$ be a function and its graph has finite length $C(f)$. Suppose that (f_n) where $f_n : [a, b] \to \mathbb{R}$ is a sequence of functions such that $f_n \xrightarrow{pw} f$. Then:*

$$C(f) \leq \liminf_{n\to\infty} C(f_n).$$

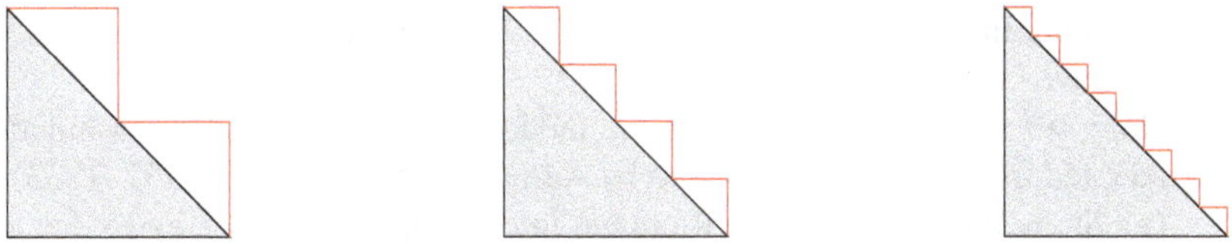

Fig. 16.13 Staircase paradox. The hypotenuse of the triangle with sidelengths 1 has length $\sqrt{2}$. This hypotenuse can be seen as the pointwise limit of the red staircase with n steps as $n \to \infty$. However, the length of the staircase remains constant $2 > \sqrt{2}$ no matter how many steps we have in the staircase!

In advanced analysis, this property is called the lower semi-continuity of the length operation. We are going to prove this proposition in several steps. WLOG, we can assume that $\liminf_{n\to\infty} C(f_n) < \infty$.

(d) For any $a, b, c \geq 0$, prove that $\sqrt{a^2 + (b+c)^2} \leq \sqrt{a^2 + b^2} + \sqrt{c^2}$.

(e) Fix $\varepsilon > 0$. Show that there exists a partition $\mathcal{P} = \{x_0, x_1, \dots, x_m\}$ of $[a, b]$ such that $C(f) - \frac{\varepsilon}{2} < C_{\mathcal{P}}$.

(f) Show that there exists an $N \in \mathbb{N}$ such that for all $n \geq N$ and $j = 0, 1, \dots, m$ we have $|f_n(x_j) - f(x_j)| < \frac{\varepsilon^2}{8m^2}$.

(g) Hence, deduce that for every $j = 1, 2, \dots, m$ and $n \geq N$ we have $|f(x_j) - f(x_{j-1})| < |f_n(x_j) - f_n(x_{j-1})| + \frac{\varepsilon^2}{4m^2}$.

(h) For all $n \geq N$, denote:

$$C_{\mathcal{P}}(f_n) = \sum_{j=1}^{m} \sqrt{(x_j - x_{j-1})^2 + (f_n(x_j) - f_n(x_{j-1}))^2}.$$

Using parts (d), (e), and (g), deduce that $C(f) < C_{\mathcal{P}}(f_n) + \varepsilon \leq C(f_n) + \varepsilon$ for every $n \geq N$.

(i) Finally, conclude that $C(f) \leq \liminf_{n\to\infty} C(f_n)$.

The lower semi-continuity property of the length operation explains the staircase paradox in Fig. 16.13.

16.19 For any $s > 0$, evaluate the following improper Riemann integrals:

(a) For $n \in \mathbb{N}$, $\int_0^\infty e^{-st} t^n \, dt = \frac{n!}{s^{n+1}}$.

(b) For $k \in \mathbb{R}$, $\int_0^\infty e^{-st} \sin(kt) \, dt = \frac{a}{a^2+s^2}$.

(c) For $k \in \mathbb{R}$, $\int_0^\infty e^{-st} \cos(kt) \, dt = \frac{s}{a^2+s^2}$.

If we extend s to the set $S = \{s \in \mathbb{C} : \mathrm{Re}(s) > 0\}$ we have the Laplace transform. The Laplace transform is a very useful concept for studying differential equations.

16.20 ($*$) In this question, we are going to define the gamma function. The idea of gamma function originated from Euler who wished to interpolate the factorial operation on the natural numbers to any real number. His original idea for the interpolating function is denoted as $[x]$ and expressed in the form of an

improper integral $[x] = \int_0^1 (-\ln(t))^x \, dt$. Using Euler's notation, for $x \in \mathbb{N}_0$ we have $[x] = x!$.
Here, we shall be working with the modern version of the gamma function adapted from Euler's formulation by Adrien-Marie Legendre (1752–1833).
(a) Show that the improper Riemann integral $\int_0^\infty t^{x-1}e^{-t}\, dt$ exists for any fixed $x > 0$.
(b) For $x > 0$, denote $\Gamma(x) = \int_0^\infty t^{x-1}e^{-t}\, dt$ which we call the gamma function. Prove that $\Gamma(x) = (x-1)\Gamma(x-1)$ for $x > 1$.
(c) Hence, deduce that $\Gamma(n) = (n-1)!$ for all $n \in \mathbb{N}$.
As intended by Euler, the gamma function extends the idea of factorials to non-integers. In fact, it can also take complex arguments $x \in \{z \in \mathbb{C} : \mathrm{Re}(z) > 0\}$.

16.21 Show, by using integral test, that the real series $\sum_{j=2}^\infty \frac{1}{j(\ln(j))^p}$ converges if and only if $p > 1$.

16.22 (∗) Find a power series of the following function centred at $x = 0$ and determine their radius of convergence:
(a) $f(x) = \ln(2-x)$.
(b) $I(x) = \int_0^x \frac{t^2}{1+t^2}\, dt$.

16.23 Recall that $\frac{d}{dx}\arctan(x) = \frac{1}{1+x^2}$.
(a) Find a power series expansion of $\arctan(x)$ for $|x| < 1$.
(b) Show, with careful reasoning, that:

$$\frac{\pi}{4} = \sum_{j=0}^{\infty} \frac{(-1)^j}{2j+1}.$$

This is called the Leibniz formula for π and can be used to approximate the value of π. In fact, this series is known way earlier by Madhava of Sangamagrama (c. 1350–1425). As a result, this series is also called the Madhava-Leibniz series for π. However, it is such a slowly converging series that the first 100,000 terms give us an approximation of π which is correct to only 5 decimal places.

(c) Let $s = \sum_{j=0}^\infty \frac{(-1)^j}{2j+1}$ and $t = \sum_{j=0}^\infty \frac{1}{(4j+1)(4j+1)}$ be two real series with partial sums (s_n) and (t_n) respectively. Show that $t_n = \frac{1}{2}s_{2n}$ for all $n \in \mathbb{N}$ and deduce that:

$$\pi = \sum_{j=0}^{\infty} \frac{8}{(4j+1)(4j+3)}.$$

16.24 (◇) Because of the slowly-converging nature of the Madhava-Leibniz series for π that we saw in Exercise 16.23, it is not very useful for computational purposes. In this question, we are going to find another series that can be used to describe π which converges faster. Let $z_j = a_j + ib_j$ for $j = 1, 2$ be two non-zero complex numbers.

(a) Prove that $\text{Arg}(z_1) + \text{Arg}(z_2) = \text{Arg}(z_1 z_2)$ if $\text{Arg}(z_1 z_2) \in (-\pi, \pi)$.

(b) Hence, prove that if $\text{Arg}(z_1), \text{Arg}(z_2), \text{Arg}(z_1 z_2) \in (-\frac{\pi}{2}, \frac{\pi}{2})$, then:

$$\arctan\left(\frac{a_1}{b_1}\right) + \arctan\left(\frac{a_2}{b_2}\right) = \arctan\left(\frac{a_1 b_2 + a_2 b_1}{b_1 b_2 - a_1 a_2}\right).$$

(c) Deduce that $\frac{\pi}{4} = \arctan(\frac{1}{2}) + \arctan(\frac{1}{3})$.

(d) Using the fact that $\arctan(x) = \int_0^x \frac{1}{1+t^2}\,dt$ and part (c), prove that:

$$\pi = \sum_{j=0}^{\infty} \frac{4(-1)^j}{2j+1}\left(\frac{1}{2^{2j+1}} + \frac{1}{3^{2j+1}}\right).$$

This is an example from a family of series for π which are called the Machin-like formula. The original formula devised by John Machin (1686–1751) in 1706 was used to compute the first 100 digits of the decimal representation for π. The formula used by him was obtained from the equation $\frac{\pi}{4} = \arctan(\frac{1}{5}) - \arctan(\frac{1}{239})$, which can be derived using the equation in part (b). Nowadays, using various choices for a_1, a_2, b_1, b_2 in part (b), one can come up with many other similar series expression for π. The year 2002 record for most number of digits of π computed (a whopping 1,241,100,000,000 digits in 600 hours) was achieved by Yasumasa Kanada (1949–2020) and his team using these Machin-like formula.

16.25 (∗) Recall the Fourier series $\sum_{j=1}^{\infty}(-1)^{j+1}\frac{2\sin(jx)}{j}$ from Exercise 11.22. We have shown that it converges pointwise for all $x \in \mathbb{R}$ and is 2π-periodic. Now we want to find its limiting function. Define the function s as this series and let $s_n(t) = \sum_{j=1}^{n}(-1)^{j+1}\frac{2\sin(jt)}{j}$ be its partial sum. Clearly, $s(n\pi) = 0$ for any $n \in \mathbb{Z}$. Now fix $x \in (0, \pi)$.

(a) For any $n \in \mathbb{N}$, show that for $t \in (0, \pi)$ we have:

$$\sum_{j=1}^{n}(-1)^{j+1} 2\cos(jt) = 1 + (-1)^{n+1}\frac{\cos((n+\frac{1}{2})t)}{\cos(\frac{t}{2})}.$$

(b) Hence, deduce that:

$$s_n(x) - x = (-1)^{n+1}\int_0^{\frac{x}{2}} \frac{2\cos((2n+1)u)}{\cos(u)}\,du.$$

(c) Using part (b), show that:

$$|s_n(x) - x| \leq \frac{2 + x\tan(\frac{x}{2})}{2(2n+1)\cos(\frac{x}{2})}.$$

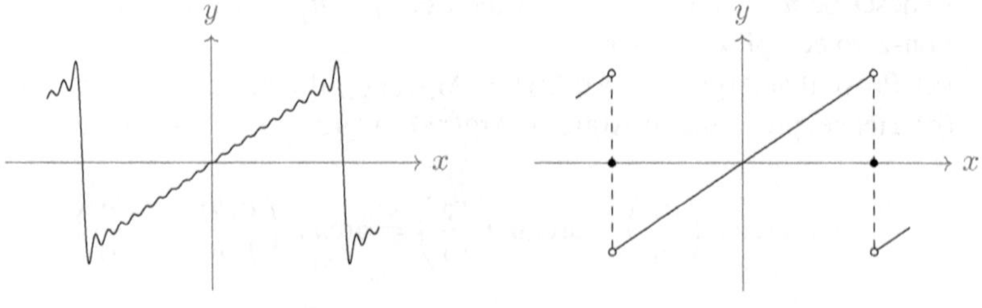

Fig. 16.14 Partial sum s_{20} and the limiting functions series s. **(a)** s_{20}. **(b)** Sawtooth function s

(d) Conclude that $s(x) = x$ on $(-\pi, \pi)$.
(e) Hence, show that $\lim_{x\uparrow(2n+1)\pi} s(x) \neq \lim_{x\downarrow(2n+1)\pi} s(x)$ for all $n \in \mathbb{Z}$.
(f) Show that for all $n \in \mathbb{Z}$ we have:

$$s((2n+1)\pi) = \frac{\lim_{x\uparrow(2n+1)\pi} s(x) + \lim_{x\downarrow(2n+1)\pi} s(x)}{2}.$$

This series is called the sawtooth function for obvious reasons from part (d) and Fig. 16.14b.

16.26 ($*$) In this question, we are going to prove that π and π^2 are irrational.
(a) In school or applied sciences, the constant π is usually approximated by the rational number $\frac{22}{7}$. By evaluating the integral $\int_0^1 \frac{x^4(1-x)^2}{1+x^2}\,dx$, show that $\pi < \frac{22}{7}$.
(b) By using suitable bounds on the integrand in part (a), show that:

$$\frac{22}{7} - \frac{1}{630} < \pi < \frac{22}{7} - \frac{1}{1260}.$$

Therefore, the approximation $\pi \approx \frac{22}{7}$ is roughly 0.0015% accurate.
There are numerous different proofs to show that π is irrational. The following proof, due to Nicolas Bourbaki which is a group of mathematicians working under a pseudonym, is one of the most elementary.
(c) For each $n \in \mathbb{N}$, let $P_n : [0, \pi] \to \mathbb{R}$ be the polynomial $P_n(x) = x^n(\pi - x)^n$. For $b \in \mathbb{N}$, define the integral $I_n(b) = \frac{b^n}{n!}\int_0^\pi P_n(x)\sin(x)\,dx$. Show that for any $n \geq 3$:

$$I_n(b) = (4n-2)bI_{n-1}(b) - (b\pi)^2 I_{n-2}(b).$$

(d) Prove that $0 < I_n(b) < \frac{2b^n}{n!}\left(\frac{\pi}{2}\right)^{2n}$ for all $b, n \in \mathbb{N}$.
(e) Hence, deduce that for any fixed $b \in \mathbb{N}$ there exists an $N \in \mathbb{N}$ such that $0 < I_n(b) < 1$ for all $n \geq N$.

Now assume for contradiction that $\pi = \frac{a}{b}$ for some $a, b \in \mathbb{N}$.

(f) By using induction and part (c), show that $I_n(b)$ are integers for any $n \in \mathbb{N}$.

(g) Deduce the contradiction and conclude the that π is irrational.

We can also prove that π^2 is irrational using the same idea. Assume for contradiction that $\pi^2 = \frac{a}{b}$ for some $a, b \in \mathbb{N}$. Then, $(b\pi)^2 = ab \in \mathbb{N}$.

(h) By using induction and part (c), show that $I_n(b)$ are integers for any $n \in \mathbb{N}$.

(i) Deduce the contradiction and conclude the that π^2 is irrational.

In fact, for any $n \in \mathbb{N}$, the number π^n is known to be irrational. The proof for $n \geq 3$ requires extra knowledge on transcendental numbers which we have defined in Exercise 4.30. However, this is beyond the scope of what we have covered in this book.

16.27 (∗) We are going to prove Wallis's formula published in 1656 by John Wallis. The formula states:

$$\lim_{n\to\infty} \frac{2^{2n}(n!)^2}{(2n)!\sqrt{n}} = \sqrt{\pi}.$$

(a) Define $I_n = \int_0^{\frac{\pi}{2}} \sin^n(x)\,dx$ for all $n \in \mathbb{N}$. Show that the sequence (I_n) satisfies the recurrence relationship $I_n = \frac{n-1}{n} I_{n-2}$ for $n \geq 2$.

(b) Hence, show that:

$$I_{2n} = \frac{2n-1}{2n}\frac{2n-3}{2n-2}\cdots\frac{3}{4}\frac{1}{2}\frac{\pi}{2} \quad \text{and} \quad I_{2n+1} = \frac{2n}{2n+1}\frac{2n-2}{2n-1}\cdots\frac{4}{5}\frac{2}{3}.$$

(c) Using part (b), show that:

$$\sqrt{\frac{\pi}{2}} = \frac{(2n)!!}{(2n-1)!!}\frac{1}{\sqrt{2n+1}}\sqrt{\frac{I_{2n}}{I_{2n+1}}},$$

where the double factorial is defined in Exercise 7.11(h).

(d) Using the definition of I_n, explain why $0 < I_{2n+1} \leq I_{2n} \leq I_{2n-1}$ for all $n \in \mathbb{N}$.

Hence, deduce the limit $\lim_{n\to\infty} \frac{I_{2n}}{I_{2n+1}} = 1$.

(e) Finally, by using algebra of limits, show that:

$$\sqrt{\frac{\pi}{2}} = \lim_{n\to\infty} \frac{(2 \cdot 4 \cdot \ldots \cdot 2n)^2}{(2n)!\sqrt{2n}},$$

and hence deduce the Wallis's formula.

(f) Using Wallis's formula, show the asymptotic equivalence $\binom{2n}{n} \sim \frac{4^n}{\sqrt{\pi n}}$.

16.28 (∗) In this question, we are going to evaluate the Gaussian integral $\int_{-\infty}^{\infty} e^{-x^2}\,dx$.

(a) We have shown in Exercise 14.2(e) that for all $x \geq 0$ we have the inequalities:

$$1 - x^2 \leq e^{-x^2} \leq \frac{1}{1+x^2}. \tag{16.23}$$

Using (16.23), deduce that the improper Riemann integral $\int_{-\infty}^{\infty} e^{-x^2}\,dx$ exists.

By raising each side of the inequalities (16.23) to the power of $n \in \mathbb{N}$ and integrating over $[0, 1]$, we have:

$$\int_0^1 (1-x^2)^n\,dx \leq \int_0^1 e^{-nx^2}\,dx \leq \int_0^1 \frac{1}{(1+x^2)^n}\,dx. \tag{16.24}$$

(b) Show that the inequalities (16.24) can be written as:

$$\int_0^{\frac{\pi}{2}} \sin^{2n+1}(y)\,dy \leq \frac{1}{\sqrt{n}}\int_0^{\sqrt{n}} e^{-y^2}\,dy \leq \int_0^{\frac{\pi}{2}} \sin^{2n-2}(y)\,dy. \tag{16.25}$$

(c) We have seen the integrals on either side of the inequalities (16.25). In Exercise 16.25, we defined them to be $I_n = \int_0^{\frac{\pi}{2}} \sin^n(x)\,dx$. Deduce that $I_{2n}I_{2n+1} = \frac{1}{2n+1}\frac{\pi}{2}$ and $I_{2n} \leq I_{2n-1} \leq \frac{2n}{2n-1}I_{2n}$.

(d) Deduce the asymptotic equivalence $I_n \sim I_{n+1}$.

(e) Hence, show that $\lim_{n\to\infty}(2n+1)I_{2n+1}^2 = \frac{\pi}{2}$.
Using the same argument, show that $\lim_{n\to\infty}(2n-2)I_{2n-2}^2 = \frac{\pi}{2}$.

(f) Using sandwiching argument and algebra of limits in the inequality in part (b), find the value of $\int_0^{\infty} e^{-y^2}\,dy$.

(g) Finally, deduce the value of the Gaussian integral $\int_{-\infty}^{\infty} e^{-x^2}\,dx$.

16.29 A real sequence (C_n) is defined as $C_n = \frac{4^{n+1}}{\pi}\int_0^1 x^{2n}\sqrt{1-x^2}\,dx$ for all $n \in \mathbb{N}$. The sequence (C_n) is called the Catalan numbers, introduced by Eugène Charles Catalan (1814–1894) for solving combinatorial problems.

(a) Show that the sequence (C_n) satisfies the recursive relation $C_{n+1} = \frac{2(2n+1)}{n+2}C_n$.

(b) Deduce that $C_n = \frac{1}{n+1}\binom{2n}{n}$ for all $n \in \mathbb{N}$.

(c) Show the asymptotic equivalence $C_n \sim \frac{4^n}{n^{\frac{3}{2}}\sqrt{\pi}}$.

Catalan numbers occur in many different mathematical problems and have many different combinatorial interpretations. One of them is that C_n is the number of ways we can place $n-1$ pairs of parentheses () in a legitimate way among $n+1$ symbols. For example, when $n = 3$, we can place the parentheses around the terms in the string of 4 symbols $abcd$ in $C_3 = 5$ distinct ways, namely $(ab)(cd)$, $(a(bc))d$, $a((bc)d)$, $((ab)c)d$, and $a(b(cd))$.

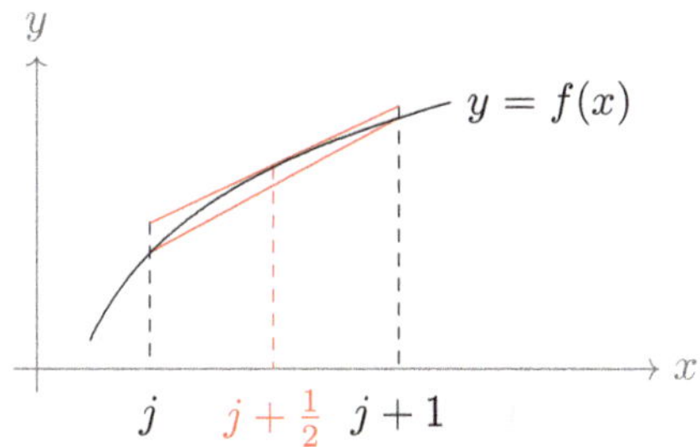

Fig. 16.15 The logarithmic graph over the interval $[j, j+1]$, its secant over this interval, and the tangent line to it at $x = j + \frac{1}{2}$

Thus, in algebra, C_n represents the number of distinct ways we can carry out a sequence of binary operations among $n + 1$ terms. On the other hand, in geometry, C_n is the number of distinct ways we can dissect a regular $(n+2)$-gon into n triangles via non-crossing diagonals.

16.30 $(*)$ In this exercise, we are going to prove Stirling's asymptotic formula which states that $n! \sim \sqrt{2\pi n}\left(\frac{n}{e}\right)^n$ or, in other words, $\lim_{n\to\infty} \frac{n!}{\sqrt{2\pi n}\left(\frac{n}{e}\right)^n} = 1$. This result was stated by De Moivre without the constant $\sqrt{2\pi}$ and was completed by James Stirling (1692–1770).

(a) Show that the logarithmic function $f : (0, \infty) \to \mathbb{R}$ where $f(x) = \ln(x)$ is concave.

Hence, at any interval $[j, j + 1]$ where $j \in \mathbb{N}$, by definition of concave functions, the secant line segment joining the points $(j, \ln(j))$ and $(j + 1, \ln(j + 1))$ lies under the graph of f. Moreover, by concavity and Exercise 14.4, the tangent line to the graph of f at the midpoint $x = j + \frac{1}{2}$ always lie above the graph of f. See Fig. 16.15 for the diagram.

Denote a_j, b_j, and c_j as the areas under the secant, logarithmic graph, and the line tangent to f at $x = j + \frac{1}{2}$ over the interval $[j, j + 1]$ respectively.

(b) Show that:

$$0 \le b_j - a_j \le c_j - a_j < \frac{1}{2}\ln\left(1 + \frac{1}{2j}\right) - \frac{1}{2}\ln\left(1 + \frac{1}{2(j+1)}\right).$$

(c) Define $A_n = \sum_{j=1}^{n-1} a_j$ and $B_n = \sum_{j=1}^{n-1} b_j$ for all $n \in \mathbb{N}$. Show that:

$$A_n = \ln(n!) - \frac{1}{2}\ln(n) \quad \text{and} \quad B_n = n\ln(n) - n + 1.$$

(d) Define $D_n = B_n - A_n$ as the difference of the areas. Show that $n! = e^{1-D_n}\sqrt{n}\left(\frac{n}{e}\right)^n$.

(e) Using the definition $D_n = B_n - A_n$, show that (D_n) is a bounded and increasing sequence.

Deduce that the sequence (D_n) has a real limit D.

(f) Determine the value of e^{1-D} by using $n! = e^{1-D_n}\sqrt{n}\left(\frac{n}{e}\right)^n$ in the Wallis's formula in Exercise 16.26.

(g) Prove that:

$$D - D_n = \sum_{j=n}^{\infty}(b_j - a_j) < \frac{1}{2}\ln\left(1 + \frac{1}{2n}\right).$$

Hence, show that $1 < e^{D-D_n} < \left(1 + \frac{1}{2n}\right)^{\frac{1}{2}}$.

(h) Deduce that:

$$e^{1-D}\sqrt{n}\left(\frac{n}{e}\right)^n < n! < e^{1-D}\sqrt{n}\left(\frac{n}{e}\right)^n\left(1 + \frac{1}{2n}\right)^{\frac{1}{2}}.$$

and conclude by showing $n! \sim \sqrt{2\pi n}\left(\frac{n}{e}\right)^n$.

16.31 (∗) We want to prove the following lemma:

Lemma 16.5.16 *Let (f_n) be a sequence of bounded non-negative functions $f_n : [a, b] \to \mathbb{R}$. If $f_n \downarrow 0$, then the sequence of lower Darboux integrals satisfies $\lim_{n\to\infty} L_{f_n} = 0$.*

Fix any $\varepsilon > 0$.

(a) Show that there exists a sequence of continuous functions (g_n) where $g_n : [a, b] \to \mathbb{R}$ such that $0 \leq g_n \leq f_n$ and:

$$L_{f_n} \leq \int_a^b g_n(x)\,dx + \frac{\varepsilon}{2^n}.$$

(b) For each $n \in \mathbb{N}$, set $h_n : [a, b] \to \mathbb{R}$ as $h_n = \min(g_1, g_2, \ldots, g_n)$. Explain why $0 \leq h_n \leq g_n \leq f_n$ and h_n is continuous for all $n \in \mathbb{N}$.

(c) Prove that $h_n \downarrow 0$.
Explain why $h_n \xrightarrow{u} 0$ on $[a, b]$ and deduce $\lim_{n\to\infty} \int_a^b h_n(x)\,dx = 0$.

(d) Prove that for each $n \in \mathbb{N}$ we have:

$$0 \leq g_n \leq h_n + \sum_{j=1}^{n-1}(\max(g_j, g_{j+1}, \ldots, g_n) - g_j).$$

(e) On the other hand, prove that for each $j \in \{1, 2, \ldots, n\}$ we have:

$$\int_a^b \max(g_j, g_{j+1}, \ldots, g_n)(x) - g_j(x)\,dx \leq L_{f_j} - \int_a^b g_j(x)\,dx \leq \frac{\varepsilon}{2^j}.$$

(f) By putting parts (d) and (e) together, show that for each $n \in \mathbb{N}$ we have:

$$\int_a^b g_n(x)\,dx \leq \int_a^b h_n(x)\,dx + \varepsilon\left(1 - \frac{1}{2^{n-1}}\right).$$

(g) Thus, using part (a), show that:

$$0 \leq L_{f_n} \leq \int_a^b h_n(x)\,dx + \varepsilon\left(1 - \frac{1}{2^n}\right),$$

and conclude the result.

16.32 (∗) Using Exercise 16.31, we are in the position to prove the DCT in Theorem 16.5.13. The following proof is due to [55].

Let (f_n) be a sequence of Riemann integrable functions $f_n : [a, b] \to \mathbb{R}$ such that $f_n \xrightarrow{pw} f$ where $f : [a, b] \to \mathbb{R}$ is a continuous function. Suppose that the sequence (f_n) is uniformly bounded. Define a sequence of functions (g_n) where $g_n : [a, b] \to \mathbb{R}$ is $g_n = |f_n - f|$. Clearly, $g_n \xrightarrow{pw} 0$.

(a) Show that the sequence of functions (g_n) is also uniformly bounded.

(b) Define a sequence of functions (h_n) where $h_n : [a, b] \to \mathbb{R}$ are $h_n(x) = \sup_{k\geq n}(g_k(x))$. Show that $0 \leq g_n \leq h_n$ for all $n \in \mathbb{N}$.

(c) Show that the sequence (h_n) is pointwise decreasing.
Deduce that $h_n \downarrow 0$.

(d) Show that $0 \leq \int_a^b g_n(x)\,dx \leq L_{h_n}$ and, using Exercise 16.31, deduce the limit $\lim_{n\to\infty} \int_a^b |f_n(x) - f(x)|\,dx = 0$.

(e) Finally, show that:

$$\lim_{n\to\infty} \int_a^b f_n(x)\,dx = \int_a^b \lim_{n\to\infty} f_n(x)\,dx = \int_a^b f(x)\,dx.$$

16.33 (a) Let $f : [0, 1] \to \mathbb{R}$ be defined as $f(0) = 0$ and $f(x) = x\ln(x)$ for $x \in (0, 1]$. Prove that f is continuous over $[0, 1]$.
Hence, deduce that f has a maximum and is Riemann integrable over $[0, 1]$.

(b) Show that $|f(x)| \leq 1$ for all $x \in [0, 1]$.

(c) For a fixed $n \in \mathbb{N}$, by using integration by parts, show that:

$$\int_0^1 f(x)^n\,dx = \frac{(-1)^n n!}{(n+1)^{n+1}}.$$

(d) Now let $g : [0, 1] \to \mathbb{R}$ be defined as $g(0) = 1$ and $g(x) = x^{-x}$ for $x \in (0, 1]$. By using parts (a)-(c) and carefully justifying your steps, show that:

$$\int_0^1 g(x)\,dx = \sum_{j=1}^{\infty} j^{-j}.$$

This integral, usually written as the improper integral $\int_0^1 x^{-x}\,dx = \sum_{j=1}^{\infty} j^{-j}$, is called "sophomore's dream" due to the fact that it looks too good to be true. But it is indeed true: we have just proven it!

16.34 ($\diamond$) We are going to prove the Picard-Lindelöf theorem. This theorem is named after Émile Picard and Ernst Lindelöf (1870–1946) and is a very important theorem in the study of differential equations as it gives us a sufficient condition for uniqueness of a solution for an ODE IVP.

Uniqueness of a solution is a very desirable thing to have for an ODE IVP because it allows us to guess for a solution and if we can find one, we do not have to worry whether there are other solutions for the ODE IVP that we might have missed. Moreover, it tells us the problem is well-behaved with no unexpected solutions. However, as we can see in Exercise 14.24, even the most unassuming looking ODE IVP might not have a unique solution. The following theorem gives us a sufficient condition on when we can expect an ODE IVP has a unique solution.

Theorem 16.5.17 (Picard-Lindelöf Theorem) *Let $I \times J \subseteq \mathbb{R}^2$ be a rectangle where I, J are closed intervals in $\mathbb{R}$. Suppose that $(x_0, y_0) \in I \times J$ not on the boundary of the rectangle and $F : I \times J \to \mathbb{R}$ is a function that is continuous over I and Lipschitz continuous over J with Lipschitz constant $K > 0$. Then, there exists an $\varepsilon > 0$ such that the initial value problem of the ODE:*

$$\frac{dy}{dx} = F(x, y(x)) \quad \textit{with} \quad y(x_0) = y_0,$$

has a unique solution $y : (x_0 - \varepsilon, x_0 + \varepsilon) \to \mathbb{R}$.

The proof relies on the fact that we can rewrite the ODE as an integral equation:

$$y(x) = y_0 + \int_{x_0}^{x} F(t, y(t))\,dt,$$

for some $x \in I$. However, this is a nested definition, namely: we want to find $y(x)$ but there is a $y(t)$ in the integral that defines it. This is similar to the situation in Exercises 6.4, 6.6, and 10.32. To get around this problem, we define a sequence of functions (y_n) where $y_n : I \to \mathbb{R}$ for all $n \in \mathbb{N}$ via the following recursively equation:

$$y_n(x) = y_0 + \int_{x_0}^{x} F(t, y_{n-1}(t))\,dt.$$

Thus, all the functions (y_n) are continuous with respect to x. Note that it is quite likely none of these (y_n) solves the ODE. The idea of the above is we put in a guess y_0 in the integral equation to get y_1, which we put back in the integral equation to get y_2, and so on. At each iteration, we would get a new function and we hope that these sequences of functions stabilise by converging to the actual solution of the ODE.

(a) Let $|x - x_0| < \varepsilon$ where $\varepsilon > 0$ will be chosen appropriately later. Since F is continuous over $I \times J$, there exists an $M > 0$ such that $|F(x, y)| \leq M$. Prove that $|y_1(x) - y_0| < M\varepsilon$.
Inductively, show that for all $n \in \mathbb{N}$:

$$|y_{n+1}(x) - y_n(x)| < MK^n\varepsilon^{n+1} = (K\varepsilon)^n M\varepsilon.$$

(b) For any $m, n \in \mathbb{N}$ with $n > m$, prove that:

$$|y_n(x) - y_m(x)| < M\varepsilon \sum_{j=m}^{n-1} (K\varepsilon)^j.$$

(c) Hence, deduce a value of $\varepsilon > 0$ for which the sequence of functions (y_n) where $y_n : (x_0 - \varepsilon, x_0 + \varepsilon) \to \mathbb{R}$ is uniformly Cauchy.

(d) By using Proposition 11.2.7, conclude that (y_n) converges uniformly to a continuous function $y : (x_0 - \varepsilon, x_0 + \varepsilon) \to \mathbb{R}$.

(e) Hence, show that for any $x \in (x_0 - \varepsilon, x_0 + \varepsilon)$ we have:

$$y(x) = y_0 + \int_{x_0}^{x} F(t, y(t))\, dt.$$

Moreover, deduce that y is differentiable with $\frac{dy}{dx} = F(x, y(x))$.

(f) Finally, prove that there is only one solution to the ODE on $(x_0 - \varepsilon, x_0 + \varepsilon)$.

16.35 (◇) Let $y : (-\infty, 1) \to \mathbb{R}$ be a function defined as $y(x) = \frac{1}{1-x}$.

(a) Show that y satisfies the ODE $\frac{dy}{dx} = y^2$.

(b) If $F : \mathbb{R} \to \mathbb{R}$ is defined as $F(y) = y^2$, show that F is Lipschitz continuous on any compact interval of the form $[-R, R] \subseteq \mathbb{R}$ for $R > 0$.

(c) Determine the value of $\varepsilon > 0$ that would allow use to use Picard-Lindelöf theorem.
Hence, by using the Picard-Lindelöf iteration, find the first three iterations y_1, y_2, and y_3 when solving the ODE IVP:

$$\frac{dy}{dx} = y^2 \quad \text{with} \quad y(0) = 1.$$

(d) Compare y_3 with the power series of the actual solution to this ODE.

Taylor and Maclaurin Series 17

Theory is the first term in the Taylor series expansion of practice.

— Thomas Cover, information theorist

In Example 12.1.9, we have seen the power series $\sum_{j=0}^{\infty} \frac{x^j}{j!}$. This series converges for every $x \in \mathbb{R}$ and it converges pointwise to the exponential function $e^x : \mathbb{R} \to \mathbb{R}$. Therefore, the exponential function that we have defined Definition 4.2.4 can also be defined via a power series definition.

Can we do this for any other functions? In other words, given a real function, can we use a power series to define or express it? The answer is: sometimes and even if we can, it probably would not be for the whole domain of definition for the original function.

In this short chapter, let us see how we can do this and thus tie up many of the questions that we had left over in the previous chapters. We shall also see how being able to express a function as a power series, albeit on a small domain, can be very useful in certain situations.

17.1 Taylor Polynomial and Series

Suppose that we start with a real function $f : \mathbb{R} \to \mathbb{R}$. Before we construct a power series for it, let us construct a polynomial that approximates the function f at a point $c \in \mathbb{R}$. Suppose further that the function f is differentiable infinitely many times at the point $x = c$.

If we want to approximate the function f with a polynomial $P_n : \mathbb{R} \to \mathbb{R}$ of some degree $n \in \mathbb{N}_0$ that agrees with the function f, we need to make sure that $P_n(c) = f(c)$ and the first n derivatives of P_n agree with f at the point $x = c$. In order to do this, we need to choose the $n + 1$ coefficients in the polynomial

S. Johar, *The Big Book of Real Analysis*,
https://doi.org/10.1007/978-3-031-30832-1_17

$P_n(x) = \sum_{j=0}^{n} a_j(x-c)^j$ in a suitable manner such that $P_n^{(j)}(c) = f^{(j)}(c)$ for all $j = 0, 1, \ldots, n$. For example, for $n = 0, 1, 2, 3$ we would have:

$$
\begin{aligned}
n = 0: &\quad P_0(x) = f(c),\\
n = 1: &\quad P_1(x) = f(c) + f'(c)(x-c),\\
n = 2: &\quad P_2(x) = f(c) + f'(c)(x-c) + \tfrac{f''(c)}{2}(x-c)^2.\\
n = 3: &\quad P_3(x) = f(c) + f'(c)(x-c) + \tfrac{f''(c)}{2}(x-c)^2 + \tfrac{f^{(3)}(c)}{6}(x-c)^3.
\end{aligned}
$$

For a general degree $n \geq 0$, by equating the coefficients of a polynomial P_n, we can deduce that $a_j = \frac{f^{(j)}(c)}{j!}$ for all $j = 0, 1, \ldots, n$. The resulting polynomial is called a Taylor polynomial, after Brook Taylor (1685–1731) who introduced them in 1715:

Definition 17.1.1 (Taylor Polynomial) For $n \in \mathbb{N}_0$ and an n-times differentiable function $f : \mathbb{R} \to \mathbb{R}$, we define the n-th order Taylor polynomial at $x = c$ as the polynomial $P_n : \mathbb{R} \to \mathbb{R}$ defined as:

$$P_n(x) = \sum_{j=0}^{n} \frac{f^{(j)}(c)}{j!}(x-c)^j.$$

Even though the polynomial P_n may not be equal to f globally over its whole domain, this is a good candidate for the power series expression for f if we take the degree of the polynomial to infinity. This is because, for starters, all the derivatives of f and P_n agree at the point c. So now we have an obvious candidate for the power series that we wanted. By taking the formal infinite sum of the Taylor polynomial as its degree goes to infinity, we define:

Definition 17.1.2 (Taylor and Maclaurin Series) Let $f : \mathbb{R} \to \mathbb{R}$ be infinitely differentiable at the point $x = c$. Then, the Taylor series of the function f expanded about the point $c \in \mathbb{R}$ is given by the power series:

$$f(x) \sim \sum_{j=0}^{\infty} \frac{f^{(j)}(c)}{j!}(x-c)^j,$$

where $f^{(j)}(c)$ is the j-th derivative of the function f evaluated at c. The point c is called the centre or point of expansion. If $c = 0$, then the power series is also called the Maclaurin series, named after Colin Maclaurin.

Remark 17.1.3 We comment the usage of the $\sim$ symbol instead of the $=$ symbol:

1. Firstly, as usual when defining a functions series, we are not sure whether the series converges for any $x \in \mathbb{R}$. The original function f may be defined for all $x \in \mathbb{R}$ but the series may not. It may not even converge for any x other than $x = c$ if the radius of convergence of the power series is $R = 0$.
2. Secondly, even if the radius of convergence R is strictly positive, meaning that the power series converges in some open ball $B_R(c)$ around the centre $c \in \mathbb{R}$, for each $x \in B_R(c)$ the series at this point may converge to values different than $f(x)$. We shall see an example of this phenomenon in Example 17.1.4(6).

These potential problems justify the use of the symbol $\sim$ instead of $=$ and so we have to treat the Taylor series with utmost care.

Example 17.1.4 Let us compute the Taylor series for some functions.

1. For the function $f : \mathbb{R} \to \mathbb{R}$ defined as $f(x) = e^x$, we have $f^{(j)}(x) = e^x$ for all $j \in \mathbb{N}$. Thus, its Taylor series centred at 0 is given by the series:

$$e^x \sim \sum_{j=0}^{\infty} \frac{f^{(j)}(0)}{j!} x^j = \sum_{j=0}^{\infty} \frac{x^j}{j!}.$$

   We have seen in Proposition 12.4.1 that these two quantities, namely the function and the power series, coincide everywhere on $\mathbb{R}$. Thus, we can safely replace the symbol $\sim$ with $=$ in the above. So this example gives us hope that the Taylor series can be a good candidate to represent a function.
2. For the function $f : \mathbb{R} \setminus \{1\} \to \mathbb{R}$ defined as $f(x) = \frac{1}{1-x}$, we have $f^{(j)}(x) = \frac{j!}{(1-x)^{j+1}}$ for all $j \in \mathbb{N}$. Therefore, $f^{(j)}(0) = j!$ for all $j \in \mathbb{N}_0$ and thus its Taylor series centred at 0 is given by the series:

$$\frac{1}{1-x} \sim \sum_{j=0}^{\infty} x^j,$$

   which converges if and only if $|x| < 1$ by the ratio test. Again, we saw earlier that this is actually an equality, so we can replace the $\sim$ with $=$. However, unlike the first example, this equality is valid only for $|x| < 1$ even though the function f is defined on a much larger domain. Thus, a Taylor series may not be able represent its original function globally.
3. For any $r \in \mathbb{R}$, we can define a generalised binomial coefficient $\binom{r}{j}$ as the product:

$$\binom{r}{j} = \prod_{k=1}^{j} \frac{r-k+1}{k} = \frac{r(r-1)(r-2)\dots(r-j+1)}{j!},$$

that we saw earlier in Exercise 12.10. Using this definition, the function $f : \mathbb{R} \setminus \{-1\} \to \mathbb{R}$ defined as $f(x) = (1+x)^r$ where $r \notin \mathbb{N}$ has j-th derivative for $j \in \mathbb{N}$ given by $f^{(j)}(x) = r(r-1)\ldots(r-j+1)(1+x)^{r-j} = j!\binom{r}{j}$. Thus, its Taylor series expanded about the point $x = 0$ is:

$$(1+x)^r \sim \sum_{j=0}^{\infty} \binom{r}{j} x^j,$$

which converges for all $|x| < 1$ by the ratio test in Exercise 12.10 (and possibly at ± 1 for some values of r). However, we do not yet know whether the function f is equal to this Taylor series other than at the centre $x = 0$.

4. For the function $f : \mathbb{R} \to \mathbb{R}$ defined as $f(x) = \sin(x)$, we can compute its j-th derivatives as:

$$f^{(j)}(x) = \begin{cases} \cos(x) & \text{for } j = 4k+1, \\ -\sin(x) & \text{for } j = 4k+2, \\ -\cos(x) & \text{for } j = 4k+3, \\ \sin(x) & \text{for } j = 4k, \end{cases}$$

so that:

$$f^{(j)}(\pi) = \begin{cases} -1 & \text{for } j = 4k+1, \\ 0 & \text{for } j = 4k+2, \\ 1 & \text{for } j = 4k+3, \\ 0 & \text{for } j = 4k, \end{cases}$$

where $k \in \mathbb{N}_0$. Therefore, the Taylor series for the function $f(x) = \sin(x)$ expanded about the point $x = \pi$ is given by the series:

$$\sin(x) \sim -(x-\pi) + \frac{(x-\pi)^3}{3!} - \frac{(x-\pi)^5}{5!} + \ldots = \sum_{j=0}^{\infty} \frac{(-1)^{j+1}}{(2j+1)!}(x-\pi)^{2j+1}. \tag{17.1}$$

If we choose to find its Taylor series expanded about the point $x = 0$ instead, we can compute its derivatives at the point $x = 0$ as:

$$f^{(j)}(0) = \begin{cases} 1 & \text{for } j = 4k+1 \\ 0 & \text{for } j = 4k+2 \\ -1 & \text{for } j = 4k+3 \\ 0 & \text{for } j = 4k, \end{cases}$$

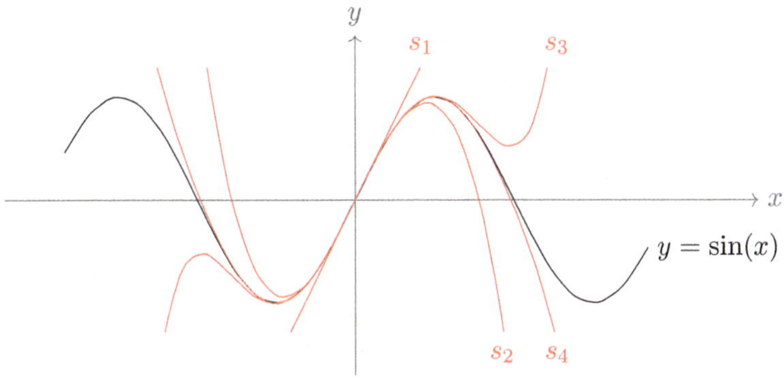

Fig. 17.1 The first four partial sums of the Taylor series (17.2) for sine centred at $x = 0$

and so this Taylor series is:

$$\sin(x) \sim x - \frac{x^3}{3!} + \frac{x^5}{5!} - \ldots = \sum_{j=0}^{\infty} \frac{(-1)^j}{(2j+1)!} x^{2j+1}. \tag{17.2}$$

Either way, it is easy to show that both of the series (17.1) and (17.2) converge for all $x \in \mathbb{R}$. We do not know whether these two functions series on $\mathbb{R}$ are equal $\sin(x)$ anywhere apart from the centre of expansion. If we refer to Fig. 17.1, we can see that the first four partial sums of the series (17.2) are close to the sine graph around a gradually larger neighbourhood of $x = 0$.
Also, notice that the Taylor series that we have computed in (17.2) is exactly the same as the series $S(x)$ that we saw in Exercises 12.11 to 12.14. In these exercises, we have seen that this power series is periodic and odd, just like the sine function. So they could be equal everywhere!

5. For the function $f : \mathbb{R}_+ \to \mathbb{R}$ defined as $f(x) = \ln(x)$, we cannot find a Taylor series expansion of this function at 0 as the function, and hence its derivatives, are not defined here. Instead, we can find its Taylor series about the point $x = 1$, for example. We find its derivatives $f^{(j)}(x) = \frac{(-1)^{j-1}(j-1)!}{x^j}$ so that $f^{(j)}(1) = (-1)^{j-1}(j-1)!$ for any $j \in \mathbb{N}$ and $f(1) = 0$. Hence, its Taylor series about the point $x = 1$ is given by:

$$\ln(x) \sim \sum_{j=1}^{\infty} \frac{(-1)^{j-1}}{j}(x-1)^j.$$

By using the ratio test, this series converges for $|x - 1| < 1$, namely for $0 < x < 2$. Clearly, by alternating series test, this series also converges at $x = 2$. So its domain of convergence is $0 < x \leq 2$.

6. Consider the function $f : \mathbb{R} \to \mathbb{R}$ defined by:

$$f(x) = \begin{cases} e^{-\frac{1}{x}} & \text{if } x > 0, \\ 0 & \text{otherwise.} \end{cases}$$

We have seen in Exercise 14.18 that this function is smooth so we can define its Taylor series. We have also computed the derivatives of f at $x = 0$ and they are given by $f^{(j)}(0) = 0$ for all $j \in \mathbb{N}$. Thus, the Taylor series centred at 0 is identically 0 and so it has a radius of convergence $R = \infty$. But clearly the function f is not identically 0, so the bump function is not equal everywhere to its Taylor series centred at 0. They are equal only for $x \leq 0$.

We have seen in Example 17.1.4 that some of the Taylor series for these functions do converge pointwise to its corresponding function, even in some small neighbourhood of the point of expansion. This would be a desirable thing to know since if this is true, we can treat the function as a power series here and vice versa.

Moreover, we can differentiate and integrate these power series term-wise within the domain of convergence, thanks to Propositions 14.2.7 and 16.5.7. This is particularly useful for the latter since it may be difficult to integrate a function in its original form from first principles or using antiderivatives. We have seen some examples on how we could do this in Example 16.5.9(2) and (3).

We have a special name for functions which agree with its Taylor series in the domain where the series converges:

Definition 17.1.5 (Analytic and Entire Functions) Let $f : \mathbb{R} \to \mathbb{R}$.

1. If the function f agrees with its Taylor series centred at c in a small ball of radius $R > 0$ around c, namely:

$$f(x) = \sum_{j=0}^{\infty} \frac{f^{(j)}(c)}{j!}(x-c)^j \quad \text{for } x \in B_R(c),$$

then we call the function f analytic in $B_R(c)$.
2. If f is equal to its Taylor series centred at c everywhere in $\mathbb{R}$, namely:

$$f(x) = \sum_{j=0}^{\infty} \frac{f^{(j)}(c)}{j!}(x-c)^j \quad \text{for } x \in \mathbb{R},$$

then we call the function an entire function.

There are many entire functions: the exponential functions and polynomials are two examples that we have seen. An example of an analytic, but not entire, function is the function $f(x) = \frac{1}{1-x}$ for $x \in \mathbb{R}\setminus\{1\}$. This function is defined on $\mathbb{R}\setminus\{1\}$ but is equal to its power series only in a finite open interval in $\mathbb{R}$ that avoids the singularity point $x = 1$.

In Example 17.1.4, we have computed the Taylor series of some functions but in some of these examples, it is not clear whether their Taylor series coincide with the original function. How do we check that the series agrees with the original function?

17.2 Taylor Remainder

To investigate this, we appeal to the definition of pointwise convergence of series. For a function f, the sequence of n-th degree Taylor polynomials (P_n) centred at $c \in \mathbb{R}$ form the sequence of partial sums for the Taylor series. So, we are now asking whether $P_n \xrightarrow{pw} f$ on the domain of convergence for the Taylor series. In other words, for all such x we want to know whether $\lim_{n\to\infty} |f(x) - P_n(x)| = 0$. Let us first give the difference $f - P_n$ in the limit above a name:

Definition 17.2.1 (Taylor Remainder) For $n \in \mathbb{N}_0$ and an n-times differentiable function $f : \mathbb{R} \to \mathbb{R}$, we define the n-th Taylor remainder $R_n : \mathbb{R} \to \mathbb{R}$ as the difference between f and its n-th order Taylor polynomial, namely $R_n = f - P_n$.

The remainder R_n measures how close the n-th Taylor polynomial is to the actual function f at any point $x \in \mathbb{R}$. Ideally, we would like this remainder at x to converge to 0 as n goes to infinity. From the discussion above, if the n-th remainder $|R_n(x)|$ converges to 0 for all x where the Taylor series is defined, then the Taylor converges to the function f here.

Since $0 \leq |R_n(x)|$, in order to show that $|R_n(x)|$ converges to 0, it is probably best to appeal to sandwich lemma to bound $|R_n(x)|$ from above by some sequence that tends to 0. Moreover, since $R_n(x)$ also depends on x, we might have to consider different sandwiching sequences for different values of x.

We have laid out the general strategy above, so now the big question is: can we find an explicit closed form for the remainder in order for us to do the analysis? We can! In fact, we have at least three expressions for R_n. The first one involves a Riemann integral:

Theorem 17.2.2 (Integral Form for Taylor Remainder) *Let $f : \mathbb{R} \to \mathbb{R}$ be an $(n+1)$-differentiable function. Then, the n-th Taylor remainder R_n at x is given by:*

$$R_n(x) = \frac{1}{n!}\int_c^x (x-t)^n f^{(n+1)}(t)\,dt.$$

Proof The integral form of R_n can be obtained by applying induction on the non-negative integers $n \in \mathbb{N}_0$. For the base case $n = 0$, we want to prove that $R_0(x) = \int_c^x (x-t)^0 f'(t)\,dt$. Indeed, from the RHS, by using the FTC, we have:

$$\int_c^x (x-t)^0 f'(t)\,dt = \int_c^x f'(t)\,dt = f(x) - f(c) = f(x) - P_0(x) = R_0(x),$$

and so the base case $n = 0$ is true.

Now assume that the statement is true for $n = k$, namely:

$$R_k(x) = \frac{1}{k!}\int_c^x (x-t)^k f^{(k+1)}(t)\,dt. \tag{17.3}$$

We want to prove the case for $n = k+1$, namely the equality $R_{k+1}(x) = \frac{1}{(k+1)!}\int_c^x (x-t)^{k+1} f^{(k+2)}(t)\,dt$. We work from the RHS of this and integrate by parts to get:

$$\begin{aligned}
&\frac{1}{(k+1)!}\int_c^x (x-t)^{k+1} f^{(k+2)}(t)\,dt \\
&= \left[\frac{(x-t)^{k+1}}{(k+1)!} f^{(k+1)}(t)\right]_{t=c}^{t=x} + \int_c^x \frac{(x-t)^k}{k!} f^{(k+1)}(t)\,dt \\
&= -\frac{(x-c)^{k+1}}{(k+1)!} f^{(k+1)}(c) + \int_c^x \frac{(x-t)^k}{k!} f^{(k+1)}(t)\,dt.
\end{aligned}$$

However, by the inductive hypothesis in (17.3), the final Riemann integral is simply $R_k(x) = f(x) - P_k(x)$. Therefore:

$$\begin{aligned}
\frac{1}{(k+1)!}\int_c^x (x-t)^{k+1} f^{(k+2)}(t)\,dt &= -\frac{(x-c)^{k+1}}{(k+1)!} f^{(k+1)}(c) + (f(x) - P_k(x)) \\
&= f(x) - P_{k+1}(x) = R_{k+1}(x),
\end{aligned}$$

by definitions of P_{k+1} and R_{k+1}. This concludes the proof. □

By applying the MVT for integrals or the IVT on continuous functions to the Riemann integral in Theorem 17.2.2, if the function $f^{(n+1)}$ is continuous, we get a less explicit but integral-free expression for the remainder. These are the second and third forms of the Taylor remainder R_n. We first prove:

Theorem 17.2.3 (Lagrange Remainder Theorem) *Let $f : \mathbb{R} \to \mathbb{R}$ be an $(n+1)$-differentiable function such that $f^{(n+1)}$ is continuous on the closed interval between the centre of expansion c and x. Then, the n-th Taylor remainder at x is given by:*

$$R_n(x) = \frac{f^{(n+1)}(\xi)}{(n+1)!}(x-c)^{n+1},$$

for some real number ξ between c and x.

Proof WLOG, suppose that $c < x$. Since $f^{(n+1)}(t)$ is continuous in $[c, x]$, it must be bounded in $[c, x]$ and attains its bounds somewhere. In other words, at all

$t \in [c, x]$, we have $m \leq f^{(n+1)}(t) \leq M$ for some constants $m, M \in \mathbb{R}$ with each equality occurring somewhere in $[c, x]$. Since $x - t \geq 0$, we then have the bounds $m(x-t)^n \leq (x-t)^n f^{(n+1)}(t) \leq M(x-t)^n$. By integrating this with respect to t from c to x, we obtain:

$$\begin{aligned}
\int_c^x m(x-t)^n\,dt \leq \int_c^x (x-t)^n f^{(n+1)}(t)\,dt &\leq \int_c^x M(x-t)^n\,dt,\\
\Rightarrow \qquad m\frac{(x-c)^{n+1}}{n+1} \leq \int_c^x (x-t)^n f^{(n+1)}(t)\,dt &\leq M\frac{(x-c)^{n+1}}{n+1},\\
\Rightarrow \qquad m \leq \frac{n+1}{(x-c)^{n+1}} \int_c^x (x-t)^n f^{(n+1)}(t)\,dt &\leq M. \qquad (17.4)
\end{aligned}$$

The quantity in the middle of the inequalities (17.4) is a continuous function of x for $x > c$. So, for every fixed $x > c$ the quantity in the middle is simply a constant value which is somewhere within $[m, M]$. Since $f^{(n+1)}$ is continuous over the interval $[c, x]$, by the IVT, there must exist some point $\xi(x, n) \in [c, x]$ such that:

$$\frac{n+1}{(x-c)^{n+1}} \int_c^x (x-t)^n f^{(n+1)}(t)\,dt = f^{(n+1)}(\xi(x,n)). \qquad (17.5)$$

Finally, by using the integral form of R_n from Theorem 17.2.2 in the equality (17.5), we obtain:

$$R_n(x) = \frac{f^{(n+1)}(\xi)}{(n+1)!}(x-c)^{n+1},$$

which completes the proof. □

Remark 17.2.4 We make some remarks regarding Theorem 17.2.3.

1. A drawback of the Lagrange remainder theorem is that we do not know the exact value of ξ in between c and x in the expression for R_n. However, we can estimate this remainder if we know the behaviour of the function $f^{(n+1)}$ between c and x.
2. A thing to note in the theorem is that the number ξ between x and c depends implicitly on the point of interest x and also the integer n. This is because the quantity $\frac{n+1}{(x-c)^{n+1}} \int_c^x (x-t)^n f^{(n+1)}(t)\,dt$ that defines it depends on both x and n. This is a very important thing to remember when we are taking limits as $n \to \infty$ since when we are increasing n, the quantity ξ also changes and cannot be treated as a constant. This can be quite troublesome as we have no idea what $\xi(x, n)$ look like!

3. Therefore, when we try to bound the remainder R_n, our main aim is to get rid of the dependency on the term ξ to avoid complicating things when we are doing the analysis.

Finally, the third expression for the Taylor remainder is given by the following theorem. The readers are invited to prove this theorem in Exercise 17.5.

Theorem 17.2.5 (Cauchy Remainder Theorem) *Let $f : \mathbb{R} \to \mathbb{R}$ be an $(n+1)$-differentiable function such that $f^{(n+1)}$ is continuous on the closed interval between the centre of expansion c and x. Then, the n-th Taylor remainder at x is given by:*

$$R_n(x) = \frac{f^{(n+1)}(\eta)}{n!}(x-\eta)^n(x-c),$$

for some real number η between c and x.

Remark 17.2.6 Similar to Lagrange remainder theorem, the value of η in Cauchy remainder theorem depends implicitly on the quantities x and n.

Example 17.2.7 We shall now show that some of the Taylor series expansions we computed in Example 17.1.4 approach their corresponding functions over their the domain of convergence.

1. We have found the Taylor expansion for the function $f(x) = \sin(x)$ around $x = 0$ to be:

$$\sin(x) \sim x - \frac{x^3}{3!} + \frac{x^5}{5!} - \ldots = \sum_{j=0}^{\infty} \frac{(-1)^j}{(2j+1)!} x^{2j+1} = S(x).$$

By using the ratio test, we can show that this series converges for all $x \in \mathbb{R}$. We now want to show that this series does converge pointwise to $\sin(x)$ for all $x \in \mathbb{R}$. Fix $x \in \mathbb{R}$. We are going to show that the n-th remainder $|R_n(x)|$ at this x converges to 0 as $n \to \infty$. We note that:

$$|R_n(x)| = |f^{(n+1)}(\xi)| \frac{|x|^{n+1}}{(n+1)!},$$

where $\xi \in (0, x)$. Since we have fixed x, this quantity ξ now depends only on the value n. This is still problematic since we want to take the limit as $n \to \infty$, so we now aim to get rid of the dependency of $R_n(x)$ on ξ altogether. We note that the derivatives of f are either the sine or cosine functions and these terms are always bounded by 1 anywhere. Therefore, we can eliminate the dependency on ξ via the following bound:

$$0 \leq |R_n(x)| = |f^{(n+1)}(\xi)|\frac{|x|^{n+1}}{(n+1)!} \leq \frac{|x|^{n+1}}{(n+1)!}. \tag{17.6}$$

Thus, if we want to find the limit of $|R_n(x)|$ as $n \to \infty$, we have to study the limit of $\frac{|x|^{n+1}}{(n+1)!}$ or equivalently the sequence (a_n) where $a_n = \frac{|x|^n}{n!}$ as $n \to \infty$. Since x is fixed, there exists an integer $m \in \mathbb{N}$ such that $|x| < m$. Therefore, for all large powers $n \geq m > |x|$, we have:

$$a_{n+1} = \frac{|x|^{n+1}}{(n+1)!} = \frac{|x|}{n+1}\frac{|x|^n}{n!} \leq \frac{|x|^n}{n!} = a_n,$$

which means the sequence (a_n) is decreasing after the m-th term. By monotone sequence theorem, since this sequence is also bounded from below by 0, it converges to some $a \geq 0$. How do we find this limit a? We know that $a_{n+1} = \frac{|x|}{n+1}\frac{|x|^n}{n!} = \frac{|x|}{n+1}a_n$. By taking the limit on both sides and using the algebra of limits, we have:

$$a = \lim_{n\to\infty} a_{n+1} = \lim_{n\to\infty}\left(\frac{|x|}{n+1}a_n\right) = \lim_{n\to\infty}\frac{|x|}{n+1}\lim_{n\to\infty} a_n = 0 \times a = 0,$$

so the sequence (a_n) converges to 0 as $n \to \infty$. Back to inequality (17.6), by sandwiching, we get:

$$0 \leq \lim_{n\to\infty}|R_n(x)| \leq \lim_{n\to\infty}\frac{|x|^{n+1}}{(n+1)!} = 0,$$

which implies that $\lim_{n\to\infty}|R_n(x)| = 0$ at x and so the Taylor series converges to $f(x) = \sin(x)$ at this x. Since x is arbitrarily fixed, we can now vary x and deduce that the Taylor series converges to $f(x) = \sin(x)$ at all $x \in \mathbb{R}$. With this, we can now write, with flourish, the equality:

$$\sin(x) = \sum_{j=0}^{\infty}\frac{(-1)^j}{(2j+1)!}x^{2j+1} = S(x) \quad \text{for all } x \in \mathbb{R}.$$

This confirms the hypothesis that we made in Exercises 12.11–12.14, and 14.14. To show that $\cos(x) = C(x)$ for all $x \in \mathbb{R}$, we leave this in Exercise 17.7.

Therefore, instead of using the geometric definition of the sine and cosine functions via the ratios of sidelengths of a right triangle, these power series $S(x)$ and $C(x)$ on $\mathbb{R}$ can instead be taken as the geometry-free definition of sine and cosine. More impressively, these series have been established by Madhava, the founder of Kerala school of astronomy and mathematics, in the fourteenth century without the language of limits or calculus.

2. Let $f : (0, \infty) \to \mathbb{R}$ be the logarithm function $f(x) = \ln(x)$. We have seen that its Taylor series about the point $x = 1$ is given by:

$$\ln(x) \sim \sum_{j=1}^{\infty} \frac{(-1)^{j-1}}{j}(x-1)^j,$$

which converges for $x \in (0, 2]$. Now we want to show that the series converges to f when $x \in (0, 2)$. Fix an x here. WLOG, suppose that $x > 1$. We can compute the derivatives f which are given by $f^{(n+1)}(x) = \frac{(-1)^n n!}{x^{n+1}}$ for $n \in \mathbb{N}_0$ so that the Taylor remainder R_n can be written as:

$$|R_n(x)| = |f^{(n+1)}(\xi)|\frac{|x-1|^{n+1}}{(n+1)!} = \frac{n!}{|\xi|^{n+1}}\frac{|x-1|^{n+1}}{(n+1)!} = \frac{1}{n+1}\left|\frac{x-1}{\xi}\right|^{n+1}, \tag{17.7}$$

for some $\xi \in (1, x)$. Since $1 < \xi < x < 2$, we must have $0 < x - \xi < 2 - 1 = 1$ so that $0 < \frac{x-1}{\xi} < 1$. Hence, when we take the limit as $n \to \infty$ in the Eq. (17.7), we would get $|R_n(x)| \to 0$ as well. Similar argument also holds for $x < 1$. Thus, the convergence is pointwise and we can conclude with the equality:

$$\ln(x) = \sum_{j=1}^{\infty} \frac{(-1)^{j-1}}{j}(x-1)^j \quad \text{for } x \in (0, 2).$$

Using Abel's theorem, we can also show the equality also holds at the endpoint $x = 2$. By relabelling $x = 1 - y$, we can also rewrite this as:

$$\ln(1-y) = \sum_{j=1}^{\infty} \frac{y^j}{j} \quad \text{for } y \in [-1, 1),$$

which is similar to the expression we obtained in Example 16.5.6(2).

17.3 Polynomial Approximation

An application of the Taylor series for a function f is that it provides us with an approximation of the function as a polynomial locally. Recall the MVT that we saw in Theorem 13.6.3: for any continuously differentiable function $f : \mathbb{R} \to \mathbb{R}$ and $c \in \mathbb{R}$, for any $x \in \mathbb{R} \setminus \{c\}$ we have $\frac{f(x)-f(c)}{x-c} = f'(\xi)$ where ξ is somewhere in between c and x. If x and c are close enough, by continuity of f', we have $f'(c) \approx f'(\xi)$, giving us the approximation:

$$f(x) = f(c) + f'(\xi)(x-c) \approx f(c) + f'(c)(x-c), \tag{17.8}$$

which is an approximation of the function f near the point $x = c$ with a linear polynomial.

The n-th Taylor polynomial generalises this. It approximates the function f around the point $c \in \mathbb{R}$ by an n-th order polynomial instead of a linear function. If $f : \mathbb{R} \to \mathbb{R}$ is $(n+1)$-times differentiable, we have the equality $f = P_n + R_n$ where P_n is the Taylor's polynomial of degree n and R_n is the error term which, by Theorem 17.2.3, can be written as $R_n(x) = \frac{f^{(n+1)}(\xi)}{(n+1)!}(x-c)^{n+1}$. Thus, $R_n \in O(|x-c|^n)$ as $x \to c$ and we can then write $f(x) = P_n(x) + O(|x-c|^n)$. This error term is very small when we are close enough to c and thus:

$$f(x) \approx P_n(x) = \sum_{j=0}^{n} \frac{f^{(j)}(c)}{j!}(x-c)^j.$$

Of course, the linear approximation of the function f obtained via the MVT in (17.8) is exactly P_1. By definition, we can then can view the Taylor's remainder R_n as the error made when we approximated the function f by an n-th order polynomial P_n. From Theorems 17.2.2 and 17.2.3, even though the remainders R_n are not explicit, we can roughly approximate the error we made from these approximations using some analysis.

Example 17.3.1 Consider the function $f(x) = \sqrt[3]{x}$ defined for $x \geq 0$. Let us find the second order Taylor polynomial centred at $x = 8$. We compute $f(8) = 2$, $f'(8) = \frac{1}{12}$, and $f''(8) = -\frac{1}{144}$. Thus, this polynomial is:

$$P_2(x) = \sum_{j=0}^{2} \frac{f^{(j)}(8)}{j!}(x-8)^j = 2 + \frac{1}{12}(x-8) - \frac{1}{288}(x-8)^2.$$

So, around the point $x = 8$, we have the approximation $\sqrt[3]{x} \approx 2 + \frac{1}{12}(x-8) - \frac{1}{288}(x-8)^2$. In other words:

$$\sqrt[3]{x} = 2 + \frac{1}{12}(x-8) - \frac{1}{288}(x-8)^2 + O(|x-8|^3).$$

How good is this approximation for $7 < x < 9$? We can see in Fig. 17.2 that the graphs are almost identical around the point $x = 8$. Now we would like to find the numerical value of the maximal error. Using the Lagrange remainder theorem, the error accrued by this approximation at $x \in (7, 9)$ is:

$$|\sqrt[3]{x} - P_2(x)| = |R_2(x)| = |f'''(\xi)|\frac{|x-8|^3}{3!} = \frac{10}{27\sqrt[3]{\xi^8}}\frac{|x-8|^3}{3!},$$

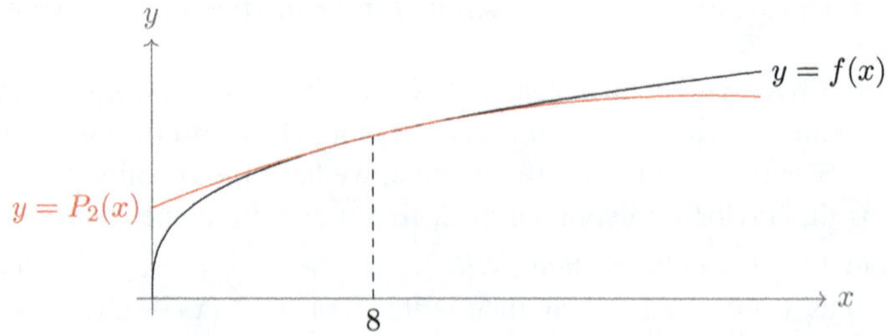

Fig. 17.2 Graph of $y = f(x) = \sqrt[3]{x}$ and its polynomial approximation $y = P_2(x)$ centred at the point $x = 8$. The approximation is close, but how close?

where ξ is between x and 8. The error is not explicit here since it depends on what x is. So we may instead ask: what is the biggest error we could have made if $7 < x < 9$? For these x we know that $|x - 8| < 1$ and so:

$$|R_2(x)| = \frac{10}{27\sqrt[3]{\xi^8}} \frac{|x-8|^3}{3!} < \frac{5}{81} \frac{1}{\sqrt[3]{\xi^8}}.$$

To find the largest possible error, we now try and minimise the value of ξ. Note that ξ is between x and 8 and x is between 7 and 9, so we can obtain an upper bound:

$$|R_2(x)| < \frac{5}{81} \frac{1}{\sqrt[3]{\xi^8}} \leq \frac{5}{81} \frac{1}{\sqrt[3]{7^8}} < \frac{5}{81 \cdot 49} < 0.0013.$$

Therefore, the polynomial P_2 gives a good approximation for f for $7 < x < 9$ since the error made by this approximation in this region is always less than 0.0013.

In Example 17.3.1, we have fixed the number of terms in the polynomial and find the error accrued by approximating f with the second order Taylor polynomial. We can also reverse engineer the problem by setting an error bound and finding how many terms in the Taylor series do we need to have for an approximation which satisfies the required error bound.

Example 17.3.2 Consider the logarithm function $f(x) = \ln(x)$ defined for $x \in \mathbb{R}_+$. We have seen in Example 17.2.7(2) that its Taylor series expansion about the point $x = 1$ is given by the series:

$$\ln(x) = \sum_{j=1}^{\infty} \frac{(-1)^{j-1}}{j}(x-1)^j \quad \text{for } x \in (0, 2].$$

We want to determine how many terms in the Taylor series we should take so that the approximation of $\ln(\frac{3}{2})$ is good to 4 decimal places. In other words, we want to find an $n \in \mathbb{N}_0$ such that $|R_n(\frac{3}{2})| < 0.0001$. Using the Lagrange remainder theorem:

$$|R_n(\tfrac{3}{2})| = |f^{(n+1)}(\xi)|\frac{|\frac{3}{2}-1|^{n+1}}{(n+1)!} = \frac{n!}{|\xi|^{n+1}}\frac{1}{2^{n+1}(n+1)!} = \frac{1}{|\xi|^{n+1}}\frac{1}{2^{n+1}(n+1)},$$

where $\xi \in (1, \frac{3}{2})$. Thus, we have the bound:

$$|R_n(\tfrac{3}{2})| \leq \frac{1}{2^{n+1}(n+1)},$$

and we want this quantity to be smaller than 0.0001. Therefore, we need to solve the inequality $\frac{1}{2^{n+1}(n+1)} < 0.0001$ or equivalently $10^4 < 2^{n+1}(n+1)$. We can then solve this by trial and error. To wit:

$$2^4 5^4 = 10^4 < 2^{n+1}(n+1) \quad \Leftrightarrow \quad 5^4 < 2^{n-3}(n+1).$$

But since $5 > 4 = 2^2$, we necessarily need $2^8 < 2^{n-3}(n+1) \Leftrightarrow 2^{11-n} < n+1$ which implies $n \geq 8$. However, we can check that $n = 8$ does not work, but $n = 9$ does! So we need the 9-th order Taylor polynomial of $\ln(x)$ expanded around $x = 1$ to approximate $\ln(\frac{3}{2})$ correct to 4 decimal places.

Exercises

17.1 Find the Taylor series of the functions $f, g : (-\infty, 1) \to \mathbb{R}$ defined as $f(x) = \cos(x)$ and $g(x) = \ln(1-x)$ centred at $x = 0$.
Hence, find the fourth order Taylor polynomial for the function $f(x)g(x) = \cos(x)\ln(1-x)$.

17.2 (∗) Find the Taylor series and its radius of convergence of the following functions.
(a) $f : \mathbb{R} \setminus \{0\} \to \mathbb{R}$ defined as $f(x) = \frac{1}{x^2}$ centred at $x = -1$.
(b) $f : \mathbb{R}_+ \to \mathbb{R}$ defined as $f(x) = \ln(x)$ centred at $x = 2$.
(c) $f : \mathbb{R} \to \mathbb{R}$ defined as $f(x) = x^3$ centred at $x = 2$.

17.3 Let $f : [-1, 1] \to \mathbb{R}$ be defined as $f(x) = \arcsin(x)$.
(a) Show that $(1-x^2)f'' - xf' = 0$ for $x \in (-1, 1)$.
(b) Hence, by using the Leibniz rule, show that for any $n \in \mathbb{N}$ and $x \in (-1, 1)$ we have:

$$(1-x^2)f^{(n+2)} - (2n+1)xf^{(n+1)} - n^2 f^{(n)} = 0.$$

(c) Show that $f^{(2n)}(0) = 0$ and $f^{(2n+1)}(0) = \frac{(2n)!}{4^n}\binom{2n}{n}$ for all $n \in \mathbb{N}_0$.
(d) Deduce the Maclaurin series of f.
(e) Find the radius of convergence of the series in part (d).
(f) Show that this series also converges at $x = \pm 1$.

17.4 Suppose that a function $f : \mathbb{R} \to \mathbb{R}$ has a power series centred at $c \in \mathbb{R}$ with positive radius of convergence $R > 0$. Prove that this is the only power series representation for f centred at c.

17.5 (∗) Prove the Cauchy remainder theorem in Theorem 17.2.5, namely:
Let $f : \mathbb{R} \to \mathbb{R}$ be an $(n+1)$-differentiable function such that $f^{(n+1)}$ is continuous on the closed interval between the centre of expansion c and x. Show that the n-th Taylor remainder at x is given by:

$$R_n(x) = \frac{f^{(n+1)}(\eta)}{n!}(x-\eta)^n(x-c),$$

for some real number η between c and x.

17.6 (∗) Suppose that $f : \mathbb{R} \to \mathbb{R}$ is a smooth function and there exists an $M > 0$ such that $|f^{(n)}(x)| \leq M$ for all $n \in \mathbb{N}_0$ and $x \in \mathbb{R}$. This is called uniform estimate on the derivatives of f. Prove that for any $c \in \mathbb{R}$ we have:

$$f(x) = \sum_{j=0}^{\infty} \frac{f^{(j)}(c)}{j!}(x-c)^j \quad \text{for all } x \in \mathbb{R}.$$

Therefore, any such function is an entire function. This is actually a really amazing fact! It says that if a smooth function has a uniform estimate on its derivatives, then all of its derivatives at any fixed point $c \in \mathbb{R}$ determines the function globally via the Taylor series. Examples of this phenomenon are the sine and cosine functions which have uniformly bounded derivatives and hence are entire functions.

17.7 Recall the bump function $\Psi : \mathbb{R} \to \mathbb{R}$ which is defined as:

$$\Psi(x) = \begin{cases} e^{-\frac{1}{1-x^2}} & \text{for } -1 < x < 1, \\ 0 & \text{otherwise,} \end{cases}$$

from Exercise 14.19. We have shown that the function Ψ is smooth. Consider the real sequence (a_n) where $a_n = \sup_{x\in\mathbb{R}} |\Psi^{(n)}(x)|$.
(a) Show that this is a well-defined sequence of real numbers.
(b) Using Exercise 17.6, prove that (a_n) is an unbounded sequence.

17.8 (∗) We have seen that $\sin(x) = S(x)$ globally in Example 17.2.7(1). Using the same idea, show that $\cos(x) = C(x)$ globally. This then confirms the hypothesis that we made in Exercises 11.3–11.6, and 13.11.
These power series are usually used as the analytic definitions of the sine and cosine functions in contrast to the geometric definitions that we have been using.

17.9 Using power series, show Euler's identity $e^{ix} = \cos(x) + i\sin(x)$ for $x \in \mathbb{R}$.

17.10 We have seen that $\lim_{x\to 0} \frac{\sin(x)}{x} = 1$ via geometrical argument in Example 13.1.9(6). Prove this instead by using power series.

17.11 (a) Using power series, evaluate the Fresnel integrals:

$$\int_0^x \sin(t^2)\,dt \quad \text{and} \quad \int_0^x \cos(t^2)\,dt.$$

(b) Show that both resulting power series converge for all $x \in \mathbb{R}$.
In fact, using tools from complex analysis, we can also deduce the values of following improper integrals $\int_0^\infty \sin(t^2)\,dt = \int_0^\infty \cos(t^2)\,dt = \frac{\sqrt{2\pi}}{4}$.

17.12 (∗) Recall that we have found a Taylor series expansion of $\ln(x+1)$ with radius of convergence $R = 1$ in Example 17.2.7(2). By carefully using this Taylor series, show that $\ln(2) = \sum_{j=1}^\infty \frac{(-1)^{j-1}}{j}$.

17.13 (∗) Recall the generalised binomial coefficient from Exercise 12.10 where for $r \in \mathbb{R}$, we define:

$$\binom{r}{j} = \prod_{k=1}^{j} \frac{r-k+1}{k} = \frac{r(r-1)(r-2)\dots(r-j+1)}{j!}.$$

We have computed the Taylor series for $(1+x)^r$ defined for $x \in \mathbb{R} \setminus \{-1\}$ in Example 17.1.4(3). It is given as:

$$(1+x)^r \sim \sum_{j=0}^{\infty} \binom{r}{j} x^j. \tag{17.9}$$

We have seen the power series on the RHS in Exercise 12.10 where we investigated its convergence for various different values of $r \in \mathbb{R}$. We now want to determine for which x are the two expressions in (17.9) equal.

(a) Suppose that $|q| < 1$ is a constant. Show that $\lim_{n\to\infty} q^n n\binom{r}{n} = 0$.

(b) Fix $|x| < 1$ and suppose that t is any number between 0 and x. Show that $\frac{|x-t|}{|1+t|} \leq q$ for some constant $0 < q < 1$.

(c) Hence, by looking at the integral form of the remainder $R_n(x)$, show that the function $(1+x)^r$ equals to the series (17.9) for $|x| < 1$.

Thus, the $\sim$ symbol in (17.9) can be safely replaced with $=$ for $|x| < 1$. This generalises the binomial theorem to exponents which are not natural numbers. Here we have proven the equality only for $|x| < 1$. How about for other values of x?

(d) Show that the equality is also true at $x = \pm 1$ if $r \geq 0$.

(e) Hence, show that for any $r \geq 0$ we have:

$$\sum_{j=0}^{\infty} \binom{r}{j} = 2^r \quad \text{and} \quad \sum_{j=0}^{\infty} (-1)^j \binom{r}{j} = 0.$$

17.14 Show that for $r > 0$ and $|x| < 1$ we have the equality:

$$\frac{1}{(1-x)^r} = \sum_{j=0}^{\infty} \frac{\Gamma(j+r)}{j!\Gamma(r)} x^j,$$

where Γ is the gamma function from Exercise 16.20.

17.15 Using Exercise 17.13, for any $p, q \in \mathbb{R}$ and $n \in \mathbb{N}$ show that the generalised binomial coefficients satisfy:

$$\binom{p+q}{n} = \sum_{j=0}^{n} \binom{p}{j}\binom{q}{n-j}.$$

17.16 (a) Using Exercise 17.13, find the Maclaurin series of the function $f : (-1, 1) \to \mathbb{R}$ defined as $f(x) = \frac{1}{\sqrt{1-x^2}}$.

(b) By using a carefully justified Riemann integration, show that the Maclaurin series in Exercise 17.3(d) is exactly equal to $\arcsin(x)$ for $|x| < 1$.

(c) Finally, by using Abel's theorem, show that the Maclaurin series in part (b) is exactly equal to $\arcsin(x)$ for $|x| \leq 1$.

17.17 We have seen some smooth functions whose Taylor series at some point c converges only within a radius so that this domain of convergence avoids any singularity of the original function. For example, the functions $\ln(1+x)$ and the $(1+x)^r$ from Exercise 17.13 defined on $(-1, \infty)$ and $\mathbb{R} \setminus \{-1\}$ respectively both have Taylor series centred at $x = 0$ that converge only for $|x| < 1$ to avoid the singularity of the functions at $x = -1$.

Find an example of a smooth function which is defined on the whole of $\mathbb{R}$ but its Maclaurin series only converges for $|x| < 1$.

17.18 Let $f : \mathbb{R} \to \mathbb{R}$ be a smooth function.

(a) Suppose that f is an even function. Show by induction that $f^{(j)}$ are odd functions for all odd $j \in \mathbb{N}$.

Hence, show that the Maclaurin series of f is of the form:

$$f(x) \sim \sum_{j=0}^{\infty} \frac{f^{(2j)}(0)}{(2j)!} x^{2j}.$$

(b) Similarly prove that if f is an odd function, then its Maclaurin series is of the form:

$$f(x) \sim \sum_{j=0}^{\infty} \frac{f^{(2j+1)}(0)}{(2j+1)!} x^{2j+1}.$$

17.19 (∗) Recall the Gaussian error function defined in Example 16.3.5(2) as the Riemann integral function $\text{erf} : \mathbb{R} \to \mathbb{R}$ as $\text{erf}(x) = \frac{2}{\sqrt{\pi}} \int_0^x e^{-t^2}\, dt$. We do not have an explicit form of this function. However, we can express it globally as a power series.

(a) Show that the Maclaurin series for this function is:

$$\text{erf}(x) = \frac{2}{\sqrt{\pi}} \sum_{j=0}^{\infty} \frac{(-1)^j}{j!(2j+1)} x^{2j+1}.$$

(b) Find a numerical value of erf(1) correct to 3 decimal places. Use a computer program to help you.

17.20 (∗) During the computation to obtain the circumference of an ellipse of eccentricity $0 \le \epsilon < 1$ in Exercise 16.13, we ran into the following integral:

$$E(\epsilon) = \int_0^{\frac{\pi}{2}} \sqrt{1 - \epsilon^2 \sin^2(\theta)}\, d\theta = \int_0^1 \sqrt{\frac{1 - \epsilon^2 t^2}{1 - t^2}}\, dt.$$

This integral is called a complete elliptic integral of the second kind and cannot be expressed exactly. However, we can express it as a real series. Show that:

$$E(\epsilon) = \frac{\pi}{2} \sum_{j=0}^{\infty} \left(\prod_{k=1}^{j} \frac{2k-1}{2k} \right)^2 \frac{\epsilon^{2j}}{1 - 2j} \quad \text{for all } 0 \le \epsilon < 1.$$

17.21 Likewise, for $0 \le \epsilon < 1$, we define a complete elliptic integral of the first kind as:

$$K(\epsilon) = \int_0^{\frac{\pi}{2}} \frac{1}{\sqrt{1 - \epsilon^2 \sin^2(\theta)}}\, d\theta = \int_0^1 \frac{1}{\sqrt{(1 - t^2)(1 - \epsilon^2 t^2)}}\, dt.$$

Show that:

$$K(\epsilon) = \frac{\pi}{2} \sum_{j=0}^{\infty} \left(\prod_{k=1}^{j} \frac{2k-1}{2k} \right)^2 \epsilon^{2j} \quad \text{for all } 0 \le \epsilon < 1.$$

17.22 (∗) In this question, we are revisiting the Basel problem in Exercise 8.12 using power series.

(a) Show that:

$$\int_0^1 \frac{x^{2n+1}}{\sqrt{1 - x^2}}\, dx = \frac{4^n (n!)^2}{(2n+1)!}.$$

(b) Using the Maclaurin series of $\arcsin(x)$ in Exercise 17.3(d) and improper integral, prove that:

$$\int_0^1 \frac{\arcsin(x)}{\sqrt{1-x^2}}\,dx = \sum_{j=0}^{\infty} \frac{1}{(2j+1)^2}.$$

Carefully justify the switching of all the sums, integrals, and limits.

(c) On the other hand, use the FTC to show that $\int_0^1 \frac{\arcsin(x)}{\sqrt{1-x^2}}\,dx = \frac{\pi^2}{8}$.

(d) Deduce that $\sum_{j=0}^{\infty} \frac{1}{(2j+1)^2} = \frac{\pi^2}{8}$.

(e) Hence, show that $\sum_{j=1}^{\infty} \frac{1}{j^2} = \frac{\pi^2}{6}$.

17.23 Using power series, find the limit $\lim_{x\to 0} \frac{1-\cos(x)}{1+x-e^x}$. Carefully justify all the steps.

17.24 Find the second order Taylor polynomial for the following functions. Hence find the maximum error when approximating the functions with these second order polynomials within the specified intervals.

(a) $f(x) = \frac{1}{x}$ centred at $x = 1$ in the interval $(0.9, 1.1)$.

(b) $f(x) = \sec(x)$ centred at $x = 0$ in the interval $(-0.2, 0.2)$.

(c) $f(x) = \ln(1+2x)$ centred at $x = 0$ in the interval $(0.5, 1.5)$.

17.25 How many terms of the Taylor series expansion of $f(x) = \ln(1+x)$ centred at $x = 0$ do we need to estimate the value of $\ln(1.4)$ with an error of less than 0.01?

17.26 For what range of values of $x \in \mathbb{R}$ can we replace the sine function $\sin(x)$ for $x \in \mathbb{R}$ by the polynomial $x - \frac{x^3}{3!}$ with an error of size no greater than 10^{-4}?

17.27 (∗) Suppose that for $|t| < 1$ we have the expression:

$$\frac{1}{\sqrt{1-2xt+x^2}} = \sum_{j=0}^{\infty} P_j(t)x^j \quad \text{for } |x| < 1,$$

where $P_j : (-1, 1) \to \mathbb{R}$ are some functions.

(a) Show that $P_0(t) = 1$ and $P_1(t) = t$.

(b) Show that P_n for $n \in \mathbb{N}$ satisfies the following recursive relationship:

$$(n+1)P_{n+1} = (2n+1)tP_n - nP_{n-1}.$$

(c) Deduce that each P_n is an n-th degree polynomial of t.
Show that P_n is an odd function if n is odd and an even function if n is even.

(d) Show that $(n+1)P'_{n+1} + nP'_{n-1} = (2n+1)(P_n + tP'_n)$ for any $n \in \mathbb{N}$ and $|t| < 1$.

(e) Using the power series, prove that that $P'_{n+1} + P'_{n-1} = 2tP'_n + P_n$ for all $n \in \mathbb{N}$ and $|t| < 1$.

(f) Using parts (d) and (e), show that for all $n \in \mathbb{N}$ and $|t| < 1$ that P_n satisfy the following second order linear ODE on $|t| < 1$:

$$(1 - t^2)P_n'' - 2tP_n' + n(n+1)P_n = 0.$$

(g) Let us introduce a differential operator D defined as $D(P_n) = \frac{d}{dt}((1 - t^2)P_n') + n(n+1)P_n$. Using this operator, we can rewrite the ODE in part (f) succinctly as $D(P_n) = 0$. By considering the equation $P_m D(P_n) - P_n D(P_m) = 0$, show that:

$$\frac{d}{dt}((1 - t)^2(P_m' P_n - P_n' P_m)) + (m - n)(m + n + 1)P_m P_n = 0,$$

for any $m, n \in \mathbb{N}$ and $|t| < 1$.

(h) Note that the polynomials P_n are only defined on the interval $(-1, 1)$. Explain why for any $m, n \in \mathbb{N}$, the improper Riemann integral $\int_{-1}^{1} P_m(t)P_n(t)\,dt$ exists.

(i) Show that for $m \neq n$ these polynomials are orthogonal, namely:

$$\int_{-1}^{1} P_m(t)P_n(t)\,dt = 0.$$

(j) Finally, for each $n \in \mathbb{N}$, show that:

$$\int_{-1}^{1} P_n(t)^2\,dt = \frac{2}{2n+1}.$$

This collection of polynomials are called the Legendre polynomials, named after Legendre. They have been studied widely on their own or as an application in some applied mathematics problems.

17.28 For $n \in \mathbb{N}_0$, let $f_n : [-1, 1] \to \mathbb{R}$ be defined as $f_n(x) = \frac{1}{2^n n!}\frac{d^n}{dx^n}(x^2 - 1)^n$. We have seen that this polynomial has n distinct roots all within the interval $(-1, 1)$ in Exercise 13.32.

(a) Show that f_n is an n-th degree polynomial with $f_0(x) = 1$ and $f_1(x) = x$.

(b) Furthermore, show that f_n satisfies the Legendre polynomial recurrence relationship in Exercise 17.27(b) and hence the Legendre polynomials can be written explicitly as $P_n = f_n$.

The functions f_n are called the Rodrigues' formula for the Legendre polynomials P_n. They were independently introduced by Olinde Rodrigues (1795–1851), James Ivory (1765–1842), and Carl Gustav Jacobi (1804–1851).

Introduction to Measure 18

I wanna know the measure from here to forever. I wanna feel the pressure of god or whatever.

— Ben Gibbard, musician

The Riemann, Darboux, and Riemann-Stieltjes integrals that we have seen in Chap. 14 were constructed by partitioning the compact domain $[a, b]$ into smaller subintervals, building an approximating step function, and approximating the area of the subgraph for this function using rectangles. In more details, we approximate a function $f : [a, b] \to \mathbb{R}$ by using a partition $\mathcal{P} = \{x_0, x_1, \ldots, x_n\}$ and a step function of the form $\phi_{\mathcal{P}} : [a, b] \to \mathbb{R}$ defined as $\phi_{\mathcal{P}}(x) = \sum_{j=1}^{n} c_j \mathbf{1}_{I'_j}(x)$ where $I'_j = (x_{j-1}, x_j]$ and some specially chosen constants $c_j \in \mathbb{R}$.

We could carry out this construction since we have declared the size of the partition intervals $(c, d]$ with either the standard length $|(c, d]| = d - c$ in the Riemann integral or using some monotone function $g : [a, b] \to \mathbb{R}$ to declare the size of the interval $(c, d]$ as $|(c, d]| = |g(d) - g(c)|$ in the Riemann-Stieltjes integral that we saw in Exercises 15.24 to 15.27.

An alternative construction for integration was proposed by Henri Lebesgue in his 1901 paper *Sur une Généralisation de l'intégrale Définie* (On a Generalisation of the Definite Integral). In this proposal, instead of approximating the function by chopping up the domain $[a, b]$ into smaller subintervals, we divide the range of the function and form a partition $\mathcal{Q} = \{y_0, y_1, \ldots y_n\}$ in the codomain. We then find the preimage of these partition subintervals, namely the sets:

$$\begin{aligned} E_j &= f^{-1}([y_{j-1}, y_j)) \\ &= \{x \in [a, b] : x = f^{-1}(y) \text{ for } y \in [y_{j-1}, y_j)\} \subseteq [a, b], \end{aligned} \tag{18.1}$$

S. Johar, *The Big Book of Real Analysis*,
https://doi.org/10.1007/978-3-031-30832-1_18

for $j = 1, 2, \ldots, n$. However, these preimage sets E_j might not be half-closed intervals as in the construction by Riemann. It could well be any subset of $[a, b]$ at all!

Using these partitions, in the same vein as the construction by Riemann, we approximate the function using a family of functions which we call simple functions. Simple functions adapted to the partition $\mathcal{Q}$ in the range is a function $\phi_{\mathcal{Q}} : [a, b] \to \mathbb{R}$ which is analogous to the step functions. The simple function approximating f adapted to the partition $\mathcal{Q}$ can be defined by using the sets in (18.1). It is given by:

$$\phi_{\mathcal{Q}}(x) = \sum_{j=1}^{n} y_{j-1} \mathbf{1}_{E_j}(x).$$

Assuming (and this is a major assumption) that we can assign "sizes" to these sets E_j just like the half-closed intervals, the approximating area of the subgraph of f by the simple function $\phi_{\mathcal{Q}}$, which we call an integral $I(\phi_{\mathcal{Q}})$, can then be defined as the sum:

$$I(\phi_{\mathcal{Q}}) = \sum_{j=1}^{n} y_{j-1} |E_j|,$$

where $|E_j|$ denotes the "size" of the set E_j. These sizes are called measure and the sets E_j which we can measure are called measurable sets.

Once we add more points in the partition $\mathcal{Q}$ of the range, if the sequence of integrals $I(\phi_{\mathcal{Q}})$ converges, we would like to call this limit the integral of f.

Example 18.0.1 We note that the function in Fig. 18.1 might be misleadingly easy and the readers might be baffled: why would this make any difference to the Riemann integral that we have defined in Chap. 15? It looks exactly the same since the shaded regions are still rectangles and we know what their width/sizes are.

As a motivating example for more complicated functions, consider the notorious Dirichlet function $f : [0, 1] \to \mathbb{R}$ which takes the value of 1 if $x \in \mathbb{Q}$ and 0 if $x \in \bar{\mathbb{Q}}$. This function is not Riemann integrable as we have seen in Example 15.3.10(3). We cannot even plot this function in a satisfactory manner as we did for the function in Fig. 18.1. Let us try and define the integral of this function using this new proposed method.

The range of this function are just the points $\{0, 1\}$. We can consider a partition $\{y_0 = 0, y_1 = 1, y_2 = 2\}$ in the codomain. Then, the preimage sets would be:

$$E_1 = \{x \in [0, 1] : x = f^{-1}([0, 1))\} = \bar{\mathbb{Q}} \cap [0, 1],$$

$$E_2 = \{x \in [0, 1] : x = f^{-1}([1, 2))\} = \mathbb{Q} \cap [0, 1].$$

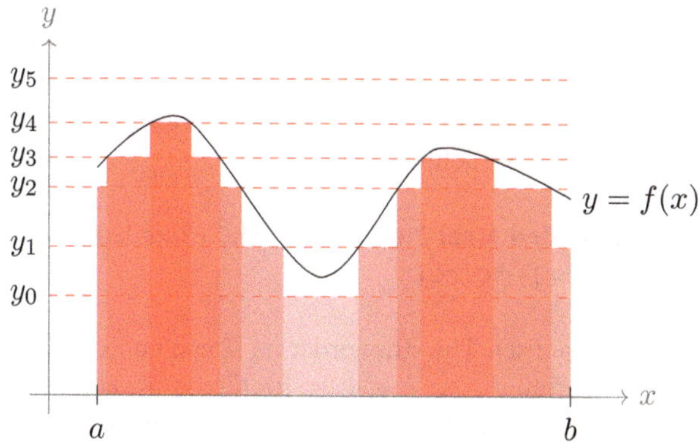

Fig. 18.1 Proposed new integral for a non-negative function $f : [a, b] \to \mathbb{R}$. We partition the codomain $\mathbb{R}_{\geq 0}$ with the points $\mathcal{Q} = \{y_0, \ldots, y_5\}$. For each subinterval $[y_{j-1}, y_j)$ we find its preimage set E_j in the domain. The total area of the regions with the same shade of red is $y_{j-1}|E_j|$. The total area of all the shaded regions is $I(\phi_{\mathcal{Q}})$

Hence, we can approximate the function f with a simple function $\phi : [0, 1] \to \mathbb{R}$ defined as $\phi(x) = \sum_{j=1}^{2} y_{j-1}\mathbf{1}_{E_j}(x) = \mathbf{1}_{E_2}(x)$. If we can assign a size to the set $E_2 = \mathbb{Q} \cap [0, 1]$, we can then approximate the area under the graph of f with an integral $I(\phi) = |E_2| = |\mathbb{Q} \cap [0, 1]|$.

However, the set E_2 is not an interval of the form $(a, b]$ that we have assigned sizes to as in the construction of the Riemann or Riemann-Stieltjes integrals. So, we need to enlarge the class of sets in $[0, 1]$ that we can assign sizes to and define their sizes properly. This is where measure theory comes in.

This chapter will be devoted to the foundations of measure theory before we continue with this integral construction in Chap. 19.

Measure theory is a rather recent branch of mathematical study. It was formally set up around early twentieth century by Borel, Lebesgue, Constantin Carathéodory (1873–1950), Maurice Fréchet (1878–1973), Nikolai Luzin (1883–1950), Johann Radon (1887–1956), and many others as a way to axiomatise how we can assign sizes to some mathematical and geometrical objects. In Exercises 18.31–18.33, we shall see that it was also adapted in the study of modern probability theory in a very natural way.

18.1 Extended Real Numbers

Before we set off towards this main goal of determining sets which we can assign sizes to and their sizes, we would like to declare some notations and conventions. So far, we have been using the set of real numbers $\mathbb{R}$ that we have constructed in Chap. 3.

We are going to extend the set of real numbers to include two new terms for convenience. This follows from the concept of actual infinity proposed by Aristotle and the concept of cardinality of sets in which the sizes of sets may be an actual infinity rather than a potential infinity. We want these infinity to interact with the real numbers algebraically. We define:

Definition 18.1.1 (Extended Real Numbers) The extended real numbers is the set $\bar{\mathbb{R}} = \mathbb{R} \cup \{-\infty, \infty\} = [-\infty, \infty]$.

This set is a totally ordered set. The strict order on $\mathbb{R}$ can be extended to include the two new quantities by defining $-\infty < a < \infty$ for all $a \in \mathbb{R}$. However, the extended real numbers is not a field. To see this, let us define the extension of the addition and multiplication on $\mathbb{R}$ to include the new quantities $\pm\infty$. For any $a \in \mathbb{R}$ we declare:

$$\begin{aligned}
a \pm \infty &= \pm\infty + a = \pm\infty, \\
\pm\infty + \pm\infty &= \pm\infty, \\
a \times \pm\infty &= \pm\infty \times a = \pm\infty \text{ if } a > 0, \\
a \times \pm\infty &= \pm\infty \times a = \mp\infty \text{ if } a < 0, \\
\frac{a}{\pm\infty} &= 0, \\
\frac{\pm\infty}{a} &= \pm\infty \text{ if } a > 0, \\
\frac{\pm\infty}{a} &= \mp\infty \text{ if } a < 0, \\
0 \times \pm\infty &= \pm\infty \times 0 = 0.
\end{aligned}$$

The rules above, except for the final one, are defined to be consistent with the algebra of limits for quantities blowing up to $\pm\infty$ that we have seen in Chap. 5. To ensure consistency, some operations, such as $\infty - \infty$, $-\infty + \infty$, and $\frac{\pm\infty}{\pm\infty}$ which we have called the indeterminate forms, are left undefined in this algebraic extension. As a result, the operations $+$ and $\times$ cannot be defined for some pairs of elements in the extended real numbers. Hence the set $\bar{\mathbb{R}}$ is not a field.

However, the final operation that we have declared above, namely $0 \times \pm\infty = \pm\infty \times 0 = 0$ which were also one of the indeterminate forms in Chap. 5, are exceptions. These are needed in defining certain quantities, such as the sum of countably infinitely many 0s, in measure theory. Indeed, this allows us to define $0 = \sum_{j=1}^{\infty} 0 = \infty \times 0$ in a natural way.

Example 18.1.2 This extension allows us to define functions with images in $\bar{\mathbb{R}}$. For example, the function $f : \mathbb{R} \setminus \{0\} \to \mathbb{R}$ with images in the real numbers defined as $f(x) = \frac{1}{x^2}$ could not be defined at $x = 0$. However, we can enlarge the domain and codomain to the extended real numbers $\bar{\mathbb{R}}$ to define an extended function $\bar{f} : \bar{\mathbb{R}} \to \bar{\mathbb{R}}$ as $\bar{f}(x) = \frac{1}{x^2}$. According to the rules on the extended real numbers, this new function then satisfies $\bar{f}(x) = f(x)$ when $x \in \mathbb{R} \setminus \{0\} \subseteq \bar{\mathbb{R}}$, $\bar{f}(\infty) = \bar{f}(-\infty) = 0$, and $\bar{f}(0) = \infty$.

18.2 π-Systems and Semirings

In this section, we interweave the discussion on the sets in $\mathbb{R}$ with definitions and results on more general universe sets X. This would give us a clear motivation on why the definitions and results are needed. Moreover, it would provide a foundation for more general results that we shall see later in Chap. 20.

As a motivating question, we first think about what class of sets in $\mathbb{R}$ that we want to assign sizes to. Of course, the largest possible collection of subsets in $\mathbb{R}$ is the power set $\mathcal{P}(\mathbb{R})$, so this would be the ultimate end goal, if it is possible. A measure should be defined on a collection of sets of our domain $\mathbb{R}$ and the measure of any set should be non-negative, but possibly ∞. This is done by requiring the measure to satisfy certain reasonable properties which we shall specify in Definition 18.5.1.

π-Systems

To reach this desired goal, let us work with what we know from the construction of Riemann integral. We start with the following special collection of subsets:

Definition 18.2.1 (π-System) Let X be a set. A non-empty collection $\mathcal{S} \subseteq \mathcal{P}(X)$ of subsets of X is called a π-system on X if:

1. $\mathcal{S}$ is non-empty, and
2. $\mathcal{S}$ is closed under finite intersection. Namely, for any $E, F \in \mathcal{S}$ we have $E \cap F \in \mathcal{S}$.

Example 18.2.2 Let us look at some examples of π-systems.

1. We have seen an example of a π-system during the construction of Riemann and Darboux integrals. The collection of sets $\mathcal{J} = \{(c, d] \subseteq [a, b] : a \leq c < d \leq b\} \cup \{\varnothing\}$ forms a π-system in $X = [a, b]$. Indeed, for any two elements in $\mathcal{J}$, they are either disjoint or intersect somewhere. For the former, their intersection would be $\varnothing \in \mathcal{J}$ and for the latter, their intersection would also be a half-open finite interval, which is also in $\mathcal{J}$.
2. Instead of looking within the compact domain $X = [a, b]$ as we had to for Riemann and Darboux integrals, let us generalise the universe X to the whole

of $\mathbb{R}$, namely we consider the sets $\mathcal{J} = \{(c, d] \subseteq \mathbb{R} : c < d\} \cup \{\varnothing\}$. Similar to the previous example, we can easily check that this collection of sets is closed under finite intersections. Thus, this is a π-system as well.

We can assign sizes to the sets in a π-system by defining the size of a set via a set function which we call a content. A content function must satisfy the following:

Definition 18.2.3 (Content) Let X be a set and $\mathcal{S} \subseteq \mathcal{P}(X)$ be any collection of subsets of X which contains $\varnothing$. A content is a set function $m : \mathcal{S} \to [0, \infty]$ such that:

1. $m(\varnothing) = 0$,
2. for any $E \in \mathcal{S}$ we have $m(E) \geq 0$, and
3. if $E, F \in \mathcal{S}$ are disjoint sets such that $E \cup F \in \mathcal{S}$, then $m(E \cup F) = m(E) + m(F)$.

Example 18.2.4 Recall from Example 18.2.2(2) that the collection $\mathcal{J} = \{(c, d] \subseteq \mathbb{R} : c < d\} \cup \{\varnothing\}$ is a π-system. A content $m : \mathcal{J} \to [0, \infty]$ can be defined via $m((c, d]) = d - c$ and $m(\varnothing) = 0$. We check that this m satisfies the axioms in Definition 18.2.3.

1. Clearly, $m(\varnothing) = 0$ by definition.
2. If $E = \varnothing$, then by $m(E) = 0$. Otherwise, if $E \in \mathcal{J}$ is non-empty, we have $E = (c, d]$ for some $c, d \in \mathbb{R}$ and so $m(E) = d - c > 0$.
3. If $E, F \in \mathcal{J}$ are disjoint sets such that $E \cup F \in \mathcal{J}$, then since $E \cup F = (a, b]$ for some $a < b$, necessarily $E = (a, c]$ and $F = (c, b]$ for some $c \in (a, b)$. Thus, we can compute $m(E) = c - a$, $m(F) = b - c$, and $m(E \cup F) = b - a$, satisfying the third condition.

Semirings

Moreover, for the π-system $\mathcal{J}$ in Example 18.2.2(2), if $(a, b] \subseteq (c, d]$, we can find more sets in $\mathcal{J}$ so that the superset $(c, d]$ can be expressed as a finite disjoint union of these sets and $(a, b]$. Namely $(c, d] = (c, a] \cup (a, b] \cup (b, d]$. Due to this, the collection $\mathcal{J}$ is also called a semiring, which we now define on a more general collection of sets as:

Definition 18.2.5 (Semiring of Sets) Let X be a set. A non-empty collection $\mathcal{S} \subseteq \mathcal{P}(X)$ of subsets of X is called a semiring if:

1. $\varnothing \in \mathcal{S}$,
2. if $E, F \in \mathcal{S}$, then $E \cap F \in \mathcal{S}$, and
3. if $E, E_1 \in \mathcal{S}$ are such that $E_1 \subseteq E$, then there exists a finite collection of pairwise disjoint non-empty sets $\{E_j\}_{j=2}^n$ where $E_j \in \mathcal{S}$ are all disjoint from E_1 such that $E = \bigcup_{j=1}^n E_j$.

Remark 18.2.6 We make some remarks regarding the third condition of Definition 18.2.5.

1. This condition is also equivalent to saying that for any sets $E, F \in \mathcal{S}$ (with F not necessarily fully contained in E), there are disjoint sets $\{E_j\}_{j=1}^m$ in $\mathcal{S}$ such that $E \setminus F = \bigcup_{j=1}^m E_j$. Indeed, if $E, F \in \mathcal{S}$, we have $E \cap F \in \mathcal{S}$ and $E \cap F \subseteq E$. Definition 18.2.5 says that there are sets $\{E_j\}_{j=1}^m$ disjoint from $E \cap F$ such that $E = (E \cap F) \cup \bigcup_{j=1}^m E_j$. Thus, since $E_j \cap F = \varnothing$ for $j = 1, 2, \ldots, m$, we have $E \setminus F = \bigcup_{j=1}^m E_j$.
2. Moreover, this condition can also be extended to more than one subset of E in a semiring $\mathcal{S}$. Namely, suppose that $\{E_j\}_{j=1}^n$ for some $n \in \mathbb{N}$ is a pairwise disjoint collection of sets in $\mathcal{S}$ and $E \in \mathcal{S}$ is such that $\bigcup_{j=1}^n E_j \subseteq E$. Then, we can find a finite collection of pairwise disjoint non-empty sets $\{F_k\}_{k=1}^m$ where $F_k \in \mathcal{S}$ are all disjoint from every E_j such that $E = \bigcup_{j=1}^n E_j \cup \bigcup_{k=1}^m F_k$. This is left for the readers to prove via induction in Exercise 18.1.

The definition of semiring implies the following lemma:

Lemma 18.2.7 *Let $\mathcal{S}$ be a semiring over a set X. If $E \subseteq X$ is such that $E = \bigcup_{j=1}^n E_j$ where $E_j \in \mathcal{S}$ (which are not necessarily pairwise disjoint), then E can be written as a union of disjoint sets in $\mathcal{S}$.*

Proof We can prove this via induction on n. For the base case when $n = 2$, assume that $E = E_1 \cup E_2$. If they are disjoint, then we are done. Otherwise, since $E_1, E_2 \in \mathcal{S}$, we can write the union as $E_1 \cup E_2 = (E_1 \setminus E_2) \cup E_2$. By Remark 18.2.6(1), we can then write $E_1 \setminus E_2 = \bigcup_{k=1}^p G_k$ where $\{G_k\}_{k=1}^p$ are pairwise disjoint collections of sets in $\mathcal{S}$ which are all disjoint from E_2. Thus, we have the decomposition $E = E_1 \cup E_2 = \bigcup_{k=1}^p G_k \cup E_2$.

Now suppose that this decomposition is true for $n = k$, namely the union $E = \bigcup_{j=1}^k E_j$ can be rewritten as a disjoint union of sets in $\mathcal{S}$. We now prove the case where $n = k + 1$. Let $E = \bigcup_{j=1}^{k+1} E_j$. By the inductive hypothesis, we have some disjoint sets $F_i \in \mathcal{S}$ where $i = 1, 2, \ldots, m$ such that $\bigcup_{j=1}^k E_j = \bigcup_{i=1}^m F_i$ and so $E = \bigcup_{i=1}^m F_i \cup E_{k+1} = \bigcup_{i=1}^m (F_i \setminus E_{k+1}) \cup E_{k+1}$. For each $i = 1, 2, \ldots, m$, using the base case, we can write $F_i \setminus E_{k+1} = \bigcup_{j=1}^{n_i} G_j^i$ where $\{G_j^i\}_{j=1}^{n_i}$ are pairwise disjoint sets in $\mathcal{S}$ which are all contained in F_i and do not intersect E_{k+1}. Therefore, the collection of sets $\bigcup_{i=1}^m \{G_j^i\}_{j=1}^{n_i}$ are pairwise disjoint and are all disjoint from E_{k+1}. Hence, we have the pairwise disjoint decomposition $E = \bigcup_{i=1}^m (F_i \setminus E_{k+1}) \cup E_{k+1} = \bigcup_{i=1}^m \bigcup_{j=1}^{n_i} G_j^i \cup E_{k+1}$. □

The content function m can also be endowed on a semiring with additional consequences. Most of the time, these additional properties can be obtained on a case-by-case basis of the underlying set and semiring. Going back to the π-system $\mathcal{J}$ of $\mathbb{R}$ that we started with in Example 18.2.2(2), we have the following results for the content m on it:

Lemma 18.2.8 *Let $(\mathcal{J}, \mathbb{R})$ be the semiring of half-closed intervals in $\mathbb{R}$ with the content m, namely $\mathcal{J} = \{(c, d] \subseteq \mathbb{R} : c < d\} \cup \{\varnothing\}$ with $m((c, d]) = d - c$ and $m(\varnothing) = 0$. Then, we have:*

1. *Additivity on $\mathcal{J}$: Let $E \in \mathcal{J}$ be a union of finitely many pairwise disjoint non-empty sets in $\mathcal{J}$, namely $E = \bigcup_{j=1}^{n} E_j$ where $E_j \in \mathcal{J}$. Then, $m(E) = \sum_{j=1}^{n} m(E_j)$.*
2. *Let $E \in \mathcal{J}$ be a union of finitely many (not necessarily pairwise disjoint) non-empty sets in $\mathcal{J}$, namely $E = \bigcup_{j=1}^{n} E_j$ where $E_j \in \mathcal{J}$. Then, $m(E) \leq \sum_{j=1}^{n} m(E_j)$.*
3. *Suppose that $\{E_j\}_{j=1}^{n}$ is a pairwise disjoint collection of sets $E_j \in \mathcal{S}$ and $E \in \mathcal{S}$ such that $\bigcup_{j=1}^{n} E_j \subseteq E$. Then, $\sum_{j=1}^{n} m(E_j) \leq m(E)$.*
4. *σ-additivity on $\mathcal{J}$: Let $E \in \mathcal{J}$ be a union of countably many pairwise disjoint non-empty sets in $\mathcal{J}$, namely $E = \bigcup_{j=1}^{\infty} E_j$ where $E_j \in \mathcal{J}$. Then, $m(E) = \sum_{j=1}^{\infty} m(E_j)$.*

Proof We prove the assertions one by one.

1. We prove this by induction on n. The case $n = 1$ is trivial. For the case $n = 2$, assume that $E = (a, b]$. We can write $E_1 = (a_1, b_1]$ and $E_2 = (a_2, b_2]$. WLOG, assume that $a_1 < a_2$. Since $E_1 \cap E_2 = \varnothing$ and $E_1 \cup E_2 = E = (a, b]$, necessarily $a_1 = a$, $b_1 = a_2$, and $b_2 = b$. Thus, $m(E) = |b - a| = b_2 - a_2 + a_2 - a_1 = |b_2 - a_2| + |b_1 - a_1| = m(E_1) + m(E_2)$.
   Now assume that the summation is true for $n = k$ disjoint sets. We now prove the case for $n = k + 1$ disjoint sets. Say $E = \bigcup_{j=1}^{k+1} E_j$. WLOG, we can write $E_j = (a_j, b_j]$ for all $j = 1, 2, \ldots, k + 1$ so that $a_1 < a_2 < a_3 < \ldots < a_{k+1}$. Necessarily, we have $b_{j-1} = a_j$ for all $j = 2, \ldots, k + 1$. Indeed:
   (a) If there is a j such that $b_{j-1} > a_j$, then $E_{j-1} \cap E_j \neq \varnothing$.
   (b) If there is a j such that $b_{j-1} < a_j$, then the union $\bigcup_{j=1}^{k+1} E_j$ is not in $\mathcal{J}$ since it is not an interval (as any $x \in (b_{j-1}, a_j]$ is not contained in E).
   Hence, the union of the first k sets in $\{E_j\}_{j=1}^{k+1}$ is $F = \bigcup_{j=1}^{k} E_j = \bigcup_{j=1}^{k} (a_j, b_j] = (a_1, b_k] \in \mathcal{J}$. Therefore, by applying the case $n = 2$ and using the inductive hypothesis, we get $m(E) = m(\bigcup_{j=1}^{k+1} E_j) = m(F \cup E_{k+1}) = m(F) + m(E_{k+1}) = \sum_{j=1}^{k} m(E_j) + m(E_{k+1})$, which is what we wanted to prove.
2. We prove this via induction over n as well. The case $n = 1$ is trivial. For the case $n = 2$, note that for any two sets $E_1, E_2 \in \mathcal{J}$, if $E_1 \cap E_2 \neq \varnothing$, by Lemma 18.2.7, we can decompose them as $E_1 = (E_1 \cap E_2) \cup \bigcup_{j=1}^{p} G_j$ and $E_2 = (E_1 \cap E_2) \cup \bigcup_{l=1}^{q} H_l$ where all the $G_j, H_l \in \mathcal{J}$ are pairwise disjoint. Then, applying the result from the first assertion, we have:

$$m(E_1) = m(E_1 \cap E_2) + \sum_{j=1}^{p} m(G_j) \quad \text{and} \quad m(E_2) = m(E_1 \cap E_2) + \sum_{l=1}^{q} m(H_l).$$

So, if $E = E_1 \cup E_2 \in \mathcal{J}$, we have the pairwise disjoint decomposition $E = (E_1 \cap E_2) \cup \bigcup_{j=1}^{p} G_k \cup \bigcup_{l=1}^{q} H_l$. Applying the previous assertion, we have:

$$\begin{aligned} m(E) = m(E_1 \cup E_2) &= m(E_1 \cap E_2) + \sum_{j=1}^{p} m(G_j) + \sum_{l=1}^{q} m(H_l) \\ &\leq 2m(E_1 \cap E_2) + \sum_{j=1}^{p} m(G_j) + \sum_{l=1}^{q} m(H_l) \\ &= m(E_1) + m(E_2). \end{aligned}$$

So, the case for $n = 2$ is true. For induction, we now assume that the case for $n = k$ is true. We prove the case for $n = k + 1$. Say $E = \bigcup_{j=1}^{k+1} E_j$. We can reorder the indices so that the union of the first k sets forms an interval in J. Indeed, by writing $E_j = (a_j, b_j]$ for $j = 1, \ldots, k + 1$, we have two cases:

(a) If $E_j \subseteq E_i$ for some indices $j \neq i$, we can reorder E_j to be the $(k + 1)$-th interval because the union of the remaining k intervals is $E \in \mathcal{J}$.

(b) If none of the E_j are contained fully within another E_i, WLOG, let $b_1 \leq b_2 \leq \ldots \leq b_k \leq b_{k+1}$. By assumption that there are no E_j fully contained in the other, we must have $a_k < a_{k+1}$. Thus, removing the set E_{k+1} results in $\bigcup_{j=1}^{k} E_j = (a, b_k] \in \mathcal{J}$.

We define $F = \bigcup_{j=1}^{k} E_j \in \mathcal{J}$. Applying the base case and the inductive hypothesis, we then have:

$$m(E) = m(F \cup E_{k+1}) \leq m(F) + m(E_{k+1}) \leq \sum_{j=1}^{k} m(E_j) + m(E_{k+1}) = \sum_{j=1}^{k+1} m(E_j),$$

which is what we wanted to prove.

3. By Remark 18.2.6(2), we can find a finite collection of pairwise disjoint non-empty sets $\{F_k\}_{k=1}^{m}$ where $F_k \in \mathcal{S}$ are all disjoint from every E_j such that $E = \bigcup_{j=1}^{n} E_j \cup \bigcup_{k=1}^{m} F_k$. By the first assertion, we then have $m(E) = \sum_{j=1}^{n} m(E_j) + \sum_{k=1}^{m} m(F_k) \geq \sum_{j=1}^{n} m(E_j)$.
4. Suppose that $E = \bigcup_{j=1}^{\infty} E_j$ where the E_j are pairwise disjoint non-empty sets in $\mathcal{J}$. We prove the desired equality in two steps.
   (a) For any $n \in \mathbb{N}$, we have $\bigcup_{j=1}^{n} E_j \subseteq E$. By the third assertion, we have the inequality $\sum_{j=1}^{n} m(E_j) \leq m(E)$ for all $n \in \mathbb{N}$. Thus, taking the limit as $n \to \infty$ we get $\sum_{j=1}^{\infty} m(E_j) \leq m(E)$.
   (b) Now we prove the opposite inequality. Since $E, E_j \in \mathcal{J}$, we can write $E = (a, b]$ and $E_j = (a_j, b_j]$ for each $j \in \mathbb{N}$. Define $\bar{E} = [a, b]$ so that $E \subseteq \bar{E}$. For a fixed $\varepsilon > 0$, define the open sets $E_j' = (a_j - \frac{\varepsilon}{2^{j+1}}, b_j + \frac{\varepsilon}{2^{j+1}})$ for all $j \in \mathbb{N}$. As a result, we have the inclusions $E_j \subseteq E_j'$ for all $j \in \mathbb{N}$. Note that $\mathcal{E} = \{E_j'\}_{j=1}^{\infty}$ forms an open cover for the set $\bar{E}$. Since $\bar{E}$ is a compact interval

in $\mathbb{R}$, there is a finite subcover of it from the collection $\mathcal{E}$, say $\{E'_{k_j}\}_{j=1}^n$ for some $n \in \mathbb{N}$.
For each $j = 1, 2, \dots, n$, define the set $F_{k_j} \in \mathcal{J}$ as $F_{k_j} = (a_{k_j} - \frac{\varepsilon}{2^{k_j+1}}, b_{k_j} + \frac{\varepsilon}{2^{k_j+1}}]$. Thus, we have the inclusion $E'_{k_j} \subseteq F_{k_j}$.
Write $F = \bigcup_{j=1}^n F_{k_j}$. Therefore, $E \subseteq \bar{E} \subseteq \bigcup_{j=1}^n E'_{k_j} \subseteq \bigcup_{j=1}^n F_{k_j} = F$. Moreover, by induction on n and using Lemma 4.5.4, the union $E \cup F = F$ is an interval. Thus, F must be of the form $(c, d]$ where $c = \min\{a_{k_j} - \frac{\varepsilon}{2^{k_j+1}} : j = 1, \dots, n\}$ and $d = \max\{b_{k_j} + \frac{\varepsilon}{2^{k_j+1}} : j = 1, \dots, n\}$ and so $F \in \mathcal{J}$.
We can then apply the third and second assertions to get the following inequality:

$$\begin{aligned} m(E) \le m(F) = m\left(\bigcup_{j=1}^n F_{k_j}\right) \le \sum_{j=1}^n m(F_{k_j}) &= \sum_{j=1}^n (b_{k_j} - a_{k_j} + \frac{\varepsilon}{2^{k_j}}) \\ &= \sum_{j=1}^n (b_{k_j} - a_{k_j}) + \sum_{j=1}^n \frac{\varepsilon}{2^{k_j}} \\ &< \sum_{j=1}^n m(E_{k_j}) + \varepsilon \\ &\le \sum_{j=1}^\infty m(E_j) + \varepsilon. \end{aligned}$$

Since $\varepsilon > 0$ is arbitrary, we arrive at the inequality $m(E) \le \sum_{j=1}^\infty m(E_j)$.
Thus, putting the two inequalities together, we obtain the desired equality. □

Lemma 18.2.8(3) says that the content function is monotone, namely as the set gets bigger, its content also gets bigger. From this lemma, we can see that the content function m on $\mathcal{J}$ behaves as we expected for "sizes", namely: it is monotone and it is both finitely and countably additive over disjoint sets of $\mathcal{J}$.

Nevertheless, we have a big word of caution here. The results in Lemma 18.2.8 hold assuming that the union E itself is contained in $\mathcal{J}$, which is not a very large collection of sets in $\mathbb{R}$. This assumption is necessary because the content function m was only defined on sets in $\mathcal{J}$. So any set in $\mathbb{R}$ that does not look like $(a, b]$ for some $a < b$ does not have a content.

18.3 Rings and Algebras

We now notice that the semiring is a very restrictive collection of sets since they are only closed under finite intersections. This is the source of limitation for the Riemann and Darboux integrals: we have only assigned sizes to a small collection of sets in $\mathbb{R}$ and any other sets which do not belong in this collection do not have assigned sizes.

To elaborate further, in Example 18.2.4, the semiring $\mathcal{J} = \{(c, d] \subseteq \mathbb{R} : c < d\} \cup \{\varnothing\}$ only consists of bounded half-closed intervals of the form $(a, b]$ for some $a < b$ and the empty set with a well-defined content m on these sets. As we know, not all subsets in $\mathbb{R}$ are necessarily half-closed intervals or empty. For example, any singleton set $\{a\}$ is not in $\mathcal{J}$, so we could not assign sizes to them using the content m. Additionally, any set of the form $(a, b] \cup [c, d]$ is also not in $\mathcal{J}$, so the content m also cannot be defined on these sets.

Rings and Algebras

Therefore, to include more sets in this collection, we want to generalise the semiring by allowing more operations on the collection of sets, such as complements and unions, and extend the content function on it appropriately. First, we define:

Definition 18.3.1 (Ring of Sets) Let X be a set. A non-empty collection $\mathcal{R} \subseteq \mathcal{P}(X)$ of subsets of X is called a ring on a set X if it is closed under union and set difference. Namely, for $A, B \in \mathcal{R}$, we have:

1. $A \cup B \in \mathcal{R}$, and
2. $A \setminus B \in \mathcal{R}$.

The two conditions above also imply that $\varnothing \in \mathcal{R}$ and $A \cap B = A \setminus (A \setminus B) \in \mathcal{R}$.

Remark 18.3.2 Let us make some remarks here:

1. We note that sometimes the definition for a ring is expressed as a collection of sets that is closed under symmetric difference and intersections. This is an equivalent condition to Definition 18.3.1 since if the collection is closed under $\triangle$ and $\cap$, then it will also be closed under $\cup$ and $\setminus$ because:
   (a) $A \cup B = (A \triangle B) \triangle (A \cap B)$.
   (b) $A \setminus B = A \triangle (A \cap B)$.
2. By induction, any finite union, intersection, and difference of sets in a ring $\mathcal{R}$ is also in $\mathcal{R}$.
3. We note that a ring is always a semiring. Indeed, the first two axioms of the semiring axioms follow immediately from the ring axioms. For the third semiring axiom, we note that if $E_1 \subseteq E$ where $E_1, E \in \mathcal{R}$, then $E \setminus E_1 \in \mathcal{R}$ and so the decomposition $E = E_1 \cup (E \setminus E_1)$ fulfills the third semiring axiom.

If the universe X is also contained in the collection $\mathcal{R}$, the ring is called an algebra:

Definition 18.3.3 (Algebra of Sets) Let X be a set. A non-empty collection $\mathcal{R} \subseteq \mathcal{P}(X)$ of subsets of X is called an algebra on X if it is a ring and contains X.

As a result, an algebra of sets is also a ring of sets which is closed under complements.

In Remark 18.3.2(3), we have proven that a ring is always a semiring. On the other hand, a semiring is not necessarily a ring. Notice that the collection $\mathcal{J}$ that we defined in Example 18.2.4 is not a ring because it is not closed under unions. Indeed, if $(a, b], (c, d] \in \mathcal{J}$, we have $(a, b] \cup (c, d] \notin \mathcal{J}$ unless they intersect.

Thus, a ring on X is a more general collection of sets than a π-system or a semiring on X. But how can we extend our π-systems or semirings to rings? Before we prove that we can always do that, let us state a lemma which the readers will prove in Exercise 18.2.

Lemma 18.3.4 *Let X be a set and $\{\mathcal{R}_j\}_{j\in J}$ be a collection of rings on X where J is some indexing set. Then, the intersection $\mathcal{R} = \bigcap_{j\in J} \mathcal{R}_j$ is also a ring on X.*

Now we prove an important result.

Lemma 18.3.5 (Ring Generated by a Collection of Subsets) *Let $S \subseteq \mathcal{P}(X)$ be an arbitrary non-empty collection of subsets of X. Then, there exists a unique ring $\mathcal{R}(S)$ containing S that is contained in every ring $\mathcal{R}$ containing S. In other words, $\mathcal{R}(S)$ is the smallest ring containing S. We call $\mathcal{R}(S)$ the ring generated by S.*

Proof We prove the existence and uniqueness of such a ring.

1. Existence: Clearly a power set $\mathcal{P}(X)$ is closed under union and set difference, so it is a ring on X as well. Moreover, $S \subseteq \mathcal{P}(X)$. So the collection of rings $C = \{\mathcal{R} : S \subseteq \mathcal{R}\}$ is non-empty and hence we can define $\mathcal{R}(S) = \cap_{\mathcal{R}\in C}\mathcal{R}$. By Lemma 18.3.4, this intersection is also a ring on X and it also contains S. Moreover, by construction, $\mathcal{R}(S)$ is contained in any other ring that contains S.
2. Uniqueness: Suppose that there are two such rings, namely $\mathcal{R}(S)$ and $\mathcal{R}(S)'$. Then, by construction, $\mathcal{R}(S)$ is contained in every ring that contains S and so $\mathcal{R}(S) \subseteq \mathcal{R}(S)'$. Likewise, we have the inclusion $\mathcal{R}(S)' \subseteq \mathcal{R}(S)$. Putting these two facts together gives us $\mathcal{R}(S) = \mathcal{R}(S)'$. □

Lemma 18.3.5 tells us that such a minimal ring that contains S exists, but how does any of the elements in this ring look like? Here we give a characterisation of a ring generated by an arbitrary collection of subsets S.

Proposition 18.3.6 *Let $S \subseteq \mathcal{P}(X)$ be a collection of subsets of X and $\mathcal{R}(S)$ is the ring generated by it. Then, any element $E \in \mathcal{R}(S)$ can be expressed as the finite* *symmetric difference* $E = E_1 \Delta E_2 \Delta \ldots \Delta E_n$ *for some* $n \in \mathbb{N}$ *where each* E_j *is a finite intersection of elements in* S*, say* $E_j = \bigcap_{k=1}^{n_j} F_k^j$ *where* $F_k^j \in S$.

The proof of this is left for the readers as Exercise 18.3. Moreover, if the collection of sets S that we wish to generate a ring from is already a semiring, the characterisation of sets in $\mathcal{R}(S)$ is so much simpler than the one in Proposition 18.3.6, namely:

Proposition 18.3.7 *Let* $\mathcal{S}$ *be a semiring on* X *and* $\mathcal{R}(\mathcal{S})$ *be the ring generated by* $\mathcal{S}$*. Then, any element* $E \in \mathcal{R}(\mathcal{S})$ *can be written as a finite union of pairwise disjoint sets in* $\mathcal{S}$.

Proof Define the collection $\mathcal{E} = \{E_1 \cup E_2 \cup \ldots \cup E_n : n \in \mathbb{N}, E_j \in \mathcal{S}\}$. Clearly $\mathcal{S} \subseteq \mathcal{E}$ since the elements in $\mathcal{S}$ correspond to the elements in $\mathcal{E}$ with $n = 1$. Moreover, since $\mathcal{R}(\mathcal{S})$ contains any element of $\mathcal{S}$ and is closed under finite unions of elements of $\mathcal{S}$, we have $\mathcal{E} \subseteq \mathcal{R}(\mathcal{S})$.

Next, we show that $\mathcal{E}$ is a ring. Pick any two elements $E, F \in \mathcal{E}$. By definition of $\mathcal{E}$, these elements can be expressed as $E = E_1 \cup \ldots \cup E_n$ and $F = F_1 \cup \ldots \cup F_m$ for some sets $E_j, F_i \in \mathcal{S}$ where $j = 1, 2, \ldots, n$ and $i = 1, 2, \ldots, m$.

1. Their union is $E \cup F = E_1 \cup \ldots \cup E_n \cup F_1 \cup \ldots \cup F_m$ which is a finite union of elements in $\mathcal{S}$. So, $E \cup F \in \mathcal{E}$.
2. We show that $E \setminus F \in \mathcal{E}$ by induction on the number of sets in the decomposition of F, namely m. For the base case $m = 1$, we have:

$$E \setminus F = (E_1 \cup \ldots \cup E_n) \setminus F_1 = \bigcup_{j=1}^{n} (E_j \setminus F_1). \tag{18.2}$$

Since $E_j, F_1 \in \mathcal{S}$ and $\mathcal{S}$ is a semiring, by Remark 18.2.6(1), the differences $E_j \setminus F_1$ for $j = 1, 2, \ldots, n$ can be written as finite unions of elements in $\mathcal{S}$. Hence, the union (18.2) is a finite union of elements in $\mathcal{S}$ and thus $E \setminus F \in \mathcal{E}$. Next, assume the inductive hypothesis for $m = k$. We now prove the case for $m = k + 1$, namely:

$$\begin{aligned} E \setminus F &= E \setminus (F_1 \cup \ldots \cup F_k \cup F_{k+1}) \\ &= E \cap (F_1 \cup \ldots \cup F_k \cup F_{k+1})^c \\ &= E \cap (F_1^c \cap \ldots \cap F_k^c \cap F_{k+1}^c) \\ &= E \cap (F_1^c \cap \ldots \cap F_k^c) \cap (E \cap F_{k+1}^c) \\ &= E \cap (F_1 \cup \ldots \cup F_k)^c \cap (E \cap F_{k+1}^c) \\ &= (E \setminus (F_1 \cup \ldots \cup F_k)) \cap (E \setminus F_{k+1}). \end{aligned} \tag{18.3}$$

By inductive hypothesis and the base case, each of the terms in (18.3) is a finite union of sets in $\mathcal{S}$. Therefore, we can write:

$$E \setminus F = \left(\bigcup_{j=1}^{p} G_j \right) \cap \left(\bigcup_{i=1}^{q} H_i \right) = \bigcup_{j=1}^{p} \bigcup_{i=1}^{q} (G_j \cap H_i),$$

for some sets $G_j, H_i \in \mathcal{S}$. Since $\mathcal{S}$ is a semiring, $G_j \cap H_i \in \mathcal{S}$ as well and so $E \setminus F$ is a union of finitely many sets in $\mathcal{S}$. This completes the induction. Hence, the collection $\mathcal{E}$ is closed under set difference as well.

Thus, $\mathcal{E}$ is a ring. Furthermore, $\mathcal{E}$ contains $\mathcal{S}$. By definition of $\mathcal{R}(\mathcal{S})$ as the smallest ring containing $\mathcal{S}$, we then have $\mathcal{R}(\mathcal{S}) \subseteq \mathcal{E}$.

By double inclusion, we have the equality $\mathcal{E} = \mathcal{R}(\mathcal{S})$. This means every element in $\mathcal{R}(\mathcal{S})$ can be expressed as a finite union of elements in $\mathcal{S}$. Finally, applying Lemma 18.2.7, we conclude that every element in $\mathcal{R}(\mathcal{S})$ can be written as a finite union of pairwise disjoint sets in $\mathcal{S}$. □

Example 18.3.8 Going back to the specific case of $X = \mathbb{R}$, by virtue of Lemma 18.3.5, there exists a unique ring generated by the semiring of half-closed finite intervals in $\mathbb{R}$, namely $\mathcal{J} = \{(c, d] \subseteq \mathbb{R} : c < d\} \cup \{\emptyset\}$. This ring $\mathcal{R}(\mathcal{J})$ is called the ring of elementary sets. Moreover, Proposition 18.3.7 tells us how we can write down any element in $\mathcal{R}(\mathcal{J})$; they are simply finite unions of half-closed intervals.

Using the characterisation in Proposition 18.3.7, let us look at examples and non-examples of elements in $\mathcal{R}(\mathcal{J})$.

1. Clearly any element in $\mathcal{J}$ is an element in $\mathcal{R}(\mathcal{J})$.
2. We note that the set $\bigcup_{j=1}^{n}(2j, 2j + 1]$ for any $n \in \mathbb{N}$ is contained in $\mathcal{R}(\mathcal{J})$ since it is a finite union of elements in $\mathcal{J}$.
3. On the other hand, the set $\bigcup_{j=1}^{\infty}(2j, 2j + 1]$ is not contained in $\mathcal{R}(\mathcal{J})$ since it cannot be decomposed into a union of finitely many sets in $\mathcal{J}$. Likewise, the universe $\mathbb{R}$ is also not contained in $\mathcal{R}(\mathcal{J})$.
4. Moreover, any singleton set $\{a\} \subseteq \mathbb{R}$ is not an element in $\mathcal{R}(\mathcal{J})$ due to the same reason.

Whilst the extension from $\mathcal{J}$ to $\mathcal{R}(\mathcal{J})$ introduced many new sets to the collection, there are still many simple subsets of $\mathbb{R}$ that are not contained in $\mathcal{R}(\mathcal{J})$.

The characterisation in Proposition 18.3.7 allows us to extend the content function on a semiring $\mathcal{S}$ to the ring $\mathcal{R}(\mathcal{S})$ generated by it. A demonstration for the case where $\mathcal{S} = \mathcal{J}\{(c, d] \subseteq \mathbb{R} : c < d\} \cup \{\emptyset\}$ is given in the following example.

Example 18.3.9 We can extend the definition of the content m from Example 18.2.4 onto the ring $\mathcal{R}(\mathcal{J})$ by setting for every $E \in \mathcal{R}(\mathcal{J})$, which has a pairwise disjoint decomposition $E = \bigcup_{j=1}^{n} E_j$ via Proposition 18.3.7, the content $m(E) = \sum_{j=1}^{n} m(E_j)$.

We make several important remarks regarding this extension.

1. This extension of the content function m from $\mathcal{J}$ to $\mathcal{R}(\mathcal{J})$ is well defined regardless of the way we decompose the set E into finite disjoint sets in $\mathcal{J}$. Indeed, suppose that we have two different pairwise disjoint decompositions of E, say $E = \bigcup_{j=1}^{n} E_j = \bigcup_{j=1}^{m} F_j$. We define $G_{ij} = E_i \cap F_j$ for $j = 1, 2, \dots, m$ and $i = 1, 2, \dots, n$. It is easy to see that the sets $\{G_{ij} : i = 1, 2, \dots, n, j = 1, 2, \dots, m\}$ are all pairwise disjoint. Moreover, $E_i = \bigcup_{j=1}^{m}(E_i \cap F_j) = \bigcup_{j=1}^{m} G_{ij}$ and $F_j = \bigcup_{i=1}^{n}(E_i \cap F_j) = \bigcup_{i=1}^{n} G_{ij}$ so that $m(E_i) = \sum_{j=1}^{m} m(G_{ij})$ and $m(F_j) = \sum_{i=1}^{n} m(G_{ij})$. Thus:
$$\sum_{i=1}^{n} m(E_i) = \sum_{i=1}^{n}\sum_{j=1}^{m} m(G_{ij}) = \sum_{j=1}^{m}\sum_{i=1}^{n} m(G_{ij}) = \sum_{j=1}^{m} m(F_j),$$
so this content extension to $\mathcal{R}(\mathcal{J})$ is well-defined regardless of the decomposition used and, by construction, is unique.
2. Another important fact that we have is if $E \in \mathcal{R}(\mathcal{J})$ is such that $E = \bigcup_{j=1}^{\infty} E_j$ for some pairwise disjoint collection of sets $E_j \in \mathcal{R}(\mathcal{J})$, then $m(E) = \sum_{j=1}^{\infty} m(E_j)$. This follows from the representation of E as a union of finite disjoint sets in $\mathcal{J}$ and Lemma 18.2.8(4). Therefore, the σ-additivity of the content m persists after its extension to $\mathcal{R}(\mathcal{J})$.

In fact, we have a name for the extension of content in Example 18.3.9, which we state for a general ring:

Definition 18.3.10 (Premeasure) Let $\mathcal{R}$ be a ring on a set X. A function $m : \mathcal{R} \to [0, \infty]$ is called a premeasure if:

1. $m(\varnothing) = 0$,
2. for any $E \in \mathcal{R}$, we have $m(E) \geq 0$, and
3. σ-additivity: For any countable collection of pairwise disjoint sets $\{E_j\}_{j=1}^{\infty}$ where $E_j \in \mathcal{R}$, if $\bigcup_{j=1}^{\infty} E_j \in \mathcal{R}$, then $m\left(\bigcup_{j=1}^{\infty} E_j\right) = \sum_{j=1}^{\infty} m(E_j)$.

The triple $(X, \mathcal{R}, m)$ is called a premeasure space.

Remark 18.3.11 One needs to be very careful with the final condition in Definition 18.3.10. It says the premeasure is defined on the countable union, provided that the countable union is itself contained in the ring $\mathcal{R}$. In general, a union of countably infinite sets in a ring may not be contained in that ring since a ring is guaranteed to be closed under finite unions only.

An example was shown in Example 18.3.8(3). In this example, for all $j \in \mathbb{N}$ the sets $(2j, 2j+1]$ are all members of $\mathcal{R}(\mathcal{J})$ with $m((2j, 2j+1]) = 1$. However, the infinite union $\bigcup_{j=1}^{\infty}(2j, 2j+1]$ is not in $\mathcal{R}(\mathcal{J})$ and thus the premeasure m on this union is undefined. Therefore, $m\left(\bigcup_{j=1}^{\infty}(2j, 2j+1]\right) \neq \sum_{j=1}^{\infty} m((2j, 2j+1])$ since the former does not even have a value in $[0, \infty]$!

We define a special kind of premeasure which we shall use several times later.

Definition 18.3.12 (σ-Finite Premeasure) Let $\mathcal{R}$ be a ring on a set X and $m : \mathcal{R} \to [0, \infty]$ be a premeasure on it. The premeasure space $(X, \mathcal{R}, m)$ is called σ-finite if there exists a sequence of sets $\{E_j\}_{j=1}^{\infty}$ where $E_j \in \mathcal{R}$ such that $m(E_j) < \infty$ for all $j \in \mathbb{N}$ and $X = \bigcup_{j=1}^{\infty} E_j$.

Remark 18.3.13 Since $\mathcal{R}$ is a ring, the sets $\{E_j\}_{j=1}^{\infty}$ in the Definition 18.3.12 can also be assumed to be pairwise disjoint.

Example 18.3.14 The premeasure space $(\mathbb{R}, \mathcal{R}(\mathcal{J}), m)$ that we have constructed from the semiring of half closed intervals $\mathcal{J}$ is a σ-finite premeasure. Indeed, for any $n \in \mathbb{N}$, we have $I_n = (-n, n] \in \mathcal{R}(\mathcal{J})$. This set has a finite premeasure since $m(I_n) = 2n < \infty$. Moreover, we can express the universe $\mathbb{R}$ as the countable union $\bigcup_{n\in\mathbb{N}} I_n = \mathbb{R}$.

σ-Rings and σ-Algebras

Usually, as we have seen in Example 18.3.8, not all subsets in a set X can be made up of finite unions of some collection of sets $S \subseteq \mathcal{P}(X)$. Therefore, working on a ring or an algebra may not be general enough for us. For example, the set of rational numbers in $\mathbb{R}$ cannot be expressed as a finite union, intersection, or differences of intervals in $\mathcal{J}$. In other words, the subset $\mathbb{Q} \subseteq \mathbb{R}$ is not contained in the ring $\mathcal{R}(\mathcal{J})$. Even worse, any singleton set $\{a\} \subseteq \mathbb{R}$ is not in this ring either!

Therefore, we would like to extend the concept of rings further to ensure closure under countably infinite set algebra operations. This is a way forward as well since we have mentioned in Remark 18.3.11 that the premeasure is countably additive if the countable union is already in the premeasure space. So, ideally we want a collection of sets that is closed under countable unions so that we can fully utilise the σ-additivity feature of premeasures. We define:

Definition 18.3.15 (σ-Ring of Sets) Let X be a set. A non-empty collection $\mathcal{R} \subseteq \mathcal{P}(X)$ of subsets of X is called a σ-ring on X if it is a ring which is closed under countable union. Namely, for $A_j, B \in \mathcal{R}$ for each $j \in \mathbb{N}$ (or any countable indexing set), we have:

1. $\bigcup_{j=1}^{\infty} A_j \in \mathcal{R}$, and
2. $A \setminus B \in \mathcal{R}$.

The two conditions above also imply that $\varnothing \in \mathcal{R}$ and $\bigcap_{j=1}^{\infty} A_j \in \mathcal{R}$.

If a σ-ring contains the universe set, we call this collection a σ-algebra.

Definition 18.3.16 (**σ-Algebra of Sets**) Let X be a set. A non-empty collection $\mathcal{F} \subseteq \mathcal{P}(X)$ of subsets of X is called a σ-algebra (or σ-field) on X if it is a σ-ring and contains X.

Remark 18.3.17 We make two remarks here:

1. Similar to algebras, a σ-algebra is also a σ-ring which is closed under complements.
2. Therefore, alternatively, instead of requiring $A \setminus B \in \mathcal{F}$ for all $A, B \in \mathcal{F}$ to establish that $\mathcal{F}$ is a σ-algebra, it is enough to have that $X \setminus C = C^c \in \mathcal{F}$ for any $C \in \mathcal{F}$ since for any $A, B \in \mathcal{F}$, we have $A \setminus B = A \cap B^c$ which is in $\mathcal{F}$ if and only if $B^c \in \mathcal{F}$.
3. The prefix σ- in σ-ring and σ-algebra indicates that these collections of sets are just rings and algebras that are closed under countably infinite set algebraic operations respectively. So, whenever we see σ- (for example, σ-additivity or σ-finite in Definitions 18.3.10 and 18.3.12), we expect some "countable infinite" features.

Example 18.3.18 For a given set X, there are many σ-algebras that we can find on X.

1. The power set $\mathcal{P}(X)$, which is the set of all subsets of X, is a σ-algebra on X because it trivially satisfies all the defining axioms of a σ-algebra. The power set $\mathcal{P}(X)$ is the largest possible σ-algebra that we can find on the set X.
2. On the other hand, the set $\{\varnothing, X\}$ is also a σ-algebra on X. This is the smallest σ-algebra we can find on the set X.
3. Let $U = \{A \subseteq X : A \text{ is finite}\}$. The collection $\mathcal{F} = U \cup \{A^c : A \in U\}$ is also a σ-algebra on X. The readers are invited to show this in Exercise 18.4(a). If X is a finite set, this σ-algebra coincides with $\mathcal{P}(X)$.
4. Let $\{E_j\}_{j=1}^{\infty}$ be a countable collection of pairwise disjoint subsets of X such that $\bigcup_{j=1}^{\infty} E_j = X$. The collection $\mathcal{F} = \{\bigcup_{j \in J} E_j : J \subseteq \mathbb{N}\}$ is a σ-algebra on X. Let us show that this set satisfies the requirements to be a σ-algebra:
   (a) Clearly $X \in \mathcal{F}$ since $\bigcup_{j \in \mathbb{N}} E_j = \bigcup_{j=1}^{\infty} E_j = X$.
   (b) For any countable collection $\{F_k\}_{k=1}^{\infty}$ where $F_k \in \mathcal{F}$ for all $k \in \mathbb{N}$, by definition, for each k we can write F_k as the union $F_k = \bigcup_{j \in J_k} E_j$ for some indexing set $J_k \subseteq \mathbb{N}$. Let $J = \bigcup_{k \in \mathbb{N}} J_k \subseteq \mathbb{N}$. Then:

$$\bigcup_{k=1}^{\infty} F_k = \bigcup_{k=1}^{\infty} \bigcup_{j \in J_k} E_j = \bigcup_{j \in J} E_j \in \mathcal{F}.$$

(c) Let $F_1, F_2 \in \mathcal{F}$. For $k = 1, 2$, we have $F_k = \bigcup_{j \in J_k} E_j$ for some indexing sets $J_k \subseteq \mathbb{N}$. We note that for any pair of sets E_j and E_k, these sets are either identical or disjoint. Thus:

$$F_1 \setminus F_2 = \bigcup_{j \in J_1} E_j \setminus \bigcup_{j \in J_2} E_j = \bigcup_{j \in J_1 \setminus J_2} E_j \in \mathcal{F}.$$

Therefore, we conclude that $\mathcal{F}$ is a σ-algebra on X.

Similar to rings and algebras, we have the following lemma for σ-algebras which allows us to construct the minimal σ-algebra that contains an initial collection of sets. The proof of the following lemma is exactly the same as the proof for rings in Lemma 18.3.5.

Lemma 18.3.19 (σ-Algebra Generated by a Collection of Subsets) *Let $S \subseteq \mathcal{P}(X)$ be an arbitrary non-empty collection of subsets of X. Then, there exists a unique σ-algebra $\mathcal{F}(S)$ containing S that is contained in every σ-algebra $\mathcal{F}$ containing S.*
In other words, $\mathcal{F}(S)$ is the smallest σ-algebra containing S. We say $\mathcal{F}(S)$ is the σ-algebra generated by S. We also denote the σ-algebra generated by S as $\sigma(S)$.

We know that the σ-algebra generated by some set S exists via a non-constructive proof of Lemma 18.3.19. However, in contrast to the concrete expression of elements in a ring generated by a collection of sets that we have seen in Proposition 18.3.6, we cannot easily get an explicit and concrete expression for any set in a σ-algebra $\sigma(S)$ generated by some set S in terms of sets in S.

Example 18.3.20 Let $X = \mathbb{R}$. Let us look at some examples of σ-algebras generated by some collections of sets in $\mathbb{R}$.

1. Let $\mathcal{O} \subseteq \mathcal{P}(\mathbb{R})$ be the collection of all open sets in $\mathbb{R}$. A Borel σ-algebra $\mathcal{B}$ on $\mathbb{R}$, named after Émile Borel, is the σ-algebra generated by open sets in $\mathbb{R}$, namely $\mathcal{B} = \sigma(\mathcal{O})$.
2. Recall the semiring $\mathcal{J} = \{(a, b] : a, b \in \mathbb{R}\} \cup \{\varnothing\}$. We claim that $\sigma(\mathcal{J}) = \mathcal{B}$ as well. To prove this, we use double inclusion.

($\subseteq$): We show that every element in $\mathcal{J}$ is contained in $\mathcal{B}$. This is true since for any $a, b \in \mathbb{R}$, we have $(a, b] = (a, \infty) \setminus (b, \infty) \in \mathcal{B}$. By definition, since $\sigma(\mathcal{J})$ is the smallest σ-algebra containing $\mathcal{J}$ and $\mathcal{B}$ is a σ-algebra containing $\mathcal{J}$, we have the inclusion $\sigma(\mathcal{J}) \subseteq \mathcal{B}$.

($\supseteq$): We show that $\sigma(\mathcal{J})$ contains all open sets in $\mathbb{R}$. To do this, by virtue of Theorem 4.5.20, it is enough to show that it contains all open intervals.
(a) For any $a, b \in \mathbb{R}$ we have $(a, b) = \bigcup_{n \geq N} (a, b - \frac{1}{n}] \in \sigma(\mathcal{J})$ for all integers n greater than or equal to $N > \frac{1}{b-a}$.

(b) Moreover, the above also implies that $(-\infty, a) = \bigcup_{n \geq a}(-n, a) \in \sigma(\mathcal{J})$ and $(b, \infty) = \bigcup_{n \geq b}(b, n) \in \sigma(\mathcal{J})$.
(c) Finally, we have $\mathbb{R} = \bigcup_{n \in \mathbb{N}}(-n, n) \in \sigma(\mathcal{J})$.
Hence, all open sets in $\mathbb{R}$ are contained in $\sigma(\mathcal{J})$, namely $\mathcal{O} \subseteq \sigma(\mathcal{J})$. Thus, by a similar argument as before, we have the inclusion $\mathcal{B} \subseteq \sigma(\mathcal{J})$.

Therefore, we have the equality of sets $\sigma(\mathcal{J}) = \mathcal{B}$.
3. First, notice that $\mathcal{J} \subseteq \mathcal{R}(\mathcal{J}) \subseteq \sigma(\mathcal{R}(\mathcal{J}))$. Since $\sigma(\mathcal{R}(\mathcal{J}))$ is a σ-algebra containing $\mathcal{J}$, by minimality of $\sigma(\mathcal{J})$, we have $\sigma(\mathcal{J}) \subseteq \sigma(\mathcal{R}(\mathcal{J}))$. On the other hand, clearly $\sigma(\mathcal{J})$ contains $\mathcal{R}(\mathcal{J})$ since $\mathcal{R}(\mathcal{J})$ is an algebra containing $\mathcal{J}$ that is only closed under finite unions and intersections. Since $\sigma(\mathcal{J})$ is a σ-algebra containing $\mathcal{R}(\mathcal{J})$, by minimality, we have $\sigma(\mathcal{R}(\mathcal{J})) \subseteq \sigma(\mathcal{J})$. Putting these inclusions together, we get the equality of σ-algebras $\sigma(\mathcal{R}(\mathcal{J})) = \sigma(\mathcal{J})$.

So, from this example, we have shown the equality of σ-algebras $\sigma(\mathcal{O}) = \mathcal{B} = \sigma(\mathcal{J}) = \sigma(\mathcal{R}(\mathcal{J}))$.

In fact, from old σ-algebras, we can also create new σ-algebras. The readers will prove the following lemma in Exercise 18.5 later.

Lemma 18.3.21 *Let $\mathcal{F}$ be a σ-algebra on a set X. If $Y \subseteq X$, then the collection of sets $\mathcal{G} = \{E \cap Y : E \in \mathcal{F}\}$ is a σ-algebra on Y.*

The σ-algebra $\mathcal{G}$ created in Lemma 18.3.21 is called the induced σ-algebra on the set Y by the σ-algebra $\mathcal{F}$.

18.4 Outer Measure

Going back to our original goal, we aimed to extend the content function m on a π-system on X to the largest collection of subsets of X possible. As we have seen In Example 18.3.9, the content m on a π-system $\mathcal{J}$ in $\mathbb{R}$ can be extended to a premeasure on the ring $\mathcal{R}(\mathcal{J})$. This can be done thanks to the explicit representation of element in $\mathcal{R}(\mathcal{J})$ in terms of elements in $\mathcal{J}$ on which we have defined m.

Now we want to extend this content to a bigger collection of set, namely to a σ-algebra of $\mathbb{R}$ which contains $\mathcal{R}(\mathcal{J})$. From Example 18.3.20, the smallest σ-algebra in $\mathbb{R}$ that contains $\mathcal{R}(\mathcal{J})$ is $\sigma(\mathcal{J}) = \mathcal{B}$, so we want to extend the premeasure m to at least this collection of sets.

How can we extend m further to a larger collection of sets? Due to the lack of expression for a general element in the σ-algebra $\sigma(\mathcal{J})$ in terms of elements of $\mathcal{J}$, extending the premeasure m to $\sigma(\mathcal{J})$ cannot be done as easily as we did for the ring $\mathcal{R}(\mathcal{J})$ in Example 18.3.9. Therefore, we have to think of another way to carry out this extension.

Let us be very ambitious and try to extend this premeasure to the largest σ-algebra in $\mathbb{R}$ that contains $\mathcal{R}(\mathcal{J})$, namely the power set $\mathcal{P}(\mathbb{R})$. For any $A \in \mathcal{P}(\mathbb{R})$, we can define the size of the set A using the premeasure m by a set function called outer measure m^*.

We do this by covering the set A with countably many sets in $\mathcal{R}(\mathcal{J})$ and defining the outer measure on A as the smallest possible sum of premeasures of covers of A, which could also be infinity. This is well-defined since we know what the premeasures of the sets in $\mathcal{R}(\mathcal{J})$ are. For a general premeasure space $(X, \mathcal{R}, m)$, this definition was proposed by Axel Harnack (1851–1888) and Carathéodory to assign a size to any set in X:

Definition 18.4.1 (Outer Measure) Let $(X, \mathcal{R}, m)$ be a premeasure space. An outer measure m^* induced by the premeasure m is a set function defined on the power set $\mathcal{P}(X)$ as:

$$m^* : \mathcal{P}(X) \to [0, \infty]$$

$$A \mapsto \inf \left\{ \sum_{j=1}^{\infty} m(I_j) : I_j \in \mathcal{R} \text{ such that } A \subseteq \bigcup_{j=1}^{\infty} I_j \right\},$$

and the infimum is taken to be ∞ if there are no such cover for A or if the infimum does not exist.

This is well-defined as the set of the sums is made up non-negative numbers, so its infimum is non-negative. It is also easy to see that since $\mathcal{R}(\mathcal{J}) \subseteq \mathcal{P}(\mathbb{R})$, the restriction of the outer measure m^* to the sets in $\mathcal{R}(\mathcal{J})$ is simply the premeasure m, namely if $A \in \mathcal{R}(\mathcal{J})$, then $m^*(A) = m(A)$.

In fact, the outer measure m^* in Definition 18.4.1 can also be endowed on any σ-algebra on X which contains $\mathcal{R}$ via restriction of m^* to said σ-algebra. We define:

Definition 18.4.2 (Outer Measure Space) Let $(X, \mathcal{R}, m)$ be a premeasure space and $\mathcal{G} \subseteq \mathcal{P}(X)$ is a σ-algebra containing $\mathcal{R}$. An outer measure space is the triple $(X, \mathcal{G}, m^*)$ where m^* is the outer measure induced by the premeasure m.

Using Definition 18.4.1, we can deduce the following properties of outer measures:

Lemma 18.4.3 *Let $(X, \mathcal{R}, m)$ be a premeasure space. The outer measure $m^* : \mathcal{P}(X) \to [0, \infty]$ induced by m satisfies:*

1. $m^*(\varnothing) = 0$.
2. *Non-negativity: $m^*(A) \geq 0$ for any $A \in \mathcal{P}(X)$.*
3. *If $A \subseteq B$, then $m^*(A) \leq m^*(B)$.*
4. *If $m^*(A) = 0$, then $m^*(A \cup B) = m^*(B)$.*

5. *σ-subadditive: For any countable collection of pairwise disjoint sets $E_j \in \mathcal{P}(X)$, we have:*

$$m^*\left(\bigcup_{j=1}^{\infty} E_j\right) \leq \sum_{j=1}^{\infty} m^*(E_j).$$

6. *If $m^*(A \triangle B) = 0$ then $m^*(A) = m^*(B)$.*

Proof The first two assertions are clearly true. Assertions 3 and 4 will be proven by the readers in Exercise 18.7. Thus, we prove only assertions 5 and 6 here.

5. If $\sum_{j=1}^{\infty} m^*(E_j) = \infty$, we are done. Otherwise, if $\sum_{j=1}^{\infty} m^*(E_j) < \infty$, then $m^*(E_j) < \infty$ for all j. Fix an arbitrary $\varepsilon > 0$. By definition of the outer measure and characterisation of infimum, for each $j \in \mathbb{N}$ there exists a countable cover $\{I_j^i\}_{i=1}^{\infty} \subseteq \mathcal{R}$ of E_j such that:

$$m^*(E_j) \leq \sum_{i=1}^{\infty} m(I_j^i) < m^*(E_j) + \frac{\varepsilon}{2^j}.$$

Since $\bigcup_{j=1}^{\infty} \{I_j^i\}_{i=1}^{\infty}$ then forms a countable cover of $\bigcup_{j=1}^{\infty} E_j$, by definition of the outer measure, we have:

$$m^*\left(\bigcup_{j=1}^{\infty} E_j\right) \leq \sum_{j=1}^{\infty}\sum_{i=1}^{\infty} m(I_j^i) < \sum_{j=1}^{\infty}\left(m^*(E_j) + \frac{\varepsilon}{2^j}\right) = \sum_{j=1}^{\infty} m^*(E_j) + \varepsilon.$$

Since $\varepsilon > 0$ is arbitrary, we conclude the proof.
6. Since $A \triangle B = (A \setminus B) \cup (B \setminus A)$, by the third assertion, we have $m^*(A \setminus B) \leq m^*(A \triangle B) = 0$ and so $m^*(A \setminus B) = 0$. Next, note that $A = (A \cap B) \cup (A \setminus B) \subseteq B \cup (A \setminus B)$. Thus, by using assertions 3 and 5, we have $m^*(A) \leq m^*(B \cup (A \setminus B)) \leq m^*(B) + m^*(A \setminus B) = m^*(B)$. By a symmetric argument, we can also deduce $m^*(B) \leq m^*(A)$. Hence, we arrive at the desired equality. □

Example 18.4.4 Let us look at some examples on how we can compute outer measures on some sets in $\mathcal{P}(\mathbb{R})$ induced by the premeasure space $(\mathbb{R}, \mathcal{R}(\mathcal{J}), m)$ from Example 18.3.9.

1. Let $\{a\} \subseteq \mathbb{R}$ be the singleton set. We can cover the set $\{a\}$ with the single interval $I = (a - \frac{1}{n}, a] \in \mathcal{J}$ for every $n \in \mathbb{N}$. Therefore, $m^*(a) \leq m((a - \frac{1}{n}, a]) = \frac{1}{n}$ for any $n \in \mathbb{N}$. This means $m^*(\{a\}) = 0$. Hence, the outer measure of any singleton set is 0.
2. For a set $[a, b] \in \mathcal{P}(\mathbb{R})$, note that $(a, b] \subseteq [a, b]$. Since $(a, b] \in \mathcal{R}(\mathcal{J})$, the premeasure m and the outer measure m^* agree on this set. Hence, we have $b - a =$

$m((a, b]) = m^*((a, b]) \leq m^*([a, b])$ where the final inequality is obtained using Lemma 18.4.3(3).
On the other hand, we can cover the set $[a, b]$ with a single set in $\mathcal{R}(\mathcal{J})$, namely $(a-\frac{1}{n}, b] \in \mathcal{R}(\mathcal{J})$ for every $n \in \mathbb{N}$. Thus, $m^*([a, b]) \leq m((a-\frac{1}{n}, b]) = b-a+\frac{1}{n}$ for any $n \in \mathbb{N}$. Hence, we have $m^*([a, b]) \leq b - a$.
Thus, we conclude that $m^*([a, b]) = b - a$.
3. Alternatively, we can compute the outer measure $m^*([a, b])$ by using the fact that $[a, b] = \{a\} \cup (a, b]$ and $m^*(\{a\}) = 0$. Using Lemma 18.4.3(4), we then have $m^*([a, b]) = m^*(\{a\} \cup (a, b]) = m^*((a, b]) = b - a$.

Lemma 18.4.3(5) implies that the outer measure m^* that we have constructed on $\mathcal{P}(\mathbb{R})$ from the content m on $\mathcal{J}$ is countably subadditive over disjoint subsets. Ideally we would like a measure to be countably additive (not just subadditive) since we expect that the sum of the measures of disjoint parts add up to be exactly equal to the measure of their union. This is what we had for the premeasure space $(\mathbb{R}, \mathcal{R}(\mathcal{J}), m)$ (albeit only for when the infinite union is contained in $\mathcal{R}(\mathcal{J})$) as we saw in Example 18.3.9(2) and we would like to preserve this property.

However, in general, this may not be possible for the outer measure m^*. Countable subadditivity in Lemma 18.4.3(5) is the best that one can guarantee for a general outer measure. Even worse, finite additivity for the outer measure m^* may not even be true! Let us look at an example for which this strict inequality for finite union occurs in $\mathbb{R}$.

Example 18.4.5 First, notice that in $\mathbb{R}$, the outer measure m^* induced by the premeasure space $(\mathbb{R}, \mathcal{R}(\mathcal{J}), m)$ is translation invariant. If we have a set $X \subseteq \mathbb{R}$ and a point $c \in \mathbb{R}$, we define the translated set $c+X = \{c+x : x \in X\}$. Translation invariant means $m^*(c + X) = m^*(X)$ for any $c \in \mathbb{R}$. This is left for the readers to prove in Exercise 18.8.

Now assume for contradiction that this outer measure m^* is finitely additive, namely we have $m^*(\bigcup_{j=1}^n A_j) = \sum_{j=1}^n m^*(A_j)$ for any pairwise disjoint collection of sets $\{A_j\}_{j=1}^n$ for any $n \in \mathbb{N}$. Consider the interval $[0, 1]$ and define an equivalence relation on the set $[0, 1]$ by $x \sim y$ iff $x - y \in \mathbb{Q}$. It is easy to see that the equivalence classes of this relation are all countable since they are of the form $[x] = \{x + r : r \in \mathbb{Q}\} \cap [0, 1]$ for some $x \in [0, 1]$.

Partition the interval $[0, 1]$ into its equivalence classes (there are uncountably many such classes since each class is countable) and choose a representative from each class. The union of these representatives is denoted as the set $A \subseteq [0, 1]$ and is called the Vitali set after Giuseppe Vitali (1875–1932). Notice that $m^*(A) \geq 0$ is a fixed constant.

The rational numbers in $[-1, 1]$ is countable, so we can enumerate them as $r_1, r_2, r_3, \ldots$. Consider the collection of countably many translated sets $\{B_j\}_{j=1}^\infty$ where $B_j = r_j + A$ for $j \in \mathbb{N}$. We state some facts about these sets:

1. These sets are pairwise disjoint by construction. Indeed, if $B_j \cap B_i \neq \varnothing$ for some $i \neq j$, then there exist $a, b \in A$ which are class representatives for the

equivalence classes of $\sim$ such that $r_j + a = r_i + b \in B_i \cap B_j$. Note that $a \neq b$ because $r_i \neq r_j$. Since A contains exactly one element from each equivalence class of $\sim$, this means $a \not\sim b$. However, $a - b = r_i - r_j \in \mathbb{Q} \cap [-1, 1]$ and so $a \sim b$ by definition, which is a contradiction.
2. Since m^* is translation invariant, we have $m^*(B_j) = m^*(r_j + A) = m^*(A)$.
3. Note that for each $j \in \mathbb{N}$, we have $B_j = r_j + A \subseteq r_j + [0, 1] \subseteq [-1, 1] + [0, 1] \subseteq [-1, 2]$.
   Moreover, note that for any $x \in [0, 1]$, the point x is related to an element in A, say $x \sim a$ for some $a \in A$. This means $x - a \in \mathbb{Q} \cap [-1, 1]$ and hence $x \in r + A$ for some $r \in \mathbb{Q} \cap [-1, 1]$. Therefore, $x \in B_j$ for some $j \in \mathbb{N}$. Thus, we have the inclusion $[0, 1] \subseteq \bigcup_{j=1}^{\infty} B_j$. Hence, we have the following set inclusions:

$$[0, 1] \subseteq \bigcup_{j=1}^{\infty} B_j \subseteq [-1, 2].$$

Now let us use these facts to obtain a contradiction. By σ-subadditivity of m^* we have:

$$1 = m^*([0, 1]) \leq m^*\left(\bigcup_{j=1}^{\infty} B_j\right) \leq \sum_{i=1}^{\infty} m^*(B_j) = \sum_{i=1}^{\infty} m^*(A).$$

So this means $m^*(A) > 0$. For any integer $n > \frac{3}{m^*(A)}$, we have $\bigcup_{j=1}^{n} B_j \subseteq \bigcup_{j=1}^{\infty} B_j \subseteq [-1, 2]$ and thus $m^*\left(\bigcup_{j=1}^{n} B_j\right) \leq m^*([-1, 2]) = 3$. Finally, by the assumption that the outer measure is additive over finitely many disjoint sets, we have:

$$3 \geq m^*\left(\bigcup_{j=1}^{n} B_j\right) = \sum_{j=1}^{n} m^*(B_j) = \sum_{j=1}^{n} m^*(A) = nm^*(A) > 3,$$

which then gives us a contradiction! Therefore, we conclude that there are sets in $\mathcal{P}(\mathbb{R})$ which are pairwise disjoint but the outer measure is not even finitely additive over their union.

Remark 18.4.6 We make some interesting remarks regarding the construction in Example 18.4.5.

1. In the construction of the Vitali set, we chose an element from each of the equivalence class. There are uncountably many such classes and our choice is non-explicit: we simply said "pick one element from each class". This construction is allowed due to the axiom of choice (AOC) in set theory.

2. The AOC, which was formulated by Ernst Zermelo (1871–1953) in axiomatic set theory, asserts that for any collection (even infinitely many) of sets, we can always construct a new set containing an element from each set in the original collection. This seems reasonable. Surely this is possible if we can do so for finitely many collections of sets? What would be the difference if there are infinitely many sets from each of which we have to select an element?
3. However, the AOC is a controversial topic that divides the mathematical community. "Constructivists" reject the AOC because they believe that mathematical objects only exist if they can be constructed explicitly. The selection imposed by AOC is arbitrary and not explicit; it simply asserts we can select one element from each set but not how. An example of such mathematician who opposed the AOC is Borel himself, who declared:

   > Any argument where one supposes an arbitrary choice to be made an uncountably infinite number of times... [is] outside the domain of mathematics.

   and thus rejects the construction of the Vitali set.
4. One of the major reasons for the rejection of AOC is that it could lead to many counterintuitive results such as the existence of the Vitali set, the Hausdorff paradox, and the Banach-Tarski paradox. The latter asserts that we can always decompose a three-dimensional ball and reassemble them into two distinct balls identical to the first one. Anything can happen in the mathematics world!
5. Others accept the AOC for convenience. Sometimes we unconsciously do this: we have done this several times in the course of this book, see if you can spot them. A joke by Jerry L. Bona (1945-) to make light of this controversy is:

   > The axiom of choice is obviously true, the well–ordering theorem is obviously false; and who can tell about Zorn's Lemma?

   which is hilarious because all three of the concepts in the quote are mathematically equivalent to each other. For further discussions on the AOC, readers are directed to [33] and [36].

On some other set and power set, the outer measure m^* induced by the premeasure space $(X, \mathcal{R}, m)$ might already be σ-additive and thus satisfies our requirement. However, as shown in Example 18.4.5, the outer measure m^* that we have constructed on the whole of power set $\mathcal{P}(\mathbb{R})$ from the premeasure m on $\mathcal{R}(\mathcal{J})$ does not behave nicely since the outer measure is not σ-additive or even finitely additive over disjoint sets.

Therefore, the power set might be too big for us to reasonably work on. Thus, we need to pare down the power set to a more well-behaved σ-algebra within $\mathcal{P}(\mathbb{R})$ so that the outer measure behaves in a way that we wanted.

18.5 Measure

Before we proceed with our construction, we define the requirements or axioms of the measure function on a generic set and show what they entail. The following is a wishlist of what we hope a measure would be. A measure is essentially what we want the outer measure to be, but with additive and σ-additive property guaranteed. We first define:

Definition 18.5.1 (Measure) Let $\mathcal{F}$ be a σ-algebra on some set X. A measure μ is a set function $\mu : \mathcal{F} \to [0, \infty]$ such that:

1. $\mu(\varnothing) = 0$,
2. Non-negativite: For any $E \in \mathcal{F}$, we have $\mu(E) \geq 0$, and
3. σ-additive: For any countable collection of pairwise disjoint sets $\{E_j\}_{j=1}^{\infty}$ where $E_j \in \mathcal{F}$, we have:

$$\mu\left(\bigcup_{j=1}^{\infty} E_j\right) = \sum_{j=1}^{\infty} \mu(E_j).$$

Remark 18.5.2 We make several remarks regarding the measure function.

1. Notice that we require the sets to be pairwise disjoint for the last assertion. For non-disjoint collection of sets, we instead have σ-subadditivity, namely for any countable collection of sets $\{E_j\}_{j=1}^{\infty}$ where $E_j \in \mathcal{F}$ for all $j \in \mathbb{N}$ which are not necessarily disjoint, we have:

$$\mu\left(\bigcup_{j=1}^{\infty} E_j\right) \leq \sum_{j=1}^{\infty} \mu(E_j).$$

   The intuition here is that for non-disjoint sets, when we sum up the measures of each individual sets, we could have possibly counted some elements in the union more than once if they lie in an intersection of the sets, thus the total sum of measures might be bigger than the measure of the union.
2. Note that from the axioms of measure, the empty set automatically has 0 measure. However, there may be some non-empty sets $E \in \mathcal{F}$ with 0 measure, namely $\mu(E) = 0$. We call such sets null or μ-null sets.

   Next, we define:

Definition 18.5.3 (Measurable Space) A set X along with a σ-algebra $\mathcal{F}$ on it, namely the pair $(X, \mathcal{F})$, is called a measurable space.

Remark 18.5.4 We make several remarks regarding Definition 18.5.3.

1. The sets in the σ-algebra $\mathcal{F}$ are called $\mathcal{F}$-measurable sets or just measurable sets when there is no ambiguity on which σ-algebra is being considered.
2. Notice the term "measurable" in Definition 18.5.3 which means on this space we are able to endow it with at least one measure. Indeed, we can always place a measure on any σ-algebra $\mathcal{F}$; an obvious measure function for any σ-algebra at all is the trivial measure, namely $\mu(E) = 0$ for all $E \in \mathcal{F}$ (one can easily check that this satisfies the conditions in Definition 18.5.1). In fact, there may be other measures that we can place on $\mathcal{F}$. In short, being a measurable space does not specify the measure.
3. Measureability is a structure on the set X. It is an additional information that we endow to the set X to make it more interesting. This is similar to how we may endow a set with a relation, an algebraic, an order, or a metric structure that we did in Chaps. 2 and 5.

As noted in Remark 18.5.4(2), a measurable space may be endowed with possibly more than one measure. If we fix a specific measure on the space, we then call it a measure space.

Definition 18.5.5 (Measure Space) A measurable space $(X, \mathcal{F})$ equipped with a measure $\mu : \mathcal{F} \to [0, \infty]$ is called a measure space. We denote it as the triple $(X, \mathcal{F}, \mu)$.

So, a measure along with the relevant σ-algebra, is a structure that we put on a set X. Instead of the boring set X with no structure at all, after endowing a σ-algebra and a measure, we have more information on it, namely the sets we can measure and the sizes of these sets.

Example 18.5.6 Let us look at an example of a measure space. Consider the set $X = \mathbb{Z}$ and consider the collection of sets $\mathcal{P}(\mathbb{Z})$. Clearly, this is a σ-algebra on $\mathbb{Z}$. We define a set function $\mu : \mathcal{P}(\mathbb{Z}) \to [0, \infty]$ as:

$$\mu(E) = \begin{cases} 0 & \text{if } E = \varnothing, \\ n & \text{if } E \text{ is a finite set with cardinality } n \in \mathbb{N}, \\ \infty & \text{if } E \text{ is an infinite set.} \end{cases}$$

We now prove that this is indeed a measure. Clearly the two first axioms in Definition 18.5.1 are satisfied. We now check the final axiom. Let $\{E_j\}_{j=1}^{\infty}$ be a collection of countably many pairwise disjoint sets in $\mathcal{P}(\mathbb{Z})$. We want to show the equality:

$$\mu\left(\bigcup_{j=1}^{\infty} E_j\right) = \sum_{j=1}^{\infty} \mu(E_j). \tag{18.4}$$

1. If at least one of the E_j is infinite, then the union $\bigcup_{j=1}^{\infty} E_j$ is infinite as well, so both of the sides of (18.4) are infinite and hence the σ-additivity of μ is satisfied here.
2. Now suppose that none of the sets E_j is infinite. We have two subcases:
   (a) If only finitely many of them is empty, then the union $\bigcup_{j=1}^{\infty} E_j$ is an infinite set. Moreover, the sum $\sum_{j=1}^{\infty} \mu(E_j)$ consists of infinitely many positive integer terms. So it must be infinite as well.
   (b) If only finitely many of the E_j are non-empty, then there exists an index $N \in \mathbb{N}$ such that $E_j = \varnothing$ for all $j > N$. Hence, $\bigcup_{j=1}^{\infty} E_j = \bigcup_{j=1}^{N} E_j$ and $\mu(E_j) = \mu(\varnothing) = 0$ for all $j > N$. We can apply induction to show σ-additivity here.
   Define $F_n = \bigcup_{j=1}^{n} E_j$ for $n = 1, \dots, N$. For $n = 1$, σ-additivity is clearly true. For $n = 2$, the σ-additivity of the measure μ is true by Lemma 3.4.8. Assume that it is true for some $n = k < N$, namely $\mu(F_k) = \sum_{j=1}^{k} \mu(E_j)$. Then, for the case $n = k + 1$, we note that $F_{k+1} = F_k \cup E_{k+1}$ is a union of two disjoint sets. Similar to the case $k = 2$, we then have:

$$\mu\left(\bigcup_{j=1}^{k+1} E_j\right) = \mu(F_k \cup E_{k+1}) = \mu(F_k) + \mu(E_{k+1}) = \sum_{j=1}^{k+1} \mu(E_j),$$

   where we used the inductive hypothesis in the second equation. This proves σ-additivity for the final case.

   In both subcases, the equality (18.4) is true.

Thus, the set function μ is σ-additive and hence defines a measure on the power set of the integers. This measure is called the counting measure on $\mathbb{Z}$, for obvious reasons.

From Definition 18.5.1, we can deduce some immediate useful features of measure spaces. This first result is similar to Lemma 18.4.3(3) for outer measures.

Proposition 18.5.7 *Let $\mathcal{F}$ be a σ-algebra on some set X along with a measure $\mu : \mathcal{F} \to [0, \infty]$. If $E \subseteq F$ where $E, F \in \mathcal{F}$, then $\mu(E) \leq \mu(F)$.*

Proof Since $F = E \cup (F \setminus E)$ where each of the sets on the RHS are disjoint, by additivity and non-negativity of measures, we have $\mu(F) = \mu(E) + \mu(F \setminus E) \geq \mu(E)$. □

Next, we have a result that is usually referred to the continuity of measures.

Proposition 18.5.8 *Let $(X, \mathcal{F}, \mu)$ be a measure space and $\{E_j\}_{j=1}^{\infty}$ is a countable collection of sets in $\mathcal{F}$.*

1. *If* $E_j \subseteq E_{j+1}$ *for all* $j \in \mathbb{N}$, *then* $\mu(\bigcup_{j=1}^{\infty} E_j) = \lim_{n\to\infty} \mu(E_n)$.
2. *If* $E_{j+1} \subseteq E_j$ *for all* $j \in \mathbb{N}$ *and* $\mu(E_1) < \infty$, *then* $\mu(\bigcap_{j=1}^{\infty} E_j) = \lim_{n\to\infty} \mu(E_n)$.

Proof We prove the assertions one by one:

1. Define $F_1 = E_1$ and $F_j = E_j \setminus E_{j-1}$ for all $j = 2, 3, \ldots$. The sets $\{F_j\}_{j=1}^{\infty}$ are pairwise disjoint. Moreover, we have $E_n = \bigcup_{j=1}^{n} F_j$ and thus $\bigcup_{j=1}^{\infty} E_j = \bigcup_{j=1}^{\infty} F_j$. Since the sets $\{F_j\}_{j=1}^{\infty}$ are pairwise disjoint, we have the equality $\mu(E_n) = \mu(\bigcup_{j=1}^{n} F_j) = \sum_{j=1}^{n} \mu(F_j)$.
   By σ-additivity of μ and definition of real series, we have:
$$\mu\left(\bigcup_{j=1}^{\infty} E_j\right) = \mu\left(\bigcup_{j=1}^{\infty} F_j\right) = \sum_{j=1}^{\infty} \mu(F_j) = \lim_{n\to\infty} \sum_{j=1}^{n} \mu(F_j) = \lim_{n\to\infty} \mu(E_n).$$
2. For each $j = 1, 2, \ldots$, define $F_j = E_1 \setminus E_j$. Then, (F_j) is an increasing nested sequence of sets. Note that $\bigcup_{j=1}^{\infty} F_j = E_1 \setminus \bigcap_{j=1}^{\infty} E_j$. Thus, by using the previous assertion, we get:
$$\lim_{n\to\infty} \mu(F_n) = \mu\left(\bigcup_{j=1}^{\infty} F_j\right) = \mu\left(E_1 \setminus \bigcap_{j=1}^{\infty} E_j\right) = \mu(E_1) - \mu\left(\bigcap_{j=1}^{\infty} E_j\right). \tag{18.5}$$
   On the other hand, since $E_1 = E_j \cup F_j$ for all $j \in \mathbb{N}$ and E_j and F_j are disjoint, we have $\mu(E_1) = \mu(F_j) + \mu(E_j)$ for all $j = 1, 2, \ldots$. Hence, (18.5) becomes:
$$\lim_{j\to\infty} (\mu(E_1) - \mu(E_j)) = \mu(E_1) - \mu\left(\bigcap_{j=1}^{\infty} E_j\right).$$
   Using the fact that $\mu(E_1) < \infty$, we can use algebra to yield the result. □

Remark 18.5.9 We note that in Proposition 18.5.8(2), the condition $\mu(E_1) < \infty$ is necessary. Finding a counterexample of the statement without this condition is left as Exercise 18.11 once the readers have seen some more concrete examples of measure spaces.

In general, we have a name for measures on which the measure of any set in $\mathcal{F}$ is finite:

Definition 18.5.10 (Finite Measure) Let $(X, \mathcal{F}, \mu)$ be a measure space. The measure μ is called a finite measure if $\mu(X) < \infty$.

Though finite measures are useful, they can be quite uncommon and restrictive. Most of the time, we are interested in a bigger space. We generalise the idea of finite measure to spaces with infinite global measure which could be broken down into countably many subsets of finite measures. This type of space allows us to work locally on subsets of finite measures in order to get a global picture. Similar to Definition 18.3.12, we define:

Definition 18.5.11 (σ-Finite Measure) Let $(X, \mathcal{F}, \mu)$ be a measure space. The measure μ is called σ-finite if there exists countably many subsets $\{E_j\}_{j=1}^{\infty}$ where $E_j \in \mathcal{F}$ for all $j \in \mathbb{N}$ such that $X \subseteq \bigcup_{j=1}^{\infty} E_j$ and $\mu(E_j) < \infty$ for all $j \in \mathbb{N}$.

In fact, many results require this condition either for convenience in proofs or simply because these results will break down under a bigger measure space. For example, refer to the results on double integral in Chap. 20 where σ-finiteness of the measure is an essential condition.

Example 18.5.12 The counting measure on $\mathbb{Z}$ that we have seen in Example 18.5.6 is a σ-finite measure since we can express the universe $\mathbb{Z}$ as a countable union of sets with finite measure. For example, if we let $E_n = \{x \in \mathbb{Z} : -n \leq x \leq n\}$, we have $\mu(E_n) = 2n + 1 < \infty$ and $\mathbb{Z} = \bigcup_{j=1}^{\infty} E_j$.

Recall that $\mathcal{P}(\mathbb{R})$ is the largest possible σ-algebra on $\mathbb{R}$, thus making the pair $(\mathbb{R}, \mathcal{P}(\mathbb{R}))$ a measurable space. On this measurable space, we can define many different kinds of measure.

Example 18.5.13 Here are some examples of measures we can put on the measurable space $(\mathbb{R}, \mathcal{P}(\mathbb{R}))$.

1. We have noted in Remark 18.5.4(2), there is a measure that we can put on any measurable space which is the trivial measure. Using this measure, the measure of any set in $\mathcal{P}(\mathbb{R})$ is 0, namely $\mu(A) = 0$ for any $A \subseteq \mathbb{R}$.
2. The counting measure μ as we have defined in Example 18.5.6 but adapted to $\mathcal{P}(\mathbb{R})$ is also a genuine measure on the σ-algebra $\mathcal{P}(\mathbb{R})$.
3. Another measure that we can define on this measurable space is the Dirac mass. Pick any $c \in \mathbb{R}$. We define the Dirac mass at $x = c$ by the set function $\delta_c : \mathcal{P}(\mathbb{R}) \to [0, \infty]$ as:

$$\delta_c(E) = \begin{cases} 1 & \text{if } c \in E, \\ 0 & \text{otherwise.} \end{cases}$$

The readers are invited to prove that this set function satisfies the conditions in Definition 18.5.1 in Exercise 18.9 and hence is a measure. As a result, this function is also called the Dirac measure, written as $\mu(A) = \delta_c(A)$ for any

$A \subseteq \mathbb{R}$. This measure was introduced by Paul Dirac (1902–1984) as a tool to study physics.

However, the measures on $(\mathbb{R}, \mathcal{P}(\mathbb{R}))$ in Example 18.5.13 are all incompatible with the premeasure m on the ring $\mathcal{R}(\mathcal{J})$ in $\mathbb{R}$ that we have constructued in Example 18.3.9. Incompatible here means if we restrict any of these measures to sets in $\mathcal{R}(\mathcal{J}) \subseteq \mathcal{P}(\mathbb{R})$, the measure μ does not coincide with the premeasure m for all sets in $\mathcal{R}(\mathcal{J})$, namely $\mu|_{\mathcal{R}(\mathcal{J})} \neq m$. For example, if μ is the trivial measure as seen in Exercise 18.5.13(1), then for $A = (a, b] \in \mathcal{R}(\mathcal{J})$ we have $\mu(A) = 0$ but $m(A) = b - a \neq 0$.

Aiming to fulfill the axioms of measure in Definition 18.5.1, let us return to where we left off with our quest on extending the premeasure m on $\mathcal{R}(\mathcal{J})$ to the whole of $\mathcal{P}(\mathbb{R})$.

18.6 Carathéodory Extension Theorem

Recall that we began with the π-system or semiring $\mathcal{J}$ on $\mathbb{R}$ made up of half-closed intervals of the form $(a, b]$ with a very basic precursor for measure called content m where $m((a, b]) = b - a$ and $m(\emptyset) = 0$. This was extended to the ring $\mathcal{R}(\mathcal{J})$ as a premeasure m in Example 18.3.9.

We then tried to extend the premeasure to the power set $\mathcal{P}(\mathbb{R})$. This extension, which we called the outer measure m^*, satisfies almost all of the measure axioms in Definition 18.5.1, except for the σ-additivity axiom. In fact, we have seen that there are sets in $\mathcal{P}(\mathbb{R})$ that give us strict inequality in the σ-subadditivity of the outer measure which is the Vitali set in Example 18.4.5.

So this suggests that the σ-algebra $\mathcal{F}$ containing $\mathcal{R}(\mathcal{J})$ that we can extend the premeasure m to must be strictly smaller than $\mathcal{P}(\mathbb{R})$. Thus, we need to throw away some sets, including the difficult Vitali set in Example 18.4.5, from $\mathcal{P}(\mathbb{R})$ so that the outer measure m^* becomes a genuine measure on the resulting σ-algebra $\mathcal{F}$. But which sets do we need to throw away?

This is where the Carathéodory condition comes in. For generality, we state the condition here for arbitrary sets X instead of $\mathbb{R}$.

Definition 18.6.1 (Carathéodory Condition) Let $(X, \mathcal{G}, m^*)$ be an outer measure space. A set $E \in \mathcal{G}$ is called m^*-measurable if it satisfies the Carathéodory condition, which is:

$$m^*(F) = m^*(F \cap E) + m^*(F \cap E^c) \quad \text{for all } F \in \mathcal{G}. \tag{18.6}$$

The subset of all elements of $\mathcal{G}$ which satisfy the Carathéodory condition is denoted $\mathcal{G}^*$ and they are called the m^*-measurable sets. Namely:

$$\mathcal{G}^* = \{E \in \mathcal{G} : \forall F \in \mathcal{G}, m^*(F) = m^*(F \cap E) + m^*(F \cap E^c)\} \subseteq \mathcal{G}.$$

Remark 18.6.2 Sometimes the Carathéodory condition (18.6) is relaxed to simply requiring:

$$m^*(F) \geq m^*(F \cap E) + m^*(F \cap E^c) \quad \text{for all } F \in \mathcal{G}, \tag{18.7}$$

since the $\leq$ inequality is trivially true by the σ-subadditivity of m^* in Lemma 18.4.3.

Note that $\mathcal{G}^*$ is non-empty since $\varnothing, X \in \mathcal{G}^*$. Other obvious sets in any $\mathcal{G}$ that satisfy the Carathéodory condition are the m^*-null sets, which are the sets with outer measure 0. Let us demonstrate this fact here:

Lemma 18.6.3 *Let X be a set and $\mathcal{G} \subseteq \mathcal{P}(X)$ is a σ-algebra on X equipped with an outer measure m^*.*

1. *Suppose that $\{E_j\}_{j=1}^{\infty}$ is a countable collection of m^*-null sets. Then, the union $\bigcup_{j=1}^{\infty} E_j$ is also m^*-null.*
2. *If E is m^*-null, then $E \in \mathcal{G}^*$.*

Proof The first assertion is clearly true by σ-subadditivity of the outer measure m^*. We prove the second assertion.

2. Pick an arbitrary $F \in \mathcal{G}$. By Lemma 18.4.3, we have $m^*(F \cap E^c) \leq m^*(F)$ and $m^*(F \cap E) \leq m^*(E)$. Since $m^*(E) = 0$, the latter implies $m^*(F \cap E) = 0$. Thus, we have $m^*(F) \geq m^*(F \cap E^c) = m^*(F \cap E^c) + m^*(F \cap E)$, which is the Carathéodory condition in (18.7). □

There are probably many other sets in $\mathcal{G}$ which satisfy the Carathéodory condition. The important reason why we are interested in these sets is the following theorem:

Theorem 18.6.4 (Carathéodory Extension Theorem) *Suppose that X is a set and $\mathcal{G} \subseteq \mathcal{P}(X)$ is a σ-algebra on X equipped with an outer measure m^*. If $X \in \mathcal{G}$, then:*

1. *$\mathcal{G}^*$ is a σ-algebra of X contained in $\mathcal{G}$.*
2. *The outer measure m^* restricted to the σ-algebra $\mathcal{G}^*$ is a measure.*

Proof We prove the assertions one by one:

1. We check the conditions in Definitions 18.3.15 and 18.3.16.
   (a) Clearly $\varnothing, X \in \mathcal{G}^*$.
   (b) First we show that for any $A, B \in \mathcal{G}^*$ we have $A \setminus B = A \cap B^c \in \mathcal{G}^*$. In other words, we aim to show that the set $A \cap B^c$ satisfies the Carathéodory condition, namely $m^*(G) = m^*(G \cap A \cap B^c) + m^*(G \cap (A^c \cup B))$ for any

$G \in \mathcal{G}$. Since $A, B \in \mathcal{G}^*$, by the Carathéodory condition, for any $F \in \mathcal{G}$ we have:

$$m^*(F) = m^*(F \cap A) + m^*(F \cap A^c), \tag{18.8}$$

$$m^*(F) = m^*(F \cap B) + m^*(F \cap B^c). \tag{18.9}$$

Substituting $F = G \cap A$ in (18.9), we get $m^*(G \cap A) = m^*(G \cap A \cap B) + m^*(G \cap A \cap B^c)$. Substituting this in (18.8) with $F = G$ yields:

$$m^*(G) = m^*(G \cap A \cap B) + m^*(G \cap A \cap B^c) + m^*(G \cap A^c). \tag{18.10}$$

Now consider (18.8) with $F = G \cap (A^c \cup B)$. This gives us:

$$\begin{aligned} m^*(G \cap (A^c \cup B)) &= m^*(G \cap (A^c \cup B) \cap A) + m^*(G \cap (A^c \cup B) \cap A^c) \\ &= m^*(G \cap B \cap A) + m^*(G \cap A^c), \end{aligned}$$

by using distributivity of the set operations. Substituting this in (18.10) yields the desired goal.

(c) Next, we want to prove that the collection $\mathcal{G}^*$ is closed under countable union. Let $\{E_j\}_{j=1}^{\infty}$ be countably many sets in $\mathcal{G}^*$. WLOG, assume that they are all pairwise disjoint. First observe that $A \cup B = X \setminus ((X \setminus A) \setminus B) \in \mathcal{G}^*$. Thus, by part (b) and induction, any finite union of elements in $\mathcal{G}^*$ also lies in $\mathcal{G}^*$. Fix an arbitrary $n < \infty$. Then, by the observation above, the finite union $\bigcup_{j=1}^n E_j$ is also m^*-measurable. By Carathéodory condition, for any $F \in \mathcal{G}$ we have:

$$m^*(F) = m^*\left(F \cap \left(\bigcup_{i=1}^n E_j\right)\right) + m^*\left(F \cap \left(\bigcup_{j=1}^n E_j\right)^c\right). \tag{18.11}$$

On the other hand, since $E_n \in \mathcal{G}^*$ and $F \cap \left(\bigcup_{i=1}^n E_j\right) \in \mathcal{G}$, the Carathéodory condition using these sets says:

$$\begin{aligned} &m^*\left(F \cap \left(\bigcup_{j=1}^n E_j\right)\right) \\ &= m^*\left(F \cap \left(\bigcup_{j=1}^n E_j\right) \cap E_n\right) + m^*\left(F \cap \left(\bigcup_{j=1}^n E_j\right) \cap E_n^c\right) \\ &= m^*(F \cap E_n) + m^*\left(F \cap \left(\bigcup_{j=1}^{n-1} E_j\right)\right). \end{aligned} \tag{18.12}$$

Thus, by induction on n, we can continue the procedure in (18.12) to eventually get:

$$m^*\left(F \cap \left(\bigcup_{j=1}^{n} E_j\right)\right) = \sum_{j=1}^{n} m^*(F \cap E_j). \tag{18.13}$$

Substituting the Eq. (18.13) in (18.11) and using the fact that $\left(\bigcup_{j=1}^{\infty} E_j\right)^c \subseteq \left(\bigcup_{j=1}^{n} E_j\right)^c$, we have:

$$\begin{aligned} m^*(F) &= \sum_{j=1}^{n} m^*(F \cap E_j) + m^*\left(F \cap \left(\bigcup_{j=1}^{n} E_j\right)^c\right) \\ &\geq \sum_{j=1}^{n} m^*(F \cap E_j) + m^*\left(F \cap \left(\bigcup_{j=1}^{\infty} E_j\right)^c\right), \end{aligned}$$

which is true for any $n \in \mathbb{N}$ since it was arbitrarily fixed. Taking the limit as $n \to \infty$ and using the σ-subadditivity of m^*, we then get:

$$\begin{aligned} m^*(F) &\geq \sum_{j=1}^{\infty} m^*(F \cap E_j) + m^*\left(F \cap \left(\bigcup_{j=1}^{\infty} E_j\right)^c\right) \\ &\geq m^*\left(\bigcup_{j=1}^{\infty} (F \cap E_j)\right) + m^*\left(F \cap \left(\bigcup_{j=1}^{\infty} E_j\right)^c\right) \\ &= m^*\left(F \cap \left(\bigcup_{j=1}^{\infty} E_j\right)\right) + m^*\left(F \cap \left(\bigcup_{j=1}^{\infty} E_j\right)^c\right) \geq m^*(F), \end{aligned} \tag{18.14}$$

where the final inequality in (18.14) is true by subadditivity of m^*. So all the inequalities in (18.14) are in fact equalities. This implies the Carathéodory condition in (18.6) which means $\bigcup_{j=1}^{\infty} E_j \in \mathcal{G}^*$.

Thus, we conclude that $\mathcal{G}^*$ is a σ-algebra.

2. Now we show that m^* is a measure on $\mathcal{G}^*$, namely it satisfies the measure axioms in Definition 18.5.1. We just have to show σ-additivity since the other conditions are already satisfied as shown in Lemma 18.4.3. In (18.14), we have seen that

all the inequalities are in fact equalities. In particular, for any countably many disjoint subsets $\{E_j\}_{j=1}^{\infty}$ where $E_j \in \mathcal{G}^*$ for all $j \in \mathbb{N}$ we must have:

$$\sum_{j=1}^{\infty} m^*(F \cap E_j) = m^*\left(F \cap \left(\bigcup_{j=1}^{\infty} E_j\right)\right) \quad \text{for any } F \in \mathcal{G}. \tag{18.15}$$

So, by setting $F = \bigcup_{j=1}^{\infty} E_j \in \mathcal{G}$ in (18.15), we have:

$$\sum_{j=1}^{\infty} m^*(E_j) = m^*\left(\bigcup_{j=1}^{\infty} E_j\right),$$

for arbitrary disjoint sets $E_j \in \mathcal{G}^*$. Thus, m^* is σ-additive on $\mathcal{G}^*$ and hence is a measure on the σ-algebra $\mathcal{G}^*$. □

Thus, the Carathéodory condition allows us to construct a genuine measure space from any outer measure space which is a very useful construction to have!

18.7 Lebesgue and Borel σ-Algebra

In this section, we are going to focus our attention on $X = \mathbb{R}$ and determine the spaces which comes from the extension of the premeasure space $(\mathbb{R}, \mathcal{R}(\mathcal{J}), m)$ in Example 18.3.9.

Lebesgue σ-Algebra

We call the sets in $\mathcal{P}(\mathbb{R})$ that satisfy the Carathéodory condition with respect to the outer measure m^* the Lebesgue σ-algebra which is denoted as $\mathcal{L} \subseteq \mathcal{P}(\mathbb{R})$. By Theorem 18.6.4, the outer measure m^* restricted to the σ-algebra $\mathcal{L}$ is a genuine measure. We denote this restriction as $m^*|_{\mathcal{L}} = \mu$. Moreover, this set is an extension of the premeasure space $(\mathbb{R}, \mathcal{R}(\mathcal{J}), m)$ that we have constructed in Example 18.3.9. Indeed, we have:

Proposition 18.7.1 *$\mathcal{R}(\mathcal{J}) \subseteq \mathcal{L}$ and $\mu|_{\mathcal{R}(\mathcal{J})} = m$.*

Proof Fix any $E \in \mathcal{R}(\mathcal{J})$. We need to show that this set satisfies the Carathéodory condition. Pick any $F \in \mathcal{P}(\mathbb{R})$. We have two cases:

1. If $F \subseteq E$ or $F \subseteq E^c$, then clearly $m^*(F) \geq m^*(F \cap E) + m^*(F \cap E^c)$ since one of the terms on the RHS is 0.

2. Otherwise, suppose that $\{I_j\}_{j=1}^\infty \subseteq \mathcal{R}(\mathcal{J})$ is a cover for the set F, namely $F \subseteq \bigcup_{j=1}^\infty I_j$. Then the collection $\{I_j \cap E\}_{j=1}^\infty \cup \{I_j \cap E^c\}_{j=1}^\infty \subseteq \mathcal{R}(\mathcal{J})$ is also a cover for the set F. Moreover, $\{I_j \cap E\}_{j=1}^\infty$ covers $F \cap E$ and $\{I_j \cap E^c\}_{j=1}^\infty$ covers $F \cap E^c$. By the additivity of premeasure m, we have $m(I_j) = m(I_j \cap E) + m(I_j \cap E^c)$ for all $j \in \mathbb{N}$. By the σ-additivity of the premeasure m and definition of outer measure, we then have:

$$\sum_{j=1}^{\infty} m(I_j) = \sum_{j=1}^{\infty} m(I_j \cap E) + \sum_{j=1}^{\infty} m(I_j \cap E^c) \geq m^*(F \cap E) + m^*(F \cap E^c). \tag{18.16}$$

   Since the inequality (18.16) is true for any cover of F, taking the infimum on the LHS over all possible covers of the set F yields $m^*(F) \geq m^*(F \cap E) + m^*(F \cap E^c)$, giving us the required Carathéodory condition in (18.7).

Finally, since $\mathcal{R}(\mathcal{J}) \subseteq \mathcal{L}$, we have $\mu|_{\mathcal{R}(\mathcal{J})} = (m^*|_{\mathcal{L}})|_{\mathcal{R}(\mathcal{J})} = m^*|_{\mathcal{R}(\mathcal{J})} = m$. □

Therefore, μ extends the premeasure m to a larger collection of sets in $\mathbb{R}$, namely to $\mathcal{L}$. This resulting measure space $(\mathbb{R}, \mathcal{L}, \mu)$ is called the Lebesgue space, the sets in $\mathcal{L}$ are called Lebesgue measurable sets, and the measure μ is called the Lebesgue measure. In fact, by construction, it is the largest possible extension of the premeasure space $(\mathbb{R}, \mathcal{R}(\mathcal{J}), m)$.

Example 18.7.2 We have seen in Proposition 18.7.1 that any element in $\mathcal{R}(\mathcal{J})$ are Lebesgue measurable. This means any half-closed intervals $(a, b]$ for any $a < b$ are also Lebesgue measurable. Also:

1. Any singleton set $\{a\}$ is Lebesgue measurable. From Example 18.4.4, we have shown $m^*(\{a\}) = 0$. By Lemma 18.6.3, since $\{a\}$ is a m^*-null set, it satisfies the Carathéodory condition. Thus any singleton set is contained in $\mathcal{L}$ and has Lebesgue measure 0, namely $\mu(\{a\}) = 0$ for all $a \in \mathbb{R}$.
2. Any finite or countable sets are also Lebesgue measurable since it is a countable union of m^*-null sets and thus is m^*-null itself by σ-subadditivity of m^*. This set also has 0 Lebesgue measure.
   On the other hand, there are also uncountable sets with measure zero. An example can be seen in Exercise 18.26.
3. Intervals of the form $[a, b]$ for $a < b$ are also Lebesgue measurable since they can be obtained via a union of a half-closed interval $(a, b]$ with a m^*-null set $\{a\}$. Moreover, by the property of outer measure in Lemma 18.4.3(4), their Lebesgue measure are:

$$\mu([a, b]) = m^*([a, b]) = m^*(\{a\} \cup (a, b]) = m^*((a, b]) = m((a, b]) = b - a.$$

Likewise, the intervals (a, b) and $[a, b)$ for $a < b$ are also Lebesgue measureable with measure $\mu((a, b)) = \mu([a, b]) = b - a$ as well.

4. In particular, any open set in $\mathbb{R}$ is Lebesgue measurable since they can be expressed as a countable union of open intervals according to Theorem 4.5.20. Consequently, any closed set in $\mathbb{R}$ are also Lebesgue measurable since they are simply complements of open sets.
5. In the album *Asphalt Meadows*, Ben Gibbard sang *"I wanna know the measure from here to forever."* Suppose that he is at the point $x = b$ in $\mathbb{R}$. Assuming that he is working with the Lebesgue measure, this can be interpreted as him asking us the measure of the set $X = [b, \infty)$ which is in $\mathcal{L}$ since it is a closed set.
The set X can be decomposed as the countable union of disjoint sets $X = \bigcup_{j=1}^{\infty}[b + j - 1, b + j)$, which are all Lebesgue measurable with measure 1 each. Hence, by σ-additivity of μ, we can compute:
$$\mu(X) = \mu\left(\bigcup_{j=1}^{\infty}[b + j - 1, b + j)\right) = \sum_{j=1}^{\infty}\mu([b+j-1, b+j)) = \sum_{j=1}^{\infty}1 = \infty.$$
6. Likewise, the universe $\mathbb{R}$ also has ∞ measure. Therefore, the Lebesgue measure space is not a finite measure space. However, it is σ-finite since the universe can be expressed as a countable union of sets of finite measures. For example, we have $\mathbb{R} = \bigcup_{n\in\mathbb{N}}(-n, n)$ where $\mu((-n, n)) = 2n < \infty$ for all $n \in \mathbb{N}$.

Apart from the Carathéodory condition, let us give another characterisation of Lebesgue measurable sets. The following result says that any Lebesgue measurable set can be approximated closely by an open set from the outside.

Proposition 18.7.3 *Let* $(\mathbb{R}, \mathcal{P}(\mathbb{R}), m^*)$ *be the outer measure space induced by the premeasure space* $(\mathbb{R}, \mathcal{R}(\mathcal{J}), m)$. *The set* $E \in \mathcal{P}(\mathbb{R})$ *is Lebesgue measurable if and only if for every* $\varepsilon > 0$ *there exists an open set* $U \subseteq \mathbb{R}$ *such that* $E \subseteq U$ *and* $m^*(U \setminus E) < \varepsilon$.

Proof We prove the implications separately:

($\Rightarrow$): Assume E is Lebesgue measurable, so we $\mu(E)$ is well-defined. We have two cases here, either $\mu(E) < \infty$ or $\mu(E) = \infty$.

1. If $\mu(E) < \infty$, using the characterisation for infimum in the definition of $m^*(E) = \mu(E)$, there exist countably many sets $\{I_j\}_{j=1}^{\infty}$ with $I_j \in \mathcal{R}(\mathcal{J})$ such that $E \subseteq \bigcup_{j=1}^{\infty} I_j$ which satisfies:
$$\sum_{j=1}^{\infty}\mu(I_j) - \frac{\varepsilon}{2} < \mu(E). \tag{18.17}$$

Since $I_j \in \mathcal{R}(\mathcal{J})$, by Proposition 18.3.7, I_j is a finite union of disjoint intervals of the form $(a, b]$. WLOG, we can assume that $I_j = (a_j, b_j]$ for all $j \in \mathbb{N}$. For each $j \in \mathbb{N}$, define the open interval $U_j = (a_j, b_j + \frac{\varepsilon}{2^{j+1}}) \supseteq I_j$ so that:

$$\mu(U_j) = b_j + \frac{\varepsilon}{2^{j+1}} - a_j = \mu(I_j) + \frac{\varepsilon}{2^{j+1}}. \tag{18.18}$$

Denote $U = \bigcup_{j=1}^{\infty} U_j$ which is an open (and hence Lebesgue measurable) set and contains E. Then, using the estimates (18.17) and (18.18), we have:

$$\mu(U) \leq \sum_{j=1}^{\infty} \mu(U_j) = \sum_{j=1}^{\infty} \left(\mu(I_j) + \frac{\varepsilon}{2^{j+1}}\right) = \sum_{j=1}^{\infty} \mu(I_j) + \frac{\varepsilon}{2} < \mu(E) + \varepsilon,$$

which implies $m^*(U \setminus E) = \mu(U \setminus E) = \mu(U) - \mu(E) < \varepsilon$.

2. If $\mu(E) = \infty$, denote $E_n = E \cap [-n, n]$ for all $n \in \mathbb{N}$ so that $E_n \subseteq E$. Then, $\mu(E_n) < \infty$ and, by the previous case, there exists an open set U_n such that $E_n \subseteq U_n$ and $\mu(U_n \setminus E_n) < \frac{\varepsilon}{2^n}$ for all $n \in \mathbb{N}$. Define $U = \bigcup_{n=1}^{\infty} U_n$. Thus, $E \subseteq U$ and $U \setminus E = \bigcup_{n=1}^{\infty}(U_n \setminus E) \subseteq \bigcup_{n=1}^{\infty}(U_n \setminus E_n)$. Hence:

$$\mu(U \setminus E) \leq \mu\left(\bigcup_{n=1}^{\infty}(U_n \setminus E_n)\right) \leq \sum_{n=1}^{\infty} \mu(U_n \setminus E_n) \leq \sum_{n=1}^{\infty} \frac{\varepsilon}{2^n} = \varepsilon.$$

($\Leftarrow$): From the assumption, for every $\varepsilon = \frac{1}{n} > 0$, we can find an open set U_n such that $E \subseteq U_n$ and $m^*(U_n \setminus E) < \frac{1}{n}$. Denote $U = \bigcap_{n=1}^{\infty} U_n$. Then, $E \subseteq U$ and $U \setminus E \subseteq U_n \setminus E$ for all $n \in \mathbb{N}$. Hence, $m^*(U \setminus E) \leq m^*(U_n \setminus E) < \frac{1}{n}$ for any $n \in \mathbb{N}$. This implies $m^*(U \setminus E) = 0$.
Lemma 18.6.3 then says $U \setminus E$ is Lebesgue measurable. We know that U is Lebesgue measurable since it is a countable intersection of open (and hence Lebesgue measurable) sets. This means $E = U \setminus (U \setminus E)$ is also Lebesgue measurable. □

A direct corollary to Proposition 18.7.3 which can be proven using complements is the following result:

Proposition 18.7.4 *Let $(\mathbb{R}, \mathcal{P}(\mathbb{R}), m^*)$ be the outer measure space induced by the premeasure space $(\mathbb{R}, \mathcal{R}(\mathcal{J}), m)$. The set $E \in \mathcal{P}(\mathbb{R})$ is Lebesgue measurable if and only if for every $\varepsilon > 0$ there exists a closed set $K \subseteq \mathbb{R}$ such that $K \subseteq E$ and $m^*(E \setminus K) < \varepsilon$.*

Analogous to Proposition 18.7.3, Proposition 18.7.4 asserts that any Lebesgue measurable set can be approximated by a closed set from the inside.

A distinctive property of the Lebesgue space is that it is a complete measure space. We define what complete measure space means first:

Definition 18.7.5 (Complete Measure Space) A measure space $(X, \mathcal{F}, \mu)$ is called a complete measure space or μ-complete if for every $E \in \mathcal{F}$ with $\mu(E) = 0$, any subset of E is also in $\mathcal{F}$.

Example 18.7.6 The Lebesgue measure space $(\mathbb{R}, \mathcal{L}, \mu)$ is a complete measure space. This is true by construction. We started with the power set $\mathcal{P}(\mathbb{R})$ which includes any m^*-null sets, along with their subsets which are also m^*-null. These null sets were not eliminated via the Carathéodory condition since they are already m^*-measurable by Lemma 18.6.3. Moreover, any subsets of these sets were also not eliminated as they are also m^*-null by σ-additivity of m^*. So, $\mathcal{L}$ contains all the m^*-null sets and their subsets.

We note that the Lebesgue space $\mathcal{L}$ is not all of $\mathcal{P}(\mathbb{R})$. Though it is quite difficult to construct or write down a set which is not in $\mathcal{L}$, we have seen that they do exist in Example 18.4.5. In fact, using the same construction, we have the following result which shows that the non-measurable subsets are abundant in $\mathbb{R}$.

Lemma 18.7.7 *Every set in $\mathcal{L}$ with positive measure contains a non-Lebesgue measurable subset.*

Borel σ-Algebra

Recall from Example 18.3.20(1) the Borel σ-algebra $\mathcal{B}$ which is the σ-algebra in $\mathbb{R}$ generated by all open sets in $\mathbb{R}$. Via construction, we can expect that the Borel σ-algebra is contained in the Lebesgue σ-algebra.

Intuitively this is true: the Borel sets were formed by generating a σ-algebra from the collection of open sets whereas the Lebesgue σ-algebra was obtained by starting with the largest possible collection of sets in $\mathbb{R}$ (namely $\mathcal{P}(\mathbb{R})$) and discarding any problematic sets via the Carathéodory condition. Moreover, the Lebesgue sets contain all of the open sets in $\mathbb{R}$ as we have remarked in Example 18.7.2(4). So we expect $\mathcal{B} \subseteq \mathcal{L}$. But do these two collections of sets somehow meet in the middle and coincide? Not quite.

Proposition 18.7.8 *Let $\mathcal{L}$ be the Lebesgue σ-algebra on $\mathbb{R}$ and $\mathcal{B}$ be the Borel σ-algebra generated by the open sets in $\mathbb{R}$. Then:*

1. *$\mathcal{B} \subseteq \mathcal{L}$.*
2. *There are Lebesgue measurable sets which are not Borel measurable.*

Proof We prove the first assertion only.

1. We have seen in Example 18.3.20 that $\sigma(\mathcal{J}) = \mathcal{B}$. The σ-algebra $\mathcal{L}$ contains all of the sets in $\mathcal{J}$ as seen in Proposition 18.7.1. Since $\mathcal{L}$ is a σ-algebra containing $\mathcal{J}$, we then have $\mathcal{B} = \sigma(\mathcal{J}) \subseteq \mathcal{L}$.

The readers shall construct a set that satisfies the second assertion in Exercise 18.27. □

As a result of Proposition 18.7.8, we can equip the Borel sets with the Lebesgue measure via the restriction $\mu|_{\mathcal{B}}$ (which we are going to denote as μ as well) to get the measure space $(\mathbb{R}, \mathcal{B}, \mu)$ which we call the Borel measure space. Furthermore, from the second assertion of Proposition 18.7.8, we can always find a Lebesgue set which is not Borel. Thus, the Borel measure space is strictly smaller than the Lebesgue measure space.

Remark 18.7.9 In contrast to the Lebesgue measure space, the Borel measure space is not complete. Here is a rough sketch of the proof: In Exercise 18.26, the readers are invited to show that the Cantor set C is Borel measurable with $\mu(C) = 0$. Using advanced set theory via transfinite induction or via an alternative method in [41], it can be shown that the cardinality of the Borel σ-algebra is equal to the cardinality of $\mathbb{R}$, namely $|\mathcal{B}| = |\mathbb{R}|$. On the other hand, Exercise 4.32(i) showed that $|C| = |\mathbb{R}|$ as well. Thus, by Theorem 3.9.6, the number of subsets of C is $|\mathcal{P}(C)| > |C| = |\mathbb{R}| = |\mathcal{B}|$. This means there are strictly more subsets of C than there are Borel sets, so there are subsets of the μ-null set C which are not Borel measurable. Thus, the Borel measure space $(\mathbb{R}, \mathcal{B}, \mu)$ is not a complete measure space.

A summary of the construction for the σ-algebras $\mathcal{L}$ and $\mathcal{B}$ from scratch that we have done in this section is given in Fig. 18.2.

A direct corollary of Propositions 18.7.3 and 18.7.4 is the following result:

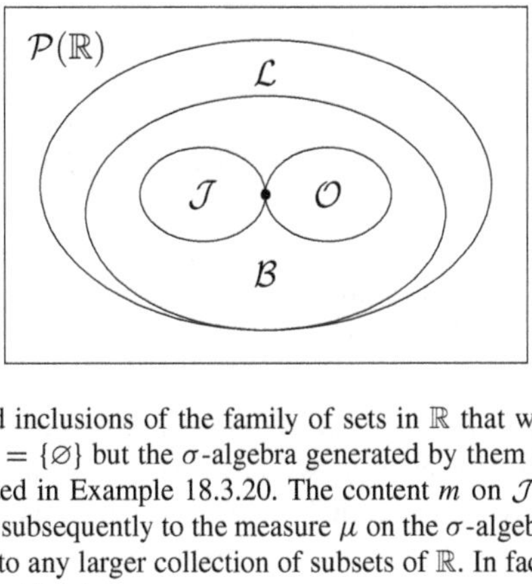

Fig. 18.2 Extensions and inclusions of the family of sets in $\mathbb{R}$ that we have constructed in this chapter. Note that $\mathcal{J} \cap \mathcal{O} = \{\emptyset\}$ but the σ-algebra generated by them are both equal to $\sigma(\mathcal{J}) = \sigma(\mathcal{O}) = \mathcal{B}$ as demonstrated in Example 18.3.20. The content m on $\mathcal{J}$ has been extended to the premeasure on $\mathcal{R}(\mathcal{J})$ and subsequently to the measure μ on the σ-algebra $\mathcal{L}$. However, we cannot extend the premeasure m to any larger collection of subsets of $\mathbb{R}$. In fact, this extension is unique as we shall see in Theorem 18.8.4

Proposition 18.7.10 *If $E \in \mathcal{L}$, then there are sets $A, B \in \mathcal{B}$ such that $A \subseteq E \subseteq B$ with $\mu(E \setminus A) = \mu(B \setminus E) = 0$.*

Proof We show only the existence of the set B. By Proposition 18.7.3, for every $n \in \mathbb{N}$, there is an open set $B_n \in \mathcal{B}$ such that $E \subseteq B_n$ and $\mu(B_n \setminus E) < \frac{1}{n}$. Define $B = \bigcap_{n=1}^{\infty} B_n \in \mathcal{B}$. Then, $E \subseteq B$ and $\mu(B \setminus E) \leq \mu(B_n \setminus E) < \frac{1}{n}$ for all $n \in \mathbb{N}$. Hence, $\mu(B \setminus E) = 0$.

The existence of the set A can be proven in a similar manner using Proposition 18.7.4. □

Proposition 18.7.10 says that any Lebesgue set differs from a Borel set by some set of zero Lebesgue measure. Therefore, the main difference between Borel and Lebesgue spaces $\mathcal{B}$ and $\mathcal{L}$ are not much: just sets of zero measure. We can state the construction of Lebesgue sets from Borel sets more rigorously via the following theorem:

Theorem 18.7.11 *Let $\mathcal{L}$ and $\mathcal{B}$ be the Lebesgue and Borel σ-algebras on $\mathbb{R}$ respectively and μ is the Lebesgue measure. Then:*

$$\mathcal{L} = \{E \cup N : E \in \mathcal{B}, N \subseteq F \in \mathcal{B} \text{ such that } \mu(F) = 0\}.$$

Proof Denote the set:

$$\mathcal{C} = \{E \cup N : E \in \mathcal{B}, N \subseteq F \in \mathcal{B} \text{ such that } \mu(F) = 0\}.$$

The collection $\mathcal{C}$ is a σ-algebra, which is left for the readers to check in Exercise 18.13. Our goal now is to show that $\mathcal{C} = \mathcal{L}$ by double inclusion.

($\subseteq$): Any $A \in \mathcal{C}$ must be of the form $A = E \cup N$ where $E \in \mathcal{B}$ and N is a subset of a μ-null set F. Since $\mathcal{B} \subseteq \mathcal{L}$ and $\mathcal{L}$ is μ-complete, necessarily $E, N \in \mathcal{L}$ and thus $A = E \cup N \in \mathcal{L}$. Therefore, $\mathcal{C} \subseteq \mathcal{L}$.

($\supseteq$): Now we show the opposite inclusion. Pick any $E \in \mathcal{L}$. We have two cases:

1. First, suppose that $\mu(E) < \infty$. By Proposition 18.7.10, there exists a Borel set $B \in \mathcal{B}$ such that $B \subseteq E$ and $\mu(E \setminus B) = 0$. Denote $N = E \setminus B$ so that $E = B \cup N$. Note that N is a μ-null set in $\mathcal{L}$. Moreover, by Proposition 18.7.10, there exists a set $A \in \mathcal{B}$ such that $N \subseteq A$ with $\mu(A \setminus N) = 0$. This implies $\mu(A) = \mu(A \cap N) + \mu(A \setminus N) = \mu(N) + \mu(A \setminus N) = 0$. Thus, N is a subset of some μ-null set A in $\mathcal{B}$ and hence $E = B \cup N \in \mathcal{C}$.
2. Otherwise, if $\mu(E) = \infty$, consider the family of sets $E_n = E \cap [-n, n] \in \mathcal{L}$ for $n \in \mathbb{N}$. For each n, we have $\mu(E_n) < \infty$ and, by repeating the argument above, we can deduce $E_n \in \mathcal{C}$ for every n. Thus, the countable union $E = \bigcup_{n=1}^{\infty} E_n$ is also in $\mathcal{C}$ since $\mathcal{C}$ is a σ-algebra. □

In mathematical language, we say that $\mathcal{L}$ is a completion of $\mathcal{B}$. In order to complete the Borel σ-algebra to form the Lebesgue σ-algebra, based on the result in Theorem 18.7.11, we simply append all the sets in the Borel σ-algebra with any subsets of any Borel set of measure 0 to each of them. This is called the completion of the Borel σ-algebra relative to the Lebesgue measure μ.

Moreover, from the Lebesgue measure space, we can create other measure spaces too. Recall Lemma 18.3.21 in which we can create new σ-algebra from old via restriction. However, in order for this new σ-algebra to inherit the Lebesgue measure μ, we need to define this restriction carefully to ensure that the measure on it exists.

Definition 18.7.12 (Subspace Measure) Let $(\mathbb{R}, \mathcal{L}, \mu)$ be the Lebesgue measure space. Let $X \subseteq \mathbb{R}$ be a Lebesgue measurable set.

1. Then, $\mathcal{G} = \{E \cap X : E \in \mathcal{L}\}$ is a σ-algebra on X.
2. Moreover this σ-algebra can be equipped with the measure $\mu_X : \mathcal{G} \to [0, \infty]$ where for any $F = E \cap X \in \mathcal{G}$ we define $\mu_X(F) = \mu(F) = \mu(E \cap X)$. We call the measure μ_X the subspace measure induced by $(\mathbb{R}, \mathcal{L}, \mu)$ on $(X, \mathcal{G})$.

The resulting measure space $(X, \mathcal{G}, \mu_X)$ is called the induced measure space on X from $(\mathbb{R}, \mathcal{L}, \mu)$. By an abuse of notation, we usually write it as $(X, \mathcal{L}, \mu)$.

The set function μ_X above is well-defined since for any $F \in \mathcal{G}$ we must have $F = E \cap X$ for some $E \in \mathcal{L}$. Moreover, since $X \in \mathcal{L}$, necessarily $F = E \cap X \in \mathcal{L}$ and thus $\mu_X(F) = \mu(F)$ has a value.

Example 18.7.13 Consider the set $X = [0, 1] \subseteq \mathbb{R}$. This is a Lebesgue set so we can turn it into a measure space with measure inherited from the Lebesgue measure μ on $\mathbb{R}$. For simplicity, by an abuse of notation, we denote the resulting measure space as $(X, \mathcal{L}, \mu)$ and call it the induced Lebesgue measure space.

1. Recall from Example 18.7.2(2) that any countable set in $\mathbb{R}$ have zero measure. The set of rational numbers in $[0, 1]$ is countable and thus $\mu(\mathbb{Q} \cap [0, 1]) = 0$.
2. On the other hand, we know that $\mu([0, 1]) = 1$ and the set of rational numbers and irrational numbers in $[0, 1]$ are disjoint. Thus:

$$\begin{aligned}1 = \mu([0, 1]) &= \mu(\mathbb{Q} \cap [0, 1]) + \mu(\bar{\mathbb{Q}} \cap [0, 1]) \\ &= 0 + \mu(\bar{\mathbb{Q}} \cap [0, 1]) = \mu(\bar{\mathbb{Q}} \cap [0, 1]),\end{aligned}$$

which implies that the set of irrational numbers in $[0, 1]$ has a very big measure compared to the rational numbers.

Note that this example answers the question posed in Example 18.0.1 on how we can assign a size to the set $\mathbb{Q} \cap [0, 1]$. Under the Lebesgue measure, its size is 0.

18.8 Uniqueness of Carathéodory Extension Theorem

One final remark that we would like to make regarding measures is the uniqueness of measure extended by the Carathéodory extension theorem.

Recall that originally we wanted to extend the premeasure m on $\mathcal{R}(\mathcal{J})$ to the σ-algebra generated by it, namely $\mathcal{B} = \sigma(\mathcal{R}(\mathcal{J}))$. We managed to construct one such measure on $\mathcal{B}$ in a very roundabout way. As depicted in Fig. 18.3, we extended it to a measure μ on a larger σ-algebra $\mathcal{L}$ and then restricted it down to $\mathcal{B} \subseteq \mathcal{L}$. We have shown that this measure μ agrees with the premeasure m on $\mathcal{R}(\mathcal{J})$. But now we might worry: is this the only measure on $\mathcal{B}$ that agrees with the premeasure m on $\mathcal{R}(\mathcal{J})$? Are we missing any other measures on $\mathcal{B}$ that might come about via a different construction?

Such a measure is unique if the premeasure space is σ-finite. In order to prove this, first, we define what a λ-system is. This system is also known as Dynkin's system, named after Eugene Dynkin (1924–2014).

Definition 18.8.1 (λ-System) Let X be a set and $\mathcal{D} \subseteq \mathcal{P}(X)$ be a collection of subsets of X. $\mathcal{D}$ is called a λ-system if:

1. $X \in \mathcal{D}$,
2. if $E, F \in \mathcal{D}$ with $E \subseteq F$, then $F \setminus E \in \mathcal{D}$, and
3. for any nested sequence of sets $E_1 \subseteq E_2 \subseteq \ldots$ in $\mathcal{D}$, we have $\bigcup_{j=1}^{\infty} E_j \in \mathcal{D}$.

We note that for any two λ-system of X, say $\mathcal{C}$ and $\mathcal{D}$, their intersection $\mathcal{C} \cap \mathcal{D}$ is also a λ-system. In fact, any arbitrary intersection of λ-systems is a λ-system.

Similar to rings and algebras, for a fixed universe X, given any collection of sets $K \subseteq \mathcal{P}(X)$, we can define the smallest λ-system generated by K, denoted by $\delta(K)$, as the intersection of all the λ-system containing K. This intersection is non-empty since $\mathcal{P}(X)$ itself is a λ-system containing K. We also leave the proof of the following result as Exercise 18.14:

Proposition 18.8.2 *Let X be a set and $\mathcal{F} \subseteq \mathcal{P}(X)$. Then, $\mathcal{F}$ is a σ-algebra on X if and only if it is both a π-system and a λ-system.*

With the definition and result above, we now state and prove Dynkin's π-λ lemma:

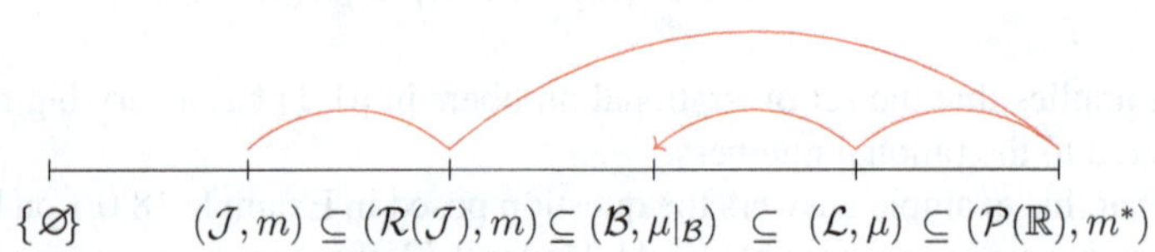

Fig. 18.3 How we constructed the Borel measure space $(\mathbb{R}, \mathcal{B}, \mu|_{\mathcal{B}})$ via a sequence of extensions and restrictions from the semiring $\mathcal{J}$ and content m

Lemma 18.8.3 (Dynkin's π-λ Lemma) *Let X be a set. Suppose that $\mathcal{S}$ and $\mathcal{D}$ are collections of subsets of X such that $\mathcal{S}$ is a π-system and $\mathcal{D}$ is a λ-system. If $\mathcal{S} \subseteq \mathcal{D}$, then $\sigma(\mathcal{S}) \subseteq \mathcal{D}$.*

Proof Denote $\delta(\mathcal{S})$ as the λ-system generated by $\mathcal{S}$. Since $\mathcal{D}$ is also a λ-system containing $\mathcal{S}$, we have the inclusion $\delta(\mathcal{S}) \subseteq \mathcal{D}$.

Now let us show that $\delta(\mathcal{S})$ is a σ-algebra. To do this, by virtue of Proposition 18.8.2, it is enough to show that $\delta(\mathcal{S})$ is also a π-system. For a fixed set $B \in \delta(\mathcal{S})$, we define the collection of sets:

$$\mathcal{D}_B = \{E \in \delta(\mathcal{S}) : E \cap B \in \delta(\mathcal{S})\} \subseteq \delta(\mathcal{S}).$$

It is a routine exercise, which we leave to the readers, to check that $\mathcal{D}_B$ is a λ-system for any $B \in \delta(\mathcal{S})$.

Next, note that if $A \in \mathcal{S}$, since $\mathcal{S}$ is a π-system, for any $C \in \mathcal{S}$ we have $A \cap C \in \mathcal{S} \subseteq \delta(\mathcal{S})$. Thus, by definition of the set $\mathcal{D}_A$, we have $C \in \mathcal{D}_A$ for all $C \in \mathcal{S}$. In other words, we have $\mathcal{S} \subseteq \mathcal{D}_A$ for any $A \in \mathcal{S}$. This means $\mathcal{D}_A$ is a λ-system which also contains $\mathcal{S}$. Since $\delta(\mathcal{S})$ is the smallest λ-system containing $\mathcal{S}$, we must have $\delta(\mathcal{S}) \subseteq \mathcal{D}_A$ for any $A \in \mathcal{S}$.

Now we define a new collection of sets:

$$\mathcal{E} = \{A \cap B : A \in \mathcal{S}, B \in \delta(\mathcal{S})\}.$$

We claim that this collection of sets is contained in $\delta(\mathcal{S})$. Indeed, pick any element $A \cap B \in \mathcal{E}$ where $A \in \mathcal{S}$ and $B \in \delta(\mathcal{S})$. Since $\delta(\mathcal{S}) \subseteq \mathcal{D}_A$, we have $B \in \delta(\mathcal{S}) \subseteq \mathcal{D}_A$. By definition of $\mathcal{D}_A$, we then have $A \cap B \in \delta(\mathcal{S})$. Since $A \cap B \in \mathcal{E}$ is arbitrary, we have the inclusion $\mathcal{E} \subseteq \delta(\mathcal{S})$.

Using the above fact, for a fixed $B \in \delta(\mathcal{S})$, for any $A \in \mathcal{S}$ we have $A \cap B \in \mathcal{E} \subseteq \delta(\mathcal{S})$. By definition of $\mathcal{D}_B$, we then have $\mathcal{S} \subseteq \mathcal{D}_B$. Since $\mathcal{D}_B$ is a λ-system containing $\mathcal{S}$, by minimality of $\delta(\mathcal{S})$, we have the inclusion $\delta(\mathcal{S}) \subseteq \mathcal{D}_B$ for any arbitrary $B \in \delta(\mathcal{S})$.

Now we show that $\delta(\mathcal{S})$ is closed under intersections. Pick any arbitrary $E, F \in \delta(\mathcal{S})$. The previous paragraph implies that $E \in \delta(\mathcal{S}) \subseteq \mathcal{D}_F$. By definition of $\mathcal{D}_F$, we have $E \cap F \in \delta(\mathcal{S})$. This shows that the set $\delta(\mathcal{S})$ is closed under intersection and hence is a π-system. Thus, by Proposition 18.8.2, we conclude that $\delta(\mathcal{S})$ is a σ-algebra.

Finally, returning to the original problem, recall that we had $\delta(\mathcal{S}) \subseteq \mathcal{D}$. Since we have shown that $\delta(\mathcal{S})$ is a σ-algebra containing $\mathcal{S}$, by minimality of $\sigma(\mathcal{S})$, we also have the inclusion $\sigma(\mathcal{S}) \subseteq \delta(\mathcal{S})$. Putting these together, we conclude that $\sigma(\mathcal{S}) \subseteq \mathcal{D}$. □

Using the Dynkin's π-λ lemma, we now prove the following unique extension theorem for σ-finite premeasure spaces:

Theorem 18.8.4 (Carathéodory Extension Theorem—Uniqueness) *Let $(X, \mathcal{R}, m)$ be a σ-finite premeasure space. If $(X, \mathcal{F})$ is a measurable space with $\mathcal{F} = \sigma(\mathcal{R})$, then there is a unique measure μ on $(X, \mathcal{F})$ such that $\mu(E) = m(E)$ for all $E \in \mathcal{R}$. Moreover, the measure space $(X, \mathcal{F}, \mu)$ is σ-finite.*

Proof The existence of such measure is given via the construction of outer measure and Theorem 18.6.4. Assume that there are two such measures, namely μ_1 and μ_2.

1. First, we assume that X has finite measure, namely $\mu_1(X) = \mu_2(X) < \infty$. Define the collection $\mathcal{E} = \{F \in \mathcal{F} : \mu_1(F) = \mu_2(F)\}$. By assumption, we have $\mathcal{R} \subseteq \mathcal{E} \subseteq \mathcal{F}$. It is routine to check that $\mathcal{E}$ is a λ-system. Thus, by Dynkin's π-λ lemma, we have $\sigma(\mathcal{R}) \subseteq \mathcal{E} \subseteq \mathcal{F} = \sigma(\mathcal{R})$ which then implies $\mathcal{E} = \mathcal{F}$. This says $\mu_1 = \mu_2$ as they agree on the whole of $\mathcal{F}$.
2. Now we prove the general case. Since the premeasure space is σ-finite, there exists a sequence of disjoint sets $\{E_j\}_{j=1}^{\infty}$ where $E_j \in \mathcal{R}$ such that $\bigcup_{j=1}^{\infty} E_j = X$ and $\mu_1(E_j) = \mu_2(E_j) = m(E_j) < \infty$ for each $j \in \mathbb{N}$. For each $j \in \mathbb{N}$, define $\mathcal{E}_j = \{F \in \mathcal{F} : \mu_1(F \cap E_j) = \mu_2(F \cap E_j)\}$.
   We can again check that this is a λ-system with $\mathcal{R} \subseteq \mathcal{E}_j$ for all $j \in \mathbb{N}$. Thus, we have the inclusions $\mathcal{R} \subseteq \mathcal{E}_j \subseteq \mathcal{F}$. By applying the finite measure case for each of the cases here, we have $\mathcal{F} = \mathcal{E}_j$ for all $j \in \mathbb{N}$. Finally, for any $A \in \mathcal{F}$ we can write it as the disjoint union $A = \bigcup_{j=1}^{\infty}(E_j \cap A)$. Using the fact that $A \in \mathcal{F} = \mathcal{E}_j$ for all $j \in \mathbb{N}$, we have:

$$
\begin{aligned}
\mu_1(A) = \mu_1\left(\bigcup_{j=1}^{\infty} E_j \cap A\right) &= \sum_{j=1}^{\infty} \mu_1(E_j \cap A) \\
&= \sum_{j=1}^{\infty} \mu_2(E_j \cap A) \\
&= \mu_2\left(\bigcup_{j=1}^{\infty} E_j \cap A\right) = \mu_2(A).
\end{aligned}
$$

   Since $A \in \mathcal{F}$ is arbitrary, we can conclude that $\mu_1 = \mu_2$.

Finally, since the premeasure space $(X, \mathcal{R}, m)$ is σ-finite, there exists countably many sets $\{E_j\}_{j=1}^{\infty}$ such that $E_j \in \mathcal{R} \subseteq \mathcal{F}$, $\mu(E_j) = m(E_j) < \infty$, and $\bigcup_{j=1}^{\infty} E_j = X$. This means the measure space $(X, \mathcal{F}, \mu)$ is also σ-finite. □

The σ-finite condition on the premeasure space is necessary for uniqueness of the extended measure. This can be seen in the following non-example:

Example 18.8.5 Let $\mathcal{J} = \{(c, d] \subseteq \mathbb{R} : c < d\} \cup \{\varnothing\}$ be a semiring on $\mathbb{R}$ with content $m(I) = \infty$ if $I \neq \varnothing$ and $m(\varnothing) = 0$. The semiring and content can then be extended to a premeasure space $(\mathbb{R}, \mathcal{R}(\mathcal{J}), m)$ which satisfies $m(E) = \infty$ if $E \neq \varnothing$ and $m(\varnothing) = 0$. This premeasure space is not σ-finite since we cannot decompose the universe $\mathbb{R}$ into a countable union of sets in $\mathcal{R}(\mathcal{J})$ with finite premeasures.

From Example 18.3.20, we have $\sigma(\mathcal{R}(\mathcal{J})) = \mathcal{B}$. We claim that there are at least two distinct measures on the measurable space $(\mathbb{R}, \mathcal{B})$ that are compatible with the premeasure m on $\mathcal{R}(\mathcal{J})$ above.

1. First, we can define a measure $\mu : \mathcal{B} \to [0, \infty]$ by defining $\mu(E) = \infty$ if $E \neq \varnothing$ and $\mu(\varnothing) = 0$. Clearly, for any $E \in \mathcal{B}$, we have $\mu(E) = \infty = m(E)$ if $E \neq \varnothing$ and $\mu(\varnothing) = 0 = m(\varnothing)$. Thus, this measure μ on $\mathcal{B}$ is compatible with m on $\mathcal{R}(\mathcal{J})$.
2. On the other hand, the counting measure $\nu : \mathcal{B} \to [0, \infty]$ similar to the one in Example 18.5.6 also agrees with the premeasure m.
   To see this, notice first that any non-empty set $E \in \mathcal{R}(\mathcal{J})$ always has infinite cardinality. This is true by virtue of Proposition 18.3.7, namely: any non-empty set in $\mathcal{R}(\mathcal{J})$ is a finite union of non-empty intervals $(c, d] \in \mathcal{J}$ which all have infinite cardinality. Thus, if $E \in \mathcal{R}(\mathcal{J})$ is non-empty, we have $\nu(E) = \infty = m(E)$. Moreover, trivially $\nu(\varnothing) = 0 = m(\varnothing)$. Thus, the counting measure ν on $\mathcal{B}$ is also compatible with m on $\mathcal{R}(\mathcal{J})$.

Therefore, there are at least two distinct measures on the measurable space $(\mathbb{R}, \mathcal{B})$ which agree with the non σ-finite premeasure space $(\mathbb{R}, \mathcal{R}(\mathcal{J}), m)$.

To conclude this section, we can now confidently say that the construction of Lebesgue measure μ from the premeasure m on $\mathcal{R}(\mathcal{J})$ in Example 18.3.9 yields a unique measure on the Borel σ-algebra $\mathcal{B}$ which agrees with the premeasure m since the premeasure space $(\mathbb{R}, \mathcal{R}(\mathcal{J}), m)$ is σ-finite.

18.9 Measurable Functions

Now that we have properly defined measures and collections of subsets which can be measured, we proceed to define functions which we want to integrate. These functions are called measurable functions, which we define as:

Definition 18.9.1 (Measurable Functions) Let $(X, \mathcal{F})$ and $(Y, \mathcal{E})$ be measurable spaces. The function $f : X \to Y$ is measurable if the preimage of any $E \in \mathcal{E}$ under f is in $\mathcal{F}$, namely $f^{-1}(E) \in \mathcal{F}$.

Remark 18.9.2 We make some remarks regarding Definition 18.9.1.

1. The measurable functions do not require measures but just measurable spaces as the domain and codomain.
2. To put emphasis on the dependence on $\mathcal{E}$ and $\mathcal{F}$, we sometimes write $f : (X, \mathcal{F}) \to (Y, \mathcal{E})$ and say that f is $(\mathcal{F}, \mathcal{E})$-measurable.

In particular, for real-valued functions, we have two special and commonly used measurable structures on the codomain, namely it could be treated as a Lebesgue space or a Borel space.

The Borel σ-algebra is very convenient to work with since we know how the Borel sets look like and how they are generated. Therefore, unless otherwise stated, when we are working with codomain $\mathbb{R}$, we are going to endow it with the Borel σ-algebra. We shall see later in Lemma 18.9.8 how this can make our lives easier. Thus, we define:

Definition 18.9.3 Let $(X, \mathcal{F})$ be a measurable space. A real-valued function $f : X \to \mathbb{R}$ is $\mathcal{F}$-measurable if the preimage of any $E \in \mathcal{B}$ under f is in $\mathcal{F}$, namely $f^{-1}(E) \in \mathcal{F}$.

In particular, we have:

Definition 18.9.4 (Borel Measurable Functions) A function $f : \mathbb{R} \to \mathbb{R}$ is called Borel measurable if for any $E \in \mathcal{B}$, its preimage under f is in $\mathcal{B}$, namely $f^{-1}(E) \in \mathcal{B}$.

Definition 18.9.5 (Lebesgue Measurable Functions) A function $f : \mathbb{R} \to \mathbb{R}$ is called Lebesgue measurable if for any $E \in \mathcal{B}$, its preimage under f is in $\mathcal{L}$, namely $f^{-1}(E) \in \mathcal{L}$.

Borel measurable functions are also Lebesgue measurable functions. However, the converse is not true. This is intuitively clear since there are more sets in a Lebesgue space than there is in a Borel space and thus one could come up with a function $f : \mathbb{R} \to \mathbb{R}$ such that $f^{-1}(E) \in \mathcal{L} \setminus \mathcal{B}$ for some $E \in \mathcal{B}$. Therefore, one has to be careful and be alert with this as warned by Barry Simon (1946-):

> Passing from Borel to Lebesgue measurable functions is the work of the devil. Don't even consider it!

Example 18.9.6 Let us look at some examples and a non-example of real-valued measurable functions.

1. Let $(X, \mathcal{F})$ be a measurable space and $E \in \mathcal{F}$. The indicator function $\mathbf{1}_E : X \to \mathbb{R}$ is $\mathcal{F}$-measurable. Indeed, let $I \in \mathcal{B}$. We have four distinct cases for I, namely:
   (a) If $1 \in I$ but $0 \notin I$, then $\mathbf{1}_E^{-1}(I) = E \in \mathcal{F}$.
   (b) If $0 \in I$ but $1 \notin I$, then $\mathbf{1}_E^{-1}(I) = E^c \in \mathcal{F}$.
   (c) If $1, 0 \in I$, then $\mathbf{1}_E^{-1}(I) = X \in \mathcal{F}$.
   (d) If $1, 0 \notin I$, then $\mathbf{1}_E^{-1}(I) = \varnothing \in \mathcal{F}$.
   Since the preimages of the indicator function $\mathbf{1}_E$ are all in $\mathcal{F}$, we conclude that the indicator function $\mathbf{1}_E$ is $\mathcal{F}$-measurable.

2. A common example of a Borel measurable function is any continuous function. Let $f : \mathbb{R} \to \mathbb{R}$ be a continuous function. By Exercise 10.16, for any open set $I \subseteq \mathbb{R}$, the preimage $f^{-1}(I)$ is open in $\mathbb{R}$. Thus, Theorem 4.5.20 says $f^{-1}(I)$ is a countable union of open intervals and hence must be contained in $\mathcal{B}$.
   To show this for any $E \in \mathcal{B}$, we note that E is generated by open intervals via countable unions, intersections, and complements of open intervals. However, as we have seen in Propositions 1.5.8 and 1.5.10, we know that inverse functions preserve countable unions, intersections, and complements. Hence, $f^{-1}(E)$ is also generated by open intervals via countable unions, intersections, and complements. Thus, it is contained in $\mathcal{B}$.
3. Similar to sets, there are also real-valued functions which are not Lebesgue measurable. An example would be the function $f : [0, 1] \to \mathbb{R}$ such that $f(x) = 1$ if $x \in A$ and $f(x) = -1$ if $x \in A^c$ where A is the Vitali set from Example 18.4.5. This is not Lebesgue measurable since for the Borel set $U = (0, \infty) \in \mathcal{B}$, we have $f^{-1}(U) = A \notin \mathcal{L}$.
   However, such functions are not very common and usually pathological in nature. Most of the functions that one would encounter are measurable. S.R. Srinivasa Varadhan (1940-) humorously quipped:

   > If you can write [a function] down, it's measurable!

   which is obviously not true since we have just written down a non-measurable function above! However, the tip from Varadhan is a good rule of thumb.

Here are some properties of measurable functions.

Proposition 18.9.7 *Let $(X, \mathcal{F})$ be a measurable space.*

1. *For $A \subseteq X$, the indicator function $\mathbf{1}_A : X \to (\mathbb{R}, \mathcal{B})$ is $\mathcal{F}$-measurable if and only if $A \in \mathcal{F}$.*
2. *Let $f : X \to (\mathbb{R}, \mathcal{B})$ be $\mathcal{F}$-measurable and $g : \mathbb{R} \to \mathbb{R}$ be a Borel measurable function. Then, the composition $g \circ f : X \to \mathbb{R}$ is $\mathcal{F}$-measurable.*

Proof We prove the assertions one by one:

1. Note that for any Borel set $E \in \mathcal{B}$, we have four cases, namely: $0, 1 \in E$, $0, 1 \notin E$, $0 \in E$ but $1 \notin E$, and $0 \notin E$ but $1 \in E$. Therefore, $f^{-1}(E)$ is either X, $\varnothing$, $X \setminus A$, or A. Thus, the assertion follows.
2. Pick any Borel set $U \in \mathcal{B}$. Then, $g^{-1}(U) \in \mathcal{B}$ and hence $(g \circ f)^{-1}(U) = f^{-1}(g^{-1}(U)) \in \mathcal{F}$. Thus, $g \circ f$ is $\mathcal{F}$-measurable. □

Example 18.9.6(2) gave us a way of checking whether a real-valued function is measurable: since Borel sets are generated by open intervals, we simply have to check that the preimage of open intervals are all measurable. In fact, we have an easier test than this which is given in the following characterisation:

Lemma 18.9.8 *Suppose that $f : (X, \mathcal{F}) \to (\mathbb{R}, \mathcal{B})$. The following are equivalent:*

1. *The function f is $\mathcal{F}$-measurable.*
2. *$f^{-1}(I) \in \mathcal{F}$ for every open interval $I \subseteq \mathbb{R}$.*
3. *$f^{-1}(J_a) \in \mathcal{F}$ for every set of the form $J_a = (-\infty, a]$ where $a \in \mathbb{R}$.*
4. *$f^{-1}(J_a) \in \mathcal{F}$ for every set of the form $J_a = (-\infty, a)$ where $a \in \mathbb{R}$.*
5. *$f^{-1}(J_a) \in \mathcal{F}$ for every set of the form $J_a = [a, \infty)$ where $a \in \mathbb{R}$.*
6. *$f^{-1}(J_a) \in \mathcal{F}$ for every set of the form $J_a = (a, \infty)$ where $a \in \mathbb{R}$.*

Proof We prove only the equivalence of some of the statements above which can then be combined together to give the full equivalence.

(1) ⇔ (2): This is true by the fact that the Borel σ-algebra is generated by open sets and the preimage operation preserves countable unions, intersections, and complements.

(2) ⇒ (3): For a fixed $a \in \mathbb{R}$, we define the open intervals $I_n = (-\infty, a+\frac{1}{n})$. Then, the set $(-\infty, a]$ can be written as the countable intersection $(-\infty, a] = \bigcap_{n\in\mathbb{N}} I_n$. Hence, we have:

$$f^{-1}(J_a) = f^{-1}\left(\bigcap_{n\in\mathbb{N}} I_n\right) = \bigcap_{n\in\mathbb{N}} f^{-1}(I_n) \in \mathcal{F}.$$

(3) ⇒ (2): We first show that we can produce any open interval (a, b) for $a, b \in \mathbb{R}$ with $a < b$ from these half-infinity intervals by countable unions, intersections, and complements. For finite $a, b \in \mathbb{R}$, we note that $(a, b] = (-\infty, b] \setminus (-\infty, a]$ and $\{b\} = \bigcap_{n\in\mathbb{N}}(b - \frac{1}{n}, b] = \bigcap_{n\in\mathbb{N}}((-\infty, b] \setminus (-\infty, b - \frac{1}{n}])$. Therefore, the set $(a, b) = (a, b] \setminus \{b\}$ can be made up of countable unions, intersections, and complements of intervals of the form $(-\infty, a]$.
Next, the intervals $(-\infty, a)$ and (a, ∞) can be obtained from the bounded open intervals by countable unions. For example, $(a, \infty) = \bigcup_{j=1}^{\infty}(a, a + j)$.
Thus, by using the fact that inverse functions preserve unions, intersections, and complements, we conclude the result.

(2) ⇒ (4): This is trivially true since $(-\infty, a)$ are all open intervals in $\mathbb{R}$.

(4) ⇒ (2): We need to show that for any $a, b \in \mathbb{R}$ with $a < b$, the interval (a, b) can be constructed by countable intersections. For finite $a, b \in \mathbb{R}$, we note that $[a, b) = (-\infty, b) \setminus (-\infty, a)$ and $\{a\} = \bigcap_{n\in\mathbb{N}}[a, a + \frac{1}{n}) = \bigcap_{n\in\mathbb{N}}((-\infty, a + \frac{1}{n}) \setminus (-\infty, a))$. Therefore, the set $(a, b) = (-\infty, b) \setminus (-\infty, a) \cap \{a\}^c$ is made up of countable unions, intersections, and

complements of intervals of the form $(-\infty, a)$. Thus, by using the fact that inverse functions preserve intersections, we conclude the result.

(3) $\Leftrightarrow$ (6): Note that for any $a \in \mathbb{R}$, we have $(a, \infty) = (-\infty, a]^c$. Thus $f^{-1}((a, \infty)) = f^{-1}((-\infty, a]^c) = X \setminus f^{-1}((-\infty, a])$. This implies $f^{-1}((a, \infty)) \in \mathcal{F}$ for all $a \in \mathbb{R}$ if and only if $f^{-1}((-\infty, a]) \in \mathcal{F}$ for all $a \in \mathbb{R}$.

(4) $\Leftrightarrow$ (5): This is similar to the previous equivalence. □

Using the characterisation in Lemma 18.9.8, we now show that measurable functions are preserved under many algebraic operations.

Proposition 18.9.9 *(Algebra of Measurable Functions) If $f, g : X \to (\mathbb{R}, \mathcal{B})$ are $\mathcal{F}$-measurable, then so are the functions:*

1. *λf for some constant $\lambda \in \mathbb{R}$,*
2. *$f \pm g$,*
3. *fg,*
4. *$\frac{f}{g}$ defined on $\{x \in X : g(x) \neq 0\} \subseteq X$,*
5. *$\max(f, g)$ and $\min(f, g)$, and*
6. *f^+, f^-, and $|f|$, where $f^+ = \max(f, 0)$ and $f^- = \max(-f, 0) = -\min(f, 0)$.*

Proof To prove these, we use Lemma 18.9.8. We only show the first four results. The rest are left to the readers as Exercise 18.17.

1. If $\lambda = 0$, then there is nothing to check. Suppose $\lambda > 0$. Since $f : X \to \mathbb{R}$ is measurable, Lemma 18.9.8 says that all the preimages of the intervals $(-\infty, b)$ for $b \in \mathbb{R}$ are contained in $\mathcal{F}$. In other words, we know that the sets $\{x \in X : f(x) < b\}$ lie in $\mathcal{F}$ for all $b \in \mathbb{R}$.
   To show that $\lambda f : X \to \mathbb{R}$ is measurable, we need to show that $\{x \in X : \lambda f(x) < c\} \in \mathcal{F}$ for all $c \in \mathbb{R}$. However, for each $c \in \mathbb{R}$, this is simply $\{x \in X : \lambda f(x) < c\} = \{x : f(x) < \frac{c}{\lambda}\} = \{x \in X : f(x) < b\}$ for $b = \frac{c}{\lambda} \in \mathbb{R}$ which we know lies in $\mathcal{F}$.
   Finally, the case for $\lambda < 0$ can also be treated using a similar argument, namely $\{x \in X : \lambda f(x) < c\} = \{x \in X : f(x) > \frac{c}{\lambda} = b\}$ which is also contained in $\mathcal{F}$ by Lemma 18.9.8(6).
2. We need to check that for all $c \in \mathbb{R}$, the sets $\{x \in X : f(x) + g(x) > c\}$ lie in $\mathcal{F}$. However, this set is the same as the countable union over the whole of rational numbers $\mathbb{Q}$, namely:

$$\begin{aligned}&\{x \in X : f(x) + g(x) > c\}\\ &= \bigcup_{q \in \mathbb{Q}} (\{x \in X : f(x) > q\} \cap \{x \in X : g(x) > c - q\}),\end{aligned}$$

and we know that all of these sets are in $\mathcal{F}$ since the functions $f, g : X \to \mathbb{R}$ are measurable. Since $\mathcal{F}$ is a σ-algebra, this countable union is also in $\mathcal{F}$. Thus, the function $f + g$ is measurable. Similarly we have:

$$\begin{aligned}&\{x \in X : f(x) - g(x) > c\}\\&= \bigcup_{q \in \mathbb{Q}} (\{x \in X : f(x) > q\} \cap \{x \in X : g(x) < q - c\}) \in \mathcal{F},\end{aligned}$$

and so $f - g$ is also measurable.

3. To show that $fg : X \to \mathbb{R}$ is measurable, it is easier to show that the function $h^2 : X \to \mathbb{R}$ is measurable first and use the previous assertions. This is equivalent to showing that the set $\{x \in X : h(x)^2 < c\}$ is in $\mathcal{F}$ for every $c \in \mathbb{R}$. We split this into two cases:
   (a) For $c < 0$, $\{x \in X : h(x)^2 < c\} = \varnothing \in \mathcal{F}$.
   (b) For $c \geq 0$, we have:

$$\begin{aligned}\{x \in X : h(x)^2 < c\} &= \{x \in X : (h(x) - \sqrt{c})(h(x) + \sqrt{c}) < 0\}\\&= (\{x \in X : h(x) > -\sqrt{c}\} \cap \{x \in X : h(x) < \sqrt{c}\})\\&\quad \cup (\{x \in X : h(x) < -\sqrt{c}\} \cap \{x \in X : h(x) > \sqrt{c}\}),\end{aligned}$$

   which lies in $\mathcal{F}$ since h is measurable.

   Thus, the square of a measurable function is also measurable. To show $fg : X \to \mathbb{R}$ is measurable, we note that:

$$fg = \frac{1}{4}(f + g)^2 - \frac{1}{4}(f - g)^2.$$

   Since f and g are measurable, necessarily $f \pm g$ are measurable by the second assertion. Hence, $\frac{1}{4}(f \pm g)^2$ are measurable by the above and the first assertion. Finally, we conclude that their difference, which is fg, is also measurable using the second assertion.

4. To show that $\frac{f}{g} : \{x \in X : g(x) \neq 0\} \to \mathbb{R}$ is measurable, it suffices to show that the function $\frac{1}{g} : \{x \in X : g(x) \neq 0\} \to \mathbb{R}$ is measurable. We check for different cases of the value c:
   (a) For $c < 0$, $\{x \in X : \frac{1}{g(x)} < c\} = \{x \in X : \frac{1}{c} < g(x)\} \cap \{x \in X : g(x) < 0\} \in \mathcal{F}$.
   (b) For $c = 0$, $\{x \in X : \frac{1}{g(x)} < 0\} = \{x \in X : g(x) < 0\} \in \mathcal{F}$.
   (c) For $c > 0$, $\{x \in X : \frac{1}{g(x)} < c\} = \{x \in X : \frac{1}{c} < g(x)\} \cup \{x \in X : g(x) < 0\} \in \mathcal{F}$.

   Thus, the function $\frac{1}{g}$ is also measurable. Since the product of measurable functions is measurable, by the third assertion, we conclude that $\frac{f}{g}$ is also measurable.

□

Remark 18.9.10 Another way of proving that the functions h^2 and $|h|$ are measurable is to note that these functions are the composition of the function h with the square and modulus functions respectively. Let $p, m : \mathbb{R} \to \mathbb{R}$ be defined as $p(x) = x^2$ and $m(x) = |x|$. These functions are continuous and hence measurable. As a result, by using Proposition 18.9.7(2), the functions $h^2 = p \circ h$ and $|h| = m \circ h$ are also measurable.

We can extend the definition of $\mathcal{F}$-measurable real-valued functions to functions with images including $\pm\infty$, namely to functions with codomain $\bar{\mathbb{R}}$.

Definition 18.9.11 A function $f : X \to \bar{\mathbb{R}}$ is $\mathcal{F}$-measurable if:

1. for any $E \in \mathcal{B}$, its preimage under f is in $\mathcal{F}$, that is $f^{-1}(E) \in \mathcal{F}$, and
2. The preimage sets $f^{-1}(\{\infty\}) = \{x \in X : f(x) = \infty\}$ and $f^{-1}(\{-\infty\}) = \{x \in X : f(x) = -\infty\}$ are $\mathcal{F}$-measurable.

In other words, a function with codomain $\bar{\mathbb{R}}$ is $\mathcal{F}$-measurable if the preimages of sets of the form $E, E \cup \{-\infty\}, E \cup \{\infty\}, E \cup \{-\infty, \infty\}$ where $E \in \mathcal{B}$ are all $\mathcal{F}$-measurable. Extending Lemma 18.9.8 using Definition 18.9.11, we have a way to check this:

Lemma 18.9.12 *Let $(X, \mathcal{F})$ be a measurable space and $f : X \to \bar{\mathbb{R}}$. The following are equivalent:*

1. *The function f is $\mathcal{F}$-measurable.*
2. *$f^{-1}(J_a) \in \mathcal{F}$ for every set of the form $J_a = [-\infty, a]$ where $a \in \mathbb{R}$.*
3. *$f^{-1}(J_a) \in \mathcal{F}$ for every set of the form $J_a = [-\infty, a)$ where $a \in \mathbb{R}$.*
4. *$f^{-1}(J_a) \in \mathcal{F}$ for every set of the form $J_a = [a, \infty]$ where $a \in \mathbb{R}$.*
5. *$f^{-1}(J_a) \in \mathcal{F}$ for every set of the form $J_a = (a, \infty]$ where $a \in \mathbb{R}$.*

Proof We prove only the equivalence of the first two statements above. The other equivalences are proven in the same way.

(1) ⇒ (2): Fix any $a \in \mathbb{R}$. Then $J_a = [-\infty, a] = \{-\infty\} \cup (-\infty, a]$. By assumption, since $(-\infty, a] \in \mathcal{B}$, both of $f^{-1}(\{-\infty\})$ and $f^{-1}((-\infty, a])$ are in $\mathcal{F}$. Hence, their union $f^{-1}(\{-\infty\}) \cup f^{-1}((-\infty, a]) = f^{-1}(J_a)$ must also be in $\mathcal{F}$.

(2) ⇒ (1): We first show that $f^{-1}(\{-\infty\}), f^{-1}(\{\infty\}) \in \mathcal{F}$.

(a) Note that $\{-\infty\} = \bigcap_{n \in \mathbb{Z}} J_n$ and thus:

$$f^{-1}(\{-\infty\}) = f^{-1}(\bigcap_{n \in \mathbb{Z}} J_n) = \bigcap_{n \in \mathbb{Z}} f^{-1}(J_n) \in \mathcal{F},$$

by assumption.

(b) Using De Morgans law, we have $\{\infty\} = \bar{\mathbb{R}} \setminus \bigcup_{n\in\mathbb{Z}} J_n = \bigcap_{n\in\mathbb{Z}} J_n^c$. Thus:

$$\begin{aligned} f^{-1}(\{\infty\}) = f^{-1}(\bigcap_{n\in\mathbb{Z}} J_n^c) &= \bigcap_{n\in\mathbb{Z}} f^{-1}(J_n^c) \\ &= \bigcap_{n\in\mathbb{Z}} (X \setminus f^{-1}(J_n)) \\ &= X \setminus \bigcup_{n\in\mathbb{Z}} f^{-1}(J_n) \in \mathcal{F}. \end{aligned}$$

This means the preimage of $(-\infty, a) = J_a \setminus \{-\infty\}$ for any $a \in \mathbb{R}$ is measurable since $f^{-1}((-\infty, a)) = f^{-1}(J_a) \setminus f^{-1}(\{-\infty\}) \in \mathcal{F}$. Using Lemma 18.9.8, we can deduce that for any $E \in \mathcal{B}$, we have $f^{-1}(E) \in \mathcal{F}$. Thus, $f : X \to \bar{\mathbb{R}}$ is $\mathcal{F}$-measurable. □

With this convention, Proposition 18.9.9 can also be extended to measurable functions $f, g : X \to \bar{\mathbb{R}}$ as long as the sums, products, and quotients $f \pm g$, fg, and $\frac{f}{g}$ are well-defined according to the rules on the extended real number set.

Limits of Measurable Functions

Measurable functions behave in a nice manner under limits. This is due to the fact that we allow countable unions, intersections, and complements in the definition of σ-algebras. We can show the following pointwise limit results:

Proposition 18.9.13 *Let (f_n) be a sequence of $\mathcal{F}$-measurable functions $f_n : X \to \bar{\mathbb{R}}$. Then, the following functions on X are also $\mathcal{F}$-measurable:*

1. $\sup_{n\in\mathbb{N}} f_n$ *and* $\inf_{n\in\mathbb{N}} f_n$.
2. $\limsup_{n\to\infty} f_n$ *and* $\liminf_{n\to\infty} f_n$.

In particular, if $f_n \xrightarrow{pw} f$ for some function $f : X \to \bar{\mathbb{R}}$, then f is $\mathcal{F}$-measurable.

Proof We prove the assertions one by one.

1. Let $G, H : X \to \bar{\mathbb{R}}$ be defined as $G(x) = \sup_{n\in\mathbb{N}} f_n(x)$ and $H(x) = \inf_{n\in\mathbb{N}} f_n(x)$. To show that these functions are measurable, we appeal to Lemma 18.9.12. Namely, we show that the preimages of the sets $[-\infty, c]$ for any $c \in \mathbb{R}$ are all measurable. Fix $c \in \mathbb{R}$. By Exercise 18.20, we have:

$$\{x\in X : G(x) \leq c\} = \{x\in X : \sup_{n\in\mathbb{N}} f_n(x) \leq c\} = \bigcap_{n\in\mathbb{N}} \{x \in X : f_n(x) \leq c\} \in \mathcal{F},$$

since each f_n are measurable. Hence, the function G is measurable. Similarly:

$$\{x\in X : H(x) \leq c\} = \{x\in X : \inf_{n\in\mathbb{N}} f_n(x) \leq c\} = \bigcup_{n\in\mathbb{N}} \{x \in X : f_n(x) \leq c\} \in \mathcal{F},$$

and hence the function H is measurable.

2. To show the result for $\liminf_{n\to\infty} f_n(x)$ and $\limsup_{n\to\infty} f_n(x)$, since these can be expressed as iterated infimum and supremum, we appeal to the first assertion applied twice.
   Consider the sequence of functions (G_n) and (H_n) where $G_n, H_n : X \to \bar{\mathbb{R}}$ are defined as $G_n(x) = \sup_{m\geq n} f_m(x)$ and $H_n(x) = \inf_{m\geq n} f_m(x)$. Using the first assertion, we note that the functions G_n and H_n are measurable for all $n \in \mathbb{N}$. Furthermore, since $\limsup_{n\to\infty} f_n(x) = \inf_{n\in\mathbb{N}}(\sup_{m\geq n} f_m(x)) = \inf_{n\in\mathbb{N}} G_n(x)$, applying the first assertion once more to the sequence of measurable functions (G_n), the function $\limsup_{n\to\infty} f_n$ is measurable. Similarly, the function $\liminf_{n\to\infty} f_n$ is measurable.

Finally if $f_n \xrightarrow{pw} f$, then for all $x \in X$ the limit superior and limit inferior coincide, namely $\limsup_{n\to\infty} f_n(x) = \liminf_{n\to\infty} f_n(x) = f(x)$. Thus, the pointwise limit function f is also measurable by the second assertion. □

Let us pause for a moment and reflect on the construction of measurable functions. In the previous chapters, we have defined many classes of functions: continuous functions, Lipschitz continuous functions, differentiable functions, smooth functions, and Riemann/Darboux integrable functions.

We have seen that these classes are closed under some mild algebraic operations such as finite addition, finite multiplication, and scalar multiplication. However, these classes might not behave well under other operations involving infinity such as pointwise limits or supremum and infimum. These classes may be preserved under limits if we place a stronger limit condition, namely uniform convergence.

On the other hand, the class of measurable functions are much more robust since they also remain measurable under pointwise limits as we have seen in Proposition 18.9.13. This makes measurable functions an unfussy and useful class of functions to work with when dealing with limits.

Almost-Everywhere Property

We end this chapter by describing another useful notation, which is the almost everywhere property. We shall be using this property many times in the later chapters. This is another example of a quantifier that we have seen in Sect. 1.4.

Definition 18.9.14 (Almost-Everywhere Property) Let $(X, \mathcal{F}, \mu)$ be a measure space and suppose that $\{P(x) : x \in X\}$ is a set of mathematical statements parametrised by points in X.

We say that the property or statement P is true almost everywhere (or written as μ-a.e. or a.e. or ae) on X if the measure of the set such that the property P does not hold is 0. In symbols:

$$P \text{ is true } \mu\text{-a.e. on } X \quad \text{if} \quad \mu\{x \in X : \neg P(x)\} = 0.$$

Since the almost everywhere condition depends on the measure space $(X, \mathcal{F}, \mu)$, we usually write it as μ-almost everywhere or μ-a.e. to emphasise the dependence on μ. However, if it is clear what measure space we are talking about, the notation a.e. is enough. Almost everywhere property allows us to ignore sets which are very small. An example is the following proposition:

Proposition 18.9.15 *Suppose that $(X, \mathcal{F}, \mu)$ is a complete measure space.*

1. *If $f : X \to \bar{\mathbb{R}}$ is $\mathcal{F}$-measurable and $f = g$ μ-a.e., then g is also $\mathcal{F}$-measurable.*
2. *If (f_n) where $f_n : X \to \bar{\mathbb{R}}$ is a sequence of $\mathcal{F}$-measurable functions and $f_n \xrightarrow{\mu-ae} f$, then f is also $\mathcal{F}$-measurable.*

Proof We prove the first assertion only. The second assertion can be proven similarly and is left as Exercise 18.29.

1. Let $E = \{x \in X : f(x) \neq g(x)\} \in \mathcal{F}$. This set has measure zero by our assumption. For each $c \in \mathbb{R}$, we have:

$$\begin{aligned} &\{x \in X : g(x) < c\} \\ &= (\{x \in X : g(x) < c\} \cap E) \cup (\{x \in X : f(x) < c\} \cap E^c). \end{aligned} \tag{18.19}$$

The first set on the RHS of (18.19) is in $\mathcal{F}$ since it is a subset of E which has measure zero and $(X, \mathcal{F})$ is a complete measure space. The second set on the RHS of (18.19) is in $\mathcal{F}$ since f is a measurable function. Thus, the union (18.19) is in $\mathcal{F}$ and hence $g : X \to \mathbb{R}$ is a measurable function by virtue of Lemma 18.9.12. □

Remark 18.9.16 A very important remark here is that Proposition 18.9.15 requires the measure space $(X, \mathcal{F}, \mu)$ to be complete. On incomplete measure spaces, these results may not be true! Here is a counterexample:

Suppose that $X = [0, 1]$. Recall from Remark 18.7.9 that the Borel space $(\mathcal{B}, \mathcal{L}, \mu)$ is not a complete measure space. We established this by showing that the Cantor set $C \subseteq X$ is Borel measurable (in Exercise 18.26) and contains subsets which are not Borel measurable. Let $A \subseteq C$ be one such subset. Define the functions $f, g : X \to \mathbb{R}$ as $f(x) = \mathbf{1}_C(x)$ and $g(x) = 2\mathbf{1}_A(x)$. Clearly, f is Borel measurable and f differs from g on the set C which has measure $\mu(C) = 0$. So $f = g$ μ-a.e.. However, the function g is not Borel measurable since $g^{-1}((1, \infty)) = A \notin \mathcal{B}$.

Therefore, one has to be careful when working with the complete Lebesgue measure space and the incomplete Borel measure space. Moving between the two can be a dangerous pitfall, as Barry Simon warned!

Exercises

18.1 (∗) Let $\mathcal{S}$ be a semiring. Suppose that $\{E_j\}_{j=1}^n$ is a pairwise disjoint collection of sets $E_j \in \mathcal{S}$ and $E \in \mathcal{S}$ is such that $\bigcup_{j=1}^n E_j \subseteq E$. Prove that there exists a finite collection of pairwise disjoint non-empty sets $\{F_k\}_{k=1}^m$ where $F_k \in \mathcal{S}$ are all disjoint from every E_j such that $E = \bigcup_{j=1}^n E_j \cup \bigcup_{k=1}^m F_k$.

18.2 (∗) Prove Lemma 18.3.4, namely:
Let X be a set and $\mathcal{R}_j$ be a collection of rings in X for $j \in J$ where J is some indexing set. Then $\mathcal{R} = \bigcap_{j \in J} \mathcal{R}_j$ is also a ring of X.

18.3 In this question, we are going to prove Proposition 18.3.6.
Let S be a collection of subsets of a set X and $\mathcal{R}(S)$ be the ring generated by it. Define $\mathcal{E}$ as the collection:

$$\mathcal{E} = \{E_1 \triangle E_2 \triangle \ldots \triangle E_n : n \in \mathbb{N},\ E_j \text{ are finite intersections of sets in } S\}.$$

(a) Show that $S \subseteq \mathcal{E} \subseteq \mathcal{R}(S)$.
(b) Prove that $\mathcal{E}$ is closed under $\triangle$ and $\cap$.
(c) Hence, deduce that $\mathcal{E}$ is closed under $\cup$ and $\setminus$ and thus is a ring.
(d) Conclude that $\mathcal{E} = \mathcal{R}(S)$.

18.4 Let X be a countably infinite set, $U = \{A \subseteq X : A \text{ is finite}\}$, and $V = \{A \subseteq X : A \text{ is countable}\}$ be the collection of finite and countable subsets of X respectively.
(a) Prove that $\mathcal{F} = U \cup \{A^c : A \in U\}$ is an algebra on X.
Is it a σ-algebra?
(b) Prove that $\mathcal{G} = V \cup \{A^c : A \in V\}$ is an algebra on X.
Is it a σ-algebra?

18.5 Prove Lemma 18.3.21, namely:
Let $\mathcal{F}$ be a σ-algebra on a set X. Suppose that $Y \subseteq X$. Prove that the collection of sets $\mathcal{G} = \{E \cap Y : E \in \mathcal{F}\}$ is a σ-algebra in Y.

18.6 Determine the σ-algebra in $\mathbb{Z}$ generated by the following collection of subsets J.
(a) $J = \{\{n\} : n \in \mathbb{Z}\}$.
(b) $J = \{\mathbb{Z}_{\leq 0}, \mathbb{Z}_{\geq 0}\}$.
(c) $J = \{\{x \in \mathbb{Z} : x \geq n\} : n \geq 0\}$.

18.7 (∗) Let m^* be an outer measure on $\mathcal{P}(\mathbb{R})$ and $A, B \subseteq \mathbb{R}$. Prove the remaining assertions in Lemma 18.4.3, namely:
(a) If $A \subseteq B$, then $m^*(A) \leq m^*(B)$.
(b) If $m^*(A) = 0$, then $m^*(A \cup B) = m^*(B)$.

18.8 (∗) For a set $A \subseteq \mathbb{R}$ and constants $c, \lambda \in \mathbb{R}$, define the translated set $c + A = \{c + x : x \in A\}$ and dilated set $\lambda A = \{\lambda x : x \in A\}$.

(a) Suppose that $\mathcal{J}$ is the π-system on $\mathbb{R}$ containing all the half-closed intervals $(a, b] \subseteq \mathbb{R}$. Show that the content m on $\mathcal{J}$ is translation-invariant and scales appropriately. In other words, show that $m(c + (a, b]) = m((a, b])$ for any $c \in \mathbb{R}$ and $m(\lambda(a, b]) = \lambda m((a, b])$ for any $\lambda > 0$.

(b) Hence, show that the outer measure m^* on $\mathcal{P}(\mathbb{R})$ adapted from m is also translation-invariant and scales appropriately, namely $m^*(c + A) = m^*(A)$ and $m^*(\lambda A) = |\lambda| m^*(A)$ for all $A \in \mathcal{P}(\mathbb{R})$ and $c, \lambda \in \mathbb{R}$.

(c) Let $E \in \mathcal{L}$. By using the Carathéodory condition, show that the sets $c + E$ and λE are also in $\mathcal{L}$ for $c, \lambda \in \mathbb{R}$.

(d) Hence, conclude that the Lebesgue measure is also translation-invariant and scales appropriately under dilation as claimed in Example 18.4.5.

18.9 (∗) For any $c \in \mathbb{R}$, the Dirac mass $\delta_c : \mathcal{P}(\mathbb{R}) \to [0, \infty]$ is defined as the set function:

$$\delta_c(E) = \begin{cases} 1 & \text{if } c \in E, \\ 0 & \text{otherwise.} \end{cases}$$

Show that this is a measure on $\mathcal{P}(\mathbb{R})$. This measure is called the Dirac measure.

18.10 (∗) Let X be an infinite set and $V = \{A \subseteq X : A \text{ is countable}\}$ be the collection of countable subsets of X. We have seen in Exercise 18.4 that $\mathcal{G} = V \cup \{A^c : A \in V\}$ is a σ-algebra in X. Define a function $\mu : \mathcal{G} \to [0, \infty]$ as:

$$\mu(E) = \begin{cases} 0 & \text{if } E \text{ is countable,} \\ 1 & \text{if } E^c \text{ is countable.} \end{cases}$$

Show that μ is a measure on $(X, \mathcal{G})$. This measure is called the co-countable measure.

18.11 Let $(X, \mathcal{F}, \mu)$ be a measure space. If $A_j \in \mathcal{F}$ such that $A_{j+1} \subseteq A_j$ for all $j \in \mathbb{N}$ and $\mu(A_1) = \infty$, is $\mu(\bigcap_{j=1}^{\infty} A_j) = \lim_{n\to\infty} \mu(A_n)$ necessarily true?

18.12 Suppose that $(X, \mathcal{F})$ is a measurable space, $\mu, \nu : \mathcal{F} \to [0, \infty]$ are measures on X, and $c \geq 0$ is a non-negative constant. Show that the functions $\mu + \nu, c\mu : \mathcal{F} \to [0, \infty]$ defined as $(\mu + \nu)(E) = \mu(E) + \nu(E)$ and $c\mu(E) = c(\mu(E))$ are also measures on $(X, \mathcal{F})$.

18.13 (∗) Let $(\mathbb{R}, \mathcal{L}, \mu)$ be the Lebesgue measure space and $\mathcal{B}$ be the Borel σ-algebra. Prove that the collection of sets:

$$\mathcal{C} = \{E \cup N : E \in \mathcal{B}, N \subseteq X \in \mathcal{B} \text{ such that } \mu(X) = 0\},$$

is a σ-algebra in $\mathbb{R}$.

18.14 Prove Proposition 18.8.2, namely:
Let X be a set and $\mathcal{F} \subseteq \mathcal{P}(X)$. Prove that $\mathcal{F}$ is a σ-algebra on X if and only if it is both a π-system and a λ-system.

18.15 $(*)$ Suppose that $(X, \mathcal{F})$ is a measurable space and $f : X \to Y$ is a function. Define the collection of sets $\mathcal{G} = \{f(X) : X \in \mathcal{F}\}$.
(a) Prove that $\mathcal{G}$ is a σ-algebra on Y if and only if f is surjective.
(b) In addition to part (a), prove that f is $(\mathcal{F}, \mathcal{G})$-measurable if and only if $f^{-1}(f(E)) \in \mathcal{F}$ for all $E \in \mathcal{F}$.

18.16 $(*)$ Let $f : \mathbb{R} \to \mathbb{R}$ be a monotone function. Show that f is Borel measurable.

18.17 $(*)$ Prove the remaining assertions in Proposition 18.9.9, namely:
If $f, g : X \to (\mathbb{R}, \mathcal{B})$ are $\mathcal{F}$-measurable, show that the following functions are $\mathcal{F}$-measurable as well:
(a) $\max(f, g)$ and $\min(f, g)$,
(b) f^+, f^-, and $|f|$, where $f^+ = \max(f, 0)$ and $f^- = \max(-f, 0) = -\min(f, 0)$.
Now let $h : X \to (\mathbb{R}, \mathcal{B})$ be a real-valued function.
(c) Prove that h is $\mathcal{F}$-measurable if and only if both the functions h^+ and h^- are $\mathcal{F}$-measurable.
(d) Suppose that $|h|$ is $\mathcal{F}$-measurable. Is it necessarily true that h is also $\mathcal{F}$-measurable?

18.18 $(*)$ Let $f : X \to (\mathbb{R}, \mathcal{B})$. Show that the collection of sets $f^{-1}(\mathcal{B}) = \{f^{-1}(E) : E \in \mathcal{B}\}$ is a σ-algebra on X.
This is called the pullback σ-algebra on X with respect to f or the σ-algebra on X generated by f, denoted as $\sigma(f)$.

18.19 (a) Let $J = \{(-\infty, a) : a \in \mathbb{R}\}$ be a collection of sets in $\mathbb{R}$. Show that the σ-algebra generated by J is the Borel σ-algebra. Namely, show $\mathcal{F}(J) = \mathcal{B}$.
(b) Repeat part (a) with the σ-algebras generated by $J = \{[a, \infty) : a \in \mathbb{R}\}$ and $J = \{(a, \infty) : a \in \mathbb{R}\}$.

18.20 $(*)$ Let (f_n) be a sequence of $\mathcal{F}$-measurable functions $f_n : (X, \mathcal{F}) \to \bar{\mathbb{R}}$ and $c \in \mathbb{R}$. Show that:
(a) $\{x \in X : \sup_{n\in\mathbb{N}} f_n(x) < c\} = \bigcap_{n\in\mathbb{N}}\{x \in X : f_n(x) < c\}$,
(b) $\{x \in X : \inf_{n\in\mathbb{N}} f_n(x) < c\} = \bigcup_{n\in\mathbb{N}}\{x \in X : f_n(x) < c\}$.

18.21 $(*)$ Let $f : \mathbb{R} \to \mathbb{R}$ be a strictly increasing continuous function. Show that if $E \in \mathcal{B}$, then $f(E) \in \mathcal{B}$.

18.22 Let $f : \mathbb{R} \to \mathbb{R}$ be a real-valued function. For each $n \in \mathbb{N}$, define the set:

$$E_n = \left\{x_0 \in \mathbb{R} : \exists\delta > 0, \forall x, y \in (x_0 - \delta, x_0 + \delta), |f(x) - f(y)| < \frac{1}{n}\right\}.$$

(a) Show that E_n is an open set for any fixed $n \in \mathbb{N}$.
(b) Prove the equality of sets $\{x_0 \in \mathbb{R} : f \text{ is continuous at } x_0\} = \bigcap_{n=1}^{\infty} E_n$.
(c) Hence, deduce that the set of points in $\mathbb{R}$ for which f is continuous is a Borel set.

18.23 Let $(X, \mathcal{F})$ be a measurable space. For a non-empty set $Y \subseteq X$, define:

$$\mathcal{G} = \{E \in \mathcal{F} : \text{ either } E \cap Y = \varnothing \text{ or } Y \subseteq E\}.$$

(a) Prove that $\mathcal{G}$ is a σ-algebra on X.
(b) Suppose that $f : X \to \mathbb{R}$ is a real-valued function. Prove that f is $\mathcal{G}$-measurable if and only if f is $\mathcal{F}$-measurable and is constant on Y.

18.24 ($\diamond$) Prove the Borel-Cantelli lemma, which was named after Borel and Francesco Paolo Cantelli (1875–1966):

Lemma 18.10.17 (Borel-Cantelli Lemma) *Let $\{E_j\}_{j=1}^{\infty}$ be Lebesgue measurable sets such that $\sum_{j=1}^{\infty} \mu(E_j) < \infty$. If $E = \bigcap_{n=1}^{\infty} \bigcup_{j=n}^{\infty} E_j$, then $\mu(E) = 0$.*

18.25 ($*$) Let $(X, \mathcal{F})$ and $(Y, \mathcal{G})$ be measurable spaces and μ is a measure on the former. Suppose that $f : X \to Y$ is a measurable function and define the set function $f_*\mu : \mathcal{G} \to [0, \infty]$ as $f_*\mu(E) = \mu(f^{-1}(E))$. Show that $f_*\mu$ is a measure on $(Y, \mathcal{G})$.
This measure is called a pushforward measure with respect to f.

18.26 Recall that the Cantor set $C \subseteq [0, 1]$ is defined as the intersection $C = \bigcap_{n \in \mathbb{N}_0} C_n$ where:

$$C_0 = [0, 1] \quad \text{and} \quad C_n = \bigcap_{m=1}^{n} \bigcup_{j=0}^{\frac{3^m-1}{2}} \left[\frac{2j}{3^m}, \frac{2j+1}{3^m}\right].$$

(a) Deduce that the set C is a Borel set.
(b) For each $n \in \mathbb{N}$, show that $\mu(C_n) = (\frac{2}{3})^n$.
(c) Hence, show that the Cantor set C is μ-null.
Therefore, the Cantor set is an example of a set that is not countable but has zero Lebesgue measure.

18.27 ($*$) Suppose that $([0, 1], \mathcal{L}, \mu)$ is the induced Lebesgue measure space. Recall the Cantor staircase $f : [0, 1] \to [0, 1]$ which was defined as the limit of the sequence of functions $f_n : [0, 1] \to [0, 1]$ defined iteratively as:

$$f_0(x) = x \quad \text{and} \quad f_n(x) = \begin{cases} \frac{f_{n-1}(3x)}{2} & \text{if } x \in \left[0, \frac{1}{3}\right], \\ \frac{1}{2} & \text{if } x \in \left[\frac{1}{3}, \frac{2}{3}\right], \\ \frac{1}{2} + \frac{f_{n-1}(3x-2)}{2} & \text{if } x \in \left[\frac{2}{3}, 1\right], \end{cases} \quad \text{for all } n \in \mathbb{N}.$$

We have shown that this function is continuous in Exercise 11.11 and has vanishing derivative on the set C^c in Exercise 13.11. By Exercise 18.19, $\mu(C^c) = 1$ so this function is constant almost everywhere. Yet it is con-

tinuous and increases from 0 to 1 over the domain. Very strange phenomenon here!
In this question, our aim is to show that there is a Lebesgue measurable set which is not Borel measurable. We have stated this fact without proof in Proposition 18.7.8.
To do this, we shall be needing the Cantor staircase function f. Define a new function $g : [0, 1] \to [0, 2]$ as $g(x) = f(x) + x$.
(a) Show that this function is strictly increasing, continuous, and bijective.
(b) Hence, show that there is a continuous inverse function $g^{-1} : [0, 2] \to [0, 1]$.
(c) Show that C^c can be written as a union of countably many pairwise disjoint open intervals.
Deduce that the image $g(C^c)$ is also a union of countably many pairwise disjoint intervals.
(d) Let μ be the Lebesgue measure on $[0, 1]$ and $[0, 2]$. For any interval $(a, b) \subseteq C^c$, show that $\mu(g((a, b))) = \mu(a, b)$.
Hence, deduce that $\mu(g(C^c)) = \mu(C^c)$.
(e) Show that $\mu(g(C)) = 1$.
Deduce that there exists a set $N \subseteq g(C)$ which is not Lebesgue measurable.
(f) Thus, prove that $g^{-1}(N)$ is Lebesgue measurable but not Borel measurable.

18.28 (∗) Let $(X, \mathcal{F}, \mu)$ be a measure space. Assume that (f_n) is a sequence of $\mathcal{F}$-measurable functions $f_n : X \to \mathbb{R}$. Suppose that $f_n \xrightarrow{pw} f$ on X to some $\mathcal{F}$-measurable function f. For any $k \in \mathbb{N}$, prove that:

$$X = \bigcup_{N\in\mathbb{N}} \bigcap_{n\geq N} \left\{x \in X : |f_n(x) - f(x)| < \frac{1}{k}\right\} = \bigcup_{N\in\mathbb{N}} \bigcap_{n\geq N} E_{n,k},$$

where $E_{n,k} = \left\{x \in X : |f_n(x) - f(x)| < \frac{1}{k}\right\}$.

18.29 Let $(X, \mathcal{F}, \mu)$ be a complete measure space. Let (f_n) where $f_n : (X, \mathcal{F}) \to \bar{\mathbb{R}}$ be a sequence of $\mathcal{F}$-measurable functions. Suppose that the function (f_n) converges almost everywhere on X to a function $f : (X, \mathcal{F}) \to \bar{\mathbb{R}}$.
(a) Explain why $f_n \xrightarrow{\mu-ae} f$ is equivalent to $\mu\{x \in X : f_n(x)$ does not converge$\} = 0$.
(b) Hence, show that:

$$\mu\left(\bigcup_{k=1}^{\infty} \bigcap_{N=1}^{\infty} \bigcup_{m,n=N}^{\infty} \left\{x \in X : |f_n(x) - f_m(x)| \geq \frac{1}{k}\right\}\right) = 0.$$

(c) Deduce that the limit function f is also measurable.

18.30 (∗) In this question, we are going to prove the monotone class theorem. We shall need this result later in Lemma 20.2.7. We first define a monotone class:

Definition 18.10.18 (Monotone Class) Let X be a set and $\mathcal{M}$ be a set of subsets of X. The set $\mathcal{M}$ is called a monotone class if:

1. for any nested $E_1 \subseteq E_2 \subseteq \ldots$ sequence of sets in $\mathcal{M}$, we have $\bigcup_{j=1}^{\infty} E_j \in \mathcal{M}$, and
2. for any nested $E_1 \supseteq E_2 \supseteq \ldots$ sequence of sets in $\mathcal{M}$, we have $\bigcap_{j=1}^{\infty} E_j \in \mathcal{M}$.

The monotone class generated by a collection of sets J, denoted as $\mathcal{M}(J)$, is the smallest monotone class containing J. The monotone class theorem states:

Theorem 18.10.19 (Monotone Class Theorem) *Let Z be a collection of subsets of X with $X \in Z$ and $\mathcal{M}(Z)$ be the monotone class generated by Z. Suppose that for all $A, B \in Z$ we have $A^c, A \cap B \in \mathcal{M}(Z)$. Then:*

1. *$\mathcal{M}(Z) = \sigma(Z)$ where $\sigma(Z)$ is the smallest σ-algebra containing Z*
2. *Moreover, if $\mathcal{M}$ is any other monotone class containing Z, then $\sigma(Z) \subseteq \mathcal{M}$.*

We prove the assertions in Theorem 18.10.19 one by one.

(a) Explain why $\mathcal{M}(Z) \subseteq \sigma(Z)$.

To prove the other inclusion $\sigma(Z) \subseteq \mathcal{M}(Z)$, we aim to show that $\mathcal{M}$ is a σ-algebra which contains Z. First we show that it is an algebra, namely it is closed under complements and finite union.

(b) Define $\mathcal{E} = \{A \in \mathcal{M}(Z) : A^c \in \mathcal{M}(Z)\}$. Show that $\mathcal{E} = \mathcal{M}(Z)$.

(c) For any fixed $B \in \mathcal{M}(Z)$, define the set $\mathcal{E}_B = \{A \in \mathcal{M}(Z) : A \cap B \in \mathcal{M}(Z)\}$. Show that $\mathcal{E}_B = \mathcal{M}(Z)$.

(d) Conclude that $\mathcal{M}(Z)$ is an algebra.

(e) Using the fact that $\mathcal{M}(Z)$ is an algebra and a monotone class, show that it is a σ-algebra.
Deduce that $\sigma(Z) \subseteq \mathcal{M}(Z)$.

(f) Hence, complete the proof for the first assertion.

(g) Explain why the second assertion of Theorem 18.10.19 is immediate.

18.31 (⋄) The theory of measures provides a solid foundation for modern probability theory. Prior to the twentieth century, there was no underlying axiomatic interpretation of probability. The early concepts of probability can be traced back to Cardano in his book *Liber de ludo aleae* (The Book on Games of Chance). Beginning in the seventeenth century probability has been studied extensively by familiar names such as Pascal, Fermat, Jacob Bernoulli, De Moivre, Thomas Bayes (1701–1761), Legendre, Gauss, Laplace, Pafnuty Chebyshev (1821–1894), and Andrey Markov (1856–1922).

The axioms of probability was finally introduced by Andrey Kolmogorov (1903–1987) in 1933 via the language of measure theory. Measure theory allow us to formalise what is a sample space, an event, and a probability of some event. Kolmogorov insisted:

> The theory of probability as mathematical discipline can and should be developed from axioms in exactly the same way as geometry and algebra.

The axiomatisation of probability is given as follows: a probability space is described by a measure space $(\Omega, \mathcal{F}, P)$ where Ω is the space of all the possible outcomes of some experiment (which we call a sample space), $\mathcal{F}$ is a σ-algebra on Ω describing the subsets of possible outcomes in Ω that we can assign a probability to (which we call events), and P is a measure on $\mathcal{F}$ which describes the probability of each event $E \in \mathcal{F}$ happening.
The probability measure P is a usual measure that satisfies Definition 18.5.1 but with an additional axiom. Here we list the three original measure axioms along with the additional fourth axiom and their probabilistic interpretation. These axioms are called the Kolmogorov axioms of probability.

1. The event of nothing happening is 0 because when we carry out the experiment, something would have happened. So, we have $P(\varnothing) = 0$.
2. $P(E) \geq 0$ for all $E \in \mathcal{F}$.
3. Two events $E, F \in \mathcal{F}$ are called mutually exclusive if $E \cap F = \varnothing$, namely these two events have no outcomes in common. The probability of a union of countably many pairwise mutually exclusive events $\{E_j\}_{j=1}^{\infty}$ is σ-additive, namely:

$$P\left(\bigcup_{i=1}^{\infty} E_i\right) = \sum_{i=1}^{\infty} P(E_i).$$

4. The probability of the event of any outcome at all happening is 1, namely $P(\Omega) = 1$.

Suppose that $A, B \in \mathcal{F}$ are two events. Show that:

(a) $P(A^c) = P(\Omega \setminus A) = 1 - P(A)$, namely the probability of the event A not happening is $1 - P(A)$.
(b) $P(A \cup B) = P(A) + P(B) - P(A \cap B)$.

18.32 (◇) Consider an experiment of tossing two fair coins with each coin landing on either a head or a tail. Fair here means the coin is equally likely to land on either head or tail. The result of each coin toss are recorded in order.

(a) List down the sample space (all possible outcomes) of the experiment, Ω.
(b) List down all the events for this experiment $\mathcal{F}$, namely all possible collections of outcomes of the experiment.
(c) If $\nu : \mathcal{F} \to [0, \infty]$ is the counting measure, define the uniform probability measure $P : \mathcal{F} \to [0, 1]$ as $P(E) = \frac{\nu(E)}{\nu(\Omega)}$. Write down the element of $\mathcal{F}$ for the following events and determine their probability:
   i. Event 1: Both of the coin tosses are heads.
   ii. Event 2: At least one of the coin tosses is heads.

iii. Event 3: Exactly one of the coin tosses is heads.
iv. Event 4: Both of the coin tosses are neither heads nor tails.

The uniform probability measure above assumes that each outcome $\omega \in \Omega$ are equally likely to happen. In other words, the probability of each outcome is uniform (hence the name).

(d) Show that Events 1 and 3 in part (c) are mutually exclusive.

18.33 ($\diamond$) We can also formalise conditional probability using measures. Suppose that the probability space for a certain experiment is given by $(\Omega, \mathcal{F}, P)$. Assume that an event $F \in \mathcal{F}$ with probability $P(F) > 0$ has occurred, so now we have to update our probability measure to take into account this given information.

(a) Define a function $\mu_F : \mathcal{F} \to [0, 1]$ where $\mu_F(E) = P(E \cap F)$. Show that this is a measure but not necessarily a probability measure.

To turn this into a probability measure, we need to normalise this measure so that the measure of Ω is 1. Define $P_F : \mathcal{F} \to [0, 1]$ as $P_F(E) = \frac{\mu_F(E)}{P(F)} = \frac{P(E\cap F)}{P(F)}$. We usually write this as $P_F(E) = P(E|F)$.

(b) Show that P_F is a probability measure.

(c) Prove Bayes' theorem, which is given in its modern interpretation as follows:

Theorem 18.10.20 (Bayes' Theorem) *Let $(\Omega, \mathcal{F}, P)$ be a probability space. If $E, F \in \mathcal{F}$ are events with non-zero probabilities, then:*

$$P(E|F) = P(F|E)\frac{P(E)}{P(F)}.$$

(d) Now suppose that $\{E_j\}_{j=1}^n$ are pairwise disjoint events each with non-zero probability such that $\bigcup_{j=1}^n E_j = \Omega$. Show that for any $m \in \{1, 2, \ldots, n\}$ we have:

$$P(E_m|F) = \frac{P(F|E_m)P(E_m)}{\sum_{j=1}^n P(F|E_j)P(E_j)}.$$

Lebesgue Integration 19

I have to pay a certain sum, which I have collected in my pocket. I take the bills and coins out of my pocket and give them to the creditor in the order I find them until I have reached the total sum. This is the Riemann integral. But I can proceed differently. After I have taken all the money out of my pocket I order the bills and coins according to identical values and then I pay the several heaps one after the other to the creditor. This is my integral.

— Henri Lebesgue, mathematician

In this chapter, we are going to define the Lebesgue integral which is the main reason why we went through the foundations of measure theory in Chap. 18. This integral was formulated by Henri Lebesgue as an alternative to the Riemann integral. The idea was encapsulated in a letter from Lebesgue to Paul Montel (1876–1975) as quoted above. We shall see later that this integration is more robust than the Riemann integral and behaves well under limits.

We have laid out the rough foundational ideas for the construction of this integral at the beginning of Chap. 18. So now let us see how to do this rigorously.

19.1 Simple Functions

After seeing how we can define a valid measure on a measurable space $(X, \mathcal{F})$, we are now ready to approximate the $\mathcal{F}$-measurable functions by an analogue of step functions we defined for Riemann and Darboux integral. These functions are called simple functions.

S. Johar, *The Big Book of Real Analysis*,
https://doi.org/10.1007/978-3-031-30832-1_19

Definition 19.1.1 (Simple Functions) Let $(X, \mathcal{F})$ be a measurable space. A function $\phi : X \to \mathbb{R}$ is called a simple function if there exists an $n \in \mathbb{N}$ such that for $j = 1, 2, \ldots, n$ there are constants $c_j \in \mathbb{R}$ and measurable sets $E_j \in \mathcal{F}$ such that:

$$\phi(x) = \sum_{j=1}^{n} c_j \mathbf{1}_{E_j}(x),$$

where $\mathbf{1}_{E_j} : X \to \mathbb{R}$ are the indicator functions on the sets E_j.

In other words, a simple function is a measurable function that attains only finitely many values in $\mathbb{R}$. Clearly, finite sums and scalar multiples of a simple function is also a simple function. WLOG (see Exercise 19.1), we can assume that all the sets $\{E_j\}$ in Definition 18.4.5 are pairwise disjoint. This assumption allows us to use the following lemma for the indicator functions, which is a special case of Lemma 15.1.5.

Lemma 19.1.2 *Let $(X, \mathcal{F})$ be a measurable space and $E, F \in \mathcal{F}$. Let $\mathbf{1}_E, \mathbf{1}_F : X \to \mathbb{R}$ be indicator functions on these sets. Then, $\mathbf{1}_E \cdot \mathbf{1}_F = \mathbf{1}_{E \cap F}$, $\mathbf{1}_E + \mathbf{1}_F = \mathbf{1}_{E \cup F} + \mathbf{1}_{E \cap F}$, and $|\mathbf{1}_E - \mathbf{1}_F| = \mathbf{1}_{E \Delta F}$.*

We note that step functions on an interval $[a, b]$ that we have seen in Definition 15.1.4 are also simple functions, but the converse is not true; the set of simple functions on $[a, b]$ is much bigger than the set of step functions because there are more measurable sets in $[a, b]$ than just half-closed intervals.

The next question is: how do we approximate a non-negative measurable function $f : (X, \mathcal{F}) \to ([0, \infty], \mathcal{B})$ using the simple functions? The following fundamental result says that this can be done by creating a partition on the subset $[0, 2^n] \subseteq \mathbb{R}$ of the codomain consisting of $2^n + 1$ equispaced points $\{y_0, y_1, \ldots, y_{2^n}\}$ from $y_0 = 0$ to $y_{2^n} = 2^n$ and finding the preimages of the sets $[y_{j-1}, y_j)$ for all $j \in \{1, \ldots, 2^n\}$. Since f is $\mathcal{F}$-measurable, these preimages are all measurable sets in X and so their indicator functions are also measurable, as we have seen in Example 18.9.6.

Using this idea, we can create a sequence of pointwise increasing simple functions which converges pointwise to f by letting $n \to \infty$ to create finer partitions on an increasingly larger subset of the codomain. We have the following result:

Proposition 19.1.3 *Let $f : X \to [0, \infty]$ be a non-negative $\mathcal{F}$-measurable function. Then, there is a pointwise increasing sequence (f_n) of simple functions $f_n : X \to \mathbb{R}$ such that $f_n \uparrow f$.*

Proof For each $n \in \mathbb{N}$, we define:

$$f_n(x) = \sum_{j=0}^{2^{2n}-1} \frac{j}{2^n} \mathbf{1}_{E_{n,j}}(x) + 2^n \mathbf{1}_{A_n}(x),$$

where $A_n = \{x \in X : f(x) \geq 2^n\}$ and:

$$E_{n,j} = \left\{x \in X : \frac{j}{2^n} \leq f(x) < \frac{j+1}{2^n}\right\} \quad \text{for } j = 0, 1, \ldots, 2^{2n} - 1.$$

First, we show that this sequence of functions is pointwise increasing. Fix $x \in X$ and $n \in \mathbb{N}$. We want to show that $f_{n+1}(x) \geq f_n(x)$. We have several cases for x.

1. If $x \in E_{n,j}$ for some $j \in \{0, \ldots, 2^{2n} - 1\}$, then $f_n(x) = \frac{j}{2^n}$ and $\frac{j}{2^n} \leq f(x) < \frac{j+1}{2^n}$. This implies $\frac{2j}{2^{n+1}} \leq f(x) < \frac{2j+2}{2^{n+1}}$. Therefore, $x \in E_{n+1,k}$ for $k = 2j$ or $k = 2j + 1$. The former means $f_{n+1}(x) = \frac{2j}{2^{n+1}} = \frac{j}{2^n} = f_n(x)$ whilst the latter means $f_{n+1}(x) = \frac{2j+1}{2^{n+1}} > \frac{j}{2^n} = f_n(x)$. In either case,we have $f_{n+1}(x) \geq f_n(x)$.
2. If $x \in A_n$ then $f(x) \geq 2^n$ and $f_n(x) = 2^n$. Thus, we either have the subcases $f(x) \geq 2^{n+1}$ or $2^n \leq f(x) < 2^{n+1}$.
   (a) For the first subcase, we then have $f_{n+1}(x) = 2^{n+1} > 2^n = f_n(x)$.
   (b) On the other hand, the second subcase implies:

$$\begin{aligned} \frac{2^n 2^{n+1}}{2^{n+1}} \leq f(x) < \frac{2^{n+1} 2^{n+1}}{2^{n+1}} \quad &\Rightarrow \quad \frac{2^{2n+1}}{2^{n+1}} \leq f(x) < \frac{2^{2n+2}}{2^{n+1}} \\ &\Rightarrow \quad x \in E_{n+1,j}, \end{aligned}$$

   for some $j \in \{2^{2n+1}, \ldots, 2^{2n+2} - 1\}$. Therefore, we have $f_{n+1}(x) \geq \frac{2^{2n+1}}{2^{n+1}} = 2^n = f_n(x)$.

   Hence, in all of these subcases we have $f_{n+1}(x) \geq f_n(x)$.

Thus, (f_n) is a sequence of pointwise increasing simple functions. Moreover, by construction, we have $0 \leq f - f_n \leq \frac{1}{2^n}$ on the set $\{x \in X : f(x) < 2^n\}$ and $f_n = 2^n$ on the set $\{x \in X : f(x) \geq 2^n\}$. So for any $x \in X$, we have two possible cases:

1. If $f(x)$ is finite, then there exists an $N \in \mathbb{N}$ such that $f(x) \leq 2^N$. Thus, for all $n \geq N$, at this point we have $0 \leq f(x) - f_n(x) \leq \frac{1}{2^n} \leq \frac{1}{2^N}$. Taking the limit as $n \to \infty$, we then have $f_n(x) \to f(x)$.
2. If $f(x) = \infty$, then $(f_n(x)) = (2^n)$ which diverges to ∞.

In either case, we have $f_n \xrightarrow{pw} f$ as $n \to \infty$. □

Remark 19.1.4 In fact, if the function $f : X \to [0, \infty]$ is bounded, we can see that the convergence $f_n \xrightarrow{pw} f$ in Proposition 19.1.3 is uniform over X.

Using the approximation by simple functions result in Proposition 19.1.3, we aim to derive an alternative definition for integration. Note that the approximation result above only works for measurable functions since we require all the preimage sets $E_{n,j}$ and A_n to be $\mathcal{F}$-measurable in order to cook up an approximating simple function f_n.

But as we have seen in Chap. 18, the class of measurable functions is a very big class of functions. Even though there are still functions which are not measurable, these functions are rather rare and pathological in nature. With this, we hope that we can integrate many more functions (despite not all functions) than we were able to using the Riemann integrals.

19.2 Integral of Simple Functions

Our aim now is to define Lebesgue integral of a function $f : X \to \bar{\mathbb{R}}$ on a measure space $(X, \mathcal{F}, \mu)$. If $X = \mathbb{R}$, we take the measure space to be the Lebesgue space.

As we did for the Riemann and Darboux integral, we first define the integral on simple functions. Recall that simple functions are functions $\phi : X \to \mathbb{R}$ on a measure space $(X, \mathcal{F}, \mu)$ of the form:

$$\phi(x) = \sum_{j=1}^{n} c_j \mathbf{1}_{E_j}(x),$$

where $E_j \in \mathcal{F}$ are pairwise disjoint and $c_j \in \mathbb{R}$ are constants. Thus, an obvious integral for the simple functions which we have discussed at the beginning of Chap. 18 would be:

$$I(\phi) = \sum_{j=1}^{n} c_j \mu(E_i). \tag{19.1}$$

We note that the function ϕ may be expressed in a different way using different collections of measurable sets E_j and constants c_j. But the value of $I(\phi)$ in (19.1) is defined to be explicitly dependent on the way it is represented.

The good news is that the value of the integral in (19.1) is well-defined regardless of the representation for ϕ. Indeed, suppose that we write ϕ in two different representations, namely: $\phi = \sum_{j=1}^{n} c_j \mathbf{1}_{E_j} = \sum_{i=1}^{n} d_i \mathbf{1}_{F_i}$ where $\{E_j\}_{j=1}^{n}$ and $\{F_i\}_{i=1}^{m}$ are pairwise disjoint collections of sets in $\mathcal{F}$ and $c_j, d_j \in \mathbb{R}$. Define $E_0 = X \setminus \bigcup_{j=1}^{n} E_j$ and $F_0 = X \setminus \bigcup_{i=1}^{m} F_i$. Furthermore, let $G_{ij} = F_i \cap E_j \in \mathcal{F}$ for all $j = 0, 1, 2, \ldots, n$ and $i = 0, 1, 2, \ldots, m$, so for each j we have $E_j =$

$\bigcup_{i=0}^{m}(F_i \cap E_j) = \bigcup_{i=0}^{m} G_{ij}$ and for each i we have $F_i = \bigcup_{j=0}^{n} G_{ij}$. Moreover, the sets $\{G_{ij} : i = 0, 1, 2, \ldots, m,\ j = 0, 1, 2, \ldots, n\}$ are pairwise disjoint.

If we set $c_0 = d_0 = 0$ we can write ϕ as:

$$\phi = \sum_{j=0}^{n} c_j \mathbf{1}_{E_j} = \sum_{j=0}^{n} c_j \sum_{i=0}^{m} \mathbf{1}_{F_i \cap E_j} = \sum_{i,j} c_j \mathbf{1}_{G_{ij}},$$

$$\phi = \sum_{i=0}^{m} d_i \mathbf{1}_{F_i} = \sum_{i=0}^{m} d_i \sum_{j=0}^{n} \mathbf{1}_{F_i \cap E_j} = \sum_{i,j} d_i \mathbf{1}_{G_{ij}}.$$

Whenever $F_i \cap E_j = G_{ij} \neq \varnothing$, evaluating the function ϕ above at any $x \in G_{ij}$, we must have $c_j = d_i$. Via contrapositive, if $c_j \neq d_i$, then $G_{ij} = \varnothing$ and hence $\mu(G_{ij}) = 0$. Using this observation, we have:

$$I\left(\sum_{j=0}^{n} c_j \mathbf{1}_{E_j}\right) = \sum_{j=1}^{n} c_j \mu(E_j) = \sum_{i,j} c_j \mu(G_{ij}) = \sum_{i,j} d_i \mu(G_{ij}) = \sum_{i=0}^{m} d_i \mu(F_i)$$
$$= I\left(\sum_{i=0}^{m} d_i \mathbf{1}_{F_i}\right),$$

and therefore, the integral $I(\phi)$ in (19.1) is independent of how the simple function ϕ is expressed as.

A thing to note here is that the value of $I(\phi)$ is also allowed to be ∞.

Example 19.2.1 Let us look at some examples of integration of simple functions.

1. Let $([0, 5], \mathcal{L}, \mu)$ be the induced Lebesgue measure space and $f : [0, 5] \to [0, \infty]$ be defined as:

$$f(x) = \begin{cases} 1 & \text{if } x \in (0, 2] \cup (4, 5], \\ 2 & \text{if } x \in (2, 3]. \end{cases}$$

The graph of this function is given in Fig. 19.1. This is an example of a step function that we saw during the construction of Riemann integral. In terms of step functions, we can write it as: $f = \mathbf{1}_{E_1} + 2\mathbf{1}_{E_2} + 0\mathbf{1}_{E_3} + \mathbf{1}_{E_4}$ where $E_1 = (0, 2]$, $E_2 = (2, 3]$, $E_3 = (3, 4]$, and $E_4 = (4, 5]$ are half-open intervals.
In terms of simple functions, we can group some of the intervals together because measurable sets can be expressed as unions of intervals. For example, we can express this function as the simple function $f(x) = \mathbf{1}_{F_1} + 2\mathbf{1}_{F_2}$ where $F_1 = (0, 2] \cup (4, 5]$ and $F_2 = (2, 3]$ are measurable sets. Moreover, this representation is not unique. We can also write $f(x) = \mathbf{1}_{G_1} + \mathbf{1}_{G_2}$ where $G_1 = (0, 3] \cup (4, 5]$ and $G_2 = (2, 3]$ are also measurable sets.

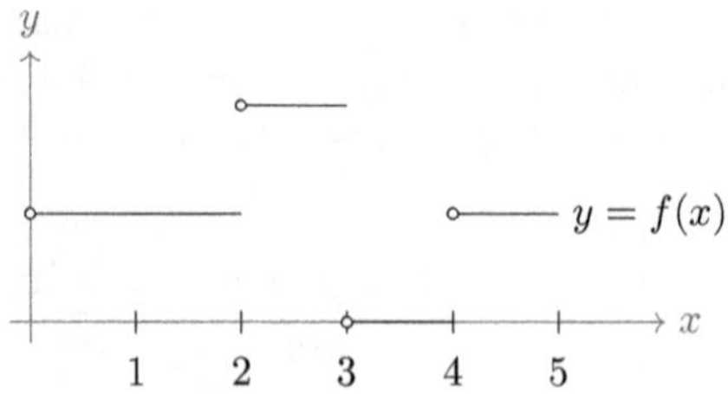

Fig. 19.1 Graph of the function f

However, as we have seen earlier, the integral is independent of the choice of representation for f, so we can compute it using the third expression as $I(f) = 1\mu(G_1) + 1\mu(G_2) = 1(3+1) + 1(1) = 5$.

2. Let $([0,1], \mathcal{L}, \mu)$ be the induced Lebesgue measure space and $f, g : [0,1] \to [0,\infty]$ be the indicator functions $f = \mathbf{1}_{\bar{\mathbb{Q}} \cap [0,1]}$ and $g = \mathbf{1}_{\mathbb{Q} \cap [0,1]}$. These functions are simple functions because $\bar{\mathbb{Q}} \cap [0,1]$ and $\mathbb{Q} \cap [0,1]$ are Lebesgue measurable sets. Furthermore, we have seen in Example 18.7.13 that the measure of these sets are $\mu(\bar{\mathbb{Q}} \cap [0,1]) = 1$ and $\mu(\mathbb{Q} \cap [0,1]) = 0$. Thus, we can compute $I(f) = 1\mu(\bar{\mathbb{Q}} \cap [0,1]) = 1$ and similarly $I(g) = 0$. We note that these functions are not Riemann integrable in Example 15.3.10(3).

From definitions, we can prove the following properties:

Proposition 19.2.2 *Let $\phi, \varphi : X \to \mathbb{R}$ be simple functions on a measure space $(X, \mathcal{F}, \mu)$.*

1. *For a constant $\lambda \in \mathbb{R}$, $I(\lambda\phi) = \lambda I(\phi)$.*
2. *$I(\phi + \varphi) = I(\phi) + I(\varphi)$.*
3. *If $0 \le \phi \le \varphi$, then $0 \le I(\phi) \le I(\varphi)$.*
4. *$|I(\phi)| \le I(|\phi|)$.*
5. *If $0 \le \phi$ and $E \in \mathcal{F}$, then $I(\mathbf{1}_E\phi) \le I(\phi)$.*

Proof The first assertion is clear. We prove assertions 2, 3, and 4 only.

2. Assume $\phi(x) = \sum_{j=1}^{n} c_j \mathbf{1}_{E_j}(x)$ and $\varphi(x) = \sum_{i=1}^{m} d_i \mathbf{1}_{F_i}(x)$ where $\{E_j\}_{j=1}^{n}, \{F_i\}_{i=1}^{m} \subseteq \mathcal{F}$ are each pairwise disjoint collections of sets in X and $c_j, d_i \in \mathbb{R}$ are constants. Define $E_0 = X \setminus \bigcup_{j=1}^{n} E_j$ and $F_0 = X \setminus \bigcup_{i=1}^{m} F_i$. Let us define $G_{ij} = F_i \cap E_j \in \mathcal{F}$ for all indices $i = 0, \ldots, m$ and $j = 0, 1, \ldots, n$. So, for each j we have $E_j = \bigcup_{i=0}^{m} (F_i \cap E_j) = \bigcup_{i=0}^{m} G_{ij}$ and for each i we have $F_i = \bigcup_{j=0}^{n} G_{ij}$. If we define $d_0 = c_0 = 0$ we can write the simple functions ϕ and φ as $\phi(x) = \sum_{i,j} c_j \mathbf{1}_{G_{ij}}(x)$ and $\varphi(x) = \sum_{i,j} d_i \mathbf{1}_{G_{ij}}(x)$

where the indices i and j run from 0 to m and n respectively. Thus $I(\phi) = \sum_{i,j} c_j \mu(G_{ij})$ and $I(\varphi) = \sum_{i,j} d_i \mu(G_{ij})$. Using the above, we have:

$$\begin{aligned} I(\phi+\varphi) = I\left(\sum_{i,j} c_j \mathbf{1}_{G_{i,j}} + \sum_{i,j} d_i \mathbf{1}_{G_{i,j}}\right) &= I\left(\sum_{i,j} (c_j + d_i)\mathbf{1}_{G_{i,j}}\right) \\ &= \sum_{i,j} (c_j + d_i)\mu(G_{i,j}) \\ &= \sum_{i,j} c_j \mu(G_{i,j}) + \sum_{i,j} d_i \mu(G_{i,j}) \\ &= I(\phi) + I(\varphi). \end{aligned}$$

3. The first inequality is clear since $\phi \geq 0$, by definition of the integral, we must have $I(\phi) \geq 0$. Now if $\phi \leq \varphi$ pointwise, then the simple function $\varphi - \phi$ is non-negative. We have several cases:
   (a) If $I(\phi) = I(\varphi) = \infty$ or $I(\phi) < \infty$ and $I(\varphi) = \infty$, then the inequality is clearly true.
   (b) The case for $I(\phi) = \infty$ and $I(\varphi) < \infty$ is impossible.
   (c) Assume that both $I(\phi)$ and $I(\varphi)$ are finite. We note that $I(\varphi - \phi) \geq 0$ and, by using the first two assertions, we have $I(\varphi) - I(\phi) \geq 0$, which implies the desired inequality.
4. Using the notation for ϕ from the proof of the second assertion, we have:

$$|I(\phi)| = \left|\sum_{j=1}^{n} c_j \mu(E_j)\right| \leq \sum_{j=1}^{n} |c_j \mu(E_j)| = \sum_{j=1}^{n} |c_j| \mu(E_j) = I(|\phi|).$$

The final assertion is left for the readers to prove in Exercise 19.4. □

The final assertion in Proposition 19.2.2 allows us to integrate a simple function over some measureable set $E \in \mathcal{F}$ rather than over the whole space X. If $E \in \mathcal{F}$, we define the integral of a simple function $\phi = \sum_{j=1}^{n} c_j \mathbf{1}_{E_j}$ over a subset $E \subseteq X$ as:

$$I_E(\phi) = I(\mathbf{1}_E \phi) = I\left(\sum_{j=1}^{n} c_j \mathbf{1}_E \mathbf{1}_{E_j}\right) = I\left(\sum_{j=1}^{n} c_j \mathbf{1}_{E \cap E_j}\right) = \sum_{j=1}^{n} c_j \mu(E \cap E_j),$$

where we used Lemma 19.1.2.

Therefore, for any two disjoint subsets $E, F \in \mathcal{F}$, we have:

$$I_{E\cup F}(\phi) = I(\mathbf{1}_{E\cup F}\phi) = I((\mathbf{1}_E + \mathbf{1}_F)\phi) = I(\mathbf{1}_E\phi + \mathbf{1}_F\phi) = I(\mathbf{1}_E\phi) + I(\mathbf{1}_F\phi)$$
$$= I_E(\phi) + I_F(\phi),$$

and so the integral of simple functions is domain additive.

19.3 Lebesgue Integral of Non-negative Functions

By virtue of Proposition 19.1.3, since we can approximate a non-negative measurable function from below by simple functions, we define:

Definition 19.3.1 (Lebesgue Integral of Non-negative Functions) Let $(X, \mathcal{F}, \mu)$ be a measure space. For a general non-negative $\mathcal{F}$-measurable function $f : X \to [0, \infty]$, we define the Lebesgue integral of f over X as:

$$\int_X f \, d\mu = \sup\{I(\phi) : \phi : X \to \mathbb{R} \text{ is a simple function with } \phi \leq f\}.$$

In fact, since f is non-negative, we can restrict the supremum to be over all non-negative simple functions ϕ instead.

Remark 19.3.2 Note the distinction between the notation between Riemann integral and Lebesgue integral:

1. The Riemann integral was defined over a compact interval $[a, b]$ so the integral is denoted with the notation $\int_a^b$. The Lebesgue integral, on the other hand, is defined over an arbitrary set X, so we use the notation $\int_X$. Even if X is an interval $X = [a, b]$, the Lebesgue integral over this set is also denoted with $\int_{[a,b]}$. This emphasises the lack of orientation in Lebesgue integral.
2. The Lebesgue integral depends heavily on the measure used on the set, hence we denote this dependence as $d\mu$ where μ is the specified measure which we are working with. On the other hand, the Riemann integral always uses the content m on the π-system of half-closed intervals in $\mathbb{R}$. Therefore, there is no ambiguity for Riemann integrals.
3. Notice that we do not require the domain X to be compact for this definition, unlike the Riemann integral. As a result, the value of this Lebesgue integral might be ∞ for some functions or domains.

From Remark 19.3.2(3), since the Lebesgue integral of any non-negative measurable function could have values in $[0, \infty]$, we want to isolate and focus our attention to the measurable functions which have finite Lebesgue integral. These functions are called Lebesgue integrable functions:

Definition 19.3.3 (Lebesgue Integrable Non-negative Function) Let $(X, \mathcal{F}, \mu)$ be a measure space. A non-negative function $f : X \to [0, \infty]$ is called Lebesgue integrable if $\int_X f\, d\mu < \infty$.
The space of Lebesgue integrable functions over the measure space $(X, \mathcal{F}, \mu)$ is denoted as $\mathcal{L}^1(X, \mathcal{F}, \mu)$ or simply $\mathcal{L}^1(X)$ if there is no confusion.

We can also define an integral of a non-negative function f over some subset $E \subseteq X$ in the σ-algebra $\mathcal{F}$. This is defined using the integral above as:

$$\begin{aligned}\int_E f\, d\mu &= \int_X \mathbf{1}_E f\, d\mu \\ &= \sup\{I(\phi) : \phi : X \to \mathbb{R} \text{ is a simple function with } \phi \leq \mathbf{1}_E f\} \\ &= \sup\{I(\mathbf{1}_E\phi) : \phi : X \to \mathbb{R} \text{ is a simple function with } \phi \leq \mathbf{1}_E f\} \\ &= \sup\{I_E(\phi) : \phi : X \to \mathbb{R} \text{ is a simple function with } \phi \leq f\}.\end{aligned}$$

From the definition of Lebesgue integral, we have the following properties:

Proposition 19.3.4 *Let $(X, \mathcal{F}, \mu)$ be a measure space and $f, g : X \to [0, \infty]$ be non-negative $\mathcal{F}$-measurable functions.*

1. *If $\phi : X \to \mathbb{R}$ is a simple function, then $\int_X \phi\, d\mu = I(\phi)$.*
2. *If $\lambda > 0$ is a real constant, then $\int_X \lambda f\, d\mu = \lambda \int_X f\, d\mu$.*
3. *If $0 \leq f \leq g$, then $\int_X f\, d\mu \leq \int_X g\, d\mu$,*
4. *If $E, F \in \mathcal{F}$ are disjoint measurable subsets of X and $f \in \mathcal{L}^1(X)$, then $\int_{E\cup F} f\, d\mu = \int_E f\, d\mu + \int_F f\, d\mu$.*
5. *If $E \in \mathcal{F}$ with $\mu(E) = 0$, then $\int_E f\, d\mu = 0$.*
6. *If $f = 0$ μ-a.e., then $\int_X f\, d\mu = 0$.*

Proof The first two assertions are left as Exercise 19.5. For brevity, we denote the sets:

$$\begin{aligned}U &= \{\phi : X \to \mathbb{R} : \phi \text{ is a simple function with } \phi \leq f\}, \\ V &= \{\phi : X \to \mathbb{R} : \phi \text{ is a simple function with } \phi \leq g\}.\end{aligned}$$

3. Since $f \leq g$ pointwise, we must have the inclusion $U \subseteq V$. Hence, we have:

$$\int_X f\, d\mu = \sup\{I(\phi) : \phi \in U\} \leq \sup\{I(\phi) : \phi \in V\} = \int_X g\, d\mu.$$

4. First, by subadditivity of supremum, we have:

$$\begin{aligned}\int_{E\cup F} f\,d\mu &= \sup\{I_{E\cup F}(\phi) : \phi \in U\}\\ &= \sup\{I_E(\phi) + I_F(\phi) : \phi \in U\}\\ &\leq \sup\{I_E(\phi) : \phi \in U\} + \sup\{I_F(\phi) : \phi \in U\}\\ &= \int_E f\,d\mu + \int_F f\,d\mu. \end{aligned} \tag{19.2}$$

To prove the opposite inequality, for any simple functions ϕ and φ both of which are smaller that f, consider the sum $\psi = \mathbf{1}_E\phi + \mathbf{1}_F\varphi$ which is also a simple function. If $x \in F$, then $\psi(x) = \mathbf{1}_E(x)\phi(x) + \mathbf{1}_F(x)\varphi(x) = \phi(x) \leq f(x)$. Likewise, for $x \in E$, we have $\psi(x) = \varphi(x) \leq f(x)$ and for $x \in (E \cup F)^c$ we have $\psi(x) = 0 \leq f(x)$.
Thus, the sum $\psi = \mathbf{1}_E\phi + \mathbf{1}_F\varphi$ is a simple function satisfying $\psi = \mathbf{1}_E\phi + \mathbf{1}_F\varphi \leq \mathbf{1}_{E\cup F} f$. Therefore, by definition of the Lebesgue integral of non-negative function, we have:

$$I_E(\phi) + I_F(\varphi) = I(\mathbf{1}_E\phi + \mathbf{1}_F\varphi) = I(\psi) \leq \int_{E\cup F} f\,d\mu.$$

Taking the supremum of $I_E(\phi)$ and $I_F(\varphi)$ over all simple functions in the set U, we have:

$$\int_E f\,d\mu + \int_F f\,d\mu \leq \int_{E\cup F} f\,d\mu. \tag{19.3}$$

Therefore, putting the two inequalities (19.2) and (19.3) together, we get the result.

5. By definition:

$$\int_E f\,d\mu = \sup\{I_E(\phi) : \phi \in U\}.$$

For any such $\phi \in U$, we can write $\phi(x) = \sum_{j=1}^n c_j \mathbf{1}_{E_j}(x)$ for some constants c_j and disjoint sets $E_j \in \mathcal{F}$. Hence, $I_E(\phi) = \sum_{j=1}^n c_j \mu(E_j \cap E) = 0$ since $\mu(E_j \cap E) \leq \mu(E) = 0$ for all $j = 1, 2, \ldots, n$. Therefore, $\int_E f\,d\mu$ is the supremum of the set $\{0\}$, which is also 0.

6. Let $E = \{x \in X : f(x) = 0\}$. Since $f = 0$ μ-a.e., we have $\mu(X \setminus E) = 0$. From the fourth and fifth assertions, we have:

$$\int_X f\,d\mu = \int_E f\,d\mu + \int_{X\setminus E} f\,d\mu = \int_E f\,d\mu.$$

Moreover, since f vanishes on E, the integral over E also vanishes since the supremum of integrals of simple function ϕ over E with $\phi \leq f = 0$ is 0. □

The converse to Proposition 19.3.4(6) is also true. Before proving this, we prove Markov's inequality, which is also known as Chebyshev's inequality.

Lemma 19.3.5 (Markov's Inequality) *Let $(X, \mathcal{F}, \mu)$ be a measure space. If $f : X \to [0, \infty]$ is $\mathcal{F}$-measurable, then for any real constant $c > 0$ we have:*

$$\mu\{x \in X : f(x) \geq c\} \leq \frac{1}{c} \int_X f \, d\mu.$$

Proof This inequality is proven by considering the function $\phi : X \to [0, \infty]$ defined as $\phi = c\mathbf{1}_{\{x \in X : f(x) \geq c\}}$. Note that, by definition, we have $\phi \leq f$ and hence:

$$\begin{aligned} c\mu\{x \in X : f(x) \geq c\} = \int_{\{x \in X : f(x) \geq c\}} c \, d\mu = \int_X c\mathbf{1}_{\{x \in X : f(x) \geq c\}} \, d\mu = \int_X \phi \, d\mu \\ \leq \int_X f \, d\mu, \end{aligned}$$

by Proposition 19.3.4(3). By algebra, we conclude with the desired inequality. □

We can then prove:

Proposition 19.3.6 *Let $(X, \mathcal{F}, \mu)$ be a measure space and $f : X \to [0, \infty]$ be an $\mathcal{F}$-measurable function.*

1. *If $f \in \mathcal{L}^1(X)$, then $\mu\{x \in X : f(x) = \infty\} = 0$.*
2. *If $\int_X f \, d\mu = 0$, then $f = 0$ μ-a.e. on X.*

Proof We prove the assertions one by one using Markov's inequality.

1. Since f is Lebesgue integrable, we know that $\int_X f \, d\mu$ is a finite constant. Let $E_n = \{x \in X : f(x) \geq n\}$ for $n \in \mathbb{N}$. Then, we have $E_{n+1} \subseteq E_n$ for all $n \in \mathbb{N}$ and $\mu(E_1) = \{x \in X : f(x) \geq 1\} \leq \int_X f \, d\mu < \infty$ by Markov's inequality. Moreover, $\{x \in X : f(x) = \infty\} = \bigcap_{j=1}^{\infty} E_j$. We can then apply Proposition 18.5.8 and Markov's inequality to get:

$$0 \leq \mu\{x \in X : f(x) = \infty\} = \mu\left(\bigcap_{j=1}^{\infty} E_j\right) = \lim_{n \to \infty} \mu(E_n) \leq \lim_{n \to \infty} \frac{1}{n} \int_X f \, d\mu = 0.$$

2. Using the assumption, Markov's inequality implies that $\mu\{x \in X : f(x) \geq c\} = 0$ for every $c > 0$. Define $E_n = \{x \in X : f(x) > \frac{1}{n}\}$ and note that $E_n \subseteq E_{n+1}$

and $\mu(E_n)$ for all $n \in \mathbb{N}$. Then, $\{x \in X : f(x) > 0\} = \bigcup_{j=1}^{\infty} E_j$. So, by Proposition 18.5.8, we have:

$$\mu\{x \in X : f(x) > 0\} = \mu\left(\bigcup_{j=1}^{\infty} E_j\right) = \lim_{n\to\infty} \mu(E_n) = \lim_{n\to\infty} 0 = 0,$$

giving us the result. □

Remark 19.3.7 Another definition of the Lebesgue integration is via the improper Riemann integration [51]. Suppose that $f : (X, \mathcal{F}, \mu) \to [0, \infty)$ is a non-negative $\mathcal{F}$-measurable function and $\mu(X) < \infty$. Then, we can also define its Lebesgue integral as the improper Riemann integral:

$$\int_X f\, d\mu = \int_0^{\infty} \mu\{x \in X : f(x) \geq t\}\, dt = \int_0^{\infty} g(t)\, dt,$$

where $g : [0, \infty) \to [0, \mu(X)]$ is defined as $g(t) = \mu\{x \in X : f(x) \geq t\}$. Since f is $\mathcal{F}$-measurable, g is a well-defined decreasing function of t with image in $[0, \mu(X)]$.

By Proposition 15.6.3, this Riemann integral makes sense over any compact interval in $[0, \infty)$. Hence, we can define the integral in an improper sense with values in $[0, \infty)$ if the improper Riemann integral converges or ∞ otherwise.

We can generalise this definition to a σ-finite domain $(X, \mathcal{F}, \mu)$. To do this, we decompose the space X into disjoint parts with finite measures and carry out the improper Riemann integral in each of the disjoint parts separately.

In Exercise 20.19, we shall show that this definition agrees with the definition of Lebesgue integral in Definition 19.3.1. This definition is usually called the layer cake representation of the Lebesgue integral since we are finding the area of the subgraph of f by "summing" up the areas of the layers of the subgraph via the Riemann integral. See Fig. 19.2 for the geometric interpretation.

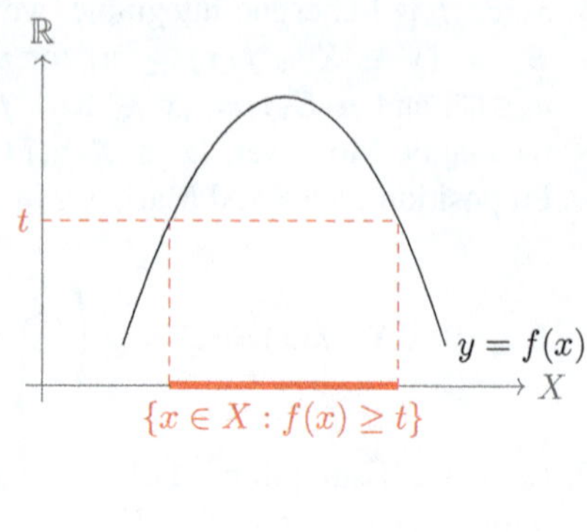

Fig. 19.2 The layer cake representation tell us that in order to find the area of the subgraph of f, we "sum" up the measures of the set $\{x \in X : f(x) \geq t\}$ for t from 0 to ∞

19.4 Monotone Convergence Theorem

A very useful result for non-negative $\mathcal{F}$-measurable functions is the monotone convergence theorem. This is a precursor to many convergence theorems which we shall prove later. This first form of the MCT for non-negative functions is also known as the Beppo Levi theorem, named after Beppo Levi (1875–1961).

Theorem 19.4.1 (Monotone Convergence Theorem, MCT) *Let (f_n) be a sequence of functions $f_n : X \to [0, \infty]$ such that $f_n \uparrow f$ to some $\mathcal{F}$-measurable function $f : X \to [0, \infty]$. Then:*

$$\int_X f \, d\mu = \int_X \lim_{n\to\infty} f_n \, d\mu = \lim_{n\to\infty} \int_X f_n \, d\mu,$$

where this integral on the LHS takes values in $[0, \infty]$.

Proof We know that $f_n \leq f_{n+1} \leq f$. Thus, we have the ordering $\int_X f_n \, d\mu \leq \int_X f_{n+1} \, d\mu \leq \int_X f \, d\mu$ for all $n \in \mathbb{N}$. Taking the limit as $n \to \infty$, we have:

$$\lim_{n\to\infty} \int_X f_n \, d\mu = \sup_{n\in\mathbb{N}} \left(\int_X f_n \, d\mu \right) \leq \int_X f \, d\mu.$$

Now we need to show the reverse inequality. Let $\phi(x) = \sum_{j=1}^m c_j \mathbf{1}_{E_j}(x)$ be a simple function such that the sets $\{E_j\}_{j=1}^m$ are pairwise disjoint and $0 \leq \phi \leq f$. Fix $\alpha \in (0, 1)$ and consider the set $B_n = \{x : f_n(x) \geq \alpha\phi(x)\}$. The set B_n is $\mathcal{F}$-measurable and, since the sequence (f_n) is increasing, we have $B_n \subseteq B_{n+1}$ for all $n \in \mathbb{N}$. Furthermore, we have $\bigcup_{n=1}^\infty B_n = X$ because $f_n(x) \to f(x) > \alpha\phi(x)$ for all $x \in X$. Since $\alpha\phi\mathbf{1}_{B_n} \leq f_n\mathbf{1}_{B_n} \leq f_n$, by integrating over the set X, we have:

$$\alpha I_{B_n}(\phi) \leq \int_X f_n \, d\mu. \tag{19.4}$$

By definition of ϕ and Proposition 18.5.8, since $E_j \cap B_n \subseteq E_j \cap B_{n+1}$ for all $j, n \in \mathbb{N}$ and $\bigcup_{n=1}^\infty (E_j \cap B_n) = E_j$, we have the following limit:

$$\begin{aligned}\lim_{n\to\infty} I_{B_n}(\phi) &= \lim_{n\to\infty} \sum_{j=1}^m c_j \mu(E_j \cap B_n) \\ &= \sum_{j=1}^m c_j \lim_{n\to\infty} \mu(E_j \cap B_n) = \sum_{j=1}^m c_j \mu(E_j) = I(\phi).\end{aligned} \tag{19.5}$$

Thus, taking the limit as $n \to \infty$ in (19.4) and using (19.5), we get:

$$\alpha I(\phi) \leq \lim_{n\to\infty} \int_X f_n \, d\mu,$$

for any arbitrary $\alpha \in (0, 1)$. Taking the limit as $\alpha \to 1$ then yields:

$$I(\phi) \leq \lim_{n\to\infty} \int_X f_n \, d\mu.$$

Since ϕ is an arbitrary simple function such that $0 \leq \phi \leq f$, by definition of the Lebesgue integral of f, taking the supremum over all such simple functions on the LHS, we have the desired inequality. Putting the two inequalities together, we obtain the desired result. □

Remark 19.4.2 We make some comments for the MCT.

1. The monotonicity condition for the MCT cannot be dropped. For example, let $X = [0, 1]$ and suppose that $(X, \mathcal{L}, \mu)$ is the induced Lebesgue measure space. Let (f_n) be a sequence of functions where $f_n : X \to \mathbb{R}$ is defined as:

$$f_n(x) = \begin{cases} n & \text{if } x \in [0, \frac{1}{n}), \\ 0 & \text{if } x \in [\frac{1}{n}, 1]. \end{cases}$$

Note that this is not a pointwise monotone sequence of functions because at any point $x \in (0, 1]$ the sequence $(f_n(x))$ is increasing up to the index $n = \lceil \frac{1}{x} \rceil$ at which the sequences goes down to 0 and remain there. Therefore, the sequence is pointwise converging to the function f which is 0 on $(0, 1]$ and ∞ at 0.
For all $n \in \mathbb{N}$, we can compute the integral $\int_X f_n \, d\mu = n\mu([0, \frac{1}{n})) + 0\mu([\frac{1}{n}, 1]) = 1$. However, the limit function is 0 a.e. so its integral is $\int_X f \, d\mu = 0$. Hence:

$$\lim_{n\to\infty} \int_X f \, d\mu = 1 \neq 0 = \int_X f \, d\mu.$$

2. Even funnier is that the MCT above does not apply to pointwise decreasing sequence of functions either. Hence, the name monotone convergence theorem can be quite misleading. An example of this is the sequence of real-valued measurable functions (f_n) defined on the induced Lebesgue measure space $(X, \mathcal{L}, \mu)$ where $X = [0, \infty)$ and:

$$f_n(x) = \begin{cases} 0 & \text{if } 0 \leq x < n, \\ 1 & \text{if } x \geq n. \end{cases}$$

We note here that the sequence of functions decreases pointwise to the zero function $f : X \to \mathbb{R}$ with $f = 0$. However, we can compute that $\int_X f_n \, d\mu = \infty$ for all $n \in \mathbb{N}$ but $\int_X f \, d\mu = 0$. So the MCT does not hold here. Therefore, probably calling the MCT the "increasing convergence theorem" would have been more suitable!

Example 19.4.3 Let $(\mathbb{N}, \mathcal{P}(\mathbb{N}), \nu)$ be the measure space on $\mathbb{N}$ with the counting measure that we saw in Example 18.5.6. For a function $f : \mathbb{N} \to [0, \infty]$, by Definition 19.3.1 we have:

$$\int_{\mathbb{N}} f \, d\nu = \sup\{I(\phi) : \phi : X \to \mathbb{R} \text{ is a simple function with } \phi \leq f\}.$$

This is the original definition of the Lebesgue integral, which can be difficult to use explicitly. Using the MCT instead, we can create a sequence of simple function (f_n) where $f_n : \mathbb{N} \to [0, \infty]$ defined as:

$$f_n(x) = \sum_{j=1}^{n} f(j)\mathbf{1}_{\{j\}}(x) \quad \Rightarrow \quad I(f_n) = \sum_{j=1}^{n} f(j)\mu(\{j\}) = \sum_{j=1}^{n} f(j).$$

The sequence (f_n) is increasing and converges pointwise to f. So, the MCT says:

$$\int_{\mathbb{N}} f \, d\nu = \lim_{n\to\infty} I(f_n) = \sum_{j=1}^{\infty} f(j).$$

The most important application of the MCT is that it gives us an explicit way of computing the Lebesgue integral of a measurable function. The Lebesgue integral by first definition is tricky to work with, thus having an explicit way to construct it is very useful. Recall from Proposition 19.1.3 that for any non-negative $\mathcal{F}$-measurable function $f : X \to [0, \infty]$, there is a sequence of simple functions (f_n) where $f_n : X \to \mathbb{R}$ such that $f_n \uparrow f$. These functions are given by:

$$f_n(x) = \sum_{j=0}^{2^{2n}-1} \frac{j}{2^n}\mathbf{1}_{E_{n,j}}(x) + 2^n\mathbf{1}_{A_n}(x),$$

where $A_n = \{x \in X : f(x) \geq 2^n\}$ and:

$$E_{n,j} = \left\{x \in X : \frac{j}{2^n} \leq f(x) < \frac{j+1}{2^n}\right\} \quad \text{for } j = 0, 1, \ldots, 2^{2n} - 1.$$

The MCT then says:

$$\int_X f \, d\mu = \lim_{n\to\infty} \int_X f_n \, d\mu = \lim_{n\to\infty} \sum_{j=0}^{2^{2n}-1} \frac{j}{2^n}\mu(E_{n,j}) + \lim_{n\to\infty} 2^n \mu(A_n). \tag{19.6}$$

Example 19.4.4 Let us compute the Lebesgue integral of some functions using the method outlined above.

1. Consider the function $f(x) = x$ defined for $x \in [0, 1]$. For any $n \in \mathbb{N}$, we have:

$$A_n = \{x \in [0, 1] : f(x) = x \geq 2^n\} = \varnothing,$$

$$E_{n,j} = \left\{x \in [0, 1] : \frac{j}{2^n} \leq f(x) = x < \frac{j+1}{2^n}\right\} = \begin{cases} \left[\frac{j}{2^n}, \frac{j+1}{2^n}\right) & \text{if } \frac{j+1}{2^n} \leq 1, \\ \varnothing & \text{if } \frac{j}{2^n} \geq 1. \end{cases}$$

Thus, by using the Eq. (19.6), we have:

$$\begin{aligned} \int_{[0,1]} f\, d\mu = \lim_{n\to\infty} \sum_{j=0}^{2^{2n}-1} \frac{j}{2^n}\mu(E_{n,j}) + \lim_{n\to\infty} 2^n\mu(A_n) &= \lim_{n\to\infty} \sum_{j=0}^{2^n-1} \frac{j}{2^n}\frac{1}{2^n} \\ &= \lim_{n\to\infty} \frac{1}{4^n}\frac{2^n(2^n-1)}{2} \\ &= \frac{1}{2}. \end{aligned}$$

2. Next, consider the function $g(x) = \frac{1}{x}$ defined for $x \in (0, 1]$. For any $n \in \mathbb{N}$, we have:

$$A_n = \{x \in (0, 1] : g(x) = \frac{1}{x} \geq 2^n\} = \left(0, \frac{1}{2^n}\right],$$

$$E_{n,j} = \left\{x \in (0, 1] : \frac{j}{2^n} \leq g(x) = \frac{1}{x} < \frac{j+1}{2^n}\right\} = \begin{cases} \varnothing & \text{if } \frac{j+1}{2^n} \leq 1, \\ \left(\frac{2^n}{j+1}, \frac{2^n}{j}\right] & \text{if } \frac{j}{2^n} \geq 1. \end{cases}$$

Thus, by using the Eq. (19.6), we have:

$$\int_{(0,1]} g\, d\mu = \lim_{n\to\infty} \sum_{j=2^n}^{2^{2n}-1} \frac{j}{2^n}\frac{2^n}{j(j+1)} + 1 = \lim_{n\to\infty} \sum_{j=2^n}^{2^{2n}-1} \frac{1}{j+1} + 1. \tag{19.7}$$

However, by using a similar inequality as in (16.17), note that:

$$\sum_{j=2^n}^{2^{2n}-1} \frac{1}{j+1} \geq \int_{2^n}^{2^{2n}} \frac{1}{x+1}\, dx = \ln\left(\frac{2^{2n}+1}{2^n+1}\right) \geq \ln\left(\frac{2^{2n}}{2(2^n)}\right) = \ln(2^{n-1}),$$

which diverges to ∞ as $n \to \infty$. This implies the limit in (19.7) is also ∞. Therefore, the Lebesgue integral of g over $(0, 1]$ is ∞.

3. Consider the function $g(x) = \frac{1}{x^2}$ defined on the set $X = [1, \infty)$. For any $n \in \mathbb{N}$, we have:

$$A_n = \{x \in X : g(x) = \frac{1}{x^2} \geq 2^n\} = \varnothing,$$

$$E_{n,j} = \left\{x \in X : \frac{j}{2^n} \leq g(x) = \frac{1}{x^2} < \frac{j+1}{2^n}\right\} = \begin{cases} \varnothing & \text{if } \frac{j}{2^n} \geq 1, \\ \left(\sqrt{\frac{2^n}{j+1}}, \sqrt{\frac{2^n}{j}}\right] & \text{if } \frac{j+1}{2^n} \leq 1. \end{cases}$$

Therefore, by using the Eq. (19.6), we have:

$$\begin{aligned} \int_X f \, d\mu &= \lim_{n\to\infty} \sum_{j=0}^{2^n-1} \frac{j}{2^n} \left(\sqrt{\frac{2^n}{j}} - \sqrt{\frac{2^n}{j+1}}\right) \\ &= \lim_{n\to\infty} \frac{1}{\sqrt{2^n}} \sum_{j=1}^{2^n-1} j \left(\frac{1}{\sqrt{j}} - \frac{1}{\sqrt{j+1}}\right). \end{aligned}$$

Let us evaluate the sum. Using summation by parts from Definition 7.8.1 with $a_j = \frac{1}{\sqrt{j}} - \frac{1}{\sqrt{j+1}}$ and $b_j = j$, we then have $s_n = \sum_{j=1}^n a_j = 1 - \frac{1}{\sqrt{n+1}}$ by telescoping and so:

$$\frac{1}{\sqrt{2^n}} \sum_{j=1}^{2^n-1} j \left(\frac{1}{\sqrt{j}} - \frac{1}{\sqrt{j+1}}\right) = \frac{1}{\sqrt{2^n}} - \sqrt{1 - \frac{1}{2^n}} + \frac{1}{\sqrt{2^n}} \sum_{j=1}^{2^n-2} \frac{1}{\sqrt{j+1}}. \tag{19.8}$$

Now we want to evaluate the limit of (19.8) as $n \to \infty$. The first two terms converge to 0 and -1 respectively. For the final term, by using similar inequalities as in (16.17), for any $n \geq 2$, we have the ordering:

$$\frac{1}{\sqrt{2^n}} \int_1^{2^n-1} \frac{1}{\sqrt{x+1}} \, dx \leq \frac{1}{\sqrt{2^n}} \sum_{j=1}^{2^n-2} \frac{1}{\sqrt{j+1}} \leq \frac{1}{\sqrt{2^n}} \int_0^{2^n-2} \frac{1}{\sqrt{x+1}} \, dx,$$

which then, by the FTC, yields:

$$2 - \frac{2\sqrt{2}}{\sqrt{2^n}} \leq \frac{1}{\sqrt{2^n}} \sum_{j=1}^{2^n-2} \frac{1}{\sqrt{j+1}} \leq 2\sqrt{1 - \frac{1}{2^n}} - \frac{2}{\sqrt{2^n}}. \tag{19.9}$$

Now note that the lower and upper bounds in the inequalities (19.9) both converge to 2 as $n \to \infty$. Hence, by sandwiching, we also have $\lim_{n\to\infty} \frac{1}{\sqrt{2^n}} \sum_{j=1}^{2^n-2} \frac{1}{\sqrt{j+1}} = 2$. Using this fact and the algebra of limits in the expression (19.8), we then get:

$$\int_X f\,d\mu = \lim_{n\to\infty} \frac{1}{\sqrt{2^n}} \sum_{j=1}^{2^n-1} j\left(\frac{1}{\sqrt{j}} - \frac{1}{\sqrt{j+1}}\right)$$

$$= \lim_{n\to\infty} \left(\frac{1}{\sqrt{2^n}} - \sqrt{1-\frac{1}{2^n}} + \frac{1}{\sqrt{2^n}} \sum_{j=1}^{2^n-2} \frac{1}{\sqrt{j+1}}\right) = 0 - 1 + 2 = 1.$$

Fatou's Lemmas

Since integrals can be thought of as limits of sums, the finite subadditivity of limit superior and finite superadditivity of limit inferior (which we have seen in Lemma 5.10.3) carry forward to Lebesgue integrals. These generalisations are called Fatou's lemmas, named after Pierre Fatou (1878–1929). They can also be seen as extensions of the MCT, but for more general sequence of functions as opposed to just pointwise increasing sequence of functions.

Theorem 19.4.5 (Fatou's Lemma) *Suppose that* (f_n) *where* $f_n : X \to [0, \infty]$ *is a sequence of* $\mathcal{F}$*-measurable functions. Then:*

$$\int_X \liminf_{n\to\infty} f_n\,d\mu \le \liminf_{n\to\infty} \int_X f_n\,d\mu.$$

In particular, if (f_n) *is a sequence of non-negative* $\mathcal{F}$*-measurable functions and* $f_n \xrightarrow{\mu-ae} f$*, then:*

$$\int_X f\,d\mu \le \liminf_{n\to\infty} \int_X f_n\,d\mu.$$

Proof Define a sequence (g_n) where $g_n : X \to [0, \infty]$ is defined as $g_n(x) = \inf_{m\ge n} f_m(x)$ for $n \in \mathbb{N}$ and $x \in X$. Thus, the sequence (g_n) is pointwise increasing and $g_n \uparrow \liminf_{n\to\infty} f_n$. We can apply the MCT to the sequence (g_n) to get:

$$\lim_{n\to\infty} \int_X g_n\,d\mu = \int_X \lim_{n\to\infty} g_n\,d\mu = \int_X \liminf_{n\to\infty} f_n\,d\mu. \tag{19.10}$$

On the other hand, $0 \le g_n \le f_m$ for all $m \ge n$ and so Proposition 19.3.4 (3) says $\int_X g_n\,d\mu \le \int_X f_m\,d\mu$ for $m \ge n$. Thus, we have:

$$\int_X g_n\,d\mu \le \inf_{m\ge n} \int_X f_m\,d\mu \quad \Rightarrow \quad \lim_{n\to\infty} \int_X g_n\,d\mu \le \liminf_{n\to\infty} \int_X f_n\,d\mu. \tag{19.11}$$

Finally, putting (19.10) and (19.11) together, we get the required inequality. □

We also have an analogous result with limit superior instead.

Corollary 19.4.6 (Reverse Fatou's Lemma) *Suppose that (f_n) where $f_n : X \to [0, \infty]$ is a sequence of $\mathcal{F}$-measurable functions such that there exists a non-negative integrable function g with $f_n \leq g$ for all $n \in \mathbb{N}$. Then:*

$$\limsup_{n\to\infty} \int_X f_n \, d\mu \leq \int_X \limsup_{n\to\infty} f_n \, d\mu.$$

Proof Apply Fatou's Lemma to the sequence (h_n) of non-negative functions $h_n : X \to \mathbb{R}$ where $h_n = g - f_n$. Note that for each $x \in X$ we have $\liminf_{n\to\infty}(-f_n(x)) = -\limsup_{n\to\infty} f_n(x)$ □

Another direct but very important corollary of the MCT is that the Lebesgue integral of non-negative functions is linear over the measurable functions:

Corollary 19.4.7 *If $f, g : X \to [0, \infty]$ are non-negative $\mathcal{F}$-measurable functions and $\lambda, \kappa \geq 0$, then:*

$$\int_X \lambda f + \kappa g \, d\mu = \lambda \int_X f \, d\mu + \kappa \int_X g \, d\mu.$$

Proof We just prove the case of $\lambda, \kappa = 1$ as the general case can then be treated by using Proposition 19.3.4(2). Consider pointwise increasing simple functions (ϕ_n) and (φ_n) where $\phi_n, \varphi_n : X \to [0, \infty]$ are such that $\phi_n \uparrow f$ and $\varphi_n \uparrow g$. It is easy to check that $\phi_n + \varphi_n \uparrow f + g$. By the MCT and the linearity of integration of simple functions, we have:

$$\begin{aligned}
\int_X f + g \, d\mu = \lim_{n\to\infty} I(\phi_n + \varphi_n) &= \lim_{n\to\infty} (I(\phi_n) + I(\varphi_n)) \\
&= \lim_{n\to\infty} I(\phi_n) + \lim_{n\to\infty} I(\varphi_n) \\
&= \int_X f \, d\mu + \int_X g \, d\mu,
\end{aligned}$$

which gives us the desired result. □

Remark 19.4.8 Corollary 19.4.7 also allows us to provide an alternative proof to Proposition 19.3.4(4) with ease. Indeed, for disjoint sets $E, F \in \mathcal{F}$ we have:

$$\begin{aligned}
\int_{E\cup F} f \, d\mu = \int_X \mathbf{1}_{E\cup F} f \, d\mu = \int_X (\mathbf{1}_E + \mathbf{1}_F) f \, d\mu &= \int_X \mathbf{1}_E f \, d\mu + \int_X \mathbf{1}_F f \, d\mu \\
&= \int_E f \, d\mu + \int_F f \, d\mu.
\end{aligned}$$

Lebesgue Integral of Non-negative Functions Series

Since the Lebesgue integral behaves well under limits on non-negative functions, we can do much more with it than we could with Riemann integral. For example, an infinite sum is defined as the limit of the finite sums as the number of terms in the sum goes to infinity. Recall that if (f_j) is a sequence of functions $f_j : X \to \mathbb{R}$, we define the infinite sum/series pointwise via the limit of partial sums $s_n = \sum_{j=1}^{n} f_j$, that is:

$$\sum_{j=1}^{\infty} f_j(x) = \lim_{n\to\infty} \sum_{j=1}^{n} f_j(x) = \lim_{n\to\infty} s_n(x).$$

Thus, we can prove the following result:

Proposition 19.4.9 (Series of Non-negative Functions) *Suppose that (f_n) is a sequence of non-negative $\mathcal{F}$-measurable functions $f_n : X \to [0, \infty]$. Then:*

$$\sum_{n=1}^{\infty} \int_X f_n \, d\mu = \int_X \sum_{n=1}^{\infty} f_n \, d\mu.$$

In particular, $\sum_{n=1}^{\infty} f_n \in \mathcal{L}^1(X, \mu)$ if and only if $\sum_{n=1}^{\infty} \int_X f_n \, d\mu < \infty$.

Proof Since the functions f_j in the series are all non-negative, the partial sums s_n are non-negative and is increasing pointwise. By using the MCT, we have:

$$\begin{aligned}\sum_{n=1}^{\infty} \int_X f_n \, d\mu &= \lim_{m\to\infty} \sum_{n=1}^{m} \int_X f_n \, d\mu = \lim_{m\to\infty} \int_X \sum_{n=1}^{m} f_n \, d\mu \\ &= \int_X \lim_{m\to\infty} \sum_{n=1}^{m} f_n \, d\mu = \int_X \sum_{n=1}^{\infty} f_n \, d\mu,\end{aligned}$$

and we are done. □

A variant of this is to split the domain of integration E into countably many disjoint subsets. In other words, we have the following domain-additive property of the Lebesgue integral of non-negative functions.

Proposition 19.4.10 *Suppose that (E_n) is a sequence of measurable sets with E_n all pairwise disjoint and $E = \bigcup_{n=1}^{\infty} E_n$. Let $f : X \to [0, \infty]$ be a non-negative $\mathcal{F}$-measurable function. Then:*

$$\int_E f \, d\mu = \sum_{n=1}^{\infty} \int_{E_n} f \, d\mu.$$

Proof For a non-negative function $f : X \to [0, \infty]$, define a sequence of non-negative functions (f_n) on X where $f_n = f\mathbf{1}_{\bigcup_{j=1}^{n} E_j}$. This sequence of functions is pointwise increasing to the limit f, so we can apply the MCT and Lemma 19.1.2 to get:

$$\begin{aligned}\int_X f\,d\mu = \int_X \lim_{n\to\infty} f_n\,d\mu = \lim_{n\to\infty}\int_X f\mathbf{1}_{\bigcup_{j=1}^{n} E_j}\,d\mu &= \lim_{n\to\infty}\int_X \sum_{j=1}^{n} f\mathbf{1}_{E_j}\,d\mu \\ &= \lim_{n\to\infty}\sum_{j=1}^{n}\int_X f\mathbf{1}_{E_j}\,d\mu \\ &= \lim_{n\to\infty}\sum_{j=1}^{n}\int_{E_j} f\,d\mu \\ &= \sum_{j=1}^{\infty}\int_{E_j} f\,d\mu,\end{aligned}$$

and we are done. □

19.5 Lebesgue Integral

The construction of Lebesgue integral can be generalised from non-negative functions to any $\mathcal{F}$-measurable functions with image in $\bar{\mathbb{R}} = [-\infty, \infty]$. This is done by breaking an arbitrary $\mathcal{F}$-measurable function $f : X \to \bar{\mathbb{R}}$ into its positive and negative parts, namely:

$$f = f^+ - f^-,$$

where the functions $f^+, f^- : X \to [0, \infty]$ defined as $f^+ = \max(f, 0)$ and $f^- = -\min(-f, 0) = \max(-f, 0)$ respectively are two non-negative functions on X. Since these two functions are also $\mathcal{F}$-measurable as we have seen in Proposition 18.9.9(6), the Lebesgue integrals $\int_X f^+\,d\mu$ and $\int_X f^-\,d\mu$ have values in in $[0, \infty]$. Moreover, if at least one of the two integrals above is finite, we can define its Lebesgue integral as:

$$\int_X f\,d\mu = \int_X f^+\,d\mu - \int_X f^-\,d\mu, \tag{19.12}$$

which takes values in $\bar{\mathbb{R}}$.

However, we note that if both of the integrals $\int_X f^+\,d\mu$ and $\int_X f^-\,d\mu$ are ∞, we would get an indeterminate case $\infty - \infty$ in (19.12), so the integral (19.12) does not exist.

On the other hand, if both of these integrals are finite and hence the value $\int_X f\,d\mu$ in (19.12) is finite, we call such functions Lebesgue integrable. We note that since $|f| = f^+ + f^-$, the integrals $\int_X f^+\,d\mu$ and $\int_X f^-\,d\mu$ are both finite if and only if the integral $\int_X f^+\,d\mu + \int_X f^-\,d\mu = \int_X |f|\,d\mu$ is finite. Hence, we define:

Definition 19.5.1 (Lebesgue Integrable Functions) Let $(X, \mathcal{F}, \mu)$ be a measure space and $f : X \to \bar{\mathbb{R}}$ be an $\mathcal{F}$-measurable function. If both of the integrals $\int_X f^+\,d\mu$ and $\int_X f^-\,d\mu$ are finite, we call the function f Lebesgue integrable. The space of Lebesgue integrable functions over X is denoted as:

$$\mathcal{L}^1(X, \mathcal{F}, \mu) = \left\{ f : X \to \bar{\mathbb{R}} : f \text{ is } \mathcal{F}\text{-measurable}, \int_X |f|\,d\mu < \infty \right\}.$$

If the measure space is clear, we sometimes write this space simply as $\mathcal{L}^1(X)$.

Example 19.5.2 Let us look at some examples.

1. Consider the measure space $(\mathbb{N}, \mathcal{P}(\mathbb{N}), \nu)$ where ν is the counting measure. For a function $f : \mathbb{N} \to \bar{\mathbb{R}}$, from Example 19.4.3, we extend the definition of the Lebesgue integral of the function f as the integral $\int_{\mathbb{N}} f\,d\nu = \sum_{j=1}^{\infty} f^+(j) - \sum_{j=1}^{\infty} f^-(j)$.
   We note that this is finite if and only if the integral $\int_{\mathbb{N}} |f|\,d\nu = \sum_{j=1}^{\infty} f^+(j) + \sum_{j=1}^{\infty} f^-(j) = \sum_{j=1}^{\infty} |f(j)|$ is finite. In other words, $f \in \mathcal{L}^1(\mathbb{N})$ if and only if the series is absolutely convergent. For the case of counting measure, purely as a convention, we denote the space of Lebesgue integrable functions l^1 instead of $\mathcal{L}^1$.
2. Consider the function $f : [-1, 1] \to \mathbb{R}$ defined as $f(x) = x$. We can split the positive and negative parts of the function as:

$$f^+(x) = \begin{cases} x & \text{if } x \geq 0, \\ 0 & \text{if } x \leq 0, \end{cases} \quad \text{and} \quad f^-(x) = \begin{cases} -x & \text{if } x \leq 0, \\ 0 & \text{if } x \leq 0. \end{cases}$$

   First, let us calculate the Lebesgue integral of the positive part. We have:

$$\int_{[-1,1]} f^+\,d\mu = \int_{[0,1]} x\,d\mu = \frac{1}{2},$$

   by Example 19.4.4(1). Likewise, we can compute the Lebesgue integral $\int_{[-1,1]} f^-\,d\mu = \frac{1}{2}$. Hence, we have the full integral $\int_{[-1,1]} f\,d\mu = \int_X f^+\,d\mu - \int_X f^-\,d\mu = \frac{1}{2} - \frac{1}{2} = 0$.

   Here are some direct consequences of Definition 19.5.1:

Proposition 19.5.3 *For $\mathcal{F}$-measurable functions $f, g : X \to \bar{\mathbb{R}}$, we have the following results:*

1. *$f \in \mathcal{L}^1(X)$ if and only if $|f| \in \mathcal{L}^1(X)$.*
2. *If $|g| \leq f$ and $f \in \mathcal{L}^1(X)$, then $g \in \mathcal{L}^1(X)$.*
3. *If $0 \leq g \leq |f|$ and $g \notin \mathcal{L}^1(X)$, then $f \notin \mathcal{L}^1(X)$.*
4. *Triangle inequality: $\left|\int_X f\,d\mu\right| \leq \int_X |f|\,d\mu$.*
5. *If $f, g \in \mathcal{L}^1(X)$ and $f \leq g$, then $\int_X f\,d\mu \leq \int_X g\,d\mu$.*
6. *If $Y \in \mathcal{F}$ is a measurable subset of X and $f \in \mathcal{L}^1(X)$, then $f \in \mathcal{L}^1(Y)$.*
7. *If $E \in \mathcal{F}$ is such that $\mu(E) = 0$, then $\int_E f\,d\mu = 0$.*
8. *If $E \in \mathcal{F}$ is such that $\mu(E) < \infty$ and f is bounded on E, then $f \in \mathcal{L}^1(E)$.*
9. *If $\mu(X) < \infty$ and f is bounded, then $f \in \mathcal{L}^1(X)$.*

Proof We leave the first four assertions as Exercise 19.6. We prove the rest here:

5. Note that if $f \leq g$, then $0 \leq f^+ \leq g^+$ and $0 \leq g^- \leq f^-$. Therefore, $\int_X f^+\,d\mu \leq \int_X g^+\,d\mu$ and $-\int_X f^-\,d\mu \leq -\int_X g^-\,d\mu$. Adding these two inequalities together yields the result.
6. By Proposition 19.3.4, we have $0 \leq \int_Y f^+\,d\mu \leq \int_X f^+\,d\mu$ and $0 \leq \int_Y f^-\,d\mu \leq \int_X f^-\,d\mu$. Since $f \in \mathcal{L}^1(X)$, the integrals of the positive and negative parts on X are finite and hence both $\int_Y f^+\,d\mu$ and $\int_Y f^-\,d\mu$ are finite as well. This then means $f \in \mathcal{L}^1(Y)$.
7. Since $\mu(E) = 0$, by Proposition 19.3.4 we have both $\int_E f^+\,d\mu = \int_E f^-\,d\mu = 0$ and hence the result.
8. Suppose that $|f| \leq M$ for some $M > 0$. Then, $\int_E |f|\,d\mu \leq \int_E M\,d\mu = M\mu(E) < \infty$ which means $|f| \in \mathcal{L}^1(E)$. By the first assertion, we deduce $f \in \mathcal{L}^1(E)$.
9. This is a corollary of the previous assertion. □

Moreover, the Lebesgue integral is a linear operation over $\mathbb{R}$, namely:

Proposition 19.5.4 *Suppose that $f, g : X \to \bar{\mathbb{R}}$ are $\mathcal{F}$-measurable functions such that $f, g \in \mathcal{L}^1(X)$ and $\lambda, \kappa \in \mathbb{R}$ are constants. Then:*

1. *$\lambda f \in \mathcal{L}^1(X)$ and $\int_X \lambda f\,d\mu = \lambda \int_X f\,d\mu$.*
2. *$\lambda f + \kappa g \in \mathcal{L}^1(X)$ and $\int_X \lambda f + \kappa g\,d\mu = \lambda \int_X f\,d\mu + \kappa \int_X g\,d\mu$.*

Proof We prove the assertions one by one.

1. Fix $\lambda \in \mathbb{R}$. If $\lambda = 0$, then there is nothing to prove. Otherwise, we have two cases:

(a) If $\lambda > 0$ then $(\lambda f)^+ = \lambda f^+$ and $(\lambda f)^- = \lambda f^-$. Therefore by definition of Lebesgue integral and Corollary 19.4.7, we have:

$$\int_X \lambda f \, d\mu = \int_X \lambda f^+ \, d\mu - \int_X \lambda f^- \, d\mu = \lambda \int_X f^+ \, d\mu - \lambda \int_X f^- \, d\mu$$
$$= \lambda \int_X f \, d\mu.$$

(b) If $\lambda < 0$ then $(\lambda f)^+ = \max(\lambda f, 0) = \max((-\lambda)(-f), 0) = -\lambda \max(-f, 0) = -\lambda f^-$ and, likewise, $(\lambda f)^- = -\lambda f^+$. Since $-\lambda > 0$, by Corollary 19.4.7, we have:

$$\int_X \lambda f \, d\mu = \int_X -\lambda f^- \, d\mu - \int_X -\lambda f^+ \, d\mu = -\lambda \int_X f^- \, d\mu + \lambda \int_X f^+ \, d\mu$$
$$= \lambda \int_X f \, d\mu.$$

Thus, we have the linearity of the integral over scalar multiplication and $\lambda f \in \mathcal{L}^1(X)$.

2. First assume $\lambda = \kappa = 1$. We note that $f + g = (f+g)^+ - (f+g)^-$ and $f+g = (f^+ - f^-) + (g^+ - g^-)$. Combining this, we have $(f+g)^+ + f^- + g^- = (f+g)^- + f^+ + g^+$, where all the terms on both sides are non-negative. Thus:

$$\int_X (f+g)^+ + f^- + g^- \, d\mu = \int_X (f+g)^- + f^+ + g^+ \, d\mu.$$

By Corollary 19.4.7 and algebra, we then get:

$$\int_X (f+g)^+ \, d\mu - \int_X (f+g)^-$$
$$= \left(\int_X f^+ \, d\mu - \int_X f^- \, d\mu \right) + \left(\int_X g^+ \, d\mu - \int_X g^- \, d\mu \right),$$

which is, by definition, the equality of integrals $\int_X f + g \, d\mu = \int_X f \, d\mu + \int_X g \, d\mu$.

We can extend the above to the general case for $\lambda, \kappa \in \mathbb{R}$ by combining the above result with the previous assertion. □

Proposition 19.5.4 shows that the space $\mathcal{L}^1(X)$ is a real vector space. Next, we have the almost everywhere properties for Lebesgue integrals, which we leave for the readers to prove as Exercise 19.7.

Proposition 19.5.5 *Suppose that $f, g : X \to \bar{\mathbb{R}}$ are $\mathcal{F}$-measurable functions on a set X with $\mu(X) > 0$. Suppose further that $f \in \mathcal{L}^1(X)$.*

1. $f(x) \in \mathbb{R}$ *μ-a.e..*
2. If $f = g$ *μ-a.e., then* $g \in \mathcal{L}^1(X)$ *with* $\int_X f\, d\mu = \int_X g\, d\mu$.
3. If $\int_X |f|\, d\mu = 0$*, then* $f = 0$ *μ-a.e..*
4. If $f > 0$ *μ-a.e., then* $\int_X f\, d\mu > 0$.

Approximations of Measurable Functions

Finally, we would like to state a useful result that says any Lebesgue integrable function defined on $\mathbb{R}$ can be approximated by a step function. However, recall that a step function was defined in Definition 15.1.4 for functions on a compact interval $[a, b]$. We need to extend this definition to the non-compact domain $\mathbb{R}$. We do this by simply declaring it to be identically 0 outside of the interval $[a, b]$.

Definition 19.5.6 (Step Function—Extended) Suppose that $[a, b] \subseteq \mathbb{R}$ is a compact interval and $\mathcal{P}$ is a partition of the interval $[a, b]$. A step function $\phi_{\mathcal{P}} : \mathbb{R} \to \mathbb{R}$ adapted to $\mathcal{P}$ is a function which is a step function on $[a, b]$ according to Definition 15.1.4 and 0 on $[a, b]^c$.

We first state and prove the following results:

Lemma 19.5.7 *Let* $(\mathbb{R}, \mathcal{L}, \mu)$ *be the Lebesgue measure space and* $E \in \mathcal{L}$ *is such that* $\mu(E) < \infty$*. For every* $\varepsilon > 0$*, there exists* $a, b \in \mathbb{R}$ *with* $a < b$*, a partition* $\mathcal{P}$ *of* $[a, b]$*, and a step function* $\phi_{\mathcal{P}} : \mathbb{R} \to \mathbb{R}$ *adapted to* $\mathcal{P}$ *such that* $\int_{\mathbb{R}} |\mathbf{1}_E - \phi_{\mathcal{P}}|\, d\mu < \varepsilon$.

Proof Fix $\varepsilon > 0$. By Proposition 18.7.3, there exists an open set $U \subseteq \mathbb{R}$ such that $E \subseteq U$ and $\mu(U \triangle E) = \mu(U \setminus E) < \frac{\varepsilon}{2}$.

Since U is open, by Theorem 4.5.20, we can write U as a countable union of disjoint open intervals. WLOG, suppose that there are infinitely many of them so that $U = \bigcup_{j=1}^{\infty} I_j$ where I_j are disjoint open intervals. By σ-additivity of μ, we have $\mu(U) = \sum_{j=1}^{\infty} \mu(I_j)$. Thus for all $j \in \mathbb{N}$, we have $\mu(I_j) \leq \mu(U) < \mu(E) + \frac{\varepsilon}{2} < \infty$ and hence all of the intervals I_j are bounded. Therefore, for $j \in \mathbb{N}$, the intervals I_j must be of the form $I_j = (a_j, b_j)$ for $a_j, b_j \in \mathbb{R}$ with $a_j < b_j$. Thus, we can write U as the union $U = \bigcup_{j=1}^{\infty}(a_j, b_j)$.

Moreover, since the series $\sum_{j=1}^{\infty} \mu(I_j) = \sum_{j=1}^{\infty} \mu((a_j, b_j))$ is increasing and converges to $\mu(U)$, there exists an $N \in \mathbb{N}$ such that $\mu(U) - \sum_{j=1}^{N} \mu((a_j, b_j)) < \frac{\varepsilon}{2}$. Writing $I = \bigcup_{j=1}^{N}(a_j, b_j)$, this is equivalent to $\mu(U \setminus I) < \frac{\varepsilon}{2}$.

Set $a = \min\{a_j : j = 1, 2, \ldots, N\}$ and $b = \max\{b_j : j = 1, 2, \ldots, N\}$. Let $\mathcal{P} = \bigcup_{j=1}^{N}\{a_j, b_j\}$ be a partition of $[a, b]$ and define $I'_j = (a_j, b_j]$ for $j = 1, 2, \ldots, N$. We claim that the step function $\phi_{\mathcal{P}}(x) = \sum_{j=1}^{N} \mathbf{1}_{I'_j}(x)$ would give us the desired estimate.

Denote $I' = \bigcup_{j=1}^{N} I'_j = \bigcup_{j=1}^{N}(a_j, b_j]$ so that $I \subseteq I'$. We have $U \triangle I' = (U \setminus I') \cup (I' \setminus U)$ First, we have $I' \setminus U = \bigcup_{j=1}^{N}(a_j, b_j] \setminus \bigcup_{j=1}^{\infty}(a_j, b_j) = \{b_1, b_2, \ldots b_N\}$ which implies $\mu(I' \setminus U) = 0$. Moreover, we also have $U \setminus I' \subseteq U \setminus I$ and so

$\mu(U \setminus I') \leq \mu(U \setminus I)$. Thus $\mu(U \Delta I') = \mu(U \setminus I') + \mu(I' \setminus U) \leq \mu(U \setminus I) < \frac{\varepsilon}{2}$. By triangle inequality, Lemma 19.1.2, and the earlier estimates, we then have:

$$\begin{aligned}
\int_{\mathbb{R}} |\mathbf{1}_E - \phi_{\mathcal{P}}| \, d\mu &= \int_{\mathbb{R}} |\mathbf{1}_E - \mathbf{1}_U + \mathbf{1}_U - \phi_{\mathcal{P}}| \, d\mu \\
&\leq \int_{\mathbb{R}} |\mathbf{1}_U - \mathbf{1}_E| \, d\mu + \int_{\mathbb{R}} |\mathbf{1}_U - \phi_{\mathcal{P}}| \, d\mu \\
&= \int_{\mathbb{R}} |\mathbf{1}_U - \mathbf{1}_E| \, d\mu + \int_{\mathbb{R}} |\mathbf{1}_U - \mathbf{1}_{I'}| \, d\mu \\
&= \int_{\mathbb{R}} \mathbf{1}_{U \Delta E} \, d\mu + \int_{\mathbb{R}} \mathbf{1}_{U \Delta I'} \, d\mu \\
&= \mu(U \Delta E) + \mu(U \Delta I') < \frac{\varepsilon}{2} + \frac{\varepsilon}{2} = \varepsilon,
\end{aligned}$$

which is the desired estimate. □

A direct corollary of this is:

Corollary 19.5.8 *Let $(\mathbb{R}, \mathcal{L}, \mu)$ be the Lebesgue measure space and $\varphi \in \mathcal{L}^1(\mathbb{R})$ is a simple function. For every $\varepsilon > 0$, there exists $a, b \in \mathbb{R}$ with $a < b$, a partition $\mathcal{P}$ of $[a, b]$, and a step function $\phi_{\mathcal{P}} : \mathbb{R} \to \mathbb{R}$ adapted to $\mathcal{P}$ such that $\int_{\mathbb{R}} |\varphi - \phi_{\mathcal{P}}| \, d\mu < \varepsilon$.*

Proof Since φ is a Lebesgue integrable simple function, we can write it as a finite sum $\varphi(x) = \sum_{j=1}^n c_j \mathbf{1}_{E_j}(x)$ where $c_j \in \mathbb{R}$ and $E_j \in \mathcal{L}$ are pairwise disjoint sets with $\mu(E_j) < \infty$. For each $j = 1, 2 \ldots, n$, by using Lemma 19.5.7, there exists a step function ϕ_j adapted to a partition $\mathcal{P}_j$ of $[a_j, b_j]$ such that $\int_{\mathbb{R}} |\mathbf{1}_{E_j} - \phi_j| \, d\mu < \frac{\varepsilon}{n|c_j|}$.

Letting $a = \min\{a_j : j = 1, 2, \ldots, n\}$, $b = \max\{b_j : j = 1, 2, \ldots, n\}$, and $\mathcal{P} = \bigcup_{j=1}^n \mathcal{P}_j$, the finite combination of step functions $\phi_{\mathcal{P}}(x) = \sum_{j=1}^n c_j \phi_j(x)$ is a step function adapted to the partition $\mathcal{P}$ of $[a, b]$ according to Proposition 15.1.6. Thus, by triangle inequality, we have:

$$\begin{aligned}
\int_{\mathbb{R}} |\varphi - \phi_{\mathcal{P}}| \, d\mu = \int_{\mathbb{R}} \left| \sum_{j=1}^n (c_j \mathbf{1}_{E_j} - c_j \phi_j) \right| \, d\mu &\leq \int_{\mathbb{R}} \sum_{j=1}^n |c_j \mathbf{1}_{E_j} - c_j \phi_j| \, d\mu \\
&= \sum_{j=1}^n |c_j| \int_{\mathbb{R}} |\mathbf{1}_{E_j} - \phi_j| \, d\mu \\
&< \sum_{j=1}^n |c_j| \frac{\varepsilon}{n|c_j|} = \varepsilon,
\end{aligned}$$

which is what we wanted to prove. □

The above two results can be used to prove the following big result that says any Lebesgue integrable function can be approximated by a step function or a continuous function. The proof of the following theorem is left for the readers to try as Exercise 19.10.

Theorem 19.5.9 *Let* $(\mathbb{R}, \mathcal{L}, \mu)$ *be the Lebesgue measure space and* $f \in \mathcal{L}^1(\mathbb{R})$.

1. *For every* $\varepsilon > 0$, *there exists* $a, b \in \mathbb{R}$ *with* $a < b$, *a partition* $\mathcal{P}$ *of* $[a, b]$, *and a step function* $\phi : \mathbb{R} \to \mathbb{R}$ *adapted to* $\mathcal{P}$ *such that* $\int_{\mathbb{R}} |f - \phi_{\mathcal{P}}| \, d\mu < \varepsilon$.
2. *For every* $\varepsilon > 0$, *there exists* $a, b \in \mathbb{R}$ *with* $a < b$ *and a continuous function* $g : \mathbb{R} \to \mathbb{R}$ *which vanishes outside of* $[a, b]$ *such that* $\int_{\mathbb{R}} |f - g| \, d\mu < \varepsilon$.

19.6 Convergence Theorems

We can improve the MCT in Theorem 19.4.1 to general Lebesgue integrable functions in X under a mild additional assumption:

Corollary 19.6.1 *Let* $(X, \mathcal{F}, \mu)$ *be a measure space and* (f_n) *be a sequence of* $\mathcal{F}$*-measurable functions* $f_n : X \to \bar{\mathbb{R}}$. *Suppose that* (f_n) *is pointwise increasing and* $f_n \uparrow f$ μ*-a.e. to some function* $f : X \to \mathbb{R}$. *Suppose further that the set of Lebesgue integrals* $\left\{\int_X f_n \, d\mu\right\}_{n=1}^{\infty}$ *is bounded. Then,* $f \in \mathcal{L}^1(X)$ *and:*

$$\int_X f \, d\mu = \int_X \lim_{n\to\infty} f_n \, d\mu = \lim_{n\to\infty} \int_X f_n \, d\mu.$$

Proof Consider the sequence of non-negative functions (g_n) where $g_n : X \to [0, \infty]$ is defined by $g_n = f_n - f_1$ for $n \in \mathbb{N}$. Clearly, this sequence of functions is increasing. We can thus apply Theorem 19.4.1 to this sequence of functions to get:

$$\begin{aligned} \lim_{n\to\infty} \int_X f_n - f_1 \, d\mu = \int_X \lim_{n\to\infty} (f_n - f_1) \, d\mu &= \int_X \lim_{n\to\infty} f_n - f_1 \, d\mu \\ &= \int_X f - f_1 \, d\mu. \end{aligned}$$

Since $\int_X f_1 \, d\mu$ is finite we can use algebra to get the desired limit. Finally, since the sequence of Lebesgue integrals $\left(\int_X f_n \, d\mu\right)$ is bounded above, since limits preserve weak inequalities, the resulting limit $\int_X f \, d\mu$ must be finite and so $f \in \mathcal{L}^1(X)$. □

The MCT is the most basic tool in studying Lebesgue integration. Despite its simplicity, there are many applications of it. One useful application of the MCT is the following result, which is usually referred to as the continuity of Lebesgue integrals:

Proposition 19.6.2 *Let $(X, \mathcal{F}, \mu)$ be a measure space and $f : X \to \bar{\mathbb{R}}$ be an $\mathcal{F}$-measurable function. For any $\varepsilon > 0$, there exists a $\delta > 0$ such that for any set $E \in \mathcal{F}$ with $\mu(E) < \delta$, we must have $\int_E |f| \, d\mu < \varepsilon$.*

Proof Let $g : X \to \bar{\mathbb{R}}$ be an $\mathcal{F}$-measurable function defined as $g = |f|$. For every $n \in \mathbb{N}$, define the sets $E_n = \{x \in X : g(x) \leq n\}$. Let (g_n) be a sequence of functions $g_n : X \to \bar{\mathbb{R}}$ where $g_n = g\mathbf{1}_{E_n}$. Note that this sequence is pointwise increasing to g, namely $g_n \uparrow g$. By the MCT, there exists an index $N \in \mathbb{N}$ such that for all $n \geq N$ we have:

$$\left| \int_X g_n \, d\mu - \int_X g \, d\mu \right| = \int_X g - g_n \, d\mu < \frac{\varepsilon}{2}.$$

For any $E \in \mathcal{F}$ with $\mu(E) < \infty$, since $g \geq g_N$ on X and $g_N = g\mathbf{1}_{E_N} \leq N$ on $E \subseteq X$, we have:

$$\int_E g \, d\mu = \int_E g - g_N \, d\mu + \int_E g_N \, d\mu \leq \int_X g - g_N \, d\mu + \int_E N \, d\mu < \frac{\varepsilon}{2} + \mu(E)N, \tag{19.13}$$

so if we pick $\delta = \frac{\varepsilon}{2N} > 0$, for any $E \in \mathcal{F}$ with $\mu(E) < \delta = \frac{\varepsilon}{2N}$, inequality (19.13) implies:

$$\int_E g \, d\mu < \frac{\varepsilon}{2} + \frac{\varepsilon}{2N} N = \varepsilon,$$

which is what we wanted to prove. □

Dominated and Bounded Convergence Theorems

The MCT in Theorem 19.4.1 and Corollary 19.6.1 require the terms in the function sequences (f_n) to be pointwise increasing. Apart from these convergence results, there are other convergence results for the Lebesgue integral that do not require the monotonicity of the functions sequence as in the MCT. These results are called the dominated and bounded convergence theorems.

Theorem 19.6.3 (Dominated Convergence Theorem, DCT) *Let $(X, \mathcal{F}, \mu)$ be a measure space and (f_n) be a sequence of $\mathcal{F}$-measurable functions where $f_n : X \to \bar{\mathbb{R}}$ and $f_n \xrightarrow{pw} f$ on X to some function $f : X \to \bar{\mathbb{R}}$. Suppose that there is a function $g \in \mathcal{L}^1(X)$ such that $|f_n| \leq g$ μ-a.e. on X for all $n \in \mathbb{N}$. Then:*

1. *f_n and f are in $\mathcal{L}^1(X)$.*
2. *$\lim_{n\to\infty} \int_X f_n \, d\mu = \int_X f \, d\mu$.*

Proof We prove the assertions separately:

1. From Proposition 18.9.13, the limiting function f is $\mathcal{F}$-measurable. By comparison, since $|f_n| \leq g$ μ-a.e., f_n are Lebesgue integrable for all $n \in \mathbb{N}$. Taking the limit as $n \to \infty$, since limits preserve weak inequalities, we have $|f| = \lim_{n\to\infty} |f_n| \leq g$ a.e.. Hence, the function f must be Lebesgue integrable as well.
2. Applying Fatou's lemma to the sequence of non-negative functions (h_n) where $h_n : X \to \mathbb{R}$ is defined as $h_n = g - f_n$, we get:

$$\begin{aligned} &\int_X \liminf_{n\to\infty} h_n \, d\mu \leq \liminf_{n\to\infty} \int_X h_n \, d\mu \\ \Rightarrow \quad &\int_X g - f \, d\mu \leq \liminf_{n\to\infty} \int_X (g - f_n) \, d\mu = \int_X g \, d\mu - \limsup_{n\to\infty} \int_X f_n \, d\mu \\ \Rightarrow \ &\limsup_{n\to\infty} \int_X f_n \, d\mu \leq \int_X f \, d\mu. \end{aligned}$$

Repeating the process with the sequence of non-negative functions $(g + f_n)$, we can deduce:

$$\int_X f \, d\mu \leq \liminf_{n\to\infty} \int_X f_n \, d\mu.$$

Thus, we have a chain of inequalities:

$$\limsup_{n\to\infty} \int_X f_n \, d\mu \leq \int_X f \, d\mu \leq \liminf_{n\to\infty} \int_X f_n \, d\mu \leq \limsup_{n\to\infty} \int_X f_n \, d\mu, \tag{19.14}$$

which implies all the inequalities in (19.14) are equalities. Since the limit superior and limit inferior coincide with $\int_X f \, d\mu$, we conclude that $\lim_{n\to\infty} \int_X f_n \, d\mu = \int_X f \, d\mu$. □

The Lebesgue integrable function g in the DCT, called the dominating or control function, is a necessary condition, which we shall see in the following example.

Example 19.6.4 Let us look at some examples:

1. A non-example for the DCT would be the Lebesgue measure space $(\mathbb{R}, \mathcal{L}, \mu)$ and the sequence of functions (f_n) with $f_n : \mathbb{R} \to \mathbb{R}$ defined as $f_n = \mathbf{1}_{(n-1,n]}$. We know that $\int_\mathbb{R} f_n \, d\mu = 1$ for all $n \in \mathbb{N}$ and $f_n \xrightarrow{pw} 0$. But:

$$1 = \lim_{n\to\infty} \int_\mathbb{R} f_n \, d\mu \quad \text{while} \quad \int_\mathbb{R} \lim_{n\to\infty} f_n \, d\mu = 0,$$

which do not agree. The DCT does not work here because there is no Lebesgue integrable function that dominates all of the functions f_n at the same time. Indeed, if $g \geq f_n$ for all $n \in \mathbb{N}$, necessarily $g_n(x) \geq 1$ for $x \in (0, \infty)$ and so g cannot be Lebesgue integrable.

2. Let $X = [-1, 1]$ and $(X, \mathcal{L}, \mu)$ be the induced Lebesgue measure space. Consider the sequence of functions (f_n) where $f_n : X \to \mathbb{R}$ is defined as $f_n(x) = \frac{n \sin(\frac{x}{n})}{x(x^2+1)}$ for $x \neq 0$ and $f_n(0) = 1$. We want to determine $\lim_{n\to\infty} \int_X f_n \, d\mu$. First, we note that for every $n \in \mathbb{N}$, the function f_n is measurable since it is continuous over X. We note that $|f_n(0)| = 1$ and, by using the estimate $|\sin(t)| \leq |t|$ for any $t \in \mathbb{R}$, we have:

$$|f_n(x)| = \left| \frac{n \sin(\frac{x}{n})}{x(x^2+1)} \right| \leq \frac{n}{|x(x^2+1)|} \frac{|x|}{n} = \frac{1}{1+x^2} \qquad \text{for } x \neq 0,$$

for all $n \in \mathbb{N}$.

Let the dominating function $g : X \to \mathbb{R}$ be $g(x) = \frac{1}{1+x^2}$. We note that since g is a bounded measurable function over a set of finite measure, we have $g \in \mathcal{L}^1(X)$. We can show that $n \sin(\frac{x}{n}) \to x$ as $n \to \infty$ (using the power series of the sine function, for example) and thus $f_n(x) \to g(x)$ for $x \neq 0$. So, we have $f_n \xrightarrow{pw} g$. Therefore, we can apply the DCT to conclude that:

$$\lim_{n\to\infty} \int_X f_n \, d\mu = \int_X \lim_{n\to\infty} f_n \, d\mu = \int_X g \, d\mu.$$

A corollary of the DCT is the bounded convergence theorem which holds on a domain of finite measure:

Corollary 19.6.5 (Bounded Convergence Theorem, BCT) *Let $(X, \mathcal{F}, \mu)$ be a finite measure space and (f_n) where $f_n : X \to \bar{\mathbb{R}}$ be a sequence of $\mathcal{F}$-measurable functions such that $f_n \xrightarrow{pw} f$ on X to some function $f : X \to \bar{\mathbb{R}}$. Suppose further that there exists a constant $M > 0$ such that $|f_n| \leq M$ μ-a.e. on X for all $n \in \mathbb{N}$. Then:*

1. *f_n and f are in $\mathcal{L}^1(X)$.*
2. $\lim_{n\to\infty} \int_X f_n \, d\mu = \int_X f \, d\mu$.

Proof Use the Lebesgue integrable function $g : X \to \mathbb{R}$ where $g = M\mathbf{1}_X$ as the dominating function and apply the DCT. □

Lebesgue Integrals of Functions Series

Limits also appear implicitly in infinite sums. Recall Proposition 19.4.9 for which we use the MCT to exchange the order of infinite sum and Lebesgue integral for positive functions. By applying the MCT this to the series of positive and negative parts separately, can deduce the full result:

Proposition 19.6.6 (Series of Functions) *Let* $(X, \mathcal{F}, \mu)$ *be a measure space and* (f_n) *be a sequence of* $\mathcal{F}$*-measurable functions where* $f_n : X \to \bar{\mathbb{R}}$*. Denote the sequence* (s_n) *where* $s_n : X \to \bar{\mathbb{R}}$ *is the partial sum* $s_n = \sum_{j=1}^{n} f_j$*.*

1. *If* $\sum_{n=1}^{\infty} \int_X |f_n| \, d\mu < \infty$*, then the sequence of partial sums* (s_n) *converges* μ*-a.e. to a Lebesgue integrable function.*
2. *If* $\sum_{n=1}^{\infty} |f_n|$ *is Lebesgue integrable, then the sequence of partial sums* (s_n) *converges* μ*-a.e. to a Lebesgue integrable function.*

In both cases, we have the equality:

$$\sum_{n=1}^{\infty} \int_X f_n \, d\mu = \int_X \sum_{n=1}^{\infty} f_n \, d\mu.$$

19.7 Comparison Between Lebesgue and Riemann Integrals

In this section, we shall look at the similarities and differences between the Riemann integral that we have constructed in Chap. 14 and the Lebesgue integral on the Lebesgue measure space $(\mathbb{R}, \mathcal{L}, \mu)$ and any of the measure spaces induced by it.

Domain of Integration

An obvious advantage of the Lebesgue integral is that the integral works for any real-valued measurable function defined on any measurable space. As long as the domain is equipped with a σ-algebra and a measure, the whole construction that we have done in this chapter could be done for measurable functions. Therefore, it allows for a more general form of integration.

For example, we have seen that the Lebesgue integral on a set $\mathbb{N}$ with the counting measure is exactly equal to a real series. So the Lebesgue integral allows us to unify some concepts in analysis and create diversity in ways we can approach a problem. Another example, recall Tannery's theorem in Theorem 8.4.5. The readers were invited to prove it in Exercise 8.9 via the ε-N argument. However, we can also prove it using Lebesgue integrals. This will be done in Exercise 19.22 using the DCT.

Moreover, in Chap. 20, we shall define double integrals as an integral on a product of two measure spaces. By induction, this would then allow for multivariable integrals of real-valued functions defined on $\mathbb{R}^n$ for $n \in \mathbb{N}$ that one might have seen in a class on introductory multivariable calculus.

Fundamental Theorem of Calculus

The most important feature of the Riemann integral is the availability of the FTC which allows us to explicitly write down the Riemann integral using antiderivatives. The Lebesgue integral in general does not have this because the Lebesgue integral deals with more complicated functions which may not even have antiderivatives.

However, we can show the following important FTC results for the Lebesgue integral. First, we have continuity of the Lebesgue integral function, which is analogous to Theorem 16.1.1:

Proposition 19.7.1 *Let $([a,b], \mathcal{L}, \mu)$ be the induced Lebesgue measure space and $f : [a,b] \to \bar{\mathbb{R}}$ be a Lebesgue integrable function. Define the Lebesgue integral function $I : [a,b] \to \mathbb{R}$ as:*

$$I(x) = \int_{[a,x]} f \, d\mu.$$

The function I is continuous on $[a,b]$.

Proof The function I is well defined since $[a,x]$ is a measurable set for any $x \in [a,b]$. Fix $\varepsilon > 0$ and $x_0 \in (a,b)$. Using Proposition 19.6.2, there always exists a $\delta > 0$ such that when $\mu(E) < \delta$ we have $\int_E |f| \, d\mu < \varepsilon$. Then, for any $x \in E = \{x \in [a,b] : |x - x_0| < \delta\}$, by setting $A = [x, x_0]$ or $[x_0, x]$ so that $\mu(A) \leq \mu(E) < \delta$, we have:

$$|I(x_0) - I(x)| = \left| \int_A f \, d\mu \right| \leq \int_A |f| \, d\mu \leq \int_E |f| \, d\mu < \varepsilon,$$

which proves continuity of I at x_0. By similar argument, we can also show left and right continuities at the endpoints $x = a$ and $x = b$ respectively. □

An analogue of the first assertion in Theorem 16.1.3 is the following result, which allows us to differentiate the Lebesgue integral function at points for which the integrand is continuous.

Theorem 19.7.2 (Fundamental Theorem of Calculus for Lebesgue Integral) *Let $([a,b], \mathcal{L}, \mu)$ be the induced Lebesgue measure space and $f : [a,b] \to \bar{\mathbb{R}}$ be a Lebesgue integrable function which is continuous and finite at $x_0 \in (a,b)$. Suppose that $I : [a,b] \to \mathbb{R}$ is the Lebesgue integral function:*

$$I(x) = \int_{[a,x]} f \, d\mu.$$

Then, $I'(x_0) = f(x_0)$.

Proof We show that $\lim_{h\to 0}\left|\frac{I(x_0+h)-I(x_0)}{h}-f(x_0)\right|=0$. Fix $\varepsilon>0$. Since f is continuous and finite at x_0, there exists a $\delta>0$ such that for any $x\in X$ with $|x-x_0|<\delta$ we have $|f(x)-f(x_0)|<\varepsilon$. Pick any $|h|<\delta$. By defining $A_h=[x_0,x_0+h]$ or $[x_0+h,x_0]$ depending on the sign of h, we note that $\mu(A_h)=|h|$. WLOG, suppose that $h\geq 0$. By using Proposition 19.5.3, we then have:

$$\begin{aligned}\left|\frac{I(x_0+h)-I(x_0)}{h}-f(x_0)\right| &= \left|\frac{\int_{[a,x_0+h]} f\,d\mu-\int_{[a,x_0]} f\,d\mu}{h}-f(x_0)\right|\\ &=\frac{1}{|h|}\left|\int_{A_h} f-f(x_0)\,d\mu\right|\\ &\leq \frac{\mu(A_h)}{|h|}\sup_{x\in A_h}|f(x)-f(x_0)|\\ &=\sup_{x\in A_h}|f(x)-f(x_0)|<\varepsilon.\end{aligned}$$

Similar estimate holds for $h\leq 0$. Therefore, this proves differentiability of I at x_0 with $I'(x_0)=f(x_0)$. □

In fact, Theorem 19.7.2 can also be extended to functions which are merely Lebesgue integrable. This allows us to differentiate a Lebesgue integrable function almost everywhere over its domain. The following is a result which we are going to state without proof as it requires more advanced techniques in measure theory to study such as the Vitali covering lemma and Hardy-Littlewood function. Readers who are interested for the proof should consult [4].

Theorem 19.7.3 (Lebesgue Differentiation Theorem) *Let $([a,b],\mathcal{L},\mu)$ be the induced Lebesgue measure space and $f:X\to\bar{\mathbb{R}}$ be a Lebesgue integrable function. Suppose that $I:[a,b]\to\mathbb{R}$ is the Lebesgue integral function:*

$$I(x)=\int_{[a,x]} f\,d\mu.$$

Then, $I'(x_0)=f(x_0)$ a.e.

Equality of Riemann and Lebesgue Integrals

A very important observation is that any Riemann integrable functions are also Lebesgue integrable. An easy case is for step functions: the readers shall prove in Exercise 19.2 that the Riemann and Lebesgue integrals for any step function agree.

In fact, they also agree for a larger class of functions. This is a very useful correspondence since computing the Lebesgue integral from scratch can be difficult.

On the other hand, we know how to evaluate some Riemann integrals using antiderivatives and the FTC. Moreover, we have additional tools in the study of Riemann integral such as integration by parts and change of variable which makes it easier to find some integrals.

Recall that Riemann integrals can only be defined for bounded functions on a compact interval. We prove:

Theorem 19.7.4 *Let $X = [a, b]$ be a compat interval in $\mathbb{R}$. Suppose that $(X, \mathcal{L}, \mu)$ is the induced Lebesgue measure space and $f : X \to \mathbb{R}$ is a bounded function. If f is Riemann integrable over X with Riemann integral $\int_a^b f(x)\,dx$, then f is also Lebesgue integrable over X with Lebesgue integral $\int_X f\,d\mu$. Moreover we have the equality of the two integrals, namely:*

$$\int_a^b f(x)\,dx = \int_X f\,d\mu.$$

Proof Since f is bounded, there exists an $M > 0$ such that $|f| \leq M$. We first show that f is measurable. Since f is Riemann integrable, by Corollary 15.3.9, there exists a sequence of refined partitions $(\mathcal{P}_n)$ of X such that $\lim_{n\to\infty} U_{f,\mathcal{P}_n} = \lim_{n\to\infty} L_{f,\mathcal{P}_n}$. Using this sequence of partitions, we can create a sequence of upper and lower approximation functions $(\overline{f}_{\mathcal{P}_n})$ and $(\underline{f}_{\mathcal{P}_n})$ for f. Observe that these approximation functions satisfy the following facts:

1. $(\overline{f}_{\mathcal{P}_n})$ and $(\underline{f}_{\mathcal{P}_n})$ are step functions which are pointwise decreasing and pointwise increasing respectively.
2. $|\overline{f}_{\mathcal{P}_n}| \leq M$ and $|\underline{f}_{\mathcal{P}_n}| \leq M$ for all $n \in \mathbb{N}$.
3. $\underline{f}_{\mathcal{P}_n}(x) \leq f(x) \leq \overline{f}_{\mathcal{P}_n}(x)$ for all $n \in \mathbb{N}$ and $x \in (a, b]$.
4. Since f is Riemann integrable, the upper and lower Darboux sums $U_{f,\mathcal{P}_n} = \int_a^b \underline{f}_{\mathcal{P}_n}(x)\,dx$ and $L_{f,\mathcal{P}_n} = \int_a^b \overline{f}_{\mathcal{P}_n}(x)\,dx$ satisfy the limits:

$$\lim_{n\to\infty} \int_a^b \underline{f}_{\mathcal{P}_n}(x)\,dx = \lim_{n\to\infty} \int_a^b \overline{f}_{\mathcal{P}_n}(x)\,dx = \int_a^b f(x)\,dx.$$

Since $(\overline{f}_{\mathcal{P}_n})$ and $(\underline{f}_{\mathcal{P}_n})$ are both sequences of step functions, they are also simple functions. Thus, each of them are also measurable. Since they are also bounded over a set of finite measure, they are Lebesgue integrable. From Exercise 19.2, for all $n \in \mathbb{N}$ we have the following equalities:

$$\int_a^b \underline{f}_{\mathcal{P}_n}(x)\,dx = \int_X \underline{f}_{\mathcal{P}_n}\,d\mu \quad \text{and} \quad \int_a^b \overline{f}_{\mathcal{P}_n}(x)\,dx = \int_X \overline{f}_{\mathcal{P}_n}\,d\mu.$$

Moreover, by observations 1, 2, and the monotone sequence theorem, the functions $(\overline{f}_{\mathcal{P}_n})$ and $(\underline{f}_{\mathcal{P}_n})$ converge pointwise to some functions $\overline{F}, \underline{F} : X \to \mathbb{R}$

respectively. By observation 3, we then have $-M \leq \underline{F} \leq f \leq \overline{F} \leq M$ on $(a, b]$. Moreover, the limit functions $\overline{F}$ and $\underline{F}$ are also measurable by Proposition 18.9.13. Thus, applying the BCT we get:

$$\lim_{n\to\infty} \int_a^b \underline{f}_{\mathcal{P}_n}(x)\,dx = \lim_{n\to\infty} \int_X \underline{f}_{\mathcal{P}_n}\,d\mu = \int_X \lim_{n\to\infty} \underline{f}_{\mathcal{P}_n}\,d\mu = \int_X \underline{F}\,d\mu, \tag{19.15}$$

$$\lim_{n\to\infty} \int_a^b \overline{f}_{\mathcal{P}_n}(x)\,dx = \lim_{n\to\infty} \int_X \overline{f}_{\mathcal{P}_n}\,d\mu = \int_X \lim_{n\to\infty} \overline{f}_{\mathcal{P}_n}\,d\mu = \int_X \overline{F}\,d\mu. \tag{19.16}$$

Since the first limits of (19.15) and (19.16) are both equal to the Riemann integral of f over $[a, b]$ by observation 4, we then have the equality:

$$\int_X \underline{F}\,d\mu = \int_a^b f(x)\,dx = \int_X \overline{F}\,d\mu.$$

This means $\int_X (\overline{F} - \underline{F})\,d\mu = 0$ and, by Proposition 19.5.5(3), since $\overline{F} - \underline{F} \geq 0$, necessarily $\overline{F} - \underline{F} = 0$ μ-a.e. on X. Moreover, since $\underline{F} \leq f \leq \overline{F}$ on $(a, b]$, we then have $\underline{F} = f = \overline{F}$ μ-a.e. on X. Since f is equal to a measurable function μ-a.e. and $(X, \mathcal{L}, \mu)$ is a complete measure space, by Proposition 18.9.15(2), the function f is also measurable.

In addition, f is a bounded function over a set of finite measure X so it must be Lebesgue integrable over X. Finally, by the MCT, we have the equality:

$$\begin{aligned}\int_a^b f(x)\,dx = \lim_{n\to\infty} \int_a^b \underline{f}_{\mathcal{P}_n}(x)\,dx &= \lim_{n\to\infty} \int_X \underline{f}_{\mathcal{P}_n}\,d\mu \\ &= \int_X \lim_{n\to\infty} \underline{f}_{\mathcal{P}_n}\,d\mu \\ &= \int_X \underline{F}\,d\mu = \int_X f\,d\mu,\end{aligned}$$

which is what we wanted to prove. □

Therefore, since any Riemann integrable function is also Lebesgue integrable, the Lebesgue integral extends the Riemann integral to a bigger class of functions along with its many consequences. This is a very important and useful outcome as pointed out by Lebesgue in his 1904 book *Leçons sur l'intégration et la Recherche des Fonctions Primitives* (Lessons on Integration and Analysis of Primitive Functions):

> It is for the resolution of these problems and not for the love of complications that I introduce in this book a definition of the integral more general than that of Riemann and containing it as a particular case.

However, Lebesgue integral ignores the orientation of the set on which a function is integrated on. Recall that for Riemann integral, we have defined in Definition 15.5.4 the following convention:

$$\int_a^b f(x)\,dx = -\int_b^a f(x)\,dx.$$

On the other hand, for Lebesgue integral, we do not have this feature since we are integrating over a set as a whole rather than a directed or oriented set. This is not a major setback and is a fairly small price to pay for the integrability of more functions.

Example 19.7.5 Recall Example 19.6.4(2) in which we looked at the sequence of functions (f_n) defined on the set $X = [-1, 1]$ where $f_n : X \to \mathbb{R}$ defined as $f_n(x) = \frac{n\sin(\frac{x}{n})}{x(x^2+1)}$ for $x \neq 0$ and $f_n(0) = 1$. We have shown that $\lim_{n\to\infty} \int_X f_n\,d\mu = \int_X \frac{1}{1+x^2}\,d\mu$.

Using the correspondence between Riemann and Lebesgue integrals on a compact domain in Theorem 19.7.4, we can compute the RHS using Riemann integrals. Indeed, $g(x) = \frac{1}{1+x^2}$ is continuous over $[-1, 1]$ and so it is Riemann integrable here. Moreover, it has an antiderivative of $\arctan(x)$ so, by using the FTC, we have:

$$\lim_{n\to\infty} \int_X f_n\,d\mu = \int_X \frac{1}{1+x^2}\,d\mu = \int_{-1}^1 \frac{1}{1+x^2}\,dx = [\arctan(x)]_{-1}^1 = \frac{\pi}{2}.$$

In fact, from Theorem 19.7.4, we can deduce a characterisation of Riemann integrable function. We have seen from Propositions 15.6.3 and 15.6.4 that any continuous functions and monotone functions on a compact interval are guaranteed to be Riemann integrable. However, there are many more Riemann integrable functions out there.

We now want to determine all the Riemann integrable functions on a compact interval $[a, b]$, namely the set $\mathcal{R}([a, b])$. This characterisation is a result called the Lebesgue's Riemann integrability criterion. We first prove the following lemma:

Lemma 19.7.6 *Let $([a, b], \mathcal{L}, \mu)$ be the induced Lebesgue measure space and $f : [a, b] \to \mathbb{R}$ be a bounded function. Let $(\mathcal{P}_n)$ be a sequence of refined partitions of $[a, b]$ such that $\|\mathcal{P}_n\| \to 0$. Suppose that $(\overline{f}_{\mathcal{P}_n})$ and $(\underline{f}_{\mathcal{P}_n})$ where $\overline{f}_{\mathcal{P}_n}, \underline{f}_{\mathcal{P}_n} : [a, b] \to \mathbb{R}$ are the upper and lower Darboux approximations for the function f with respect to the partitions $(\mathcal{P}_n)$. If $\overline{f}_{\mathcal{P}_n} \downarrow \overline{F}$ and $\underline{f}_{\mathcal{P}_n} \uparrow \underline{F}$ to some functions $\overline{F}, \underline{F} : [a, b] \to \mathbb{R}$ respectively, then we have the equality:*

$$\mu\{x \in [a, b] : \overline{F} = \underline{F}\} = \mu\{x \in [a, b] : f \text{ is continuous at } x\}.$$

Proof Write $A = \{x \in [a, b] : \overline{F} = \underline{F}\}$ and $B = \{x \in [a, b] : f \text{ is continuous at } x\}$. The set A is measurable since it is the preimage of the set

$\{0\}$ for the measurable function $\overline{F} = \underline{F} : [a, b] \to \mathbb{R}$ and the set B is measurable by Exercise 18.22. Our goal is to show that $\mu(A) = \mu(B)$.

Write also $P = \bigcup_{n=1}^{\infty} \mathcal{P}_n$. Since $\mathcal{P}_n$ is finite for each n, this union is countable and hence $\mu(P) = 0$. We now show the desired equality $\mu(A) = \mu(B)$ in two steps.

1. Since $A = (A \cap P) \cup (A \setminus P)$ and $\mu(A \cap P) \leq \mu(P) = 0$, we have $\mu(A) = \mu(A \setminus P)$. We now show that $A \setminus P \subseteq B$. Pick any $x_0 \in A \setminus P \subseteq (a, b)$. Since $\underline{F} \leq f \leq \overline{F}$ on $(a, b]$ as noted in the proof of Theorem 19.7.4, necessarily $\overline{F}(x_0) = \underline{F}(x_0) = f(x_0)$.
   Fix $\varepsilon > 0$. Since the sequence $(\overline{f}_{\mathcal{P}_n}(x_0))$ decreases to $\overline{F}(x_0) = f(x_0)$, there exists an index $N_1 \in \mathbb{N}$ such that $\overline{f}_{\mathcal{P}_n}(x_0) - f(x_0) < \varepsilon$ for all $n \geq N_1$. Likewise, there exists an index $N_2 \in \mathbb{N}$ such that $f(x_0) - \underline{f}_{\mathcal{P}_n}(x_0) < \varepsilon$ for all $n \geq N_2$.
   Set $N = \max\{N_1, N_2\}$. Then, the point x_0 is contained in some interval $I = [x_{j-1}, x_j]$ for the partition $\mathcal{P}_N$. Since $x_0 \notin P$, necessarily $x_0 \in (x_{j-1}, x_j)$. Set $\delta = \min\{x_0 - x_{j-1}, x_j - x_0\} > 0$. Then, for any $x \in [a, b]$ such that $|x - x_0| < \delta$ we have $x \in (x_0 - \delta, x_0 + \delta) \subseteq I$. By definition of $\overline{f}_{\mathcal{P}_N}$ and $\underline{f}_{\mathcal{P}_N}$, we then have:
   $$\underline{f}_{\mathcal{P}_N}(x_0) = \inf_{x \in I} f(x) \leq f(x) \leq \sup_{x \in I} f(x) = \overline{f}_{\mathcal{P}_N}(x_0).$$
   Using the inequalities earlier, for any $x \in [a, b]$ with $|x - x_0| < \delta$ we have:
   $$\begin{aligned} & -\varepsilon + f(x_0) < \underline{f}_{\mathcal{P}_N}(x_0) \leq f(x) \leq \overline{f}_{\mathcal{P}_N}(x_0) < f(x_0) + \varepsilon \\ \Rightarrow \quad & |f(x) - f(x_0)| < \varepsilon, \end{aligned}$$
   proving that f is continuous at x_0 and so $x_0 \in B$. Since $x_0 \in A \setminus P$ is arbitrary, we have the inclusion $A \setminus P \subseteq B$ and hence $\mu(A \setminus P) \leq \mu(B)$.
2. Now we show $B \subseteq A$. Pick any $x_0 \in B$ so that f is continuous at x_0. We prove that the real sequences $(\overline{f}_{\mathcal{P}_n}(x_0))$ and $(\underline{f}_{\mathcal{P}_n}(x_0))$ both converge to $f(x_0)$. By continuity of f, for every $\varepsilon > 0$ there exists a $\delta > 0$ such that when $x \in [a, b]$ satisfies $|x - x_0| < \delta$ we have $|f(x) - f(x_0)| < \varepsilon$. This means:
   $$f(x) < f(x_0) + \varepsilon \quad \text{for all } x \in (x_0 - \delta, x_0 + \delta). \tag{19.17}$$
   Since $(\mathcal{P}_n)$ satisfies $||\mathcal{P}_n|| \to 0$, we can find an $N \in \mathbb{N}$ such that the partition interval in $\mathcal{P}_N$ that contains x_0, which we denote as I_N, is of length smaller than $\frac{\delta}{2}$.
   Since $(\mathcal{P}_n)$ is a sequence of refined partitions, for all $n \geq N$ the partition in $\mathcal{P}_n$ containing the point x_0, which we denote as I_n, also has length smaller than $\frac{\delta}{2}$. From the estimate (19.17), we have $f(x_0) \leq \overline{f}_{\mathcal{P}_n}(x_0) = \sup_{x \in I_n} f(x) \leq f(x_0) + \varepsilon$ for all $n \geq N$.
   Taking the limit as $n \to \infty$, since limits preserve weak inequalities, we have $f(x_0) \leq \overline{F}(x_0) \leq f(x_0) + \varepsilon$. But since $\varepsilon > 0$ is arbitrary, necessarily $f(x_0) =$

$\overline{F}(x_0)$. In a similar manner, we can show $f(x_0) = \underline{F}(x_0)$. Thus, $\overline{F}(x_0) = \underline{F}(x_0)$ and hence $x_0 \in A$. Since $x_0 \in B$ is arbitrary, we have the inclusion $B \subseteq A$ and hence $\mu(B) \leq \mu(A)$.

Putting the two measure inequalities together, we have $\mu(B) \leq \mu(A) = \mu(A \backslash P) \leq \mu(B)$, giving us the desired equality. □

Now we prove the theorem characterising all Riemann integrable functions:

Theorem 19.7.7 (Lebesgue's Riemann Integrability Criterion) *Let $X = [a, b]$ and $(X, \mathcal{L}, \mu)$ be the induced Lebesgue measure space. Suppose that $f : X \to \mathbb{R}$ is a bounded function. Then, the function f is Riemann integrable over $[a, b]$ if and only if it is continuous μ-a.e. on $[a, b]$.*

Proof From Corollary 15.3.9, the function f is Riemann integrable over X if and only if there is a sequence of refined partitions $(\mathcal{P}_n)$ of X for which the upper and lower Darboux sums converge to the same number, namely $\lim_{n\to\infty} \int_a^b \underline{f}_{\mathcal{P}_n}(x)\,dx = \lim_{n\to\infty} \int_a^b \overline{f}_{\mathcal{P}_n}(x)\,dx$. From the equalities (19.15) and (19.16), this is equivalent to having the limiting step functions $\overline{F} = \lim_{n\to\infty} \overline{f}_{\mathcal{P}_n}$ and $\underline{F} = \lim_{n\to\infty} \underline{f}_{\mathcal{P}_n}$ satisfying the equality:

$$\int_X \overline{F}\,d\mu = \int_X \underline{F}\,d\mu,$$

or in other words, $\overline{F} = \underline{F}$ μ-a.e.. Thus, by Lemma 19.7.6, we have $\mu([a, b]) = \mu\{x \in [a, b] : \overline{F} = \underline{F}\} = \mu\{x \in [a, b] : f \text{ is continuous at } x\}$. This is the same as saying that f is continuous μ-a.e. which completes the proof. □

Therefore, Riemann integrals can only exist for μ-a.e. continuous functions defined on a compact interval $[a, b]$, which is a small class of functions compared to the class of measurable functions.

Improper Integrals

Unlike the Riemann integral, the definition of Lebesgue integral allows us to carry out the integral construction on unbounded domain as long as we have a measure on the underlying σ-algebra and the integrand is a measurable function.

However, even though the construction allows it, there is no guarantee that such Lebesgue integrals are finite. For example, the function $f : \mathbb{R} \to \mathbb{R}$ defined as $f \equiv 1$ is measurable and we can (Lebesgue) integrate it over $\mathbb{R}$, but it is not Lebesgue integrable since the value of the integral is ∞.

For Riemann integrals, the definition and construction depends heavily on the fact that the domain of the integral is compact. For unbounded domains or non-compact domains, the Riemann integrals can be extended via improper Riemann integrals which we saw in Definition 16.4.2. We have to be very careful with this definition since it is defined using limits. Even though the Riemann integral and Lebesgue integral of a Riemann integrable function agree on compact domains thanks to Theorem 19.7.4, we have no guarantee that they remain so over other domains.

Most of the time, improper Riemann integrable functions are also Lebesgue integrable. But sometimes we might not be so lucky.

Example 19.7.8 Let us look at some improper Riemann integrals and see that some agrees with the Lebesgue integral whilst the other does not.

1. Let $((0, 1], \mathcal{L}, \mu)$ be the induced Lebesgue measure space. Consider the function $f : (0, 1] \to \mathbb{R}$ defined as $f(x) = \frac{1}{\sqrt{x}}$. We have seen in Example 16.4.4(1) that this function is improperly Riemann integrable over $(0, 1]$ with value of 2. Let us show that it is also Lebesgue integrable over the same region with the same value. Clearly, this function f is measurable since it is continuous, so asking whether it is Lebesgue integrable is a valid question.
   Computing the Lebesgue integral from first principle is fiddly, so let us use some technology here. We define a sequence of functions (f_n) where $f_n : (0, 1] \to \mathbb{R}$ is defined as $f_n = f\mathbf{1}_{[\frac{1}{n},1]}$. Note that this sequence of function is pointwise increasing and its pointwise limit is the function f which is what we want to integrate. Thus, by using the MCT, we have:
$$\int_{(0,1]} f \, d\mu = \int_{(0,1]} \lim_{n\to\infty} f_n \, d\mu = \lim_{n\to\infty} \int_{(0,1]} f_n \, d\mu = \lim_{n\to\infty} \int_{[\frac{1}{n},1]} f \, d\mu. \tag{19.18}$$
   Now notice that for any $n \in \mathbb{N}$, the function f is continuous over the compact domain $[\frac{1}{n}, 1]$ so the Lebesgue integral over this domain is equal to its Riemann integral by Theorem 19.7.4. We can then compute it using the FTC as:
$$\int_{[\frac{1}{n},1]} f \, d\mu = \int_{\frac{1}{n}}^{1} \frac{1}{\sqrt{x}} \, dx = 2 - 2\frac{1}{\sqrt{n}}.$$
   Putting this in (19.18) we then have:
$$\int_{(0,1]} f \, d\mu = \lim_{n\to\infty} \int_{[\frac{1}{n},1]} f \, d\mu = \lim_{n\to\infty} \left(2 - 2\frac{1}{\sqrt{n}}\right) = 2 = \int_{0}^{1} f(x) \, dx.$$
   Thus, for this function, we have the equality of the improper Riemann integral with the Lebesgue integral.
2. In Example 19.4.4(3), we have seen that the Lebesgue integral of the function $f : [1, \infty) \to \mathbb{R}$ defined as $f(x) = \frac{1}{x^2}$ over the set $[1, \infty)$ exists and is equal to

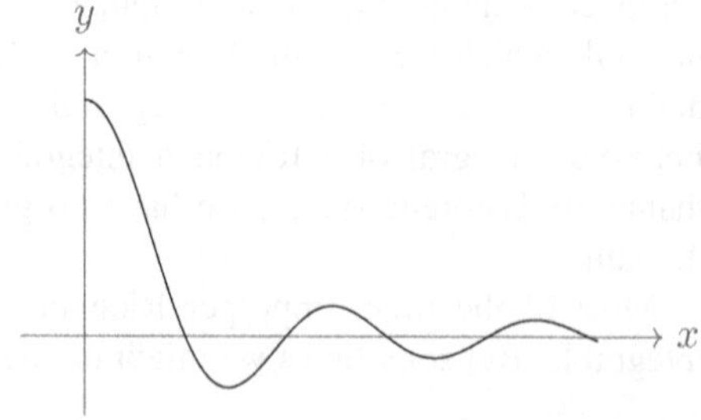

Fig. 19.3 The graph for $f(x) = \frac{\sin(x)}{x}$

1. This value is agreement with the improper Riemann integral $\int_1^\infty \frac{1}{x^2}\,dx$ which we have noted in Example 16.4.4(3).
3. Define the function $f : [0,\infty) \to \mathbb{R}$ by $f(x) = \frac{\sin(x)}{x}$ for $x \neq 0$ and $f(0) = 1$. Figure 19.3 is the graph of this function.
   This function is continuous and hence measurable. To compute its Lebesgue integral, we split this into the positive and negative parts. Define the sets E and F where f is non-negative and non-positive respectively, namely:
   $$E = \bigcup_{\substack{n\in\mathbb{N} \\ n \text{ odd}}} [(n-1)\pi, n\pi] \quad \text{and} \quad F = \bigcup_{\substack{n\in\mathbb{N} \\ n \text{ even}}} [(n-1)\pi, n\pi].$$
   Then, the positive and negative parts of f are given as:
   $$f^+(x) = \frac{\sin(x)}{x}\mathbf{1}_E(x) \quad \text{and} \quad f^-(x) = -\frac{\sin(x)}{x}\mathbf{1}_F(x).$$
   To find the whole Lebesgue integral, we find the Lebesgue integral of each parts above. Consider first the positive part of f. We split the domain of integration into smaller compact intervals. Due to the fact that the function is continuous over each compact interval and hence Riemann integrable here, we can carry out the Lebesgue integral as a Riemann integral. Since $\sin(x)$ is non-negative over each of the intervals in E, we have:
   $$\begin{aligned}\int_{[0,\infty)} f^+\,d\mu = \int_E \frac{\sin(x)}{x}\,dx &= \sum_{n\in\mathbb{N} n \text{ odd}} \int_{(n-1)\pi}^{n\pi} \frac{\sin(x)}{x}\,dx \\ &\geq \sum_{\substack{n\in\mathbb{N} \\ n \text{ odd}}} \int_{(n-1)\pi}^{n\pi} \frac{\sin(x)}{n\pi}\,dx \\ &= \frac{2}{\pi}\sum_{\substack{n\in\mathbb{N} \\ n \text{ odd}}} \frac{1}{n} = \frac{2}{\pi}\sum_{n=1}^{\infty} \frac{1}{2n-1}.\end{aligned}$$

But this sum diverges to ∞ by comparison with the harmonic series. Similarly the integral of the negative part of f, namely $\int_{[0,\infty)} f^- \, d\mu$, also diverges to ∞. Therefore, we have the indeterminate $\infty - \infty$ case for the Lebesgue integral. Hence, this function is not Lebesgue integrable.

On the other hand, if we were to use the Riemann integral, since the domain of integration is unbounded, we have to use the improper Riemann integral. The improper integral can be defined as the limit of the integral function $I : [0, \infty) \to \mathbb{R}$ defined as $I(t) = \int_0^t f(x)\, dx$ as $t \to \infty$. For any $t > 1$, we can apply integration by parts to get:

$$\begin{aligned} I(t) &= \int_0^1 f(x)\,dx + \int_1^t \frac{\sin(x)}{x}\,dx \\ &= \int_0^1 f(x)\,dx + \left[-\frac{\cos(x)}{x}\right]_1^t - \int_1^t \frac{\cos(x)}{x^2}\,dx \\ &= \int_0^1 f(x)\,dx - \frac{\cos(t)}{t} + \cos(1) - \int_1^t \frac{\cos(x)}{x^2}\,dx. \end{aligned}$$

We note that the Riemann integral $\int_0^1 f(x)\,dx$ exists since the integrand is continuous over $[0, 1]$. Moreover, the limits $\lim_{t\to\infty} \frac{\cos(t)}{t}$ and $\lim_{t\to\infty} \int_1^t \frac{\cos(x)}{x^2}\,dx$ both exist. The latter has been shown in Example 16.4.13(5). Thus, by the algebra of limits, the improper integral $\lim_{t\to\infty} I(t) = \int_0^\infty f(x)\,dx$ exists. Hence, the function f is Riemann integrable over $\mathbb{R}$ in the improper sense.

Despite Example 19.7.8(3), we have a very good news: if the function does not change sign over an unbounded domain, the improper Riemann integral (if it exists) coincides with the Lebesgue integral. The proof of this result is left as Exercise 19.20.

Proposition 19.7.9 *Let $X = [0, \infty)$, $(X, \mathcal{L}, \mu)$ be the induced Lebesgue measure space and $f : X \to [0, \infty)$ be a non-negative function. If the improper Riemann integral $\int_0^\infty f(x)\,dx = \lim_{k\to\infty} \int_0^k f(x)\,dx$ exists, then the Lebesgue integral $\int_{[0,\infty)} f\,d\mu$ exists and coincides with the improper Riemann integral.*

By the correspondence between Riemann integrals and Lebesgue integral for compact domains, we still could utilise the FTC for Riemann integration for nice enough functions. In fact, Proposition 19.7.9 says that we do have the correspondence for certain improper integrals too.

Example 19.7.10 Recall Example 19.7.5 in which we looked at the sequence of functions (f_n) defined on the set $X = [-1, 1]$ where $f_n : X \to \mathbb{R}$ defined as $f_n(x) = \frac{n \sin(\frac{x}{n})}{x(x^2+1)}$ for $x \neq 0$ and $f_n(0) = 1$. We have shown that $\lim_{n\to\infty} \int_X f_n\,d\mu = \frac{\pi}{2}$.

Suppose now that we change the domain of integration to the unbounded set $\mathbb{R}$ with the usual Lebesgue measure. We want to evaluate $\lim_{n\to\infty}\int_{\mathbb{R}} f_n\,d\mu$. For every $n \in \mathbb{N}$, the function f_n is measurable since it is a continuous function. We also have the estimate $|f_n(x)| \leq \frac{1}{1+x^2} = g(x)$ for all $x \in \mathbb{R}$. In order to use the DCT, we need to show that the function g, which is a dominating function for the sequence, itself is Lebesgue integrable over the whole of $\mathbb{R}$.

In Example 19.6.4 this is immediate since the domain of integration was bounded. For this case, since the domain $\mathbb{R}$ is unbounded, we can use Proposition 19.7.9. Indeed, we compute the improper Riemann integral:

$$\int_{-\infty}^{\infty} g(x)\,dx = \lim_{k\to\infty}\int_{-k}^{k}\frac{1}{1+x^2}\,dx = \lim_{k\to\infty}[\arctan(x)]_{-k}^{k} = \pi < \infty.$$

By Proposition 19.7.9, since $g \geq 0$, we have the equality $\int_{\mathbb{R}}|g|\,d\mu = \int_{\mathbb{R}} g\,d\mu = \int_{-\infty}^{\infty} g(x)\,dx = \pi < \infty$. Thus, $g \in \mathcal{L}^1(X)$ and we can apply the DCT to conclude that:

$$\lim_{n\to\infty}\int_{\mathbb{R}} f_n\,d\mu = \int_{\mathbb{R}}\lim_{n\to\infty} f_n\,d\mu.$$

In fact, we can show that $f_n \xrightarrow{pw} g$. Thus, by Proposition 19.5.5, we have:

$$\lim_{n\to\infty}\int_{\mathbb{R}} f_n\,d\mu = \int_{\mathbb{R}}\lim_{n\to\infty} f_n\,d\mu = \int_{\mathbb{R}} g\,d\mu = \pi.$$

Limits of Integrals

The Lebesgue integral behaves very well under limits. This is because we have defined it on the class of measurable functions, which is a large class of functions that behave well under limits and infinite processes. We saw in the previous section that we have monotone convergence theorem, Fatou's lemma, dominated convergence theorem, and bounded convergence theorem which could be used to deal with these limits.

On the other hand, we also have some convergence results for Riemann integral, namely integrable limit theorem (Theorems 16.5.3 and 16.5.5), monotone convergence theorem (Theorem 16.5.11), and dominated convergence theorem (Theorem 16.5.13). However, they are much more restrictive in nature. For example, the integrable limit theorem for Riemann integral requires uniform convergence of the functions sequence and all three of the results require the limit function f to be continuous over $[a, b]$. The convergence theorems for Lebesgue integration are much less restrictive, which makes them easier to work with.

Exercises

19.1 (∗) Let $(X, \mathcal{F})$ be a measurable space. A simple function $\phi : X \to \mathbb{R}$ is a function of the form:

$$\phi(x) = \sum_{j=1}^{n} c_j \mathbf{1}_{E_j}(x),$$

where $c_j \in \mathbb{R}$ and $\{E_j\}_{j=1}^n$ is a collection of measurable sets $E_j \in \mathcal{F}$. Prove that we can write ϕ as:

$$\phi(x) = \sum_{j=1}^{m} d_j \mathbf{1}_{F_j}(x),$$

for some constants $d_j \in \mathbb{R}$ and pairwise disjoint measurable sets $\{F_j\}_{j=1}^n$.

19.2 (∗) Let $X = [a, b]$ and $(X, \mathcal{L}, \mu)$ be the induced Lebesgue measure space. Let $\phi : X \to \mathbb{R}$ be a step function with respect to some partition $\mathcal{P}$ of $[a, b]$ as in Definition 15.1.4. Show that:
(a) ϕ is a measurable function.
(b) $\int_X \phi \, d\mu = \int_a^b \phi(x) \, dx$ where the former is the Lebesgue integral and the latter is the Riemann integral.

19.3 Let $f : X \to \mathbb{R}$ be an $\mathcal{F}$-measurable function. Show that there exists a sequence of simple functions (f_n) where $f_n : X \to \mathbb{R}$ such that $|f_n(x)| \leq |f_{n+1}(x)|$ for all $n \in \mathbb{N}$ and $f_n \xrightarrow{pw} f$ on X.

19.4 (∗) Suppose that $\phi : X \to \mathbb{R}$ is a simple function. If $0 \leq \phi$ and $E \in \mathcal{F}$, show that $I(\mathbf{1}_E \phi) \leq I(\phi)$.

19.5 (∗) Prove the first two assertions in Proposition 19.3.4, namely:
(a) If $\phi : X \to \mathbb{R}$ is a simple function, then $\int_X \phi \, d\mu = I(\phi)$.
(b) Let $\lambda > 0$ and $f : X \to [0, \infty]$ is a measurable function. Prove that $\int_X \lambda f \, d\mu = \lambda \int_X f \, d\mu$.

19.6 (∗) Prove the first four assertions of Proposition 19.5.3, namely:
For $\mathcal{F}$-measurable functions $f, g : X \to \bar{\mathbb{R}}$, prove the following results:
(a) $f \in \mathcal{L}^1(X)$ if and only if $|f| \in \mathcal{L}^1(X)$.
(b) If $|g| \leq f$ and $f \in \mathcal{L}^1(X)$, then $g \in \mathcal{L}^1(X)$.
(c) If $0 \leq g \leq |f|$ and $g \notin \mathcal{L}^1(X)$, then $f \notin \mathcal{L}^1(X)$.
(d) Triangle inequality: $\left|\int_X f \, d\mu\right| \leq \int_X |f| \, d\mu$.

19.7 (∗) Prove Proposition 19.5.5, namely:
Suppose that $f, g : X \to \bar{\mathbb{R}}$ are $\mathcal{F}$-measurable functions on a set X with $\mu(X) > 0$. Suppose further that $f \in \mathcal{L}^1(X)$. Prove that:
(a) $f(x) \in \mathbb{R}$ μ-a.e..
(b) If $f = g$ μ-a.e., then $g \in \mathcal{L}^1(X)$ with $\int_X f \, d\mu = \int_X g \, d\mu$.
(c) If $\int_X |f| \, d\mu = 0$, then $f = 0$ μ-a.e..
(d) If $f > 0$ μ-a.e., then $\int_X f \, d\mu > 0$.

19.8 (∗) Let $(X, \mathcal{P}(X))$ be a measureable space. For a fixed $c \in X$, we have seen in Exercise 18.9 the Dirac measure $\delta_c : \mathcal{P}(X) \to [0, \infty]$ where:

$$\delta_c(E) = \begin{cases} 1 & \text{if } c \in E, \\ 0 & \text{otherwise.} \end{cases}$$

Suppose that $f : X \to [0, \infty]$ is a measurable function. Show that $\int_X f \, d\delta_c = f(c)$.

19.9 (∗) Let $([0, 1], \mathcal{L}, \mu)$ be the induced Lebesgue measure space. Recall the Cantor's staircase $f : [0, 1] \to [0, 1]$ from Exercises 11.11, 13.11, and 18.27 which is the limit of the sequence of functions $f_n : [0, 1] \to [0, 1]$ defined iteratively as:

$$f_0(x) = x \quad \text{and} \quad f_n(x) = \begin{cases} \frac{f_{n-1}(3x)}{2} & \text{if } x \in \left[0, \frac{1}{3}\right], \\ \frac{1}{2} & \text{if } x \in \left[\frac{1}{3}, \frac{2}{3}\right], \\ \frac{1}{2} + \frac{f_{n-1}(3x-2)}{2} & \text{if } x \in \left[\frac{2}{3}, 1\right], \end{cases} \quad \text{for all } n \in \mathbb{N}.$$

(a) Show that the function f is Lebesgue measurable.
Explain why it is also Riemann integrable and hence Lebesgue integrable.
(b) Show the following symmetries of f, namely for $x \in [0, 1]$ we have:
i. $f(x) = 1 - f(1 - x)$.
ii. $f(\frac{x}{3}) = \frac{f(x)}{2}$.
iii. $f(\frac{x+2}{3}) = \frac{1+f(x)}{2}$.
(c) Using part (b), find the integral $\int_X f \, d\mu$.

19.10 (∗) Prove Theorem 19.5.9, namely:
Let $(\mathbb{R}, \mathcal{L}, \mu)$ be the Lebesgue measure space and $f \in \mathcal{L}^1(\mathbb{R})$. Prove that:
(a) For every $\varepsilon > 0$, there exists $a, b \in \mathbb{R}$ with $a < b$, a partition $\mathcal{P}$ of $[a, b]$, and a step function $\phi_{\mathcal{P}} : \mathbb{R} \to \mathbb{R}$ adapted to $\mathcal{P}$ such that $\int_{\mathbb{R}} |f - \phi_{\mathcal{P}}| \, d\mu < \varepsilon$.
(b) For every $\varepsilon > 0$, there exists $a, b \in \mathbb{R}$ with $a < b$ and a continuous function $g : \mathbb{R} \to \mathbb{R}$ which vanishes outside of $[a, b]$ such that $\int_{\mathbb{R}} |f - g| \, d\mu < \varepsilon$.

19.11 (∗) Let $(\mathbb{R}, \mathcal{L}, \mu)$ be the Lebesgue measure space and $f : \mathbb{R} \to [0, \infty]$ be a non-negative Lebesgue measurable function.
(a) Show that the set function on $\mathcal{L}$ defined as:

$$\nu : \mathcal{L} \to [0, \infty]$$
$$A \mapsto \int_A f \, d\mu,$$

is also a measure on $\mathcal{L}$.
(b) Show that if $f \in \mathcal{L}^1(\mathbb{R})$, then ν is a finite measure.

(c) Show that if $\mu(A) = 0$, then $\nu(A) = 0$.
(d) Prove: $f > 0$ μ-a.e. if and only if $\nu(A) = 0$ implies $\mu(A) = 0$.
(e) Suppose that $g : \mathbb{R} \to [0, \infty]$. Prove that $g \in \mathcal{L}^1(\mathbb{R}, \nu)$ if and only if $gf \in \mathcal{L}^1(\mathbb{R}, \mu)$ and we have:

$$\int_{\mathbb{R}} g \, d\nu = \int_{\mathbb{R}} gf \, d\mu.$$

We call the measure ν has density f with respect to μ. Due to the result in part (e), we usually write this formally as $d\nu = f \, d\mu$. This generalises the change of variables result that we have seen for Riemann integrals in Theorem 16.3.2.

A more general result for this is called the Radon-Nikodym theorem, which one can find in any measure theory literature. The readers can choose to either consult [12] for a lengthy but basic proof or [4] for a proof which requires additional knowledge on the concept of signed measures.

19.12 (∗) Let $(X, \mathcal{F}, \mu)$ be a finite measure space and $f : X \to \mathbb{R}$ is an $\mathcal{F}$-measurable function.
(a) Prove that if f is bounded a.e. on X, then $f \in \mathcal{L}^1(X)$.
(b) Suppose that $\alpha \leq f(x) \leq \beta$ a.e. for some $\alpha, \beta \in \mathbb{R}$. Prove that:

$$\alpha\mu(X) \leq \int_X f \, d\mu \leq \beta\mu(X).$$

19.13 Suppose that $(\mathbb{R}, \mathcal{L}, \mu)$ is the Lebesgue measure space. Let $f : \mathbb{R} \to \bar{\mathbb{R}}$ be an $\mathcal{L}^1(\mathbb{R})$ function and $c \in \mathbb{R} \setminus \{0\}$ be a constant. Show that:
(a) If $f_c : \mathbb{R} \to \bar{\mathbb{R}}$ is defined as $f_c(x) = f(x + c)$, then:

$$\int_{\mathbb{R}} f_c \, d\mu = \int_{\mathbb{R}} f \, d\mu.$$

(b) If $f_c : \mathbb{R} \to \bar{\mathbb{R}}$ is defined as $f_c(x) = f(cx)$, then:

$$\int_{\mathbb{R}} f_c \, d\mu = \frac{1}{|c|} \int_{\mathbb{R}} f \, d\mu.$$

19.14 Suppose that $(X, \mathcal{F}, \mu)$ is a finite measure space. Let $f : X \to \mathbb{R}$ be an $\mathcal{F}$-measurable function such that:

$$\int_X f^2 \, d\mu = \int_X f^3 \, d\mu = \int_X f^4 \, d\mu.$$

Show that $f = \mathbf{1}_E$ μ-a.e. for some $E \in \mathcal{F}$.

19.15 (◊) Let $([0, \infty), \mathcal{L}, \mu)$ be the induced Lebesgue measure space and $f \in \mathcal{L}^1([0, \infty))$. Prove that if $\int_{[0,c]} f \, d\mu = 0$ for any $c > 0$, then $f \equiv 0$ a.e.

19.16 ($*$) Let $(\mathbb{R}, \mathcal{L}, \mu)$ be the Lebesgue space and $f \in \mathcal{L}^1(\mathbb{R})$. Prove that $\lim_{h\to 0} \int_{\mathbb{R}} |f(x+h) - f(x)|\, d\mu = 0$.

19.17 ($\diamond$) In this question, we are going to prove the Riemann-Lebesgue lemma, which is an important result in Fourier analysis.

Theorem 19.8.11 (Riemann-Lebesgue Lemma) *Suppose that $(\mathbb{R}, \mathcal{L}, \mu)$ is the Lebesgue measure space and $f : \mathbb{R} \to \mathbb{R}$ is a Lebesgue integrable function. Then, we have the following limits:*

$$\lim_{h\to\infty} \int_{\mathbb{R}} f(x)\sin(hx)\, d\mu = 0 \quad \text{and} \quad \lim_{h\to\infty} \int_{\mathbb{R}} f(x)\cos(hx)\, d\mu = 0.$$

We are going to prove the first equality in Theorem 19.8.11 in three steps. The second equality can be proven in a similar manner.

(a) Prove the Riemann-Lebesgue lemma for $f = \mathbf{1}_I$ where $I = (a, b]$.
(b) Using part (a), deduce the Riemann-Lebesgue lemma for any step function f.
(c) Using Exercise 19.10 and part (b), show that for all $\varepsilon > 0$ there exists an $M > 0$ such that if $h > M$ we have $|\int_{\mathbb{R}} f(x)\sin(hx)\, d\mu| < \varepsilon$ for any $f \in \mathcal{L}^1(\mathbb{R})$.
(d) Conclude the result.

19.18 ($\diamond$) We are going to prove the generalised DCT:

Theorem 19.8.12 (Generalised Dominated Convergence Theorem) *Suppose that $(X, \mathcal{F}, \mu)$ is a measure space. Let (g_n) be a sequence of Lebesgue integrable functions $g_n : X \to [0, \infty)$ which converges pointwise to a Lebesgue integrable function $g : X \to [0, \infty)$ such that:*

$$\lim_{n\to\infty} \int_X g_n\, d\mu = \int_X g\, d\mu < \infty.$$

Suppose further that (f_n) be a sequence of Lebesgue integrable functions $f_n : X \to \mathbb{R}$ such that it converges pointwise to a function $f : X \to \mathbb{R}$ and $|f_n(x)| \leq g_n(x)$ for all $x \in X$ and $n \in \mathbb{N}$. Then, $f \in \mathcal{L}^1(X)$ and:

$$\lim_{n\to\infty} \int_X f_n\, d\mu = \int_X f\, d\mu.$$

(a) First, prove that $f \in \mathcal{L}^1(X)$.
(b) By considering the sequence of functions (h_n) where $h_n : X \to \mathbb{R}$ are $h_n = g_n - f_n$, show that:

$$\limsup_{n\to\infty} \int_X f_n\, d\mu \leq \int_X f\, d\mu.$$

(c) Likewise, by considering the functions sequence (h_n) where $h_n = g_n + f_n$, show:

$$\int_X f \, d\mu \leq \liminf_{n\to\infty} \int_X f_n \, d\mu.$$

(d) Conclude the result.

19.19 ($*$) Let $(X, \mathcal{F}, \mu_1)$ and $(Y, \mathcal{G}, \mu_2)$ be measure spaces. Suppose that there exists a function $f : X \to Y$ such that $\mu_2 = f_*\mu_1$ is the pushforward measure from Exercise 18.25. Let $g : Y \to [0, \infty]$.

(a) Show that if $g : Y \to \mathbb{R}$ is the indicator function $g = \mathbf{1}_E$ for some $E \in \mathcal{G}$, then:

$$\int_Y g \, d\mu_2 = \int_X (g \circ f) \, d\mu_1. \tag{19.19}$$

(b) Show that the integral (19.19) holds for simple function g as well.

(c) Finally deduce that the integral (19.19) holds for any $\mathcal{G}$-measurable function g.

19.20 ($*$) Prove Proposition 19.7.9, namely:
Let $X = [0, \infty)$, $(X, \mathcal{L}, \mu)$ be the induced Lebesgue measure space and $f : X \to [0, \infty)$ be a non-negative function. If the improper Riemann integral $\int_0^\infty f(x) \, dx = \lim_{k\to\infty} \int_0^k f(x) \, dx$ exists, then the Lebesgue integral $\int_{[0,\infty)} f \, d\mu$ exists and coincides with the improper Riemann integral.

19.21 ($*$) Let μ be the usual Lebesgue measure on $\mathbb{R}$. Find the limit of the following integrals:

(a) $\lim_{n\to\infty} \int_{[0,1]} (1 - \frac{x}{n})^n \, d\mu$.

(b) $\lim_{n\to\infty} \int_{[0,1]} \frac{n \sin(x)}{1+n^2\sqrt{x}} \, d\mu$.

(c) $\lim_{n\to\infty} \int_{[0,\infty)} \frac{1}{(1+\frac{x}{n})^n x^{\frac{1}{n}}} \, d\mu$.

(d) $\lim_{n\to\infty} \int_{[0,1]} \cos(x^n) \, d\mu$.

(e) $\lim_{n\to\infty} \int_{[1,\infty)} \frac{\ln(nx)}{x+x^2 \ln(n)} \, d\mu$.

(f) $\lim_{n\to\infty} \int_{[1,\infty)} \frac{\ln(1+nx)}{1+x^2 \ln(n)} \, d\mu$.

(g) $\lim_{n\to\infty} \int_{[0,n)} \left(1 + \frac{x}{n}\right)^n e^{-\pi x} \, d\mu$.

(h) $\lim_{n\to\infty} \int_{(0,n)} \left(\frac{\sin(x)}{x}\right)^n \, d\mu$.

(i) $\lim_{n\to\infty} \int_{[0,\infty)} \frac{\sin(nx)}{1+x^2} \, d\mu$.

(j) $\lim_{n\to\infty} \int_{[0,1]} \frac{nx^{n-1}}{1+x} \, d\mu$.

19.22 Recall Tannery's theorem in Theorem 8.4.5. By considering the measure space $(\mathbb{N}, \mathcal{P}(\mathbb{N}), \nu)$ where ν is the counting measure and the doubly indexed real sequence $(a_{m,n})$ as a sequence of functions $a_n : \mathbb{N} \to \mathbb{R}$ where $a_n(m) = a_{m,n}$, prove Tannery's theorem using the DCT.

19.23 Using a counting measure space, show that:

$$\lim_{n\to\infty}\sum_{m=1}^{\infty}\frac{(-1)^m n}{n+nm^2+1}=\sum_{m=1}^{\infty}\frac{(-1)^m}{1+m^2}.$$

19.24 (◇) In this question, we are going to prove the inclusion-exclusion principle. This principle gives us a formula for measuring or counting the elements in a set obtained by the union of n sets. For finite sets with the counting measure, this has been proven for the case of two sets in Lemma 3.4.8. Here we shall prove the general case for higher number of sets equipped with arbitrary measure. As an example, for $n = 2$, similar to Lemma 3.4.8, we have $\mu(E_1 \cup E_2) = \mu(E_1) + \mu(E_2) - \mu(E_1 \cap E_2)$. For $n = 3$ we have $\mu(E_1 \cup E_2 \cup E_3) = \mu(E_1) + \mu(E_2) + \mu(E_3) - \mu(E_1 \cap E_2) - \mu(E_1 \cap E_3) - \mu(E_2 \cap E_2) + \mu(E_1 \cap E_2 \cap E_3)$. The idea behind this equality is demonstrated in Fig. 19.4.

In Fig. 19.4, if we consider the sum of the measures of each set E_1, E_2, and E_3, we would have measured some subsets twice or thrice as denoted in Fig. 19.4a. Therefore, we need to exclude all the intersections of any two sets $E_j \cap E_i$. This then results in the intersection of all three sets $E_1 \cap E_2 \cap E_3$ not being counted in Fig. 19.4b. So, we include this intersection in Fig. 19.4c to get the measure of $E_1 \cup E_2 \cup E_3$ where each part is only measured once. This is where the term inclusion-exclusion comes from: we include, exclude, include, exclude, include, ... et cetera.

For higher number of sets, we have the following general formula:

Proposition 19.8.13 (Inclusion-Exclusion Principle) *Let $(X, \mathcal{F}, \mu)$ be a measure space and $E_j \in \mathcal{F}$ for $j = 1, 2, \ldots, n$. Define $\mathcal{N} = \{1, 2, \ldots, n\}$ and $E = \bigcup_{j=1}^{n} E_j$. Suppose that $\mathcal{E}^p = \{E_{k_1} \cap E_{k_2} \cap \ldots \cap E_{k_p} : k_j \in$*

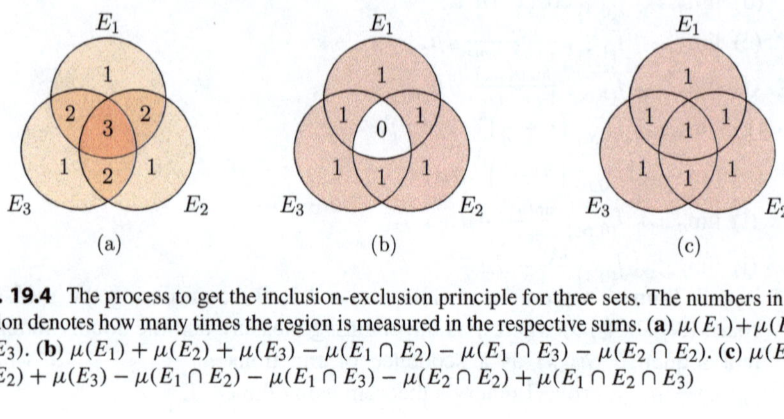

Fig. 19.4 The process to get the inclusion-exclusion principle for three sets. The numbers in each region denotes how many times the region is measured in the respective sums. (**a**) $\mu(E_1)+\mu(E_2)+\mu(E_3)$. (**b**) $\mu(E_1) + \mu(E_2) + \mu(E_3) - \mu(E_1 \cap E_2) - \mu(E_1 \cap E_3) - \mu(E_2 \cap E_2)$. (**c**) $\mu(E_1) + \mu(E_2) + \mu(E_3) - \mu(E_1 \cap E_2) - \mu(E_1 \cap E_3) - \mu(E_2 \cap E_2) + \mu(E_1 \cap E_2 \cap E_3)$

$\mathcal{N}$ distinct} for $p \in \mathcal{N}$ be the collection of intersections of p sets from $\{E_j\}_{j=1}^n$. Then:

$$\mu(E) = \sum_{p=1}^{n} (-1)^{p-1} \sum_{F \in \mathcal{E}^p} \mu(F). \tag{19.20}$$

(a) Define the function $f : X \to \mathbb{R}$ as $f = \prod_{j=1}^n (\mathbf{1}_E - \mathbf{1}_{E_j})$. Show that f is identically 0.

(b) Using the function f from part (a), prove that:

$$\mathbf{1}_E = \sum_{p=1}^{n} (-1)^{p-1} \sum_{F \in \mathcal{E}^p} \mathbf{1}_F.$$

(c) Hence, deduce the inclusion-exclusion principle.

19.25 (◇) Continuing from Exercise 19.24, we now prove the Bonferroni inequalities which tell us what happens if we truncate the sum in the inclusion-exclusion principle in (19.20).

Proposition 19.8.14 (Bonferroni Inequalities) *Let $(X, \mathcal{F}, \mu)$ be a measure space, $E_j \in \mathcal{F}$ for $j = 1, 2, \ldots, n$, and $E = \bigcup_{j=1}^n E_j$. Suppose $0 \le m \le n$. For odd m, we have:*

$$\mu(E) \le \sum_{p=1}^{m} (-1)^{p-1} \sum_{F \in \mathcal{E}^p} \mu(F), \tag{19.21}$$

and for even m we have:

$$\mu(E) \ge \sum_{p=1}^{m} (-1)^{p-1} \sum_{F \in \mathcal{E}^p} \mu(F). \tag{19.22}$$

These inequalities are named after Carlo Emilio Bonferroni (1892–1960). For the specific case of $m = 1$, this is called Boole's inequality after George Boole (1815–1864) and when $m = n$, we have equality by the inclusion-exclusion principle. We will prove the inequalities (19.21) and (19.22) via the following steps.

(a) Let (a_j) for $j = 0, \ldots, n$ be a sequence of positive real numbers such that there exists a $0 \le k \le n$ where (a_j) is increasing between $0 \le j \le k$ and decreasing between $k \le j \le n$. Assume that $\sum_{j=0}^n (-1)^j a_j = 0$. Prove that $\sum_{j=0}^q (-1)^j a_j \ge 0$ for even q and $\sum_{j=0}^q (-1)^j a_j \le 0$ for odd q.

(b) Hence, deduce that for any $r \in \mathbb{N}$ and $0 \le q \le r$, we have $\sum_{j=0}^q (-1)^j \binom{r}{j} \ge 0$ for even q and $\sum_{j=0}^q (-1)^j \binom{r}{j} \le 0$ for odd q.

(c) For any $x \in E$, let $I_x = \{j \in \mathbb{N} : x \in E_j\}$. For any $p \in \mathcal{N} = \{1, 2, \ldots, n\}$ prove that:

$$\sum_{F \in \mathcal{E}^p} \mathbf{1}_F(x) = \binom{|I_x|}{p},$$

where $\mathcal{E}^p = \{E_{k_1} \cap E_{k_2} \cap \ldots \cap E_{k_p} : k_j \in \{1, 2, \ldots, n\} \text{ are distinct}\}$.

(d) Fix $1 \leq m \leq n$. Using part (c), show that:

$$\int_E \sum_{p=0}^{m} (-1)^p \binom{|I_x|}{p} \, d\mu = \mu(E) + \sum_{p=1}^{m} (-1)^p \sum_{F \in \mathcal{E}^p} \mu(F).$$

(e) Hence, obtain the Bonferroni inequalities (19.21) and (19.22).

19.26 (∗) Let $(X, \mathcal{F}, \mu)$ be a measure space and recall that $\mathcal{L}^1(X)$ is a real vector space. Define an operation $||\cdot||_1 : \mathcal{L}^1(X) \to \mathbb{R}$ as:

$$||f||_1 = \int_X |f| \, d\mu.$$

(a) Show that if $f, g \in \mathcal{L}^1(X)$ and $\lambda \in \mathbb{R}$, then we have $||f|| \geq 0$, $||\lambda f||_1 = |\lambda| ||f||_1$, and $||f + g||_1 \leq ||f||_1 + ||g||_1$.

(b) Explain why $||\cdot||_1$ is not a norm on the space $\mathcal{L}^1(X)$.

We define a relation $\sim$ on the space of $\mathcal{F}$-measurable functions by $f \sim g$ iff $f = g$ μ-a.e.

(c) Show that $\sim$ is an equivalence relation.

(d) Show that if $f \sim g$, then $\int_X f \, d\mu = \int_X g \, d\mu$.

The L^1 function space is denoted as the quotient space $L^1(X) = \mathcal{L}^1(X)/\sim$.

(e) Show that $L^1(X)$ is also a real vector space.

Define an operation $||\cdot||_1 : L^1(X) \to \mathbb{R}$ as:

$$||[f]||_1 = \int_X |f| \, d\mu,$$

for any $[f] \in L^1(X)$.

(f) Show that the function $||\cdot||_1$ is well-defined on $L^1(X)$. Namely, if $f \sim g$, then $||[f]||_1 = ||[g]||_1$.

(g) Show that $||\cdot||_1$ is a norm on the space $L^1(X)$.

19.27 (∗) For any $1 \leq p < \infty$, we define the space:

$$\mathcal{L}^p(X) = \left\{ f : X \to \mathbb{R} : f \text{ is } \mathcal{F}\text{-measurable}, \int_X |f|^p \, d\mu < \infty \right\}.$$

The $L^p(X)$ space is then defined by taking the quotient of the $\mathcal{L}^p(X)$ space by the equivalence relation $\sim$ from Exercise 19.27(c). For brevity, instead of writing $[f]$ for the equivalence class, we write it simply as f. We define an L^p-norm on this space as:

$$||f||_p = \left(\int_X |f|^p \, d\mu \right)^{\frac{1}{p}} \geq 0.$$

Prove the following results:

(a) Hölder's inequality: If $p, q > 1$ are such that $\frac{1}{p} + \frac{1}{q} = 1$, $f \in L^p(X)$, and $g \in L^q(X)$, then $fg \in L^1(X)$ with:

$$||fg||_1 \leq ||f||_p ||g||_q.$$

For the case $p = q = 2$, the inequality is called the Cauchy-Bunyakovsky-Schwarz inequality.

(b) Minkowski's inequality: If $p \geq 1$ and $f, g \in L^p(X)$, then:

$$||f + g||_p \leq ||f||_p + ||g||_p.$$

(c) Suppose that $\mu(X) < \infty$. If $1 \leq p < q < \infty$, then $L^q(X) \subseteq L^p(X)$.

(d) Generalised Hölder's inequality: Suppose that $r > 0$ and $p_1, p_2, \ldots, p_n > 0$ are real numbers such that $\sum_{j=1}^{n} \frac{1}{p_j} = \frac{1}{r}$. If $f_j : X \to \mathbb{R}$ for $j = 1, 2, \ldots, n$, then:

$$||f_1 f_2 \ldots f_n||_r \leq ||f_1||_{p_1} ||f_2||_{p_2} \ldots ||f_n||_{p_n}.$$

19.28 We can also define the space $\mathcal{L}^\infty(X)$. Let $(X, \mathcal{F}, \mu)$ be a measure space. We first define the essential supremum of a function $f : X \to \mathbb{R}$ as $\text{ess sup}(f) = \inf\{C > 0 : |f(x)| < C \text{ for } \mu\text{-a.e.}\}$. Define:

$$\mathcal{L}^\infty(X) = \{f : X \to \mathbb{R} : \text{ess sup}(f) < \infty\}.$$

In words, the set $\mathcal{L}^\infty(X)$ is the set of functions that is bounded almost everywhere on X. As in Exercises 18.26 and 18.27, the $L^\infty(X)$ space is then defined by taking the quotient of the $\mathcal{L}^\infty(X)$ space by the equivalence relation $\sim$. We define the L^∞ norm on this space as $||f||_\infty = \text{ess sup}(f)$. Suppose that $\mu(X) < \infty$.

(a) Show that $L^\infty(X) \subseteq L^p(X)$ for all $1 \leq p < \infty$.

(b) Let $f \in L^\infty(X)$. By considering the set $E = \{x \in X : f(x) \geq ||f||_\infty - \varepsilon\}$ for small $\varepsilon > 0$, show that $\liminf_{p \to \infty} ||f||_p \geq ||f||_\infty$.

(c) On the other hand, show that $||f||_\infty \geq \limsup_{p \to \infty} ||f||_p$.

(d) Conclude that $\lim_{p \to \infty} ||f||_p = ||f||_\infty$.

19.29 ($\diamond$) In Exercises 18.31, 18.32, and 18.33, we have seen that measure theory forms an axiomatic foundation for the theory of probability. We continue with some introductory concepts from probability theory here. Let $(\Omega, \mathcal{F}, P)$ be a probability space.

Definition 19.8.15 (Random Variable) Let $(G, \mathcal{G})$ be a measurable space. A G-valued random variable is an $\mathcal{F}$-measurable function $X : \Omega \to G$.

The measurable codomain space $(G, \mathcal{G})$ is usually chosen as the Borel space $(\mathbb{R}, \mathcal{B})$. This choice allows us to work on the field of real numbers which we are more familiar with as opposed to the abstract space of outcomes Ω. It also allows us to combine some outcomes together if X is not injective.

(a) Recall the experiment of tossing two fair coins in Exercise 18.32. Suppose that we are interested with the number of heads in each experiment. Describe the random variable $X : \Omega \to \mathbb{R}$ where $X(\omega) =$ total number of heads in the outcome ω.

In part (a), we have combined some of the outcomes in the experiment together with a non-injective random variable: for example the outcomes HT and TH are combined as their images under the random variable X are the same.

We have a σ-algebra structure $\mathcal{B}$ in the codomain $\mathbb{R}$, so now we need to define a measure on the codomain measurable space $(\mathbb{R}, \mathcal{B})$. There are many measures that can be endowed on this codomain, for example the Lebesgue measure, counting measure, or trivial measure.

However, for this random variable mapping to be meaningful, we want to choose the one that remembers the probability measure in the original probability space $(\Omega, \mathcal{F}, P)$. We define $P_X : \mathcal{B} \to [0, \infty]$ as the measures of the preimages of sets in $\mathcal{B}$ under X, namely for any $E \in \mathcal{B}$:

$$P_X(E) = P(X^{-1}(E)) = P\{\omega \in \Omega : X(\omega) \in E\}.$$

We have seen that this is called the pushforward measure P with respect to the function X in Exercises 18.25 and 19.15. See Fig. 19.5.

(b) For every $a \in \mathbb{R}$, determine $P_X(\{a\})$.

(c) Thus, for any $E \in \mathcal{B}$, determine $P_X(E)$.

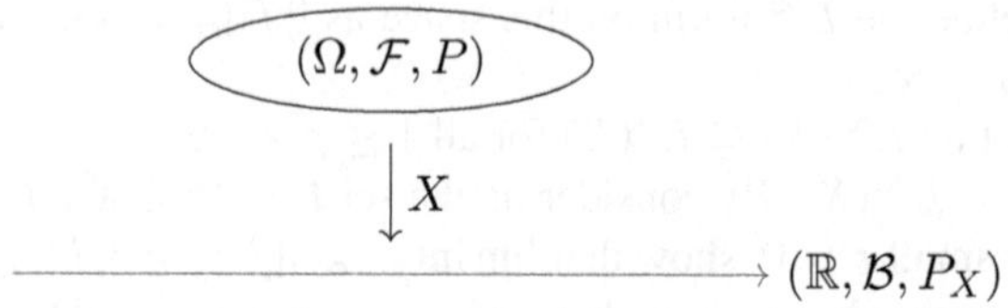

Fig. 19.5 Probability space $(\Omega, \mathcal{F}, P)$ contains all the raw information and data about the experiment. The random variable X maps the abstract space Ω to a more familiar space of $(\mathbb{R}, \mathcal{B})$. To carry the information from the original probability space, we endow the codomain with the pushforward probability measure P_X

(d) Finally, write the probability measure $P_X : \mathcal{B} \to [0, 1]$ using Dirac delta measures which we saw in Exercise 18.9.

Now note that for every $x \in \mathbb{R}$, the set $(-\infty, x]$ is a Borel set. Therefore, the quantity $P_X((-\infty, x])$ for each $x \in \mathbb{R}$ is well-defined and unique. The assignment $x \mapsto P_X((-\infty, x])$ defines a function $F_X : \mathbb{R} \to \mathbb{R}$. This is called the cumulative distribution function (or cdf for short), denoted as:

$$F_X(x) = P_X((-\infty, x]) = P(X^{-1}((-\infty, x])) = P\{\omega \in \Omega : X(\omega) \leq x\}.$$

(e) Write fown the cumulative distribution function for the random variable X.

Now suppose that for each heads we throw in an experiment, we will win \$5. Define a function $g : \mathbb{R} \to \mathbb{R}$ as $g(x) = 5x$.

(f) Show that $g \circ X : \Omega \to \mathbb{R}$ is also a random variable. List down the images of the elements in Ω under the random variable $g \circ X$.

(g) Hence, determine $P_{g \circ X}$.

(h) Discuss the relevance of the function g with the random variable of winnings.

Therefore, from the above, we can see that there are many random variables that we can have on a probability space. Moreover, the final discussion tells us that we can compose any Borel measurable function with a random variable to create another random variable.

19.30 (◇) Roughly speaking, the expected value of a random variable is the average value for the random variable as we carry out a large number of repeated identical experiments. More concretely, we define:

Definition 19.8.16 (Expectation of a Random Variable) Let $(\Omega, \mathcal{F}, P)$ be a probability space and $X : \Omega \to \mathbb{R}$ be a random variable. The expected value or the expectation of the random variable X is the integral of the function X over Ω, denoted as:

$$\mathbb{E}[X] = \int_\Omega X \, dP.$$

(a) Using Exercise 19.19, show that we can also write the expectation as:

$$\mathbb{E}[X] = \int_\mathbb{R} \mathrm{id}_\mathbb{R} \, dP_X = \int_\mathbb{R} x \, dP_X(x),$$

if x is the variable used to denote $\mathbb{R}$.

(b) In general, for any function $g : \mathbb{R} \to \mathbb{R}$, show that $\mathbb{E}[g(X)] = \int_\Omega X \, dP = \int_\mathbb{R} g \, dP_X$.

(c) Using the random variables from Exercise 19.30, compute $\mathbb{E}[X]$ and $\mathbb{E}[g \circ X]$.

19.31 (◇) Consider the experiment of rolling two regular fair dice.

(a) Write down the sample space Ω of this experiment. Assuming uniform probability measure P (see Exercise 18.32(c) for the definition) on $(\Omega, \mathcal{F})$, state the probabilities for each outcome.

(b) Without listing them down, compute the size of the events space $\mathcal{F}$.

(c) Suppose that we are interested in with the sum of the numbers on the two dice.

   i. Write down in full the function $X : \Omega \to \mathbb{R}$ where $X(\omega) =$ sum of the numbers obtained in the outcome ω.

   ii. For every $a \in \mathbb{R}$, determine $P_X(\{a\})$. Hence, for any $E \in \mathcal{B}$, determine $P_X(E)$.

   iii. Find $\mathbb{E}[X]$.

(d) Suppose that we are interested in with the maximum number on the two dice.

   i. Write down in full the function $Y : \Omega \to \mathbb{R}$ where $Y(\omega) =$ maximum number of the dice in the outcome ω.

   ii. For every $a \in \mathbb{R}$, determine $P_Y(\{a\})$. Hence, for any $E \in \mathcal{B}$, determine $P_Y(E)$.

   iii. Find $\mathbb{E}[Y]$.

(e) i. From part (c), write a computer program simulating the outcomes for the random variable X for 1000 experiments.
   For each $n = 1, 2, \ldots, 1000$, compute the average value a_n of the first n random variable outcomes of the experiments.
   Plot these values in a graph and compare them with the computed expected value $\mathbb{E}[X]$ in part (c).

   ii. Repeat the activity above for the random variable Y.

These should justify the interpretation of expected value as we have mentioned in Exercise19.31.

Double Integrals

20

Prepare for trouble! Make it double!

— Jessie and James, Team Rocket

In this chapter, we want to extend the Lebesgue integral from two distinct σ-finite measure spaces $(X, \mathcal{F}, \mu_1)$ and $(Y, \mathcal{G}, \mu_2)$ to their Cartesian product $X \times Y$. In other words, we are going to address how can we integrate a real-valued function $f : X \times Y \to \mathbb{R}$. Towards the end of this chapter, we shall give a generalisation to higher number of Cartesian products of measure spaces.

20.1 Product Measure Space

In order to define a Lebesgue integral of a function $f : X \times Y \to \mathbb{R}$, we first need to determine a candidate for σ-algebra and measure μ on the Cartesian product $X \times Y$ which are compatible with the σ-algebras $\mathcal{F}$ and $\mathcal{G}$ and the measures μ_1 and μ_2. Due to the generality of the results we have proven in Chap. 18, this task can be done with all the machineries that we have established.

Product Measurable Space

An obvious candidate for the collection of sets in $X \times Y$ that we want to work on is the collection of the Cartesian products of the sets in each constituent σ-algebra. Namely, we consider the collection of sets $\mathcal{J} = \mathcal{F} \times \mathcal{G} = \{A \times B : A \in \mathcal{F}, B \in \mathcal{G}\}$. We can clearly define a content on the subsets of $X \times Y$ in $\mathcal{J}$ via $m(A \times B) = \mu_1(A)\mu_2(B)$. This content is called the product content on $\mathcal{J}$ induced by the measures μ_1 and μ_2.

S. Johar, *The Big Book of Real Analysis*,
https://doi.org/10.1007/978-3-031-30832-1_20

However, the collection $\mathcal{J}$ does not satisfy the ring axioms. In particular, a union of two nonempty sets in $\mathcal{J}$ is not necessarily in $\mathcal{J}$. The collection $\mathcal{J}$ is instead called the rectangular sets and they form a π-system and semiring. By virtue of Proposition 18.3.7, the elements in the ring generated by the rectangular sets $\mathcal{R}(\mathcal{J})$ can be expressed as a finite union of pairwise disjoint rectangular sets in $\mathcal{J}$. This allows us to extend the product content m that we have defined on $\mathcal{J}$ above to create a premeasure space $(X \times Y, \mathcal{R}(\mathcal{J}), m)$. This is left as Exercise 20.1 for the readers to verify. The premeasure space above is called a product premeasure space.

Moreover, if the constituent measure spaces are σ-finite, the product premeasure space must also be σ-finite. The following lemma will be proven by the readers in Exercise 20.2.

Lemma 20.1.1 *Let $(X, \mathcal{F}, \mu_1)$ and $(Y, \mathcal{G}, \mu_2)$ be σ-finite measure spaces. Let $\mathcal{J}$ be the semiring of rectangles $\mathcal{J} = \mathcal{F} \times \mathcal{G}$ with the product content m and $\mathcal{R}(\mathcal{J})$ be the ring generated by $\mathcal{J}$. Then, the premeasure space $(X \times Y, \mathcal{R}(\mathcal{J}), m)$ is also σ-finite.*

However, as we have observed in Chap. 18, this is not good enough for a measure since the underlying space is not a σ-algebra. By Lemma 18.3.19, we can thus construct the smallest σ-algebra containing $\mathcal{J}$, which we call $\mathcal{H} = \sigma(\mathcal{J}) = \sigma(\mathcal{R}(\mathcal{J}))$. Usually we denote this σ-algebra as $\mathcal{H} = \mathcal{F} \otimes \mathcal{G}$ and call it the product σ-algebra of $\mathcal{F}$ and $\mathcal{G}$. In fact, we can also characterise the σ-algebra $\mathcal{H}$ using projection maps. We first define:

Definition 20.1.2 (Projection Maps) Let X and Y be non-empty sets and $X \times Y$ be their Cartesian product.

1. A projection map onto X is the function $\pi_X : X \times Y \to X$ defined as $\pi_X(x, y) = x$ for all $(x, y) \in X \times Y$.
2. A projection map onto Y is the function $\pi_Y : X \times Y \to Y$ defined as $\pi_Y(x, y) = y$ for all $(x, y) \in X \times Y$.

Note that these projection maps are surjective. Now we prove a characterisation of the product σ-algebra using the projection maps:

Proposition 20.1.3 *Let $(X, \mathcal{F})$ and $(Y, \mathcal{G})$ be two measurable spaces. The product σ-algebra $\mathcal{H} = \mathcal{F} \otimes \mathcal{G}$ is the smallest σ-algebra on the Cartesian product $X \times Y$ such that both the projection maps $\pi_X : (X \times Y, \mathcal{H}) \to (X, \mathcal{F})$ and $\pi_Y : (X \times Y, \mathcal{H}) \to (Y, \mathcal{G})$ are measurable.*

Proof Let $\mathcal{K}$ be the smallest σ-algebra on $X \times Y$ for which the projection maps π_X and π_Y are $\mathcal{F}$-measurable and $\mathcal{G}$-measurable respectively. In other words, for every $A \in \mathcal{F}$ we must have $\pi_X^{-1}(A) \in \mathcal{K}$ and for every $B \in \mathcal{G}$ we must have $\pi_Y^{-1}(B) \in \mathcal{K}$. We want to show the equality of sets $\mathcal{K} = \mathcal{H}$ via double inclusion.

($\subseteq$): For any $A \in \mathcal{F}$, we have $\pi_X^{-1}(A) = A \times Y \in \mathcal{J} \subseteq \mathcal{H}$. Thus, π_X is $\mathcal{H}$-measurable. Likewise, for any $B \in \mathcal{G}$, we have $\pi_Y^{-1}(B) = X \times B \in \mathcal{J} \subseteq \mathcal{H}$. Therefore, π_Y is $\mathcal{H}$-measurable. So the projection maps are also measurable on the measure space $(X \times Y, \mathcal{H})$. Since $\mathcal{K}$ is the smallest such measure space, we must have the inclusion $\mathcal{K} \subseteq \mathcal{H}$.

($\supseteq$): For any element $A \times B \in \mathcal{J}$, we have $A \times B = (A \times X) \cap (Y \times B) = \pi_X^{-1}(A) \cap \pi_Y^{-1}(B) \in \mathcal{K}$. Thus, we have the inclusion $\mathcal{J} \subseteq \mathcal{K}$. By minimality of $\sigma(\mathcal{J})$, we have the inclusion $\mathcal{H} = \sigma(\mathcal{J}) \subseteq \mathcal{K}$. □

The characterisation in Proposition 20.1.3 can help us build more general product measurable spaces. In particular, this helps us build a unique measure space for higher number of factors. Let us demonstrate this.

Suppose that we have three measurable spaces $(X, \mathcal{F})$, $(Y, \mathcal{G})$, and $(Z, \mathcal{H})$. We would like to construct a σ-algebra on the Cartesian product $W = X \times Y \times Z$. There are several ways we can do this, namely:

1. construct the product σ-algebra $\mathcal{F} \otimes \mathcal{G}$ first, then construct the σ-algebra $(\mathcal{F} \otimes \mathcal{G}) \otimes \mathcal{H}$, or
2. construct the product σ-algebra $\mathcal{G} \otimes \mathcal{H}$ first, then construct the σ-algebra $\mathcal{F} \otimes (\mathcal{G} \otimes \mathcal{H})$, or
3. consider the cubical sets $\mathcal{J} = \{A \times B \times C : A \in \mathcal{F}, B \in \mathcal{G}, C \in \mathcal{H}\}$ and generate $\mathcal{F} \otimes \mathcal{G} \otimes \mathcal{H} = \sigma(\mathcal{J})$.

These constructions actually yield the same σ-algebra, according to the following result:

Proposition 20.1.4 *Suppose that $(X, \mathcal{F})$, $(Y, \mathcal{G})$, and $(Z, \mathcal{H})$ are three measurable spaces. Then, we have the equality of the σ-algebras $(\mathcal{F} \otimes \mathcal{G}) \otimes \mathcal{H} = \mathcal{F} \otimes (\mathcal{G} \otimes \mathcal{H}) = \mathcal{F} \otimes \mathcal{G} \otimes \mathcal{H}$ on $W = X \times Y \times Z$.*

Proof We shall only prove the equality $\mathcal{F} \otimes (\mathcal{G} \otimes \mathcal{H}) = \mathcal{F} \otimes \mathcal{G} \otimes \mathcal{H}$ as the other equality can be proven similarly. We prove this via double inclusion. Denote the cubical sets $\mathcal{J} = \{A \times B \times C : A \in \mathcal{F}, B \in \mathcal{G}, C \in \mathcal{H}\}$ and the σ-algebra generated by it as $\mathcal{K} = \sigma(\mathcal{J}) = \mathcal{F} \otimes \mathcal{G} \otimes \mathcal{H}$ for brevity.

($\supseteq$): For any $A \in \mathcal{F}$, $B \in \mathcal{G}$, and $C \in \mathcal{H}$, we have $B \times C \in \mathcal{G} \otimes \mathcal{H}$ and hence $A \times B \times C \in \mathcal{F} \otimes (\mathcal{G} \otimes \mathcal{H})$. This means $\mathcal{J} \subseteq \mathcal{F} \otimes (\mathcal{G} \otimes \mathcal{H})$ and so $\mathcal{K} = \sigma(\mathcal{J}) \subseteq \mathcal{F} \otimes (\mathcal{G} \otimes \mathcal{H})$.

($\subseteq$): Define the projection maps $\pi_1 : W \to X$ and $\pi_2 : W \to Y \times Z$. By Proposition 20.1.3, the σ-algebra $\mathcal{F} \otimes (\mathcal{G} \otimes \mathcal{H})$ is the smallest σ-algebra on W such that the projection maps π_1 and π_2 are measurable. We now show that π_1 and π_2 are also $\mathcal{K}$-measurable.

1. For π_1, for any $A \in \mathcal{F}$ we have $\pi^{-1}(A) = A \times Y \times Z \in \mathcal{J} \subseteq \mathcal{K}$.
2. For π_2, first denote the rectangular sets $\mathcal{S} = \{B \times C : B \in \mathcal{G}, C \in \mathcal{H}\}$. Note that $\mathcal{G} \otimes \mathcal{H} = \sigma(\mathcal{S})$. For arbitrary sets $B \in \mathcal{G}$ and $C \in \mathcal{H}$, we have $\pi_2^{-1}(B \times C) = X \times B \times C \in \mathcal{K}$. Thus the preimages of all the rectangular sets $\mathcal{S}$ are contained in $\mathcal{K}$ which implies that $\mathcal{S} \subseteq \{\pi_2(D) : D \in \mathcal{K}\}$. Since π_2 is surjective, by Exercise 18.15(a), the collection $\{\pi_2(D) : D \in \mathcal{K}\}$ is a σ-algebra that contains the rectangular sets $\mathcal{S}$. Thus, we have the inclusion $\mathcal{G} \otimes \mathcal{H} = \sigma(\mathcal{S}) \subseteq \{\pi_2(D) : D \in \mathcal{K}\}$ which means the preimage of any set in $\mathcal{G} \otimes \mathcal{H}$ under the projection map π_2 is in $\mathcal{K}$. Therefore, π_2 is also $\mathcal{K}$-measurable.

Hence, since π_1 and π_2 are also $\mathcal{K}$-measurable, by minimality of $\mathcal{F}\otimes(\mathcal{G}\otimes\mathcal{H})$, we have the inclusion $\mathcal{F} \otimes (\mathcal{G} \otimes \mathcal{H}) \subseteq \mathcal{K}$. □

By Proposition 20.1.4, the product operation $\otimes$ on σ-algebras are associative. Thus, we can write the unique product of n measurable spaces $(X_j, \mathcal{F}_j)$ for $j = 1, 2, \ldots, n$ as $(X_1 \times \ldots \times X_n, \mathcal{F}_1 \otimes \ldots \otimes \mathcal{F}_n)$ without any ambiguity.

Product Measure

Next, we want to define a measure on the measurable product space $(X \times Y, \mathcal{F} \otimes \mathcal{G})$ using the product premeasure m. This measure will be called the product measure on $X \times Y$ induced by μ_1 and μ_2.

A way to construct the product measure is to first define an outer measure m^* induced by the premeasure m on the largest σ-algebra on $X \times Y$, namely the power set $\mathcal{P}(X \times Y)$, by the countable covering argument that we have seen in Definition 18.4.1. More specifically, for any $E \in \mathcal{P}(X \times Y)$, we define:

$$m^*(E) = \inf \left\{ \sum_{j=1}^{\infty} m(I_j) : I_j \in \mathcal{R}(\mathcal{J}) \text{ such that } E \subseteq \bigcup_{j=1}^{\infty} I_j \right\},$$

where $\mathcal{J} = \mathcal{F} \times \mathcal{G} = \{A \times B : A \in \mathcal{F}, B \in \mathcal{G}\}$ is the set of rectangular sets and $(\mathcal{R}(\mathcal{J}), m)$ is the ring generated by it along with the premeasure m that we saw previously. If this infimum does not exist, we set $m^*(E) = \infty$.

Similar to the construction in Chap. 18, this outer measure m^* is probably not a genuine measure on the measurable space $(X \times Y, \mathcal{P}(X \times Y))$ since it might lack the σ-additivity property. Therefore, in order to turn it into a genuine measure, we proceed by removing the sets which does not satisfy the Carathéodory condition (see Definition 18.6.1) from $\mathcal{P}(X \times Y)$. The resulting collection of sets, which we now call $\mathcal{K}$, is a complete σ-algebra and the outer measure restricted to this σ-algebra $m^*|_{\mathcal{K}}$, which we now call μ, is a genuine measure on $\mathcal{K}$. This results in a measure space $(X \times Y, \mathcal{K}, \mu)$.

Clearly $\mathcal{F}\otimes\mathcal{G} \subseteq \mathcal{K}$ since $\mathcal{K}$ is a σ-algebra that contains all of the sets in $\mathcal{J}$. Thus, the measure μ that we have constructed on $\mathcal{K}$ can also be endowed on the σ-algebra $\mathcal{F}\otimes\mathcal{G}$ via the restriction $\mu|_{\mathcal{F}\otimes\mathcal{G}}$, which we also denote as μ. The resulting measure is called the product measure on $(X \times Y, \mathcal{F}\otimes\mathcal{G})$ induced by the measure spaces $(X, \mathcal{F}, \mu_1)$ and $(Y, \mathcal{G}, \mu_2)$.

Furthermore, for any $A \in \mathcal{F}$ and $B \in \mathcal{G}$, we have $A \times B \in \mathcal{J} \subseteq \mathcal{K}$ and $\mu(A \times B) = \mu_1(A)\mu_2(B) = m(A \times B)$. This means the measure μ on $\mathcal{F}\otimes\mathcal{G}$ is compatible with the premeasure m on $\mathcal{R}(\mathcal{J})$. Moreover, if each of the measure spaces $(X, \mathcal{F}, \mu_1)$ and $(Y, \mathcal{G}, \mu_2)$ are σ-finite, Lemma 20.1.1 implies that the premeasure space $(X \times Y, \mathcal{R}(\mathcal{J}), m)$ is also σ-finite. Thus, by applying Theorem 18.8.4, the induced product measure μ is unique and also σ-finite.

We note that, similar to the inclusion of the Borel σ-algebra in the Lebesgue σ-algebra on $\mathbb{R}$ that we have seen in Proposition 18.7.8, the inclusion $\mathcal{F}\otimes\mathcal{G} \subseteq \mathcal{K}$ might be strict. Clearly the σ-algebra $\mathcal{K}$ is complete since all the m^*-null sets and their subsets were not eliminated from $\mathcal{P}(X \times Y)$ during the construction using the Carathéodory condition. On the other hand, the product σ-algebra $\mathcal{F}\otimes\mathcal{G}$ might not be complete as they possibly lack some of these m^*-null sets.

Example 20.1.5 Consider the Lebesgue measure space $(\mathbb{R}, \mathcal{L}, \mu)$ on the real numbers. We can construct two different measure spaces on the Cartesian product $\mathbb{R}^2$.

The rectangular sets $\mathcal{J} = \{A \times B : A, B \in \mathcal{L}\}$ is a π-system with content $m(A \times B) = \mu(A)\mu(B)$. This π-system could then be extended to a semiring $\mathcal{R}(\mathcal{J})$ with premeasure induced by m.

1. Using the premeasure m, we could define an outer measure m^* on $\mathcal{P}(\mathbb{R}^2)$. Next, by using the Carathéodory condition, we can extract a σ-algebra $\mathcal{L}(\mathbb{R}^2) \subseteq \mathcal{P}(\mathbb{R}^2)$ such that the outer measure m^* restricted to the σ-algebra $\mathcal{L}(\mathbb{R}^2)$, which we call $m^*|_{\mathcal{L}(\mathbb{R}^2)} = \tilde{\mu}$, is a genuine measure. We call the σ-algebra $\mathcal{L}(\mathbb{R}^2)$ the Lebesgue σ-algebra of $\mathbb{R}^2$ and the measure $\tilde{\mu}$ the Lebesgue measure on $\mathbb{R}^2$.
2. Alternatively, we could construct the product σ-algebra $\mathcal{L}\otimes\mathcal{L} = \sigma(\mathcal{J})$ on $\mathbb{R}^2$. By Theorem 18.8.4, since $(\mathbb{R}, \mathcal{L}, \mu)$ is σ-finite and hence the premeasure space $(\mathbb{R}^2, \mathcal{R}(\mathcal{J}), m)$ is also σ-finite, the unique measure on this σ-algebra induced by the premeasure m on $\mathcal{R}(\mathcal{J})$ is also the Lebesgue measure $\tilde{\mu}$ as defined above.

Thus, we have two possible measure space structures on $\mathbb{R}^2$ induced by $(\mathbb{R}, \mathcal{L}, \mu)$, namely $(\mathbb{R}^2, \mathcal{L}(\mathbb{R}^2), \tilde{\mu})$ and $(\mathbb{R}^2, \mathcal{L}\otimes\mathcal{L}, \tilde{\mu})$. We note that $\mathcal{L}(\mathbb{R}^2)$ and $\mathcal{L}\otimes\mathcal{L}$ are two distinct σ-algebras on $\mathbb{R}^2$. The former is called the Lebesgue σ-algebra on $\mathbb{R}^2$ whilst the latter is the product σ-algebra of Lebesgue sets. The difference is very subtle but crucial, namely: we have the strict inclusion $\mathcal{L}\otimes\mathcal{L} \subsetneq \mathcal{L}(\mathbb{R}^2)$. In Exercise 20.9, the readers are invited to construct an example of a set which is in $\mathcal{L}(\mathbb{R}^2) \setminus \mathcal{L}\otimes\mathcal{L}$. Moreover, $\mathcal{L}(\mathbb{R}^2)$ is complete whereas $\mathcal{L}\otimes\mathcal{L}$ is not.

Of course, this construction of product measures also works for products of measurable spaces with more than two factors by induction.

20.2 Iterated Integrals

Lebesgue Integral over $X \times Y$

Using a measure space structure on $X \times Y$, we can define the Lebesgue integral of non-negative functions on the space $X \times Y$ as we did before in Definition 19.3.1 using simple functions. Namely:

Definition 20.2.1 (Lebesgue Integral of Non-negative Functions) Let $(X \times Y, \mathcal{H}, \mu)$ be a measure space. For a general non-negative $\mathcal{H}$-measurable function $f : X \times Y \to [0, \infty]$, we define:

$$\int_{X\times Y} f \, d\mu = \sup\{I(\phi) : \phi : X \times Y \to \mathbb{R} \text{ is a simple function with } \phi \leq f\}.$$

Likewise, the integral of a general $\mathcal{H}$-measurable function to a larger codomain $f : X \times Y \to \bar{\mathbb{R}}$ can be done by first breaking the function f into its positive and negative parts. Namely, we write $f = f^+ - f^-$ where $f^+, f^- : X \times Y \to [0, \infty]$ are defined as $f^+(x, y) = \max\{f(x, y), 0\}$ and $f^-(x, y) = -\min\{f(x, y), 0\}$ for every $(x, y) \in X \times Y$ respectively. The functions f^+ and f^- are thus two non-negative functions on $X \times Y$ for which we can compute their Lebesgue integral using Definition 20.2.1.

Therefore, if at least one of the integrals of the non-negative functions f^+ or f^- above is finite, we can define the Lebesgue integral for the function f as the sum:

$$\int_{X\times Y} f \, d\mu = \int_{X\times Y} f^+ \, d\mu - \int_{X\times Y} f^- \, d\mu,$$

which takes values in $\bar{\mathbb{R}}$.

Moreover, if both of the integrals for f^+ and f^- are finite (which then implies that $\int_{X\times Y} f \, d\mu = \int_{X\times Y} f^+ \, d\mu - \int_{X\times Y} f^- \, d\mu < \infty$), we call such functions Lebesgue integrable. This is similar to Definition 19.5.1.

Definition 20.2.2 (Lebesgue Integrable Functions) Let $(X \times Y, \mathcal{H}, \mu)$ be a measure space and $f : X \times Y \to \bar{\mathbb{R}}$ be an $\mathcal{H}$-measurable function. If both of the integrals $\int_{X\times Y} f^+ \, d\mu$ and $\int_{X\times Y} f^- \, d\mu$ are finite, we call the function f Lebesgue integrable.
The space of Lebesgue integrable functions over $X \times Y$ is denoted as:

$$\mathcal{L}^1(X \times Y, \mathcal{H}, \mu) = \left\{ f : X \times Y \to \bar{\mathbb{R}} : f \text{ is } \mathcal{H}\text{-measurable}, \int_{X\times Y} |f| \, d\mu < \infty \right\}.$$

What we have defined so far are simply the exact same definitions for the measure and integral for a general space X that we have stated in Chaps. 18 and 19, but with the measure space $(X, \mathcal{F}, \mu)$ replaced with the product space $X \times Y$, its σ-algebra $\mathcal{H}$, and a measure μ defined on the σ-algebra $\mathcal{H}$.

In fact, by the generality of definitions that we had used in the previous chapters, all the properties and results of integrals in Propositions 19.3.4, 19.3.6, 19.5.3, 19.5.4, and 19.5.5 as well as the convergence theorems carry forward to this new measure space.

Now suppose that we are working with the product measure space on $X \times Y$ induced by the constituent measure spaces $(X, \mathcal{F}, \mu_1)$ and $(Y, \mathcal{G}, \mu_2)$. Namely, our measure space is $(X \times Y, \mathcal{F} \otimes \mathcal{G}, \mu)$ which was constructed in Sect. 20.1. We would like to know how the Lebesgue integral over the product measure space $(X \times Y, \mathcal{F} \otimes \mathcal{G}, \mu)$ relates to the integral over its constituent measure spaces $(X, \mathcal{F}, \mu_1)$ and $(Y, \mathcal{G}, \mu_2)$. We shall devote the remaining part of this chapter to study this relationship.

Sections and Section Functions

First, we need to define a new terminology, namely cross-sections or sections.

Definition 20.2.3 (Sections) Let $W \subseteq X \times Y$ be a subset of the Cartesian product $X \times Y$.

1. The section of W at a fixed $x \in X$ is the set $W^x = \{y \in Y : (x, y) \in W\} \subseteq Y$.
2. The section of W at a fixed $y \in Y$ is the set $W_y = \{x \in X : (x, y) \in W\} \subseteq X$.

See Fig. 20.1 for an example of a section. One crucial result regarding the sections is that if a set $W \subseteq X \times Y$ is $(\mathcal{F} \otimes \mathcal{G})$-measurable, then any of its sections is measurable with respect to $\mathcal{F}$ or $\mathcal{G}$, namely:

Lemma 20.2.4 *Let $(X, \mathcal{F})$ and $(Y, \mathcal{G})$ be two non-empty measurable spaces. If $W \in \mathcal{H} = \mathcal{F} \otimes \mathcal{G}$, then $W_y \in \mathcal{F}$ for every $y \in Y$ and $W^x \in \mathcal{G}$ for every $x \in X$.*

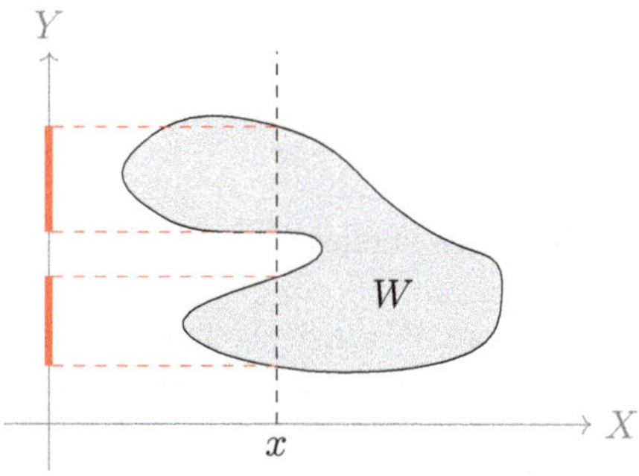

Fig. 20.1 The section W^x for the set W at $x \in X$ is highlighted in red on the Y axis

Proof Denote the rectangular sets $\mathcal{J} = \mathcal{F} \times \mathcal{G}$. Define the collection of sets $\mathcal{W} = \{W \in \sigma(\mathcal{J}) : W_y \in \mathcal{F} \text{ for all } y \in Y\} \subseteq \sigma(\mathcal{J}) = \mathcal{F} \otimes \mathcal{G}$. We now aim to show the opposite inclusion, namely $\mathcal{F} \otimes \mathcal{G} \subseteq \mathcal{W}$, to deduce the equality of the two σ-algebras. To do this, we prove that that $\mathcal{W}$ is a σ-algebra containing the rectangles $\mathcal{J}$ and use the minimality of $\sigma(\mathcal{J}) = \mathcal{F} \otimes \mathcal{G}$.

We first show that all the rectangular sets are contained in $\mathcal{W}$. Pick a rectangle $W = A \times B \in \mathcal{J}$ where $A \in \mathcal{F}$ and $B \in \mathcal{G}$. Then:

$$W_y = \begin{cases} A & \text{if } y \in B, \\ \varnothing & \text{if } y \notin B. \end{cases}$$

Either way, $W_y \in \mathcal{F}$ and so, by definition, we have $W \in \mathcal{W}$ and thus $\mathcal{J} \subseteq \mathcal{W}$. Next, we show that $\mathcal{W}$ satisfies the axioms of σ-algebra.

1. Clearly, $X \times Y \in \mathcal{W}$.
2. If $W \in \mathcal{W}$, we need to show that its complement is also in $\mathcal{W}$, namely $(X \times Y) \setminus W \in \mathcal{W}$. We just need to check that for any $y \in Y$, the section of this set is in $\mathcal{F}$. For a fixed $y \in Y$, we have:

$$\begin{aligned} ((X \times Y) \setminus W)_y &= \{x \in X : (x, y) \in (X \times Y) \setminus W\} \\ &= \{x \in X : (x, y) \in X \times Y\} \cap \{x \in X : (x, y) \notin W\} \\ &= X \cap \{x \in X : (x, y) \in W\}^c \\ &= X \setminus \{x \in X : (x, y) \in W\} = X \setminus W_y. \end{aligned}$$

   However, since $W \in \mathcal{W}$, we must have $W_y \in \mathcal{F}$ and hence $((X \times Y) \setminus W)_y = X \setminus W_y \in \mathcal{F}$. Thus, we also have $(X \times Y) \setminus W \in \mathcal{W}$.
3. We need to check that the collection of sets $\mathcal{W}$ is closed under countable unions. Pick countably many elements $\{W_j\}_{j=1}^{\infty}$ in $\mathcal{W}$. By definition of $\mathcal{W}$, for every $y \in Y$ and for all $j \in \mathbb{N}$ we have $(W_j)_y \in \mathcal{F}$. Thus, for a fixed $y \in Y$, we have:

$$\begin{aligned} \left(\bigcup_{j=1}^{\infty} W_j\right)_y = \left\{x \in X : (x, y) \in \bigcup_{j=1}^{\infty} W_j\right\} &= \bigcup_{j=1}^{\infty} \{x \in X : (x, y) \in W_j\} \\ &= \bigcup_{j=1}^{\infty} (W_j)_y \in \mathcal{F}, \end{aligned}$$

   which then implies $\bigcup_{j=1}^{\infty} W_j \in \mathcal{W}$.

Therefore, $\mathcal{W}$ is a σ-algebra which contains $\mathcal{J}$ and thus, by minimality of $\sigma(\mathcal{J})$, we have $\mathcal{F} \otimes \mathcal{G} = \sigma(\mathcal{J}) \subseteq \mathcal{W}$. This implies the equality $\mathcal{W} = \mathcal{F} \otimes \mathcal{G} = \sigma(\mathcal{J})$. From this equality, we conclude that for any $W \in \mathcal{F} \otimes \mathcal{G}$ and $y \in Y$ we must have $W_y \in \mathcal{F}$.

Similar argument can be employed to prove that for every $x \in X$, $W^x \in \mathcal{G}$. □

Now we define section functions as functions defined on two or more variables but with one of the arguments fixed. Specifically, for functions of two variables, we define:

Definition 20.2.5 (Section Functions) Let $f : W \to \bar{\mathbb{R}}$ be a function where $W = X \times Y$ is a non-empty set.

1. For any fixed $x \in X$, the section function $f^x : Y \to \bar{\mathbb{R}}$ is defined as the restriction of the function f to the section $W^x = Y$, namely $f^x = f|_{W^x}$.
2. For any fixed $y \in Y$, the section function $f_y : X \to \bar{\mathbb{R}}$ is defined as the restriction of the function f to the section $W_y = X$, namely $f_y = f|_{W_y}$.

Suppose now that $f : X \times Y \to [0, \infty]$ so that for any $x \in X$ the section function $f^x : Y \to [0, \infty]$ is also non-negative. If f^x is $\mathcal{G}$-measurable for each $x \in X$, then for any fixed $x \in X$ we can find its Lebesgue integral, which depends on the fixed x and is denoted as $F(x) = \int_Y f^x \, d\mu_2$. By varying x, this defines a function $F : X \to [0, \infty]$. Moreover, if the Lebesgue integral function F is $\mathcal{F}$-measurable, we can integrate it further over the set X to get:

$$\int_X F \, d\mu_1 = \int_X \left(\int_Y f^x \, d\mu_2 \right) d\mu_1. \tag{20.1}$$

This integral is called an iterated Lebesgue integral and its construction hinges on two crucial assumptions that we made above, namely:

1. whether the section functions f^x are $\mathcal{G}$-measurable for all $x \in X$, and
2. whether the integral function F is $\mathcal{F}$-measurable.

In general, the two assumptions above may not hold. However, under some mild conditions, these may be true. We shall later verify these assumptions for some cases in Lemma 20.2.10.

Similarly, with analogous assumptions as above, namely the section functions f_y are $\mathcal{F}$-measurable for each $y \in Y$ and the Lebesgue integral function $G : Y \to [0, \infty]$ defined as $G(y) = \int_X f_y \, d\mu_1$ is $\mathcal{G}$-measurable, we can construct another iterated integral:

$$\int_Y G \, d\mu_2 = \int_Y \left(\int_X f_y \, d\mu_1 \right) d\mu_2. \tag{20.2}$$

Now we ask ourselves: if these two iterated integrals (20.1) and (20.2) are well-defined, is it necessary that they are equal? In addition, we also have the Lebesgue integral which was constructed over $X \times Y$ without iterations, namely $\int_{X \times Y} f \, d\mu$ in Definition 20.2.2, using the product measure μ on $(X \times Y, \mathcal{F} \otimes \mathcal{G})$. As a result, we have three different possible Lebesgue integral expressions of the function $f : X \times Y \to [0, \infty]$, namely:

$$\int_X \left(\int_Y f^x \, d\mu_2 \right) d\mu_1 \quad \text{and} \quad \int_Y \left(\int_X f_y \, d\mu_1 \right) d\mu_2 \quad \text{and} \quad \int_{X \times Y} f \, d\mu.$$

In general, these integrals may not coincide.

Example 20.2.6 Consider two sets $X = Y = [0, 1]$. Define the measure spaces $(X, \mathcal{F}, \mu_1)$ and $(Y, \mathcal{G}, \mu_2)$ where $\mathcal{F} = \mathcal{P}(X)$ with the counting measure μ_1 and $\mathcal{G} = \mathcal{L}$ with the Lebesgue measure μ_2.

Let $(W, \mathcal{H}, \mu)$ where $W = X \times Y$ and $\mathcal{H} = \mathcal{F} \otimes \mathcal{G}$ be the product measure space. Consider the function $f : W \to [0, \infty]$ which is 1 on the diagonal and 0 everywhere else, namely $f(x, y) = \mathbf{1}_E(x, y)$ where $E = \{(x, x) : x \in [0, 1]\}$ is the diagonal set. This function is $\mathcal{H}$-measurable, which we left for the readers to check in Exercise 20.3.

1. For any $x \in X$, the section function $f^x(y) = \mathbf{1}_E|_{W^x}(y) = \mathbf{1}_{\{x\}}(y)$ is $\mathcal{G}$-measurable since $\{x\} \in \mathcal{L}$ and by virtue of Proposition 18.9.7. The construction of this section function can be seen in Fig. 20.2.
   For any fixed $x \in X$, we can compute its Lebesgue integral as:
   $$F(x) = \int_Y f^x \, d\mu_2 = \int_Y \mathbf{1}_{\{x\}} \, d\mu_2 = \mu_2(\{x\}) = 0,$$
   since μ_2 is the Lebesgue measure. Note that the function $F : X \to [0, \infty]$ is $\mathcal{F}$-measurable since it is a constant. Thus, we can compute the iterated integral (20.1) as:
   $$\int_X \left(\int_Y f^x \, d\mu_2 \right) d\mu_1 = \int_X F(x) \, d\mu_1 = \int_X 0 \, d\mu_1 = 0.$$
2. On the other hand, for all $y \in Y$ the section function $f_y(x) = \mathbf{1}_{\{y\}}(x)$ is $\mathcal{F}$-measurable since $\{y\} \in \mathcal{P}(X) = \mathcal{F}$. For any fixed $y \in Y$, since μ_1 is the counting

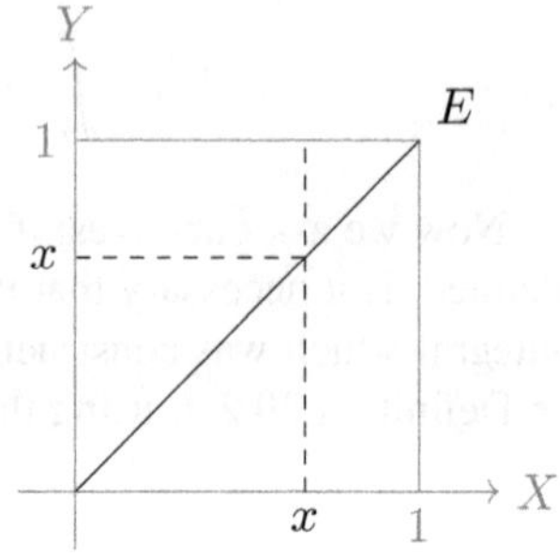

Fig. 20.2 The set $[0, 1]^2$ is the domain of the function f which is 1 on the diagonal E and 0 elsewhere. For a fixed $x \in [0, 1]$, the section function $f^x(y)$ has value 1 at $y = x$ and 0 elsewhere. Thus, $f^x(y) = \mathbf{1}_{\{x\}}(y)$

measure, we have:

$$G(y) = \int_X f_y \, d\mu_1 = \int_X \mathbf{1}_{\{y\}} \, d\mu_1 = \mu_1(\{y\}) = 1.$$

Thus, the function $G : Y \to [0, \infty]$ is $\mathcal{G}$-measurable and so the iterated integral (20.2) is given by:

$$\int_Y \left(\int_X f_y \, d\mu_1 \right) d\mu_2 = \int_Y G(y) \, d\mu_2 = \int_Y 1 \, d\mu_2 = 1.$$

Therefore, we have:

$$\int_X \left(\int_Y f^x \, d\mu_2 \right) d\mu_1 = 0 \neq 1 = \int_Y \left(\int_X f_y \, d\mu_1 \right) d\mu_2.$$

Hence, this is an example that the iterated integrals of the same function are not necessarily equal to each other.

The problem with the product measure space in Exercise 20.2.6 is that one of the constituent spaces is not σ-finite. Our main goal now is to prove a theorem by Tonelli, which allows us to swap the order of integration iteration of non-negative functions under some σ-finite conditions on the constituent spaces.

So from now on, we will be working mostly on σ-finite measure spaces. Thus, by Lemma 20.1.1 and Theorem 18.8.4, the product measure space is also σ-finite.

Measurability of Section Functions

To move forward from these σ-finite assumptions, we need to first answer the questions that we posed regarding the $(\mathcal{F} \otimes \mathcal{G})$-measurable function $f : (X \times Y, \mathcal{F} \otimes \mathcal{G}, \mu) \to [0, \infty]$ earlier, namely: whether the section functions f^x are $\mathcal{G}$-measurable for all $x \in X$, whether the section functions f_y are $\mathcal{F}$-measurable for all $y \in Y$, and whether the Lebesgue integral functions $F(x) = \int_Y f^x \, d\mu_2$ and $G(y) = \int_Y f_y \, d\mu_1$ are $\mathcal{F}$-measurable and $\mathcal{G}$-measurable respectively.

To address these important questions, we need to prove two technical results first:

Lemma 20.2.7 *Let $(X, \mathcal{F}, \mu_1)$ and $(Y, \mathcal{G}, \mu_2)$ be finite measure spaces and $(X \times Y, \mathcal{H}, \mu)$ where $\mathcal{H} = \mathcal{F} \otimes \mathcal{G}$ be its product measure space. For any $E \in \mathcal{H}$, we have:*

1. *the function $F_E : X \to [0, \infty]$ defined as $F_E(x) = \mu_2(E^x)$ is $\mathcal{F}$-measurable, and*
2. *the function $G_E : Y \to [0, \infty]$ defined as $G_E(y) = \mu_1(E_y)$ is $\mathcal{G}$-measurable.*

Proof We shall only prove the first assertion since the second can be proven similarly.

1. Let $\mathcal{E} = \{E \in \mathcal{H} : F_E \text{ is } \mathcal{F}\text{-measurable}\} \subseteq \mathcal{H}$. Our goal is to show that $\mathcal{E} = \mathcal{H}$ by proving the opposite inclusion $\mathcal{H} \subseteq \mathcal{E}$.
   Note that $\mathcal{H}$ is the smallest σ-algebra containing the rectangles $\mathcal{J} = \mathcal{F} \times \mathcal{G} = \{A \times B : A \in \mathcal{F}, B \in \mathcal{G}\}$. Thus, it is enough to show that $\mathcal{E}$ is a σ-algebra containing $\mathcal{J}$. We aim to show this via the monotone class theorem in Theorem 18.10.19 (see Exercise18.30). First, we check that the conditions for this theorem are satisfied, namely: $\mathcal{J} \subseteq \mathcal{E}$, $\mathcal{E}$ is a monotone class, and $A^c, A \cap B \in \mathcal{E}$ for all $A, B \in \mathcal{J}$.
   (a) Let $A \in \mathcal{J}$ be $A = C \times D$ for some $C \in \mathcal{F}$ and $D \in \mathcal{G}$. By an easy check in Exercise 20.6, we have $F_A(x) = \mu_2(A^x) = \mu_2((C \times D)^x) = \mu_2(D)\mathbf{1}_C(x)$. Note that this is $\mathcal{F}$-measurable as it is a constant scale of an indicator function over a measurable set $C \in \mathcal{F}$. Thus, we have $A \in \mathcal{E}$. Since $A \in \mathcal{J}$ is arbitrary, we conclude that $\mathcal{J} \subseteq \mathcal{E}$.
   (b) We now show that $\mathcal{E}$ is a monotone class. Pick any nested sequence of sets $\{E_j\}_{j=1}^{\infty}$ in $\mathcal{E}$ where $E_j \subseteq E_{j+1}$ for all $j \in \mathbb{N}$. We want to first show that $E = \bigcup_{j=1}^{\infty} E_j \in \mathcal{E}$ as well, namely F_E is $\mathcal{F}$-measurable. We note that for a fixed $x \in X$, we have $E^x = (\bigcup_{j=1}^{\infty} E_j)^x = \bigcup_{j=1}^{\infty} E_j^x$. Since $E_j \subseteq E_{j+1}$ for all $j \in \mathbb{N}$, we must have the inclusion of sections $E_j^x \subseteq E_{j+1}^x$. By Proposition 18.5.8, we then have:

$$F_E(x) = \mu_2(E^x) = \mu_2\left(\bigcup_{j=1}^{\infty} E_j^x\right) = \lim_{n\to\infty} \mu_2(E_n^x) = \lim_{n\to\infty} F_{E_n}(x),$$

   which is also $\mathcal{F}$-measurable as it is a limit of a sequence of $\mathcal{F}$-measurable functions (F_{E_n}). Thus, we have $E \in \mathcal{E}$.
   Similarly, by using Proposition 18.5.8, since $\mu_2(Y) < \infty$, we can show that for any nested sequence of sets $\{E_j\}_{j=1}^{\infty}$ in $\mathcal{E}$ where $E_j \supseteq E_{j+1}$ for all $j \in \mathbb{N}$, if $E = \bigcap_{j=1}^{\infty} E_j$, then $F_E(x) = \mu_2(E^x) = \lim_{n\to\infty} \mu_2(E_n^x) = \lim_{n\to\infty} F_{E_n}(x)$. Thus, the function F_E is also $\mathcal{F}$-measurable since it is a limit of a sequence of $\mathcal{F}$-measurable functions and so $E \in \mathcal{E}$. Thus, we conclude that $\mathcal{E}$ is a monotone class.
   (c) Now we show that complements and intersection of elements in $\mathcal{J}$ are in $\mathcal{E}$. Fix $A = C \times D \in \mathcal{J}$. From Exercise 1.24, we can express $A^c = (C \times D)^c = (C^c \times D) \cup (C \times D^c) \cup (C^c \times D^c)$ as a union of pairwise disjoint sets. Thus we have:

$$\begin{aligned}
F_{A^c}(x) &= \mu_2(((C \times D)^c)^x) \\
&= \mu_2((C^c \times D)^x \cup (C \times D^c)^x \cup (C^c \times D^c)^x) \\
&= \mu_2((C^c \times D)^x) + \mu_2((C \times D^c)^x) + \mu_2((C^c \times D^c)^x) \\
&= \mu_2(D)\mathbf{1}_{C^c}(x) + \mu_2(D^c)\mathbf{1}_C(x) + \mu_2(D^c)\mathbf{1}_{C^c}(x),
\end{aligned}$$

which is an $\mathcal{F}$-measurable function since $C, C^c \in \mathcal{F}$. Thus, we have $A^c \in \mathcal{E}$. Additionally, if $B = P \times Q \in \mathcal{J}$, then $A \cap B = (C \times D) \cap (P \times Q) = (C \cap P) \times (D \cap Q)$. Thus, we have $F_{A \cap B}(x) = \mu_2(((C \cap P) \times (D \cap Q))^x) = \mu_2(D \cap Q)\mathbf{1}_{C \cap P}(x)$ which is also $\mathcal{F}$-measurable. This means $A \cap B \in \mathcal{E}$.

Therefore, by monotone class theorem, we deduce that $\mathcal{H} = \sigma(\mathcal{J}) \subseteq \mathcal{E}$. Hence, we conclude that $\mathcal{H} = \mathcal{E}$. This means for every $E \in \mathcal{H}$ the function F_E is $\mathcal{F}$-measurable. □

We now extend Lemma 20.2.7 to σ-finite measure spaces.

Corollary 20.2.8 *Let $(X, \mathcal{F}, \mu_1)$ and $(Y, \mathcal{G}, \mu_2)$ be σ-finite measure spaces and $(X \times Y, \mathcal{H}, \mu)$ where $\mathcal{H} = \mathcal{F} \otimes \mathcal{G}$ be its product measure space. For any $E \in \mathcal{H}$, we have:*

1. *the function $F_E : X \to [0, \infty]$ defined as $F_E(x) = \mu_2(E^x)$ is $\mathcal{F}$-measurable, and*
2. *the function $G_E : Y \to [0, \infty]$ defined as $G_E(y) = \mu_1(E_y)$ is $\mathcal{G}$-measurable.*

Proof We shall only prove the first assertion since the second can be proven similarly.

1. For the case where $(Y, \mathcal{G}, \mu_2)$ is σ-finite, we could not follow the argument in Lemma 20.2.7 since in the proof, we explicitly require $\mu_2(Y) < \infty$ in step (b) so that we can use Proposition 18.5.8.
   However, we can apply a limiting argument to extend Lemma 20.2.7. Since $(Y, \mathcal{F}, \mu_2)$ is σ-finite, there exists an increasing sequence of sets $\{Y_j\}_{j=1}^{\infty}$ in $\mathcal{G}$ such that $\mu_2(Y_j) < \infty$ and $Y_j \subseteq Y_{j+1}$ for all $j \in \mathbb{N}$ with $\bigcup_{j=1}^{\infty} Y_j = Y$. For each $j \in \mathbb{N}$, we define $Z_j = X \times Y_j$. We note that for any $E \in \mathcal{H}$, the functions $F_{E \cap Z_j} : X \to [0, \infty]$ are all $\mathcal{F}$-measurable since $\mu_2(Y_j) < \infty$ and by using Lemma 20.2.7 for finite measure space.
   The collection $\{E \cap Z_j\}_{j=1}^{\infty}$ is an increasing sequence of sets such that $E = \bigcup_{j=1}^{\infty}(E \cap Z_j)$. Thus, by Proposition 18.5.8, we have:

$$F_E(x) = \mu_2(E^x) = \mu_2\left(\bigcup_{j=1}^{\infty}(E \cap Z_j)^x\right) = \lim_{n \to \infty} \mu_2((E \cap Z_n)^x)$$

$$= \lim_{n \to \infty} F_{E \cap Z_n}(x),$$

   which is $\mathcal{F}$-measurable since it is a limit of a sequence of $\mathcal{F}$-measurable functions. □

Next, we prove a very simple lemma which we shall need later:

Lemma 20.2.9 *Let $(X, \mathcal{F})$ and $(Y, \mathcal{G})$ be measurable spaces and $(X \times Y, \mathcal{H})$ where $\mathcal{H} = \mathcal{F} \otimes \mathcal{G}$ be its product measurable space. Suppose that $E \in \mathcal{H}$. Then, the sections of the indicator function $\mathbf{1}_E : X \times Y \to \mathbb{R}$ satisfy $(\mathbf{1}_E)^x = \mathbf{1}_{E^x}$ for all $x \in X$ and $(\mathbf{1}_E)_y = \mathbf{1}_{E_y}$ for all $y \in Y$.*

Proof We prove only the second equality. Fix $y \in Y$. For any $x \in X$ we have two cases:

1. $(x, y) \in E$, namely $x \in E_y$. Then, $(\mathbf{1}_E)_y(x) = \mathbf{1}_E(x, y) = 1 = \mathbf{1}_{E_y}(x)$.
2. $(x, y) \notin E$, namely $x \notin E_y$. Then, $(\mathbf{1}_E)_y(x) = \mathbf{1}_E(x, y) = 0 = \mathbf{1}_{E_y}(x)$.

Either way, we have $(\mathbf{1}_E)_y(x) = \mathbf{1}_{E_y}(x)$ for all $x \in X$. Since $y \in Y$ is arbitrary, we conclude that $(\mathbf{1}_E)_y = \mathbf{1}_{E_y}$ for all $y \in Y$. □

Now that we have the required tools, we prove the important lemma for measurability of section functions and Lebesgue integral functions. This is necessary to ensure that the iterated integrals make sense in the first place. We do this for non-negative functions on a σ-finite measure space.

Lemma 20.2.10 *Let $(X, \mathcal{F}, \mu_1)$ and $(Y, \mathcal{G}, \mu_2)$ be σ-finite measure spaces and $(X \times Y, \mathcal{H}, \mu)$ where $\mathcal{H} = \mathcal{F} \otimes \mathcal{G}$ be its product measure space. Suppose that $f : X \times Y \to [0, \infty]$ is an $\mathcal{H}$-measurable non-negative function.*

1. *For any fixed $y \in Y$, the section function $f_y : X \to [0, \infty]$ defined as $f_y(x) = f(x, y)$ is $\mathcal{F}$-measurable.*
2. *For any fixed $x \in X$, the section function $f^x : Y \to [0, \infty]$ defined as $f^x(y) = f(x, y)$ is $\mathcal{G}$-measurable.*
3. *The function $G : Y \to [0, \infty]$ defined as $G(y) = \int_X f_y \, d\mu_1$ is non-negative and $\mathcal{G}$-measurable.*
4. *The function $F : X \to [0, \infty]$ defined as $F(x) = \int_Y f^x \, d\mu_2$ is non-negative and $\mathcal{F}$-measurable.*

Proof We prove only two of the assertions as the others are similarly done.

1. Fix $y_0 \in Y$. By virtue of Lemma 18.9.8, we simply need to show that for every $c \in \mathbb{R}$, the set $\{x \in X : f_{y_0}(x) = f(x, y_0) < c\}$ is in $\mathcal{F}$. Indeed, since f itself is $\mathcal{H}$-measurable, the set $W = \{(x, y) \in X \times Y : f(x, y) < c\} \in \mathcal{H}$. By taking the section at y_0 and using Lemma 20.2.4, we have $\{x \in X : f(x, y_0) < c\} = \{x \in X : (x, y_0) \in W\} = W_{y_0} \in \mathcal{F}$. Since $y_0 \in Y$ was arbitrary fixed, we are done.
3. Non-negativity is clear since $f_y \geq 0$. We now prove that the Lebesgue integral function G is $\mathcal{G}$-measurable in three steps:

(a) First, let $f(x, y) = \mathbf{1}_E(x, y)$ be an indicator function on an $\mathcal{H}$-measurable set $E \in \mathcal{H}$. By Lemma 20.2.9, we have $G(y) = \int_X (\mathbf{1}_E)_y \, d\mu_1 = \int_X \mathbf{1}_{E_y} \, d\mu_1 = \mu_1(E_y)$. From Corollary 20.2.8, this is a measurable function of y.

(b) For a simple function $f : X \times Y \to [0, \infty)$, we can write $f(x, y) = \sum_{j=1}^{n} c_j \mathbf{1}_{E_j}(x, y)$ where $c_j \geq 0$ and $E_j \in \mathcal{H}$ are pairwise disjoint $\mathcal{H}$-measurable sets. By linearity of Lebesgue integrals, we have:

$$G(y) = \int_X f_y \, d\mu_1 = \int_X \sum_{j=1}^{n} (c_j \mathbf{1}_{E_j})_y \, d\mu_1 = \sum_{j=1}^{n} c_j \int_X (\mathbf{1}_{E_j})_y \, d\mu_1$$

$$= \sum_{j=1}^{n} c_j \mu_1((E_j)_y),$$

which is also measurable since it is a linear combination of functions from part (a).

(c) Finally, for a general $\mathcal{H}$-measurable function $f : X \times Y \to [0, \infty]$, by virtue of Proposition 19.1.3, we can find a sequence of simple functions (f_n) where $f_n : X \times Y \to [0, \infty)$ such that $f_n \uparrow f$ on $X \times Y$. For any fixed $y \in Y$, the sequence of section functions $((f_n)_y)$ where $(f_n)_y : X \to [0, \infty)$ is also a sequence of simple functions with respect to the measure space $(X, \mathcal{F}, \mu_1)$ which is pointwise increasing and converges to $f|_{X_y} = f_y$, namely $(f_n)_y \uparrow f_y$.

By part (b), the sequence of functions (G_n) defined as $G_n : Y \to [0, \infty]$ where $G_n(y) = \int_X (f_n)_y \, d\mu_1$ are all $\mathcal{G}$-measurable. By the MCT, we then have:

$$G(y) = \int_X f_y \, d\mu_1 = \int_X \lim_{n\to\infty} (f_n)_y \, d\mu_1 = \lim_{n\to\infty} \int_X (f_n)_y \, d\mu_1 = \lim_{n\to\infty} G_n(y).$$

Since (G_n) are all $\mathcal{G}$-measurable functions, the limit G is also a $\mathcal{G}$-measurable function. □

Alternative Formulation of Product Measure

A very useful and important note that we want to highlight is that Lemma 20.2.10 also gives us another way to describe the product measure of σ-finite measure spaces which we have constructed in Sect. 20.1. This characterisation is given by:

Proposition 20.2.11 *Let $(X, \mathcal{F}, \mu_1)$ and $(Y, \mathcal{G}, \mu_2)$ be σ-finite measure spaces and $(X \times Y, \mathcal{H}, \mu)$ where $\mathcal{H} = \mathcal{F} \otimes \mathcal{G}$ be its product measure space. Then, the product measure of a set $E \in \mathcal{H}$ satisfies:*

$$\mu(E) = \int_{X\times Y} \mathbf{1}_E \, d\mu = \int_X \left(\int_Y (\mathbf{1}_E)^x \, d\mu_2 \right) d\mu_1 = \int_Y \left(\int_X (\mathbf{1}_E)_y \, d\mu_1 \right) d\mu_2.$$

Proof Since $(X, \mathcal{F}, \mu_1)$ and $(Y, \mathcal{G}, \mu_2)$ are σ-finite measure spaces, by Lemma 20.1.1, the premeasure space $(W, \mathcal{R}(\mathcal{J}), m)$ is also σ-finite.

Define a set function $\mu' : \mathcal{H} \to [0, \infty]$ as $\mu'(E) = \int_X \left(\int_Y (\mathbf{1}_E)^x \, d\mu_2\right) d\mu_1$ for all $E \in \mathcal{H}$. By a routine check in Exercise 20.8, this is a measure on $(X \times Y, \mathcal{H})$.

Now we check that the measures μ and μ' agree on the premeasure space $(X \times Y, \mathcal{R}(\mathcal{J}), m)$. Pick an arbitrary rectangular set $A \times B$ from the semiring of rectangles $\mathcal{J} = \{A \times B : A \in \mathcal{F}, B \in \mathcal{G}\}$. Then, by using Exercise 20.6, we have:

$$\begin{aligned}
\mu'(A \times B) = \int_X \left(\int_Y (\mathbf{1}_{A\times B})^x \, d\mu_2\right) d\mu_1 &= \int_X \mu_2((A \times B)^x) \, d\mu_1 \\
&= \int_X \mu_2(B)\mathbf{1}_A \, d\mu_1 \\
&= \mu_2(B) \int_X \mathbf{1}_A \, d\mu_1 \\
&= \mu_2(B)\mu_1(A) \\
&= m(A \times B) = \mu(A \times B).
\end{aligned}$$

Hence, we have the equality $\mu = \mu'$ on $\mathcal{J}$. Furthermore, by unique extension to $\mathcal{R}(\mathcal{J})$, we then have $\mu = \mu'$ on $\mathcal{R}(\mathcal{J})$. Finally, since the premeasure space $(X \times Y, \mathcal{R}(\mathcal{J}), m)$ is σ-finite, by Theorem 18.8.4, such a measure on $\sigma(\mathcal{R}(\mathcal{J})) = \mathcal{H}$ must be unique. Therefore, we conclude that $\mu = \mu'$ everywhere on $\mathcal{H}$.

An identical argument works for the other iterated integral. □

At the beginning of this chapter in Sect. 20.1, we went through a whole rigmarole of constructing the product measure on the measurable space $(X \times Y, \mathcal{F} \otimes \mathcal{G})$ via outer measures induced by the product premeasure m on $\mathcal{R}(\mathcal{J})$. This was an established, albeit lengthy and involved, routine which was introduced and carried out in Chap. 18. Therefore, an important consequence of the Proposition 20.2.11 is that it gives us an unfussy and easier way of constructing the product measure on σ-finite measure spaces.

Remark 20.2.12 Since μ' in the proof of Proposition 20.2.11 is also a measure and coincides with the construction that we did, many literature refers to this as the primary definition of product measure for convenience and brevity.

20.3 Fubini's and Tonelli's Theorems

Now we prove that the two different iterated integrals for non-negative functions agree when we carry them out over σ-finite measure spaces. The following result is due to Leonida Tonelli (1885–1946).

Theorem 20.3.1 (Tonelli's Theorem) *Let $(X, \mathcal{F}, \mu_1)$ and $(Y, \mathcal{G}, \mu_2)$ be σ-finite measure spaces and $(X \times Y, \mathcal{H}, \mu)$ where $\mathcal{H} = \mathcal{F} \otimes \mathcal{G}$ be its product measure space.*

Let $f : X \times Y \to [0, \infty]$ be an $\mathcal{H}$-measurable non-negative function. Then, we have the equality:

$$\int_{X\times Y} f\, d\mu = \int_X \left(\int_Y f^x\, d\mu_2 \right) d\mu_1 = \int_Y \left(\int_X f_y\, d\mu_1 \right) d\mu_2. \tag{20.3}$$

Proof By Lemma 20.2.10, both of the iterated integrals in (20.3) are well-defined. We only prove the first equality in (20.3) as the other can be similarly proven. This is done in three steps:

1. First, if $f(x, y) = \mathbf{1}_E(x, y)$ is an indicator function on a measurable set $E \in \mathcal{H}$, the equality holds by Proposition 20.2.11.
2. Next, for a simple function $f : X \times Y \to [0, \infty)$ we can write $f(x, y) = \sum_{j=1}^n c_j \mathbf{1}_{E_j}(x, y)$ where $c_j \geq 0$ and $E_j \in \mathcal{H}$ are pairwise disjoint sets. By linearity of integrals and the first case above, we then have:

$$\begin{aligned}\int_X \left(\int_Y f^x\, d\mu_2 \right) d\mu_1 &= \int_X \left(\int_Y \sum_{j=1}^n c_j \mathbf{1}_{E_j}^x\, d\mu_2 \right) d\mu_1 \\ &= \sum_{j=1}^n c_j \int_X \left(\int_Y \mathbf{1}_{E_j}^x\, d\mu_2 \right) d\mu_1 \\ &= \sum_{j=1}^n c_j \int_{X\times Y} \mathbf{1}_{E_j}\, d\mu \\ &= \sum_{j=1}^n c_j \mu(E_j) = \int_{X\times Y} f\, d\mu.\end{aligned}$$

3. Finally, for a general $\mathcal{H}$-measurable function $f : X \times Y \to [0, \infty]$, by Proposition 19.1.3 we can find a sequence of simple functions (f_n) where $f_n : X \times Y \to [0, \infty)$ such that $f_n \uparrow f$ on $X \times Y$. For any fixed $x \in X$ we have the ordering $f_n^x(y) = f_n(x, y) \leq f_{n+1}(x, y) = f_{n+1}^x(y)$ for all $y \in Y$ and $n \in \mathbb{N}$.
   Define a sequence of function (F_n) where $F_n : X \to [0, \infty]$ is given by $F_n(x) = \int_Y f_n^x\, d\mu_2$. The functions sequence (F_n) is also increasing pointwise since $f_n^x(y) \leq f_{n+1}^x(y)$ for all $y \in Y$ and $n \in \mathbb{N}$ and hence, for any fixed $x \in X$, we have the ordering $F_n(x) \leq \int_Y f_n^x(y)\, d\mu_2 \leq \int_Y f_{n+1}^x(y)\, d\mu_2 = F_{n+1}(x)$. Thus, we have three sequences of functions which are pointwise increasing over their domain of definitions, namely (f_n), (f_n^x), and (F_n).
   For any fixed $n \in \mathbb{N}$, the previous case implies:

$$\int_X F_n\, d\mu_1 = \int_X \left(\int_Y f_n^x\, d\mu_2 \right) d\mu_1 = \int_{X\times Y} f_n\, d\mu.$$

Taking the limit as $n \to \infty$ on both sides and applying the MCT several times, we have:

$$\begin{aligned}
&\lim_{n\to\infty}\int_X F_n\,d\mu_1 = \lim_{n\to\infty}\int_{X\times Y} f_n\,d\mu \\
\Rightarrow &\int_X \lim_{n\to\infty} F_n\,d\mu_1 = \int_{X\times Y}\lim_{n\to\infty} f_n\,d\mu \\
\Rightarrow &\int_X \lim_{n\to\infty}\left(\int_Y f_n^x\,d\mu_2\right)d\mu_1 = \int_{X\times Y} f\,d\mu \\
\Rightarrow &\int_X \left(\int_Y \lim_{n\to\infty} f_n^x\,d\mu_2\right)d\mu_1 = \int_{X\times Y} f\,d\mu \\
\Rightarrow &\int_X \left(\int_Y f^x\,d\mu_2\right)d\mu_1 = \int_{X\times Y} f\,d\mu,
\end{aligned}$$

which is the desired equality.

The proof for the other iterated integral is also done in the same way. □

Example 20.3.2 Let $([0,1],\mathcal{L},\mu)$ be the induced Lebesgue measure space and $X = [0,1]^2 \subseteq \mathbb{R}^2$ be a set with the product σ-algebra $\mathcal{H}$ and product measure $\tilde{\mu}$. Let $f : X \to \mathbb{R}$ be defined as $f(x,y) = xe^y$. We want to integrate this function over X. This is a non-negative $\mathcal{H}$-measurable function, so we can simply apply Tonelli's theorem to find the integral.

Picking one of the iterated integrals, we choose $\int_X f\,d\tilde{\mu} = \int_{[0,1]}\left(\int_{[0,1]} f^x\,d\mu(y)\right)d\mu(x)$ where we denoted the Lebesgue measures μ on $[0,1]$ with $\mu(x)$ and $\mu(y)$ to distinguish which variable we are integrating with respect to. Let us evaluate the inner integral first. For a fixed $x \in [0,1]$ we have:

$$\int_{[0,1]} f^x\,d\mu(y) = \int_{[0,1]} xe^y\,d\mu(y) = x\int_{[0,1]} e^y\,d\mu(y).$$

Now notice that the exponential function e^y is Riemann integrable over $[0,1]$ since it is continuous. Therefore, Theorem 19.7.4 says that the Lebesgue integral of e^y over $[0,1]$ is the same as its Riemann integral. By the FTC, we can compute:

$$\int_{[0,1]} f^x\,d\mu(y) = x\int_{[0,1]} e^y\,d\mu(y) = x\int_0^1 e^y\,dy = x(e-1),$$

and so the iterated integral becomes:

$$\int_X f\,d\tilde{\mu} = \int_{[0,1]}\left(\int_{[0,1]} f^x\,d\mu(y)\right)d\mu(x) = \int_{[0,1]} x(e-1)\,d\mu(x)$$
$$= (e-1)\int_{[0,1]} x\,d\mu(x).$$

Again, the integrand x is Riemann integrable over $[0, 1]$. So, by using Theorem 19.7.4 and the FTC, we have:

$$\int_X f\,d\mu = (e-1)\int_{[0,1]} x\,d\mu_1 = (e-1)\int_0^1 x\,dx = \frac{e-1}{2}.$$

We can extend Tonelli's theorem to general Lebesgue integrable functions by considering the positive and negative parts separately. However, unlike Tonelli's theorem, the commutativity of the iterated integrals on functions with images with mixed signs requires an additional prior knowledge that $f \in \mathcal{L}^1(X \times Y)$ to work. This result was introduced by Guido Fubini (1879–1943).

Theorem 20.3.3 (Fubini's Theorem) *Let $(X, \mathcal{F}, \mu_1)$ and $(Y, \mathcal{G}, \mu_2)$ be σ-finite measure spaces and $(X \times Y, \mathcal{H}, \mu)$ where $\mathcal{H} = \mathcal{F} \otimes \mathcal{G}$ be its product measure space. Let $f \in \mathcal{L}^1(X \times Y, \mathcal{H})$. Then:*

1. *$f_y \in \mathcal{L}^1(X)$ for μ_2-a.e. on Y and $f^x \in \mathcal{L}^1(Y)$ for μ_1-a.e. on X.*
2. *The integral function $G(y) = \int_X f_y\,d\mu_1$ is in $\mathcal{L}^1(Y)$ and we have the equality:*

$$\int_Y G\,d\mu_2 = \int_{X\times Y} f\,d\mu.$$

3. *The integral function $F(x) = \int_Y f^x\,d\mu_2$ is in $\mathcal{L}^1(X)$ and we have the equality:*

$$\int_X F\,d\mu_1 = \int_{X\times Y} f\,d\mu.$$

4. *We have the equality of iterated integrals:*

$$\int_{X\times Y} f\,d\mu = \int_Y\left(\int_X f_y\,d\mu_1\right)d\mu_2 = \int_X\left(\int_Y f^x\,d\mu_2\right)d\mu_1.$$

Proof We shall prove assertions 1 and 2 only. To set notations, let $f = f^+ - f^-$ be the decomposition of the function f into its positive and negative parts. We note that for a fixed $y \in Y$, we have $(f^+)_y = \max(f, 0)_y = \max(f_y, 0) = (f_y)^+$ and likewise $(f^-)_y = (f_y)^-$. So we can write them unambiguously as f_y^+

and f_y^- respectively. This also means $|f_y| = f_y^+ - f_y^- = (f^+ - f^-)_y = |f|_y$. Moreover, by Lemma 20.2.10, these functions and their respective integral functions are measurable with respect to the measures on their domains.

1. First, denote the function $H : Y \to [0, \infty]$ as $H(y) = \int_X |f_y| \, d\mu_1$. Note that $f \in \mathcal{L}^1(X \times Y, \mathcal{H})$ means $\int_{X\times Y} |f| \, d\mu < \infty$. Since $|f| \geq 0$, we can apply the fact that $|f_y| = |f|_y$ and Tonelli's theorem as thus:

$$\begin{aligned}\int_Y H(y) \, d\mu_2 = \int_Y \left(\int_X |f_y| \, d\mu_1 \right) d\mu_2 &= \int_Y \left(\int_X |f|_y \, d\mu_1 \right) d\mu_2 \\ &= \int_{X\times Y} |f| \, d\mu < \infty.\end{aligned}$$

   This says the integral function H is in $\mathcal{L}^1(Y)$. By Proposition 19.5.5, this implies $H(y) = \int_X |f_y| \, d\mu_1$ is finite for μ_2-a.e. in Y. In other words, we have $f_y \in \mathcal{L}^1(X)$ for μ_2-a.e. in Y. In a similar manner, we can show that $f^x \in \mathcal{L}^1(Y)$ for μ_1-a.e. in X.
2. Using the first assertion, for μ_2-a.e. in Y we have $G(y) = \int_X f_y \, d\mu_2 = \int_X f_y^+ \, d\mu_2 - \int_X f_y^- \, d\mu_2 < \infty$. Denote the set on which this is true as $E \subseteq Y$ so that $\mu_2(Y \setminus E) = 0$. Then, by using definition of the integral for f and Tonelli's theorem, we have:

$$\begin{aligned}\int_{X\times Y} f \, d\mu &= \int_{X\times Y} f^+ \, d\mu - \int_{X\times Y} f^- \, d\mu \\ &= \int_Y \left(\int_X f_y^+ \, d\mu_1 \right) d\mu_2 - \int_Y \left(\int_X f_y^- \, d\mu_1 \right) d\mu_2 \\ &= \int_Y \left(\int_X f_y^+ \, d\mu_1 - \int_X f_y^- \, d\mu_1 \right) d\mu_2 \\ &= \int_E \left(\int_X f_y^+ \, d\mu_1 - \int_X f_y^- \, d\mu_1 \right) d\mu_2 \\ &= \int_E \left(\int_X f_y \, d\mu_1 \right) d\mu_2 = \int_E G \, d\mu_2 = \int_Y G \, d\mu_2.\end{aligned}$$

   Thus, we have $G \in \mathcal{L}^1(Y)$ with the desired equality.

Assertion 3 can be proven in the similar manner. Putting together assertions 2 and 3 then yields the equality of iterated integrals in assertion 4. □

As we have mentioned earlier, Fubini's theorem requires the prior knowledge that $f \in \mathcal{L}^1(X \times Y, \mathcal{H})$ before we can use it. This can be annoying or difficult for us to show. Luckily, Tonelli's theorem could help us show this fact. Putting Theorems 20.3.1 and 20.3.3 together, we have the Fubini-Tonelli theorem:

Theorem 20.3.4 (Fubini-Tonelli Theorem) *Let $(X, \mathcal{F}, \mu_1)$ and $(Y, \mathcal{G}, \mu_2)$ be σ-finite measure spaces and $(X \times Y, \mathcal{H}, \mu)$ where $\mathcal{H} = \mathcal{F} \otimes \mathcal{G}$ be its product measure space. Let $f : X \times Y \to \bar{\mathbb{R}}$ be $\mathcal{H}$-measurable and suppose that at least one of the following:*

$$\int_X \left(\int_Y |f^x| \, d\mu_2 \right) d\mu_1 < \infty \quad \text{or} \quad \int_Y \left(\int_X |f_y| \, d\mu_1 \right) d\mu_2 < \infty,$$

is true. Then, $f \in \mathcal{L}^1(X \times Y, \mathcal{H})$ and we have the equality:

$$\int_{X \times Y} f \, d\mu = \int_X \left(\int_Y f^x \, d\mu_2 \right) d\mu_1 = \int_Y \left(\int_X f_y \, d\mu_1 \right) d\mu_2.$$

Proof WLOG, assume that $\int_X \left(\int_Y |f^x| \, d\mu_2 \right) d\mu_1 < \infty$. By Tonelli's theorem, we have $\int_{X \times Y} |f| \, d\mu < \infty$ and so $f \in \mathcal{L}^1(X \times Y, \mathcal{H})$. Applying Fubini's theorem then yields the desired equality. □

We note that the σ-finite conditions on the constituent measure spaces of the product space in Theorems 20.3.1, 20.3.3, and 20.3.4 cannot be removed. Example 20.2.6 showed us that if we drop the σ-finite condition, the iterated integrals might not commute.

Remark 20.3.5 Historically, Fubini's theorem was formulated first in 1907. Tonelli's theorem was introduced two years later in 1909 as a special case for Fubini's theorem. We presented them here in reverse time order in increasing generality.

Example 20.3.6 Let us look at some examples on how to use these results:

1. Let $([0, 1], \mathcal{L}, \mu)$ be the induced Lebesgue measure space and $X = [0, 1]^2 \subseteq \mathbb{R}^2$ be a set with the product σ-algebra $\mathcal{H}$ and product measure $\tilde{\mu}$. Consider the function $f : X \to \mathbb{R}$ defined as $f(x, y) = \sin(y^2)$. This function is continuous and hence $\mathcal{H}$-measurable over X. We wish to integrate this over a measurable triangle $A = \{(x, y) : 0 \leq x \leq y \leq 1\} \in \mathcal{H}$.
   Note that A is a set of finite measure and $|f| \leq 1$ on A. So, by Proposition 19.5.3, f must be Lebesgue integrable over A, namely $\mathbf{1}_A f \in \mathcal{L}^1(X)$. Now we want to compute its value. In other words, we want to evaluate $\int_A f \, d\tilde{\mu} = \int_X \mathbf{1}_A f \, d\tilde{\mu}$. We can use Fubini's theorem which says there are two possible ways that we can evaluate this via iterated integrals, namely either by:

$$\int_{[0,1]} \left(\int_{[0,1]} (\mathbf{1}_A f)^x \, d\mu(y) \right) d\mu(x) \quad \text{or} \quad \int_{[0,1]} \left(\int_{[0,1]} (\mathbf{1}_A f)_y \, d\mu(x) \right) d\mu(y).$$

(a) Let us try the first integral. For a fixed $x \in [0, 1]$, the inner integral becomes:

$$\begin{aligned}\int_{[0,1]} (\mathbf{1}_A f)^x \, d\mu(y) &= \int_{[0,1]} \mathbf{1}_A^x f^x \, d\mu(y) \\ &= \int_{[0,1]} \mathbf{1}_{A^x} \sin(y^2) \, d\mu(y) \\ &= \int_{A^x} \sin(y^2) \, d\mu(y) = \int_{[x,1]} \sin(y^2) \, d\mu(y).\end{aligned}$$

We know that $\sin(y^2)$ is continuous over $[x, 1]$ so, by Proposition 19.7.4, this is equal to the Riemann integral with value $\int_x^1 \sin(y^2) \, dy$.
However, there is no explicit antiderivative for $\sin(y^2)$ in terms of elementary functions (we have a power series form of it, which is the Fresnel integral seen in Exercise 17.11). So, even though we know that it is Riemann integrable, we do not have an explicit expression for it for us to continue to get a concrete value.

(b) Let us try the second iteration instead. For a fixed $y \in [0, 1]$, the inner integral is:

$$\begin{aligned}\int_{[0,1]} (\mathbf{1}_A f)_y \, d\mu(x) &= \int_{[0,1]} \mathbf{1}_{A_y} f_y \, d\mu(x) \\ &= \int_{A_y} \sin(y^2) \, d\mu(x) \\ &= \sin(y^2) \int_{[0,y]} 1 \, d\mu(x) = y \sin(y^2),\end{aligned}$$

which looks more promising. Note that this is continuous and hence Riemann integrable over $[0, 1]$. Putting this in the iterated integral and applying Proposition 19.7.4, we have:

$$\int_A f \, d\tilde{\mu} = \int_{[0,1]} y \sin(y^2) \, d\mu(y) = \int_0^1 y \sin(y^2) \, dy = \frac{1 - \cos(1)}{2},$$

where we used the change of variable $z = y^2$ followed by the FTC to evaluate the final Riemann integral.

2. Let $((0, \infty), \mathcal{L}, \mu)$ be the induced Lebesgue measure space and $X = (0, \infty)^2 \subseteq \mathbb{R}^2$ be a set with the the product σ-algebra $\mathcal{H}$ and product measure $\tilde{\mu}$. Consider the function $f : X \to \mathbb{R}$ defined as $f(x, y) = \sin(\frac{1}{x^2+y^4}) \cos(x^2 + y^4)$. Note that this function is continuous and hence $\mathcal{H}$-measurable. We want to show that $f \in \mathcal{L}^1(X)$.
To do this, we split the region X into the union of three disjoint regions $X_1, X_2, X_3 \subseteq X$ where $X_1 = (0, 1) \times (0, 1)$, $X_2 = [1, \infty) \times (0, \infty)$, and

$X_3 = (0, 1) \times [1, \infty)$. Before we begin with the analysis, we note the following useful bounds $|\sin(t)| \leq t$ and $|\cos(t)| \leq 1$ for any $t \geq 0$.

(a) On $X_1 = (0, 1) \times (0, 1)$, using the bound on sine and cosine functions above, we have $|f| \leq 1$ on X_1. Therefore, by Proposition 19.5.3(8), since the set X_1 has finite measure, we have $f \in \mathcal{L}^1(X_1)$.

(b) On $X_2 = [1, \infty) \times (0, \infty)$, aiming to utilise Fubini-Tonelli's theorem, we first show that $\int_{[1,\infty)} \left(\int_{(0,\infty)} |f^x| \, d\mu(y) \right) d\mu(x) < \infty$. For a fixed $x \in [1, \infty)$, the inner integral can be bound as such:

$$\int_{(0,\infty)} |f^x| \, d\mu(y) = \int_{(0,\infty)} \left| \sin\left(\frac{1}{x^2 + y^4} \right) \cos(x^2 + y^4) \right| d\mu(y)$$
$$\leq \int_{(0,\infty)} \frac{1}{x^2 + y^4} \, d\mu(y).$$

The final integrand is non-negative. Moreover, it is improperly Riemann integrable over $(0, \infty)$. To evaluate the improper integral, we carry out a change of variable $\frac{y^2}{x} = z$. Since $x \geq 1$ is a fixed constant, for any $s, t > 0$ such that $0 < s^2 < x < t^2$, we have:

$$\int_s^t \frac{1}{x^2 + y^4} \, dy = \frac{1}{x^2} \int_s^t \frac{1}{1 + \frac{y^4}{x^2}} \, dy = \frac{1}{2x^{\frac{3}{2}}} \int_{\frac{s^2}{x}}^{\frac{t^2}{x}} \frac{1}{z^{\frac{1}{2}} + z^{\frac{5}{2}}} \, dz$$
$$\leq \frac{1}{2x^{\frac{3}{2}}} \left(\int_{\frac{s^2}{x}}^{1} \frac{1}{z^{\frac{1}{2}}} \, dz + \int_1^{\frac{t^2}{x}} \frac{1}{z^{\frac{5}{2}}} \, dz \right)$$
$$= \frac{1}{2x^{\frac{3}{2}}} \left(2 - \frac{2s}{x^{\frac{1}{2}}} + \frac{2}{3} - \frac{2x^{\frac{3}{2}}}{3t^3} \right).$$

Taking the limit as $s \downarrow 0$ and $t \uparrow \infty$, we get the following bound on the improper integral:

$$\int_0^\infty \frac{1}{x^2 + y^4} \, dy \leq \frac{4}{3x^{\frac{3}{2}}},$$

and thus, by Proposition 19.7.9, the improper Riemann integral equals the Lebesgue integral. Therefore, we have the bound:

$$\int_{(0,\infty)} |f^x| \, d\mu(y) \leq \int_{(0,\infty)} \frac{1}{x^2 + y^4} \, d\mu(y) = \int_0^\infty \frac{1}{x^2 + y^4} \, dy \leq \frac{4}{3x^{\frac{3}{2}}}$$

Notice also that $\frac{4}{3x^{\frac{3}{2}}}$ is a non-negative function that is improperly Riemann integrable over $[1,\infty)$. Again, by using Proposition 19.7.9, we have:

$$\int_{[1,\infty)}\left(\int_{(0,\infty)}|f^x|\,d\mu_2\right)d\mu_1 \le \int_{[1,\infty)}\frac{4}{3x^{\frac{3}{2}}}\,d\mu_1 = \int_1^\infty \frac{4}{3x^{\frac{3}{2}}}\,dx = \frac{8}{3} < \infty.$$

Thus, by Fubini-Tonelli theorem, we have $f \in \mathcal{L}^1(X_2)$.

(c) On $X_3 = (0,1)\times[1,\infty)$, we repeat the same argument but with the other iterated integral for ease. For a fixed $y \in [1,\infty)$, we compute the improper Riemann integral:

$$\int_0^1 |f_y|\,dx \le \int_0^1 \frac{1}{x^2+y^4}\,dx \le \int_0^1 \frac{1}{y^4}\,dx = \frac{1}{y^4},$$

and so, by using Proposition 19.7.9 to evaluate the Lebesgue integral as an improper Riemann integral, we have:

$$\int_{(1,\infty)}\left(\int_{(0,1)}|f_y|\,d\mu_1\right)d\mu_2 \le \int_{(1,\infty)}\frac{1}{y^4}\,d\mu_2 = \int_1^\infty \frac{1}{y^4}\,dy = \frac{1}{3} < \infty.$$

Thus, by Fubini-Tonelli theorem, we have $f \in \mathcal{L}^1(X_3)$.

By additivity of Lebesgue integral over disjoint domains, we conclude that $f \in \mathcal{L}^1(X)$.

3. If we consider the space $(\mathbb{N},\mathcal{P}(\mathbb{N}),\nu)$ where ν is the counting measure on $\mathcal{P}(\mathbb{N})$, The product space $(\mathbb{N}\times\mathbb{N},\mathcal{P}(\mathbb{N}\times\mathbb{N}),\mu)$ is also a measure space with counting measure. Suppose that $a : \mathbb{N}\times\mathbb{N}\to\mathbb{R}$. Fubini-Tonelli's theorem says that the infinite sum $\int_{\mathbb{N}\times\mathbb{N}}|a|\,d\mu = \sum_{m,n\in\mathbb{N}}|a(m,n)|$ is finite if either one of the iterated integrals:

$$\int_{\mathbb{N}}\left(\int_{\mathbb{N}} a^m\,d\nu\right)d\nu = \sum_{m=1}^{\infty}\left(\sum_{n=1}^{\infty}|a^m|\right)$$

or

$$\int_{\mathbb{N}}\left(\int_{\mathbb{N}} a^m\,d\nu\right)d\nu = \sum_{n=1}^{\infty}\left(\sum_{m=1}^{\infty}|a_n|\right),$$

is finite.

Remark 20.3.7 One final remark is that in iterated integrals it is common to omit the subscript y and superscript x to indicate section functions. For example, we usually write $\int_X\int_Y f^x\,d\mu(x)\,d\mu(y)$ as $\int_X\int_Y f\,d\mu(y)d\mu(x)$ with the implicit assumption that the inner integral is carried out for a fixed $x \in X$. This would help us declutter the notation and make room for any other important subscripts and superscripts on the function.

20.4 Multiple Integrals

Finally, we can easily see that by induction, the whole construction that we have done in this chapter can be extended to n-fold products of σ-finite measure spaces. For completeness, we list the results here.

Suppose that $(X_j, \mathcal{F}_j, \mu_j)$ be σ-finite measure spaces for $j = 1, 2, \ldots, n$. Let $\mathcal{N} = \{1, 2, \ldots, n\}$ be the set of the integers 1 to n. We state without proof the following facts:

1. (Product σ-algebra) There exists a unique product σ-algebra on the Cartesian product $W = X_1 \times \ldots \times X_n$ denoted as $\mathcal{H} = \mathcal{F}_1 \otimes \ldots \otimes \mathcal{F}_n$. This construction is well-defined by Proposition 20.1.4.
2. (Product measure) Furthermore, on this measurable space $(W, \mathcal{H})$, there is a unique product measure $\tilde{\mu}$ such that $\tilde{\mu}(E_1 \times \ldots \times E_n) = \mu_1(E_1)\mu_2(E_2)\ldots \mu_n(E_n)$ where $E_j \in \mathcal{F}_j$ for each $j = 1, 2, \ldots, n$.
3. (Tonelli's theorem) If $f : W \to [0, \infty]$ is an $\mathcal{H}$-measurable non-negative function, then:

$$\int_W f\, d\tilde{\mu} = \int_{X_{\sigma(1)}} \int_{X_{\sigma(2)}} \cdots \int_{X_{\sigma(n)}} f\, d\mu_{\sigma(n)} \ldots d\mu_{\sigma(2)}\, d\mu_{\sigma(1)},$$

where $\sigma : \mathcal{N} \to \mathcal{N}$ is any permutation of the integers 1 to n.
4. (Fubini's theorem) If $f : W \to \bar{\mathbb{R}}$ is an $\mathcal{H}$-measurable function such that $f \in \mathcal{L}^1(W, \mathcal{H})$, then:

$$\int_W f\, d\tilde{\mu} = \int_{X_{\sigma(1)}} \int_{X_{\sigma(2)}} \cdots \int_{X_{\sigma(n)}} f\, d\mu_{\sigma(n)} \ldots d\mu_{\sigma(2)}\, d\mu_{\sigma(1)},$$

where $\sigma : \mathcal{N} \to \mathcal{N}$ is any permutation of the integers 1 to n.
5. (Fubini-Tonelli theorem) If $f : W \to \bar{\mathbb{R}}$ is an $\mathcal{H}$-measurable such that for some permutation $\tau : \mathcal{N} \to \mathcal{N}$ we have:

$$\int_{X_{\tau(1)}} \int_{X_{\tau(2)}} \cdots \int_{X_{\sigma(n)}} |f|\, d\mu_{\tau(n)} \ldots d\mu_{\tau(2)}\, d\mu_{\tau(1)} < \infty,$$

then $f \in \mathcal{L}^1(W, \mathcal{H})$ with the equality:

$$\int_W f\, d\tilde{\mu} = \int_{X_{\sigma(1)}} \int_{X_{\sigma(2)}} \cdots \int_{X_{\sigma(n)}} f\, d\mu_{\sigma(n)} \ldots d\mu_{\sigma(2)}\, d\mu_{\sigma(1)},$$

where $\sigma : \mathcal{N} \to \mathcal{N}$ is any permutation of the integers 1 to n.

Exercises

20.1 (∗) Let $(X, \mathcal{F}, \mu_1)$ and $(Y, \mathcal{G}, \mu_2)$ be two σ-finite measure spaces. Define the collection $\mathcal{J} = \mathcal{F} \times \mathcal{G} = \{A \times B : A \in \mathcal{F}, B \in \mathcal{G}\}$ of rectangular sets with a product content m on $\mathcal{J}$ defined via $m(A \times B) = \mu_1(A)\mu_2(B)$.
(a) Show that $\mathcal{J}$ is a semiring.
(b) Let $\mathcal{R}(\mathcal{J})$ be the ring generated by $\mathcal{J}$. Extend the definition of m to $\mathcal{R}(\mathcal{J})$.
(c) Show that $(X \times Y, \mathcal{R}(\mathcal{J}), m)$ is a premeasure space.

20.2 (∗) Prove Lemma 20.1.1, namely:
Let $(X, \mathcal{F}, \mu_1)$ and $(Y, \mathcal{G}, \mu_2)$ be σ-finite measure spaces. Let $\mathcal{J}$ be the semiring of rectangles $\mathcal{J} = \mathcal{F} \times \mathcal{G}$ with the product content m and $\mathcal{R}(\mathcal{J})$ be the ring generated by $\mathcal{J}$. Prove that the premeasure space $(X \times Y, \mathcal{R}(\mathcal{J}), m)$ is σ-finite.

20.3 Consider two measure spaces $(X, \mathcal{F}, \mu_1)$ and $(Y, \mathcal{G}, \mu_2)$ with $X = Y = [0, 1]$. Let $\mathcal{F} = \mathcal{P}(X)$ with the counting measure μ_1 and $\mathcal{G} = \mathcal{L}$ with the Lebesgue measure μ_2. Let $(W, \mathcal{H}, \mu)$ where $W = X \times Y$ and $\mathcal{H} = \mathcal{F} \otimes \mathcal{G}$ be the product measure space.
Define $f : W \to \mathbb{R}$ to be the function which is 1 on the diagonal and 0 everywhere else, namely $f(x, y) = \mathbf{1}_E$ where $E = \{(x, x) : x \in [0, 1]\}$ is the diagonal. Show that the function f is $\mathcal{H}$-measurable.

20.4 (∗) Let $(X, \mathcal{F})$ be a measurable space and $f : X \to [0, \infty)$ is a non-negative function. Suppose that $U = \{(x, y) \in X \times [0, \infty) : 0 \leq y \leq f(x)\}$ is the subgraph of f.
(a) Prove that f is an $\mathcal{F}$-measurable function if and only if the subgraph U is measurable in the product space $X \times [0, \infty)$ with the product σ-algebra $\mathcal{F} \otimes \mathcal{B}$.
(b) Suppose that μ is the measure on $(X, \mathcal{F})$ and $([0, \infty), \mathcal{B})$ are the induced Lebesgue measure. Let $\tilde{\mu}$ be the product measure on $(X \times [0, \infty), \mathcal{F} \otimes \mathcal{B})$. Prove that:

$$\tilde{\mu}(U) = \int_X f \, d\mu.$$

20.5 (∗) Let $(X, \mathcal{F})$ be a measurable space and $f : X \to \mathbb{R}$ be a measurable function.
(a) Suppose that $(X \times \mathbb{R}, \mathcal{F} \otimes \mathcal{B})$ is the product measurable space. Prove that the function $h : X \times \mathbb{R} \to \mathbb{R}$ defined as $h(x, y) = f(x) - y$ is $(\mathcal{F} \otimes \mathcal{B})$-measurable.
(b) Hence, show that the graph of the function $G_f = \{(x, f(x)) : x \in X\} \subseteq X \times \mathbb{R}$ is a measurable set.
(c) Suppose that μ_1 and μ_2 are measures on the spaces $(X, \mathcal{F})$ and $(\mathbb{R}, \mathcal{B})$ respectively (so μ_2 is the Lebesgue measure). Assume that $\mu_1(X) < \infty$. If $\tilde{\mu}$ is their product measure, show that $\tilde{\mu}(G_f) = 0$.
(d) Extend the result in part (c) for σ-finite domain $(X, \mathcal{F}, \mu_1)$.

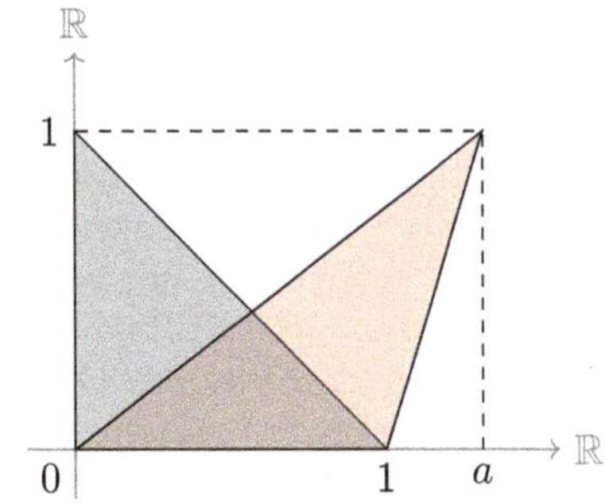

Fig. 20.3 The triangle A is in grey and the triangle B is in red

20.6 $(*)$ Consider two measure spaces $(X, \mathcal{F}, \mu_1)$ and $(Y, \mathcal{G}, \mu_2)$. Let $A \in \mathcal{J}$ be a rectangular set $A = C \times D$ for some $C \in \mathcal{F}$ and $D \in \mathcal{G}$. Prove that $\mu_2(A^x) = \mu_2(D)\mathbf{1}_C(x)$.

20.7 Let $f : X \times Y \to \mathbb{R}$ be a function defined on the product set. Prove that for any $E \subseteq \mathbb{R}$ we have $(f^x)^{-1}(E) = (f^{-1}(E))^x$ and $(f_y)^{-1}(E) = (f^{-1}(E))_y$.

20.8 $(*)$ Let $(X, \mathcal{F}, \mu_1)$ and $(Y, \mathcal{G}, \mu_2)$ be σ-finite measure spaces and $(X \times Y, \mathcal{H}, \mu)$ where $\mathcal{H} = \mathcal{F} \otimes \mathcal{G}$ be its product measure space.

(a) Define a set function $\mu' : \mathcal{H} \to [0, \infty]$ by $\mu'(E) = \int_X \left(\int_Y (\mathbf{1}_E)^x \, d\mu_2\right) d\mu_1$. Show that $\mu'(E)$ is a measure.

(b) Deduce that the set function $\mu'' : \mathcal{H} \to [0, \infty]$ defined as $\mu''(E) = \int_Y \left(\int_X (\mathbf{1}_E)_y \, d\mu_1\right) d\mu_2$ is also a measure.

20.9 $(\diamond)$ Consider the Lebesgue measure space $(\mathbb{R}, \mathcal{L}, \mu)$. In Example 20.1.5, we have constructed two different measure spaces on the Cartesian product $\mathbb{R}^2$, namely the Lebesgue σ-algebra $\mathcal{L}(\mathbb{R}^2)$ and the product σ-algebra $\mathcal{L} \otimes \mathcal{L}$. Moreover, we claimed that we have the strict inclusion $\mathcal{L} \otimes \mathcal{L} \subsetneq \mathcal{L}(\mathbb{R}^2)$. Construct an example of a set which is in $\mathcal{L}(\mathbb{R}^2) \setminus \mathcal{L} \otimes \mathcal{L}$.

20.10 $(*)$ Suppose that $(X, \mathcal{F}, \mu_1)$ and $(Y, \mathcal{G}, \mu_2)$ are σ-finite measure spaces and $(X \times Y, \mathcal{H}, \mu)$ where $\mathcal{H} = \mathcal{F} \otimes \mathcal{G}$ is its product measure space. Let $E, F \in \mathcal{H}$ be such that for every $y \in Y$, we have $\mu_1\{x \in X : (x, y) \in E\} = \mu_1\{x \in X : (x, y) \in F\}$. Show that $\mu(E) = \mu(F)$.

This is called the Cavalieri's principle. In two dimensions, this principle states that if two regions in $\mathbb{R}^2$ are contained within two parallel lines and if every line parallel to these lines intersect the two regions in line segments of equal measure, then the total areas of the regions are equal. Let us demonstrate this principle with a concrete case. Suppose that we have two triangles A and B on the product measure space $(\mathbb{R}^2, \mathcal{L} \otimes \mathcal{L}, \tilde{\mu})$. Let A be the solid triangle with vertices $(0, 0)$, $(0, 1)$, and $(1, 0)$ while B be the solid triangle with vertices $(0, 0)$, $(1, 0)$, and $(a, 1)$ for some $a \in \mathbb{R}$. See Fig. 20.3.

(a) Find the equations of the non-horizontal sides of the triangles A and B.

(b) For any fixed $y \in [0, 1]$, show that $\mu\{x \in \mathbb{R} : (x, y) \in A\} = \mu\{x \in \mathbb{R} : (x, y) \in B\}$ where μ is the Lebesgue measure on $\mathbb{R}$.

(c) Conclude that the two triangles have the same area, namely $\tilde{\mu}(A) = \tilde{\mu}(B)$ and show that this agrees with the classical geometrical interpretation of it.

In fact, this principle also generalises to higher dimensions. This principle can be used to prove many interesting mathematical phenomenon and results such as the napkin ring problem and the quadrature of a cycloid by Gilles de Roberval (1602–1675).

20.11 (∗) Let $(X, \mathcal{F}, \mu_1)$ and $(Y, \mathcal{G}, \mu_2)$ be two measure spaces and $(X \times Y, \mathcal{H}, \mu)$ be their product measure space. Suppose that $f \in \mathcal{L}^1(X)$ and $g \in \mathcal{L}^1(Y)$.

(a) Show that the functions $h_1, h_2 : X \times Y \to \mathbb{R}$ defined as $h_1(x, y) = f(x)$ and $h_2(x, y) = g(y)$ are $\mathcal{H}$-measurable.

(b) Hence, show that the function $h : X \times Y \to \mathbb{R}$ defined as the product $h(x, y) = f(x)g(y)$ is $\mathcal{H}$-measurable.

(c) Prove that $fg \in \mathcal{L}(X \times Y)$ and:

$$\int_{X\times Y} f(x)g(y)\, d\mu = \int_X f(x)\, d\mu_1 \int_Y g(y)\, d\mu_2.$$

20.12 Let $([-1, 1], \mathcal{L}, \mu)$ be the induced Lebesgue measure space. Suppose that $X = [-1, 1]^2$ and $(X, \mathcal{L} \otimes \mathcal{L}, \tilde{\mu})$ is the product measure space. Explain why the following functions are Lebesgue integrable over X and determine their values.

(a) $f : X \to \bar{\mathbb{R}}$ defined as $f(x, y) = xy^2$.

(b) $g : X \to \bar{\mathbb{R}}$ defined as $f(x, y) = x + 2y$.

20.13 (∗) Recall the gamma function from Exercise 16.20 which was defined as the improper Riemann integral $\Gamma(x) = \int_0^\infty t^{x-1}e^{-t}\, dt$ for $x > 0$.

(a) Let $((0, \infty), \mathcal{L}, \mu)$ be the induced Lebesgue measure space. For a fixed $x > 0$, explain why $t^{x-1}e^{-t} \in \mathcal{L}^1((0, \infty))$.

(b) Using Exercise 20.11, show that $\Gamma(x)\Gamma(y) = \int_{(0,\infty)} \int_{(0,\infty)} t^{x-1}s^{y-1} e^{-(t+s)}\, d\mu(t)\, d\mu(s)$.

(c) Using part (b), by carrying out some change of variables, show that:

$$\int_0^1 u^{x-1}(1-u)^{y-1}\, du = \frac{\Gamma(x)\Gamma(y)}{\Gamma(x+y)}.$$

The function $(x, y) \mapsto \int_0^1 u^{x-1}(1-u)^{y-1}\, du$ for $x, y > 0$ is called the beta function and is denoted as $B(x, y)$.

(d) Using the beta function, deduce the value of $\Gamma(\frac{1}{2})$.

(e) Show that for any $x, y > 0$ we have $\int_0^1 u^{x-1}(1-u)^{y-1}\, du = \int_0^1 u^{y-1}(1-u)^{x-1}\, du$.

20.14 (∗) Let $(0, \infty)$ and $[0, 1]$ be intervals in $\mathbb{R}$ with the induced Lebesgue measure μ. Let $X = (0, \infty) \times [0, 1]$ be the Cartesian product with product σ-algebra and product measure $\tilde{\mu}$. Define a function $f : X \to \mathbb{R}$ as $f(x, y) = e^{-x} \sin(2xy)$.

(a) Show that $f \in \mathcal{L}^1(X)$.
(b) By carefully justifying all the steps, determine the value $\int_X f \, d\tilde{\mu}$.

20.15 Let $([0,1], \mathcal{L}, \mu)$ be the induced Lebesgue measure space. Suppose that $X = [0,1]^2$ and $(X, \mathcal{L} \otimes \mathcal{L}, \tilde{\mu})$ is the product measure space. Define a function $f : \mathbb{R}^2 \to \mathbb{R}$ as:

$$f(x,y) = \begin{cases} \frac{x-y}{(x+y)^3} & \text{if } (x,y) \neq (0,0), \\ 0 & \text{if } (x,y) = (0,0). \end{cases}$$

Show that f is not Lebesgue integrable over X.

20.16 Let $(\mathbb{N}, \mathcal{P}(\mathbb{N}), \nu)$ be the measure space where ν is the counting measure and $(\mathbb{N}^2, \mathcal{P}(\mathbb{N}) \otimes \mathcal{P}(\mathbb{N}), \tilde{\nu})$ be the product measure space. Consider the function $f : X \to \mathbb{R}$ defined as:

$$f(x,y) = \begin{cases} 1 & \text{if } x = y, \\ -1 & \text{if } x = y+1, \\ 0 & \text{otherwise.} \end{cases}$$

(a) Show that $\int_{\mathbb{N}} \left(\int_{\mathbb{N}} f^x \, d\nu(y) \right) d\nu(x)$ and $\int_{\mathbb{N}} \left(\int_{\mathbb{N}} f_y \, d\nu(x) \right) d\nu(y)$ both exist as iterated integrals but are not equal.
(b) Explain why the function f does not satisfy Fubini's theorem.

20.17 $(*)$ Let $([-1,1], \mathcal{L}, \mu)$ be the induced Lebesgue measure space. Suppose that $X = [-1,1]^2$ and $(X, \mathcal{L} \otimes \mathcal{L}, \tilde{\mu})$ is the product measure space. Consider the function $f : X \to \bar{\mathbb{R}}$ defined as:

$$f(x,y) = \begin{cases} \frac{xy}{(x^2+y^2)^2} & \text{if } (x,y) \neq (0,0), \\ 0 & \text{if } (x,y) = (0,0). \end{cases}$$

(a) Show that:

$$\int_{[-1,1]} \left(\int_{[-1,1]} f^x \, d\mu(y) \right) d\mu(x) = \int_{[-1,1]} \left(\int_{[-1,1]} f_y \, d\mu(x) \right) d\mu(y),$$

as iterated integrals.
(b) Show that f is not integrable over X.
(c) Explain why parts (a) and (b) do not violate Fubini's theorem.

20.18 $(*)$ Let $(\mathbb{R}, \mathcal{L}, \mu)$ be the Lebesgue measure space and $(\mathbb{R}^2, \mathcal{L} \otimes \mathcal{L}, \tilde{\mu})$ be the product measure space. Fix $k \in (0,1)$ and define a function $f : \mathbb{R}^2 \to \mathbb{R}$ as:

$$f(x,y) = \begin{cases} \frac{(1-y)^k}{(x-y)^k} & \text{if } 0 < x < 1 \text{ and } 0 \leq y < x, \\ 0 & \text{otherwise.} \end{cases}$$

Find the value of the integral $\int_{\mathbb{R}^2} f \, d\tilde{\mu}$.

20.19 (∗) In this question, we will justify the layer cake representation of the Lebesgue integral as seen in Remark 19.3.7.
Let $(X, \mathcal{F}, \mu)$ be a measure space and $f : X \to [0, \infty)$ be an $\mathcal{F}$-measurable function. Denote the set $L_f = \{(y, s) \in X \times [0, \infty) : 0 \leq s \leq f(y)\}$. We have seen in Exercise 20.4(a) that the set L_f is $(\mathcal{F} \otimes \mathcal{B})$-measurable. For any $t \geq 0$, denote $L_f^t = \{y \in X : f(y) \geq t\}$.
(a) Show that $\mathbf{1}_{L_f^t}(x) = \mathbf{1}_{[0,f(x)]}(t)$ for all $(x, t) \in X \times [0, \infty)$.
(b) Hence, deduce that for all $x \in \mathbb{R}$, we have $f(x) = \int_0^\infty \mathbf{1}_{L_f^t}(x)\, dt$.
(c) Finally, derive the layer cake representation of the Lebesgue integral, namely:

$$\int_X f \, d\mu = \int_0^\infty \mu\{y \in X : f(y) \geq t\} \, dt.$$

20.20 (∗) Let $(\mathbb{R}, \mathcal{L}, \mu)$ be the Lebesgue space and $(\mathbb{R}^2, \mathcal{H}, \tilde{\mu})$ where $\mathcal{H} = \mathcal{L} \otimes \mathcal{L}$ be its product measure space. Fix any $E \in \mathcal{H}$. For $(a, b) \in \mathbb{R}^2$ and $\lambda \in \mathbb{R}$, define the translated and scaled sets:

$$(a, b) + E = \{(a, b) + (x, y) : (x, y) \in E\},$$
$$\lambda E = \{\lambda(x, y) : (x, y) \in E\}.$$

(a) Prove that the sets $(a, b) + E$ and λE are also in $\mathcal{H}$.
(b) Show that $((a, b) + E)^x = E^{x-a} + b \subseteq \mathbb{R}$.
(c) Hence, deduce that $\tilde{\mu}((a, b) + E) = \tilde{\mu}(E)$.
(d) For $\lambda \neq 0$, show that $(\lambda E)^x = \lambda E^{\frac{x}{\lambda}} \subseteq \mathbb{R}$.
(e) Hence, deduce that $\tilde{\mu}(\lambda E) = |\lambda|^2 \tilde{\mu}(E)$.

We can extend the results above for n products of the Lebesgue spaces, which we denote as $(\mathbb{R}^n, \overset{n}{\otimes}\mathcal{L}, \mu^n)$. For $E \in \overset{n}{\otimes}\mathcal{L}$, the measures of translated and scaled sets are $\mu^n((a_1, \ldots, a_n) + E) = \mu^n(E)$ for any $(a_1, a_2, \ldots, a_n) \in \mathbb{R}^n$ and $\mu^n(\lambda E) = |\lambda|^n \mu^n(E)$ for any $\lambda \in \mathbb{R}$.

20.21 (◇) In this question, we are going to compute the volume of an open ball in the real n-space $\mathbb{R}^n$. We know that the volume of n-balls for $n = 1$ and $n = 2$, which are 2 and π respectively. Let $(\mathbb{R}, \mathcal{L}, \mu)$ be the Lebesgue space and for any $k \in \mathbb{N}$, denote the product space of k copies of the Lebesgue space as $(\mathbb{R}^k, \overset{k}{\otimes}\mathcal{L}, \mu^k)$. For any $k \in \mathbb{N}$, define the open unit ball as the set:

$$B_1^k = \{(x_1, \ldots, x_k) \in \mathbb{R}^k : x_1^2 + \ldots + x_k^2 < 1\}.$$

(a) Show that $B_1^k \in \overset{k}{\otimes}\mathcal{L}$ for any $k \in \mathbb{N}$.
(b) Fix $n \in \mathbb{N}$ where $n \geq 3$. Explain why the indicator function $\mathbf{1}_{B_1^n} : \mathbb{R}^n \to \mathbb{R}$ is Lebesgue integrable.

(c) Show that:

$$\mu^n(B_1^n) = \int_{\mathbb{R}^n} \mathbf{1}_{B_1^n}\, d\mu^n = \int_{(-1,1)} \left(\int_{\mathbb{R}^{n-1}} \mathbf{1}_{B_1^n}\, d\mu^{n-1} \right) d\mu(x_n).$$

(d) For any fixed $x_n \in (-1, 1)$, show that:

$$\begin{aligned} B_1^n &= \sqrt{1 - x_n^2}\{(y_1, \ldots, y_n) : y_1^2 + \ldots + y_{n-1}^2 < 1\} \\ &= (\sqrt{1 - x_n^2} B_1^{n-1}, x_n) \subseteq \mathbb{R}^n. \end{aligned}$$

(e) Using Exercise 20.20 and part (d), deduce that:

$$\mu^n(B_1^n) = 2\mu^{n-1}(B_1^{n-1}) \int_{[0,1)} (1 - x_n^2)^{\frac{n-1}{2}}\, d\mu(x_n).$$

(f) Derive the recursive formula $\mu^n(B_1^n) = \sqrt{\pi}\mu^{n-1}(B_1^{n-1})\frac{\Gamma(\frac{n+1}{2})}{\Gamma(\frac{n+2}{2})}$ for $n \geq 3$.

(g) Using induction, show that $\mu^n(B_1^n) = \frac{\pi^{\frac{n}{2}}}{\Gamma(\frac{n+2}{2})}$ for any $n \in \mathbb{N}$.

20.22 Recall from Exercise 16.28 that we have computed the Gaussian integral $\int_{-\infty}^{\infty} e^{-x^2}\, dx = \sqrt{\pi}$. In this question, we are going to compute it using Tonelli's theorem. Let $X = [0, \infty)$ and $(X, \mathcal{L}, \mu)$ is the induced Lebesgue space.

(a) Denote $I = \int_X e^{-x^2}\, d\mu(x)$. Show that $e^{-x^2} \in \mathcal{L}^1(X)$, namely $I < \infty$.

(b) Suppose that $(X \times X, \mathcal{L} \otimes \mathcal{L}, \tilde{\mu})$ is the product measure space. Show that:

$$\int_{X \times X} e^{-(x^2+y^2)}\, d\tilde{\mu} = I^2.$$

(c) Using the change of variable $u = \frac{x}{\sqrt{x^2+y^2}}$ for a fixed $y \in [0, \infty)$ and Tonelli's theorem, show that $I^2 = \frac{\pi}{4}$.
Hence, conclude the result.

20.23 (◇) Suppose that $f, g \in \mathcal{L}^1(X)$. Let $(\mathbb{R}, \mathcal{L}, \mu)$ be the Lebesgue space and $(\mathbb{R}^2, \mathcal{L} \otimes \mathcal{L}, \tilde{\mu})$ be the product measure space.

(a) Show that the function $h : \mathbb{R}^2 \to \mathbb{R}$ defined as $h(x, y) = f(x - y)g(y)$ is measurable and in $\mathcal{L}^1(\mathbb{R}^2)$.

(b) Define a function $f * g : \mathbb{R} \to \mathbb{R}$ as $(f * g)(x) = \int_{\mathbb{R}} f(x-y)g(y)\, d\mu(y)$. Show that $f * g < \infty$ a.e..

(c) Show that $f * g = g * f$ on $\mathbb{R}$.

(d) Prove that $||f * g||_1 \leq ||f||_1||g||_1$ where $|| \cdot ||_1$ was defined in Exercise 19.27.

The function $f * g$ is called the convolution of the functions f and g. This function is used commonly in the study of partial differential equations and Fourier analysis. This function is also used in many different areas of science such as geophysics, signal and image processing, engineering, and acoustics.

20.24 ($\diamond$) An application of the convolution process in Exercise 20.23 is that it allows us to "smoothen" any Lebesgue integrable function. Suppose that $\phi \in C^\infty(\mathbb{R})$ is a smooth continuous function with bounded derivatives and $f \in \mathcal{L}^1(\mathbb{R})$. Let $\phi * f : \mathbb{R} \to \mathbb{R}$ be their convolution.
(a) Show that $\phi * f$ is continuous.
(b) Prove that $\phi * f$ is differentiable with $\frac{d}{dx}(\phi * f) = (\frac{d}{dx}\phi) * f$.
(c) Deduce that $\phi * f$ is also smooth.

20.25 ($\diamond$) By a suitable choice of the function ϕ in Exercise 20.24, we can also show that the convolution $\phi * f$ is "close" to f. Close here means the function f can be approximated by the convolution $\phi * f$ to some degree of accuracy. Recall the bump function $\Psi : \mathbb{R} \to \mathbb{R}$ in Exercise 14.19 defined as $\Psi(x) = e^{-\frac{1}{1-x^2}}$ for $x \in (-1, 1)$ and 0 elsewhere. Let $\varphi : \mathbb{R} \to \mathbb{R}$ be defined as its normalisation, namely $\varphi(x) = c\Psi(x)$ where $c = (\int_{\mathbb{R}} \Psi(x)\, dx)^{-1}$, so that $\int_{\mathbb{R}} \varphi(x)\, dx = 1$.
(a) Show that $\varphi \geq 0$, $\varphi \in C^\infty(\mathbb{R})$, and $\text{supp}(\varphi) \subseteq [-1, 1]$. Recall that the support of a function is defined in Exercise 15.10.

The function φ in part (a) is called a standard mollifier function, a name coined by Kurt Otto Friedrichs (1901–1982). This function was independently introduced for the study of PDEs by Friedrichs and Sergei Sobolev (1908–1989) who used the mollifiers to prove the Sobolev embedding theorem in the study of functions spaces.

For any $\varepsilon > 0$, we now define the scaled function $\varphi_\varepsilon(x) = \frac{1}{\varepsilon}\varphi(\frac{x}{\varepsilon})$.
(b) Show that $\varphi_\varepsilon \geq 0$, $\varphi_\varepsilon \in C^\infty(\mathbb{R})$, $\text{supp}(\varphi_\varepsilon) \subseteq [-\varepsilon, \varepsilon]$, and $\int_{\mathbb{R}} \varphi_\varepsilon(x)\, dx = 1$.

Suppose that $f \in \mathcal{L}^1(\mathbb{R})$.
(c) Show that $\text{supp}(\varphi_\varepsilon * f) \subseteq \text{supp}(f) + [-\varepsilon, \varepsilon]$.
(d) Show that $\varphi_\varepsilon * f \in C^\infty(\mathbb{R})$ for any $\varepsilon > 0$.
(e) Prove that $||\varphi_\varepsilon * f - f||_1 \to 0$ as $\varepsilon \to 0$.
(f) Now assume that $f \in C^0(\mathbb{R})$. Show that $\varphi_\varepsilon * f \xrightarrow{u} f$ as $\varepsilon \to 0$ over any compact subset $K \subseteq \mathbb{R}$.

The result in part (e) can also be extended to functions in the function space $\mathcal{L}^p$ for any other $p > 1$. Thus, from this question, we can see that the functions $\varphi_\varepsilon * f$ (which are called mollified functions) are smooth and can be used to approximate less regular functions such as continuous functions or even $\mathcal{L}^p$ functions.

20.26 ($\diamond$) Recall that for a probability space $(\Omega, \mathcal{F}, P)$ and a random variable $X : \Omega \to \mathbb{R}$, its cumulative distribution function and expectation are given as:

$$F_X(x) = P\{\omega \in \Omega : X(\omega) \leq x\},$$

$$\mathbb{E}[X] = \int_\Omega X\, dP.$$

Suppose that X is a non-negative random variable. Consider the product space $\Omega \times \mathbb{R}$ with the σ-algebra $\mathcal{F} \otimes \mathcal{B}$ and product measure $\tilde{\mu}$.

(a) Let π_Ω and $\pi_\mathbb{R}$ be the projection maps from the space $\Omega \times \mathbb{R}$ to Ω and $\mathbb{R}$ respectively. Prove that the function $g : \Omega \times \mathbb{R} \to \mathbb{R}$ defined as $g = \pi_\mathbb{R} - X \circ \pi_\Omega$ is a measurable.

(b) Using part (a), deduce that the set $E = \{(\omega, x) \in \Omega \times \mathbb{R} : x < X(\omega)\}$ is in $\mathcal{F} \otimes \mathcal{B}$.

(c) Hence prove that the function $f : \Omega \times \mathbb{R} \to \mathbb{R}$ defined as $f(\omega, x) = \mathbf{1}_{X(\omega) > x \geq 0}$ is measurable.

(d) Show that for a fixed $\omega \in \Omega$, we have $X(\omega) = \int_\mathbb{R} \mathbf{1}_{X(\omega) > x \geq 0} \, d\mu(x)$.

(e) Hence, by using Tonelli's theorem, deduce that $\mathbb{E}[X] = \int_\mathbb{R} 1 - F_X(x) \, d\mu(x)$.

Suppose that X is a non-negative random variable. Consider the product space $\Omega \times \mathbb{R}$ with the σ-algebra $\mathcal{F} \otimes \mathcal{B}$ and product measure $\mathbb{P} \otimes \lambda$.

(a) Let π_Ω and $\pi_\mathbb{R}$ be the projection maps from the space $\Omega \times \mathbb{R}$ to Ω and $\mathbb{R}$ respectively. Prove that the function $g: \Omega \times \mathbb{R} \to \mathbb{R}$ defined as $g = \pi_\mathbb{R} - X \circ \pi_\Omega$ is measurable.

(b) Using part (a), deduce that the set $E = \{(\omega, x) \in \Omega \times \mathbb{R} : x < X(\omega)\}$ is in $\mathcal{F} \otimes \mathcal{B}$.

(c) Hence prove that the function $f: \Omega \times \mathbb{R} \to \mathbb{R}$ defined as $f(\omega, x) = \mathbb{1}_E(\omega, x)$ is measurable.

(d) Show that for a fixed $\omega \in \Omega$, we have $X(\omega) = \int_{\mathbb{R}} f(\omega, x) \, d\lambda(x)$.

(e) Hence, by using Tonelli's theorem, deduce that $\mathbb{E}[X] = \int_{\mathbb{R}} (1 - F_X(x)) \, d\lambda(x)$.

Hints for Exercises

Chapter 1: Logic and Sets

1.6 You may use a truth table, but this is not necessary.
1.12 Prove the implications **(1)** $\Rightarrow$ **(2)**, **(2)** $\Rightarrow$ **(3)**, and **(3)** $\Rightarrow$ **(1)** separately.
1.21 Write all the Δ in terms of the set operations $\setminus$, $\cup$, and $\cap$. Then use set algebra.
1.22 (a) Prove that both sides are equal to $(X \cap Y) \cup (X^c \cap Y^c)$.
1.23 Use set algebra.
1.24 (a) Use Proposition 1.3.26(1) and (4). Then show the inclusions $(A \times C) \cap (B \times D) \subseteq B \times C$ and $(A \times C) \cap (B \times D) \subseteq A \times D$. Finally, apply Exercise 1.18(b).
(b) Use Proposition 1.3.26(2) and (5) to show $X \times Y = (A \cup A^c) \times (C \cup C^c) = (A \times C) \cup (A^c \times C) \cup (A \times C^c) \cup (A^c \cup C^c)$. Show that any two sets on the RHS are disjoint by using part (a) and then find $(A \times C)^c$.
(c) Write $A = (A \cap B) \cup (A \setminus B)$ and $B = (B \cap A) \cup (B \setminus A)$. Then use Proposition 1.3.26(2) and (5).
(d) Use parts (a) and (b), the distributivity of set operations, and Proposition 1.3.26(2) and (5).
1.25 Use the definition of image and preimage of sets.
1.27 Use the results from Exercise 1.25 and 1.26.
1.28 Prove each implication via contrapositive.
1.31 Prove parts (c) and (d) via contrapositive.
1.32 Construct a suitable left-inverse for injective f and right-inverse for surjective f. Exercise 1.31 might be useful for the converses.

Chapter 2: Integers

2.4 For parts (e) and (f), recall the fundamental theorem of equivalence relation.
2.8 Prove only one implication as the other is similarly done by noting that $(\mathcal{R}^{-1})^{-1} = \mathcal{R}$.
2.13 (d) Use part (b).
2.15 (k) Use induction on p and with part (j) as the base case.

S. Johar, *The Big Book of Real Analysis*,
https://doi.org/10.1007/978-3-031-30832-1

2.16 Prove the contrapositive of the statements.

2.20 (a) Suppose for contradiction that $\gcd(p, q) = k > 1$.
(c) Show both $\text{lcm}(p, q) \leq pq$ and $\text{lcm}(p, q) \geq pq$. Part (a) might be useful.
(d) Show that $\text{lcm}(a, b) = gpq$. Start by noting that $\text{lcm}(a, b) = gk$ for some $k \in \mathbb{N}$

2.22 Assume that there are two of these objects. Show that these two objects are identical.

2.23 (a) Show that $a \times (-1)$ is an additive inverse of a and use Exercise 2.22(c).
(c) Prove this via contradiction.

2.24 (a) Prove this by contradiction.
(b) Use induction on the set of negative and positive integers separately.

2.26 (a) Show that the set $\{n : n \in \mathbb{N}, nb > a\}$ is non-empty and use the well-ordering principle. Define q_1 appropriately.
(e) For the second part, successively substitute the equations in the algorithm backwards and use part (d).

2.27 (a) Multiply the equations obtained from Bézout's identity.

2.28 Use Bézout's identity from Exercise 2.26 for both implications.

2.29 (a) Use Bézout's identity from Exercise 2.26.
(b) Show that that $\gcd(a, c)$ can only be 1 or c. Then part (a) might be useful.

2.30 (b) Exercise 2.29(b) and cancellation law on $\mathbb{N}$ might be useful.
(c) Assume for contradiction that there are only finitely many primes $P = \{p_j\}_{j=1}^n$. Now consider the number $q = p_1 p_2 \dots p_n + 1$ and use the fundamental theorem of arithmetic.

2.32 Show that $\min\{x, y\} + \max\{x, y\} = x + y$ for any $x, y \in \mathbb{N}_0$ and use this result.

2.33 Prove the results by contradiction.

Chapter 3: Construction of Real Numbers

3.4 Prove parts (a) and (b) by contradiction. For parts (c) and (d), prove the statements via induction on n.

3.5 (a) To construct an injection $\phi : \mathbb{N}^2 \to \mathbb{N}$, recall how we constructed an injection from $\mathbb{Q}_+$ to $\mathbb{N}$.
(b) Use part (a), composition of functions, and Exercise 1.31(a).

3.6 To show that h is surjective, pick any $y \in Y$. Then, either $g(y) \in A$ or $g(y) \in B$. Find a suitable preimage of y from these two cases.
To show that h is injective, suppose $h(x) = h(z)$ for some $x, z \in X$. Then, either both x and z are in A, both are in B, or each in separate subsets. Check these cases separately.

3.7 (b) Suppose for contradiction that B is countable.

3.10 (b) Show that every element $r \in \mathbb{Q}$ is in PQ. Split into cases of $r = 0$, $r > 0$ and $r < 0$.

(d) Split into two cases: at least one of p, q is zero or $p, q > 0$. For the latter case, assume for contradiction that there is a maximal element $k \in \{x \in \mathbb{Q} : 0 \leq x < p\}\{x \in \mathbb{Q} : 0 \leq x < q\} \cup \{x \in \mathbb{Q} : x < 0\}$.

3.12 Use the characterisation in Exercise 3.9.

3.14 (b) For the second half of each question, use part (a)(i).
(c) Use the identity in part (a)(ii) during the inductive step.
(f) Consider the binomial expansion of $(1+x)^{2n} = (1+x)^n(1+x)^n$.
(g) ii. Use part (g)(i) to deduce the final inequality.
iii. Use parts (g)(i) and (g)(ii) to deduce the final inequality.

3.16 (a) Use the inequality in Exercise 2.13(a) during the inductive step.

3.17 (a) Show that L does not have a maximal element via contradiction. Recall Example 3.6.9 and binomial expansion.
(b) For the inclusion $M \subseteq L \otimes \ldots \otimes L$, use Proposition 3.3.10.
(d) Suppose that $c^p = b^p = a$ with $c, b \geq 0$. Then use part (c).

3.18 Remember poor Hippasus.

3.19 (a) Suppose that $\sqrt{n}$ not irrational. Show that it must be an integer.
(c) From part (b), the number $\sqrt[n]{n}$ is either an integer or irrational. Assume for contradiction that it is an integer. Bernoulli's inequality from Exercise 3.15 may be helpful.

3.20 (f) Prove this by contradiction and consider the leading coefficients.

3.21 (a) Note that $P(x) = P(x) - P(c)$ for all $x \in \mathbb{R}$. Then use Exercise 3.17(c).

3.22 (a) Use Exercise 3.21(b).
(b) For a fixed $k \in \mathbb{R}$, define a polynomial $Q : \mathbb{R} \to \mathbb{R}$ as $Q(x) = P(x) - k$. Use part (a).

3.23 (a) Complete the square.

3.26 The quadratic polynomial $P(x) = \sum_{j=1}^{n}(a_j x + b_j)^2$ is non-negative for all $x \in \mathbb{R}$ so it has at most one real root. Then use Exercise 3.23(c).

3.27 (a) The square of any real number is non-negative.
(b) At the inductive step, let the arithmetic mean of the $k+1$ numbers in the set $A = \{a_1, a_2, a_3, \ldots, a_{k+1}\}$ be $\bar{a}$. If all the numbers in the list are equal to the $\bar{a}$, then we are done. Otherwise, we can find two numbers in the list, WLOG, a_1, a_2 such that $a_1 < \bar{a} < a_2$. Consider the number $y = a_1 + a_2 - \bar{a}$. Show that $\bar{a}y \geq a_1 a_2$. Next consider the set of k numbers $A' = \{y, a_3, \ldots, a_{k+1}\}$. Apply the inductive hypothesis on this set.

3.28 To show HM $\leq$ GM, apply AM-GM on the set $\{\frac{1}{a_1}, \ldots, \frac{1}{a_n}\}$. To show AM $\leq$ QM, apply the Cauchy-Schwarz inequality in Exercise 3.26.

3.29 (b) Use the fact that $0 \prec i$ and $-1 \prec 0$ to get a contradiction.
(c) Suppose for contradiction that $\lhd$ is a total order which is compatible with the field structure. By Lemma 3.3.3, the multiplicative identity must be strictly bigger than the additive identity. Derive a contradiction.

3.30 (c) Find an irrational number in $\mathbb{R}$ which is not in $\mathbb{Q}[\sqrt{2}]$.
(d) Assume for contradiction that $\prec$ is compatible with the field structure. Note that, via this order, we have $-1 \prec 0$, $0 \prec 1 + \sqrt{2}$, and $0 \prec 1 - \sqrt{2}$.

3.31 (a) Apply Bézout's identity from Exercise 2.26 to a and r.

(c) Suppose that there is such a field-compatible order. By Lemma 3.3.3, the multiplicative identity must be strictly bigger than the additive identity. Derive a contradiction.

Chapter 4: Real Numbers

4.1 Use the floor or ceiling function.

4.3 (a) Recall Exercise 2.29(a).
(b) This can be shown via induction or using Bernoulli's inequality.
(c) Write down $r = \frac{a}{b}$ and $s = \frac{c}{d}$ as ratios of natural numbers in their lowest forms. For contradiction, assume that $b \neq 1$. Use Exercise 2.27 and parts (a) and (b).

4.4 Show $\sup(Y) \leq \sup(X)$ and $\sup(Y) \geq \sup(X)$.

4.5 Denote $\sup\{f(x, y) : x \in X, y \in Y\} = M$, $M(x) = \sup\{f(x, y) : y \in Y\}$ for any $x \in X$, and $N(y) = \sup\{f(x, y) : x \in X\}$ for any $y \in Y$. Show that $\sup\{M(x) : x \in X\} = M = \sup\{N(y) : y \in Y\}$.

4.6 (a) Similar idea to Exercise 4.5.
(b) A very simple example can be obtained with $|X| = |Y| = 2$ and $f : X \times Y \to \{0, 1\}$.

4.7 (a) For $n \geq 2$, split into two cases, namely $-1 < x \leq -\frac{1}{n}$ or $x > -\frac{1}{n}$. For the latter, use AM-GM on 1 copy of $1 + nx$ and $n - 1$ copies of 1.
(b) Use AM-GM with some copies of $1 + y$ and some copies of 1.
(c) Split into two cases, namely $-1 < x \leq -\frac{1}{r}$ or $x > -\frac{1}{r}$. For the latter, use part (b).

4.8 Prove this by contradiction.

4.9 (c) Use the characterisation of supremum and Exercise 4.7(c).
(d) Use Exercise 4.8.

4.11 Let $M = \sup(X)$ and $N = \sup(Y)$
(a) Show that $\sup(X \cup Y) \leq \max\{M, N\}$ and $\sup(X \cup Y) \geq \max\{M, N\}$. Proposition 4.1.10(2) might be useful for the latter. Similar for infimum.
(b) For $\lambda > 0$ show that $\sup(\lambda X) \leq \lambda M$ and $\sup(\lambda X) \geq \lambda M$. For the latter, consider the set $Z = \lambda X$. Similar for infimum and the results for $\lambda < 0$.
(c) Show that $\sup(X + Y) \leq M + N$ and $\sup(X + Y) \geq M + N$. Characterisation of supremum and Exercise 4.8 may be useful for the latter. Similar for infimum.

4.12 (a) Use induction and Exercise 4.11(a).

4.13 For $a > 1$, write $M = \sup\{a^p : p \in \mathbb{Q}, p \leq x\}$ and $N = \inf\{a^r : r \in \mathbb{Q}, r \geq x\}$. Show $M \leq N$ first. To show the equality, split into two cases: $x \in \mathbb{Q}$ and $x \notin \mathbb{Q}$. For the latter, assume for contradiction that $M < N$. Show that for every $n \in \mathbb{N}$, there exists a $k \in \mathbb{Z}$ such that $a^{\frac{k-1}{n}} < M < N < a^{\frac{k}{n}}$ and so $1 < \frac{N}{M} \leq a^{\frac{1}{n}}$. Then apply Bernoulli's inequality to get a contradiction.
For $0 < a < 1$, write $b = \frac{1}{a} > 1$ and apply Proposition 4.1.10 to the case above.

4.14 Use Proposition 4.2.6.

4.15 (a) If $x > 0$, then $(1 + x)^p$ is increasing with p. Use the definition of exponentiation with irrational exponent and Exercise 4.7(c).
If $-1 < x < 0$, then $(1 + x)^p$ is decreasing with p. Use similar argument as the previous case.
(b) Similar to part (a).
4.17 (h) Recall Exercise 2.13(f).
4.18 Split into cases of different sign combinations of a and b.
4.19 (a) Show that any real number smaller than $\sup(I)$ is in I.
4.20 (b) Pick any two $x, y \in I \cup J$ such that $x < y$. Let $z \in I \cap J$. Then, there are three possible locations of z relative to x and y.
4.21 (b) Use De Morgan's laws.
4.24 (c) Show that for any $x \in A$, $x \in B_{n_x}$ for exactly one $n_x \in \mathbb{N}$. Define a suitable injection from A to $\mathbb{N}^2$ using this fact. Recall also Exercise 3.5.
4.25 Use Exercise 3.7 or Exercise 4.24.
4.26 (a) Use induction. Alternatively, for a fixed $n \in \mathbb{N}$ and every $m \in \mathbb{N}$, consider the sets $B_m^n = \{X \subseteq \mathbb{N} : |X| = n, \sum_{x \in X} x = m\}$. Show that B_m^n are all finite, $B_m^n \cap B_k^n = \varnothing$ for $m \neq k$, and $A_n = \bigcup_{m=1}^{\infty} B_m^n$.
(b) Use Exercise 4.24.
4.27 (c) Use Cantor-Bernstein-Schröder theorem.
4.28 (a) Use Exercise 4.27(b) and (c).
(b) Consider the decimal representation of a number $x \in [0, 1)$ and map it to an appropriate subset of $\mathbb{N}$.
(c) For any non-empty subset $X \subseteq \mathbb{N}$, construct a suitable corresponding decimal representation consisting of the digits 0s and 1s only. Do not forget to find the suitable images of $\varnothing, \mathbb{N} \in \mathcal{P}(\mathbb{N})$ as well.
(d) Use Cantor-Bernstein-Schröder theorem.
4.30 (b) For each $n \in \mathbb{N}$, show that the set of polynomials of degree n with integer coefficients are countable. Then use Lemma 3.4.11.
4.31 (a) Clear out the denominators by algebra and use divisibility arguments.
4.32 (a) See Fig. 4.11 to visualise the process of getting C_1 from C_0 and getting C_2 from C_1.
(c) Use Exercise 4.22.
(d) Use part (a). Similar to decimal representations, we can identify 0.1 with 0.0222... in base-3.
(i) Exercise 4.28 is useful here.
(j) Suppose for contradiction that there are $a, b \in C$ with $a < b$ and $[a, b] \subseteq C$. Write $a \sim 0.a_1a_2\ldots$ and $b \sim 0.b_1b_2\ldots$ in ternary representation and derive a contradiction.

Chapter 5: Real Sequences

5.5 Prove this via contradiction.
5.6 (b) Use Example 5.4.4.

5.7 For a fixed $\varepsilon > 0$, find a suitable N that would work for the whole sequence using the indices from the subsequences.

5.8 (a) Use the triangle inequality.

5.10 Use monotone sequence theorem.

5.14 (a) Prove by induction and the AM-GM inequality.
(b) Use part (a). Consider the cases $b_1 = 0$ and $b_1 > 0$.

5.15 (a) Since $\frac{1}{r} > 1$, write $\frac{1}{r} = 1 + x$ for some $x > 0$. Then apply Bernoulli's inequality.
(b) Use Exercise 5.3 or Lemma 5.9.3.
(c) Use part (a) and Proposition 5.9.5.

5.16 (a) Show that $\exists k \in (0, 1), \exists N \in \mathbb{N} : \forall n \geq N, |a_{n+1}| < k|a_n|$. Then Exercise 5.15 might be useful.
Use part (a) for the rest of the question.

5.17 (a) Clearly true for $r = 1$. For $r > 1$, write $r = 1 + x$ for some $x > 0$. Then apply Bernoulli's inequality and sandwiching.
(c) Show that $\exists N \in \mathbb{N}, \exists \alpha, \beta > 0 : \forall n \geq N, \alpha \leq a_n \leq \beta$. Then use sandwiching along with parts (a) and (b).
(d) Use sandwiching and parts (a) and (b).

5.18 (a) Use Exercise 5.17(c) and the AOL.
(b) To find the limit of $1 - \frac{n}{2^n}$, recall Exercise 2.13(f).

5.19 (a) Split into two cases, namely $x \geq 0$ and $x < 0$.
(b) Fix $\varepsilon > 0$. Show that $\exists N \in \mathbb{N} : \forall n \geq N, |a_n - a| < \log_r(\frac{\varepsilon}{r^a} + 1)$. Then use part (a).

5.20 (c) Show that $0 < a_n \leq e \leq b_n$ for all $n \in \mathbb{N}$.
(d) Use part (c).
(e) Use Exercise 5.17(c).
(f) Use Bernoulli's inequality.

5.21 Assume for contradiction that $a_n \to \infty$.

5.22 (b) Show that there is an $N \in \mathbb{N}$ such that $a_{n+1} - a_n < 0$ for all $n \geq N$.
(d) Use the fact that $2 < e$ and part (c).
(f) Use part (e) and Proposition 5.5.4.

5.24 Use sandwiching and Exercise 5.23.

5.28 (a) First bound $|a_j - a_{j-1}|$ for any $j \geq 3$ in terms of r and $|a_2 - a_1|$. Next, use the triangle inequality repeatedly on $|a_n - a_m|$ to bound it in terms of r and $|a_2 - a_1|$ and r. Use Exercise 3.13.
(b) Use Exercise 5.15.

5.29 Assume for contradiction that (a_n) converges. Then, the sequence (a_n) is Cauchy and, in particular, we have $|a_{n+2} - a_n| \to 0$. Derive a contradiction using the product-to-sum rule, angle summation formula, and the Pythagorean trigonometric identity.

5.31 (a) Denote $r = \frac{1+L}{2} < 1$. Using the definition of limit superior, show that $\exists N \in \mathbb{N} : \forall n \geq N, a_{n+1} < ra_n$. Then bound all a_n for $n \geq N$ in terms of r and a_N and use Exercise 5.15.

5.33 (a) To show the former, let $A = \{a \in \mathbb{R} : a_n > a \text{ for infinitely many } n\}$ be the set in Proposition 5.10.5. Show that A is bounded from above and closed downwards. Then Exercise 4.19 might be useful here.
(b) Prove each inequality via contradiction and using part (a).

5.34 (a) Since the sequence (b_n) converges, $\limsup_{n\to\infty} b_n = \liminf_{n\to\infty} b_n = b$. Prove the identity using subadditivity of limit superior in Lemma 5.10.3.
(c) There is a tail of the sequence for which b_n are all positive. WLOG, we can assume all of b_n are positive. Use Lemma 5.10.4.

Chapter 6: Some Applications of Real Sequences

6.1 (b) Each of the interval in part (a) has length $\frac{4\pi}{6} > 1$, so each of them must contain at least one integer.
(c) Use sandwiching in part (b) and use Proposition 5.10.9.

6.2 (a) Use the rational root theorem from Exercise 4.31.
(d) Use part (c).
(e) The terms in the sum is the first n terms of geometric sequence with first term φ^{n+1} and common ratio $\frac{\psi}{\varphi}$.

6.3 (b) For the inequality, use part (a).
(d) For $n > m$, use triangle inequality repeatedly on $|r_n - r_m|$ and apply part (c).
(e) Recall Exercise 6.2(b).

6.4 (b) Similar to Exercise 5.28(a).
(c) Use induction.

6.5 (d) Use the HM-AM inequality from Exercise 3.28 to get the first inequality.
(e) Take the limit as $n \to \infty$ in part (d).

6.6 (b) Write $x_n^2 - 2$ in terms of x_{n-1}.
(c) Use part (b).

6.9 (a) One can either construct this infinite set of points or prove the statement via contradiction.
(b) Use part (a).

6.10 Suppose for contradiction that there are countably many such sequences. List them all down and construct a new sequence in $X \setminus \{x_0\}$ converging to the point x_0 which is not in the list using a "modified" Cantor diagonal argument and Exercise 6.9(a).

6.11 For each $x \in X \setminus X'$, find a pair of rational numbers p_x and q_x such that $p_x < x < q_x$ and $[p_x, q_x] \subseteq X^c$. Using this, define an appropriate injective function $f : X \setminus X' \to \mathbb{Q}^2$.

6.12 (b) Use Lemma 6.2.6 and part (a).
(c) Prove this via double inclusion. For the $\subseteq$ inclusion, prove that if $x \notin A' \cup B'$, then $x \notin (A \cup B)'$.
(d) Use parts (a) and (c) and Lemma 6.2.6.

6.13 (a) Prove the forward implication via contradiction. For the converse, Bolzano-Weierstrass theorem might be useful.

6.14 (a) Use the fact that $|z_n| = (a_n^2 + b_n^2)^{\frac{1}{2}}$.

6.18 (c) Show that $\exists N \in \mathbb{N}$ such that $\forall n \geq N$ we have $|a_n - L| < \frac{1}{2}$. Part (b) might be useful.

6.21 Use Exercise 6.21(a) for parts (a) and (b).

6.22 (a) Choose a very small $0 < \varepsilon < 1$ in the definition of Cauchy sequence.

6.23 To show that $\mathbb{R}^k$ is complete, let $(\mathbf{x}_n)$ be any Cauchy sequence in $\mathbb{R}^k$. Write the term $\mathbf{x}_n = (x_{n,1}, x_{n,2}, \ldots, x_{n,k})$ for $x_{n,j} \in \mathbb{R}$. Show that the sequences $(x_{n,j})$ for $j = 1, 2, \ldots, k$ are all Cauchy sequences in $\mathbb{R}$ and hence are all convergent.

6.25 (a) Let $\{I_n\}$ be a sequence of nested closed intervals $I_n = [a_n, b_n]$ for $a_n \leq b_n$. Show that (a_n) and (b_n) converge to the same point.
(b) Let $X \subseteq \mathbb{R}$ be a bounded subset with an upper bound M. Pick a point $a_1 \in X$. If X is an upper bound of X and so is its supremum, then we are done. Otherwise, let $b_1 = M$ and recursively choose c_1 to be the midpoint of the interval $[a_1, b_1]$. If $c_1 \in X$, then define $a_2 = c_1$ and $b_2 = M$. Otherwise, define $a_2 = a_1$ and $b_2 = c_1$. Repeat the procedure to create the points a_j and b_j (so the a_j are always in X and the b_j are always outside of X) using the midpoint c_{j-1} until either we get an $a_k \in X$ as an upper bound (and hence supremum) of X or we get infinite sequences of points (a_n) and (b_n). For the latter case, use the nested interval property to deduce that X has a supremum.

6.26 (a) Similar to the proof of Lemma 5.8.2.
(b) To show (d_n) is Cauchy, remember that (a_n) and (b_n) are Cauchy and so are bounded.
(f) Use the fact that (a_n) does not converge to 0 in $\mathbb{Q}$ (warning: it might not even converge in $\mathbb{Q}$) and is a Cauchy sequence in $\mathbb{Q}$.
(g) Use the notation and results from part (f).
(l) Showing (u_n) is decreasing is straightforward. To show it is Cauchy, show that $u_n - l_n = \frac{u_{n-1} - l_{n-1}}{2}$ for all $n \in \mathbb{N}$ and use this to estimate the value of $|u_{n+1} - u_n|$ in terms of the length $|u - l|$. Now apply triangle inequality repeatedly on $|u_n - u_m|$ for $n > m$. Similar approach for (l_n).
(m) Use the estimate $|u_n - l_n| \leq \frac{1}{2^{n-1}}|u - l|$ from part (l) and sandwiching.
(n) Prove this via contradiction.
(o) Prove this via contradiction. Note that, by construction, $[(l_n)]$ are not upper bounds of the set S for any $n \in \mathbb{N}$.

6.27 (a) Use the Archimedean property and unboundedness of the rational numbers.
(c) Find the value $x_{n+1} - x_n$ explicitly to show that (x_n) is increasing. Likewise for (y_n).
(d) Use parts (b) and (c).
(f) Note that $x_n, y_n \to b$. Now use the recursive relation and the AOL to get $b^p = a$.
(g) Similar to Exercise 3.17(d).

6.28 (a) See Exercise 5.19(a).

(b) Similar idea to Exercise 5.19(b).
(c) By the AOL, this is equivalent to showing $\lim_{n\to\infty} \frac{a^{x_n}}{a^{y_n}} = \lim_{n\to\infty} a^{x_n - y_n} = 1$. Use part (a) to show this.
(d) For $0 < a < 1$, since $b = \frac{1}{a} > 1$, we apply part (b) to b and then use the AOL.

6.29 (a) For $a \in \bar{\mathbb{Q}}$ with $a > 1$, there exists a rational sequence (a_n) such that $a_n \to a$. WLOG, assume $a_n > a$ for all n. Use Exercises 4.7(c) and 6.29.

Chapter 7: Real Series

7.1 Find the explicit form of a_n. Then use Exercise 5.19.
7.2 Write the partial sums of the series $\sum_{j=1}^{\infty} b_j$ in terms of the partial sums of the series $\sum_{j=1}^{\infty} a_j$.
7.4 Use the ε-N definition to prove the convergence $c_n \to a$. Note that $|c_n - a| = \left|\frac{1}{n}\sum_{j=1}^{n}(a_n - a)\right|$.
7.5 Write the repeating string in the decimal representation as a geometric series.
7.6 Use the AOL.
7.7 (d) Use Cauchy-Schwarz inequality from Exercise 3.25.
7.8 (a) Use the fact that the sequence of partial sums of $\sum_{j=1}^{\infty} a_j$ is Cauchy.
7.12 (a) Split the partial sum S_{2n} into sums of terms with odd and even indices. Bound the sum of terms with odd indices with the sum of terms with even indices.
7.13 (a) Consider the partial sums of $\sum_{j=1}^{n} a_n b_n$.
7.14 Consider the partial sums.
7.15 The AM-GM inequality in Exercise 3.28 might help.
7.16 Recall the algebraic expressions for $\min\{p, q\}$ and $\max\{p, q\}$ for $p, q \in \mathbb{R}$ in Exercise 5.11. Use comparison test.
7.17 Start with two sequences of 1s and interlace with terms from a convergent sequence.
7.18 Exercise 7.12 might be useful. Split into three cases, namely: $p > q > 1$, $0 < q < p \leq 1$, and $q \leq 1 < p$.
7.19 Use Raabe's test.
7.20 Consider a suitable subsequence of the partial sums for the absolute series.
7.21 (a) Show that the even-indexed and odd-indexed subsequences of (na_n) both converge to 0 using Exercise 7.8. Then combine the limits using Exercise 5.7.
(b) Use part (a).
7.22 Bound the terms of one sequence with the terms in another using the given limits and use direct comparison test.
7.23 Write $s_n = \sum_{j=1}^{n} ja_j$ and $t_n = \sum_{j=1}^{n} ja_{j+1}$. Prove that $t_n = \frac{n}{n+1}s_n - \frac{1}{2}s_1 - \sum_{j=2}^{n} \frac{1}{j(j+1)} s_j$. Show that the RHS converges. Exercise 7.13(a) might be helpful.
7.24 (a) Prove this via contradiction.

(b) If $\limsup_{j\to\infty} |\frac{a_{j+1}}{a_j}| = \infty$, we are done. Otherwise, pick any $r \geq \limsup_{j\to\infty} |\frac{a_{j+1}}{a_j}|$. Use part (a) to get the result.
(c) Use part (b) and apply generalised root test.

7.25 Put $a_n = n$ in the inequalities from Exercise 7.24.

7.26 Show that $|a_{n+1}| \geq \frac{(N-1)|a_N|}{n}$ for all $n \geq N$.

7.27 (a) Utilise Raabe's test.
(b) Show that there exists an $N_1 \in \mathbb{N}$ such that $\forall n \geq N_1$, $|\frac{a_n}{a_{n+1}}| > 1 + \frac{L+1}{2}$. Proceed as the proof of Theorem 7.7.3.
(c) Show that there exists an $N_1 \in \mathbb{N}$ such that $\forall n \geq N_1$, $|\frac{a_n}{a_{n+1}}| < \frac{1+n}{n}$. Proceed as the proof of Theorem 7.7.3.

7.29 For the sum of fourth powers, consider the binomial expansion for $(x+1)^5$ and the telescoping sum $\sum_{j=1}^{n}((j+1)^5 - j^5)$.

7.30 Use summation by parts.

7.31 Show that $\frac{|\cos(j)|}{j} \geq \frac{\cos(2j)}{2j} + \frac{1}{2j}$ for all $j \in \mathbb{N}$. Then recall Example 7.8.6.

7.32 Prove the forward implication via contrapositive. To show the converse, let $t_n = \sum_{j=1}^{n} a_j$ be the n-th partial sum. Use Exercise 5.21 with the subsequence (t_{2^n}).

7.35 (a) Use Exercise 5.20 and sandwiching.
(b) Use Proposition 5.7.5 and part (a).
(c) For convergence, use part (b). For divergence, use part (b) and Exercise 7.33(b).

Chapter 8: Additional Topics in Real Series

8.3 (a) Show that the partial sums of the series $\sum_{j=1}^{\infty} |c_j|$ are bounded.

8.4 To prove that the series diverges to ∞, we carry out the same arrangement argument as in the proof for Theorem 8.1.5 but at the n-th iteration, we add positive terms until we exceed $n+1$ and then add negative terms before it reaches n. Repeat this construction so that the partial sums of the rearranged sum diverge to ∞. Similar argument works for finding a rearrangement that diverges to $-\infty$.

8.5 (f) Show that $u_{3n} = s_{4n} + \frac{s_{2n}}{2}$, $u_{3n+1} = u_{3n} + \frac{1}{4n+1}$, and $u_{3n+2} = u_{3n+1} + \frac{1}{4n+3}$.

8.6 (b) If $x_1 \neq 0$, then show that $0 < x_1 \leq 1$. Use Archimedean property on the rationals and the well-ordering principle.
(c) To show the inequality, use part (b).
(e) Backwards substitution.

8.7 (a) Use Exercise 7.8(b).
(b) For any $m = yz^2 \in M(x)$, explain why there are 2^{k-1} possibilities for the value of y and $\sqrt{x}$ possibilities for the value of z.
(e) Use parts (a), (c), and (d).
(f) Use parts (b) and (e).

8.8 (a) First, count the number of natural numbers with m digits and no 9 in its decimal representation.
(c) For a fixed $j \in \mathbb{N}$ show that $c_j \leq 8(0.9)^{j-1}$.
(d) Use Exercise 5.21.

8.9 (a) Use the third assumption and the fact that limits preserve weak inequalities.
(b) Show that for a fixed $n \in \mathbb{N}$, for all $j \in \mathbb{N}$ we have $|a_{j,n} - a_j| \leq 2M_j$.
(c) Use Exercise 7.8(b).
(d) Simply follow the inequalities and use parts (b) and (c).
(e) Use the second assumption in the theorem.

8.10 Fix $x \in \mathbb{R}$. $(1 + \frac{x}{n})^n = \sum_{j=0}^{n} \binom{n}{j} \frac{x^j}{n^j}$. Choose $a_{j,n} = \binom{n}{j} \frac{x^j}{n^j}$ if $0 \leq j \leq n$ and $a_{j,n} = 0$ if $j > n$. Check that this double sequence satisfies the conditions of Tannery's theorem.

8.11 (b) $\sum_{m=1}^{\infty} \sum_{n=1}^{\infty} b_{m,n} = \sum_{m=1}^{\infty} (\lim_{q\to\infty} B_{m,q})$ where $B_{m,q} = \sum_{n=1}^{q} b_{m,n}$. Show that the double sequence $(B_{m,q})$ satisfies the conditions in Tannery's theorem. Then use Tannery's theorem.

8.12 (a) Recall the construction in Sect. 6.1.
(c) Use part (b).
(d) The first part should be $\sin(mx) = \sum_{j=0,\, j \text{ odd}}^{m} \binom{m}{j} i^{j-1} \sin^j(x) \cos(x)^{m-j}$ where i is the imaginary unit.
(e) For each $j = 1, 2, \ldots, n$, put in $x = \frac{j\pi}{2n+1}$ in the identity in part (d).
(f) Explain why the roots of P are $y = \cot^2(\frac{j\pi}{2n+1})$ for $j = 1, 2, \ldots, n$ and are all distinct.
(g) From part (f), we can write $P(y) = a(y - a_1)\ldots(y - a_n)$ where a is the leading coefficient and a_j are all the roots of P. Multiply out to determine the coefficient of y^{n-1}.
(h) Put the identities from part (g) in part (c).

8.13 (c) Use part (b).
(d) Show that $f(X) - f(Y) = \sum_{n\in X\setminus Y} \frac{1}{3^n} - \sum_{n\in Y\setminus X} \frac{1}{3^n}$. Using part (c), find suitable lower and upper bounds of these sums respectively.
(e) Use Theorem 3.9.6 from Exercise 3.8.

Chapter 9: Functions and Limits

9.1 Use Lemma 4.1.13.

9.2 (e) Bernoulli's inequality might help here.

9.4 (a) For the inductive step when $n = k + 1$, set $y = \sum_{j=1}^{k} \frac{t_j}{1-t_{k+1}} a_j$ so that $\sum_{j=1}^{k+1} t_j a_j = (1 - t_{k+1})y + t_{k+1} a_{k+1}$. Then use the convexity of f. Notice that $\sum_{j=1}^{k} \frac{t_j}{1-t_{k+1}} = 1$. Use this in the inductive step.
(b) Recall from Exercise 9.2(e) that the exponential function $f(x) = e^x$ on $\mathbb{R}$ is a strictly increasing convex function. By Exercise 9.3(b), its inverse $\ln(x)$ is then strictly increasing concave function. Use this fact in part (a).

(c) Use the fact that $ab = e^{\ln(ab)}$ and the logarithm is a concave function.
(d) Write $A = \sqrt[p]{|a_1|^p + \ldots + |a_n|^p}$, $B = \sqrt[q]{|b_1|^q + \ldots + |b_n|^q}$, $c_j = \frac{a_j}{A}$, and $d_j = \frac{b_j}{B}$ for all $j = 1, 2, \ldots, n$. Then apply triangle and Young's inequalities.

9.7 (a) Pick two sequences of points converging to 0, one with rational terms and one with irrational terms.
(b) At $x_0 = 0$, use the ε-δ definition. For $x_0 \neq 0$, use a similar argument from part (a).

9.9 Use the fact that $|\sin(x)| \leq |x|$ for any $x \in (0, \frac{\pi}{2})$ which was proven in Exercise 8.12(a).

9.10 (a) Pick any sequence (x_n) with $x_n \neq x_0$ such that $x_n \to x_0$. WLOG, suppose that (x_n) are all contained in the punctured ball $B_\delta(x_0) \setminus \{x_0\}$. Use the fact that limits preserve weak inequalities.

9.11 If $L < M$, use $\varepsilon = \frac{M-L}{2} > 0$ in the definition of limits for f and g.

9.12 (a) Use triangle inequality.
(b) Show this using the sequence definition of limits. Pick any two arbitrary sequences (x_n) and (y_n) in $X \setminus \{x_0\}$ such that both of them converge to x_0. Show that the image sequences $(f(x_n))$ and $(f(y_n))$ both converge and their limits are the same.

9.14 (c) Consider the cases of even and odd degree polynomials separately.

9.15 (c) Show that $f(x) = \frac{2}{\sqrt{1+\frac{1}{\sqrt{x}}}+\sqrt{1-\frac{1}{\sqrt{x}}}}$.
(d) For maximum and supremum, show that $f(x)^2 \leq 2$ on X.

9.16 Recall the algebraic expressions for $\min\{p, q\}$ and $\max\{p, q\}$ for $p, q \in \mathbb{R}$ in Exercise 5.11.

9.18 Find the one-sided limits.

9.23 (a) Prove this via contradiction.
(b) Use the AM-GM inequality from Exercise 3.27(b) with some copies of 1 and some copies of $\sqrt{x}$.
(c) Use parts (a) and (b).

9.24 Follow the steps in Exercise 9.23. Use the AM-GM inequality to get $\frac{(n-m)+mx^{\frac{k}{2}}}{n} \geq x^{\frac{mk}{2n}}$ where $m, n \in \mathbb{N}$ are chosen appropriately.

9.25 Prove this via contradiction.

9.26 (a) The set $\{f(x) : x \in [a, \infty)\}$ has a supremum. Prove that the limit $\lim_{x\to\infty} f(x)$ is this supremum.
(b) Prove that the set $\{f(x) : x \in [a, \infty)\}$ is bounded from above by the limit $\lim_{n\to\infty} f(x_n)$. Then use part (a).

9.27 (b) Recall Exercise 9.7(b).

9.29 Use the characterisation of supremum and infimum.

9.30 (b) Use Proposition 4.1.10 to show that S is increasing and then apply Exercise 9.29(a).
(c) Use Proposition 4.1.10 and Exercise 9.10.
(d) For the right-most inequality, show that $\forall h > 0, \exists N \in \mathbb{N} : \forall n \geq N, f(x_n) \leq S(h)$.

(e) Using the characterisation of supremum in the definition of S, construct a sequence (x_n) such that for each $n \in \mathbb{N}$, we have $S(\frac{1}{n}) - \frac{1}{n} < f(x_n) \leq S(\frac{1}{n})$.
(f) Use parts (d) and (e).
(g) Use parts (d) and (e).

9.31 For $K \geq a$, write $S(K) = \sup\{f(x) : x \in [K, \infty)\}$, $T(K) = \inf\{f(x) : x \in [K, \infty)\}$, $S = \limsup_{x\to\infty} f(x)$, and $T = \liminf_{x\to\infty} f(x)$.
(a) Show that $S(K)$ is decreasing and bounded from below to deduce that S exists. Next, note that $\forall K \geq a$, $T(K) \leq S(K)$. Use Exercise 9.10 for limit at ∞.
(b) Use ε-K definition for limits at ∞.

Chapter 10: Continuity

10.3 Prove the forward implication by constructing a sequence (x_n) with $x_n \to x_0$ and $f(x_n) \to f(x_0)$ using the ε-δ definition. Prove the converse via contrapositive.

10.4 Recall the algebraic expressions for $\min\{p, q\}$ and $\max\{p, q\}$ for $p, q \in \mathbb{R}$ in Exercise 5.11.

10.5 To show $\tilde{f}$ is continuous at a, use Definition 10.1.4.

10.6 To show f is not continuous at $x = 0$, pick a suitable sequence of points (x_n) with $x_n \to 0$ but $f(x_n) \not\to f(0) = 0$.

10.7 (a) Use the ε-δ definition of continuity.
(b) Use the sequence definition to disprove continuity.

10.8 (a) Break into two cases, namely: $x \in \mathbb{Q}$ and $x \in \bar{\mathbb{Q}}$. For the former case, prove first that if $\gcd(p, q) = 1$, then $\gcd(p + q, q) = 1$ as well.
(b) For any $m \in \mathbb{N}$, let $R_m = \{\frac{p}{m} \in [0, 1] \cap \mathbb{Q} : \gcd(p, m) = 1\}$ and so $Q_n = \bigcup_{m=1}^{n} R_m$.
(d) Use the ε-δ definition and Archimedean property. Use part (c).
(f) Fix $x_0 = \frac{p}{q}$ so that $f(x_0) = \frac{1}{q}$. Show that the set $\{|x_0 - r| : r \in Q_{q-1}\}$ has a positive minimum. Hence, find a neighbourhood of x_0 which does not contain any elements from Q_{q-1}. Show that $f(x_0) \geq f(x)$ for every x in this neighbourhood.

10.9 Use the IVT.

10.10 (a) Prove this via contradiction and the IVT.
(b) Prove this via contradiction. Part (a) is useful.
(c) Prove this via contradiction. Part (a) is useful.
(d) Show f is not continuous at $x = 1$ by picking a suitable sequence (x_n) with $x_n \to 1$.
(e) Define $g : Y \to X$ as the inverse of the function f in part (d). Show it is not continuous at $x = \frac{3}{4}$.

10.11 (a) Recall Exercise 9.29.

(b) By using part (a), eliminate the possibility of essential and removable discontinuity.
(c) Construct an injective mapping from the set of jump points to the rational numbers.

10.12 (a) Define $g : [0, \frac{1}{2}] \to \mathbb{R}$ as $g(x) = f(x) - f(x + \frac{1}{2})$. Then use the IVT.
(b) Define $g : [0, 1 - \frac{1}{n}] \to \mathbb{R}$ as $g(x) = f(x) - f(x + \frac{1}{n})$. Show that $g(0) + g(\frac{1}{n}) + g(\frac{2}{n}) + \ldots + g(\frac{n-1}{n}) = 0$. Then use the IVT.

10.13 (a) Use the ε-δ definition. Split into cases to check continuity at $x_0 = 0$ and $x_0 > 0$.
(b) Use part (a) and an overlapping argument.

10.16 For an open set $U \subseteq \mathbb{R}$, consider two cases, namely: $f^{-1}(U) = \varnothing$ and $f^{-1}(U) \neq \varnothing$.

10.17 Use induction and the IVT.

10.18 Use the IVT on the function $f : \mathbb{R} \setminus \{m\pi + \frac{\pi}{2} : m \in \mathbb{Z}\} \to \mathbb{R}$ defined as $f(x) = \tan(x) - kx$.

10.19 (a) Consider a function $g : \mathbb{R} \to \mathbb{R}$ defined as $g(x) = f(x) - f(0)$. Show that $g(x) > 1$ over some set $(-\infty, p) \cup (q, \infty)$ and use the EVT on the complement of this set.
(d) WLOG, consider $K = 0$. If f is not identically 0, split into several cases: $f \geq 0$, $f \leq 0$, and f has mixed signs. Use the EVT to study these cases. For the mixed sign, split the function f into $f = f^+ + f^-$ where $f^+ = \max(f, 0) \geq 0$ and $f^- = \min(f, 0) \leq 0$.

10.20 Exercise 10.10 might be useful.

10.22 (a) Use Proposition 10.6.14.
(b) Use Proposition 10.2.6 to extend the function continuously to 0. Then apply an overlapping argument and part (a).

10.23 (a) Consider the collection of closed intervals $A = \{I_n = [nP, (n+2)P] : n \in \mathbb{Z}\}$ each of length $2P$ which covers the whole of $\mathbb{R}$. Use an overlapping argument and note that the function is periodic.
(b) Use Proposition 10.6.14 and overlapping argument.

10.24 (a) For the equality case, use the characterisation of infimum.

10.25 (a) Use the factorisation in Exercise 3.17(c) and triangle inequality.
(b) Prove this via contradiction.

10.26 (a) Use Exercise 9.14(b)(c) and Exercise 10.19(a).
(c) One direction is done in part (b). For the other implication, use Exercise 9.14(b)(c) and the IVT.
(e) Prove the contrapositive by using Exercise 9.14.

10.27 (a) Apply the IVT on the interval $[-c, c]$.

10.28 (a) Write $x = \frac{a}{a+b}$ and $y = \frac{b}{a+b}$. Show that $x^p + y^p \geq 1$.
(b) Use part (a).
(c) For the backwards implication, use part (b). For the forward implication, suppose for contradiction that g is α-Hölder for $\alpha > \beta$.

(d) For $x < y$ in $\mathbb{R}$, split the interval $[x, y]$ into n subintervals of equal length and apply the α-Hölder condition on each subinterval. Use triangle inequality on $|h(y) - h(x)|$ and take the limit as n goes to ∞.

10.29 (d) For any $y \in \mathbb{R}$, we have $-m\delta < y < m\delta \Leftrightarrow -\delta < \frac{y}{m} < \delta$ for some $m \in \mathbb{N}$ and δ is as in part (a).

(e) Use the fact that f is continuous at 0 and the identity $f(x + y) = f(x)f(y)$.

(f) Show this for $r = \frac{1}{m}$ where $m \in \mathbb{N}$ first. Then use part (c).

(g) Use part (f).

(h) Use part (e) and Exercise 5.19.

10.30 (a) Prove each inequality separately. Write $v = tu + (1 - t)w$ for some $t \in [0, 1]$.

(b) Use the first inequality in part (a) with $u = c$ and $w = d$.

(c) Write any arbitrary point $x \in [c, d]$ as $x = \frac{c+d}{2} + y$ for some $y \in \left[\frac{c-d}{2}, \frac{d-c}{2}\right]$ and use the upper bound of f from part (b).

(d) Pick any $c', d' \in (a, b)$ with $a < c' < c$ and $d < d' < b$ so that $[c, d] \subsetneq [c', d'] \subseteq (a, b)$. Let $x, y \in [c, d]$ and WLOG $y > x$. Look at the set of points $\{c', c, x, y\}$ and $\{x, y, d, d'\}$. Use the inequalities in part (a) to bound $\frac{f(y)-f(x)}{y-x}$ from above and below with the terms $f(c), f(c'), f(d), f(d'), c, c', d, d'$. Using the fact that f is bounded over $[c', d']$ from parts (b) and (c), deduce Lipschitz continuity.

10.31 (a) Use Exercises 9.2(e) and 10.30(e).

(b) Use Theorem 10.5.4.

(c) Use Exercise 10.30(a) with suitable choices of u, v, w.

(d) Fix $x < 1$. Split into four cases, namely: $x = 0$, $x \leq -1$, $x \in (0, 1)$, and $x \in (-1, 0)$. For $x \in (0, 1)$, use Exercise 10.30(a) with $u = 0$, $v = \frac{1}{n}$ for integers $n \geq \frac{1}{x}$, and $w = x$. Recall also from Example 5.4.4 that e is the limit of the increasing real sequence (a_n) where $a_n = (1 + \frac{1}{n})^n$. For $x \in (-1, 0)$, a similar argument may be employed by recalling Exercise 5.20.

(f) Use part (e) to show the Lipschitz continuity on $[1, \infty)$.

(g) Recall from Exercise 9.24 that $\lim_{x\to\infty} x^{\frac{1}{x^k}} = 1$ and the continuity of the logarithm function. Exercise 10.15 might be helpful.

10.32 (b) Let $n > m$. Use triangle inequality repeatedly on $|x_n - x_m|$ and apply part (a).

(c) Use the recursive relation and continuity of f.

(e) Use parts (c) and (d).

Chapter 11: Functions Sequence and Series

11.2 Split into two cases, namely: $x \in [0, 1)$ and $x = 1$.

11.3 (a) Use Theorem 11.3.4.

(b) Use Proposition 11.2.5.

(c) Use Theorem 11.3.4.

11.4 Use Theorem 5.4.2.

11.6 (a) Use the AM-GM inequality.

(b) To show uniform convergence, fix $\varepsilon > 0$. We split into two cases, namely: $|x| < \varepsilon^2$ and $|x| \geq \varepsilon^2$. For the former, use the AM-GM inequality and the estimate from Example 10.6.4. Find an N that works for both cases.

11.7 Use Proposition 11.2.5 to determine whether the convergence is uniform.

11.8 (d) Prove this using the ε-N definition.

11.9 (a) Assume for contradiction that f has $K+1$ discontinuities $\{p_1, p_2, \ldots, p_{K+1}\}$. Show that for every $j = 1, 2, \ldots, K+1$ there exists $\varepsilon > 0$ so that $\forall \delta > 0, \exists x_j \in X : |x_j - p_j| < \delta$ and $|f(x_j) - f(p_j)| \geq \varepsilon$.

Next, using the fact that $f_n \xrightarrow{u} f$, show that there exists an $N \in \mathbb{N}$ such that for $j = 1, 2, \ldots, K+1$, we have $|f_N(x_j) - f(x_j)| < \frac{\varepsilon}{4}$ and $|f_N(p_j) - f(p_j)| < \frac{\varepsilon}{4}$. Deduce that f_N is discontinuous at $\{p_1, p_2, \ldots, p_{K+1}\}$ and get a contradiction.

(b) Refer to Exercise 11.7.

11.11 (a) At the inductive step, split the region $[0, 1]$ into three separate regions $[0, \frac{1}{3}]$, $[\frac{1}{3}, \frac{2}{3}]$, and $[\frac{2}{3}, 1]$ to use the recursive formulation.

(b) Use the Cauchy criterion for uniform convergence by using the result in part (a).

(c) Apply Theorem 11.3.4.

(d) Use the IVT.

(e) Prove by induction that f_n is increasing for all $n \in \mathbb{N}$. Then apply Exercise 11.10.

11.12 Prove this using the ε-N definition for uniform convergence of the functions sequence (f_n) as well as the uniform continuity definition of the functions f_n.

11.14 (a) Find the negation of the definition for $f_n \xrightarrow{u} f$ on $[a, b]$ via Proposition 11.2.5.

(b) Use the characterisation of supremum and Bolzano-Weierstrass theorem.

(c) Fix k_n and consider the quantity $f_{k_n}(x_{k_m}) - f(x_{k_m})$ for $m \geq n$. Use the assumption that (f_n) is pointwise decreasing, the estimate in part (b), and the assumption that f and f_n are all continuous to deduce that f_n cannot converge pointwise to f.

11.16 (a) At the inductive step, showing the lower inequality $0 \leq 0f_k(x)$ is straightforward. For the other inequality, show first that $f_{k+1}(x) = \frac{1+x^2-(f_k(x)^2-1)^2}{2}$ and prove that the numerator is smaller than $2|x|$.

(b) Use part (a) and the recursive relation to deduce that the function is pointwise increasing. Use Theorem 5.4.2 and the recursive relation to find the limiting function.

(c) Use Dini's theorem.

11.17 Use root test and Proposition 11.3.1.

11.18 (a) Multiply the partial sum with $2\sin(\frac{1}{2})$ and use the product-to-sum formula.

(b) Using angle addition formula, write the partial sum of the series in terms of the partial sums of the series $\sum_{j=1}^{\infty} \frac{\sin(j)}{j}$ and $\sum_{j=1}^{\infty} \frac{\cos(j)}{j}$. These two series converge, so they satisfy the Cauchy criterion.

11.19 (a) Use the inequality derived in Exercise 10.31(c)(d).
(b) Use the estimate in part (a) and Weierstrass M-test.

11.21 (b) Use geometric series and part (a). Proposition 11.3.1 might help.

11.22 (a) Multiply t_n with $2\cos(\frac{x}{2})$ and use the product-to-sum formula.
(b) Use part (a) and Dirichlet's test for real series.
(c) Note that $(-1)^{j+1} = -\cos(j\pi)$. Show that for any $n \in \mathbb{N}$ the supremum of the quantity $|s_{2n}(x) - s_n(x)|$ over $[-\pi, \pi]$ is bounded away from 0 by a positive constant independent of n by evaluating it at $x = \pi - \frac{\pi}{4n}$.

11.23 Denote (t_n) as the sequence of partial sums of the series $\sum_{j=1}^{\infty} f_j g_j$. Prove pointwise convergence using the Cauchy criterion and summation by parts. Uniform convergence can then be proven using the final condition.

11.25 Check the cases $x = \pm 1$, $x = 0$, $|x| > 1$, and $0 < |x| < 1$.

11.27 (b) Show that the series cannot converge uniformly over $(0, \infty)$ using contradiction and Cauchy criterion.

11.28 Use the alternating series test and Dirichlet's test for uniform convergence.

11.29 (c) Recall Cantor's diagonal argument from Sect. 4.4.

11.30 (b) The set I is compact.
(d) Use parts (b) and (c).

11.31 (e) Use part (d) and complete the square.
(f) Use definition of $B_n^f(x)$ and part (a).
(g) i. Use the uniform continuity estimate for f and part (b).
ii. Use the uniform bound on the function f, the estimate $\frac{(nx-j)^2}{n^2\delta^2} \geq 1$ for any $j \in I$ (obtained from the definition of the set I), and part (e).

Chapter 12: Power Series

12.1 (g) To investigate the convergence/divergence at the boundary points, recall Exercise 3.14(g).

12.3 The domain of convergence depends on the value of p.

12.4 (b) Use Raabe's test.

12.5 Consider the value of the coefficients for $j \equiv 0, 1, \ldots, 11 \pmod{12}$ and use Cauchy-Hadamard theorem.

12.6 (b) Use partial fractions to decompose the reciprocal function in part (a).

12.8 Prove that if a series is 0 on the punctured disc $B_r(c) \setminus \{c\}$, then the series is also 0 at $x = 0$. Use strong induction to deduce $a_j = b_j$ for all $j \in \mathbb{N}_0$.

12.10 (b) Use the ratio test.
(c) Use Raabe's test (also done in Exercise 7.11(d)).
(d) Use Exercise 7.11(f) and Raabe's test.
(e) Use Exercise 7.11(e).

12.11 (d) Use Mertens' theorem.

(g) Prove this via contradiction. Use part (f). Note that the partial sums are polynomials and recall Exercise 9.14.

12.12 (b) Use Mertens' theorem.

(c) Use part (b).

12.13 (a) Split the sum for $C(x)$ into the sum of terms with odd and even indices.

(b) Split the sum into terms with indices $j \leq 2$ and $j \geq 3$. For the latter, split the sum further into the sum of terms with odd and even indices.

(c) Use the IVT.

(d) Use Exercise 12.12(c).

(e) Use Exercise 12.12(b).

(f) Show first $S(x) = \sum_{j=1}^{\infty} \frac{x^{2j+1}}{(2j+1)!}\left(1 - \frac{x^2}{(2j+2)(2j+3)}\right)$.

(g) Use Exercise 12.11(e) to find $S(\tau)$. Then use Exercise 12.12(b)(c).

12.14 (a) Exercises 12.12(c) and 12.13(f) might help.

(b) Use Exercise 12.12(c).

12.16 (b) Show that the series is $\sum_{j=0}^{\infty} x^j(x^j + 2x^{j-1} + 2^2x^{j-2} + \ldots + 2^j)$. For each $n \in \mathbb{N}_0$, find the coefficient a_n of x^n in this series. Split into cases of n even and n odd. When n is even, the coefficient comes from the terms with indices $j = \frac{n}{2}$ to $j = n$. When n is odd, the coefficient comes from the terms with indices $j = \lceil \frac{n}{2} \rceil$ to $j = n$. Then we can write the series in the standard power series form $\sum_{n=0}^{\infty} a_n x^n$.

12.19 (a) The first inequality is clear. For the other inequality, bound the difference with a geometric series.

12.20 Find a power series expression for f and use Theorem 11.4.17.

12.21 (a) Start with $e^n n!$ and use the power series for e^n.

12.22 (d) Showing sinh is increasing over $x \geq 0$ is straightforward. Use part (a) to show that cosh is increasing over $x \geq 0$. Use part (c) to answer the remaining part of the question.

12.26 This can be written as $\lim_{n\to\infty} \sum_{j=0}^{n} (\frac{n-j}{n})^n$. To use Tannery's theorem, we set $(a_{j,n})$ for $j \in \mathbb{N}_0$ and $n \in \mathbb{N}$ with $a_{j,n} = (\frac{n-j}{n})^n = (1 - \frac{j}{n})^n$ for $0 \leq j \leq n-1$ and $a_{j,n} = 0$ otherwise. Check the conditions in Tannery's theorem.

12.27 (a) Exercise 11.19(a) might help.

(b) Similar argument to part (a).

12.28 (a) Show that there exists an $N \in \mathbb{N}$ such that $\ln|1 - \frac{r}{n}| \leq -\frac{r}{2n}$ for all $n \geq N$. Exercise 11.19(a) might be helpful.

(b) Let (a_n) be the sequence defined as the product $a_n = |1 - \frac{r}{1}| \ldots |1 - \frac{r}{n}|$. Use part (a).

(c) Use part (b).

12.29 (a) Test $x = \frac{1}{4}$ and use Corollary 12.1.3.

Chapter 13: Differentiation

13.2 $f'(x) = \lim_{h \to 0} \frac{f(x+h)-f(x)}{h} = \lim_{h \to 0} \frac{f(x-h)-f(x)}{-h}$.

13.5 Use chain rule.

13.7 Use chain rule.

13.8 Split into three regions, namely: $x \in (-1, 0) \cup (1, \infty)$, $x \in (-\infty, -1) \cup (0, 1)$, and $x \in \{-1, 0, 1\}$.

13.9 Use one sided limits to study the continuity and differentiability.

13.10 (a) Recall Exercise 10.8(e).

(b) Let $x_0 \in \bar{\mathbb{Q}} \cap (0, 1)$. Suppose for contradiction that it is differentiable at x_0. Show that 0 is a candidate for the derivative of f at x_0. Then, by using the ε-δ definition for differentiability with $f'(x_0) = 0$ with $\varepsilon = 1$, show that for any $\delta > 0$ there exists a point $y = \frac{p}{q} \in \mathbb{Q}$ such that $|x_0 - y| < \delta$ but $|\frac{f(x_0)-f(y)}{x_0-y} - 0| \geq 1$. Prove this by looking at the subcases $\delta > 1$ and $0 < \delta \leq 1$.

13.11 (c) Show $E_1^c = C_1$. Then use the recurrence relation in part (b) and induction. The final part is obtained by recalling that $C = \bigcap_{n \in \mathbb{N}_0} C_n$.

(d) Use induction.

(e) Use part (c) to show that the interval must be contained in some C_N^c for $N \in \mathbb{N}$. Then use part (d) to show that it is constant on C_k^c for all $k \geq N$. Show that it is the same constant for all $k \geq N$ by induction.

(f) Note that $x \in I$ for some interval I as in part (e). Show that the limiting function is also a constant on I.

13.15 Prove this by contrapositive.

13.16 Consider the function $h(x) = e^{g(x)} f(x)$ for $x \in [a, b]$.

13.17 Consider the function $g(x) = f(x) - x$ for $x \in [-R, R]$. Show that g cannot be positive and negative anywhere within $(-R, R)$ by using the MVT.

13.18 Use the IVT to show that there is at least one solution. Show that there is only one solution using Exercise 13.15.

13.19 (b) Assume for contradiction that g never vanishes in (c, d). Consider the function $\frac{f}{g} : [c, d] \to \mathbb{R}$. Explain why this function exists and differentiable. Get a contradiction using part (a) and Rolle's theorem.

13.20 Consider the differentiable function $g : [a, b] \to \mathbb{R}$ defined as $g(x) = f(x) - kx$. Show that its global minimum cannot be at a or b.

13.21 Suppose for contradiction that there are more than one fixed point of f. Use the MVT.

13.23 Show that $f'(x) = 0$ for any $x \in \mathbb{R}$.

13.24 The identities in Exercise 3.14(a) might be useful.

13.27 (a) Let $f' = g$ so that $g' = 0$.

(b) Use induction.

13.28 (a) Use Rolle's theorem.

13.29 (a) Write $P(x) = (x - x_0)^m Q(x)$ where $Q(x)$ is a polynomial of degree $n - m$ such that $Q(x_0) \neq 0$.

(b) Prove the forward implication via induction and part (a). For the converse, since P has root x_0, we must have $P(x) = (x - x_0)^k Q(x)$ for some $k \in \{1, 2, \ldots, n\}$ and $Q(x_0) \neq 0$. Determine the value of k.
(c) Consider $P^{(n-2)}$. This is a quadratic equation with how many roots?

13.30 (a) Use the IVT.
(b) For the latter, prove via contradiction and Rolle's theorem.
(c) Use Exercise 13.29(c) and part (a).

13.31 (a) Use induction. In the inductive step, show that Q'_{k+1} has at most $k+1$ distinct zeroes and deduce the maximum possible number of distinct zeroes of Q_{k+1} by Rolle's theorem.

13.32 (a) Let $h_n(x) = (x^2 - 1)^n$ so that $f_n(x) = \frac{d^n}{dx^n} h_n(x)$. Use Exercise 13.29 on h_n.
(b) Use Exercise 13.29(b) and Rolle's theorem.
(c) Use part (b), Exercise 13.29(b), and Rolle's theorem.

Chapter 14: Some Applications of Differentiation

14.1 Break into cases, namely $x > x_0$ and $x < x_0$. Exercise 10.30(a) might be useful.

14.2 Consider the function $h(x) = g(x) - f(x)$ on $[0, \infty)$. For parts (b) and (c), prove the inequalities first for $x \geq 0$ and use the parity of the functions to deduce the equalities for $x < 0$. For part (c), the inequality in part (b) might be useful. For parts (d) and (e), prove the inequalities separately.

14.4 (b) For the forward implication, use Exercise 10.30(a) with the points x_0, x, and a point between them. Take the limit as this point approaches x_0. For the converse, show that for any $x, y \in I$ with $x < y$ we have $(f'(y) - f'(x))(y - x) \geq 0$ so that f' is increasing. Pick three points $x < y < z$ and use the MVT.

14.6 (c) Fix $x_0 \in \mathbb{R}$. Explain why $\exists! s, t \in \mathbb{R}$ such that $x_0 = t - \sin(t)$ and $x_0 + 2\pi = s - \sin(s)$. Show $y(x_0) = y(x_0 + 2\pi)$ using this.
(e) Use part (d).
(g) Use part (f) to get $y'(\sqrt{y(2-y)} - (1-y)\sin(x + \sqrt{y(2-y)})) = \sqrt{y(2-y)}\sin(x + \sqrt{y(2-y)})$. Show that the bracketed terms on the LHS is never 0 for $y \neq 0, 2$ by using part (d).
(h) Prove the first part via contradiction. Use parts (d) and (f) to do this.
(i) From part (g), show that $y'\sqrt{y(2-y)} = \sin(x + \sqrt{y(2-y)})(\sqrt{y(2-y)} + y'(1-y))$. Square both sides and use part (d).
(j) Use the AOL, the fact that $y(x)$ is continuous, and the signs of $y'(x)$ from part (h).
(k) Differentiate the equation in part (i) implicitly.

14.7 (e) Use the sequence constructed in part (d).

14.8 (a) Use the AM-GM inequality.
(b) Recall Theorem 11.3.4.

14.13 Recall that a power series and its derivative converge uniformly over any closed interval.

14.17 Use induction and L'Hôpital's rule.

14.18 (a) At $x = 0$, find the left- and right-derivatives. Carry out a change of variable to utilise Exercise 14.17.
(c) Show that the left- and right- n-th derivatives at $x = 0$ are 0. Use Exercise 14.17 and part (b).

14.19 Write Ψ as a product of two transformed copies of F from Exercise 14.18. Use Exercise 13.26.

14.21 (c) Use L'Hôpital's rule.
(d) Use Proposition 9.5.6.

14.22 (c) Add $+1 - 1$ to the function and use reverse product rule.

14.26 Recall the bump function from Exercise 14.19.

14.28 (a) Differentiate the Wronskian and substitute in the relevant quantities.

14.29 (b) Use Exercise 14.28(a).
(d) Set $y_1(x) = e^{2x}$ in the equation from part (c).

14.30 (b) Derive an ODE for W using Abel's identity and solve it by using an integrating factor.
(d) Solve the ODE for y_2 in part (c) by using an integrating factor.

14.31 Use Abel's identity to get another linearly independent solution if necessary.

14.33 (a) Use sandwiching.
(c) Use the MVT and triangle inequality.
(d) Substitute in the expressions of x_0 and y_m in terms of x_m, α_m, and b and use the angle summation formula. Note that b is a positive odd integer. To show the inequality, use the fact that $x_m \in [-\frac{1}{2}, \frac{1}{2})$.
(e) Use reverse triangle inequality and the assumption that $ab > 1 + \frac{3\pi}{2}$.

14.34 (c) Use parts (a) and (b).

Chapter 15: Riemann and Darboux Integrals

15.2 (a) Use the density of irrational numbers in the real numbers.

15.3 (b) Use part (a) to bound $U_{f,\mathcal{P}}$.

15.4 By Riemann integrability of f over $[a, b]$, $\forall \varepsilon > 0, \exists$ partition $\mathcal{P}$ of $[a, b]$: $U_{f,\mathcal{P}} - L_{f,\mathcal{P}} < \varepsilon$. We can assume that $c, d \in \mathcal{P}$. Now consider $\mathcal{P}' = \mathcal{P} \cap [c, d]$.

15.5 (a) Denote $f(t) = |t|$. WLOG, suppose that $x > 0$. Denote $\mathcal{U} \subseteq \mathbb{R}$ as the set of upper Darboux sums of f over all partitions of $[-x, 0]$ and $\mathcal{U}' \subseteq \mathbb{R}$ as set of upper Darboux sums of f over all partitions of $[0, x]$. Show that $\mathcal{U} = \mathcal{U}'$. This can be done by using the fact that if $\mathcal{P}$ is a partition of $[0, x]$, then $-\mathcal{P}$ is a partition of $[-x, 0]$.
(b) Find the expression for $x \geq 0$ first. Then use part (a).

15.6 For all $t \in \mathbb{R}$ and $x \in [a, b]$ we have the inequality $0 \leq (tf(x) + g(x))^2$. Take the integral and get a quadratic polynomial in terms of t.

15.7 Recall the algebraic expressions for $\min\{p, q\}$ and $\max\{p, q\}$ for $p, q \in \mathbb{R}$ in Exercise 5.11.

15.8 (b) Use the ε-N definition of uniform convergence for a suitable $\varepsilon > 0$ and part (a).

15.9 (b) For any partition $\mathcal{P}$, the supremum of f over each subinterval is non-negative with at least one is strictly positive by part (a).
(c) Use contrapositive and parts (a) and (b).

15.10 Suppose for contradiction that $f(c) \neq 0$ at some $c \in [a, b]$. WLOG, suppose that $f(c) > 0$. Use Exercise 15.9 and pick an appropriate function g from Exercise 14.19.

15.11 By Exercise 11.31, we can find a sequence of polynomials (P_n) of degree n such that $P_n \xrightarrow{u} f$ over $[0, 1]$. So, for any $\varepsilon > 0, \exists N \in \mathbb{N} : \forall n \geq N, \sup_{x\in[0,1]} |P_n(x) - f(x)| < \varepsilon$. Use the assumption and the Cauchy-Bunyakovsky-Schwarz inequality in Exercise 15.6 to show $(\int_0^1 f(x)^2\, dx)^{\frac{1}{2}} < \varepsilon$. Conclude by using Exercise 15.9(c).

15.12 (a) Prove this via contradiction. Use Exercise 15.9.

15.13 (a) Similar to Exercise 15.5(a).
(b) Define a function $g : [-a, a] \to \mathbb{R}$ such that $g = -f$ on $[-a, 0]$ and $g = f$ on $[0, a]$ and use part (a).

15.14 (a) Fix $\varepsilon > 0$. Since f is Riemann integrable, $\exists$ a partition $\mathcal{P} = \{x_0, \ldots, x_n\}$ of $[0, 1]$ such that $U_{f,\mathcal{P}} - L_{f,\mathcal{P}} < \frac{\varepsilon}{2}$. Set $\mathcal{P}' = \{y_0, \ldots, y_n\}$ such that $y_j = x_j^2$ for all j.
(b) Use Exercise 15.13.

15.15 (a) Use the ε-δ definition of Riemann integrability and the Archimedean property. Then pick an appropriate tagged partition $\mathcal{P}_\tau$ that allows us to write $R_{f,\mathcal{P}_\tau}$ in terms of S_n.

15.16 Use the characterisation of supremum and infimum.

15.17 (a) Use the EVT on the function f and Proposition 15.5.5.

15.18 (a) For the function g, recall Exercise 15.3.
(b) Recall Theorem 10.6.10.
(d) i. Use part (b).
ii. Use part (c).

15.19 (a) Let $F = (\int_a^b |f(x)|^p\, dx)^{\frac{1}{p}}$ and $G = (\int_a^b |g(x)|^q\, dx)^{\frac{1}{q}}$. If one of F or G is 0, the inequality is trivial. Otherwise, let $\tilde{f}(x) = \frac{f(x)}{F}$ and $\tilde{g}(x) = \frac{g(x)}{G}$ and recall Young's inequality from Exercise 9.4(c).
(b) Let $M = (\int_a^b |f(x) + g(x)|^p\, dx)^{\frac{1}{p}}$. Show $M^p \leq \int_a^b |f(x) + g(x)|^{p-1}|f(x)|\, dx + \int_a^b |f(x) + g(x)|^{p-1}|g(x)|\, dx$ and use part (a) with a suitable choice of q.

15.20 (a) Use the same notation as Exercise 15.19(a). Clearly, the inequality is true if $F = 0$ or $G = 0$. Else, note that the function $h : [0, \infty) \to \mathbb{R}$ defined as $h(x) = x^p$ is concave. Using the function h, bound $|f(x) + g(x)|^p$ from below and integrate.

15.21 The forward implication is straightforward. To show the converse, show that if we start with the given estimate, for some sequence of tagged partitions $(\mathcal{P}^n_{\tau_n})$, the sequence of Riemann sums $(R_{f,\mathcal{P}^n_{\tau_n}})$ is Cauchy.

15.23 This is essentially an easy but fiddly construction. Use the definition of L to find a step function close to f. Then "shave off" a small bit of the sides of rectangles to create graphs in the shape of trapeziums/triangles with touching bases, so the graph is continuous with no jumps.

15.24 (a) Identical argument that leads to Eq. (15.2).
(b) Same as the proof for Proposition 15.3.5.

15.25 (a) Same as the proof for Theorem 15.3.8.
(b) Same as the proof for Corollary 15.3.9.

15.26 (a) Fix $\varepsilon > 0$. Since f is continuous at 0, $\exists \delta > 0 : \forall x \in [-1, 1], |x - 0| < \delta \Rightarrow |f(x) - f(0)| < \varepsilon$. Use this in the definition of Riemann-Stieltjes integral.
(b) Use an equispaced partition of $[0, 1]$ with $n + 1$ points.

15.27 Prove these results using the ε-δ definition of Riemann-Stieltjes integral given in the question.
(b) Use triangle inequality.
(c) Use the facts that f is bounded, g' is uniformly continuous over $[a, b]$, and the MVT.

15.28 (a) Find the expressions for $U_{f,\mathcal{P},g}$ and $L_{g,\mathcal{P},f}$ first. Then show $U_{f,\mathcal{P},g} - L_{g,\mathcal{P},f} = f(b)g(b) - f(a)g(a)$. Same for the other equality.
(b) Use part (a) to show that $\int_a^b g(x)\,df \leq f(b)g(b) - f(a)g(a) - \int_a^b f(x)\,dg$ and the opposite inequality.
(c) Use Exercise 15.27(c).

15.29 (b) Find a tagged partition $\mathcal{P}_\tau$ such that each subinterval contains at most one of the points d_j and use the ε-δ definition of Riemann-Stieltjes integral. Use the fact that f is uniformly continuous on $[0, z]$ by Theorem 10.6.10.

15.30 Use Exercise 15.29.

15.31 Recall Exercise 15.26(a).

Chapter 16: Fundamental Theorem of Calculus

16.1 (a) Use the fact that if f is Riemann integrable, then it is a bounded function over $[a, b]$.
(b) Define the function $J(x) = f(a)\int_a^x g(t)\,dt + f(b)\int_x^b g(t)\,dt$ for $x \in [a, b]$ and use the IVT.
(c) Define the function $J(x) = f(a)\int_a^x g(t)\,dt$ for $x \in [a, b]$ and use the IVT.

16.3 Differentiate and use the FTC to get $F'(x) = 0$.

16.5 Define $\bar{f} : [a, b] \to \mathbb{R}$ as the continuous extension of f. Show that for any $t \in (a, b)$ we have $\int_a^b \bar{f}(x)\,dx = \int_a^t f(x)\,dx + \int_t^b \bar{f}(x)\,dx$. Take the limit as $t \uparrow b$.

16.8 Fix $t \in [0, \infty)$ and let $n = \lfloor t \rfloor$. Find the integral $s \int_0^t \frac{\lfloor x \rfloor}{x^{s+1}}\, dx$ in terms of n and $t - n$. Find the limit as $t \to \infty$.
For the second identity, show that $s \int_1^\infty \frac{x}{x^{s+1}}\, dx = \frac{s}{s-1}$ and use the previous part.

16.9 (a) Find an antiderivative of f, use the FTC, and take limits.
(b) The quantity $\frac{\sin(x)}{x}$ (extended continuously to $x = 0$) is bounded from above and from below by some positive constants over $[0, 1]$. Then use part (a).

16.10 (c) Start with the integral $\int_{f(a)}^{f(b)} f^{-1}(x)\, dx$ and apply a change of variable.
(d) Draw a picture.

16.11 (c) The arclength of an increasing function f over the interval $[a, b]$ is the same as the arclength of the inverse function f^{-1} over the interval $[f^{-1}(a), f^{-1}(b)]$. This is because the graph for the inverse is just a reflection of the graph for the original function about the diagonal line $y = x$.

16.13 (b) By symmetry, compute the arclength D in the first quadrant and multiply by 4.

16.17 To deduce the equality, show the inequalities $C(f) \leq \sup_{\mathcal{P}}\{C_{\mathcal{P}}\}$ and $C(f) \geq \sup_{\mathcal{P}}\{C_{\mathcal{P}}\}$ separately. For the former, use the fact that $\sqrt{1 + f'(x)^2}$ is Riemann integrable and the MVT. For the latter, pick any partition $\mathcal{P}$ of $[a, b]$ and use the reverse Minkowski inequality in Exercise 15.20 to show that $C(f) \geq C_{\mathcal{P}}$.

16.20 (a) Fix $x > 0$. Prove that $\lim_{t\to\infty} t^{x-1} e^{-\frac{t}{2}} = 0$. Hence, deduce that there $\exists K > 0$ such that $t^{x-1}e^{-t} \leq e^{-\frac{t}{2}}$ for all $t \geq K$. The latter is improperly Riemann integrable over some unbounded interval of $[0, \infty)$. Exercise 9.26 might be useful.
(b) Use integration by parts.

16.22 (a) Note $f'(x) = -\frac{1}{2-x}$. Find its power series and integrate term-by-term (justify this).

16.23 (a) Similar to Exercise 16.22.
(b) Use Abel's theorem.

16.24 (a) Use the exponential form of complex numbers.
(c) Use Proposition 16.5.7 to switch the order of integration and infinite summation.

16.25 (a) Multiply the LHS with $\cos(\frac{t}{2})$ and use the product-to-sum formula.
(b) Integrate the sum in part (a) with respect to t on $[0, x]$.
(c) Use integration by parts and triangle inequality on part (b).
(d) For any $x \in (0, \pi)$, take the limit as $n \to \infty$ on the inequality in part (c). For $x \in (-\pi, 0)$, use the fact that the series is odd.
(e) Use part (d). Note that the series is 2π-periodic.

16.26 (a) Expand the numerator and use polynomial long division.
(b) Note that $0 \leq x \leq 1$.
(c) Integrate by parts twice. Note that $(\pi - 2x)^2 = \pi^2 - 4x(\pi - x)$.
(d) Find the maximum of P_n over $[0, \pi]$.

(e) Use Exercise 5.16.
(f) $I_1(b) = 2$ and $I_2(b) = 4b$. Use induction on the result in part (c). Note that, by the assumption that $\pi = \frac{a}{b}$, we have $(b\pi)^2 = a^2 \in \mathbb{N}$.

16.27 (a) Integrate by parts.
(b) $I_1 = 1$ and $I_0 = \frac{\pi}{2}$. Use the recursive relation in part (a).
(c) Use part (b) to evaluate $\frac{I_{2n}}{I_{2n+1}}$.
(d) Note that for $x \in [0, \frac{\pi}{2}]$ we have $0 \leq \sin(x) \leq 1$.
(e) Note that the identity in part (c) can be written as $\sqrt{\frac{2n+1}{2n}}\sqrt{\frac{I_{2n+1}}{I_{2n}}}\sqrt{\frac{\pi}{2}} = \frac{(2n)!!(2n)!!}{(2n)!\sqrt{2n}}$.

16.28 (a) Since the integrand is even, it is enough to show that $\int_0^\infty e^{-x^2}\,dx$ exists. Recall the identity in Exercises 16.23 and 9.26.
(b) In the first integral, change the variable $x = \cos(y)$. In the second integral, change the variable $y = \sqrt{n}x$. In the third integral, change the variable $x = \cot(y)$.
(c) Use Exercise 16.27(b)(d).
(d) Use part (c).
(e) Use parts (c) and (d).
(f) The inequality in part (b) can be written as $I_{2n+1} \leq \frac{1}{\sqrt{n}} \int_0^{\sqrt{n}} e^{-y^2}\,dy \leq I_{2n-2}$. Use part (e) and sandwiching.

16.29 (a) Use integration by parts twice. The antiderivatives of $\sqrt{1-x^2}$ and $\arcsin(x)$ from Exercise 16.4 are useful here.
(b) Prove this by induction and the recursive formula in part (a).
(c) Use Exercise 16.27(f).

16.30 (b) Compute $c_j - a_j$ using geometry.
(e) To show that D_n is bounded, use the estimate in Exercise 10.31(c) and the series in Exercise 8.12.
(f) Use the $n!$ from part (d) in Wallis's formula and the AOL.
(g) Use part (b) and telescoping sum.
(h) Use parts (d), (f), (g), and sandwiching.

16.31 (a) Use Exercise 15.23.
(c) Use Dini's theorem (Theorems 11.2.4) and 16.5.3.
(d) Fix $n \in \mathbb{N}$. Note that $\max(g_j, \ldots, g_n) - g_j \geq 0$ for all $j = 1, 2, \ldots, n-1$. Then use the fact that $0 \leq g_n \leq g_j + (g_n - g_j)$ for all $j = 1, 2, \ldots, n-1$ to show $0 \leq g_n \leq g_j + \sum_{j=1}^{n-1}(\max(g_j, \ldots, g_n) - g_j)$ for $j = 1, 2, \ldots, n-1$.
(e) Fix $j \in \{1, \ldots, n\}$. Note that $G_j = \max(g_j, \ldots, g_n)$ is Riemann integrable. Prove that $G_j = \max(g_j, \ldots, g_n) \leq f_j$.
(f) Combine parts (d) and (e).
(g) Combine parts (a) and (f).

16.32 (a) Use the fact that the functions sequence (f_n) is uniformly bounded and f is continuous.

(d) Use part (b) to show the inequality. Use Exercise 16.31 and part (c) to deduce the limit.

16.33 (b) Use Proposition 13.5.7.

(c) Fix n and use induction on m for the improper integral $\int_0^1 x^n(\ln(x))^m\,dx$ with the base case for $m = 0$ given by $\int_0^1 x^n\,dx = \frac{1}{n+1}$.

(d) Show first that $g(x) = e^{-f(x)}$ and write it as a power series. Integrate it and justify switching the order of integral and infinite sum. Use part (c).

16.34 (b) Use the triangle inequality repeatedly on $|y_n(x) - y_m(x)|$ and part (a).

(e) Take the limit as $n \to \infty$ in the recursive relation. Use the FTC to show that the limit satisfies the ODE.

(f) Assume that there are two solutions y and z. Suppose that $\sup_{|x-x_0|<\varepsilon} |y(x) - z(x)| = C \geq 0$. Show that for all $n \in \mathbb{N}$ we have $\sup_{|x-x_0|<\varepsilon} |y(x) - z(x)| = C(K\varepsilon)^n$ and conclude that $y = z$.

Chapter 17: Taylor and Maclaurin Series

17.3 (b) Differentiate the equation in part (a) n times by applying the Leibniz rule.

(c) Prove each equality separately using induction and the recursive formula in part (b).

(f) Use the bounds of the central binomial coefficients in Exercise 3.14(g). Alternatively, use the asymptotics of the central binomial coefficient as seen in Exercise 16.27.

17.4 Suppose that there are two such power series, namely $\sum_{j=0}^{\infty} a_j(x-c)^j = \sum_{j=0}^{\infty} b_j(x-c)^j$. Differentiate term-wise (justify this) and substitute $x = c$ to show that the coefficients are identical.

17.5 WLOG, suppose that $c < x$. Use $R_n(x) = \frac{1}{n!}\int_c^x (x-t)^n f^{(n+1)}(t)\,dt$. Note that the integrand is continuous on $[c, x]$ and hence attains its minimum m and maximum M somewhere. Use the IVT on an appropriate function.

17.6 Show that $R_n \xrightarrow{pw} 0$ using Lagrange's remainder form and Exercise 5.16.

17.7 (a) Apply the EVT on $\Psi^{(n)}$.

(b) Assume for contradiction that the sequence (a_n) is bounded. Use Exercise 17.6.

17.8 Similar idea to Example 17.2.7.

17.11 (a) WLOG, since the integrand is an even function, suppose $x > 0$. Write the integrands in terms of power series and justify switching the integral and infinite sum.

17.12 Use Abel's theorem.

17.13 (a) Let $a_n = q^n n\binom{r}{n}$ for all $n \in \mathbb{N}$. Show that the series $\sum_{j=1}^{\infty} a_j$ converges.

(b) Consider the two cases, namely $x < 0$ and $x > 0$. Show $\frac{|x-t|}{|1+t|} < 1$ for all t in the closed interval with 0 and x as endpoints. Then apply the EVT.

(c) Use parts (a) and (b) to bound and show that for any fixed $|x| < 1$ we have $|R_n(x)| \to 0$.

(d) Use Exercise 12.10 and Abel's theorem.
(e) Use Abel's theorem.

17.14 Use Exercise 17.13, definition of generalised binomial coefficients, and gamma function.

17.15 Note that $(1+x)^{p+q} = (1+x)^p(1+x)^q$. Use Exercise 17.13 and Mertens' theorem.

17.16 (b) Integrate the series in part (a).
(c) Use Raabe's test and Abel's theorem.

17.18 (a) Let f be an even function. For any $n \in \mathbb{N}_0$ prove that $f^{(2n)}(x)$ is even and $f^{(2n+1)}(x)$ is odd by induction.

17.19 (a) Fix an $x \in \mathbb{R}$. Since the function erf is even, WLOG, suppose that $x > 0$. Write the integrand in terms of a power series and justify the switching of integral and infinite sum.

17.20 Fix $0 \leq \epsilon < 1$. Use Exercise 17.13 and justify switching the order of integration and infinite sum. Recall the integral $I_{2n} = \int_0^{\frac{\pi}{2}} \sin(x)^{2n}\, dx$ from Exercise 16.27.

17.21 Similar to Exercise 17.20.

17.22 (a) Change the variable to $x = \sin(\theta)$ and use Exercise 16.27

17.27 (a) To get $P_1(t)$, we can differentiate term-by-term with respect to x since the series converges uniformly over some closed interval in $(-1, 1)$.
(b) Differentiate the equation with respect to x, rewrite the LHS in terms of f, use algebraic manipulations so that both sides of the equation are power series, and equate coefficients of x^n.
(c) Prove by induction and the recursive relation in part (b).
(d) Differentiate the recursive relation in part (b).
(e) Differentiate $\frac{1}{\sqrt{1-2xt+x^2}} = \sum_{j=0}^{\infty} P_j(t)x^j$ with respect to t and equate coefficients.
(f) First get the equations $P_n' = (n-1)P_{n-1} + tP_n'n - 1$ and $P_{n-1}' = -nP_n + tP_n'$. Combine them to get $(1-t)^2 P_n' = -ntP_n + nP_{n-1}$ and deduce the desired ODE.
(g) Compute $\frac{d}{dt}((1-t^2)(P_m' P_n - P_m' P_n))$.
(h) Recall Proposition 16.4.5.
(i) Integrate the equation in part (g) from $t = -1$ to 1.
(j) Let $I_n = \int_{-1}^{1} P_n(t)^2\, dt$. Show that $I_n = \frac{2n-1}{2n+1} I_{n-1}$ for $n \in \mathbb{N}$. This can be done by recalling the recursive formula $nP_n = (2n-1)tP_{n-1} - (n-1)P_{n-2}$ and the orthogonal relationship in part (i).

17.28 (a) Prove this via induction on n. Use the Leibniz rule.

Chapter 18: Introduction to Measure

18.1 Prove this via induction on n and use Remark 18.2.6(1).

18.3 (a) To show $\mathcal{E} \subseteq \mathcal{R}(S)$, show by induction that $E_1 \Delta \ldots \Delta E_n \in \mathcal{R}(S)$ for any $n \in \mathbb{N}$.

(b) To show that $\mathcal{E}$ is closed under $\cap$, use Exercise 1.21(b).
(c) Recall Remark 18.3.2.

18.7 (a) Let $\{I_j\}_{j=1}^{\infty}$ be any countable cover of the set B for which $I_j \in \mathcal{R}$.
(b) Show that $m^*(B) \leq m^*(A \cup B)$ and $m^*(A \setminus B) \leq m^*(A) = 0$ by using part (a). Using the latter and the fact that $A \cup B = B \cup (A \setminus B)$, derive $m^*(A \cup B) \leq m^*(B)$ using the definition of outer measure and assumption.

18.8 (b) Use the definition of outer measure and part (a). For the scaling of the set, split into cases of $\lambda > 0$, $\lambda = 0$, and $\lambda < 0$. The first case follows from part (a). The second case is trivial. For the final case of $\lambda < 0$, WLOG assume that $\lambda = -1$ and show $m^*(-A) = m^*(A)$. Note that if $I = (a, b] \in \mathcal{R}$, then $-I = [b, a) \notin \mathcal{R}$ so the content m cannot be defined on $-I$. However, for any $\varepsilon > 0$ the slightly enlarged set $I' = (b - \varepsilon, a]$ is in $\mathcal{R}$ (hence has a content) and contains $-I$.
(c) Let $G \in \mathcal{P}(\mathbb{R})$ be arbitrary. Then $G = c + F$ where $F \in \mathcal{P}(\mathbb{R})$ is defined as $F = -c + G$. Show that $G \cap (c + E) = c + (F \cap E)$ and $G \cap (c + E)^c = c + (F \cap E^c)$. Use these in the Carathéodory condition.

18.13 To show that $\mathcal{C}$ is closed under set difference, prove that it is closed under intersection and complement. To show the latter, note that $A \in \mathcal{C}$ can be written as $A = E \cup M$ where $M \subseteq X \in \mathcal{B}$ with $\mu(X) = 0$. Then show that $A^c = (E^c \cap X^c) \cup ((E^c \cap M^c) \setminus (E^c \cap X^c))$ and $(E^c \cap M^c) \setminus (E^c \cap X^c) \subseteq X$.

18.16 Use Lemma 18.9.8.

18.17 (a) Let $h = \max(f, g)$. Explain why for any $c \in \mathbb{R}$, we have $h(x) < c$ if and only if $f(x) < c$ and $g(x) < c$.
(b) The zero function is $\mathcal{F}$-measurable. Then use part (a). Note also that $|f| = f^+ + f^-$.

18.18 Use Exercises 1.26 and 1.27.

18.19 (a) To show the inclusion $\mathcal{B} \subseteq \mathcal{F}(J)$, show that $\mathcal{F}(J)$ contains any open set. Theorem 4.5.20 might be useful.

18.21 Use Theorem 10.5.4.

18.22 (a) Fix $n \in \mathbb{N}$ and $x_0 \in E_n$ so that $\exists \delta > 0 : \forall x, y \in B_\delta(x_0), |f(x) - f(y)| < \frac{1}{n}$. Show that $B_\delta(x_0) \subseteq E_n$ so that E_n is an open set.
(b) Use double inclusion.

18.23 (a) To show that the set difference is closed, pick any two $E, F \in \mathcal{G}$ to consider $E \setminus F$. We have four cases: both contain Y, both do not contain Y, E contains Y but F does not, and E does not contain Y but F does. For the countable union, consider $\{E_j\}_{j=1}^{\infty}$. We have two cases: all of them do not contain Y and at least one of them contains Y.

18.24 Use Proposition 18.5.8 and Exercise 7.8(a).

18.26 (b) Prove via induction.
(c) Use Proposition 18.5.8.

18.27 (b) Use Theorem 10.5.4.
(c) Recall the construction of C from $[0, 1]$ by removing open intervals in Exercise 4.32. To show the results for the image $g(C^c)$, use part (a).

(d) Use the facts on the Cantor staircase function f from Exercises 11.11 and 13.11 as well as part (c).
(e) Use Exercise 18.26(c) and Lemma 18.7.7.
(f) Use Exercise 18.21.

18.28 Prove this via double inclusion.

18.29 (b) If the functions sequence (f_n) does not converge at the point $x \in X$, then the sequence $(f_n(x))$ must not be Cauchy. Use the negation of the definition for Cauchy sequence and put $\varepsilon = \frac{1}{k} > 0$ for any $k \in \mathbb{N}$.
(c) Let $Y = \{x : f_n(x) \text{ does not converge}\}$. Then, the sequence of restricted functions $(f_n|_{Y^c})$ (which are still measurable) converges pointwise. Use Proposition 18.9.15.

18.30 (b) Show first that $Z \subseteq \mathcal{E} \subseteq \mathcal{M}(Z)$. Then show that $\mathcal{E}$ is a monotone class.
(c) Similar strategy to part (b).
(d) Use parts (b) and (c).
(e) Pick any countable collection of sets $\{E_j\}_{j=1}^{\infty}$ in $\mathcal{M}(Z)$. Define $F_n = \bigcup_{j=1}^{n} E_j \in \mathcal{M}(Z)$. Use the fact that $\mathcal{M}(Z)$ is a monotone class.
(f) Put parts (a) and (e) together.

18.31 (b) Write $A \cup B = (A \setminus B) \cup (B \setminus A) \cup (A \cap B)$, $A = A \setminus B \cup (A \cap B)$, and $B = (B \setminus A) \cup (A \cap B)$ where the sets in each union are disjoint.

18.33 (d) Note that $F = \bigcup_{j=1}^{\infty}(F \cap E_j)$ and the sets in the union are pairwise disjoint. Use part (c) to find what $P(E_m|F)$ is.

Chapter 19: Lebesgue Integration

19.1 Prove this via induction on n.

19.2 (a) Use Lemma 18.9.8.
(b) Consider first $\phi \geq 0$. For the general ϕ, split the function ϕ into its positive and negative parts.

19.3 Use Proposition 19.1.3 on the positive and negative parts of the function f.

19.7 (d) Define $E_n = \{x \in X : f(x) \geq \frac{1}{n}\} \subseteq X$. Show that there exists an $N \in \mathbb{N}$ such that $\int_X f \, d\mu \geq \int_{E_N} f \, d\mu > 0$.

19.8 Show that for any simple function $\phi : X \to \mathbb{R}$ we have $I(\phi) = \phi(c)$. Then use Proposition 19.1.3.

19.9 (a) Use Proposition 18.9.13 for the first part. For the second part, recall the properties of the function f from Exercise 11.11 and Theorem 19.7.4.
(c) Use part (b) to evaluate the integral of f over $[0, \frac{1}{3}]$, $[\frac{1}{3}, \frac{2}{3}]$, and $[\frac{2}{3}, 1]$.

19.10 First, suppose that $f \geq 0$. For the general case, split into positive and negative parts.
(a) Use the definition of Lebesgue integral and Corollary 19.5.8.
(b) Use part (a). Then recall Exercise 15.23 in which we created a continuous function from a step function.

19.11 (d) Prove both implications via contradiction. For the forward implication, recall Proposition 19.5.5.

(e) Prove this in three steps: g is an indicator function, g is a simple function, and g is a general measurable function.

19.13 Use Exercise 18.8 and definition of Lebesgue integral.

19.14 Show first $\int_X f^2(f-1)^2\,d\mu = 0$.

19.15 Show first that $f \equiv 0$ a.e. on any bounded interval in $[0,\infty)$. Next, consider the subset $[0,1] \subseteq [0,\infty)$. Assume for contradiction that $f > 0$ on a subset $E \subseteq [0,1]$ with positive measure. Then use Proposition 18.7.4, Theorem 4.5.20, and Proposition 19.5.5 to get a contradiction. Repeat to show that f cannot be negative on a subset $E \subseteq [0,1]$ with positive measure. Conclude that $f \equiv 0$ a.e. on $[0,1]$ and hence on $[0,\infty)$.

19.16 Prove this first when f is step function and then use Exercise 19.10(a) for the general f.

19.18 (b) Use Fatou's lemma.

19.19 (c) First, suppose that $g \geq 0$. Use the MCT and part (b). For a general function g, split into its positive and negative parts.

19.20 For any $n \in \mathbb{N}$, define $f_n(x) = f(x)\mathbf{1}_{[0,n]}$. Use the MCT.

19.21 (b) Show that $\frac{n\sin(x)}{1+n^2\sqrt{x}}$ is uniformly bounded over $[0,1]$.

(c) Split the domain of integration into two, namely $[0,1]$ and $[1,\infty)$. Exercise 19.14 might be helpful.

(e) Use the DCT with dominating function $g(x) = \frac{1}{x^2} + \frac{\ln(x)}{x^2}$.

(f) Use the bounds $nx < 1+nx < 2nx$ and sandwiching to show that $\frac{\ln(1+nx)}{1+x^2\ln(n)} \xrightarrow{pw} \frac{1}{x^2}$. Use the DCT with an appropriate dominating function.

(h) Use the DCT with the dominating function $g(x) = 1$ for $x \in (0,1]$ and $g(x) = \frac{1}{x^2}$ for $x \in (1,\infty)$.

(j) Justify the change of variable $y = x^n$.

19.22 Let $(a_{m,n}) = (a_n(m))$ be the doubly indexed sequence. Treat this as a sequence of functions (a_n) where $a_n : \mathbb{N} \to \mathbb{R}$.

19.25 (a) Denote $S_q = \sum_{j=1}^{q}(-1)^j a_j$. Suppose first that q is even. We have two cases: either $q \leq k$ or $q \geq k$. For the latter case, show that $S_q = \sum_{j=1}^{q}(-1)^j a_j = \sum_{j=q+1}^{n}(-1)^{j+1}a_j$ and study the cases for n is even and n is odd. Similar argument for odd q.

(b) Recall Exercise 3.14(b). Then use part (a).

(d) Use part (c) and integrate.

(e) Combine parts (b) and (d).

19.27 (a) Refer to Exercise 15.19(a).

(b) Refer to Exercise 15.19(b).

(c) Since $p < q$, we have $\frac{1}{p} > \frac{1}{q}$ so there exists some $r > 0$ such that $\frac{1}{q} + \frac{1}{r} = \frac{1}{p} \Leftrightarrow \frac{1}{\frac{q}{p}} + \frac{1}{\frac{r}{p}} = 1$. Use this and Hölder's inequality on $f \in L^q(X)$ to show that it lies in $L^p(X)$ as well.

(d) To prove the inductive step, suppose $\sum_{j=1}^{k+1}\frac{1}{p_j} = \frac{1}{r}$. Necessarily $p_j > r$ for all $j = 1, 2, \ldots, k+1$. Apply part (a) with $p = \frac{p_{k+1}}{p_{k+1}-r}$, $q = \frac{p_{k+1}}{r}$, $f = (f_1 \ldots f_k)^r$, and $g = f_{k+1}^r$. Then apply the inductive hypothesis.

19.28 (b) If $||f||_\infty = 0$, then we have nothing to show. Else, fix $\varepsilon > 0$ and show that $||f||_p \geq (||f||_\infty - \varepsilon)\mu(E)^{\frac{1}{p}}$ for all $p \geq 1$. Then apply limit inferior as $p \to \infty$ on both sides.

(c) Show first that $\forall p > 1$, we have $||f||_p^p \leq ||f||_\infty^{p-1}||f||_1$.

19.29 (b) $P_X(\{a\})$ is non-zero only for three values of a.

(d) Use part (c).

(e) F_X is a piecewise constant function taking four values.

19.30 (c) To evaluate $\mathbb{E}[X]$, use the integral with respect to the Dirac delta measure that we have derived in Exercise 19.8. For $\mathbb{E}[g \circ X]$, either compute the integral using the probability measure $P_{g \circ X}$ obtained in Exercise 19.29(f) or the formula in part (b).

Chapter 20: Double Integrals

20.3 Show that the set $A = \{(x, y) : y > x\}$ is measurable by tiling it with countably many squares which get smaller in size. Likewise, show that the set $B = \{(x, y) : y < x\}$ is also measurable.

20.4 (a) For the forward implication, use Proposition 19.1.3. For the converse, recall Lemma 18.9.8.

(b) Use part (a) and the MCT.

20.5 (a) Show that the functions $(x, y) \mapsto f(x)$ and $(x, y) \mapsto -y$ are both $(\mathcal{F} \otimes \mathcal{B})$-measurable.

(b) Write G_f using the function h.

(c) Suppose first that $f \geq 0$. Fix $\varepsilon > 0$. For every $n \in \mathbb{N}$, define the sets $E_n = f^{-1}([\varepsilon(n-1), \varepsilon n))$ and $F_n = E_n \times [\varepsilon(n-1), \varepsilon n)$. Show that $G_f \subseteq \bigcup_{n=1}^\infty F_n$. For the general function f with mixed signs, break f into its positive and negative parts.

20.8 (a) To show σ-additivity, use the MCT.

(b) Use Proposition 20.2.11.

20.9 Recall the Vitali set in Example 18.4.5.

20.11 (c) Use Fubini-Tonelli theorem.

20.13 (a) Use Proposition 19.7.9.

(c) For a fixed $s \in (0, \infty)$, write $w = t + s$ in the inner integral. Then for a fixed $w > 0$, write $u = \frac{s}{w}$.

20.14 (a) Show that f is measurable and Lebesgue integrable using Tonelli's theorem, comparison, and Proposition 19.7.9.

(b) To find the actual value of the integral, consider the cases where $y = 0$ and $y > 0$. For the latter, break the integrand into positive and negative parts.

20.15 Use the contrapositive to Fubini's theorem.

20.18 For $n \in \mathbb{N}$ let $E = \{(x, y) : y + \frac{1}{n} \leq x < 1, 0 \leq y < 1 - \frac{1}{n}\}$. Use Tonelli's theorem to evaluate $\int_E f \, d\tilde{\mu}$. Then take the limit as $n \to \infty$ by using the MCT.

20.19 (a) Use the definition of indicator functions.
(b) Note that $f(x) = \int_0^{f(x)} dt$.

20.20 (a) Let $f : \mathbb{R}^2 \to \mathbb{R}^2$ be defined as $f(x, y) = (-a, -b) + (x, y)$ for some constants $a, b \in \mathbb{R}$. Show that the preimage of E is measurable. Similar for the scale: define a suitable function (break into cases of $\lambda = 0$ and $\lambda \neq 0$).
(c) Use Proposition 20.2.11 for the definition of $\tilde{\mu}$. Use part (b) and Lemma 20.2.9 to show the equality. Note that the Lebesgue measure on $\mathbb{R}$ is translation invariant as shown in Exercise 18.8.

20.21 (a) Let $f : \mathbb{R}^k \to \mathbb{R}$ be defined as $f(x_1, \ldots, x_k) = \sum_{j=1}^k x_j^2 = \sum_{j=1}^k \pi_j(x_1, \ldots, x_k)^2$. Note that $B_1^k = f^{-1}([0, 1))$.
(c) $\mathbb{R}^n = \mathbb{R} \times (\mathbb{R}^{n-1})$ with the σ-algebras $\mathcal{L}$ and $\overset{n-1}{\otimes} \mathcal{L}$ respectively. The latter has product measure μ^{n-1}.
(d) $B_1^n = \{(x_1, \ldots, x_n) : x_1^2 + \ldots + x_n^2 < 1\} = \{(x_1, \ldots, x_n) : x_1^2 + \ldots + x_{n-1}^2 < 1 - x_n^2\}$. Write $y_j = \frac{x_j}{\sqrt{1-x_n^2}}$ for $j = 1, \ldots, n-1$.
(f) Use the substitution $x^2 = u$ in the integral from part (e). Recall the gamma and beta functions in Exercises 16.20 and 20.13.

20.22 (a) Use Exercise 14.2(e) and comparison.
(b) Use Exercise 20.11(c).
(c) Apply a second change of variable, namely: for a fixed u, define $v = \frac{y^2}{1-u^2}$.

20.23 (a) Show that the function $(x, y) \mapsto x - y$ is $(\overset{2}{\otimes}\mathcal{L}, \mathcal{L})$-measurable and thus $f(x - y)$ is $\overset{2}{\otimes}\mathcal{L}$-measurable. Then use Tonelli's theorem and the fact the Lebesgue measure on $\mathbb{R}$ is translation invariant in Exercise 18.8.
(b) Use Theorem 20.3.3 and Proposition 19.5.5.
(c) Use the translation invariance of the Lebesgue integrals on $\mathbb{R}$.
(d) Use Fubini's theorem and similar argument to part (a).

20.24 (a) For any $x_0 \in \mathbb{R}$ show that for any sequence (x_n) converging to x_0 we have $(\phi * f)(x_n) \to (\phi * f)(x_0)$ via the DCT.
(b) First show that $\frac{d}{dx}(\phi * f)(x) = \lim_{h \to 0} \int_{\mathbb{R}} f(y) \frac{\phi(x-y+h)-\phi(x-y)}{h} \, d\mu(y)$. Then use the MVT, part (a), and the assumption that the derivatives of ϕ are bounded.

20.25 (c) Since $(\varphi_\varepsilon * f)(x) = \int_{\mathbb{R}} \varphi_\varepsilon(x - y) f(y) \, d\mu(y)$, for the integral to be non-vanishing, necessarily both of the integrands are non-zero.
(d) Use Exercise 20.24.
(e) Use Fubini's theorem, Exercise 19.16, and the BCT. A change of variable to remove the dependency of φ_ε on ε might be helpful too.
(f) Show this via the first principle, namely show that $\forall x \in K, \forall \eta > 0, \exists \nu > 0 : \forall 0 < \varepsilon < \nu, |(\varphi_\varepsilon * f)(x) - f(x)| < \eta$.

20.26 (b) Note $g(x, \omega) = x - X(\omega)$ so that $E = g^{-1}((-\infty, 0))$.
(c) Denote $A = \{(\omega, x) : 0 \leq x < X(\omega)\}$. Then $A = E \cap (\Omega \times [0, \infty)) \in \mathcal{F} \otimes \mathcal{B}$.

Reference

1. Abbott, S. *Understanding Analysis*. Springer Science+Business Media, New York (2016).
2. Aksoy, A.G., Khamsi, M.A. *A Problem Book in Real Analysis*. Springer Science+Business Media, New York (2010).
3. Asmar, N.H., Grafakos, L. *Complex Analysis with Applications*. Springer Nature, Switzerland (2018).
4. Axler, S. *Measure, Integration and Real Analysis*. Springer Open, New York (2019).
5. Bartle, R.G., Sherbert, D.R. *Introduction to Real Analysis*. John Wiley & Sons, United States of America (2010).
6. Baritompa, B., Lowen, R., Polster, B. M., Ross, M. *Mathematical Table-Turning Revisited*. Mathematical Intelligencer, Vol. 29, Issue 2 (2007): 49–58.
7. Batchelor, G.K. *An Introduction to Fluid Mechanics*. Cambridge University Press, Cambridge (2000).
8. Biggs, N.L. *Discrete Mathematics*. Oxford University Press, New York (2002).
9. Billingsley, P. *Probability and Measure*. John Wiley & Sons, Canada (1995).
10. Bromwich, T.J. *An Introduction To The Theory Of Infinite Series*. Macmillan and Co. Limited, London (1908).
11. Cajori, F. *A History of Mathematics*. (5th edition). American Mathematical Society, Providence (1999).
12. Capiński, M., Kopp, E. *Measure, Integral and Probability*. Springer-Verlag, London (2004).
13. Cheney, W., Kincaid, D. *Numerical Mathematics and Computing*. (6th edition). Thomson Brooks/Cole, Belmont (2008).
14. Chrisomalis, S. *Numerical Notation: A Comparative History*. Cambridge University Press, Cambridge (2010).
15. Collins, P.J. *Differential and Integral Equations*. Oxford University Press, New York (2006).
16. Cunningham, D.W. *Elementary Analysis with Proof Strategies*. CRC Press, Boca Raton (2021).
17. Crossley, M.D. *Essential Topology*. Springer-Verlag, London (2010).
18. Dineen, S. *Multivariate Calculus and Geometry*. Springer-Verlag, London (2014).
19. Dunham, W. *The Calculus Gallery*. Princeton University Press, Princeton (2005).
20. Dyer, R.H., Edmunds, D.E. *From Real to Complex Analysis*. Springer International Publishing, Switzerland (2014).
21. Elaydi, S.N. *Discrete Chaos with Applications in Science and Engineering*. (2nd edition). CRC Press, Boca Raton (2007).
22. Epperson, J.F. *An Introduction to Numerical Methods and Analysis*. John Wiley & Sons, New Jersey (2013).
23. Euclid, Heath, T.L., Densmore, D. *Euclid's Elements: All Thirteen Books Complete in One Volume*. Green Lion Press, Santa Fe (2007).
24. Gelbaum, B.R., Olmsted, J.M.H. *Counterexamples in Analysis*. Dover Publications, New York (1992).

S. Johar, *The Big Book of Real Analysis*,
https://doi.org/10.1007/978-3-031-30832-1

25. Gowers, T., Barrow-Green, J., Leader, I. *The Princeton Companion to Mathematics*. Princeton University Press, Princeton (2008).
26. Griffiths, D.J. *Introduction to Quantum Mechanics*. Prentice Hall, New Jersey (1995).
27. Hamkins, J.D. *Lectures on Philosophy of Mathematics*. MIT Press (2021).
28. Hammack, R. *Book of Proof*. Self-published (2018).
29. Hamming, R.W. *Numerical Methods for Scientists and Engineers*. (2nd edition). Dover Publications, New York (1987).
30. Hansheng, Y., Lu, B. *Another Proof for the p-series Test*. The College Mathematics Journal, Vol. 36, No. 3 (2005): 235–237.
31. Hata, M. *Problems and Solutions in Real Analysis*. World Scientific, Singapore (2007).
32. Harlan, J.B. *Math Bite: Finding e in Pascal's Triangle*. Mathematics Magazine, Vol. 85, No. 1 (2012): 51.
33. Hodel, R.E. *An Introduction to Mathematical Logic*. Dover Publications, New York (1995).
34. Houston, K. *How to Think Like a Mathematician: A Companion to Undergraduate Mathematics*. Cambridge University Press, Cambridge (2009).
35. Howie, J.M. *Real Analysis*. Springer-Verlag, Berlin (2001).
36. Jech, T.J. *The Axiom of Choice*. Dover Publications, New York (2008).
37. Jones, G.A., Jones, J.M. *Elementary Number Theory*. Springer-Verlag, London (1998).
38. Kaczor, W.J., Nowak, M.T. *Problems in Mathematical Analysis I: Real Numbers, Sequences, and Series*. American Mathematical Society, Providence (2000).
39. Kaczor, W.J., Nowak, M.T. *Problems in Mathematical Analysis II: Continuity and Differentiation*. American Mathematical Society, Providence (2000).
40. Kaczor, W.J., Nowak, M.T. *Problems in Mathematical Analysis III: Integration*. American Mathematical Society, Providence (2000).
41. Kánnai, Z. *An Elementary Proof That The Borel Class of the Reals has Cardinality Continuum*. Acta Math. Hungar, Vol. 159, No. 1 (2019): 124–130.
42. Katz, V.J. *A History of Mathematics: An Introduction*. (3rd edition). Addison-Wesley, Boston (2009).
43. Kay, A. *Number Systems: A Path into Rigorous Mathematics*. CRC Press, Boca Raton (2022).
44. Kleppner, D., Kolenkow, R. *An Introduction to Mechanics*. (2nd edition). Cambridge University Press, Cambridge (2014).
45. Kolmogorov, A.N., Fomin, S.V. *Introductory Real Analysis*. Dover Publications, New York (1975).
46. Krantz, S.G. *Real Analysis and Foundations*. CRC Press, Boca Raton (2005).
47. Kreyszig, E. *Introductory Real Analysis with Applications*. John Wiley & Sons, Canada (1978).
48. Laczkovich, M., Sós, V.T. *Real Analysis: Foundations and Functions of One Variable*. Springer Science+Business Media, New York (2015).
49. Lax, P.D., Terrell, M.S. *Calculus with Applications*. Springer Science+Business Media, New York (2014).
50. Liebeck, M. *A Concise Introduction to Pure Mathematics*. CRC Press, Boca Raton (2011).
51. Lieb, E.H., Loss, M. *Analysis*. (2nd edition). American Mathematical Society, Providence (2001).
52. Liesen, J., Mehrmann, V. *Linear Algebra*. Springer International Publishing, Switzerland (2015).
53. Little, C.H.C., Teo, K.L., Brunt, B. *Real Analysis via Sequences and Series*. Springer Science+Business Media, New York (2015).
54. Lovett, S. *Abstract Algebra: Structures and Applications*. CRC Press, Boca Raton (2016).
55. Luxemburg, W.A.J. *Arzelá's Dominated Convergence Theorem for the Riemann Integral*. The American Mathematical Monthly, Vol. 78, Issue 9 (1971): 2970–979.
56. Muscat, J. *Functional Analysis: An Introduction to Metric Spaces, Hilbert Spaces, and Banach Algebras*. Springer International Publishing, Switzerland (2014).
57. Perko, L. *Differential Equations and Dynamical Systems*. Springer Science+Business Media, New York (2001).
58. Pinter, C.C. *A Book of Abstract Algebra*. McGraw-Hill, Singapore (1990).

59. Priestley, H.A. *Introduction to Complex Analysis*. Oxford University Press, Oxford (2003).
60. Radulescu, T.-L.T., Radulescu, V.D., Andreescu, T. *Problems in Real Analysis: Advanced Problems on the Real Axis*. Springer Science+Business Media, New York (2009).
61. Rosentrater, C.R. *Varieties of Integration*. MAA Press (2015).
62. Ross, K.A. *Elementary Analysis: The Theory of Calculus*. Springer Science+Business Media, New York (2013).
63. Rudin, W. *Principles of Mathematical Analysis*. McGraw-Hill, United States of America (1976).
64. Rynne, B.P., Youngson, M.A. *Linear Functional Analysis*. Springer-Verlag, London (2008).
65. Salsa, S. *Partial Differential Equations in Action: From Modelling to Theory*. Springer International Publishing, Switzerland (2015).
66. Saoub, K.R. *A Tour Through Graph Theory*. CRC Press, Boca Raton (2018).
67. Schmelzer, T., Baillie, R. *Summing a Curious, Slowly Convergent Series*. The American Mathematical Monthly, Vol. 115, No. 6 (2008): 525–540.
68. Sonar, T. *3000 Years of Analysis*. Springer Nature, Switzerland (2021).
69. Stillwell, J. *Mathematics and its History*. (3rd edition). Springer Science+Business Media, New York (2010).
70. Stolz, O. *Ueber die Grenzwerthe der Quotienten*. Mathematische Annalen, Vol. 15 (1879): 556–559.
71. Strogatz, S. *Nonlinear Dynamics and Chaos with Applications to Physics, Biology, Chemistry, and Engineering*. CRC Press, Boca Raton (2018).
72. Strogatz, S. *Infinite Powers: How Calculus Reveals the Secrets of the Universe*. Houghton Mifflin Harcourt, Boston (2019).
73. Sutherland, W.A. *Introduction to Metric and Topological Spaces*. Oxford University Press, New York (2009).
74. Tao, T. *Analysis I*. Springer Science+Business Media and Hindustan Book Agency, Singapore (2016).
75. Wallace, D.A.R. *Groups, Rings and Fields*. Springer-Verlag, London (1998).

Index

S. Johar, *The Big Book of Real Analysis*,
https://doi.org/10.1007/978-3-031-30832-1

R

S

T

GPSR Compliance
The European Union's (EU) General Product Safety Regulation (GPSR) is a set of rules that requires consumer products to be safe and our obligations to ensure this.

If you have any concerns about our products, you can contact us on

ProductSafety@springernature.com

In case Publisher is established outside the EU, the EU authorized representative is:

Springer Nature Customer Service Center GmbH
Europaplatz 3
69115 Heidelberg, Germany

www.ingramcontent.com/pod-product-compliance
Lightning Source LLC
Chambersburg PA
CBHW072338060126
37739CB00028B/118
* 9 7 8 3 0 3 1 3 0 8 3 1 4 *